# Range Rover
## Workshop Manual
## 1986-1989 Model Years

RR2901

All Petrol and Diesel Models from 1986-1989 MY
3.5 V8 Petrol and the 2.4 Diesel Engine

Publication Number SRR660ENWM
(Includes Pub. No. LSM 180WM)

Land Rover
Lode Lane,
Solihull
West Midlands
B92 8NW
England

© Content Copyright of Jaguar Land Rover Limited 1986, 1987, 1988, 1989 and 1991
Brooklands Books Limited 1992 and 2016

This book is published by Brooklands Books Limited and based upon text and illustrations
protected by copyright and first published in 1986 by Jaguar Land Rover Limited and
may not be reproduced transmitted or copied by any means without the prior written
permission of Jaguar Land Rover Limited and Brooklands Books Limited.

Whilst every effort is made to ensure the accuracy of the particulars contained in
this book the Manufacturing Companies will not, in any circumstances,
be held liable for any inaccuracies or the consequences thereof.

Brooklands Books Ltd., PO Box 146, Cobham,
Surrey KT11 1LG, England.
E-mail: sales@brooklands-books.com    www.brooklands-books.com

Part Number: SSR660ENWM

ISBN 9781783180707        Ref: RR29WH        10T6/2484

# CONTENTS

| **Workshop Manual 1986 Onwards - Part 1** | Section | Page |
|---|---|---|
| Introduction | 01 | 6 |
| General Specification Data | 04 | 9 |
| Engine Tuning Data | 05 | 12 |
| General Fitting Instructions | 07 | 14 |
| Recommended Lubricants, Fluids & Capacities | 09 | 17 |
| Maintenance | 10 | 19 |

| **Workshop Manual 1986 Onwards - Part 2** | Section | Page |
|---|---|---|
| General Specification Data | 04 | 31 |
| Torque Wrench Settings | 06 | 33 |
| Engine | 12 | 34 |
| Emission Control | 17 | 50 |
| Fuel System (including Fuel Injection) | 19 | 55 |
| Cooling Systems | 26 | 78 |
| Manifold & Exhaust System | 30 | 80 |
| Clutch | 33 | 82 |

| **Workshop Manual 1986 Onwards - Part 3** | Section | Page |
|---|---|---|
| General Specification Data | 04 | 88 |
| Torque Wrench Settings | 06 | 89 |
| Manual Gearbox and Transfer Box | 37 | 91 |
| Automatic Gearbox | 44 | 114 |
| Propeller Shafts | 47 | 142 |
| Rear Axles & Final Drive | 51 | 143 |
| Front Axles & Final Drive | 54 | 149 |

| **Workshop Manual 1986 Onwards - Part 4** | Section | Page |
|---|---|---|
| General Specification Data | 04 | 156 |
| Torque Wrench Settings | 06 | 157 |
| Steering | 57 | 158 |
| Front Suspension | 60 | 171 |
| Rear Suspension | 64 | 173 |
| Brakes | 70 | 176 |
| Body | 76 | 189 |

| **Workshop Manual 1986 Onwards - Part 5** | Section | Page |
|---|---|---|
| General Specification Data | 04 | 206 |
| Torque Wrench Settings | 06 | 207 |
| Heating & Ventilation | 80 | 208 |
| Air Conditioning | 82 | 209 |
| Wipers & Washers | 84 | 219 |
| Electrical | 86 | 223 |

| **1987 Model Year Supplement** | Section | Page |
|---|---|---|
| Introduction | 01 | 254 |
| General Specification Data | 04 | 255 |
| Engine Tuning Data | 05 | 256 |
| Lubricants and Fluids | 09 | 258 |
| Maintenance | 10 | 260 |
| Fuel System | 19 | 262 |
| Clutch | 33 | 268 |
| Steering | 57 | 269 |
| Brakes | 70 | 273 |
| Body | 76 | 275 |
| Heating & Ventilation | 80 | 278 |
| Air Conditioning | 82 | 279 |
| Wipers & Washers | 84 | 280 |
| Electrical | 86 | 281 |

| **1988 Model Year Supplement** | Section | Page |
|---|---|---|
| Introduction | 01 | 306 |
| General Specification Data | 04 | 307 |
| Engine Tuning Data | 05 | 309 |

## CONTENTS - Continued

| | Section | Page |
|---|---|---|
| Torque Wrench Settings | 06 | 310 |
| Lubricants and Fluids | 09 | 311 |
| Maintenance | 10 | 312 |
| Emission Control | 17 | 319 |
| Fuel System | 19 | 322 |
| Manifold & Exhaust System | 30 | 342 |
| Propeller Shafts | 47 | 343 |
| Electrical | 86 | 344 |

### Lucas A127/65 Amp Alternator Bulletin

| | Section | Page |
|---|---|---|
| Introduction | 01 | 362 |
| General Specification Data | 04 | 363 |
| Electrical | 86 | 363 |

### Lucas Girling-Type 115 Brake Bulletin

| | Section | Page |
|---|---|---|
| Introduction | 01 | 371 |
| General Specification Data | 04 | 372 |
| Maintenance | 10 | 372 |
| Brakes | 70 | 373 |

### Tudor Webasto Sunshine Roof Bulletin

| | Section | Page |
|---|---|---|
| Introduction | 01 | 381 |
| Body | 76 | 382 |
| Electrical | 86 | 385 |

### Vogue SE Bulletin

| | Section | Page |
|---|---|---|
| Introduction | 01 | 387 |
| General Specification Data | 04 | 388 |
| Torque Wrench Settings | 06 | 388 |
| Body | 76 | 388 |
| Electrical | 86 | 389 |

### 1989 Model Year Supplement

| | Section | Page |
|---|---|---|
| Introduction | 01 | 404 |
| General Specification Data | 04 | 405 |
| Torque Wrench Settings | 06 | 406 |
| Lubricants and Fluids | 09 | 407 |
| Maintenance | 10 | 408 |
| Manual Gearbox and Transfer Box | 37 | 410 |
| Automatic Gearbox | 44 | 440 |
| Brakes | 70 | 443 |
| Body | 76 | 444 |
| Air Conditioning | 82 | 448 |
| Wipers & Washers | 84 | 449 |
| Electrical | 86 | 450 |

### Concealed Door Hinges Bulletin

| | Section | Page |
|---|---|---|
| Introduction | 01 | 460 |
| Torque Wrench Settings | 06 | 460 |
| Lubricants and Fluids | 09 | 461 |
| Maintenance | 10 | 462 |
| Body | 76 | 463 |

### 2.4 & 2.5 Turbo Diesel Engine Supplement

| | Section | Page |
|---|---|---|
| Introduction | 01 | 467 |
| General Specification Data | 04 | 468 |
| Engine Tuning Data | 05 | 474 |
| Torque Wrench Settings | 06 | 475 |
| Lubricants and Fluids | 09 | 476 |
| Maintenance | 10 | 478 |
| Service Tools | 99 | 483 |
| VW Diesel Engine | 12 | 484 |
| Fuel System | 19 | 511 |
| Cooling Systems | 26 | 514 |
| Electrical | 86 | 515 |

# RANGE ROVER WORKSHOP MANUAL 1986 ONWARDS

# Part 1

|  | Section | Page |
|---|---|---|
| Introduction | 01 | 6 |
| General Specification Data | 04 | 9 |
| Engine Tuning Data | 05 | 12 |
| General Fitting Instructions | 07 | 14 |
| Recommended Lubricants, Fluids & Capacities | 09 | 17 |
| Maintenance | 10 | 19 |

# 01

## INTRODUCTION

This Workshop Manual covers the Range Rover range of vehicles. It is primarily designed to assist skilled technicians in the efficient repair and maintenance of Range Rover vehicles.

WARNINGS and CAUTIONS are given throughout this manual in the following form:

**WARNING: Procedures which must be followed precisely to avoid the possibility of personal injury.**

**CAUTION: This calls attention to procedures which must be followed to avoid damage to components.**

NOTE: This calls attention to methods which make a job easier to perform.

### REFERENCES

References to the left- or right-hand side in the manual are made when viewing the vehicle from the rear. With the engine and gearbox assembly removed, the water pump end of the engine is referred to as the front.

To reduce repetition, operations covered in this manual do not include reference to testing the vehicle after repair. It is essential that work is inspected and tested after completion and if necessary a road test of the vehicle is carried out particularly where safety related items are concerned.

### DIMENSIONS

The dimensions quoted are to design engineering specification. Alternative unit equivalents, shown in brackets following the dimensions, have been converted from the original specification.

During the period of running-in from new, certain adjustments may vary from the specification figures given in this Manual. These adjustments will be re-set by the Distributor or Dealer at the After Sales Service, and thereafter should be maintained at the figures specified in the Manual.

### REPAIRS AND REPLACEMENTS

When replacement parts are required it is essential that only Land Rover parts are used.

Attention is particularly drawn to the following points concerning repairs and the fitting of replacement parts and accessories:

Safety features embodied in the vehicle may be impaired if other than Land Rover parts are fitted. In certain territories, legislation prohibits the fitting of parts not to the vehicle manufacturer's specification. Torque wrench setting figures given in the Repair Operation Manual must be strictly adhered to. Locking devices, where specified, must be fitted. If the efficiency of a locking device is impaired during removal it must be renewed. Owners purchasing accessories while travelling abroad should ensure that the accessory and its fitted location on the vehicle conform to mandatory requirements existing in their country of origin. The terms of the Owners Service Statement may be invalidated by the fitting of other than Land Rover parts.

All Land Rover parts have the full backing of the Owners Service Statement.

Land Rover Distributors and Dealers are obliged to supply only Land Rover service parts.

### POISONOUS SUBSTANCES

Many liquids and other substances used in motor vehicles are poisonous and should under no circumstances be consumed and should as far as possible be kept away from open wounds. These substances among others include antifreeze, brake fluid, fuel, windscreen washer additives, lubricants and various adhesives.

### FUEL HANDLING PRECAUTIONS

The following information provides basic precautions which must be observed if petrol (gasoline) is to be handled safely. It also outlines the other areas of risk which must not be ignored.

This information is issued for basic guidance only, and in any case of doubt appropriate enquiries should be made of your local Fire Officer.

### GENERAL

Petrol/gasoline vapour is highly flammable and in confined spaces is also very explosive and toxic.

When petrol/gasoline evaporates it produces 150 times its own volume in vapour, which when diluted with air becomes a readily ignitable mixture. The vapour is heavier than air and will always fall to the lowest level. It can readily be distributed throughout a workshop by air current, consequently, even a small spillage of petrol/gasoline is potentially very dangerous.

*Continued*

---

Section Number

| 01 | INTRODUCTION | |
|---|---|---|
| | —Repairs and replacement parts | 1 |
| | —Poisonous substances | 1 |
| | —Fuel handling precautions | 2 |
| | —Fuel tank draining and repair | 3 |
| | —Service tools | 4 |
| | —Abbreviations and symbols | 4 |
| | —Location of vehicle identification numbers | 5 |

| 04 | GENERAL SPECIFICATION DATA | |
|---|---|---|
| | —V8 engine data | 1 |
| | —General data—all models | 2 |
| | —Electrical data | 3 |
| | —Vehicle dimensions and weights | 4 |
| | —Tyre pressures | 5 |

| 05 | ENGINE TUNING DATA | |
|---|---|---|
| | —V8 engine tuning data | 1 |
| | —Fuel injection models | 1 |
| | —Carburetter models | 2 |

| 07 | GENERAL FITTING INSTRUCTIONS | |
|---|---|---|
| | —Precautions against damage | 1 |
| | —Safety precautions | 1 |
| | —Preparation and dismantling | 1 |
| | —Inspection of components | 2 |
| | —Ball and roller bearings | 2 |
| | —Oil seals | 2 |
| | —Joints and joint faces | 3 |
| | —Flexible hydraulic pipes and hoses | 3 |
| | —Metric bolt identification | 4 |
| | —Metric nut identification | 4 |
| | —Keys and keyways | 4 |
| | —Tab washers, split pins, nuts and locking wire | 4 |
| | —Screw threads | 4 |
| | —Unified thread identification | 5 |

| 09 | RECOMMENDED LUBRICANTS, FLUIDS AND CAPACITIES | |
|---|---|---|
| | —Recommended lubricants | 1 |
| | —Corrosion inhibitor | 2 |
| | —Anti-freeze proportions | 3 |
| | —Capacities | 3 |

| 10 | MAINTENANCE | |
|---|---|---|
| | —Maintenance Schedules | 1 |
| | —Maintenance—lubrication | 3 |
| | —Maintenance—general maintenance and adjustments | 7 |

# Special Service Tools

The use of approved special service tools is important. They are essential if service operations are to be carried out efficiently, and safely. The amount of time which they save can be considerable.

Every special tool is designed with the close co-operation of Land Rover Ltd., and no tool is put into production which has not been tested and approved by us. New tools are only introduced where an operation cannot be satisfactorily carried out using existing tools or standard equipment. The user is therefore assured that the tool is necessary and that it will perform accurately, efficiently and safely.

Special tools bulletins will be issued periodically giving details of new tools as they are introduced.

All orders and enquiries from the United Kingdom should be sent direct to V. L. Churchill. Overseas orders should be placed with the local V. L. Churchill distributor, where one exists. Countries where there is no distributor may order direct from V. L. Churchill Limited.

The tools recommended in this Repair Operation Manual are listed in a multi-language, illustrated catalogue obtainable from Messrs. V. L. Churchill at the above address under publication number 2217/2/84 or from Land Rover Ltd., under part number LSM0052TC from the following address: Land Rover Limited, Service Department, Lode Lane, Solihull, West Midlands, England B92 8NW.

---

Always have a fire extinguisher containing FOAM $CO_2$ GAS, or POWDER close at hand when handling or draining fuel, or when dismantling fuel systems and in areas where fuel containers are stored.
Always disconnect the vehicle battery BEFORE carrying out dismantling or draining work on a fuel system.
Whenever petrol/gasoline is being handled, drained or stored, or when fuel systems are being dismantled all forms of ignition must be extinguished or removed, any head-lamps used must be flameproof and kept clear of spillage.

NO ONE SHOULD BE PERMITTED TO REPAIR COMPONENTS ASSOCIATED WITH PETROL/GASOLINE WITHOUT FIRST HAVING HAD SPECIALIST TRAINING.

## FUEL TANK DRAINING

WARNING: PETROL/GASOLINE MUST NOT BE EXTRACTED OR DRAINED FROM ANY VEHICLE WHILST IT IS STANDING OVER A PIT.

Draining or extracting petrol/gasoline from vehicle fuel tank must be carried out in a well ventilated area.
The receptacle used to contain the petrol/gasoline must be more than adequate for the full amount of fuel to be extracted or drained. The receptacle should be clearly marked with its contents, and placed in a safe storage area which meets the requirements of local authority regulations.
WHEN PETROL/GASOLINE HAS BEEN EXTRACTED OR DRAINED FROM A FUEL TANK THE PRECAUTIONS GOVERNING NAKED LIGHTS AND IGNITION SOURCES SHOULD BE MAINTAINED.

## FUEL TANK REMOVAL

On vehicles where the fuel line is secured to the fuel tank outlet by a spring steel clip, it is recommended that such clips are released before the fuel line is disconnected or the fuel tank unit is removed. This procedure will avoid the possibility of residual petrol fumes in the fuel tank being ignited when the clips are released.
As an added precaution fuel tanks should have a PETROL/GASOLINE VAPOUR warning label attached to them as soon as they are removed from the vehicle.

## FUEL TANK REPAIR

Under no circumstances should a repair to any tank involving heat treatment be carried out without first rendering the tank SAFE, by using one of the following methods:

STEAMING: With the filler cap and tank unit removed, empty the tank. Steam the tank for at least two hours with low pressure steam. Position the tank so that condensation can drain away freely, ensuring that any sediment and sludge not volatised by the steam, is washed out during the steaming process.
BOILING: With the filler cap and tank unit removed, empty the tank. Immerse the tank completely in boiling water containing an effective alkaline degreasing agent or a detergent, with the water filling and also surrounding the tank for at least two hours.
After steaming or boiling a signed and dated label to this effect should be attached to the tank.

## SPECIFICATION

Purchasers are advised that the specification details set out in this Manual apply to a range of vehicles and not to any specific one. For the specification of a particular vehicle, purchasers should consult their Distributor or Dealer.

The Manufacturers reserve the right to vary their specifications with or without notice, and at such times and in such manner as they think fit. Major as well as minor changes may be involved in accordance with the Manufacturer's policy of constant product improvement.

Whilst every effort is made to ensure the accuracy of the particulars contained in this Manual, neither the Manufacturer nor the Distributor or Dealer, by whom this Manual is supplied, shall in any circumstances be held liable for any inaccuracy or the consequences thereof.

## COPYRIGHT

© Land Rover Limited 1986

All rights reserved. No part of this publication may be reproduced, stored in a retrieval system or transmitted in any form, electronic, mechanical, photocopying, recording or other means without prior written permission of Land Rover Limited, Service Department, Solihull, England.

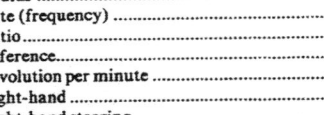

# 01 | LOCATION OF VEHICLE IDENTIFICATION AND UNIT NUMBERS

## VEHICLE IDENTIFICATION NUMBER (VIN)

The Vehicle Identification Number and the recommended maximum vehicle weights are stamped on a plate located under the bonnet riveted to the top of the front grille at the front of the engine compartment.
The number is also stamped on the right side of the chassis forward of the spring mounting turret.

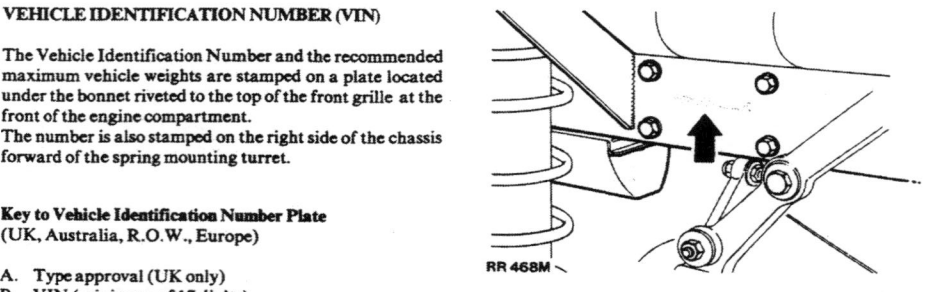

**Key to Vehicle Identification Number Plate**
(UK, Australia, R.O.W., Europe)

A. Type approval (UK only)
B. VIN (minimum of 17 digits)
C. Maximum permitted laden weight for vehicle
D. Maximum vehicle and trailer weight
E. Maximum road weight—front axle
F. Maximum road weight—rear axle

**Key to Vehicle Identification Number**
(Saudi Arabia)

A. Year of manufacture
B. Month of manufacture
C. Maximum vehicle weight
D. Maximum road weight—front axle
E. Maximum road weight—rear axle
F. VIN (minimum of 17 digits)

## ENGINE SERIAL NUMBER—V8 ENGINE

The V8 engine serial number is stamped on a cast pad on the cylinder block between numbers 3 and 5 cylinders. The appropriate engine compression ratio is stamped above the serial number.

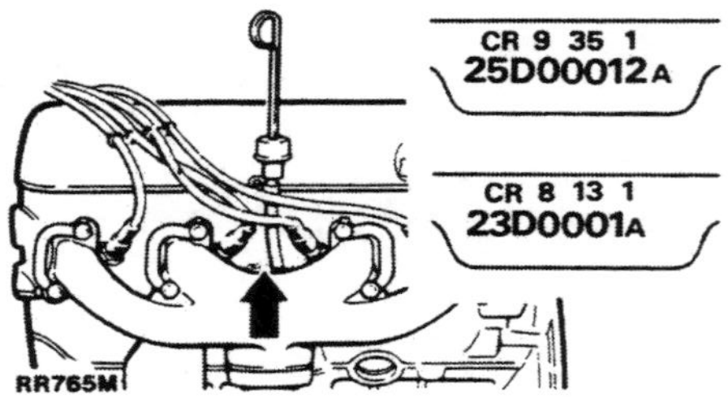

NOTE:
CR 9 35 1
25D00012A
CR 8 13 1
23D0001A

## MAIN GEARBOX LT77—5 SPEED

The serial number is stamped on a cast pad on the bottom right-hand side of the gearbox.

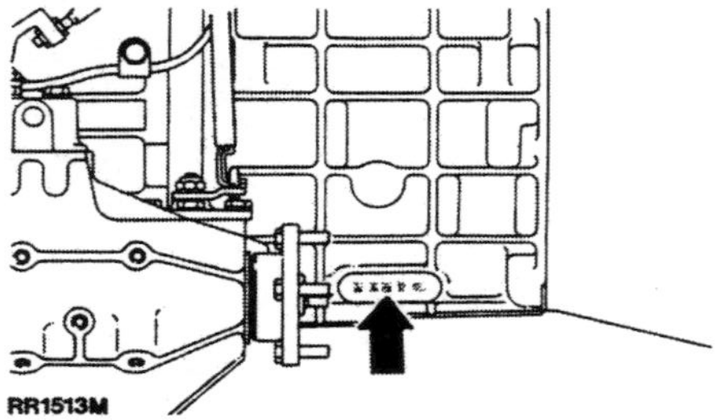

---

# 01 | ABBREVIATIONS AND SYMBOLS USED IN THIS MANUAL

| | | | |
|---|---|---|---|
| Across flats (bolt size) | AF | Midget edison screw | MES |
| After bottom dead centre | ABDC | Millimetre | mm |
| After top dead centre | ATDC | Miles per gallon | mpg |
| Alternating current | a.c. | Miles per hour | mph |
| Ampere | amp | Minimum | min |
| Ampere-hour | amp hr | Minute (angle) | — |
| Atmospheres | Atm | Minus (of tolerance) | — |
| Before bottom dead centre | BBDC | Negative (electrical) | — |
| Before top dead centre | BTDC | Number | No. |
| Bottom dead centre | BDC | Ohms | ohm |
| Brake mean effective pressure | BMEP | Ounces (force) | ozf |
| Brake horse power | bhp | Ounces (mass) | oz |
| British Standards | BS | Ounce inch (torque) | ozf.in. |
| Carbon monoxide | CO | Outside diameter | o.dia. |
| Centimetre | cm | Paragraphs | para. |
| Centigrade (Celsius) | C | Part number | Part No. |
| Cubic centimetre | cm³ | Percentage | % |
| Cubic inch | in³ | Pints | pt |
| Degree (angle) | deg or ° | Pints (US) | US pt |
| Degree (temperature) | deg or ° | Plus (tolerance) | + |
| Diameter | dia. | Positive (electrical) | + |
| Direct current | d.c. | Pound (force) | lbf |
| Fahrenheit | F | Pounds feet (torque) | lbf.ft. |
| Feet | ft | Pounds inches (torque) | lbf.in. |
| Feet per minute | ft/min | Pound (mass) | lb |
| Fifth | 5th | Pounds per square inch | lb/in² |
| Figure (illustration) | Fig. | Radius | r |
| First | 1st | Rate (frequency) | c/min |
| Fourth | 4th | Ratio | — |
| Gramme (force) | gf | Reference | ref. |
| Gramme (mass) | g | Revolution per minute | rev/min |
| Gallons | gal | Right-hand | RH |
| Gallons (US) | US gal | Right-hand steering | RHStg |
| High compression | h.c. | Second (angle) | — |
| High tension (electrical) | H.T. | Second (numerical order) | 2nd |
| Hundredweight | cwt | Single carburetter | SC |
| Independent front suspension | i.f.s. | Specific gravity | sp.gr. |
| Internal diameter | i.dia. | Square centimetres | cm² |
| Inches of mercury | in.Hg | Square inches | in² |
| Inches | in | Standard | std. |
| Kilogramme (force) | kgf | Standard wire gauge | s.w.g. |
| Kilogramme (mass) | kg | Synchroniser/synchromesh | synchro. |
| Kilogramme centimetre (torque) | kgf.cm | Third | 3rd |
| Kilogramme per square centimetre | kg/cm² | Top dead centre | TDC |
| Kilogramme metres (torque) | kgf.m | Twin carburetters | TC |
| Kilometres | km | United Kingdom | UK |
| Kilometres per hour | km/h | Vehicle Identification Number | VIN |
| Kilovolts | kV | Volts | V |
| King pin inclination | k.p.i. | Watts | W |
| Left-hand steering | LHStg | **SCREW THREADS** | |
| Left-hand thread | LHThd | American Standard Taper Pipe | NPTF |
| Litres | litre | British Association | BA |
| Low compression | l.c. | British Standard Fine | BSF |
| Low tension | l.t. | British Standard Pipe | BSP |
| Maximum | max. | British Standard Whitworth | Whit. |
| Metre | m | Unified Coarse | UNC |
| Microfarad | mfd | Unified Fine | UNF |

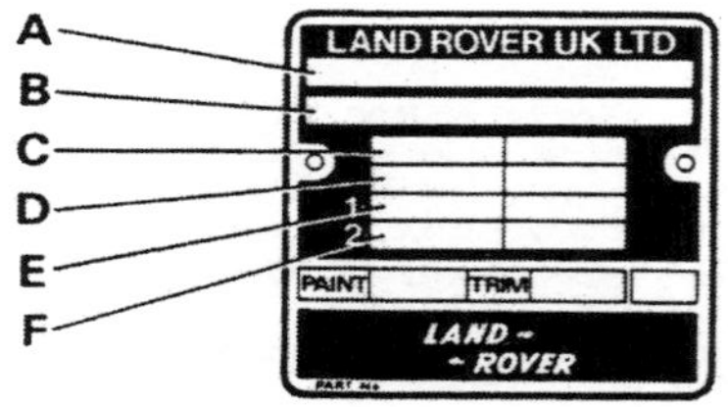

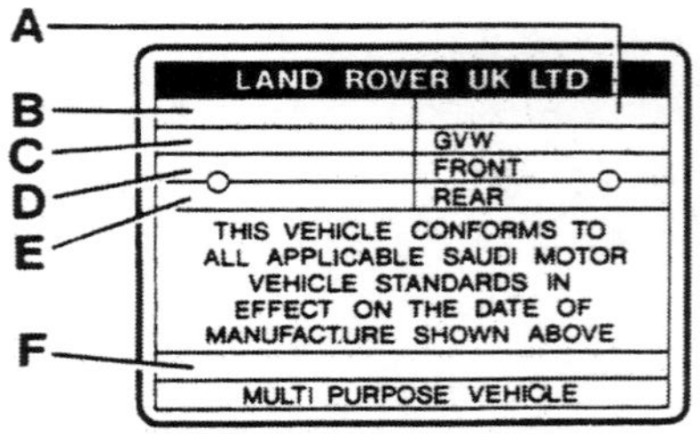

# GENERAL SPECIFICATION DATA

## GENERAL SPECIFICATION DATA

### V8 ENGINE

**ENGINE**
- Type: V8
- Number of cylinders: Eight, two banks of four
- Bore: 88.90 mm (3.500 in)
- Stroke: 71.12 mm (2.800 in)
- Capacity: 3528 cc (215 in³)
- Valve operation: Overhead by push-rod

**Fuel-injection models**
- Maximum power—bhp—9.35:1: 165 at 4750 rev/min
- Maximum power—bhp—8.13:1: 150 at 4750 rev/min
- Maximum torque—9.35:1: 206 lbf.ft. (279 Nm) at 3200 rev/min
- Maximum torque—8.13:1: 190 lbf.ft. (258 Nm) at 2500 rev/min

**Carburetter models**
- Maximum power—bhp—9.35:1: 127 at 4000 rev/min
- Maximum power—bhp—8.13:1: 132 at 5000 rev/min
- Maximum torque—9.35:1: 194 lbf.ft. (263 Nm) at 2500 rev/min
- Maximum torque—8.13:1: 185 lbf.ft. (251 Nm) at 2500 rev/min

**Main bearings**
- Number and type: 5 Vandervell shells
- Material: Lead–indium

**Connecting rods**
- Type: Horizontally split big-end, plain small-end

**Big-end bearings**
- Number and type: 8 Vandervell VP shells
- Material: Lead–indium

**Gudgeon pins**
- Type: Press fit in connecting rod. Clearance on piston

**Pistons**
- Type: Aluminium alloy, concave crown

**Camshaft**
- Location: Central
- Number of bearings: 5
- Bearings: Non-serviceable

**Tappets**: Hydraulic—self-adjusting

**Lubrication**
- System type: Wet sump, pressure fed
- Oil pump type: Gear
- Oil filter: Full flow, self-contained cartridge

**Cooling system**
- Type: Pressurised spill-return system with thermostat control, pump and fan assisted
- Type of pump: Centrifugal

---

# LOCATION OF VEHICLE IDENTIFICATION AND UNIT NUMBERS

## AUTOMATIC GEARBOX ZF4HP22

The serial number is stamped on a plate riveted to the bottom left-hand side of the gearbox casing.

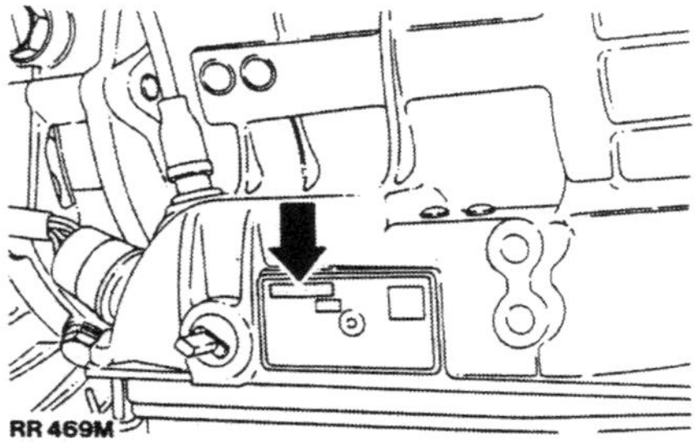

RR 469M

## TRANSFER GEARBOX LT230T

The serial number is stamped on the left-hand side of the gearbox casing below the mainshaft rear bearing housing adjacent to the bottom cover.

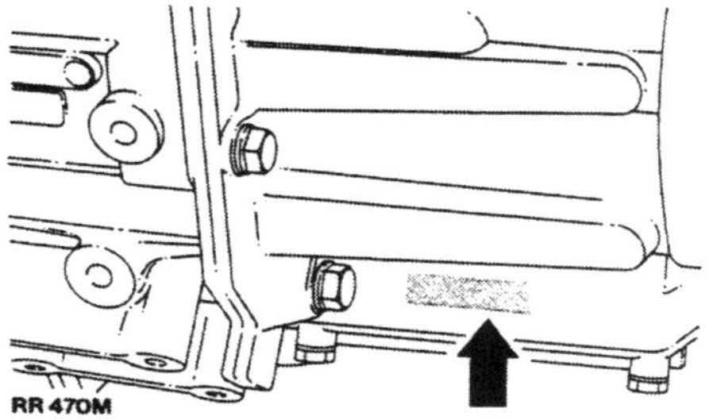

RR 470M

## FRONT AXLE

The serial number is stamped on top of the left-hand axle tube.

## REAR AXLE

The serial number is stamped on the left-hand rear axle tube.

## GENERAL SPECIFICATION DATA

**Fuel system—carburetter**
- 8.13 :1, 9.35 :1 Emission controlled ............ 2 × Solex 175 CDSE
- 8.13 :1 Non emission ............................ 2 × Solex 175 CD3
- Fuel pump ........................................ AC Delco—low pressure (electrical), immersed in the fuel tank
- Fuel filter ...................................... AC Delco CD600, element ACD60

**Fuel system—fuel injection**
- Type ............................................. Lucas 'L' system electronically controlled
- Fuel pump—make and type ......................... AC Delco—high pressure (electrical), immersed in the fuel tank
- Fuel filter—make and type ....................... Bosch in-line filter, cannister type

### TRANSMISSION

**Clutch**
- Make and type ................................... Borg & Beck, diaphragm type
- Clutch plate diameter ........................... 266.5 mm (10.5 in)

**Gearbox**
- Model ........................................... LT77 (manual)
- Type ............................................ Five speed, single helical constant mesh with synchromesh on all forward gears

**Gearbox**
- Model ........................................... ZF 4HP22 (automatic)
- Type ............................................ Automatic four speed and reverse, epicyclic gearbox with fluid torque converter and lock up

**Transfer gearbox**
- Model ........................................... LT230T
- Type ............................................ Two-speed reduction on gearbox output, front and rear drive permanently engaged via a lockable differential

**Propellor shafts**
- Type ............................................ Open type 50.8 mm (2 in) diameter
- Universal joints ................................ 03EHD standard shafts

**Rear axle**
- Type ............................................ Spiral bevel, fully floating shafts
- Ratio ........................................... 3.54 :1

**Front axle**
- Type ............................................ Spiral bevel enclosed constant velocity joints
- Ratio ........................................... 3.54 :1

### SUSPENSION
- Front ........................................... Coil springs, radius arms and panhard rod
- Rear ............................................ Coil springs, radius arms, 'A' frame location arms with 'Boge' hydromat self-energising levelling device

### Shock absorbers
- Type ............................................ Telescopic, double acting—non adjustable

### STEERING
- Power assisted type ............................. Adwest varamatic

## GENERAL SPECIFICATION DATA

### BRAKES

**Foot Brake**
- Type ............................................ Disc
- Operation ....................................... Hydraulic, servo assisted, self adjusting

**Front Brake**
- Type ............................................ Outboard discs with four piston calipers
- Pad wear indicator .............................. Right-hand caliper, inboard pad

**Rear Brake**
- Type ............................................ Outboard disc with two piston calipers
- Pad wear indicator .............................. Left-hand caliper, inboard pad

**Transmission Brake**
- Type ............................................ Mechanical-cable operated drum brake on the rear of the transfer gearbox output shaft

### WHEELS AND TYRES
- Type and size ................................... Steel 6.00 JK × 16 (tubed)
  Alloy 7.00 J × 16 (tubed or tubeless)
- Tyre size ....................................... 205 R16 or 215/75 R16

NOTE: Fuel injection vehicles must be fitted with 'S' rated high speed tyres.

### ELECTRICAL EQUIPMENT
- System .......................................... 12 volt negative earth
- Fuses ........................................... 'Autofuse' (blade type)
  Blow ratings to suit individual circuits

**Battery**
- Make ............................................ Chloride maintenance free
- Type ............................................ 9 plate 210/85/90
  13 plate 380/120/90

**Starter Motor**
- Make and type ................................... Lucas 3M100 pre-engaged

**Alternator**
- Make and type ................................... Lucas 133/65

**Wiper Motor**
- Make and type ................................... Front—Lucas 28W 2-speed
  Rear—Lucas 14W PM

**Horns**
- Make and type ................................... Klaxix (Mixo) TR99

**Distributor**
- Make and type ................................... Lucas 35DM8 electronic

# GENERAL SPECIFICATION DATA

## VEHICLE WEIGHTS AND PAYLOAD

When loading a vehicle to its maximum (Gross Vehicle Weight), consideration must be taken of the unladen vehicle weight and the distribution of the payload to ensure that axle loadings do not exceed the permitted maximum values.

It is the customer's responsibility to limit the vehicle's payload in an appropriate manner such that neither maximum axle loads nor Gross Vehicle Weight are exceeded.

### Range Rover

| Model | | Manual | | | Automatic | | |
|---|---|---|---|---|---|---|---|
| | Front Axle kg (lb) | Rear Axle kg (lb) | Total kg (lb) | Front kg (lb) | Rear kg (lb) | Total kg (lb) |
| 2 DOOR Unladen Weight | 893 (1969) | 867 (1911) | 1760 (3880) | 920 (2028) | 871 (1920) | 1791 (3948) |
| EEC Kerb Weight | 912 (2011) | 983 (2167) | 1895 (4178) | 939 (2070) | 987 (2176) | 1926 (4246) |
| Gross Vehicle Weight | 1100 (2425) | 1510 (3329) | 2510 (5534) | 1100 (2425) | 1510 (3329) | 2510 (5534) |
| 4 DOOR Unladen Weight | 909 (2004) | 883 (1947) | 1792 (3951) | 925 (2039) | 911 (2008) | 1836 (4048) |
| EEC Kerb Weight | 928 (2046) | 999 (2202) | 1927 (4248) | 944 (2081) | 1027 (2264) | 1971 (4345) |
| Gross Vehicle Weight | 1100 (2425) | 1510 (3329) | 2510 (5534) | 1100 (2425) | 1510 (3329) | 2510 (5534) |

Note: UNLADEN WEIGHT is the minimum vehicle specification, excluding fuel and driver.
EEC KERB WEIGHT is the minimum vehicle specification, plus full fuel tank and 75 kg (165 lb) driver.
GROSS VEHICLE WEIGHT is the maximum all-up weight of the vehicle including driver, passengers, payload and equipment. This figure is liable to vary according to legal requirements in certain countries.

### Maximum permissible towed weights

| | On-road | Off-road |
|---|---|---|
| Trailers without brakes | 750 kg / 1650 lb | 750 kg / 1650 lb |
| Trailers with overrun brakes | 3500 kg / 7700 lb | 1000 kg / 2200 lb |
| 4-wheel trailers with continuous or semi-continuous brakes, i.e. coupled brakes | 4000 kg / 8800 lb | 1000 kg / 2200 lb |

NOTE: It is the Owner's responsibility to ensure that all regulations with regard to towing are complied with. This applies also when towing abroad. All relevant information should be obtained from the appropriate motoring organisation.

### Tyre Pressures

Pressures: Check with tyres cold

| | Normal on- and off-road use. All speeds and loads | | | Off-road 'emergency' soft use maximum speed of 40 kph (25 mph) | | |
|---|---|---|---|---|---|---|
| | | Front | Rear | | Front | Rear |
| | bars | 1,9 | 2,4 | bars | 1,1 | 1,6 |
| | lbf/in² | 28 | 35 | lbf/in² | 16 | 23 |
| | kgf/cm² | 2,0 | 2,5 | kgf/cm² | 1,1 | 1,6 |

For extra ride comfort rear tyre pressures may be reduced to 2,1 bars (31 lbf/in²) 2,2 kgf/cm² when the rear axle weight does not exceed 1250 kg (2755 lb).

These pressures may be increased for rough off-road usage where the risk of tyre cutting or penetration is more likely.

Pressures may also be increased for high speed motoring near the vehicle's maximum speed. Any such increase in pressures may be up to an absolute maximum pressure of 2,9 bars (42 lbf/in²) 3,0 kgf/cm².

Normal operating pressures should be restored as soon as reasonable road conditions or hard ground is reached.

After any usage off the road, tyres and wheels should be inspected for damage particularly if high cruising speeds are subsequently to be used.

Towing: When the vehicle is used for towing, the reduced rear tyre pressures for extra ride comfort are not applicable.

*WARNING: Wheels and tyres. Unless both wheel rim and tyre are marked 'TUBELESS', an inner tube MUST be fitted.*

---

# GENERAL SPECIFICATION DATA

**Ignition Module**
Make and type .................. Lucas 2CE 12 volt electronic

**Airflow Meter**
Make and type .................. Lucas 2AM
Serial number .................. 73243A

**Injectors**
Make and type .................. Lucas 6NJ
Serial number .................. 73178A

**Electronic Control Unit**
Make and type .................. Lucas 4CU
Serial number .................. 84477A

**Power Resistor Pack**
Make and type .................. 2PR
Serial number .................. 73184A

## VEHICLE DIMENSIONS

| | |
|---|---|
| Overall length | 4.45 m (175 in) |
| Overall width | 1.82 m (71.6 in) |
| Overall height | 1.90 m (74.8 in) |
| Wheelbase | 2.54 m (100 in) |
| Track: front and rear | 1.49 m (58.5 in) |
| Ground clearance: under differential | 190 mm (7.5 in) |
| Turning circle | 11.89 m (39 ft) |
| Loading height | 686 mm (27 in) |
| Maximum cargo height | 1.01 m (40 in) |
| Rear opening height | 1.01 m (40 in) |
| Usable luggage capacity, rear seat folded | 2.00 m³ (70 ft³) |
| Usable luggage capacity, rear seat in use: | |
| —four door vehicles | 1.25 m³ (44.14 ft³) |
| —two door vehicles | 1.17 m³ (41.31 ft³) |
| Maximum roof rack load | 75 kg (165 lb) |

# ENGINE TUNING DATA

## ENGINE TUNING DATA

| | |
|---|---|
| Type | V8 Cylinder |
| Firing order | 1—8—4—3—6—5—7—2 |
| Cylinder Numbers—Left bank | 1—3—5—7 |
| Right bank | 2—4—6—8 |
| No 1 Cylinder location | Pulley end of left bank |
| Timing marks | On crankshaft vibration damper |
| Spark plugs Make/type | Champion N9YC |
| Gap | 0.70–0.80 mm (0.030–0.032 in) |
| Ignition module Make/type | Lucas 2CE 12 volt electronic |
| Distributor Make/type | Lucas 35DM8 electronic |
| Air gap | 0.20–0.35 mm (0.008–0.014 in) |

## FUEL INJECTION MODELS

| | |
|---|---|
| Engine | V8 Cylinder |
| Compression ratio | 9.35:1 and 8.13:1 |
| Fuel injection system | Lucas 'L' system electronically controlled |

Valve Timing
| | Inlet | Exhaust |
|---|---|---|
| Opens | 24° BTDC | 62° BBDC |
| Closes | 52° ABDC | 14° ATDC |
| Duration | 256° | 256° |
| Valve peak | 104° ATDC | 114° BTDC |

| | |
|---|---|
| Idle Speed | 700 to 800 rev/min |
| Ignition Timing at | 600 rev/min |
| Ignition Timing Dynamic or static | TDC ±1° |
| Exhaust Gas CO content at idle | 0.5–1.0% max |
| Distributor Make/type | Lucas 35 DM8 electronic |
| Rotation | Clockwise |
| Air gap | 0.20–0.35 mm (0.008–0.014 in) |
| Serial number | 42608 |

Centrifugal Advance
| | |
|---|---|
| Decelerating check—vacuum pipe disconnected | Distributor advance 12° to 14° |
| Distributor decelerating speeds—1600 | |
| 1100 | 9° 30' to 11° 30' |
| 600 | 1° 30' to 4° |
| No centrifugal advance below | 600 rev/min |

| | |
|---|---|
| Fuel | 96 octane—9.35:1 |
| | 93 octane—8.13:1 |

## Notes

## ENGINE TUNING DATA 05

### EUROPE — Emission Controlled

#### CARBURETTER MODELS

Engine ................... V8 Cylinder

Compression ratio ................... 9.35:1

**Valve timing**

| | Inlet | Exhaust |
|---|---|---|
| Opens | 36° BTDC | 74° BBDC |
| Closes | 64° ABDC | 26° ATDC |
| Duration | 280° | 280° |
| Valve peak | 99° ATDC | 119° BTDC |

**Carburetters**

Type ................... 2 × Solex 175 CDSE
Solex specification number ................... 4187
Needle ................... B1GG
Idle speed (engine hot) ................... 650–750 rev/min
Fast idle speed (engine hot) ................... 1050–1150 rev/min
Mixture setting — CO at idle ................... 0.5%–2.5% pulsair connected

**Ignition**

Distributor make/type ................... Lucas 35DM8 electronic
Direction of rotation ................... Clockwise
Air gap ................... 0.20–0.35 mm (0.008–0.014 in)
Distributor serial number ................... 41980

**Centrifugal advance**

Decelerating check with vacuum pipe disconnected
Distributor decelerating speeds — 2800 ................... Distributor advance 5° 30' to 9°
1750 ................... 6° 30' to 9°
1000 ................... 2° 30' to 4° 30'
No advance below ................... 250 rev/min

**Ignition timing**

Dynamic or static ................... 6° BTDC at 650–750 rev/min (vacuum pipe disconnected)

Fuel ................... 96 octane

---

## ENGINE TUNING DATA 05

### GULF STATES — Emission Controlled

#### CARBURETTER MODELS

Engine ................... V8 Cylinder

Compression ratio ................... 8.13:1

**Valve timing**

| | Inlet | Exhaust |
|---|---|---|
| Opens | 30° BTDC | 68° BBDC |
| Closes | 75° ABDC | 37° ATDC |
| Duration | 285° | 285° |
| Valve peak | 106° ATDC | 112° BTDC |

**Carburetters**

Type ................... 2 × Solex 175 CDSE
Solex specification number ................... 4186
Needle ................... B1FC
Idle speed (engine hot) ................... 650–750 rev/min
Fast idle speed (engine hot) ................... 1050–1150 rev/min
Mixture setting — CO at idle ................... 1.5%–3.5% pulsair connected

**Ignition**

Distributor make/type ................... Lucas 35DM8 electronic
Direction of rotation ................... Clockwise
Air gap ................... 0.20–0.35 mm (0.008–0.014 in)
Distributor serial number ................... 42609

**Centrifugal advance**

Vacuum pipe disconnected
Distributor decelerating speeds — 2300 ................... Distributor advance 10° 30' to 13° 30'
1800 ................... 8° to 10°
1200 ................... 3° 30' to 5° 30'
No advance below ................... 450 rev/min

**Ignition timing**

Dynamic or static ................... 6° BTDC at 600 rev/min (vacuum pipe disconnected)

Fuel ................... 90–93 octane

## 05 ENGINE TUNING DATA

### REST OF THE WORLD—Non Emission

#### CARBURETTER MODELS

| | |
|---|---|
| Engine | V8 Cylinder |
| Compression ratio | 8.13:1 |

**Valve timing**

| | Inlet | Exhaust |
|---|---|---|
| Opens | 30° BTDC | 68° BBDC |
| Closes | 75° ABDC | 37° ATDC |
| Duration | 285° | 285° |
| Valve peak | 106° ATDC | 112° BTDC |

**Carburetters**

| | |
|---|---|
| Type | 2 × Solex 175 CD3 |
| Solex specification number | 4185 |
| Needle | B1FF |
| Idle speed (engine hot) | 550—650 rev/min |
| Fast idle speed (engine hot) | 1050—1150 rev/min |

**Ignition**

| | |
|---|---|
| Distributor make/type | Lucas 35DM8 electronic |
| Direction of rotation | Clockwise |
| Air gap | 0.20–0.35 mm (0.008–0.014 in) |
| Distributor serial number | 41981 |

**Centrifugal advance**

| | Vacuum pipe disconnected |
|---|---|
| Distributor decelerating speeds—2100 | Distributor advance 11° 30' to 13° 30' |
| 1500 | 8° to 10° |
| 1000 | 4° to 6° |
| No advance below | 450 rev/min |

**Ignition timing**

| | |
|---|---|
| Dynamic or static | 6° BTDC at 650–750 rev/min (vacuum pipe disconnected) |
| Fuel | 90–93 octane |

---

## 07 GENERAL FITTING INSTRUCTIONS

### GENERAL FITTING INSTRUCTIONS

#### PRECAUTIONS AGAINST DAMAGE

1. Always fit covers to protect wings before commencing work in engine compartment.
2. Cover seats and carpets, wear clean overalls and wash hands or wear gloves before working inside car.
3. Avoid spilling hydraulic fluid or battery acid on paint work. Wash off with water immediately if this occurs. Use Polythene sheets in boot to protect carpets.
4. Always use a recommended Service Tool, or a satisfactory equivalent, where specified.
5. Protect temporarily exposed screw threads by replacing nuts or fitting plastic caps.

#### SAFETY PRECAUTIONS

1. Whenever possible use a ramp or pit when working beneath vehicle, in preference to jacking. Chock wheels as well as applying hand brake.
2. Never rely on a jack alone to support vehicle. Use axle stands or blocks carefully placed at jacking points to provide rigid location.
3. Ensure that a suitable form of fire extinguisher is conveniently located.
4. Check that any lifting equipment used has adequate capacity and is fully serviceable.
5. Inspect power leads of any mains electrical equipment for damage and check that it is properly earthed.
6. Disconnect earth (grounded) terminal of vehicle battery.
7. Do not disconnect any pipes in air conditioning refrigeration system, if fitted, unless trained and instructed to do so. A refrigerant is used which can cause blindness if allowed to contact eyes.
8. Ensure that adequate ventilation is provided when volatile degreasing agents are being used.

**CAUTION: Fume extraction equipment must be in operation when trachloride, methylene chloride, chloroform or perchlorethylene are used for cleaning purposes.**

9. Do not apply heat in an attempt to free stiff nuts or fittings; as well as causing damage to protective coatings, there is a risk of damage to electronic equipment and brake lines from stray heat.
10. Do not leave tools, equipment, spilt oil, etc., around or on work area.
11. Wear protective overalls and use barrier creams when necessary.

#### PREPARATION

1. Before removing a component, clean it and its surrounding areas as thoroughly as possible.
2. Blank off any openings exposed by component removal, using greaseproof paper and masking tape.
3. Immediately seal fuel, oil or hydraulic lines when separated, using plastic caps or plugs, to prevent loss of fluid and entry of dirt.
4. Close open ends of oilways, exposed by component removal, with tapered hardwood plugs or readily visible plastic plugs.
5. Immediately a component is removed, place it in a suitable container; use a separate container for each component and its associated parts.
6. Before dismantling a component, clean it thoroughly with a recommended cleaning agent; check that agent is suitable for all materials of component.
7. Clean bench and provide marking materials, labels, containers and locking wire before dismantling a component.

#### DISMANTLING

1. Observe scrupulous cleanliness when dismantling components, particularly when brake, fuel or hydraulic system parts are being worked on. A particle of dirt or a cloth fragment could cause a dangerous malfunction if trapped in these systems.
2. Blow out all tapped holes, crevices, oilways and fluid passages with an air line. Ensure that any O-rings used for sealing are correctly replaced or renewed, if disturbed.
3. Mark mating parts to ensure that they are replaced as dismantled. Whenever possible use marking ink, which avoids possibilities of distortion or initiation of cracks, liable if centre punch or scriber are used.
4. Wire together mating parts where necessary to prevent accidental interchange (e.g. roller bearing components).
5. Wire labels on to all parts which are to be renewed, and to parts requiring further inspection before being passed for reassembly; place these parts in separate containers from those containing parts for rebuild.
6. Do not discard a part due for renewal until after comparing it with a new part, to ensure that its correct replacement has been obtained.

## GENERAL FITTING INSTRUCTIONS

### INSPECTION—GENERAL

1. Never inspect a component for wear or dimensional check unless it is absolutely clean; a slight smear of grease can conceal an incipient failure.
2. When a component is to be checked dimensionally against figures quoted for it, use correct equipment (surface plates, micrometers, dial gauges, etc.) in serviceable condition. Makeshift checking equipment can be dangerous.
3. Reject a component if its dimensions are outside limits quoted, or if damage is apparent. A part may, however, be refitted if its critical dimension is exactly limit size, and is otherwise satisfactory.
4. Use 'Plastigauge' 12 Type PG-1 for checking bearing surface clearances; directions for its use, and a scale giving bearing clearances in 0.0025 mm (0.0001 in) steps are provided with it.

### BALL AND ROLLER BEARINGS

**NEVER REPLACE A BALL OR ROLLER BEARING WITHOUT FIRST ENSURING THAT IT IS IN AS-NEW CONDITION**

1. Remove all traces of lubricant from bearing under inspection by washing in petrol or a suitable degreaser; maintain absolute cleanliness throughout operations.
2. Inspect visually for markings of any form on rolling elements, raceways, outer surface of outer rings or inner surface of inner rings. Reject any bearings found to be marked, since any marking in these areas indicates onset of wear.
3. Holding inner race between finger and thumb of one hand, spin outer race and check that it revolves absolutely smoothly. Repeat, holding outer race and spinning inner race.
4. Rotate outer ring gently with a reciprocating motion, while holding inner ring; feel for any check or obstruction to rotation, and reject bearing if action is not perfectly smooth.
5. Lubricate bearing generously with lubricant appropriate to installation.
6. Inspect shaft and bearing housing for discolouration or other marking suggesting that movement has taken place between bearing and seatings. (This is particularly to be expected if related markings were found in operation 2.) If markings are found, use 'Loctite' in installation of replacement bearing.
7. Ensure that shaft and housing are clean and free from burrs before fitting bearing.
8. If one bearing of a pair shows an imperfection it is generally advisable to renew both bearings; an exception could be made if the faulty bearing had covered a low mileage, and it could be established that damage was confined to it only.
9. When fitting bearing to shaft, apply force only to inner ring of bearing, and only to outer ring when fitting into housing.
10. In the case of grease-lubricated bearings (e.g. hub bearings) fill space between bearing and outer seal with recommended grade of grease before fitting seal.
11. Always mark components of separable bearings (e.g. taper roller bearings) in dismantling, to ensure correct reassembly. Never fit new rollers in a used cup.

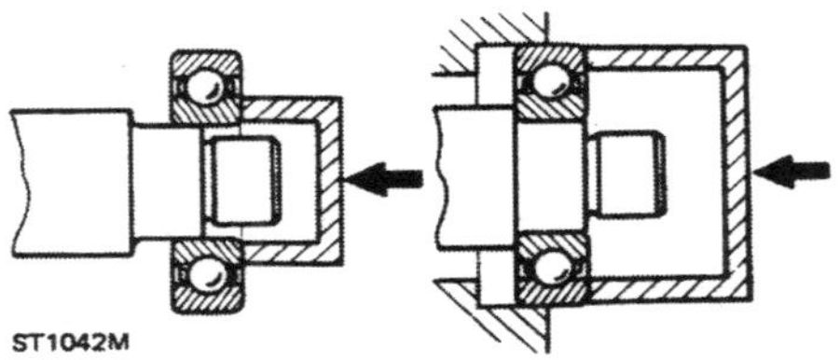

### OIL SEALS

1. Always fit new oil seals when rebuilding an assembly. It is not physically possible to replace a seal exactly when it has bedded down.
2. Carefully examine seal before fitting to ensure that it is clean and undamaged.
3. Smear sealing lips with clean grease; pack dust excluder seals with grease, and heavily grease duplex seals in cavity between sealing lips.
4. Ensure that seal spring, if provided, is correctly fitted.
5. Place lip of seal towards fluid to be sealed and slide into position on shaft, using fitting sleeve when possible to protect sealing lip from damage by sharp corners, threads or splines. If fitting sleeve is not available, use plastic tube or adhesive tape to prevent damage to sealing lip.
6. Grease outside diameter of seal, place square to housing recess and press into position, using great care and if possible a 'bell piece' to ensure that seal is not tilted. (In some cases it may be preferable to fit seal to housing before fitting to shaft.) Never let weight of unsupported shaft rest in seal.
7. If correct service tool is not available, use a suitable drift approximately 0.4 mm (0.015 in) smaller than outside diameter of seal. Use a hammer VERY GENTLY on drift if a press is not suitable.
8. Press or drift seal in to depth of housing if housing is shouldered, or flush with face of housing where no shoulder is provided. Ensure that the seal does not enter the housing in a tilted position.

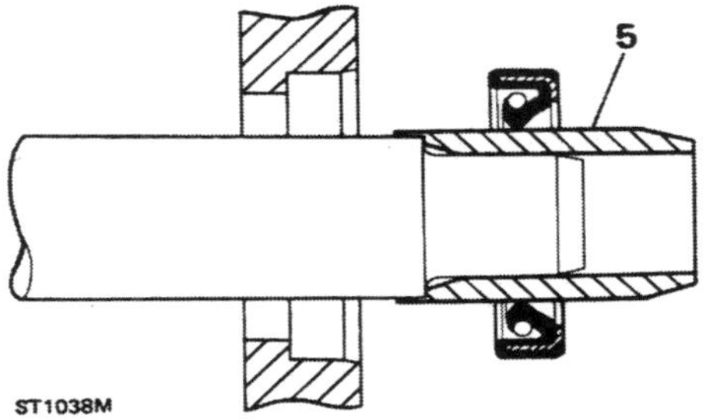

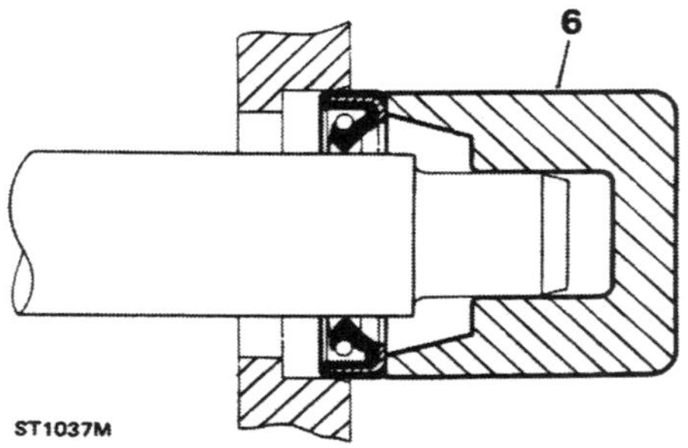

**NOTE: Most cases of failure or leakage of oil seals are due to careless fitting, and resulting damage to both seals and sealing surfaces. Care in fitting is essential if good results are to be obtained.**

### FLEXIBLE HYDRAULIC PIPES, HOSES

1. Before removing any brake or power steering hose, clean end fittings and area surrounding them as thoroughly as possible.
2. Obtain appropriate blanking caps before detaching hose end fittings, so that ports can be immediately covered to exclude dirt.
3. Clean hose externally and blow through with airline. Examine carefully for cracks, separation of plies, security of end fittings and external damage. Reject any hose found faulty.
4. When refitting hose, ensure that no unnecessary bends are introduced, and that hose is not twisted before or during tightening of union nuts.
5. Containers for hydraulic fluid must be kept absolutely clean.
6. Do not store hydraulic fluid in an unsealed container. It will absorb water, and fluid in this condition would be dangerous to use due to a lowering of its boiling point.
7. Do not allow hydraulic fluid to be contaminated with mineral oil, or use a container which has previously contained mineral oil.
8. Do not re-use fluid bled from system.
9. Always use clean brake fluid to clean hydraulic components.
10. Fit a blanking cap to a hydraulic union and a plug to its socket after removal to prevent ingress of dirt.
11. Absolute cleanliness must be observed with hydraulic components at all times.
12. After any work on hydraulic systems, inspect carefully for leaks underneath the vehicle while a second operator applies maximum pressure to the brakes (engine running) and operates the steering.

### METRIC BOLT IDENTIFICATION

1. An ISO metric bolt or screw, made of steel and larger than 6 mm in diameter can be identified by either of the symbols ISO M or M embossed or indented on top of the head.
2. In addition to marks to identify the manufacture, the head is also marked with symbols to indicate the strength grade, e.g. 8.8, 10.9, 12.9 or 14.9, where the first figure gives the minimum tensile strength of the bolt material in tens of kg/sq mm.
3. Zinc plated ISO metric bolts and nuts are chromate passivated, a greenish-khaki to gold-bronze colour.

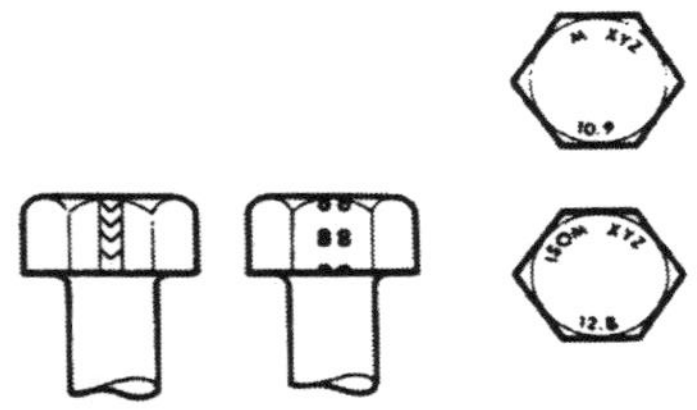

### JOINTS AND JOINT FACES

1. Always use correct gaskets where they are specified.
2. Use jointing compound only when recommended. Otherwise fit joints dry.
3. When jointing compound is used, apply in a thin uniform film to metal surfaces; take great care to prevent it from entering oilways, pipes or blind tapped holes.
4. Remove all traces of old jointing materials prior to reassembly. Do not use a tool which could damage joint faces.
5. Inspect joint faces for scratches or burrs and remove with a fine file or oil stone; do not allow swarf or dirt to enter tapped holes or enclosed parts.
6. Blow out any pipes, channels or crevices with compressed air, renewing any O-rings or seals displaced by air blast.

# GENERAL SPECIFICATION DATA

## GENERAL FITTING INSTRUCTIONS

### METRIC NUT IDENTIFICATION

1. A nut with an ISO metric thread is marked on one face or on one of the flats of the hexagon with the strength grade symbol 8, 12 or 14. Some nuts with a strength 4, 5 or 6 are also marked and some have the metric symbol M on the flat opposite the strength grade marking.
2. A clock face system is used as an alternative method of indicating the strength grade. The external chamfers or a face of the nut is marked in a position relative to the appropriate hour mark on a clock face to indicate the strength grade.
3. A dot is used to locate the 12 o'clock position and a dash to indicate the strength grade. If the grade is above 12, two dots identify the 12 o'clock position.

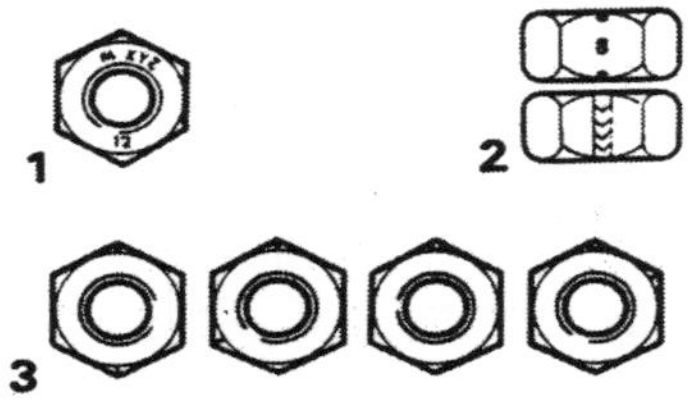

### NUTS

1. When tightening a slotted or castellated nut never slacken it back to insert split pin or locking wire except in those recommended cases where this forms part of an adjustment. If difficulty is experienced, alternative washers or nuts should be selected, or washer thickness reduced.
2. Where self-locking nuts have been removed it is advisable to replace them with new ones of the same type.

NOTE: Where bearing pre-load is involved nuts should be tightened in accordance with special instructions.

### KEYS AND KEYWAYS

1. Remove burrs from edges of keyways with a fine file and clean thoroughly before attempting to refit key.
2. Clean and inspect key closely; keys are suitable for refitting only if indistinguishable from new, as any indentation may indicate the onset of wear.

### TAB WASHERS

1. Fit new washers in all places where they are used. Always renew a used tab washer.
2. Ensure that the new tab washer is of the same design as that replaced.

### SPLIT PINS

1. Fit new split pins throughout when replacing any unit.
2. Always fit split pins where split pins were originally used. Do not substitute spring washers: there is always a good reason for the use of a split pin.
3. All split pins should be fitted as shown unless otherwise stated.

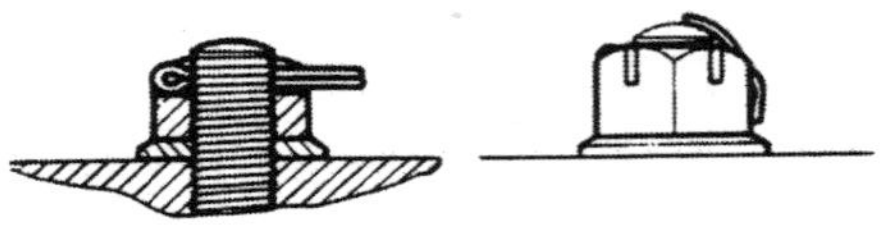

### LOCKING WIRE

1. Fit new locking wire of the correct type for all assemblies incorporating it.
2. Arrange wire so that its tension tends to tighten the bolt heads, or nuts, to which it is fitted.

### SCREW THREADS

1. Both UNF and Metric threads to ISO standards are used. See below for thread identification.
2. Damaged threads must always be discarded. Cleaning up threads with a die or tap impairs the strength and closeness of fit of the threads and is not recommended.
3. Always ensure that replacement bolts are at least equal in strength to those replaced.
4. Do not allow oil, grease or jointing compound to enter blind threaded holes. The hydraulic action on screwing in the bolt or stud could split the housing.
5. Always tighten a nut or bolt to the recommended torque figure. Damaged or corroded threads can affect the torque reading.
6. To check or re-tighten a bolt or screw to a specified torque figure, first slacken a quarter of a turn, then re-tighten to the correct figure.
7. Always oil thread lightly before tightening to ensure a free running thread, except in the case of self-locking nuts.

### UNIFIED THREAD IDENTIFICATION

1. **Bolts**
   A circular recess is stamped in the upper surface of the bolt head.
2. **Nuts**
   A continuous line of circles is indented on one of the flats of the hexagon, parallel to the axis of the nut.
3. **Studs, Brake Rods, etc.**
   The component is reduced to the core diameter for a short length at its extremity.

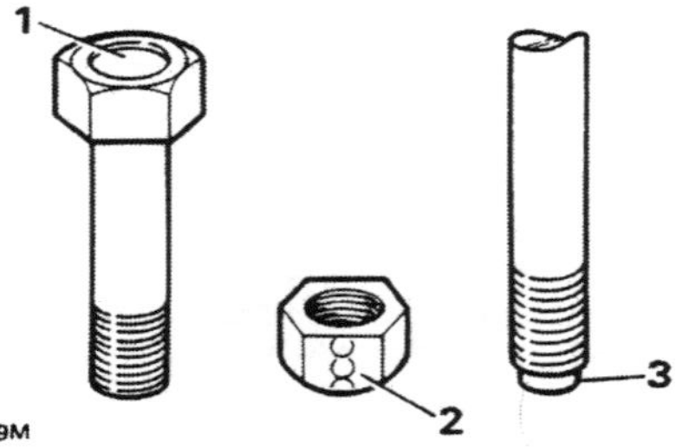

## LUBRICANTS, FLUIDS AND CAPACITIES

### RECOMMENDED LUBRICANTS, FLUIDS AND CAPACITIES

These recommendations apply to temperate climates where operational temperatures may vary between -10C (14F) and 35C (95F)

| COMPONENTS | BP | CASTROL | DUCKHAMS | ESSO | MOBIL | PETROFINA | SHELL | TEXACO |
|---|---|---|---|---|---|---|---|---|
| Engine V8 Sump Dashpots (Carburetter models only) | BP Super Viscostatic (20/50) or BP VF7 (10W/30) | Castrol GTX (15W/50) or Castrolite (10W/40) | Duckhams 15W/50 Hypergrade Motor Oil | Esso Superlube (15W/40) | Mobil Super 10W/40 or Mobil 1 Rally Formula 5W/50 | Fina Supergrade Motor Oil 15W/40 10W/40 | Shell Super Motor Oil 15W/40 or 10W/40 | Havoline Motor Oil 15W/40 or Eurotex HC (10W/30) |
| LT77—five-speed gearbox ZF4HP22 Automatic Gearbox | BP Autran DX2D | Castrol TQ Dexron IID | Duckhams Fleetmatic CD or Duckhams D-Matic | Esso ATF Dexron IID | Mobil ATF 220D | Fina Dexron IID | Shell ATF Dexron IID | Texamatic Fluid 9226 |
| Front differential Rear differential Swivel pin housings and LT230T Transfer box* | BP Gear Oil SAE 90EP | Castrol Hypoy SAE 90EP | Duckhams Hypoid 90 | Esso Gear Oil GX 85W/90 | Mobil Mobilube HD90 | Fina Pontonic MP SAE 80W/90 | Shell Spirax 90 EP | Texaco Multigear Lubricant EP 85W/90 |
| Brake and clutch reservoirs | Brake Fluids having a minimum boiling point of 260°C (500°F) and complying with FMVSS 116 DOT3 or DOT4 ||||||||
| Prop. shaft Front and rear | BP Energrease L2 | Castrol LM Grease | Duckhams LB 10 | Esso Multi-purpose Grease H | Mobilgrease MP | Fina Marson HTL 2 | Shell Retinax A | Marfak All purpose Grease |
| Power steering box and fluid reservoir as applicable | BP Autran DX2D | Castrol TQ Dexron IID | Duckhams Fleetmatic CD or Duckhams D-Matic | Esso ATF Dexron IID | Mobil ATF 220D | Fina Dexron II | Shell ATF Dexron IID | Texamatic Fluid 9226 |
| Lubrication nipples (hubs, ball joints, etc.) | BP Energrease L2 | Castrol LM Grease | Duckhams LB 10 | Esso Multi-purpose Grease H | Mobilgrease MP | Fina Marson HTL 2 | Shell Retinax A | Marfak All purpose Grease |
| Ball joint assembly Top link | Dextragrease Super GP ||||||||
| Seat slides Door lock striker | BP Energrease L2 | Castrol LM Grease | Duckhams LB 10 | Esso Multi-purpose Grease H | Mobil Mobilgrease MP | Fina Marson HTL2 | Shell Retinax A | Marfak All purpose Grease |
| Engine cooling system | Use an ethylene glycol based anti-freeze (containing no methanol) with non-phosphate corrosion inhibitors suitable for use in aluminium engines to ensure the protection of the cooling system against frost and corrosion in all seasons. Use one part anti-freeze to one part water for protection down to –36°C (–65°F). IMPORTANT: Coolant solution must not fall below proportions one part anti-freeze to three parts water, i.e. minimum 25% anti-freeze in coolant otherwise damage to engine is liable to occur. When anti-freeze is not required, the cooling system must be flushed out with clean water and refilled with a solution of one part Marstons SQ36 inhibitor to nine parts water, i.e. minimum 10% inhibitor in coolant ||||||||

* Engine oil or gearbox oil or a mixture of both may be used as an alternative to the gear oil specified for the transfer box.

# LUBRICANTS, FLUIDS AND CAPACITIES

## RECOMMENDED LUBRICANTS AND FLUIDS  SERVICE INSTRUCTIONS ALL MARKETS

| COMPONENTS | BP | CASTROL | DUCKHAMS | ESSO | MOBIL | PETROFINA | SHELL | TEXACO | SPEC. REF. ALL BRANDS |
|---|---|---|---|---|---|---|---|---|
| Seat slides Door lock striker | BP Energrease L2 | Castrol LM Grease | Duckhams LB 10 | Esso Multi-purpose Grease H | Mobil Mobil-grease MP | Fina Marson HTL2 | Shell Retinax A | Marfak All purpose Grease | NGLI-2 Multi-purpose Lithium based Grease |
| Windscreen Washers | Screen Washer Fluid | | | | | | | | |
| Bonnet pintle | Graphite Lock Grease Type 'B' | | | | | | | | |
| Door locks (anti-burst) Inertia reels | DO NOT LUBRICATE. These components are 'life' lubricated at the manufacturing stage | | | | | | | | |
| | NOTE: The above lubricants are considered to be suitable for ambient temperatures in the range of −40°C to +35°C. For extreme ambient temperatures, outside the above range, refer to local Distributor. | | | | | | | | |
| Battery lugs Earthing surfaces Where paint has been removed | Petroleum jelly. NOTE: Do not use Silicone Grease | | | | | | | | |
| Air Conditioning System Refrigerant | METHYL CHLORIDE REFRIGERANTS MUST NOT BE USED Use only with refrigerant 12. This includes 'Freon 12' and 'Arcton 12' | | | | | | | | |
| Compressor Oil | Shell Clavus 68   BP Energol LPT 68   Sunisco 4GS   Texaco Capella E Wax Free 68 | | | | | | | | |

## RECOMMENDED LUBRICANTS AND FLUIDS
### SERVICE INSTRUCTIONS FOR AMBIENT CONDITIONS OUTSIDE TEMPERATE CLIMATE LIMITS OR FOR MARKETS WHERE THE PRODUCTS LISTED ARE NOT AVAILABLE

| | Service Classification | | Ambient Temperature °C |
|---|---|---|---|
| | Specification | SAE Classification | −30 −20 −10 0 10 20 30 40 50 |
| Engine sump Dashpots (carburetter models only) oil can | Oils must meet BLS.22.OL.07 or CCMC G3 or API service levels SF | 5W/30 5W/40 5W/50 10W/30 10W/40 10W/50 | |
| | Oils must meet BLS.22.OL.02 or CCMC G1 or G2 or API service levels SE or SF | 15W/40 15W/50 20W/40 20W/50 25W/40 25W/50 | |
| Main gearbox, manual or automatic | ATF Dexron IID | | |
| Transfer gearbox Final drive units Swivel pin housings Steering box | API GL4 or GL5 MIL-L-2105 or MIL-L-2105B | 90 EP 80W EP | |
| Power steering | ATF M2C 33 (F or G) ATF Dexron IID | | |
| Brake and clutch reservoirs | Brake fluid must have a minimum boiling point of 260°C (500°F) and comply with FMVSS 116 DOT 3 | | |
| Lubrication nipples (hubs, ball joints, etc.) | NLGI-2 multipurpose lithium based grease | | |
| Engine cooling system | Use an ethylene glycol based anti-freeze (containing no methanol) with non-phosphate corrosion inhibitors suitable for use in aluminium engines to ensure the protection of the cooling system against frost and corrosion in all seasons. Use one part anti-freeze to one part water for protection down to −36°C (−65°F). IMPORTANT: Coolant solution must not fall below proportions one part anti-freeze to three parts water, i.e. minimum 25% anti-freeze in coolant otherwise damage to engine is liable to occur. When anti-freeze is not required, the cooling system must be flushed out with clean water and refilled with a solution of one part Marstons SQ36 inhibitor to nine parts water, i.e. minimum 10% inhibitor in coolant. | | |

## 09 LUBRICANTS, FLUIDS AND CAPACITIES

### ANTI-FREEZE

| ENGINE | MIXTURE | PERCENTAGE CONCENTRATION | PROTECTION |
|---|---|---|---|
| V8 (aluminium) | One part anti-freeze<br>One part water | 50% | −33°C<br>−36°C<br>−41°C<br>−42°F<br>−47°C<br>−53°F |

**Complete protection**
Vehicle may be driven away immediately from cold

**Safe limit protection**
Coolant in mushy state. Engine may be started and driven away after short warm-up period

**Lower protection**
Prevents frost damage to cylinder head, block and radiator. Thaw out before starting engine

### RECOMMENDED FUEL

With the exception of the 9.35:1 high compression (emission) engine which is designed to operate on 97 octane fuel (British 4-star rating) all other Range Rover engines are designed for fuel having a minimum octane rating of 91 to 93 (the British 2-star rating).

Where these fuels are not available and it is necessary to use fuels of lower or unknown rating, the ignition timing must be retarded from the specified setting, just sufficiently to prevent audible detonation (pinking) under all operating conditions, otherwise damage to the engine may occur. Use exhaust gas analysis equipment to check the final engine exhaust emissions after resetting. (See 'Engine Tuning' data).

The use of lower octane fuels will result in the loss of engine power and efficiency.

**CAUTION:** Do not use oxygenated fuels such as blends of methanol/gasolene or ethanol/gasoline (e.g. 'GASOHOL').

In the interests of public health, and to assist in keeping undesirable exhaust emissions as low as possible, fuels of an octane rating higher than that recommended should not be used.

**NOTE:** Vehicles fitted with 8.13 :1 low compression engines will operate on unleaded fuel with an octane rating of 91 octane without any engine tune adjustments.

**Capacities (approx.)*** 

| | Litres | Imperial unit | US unit |
|---|---|---|---|
| Engine sump and filter from dry | 5.66 | 10 pints | 12.8 pints |
| Gearbox from dry—manual LT77 | 2.7 | 4.7 pints | 5.9 pints |
| Gearbox from dry—automatic ZF | 9.1 | 16 pints | 20 pints |
| Transfer gearbox from dry | 2.5 | 4.4 pints | 5.5 pints |
| Front axle from dry | 1.7 | 3.0 pints | 3.75 pints |
| Front axle swivel pin housing (each) | 0.35 | 0.6 pints | 0.75 pints |
| Rear axle from dry | 1.7 | 3.0 pints | 3.75 pints |
| Power steering box and reservoir | 2.9 | 5.0 pints | 6.25 pints |
| Cooling system | 11.4 | 20 pints | 24 pints |
| Fuel tank | 80.0 | 17.5 gallons | 22 gallons |

**NOTE:** * All levels must be checked by dipstick or level plugs as applicable.

When draining oil from the ZF automatic gearbox, oil will remain in the torque converter, refill to high level on dipstick only.

---

## MAINTENANCE 10

### MAINTENANCE SCHEDULES

Efficient maintenance is one of the biggest factors in ensuring continuing reliability and efficiency. For this reason the following detailed schedules are included so that at the appropriate intervals owners and operators may know what is required. The Maintenance Schedules are based upon intervals of 10,000 km (6,000 miles) or 6 months unless otherwise stated.

A = 10,000 km (6,000 miles)  B = 20,000 km (12,000 miles)
30,000 km (18,000 miles)   40,000 km (24,000 miles)
50,000 km (30,000 miles)   60,000 km (36,000 miles)
70,000 km (42,000 miles)   80,000 km (48,000 miles)

CORRECT = ☑   INCORRECT = ☒

| No. | A | B | Operations |
|---|---|---|---|
| 1 | ☐ | | Check condition and security of seats, seat belt mountings, seat belts and buckles |
| 2 | ☐ | | Check operation of footbrake and clutch with engine running; stop engine |
| 3 | ☐ | | Check operation of; all lamps, horns, warning indicators |
| 4 | ☐ | | Check operation of front/rear screen wipers and washers and condition of wiper blades |
| 5 | ☐ | ☐ | Check security and operation of hand brake; release fully after checking |
| 6 | ☐ | | Check rear view mirrors for security, cracks and crazing |
| 7 | ☐ | | Remove road wheels |
| 8 | ☐ | ☐ | Check tyres for: compliance with manufacturers specification; visually for cuts, lumps, bulges, uneven tread wear and depth; tyre pressures (including spare) adjust if required—see Owners Manual |
| 9 | ☐ | | Inspect brake pads for wear, calipers for leaks and discs for condition |
| 10 | ☐ | | Check for oil/fluid leaks from steering and suspension systems |
| 11 | ☐ | | Check condition and security of steering, joints and gaiters |
| 12 | ☐ | | Refit road wheels to original position |
| 13 | ☐ | | Drain flywheel housing if drain plug is fitted for wading (refit) |
| 14 | ☐ | | Check/top up engine oil |
| 15 | ☐ | | Renew engine oil and filter |
| 16 | ☐ | | Check/top up automatic transmission fluid |
| 17 | 40,000 km (24,000 miles) | | Renew automatic transmission fluid and filter |
| 18 | ☐ | | Check/top up gearbox oil |
| 19 | ☐ | | Renew gearbox oil |
| 20 | ☐ | | Check/top up transfer box oil |
| 21 | ☐ | | Renew transfer box oil |
| 22 | 40,000 km (24,000 miles) | | Check/top up front axle oil |
| 23 | ☐ | | Renew front axle oil |
| 24 | 40,000 km (24,000 miles) | | Check/top up swivel pin housing oil |
| 25 | 40,000 km (24,000 miles) | | Renew swivel pin housing oil |
| 26 | 40,000 km (24,000 miles) | | Check/top up rear axle oil |
| 27 | | | Renew rear axle oil |
| 28 | ☐ | | Check visually brake, fuel, clutch pipes/unions for chafing, leaks and corrosion |
| 29 | ☐ | ☐ | Check exhaust system for leakage and security |
| 30 | ☐ | | Lubricate hand brake mechanical linkage and adjust to manufacturers instructions if required |
| 31 | | | Lubricate propeller shaft universal joints |
| 32 | 40,000 km (24,000 miles) | | Lubricate propeller shaft sliding joints |
| 33 | ☐ | | Check tightness of propeller shaft coupling bolts |
| 34 | ☐ | | Ensure front and rear axle breathers are free from obstruction |
| 35 | ☐ | | Check security and condition of suspension fixings |
| 36 | ☐ | | Check for oil leaks from engine and transmission |
| 37 | 80,000 km (48,000 miles) | | Clean fuel pump filter |
| 38 | 80,000 km (48,000 miles) | | Check suspension self levelling unit for leaks |
| 39 | ☐ | | Renew fuel filter (EFi engines) |
| 40 | ☐ | | Renew fuel filter element (carburetter engines) |
| 41 | ☐ | | Renew air cleaner element(s) |
| 42 | ☐ | | Check air cleaner dump valve, clean or renew |
| 43 | ☐ | | Renew engine breather filter |
| 44 | ☐ | | Clean or renew engine flame trap(s) |
| 45 | ☐ | | Check condition of driving belts—adjust if required |
| 46 | 80,000 km (48,000 miles) | | Renew charcoal canister |
| 47 | ☐ | | Check crankcase breathing system for leaks, hoses for security and condition |
| 48 | ☐ | | Clean/adjust spark plugs |
| 49 | ☐ | | Renew spark plugs |
| 50 | ☐ | | Top up carburetter piston dampers |
| 51 | ☐ | | Check/top up cooling system |
| 52 | ☐ | | Check brake servo hose for security and condition |
| 53 | ☐ | | Check ignition wiring and H.T. leads for fraying, chafing and deterioration |
| 54 | ☐ | | Clean distributor cap, check for cracks and tracking |
| 55 | ☐ | | Check/adjust ignition timing (refer to the repair operation manual for details) |
| 56 | ☐ | | Lubricate accelerator control linkages and pedal pivot |

*Continued*

# MAINTENANCE 10

| No. | A | B | Operations |
|---|---|---|---|
| 57 | ☐ | ☐ | Check throttle operation |
| 58 | ☐ | ☐ | Lubricate all locks (not steering lock), hinges and door—check mechanisms |
| 59 | ☐ | ☐ | Check operation of all doors, bonnet and tailgate locks |
| 60 | ☐ | ☐ | Check/adjust carburetter mixture settings, fuel injection idle air mixture and engine idle speed with engine at normal running temperature |
| 61 | ☐ | ☐ | Check operation of air intake temperature control system |
| 62 | ☐ | ☐ | Check/adjust steering box |
| 63 | ☐ | ☐ | Check power steering system for leaks, hydraulic pipes and unions for chafing and corrosion |
| 64 | ☐ | ☐ | Check/top up fluid in power steering reservoir |
| 65 | ☐ | ☐ | Check/top up clutch fluid reservoir |
| 66 | ☐ | ☐ | Check/top up brake fluid reservoir |
| 67 | ☐ | ☐ | Check/top up windscreen washer reservoir |
| 68 | ☐ | ☐ | Check cooling and heater system for leaks, hoses for security and condition |
| 69 | ☐ | ☐ | Check/top up battery electrolyte |
| 70 | ☐ | ☐ | Remove battery connections; clean and grease—refit |
| 71 | ☐ | ☐ | Check/adjust headlamp alignment |
| 72 | ☐ | ☐ | Check front wheel alignment |

| No. | A | B | Operations |
|---|---|---|---|
| 73 | ☐ | ☐ | Road test—Check: |
|  | ☐ | ☐ | And ensure automatic gearbox starter/isolator switch will only operate in 'P' and 'N'. |
|  | ☐ | ☐ | For excessive engine noise |
|  | ☐ | ☐ | Clutch for slipping/judder/spinning |
|  | ☐ | ☐ | Gear selection/noise—high and low range |
|  | ☐ | ☐ | Automatic gear selection/shift speeds |
|  | ☐ | ☐ | Steering for noise/abnormal effort required |
|  | ☐ | ☐ | All instruments, pressure, fuel and temperature gauges, warning indicators |
|  | ☐ | ☐ | Heater and air conditioning systems |
|  | ☐ | ☐ | Heated rear screen |
|  | ☐ | ☐ | Shock absorbers (irregularities in ride) |
|  | ☐ |  | Footbrake, on emergency stop, pulling to one side, binding, pedal effort |
|  | ☐ | ☐ | Handbrake efficiency |
|  | ☐ | ☐ | Fully extend seat belt, check for correct operation of retraction and latching. Inertia belts lock when snatched and when car is on slope |
|  | ☐ | ☐ | Road wheel balance |
|  | ☐ | ☐ | Transmission for vibrations |
|  | ☐ | ☐ | For body noises (squeaks and rattles) |
| 74 | ☐ | ☐ | Endorse service record |
| 75 | ☐ | ☐ | Report any additional work required |

It is recommended that:
At 30,000 km (18,000 mile) intervals or every 18 months, whichever is the sooner, the hydraulic brake fluid should be completely renewed.

At 60,000 km (36,000 mile) intervals or every 3 years, whichever is the sooner, all hydraulic brake fluid, seals and flexible hoses should be renewed, all working surfaces of the master cylinder, wheel cylinders and caliper cylinders should be examined and renewed where necessary.

At 60,000 km (36,000 mile) intervals remove all suspension dampers, test for correct operation, refit or renew as necessary.

## MAINTENANCE SUMMARY

**The following should be checked weekly or before a long journey.**

| | |
|---|---|
| Engine oil level | Windscreen/tailgate/headlamp washer. Reservoir level(s) |
| Brake fluid level | Battery electrolyte level(s) |
| Radiator coolant level | All tyres for pressure and condition |
| | Operation of horn |
| | Operation of washers and wipers |
| | Operation of all lights |

---

# MAINTENANCE 10

## LUBRICATION

This first part of the maintenance section covers renewal of lubricating oils for the major units of the vehicle and other components that require lubrication, as detailed in the 'Maintenance Schedules'. Refer to the 'General Specification Data' for capacities and recommended lubricants.

Vehicles operating under severe conditions of dust, sand, mud and water should have the oils changed and lubrication carried out at more frequent intervals than that recommended in the maintenance schedules.

Draining of used oil should take place after a run when the oil is warm. Always clean the drain and filler-level plugs before removing. In the interests of safety disconnect the vehicle battery to prevent the engine being started and the vehicle moved inadvertently, while oil changing is taking place.

Allow as much time as possible for the oil to drain completely except where blown sand or dirt can enter the drain holes. In these conditions clean and refit the drain plugs immediately while the main bulk of oil has drained. Where possible, always refill with oil of the make and specification recommended in the lubrication charts and from sealed containers.

## RENEW ENGINE OIL AND FILTER

### DRAIN THE OIL

1. Drive vehicle to level ground.
2. Run the engine to warm the oil; switch off the ignition and disconnect the battery for safety.
3. Place an oil tray under the drain plug.
4. Remove the drain plug in the bottom of the sump at the left-hand side. Allow oil to drain away completely and replace the plug and tighten to the correct torque.

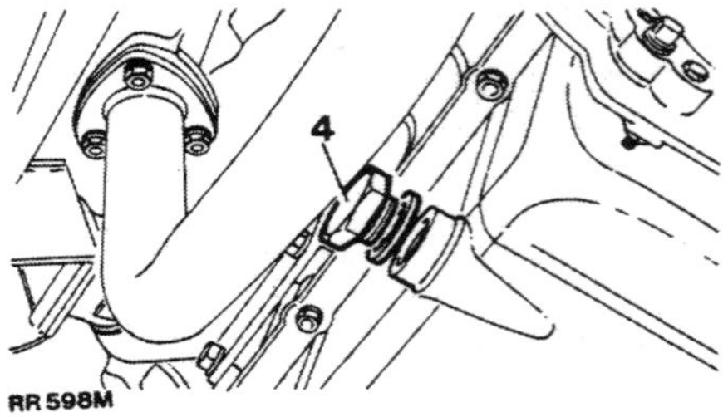

### Renew oil filter

5. Place an oil tray under the engine.
6. Unscrew the filter anti-clockwise, using a strap spanner as necessary.
7. Smear a little clean engine oil on the rubber washer of the new filter, then screw the filter on clockwise until the rubber sealing ring touches the machined face, then tighten a further half turn by hand only. Do not overtighten.

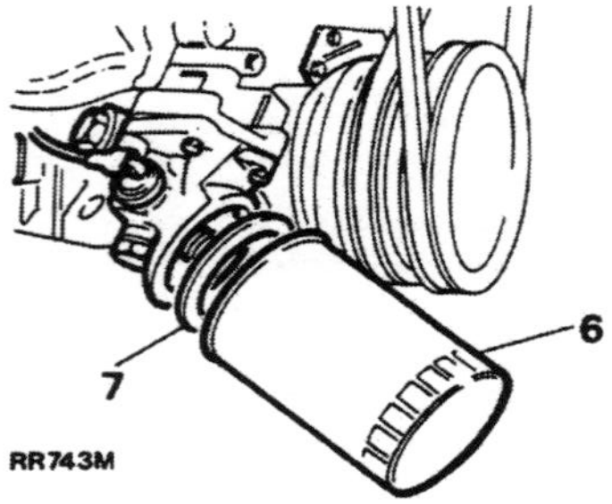

### Refill sump with oil

8. Check that the drain plug is tight.
9. Clean the outside of the oil filler cap, remove it from the rocker cover or extension filler neck and clean the inside.
10. Pour in the correct quantity of new oil of the correct grade from a sealed container to the high mark on the dipstick and firmly replace the filler cap. DO NOT FILL ABOVE 'HIGH MARK'.
11. Run the engine and check for leaks from the filter. Stop the engine, allow the oil to run back into the sump for a few minutes, then check the oil level again and top up if necessary.

# MAINTENANCE

## RENEW MAIN AND TRANSFER GEARBOX OILS

### DRAIN AND REFILL 77 mm MAIN GEARBOX

1. Drive the vehicle to level ground and place a suitable container under the gearbox to catch the old oil.
2. Remove the gearbox and extension case drain plugs and allow the oil to drain completely. Wash the extension case filter in kerosene and refit the plugs using new washers, if necessary, and tighten to the correct torque.
3. Remove the oil filler-level plug and inject the approximate quantity of new oil of the correct make and grade until it begins to run out of the filler-level hole. Fit the plug and tighten to the correct torque. Since the plug has a tapered thread it must not be overtightened. Wipe away any surplus oil.

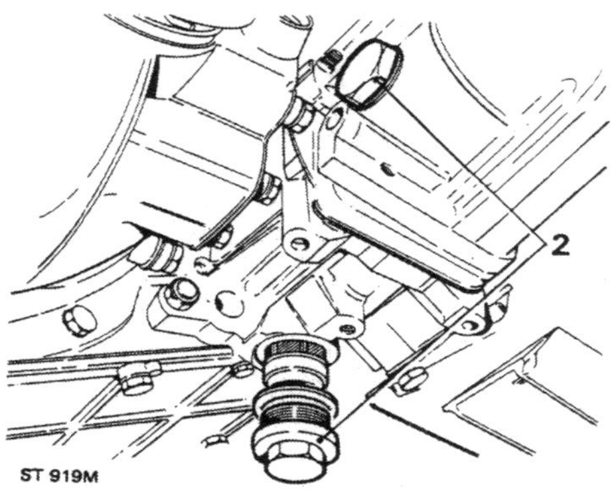

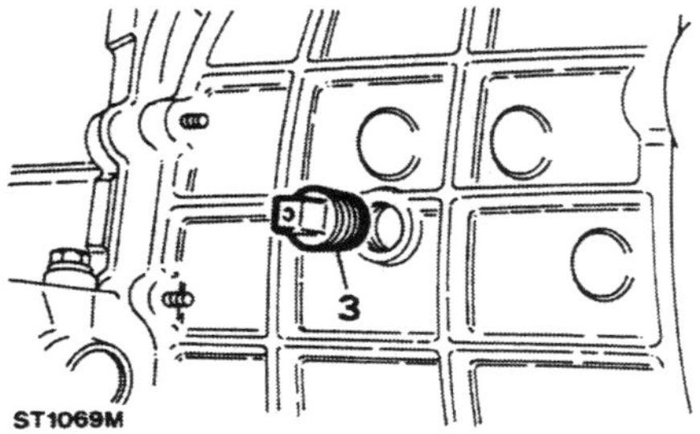

### DRAIN AND RENEW 230T TRANSFER GEARBOX

1. Drive the vehicle to level ground and place a container under the gearbox to catch the old oil.
2. Remove the drain plug and allow the oil to drain. Fit the plug using a new washer, if necessary, and tighten to the correct torque.
3. Remove the filler-level plug and inject the approximate quantity of the recommended oil until it begins to run from the plug hole. Fit the level plug and tighten only to the correct torque, do not overtighten, wipe away any surplus oil.

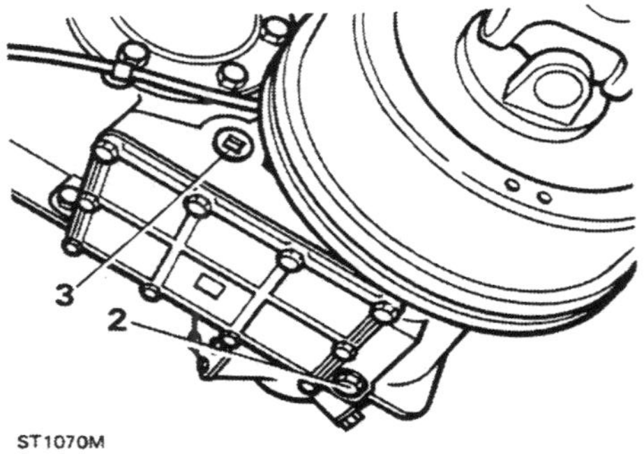

### RENEW ZF AUTOMATIC GEARBOX FLUID

1. Drive the vehicle onto a suitable hydraulic ramp. Obtain a suitable container to drain the gearbox fluid into.
2. Remove the drain plug from the bottom of the sump and allow time for the fluid to drain.
3. Refit the plug using a new sealing washer and tighten to the correct torque.

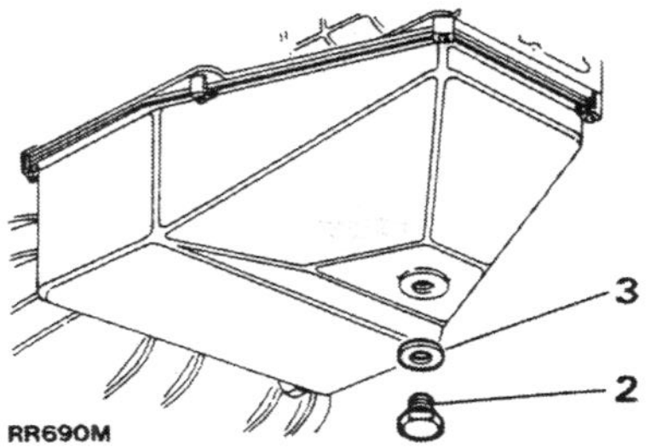

*Continued*

---

## REFILL ZF AUTOMATIC GEARBOX AND CHECK FLUID LEVEL

**NOTE: The fluid level in the ZF automatic gearbox is checked when the engine is at normal ambient temperature, and idling in neutral gear.**

4. Ensure the vehicle is on level ground.
5. Lift the bonnet and remove the gearbox dipstick located at the rear of the right-hand rocker cover.
6. Refill or top-up the fluid with the correct quantity and grade of fluid (see 'Data' section).
7. Ensure that the fluid level registers between the minimum and maximum level markings on the dipstick.

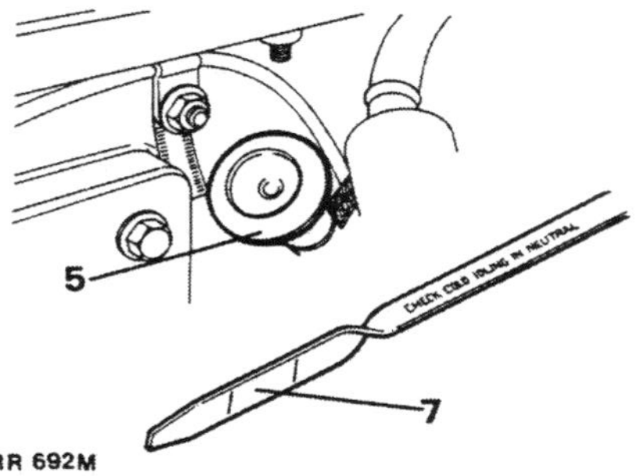

## OIL SCREEN REPLACEMENT ZF AUTOMATIC GEARBOX

### Removing

1. Place the vehicle on a ramp or over a pit, open the bonnet and disconnect the battery leads.
2. From underneath the vehicle drain the gearbox using a suitable container.
3. Discard the oil pan plug seal ring.
4. Remove the filler/level tube from the oil pan.
5. Remove the six retaining plates and bolts.
6. Remove the oil pan and discard the gasket.
7. Using TX27 torx bit undo the three screws which hold the oil screen.
8. Remove the oil screen and discard the 'O' rings.
9. Separate the oil screen from the suction tube and discard the 'O' ring and oil screen.

### Refitting

10. Fit two new 'O' rings to the oil screen using a light grease for ease of assembly.
11. Fit the suction tube to the oil screen.
12. Fit the oil screen to the control unit and secure with three bolts using TX27 torx bit tighten to the specified torque.
13. Refit the oil pan using a new gasket.

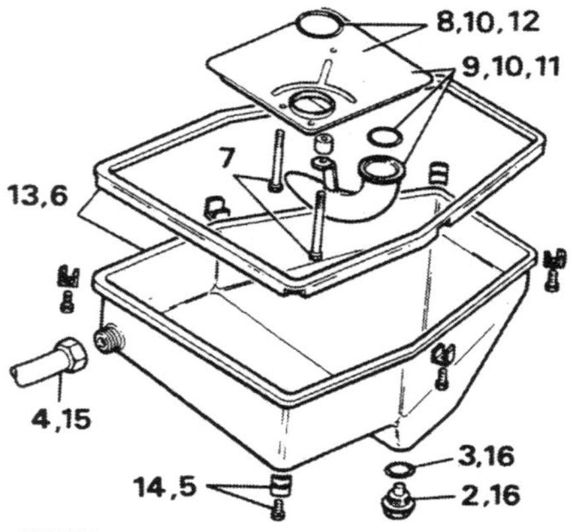

14. Secure using the six retaining plates and bolts (two straight and four corner plates), tighten to the specified torque.
15. Reconnect the oil level/filler tube.
16. Fit oil pan plug using a new seal.
17. Connect the battery leads.
18. Fill the gearbox with the correct oil through the filler/level tube located within the engine bay (see 'Data' section).
19. Ensuring the vehicle is on level ground with the handbrake applied, check oil level while engine is running at idle with neutral selected.

## POWER STEERING—FLUID RESERVOIR

1. Clean and remove the reservoir cap, observe the fluid level in relation to the level mark on the side of the reservoir.
2. If necessary top-up with a recommended fluid (see 'Data' section), until the fluid is 12 mm (½ in) above the filter.
3. Refit the cap.

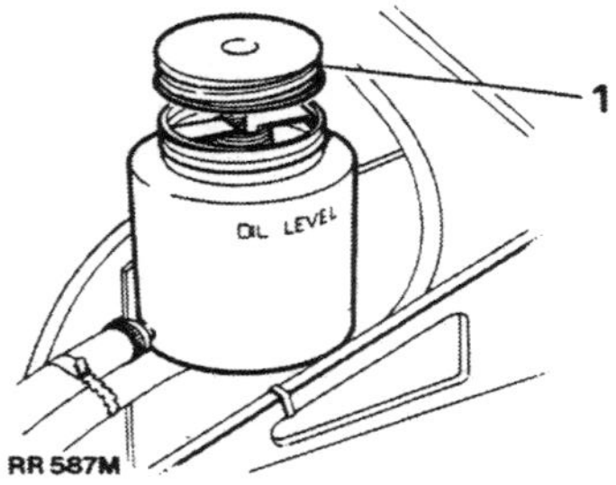

# MAINTENANCE

## RENEW FRONT AND REAR AXLE OIL

1. Drive the vehicle to level ground and place a container under the axle to be drained.
2. Using a spanner with a 13 mm (0.5 in) square drive remove the drain plug and allow the oil to drain completely. Clean and refit the drain plug.
3. Remove the oil filler-level plug and inject new oil of a recommended make and grade until it begins to run from the hole. Clean and fit the filler-level plug and wipe away any surplus oil.

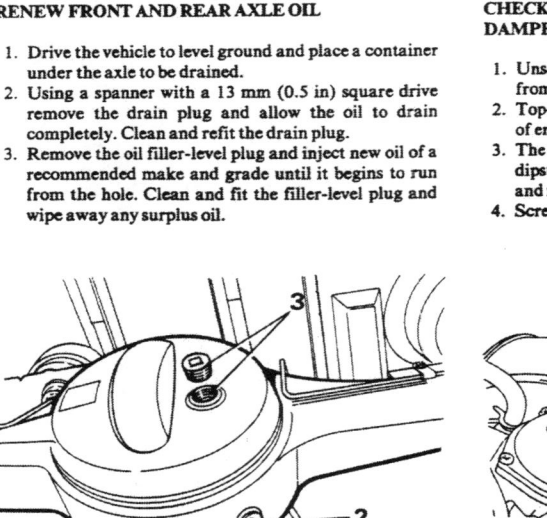

ST 922M

## RENEW SWIVEL PIN HOUSING OIL

1. Drive the vehicle to level ground and place a container under each swivel housing to catch the old oil.
2. Remove the drain plug and allow the oil to drain completely, clean and refit the plug.
3. Remove the level plug.
4. Remove the filler plug and inject the recommended make and grade of oil until oil begins to run from the level hole.
5. Clean and refit the level and filler plugs, wipe away any surplus oil.

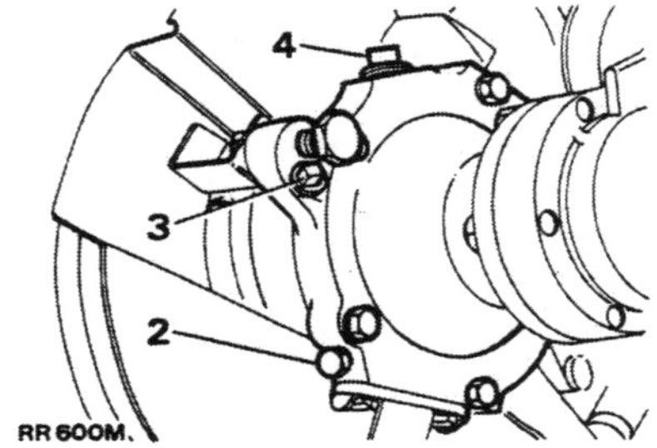

RR 600M.

## CHECK/TOP UP CARBURETTER PISTON DAMPERS

1. Unscrew and withdraw the plug and damper assembly from the top of each carburetter.
2. Top-up the damper chambers with the seasonal grade of engine oil.
3. The oil level is correct, when utilising the damper as a dipstick its threaded plug is 6 mm above the dash pots and resistance is felt.
4. Screw down the damper plugs.

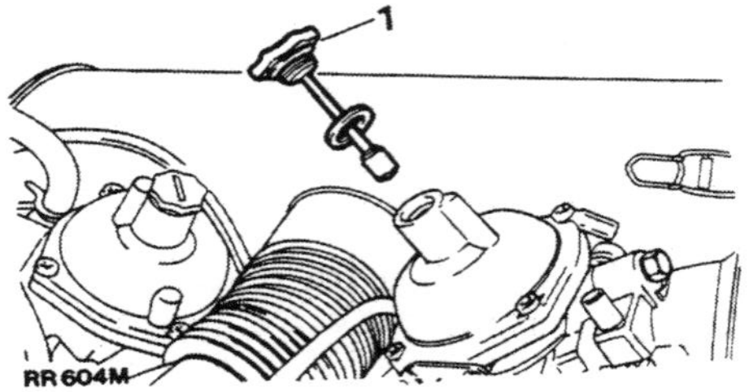

RR 604M

## LUBRICATE PROPELLER SHAFTS

1. Clean all grease nipples on the front and rear propshaft universal joints and sliding portion of the rear shaft.
2. Charge a low pressure hand-grease gun with grease of a recommended make and grade and apply to the grease nipples.

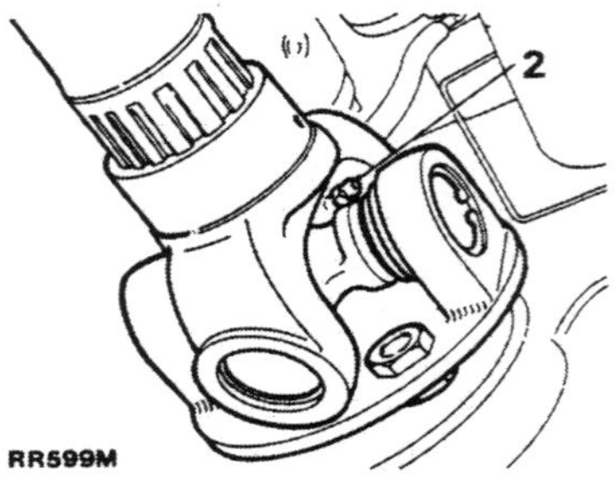

RR599M

*Continued*

## Lubricate the propeller shaft sliding joint

3. Disconnect one end of the front propeller shaft and compress the sliding portion whilst applying grease. It is necessary to compress the shaft to prevent overfilling with grease. It should be noted that this sliding joint must only be lubricated at 40,000 km (24,000 mile) intervals.

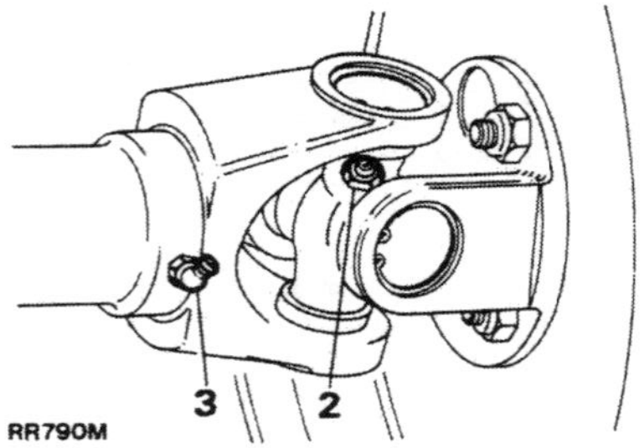

RR790M

## GENERAL MAINTENANCE AND ADJUSTMENT

## FILTERS

### RENEW THE AIR CLEANER ELEMENT — carburetter type

1. Slacken the clip retaining the advance/retard vacuum pipes from the air cleaner and release pipes from intake.
2. Slacken the clip retaining hose air cleaner to temperature sensing device from air intake and remove pipe from flap valve on air intake.
3. Slacken the hose clip attaching warm air intake hose to air intake.
4. Withdraw air intake from steady post and hoses.
5. Slacken the clips and remove the air cleaner elbows.
6. Emission pipes — slacken the clips and withdraw 'Pulsair' hoses.
7. Remove the air cleaner from the retaining posts by lifting and easing forward.
8. At the same time disconnect the hose engine breather filter to air cleaner. Place air cleaner to one side.

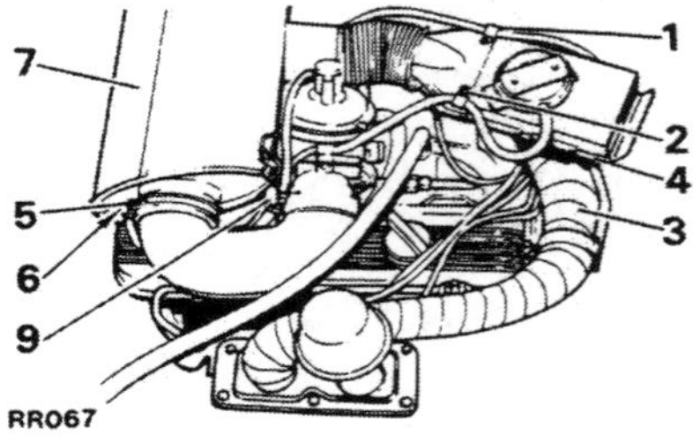

RR067

9. Remove hose with the non-return valve from the manifold.
   The air cleaner can now be completely removed.
10. Release the two clips at each side of air cleaner casing and withdraw the frames and elements.
11. Release wing nuts and withdraw plate and sealing washers, and discard old elements.
12. Discard any faulty rubber seals.
13. Assemble new elements into air cleaner frames and secure with seals, end plate and wing nuts.
14. Fit the carrier frames and elements into the air cleaner body and secure with clips.

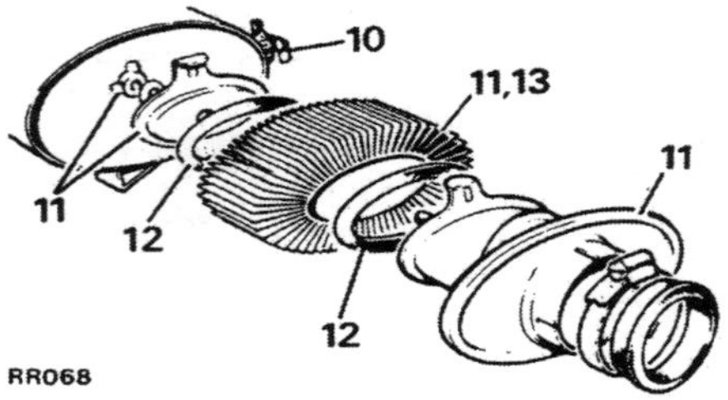

RR068

### Check air cleaner dump valve

15. Squeeze open the dump valve and check that the interior is clean. Also check that the rubber is flexible and in a good condition.
16. If necessary, remove the dump valve to clean the interior. Fit a new valve if the original is in a poor condition.

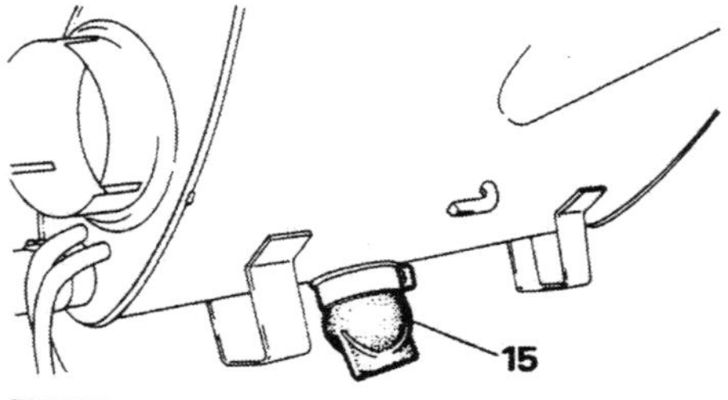

### Fit air cleaner

17. Reconnect hose with non-return valve to the manifold connection.
18. Place air cleaner on to the retaining posts.

RR597M

*Continued*

# MAINTENANCE

19. At the same time reconnect the breather filter hose to the base of the air cleaner.
20. Emission engines — reconnect and secure 'Pulsair' hoses.
21. Refit the air cleaner elbows and tighten clips.
22. Replace air intake on to steady post and reconnect air cleaner and warm air intake hoses. Tighten clip.
23. Reconnect pipe, air cleaner to temperature sensor and vacuum advance/retard pipe to air intake. Position pipes in retaining clips on air intake and tighten clips.

## RENEW THE AIR CLEANER ELEMENT — fuel injection type

1. Release the clip securing the hose to the rear of the air cleaner case.
2. Remove the two nuts and bolts securing the air cleaner retaining bracket to the left-hand valance. Remove air cleaner case from the hose, and remove from the vehicle.
3. Unclip the three catches securing the inlet tube to the air cleaner body and remove the inlet tube.
4. Remove the nut and end plate securing the air cleaner element in position.
5. Withdraw and discard the old element.

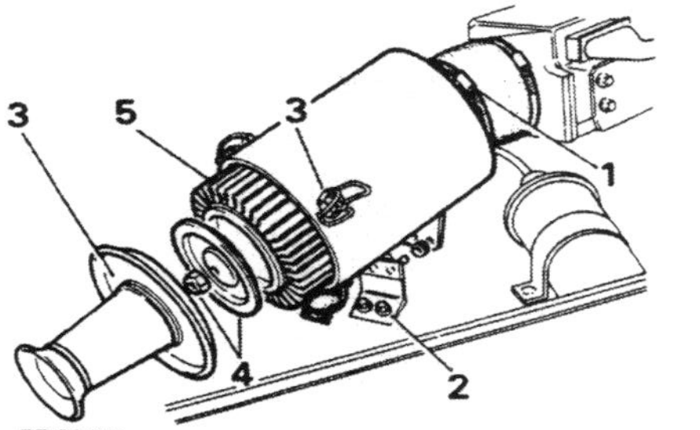

RR596M

6. Reverse the removal instructions 1 to 5 ensuring that all hose connections are secure.

## RENEW FUEL LINE FILTER — fuel injection type

WARNING: The spilling of fuel is unavoidable during this operation. Ensure that all necessary precautions are taken to prevent fire and explosion.

1. Depressurise the fuel system.
2. The fuel line filter is located on the right-hand chassis side member forward of the fuel tank filler neck. Access to the filter is gained through the right-hand rear wheel arch.
3. Clamp the inlet and outlet hoses to prevent the minimum of fuel spillage when disconnecting the hoses. (Denoted by the arrows).
4. Slacken the two securing clip screws nearest the filter to enable the hoses to be removed from the filter canister.
5. Remove the filter from the chassis side member by releasing the securing bolt and bracket.

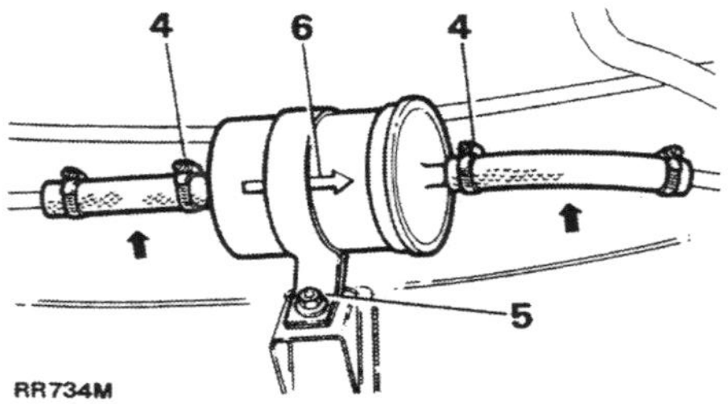

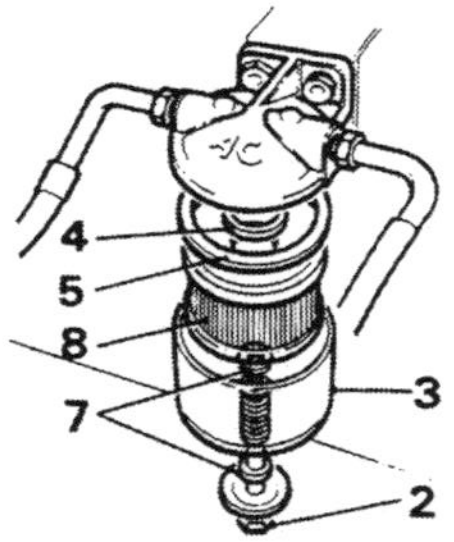

ST1075M

7. Ensure that the centre and top sealing rings are in good condition and replace as necessary.
8. Fit new element, small hole downwards.
9. Refit sealing rings (small and large).
10 Replace filter bowl and tighten the centre bolt.

## RENEW ADSORPTION CANISTER

1. Disconnect from the canister:
    (i) Canister line to fuel tank.
    (ii) Canister purge line.
    (iii) Carburetter vent pipe (blanked off on 'fuel injection models').
2. Slacken the clamp nut screw.
3. Remove the canister.

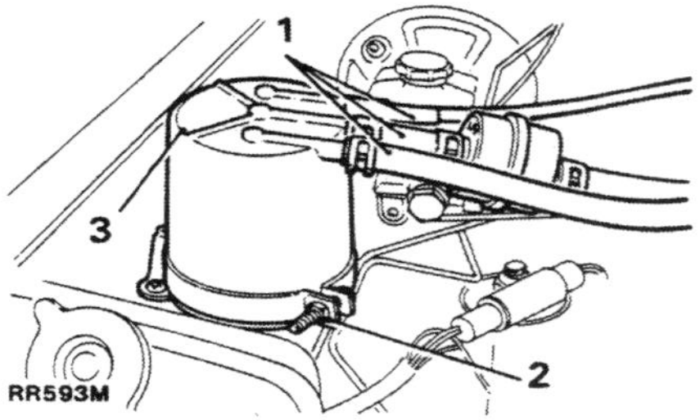

RR593M

### Fit new canister

4. Secure the canister in the clamp.
5. Reverse instructions 1 and 2 above.

WARNING: The use of compressed air to clean an adsorption canister or clear a blockage in the evaporative system is very dangerous. An explosive gas present in a fully saturated canister may be ignited by the heat generated when compressed air passes through the canister.

## RENEW CRANKCASE AIR INTAKE FILTER — fuel injection type

1. Prise the filter holder upwards to release it from the rocker cover.

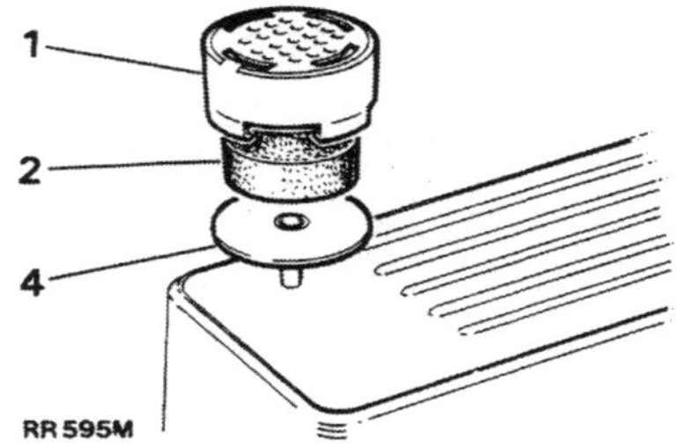

RR595M

2. Discard the sponge filter.

### Fit new filter

3. Insert a new filter into the plastic body.
4. Push filter holder onto the rocker cover until it clips firmly into place.

RR734M

### Fit new filter

6. Fit a new filter observing the direction of flow arrows stamped on the canister.
7. Start the engine and inspect for fuel leaks around the hose connections.

## RENEW FUEL FILTER ELEMENT — carburetter type

### Renew fuel filter element

The element provides a filter between the pump and carburetter and is located on the front LH wing.
Replace as follows:

1. Remove all dirt, grit, grease from around the filter body before dismantling.
2. Unscrew the centre bolt.
3. Withdraw the filter bowl.
4. Remove the small sealing ring and remove the element.
5. Withdraw the large sealing ring from the underside of the filter body.
6. Discard the old element and replace with a new unit.

*Continued*

# MAINTENANCE

## CLEAN CRANKCASE FLAME TRAP/BREATHER FILTER—fuel injection type

1. Release the hose clip and pull the hose off the canister.
2. Unscrew the canister and remove it from the rocker cover.
3. Remove the large 'O' ring from the screwed end of the canister.

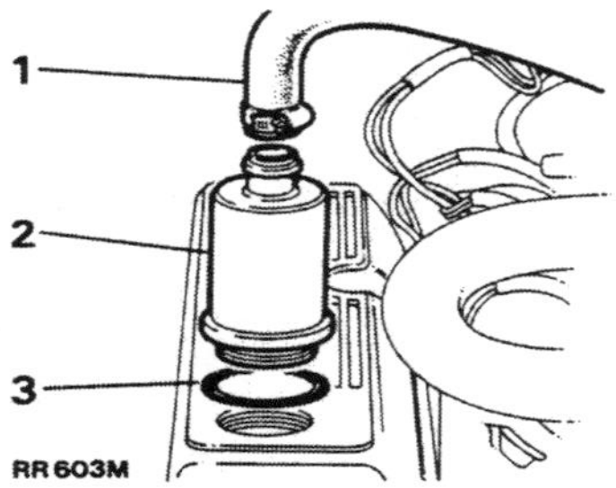

4. Visually inspect the condition of the wire gauze within the canister, if in poor condition renew the whole assembly, if in an acceptable condition clean the gauze as follows:
5. Immerse canister in a small amount of petrol and allow time for the petrol to dissolve and loosen any engine fume debris within the canister.
6. Remove canister from petrol bath and allow to dry out in still air.

**WARNING: Do not use a compressed air line to remove any remaining petrol or particles of debris within the canister as this could cause fire or personal injury.**

### Refitting the flame trap/breather

7. Fit a new rubber 'O' ring.
8. Screw the flame trap canister into the rocker cover, hand tight only.
9. Refit hose and tighten hose clip securely.

## CLEAN OR RENEW ENGINE FLAME TRAP(S)—carburetter type

1. Pull the flame trap hoses out of the retaining clips.
2. Pull the hoses from the flame trap.
3. Withdraw the flame trap.

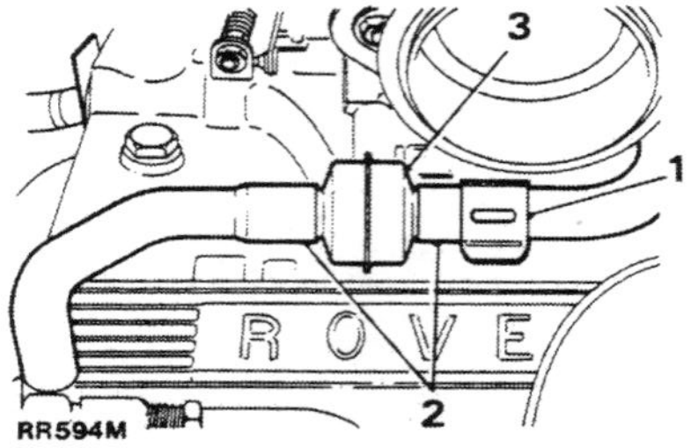

4. Visually inspect wire gauze inside the flame trap, if in poor condition renew the unit. If the gauze is in a satisfactory condition clean as follows:
5. Emerse flame traps in a small amount of petrol, allow time for the petrol to dissolve and loosen any debris within the flame trap.
6. Allow the flame traps to dry in still air.

**WARNING: Do not use a compressed air line to dry or clean the flame traps as this could cause fire or personal injury.**

### Fit engine flame traps

7. Push the hoses onto the flame trap and ensure that they are secure.
8. Locate the hoses in their respective retaining clips.

## CHECK/ADJUST OPERATION OF ALL WASHERS AND TOP-UP RESERVOIR

1. Check the operation of windscreen, tailgate and headlamp washers.
2. Adjust jets if necessary by inserting a needle or very fine sharp implement into the jet orifice and manoeuvring to alter the jet direction.
3. Unclip the reservoir cap.
4. Top up reservoir to within 25 mm (1 in) below the bottom of the filler neck.
   Use a screen washer solvent in the reservoir, this will assist in removing mud, flies and road film.
5. In cold weather to prevent freezing of the water add 'Isopropyl Alcohol' to the reservoir.

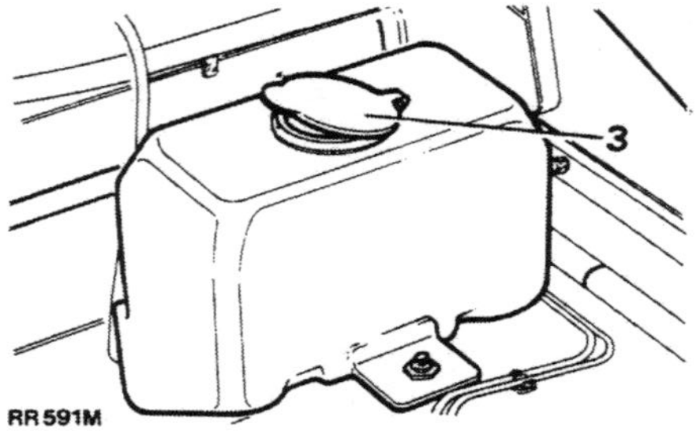

## RENEW EVAPORATIVE LOSS FILTER—carburetter models

NOTE: **Engine breather filters may have alternative locations according to engine build specification.**

1. Squeeze the short centre hose retaining clip and pull the filter from the hose.
2. Squeeze the hose retaining clip on the opposite end of the filter and remove the hose.

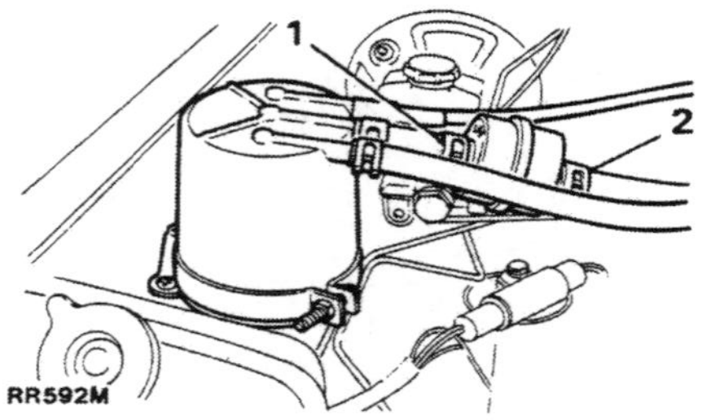

3. Fit a new filter, ensuring that the end of the filter marked 'IN' is fitted to the short hose from the adsorption cannister.

## CHECK CLUTCH FLUID RESERVOIR

1. Check the fluid level in the reservoir mounted on the bulkhead adjacent to the brake servo.
2. Remove the cap, top-up if necessary to the bottom of the filler neck. (Use the correct fluid specified in the 'Data' section).

NOTE: **If significant topping-up is required check for leaks at master cylinder, slave cylinder and connecting pipes.**

3. Check clutch pipes for chafing and corrosion.

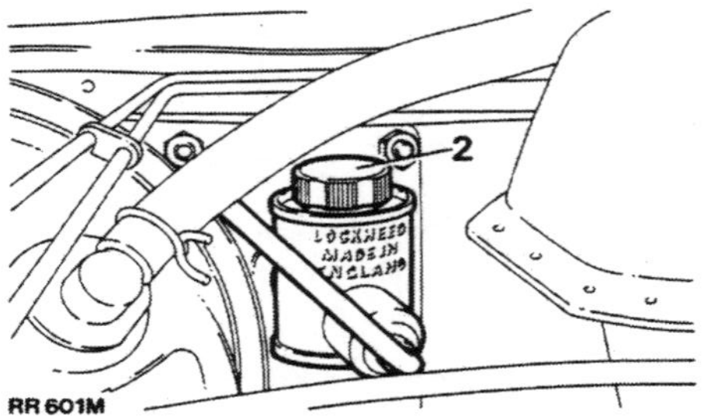

### CHECK

Check ignition wiring and high tension leads for fraying, chafing and deterioration.

### CHECK

Check/adjust ignition timing, (see engine tuning data/procedure) using suitable electronic equipment.

## FUEL SYSTEM (All models)

Check all hose connections for leaks and hose deterioration, renew hoses or tighten hose clips as necessary.

# MAINTENANCE

## IGNITION

### Clean/adjust or renew spark plugs

1. Use the special spark plug spanner and tommy bar supplied in the tool kit when removing or refitting spark plugs.
2. Take great care when fitting spark plugs not to cross-thread the plug, otherwise costly damage to the cylinder head will result.
3. Check or replace the spark plugs as applicable. If the plugs are in good condition, clean and reset the electrode gaps, refer to engine tuning data. At the same time file the end of the central electrode until bright metal can be seen.
4. It is important that only the correct type of spark plugs are used for replacements.
5. Incorrect grades of plugs may lead to piston overheating and engine failure.

To remove spark plugs proceed as follows:

6. Remove the leads from the spark plugs.
7. Remove the plugs and washers.

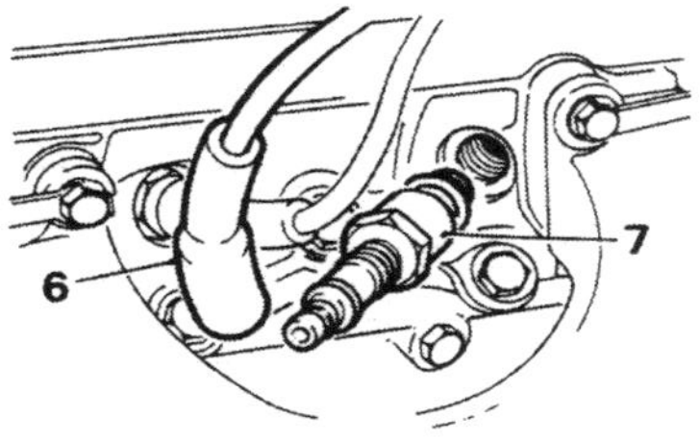

### Clean the spark plugs

8. (a) Fit the plug into a 14 mm adaptor of an approved spark plug cleaning machine.
   (b) Wobble the plug in the adaptor with a circular motion for three or four seconds only with the abrasive blast in operation. Important: Excessive abrasive blasting will lead to severe erosion of the insulator nose. Continue to wobble the plug in its adaptor with air only, blasting the plug for a minimum of 30 seconds: this will remove abrasive grit from the plug cavity.
   (c) Wire-brush the plug threads; open the gap slightly, and vigorously file the electrode sparking surfaces using a point file. This operation is important to ensure correct plug operation by squaring the electrode sparking surfaces.
   (d) Wash new plugs in petrol to remove protective coating.
9. Set the electrode gap to the recommended clearance.

10. Shows dirty plug.
11. Filing plug electrodes.
12. Clean plug set to correct gap.

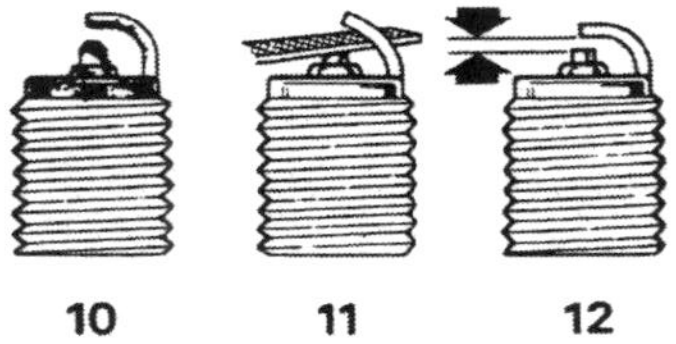

13. Test the plugs in accordance with the plug cleaning machine manufacturers' recommendations.
14. If satisfactory the plugs can be refitted.
15. When pushing the leads onto the plugs, ensure that the shrouds are firmly seated on the plugs.

### Fitting H.T. leads

16. Ensure that replacement H.T. leads are refitted in their spacing cleats in accordance with the correct layout illustrated.
Failure to observe this instruction may result in cross-firing between two closely fitted leads which are consecutive in the firing order.

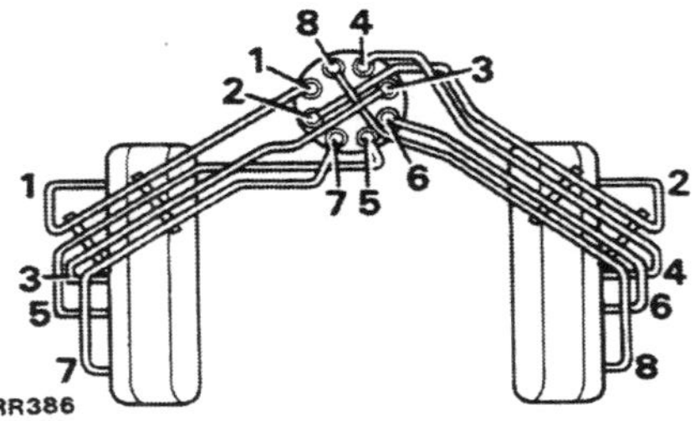

---

# MAINTENANCE

## DISTRIBUTOR – LUCAS 35DM8

The electronic ignition employs a Lucas 35DM8 distributor.
The internal operating parts of the distributor are pre-set at the factory and should not normally require resetting. Adjustments should only be made, if the unit is known to be faulty or damaged.
Maintenance of the distributor consists of the following items.

1. Clean outer surfaces of distributor cap to remove dirt, grease etc.
2. Unclip the cap, check cap for signs of cracking.
3. Wipe inside cap with nap free cloth.
4. Check rotor arm, cap and flash shield for signs of tracking.
5. Apply three drops of clean engine oil to the felt pad in the rotor shaft.
**DO NOT DISTURB the clear plastic insulating cover (flash shield) which protects the magnetic pick-up module.**

## CHECK AIR CONDITIONING SYSTEM (where fitted)

**WARNING: Adjustments or rectification operations should be carried out by a Range Rover dealer or an approved automotive air conditioning specialist. Under no circumstances should non-qualified personnel attempt repair or servicing of air conditioning equipment.**

The following items should be checked,

1. **Condenser:** Clean the exterior of the condenser matrix using a water hose or compressed air-line.
2. Check pipe connections for signs of fluid leakage.
3. **Evaporator:** Examine the pipe connections for signs of fluid leakage.
4. **Receiver/drier sight glass:** After running the engine for five minutes with the air conditioning system in operation, examine the sight glass, there should be no sign of bubbles.
5. Check pipe connections for signs of fluid leakage.
6. **Compressor:** Check the pipe connections for fluid leakage and the hoses for swellings.

---

# MAINTENANCE

## CHECK/TOP UP COOLING SYSTEM

1. To prevent corrosion of the aluminium alloy engine parts it is imperative that the cooling system is filled with a solution of water and anti-freeze, winter or summer, or water and inhibitor during the summer only. Never fill or top up with plain water.
2. The expansion tank filler cap is under the bonnet.
3. With a cold engine, the correct coolant level should be up to the 'Water Level' plate situated inside the expansion tank below the filler neck.

**WARNING: Do not remove the filler cap when engine is hot because the cooling system is pressurised and personal scalding could result.**

4. When removing the filler cap, first turn it anti-clockwise a quarter of a turn and allow all pressure to escape, before turning further in the same direction to lift it off.

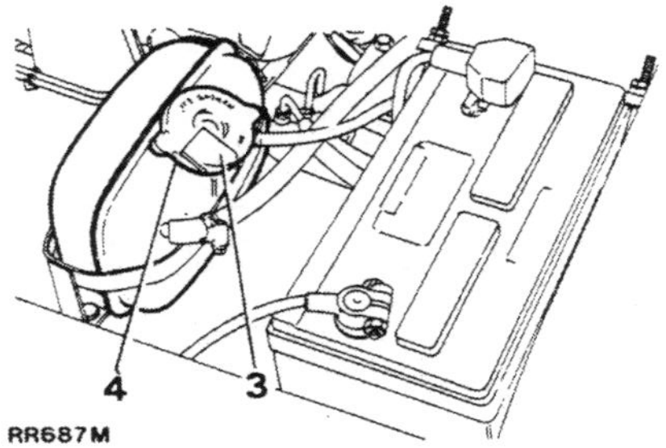

5. When replacing the filler cap it is important that it is tightened down fully, not just to the first stop. Failure to tighten the filler cap properly may result in water loss, with possible damage to the engine through overheating. Use soft water whenever possible, if local water supply is hard, rainwater should be used.

**Check cooling/heater systems for leaks and hoses for security and condition.**

Cooling system hoses should be changed at the first signs of deterioration.

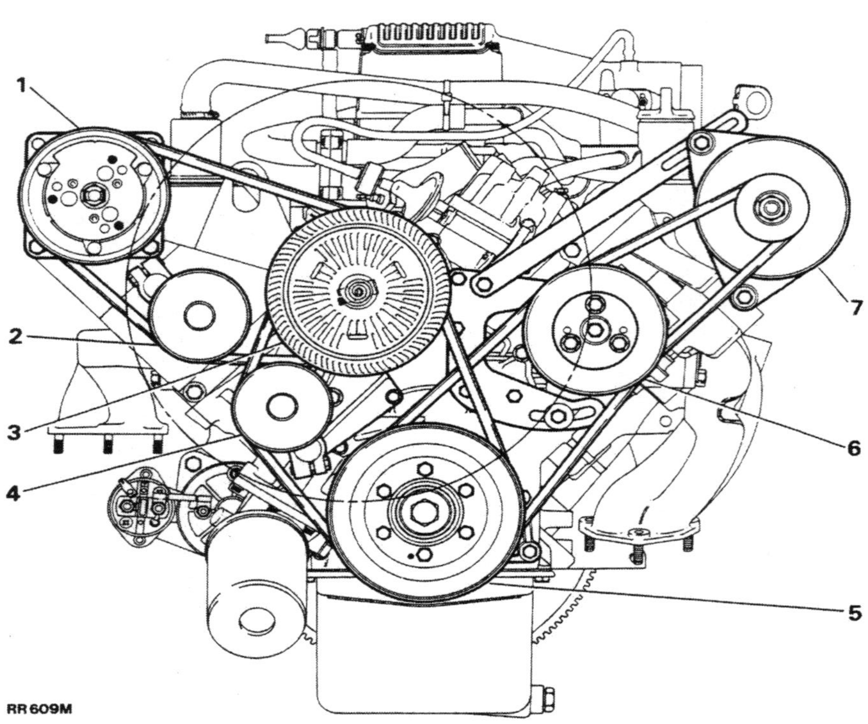

# MAINTENANCE 10

## Check driving belts, adjust or renew as necessary

1. Examine the following belts for wear and condition and renew if necessary:
   (A) Crankshaft—Jockey Pulley—Water Pump
   (B) Crankshaft—Steering Pump
   (C) Steering Pump—Alternator

1. Air conditioning compressor
2. Jockey wheel
3. Viscous fan—water pump unit
4. Jockey wheel
5. Crankshaft
6. Power steering pump
7. Alternator

RR 609M

## DRIVE BELTS—adjust or renew

### COMPRESSOR DRIVE BELT

The belt must be tight with not more than 4 to 6 mm (0.19 to 0.25 in) total deflection when checked by hand midway between the pulleys on the longest run.

Where a belt has stretched beyond the limits, a noisy whine or knock will often be evident during operation, if necessary adjust as follows:

14

1. Slacken the jockey wheel securing bolt.
2. Adjust the position of the jockey wheel until the correct tension is obtained.
3. Tighten the securing bolt and re-check the belt tension.

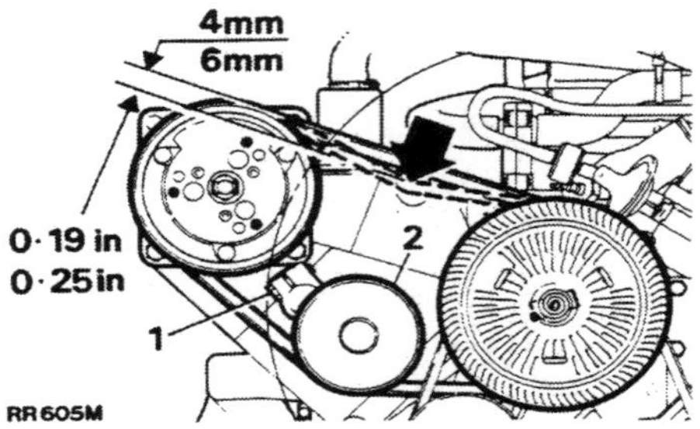

4mm  0·19 in
6mm  0·25 in

RR 605M

---

# MAINTENANCE 10

2. Each belt should be sufficiently tight to drive the appropriate auxiliary without undue load on the bearings.
3. Slacken the bolts securing the unit to its mounting bracket.
4. Slacken the appropriate pivot bolt or jockey wheel and the fixing at the adjustment link where applicable.
5. Pivot the unit inwards or outwards as necessary and adjust until the correct belt tension is obtained.
6. Belt tension should be approximately 11 to 14 mm (0.437 to 0.562 in) at the points denoted by the bold arrows.
7. Tighten all unit adjusting bolts. Check adjustment again, when a new belt is fitted, after approximately 1,500 km (1,000 miles) running.

## CHECK AIR INTAKE TEMPERATURE CONTROL SYSTEM—carburetter type

1. Check operation of the mixing flap valve in the air cleaner by starting the engine from cold and observing the flap valve as the engine temperature rises.
2. The valve should start to open slowly within a few minutes of starting and continue to open until a stabilised position is achieved. This position and the speed of operation will be entirely dependent on prevailing ambient conditions.
3. Failure to operate indicates failure of flap valve vacuum capsule or thermostatically controlled vacuum switch or both.
4. Check by connecting a pipe directly from the banjo on No. 8 point inlet manifold to the flap valves, thus by-passing the temperature sensor.
5. If movement of the flap valve is evident the temperature sensor is faulty. If no movement is detected, the vacuum capsule is faulty.
6. Fit new parts where necessary.

Illustration A  RR 606M

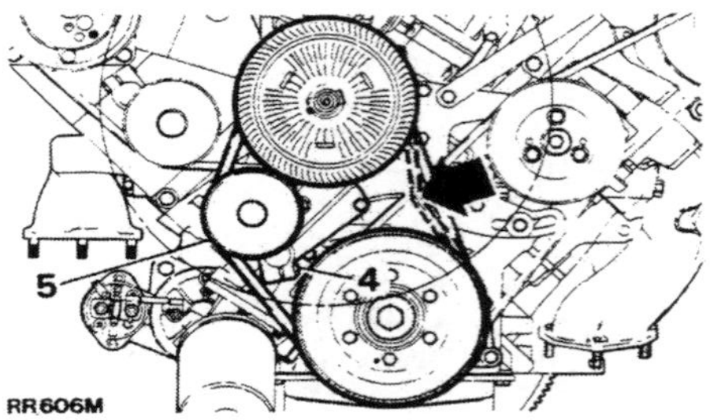

Illustration B  RR 607M

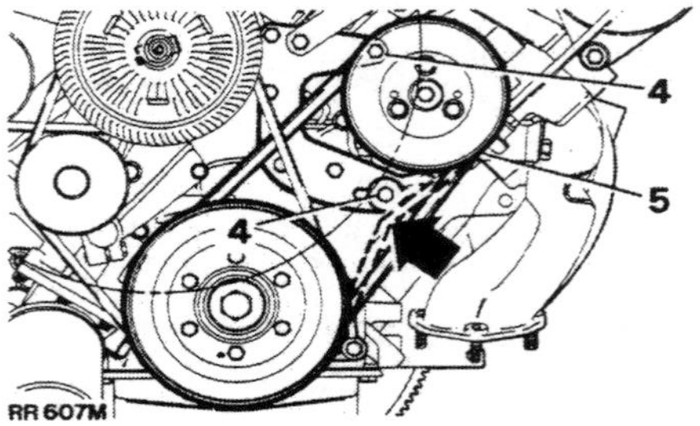

Illustration C  RR 608M

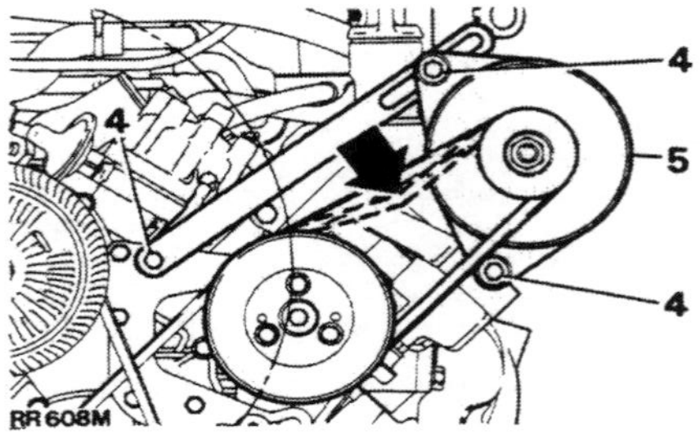

## STEERING AND SUSPENSION

Check condition and security of steering unit, joints, relays and gaiters

Check steering box for oil/fluid leaks

Check shock absorbers for fluid leaks

Check power steering system for leaks, hydraulic pipes and unions for chafing and corrosion

Check security of suspension fixings

15

# MAINTENANCE

## CHECK STEERING BALL JOINTS

Ball joints are lubricated for the normal life of ball joints during manufacture and require no further lubrication. This applies only if the rubber gaiter has not become dislodged or damaged. The joints should be checked at the specified mileage intervals but more frequently if the vehicle is used under arduous conditions.

1. Check for wear in the joints by moving the ball joint up and down vigorously. If free movement is apparent renew the complete joint assembly.

### Check/adjust front wheel alignment

Use recognised wheel alignment equipment to perform this check and adjustment.
See 'Data' section for the correct alignment.

### To adjust

1. Set the vehicle on level ground, with the road wheels in the straightforward position, and push it forward a short distance.
2. Slacken the clamps securing the adjusting shaft to the track rod.
3. Turn the adjusting shaft to decrease or increase the effective length of the track rod, as necessary, until the toe-out is correct.
4. Re-tighten the clamps.
5. Push the vehicle rearwards, turning the steering wheel from side to side to settle the ball joints. Then with the road wheels in the straight ahead position, push the vehicle forward a short distance.
6. Re-check the toe-out. If necessary carry out further adjustment.

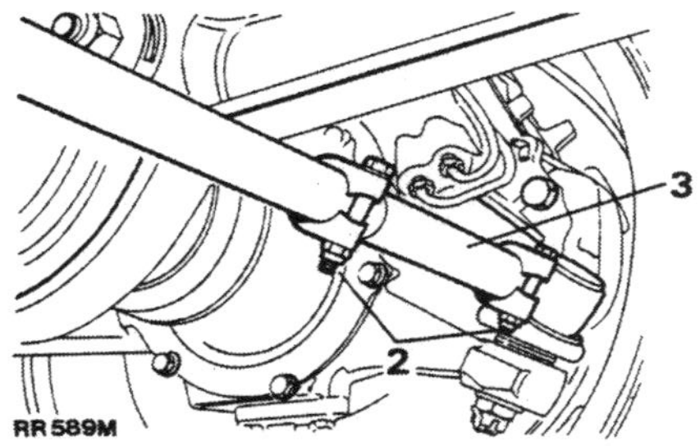

### Drain flywheel housing if drain plug is fitted for wading

1. The flywheel housing can be completely sealed to exclude mud and water under severe wading conditions, by means of a plug fitted in the bottom of the housing.
2. The plug is screwed into the housing adjacent to the drain hole, and should only be fitted when the vehicle is expected to do wading or very muddy work.
3. When the plug is in use it must be removed periodically and all oil allowed to drain off before the plug is replaced.

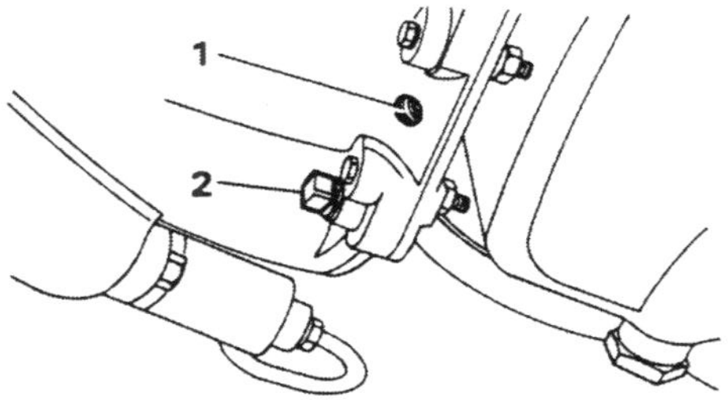

RR105M

### CHECK ROAD SPRINGS

Verify that the vehicle is being operated within the specified maximum loading capabilities. Drive the vehicle onto level ground and remove all loads. Should the vehicle lean to one side it indicates a fault with the springs or shock absorbers, not the self-levelling unit. If the levelling unit is believed to be at fault, the procedure below should be followed:

1. Check the levelling unit for excessive oil leakage and if present the unit must be changed. Slight oil seepage is permissible.
2. Remove any excessive mud deposits and loose items from the rear seat and load area.
3. Measure the clearance between the rear axle bump pad and the bump stop rubber at the front outer corner on both sides of the vehicle. The average clearance should be in excess of 67 mm (2.8 in). If it is less than this figure remove the rear springs and check their free length against the 'Road Spring Data'. Replace any spring whose free length is more than 20 mm (0.787 in) shorter than the figure given. If after replacing a spring the average bump clearance is still less than 67 mm (2.8 in), replace the levelling unit.

*Continued*

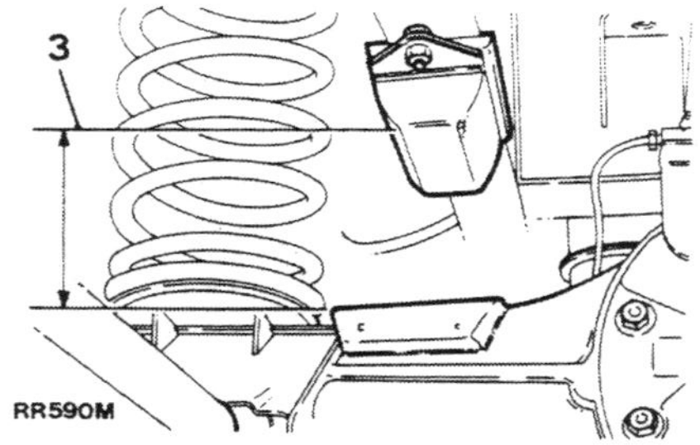

RR588M

---

# MAINTENANCE

CAUTION: When topping-up the reservoir, care should be taken to ensure that brake fluid does not come into contact with any paintwork on the vehicle.

## CHECK AND ADJUST TRANSMISSION BRAKE (Handbrake)

The handbrake lever acts on a transmission brake at the rear of the transfer box.

1. Set the vehicle on level ground.
2. Fully release the handbrake.
3. Disconnect the handbrake cable linkage at the transmission brake drum by removing the split pin, plain and spring washer and clevis pin.
4. Turn the adjuster on the back plate clockwise until the shoes are fully expanded against the drum.
5. Slacken the four locknuts on the handbrake adjustment link.

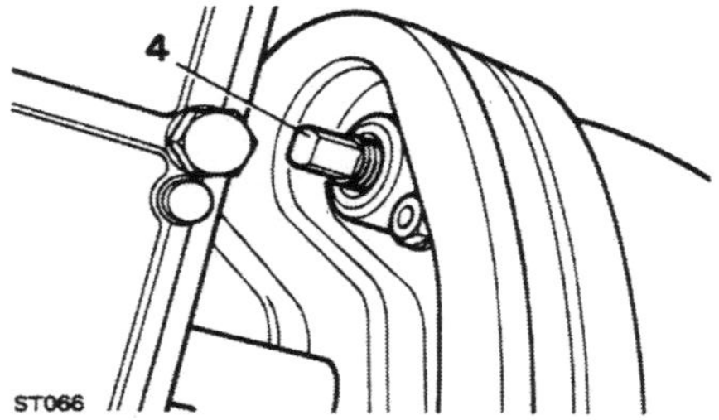

ST066

4. With the rear seat upright, load 450 kg (992 lb) into the rear of the vehicle, distributing the load evenly over the floor area. Check the bump stop clearance, with the driving seat occupied.
5. Drive the vehicle for approximately 5 km (3 miles) over undulating roads or graded tracks. Bring the vehicle to rest by light brake application so as not to disturb the vehicle loading. With the driving seat occupied, check the bump stop clearance again.
6. If the change in clearance is less than 20 mm (0.787 in) the levelling unit must be replaced.

## BRAKES

Check visually, hydraulic pipes and unions for chafing, leaks and corrosion.

### Check/top up brake fluid reservoir

The tandem brake reservoir is integral with the servo unit and master cylinder.

1. Remove cap to check fluid level; top up if necessary until the fluid reaches the bottom of the filter neck. See 'Data' section for recommended fluids.
2. If significant topping up is required check master cylinder, brake disc cylinders and brake pipes and connections for leakage; any leakage must be rectified immediately.

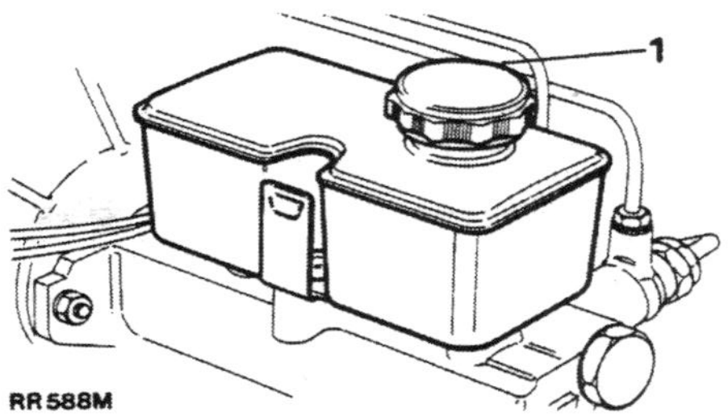

RR588M

6. Rotate the link clockwise or anti-clockwise until the clevis pin holes in the link bracket and brake drum lever line up.
7. Fit the clevis pin, plain and spring washer and a new split pin, lightly grease the assembly.
8. Fully tighten the four link locknuts.

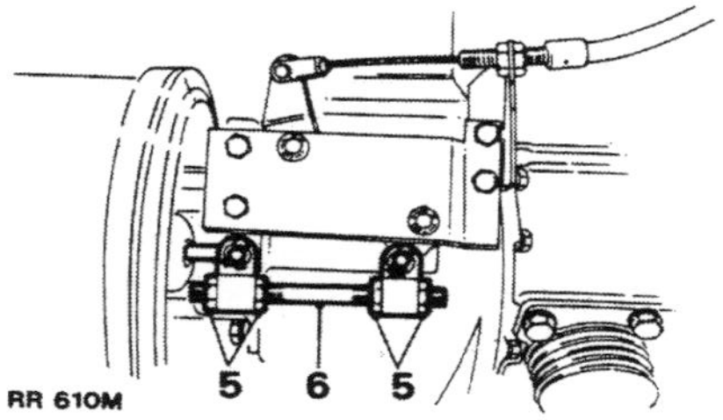

RR610M

9. Slacken the brake drum adjuster, until the handbrake becomes fully operational on the third or fourth notch of the handbrake lever quadrant.

CAUTION: DO NOT over-adjust the handbrake, the drum must be free to rotate when the handbrake is released, otherwise serious damage will result.

*Continued*

# MAINTENANCE

## CHECK FOOTBRAKE OPERATION

If the footbrake is 'spongy' bleed the brake system. Check all hoses and pipes for security, fractures and leaks. Renew as necessary.

## RENEW FRONT AND REAR BRAKE PADS

Brake pad wear is indicated by a pad wear warning light incorporated into the instrument binnacle. The warning lamp is illuminated when pad wear is reduced to approximately 3.0 mm (0.118 in). The system is operated by an electrical sensor incorporated into the front and rear right-hand side inboard brake pads.

When pad wear is sufficient in either front or rear pads allowing the sensor within the pads to complete a circuit to earth through the disc, thus illuminating the warning lamp in the instrument binnacle.

### Renew front brake pads

1. Slacken both front wheel nuts, jack up the vehicle and lower onto axle stands.
2. Disconnect the battery.
3. Disconnect the two-pin electrical plug at the rear of the disc mudshield (front right-hand side only).
4. Clean the exterior of the calipers.
5. Remove the split pins from the brake caliper.
6. Remove the retaining springs.
7. Withdraw the brake pads.
8. Clean the exposed parts of the pistons, using new brake fluid.

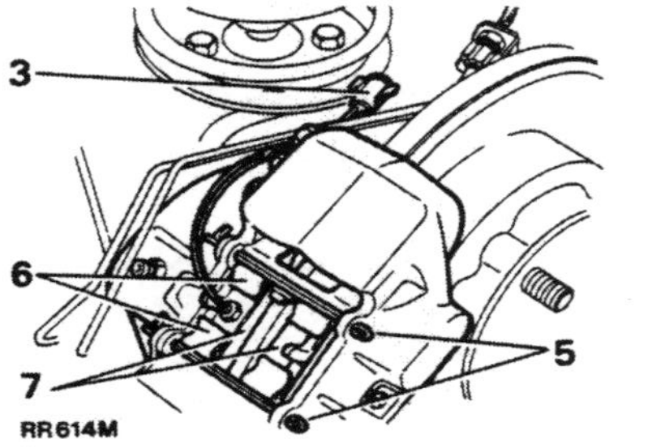

RR614M

9. Using piston clamp 18G672 press each piston back into its bore, whilst ensuring that the displaced brake fluid does not overflow from the reservoir.
10. Smear the faces of the pistons with Lockheed disc brake lubricant taking care not to let any reach the lining material.

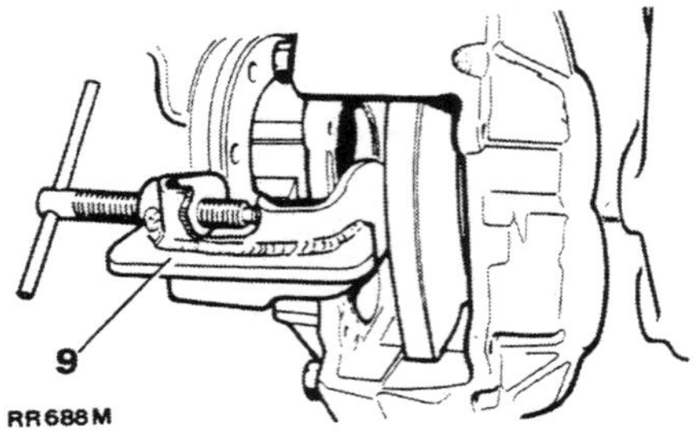

RR688M

11. Insert the new brake pads.
12. Place the brake pad retaining springs in position, fit new split pins and splay the ends.
13. Apply the footbrake several times to locate the pads.
14. Check the fluid reservoir and top-up if necessary.
15. Fit the road wheels, lower the vehicle and finally tighten the road wheels.

### Renew rear brake pads

Jacking up the rear of the vehicle, follow the procedure as for front pads.

1. The two-pin electrical plug for rear brake pad wear indication is located on the rear left-hand inboard pad.

## CHECK BRAKE SERVO HOSE(S)

Visually inspect all servo hoses and connections for condition and security.

## RENEW BRAKE FLUID

Brake fluid absorbs water and in time the boiling point of the fluid will be lowered sufficiently to cause the fluid to be vapourised by the heat generated when the vehicle brakes are applied. This will result in loss of braking efficiency or in extreme cases brake failure.

Therefore, all fluid in the brake system should be changed every eighteen months or 30,000 km (18,000 miles), whichever is the sooner. It should also be changed before touring in mountainous areas if not done in the previous nine months.

Care must be taken always to observe the following points:
(a) At all times use the recommended brake fluid.
(b) Never leave fluid in unsealed containers. It absorbs moisture quickly and can be dangerous if used in the braking system in this condition.

*Continued*

(c) Fluid drained from the system or used for bleeding is best discarded.
(b) The necessity for absolute cleanliness throughout cannot be over emphasised.

## CLEAN THE THROTTLE BUTTERFLY HOUSING—plenum chamber

### Fuel injection

At regular service intervals it is recommended that any carbon/oil build up around the throttle butterfly seat (plenum chamber bore), should be removed using a suitable solvent.

1. Remove the inlet pipe from the plenum chamber inlet neck.
2. Open the throttle butterfly and remove any carbon/oil deposits around the butterfly seating.
3. Refit inlet pipe.
4. Check idle speed (see 'Engine Tuning Data').
5. Check and reset CO levels as necessary (see 'Engine Tuning Data').

## BATTERY

A low maintenance battery is installed in the vehicle. Dependent upon climate conditions the electrolyte levels should be checked as follows:

Temperate climates every three years.
Hot climates every year.

The exterior of the battery should be occasionally wiped clean to remove any dirt or grease.

Periodically remove the battery terminals to clean and coat with petroleum jelly.

**NOTE: If a new battery is fitted to the vehicle it should be the same type as fitted to the vehicle when new. Alternative batteries may vary in size and terminal positions and this could be a possible fire hazard if the terminals or leads come into contact with the battery clamp assembly. When fitting a new battery ensure that the terminals and leads are clear of the battery clamp assembly.**

# RANGE ROVER WORKSHOP MANUAL 1986 ONWARDS

# Part 2

|  | Section | Page |
|---|---|---|
| General Specification Data | 04 | 31 |
| Torque Wrench Settings | 06 | 33 |
| Engine | 12 | 34 |
| Emission Control | 17 | 50 |
| Fuel System (including Fuel Injection) | 19 | 55 |
| Cooling Systems | 26 | 78 |
| Manifold & Exhaust System | 30 | 80 |
| Clutch | 33 | 82 |

Section
Number

## 04 GENERAL SPECIFICATION DATA

— V8 engine data .................................... 1
— Fuel system ....................................... 3
— Cooling system .................................... 3
— Clutch ............................................ 3

## 06 TORQUE WRENCH SETTINGS

— Engines ........................................... 1
— Clutch ............................................ 1

## 12 V8 CYLINDER ENGINE

— Overhaul engine ................................... 1
— Cylinder head ..................................... 3
— Flywheel .......................................... 6
— Timing gear cover and water pump .................. 7
— Oil pump .......................................... 8
— Timing chain gears and camshaft ................... 9
— Connecting rods and pistons ...................... 10
— Crankshaft ....................................... 16
— Assembling engine ................................ 18
— Fault diagnosis .................................. 30

## 17 EMISSION CONTROL

— Description ....................................... 1
— Engine breather filter — remove and refit ........ 1
— Engine flame traps — remove and refit ............ 2
— Adsorption canister — remove and refit ........... 3
— Pulsair air injection ............................. 3
— Pulsair manifold — remove and refit .............. 4
— Pulsair check valve — remove and refit ........... 5
— Air intake temperature control system — description 6
— Control valve — check — remove and refit ........ 6
— Temperature sensor — remove and refit ............ 7
— Vacuum delay valve ............................... 7
— Air intake temperature control — fault diagnosis . 9

*Continued*

## 19 FUEL SYSTEM—ELECTRONIC FUEL INJECTION

— Components ........................................ 1
— Wiring diagram .................................... 2
— Description ....................................... 3
— Engine setting procedure .......................... 4
— Continuity tests AVO meter ....................... 12
— Fault diagnosis — hot start ...................... 13
— Fault diagnosis — general ........................ 14
— Engine tuning procedure .......................... 14
— Check and adjust ignition timing ................. 14
— Check and adjust idle speed ...................... 14
— Check and adjust idle CO level ................... 15
— Air cleaner — remove and refit .................. 15
— Resetting throttle levers ........................ 16
— Renew throttle cable ............................. 16
— Relays — remove and refit ....................... 17
— Electronic control unit .......................... 18
— Throttle potentiometer — remove, refit and setting 18
— Thermotime switch — test — remove and refit ..... 19
— Coolant temperature sensor — test — remove and refit 20
— Air temperature sensor — test .................... 20
— Airflow meter — remove and refit ................ 20
— Power resistor — remove and refit ............... 21
— Over-run fuel shut-off valve — remove and refit . 22
— Solenoid operated air valve — remove and refit .. 22
— Depressurise fuel system ......................... 23
— Fuel pressure regulator — test — remove and refit 24
— Plenum chamber — remove and refit ............... 24
— Ram housing — remove and refit .................. 26
— Extra air valve — test — remove and refit ...... 27
— Fuel rails — remove and refit ................... 27
— Injectors — test — remove and refit ............. 28
— Intake manifold — remove and refit
— Fuel line filter — remove and refit
— Fuel pump — remove and refit
— Fuel pipes

## 19 FUEL SYSTEM—CARBURETTER

— Air filter — remove and refit ..................... 1
— Carburetter description — tamperproofing ......... 2
— Carburetters — tune and adjust ................... 2
— Slow running — adjust ............................ 3
— Fast idle — adjust ............................... 4
— Idle mixture — adjust ............................ 4
— Carburetter — remove and refit ................... 6
— Carburetter overhaul ............................. 8
— Throttle cable — remove and refit ............... 11
— Throttle linkage — remove and refit ............. 12
— Choke cable — remove and refit .................. 13
— Throttle pedal — remove and refit ............... 14
— Fuel filter — remove and refit .................. 14
— Fuel pipes ....................................... 15
— Fuel tank and pump — remove and refit ........... 15
— Fault diagnosis .................................. 16

*Continued*

30

# GENERAL SPECIFICATION DATA | 04

| Section Number | | |
|---|---|---|
| 26 | **COOLING SYSTEM** | |
| | —Coolant — drain and refill | 1 |
| | —Coolant requirements | 2 |
| | —Expansion tank — remove and refit | 2 |
| | —Fan belt — check and adjust | 3 |
| | — remove and refit | 3 |
| | —Viscous coupling, fan blades, pulley and fan cowl | 3 |
| | —Radiator — remove and refit | 4 |
| | —Thermostat — remove and refit | 4 |
| | — test | 5 |
| | —Water pump — remove and refit | 5 |
| | —Fault diagnosis | 6 |
| 30 | **MANIFOLD AND EXHAUST SYSTEM** | |
| | —Induction manifold — remove and refit | 1 |
| | —Exhaust system complete — remove and refit | 2 |
| | —Exhaust manifold — remove and refit | 4 |
| 33 | **CLUTCH** | |
| | —Clutch assembly — overhaul | 1 |
| | — remove and refit | 1 |
| | —Hydraulic system — bleed | 1 |
| | —Master cylinder — remove and refit | 2 |
| | — overhaul | 2 |
| | —Release bearing assembly — remove and refit | 3 |
| | —Clutch pedal — remove and refit | 4 |
| | —Slave cylinder — remove and refit | 5 |
| | — overhaul | 5 |

## GENERAL SPECIFICATION DATA

### ENGINE

**Crankshaft**
Main journal diameter ............... 58.409–58.422 mm (2.2996–2.3001 in)
Minimum regrind diameter ......... 57.393–57.406 mm (2.2596–2.2601 in)
Crankpin journal diameter .......... 50.800–50.812 mm (2.0000–2.0005 in)
Minimum regrind diameter ......... 49.784–49.797 mm (1.9600–1.9605 in)
Crankshaft end thrust ............... Taken on thrust washers of centre main bearing
Crankshaft end-float ................. 0.10–0.20 mm (0.004–0.008 in)

**Main bearings**
Number and type ...................... 5, Vandervell shells
Material ................................. Lead–indium
Diametrical clearance ............... 0.010–0.048 mm (0.0004–0.0019 in)
Undersizes ............................. 0.254 mm, 0.508 mm (0.010 in, 0.020 in)

**Connecting rods**
Type ..................................... Horizontally split big-end, plain small-end
Length between centres ........... 143.81–143.71 mm (5.662–5.658 in)

**Big-end bearings**
Type and material .................... Vandervell VP lead–indium
Diametrical clearance ............... 0.015–0.055 mm (0.006–0.0022 in)
End-float on crankpin ............... 0.15–0.36 mm (0.006–0.014 in)
Undersizes ............................. 0.254 mm, 0.508 mm (0.010 in, 0.020 in)

**Gudgeon pins**
Length ................................... 72.67–72.79 mm (2.861–2.866 in)
Diameter ................................ 22.215–22.220 mm (0.8746–0.8749 in)
Fit-in connecting rod ................ Press fit
Clearance in piston .................. 0.002–0.007 mm (0.0001–0.0003 in)

**Pistons**
Clearance in bore, measured at bottom of skirt at right angles to gudgeon pin ............... 0.018–0.033 mm (0.0007–0.0013 in)

**Piston rings**
Number of compression ............ 2
Number of oil .......................... 1
No. 1 compression ring ............. Chrome parallel faced
No. 2 compression ring ............. Stepped to 'L' shaped and marked 'T or 'TOP'
Width of compression rings ....... 1.56–1.59 mm (0.0615–0.0625 in)
Compression ring gap .............. 0.44–0.57 mm (0.017–0.022 in)
Oil ring type ........................... Perfect circle, type 98-6
Oil ring width .......................... 4.811 mm (0.1894 in) max
Oil ring gap ............................ 0.38–1.40 mm (0.015–0.055 in)

**Camshaft**
Location ................................ Central
Bearings ................................ Non serviceable
Number of bearings ................. 5
Drive ..................................... Chain 9.52 mm (0.375 in) pitch × 54 pitches

31

# GENERAL SPECIFICATION DATA

**Tappets**
Type ............................................. Hydraulic, non adjustable

**Valves**
Length:
  Inlet ........................................... 116.59–117.35 mm (4.590–4.620 in)
  Exhaust ....................................... 116.59–117.35 mm (4.590–4.620 in)
Seat angle:
  Inlet ........................................... 45°–45½°
  Exhaust ....................................... 45°–45½°
Head diameter:
  Inlet ........................................... 39.75–40.00 mm (1.565–1.575 in)
  Exhaust ....................................... 34.226–34.480 mm (1.3475–1.3575 in)
Stem diameter:
  Inlet ........................................... 8.664–8.679 mm (0.3411–0.3417 in)
  Exhaust ....................................... 8.651–8.666 mm (0.3406–0.3412 in)
Stem to guide clearance:
  Inlet ........................................... 0.025–0.066 mm (0.0010–0.0026 in)
  Exhaust ....................................... 0.038–0.078 mm (0.0015–0.0031 in)
Valve lift (inlet and exhaust) ............. 9.49 mm (0.374 in)
Valve spring length fitted ................. 40.4 mm (1.590 in) at pressure of 29.5 kg (65 lb)

**Lubrication**
System ......................................... Wet sump, pressure fed
System pressure, engine warm at 2400 rpm .... 2.1–2.8 kgf/cm² (30–40 lbf/in²)
Oil filter (external) ........................... Full-flow, self-contained cartridge
Oil filter (internal) ........................... Gauze. Pump intake filter
Oil pump type ................................. Gear

**Oil pressure relief valve**
Type ............................................. Non adjustable
Relief valve spring:
  Free length ................................... 81.2 mm (3.200 in)
  Compressed length at 4.2 kg (9.3 lb) load ... 45.7 mm (1.800 in)

**Oil filter by-pass valve**
Type ............................................. Non adjustable
By-pass valve spring:
  Free length ................................... 37.5 mm (1.48 in)
  Compressed length at 0.34 kg (0.75 lb) ..... 22.6 mm (0.89 in)

# GENERAL SPECIFICATION DATA

**FUEL SYSTEM—carburetter**
Carburetter type ............................. See 'Engine Tuning Data', in Book I
Fuel pump—make/type ..................... AC Delco—low pressure (electrical) immersed in the fuel tank
Pump delivery pressure .................... 0.28–0.42 kgf/cm² (4–6 lbf/in²)
Fuel filter ....................................... AC Delco CD600—element ACD60

**FUEL SYSTEM—fuel injection**
Fuel system type ............................. See 'Engine Tuning Data', in Book I
Fuel pump—make/type ..................... AC Delco—high pressure (electrical) immersed in the fuel tank
Pump delivery pressure .................... 1.83–2.5 kgf/cm² (26–36 lbf/in²)
Fuel filter ....................................... Bosch in-line filter 'canister' type

**COOLING SYSTEM**
Type ............................................. Pressurized spill return system with thermostat control, pump and fan assisted
Thermostat .................................... 88°C
Type of pump ................................. Centrifugal

**CLUTCH**
Type ............................................. Borg and Beck diaphragm spring
Centre plate diameter ...................... 267 mm (10.5 in)
Facing material ............................... Raybestos 1488-05. Grooved
Damper spring colour ...................... Light blue/dark blue
Release bearing .............................. Ball journal
Number of damper springs ................ 6

## TORQUE WRENCH SETTINGS

| | Nm | lbf ft |
|---|---|---|
| **ENGINE—V8 Petrol engine** | | |
| Air intake adaptor to carbs | 24 | 17 |
| Alternator mounting bracket to cylinder head | 34 | 25 |
| Alternator to mounting bracket | 24 | 17 |
| Alternator to adjusting link | 24 | 17 |
| Chainwheel to camshaft | 54–61 | 40–45 |
| Connecting rod bolt | 47–54 | 35–40 |
| Clutch attachment to flywheel | 24–30 | 18–22 |
| Cylinder head: | | |
| Outer row | 54–61 | 40–45* |
| Centre row | 88–95 | 65–70* |
| Inner row | 88–95 | 65–70* |
| Distributor clamp bolt | 19–22 | 14–16 |
| Exhaust manifold to cylinder heads | 19–22 | 14–16 |
| Fan to viscous unit | 26–32 | 19–24 |
| Flexible drive plate to crankshaft | 74–81 | 55–60 |
| Flywheel to crankshaft | 74–81 | 55–60 |
| Inlet manifold to cylinder heads | 47–54 | 35–40 |
| Lifting eye to cylinder heads | 24 | 17 |
| Main bearing cap bolts | 68–75 | 50–55** |
| Main bearing cap rear bolts | 88–95 | 65–70*** |
| Manifold gasket clamp bolt | 13.5–20 | 10–15 |
| Oil pump cover to timing cover | 11–14 | 8–10 |
| Oil plug | 24–30 | 18–22 |
| Oil relief valve plug | 40–47 | 30–35 |
| Oil sump drain plug | 40.6–47 | 30–35 |
| Oil sump to cylinder block | 8–11 | 5–8 |
| Oil sump rear to cylinder block | 17.6–20.3 | 13–15 |
| Rocker cover to cylinder head | 7 | 5 |
| Rocker shaft bracket to cylinder head | 34–40 | 25–30 |
| Spark plug | 13.8–16.2 | 10–12 |
| Starter motor attachment | 40.6–47.4 | 30–35 |
| Damper to crankshaft | 257–285 | 190–210 |
| Timing cover to cylinder block | 24–30 | 18–22*** |
| Viscous unit to water pump hub | 36–40 | 27–30 |
| Water pump pulley to water pump hub | 23 | 17*** |
| Water pump timing cover to cylinder block | 24–30 | 18–22 |
| Water jacket to plenum chamber | 10–14 | 8–10*** |
| Plenum chamber to ram housing | 20–24 | 15–18 |
| Ram housing to intake manifold | 20–24 | 15–18 |
| Thermostat housing to intake manifold | 24–30 | 18–22 |

Lubricants/sealants have been specified in certain applications for assembly purposes.
* These bolts must have threads coated with Loctite 572 prior to assembly. For this purpose it is necessary to use an approved dispenser to apply the sealant/lubricant to the first three threads of the bolts.
** These bolts must have threads coated in lubricant EXP16A (Marston Lubricants) prior to assembly.
*** These bolts must have threads coated in sealant Loctite 572 prior to assembly.
It is essential that all bolts are securely tightened and it is imperative that the correct torque is adhered to.

## Notes

# ENGINE 12

## TORQUE WRENCH SETTINGS

Charts below give torque settings for all screws and bolts used except for those that are specified otherwise

| SIZE | METRIC Nm | lbf ft |
|---|---|---|
| M5 | 3.7–5.2 | 5–7 |
| M6 | 5.2–7.4 | 7–10 |
| M8 | 16.2–20.7 | 22–28 |
| M10 | 29.5–36.9 | 40–50 |
| M12 | 59.0–73.8 | 80–100 |
| M14 | 66.4–88.5 | 90–120 |
| M16 | 118.0–147.5 | 160–200 |

| SIZE UNC | Nm | lbf ft |
|---|---|---|
| ¼ | 6.8–9.5 | 5–7 |
| ⁵⁄₁₆ | 20.3–27.1 | 15–20 |
| ⅜ | 35.2–43.4 | 26–32 |
| ⁷⁄₁₆ | 67.8–88.1 | 50–65 |
| ½ | 81.3–101.7 | 60–75 |
| ⅝ | 122.0–149.1 | 90–110 |

| UNF | Nm | lbf ft |
|---|---|---|
|  | 8.1–12.2 | 6–9 |
|  | 20.3–27.1 | 15–20 |
|  | 35.2–43.4 | 26–32 |
|  | 67.8–88.1 | 50–65 |
|  | 81.3–101.7 | 60–75 |
|  | 122.0–149.1 | 90–110 |

## V8 CYLINDER ENGINE

The following V8 Engine overhaul is applicable to all Range Rovers.

Before removing the fuel injection engine from the vehicle, the following equipment and bracketry must be disconnected or removed.

Air conditioning compressor (if fitted, do not evacuate or disconnect any hoses. Release the compressor from its mounting and lay to one side)
Alternator
Radiator
Power steering pump (disconnect pipes and hoses)
Fan blades
Automatic transmission fluid pipes (from side of block)
Any electrical wiring or harnesses

For the removal procedure of the following items see the 'Fuel Injection' section of manual.

Air cleaner
Air flow meter
Throttle lever bracketry
Plenum chamber
Ram housing
Any electrical wiring or harnesses

## DISMANTLE AND OVERHAUL

Remove the engine from the vehicle and clean the exterior. In the interests of safety and efficient working secure the engine to a recognised engine stand. Drain and discard the sump oil.

**Special tools:**
RO605351—Guide bolts
18G 537—Torque wrench
18G 79—Clutch centralising tool
18G 1150—Gudgeon pin remover/replacer—Basic tool
18G 1150E—Adaptor remover/replacer—gudgeon pin.
18G 106A—Spring compressor
600959—Valve guide drift exhaust
MS76—Valve cutter handle set
MS150—8.5 Adjustable pilot
MS621—Valve seat cutter
RO605774—Distance piece for valve guide
RO274401—Drift for guide removal—inlet and exhaust
RO1014—Camshaft rear seal sleeve

## REMOVE ANCILLARY EQUIPMENT

Before commencing, and whilst dismantling, make a careful note of the position of brackets, clips, harnesses, pipes, hoses, filters and other miscellaneous and non-standard items to facilitate reassembly.

1. Remove the following items of equipment:
   Starter motor.
   Alternator and mounting bracket.
   Power steering pump.
   Disconnect spark plug H.T. leads and remove the distributor.
   Clutch.
   Fan blades, pulley and drive belt.
   Remove pulse air rails from cylinder heads.
   Dipstick and engine mounting brackets.

### Remove exhaust manifolds

2. Bend back the lock tabs, and remove the eight bolts securing each manifold, and withdraw the manifolds.

### Remove intake manifold—Carburetter versions only

NOTE: The removal procedure for the intake manifold on fuel injection models is incorporated in the 'Fuel Injection' section of the manual.

3. Disconnect miscellaneous pipes and hoses from the intake manifold and the carburetters.
4. Evenly slacken and remove the twelve bolts and lift off the intake manifold complete with carburetters.
5. Wipe away any surplus coolant lying on the manifold gasket and remove the gasket clamp bolts and remove the clamps.
6. Lift off the manifold gasket and seals.

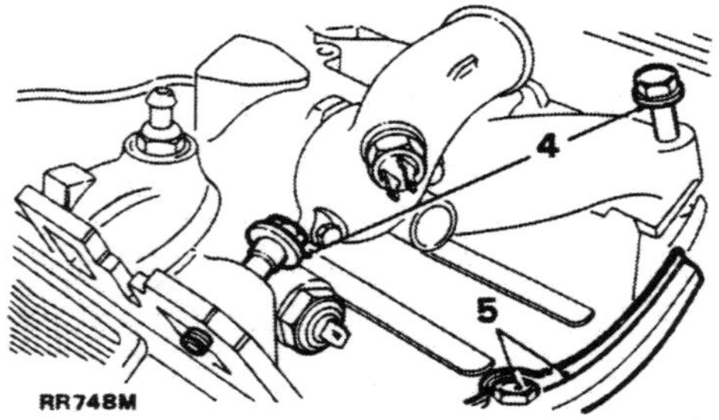

RR748M

*Continued*

# ENGINE

## Remove water pump

7. Remove the fifteen bolts and withdraw the water pump and joint washer.

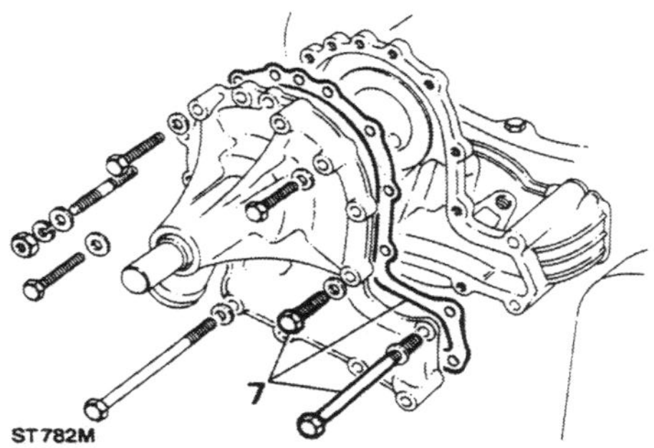

## REMOVE AND OVERHAUL ROCKER SHAFTS AND VALVE GEAR

1. Remove the four screws and lift off the rocker covers.
2. Remove the four rocker shaft retaining bolts and lift off the assembly complete with baffle plate.
3. Withdraw the pushrods and retain in the sequence removed.
4. Remove the hydraulic tappets and place to one side with their respective pushrods. If a tappet cannot be removed leave in position until the camshaft is removed.

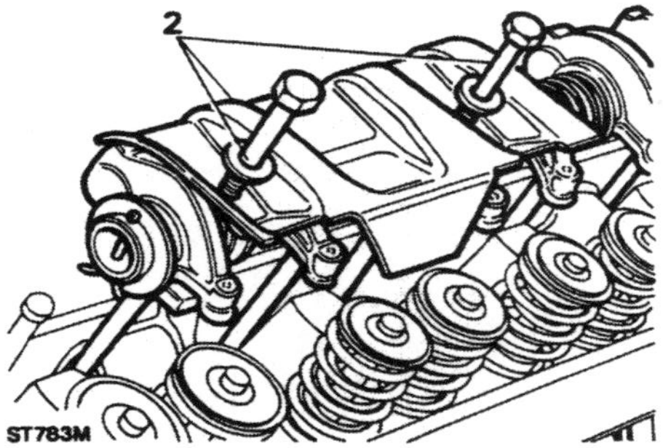

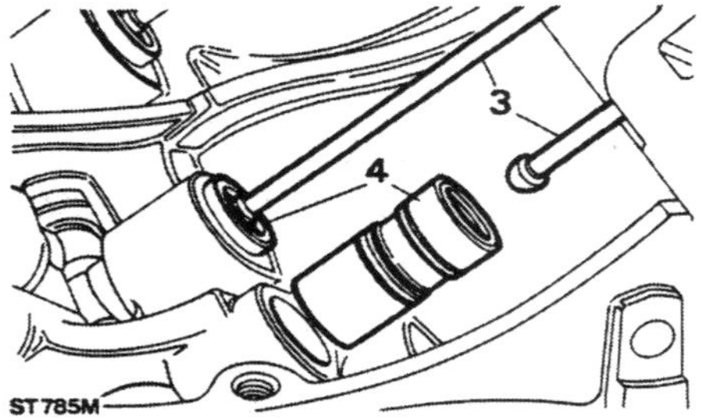

## Dismantle rocker shafts

5. Remove the split pin from one end of the rocker shaft.
6. Withdraw the following components and retain them in the correct sequence for reassembly:
7. A plain washer.
8. A wave washer.
9. Rocker arms.
10. Brackets.
11. Springs.
12. Examine each component for wear, in particular the rockers and shafts. Discard weak or broken springs.

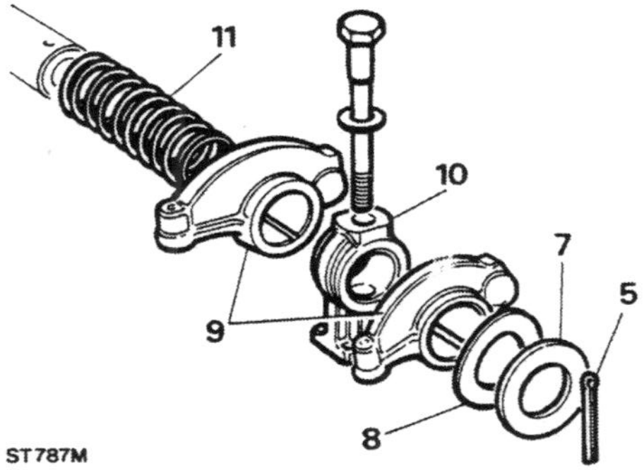

## Inspect tappets and pushrods

13. Hydraulic tappet: inspect inner and outer surfaces of body for blow holes and scoring. Replace hydraulic tappet if body is roughly scored or grooved, or has a blow hole extending through the wall in a position to permit oil leakage from lower chamber.
14. The prominent wear pattern just above lower end of body should not be considered a defect unless it is definitely grooved or scored. It is caused by side thrust of the cam against the body while the tappet is moving vertically in its guide.
15. Inspect the cam contact surface of the tappets. Fit new tappets if the surface is excessively worn or damaged.
16. A hydraulic tappet body that has been rotating will have a round wear pattern and a non-rotating tappet body will have a square wear pattern with a very slight depression near the centre.
17. Tappets MUST rotate and a circular wear condition is normal. Tappets with this wear pattern can be refitted provided there are no other defects.
18. In the case of a non-rotating tappet, fit a new replacement and check camshaft lobes for wear; also ensure the new tappet rotates freely in the cylinder block.
19. Fit a new hydraulic tappet if the area where the pushrod contacts is rough or otherwise damaged.
20. Renew any pushrod having a rough or damaged ball end or seat. Also bent rods must be renewed.

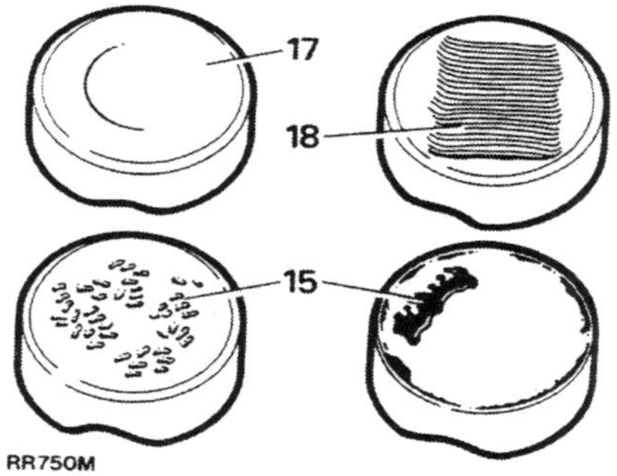

## Assemble rocker shafts

21. Fit a split pin to one end of the rocker shaft.
22. Slide a plain washer over the long end of the shaft to abut the split pin.
23. Fit a wave washer to abut the plain washer.

**NOTE:** Two different rocker arms are used and must be fitted so that the valve ends of the arms slope away from the brackets, as indicated by the dotted lines 'A' on the illustration.

24. Assemble the rocker arms, brackets and springs to the rocker shaft.
25. Compress the springs, brackets and rockers, and fit a wave washer, plain washer and split pin to the end of the rocker shaft.
26. Locate the oil baffle plate in place over the rockers furthest from the notched end of the rocker shaft and fit the bolts through the brackets and place the assemblies to one side.

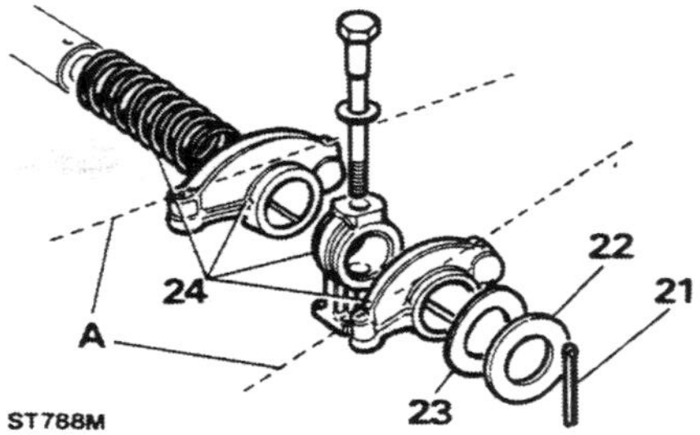

## REMOVE AND OVERHAUL THE CYLINDER HEADS

1. Evenly slacken the fourteen cylinder head bolts reversing the tightening order.
2. Before removing the heads mark them relative to the LH and RH side of the engine.
3. Lift off the cylinder heads and discard the gasket.

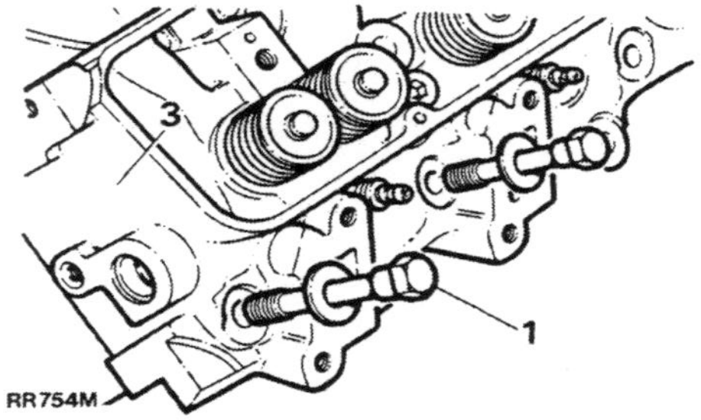

## Dismantle cylinder heads

4. Remove the spark plugs.
5. Using the valve spring compressor 18G 106A or a suitable alternative, remove the valves and springs and retain in sequence for refitting.

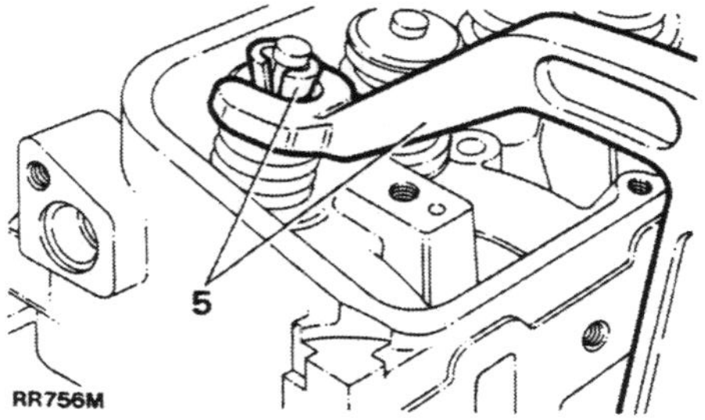

*Continued*

# ENGINE 12

6. Clean the combustion chambers with a soft wire brush.
7. Clean the valves.
8. Clean the valve guide bores.
9. Regrind or fit new valves as necessary.
10. If a valve must be ground to a knife-edge to obtain a true seat, fit a new valve.
11. The correct angle for the valve face is 45 degrees.
12. The correct angle for the seat is 46 ± ¼ degrees and the seat witness should be towards the outer edge.
13. Check the valve guides and fit replacements as necessary. Using the valve guide remover 274401, drive out the old guides from the combustion chamber side.

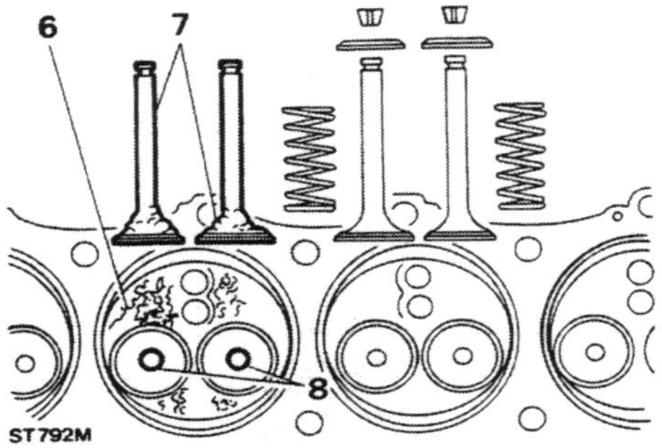

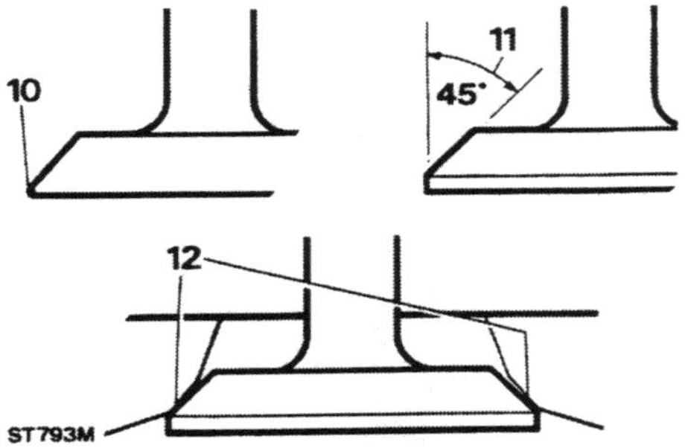

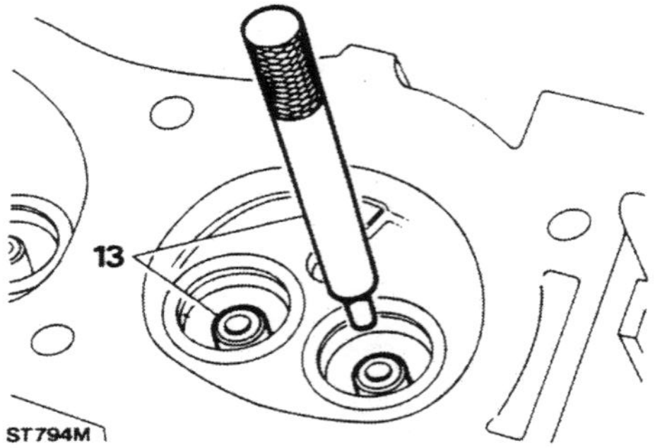

## Fit new valve guides

14. Lubricate the new valve guide and place in position. Using guide drift 600959 drive the guide into the cylinder head until it protrudes 19 mm (¾ in) above the valve spring recess in the head.

NOTE: Service valve guides are 0.02 mm (0.001 in) larger on the outside diameter than the original equipment to ensure interference fit.

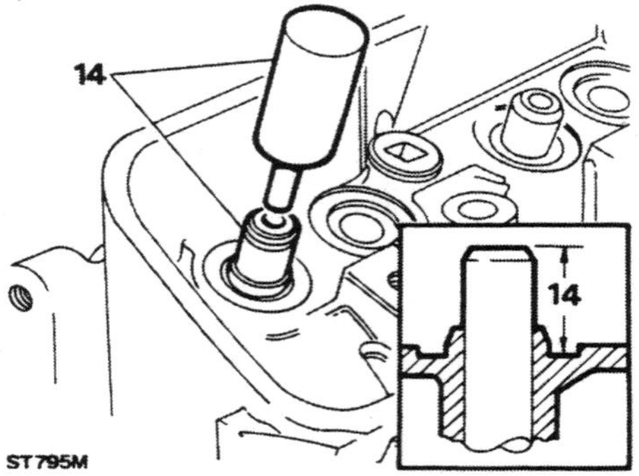

## Examine and fit new valve seats

15. Check the valve seats for wear, pits and burning and renew the inserts if necessary.

16. Remove the old seat inserts by grinding them away until they are thin enough to be cracked and prised out.
17. Heat the cylinder head evenly to approximately 65° C (150° F).
18. Press the new insert into the recess in the cylinder head.

NOTE: Service valve seat inserts are available in two over-sizes 0.25 and 0.50 mm (0.010 and 0.020 in) larger on the outside diameter to ensure interference fit.

19. If necessary, cut the valve seats to 46 ± ¼ degrees.
20. The nominal seat width is 1.5 mm (0.031 in). If the seat exceeds 2.0 mm (0.078 in) it should be reduced to the specified width by the use of 20° and 70° cutters.
21. The inlet valve seat diameter: 'A' is 37.03 mm (1.458 in) and the exhaust valve seat is 31.50 mm (1.240 in).

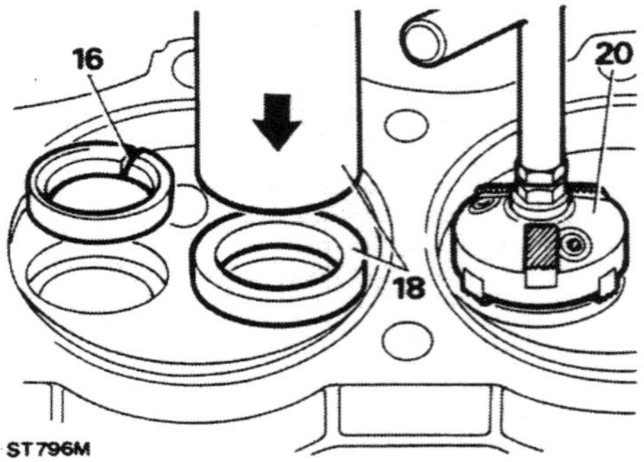

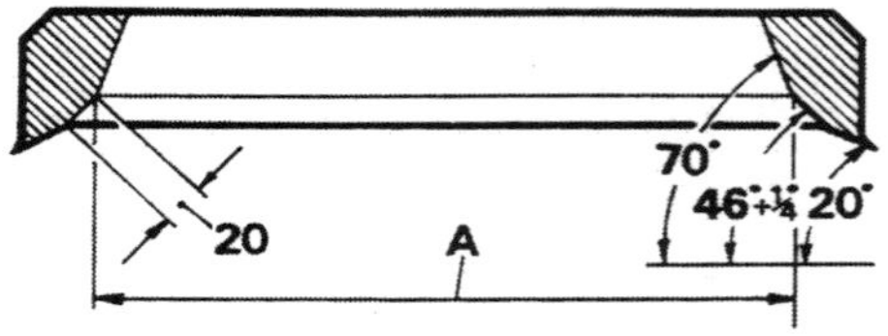

22. Ensure that the cutter blades are correctly fitted to the cutter head with the angled end of the blade downwards facing the work, as illustrated. Check that the cutter blades are adjusted so that the middle of the blade contacts the area of material to be cut. Use the key provided in the hand set MS76. Use light pressure and remove only the minimum material necessary.

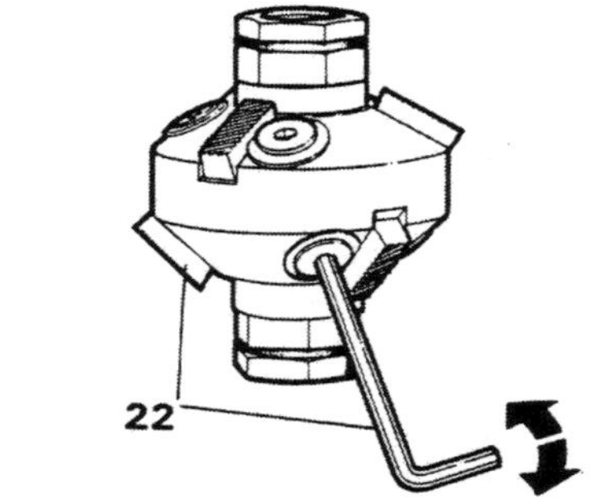

23. Smear a small quantity of engineers' blue round the valve seat and revolve a properly ground valve against the seat. A continuous fine line should appear round the valve. If there is a gap of not more than 12 mm it can be corrected by lapping.
24. Alternatively, insert a strip of cellophane between the valve and seat, hold the valve down by the stem and slowly pull out the cellophane. If there is a drag the seal is satisfactory in that spot. Repeat this in at least eight places. Lapping-in will correct a small open spot.

## Assemble valves to cylinder head

25. Before fitting the valves and springs the height of each valve above the head must be checked. Insert each valve in turn in its guide and whilst holding the head firmly against the seat, measure the height of the stem above the valve spring seat surface. This dimension must not exceed 47.63 mm (1.875 in). If necessary renew the valve or grind the end of the valve stem.
26. Lubricate the valve stems and assemble the valves, springs and caps and secure with the collets using valve spring compressor 18G 106A.

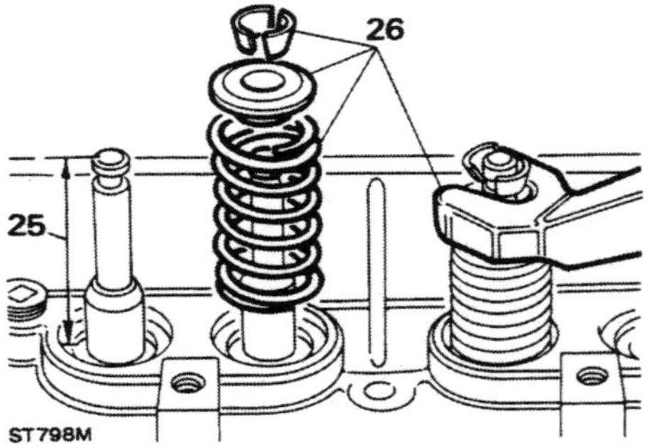

Continued

## ENGINE 12

### Reclaiming cylinder head threads

Damaged or stripped threads in the cylinder head can be salvaged by fitting Helicoils as follows:

Holes A — These three holes may be drilled 0.3906 in dia. X 0.937 + 0.040 in deep. Tapped with Helicoil Tap No. 6 CPB or 6CS × 0.875 in (min) deep (⅜ UNC 1½D insert).

Holes B — These eight holes may be drilled 0.3906 in dia. X 0.812 + 0.040 in deep. Tapped with Helicoil Tap No. 6 CBB 0.749 in (min).

Holes C — These four holes may be drilled 0.3906 in dia. X 0.937 + 0.040 in deep. Tapped with Helicoil Tap No. 6 CPB or 6CS × 0.875 in (min) deep (⅜ UNC 1½D insert).

Holes D — These four holes may be drilled 0.261 in dia. X 0.675 + 0.040 in deep. Tapped with Helicoil Tap No. 4 CPB or 4CS × 0.625 in (min) deep (¼ UNC 1½D insert).

Holes E — These six holes may be drilled 0.3906 in dia. X 0.937 + 0.040 in deep. Tapped with Helicoil Tap No. 6 CPB or 6CS × 0.875 in (min) deep (⅜ UNC 1½D insert).

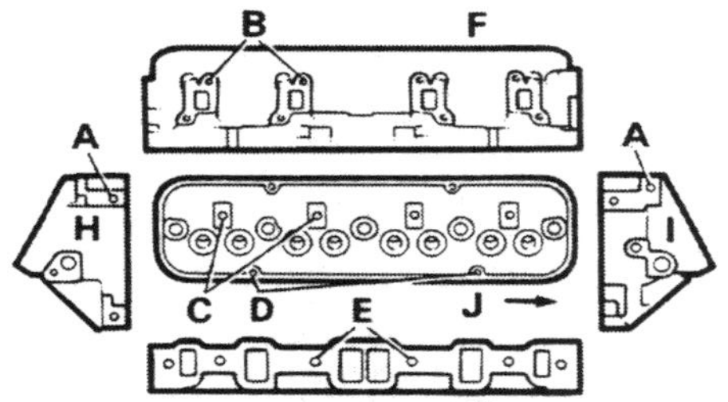

**NOTE: Right-hand cylinder head illustrated. American projection.**

F Exhaust manifold face
G Inlet manifold face
H Front face
I Rear face
J Front of engine

### REMOVE AND OVERHAUL FLYWHEEL

1. Remove the retaining bolts and withdraw the flywheel from the crankshaft.

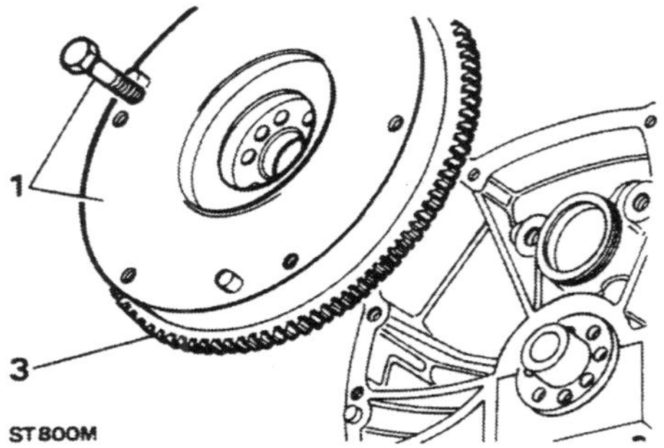

2. Examine the flywheel clutch face for cracks, scores and overheating. If the overall thickness of the flywheel is in excess of the minimum thickness i.e. 39.93 mm (1.572 in) it can be refaced provided that after machining it will not be below the minimum thickness. Remove the three dowels before machining.

3. Examine the ring gear and if worn or the teeth are chipped and broken it can be renewed as follows:

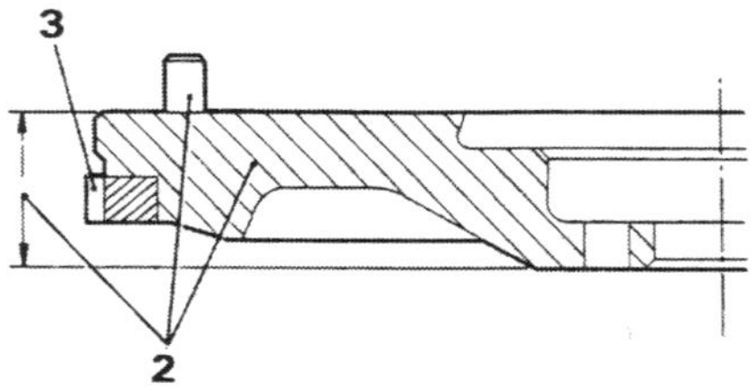

4. Drill a 10 mm (0.393 in) diameter hole axially between the root of any tooth and the inner diameter of the starter ring sufficiently deep to weaken the ring. Do NOT allow the drill to enter the flywheel.

5. Secure the flywheel in a vice fitted with soft jaws and place a cloth over the flywheel to protect the operator from flying fragments.

**WARNING: Take adequate precautions against flying fragments when splitting the ring gear.**

6. Place a chisel immediately above the drilled hole and strike it sharply to split the starter ring gear.

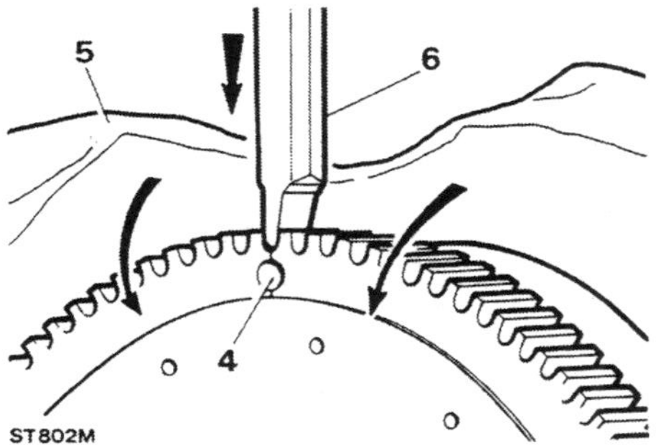

7. Heat the new ring gear uniformly to between 170° and 175°C (338° to 347°F) but do not exceed the higher temperature.
8. Place the flywheel, clutch side down, on a flat surface.
9. Locate the heated starter ring gear in position on the flywheel, with the chamfered inner diameter towards the flywheel flange. If the starter ring gear is chamfered both sides, it can be fitted either way round.
10. Press the starter ring gear firmly against the flange until the ring contracts sufficiently to grip the flywheel.
11. Allow the flywheel to cool gradually. Do NOT hasten cooling in any way or distorting may occur.
12. Fit new clutch assembly location dowels to the flywheel.

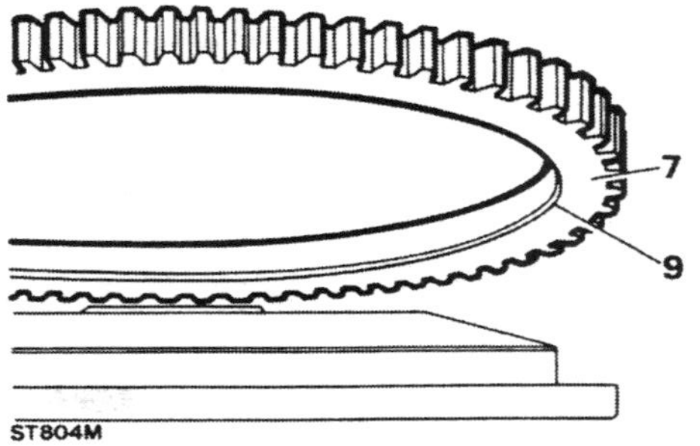

### REMOVE TIMING GEAR COVER AND WATER PUMP

1. Place an oil drip-tray beneath the timing cover and remove the oil filter element.
2. Remove the crankshaft pulley bolt and special washer and withdraw the pulley.
3. Remove the two bolts securing the sump to the bottom of the timing cover.
4. Remove the remaining timing cover retaining bolts and withdraw the cover complete with oil pump.

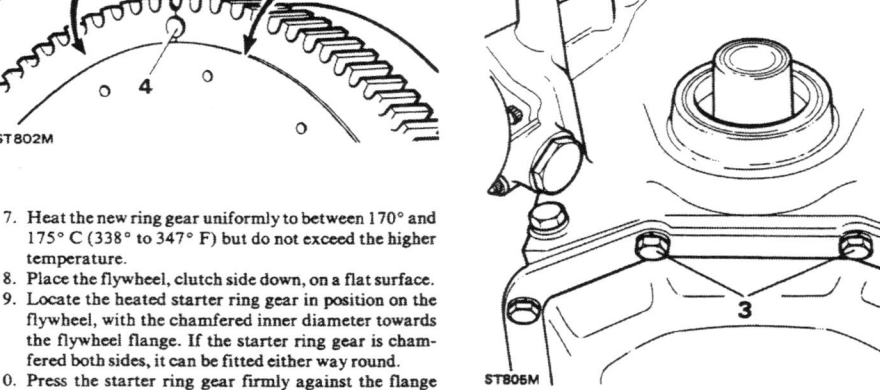

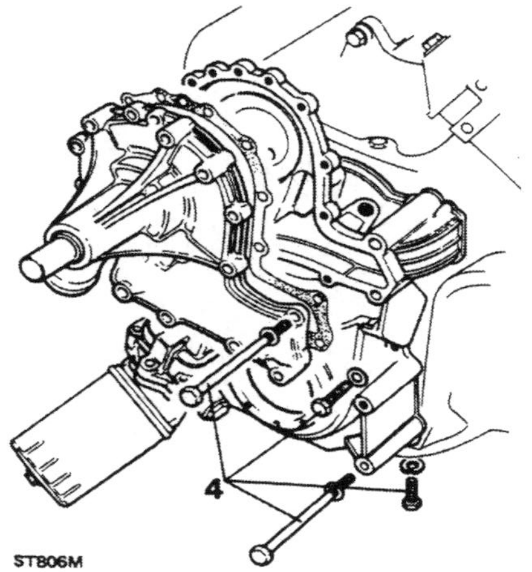

*Continued*

## ENGINE 12

### Renewing timing cover oil seal

5. Remove the seven drive screws and withdraw the mud shield and the oil seal.
6. Position the gear cover with the front face uppermost and the underside supported across the oil seal housing bore on a suitable wooden block.
7. Enter the oil seal, lip side leading, into the housing bore.
8. Press in the oil seal until the plain face is 1.5 mm (0.062 in) approximately below the gear cover face.
9. Fit the mud shield and secure with the screws.

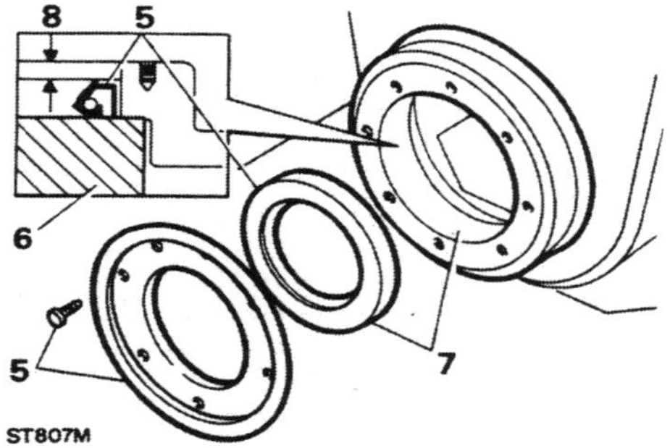

### REMOVE AND OVERHAUL THE OIL PUMP

1. Remove the bolts from the oil pump cover.
2. Withdraw the oil pump cover.
3. Lift off the cover gasket.
4. Withdraw the oil pump gears.

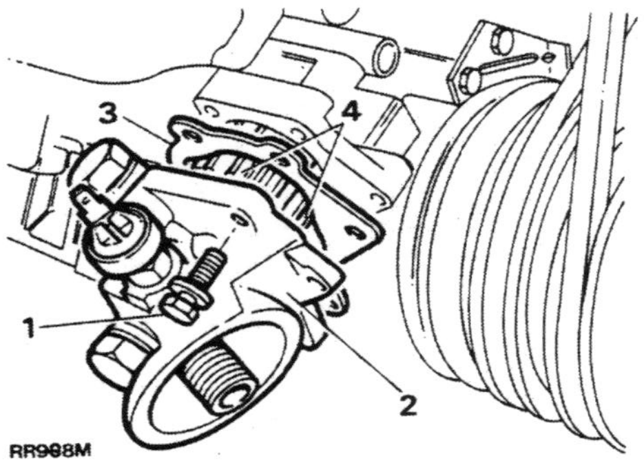

### Dismantle pump

5. Unscrew the plug from the pressure relief valve.
6. Lift off the joint washer for the plug.
7. Withdraw the spring from the relief valve.
8. Withdraw the pressure relief valve.

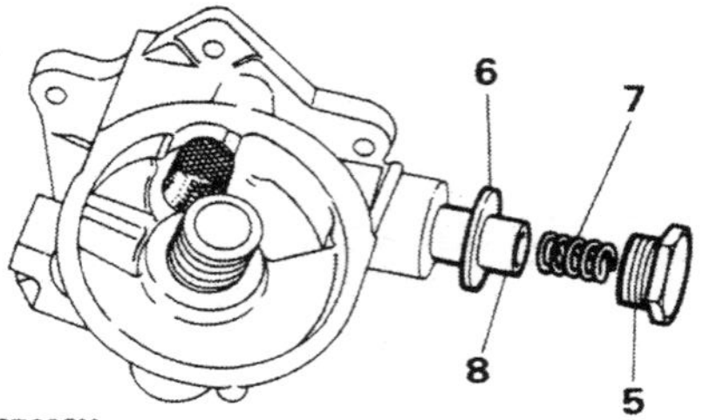

### Examine pump

9. Check the oil pump gears for wear or scores.
10. Fit the oil pump gears and shaft into the front cover.
11. Place a straight-edge across the gears.
12. Check the clearance between the straight-edge and the front cover. If less than 0.05 mm (0.0018 in), check the front cover gear pocket for wear.
13. Check the oil pressure relief valve for wear or scores.
14. Check the relief valve spring for wear at the sides or signs of collapse.
15. Clean the gauze filter for the relief valve.
16. Check the fit of the relief valve in its bore. The valve must be an easy slide fit with no perceptible side movement.

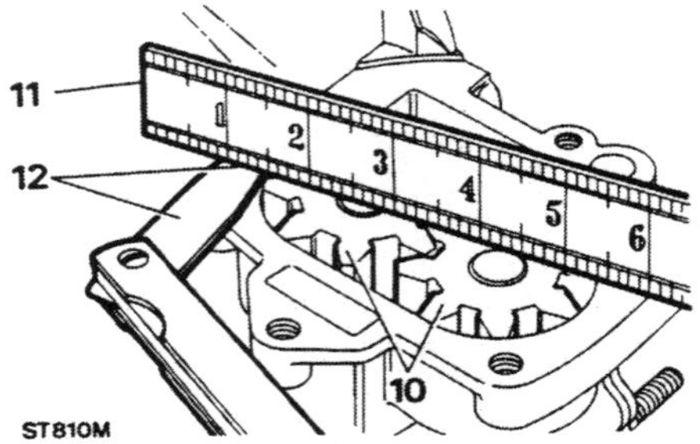

### Assemble pump

17. Insert the relief valve spring.
18. Locate the sealing washer on to the relief valve plug.
19. Fit the relief valve plug and tighten to correct torque —see data section.
20. Fully pack the oil pump gear housing with petroleum jelly. Use only petroleum jelly; no other grease is suitable.
21. Fit the oil pump gears so that the petroleum jelly is forced into every cavity between the teeth of the gears.

**IMPORTANT:** Unless the pump is fully packed with petroleum jelly it may not prime itself when the engine is started.

22. Place a new gasket on the oil pump cover.
23. Locate the oil pump cover in position.
24. Fit the special fixing bolts and tighten alternately and evenly to the correct torque.

### REMOVE TIMING CHAIN GEARS AND CAMSHAFT

**CAUTION: If this operation is carried out with cylinder heads and rocker shafts in position the engine must NOT be rotated once the timing chain has been removed otherwise the pistons and valves will be damaged.**

1. Remove the retaining bolt and washer and withdraw the distributor drive gear and spacer.
2. Withdraw the chain wheels complete with timing chain.
3. Withdraw the camshaft whilst taking particular care not to damage the bearings in the cylinder block.

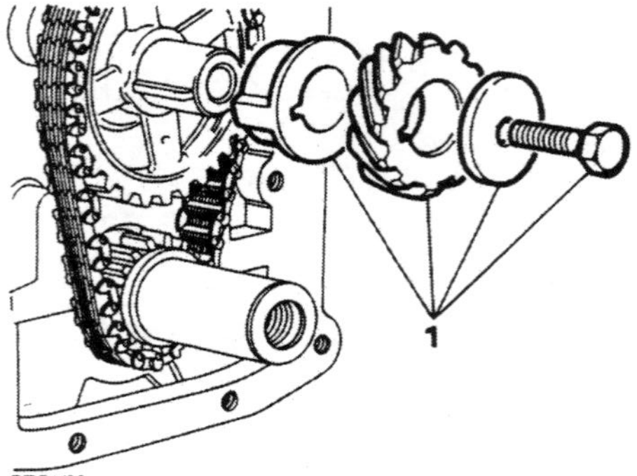

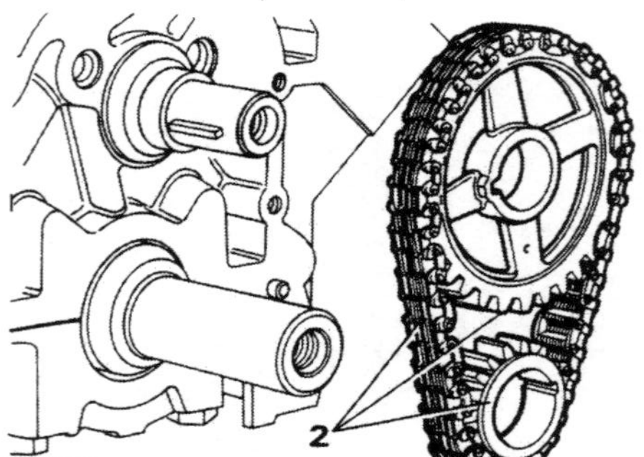

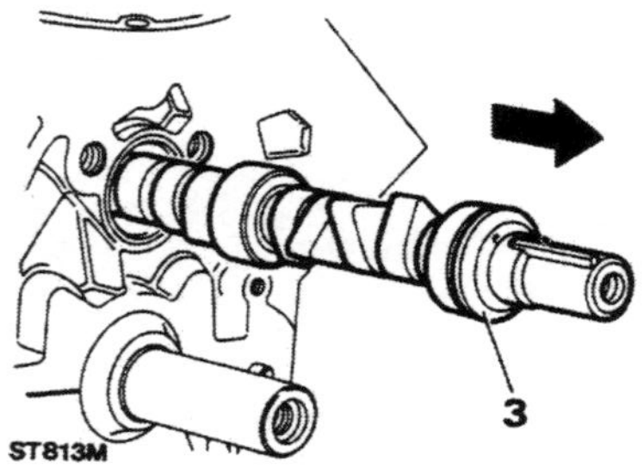

### Examine components

4. Visually examine all parts for wear. Check the camshaft bearing journals and cams for wear, pits, scores and overheating. Should any of these conditions be present the shaft should be renewed.
5. Examine the links and pins of the timing chain for wear and compare its condition with that of a new chain. Similarly the teeth of the chain wheels should be inspected and if necessary the wheels should be renewed.

*Continued*

# ENGINE 12

6. Measure the camshaft journals for overall wear, ovality and taper. The diameters of the five journals are as follows commencing from the front of the shaft:

   Number 1 journal 1.786 to 1.785 in
   Number 2 journal 1.750 to 1.755 in
   Number 3 journal 1.726 to 1.725 in
   Number 4 journal 1.696 to 1.695 in
   Number 5 journal 1.666 to 1.665 in.

7. To check the camshaft for bow, rest the two end journals i.e. numbers 1 and 5 on 'V' blocks and mount a dial gauge on the centre journal. Rotate the shaft and note the reading. If the run out is more than 0.05 mm (0.002 in) it should be renewed.

## REMOVE AND OVERHAUL CONNECTING-RODS AND PISTONS

1. Withdraw the remaining bolts and remove the sump.
2. Remove the sump oil strainer.
3. Remove the connecting-rod caps and retain them in sequence for reassembly.
4. Screw the guide bolts 605351 onto the connecting-rods.
5. Push the connecting-rod and piston assembly up the cylinder bore and withdraw it from the top. Retain the connecting-rod and piston assemblies in sequence with their respective caps.
6. Remove the guide bolts 605351 from the connecting-rod.

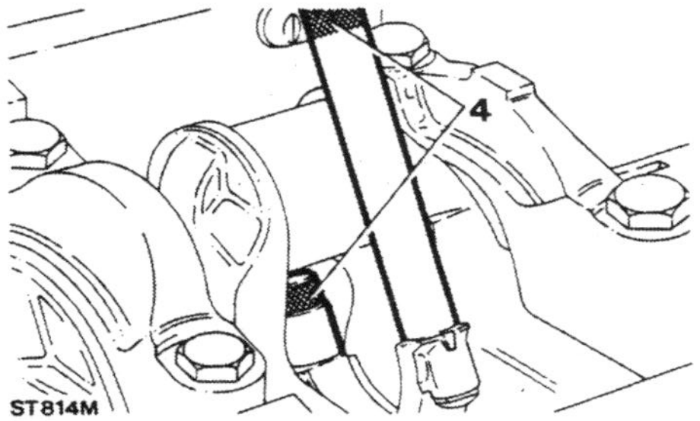

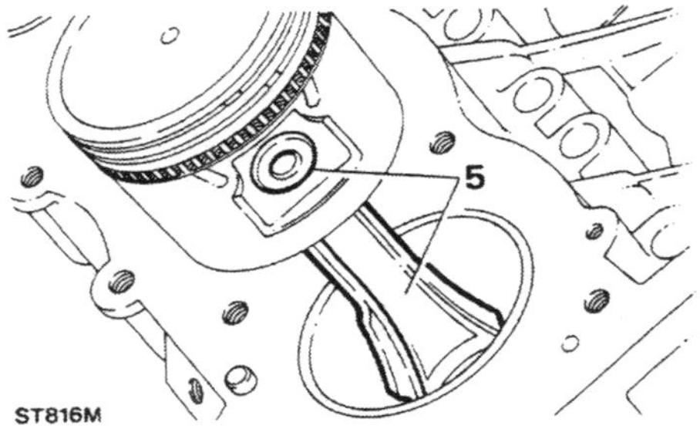

## Overhaul

NOTE: **The connecting-rods, caps and bearing shells must be retained in sets, and in the correct sequence. Remove the piston rings over the crown of the piston. If the same piston is to be refitted, mark it relative to its connecting-rod to ensure that the original assembly is maintained.**

7. Withdraw the gudgeon pin, using tool 18G 1150 as follows:
   a. Clamp the hexagon body of 18G 1150 in a vice.
   b. Position the large nut flush with the end of the centre screw.
   c. Push the screw forward until the nut contacts the thrust race.
   d. Locate the piston adaptor 18G 1150 E with its long spigot inside the bore of the hexagon body.
   e. Fit the remover/replacer bush of 18G 1150 on the centre screw with the flanged end away from the gudgeon pin.
   f. Screw the stop-nut about half-way onto the smaller threaded end of the centre screw, leaving a gap 'A' of 3 mm (⅛ in) between this nut and the remover/replacer bush.
   g. Lock the stop-nut securely with the lock screw.
   h. Check that the remover/replacer bush is correctly positioned in the bore of the piston.
   i. Push the connecting-rod to the right to expose the end of the gudgeon pin, which must be located in the end of the adaptor 'd'.
   j. Screw the large nut up to the thrust race.
   k. Hold the lock screw and turn the large nut until the gudgeon pin has been withdrawn from the piston. Dismantle the tool.

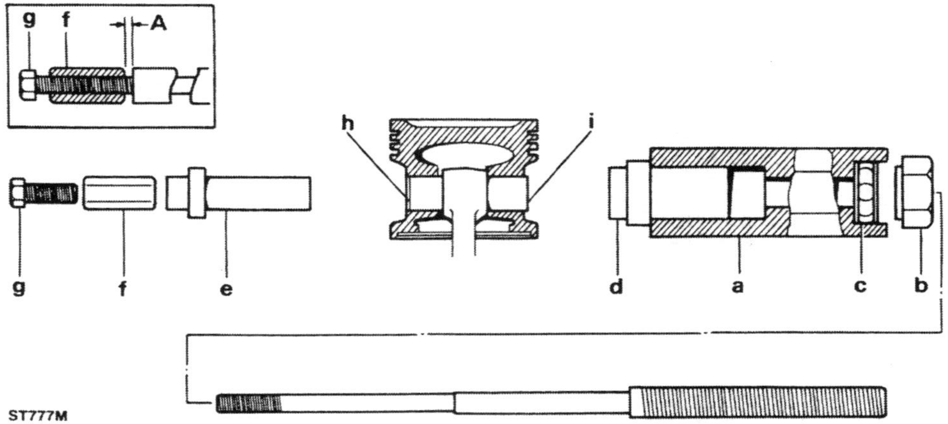

8. As an alternative to tool 18G 1150, press the gudgeon pin from the piston using an hydraulic press and the components which comprise tool 605350 as follows:
   A. Place the base of tool 605350 on the bed of an hydraulic press which has a capacity of 8 tons (8 tonnes).
   B. Fit the guide tube into the bore of the base with its countersunk face uppermost.
   C. Push the piston to one side so as to expose one end of the gudgeon pin and locate this end in the guide tube.
   D. Fit the spigot end of the small diameter mandrel into the gudgeon pin.
   E. Press out the gudgeon pin, using the hydraulic press.

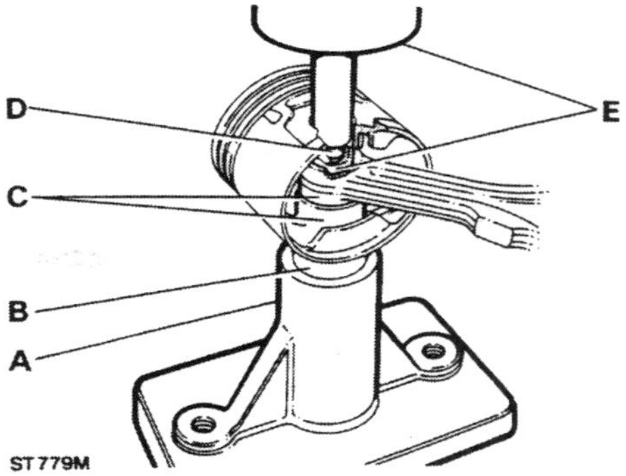

### Original pistons

Remove the carbon deposits, particularly from the ring grooves.
Examine the pistons for signs of damage or excessive wear; refer to 'new pistons' for the method of checking the running clearance. Fit new pistons if necessary.

### New pistons

Pistons are available in service standard size and in oversizes of 0.25 mm (0.010 in) and 0.50 mm (0.020 in). Service standard size pistons are supplied 0.0254 mm (0.001 in) oversize. When fitting new service standard size pistons to a cylinder block, check for correct piston to bore clearance, honing the bore if necessary. Bottom of piston skirt/bore clearance should be 0.018 to 0.033 mm (0.0007 to 0.0013 in).

NOTE: **The temperature of the piston and cylinder block must be the same to ensure accurate measurement. When reboring the cylinder block, the crankshaft main bearing caps must be fitted and tightened to the correct torque.**

9. Check the cylinder bore dimension at right angles to the gudgeon pin, 40 to 50 mm (1½ to 2 in) from the top.

*Continued*

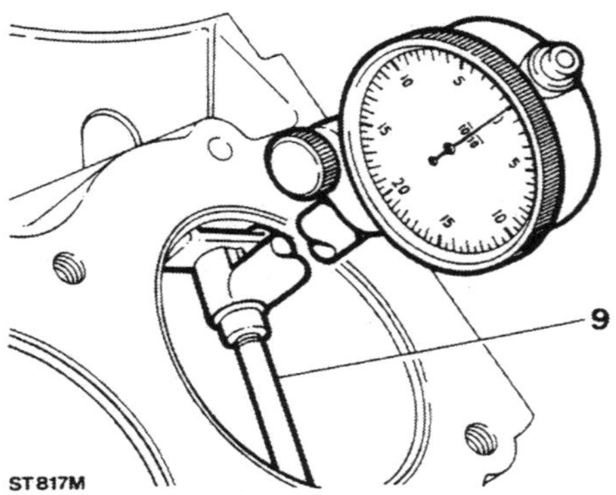

# ENGINE 12

10. Check the piston dimension at right angles to the gudgeon pin, at the bottom of the skirt.
11. The piston dimension must be 0.018 to 0.033 mm (0.0007 to 0.0013 in) smaller than the cylinder.
12. If new piston rings are to be fitted without reboring, deglaze the cylinder walls with a hone, without increasing the bore diameter to provide a cross-hatch finish.
13. Check the compression ring gaps in the applicable cylinder, held square to the bore with the piston. Gap limits: 0.44 to 0.56 mm (0.017 to 0.022 in). Use a fine-cut flat file to increase the gap if required. Select a new piston ring if the gap exceeds the limit.

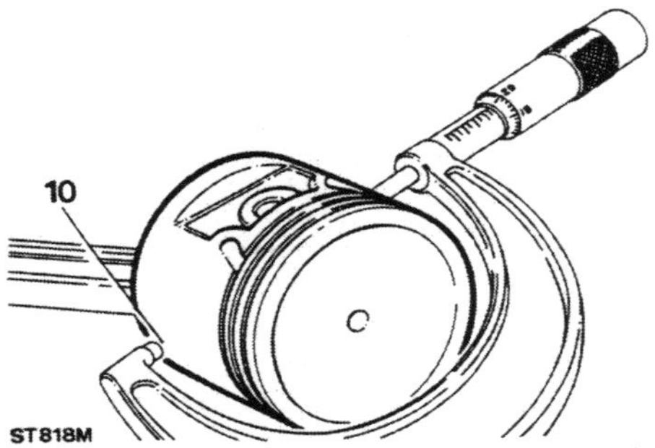

NOTE: Gapping does not apply to oil control rings.

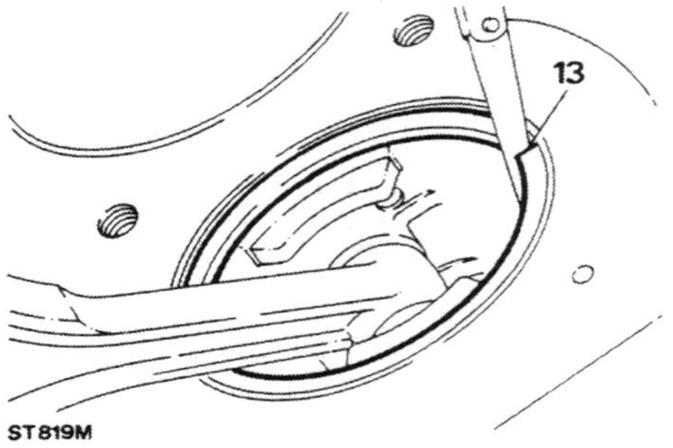

14. Temporarily fit the compression rings to the piston.
15. The ring marked 'TOP' must be fitted with the marking uppermost and into the second groove. The chrome ring is for the top groove and can be fitted either way round.
16. Check the compression ring clearance in the piston groove. Clearance limits: 0.05 to 0.10 mm (0.002 to 0.004 in).

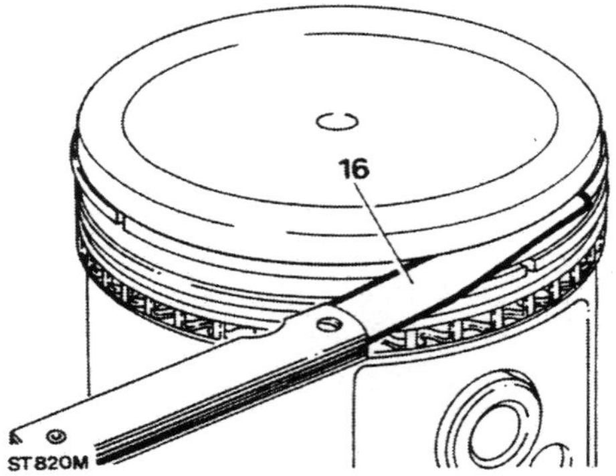

## Fit piston rings

17. Fit the expander ring into the bottom groove making sure that the ends butt and do not overlap.
18. Fit two ring rails to the bottom groove, one above and one below the expander ring.
19. Fit the second compression ring with the marking 'TOP' uppermost and the chrome compression ring in the top groove, either way round.

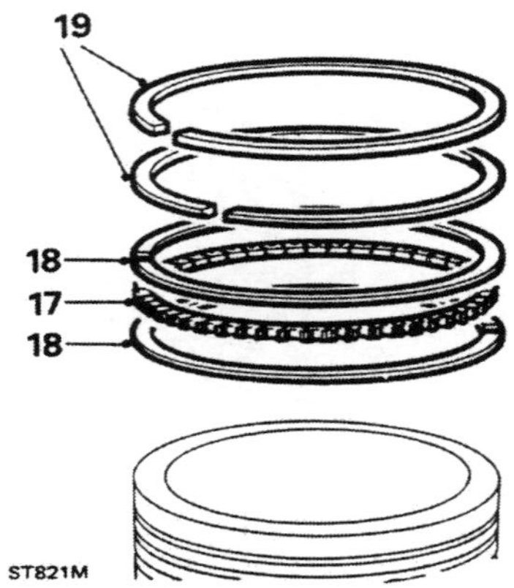

## Examine connecting-rods

20. Check the alignment of the connecting-rod.
21. Check the connecting-rod small end, the gudgeon pin must be a press fit.

## Check crankshaft bearings

22. Locate the bearing upper shell into the connecting-rod.
23. Locate the connecting-rod and bearing onto the applicable crankshaft journal, noting that the domed shape boss on the connecting-rod must face towards the front of the engine on the right-hand bank of cylinders and towards the rear on the left-hand bank.
24. When both connecting-rods are fitted, the bosses will face inwards towards each other.
25. Place a piece of Plastigauge across the centre of the lower half of the crankshaft journal.
26. Locate the bearing lower shell into the connecting-rod cap.
27. Locate the cap and shell onto the connecting-rod. Note that the rib on the edge of the cap must be the same side as the domed shape boss on the connecting-rod.
28. Secure the connecting-rod cap. Tighten to the correct torque, see data section.
29. Do not rotate the crankshaft or connecting rod while the Plastigauge is in use.
30. Remove the connecting-rod cap and shell.
31. Using the scale printed on the Plastigauge packet, measure the flattened Plastigauge at its widest point.
32. The graduation that most closely corresponds to the width of the Plastigauge indicates the bearing clearance.
33. The correct bearing clearance with new or overhauled components is 0.013 to 0.06 mm (0.0006 to 0.0022 in).
34. If a bearing has been in service, it is advisable to fit a new bearing if the clearance exceeds 0.08 mm (0.003 in).
35. If a new bearing is being fitted, use selective assembly to obtain the correct clearance.
36. Wipe off the Plastigauge with an oily rag. DO NOT scrape it off.

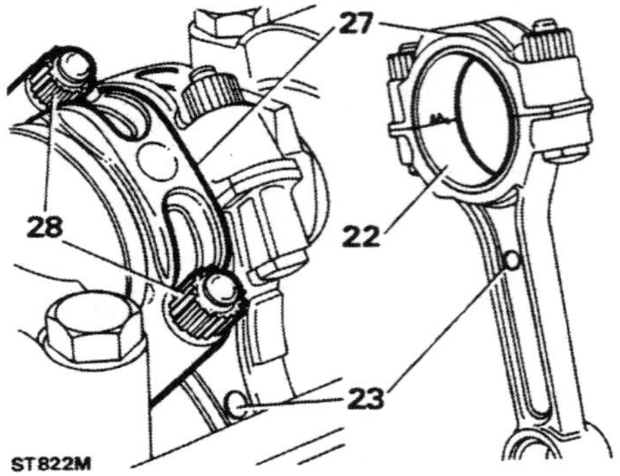

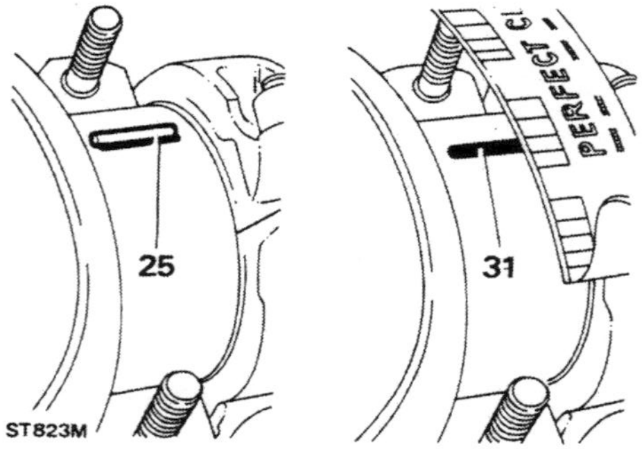

IMPORTANT: The connecting-rods, caps and bearing shells must be retained in sets, and in the correct sequence.

*Continued*

# ENGINE 12

j. Set the torque wrench 18G 537 to 12 lbf/ft. This represents the minimum load for an acceptable interference fit of the gudgeon pin in the connecting-rod.

k. Using the torque wrench and socket 18G 587 on the large nut, and holding the lock screw, pull the gudgeon pin in until the flange of the remover/replacer bush is 4 mm (0.160 in) 'B' from the face of the piston. Under no circumstances must this flange be allowed to contact the piston.

**CAUTION: If the torque wrench has not broken throughout the pull, the fit of the gudgeon pin to the connecting-pin is not acceptable and necessitates the renewal of components. The large nut and centre screw of the tool must be kept well-oiled.**

39. Remove the tool and check that the piston moves freely on the gudgeon pin and that no damage has occurred during pressing.

---

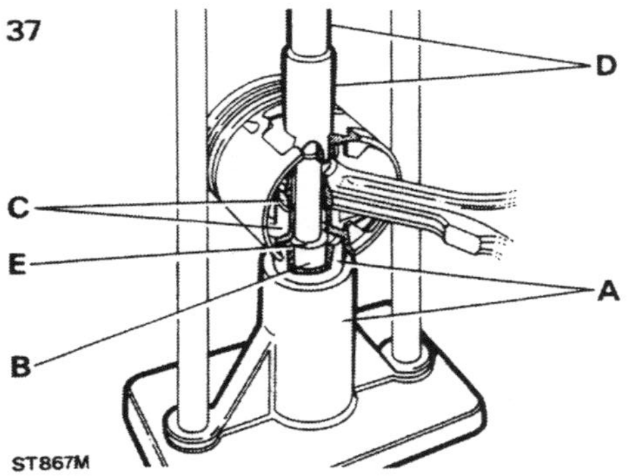

## Assembling pistons to connecting-rods

37. If an hydraulic press and tool 605350 was used for dismantling, refit each piston to its connecting-rod as follows:

    A. Check that the base of tool 605350 and the guide tube are fitted as follows:
       Place the base of tool 605350 on the bed of an hydraulic press which has a capacity of 8 tons (8 tonnes).
       Fit the guide tube into the bore of the base with its countersunk face uppermost.
    B. Fit the long mandrel inside the guide tube.
    C. Fit the connecting-rod into the piston with the markings together if the original pair are being used, then place the piston and connecting rod assembly over the long mandrel until the gudgeon pin boss rests on the guide tube.
    D. Fit the gudgeon pin into the piston up to the connecting-rod, and the spigot end of the small diameter mandrel into the gudgeon pin.
    E. Press in the gudgeon pin until it abuts the shoulder of the long mandrel.

38. If tool 18G 1150 was used for dismantling, refit each piston to its connecting-rod as follows:

    a. Clamp the hexagon body of 18G 1150 in a vice, with the adaptor 18G 1150 E positioned as in 7d.
    b. Remove the large nut of 18G 1150 and push the centre screw approximately 50 mm (2 in) into the body until the shoulder is exposed.
    c. Slide the parallel guide sleeve, grooved end last, onto the centre screw and up to the shoulder.
    d. Lubricate the gudgeon pin and bores of the connecting-rod and piston with graphited oil (Acheson's Colloids 'Oildag'). Also lubricate the ball race and centre screw of 18G 1150.
    e. Fit the connecting-rod and the piston together onto the tool with the markings together if the original pair are being used and with the connecting-rod around the sleeve up to the groove.
    f. Fit the gudgeon pin into the piston bore up to the connecting-rod.
    g. Fit the remover/replacer bush 18G 1150/3 with its flanged end towards the gudgeon pin.
    h. Screw the stop-nut onto the centre screw and adjust this nut to obtain a 1 mm (1/32 in) end-float 'A' on the whole assembly, and lock the nut securely with the screw.
    i. Slide the assembly back into the hexagon body and screw on the large nut up to the thrust race.

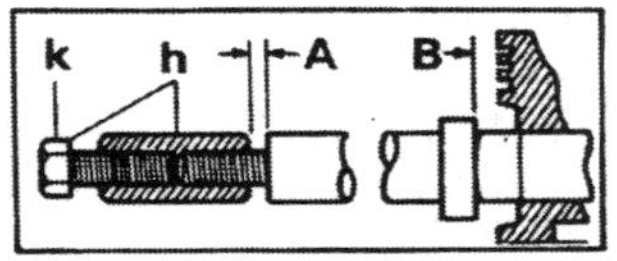

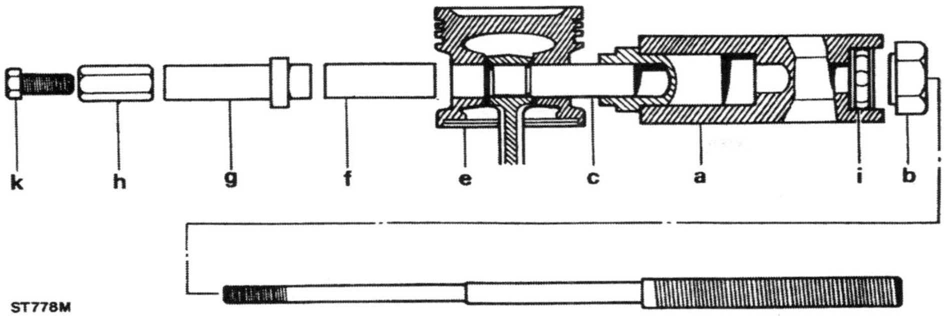

## ENGINE 12

### REMOVE AND OVERHAUL CRANKSHAFT

1. Remove the main bearing caps and lower bearing shells and retain in sequence. It is important to keep them in pairs and mark them with the number of the respective journal until it is decided if the bearing shells are to be refitted.
2. Lift out the crankshaft and rear oil seal.

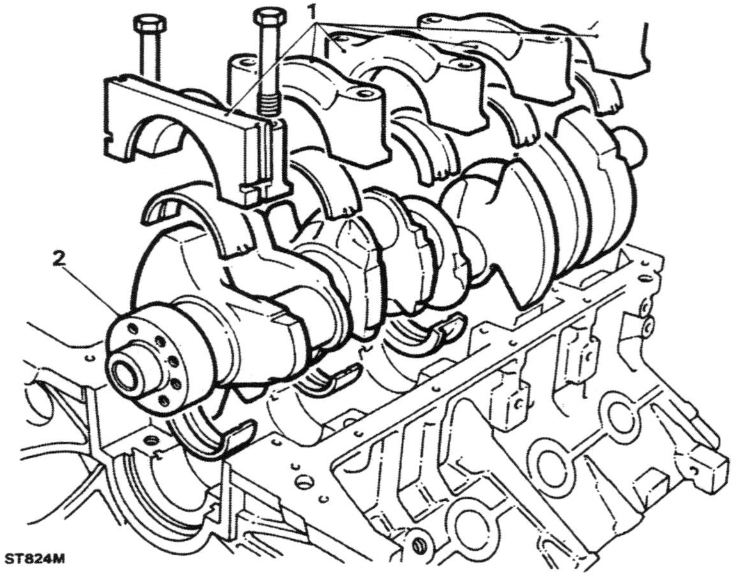

### Inspect and overhaul crankshaft

3. Rest the crankshaft on vee-blocks at numbers one and five main bearing journals.
4. Using a dial test indicator, check the run-out at numbers two, three and four main bearing journals. The total indicator readings at each journal should not exceed 0.08 mm (0.003 in).
5. While checking the run-out at each journal, note the relation of maximum eccentricity on each journal to the others. The maximum on all journals should come at very near the same angular location.
6. If the crankshaft fails to meet the foregoing checks it is bent and is unsatisfactory for service.
7. Check each crankshaft journal for ovality. If ovality exceeds 0.040 mm (0.0015 in), a reground or new crankshaft should be fitted.

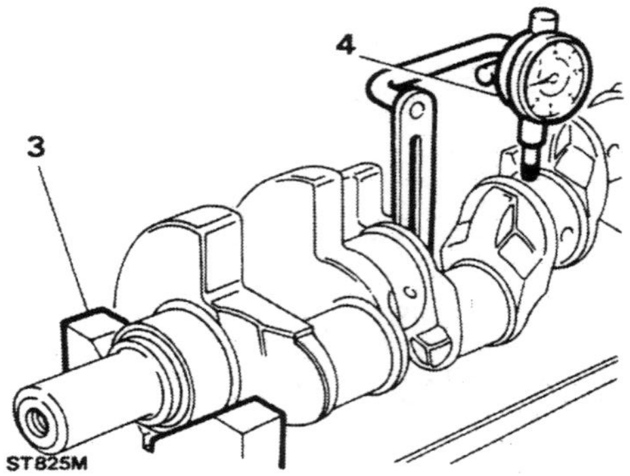

8. Bearings for the crankshaft main journals and the connecting-rod journals are available in the following undersizes:
   0.25 mm (0.010 in)
   0.50 mm (0.020 in)
9. The centre main bearing shell, which controls crankshaft thrust, has the thrust faces increased in thickness when more than 0.25 mm (0.010 in) undersize, as shown on the following chart.

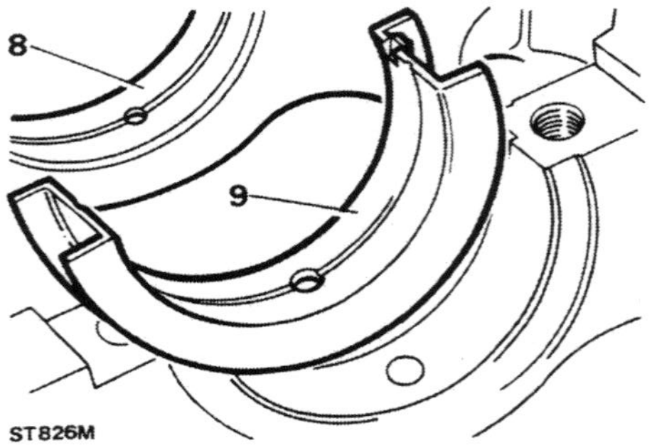

### Crankshaft dimensions — millimetres

| Crankshaft Grade | Diameter '12' | Width '13' | Diameter '14' |
|---|---|---|---|
| Standard | 58.400–58.413 | 26.975–27.026 | 50.800–50.812 |
| 0.254 U/S | 58.146–58.158 | 26.975–27.026 | 50.546–50.559 |
| 0.508 U/S | 57.892–57.904 | 27.229–27.280 | 50.292–50.305 |

### Crankshaft dimensions — inches

| Crankshaft Grade | Diameter '12' | Width '13' | Diameter '14' |
|---|---|---|---|
| Standard | 2.2992–2.2997 | 1.062–1.064 | 2.0000–2.0005 |
| 0.010 U/S | 2.2892–2.2897 | 1.062–1.064 | 1.9900–1.9905 |
| 0.020 U/S | 2.2792–2.2797 | 1.072–1.074 | 1.9800–1.9805 |

### Check main bearing clearance

16. Remove the oil seals from the cylinder block and the rear main bearing cap.
17. Locate the upper main bearing shells into the cylinder block. These must be the shells with the oil drilling and oil grooves.
18. Locate the flanged upper main bearing shell in the centre position.
19. Place the crankshaft in position on the bearings.

*Continued*

10. When a crankshaft is to be reground, the thrust faces on either side of the centre main journal must be machined in accordance with the dimensions on the following charts.

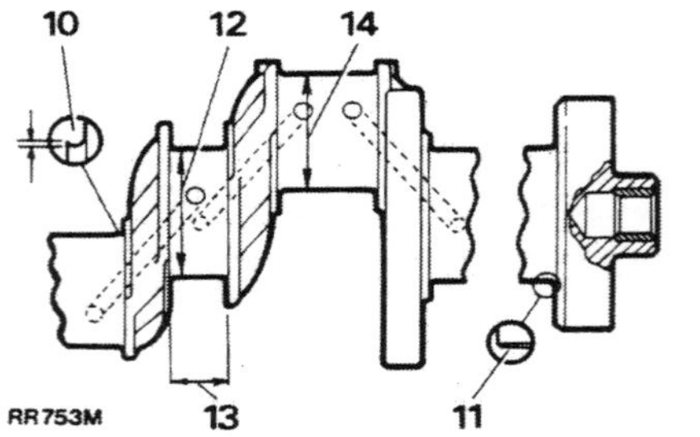

| Main bearing journal size | Thrust face width |
|---|---|
| Standard | Standard |
| 0.25 mm (0.010 in) undersize | Standard |
| 0.50 mm (0.020 in) undersize | 0.25 mm (0.010 in) oversize |

11. For example: If a 0.50 mm (0.020 in) undersize bearing is to be fitted, then 0.12 mm (0.005 in) must be machined off each thrust face of the centre journal, maintaining the correct radius.

### Crankshaft dimensions

12. The radius for all journals except the rear main bearing is 1.90 to 2.28 mm (0.075 to 0.090 in).
13. The radius for the rear main bearing journal is 3.04 mm (0.120 in).
14. Main bearing journal diameter, see the following charts.
15. Thrust face width, and connecting-rod journal diameter, see the following charts.

## ENGINE 12

20. Place a piece of Plastigauge across the centre of the crankshaft main bearing journals.
21. Locate the bearing lower shell into the main bearing cap.
22. Fit numbers one to four main bearing caps and shells, tighten to the correct torque, see data section.
23. Fit the rear main bearing cap and shell and tighten to the correct torque, see data section. Do not allow the crankshaft to be rotated while the Plastigauge is in use.
24. Remove the main bearing caps and shells.
25. Using the scale printed on the Plastigauge packet, measure the flattened Plastigauge at its widest point.

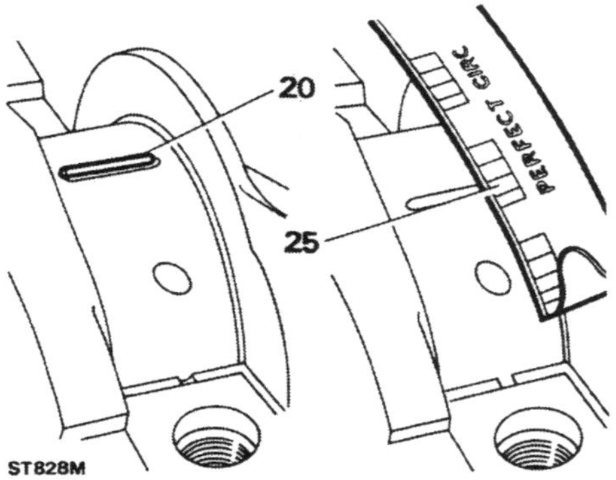

26. The graduation that most closely corresponds to the width of the Plastigauge indicates the bearing clearance.
27. The correct bearing clearance with new or overhauled components is 0.023 to 0.065 mm (0.0009 to 0.0025 in).
28. If the correct clearance is not obtained initially, use selective bearing assembly.
29. Wipe off the Plastigauge with an oily rag. Do NOT scrape it off.
30. Maintain the bearing shells and caps in sets and in the correct sequence.

### Renew spigot bearing

31. Carefully remove the old bearing.
32. Fit the spigot bearing flush with, or to a maximum of 1.6 mm (0.063 in) below the end face of the crankshaft.
33. Ream the spigot bearing to 19.177 + 0.025 mm (0.7504 + 0.001 in) inside diameter. Ensure all swarf is removed.

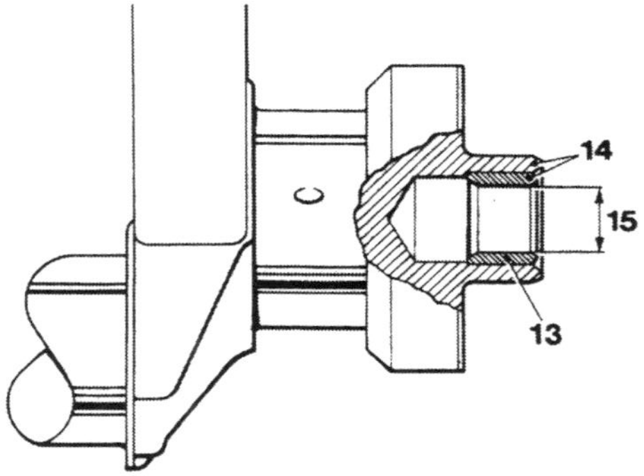

### ASSEMBLING ENGINE

### FIT CRANKSHAFT AND MAIN BEARINGS

1. Locate the upper main bearing shells into the cylinder block; these must be the shells with the oil drilling and oil grooves.
2. Locate the flanged upper main bearing shell in the centre position.
3. Lubricate the crankshaft main bearing journals and bearing shells with clean engine oil and lower the crankshaft into position.

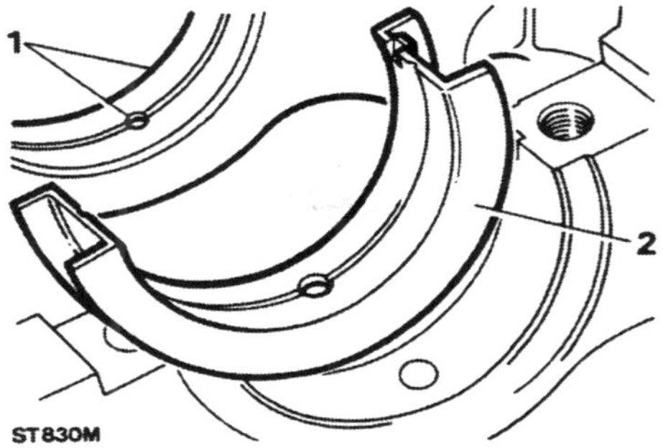

4. Lubricate the lower main bearing shells and fit numbers one to four main bearing caps and shells only, leaving the fixing bolts finger-tight at this stage.
5. Fit the cruciform side seals to the grooves each side of the rear main bearing cap. Do not cut the side seals to length, they must protrude 1.5 mm (0.062 in) approximately above the bearing cap parting face.
6. Apply Hylomar PL32M jointing compound to the rearmost half of the rear main bearing cap parting face or, if preferred, to the equivalent area on the cylinder block as illustrated.
7. Lubricate the bearing half and bearing cap side seals with clean engine oil.
8. Fit the bearing cap assembly to the engine. Do not tighten the fixings at this stage but ensure that the cap is fully home and squarely seated on the cylinder block.
9. Tension the cap bolts equally by one-quarter turn approximately, then back off one complete turn on each fixing bolt.

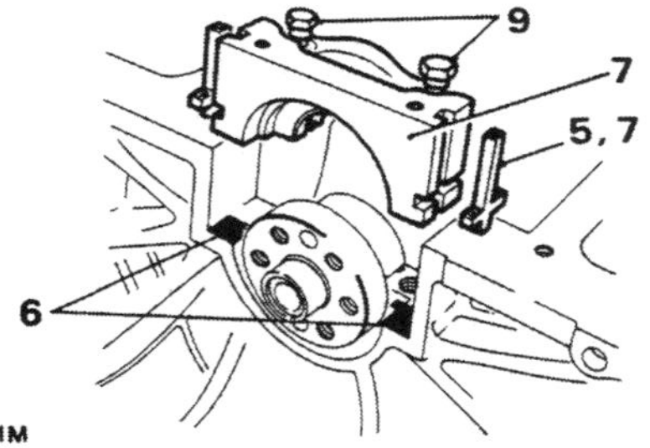

NOTE: The lubricant coating must cover the seal guide outer surface completely to ensure that the oil seal lip is not turned back during assembly.

Position the oil seal, lipped side towards the engine, onto the seal guide. The seal outside diameter must be clean and dry.

12. Push home the oil seal fully and squarely by hand into the recess formed in the cap and block until it abuts against the machined step in the recess. Withdraw the seal guide.

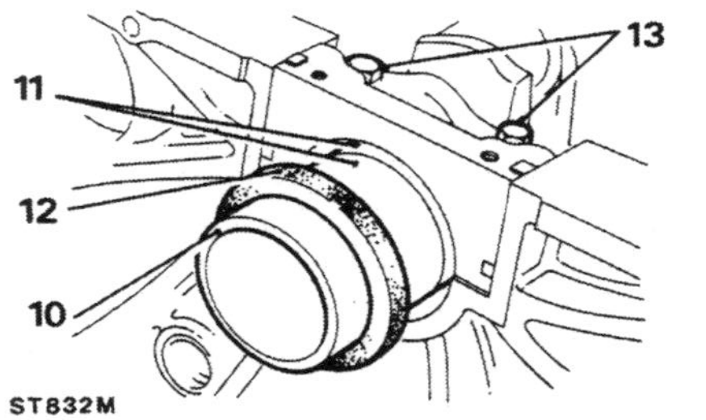

13. Tighten the main bearing cap bolts to the correct torque noting that the bolts for numbers one to four bearings have a different torque to number five bearing cap bolts.
14. Using a feeler gauge or a dial indicator check the crankshaft end-float, 0.10 to 0.20 mm (0.004 to 0.008 in).

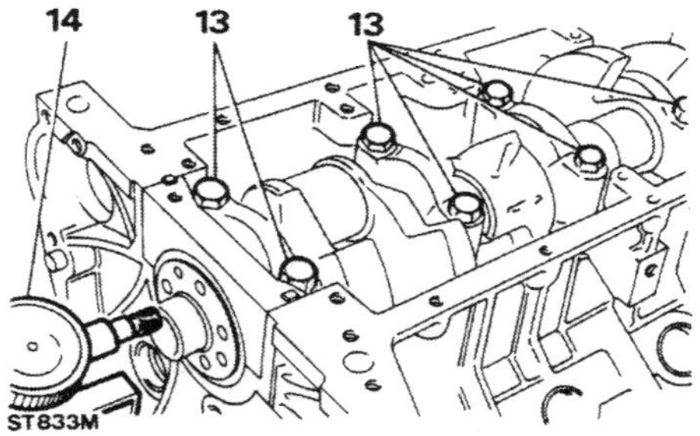

CAUTION: Do not exceed 1,000 engine rev/min when first starting the engine, otherwise the crankshaft rear oil seal will be damaged.

CAUTION: Do not handle the seal lip, visually check that it is not damaged and ensure that the outside diameter remains clean and dry.

10. Position the oil seal guide RO 1014 on the crankshaft flange.
11. Ensure that the oil seal guide and the crankshaft journal are scrupulously clean, then coat the seal guide and oil seal journal with clean engine oil.

# ENGINE 12

## FIT CONNECTING-RODS AND PISTONS

1. Locate the applicable crankshaft journal at BDC.
2. Place the bearing upper shell in the connecting-rod.
3. Retain the upper shell by screwing the guide bolts 605351 onto the connecting-rods.
4. Insert the connecting-rod and piston assembly into its respective bore, noting that the domed shape boss on the connecting-rod must face towards the front of the engine on the right-hand bank of cylinders and towards the rear on the left-hand bank. When both connecting-rods are fitted, the bosses will face inwards towards each other.
5. Position the oil control piston rings so that the ring gaps are all at one side, between the gudgeon pin and piston thrust face. Space the gaps in the ring rails approximately 25 mm (1 in) each side of the expander ring joint.
6. Position the compression rings so that their gaps are on opposite sides of the piston between the gudgeon pin and piston thrust face.
7. Using a piston ring compressor, locate the piston into the cylinder bore, until the piston crown is just below the cylinder block top face.
8. Pull the connecting rods on to the crankpins using the guide rods.
9. Place the bearing lower shell in the connecting-rod cap.
10. Locate the cap and shell onto the connecting-rod, noting that the rib on the edge of the cap must be towards the front of the engine on the right-hand bank of cylinders and towards the rear on the left-hand bank.
11. Check that the connecting-rods move freely sideways on the crankshaft. Tightness indicates insufficient bearing clearance or a misaligned connecting-rod.
12. Check the end-float between the connecting-rods on each crankshaft journal. Clearance limits: 0.15 to 0.37 mm (0.006 to 0.014 in)
13. Tighten the connecting-rod nuts to the correct torque. Fit the oil strainer and joint washer.

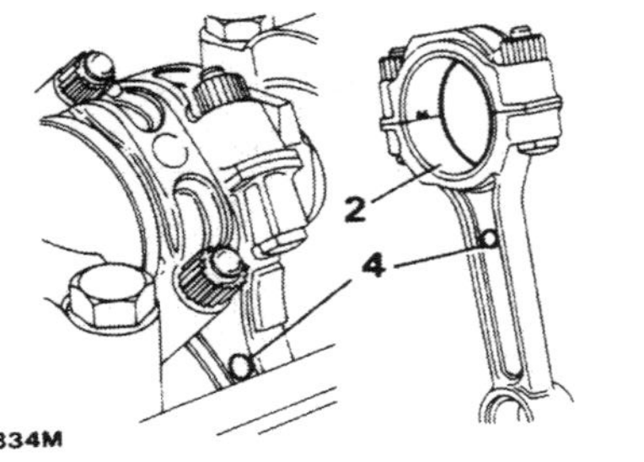

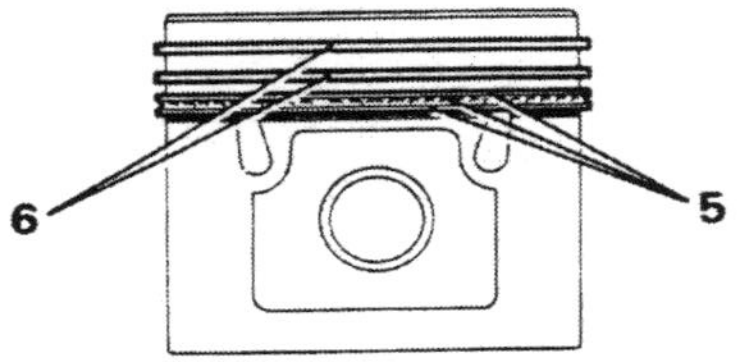

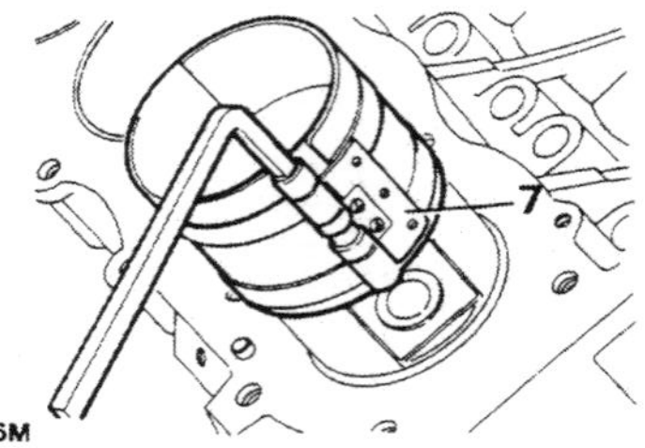

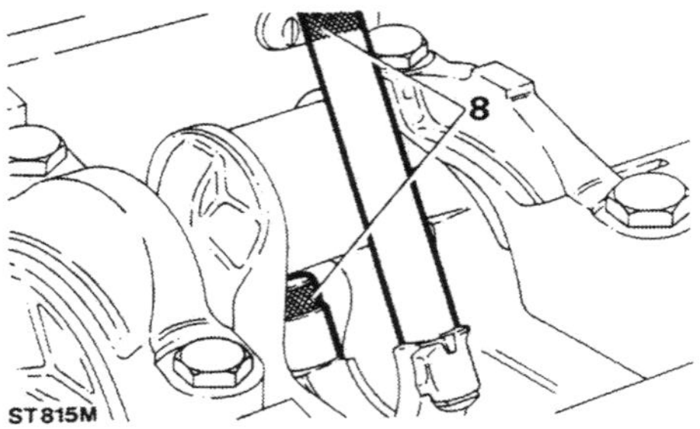

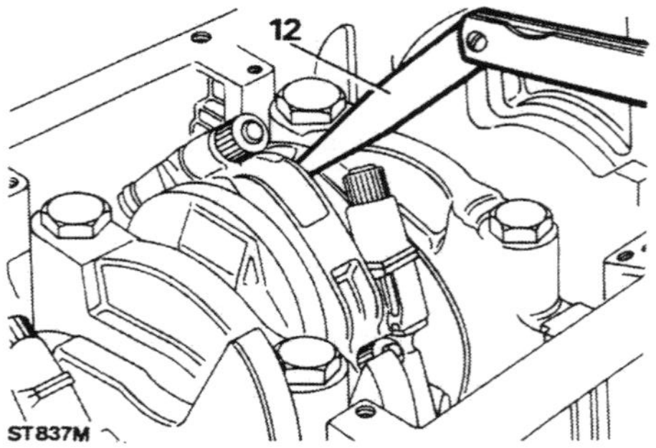

## FIT CAMSHAFT TIMING GEARS AND CHAIN

1. Lubricate the camshaft journals and carefully insert the camshaft into the cylinder block.
2. Turn the crankshaft to bring number one piston to TDC.
3. Temporarily fit the camshaft chain wheel with the marking 'FRONT' outward.
4. Turn the camshaft until the mark on the camshaft chain wheel is at the six o'clock position, then remove the chain wheel without disturbing the camshaft.
5. Encircle the chain wheels with the chain keeping the timing marks aligned.
6. Engage the chain wheel assembly on the camshaft and crankshaft key locations and check that the camshaft key is parallel to the shaft axis to ensure adequate lubrication of the distributor drive gear.

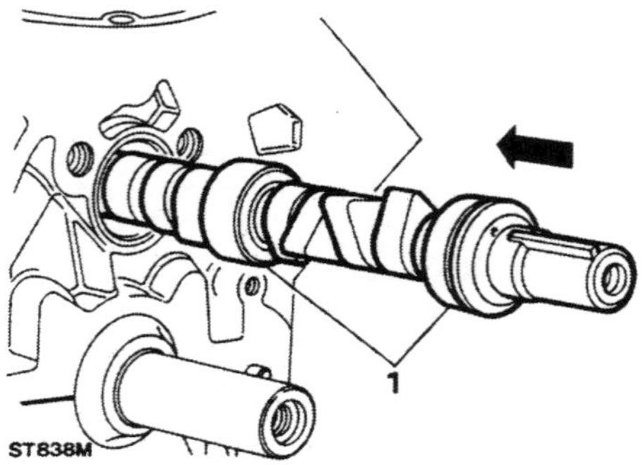

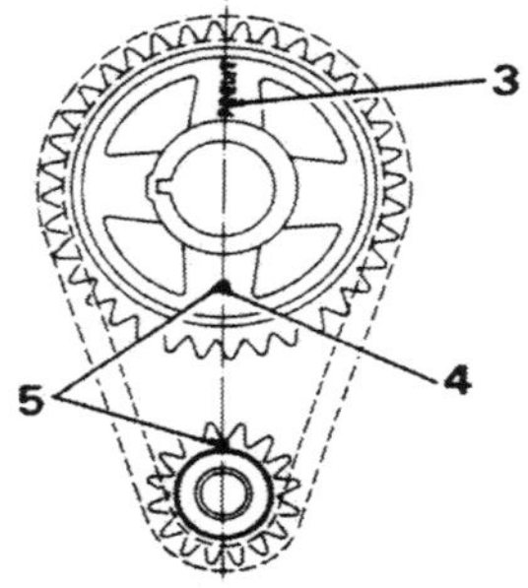

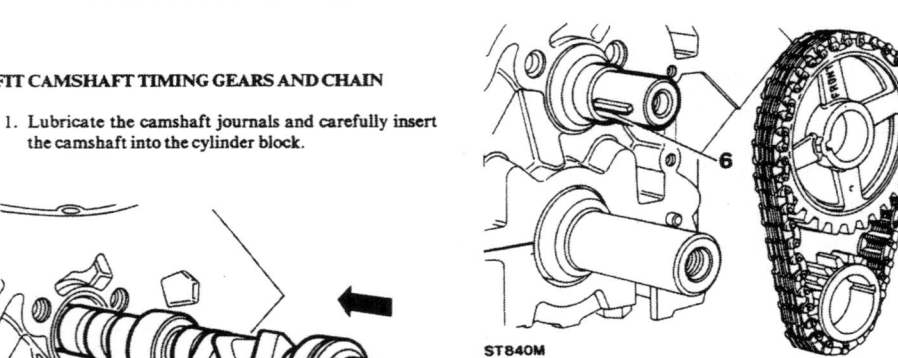

*Continued*

… ENGINE

**CAUTION:** The space between the key and keyway acts as an oilway for lubrication of the drive gear. Ensure that the key is seated to the full depth of the keyway. The overall dimension of shaft and key must not exceed 30.15 mm (1.187 in). Dimension A below.

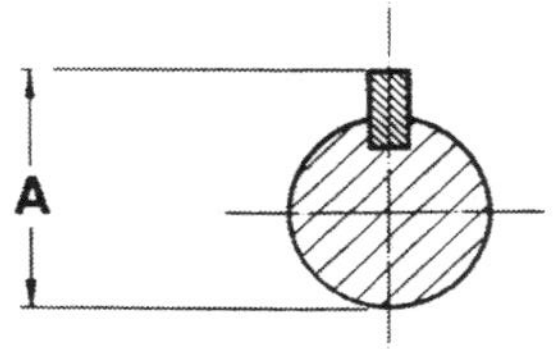

7. Check that the timing marks line-up and fit the spacer with the flange to the front.
8. Fit the distributor drive gear ensuring that the annular grooved side is fitted to the rear, that is towards the spacer.
9. Secure the drive gear and camshaft chain wheel assembly with the bolt and washer and tighten to the correct torque.

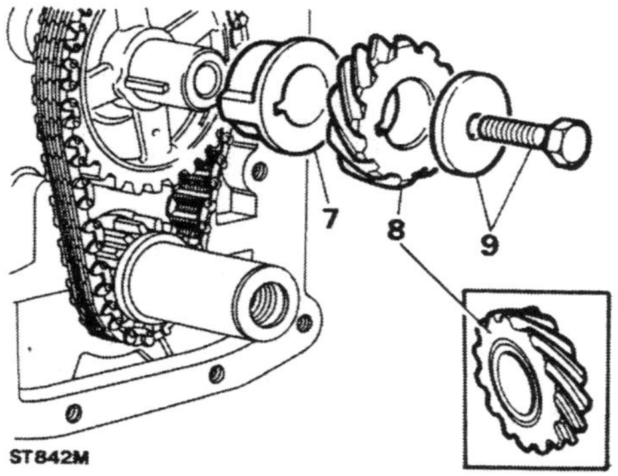

2. Locate the cylinder heads on the block dowel pins.
3. Clean the threads of the cylinder head bolts then coat them with Thread Lubricant-Sealant Loctite 572.

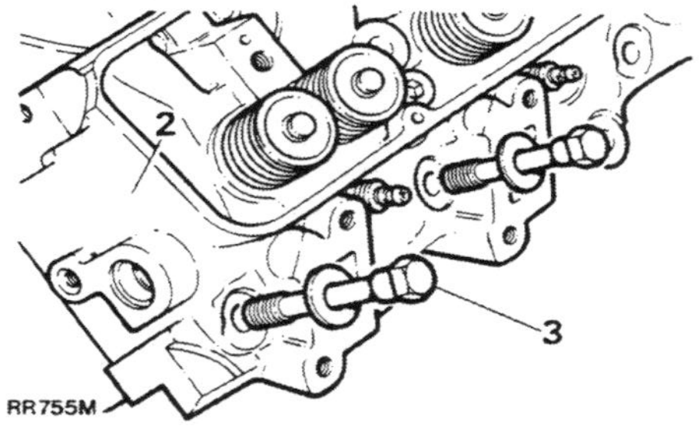

### FIT THE FLYWHEEL

1. Locate the flywheel in position on the crankshaft spigot, with the ring gear towards the engine.
2. Align the flywheel fixing bolt holes which are off-set to prevent incorrect assembly.
3. Fit the flywheel fixing bolts and before finally tightening, take up any clearance by rotating the flywheel against the direction of engine rotation. Tighten the bolts evenly to the correct torque.

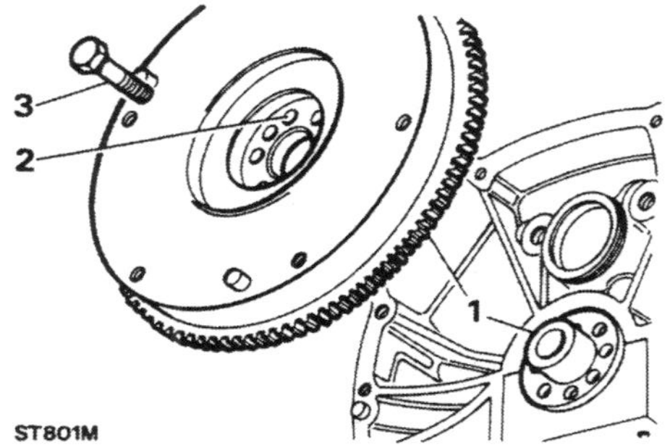

### FIT CYLINDER HEADS

1. Fit new cylinder head gaskets with the word 'TOP' uppermost. Do NOT use sealant.

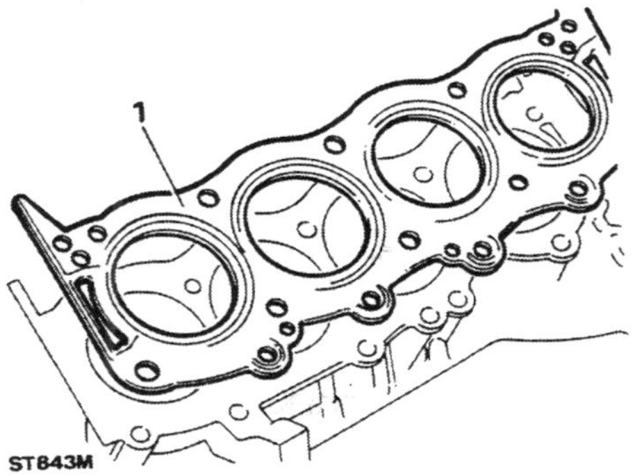

4. Locate the cylinder head bolts in position as illustrated and fit dipstick tube.

  Long bolts—1, 3 and 5.
  Medium bolts—2, 4, 6, 7, 8, 9 and 10.
  Short bolts—11, 12, 13 and 14.

5. Tighten the cylinder head bolts a little at a time in the sequence shown. See data for correct tightening torque.
6. When all bolts have been tightened, re-check the torque settings.

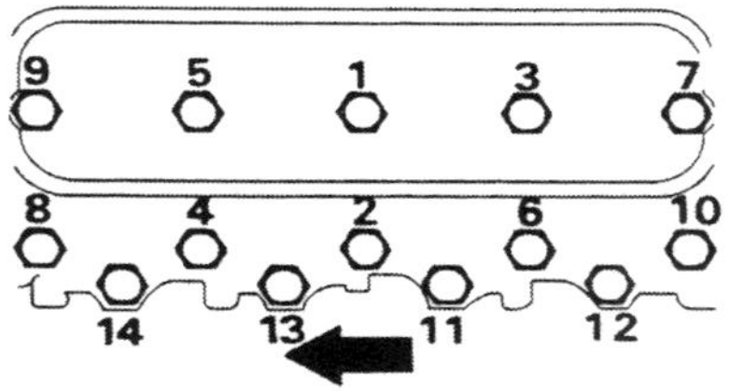

**Note:** Left-hand cylinder head illustrated. Arrow points to front of vehicle.

### FIT TAPPETS, PUSH RODS AND ROCKER ASSEMBLIES

#### Fit tappets and push rods

1. Fit the tappets and push rods to their original locations. Ensure that the tappets move freely in their respective bores. Before fitting the tappets immerse them in clean engine oil to reduce tappet noise when the engine is first started after the overhaul.

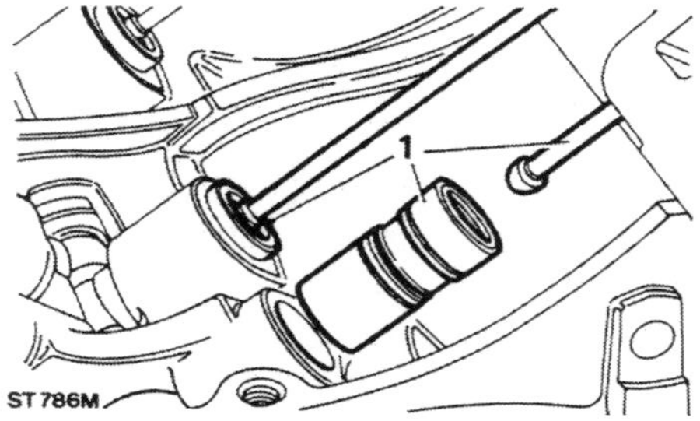

#### Fit the rocker assemblies

2. The rocker shafts are handed and must be fitted correctly to align the oilways.
3. Each rocker shaft is notched at one end and on one side only. The notch must be uppermost and towards the front of the engine on the right-hand side, and towards the rear on the left-hand side.

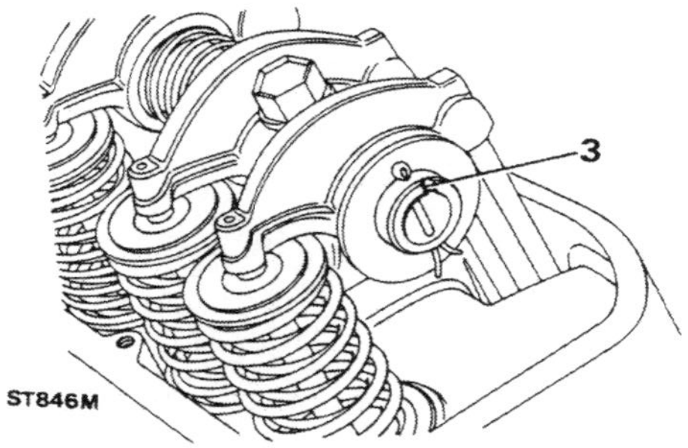

*Continued*

# ENGINE

4. Fit the rocker shaft assemblies. Ensure that the push-rods engage the rocker cups and that the baffle plates are fitted to the front on the left-hand side, and to the rear on the right-hand side. Tighten the bolts.

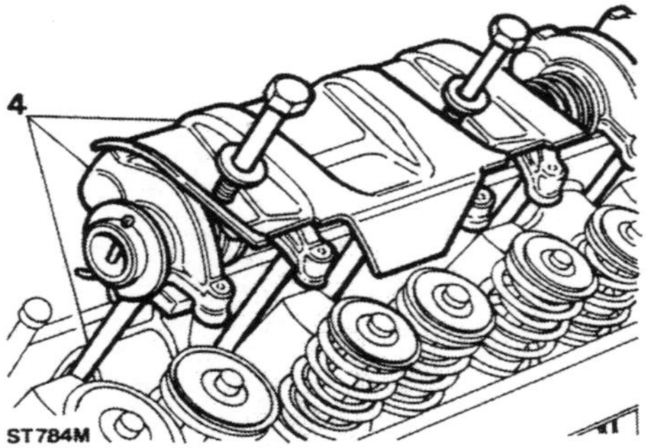

It should be noted that tappet noise can be expected on initial starting-up after an overhaul due to oil drainage from the tappet assemblies or indeed if the vehicle has been standing over a very long period. If excessive noise should be apparent after an overhaul, the engine should be run at approximately 2,500 rev/min for a few minutes (subject to the following caution) when the noise should be eliminated.

**CAUTION: Do not exceed 1,000 engine rev/min when first starting the engine, otherwise the crankshaft rear oil seal will be damaged.**

## FIT THE INTAKE MANIFOLD

1. Coat both sides of new manifold gasket seals with silicon grease.
2. Locate the seals in position with their ends engaged in the notches formed between the cylinder head and block.
3. Apply 'Hylomar' sealing compound SQ32M on the corners of the cylinder head, manifold gasket and manifold, around the water passage joints.
4. Fit the manifold gasket with the word 'FRONT' to the front and the open bolt hole at the front R.H. side.
5. Fit the gasket clamps but do not fully tighten the bolts at this stage.
6. Locate the manifold onto the cylinder head.
7. Clean the threads of the manifold securing bolts.
8. Fit all the manifold bolts and tighten them a little at a time, evenly, alternate sides working from the centre to each end and finally tighten to the correct torque.
9. Tighten the gasket clamp bolts to the correct torque.

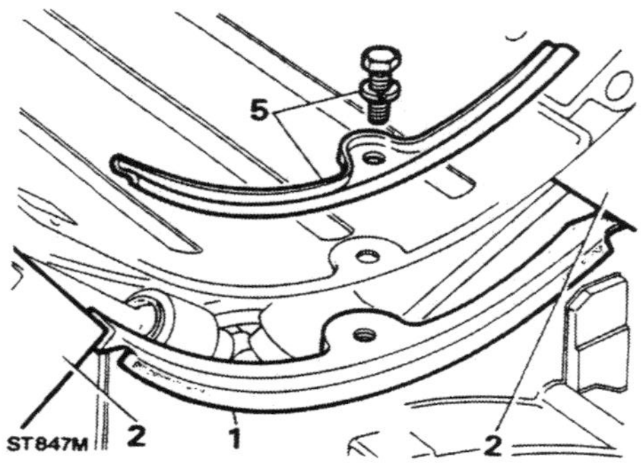

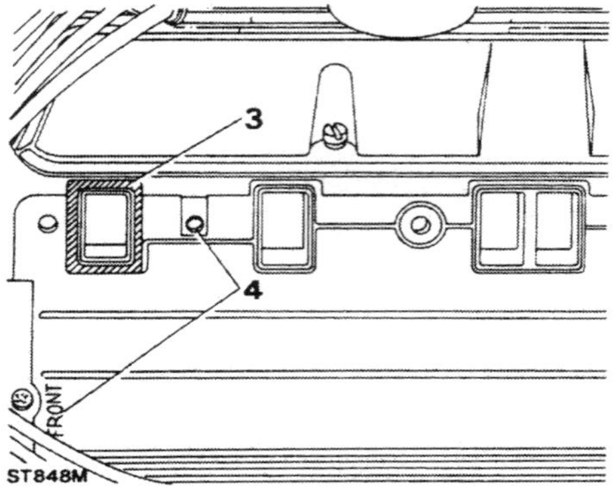

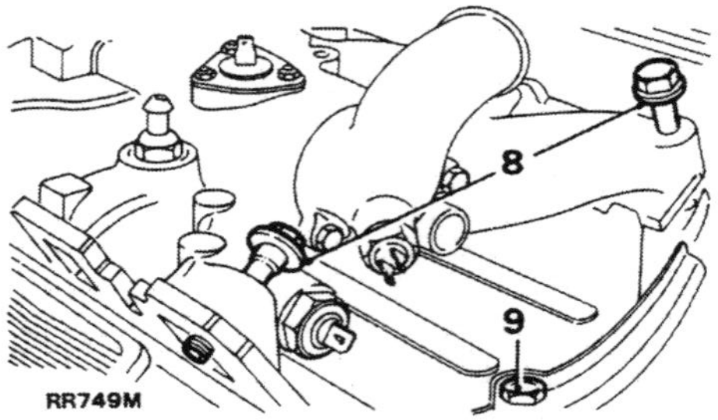

## FIT THE TIMING COVER AND CRANKSHAFT PULLEY

1. Place a new timing cover joint washer in position and fit the timing cover locating it on the two dowels.
2. Clean the threads of the timing cover securing bolts, then coat them with Thread Lubricant-Sealant Loctite 572.
3. Fit and evenly tighten the timing cover bolts to the correct torque figure.
4. Fit the crankshaft pulley and tighten the retaining bolt to the correct torque.
5. Fit timing pointer.

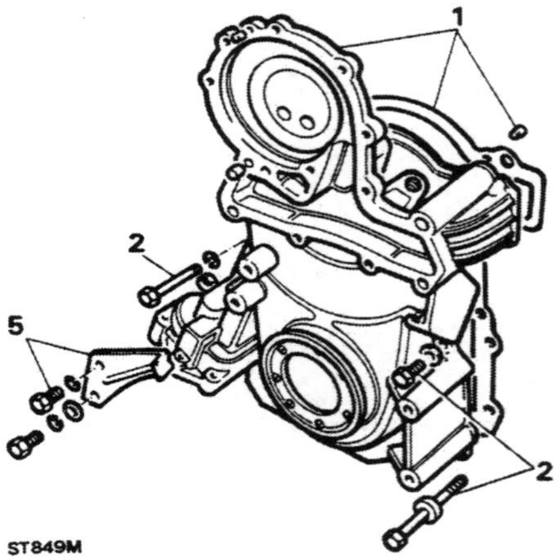

## FIT THE SUMP

1. Clean the sump mating faces and at the joint between the timing cover and crankcase apply a coating of a universal jointing compound about 13 to 19 mm (½ to ¾ in) wide in the area illustrated.
2. Place the sump gasket in position, fit the sump and evenly tighten the retaining bolts to the correct torque.

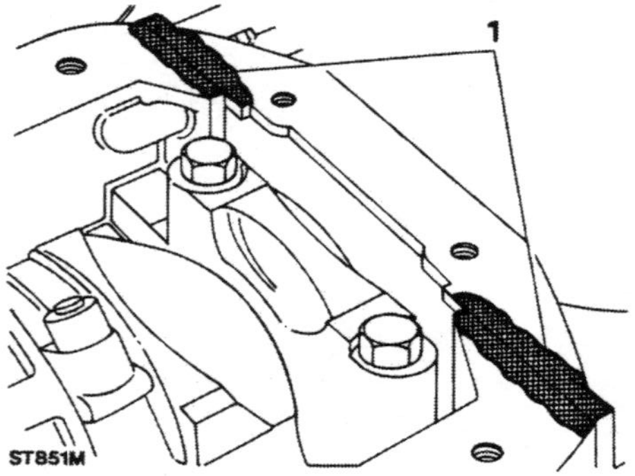

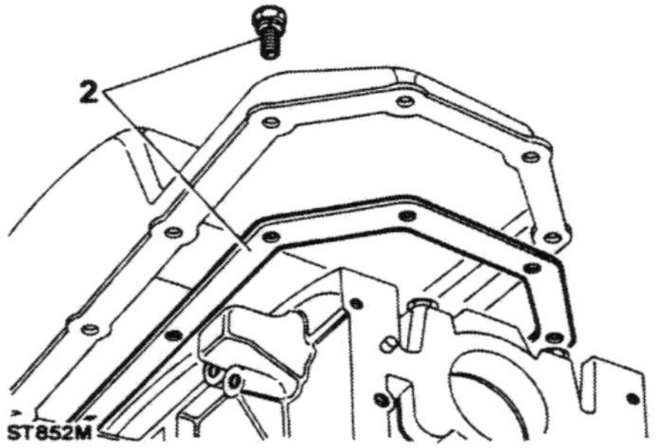

## FIT THE WATER PUMP

1. Lightly grease a new joint washer and place it in position on the timing cover.
2. Clean the threads of the four long bolts and smear them with Loctite 572 thread lubricant sealant. Locate the water pump in position.
3. Locate the alternator adjusting link on the water pump.
4. Leave the alternator adjusting link loose and tighten the remaining water pump housing bolts evenly and to the correct torque.

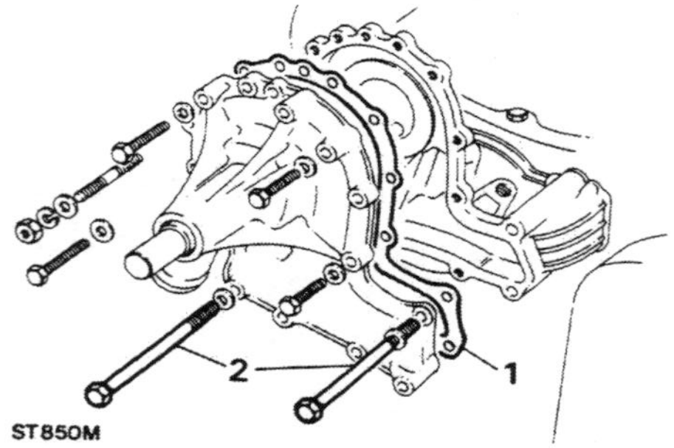

# ENGINE

## TEST AND FIT THERMOSTAT

1. Remove the two bolts securing the thermostat housing to the intake manifold.
2. Remove the housing gasket.
3. Withdraw the thermostat.

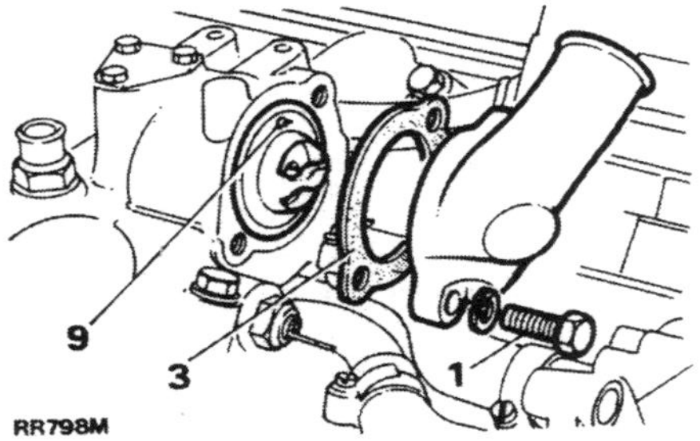

*RR798M*

4. Note the temperature stamped on the thermostat at which it should be fully open.

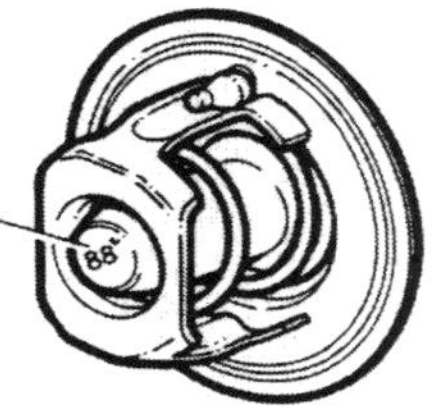

*RR799M*

5. Place the thermostat and a Centigrade thermometer in a laboratory beaker, or a suitable alternative, half full of water.
6. Heat the water and observe the temperature at which the thermostat opens.
7. If faulty discard the thermostat.

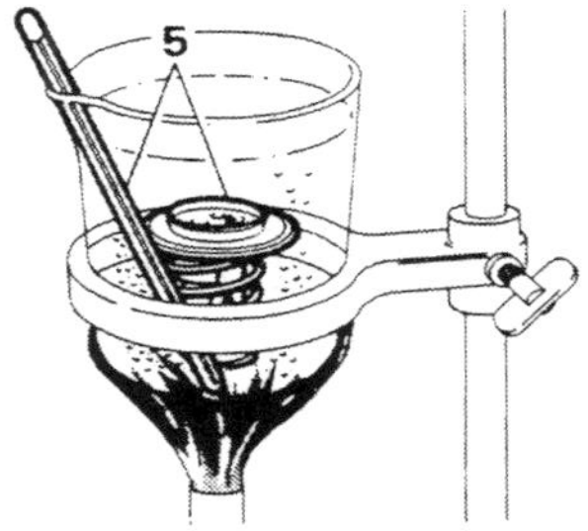

*ST858M*

8. Clean the intake manifold and thermostat housing mating faces.
9. Fit the thermostat with the jiggle pin uppermost at 12 o'clock.
10. Fit the housing using a new gasket, and tighten the two bolts to the correct torque, see 'data section'.

## FIT THE DISTRIBUTOR

1. Turn the crankshaft to bring number one piston to TDC on the compression stroke (both valves closed number one cylinder).
2. Turn distributor drive until rotor arm is approximately 30° anti-clockwise from number one sparking plug lead position on cap.
3. Turn the oil pump and distributor common drive shaft so that the tongue is in the approximate position as illustrated.

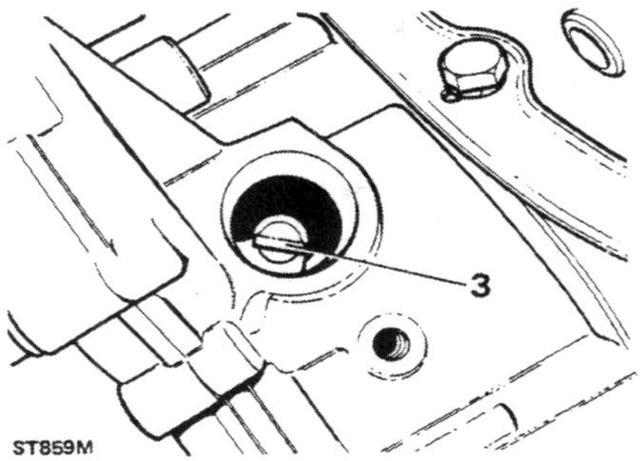

*ST859M*

## FIT ROCKER COVERS

1. Remove all traces of old gasket on the covers and cylinder heads.
2. Clean and dry the gasket mounting surface, using Bostik cleaner 6001.
3. Apply Bostik 1775 impact adhesive to the seal face and the gasket, using a brush to ensure an even film. Allow the adhesive to become touch-dry, approximately fifteen minutes.

**NOTE: The gasket fits one way round only and must be fitted accurately first time; any subsequent movement would destroy the bond.**

4. Place one end of the gasket into the cover recess with the edge firmly against the recess wall: at the same time hold the remainder of the gasket clear; then work around the cover, pressing the gasket into place ensuring that the outer edge firmly abuts the recess wall.

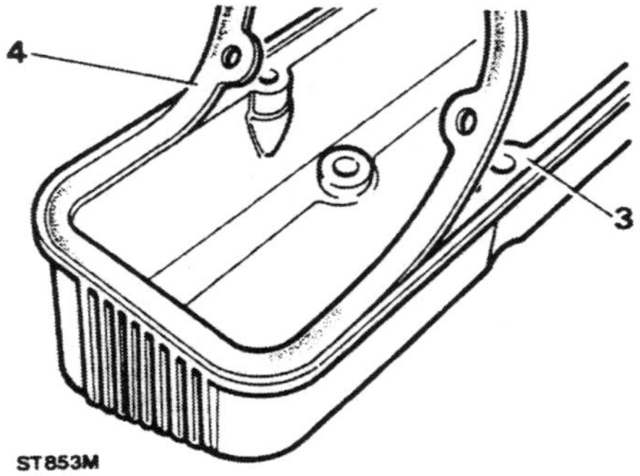

*ST853M*

4. Fit distributor to engine and check that centre line of rotor arm is now in line with number one sparking plug lead in cap. Reposition distributor if necessary. The vacuum capsule should be at 90° to the camshaft.
5. If distributor does not seat correctly in front cover, oil pump drive is not engaged. Engage by lightly pressing down distributor while turning engine.
6. Fit clamp and bolt leaving both loose at this stage.

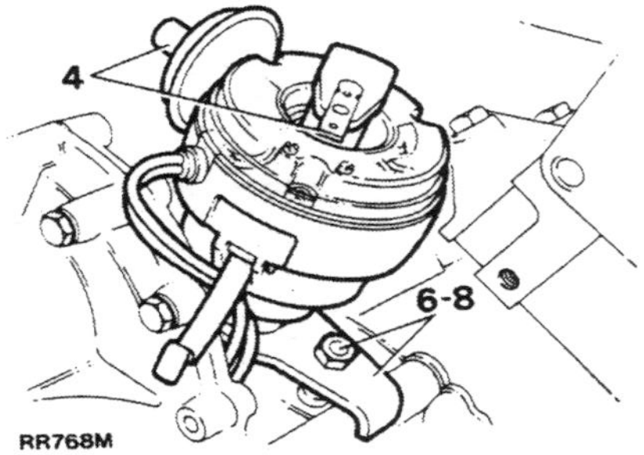

*RR768M*

7. Rotate distributor until the static timing is within 2-3° of TDC.

**CAUTION: On no account must the engine be started before this operation is carried out.**

8. Secure distributor in this position by tightening clamp bolt.
9. Fit the distributor cap and spark plugs and connect the H.T. leads in accordance with the illustration below.

**NOTE: The above distributor setting is to enable the engine to be run so that the correct setting given in 'Engine Tuning Data' can be achieved once the engine is refitted to the vehicle.**

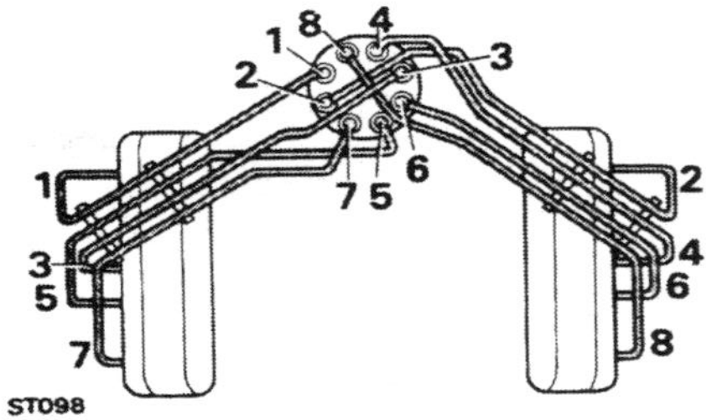

5. Allow the cover to stand for thirty minutes before fitting it to the engine.
6. Secure the rocker covers to the engine with the four screws. Short screws—in board, long screws—out board.

*ST098*

# ENGINE 12

## FIT ENGINE OIL FILTER

1. Clean oil pump mating face with filter.
2. Smear clean engine oil on the rubber washer of the new filter.
3. Fill the filter with new oil as far as possible, noting the angle at which the filter is to be fitted.
4. Screw on the filter until the sealing ring touches the oil pump cover face, then tighten it a further half turn by hand only. **Do not overtighten.**

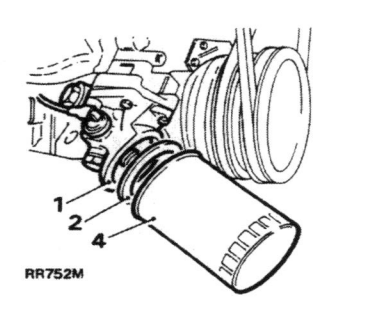

## FIT THE CLUTCH

1. Clean the flywheel and clutch assembly pressure plate.
2. Fit the centre plate and the clutch assembly and loosely secure to the flywheel with the retaining bolts.
3. Insert centralising tool 18G 79 or a slave primary shaft and finally tighten the clutch assembly retaining bolts in a diagonal sequence, to the correct torque.
4. Smear the centre plate splines with Rocol MV3 or Rocol MTS 1000 grease.

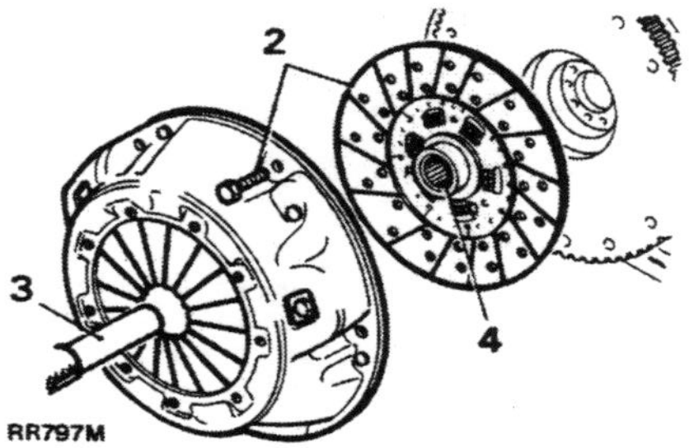

## FIT THE FAN PULLEY, VISCOUS COUPLING AND FAN

1. Secure the fan to the viscous coupling with the four bolts and tighten evenly.
2. If removed, fit the pulley to hub assembly adaptor and secure with the three bolts and tighten to the correct torque.
3. Screw the viscous coupling onto the adaptor thread tighten to the correct torque—see data section.
4. Fit the fan belt and compressor belt if fitted.

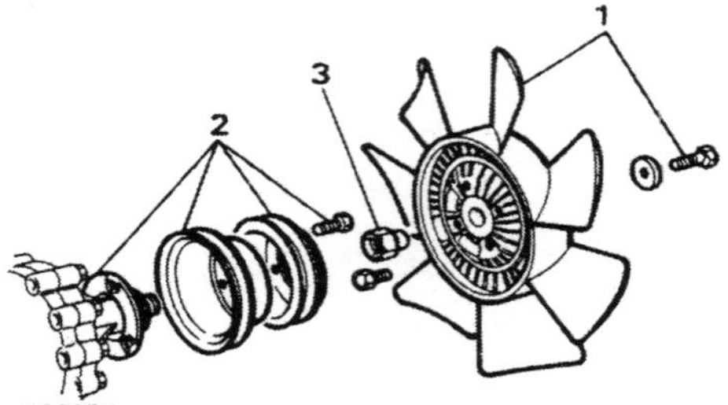

## FIT THE ALTERNATOR

1. Offer up the alternator to the mounting bracket and locate the pivot nuts and bolts noting that the fan guard is attached to the front nut and bolts.
2. Slacken the alternator adjustment bracket and alternator bracket. Note that the fan guard is attached to the adjustment bracket bolt.
3. Fit and tension the fan belt as follows:
   Pivot the alternator away from the engine but in doing so, do not apply any pressure to the slip ring end bracket. Tighten the pivot nuts and bolts and adjustment bolt. The tension is correct when, with thumb pressure, the belt deflection is approximately 11 to 14 mm (0.437 to 0.562 in) between alternator and the power steering pulley. Finally, tighten the fixings and connect the wiring plug to the alternator.

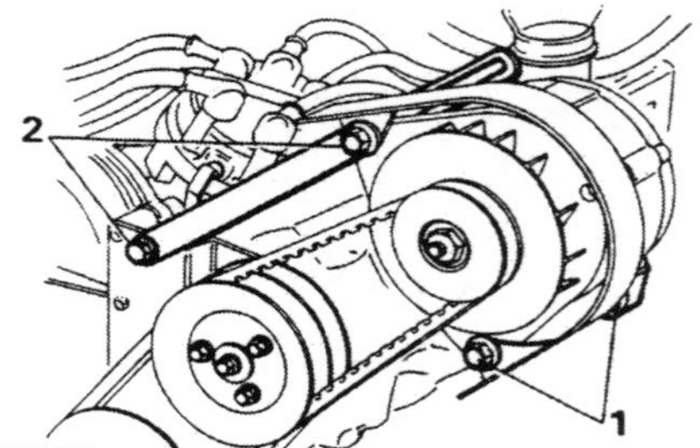

## FIT THE CARBURETTERS

If removed, fit the carburetters to the induction manifold, using new joint washers in the correct sequence as illustrated.

Illustration A
1. Fit the liner.
2. Fit the two joint washers.
3. In between the washers fit the insulator.

Illustration B
NOTE: Ensure the teeth are pointing towards the manifold penthouse when the deflector is refitted.

4. Fit the three joint washers.
5. Fit the saw toothed deflector.
6. Fit the insulator.

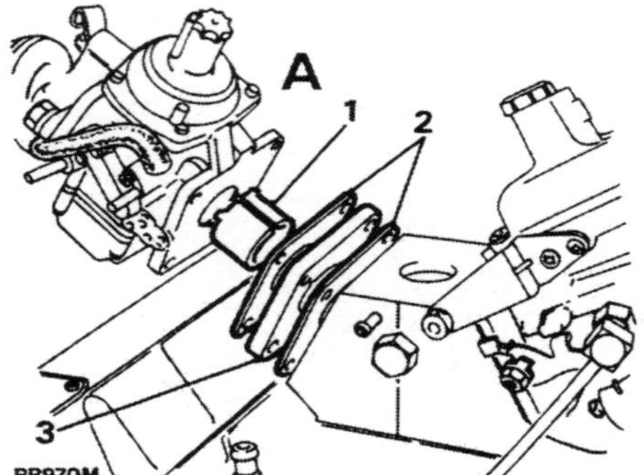

## FIT EXHAUST MANIFOLD AND AIR RAILS

1. Ensure that the mating surfaces of the cylinder head and exhaust manifold are clean and smooth and coat the faces with 'Foliac J 166' or 'Moly Paul' anti-seize compound. 'Foliac J 166' is manufactured by Rocol Ltd., Rocol House, Swillington, Leed, England. 'Moly Paul' is manufactured by K.S. Paul Products Ltd., Nobel Road, London N18.
2. Place the manifold in position on the cylinder head and fit the securing bolts, lockplates and plain washers. The plain washers are fitted between the manifold and lockplates. Evenly tighten the manifold bolts to the correct torque figure and bend over the lock tabs.
3. Fit pulse-air rails to cylinder heads.

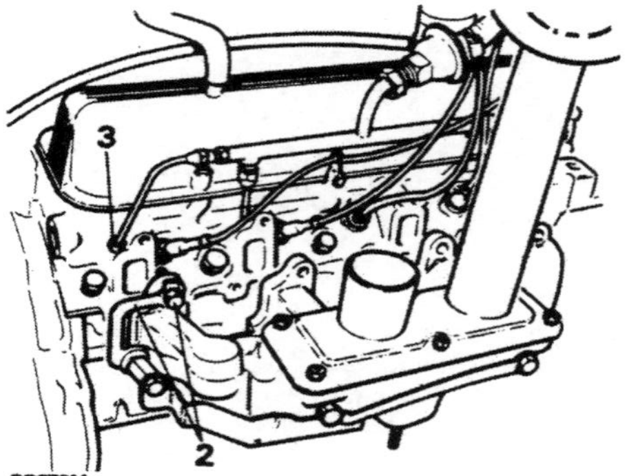

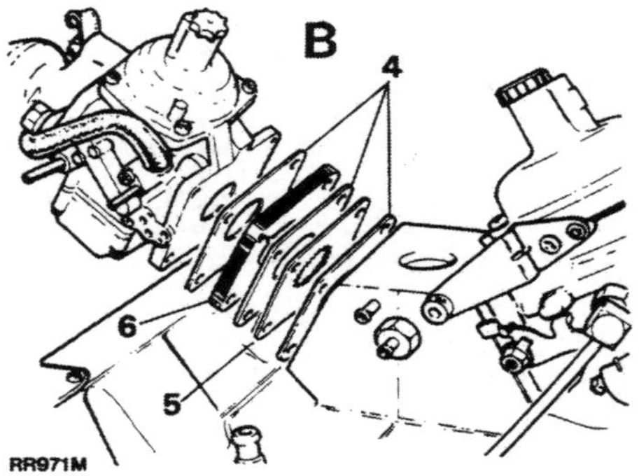

## MISCELLANEOUS AND NON-STANDARD ITEMS

Fit any other items of equipment and miscellaneous hoses, pipes, filters, clips and brackets to the positions noted during dismantling.

## ENGINE

### FAULT DIAGNOSIS

| SYMPTOM | POSSIBLE CAUSE | CURE |
|---|---|---|
| ENGINE FAILS TO START | 1. Incorrect starting procedure<br>2. Starter motor speed too slow<br>3. Faulty ignition system<br>4. Water or dirt in fuel system<br>5. Carburetter(s) flooding<br>6. Defective fuel pump<br>7. Defective starter motor<br>8. Starter pinion not engaging | See owners handbook<br>Check battery and connections<br>Check each component in system<br>Flush out system with clean fuel<br>Check float chamber needle valve<br>Remove, overhaul or renew<br>Remove starter motor and overhaul |
| ENGINE STALLS | 1. Low idling speed<br>2. Faulty sparking plugs<br>3. Faulty coil<br>4. Faulty reluctor<br>5. Incorrect mixture<br>6. Foreign matter in fuel system | Adjust carburetter(s)<br>Clean and test; renew if necessary<br>Renew<br>Renew<br>Adjust carburetter(s)<br>Investigate source of foreign matter and clean as necessary |
| LACK OF POWER | 1. Poor compression<br><br>2. Badly seating valves<br>3. Faulty exhaust silencer<br>4. Incorrect ignition timing<br>5. Leaks or restriction in fuel system<br>6. Faulty sparking plugs<br>7. Sticking valves<br>8. Defective valve springs<br>9. Faulty coil or battery<br>10. Worn valve guides or valves<br>11. Faulty cylinder head gaskets<br>12. Damaged exhaust system<br>13. Vacuum pipes, disconnected at inlet manifold, distributor or gearbox | If the compression is appreciably less than the correct figure, the piston rings or valves are faulty. Low pressure in adjoining cylinders indicates a faulty cylinder head gasket<br>Overhaul cylinder head(s)<br>Renew<br>Check and adjust using electronic equipment<br>Check through system<br>Clean, test and renew if necessary<br>Decarbonize<br>Renew<br>Adjust brakes or overhaul<br>Overhaul cylinder head(s)<br>Renew gaskets<br>Rectify or renew<br>Refit pipes |
| ENGINE RUNS ERRATICALLY | 1. Faulty electrical connections<br>2. Defective sparking plugs<br>3. Low battery charge<br>4. Defective distributor<br>5. Foreign matter in fuel system<br>6. Faulty fuel pump<br>7. Sticking valves<br>8. Defective valve springs<br>9. Incorrect ignition timing | Check security of all ignition connections<br>Clean, test and renew if necessary<br>Recharge battery and test for condition<br>Remove and overhaul<br>Determine source of dirt and clean system<br>Remove and overhaul or renew<br>Overhaul cylinder head(s)<br>Renew<br>Check timing with electronic equipment, if possible |
| ENGINE STARTS, BUT STOPS IMMEDIATELY | 1. Faulty electrical connections<br>2. Foreign matter in fuel system<br>3. Faulty fuel pump<br>4. Low fuel level in tank | Check HT leads for cracked insulation, check low tension circuit<br>Determine source of matter and clean system<br>Remove, overhaul or renew<br>Replenish |
| ENGINE FAILS TO IDLE | 1. Incorrect carburetter setting<br>2. Faulty fuel pump<br>3. Sticking valves<br>4. Faulty cylinder head gasket(s) | Adjust as necessary<br>Remove, overhaul or renew<br>Overhaul cylinder head(s)<br>Renew |
| ENGINE MISFIRES ON ACCELERATION | 1. Distributor incorrectly set<br>2. Faulty coil<br>3. Faulty sparking plugs<br>4. Faulty carburetter(s)<br>5. Vacuum pipes disconnected at inlet manifold | Adjust<br>Renew<br>Clean, test or renew<br>Overhaul<br>Check all vacuum connections. Renew faulty pipes |
| ENGINE KNOCKS | 1. Ignition timing advanced<br>2. Excessive carbon deposit<br>3. Incorrect carburetter setting<br>4. Unsuitable fuel<br>5. Worn pistons or bearings<br>6. Distributor advance mechanism faulty<br>7. Defective sparking plugs | Adjust using electronic equipment<br>Decarbonise<br>Adjust<br>Adjust ignition timing to suit octane rating<br>Overhaul engine<br>Renew capsule and re-check<br>Clean, test and renew if necessary |
| ENGINE BACKFIRES | 1. Ignition defect<br>2. Carburetter defect<br>3. Sticking valve<br>4. Weak valve springs<br>5. Badly seating valves<br>6. Excessively worn valve stems and guides<br>7. Excessive carbon deposit<br>8. Incorrect sparking plug gap<br>9. Air leak in induction or exhaust systems | Check all ignition components and timing<br>Overhaul carburetter(s)<br>Overhaul cylinder head<br>Clean and reset<br>Renew faulty gaskets or components |

### FAULT DIAGNOSIS

| SYMPTOM | POSSIBLE CAUSE | CURE |
|---|---|---|
| BURNED VALVES | 1. Sticking valves<br>2. Weak valve springs<br>3. Excessive deposit on valve seats<br>4. Distorted valves<br>5. Excessive mileage between overhauls | Overhaul cylinder head |
| NOISY VALVE MECHANISM | 1. Excessive oil in sump, causing air bubbles in hydraulic tappets<br>2. Worn or scored parts in valve operating mechanism<br>3. Valves and seats cut down excessively, raising end of valve stem, 1.27 mm (0.050 in) above normal position<br>4. Sticking valves<br>5. Weak valve springs<br>6. Worn timing chain or chain wheels | Drain and refill to correct level on dipstick<br>Replace faulty parts<br>Grind off end of valve stem or replace parts<br>Overhaul cylinder head<br>Renew worn parts |
| NOISE FROM HYDRAULIC TAPPETS<br>1. Rapping noise only when engine is started<br>2. Intermittent rapping noise<br>3. Noise on idle and low speed<br>4. General noise at all speeds<br>5. Loud noise at normal operating temperature only | 1. Oil too heavy for prevailing temperature<br>2. Excessive varnish in tappet<br>3. Leakage at check ball<br>4. High oil level in sump<br>  Leakage at check ball<br>  Worn tappet body<br>5. Excessive leak-down rate or scored lifter plunger | Drain and refill with correct grade<br>Replace tappet<br>Replace tappet<br>Drain and refill to correct level on dipstick<br>Replace tappet<br>Replace camshaft<br>Replace tappet |
| MAIN BEARING RATTLE | 1. Low oil level in sump<br>2. Low oil pressure<br>3. Excessive bearing clearance<br>4. Burnt-out bearings<br>5. Loose bearing caps | Replenish as necessary to high mark on dipstick<br>Worn bearings<br>Renew bearings; grind crankshaft<br>Renew and investigate reason for failure<br>Tighten to correct torque |
| LOW OIL PRESSURE WARNING LIGHT REMAINS ON, ENGINE RUNNING | 1. Thin or diluted oil<br>2. Low oil level<br>3. Choked pump strainer<br>4. Faulty release valve<br>5. Excessive bearing clearance<br>6. Oil pressure switch unserviceable<br>7. Electrical fault<br>8. Relief valve plunger sticking<br>9. Weak relief valve spring<br>10. Pump rotors excessively worn<br>11. Excessively worn bearings: main connecting rod, big end, camshaft, etc | Drain and refill with correct oil and renew filter<br>Replenish to high mark on dipstick<br>Clean<br>Rectify<br>Rectify<br>Renew<br>Check circuit<br>Remove and ascertain cause<br>Renew<br>Overhaul oil pump<br>Ascertain which bearings and rectify |
| RATTLE IN LUBRICATION SYSTEM | 1. Oil pressure relief valve plunger sticking | Remove and clean |
| ENGINE OVERHEATING | 1. Low coolant level<br>2. Faulty cooling system<br>3. Faulty thermostat<br>4. Incorrect timing<br>5. Defective lubrication system | Check for leaks. Check expansion tank level<br>Check fan and belt, pump, radiator fins not blocked<br>Test and renew if necessary<br>Check and adjust using electronic equipment<br>Renew filter. Check pump. Clean strainer.<br>Check oil circulation |
| MECHANICAL NOISES:<br>Medium low pitch knock<br>Low pitch thud<br>High pitch tap<br>Intermittent thuds<br>Continual slapping | Big end bearing slack or run<br>Main bearing slack or run<br>Worn gudgeon pins<br>Loose flywheel or excessive crankshaft endfloat<br>Piston clearance excessive—more apparent when engine cold, may disappear when engine hot | |

# EMISSION CONTROL 17

## EMISSION AND EVAPORATIVE CONTROL

Range Rover—carburetter and electronic fuel injection models are fitted during manufacture with various items of emission and evaporative control equipment to meet individual territory requirements. Therefore, some operations in this section may not be applicable to all vehicles. Where possible, items have been territorised to help identify individual requirements.

Unauthorised replacement or modification of the emission or evaporative control equipment may contravene local territory legislation and render the vehicle user and/or repairer liable to legal penalties. Any repairs or adjustments to the systems must be undertaken by an approved dealer. Refer to 'Engine Tuning Data' for the appropriate CO levels applicable to the vehicle specification.

## CRANKCASE BREATHING SYSTEM

### Description

The 'blow-by' gases from the crankcase are vented into the combustion system to be burned with the fuel/air mixture. The system provides positive emission control under all conditions. During engine running, crankcase fumes which may collect in the crankcase are vented to the carburetter or plenum chamber via hoses and flame traps.

**NOTE: Filters may have alternative locations according to engine build specification.**

## ENGINE BREATHER FILTER—carburetter models

### Remove and refit

#### Removing

1. Remove the air cleaner.
2. Withdraw the filter top hose.
3. Slacken the filter clip.
4. Withdraw the filter from the bottom hose.

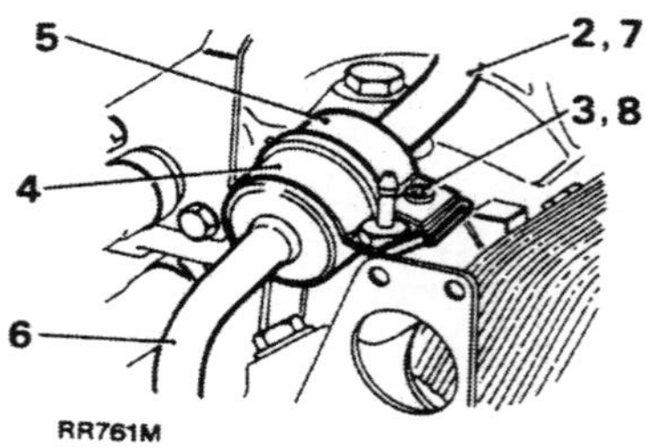

#### Refitting

5. Fit the filter with the end marked 'IN' facing forward.
6. Connect the filter bottom hose.
7. Connect the filter top hose.
8. Secure the filter retaining clip.
9. Fit the air cleaner.

## ENGINE BREATHER FILTER—fuel injection models

The engine breather filter is located at the rear of the left-hand rocker cover, beneath the throttle linkage bracketry.

### Remove and refit

#### Removing

1. Prise the filter outer cover upwards to release it from its mounting.
2. Remove the sponge filter from the cover and discard the sponge.

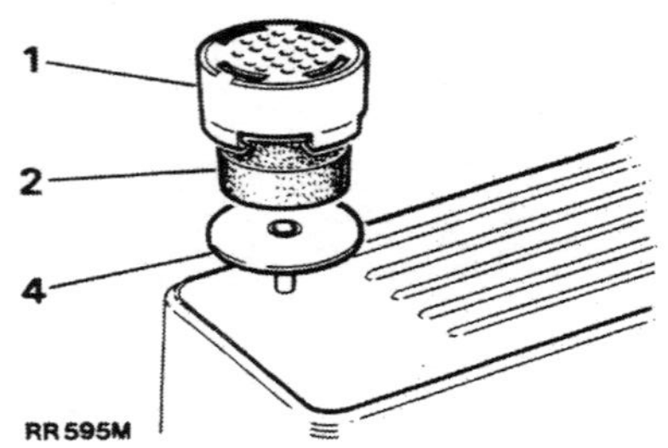

#### Refitting

3. Insert a new filter into the filter cover.
4. Press the filter onto its mounting until it clips firmly into position.

# EMISSION CONTROL

## ENGINE FLAME TRAPS—carburetter models

NOTE: Flame traps may have alternative locations according to engine build specification.

### Remove and refit

#### Removing

1. Pull the flame trap hoses out of the retaining clips.
2. Pull the hoses from the flame trap.
3. Withdraw the flame trap.

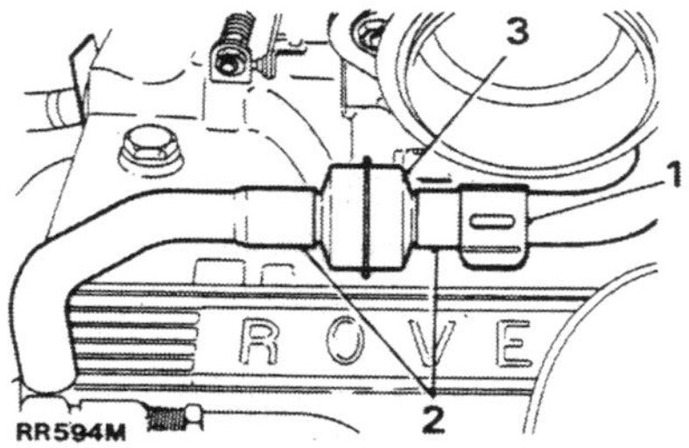

4. Visually inspect the wire gauze inside the flame trap, if in poor condition, renew the unit. If the unit is in a satisfactory condition, clean as follows:
5. Immerse the flame trap in a small amount of petrol, allow time for the petrol to dissolve and loosen any debris.
6. Remove the flame trap from the petrol bath and allow it to dry in still air.

WARNING: Do not use a compressed air line to dry or clean the flame traps as this could cause fire or personnel injury.

#### Refitting

7. Push the hoses onto the flame trap and ensure that they are secure.
8. Locate the hoses in their respective retaining clips.

## ENGINE FLAME TRAP—fuel injection models

### Remove and refit

#### Removing

1. Release the hose clip and pull the hose off the canister.
2. Unscrew the canister and remove it from the rocker cover.
3. Remove the large rubber 'O' ring and inspect for deterioration.

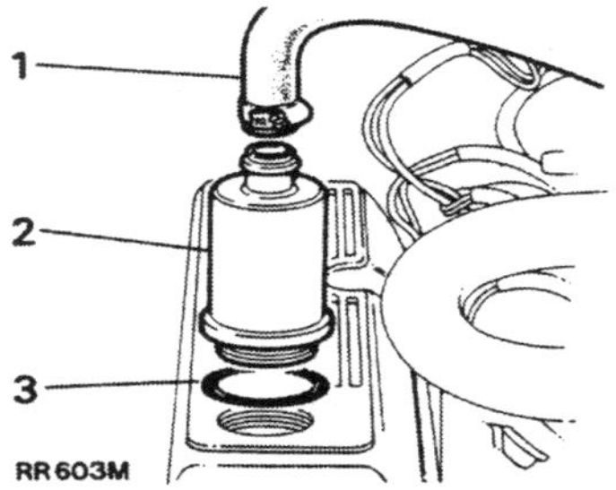

4. Visually inspect the condition of the wire gauze within the canister, if in poor condition, renew the whole assembly, if the flame trap is in an acceptable condition, clean as follows:
5. Immerse the canister in a small amount of petrol and allow time for the petrol to dissolve or loosen any debris.
6. Remove the canister from the petrol bath and allow to dry in still air.

WARNING: Do not use a compressed air line to dry; clean or remove any remaining particles of debris within the canister as this could cause fire or personnel injury.

#### Refitting

7. If the original canister is being refitted, fit a new 'O' ring.
8. Screw the canister into the rocker cover securely—hand tight only.
9. Refit the hose and tighten the hose clip securely.

## EVAPORATIVE LOSS CONTROL SYSTEM
(Saudi Arabia carburetter models only)

The evaporative loss control system reduces the amount of fuel vapour vented to the atmosphere. An adsorption canister receives vapour from the carburetter float chamber and from the fuel tank. The canister is purged of these vapours by inlet manifold or carburetter depression.

## ADSORPTION CANISTER

### Remove and refit

#### Removing

1. Disconnect from the canister:
   (i) Canister line to fuel tank.
   (ii) Canister purge line.
   (iii) Carburetter vent pipe.
2. Slacken the clamp nut screw.
3. Remove the canister.

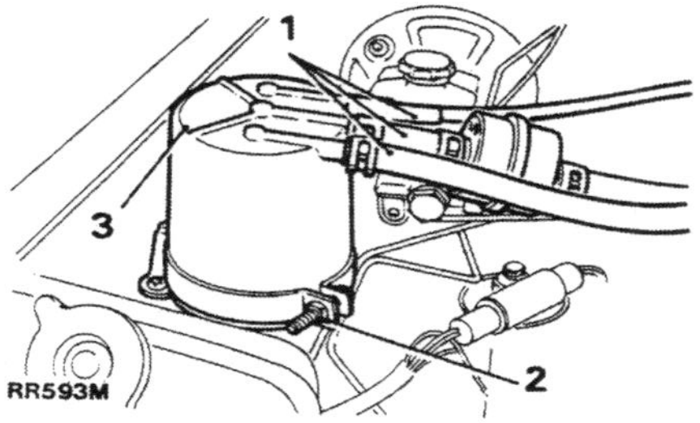

#### Refitting

4. Secure the canister in the clamp.
5. Reverse instructions 1 and 2 above.

## EVAPORATIVE LOSS CONTROL SYSTEM
(Australian and Saudi Arabia fuel injection models only)

The evaporative loss control system reduces the amount of fuel vapour vented to the atmosphere. An adsorption canister receives vapour from the fuel tank. The canister is purged of these vapours by plenum chamber depression.

The remove and refit of the canister is as described for carburetter models, noting that the carburetter vent pipe marked *C* is blanked off.

WARNING: The use of compressed air to clean an adsorption canister or clear a blockage in the evaporative system is very dangerous. An explosive gas present in a fully saturated canister may be ignited by the heat generated when compressed air passes through the canister.

## PULSAIR AIR INJECTION
(UK, Europe and Saudi Arabia carburetter models only)

'Pulsair' is a system of self-induced air injection. The induced air which is taken from the air cleaner (UK and Europe models) or from the air intake elbows (Saudi Arabia models) at high manifold depressions, passes through one-way valves and a configuration of pipes which inject clean air into the exhaust gases via the manifold, thereby reducing carbon monoxide emission to the atmosphere.

### UK and Europe Version

1. Air cleaner.
2. Pulsair manifolds.
3. Check valve housings.
4. Connecting hoses.

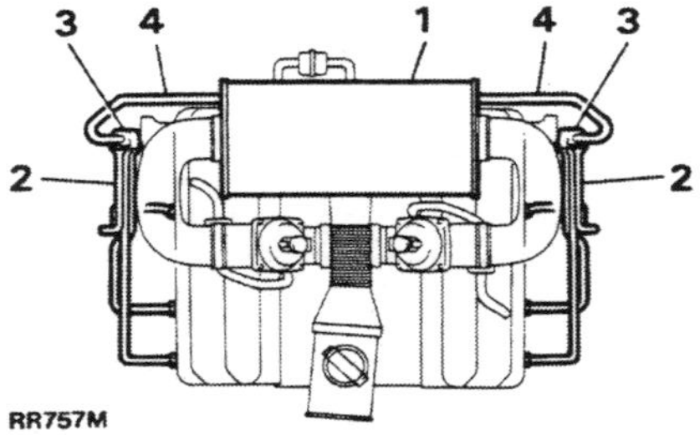

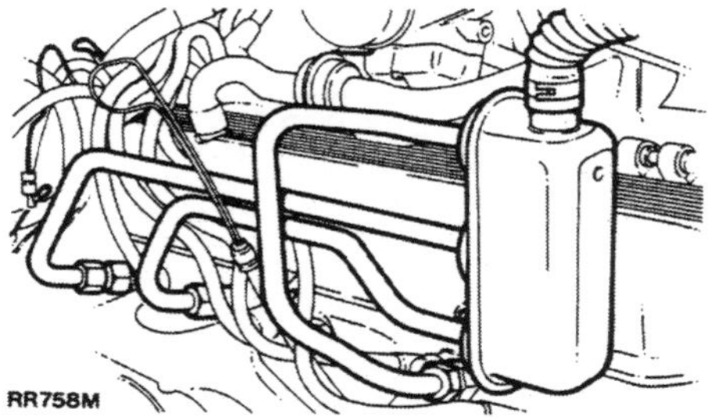

Continued

# EMISSION CONTROL

## Saudi Arabia Version

1. Air cleaner.
2. Connecting pipes.
3. Connecting hoses.
4. Pulsair valves.
5. Air manifolds.

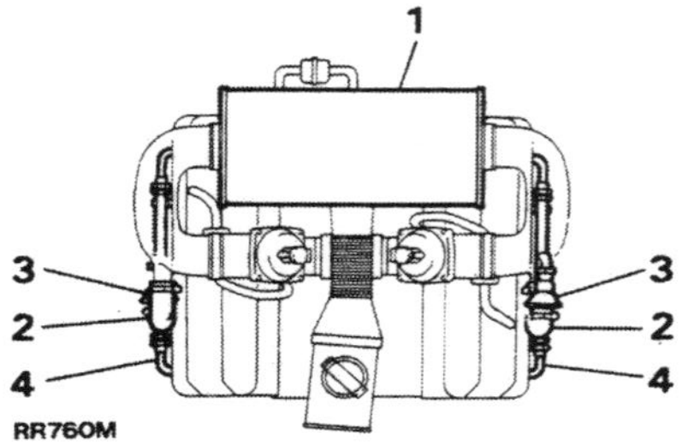

RR760M

## PULSAIR MANIFOLD (UK and Europe carburetter models)

### Remove and refit

### Removing

1. Release the clip securing the hose to the check valve housing.
2. Release the manifold from the cylinder head.
3. Release the single screw securing the check valve housing to the rocker cover.
4. Withdraw the manifold.
5. Remove the threaded inserts from the cylinder head.
6. Using a piece of bent wire, retrieve the spargas tube from the bottom of the insert bore.

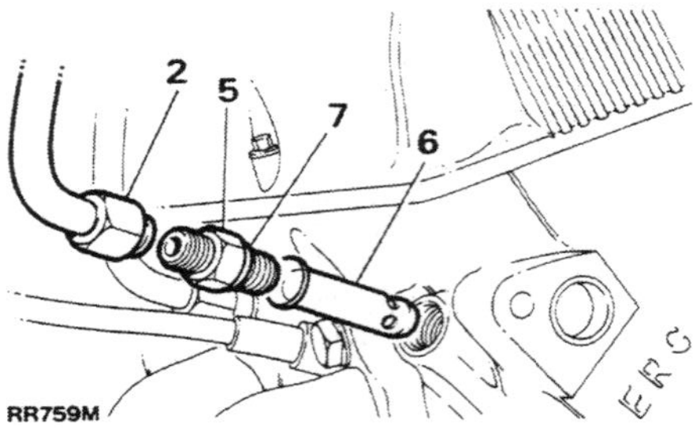

RR759M

### Refitting

7. Reverse the removal procedure ensuring that the threaded double ended insert is fitted with the plain shank nearest the cylinder head.
8. Run the engine and check for air leaks at the manifold.

## PULSAIR MANIFOLD (Saudi Arabia carburetter models)

### Remove and refit

### Removing

1. Release the hose clip securing the hose at the air intake elbow.
2. Pull the hose from the elbow.
3. Release the check valve from the manifold.
4. Release the manifold from the cylinder head.
5. Lift off the manifold.

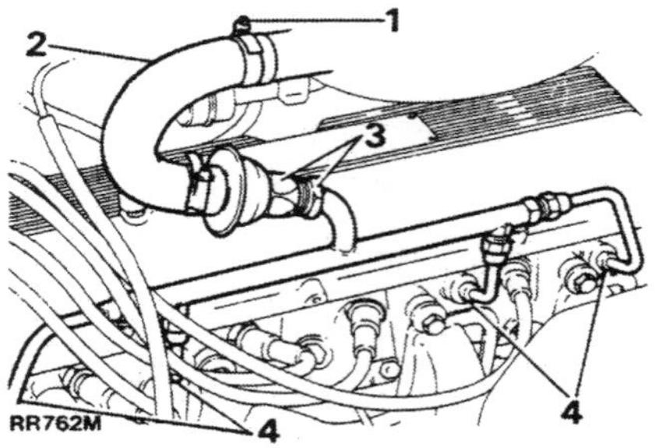

## PULSAIR CHECK VALVE (Saudi Arabia carburetter models only)

### Remove and refit

### Removing

1. Disconnect the hose from the check valve/pulse air valve.
2. Using two open-ended spanners—one on the air distribution manifold hexagon, to support the manifold, and the other to remove the valve anti-clockwise.

CAUTION: Do not impose any strain on the air manifold.

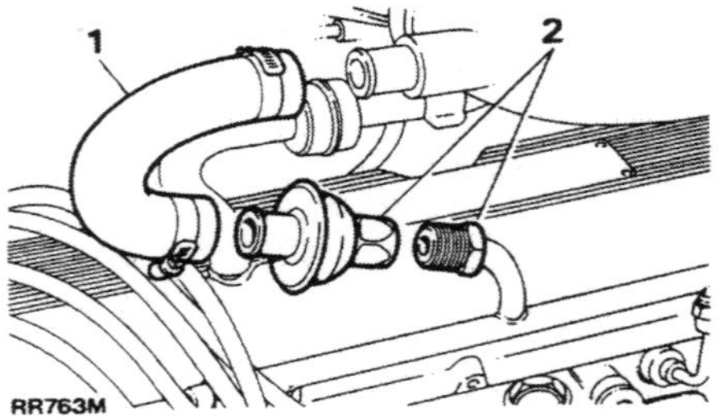

RR763M

### Refitting

3. Reverse instructions 1 and 2.

NOTE: The pulsair valve is identified by a pink paint spot and the part number 4974-196 on the face.

## PULSAIR CHECK VALVE

### Test

The check valve is a one-way valve positioned to prevent the back-flow of exhaust gases.

### Testing

CAUTION: Do not use a pressure air supply for this test.

1. Remove the check valve.
2. Blow through the valve orally in both directions in turn. Air should only pass through the valve when blown from the hose connection end. Should air pass through the valve when blown from the air manifold end, renew the valve.

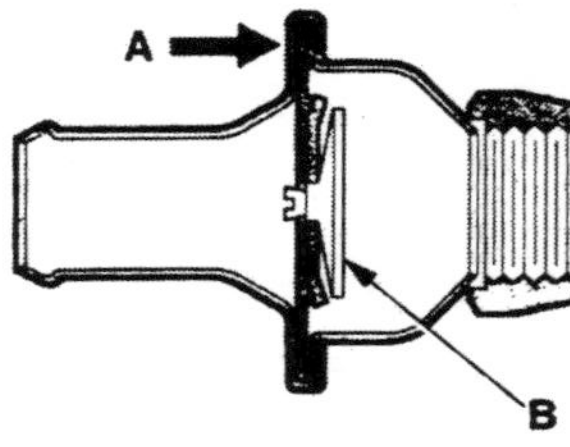

RR762M

### Refitting

6. Reverse the removal procedure.
7. Run the engine and check for air leaks at the manifold.

RR764M

A. Direction of flow.
B. Valve diaphragm.

3. Refit the pulsair check valve.

## PULSAIR CHECK VALVE (UK and Europe carburetter models only)

The pulsair check valves are an integral part of the pulsair manifold assembly and are located in the check valve housing. If a single valve within that housing fails it will be necessary to renew the complete pulsair manifold assembly.

To remove the pulsair assembly see pulsair manifold, remove and refit.

### TEST

1. Remove the pulsair manifold.
2. Blow through the valve orally in both directions in turn. Air should only pass through the valve or valves when blown from the check valve housing. Should air pass through the valves when blown from the manifold end, renew the complete assembly.

# EMISSION CONTROL

## AIR INTAKE TEMPERATURE CONTROL SYSTEM
(UK, Europe and Saudi Arabia carburetter models only)

### Description

To maintain an efficient air intake temperature, a sensor is incorporated in the air cleaner. The sensor allows inlet manifold vacuum to operate a flap in the air cleaner intake. The flap permits cold air, from forward of the radiator and/or hot air, from a scoop around the exhaust manifold, to mix in varying amounts to provide the required air temperature.

The specification comprises:

1. A hot box surrounding the exhaust manifold.
2. A vacuum operated thermostatically controlled flap valve in the air cleaner.
3. The flap valve controls the source of the intake air supply which may be warm air drawn entirely from the hot box or cold air drawn from the under bonnet area or a combination of both.
4. The hot box is connected via a hose to the flap valve in the air intake.
5. The temperature sensing device is situated in the air cleaner on the clean side of the element.
6. A pipe from the manifold is attached to the temperature sensing device via a non-return valve.
7. From the other side of the temperature sensing device is a pipe connecting the vacuum capsule operating the flap valve.

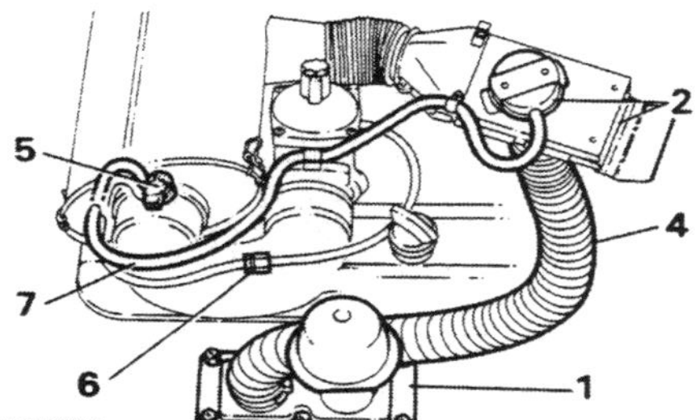

RR770M

### CONTROL VALVE

#### Check operation

1. Check operation of the mixing flap valve in the air cleaner by starting the engine from cold and observing the flap valve as the engine temperature rises.
2. The valve should start to open slowly within a few minutes of starting and continue to open until a stabilised position is achieved. This position and the speed of operation will be entirely dependent on prevailing ambient conditions.
3. Failure to operate indicates failure of the flap valve vacuum capsule or the thermostatically controlled vacuum switch or both.
4. Check by connecting a pipe directly to the flap valve, thus by-passing the temperature sensor.
5. If movement of the flap valve is evident the temperature sensor is faulty. If no movement is detected, the vacuum capsule is faulty.
6. Fit new parts where necessary.

### AIR INTAKE TEMPERATURE CONTROL VALVE

#### Remove and refit

**NOTE: Alternative valves may be fitted according to engine build specifications.**

#### Removing

1. Release the two clips securing the pipes at either side of the valve unit.
2. Pull the pipe from the vacuum capsule.
3. Release the hose clip and remove the hot box hose from the bottom of the valve unit.
4. Disconnect the hose between the valve unit and air cleaner.
5. Release the valve unit from its support and lift it from the engine compartment.

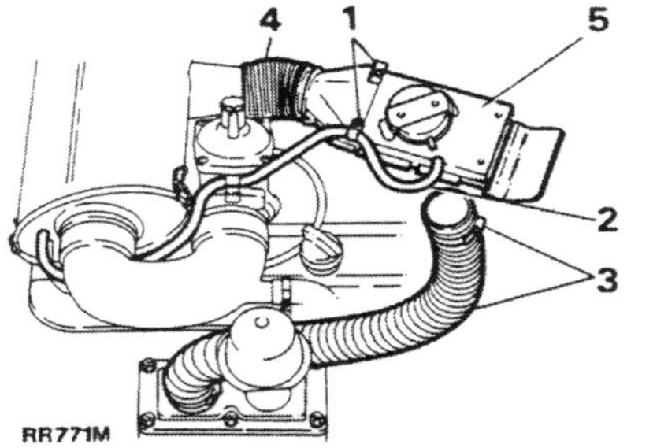

#### Refitting

6. Reverse the removal procedure. Ensure that all hoses and pipes are fitted securely.

RR771M

## TEMPERATURE SENSOR

The temperature sensor is incorporated into the clean air side of the air cleaner case. The sensor is located in the right-hand end cover of the air cleaner. Access to the unit is gained as follows:

### Remove and refit

#### Removing

1. Remove the right-hand air cleaner elbow.

**NOTE: Saudi Arabia versions only—disconnect the hose from the check valve at the air cleaner elbow.**

**Care should be taken not to damage any seals or 'O' rings when removing the elbows.**

2. Remove the air cleaner right-hand end cover.

**NOTE: UK and Europe versions only—remove the hose from the pulsair manifold at the right-hand air cleaner end cover.**

3. Remove the wing nut, fibre washer and end plate securing the air cleaner element.
4. Remove the element.
5. Pull the two pipes off the sensor, noting their position for re-assembly.
6. Prise the spring retaining clip of the sensor.
7. Manoeuvre the sensor and remove it from the end cover.

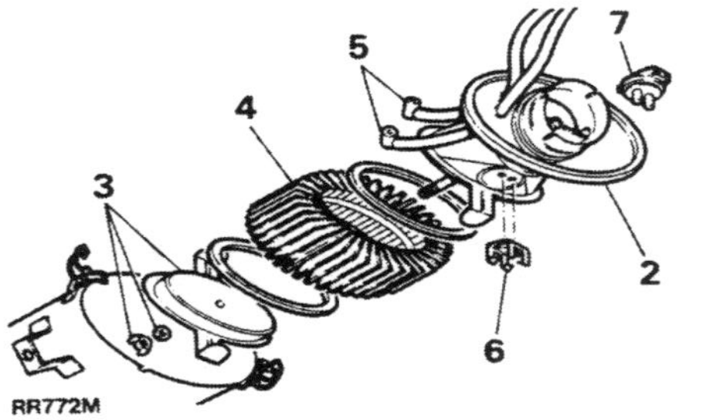

#### Refitting

8. Reverse the removal procedure.

**NOTE: When refitting the elbows coat the air cleaner seal and carburetter 'O' ring with petroleum jelly. Ensure that all hoses and connections are secure.**

RR772M

## EMISSIONS LABEL (Australian and Saudi Arabian fuel injection models only)

An exhaust emission control information label is attached to the right-hand rocker cover. The label is fitted to comply with local territory emission requirements.

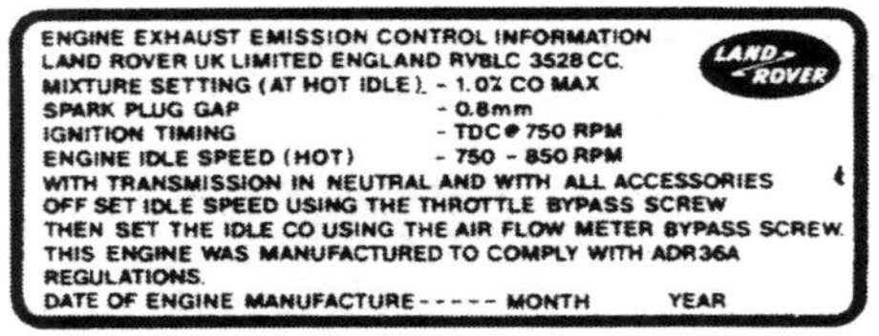

```
ENGINE EXHAUST EMISSION CONTROL INFORMATION
LAND ROVER UK LIMITED ENGLAND RVBLC 3528 CC.
MIXTURE SETTING (AT HOT IDLE):    - 1. 0% CO MAX
SPARK PLUG GAP                    - 0.8mm
IGNITION TIMING                   - TDC @ 750 RPM
ENGINE IDLE SPEED (HOT)           - 750 - 850 RPM
OFF SET IDLE SPEED IN NEUTRAL WITH ALL ACCESSORIES
OFF SET THE IDLE CO USING THE THROTTLE BYPASS SCREW.
THEN SET THE IDLE CO USING THE AIR FLOW METER BYPASS SCREW.
THIS ENGINE WAS MANUFACTURED TO COMPLY WITH ADR36A
REGULATIONS.
DATE OF ENGINE MANUFACTURE ----- MONTH    YEAR
```

RR769M

## VACUUM DELAY VALVE—all carburetter emission controlled models

The coloured end of the vacuum delay valve should always be fitted to the short hose from the distributor.

### Test check valve air flow

1. Attach a 10.00 ± .250 in³ vacuum tank to the coloured side of the valve.
2. Expose the black side to atmospheric pressure.
3. Expected result: The time required for the vacuum to drop from minus 20 in Hg to minus 2 in Hg will be 0.5 seconds maximum.

### Test

4. **Vacuum recovery air flow:** Attach a 22.75 ± .5 in³ vacuum tank to the black side of the valve.
5. Expose the coloured side to atmospheric pressure.
6. Expected result: The time required for the vacuum to drop from 16 in Hg to 8 in Hg will be 240 to 360 seconds.

### Test external leakage check

1. Seal the coloured side of the valve and attach a short flexible pipe to the other end.
2. Submerge the valve in water and orally blow through the valve.
3. If any external leakage is noticeable, renew the valve.

If the delay valve does not comply with any of the test results, renew the unit.

*Continued*

## EMISSION CONTROL

**Remove and refit**

**Removing**

1. Pull the two flexible hoses from the delay unit.
2. Remove the unit from the engine compartment.

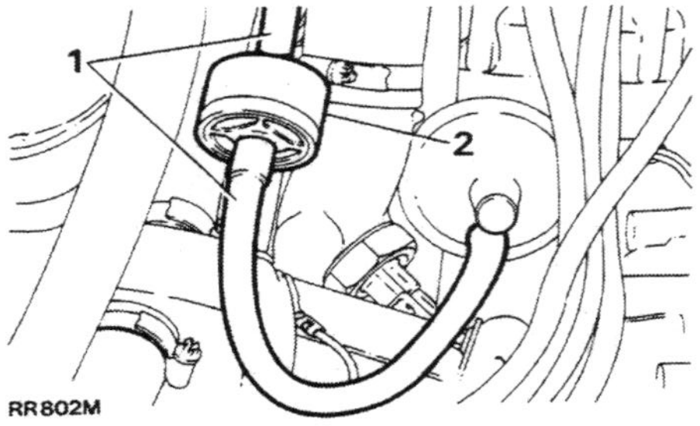

RR802M

**Refitting**

3. Reverse the removal instructions ensuring that the coloured side of the valve is fitted to the short hose from the distributor.

## AIR INTAKE TEMPERATURE CONTROL — FAULT FINDING

| SYMPTOM | POSSIBLE CAUSE | ACTION |
|---|---|---|
| Poor or erratic idle<br>Hesitation or flat spot (cold engine)<br>Excessive fuel consumption<br>Lack of power<br>Engine overheating | Hot air inlet hose loose, adrift or blocked | Check hot air inlet hose for condition and security. Renew if necessary. |
| Poor or erratic idle<br>Hesitation or flat spot (cold engine)<br>Excessive fuel consumption<br>Lack of power<br>Engine cuts out or stalls (at idle)<br>Engine 'runs-on'<br>Engine 'knocks or pinks'<br>Rich running (excess CO) | Flap valve jammed | Check operation of flap valve. If fault cannot be rectified renew air cleaner outer cover which includes flap valve. |
| Poor or erratic idle<br>Hesitation or flat spot (cold engine)<br>Excessive fuel consumption<br>Lack of engine power<br>Engine cuts out or stalls (at idle)<br>Engine misfires<br>Lean running (low CO) | Vacuum pipes disconnected or leaking | Check the vacuum pipes for security and deterioration. Renew if necessary. |
| Hesitation or flat spot (cold engine)<br>Excessive fuel consumption<br>Lack of power<br>Engine overheating<br>Engine cuts out or stalls (at idle)<br>Engine 'runs-on'<br>Engine 'knocks or pinks'<br>Rich running (excess CO) | One-way valve faulty | Blow through valve to check 'one-way' action. If the valve leaks fit a new valve. |
| Poor or erratic idle<br>Hesitation or flat spot (cold engine)<br>Excessive fuel consumption<br>Lack of power<br>Engine overheating<br>Engine cuts out or stalls (at idle)<br>Engine 'runs-on'<br>Engine 'knocks or pinks'<br>Rich running (excess CO) | Temperature sensor faulty, leaking or jammed | Check and renew if necessary. |
| Poor or erratic idle<br>Hesitation or flat spot (cold engine)<br>Excessive fuel consumption<br>Lack of power<br>Engine cuts out and stalls<br>Rich running | Flap valve diaphragm leaking | Check with a distributor vacuum test unit. If leakage is apparent renew air cleaner outer cover which includes the servo motor. |

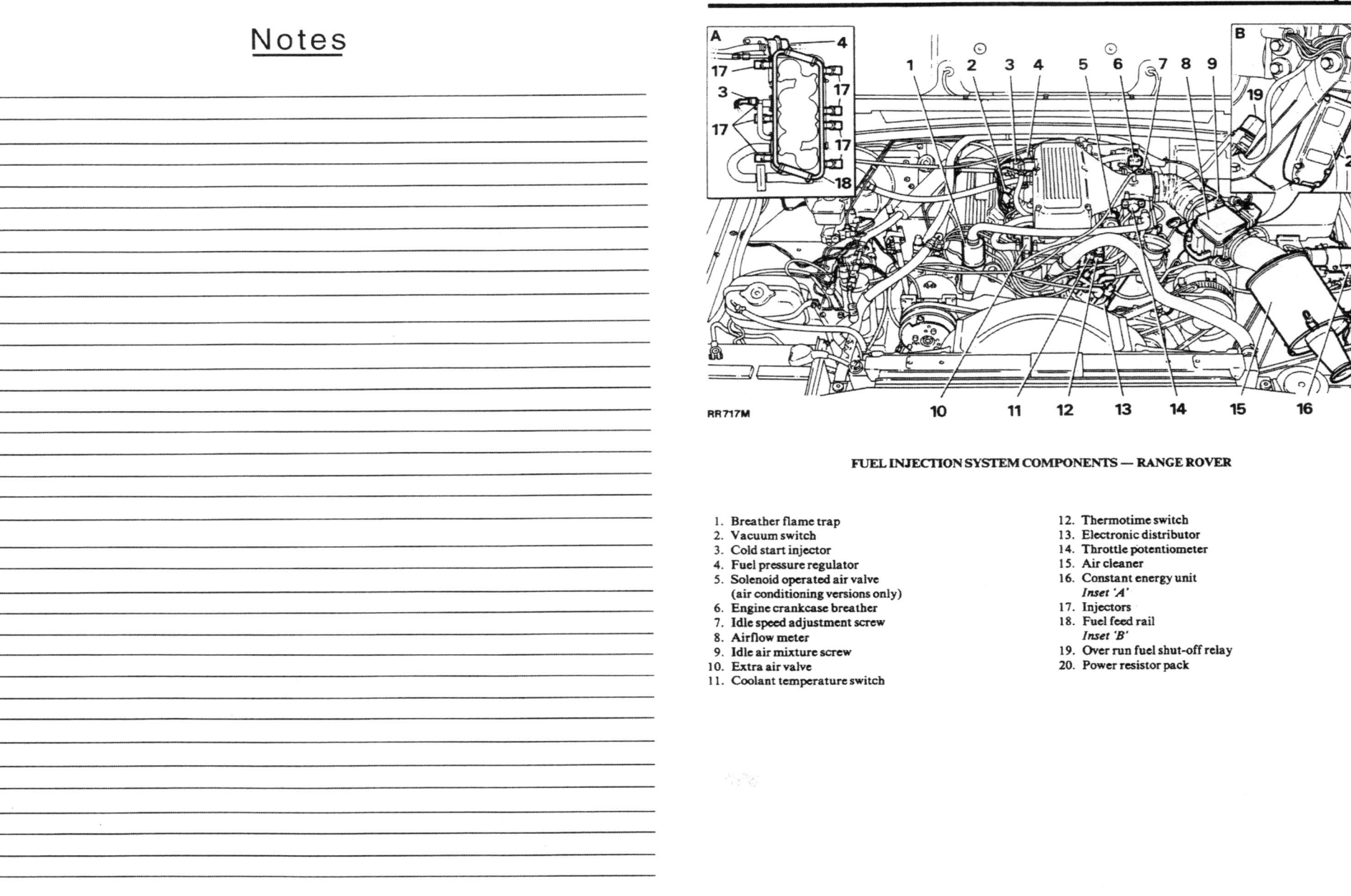

## FUEL INJECTION SYSTEM [19]

**FUEL INJECTION SYSTEM COMPONENTS — RANGE ROVER**

1. Breather flame trap
2. Vacuum switch
3. Cold start injector
4. Fuel pressure regulator
5. Solenoid operated air valve (air conditioning versions only)
6. Engine crankcase breather
7. Idle speed adjustment screw
8. Airflow meter
9. Idle air mixture screw
10. Extra air valve
11. Coolant temperature switch
12. Thermotime switch
13. Electronic distributor
14. Throttle potentiometer
15. Air cleaner
16. Constant energy unit
   *Inset 'A'*
17. Injectors
18. Fuel feed rail
   *Inset 'B'*
19. Over run fuel shut-off relay
20. Power resistor pack

## Notes

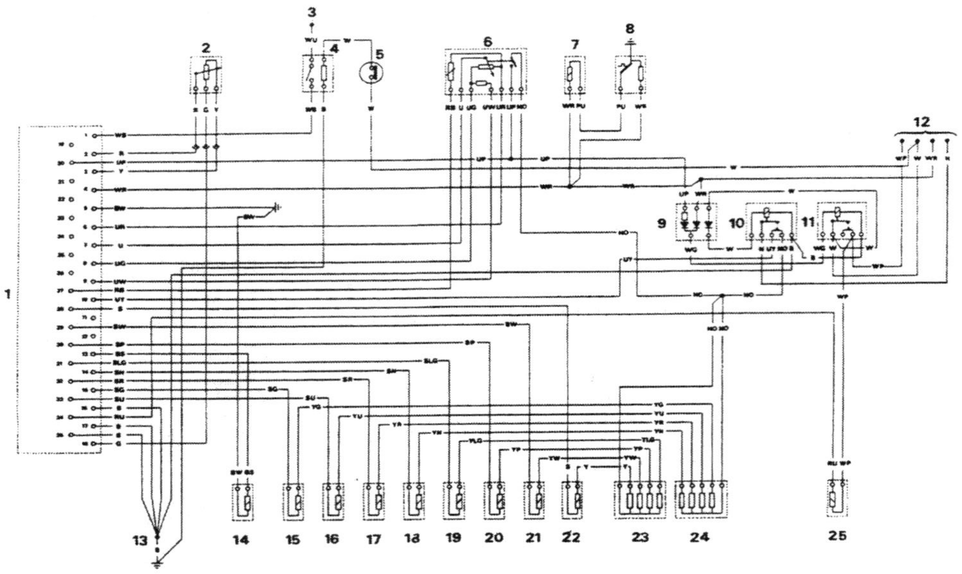

# FUEL INJECTION SYSTEM

## FUEL INJECTION—Circuit Diagram

1. Electronic control unit (ECU)
2. Throttle potentiometer
3. Ignition pick-up point (142 on main circuit diagram)
4. Over-run fuel shut-off relay
5. Vacuum switch
6. Air flow meter
7. Cold start injector
8. Thermotime switch
9. Current steering module for fuel injection relays
10. Main relay
11. Fuel pump relay
12. Pick-up point fuel injection circuit (111 on main circuit diagram)
13. Clinch
14. Temperature sensor
15. Injector (No. 1 cylinder)
16. Injector (No. 3 cylinder)
17. Injector (No. 5 cylinder)
18. Injector (No. 7 cylinder)
19. Injector (No. 2 cylinder)
20. Injector (No. 4 cylinder)
21. Injector (No. 6 cylinder)
22. Injector (No. 8 cylinder)
23. Power resistors
24. Power resistors
25. Extra air valve

**Key to cable colours**

- B — Black
- G — Green
- K — Pink
- L — Light
- N — Brown
- O — Orange
- P — Purple
- R — Red
- S — Slate
- U — Blue
- W — White
- Y — Yellow

The last letter of a colour code denotes the tracer colour

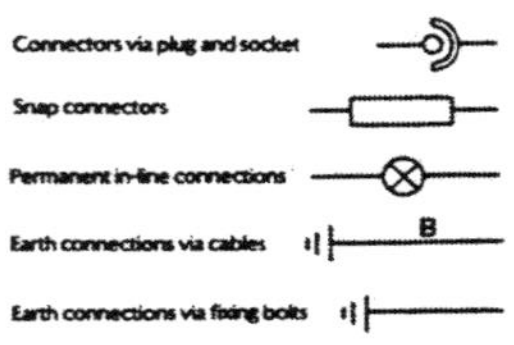

## ELECTRONIC FUEL INJECTION

### Description

### Fuel Injection System

Fuel is drawn from the fuel tank by a high pressure electric pump located in the fuel tank. The fuel pump operates only when the pump relay and/or the starter motor circuits are energised. Fuel passes to the pressure regulator via an in-line filter. The pressure regulator varies pressure in direct proportion to manifold depression and thus varies injection pressure between 1.8 to 2.5 kgf/cm², 26 to 36 lbf/in². Excess fuel is returned to the fuel tank.

A fuel rail links the pressure regulator with the fuel injectors, one injector being fitted to each inlet manifold spur. The injectors are solenoid operated being pulse energised by the electronic control unit, ECU, completing a circuit to 'earth'. When 'open' the injectors spray fuel into the inlet manifold to be drawn into the engine cylinders at the next induction stroke of each piston.

Therefore there needs to be no fixed relationship between the injector timing and the engine ignition or valve timing.

The injectors are programmed to 'open' alternately in two banks of four, twice per engine operating cycle. The time that the injectors are 'open' governs the amount of fuel supplied to the engine and this 'open' time is computed by the ECU from the inputs it receives from the various sensors.

To assist cold starting, a separate cold start injector sprays a fine jet of fuel against the air stream entering the plenum chamber. The cold start injector is energised from the engine starter motor circuit and is in series with a thermotime switch which is dual activated by the engine coolant temperature (heat) and a heater coil around a bi-metal strip (time) the coil being energised from the starter motor circuit. The thermotime switch ensures that the cold start injector will not be energised when the engine is at normal operating temperature or during prolonged operation of the starter motor when the engine is below normal operating temperature. The switch will isolate the cold start injector after approximately 8 to 12 seconds at −20°C, −4°F, this time is decreased as the engine approaches its normal operating temperature.

### Fuel Pump

The fuel pump is energised initially via a relay during operation of the starter motor solenoid and then by a switch operated by the air flow meter and is independent of the ECU.

### Electronic Control Unit — ECU.

The ECU is a sealed unit and receives input signals from various sensors and computes from these an output signal to the fuel injector solenoid circuits. When activated the solenoids 'open' the injectors to spray fuel into the engine inlet manifold spurs, the injectors remaining open for between 1.5 and 10 milli-seconds depending on engine running requirements.

The ECU is protected by various devices: the diode pack protects against reversed battery connection. The main relay is controlled by the ignition switch and connects battery voltage directly to the ECU.

### Engine Speed

Low tension circuit pulses from the ignition coil negative terminal, are passed to the ECU to be computed into an engine speed input.

### Air Flow Meter

The air flow meter measures induction air flow mass. The plenum chamber absorbs any rapid fluctuations in air flow that might upset the air flow meter signals.

The movement of the measuring flap is dampened by a compensating flap which prevents flutter. The position of the flap is controlled by the air drawn into the engine and the action of a return spring. The mass of air drawn into the engine at any time is indicative of the engine load and a signal from a potentiometer, variable resistance, proportional to the flap position, is passed to the ECU. However, the air mass and air density is dependent upon air temperature. Therefore, an air temperature sensor is incorporated into the air flow meter and this sends a separate signal to the ECU.

Due to the action of the return spring, the measuring flap is almost closed when the engine is idling and an idle air by-pass channel is provided to assist the engine to breath at low speed. Air passing through the by-pass channel is not registered by the air flow meter measuring flap. The idle air mixture screw is fitted into the by-pass channel to regulate the air flow to adjust the air to fuel ratio CO content at idle speed.

### Coolant Temperature Sensor

The sensor provides coolant temperature information to the ECU.

The sensor causes the ECU to lengthen slightly, the time that the main injectors are 'open' reducing this time as the engine warms up and cutting it off when normal engine operating temperature is reached.

*Continued*

# FUEL INJECTION SYSTEM

## Extra Air Valve

This valve is mounted above a water passage in the inlet manifold and registers engine coolant temperature. The valve provides the additional air required to maintain satisfactory cold start mixture until the engine reaches normal operating temperature. This air is taken after it has passed through the air flow meter, so that the air is registered by the ECU, and returned to the plenum chamber after the throttle butterfly.

The valve allows extra air to pass under cold start conditions, the extra air source is reduced and finally terminated as normal engine operating temperature is reached.

The valve is controlled by a bi-metal strip which is heated from two sources; the coolant and a heater coil around the strip. The heater coil is energised from the fuel pump circuit and comes into operation when the engine is under crank or running.

## Throttle Potentiometer

The electrical signal from the potentiometer to the ECU depends upon the position of the throttle butterfly spindle and hence the position of the accelerator pedal. By using the variable voltage output in conjunction with the information from the other sensors, the ECU adjusts fuel input to accommodate requirements for acceleration, deceleration and constant engine speed. When sudden acceleration is signalled to the ECU by the throttle potentiometer, all injectors are instantly pulsed to operate once simultaneously to ensure adequate engine response.

## Over-run Fuel Cut-off — Vacuum Switch

The manifold depression switch senses manifold depression above 24 in Hg ± 1 in Hg. The switch operates the over-run relay which interrupts the ignition signal to the ECU.

## Air Temperature Sensor

The air temperature sensor is an integral part of the air flow meter and cannot be replaced as a separate item.

The sensor sends a separate signal to the ECU, the ECU then alters the length of time the injectors remain open, correcting the air/fuel mixture.

## FUEL INJECTION SYSTEM

CAUTION: **The fuel system incorporates fine metering components that would be affected by any dirt in the system; therefore it is essential that working conditions are scrupulously clean.**

If it is necessary to disconnect any part of the fuel injection system i.e. pipes, hoses, etc., these must be blocked off to prevent ingress of dirt.

### ENGINE SETTING PROCEDURE — FUEL INJECTION

If a major overhaul has been undertaken on the fuel injection/engine system, the following check and adjustments must be carried out before attempting to start the engine.

A. Throttle potentiometer setting — see 'Throttle switch —potentiometer' setting procedure.

B. Spark plug gaps—see 'Data section'.

C. Throttle levers—see 'Throttle lever setting procedure'.

D. Ignition timing—static—see 'Engine tuning data'.

E. By-pass idle screw—see 'Engine tuning procedure'.

NOTE: If the previous checks and adjustments are satisfactory but the engine will not start the fuel injection electrical circuitory must be checked using the appropriate recommended equipment.

Recommended Equipment:

Lucas 'Electronic Ignition Analyser'
Lucas Part Number—TWB 119.

Lucas 'E.F.I. Throttle Potentiometer Adjustment Gauge'
Lucas Part Number—YWB 121.

Lucas 'Epitest' diagnostic system
Lucas Part Number—YWB 105.

Use in conjunction with the Lucas Operating Instruction Manuals.

If the above equipment is unavailable the tests can be carried out using an AVO meter, following the instructions given in the charts.

CAUTION: Ensure the AVO is correctly set to volts or ohms, dependent upon which test is being undertaken.

## CONTINUITY TESTS—Using an AVO meter

The following continuity tests are intended as a guide to identifying where a fault may occur within a circuit; reference should be made to the fuel injection circuit diagram for full circuit information.

### Key to Symbols

| | | | |
|---|---|---|---|
| Cold Start Injector | 35-Way Harness Multiplug | P – Pump Relay  M – Main Relay  S – Steering Module | Airflow Meter Harness Plug | Throttle Switch (potentiometer type) |
| Temporary Connection | Resistor Box | Injector | Thermotime Switch | Extra Air Valve |
| Coolant Temp. Sensor | Ignition | Pressure Gauge | Electrical Multiplug | Constant Energy Unit |

RR 716M

## CONTINUITY TEST—Using an AVO meter

NOTE: **All tests are carried out from the electronic control unit (ECU) harness multi-plug unless stated otherwise in the test procedure.**

*Continued*

# FUEL INJECTION SYSTEM

| TEST | CIRCUIT TESTING | EXPECTED RESULTS | POSSIBLE FAULTS AND REMEDIES |
|---|---|---|---|
| **1. ECU SUPPLY** Disconnect the multi-plug from the ECU. Switch on the ignition. Connect a voltmeter between pin 10 and earth. | RR670M | 11–12.5 volts | No reading; check all wiring to main relay, check main relay by substitution. Below 11 volts; check battery. Check circuit for high resistance connection. |
| **2. FUEL PUMP CONTACTS** Switch on the ignition and connect a voltmeter between pin 20 and earth. Airflow meter flap closed. | RR671M | 0 volts | If reading registered, check airflow meter switch action. |
| Operate the flap in the airflow meter. | | 11–12.5 volts | No reading, check wiring from main relay to airflow meter. Check wiring from airflow meter to fuel pump relay. Check pump relay by substitution. Check fuel pump operation by connecting a direct supply to the pump terminals. |
| **3. CRANKING SIGNAL** Connect a voltmeter between pin 4 and earth. Crank the engine. | RR672M | 8–12 volts | No reading but starter motor operates; check wiring from starter relay to steering module and from steering module to ECU. No reading starter motor does not operate; check starter relay and starter motor. Below 8 volts check battery, check starter motor. |

| TEST | CIRCUIT TESTING | EXPECTED RESULTS | POSSIBLE FAULTS AND REMEDIES |
|---|---|---|---|
| **4. AIRFLOW METER** Connect ohmmeter between pin 6 and 8. | RR673M | 360 Ω ±10 Ω | Should readings be different from expected results; check harness continuity between ECU plug and airflow meter plug, i.e. pins 6–6, 8–8, 9–9. Check variable resistor circuit in airflow meter with ohmmeter, check that meter flap is closed before substituting airflow meter. |
| Connect ohmmeter between pins 6 and 9. | | 560 Ω ±10 Ω | |
| Connect ohmmeter between pins 8 and 9. | | 200 Ω ±10 Ω | |
| **5. WATER TEMPERATURE SENSOR** Connect ohmmeter between pin 13 and earth. | RR674M | Temp °C   KΩ<br>−10 ±1°C   7.0 to 11.6<br>+20 ±1°C   2.1 to 2.9<br>+80 ±1°C   0.27 to 0.39 | Incorrect reading; check wiring and electrical plug, renew sensor if readings do not comply with results.<br><br>NOTE: Connect the AVO measuring leads to the sensor for short periods only, to minimise the effect of self heating due to the measuring current. |
| **6. SPEED SIGNAL** (Constant energy ignition) Fit a jump lead between the negative terminal on the ignition coil and pin 1. Connect voltmeter between pin 1 and earth. Crank engine. | RR675M | 6–9 volts fluctuating | No reading; check harness connections between coil and ECU. |

Continued

# FUEL INJECTION SYSTEM 19

| TEST | CIRCUIT TESTING | EXPECTED RESULTS | POSSIBLE FAULTS AND REMEDIES |
|---|---|---|---|
| **7. INJECTOR CHECK Numbers 7 and 8**<br>**Injector 7**<br>Connect ohmmeter between pin 14 and 87 on main relay.<br><br>**Injector 8**<br>Connect ohmmeter between pin 28 and 87 on main relay. | No 7, No 8 (RR676M) | 7–10 KΩ<br><br>7–10 KΩ | If reading is below expected results, disconnect each injector in turn to find injector with '00' or low reading; renew injector. If winding resistance of injector is satisfactory, check wiring circuits and resistor pack for open circuit condition. |
| **7A. INJECTOR CHECK Numbers 2 and 4**<br>**Injector 2**<br>Connect ohmmeter between pin 31 and 87 on main relay.<br><br>**Injector 4**<br>Connect ohmmeter between pin 30 and 87 on main relay. | No 2, No 4 (RR677M) | 7–10 KΩ<br><br>7–10 KΩ | See injectors 7 and 8. |
| **7B. INJECTOR CHECK Numbers 3 and 5**<br>**Injector 3**<br>Connect ohmmeter between pin 15 and 87 on main relay.<br><br>**Injector 5**<br>Connect ohmmeter between pin 29 and 87 on main relay. | No 3, No 5 (RR678M) | 7–10 KΩ<br><br>7–10 KΩ | See injectors 7 and 8. |
| **7C. INJECTOR CHECK Numbers 1 and 6**<br>**Injector 1**<br>Connect ohmmeter between pin 33 and 87 on main relay.<br><br>**Injector 6**<br>Connect ohmmeter between pin 32 and 87 on main relay. | No 1, No 6 (RR679M) | 7–10 KΩ<br><br>7–10 KΩ | See injectors 7 and 8. |
| **8. PRESSURISED FUEL RAIL**<br>Fit pressure gauge to cold start injector fuel hose. Switch the ignition on and operate the airflow meter spring return switch to energise the fuel pump. | (RR680M) | 2.4–2.6 kgf/cm²<br>34–37 lbf in² | No pressure build up; check fuel pump circuit with voltmeter. If a reading of 12 volts obtained, check fuel pump earth circuit, satisfactory. If earth circuit check pump. Check operation of fuel pump relay and main relay by substitution if pressure reading is zero. Fuel pressure above or below limits; check pipe work and regulator for blockages and falling pressure reading, check pipe work for leaks, check for leaking injectors, pressure regulator and fuel pump non-return valve. |
| **9. EXTRA AIR VALVE**<br>Connect ohmmeter between pin 34 and 87 on fuel pump relay. | (RR681M) | 30–40 Ω | No reading; check wiring and connections between the pump relay, extra air valve and ECU. Check air valve for continuity with an ohmmeter. |
| **10. COLD START INJECTOR**<br>Disconnect the thermotime switch. With each of the leads being connected to earth in turn. Connect ohmmeter between pin 4 and earth. | (RR682M) | 0–5.0 Ω | No reading; check wiring and connections between the ECU cold start injector and thermotime switch. Open circuit; check wiring and connections and cold start injector windings. |

*Continued*

# FUEL INJECTION SYSTEM

| | | | |
|---|---|---|---|
| **11. AIR TEMPERATURE SENSOR** (Airflow meter)<br>Connect ohmmeter between pin 6 and 27.<br>**NOTE:** Connect the AVO measuring leads to the sensor for short periods only, to minimise the effect of self heating due to the measuring current. | RR683M | Temp °C  KΩ<br>−10      8.26<br>±0.5°C   to<br>         10.56<br>+20      2.28<br>±0.5°C   to<br>         2.72<br>+50      0.76<br>±0.5     to<br>         0.91 | If reading is infinity; disconnect airflow meter, bridge terminals 6 and 27. If ohmmeter reads zero air temperature, sensor is faulty. If after replacement ohmmeter shows infinity, check wiring and connections to the ECU. |
| **12. THROTTLE POTENTIOMETER**<br>**CAUTION:** Ensure the AVO is set to volts.<br>Reconnect the ECU switch ignition on. Measure voltage between green −VE lead and yellow +VE lead by inserting the meter probes into the rear of the multi plug.<br>With ECU connected insert meter −VE lead to green wire and meter +VE lead to red wire measure voltage.<br>ECU connected and with leads connected as above; open throttle voltage should steadily increase. | RR684M | 4.3 ± 0.2 volts<br><br>0.3–0.36 volts<br><br>Smooth swing within 0.3 to 4.5 volt range | No reading or low reading; check wiring and connections<br><br><br><br>If meter reading drops and suddenly picks up through the voltage range—indicates faulty track—renew potentiometer. |
| **13. OVER-RUN RELAY**<br>Disconnect the negative lead from coil to relay. Ignition off, connect ohmmeter between pin 1 and 30 on relay.<br>Ignition on, connect ohmmeter between pin 1 and 30 on relay.<br>Disconnect the vacuum switch and repeat the above test. | RR685M | Infinity Ω<br><br>0 Ω<br><br>Infinity Ω | Reading other than infinity, check wiring and connections for security. Substitute relay.<br>Readings other than zero; check wiring and connections, renew vacuum switch if necessary.<br>Readings other than infinity; renew relay. |
| **14. AIRFLOW METER** (Potentiometer)<br>Reconnect the ECU switch ignition on. Peel back rubber boot on plug. Insert +VE meter probe to pin 6 and −VE lead to pin 9<br>Connect −VE meter probe to pin 9 and +VE probe to pin 7 measure voltage<br>With leads connected as above gradually open air flap. Voltage should decrease. | RR714M | 1.55 ± 0.1 volts<br><br>3.7 ± 0.1 volts<br><br>1.6 ± 0.1 volts | No reading or low reading; check wiring and connections<br><br><br><br>Renew airflow meter if results are not within expected results. |
| **15. AIRFLOW METER** (Potentiometer)<br>Disconnect the ECU. Switch ignition on. Peel back electrical plug rubber boot. Insert −VE meter probe to pin 9 and +VE probe to pin 8. | RR715M | 4.3 ± 0.2 volts | If actual results do not meet expected results, renew the airflow meter. |

After completing the tests with either the 'Epitest' equipment or AVO meter, retest the vehicle to ensure the faults have been rectified.
If faults still persist, check the ECU by substitution.

# FUEL INJECTION SYSTEM

## FAULT DIAGNOSIS—FUEL INJECTION—HOT START

| SYMPTOM | POSSIBLE CAUSE | CURE |
|---|---|---|
| A. Engine will not start. | 1. Battery discharged. | 1. Remove battery from vehicle and recharge. |
| | **WARNING: BEFORE CARRYING OUT THE NEXT TWO OPERATIONS, ENSURE THAT THE RECOMMENDED INSULATED EQUIPMENT IS USED WHEN HANDLING THE HIGH TENSION LEAD.** | |
| | 2. Ignition—coil. | 2. Remove HT lead from distributor, turn ignition on and check for spark at HT to earth—no spark—renew ignition coil and amplifier. |
| | 3. Ignition—distributor. | 3. Remove HT lead from spark plug, turn ignition on and crank engine—no spark—rectify or renew distributor. |
| | 4. Injectors failing to operate. | 4. Turn ignition on and operate the throttle briskly; injectors should be heard to click. If no clicking audible, check throttle potentiometer connections. |
| | 5. Fuel tank ventilation. | 5. Check ventilation pipe work and that a non-vented petrol cap is fitted, rectify or renew. |
| | 6. Cold start system operating. | 6. Cold start system should be inoperative at coolant temperatures above 35°C. If the cold start injector is operating it will cause over-richness of fuel and the vehicle will not start. Disconnect cold start injector, if vehicle starts renew thermotime switch and or cold start injector if injector is leaking. |
| | 7. Loss of fuel pressure. | 7. Disconnect the high tension lead from the coil. Insert a pressure gauge between fuel pump and fuel rail (between fuel rail and cold start injector), crank engine to pressurise fuel system, pressure under cranking should be 2.5 kgf/cm² (36 lbf/in²). Pressure should not drop by more than 0.7 kgf/cm² (10 lbf/in²) in first hour, if greater check for fuel leaks in system. Clamp fuel pipe between gauge and fuel rail, if pressure drops renew fuel pump. |
| | 8. Loss of pressure due to faulty fuel pressure regulator. | 8. Insert a pressure gauge between fuel pump and fuel rail. (Between fuel rail and cold start injector). Crank engine to pressurise fuel system, and clamp fuel pipe between pressure regulator and fuel rail; if pressure still drops rectify or renew pressure regulator. If regulator leaking it may be due to ingress of dirt or grit lodged on the valve seat. It is possible to clear this by clamping fuel return pipe after the regulator, run engine for no more than two seconds; release the clamp quickly. The resultant fuel flow may be sufficient to move the dirt or grit off the valve seat. If valve still loses pressure after re-testing—renew the pressure regulator. |
| | 9. Loss of fuel pressure due to leaking injectors. | 9. Insert a pressure gauge between fuel pump and fuel rail; crank engine to pressurise fuel system. Clamp fuel pipe between pressure regulator and fuel rail; if pressure drops locate and renew a faulty injector or cold start injector. |

## FAULT DIAGNOSIS CHART—GENERAL—FUEL INJECTION

| SYMPTOM | POSSIBLE CAUSE | CURE |
|---|---|---|
| 1. Fuel/air mixture weak. | Air leaks (unmetered air entering engine). | Check engine for air leaks, renew 'O' rings and gaskets as necessary. |
| | Low fuel pressure. | See fault diagnosis 'Hot start' tests 7, 8, 9. |
| | Blocked fuel injectors. | Remove injectors from manifold; do not disconnect from fuel rail. Using Lucas 'Epitest' diagnostic equipment locate faulty injector and renew. |
| 2. Fuel/air mixture rich. | Fuel pressure high. | See fault diagnosis 'Hot start' test 8. Check for faulty pressure regulator; check fuel return pipes for blockages. |
| | Cold start system operating continuously. | Disconnect cold start injector, if mixture returns to correct levels, renew the thermotime switch or cold start injector. |
| | Exhaust leaks. | Check all exhaust, joints, pipes, silencers for security and leaks; retighten or renew as necessary. |
| | Leaking injector. | Locate and renew faulty injector. |
| 3. Engine erratic (hesitating/intermittent). | Engine compressions. | If pressure is considerably low, this indicates faulty valves or piston rings. Low pressure in other cylinders indicates faulty cylinder head gasket. |
| | Fuel injection electrical earth. | Check earth connection on rear of block, clean and check for security. |
| | Ignition system. | See fault diagnosis 'Hot start' test 2 and 3 for H.T. checks. Check low tension side of ignition using 'Lucas Electronic Ignition Analyser' Part number YWB 119. |
| | Electrical connections. | Check all connections for security. |
| | Air filter blocked. | Check filter—renew if necessary. |
| | Inadequate fuel pressure. | Insert pressure gauge into fuel system; check for correct operating pressure. If pressure low, investigate for blockages on fuel feed line. |

# FUEL INJECTION SYSTEM

## ENGINE TUNING PROCEDURE

Before carrying out 'Engine Tuning' on fuel injection vehicles, it is important that all other engine related setting procedures are undertaken first; air meter and air cleaner correctly fitted, ignition and throttle potentiometer correctly set; all hoses correctly fitted and secured.

When the engine is running at its normal operating temperature; thermostat open, the following additional checks and adjustments can be made.

## CHECK AND ADJUST IGNITION TIMING

1. Timing to be checked at not more than 600 rev/min from number 1 cylinder using a stroboscopic lamp.
2. If adjustment is necessary, slacken the distributor clamp bolt and rotate clockwise to retard or anti-clockwise to advance. When the required setting has been attained, tighten the clamp bolt and re-check the setting.

NOTE: Timing must be checked with vacuum pipe disconnected.

## CHECK AND ADJUST IDLE SPEED

1. Remove tamperproof blanking plug from the plenum chamber.
2. Rotate the idle adjustment screw to set the engine idle speed at 700 to 800 rev/min.
   (Clockwise decreases speed — anti-clockwise increases speed.)

## CHECK AND ADJUST IDLE CO LEVEL

The following measurements must be taken with the air cleaner connected, exhaust system correctly fitted and checked for leaks.

1. Remove the tamperproof blanking plug from the top of the air flow meter.
2. Adjust the screw until the CO level reading is to specification; 1% maximum.

NOTE: Do not allow the engine to idle for longer than 3 minutes when setting.

3. Give engine a 'clear out' burst of 30 seconds at 2000 rev/min light load or longer if necessary to maintain normal running temperature.
4. Re-check CO level and idle speed; adjust if necessary.
5. Fit new tamperproof plugs to air flow meter and idle adjustment screws.

## AIR CLEANER

### Remove

1. Release the hose clip securing the hose to the rear of the air cleaner canister.
2. Remove the two nuts and bolts securing the air cleaner to the left-hand valance. Pull the canister from the hose and remove the air filter canister from the engine compartment.
3. Unclip the three catches securing the inlet tube to the air cleaner canister and remove the inlet tube.
4. Remove the nut and end plate securing the air cleaner element in position.
5. Withdraw the air cleaner element and discard.

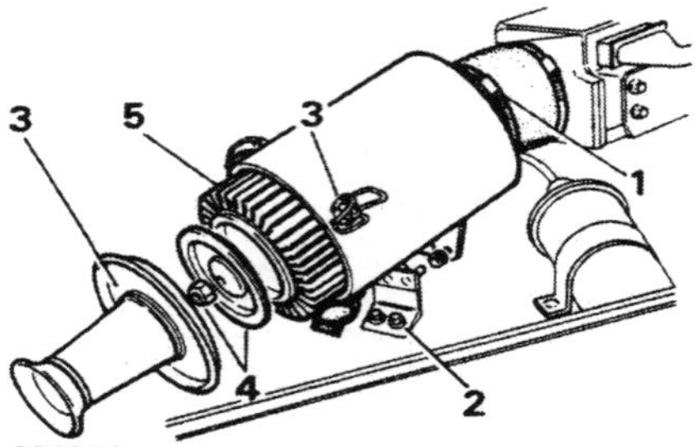

### Refit

6. Fit new element and secure in position.
7. Refit the inlet tube to the air cleaner canister.
8. Refit the air cleaner to the mounting bracket and tighten the two nuts.
9. Fit the hose to the air cleaner outlet aperture and tighten the hose clip.

## RESETTING THROTTLE LEVERS—fuel injection

### Manual/Automatic Transmission

NOTE: The setting procedure outlined is applicable at minimum throttle condition only.

1. Set the throttle butterfly to a maximum of 0.05 mm (0.002 in) clearance, measured at the vertical centre line within ± 10°.

2. Ensuring that the butterfly is retained at its setting, remove the stop adjustment screw tamperproof cap (if fitted).
3. Rotate the screw until contact is made with the stop lever, refit the tamperproof cap (if fitted).

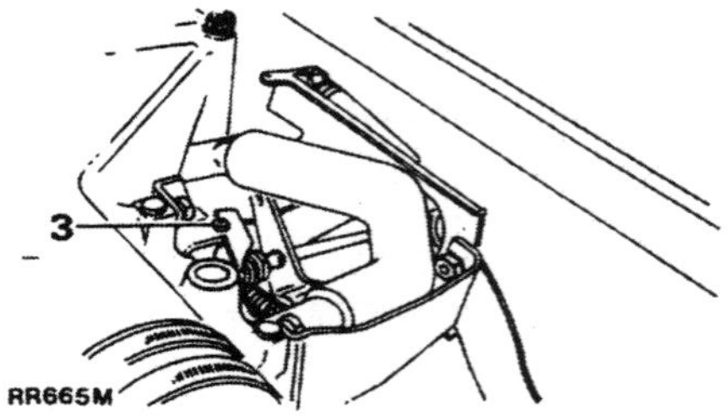

4. Release the throttle operating lever securing screw and adjust the lever until contact is made with the top end of the slot in the throttle lever mounting bracket; retaining the lever in this position re-tighten the screw.

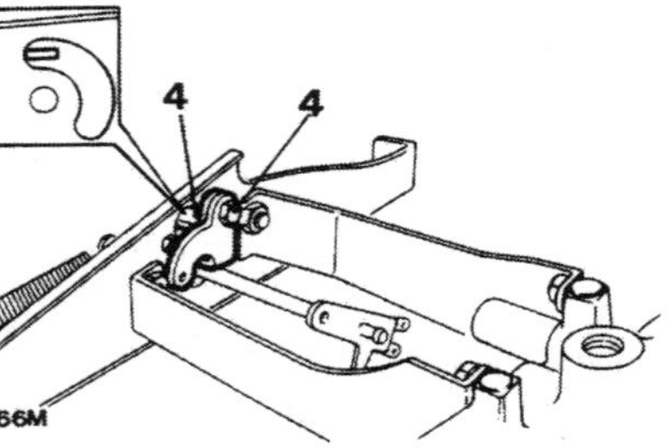

NOTE: Re-check the throttle potentiometer setting after adjusting the throttle levers.

## RENEW THE THROTTLE CABLE

### Removing

1. Remove the split pin, washer and clevis pin securing the throttle linkage end of the cable.
2. Carefully prise the cable adjustment nut out of the linkage mounting bracket.
3. Withdraw the cable from the mounting bracket.
4. Release the outer cable from the retaining clips within the engine compartment.
5. Remove the lower fascia panel from beneath the steering column.
6. Disconnect the cable from the throttle pedal.
7. Feed the cable through the bulkhead grommet and into the engine compartment.

### FIT NEW THROTTLE CABLE

8. Feed the new cable from the engine compartment through the bulkhead grommet.
9. Connect the cable to the throttle pedal.
10. Connect the cable to the throttle linkage, fit a new split pin and secure in position.
11. Clip the outer cable adjustment nut into the mounting bracket.
12. Adjust the outer cable to give 1.57 mm (0.062 in) free play in the throttle cable and check the throttle operation.

# FUEL INJECTION SYSTEM | 19

## ELECTRONIC FUEL INJECTION RELAYS

Incorporated into the fuel injection electrical circuits are three relays and a current steering diode pack. Two of the relays and the steering diode pack are located beneath the front right-hand seat protected by a black cover. The remaining relay is located in the engine compartment attached to the air flow meter mounting bracket.

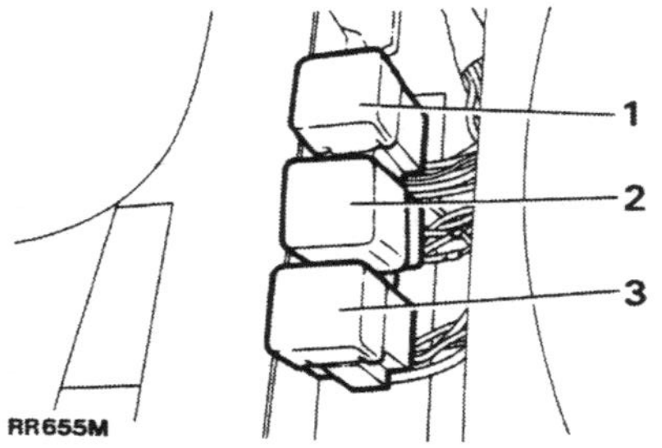

RR 655M

Relays viewed from within the vehicle.

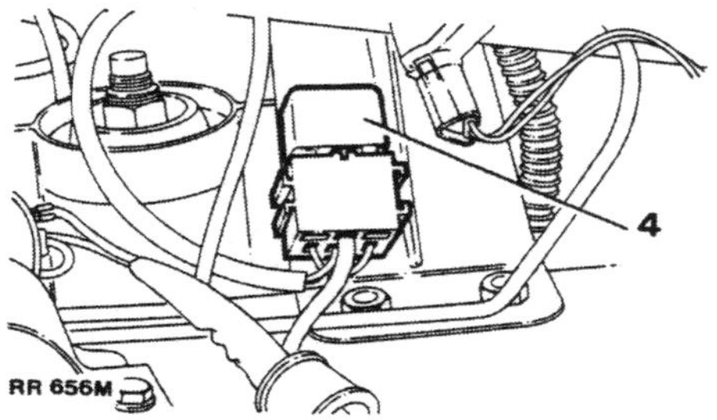

RR 656M

Relay viewed from engine compartment

1. Fuel pump relay (Item 11 on fuel injection circuit diagram).
2. Steering module (red case) (Item 9 on fuel injection circuit diagram).
3. Main relay (Item 10 on fuel injection circuit diagram).
4. Over-run fuel shut-off relay (Item 4 on fuel injection circuit diagram).

### Removing

1. Disconnect the battery.
2. Remove the black protective cover (applicable only to the relays located under the right-hand seat).
3. Pull the relay(s) from the multi-plug(s).

### Refitting

4. Reverse the removal procedure.

## ELECTRONIC CONTROL UNIT — ECU

### Remove and refit

### Removing

1. Disconnect the battery.
2. The ECU is located under the front right-hand seat and is accessible through the front aperture of the seat plinth.
3. Release the quarter turn screw and lift off the black protection cover.
4. Remove the three relays, noting their position for re-assembly.
5. Pull the ECU multi-plug retaining clip towards the rear of the seat.
6. Pull the rear of the multi-plug out of the ECU.
7. Manoeuvre the rear of the plug towards the gearbox tunnel to release the hooked front end of the plug from the retaining peg.

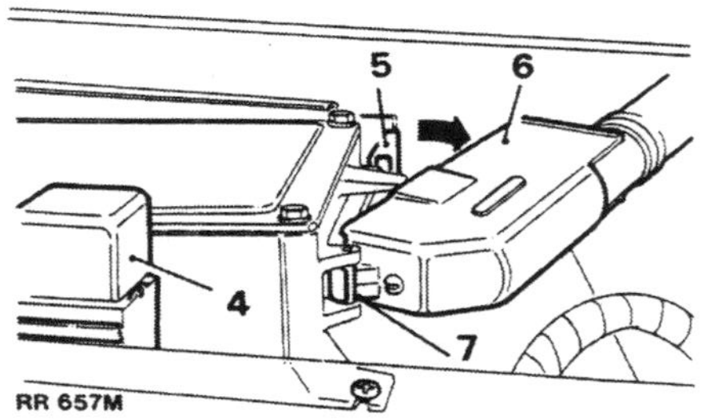

RR 657M

8. Release the three screws securing the ECU to the mounting bracket.

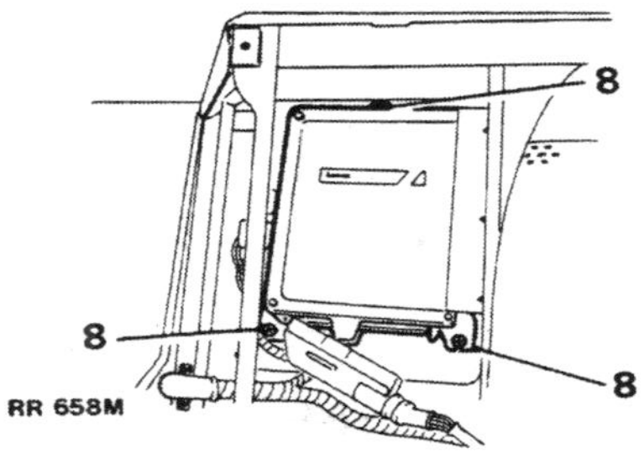

RR 658M

9. Withdraw the ECU from beneath the seat.

NOTE: The ECU is not itself a serviceable item, in the event of a unit failure, the ECU must be renewed.

### Refitting

10. Secure the ECU in position.
11. Push the multi-plug into the ECU until an audible click is heard denoting that the plug is fitted securely.
12. Refit the relays.
13. Refit the ECU protective cover.
14. Re-connect the battery.

## THROTTLE POTENTIOMETER

### Remove

1. Disconnect the battery.
2. Disconnect the electrical three pin plug.
3. Remove the two screws securing the switch to the plenum chamber and carefully pull the switch off the throttle butterfly spindle.

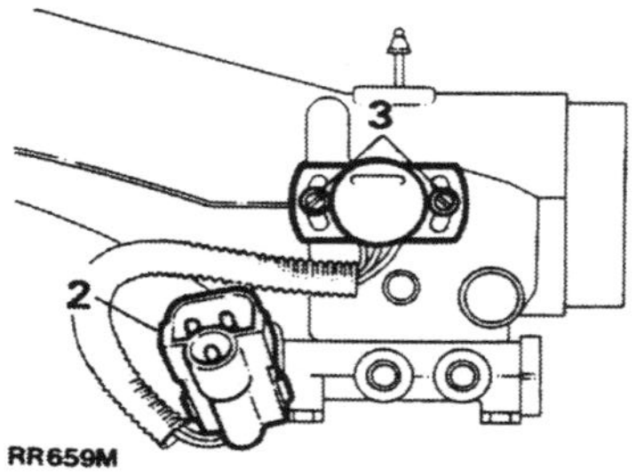

RR659M

4. Remove the old gasket.

### Refit

5. Fit a new gasket between the throttle switch and plenum chamber.
6. Align the switch and spindle flats; slide the switch onto the throttle spindle and secure the switch to the plenum chamber.
7. The throttle potentiometer must be reset using a potentiometer adjustment gauge.

### Setting the potentiometer

Equipment required:—

Lucas electronic fuel injection throttle potentiometer adjustment gauge — Lucas Part Number YWB 121.

8. Slacken the potentiometer securing screws.
9. Disconnect the three-pin plug from the potentiometer electrical lead. Connect the adjustment gauge plug to the potentiometer.
10. Connect the two crocodile clips from the throttle potentiometer gauge to the appropriate battery terminals.
11. Rotate the potentiometer clockwise or anti-clockwise until the middle lamp of the three indication lamps remains illuminated.
12. Tighten the potentiometer securing screws.
13. Re-check the potentiometer setting.
14. Disconnect the adjustment gauge from the potentiometer and battery terminals.
15. Re-connect the harness three-pin plug to the potentiometer.

NOTE: If a potentiometer adjustment gauge is unavailable, the setting procedure can be carried out using a voltmeter.

IF AN AVO METER IS USED TO CARRY OUT THIS CHECK — ENSURE THE AVO IS SET TO VOLTS. AN AVO METER SETTING OTHER THAN VOLTS WILL RESULT IN DAMAGE TO THE POTENTIOMETER.

### Setting the Potentiometer using a voltmeter

16. Slacken the potentiometer securing screws.
17. Switch on the ignition.
18. Connect a voltmeter between the red and green leads at the potentiometer electrical plug.
19. Rotate the potentiometer clockwise or anti-clockwise, until the volt meter reads 290 ± 20 Mv.
20. Tighten the potentiometer securing screws.
21. Re-check the voltmeter reading.

# FUEL INJECTION SYSTEM

## THERMOTIME SWITCH

### Test

**WARNING: When the cooling system is hot take care to avoid scalding.**

1. Remove the pressure relief cap from the coolant expansion tank and remove the filler plug from the radiator. Use a thermometer and note the coolant temperature.
2. Disconnect the battery and pull the electrical connector from the thermotime switch.

Note the rated value stamped on body of the switch.

3. Connect ohmmeter between switch terminal 'W' and earth:

    (a) Coolant temperature higher than the switch rated value; a very high resistance reading, (approximately 300 ohms at temperature greater than 40°C) open circuit, should be obtained. Renew the switch if a low resistance, short circuit, is shown.

    (b) Coolant temperature lower than the switch rated value; a very low resistance reading, closed circuit, (approximately zero ohms at temperature less than 40°C) should be obtained. Renew the switch if a high reading, open circuit, is shown.

4. Connect a 12v supply via an isolating switch to terminal 'G' of the thermotime switch.

Use a stopwatch, check the time delay between making the isolating switch, and the ohmmeter showing the change from low to high resistance. The delay period must closely approximate to time, according to temperature:

| Coolant Temperature °C | Delay in Seconds |
|---|---|
| −10 | 8 |
| 0 | 4.5 |
| 10 | 3.5 |
| 35 | 0 |

Renew the thermotime switch if necessary.

5. Re-connect the plug to the switch and connect the battery.

### Removing

**WARNING: When the cooling system is hot take care to avoid scalding**

6. Remove the pressure relief cap from the coolant expansion tank.
7. Remove the radiator bottom hose and partially drain the cooling system.
8. Disconnect the battery.
9. Remove the electrical plug from the switch.
10. Unscrew the switch and remove it from the inlet manifold.
11. Remove the copper seating washer.

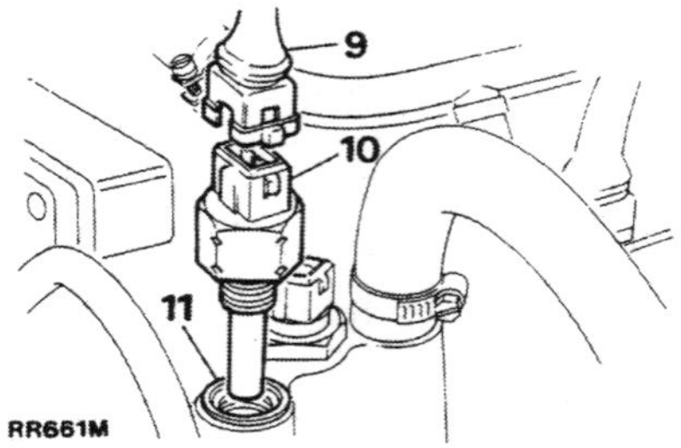

RR661M

### Refitting

12. Fit a new copper washer.
13. Reverse the removal instructions ensuring that the switch is firmly screwed into position.
14. Re-connect the electrical plugs.
15. Visually inspect for water leaks.

## COOLANT TEMPERATURE SENSOR

### Test

NOTE: When using an AVO connect the measuring leads to the sensor for short periods of time only to minimise the effects of self heating due to the measuring current.

1. Disconnect the battery and remove the electrical plug from the temperature sensor.
2. Connect an ohmmeter between the sensor terminals and note the resistance reading, disconnect the ohmmeter.

The reading should closely approximate the following according to temperature:

| Coolant Temperature °C | Resistance Kilohms |
|---|---|
| −10 | 9.2 |
| 0 | 5.9 |
| 20 | 2.5 |
| 40 | 1.18 |
| 60 | 0.60 |
| 80 | 0.33 |

3. Check the resistance between each terminal in turn against the body of the sensor. A very high resistance reading, open circuit, must be obtained.
4. Re-connect the sensor and the battery.

### Removing

**WARNING: When the cooling system is hot take care to avoid scalding.**

5. Remove the pressure relief cap from the coolant expansion tank.
6. Remove the radiator bottom hose and partially drain the cooling system.
7. Disconnect the battery.
8. Remove the electrical plugs from the coolant sensor and thermotime switch.
9. Unscrew the thermotime switch to give access to the coolant sensor.
10. Unscrew the coolant sensor and remove it from the inlet manifold.
11. Remove the copper washer from the manifold.

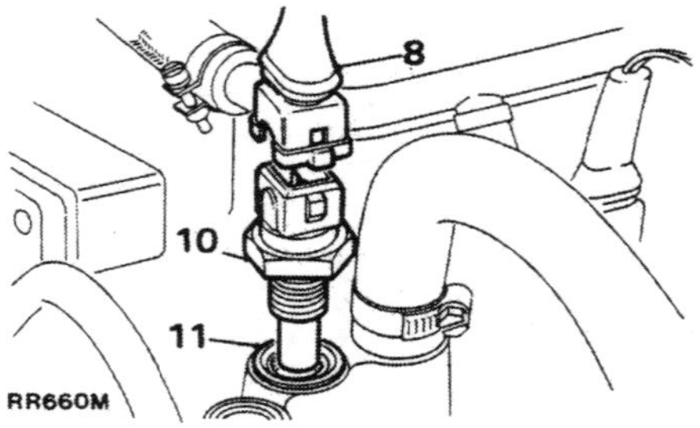

RR660M

### Refitting

12. Reverse the removal procedure ensuring that the switches are securely fitted. Fit new copper washers.
13. Re-connect the electrical plugs.
14. Visually inspect for water leaks.

## AIR TEMPERATURE SENSOR

### Test

NOTE: To prevent self heating of the sensor during the test procedure connect the measuring leads to the sensor for short periods of time only.

1. Disconnect the battery and the electrical multiplug from air flow meter.
2. Connect the ohmmeter between terminals 6 and 27 of the air flow meter and note the resistance reading. The reading should closely approximate the following, according to temperature:

| Ambient air temperature (°C) | Resistance kilohms |
|---|---|
| −10 | 9.2 |
| 0 | 5.9 |
| 20 | 2.5 |
| 40 | 1.18 |
| 60 | 0.60 |

3. Disconnect the ohmmeter. Re-connect the multi-plug and the battery.

The air temperature sensor is not a serviceable item. If it does not meet the test requirements, the complete air flow meter must be renewed.

## AIR FLOW METER

### Remove and Refit

### Removing

1. Disconnect the battery.
2. Disconnect the electrical multi-plug from the plenum chamber hose from the rear of the meter.
3. Release the hose clip and remove the plenum chamber hose from the rear of the meter.
4. Release the hose clip at the air intake side of the air flow meter.
5. Remove the three securing screws and plain washers retaining the meter to the mounting bracket.
6. Detach the air flow meter from the hose and withdraw it from the engine compartment.

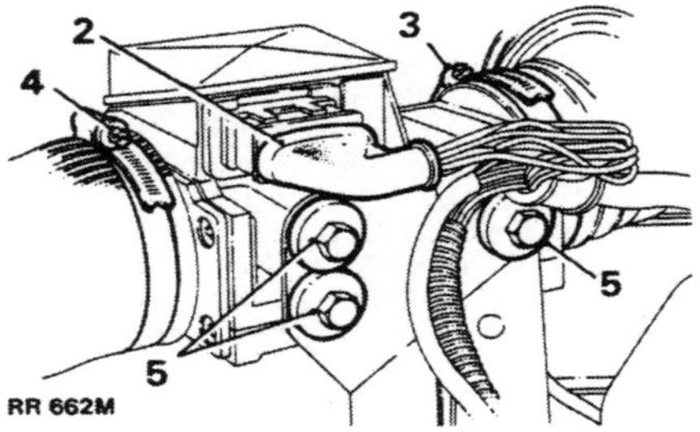

RR662M

### Refitting

7. Reverse the removal operations ensuring the multi-plug is firmly re-connected on re-assembly.

NOTE: The air flow meter is not a serviceable item. In the event of failure or damage the complete unit is to be renewed.

*Continued*

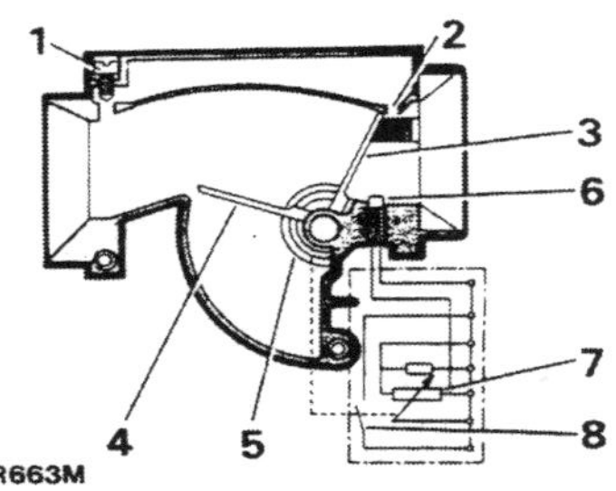

# FUEL INJECTION SYSTEM

1. Idle mixture adjustment screw
2. Air by-pass channel
3. Measuring flap
4. Compensating flap
5. Coil spring—flap return
6. Air temperature sensor
7. Potentiometer
8. Fuel pump switch

## POWER RESISTOR RACK

The power resistor pack is located under the air flow meter attached to the meter mounting bracket.

The resistor pack is not a serviceable item, in the event of failure or damage the unit must be renewed.

### Removing

1. Disconnect the battery.
2. Disconnect the multi-plug from the bottom of the resistor pack.
3. Remove the two nuts, bolts and spring washers securing the resistor pack to the air flow meter mounting bracket.
4. Withdraw the resistor pack from the engine compartment.

### Refitting

5. Reverse the removal instructions ensuring that the multi-plug is fitted securely.

## OVER-RUN FUEL SHUT-OFF VALVE—VACUUM SWITCH

1. Disconnect the battery.
2. Disconnect the two electrical leads.
3. Remove the vacuum hose from the valve.
4. Remove the single nut and spring washer securing the valve to the injector retaining plate.
5. Withdraw the valve.

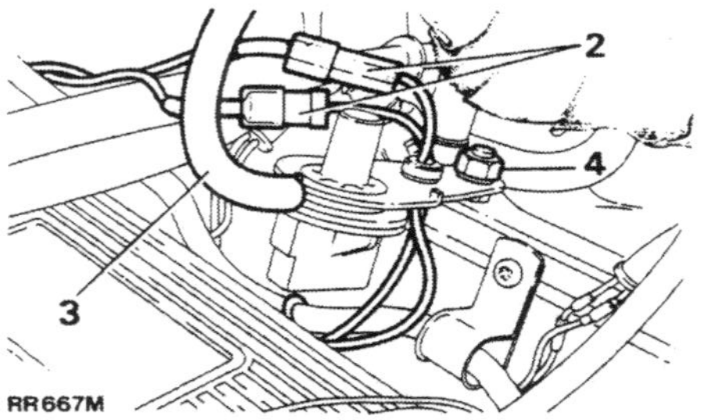

## DEPRESSURISING THE FUEL SYSTEM

**WARNING:** Under normal operating conditions the fuel injection system is pressurised by a high pressure fuel pump, operating at 1.8 to 2.5 kgf/cm² (26 to 36 lbf/in²). When the engine is stationary this pressure is maintained within the system.

To prevent pressurised fuel escaping and to avoid personal injury it is necessary to depressurise the fuel injection system before any service operations are carried out.

1. Remove the ECU protective cover located under the front right hand seat.
2. Pull the fuel pump relay off its multi-plug (see Electronic fuel injection relays).
3. Start and run the engine.
4. When sufficient fuel has been used up causing the fuel line pressure to drop, the injectors will become inoperative, resulting in engine stall. Switch ignition off.

NOTE: Fuel at low pressure will still remain in the system. This low pressure can be removed by releasing the cold start injector from the plenum chamber and then placing the injector with hose still attached into a suitable container. Release the hose clip and carefully remove the hose from the injector to release any remaining pressurised fuel.

5. Disconnect the battery.

### Refitting

6. Refit the cold start injector.
7. Refit the fuel pump relay, re-connect the battery.
8. Crank the engine (engine will fire within approximately 6 to 8 seconds).

## FUEL PRESSURE REGULATOR

### Test

1. Depressurise the fuel system.
2. Release the clip and pull the cold start injector supply hose from the fuel rail. Connect a pressure gauge to the fuel rail.
3. Switch the ignition on and operate the flap in the air flow meter to energise the fuel pump.
4. Check that the pressure gauge reading is between 2.5 to 2.6 kgf/cm² (35 to 37 lbf/in²).
5. Switch off the ignition. The fuel pressure should be maintained between 2.1 to 2.6 kgf/cm² (30 to 37 lbf/in²).

NOTE: The pressure reading may slowly drop through either the regulator valve or the fuel pump non-return valve. A slow steady drop is permissable; a rapid fall must be investigated.

6. If the pressure reading is unsatisfactory renew the pressure regulator.
7. After fitting a new regulator re-test the system, if the pressure continues to drop off, the fuel injectors, fuel pump, non-return valve and fuel system pipework should be checked for leaks.
8. Depressurise the fuel system. Remove the gauge and connect the cold start injector supply hose.
9. Check for fuel leaks around the hose connection.

### Remove and Refit

### Removing

10. Depressurise the fuel system.
11. Disconnect the battery.
12. Pull the flexible rubber vacuum hose from the bottom of the regulator.
13. Release the spill return hose from the side of the regulator.
14. Release the clip and remove the spill return hose from the top of the regulator.
15. Release the single large nut securing the regulator to the regulator mounting bracket.
16. Withdraw the regulator.

### Refitting

17. Reverse the removal instructions ensuring that all hose connections are secure.
18. Energise the fuel pump and visually inspect all hose connections for fuel leaks.

NOTE: In the event of failure or damage the valve is not serviceable.

### Refitting

6. Reverse the removal instructions.

## SOLENOID OPERATED AIR VALVE

—fitted to vehicles with air conditioning only.

1. Disconnect the battery.
2. Disconnect the electrical lead.
3. Pull the front air valve pipe from the plenum chamber.
4. Pull the rear air valve pipe from the extra air valve rail.
5. Remove the single nut and spring washer securing the valve to the injector retaining plate.
6. Withdraw the valve.
7. Remove the hoses from the valve.

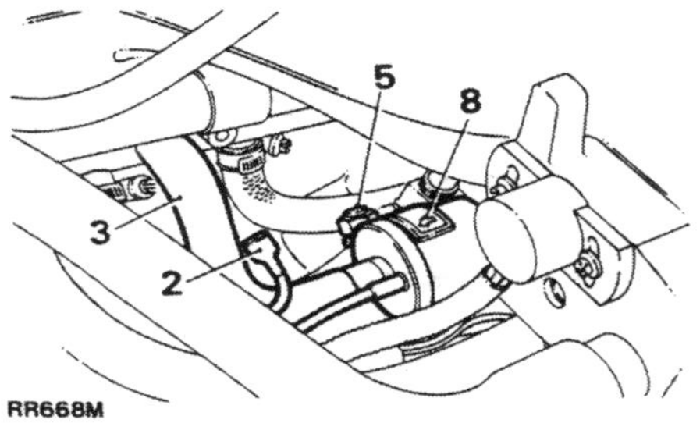

### Refitting

8. Reverse the removal procedure, observing the direction of flow arrow on the air valve body. The direction of flow is from air rail to plenum chamber.

## FUEL INJECTION SYSTEM

### PLENUM CHAMBER

**Remove and refit**

**Removing**

1. Disconnect the battery.
2. Release the hose clip at the rear of the air flow meter and remove the hose from its location.
3. Release the radiator bottom hose and partially drain the cooling system.
4. Release the two coolant hoses from the bottom of the plenum chamber inlet neck.
5. Disconnect the breather hose and vacuum pipe.
6. Disconnect the throttle potentiometer electrical multi-plug.
7. Remove the two screws securing the cold start injector to the plenum chamber.
8. Withdraw the cold start injector and gasket.
9. Disconnect the flexible rubber hose from the extra air valve rail and plenum chamber.
10. Remove the two bolts (with spring washers) securing the throttle cable anchor bracket to the throttle lever support bracket.
11. Unhook the small return spring from the throttle levers.
12. Remove the small vacuum hose from the rear of the plenum chamber.
13. Remove the six socket head bolts securing the plenum chamber to the ram housing.
14. Manoeuvre the plenum chamber and lift it off the ram housing.

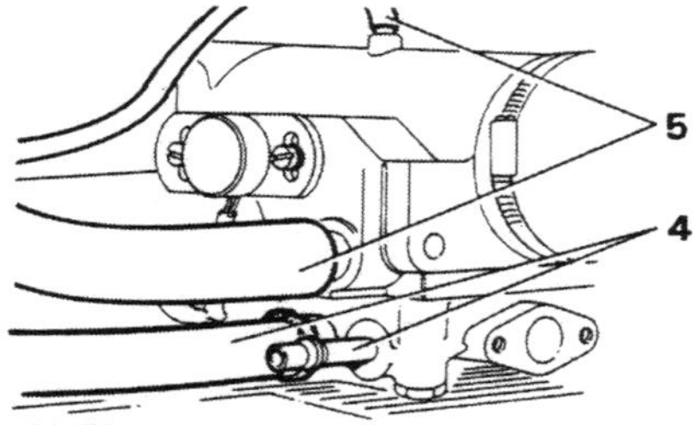

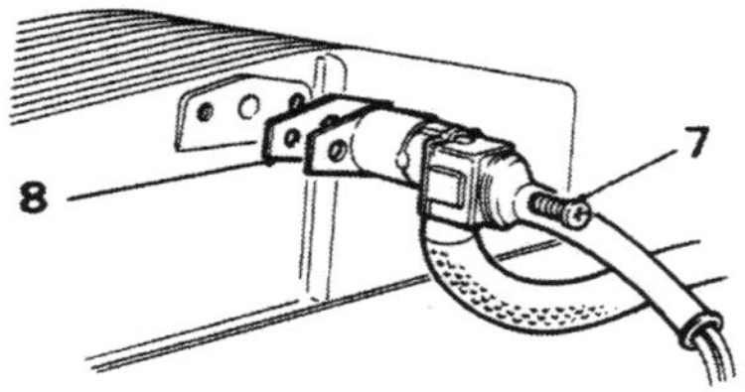

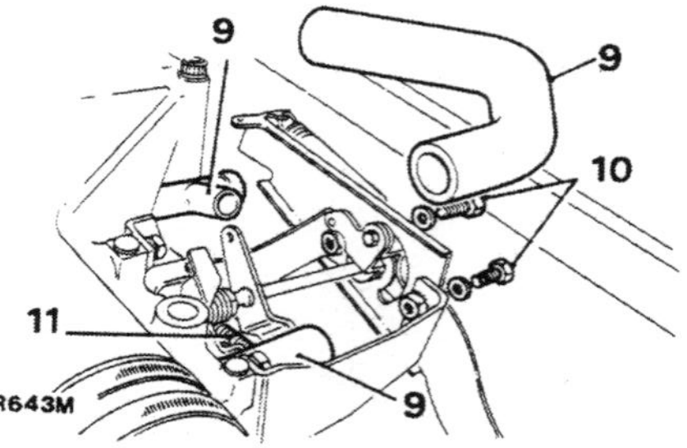

**NOTE: To prevent dirt entering the ram tubes, place a protective cover over the ram tube apertures.**

**Refitting**

15. Ensure that all mating surfaces are free from any previous sealing compounds.
16. Smear the mating surfaces of the plenum chamber and ram housing with 'Hylomar' sealant.
17. Reverse the removal procedure.

### RAM HOUSING

**Remove and refit**

**Removing**

1. Depressurise the fuel system.
2. Disconnect the battery.
3. Remove the plenum chamber (see plenum chamber remove and refit).
4. Release the brake servo hose from the ram housing.
5. Release the extra air valve hose at the ram housing.
6. Remove the solenoid operated air valve hoses from the front left-hand side of the ram housing and from the rear of the extra air valve rail.
7. Remove the extra air valve hose from the air valve rail.
8. Remove the six through bolts (with plain washers) securing the ram housing to the intake manifold. 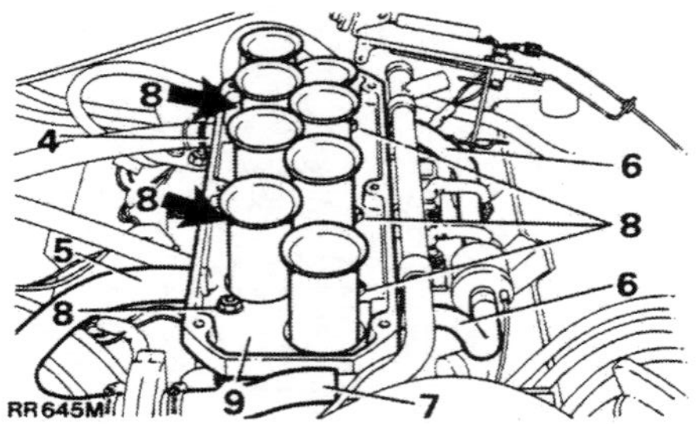
9. Lift the ram housing off the intake manifold face and remove it from the engine compartment.
10. Place a protective cover over the top of the intake manifold inlet bores to prevent ingress of dirt.

**Refitting**

11. Ensure that all mating faces are clean and free from dirt and any previous sealing compounds.
12. Apply 'Hylomar' sealant to the intake manifold face before refitting the ram housing.
13. Fit the ram housing and retighten the bolts, working from the two centre bolts, diagonally towards the outer four bolts.
14. See 'Data section' for correct tightening torque.

5. Disconnect the electrical multi-plug.
6. Remove the two bolts securing the air valve to the intake manifold.
7. Move the plug lead support bracket and earth lead aside.
8. Withdraw the air valve from the engine compartment.

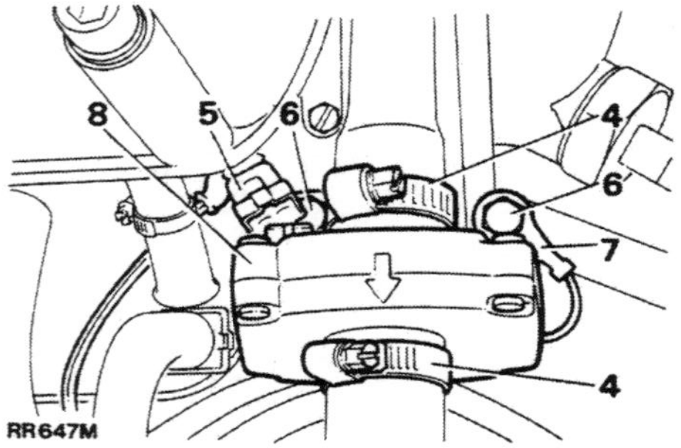

**Refitting**

9. Refit the hoses securely.
10. Refit the plug lead support bracket and earth lead.
11. Reverse the removal operations.

### EXTRA AIR VALVE

**Test**

1. Disconnect the electrical multi-plug from the valve and connect a voltmeter across the terminals of the connector.
2. Operate the starter motor—battery voltage should be obtained on the voltmeter.

No voltage—check wiring for continuity and condition; Battery voltage—check the resistance of the heating coils.

3. Connect an ohmmeter between the terminals of the air valve. A resistance of 33 ohm should be obtained.

No resistance—renew the air valve.

**Remove and refit**

**Removing**

4. Release the two hose clips and detach the hoses from the air valve.

# FUEL INJECTION SYSTEM

## FUEL RAILS

### Remove

1. Depressurise the fuel system.
2. Remove the plenum chamber.
3. Remove the ram housing.
4. Disconnect the fuel feed pipe at the fuel rail.
5. Disconnect the spill return pipe from the pressure regulator at the fuel rail.
6. Release the eight injector hose clips at the fuel rail.
7. Remove the fuel rail from the hoses.

### Refitting the fuel rails

8. Push the fuel rail outlet pipes onto the injector hoses.
9. Tighten the eight hose clips securely.
10. Re-connect the spill return and fuel feed pipes.
11. Refit the ram housing and plenum chamber, tighten bolts to correct torque. See 'Data section'.
12. Start the engine and visually check the fuel rails and connections for leaks.

## INJECTORS—RH AND LH BANKS—COLD START INJECTOR

### Test

1. Use an ohmmeter to measure resistance value of each injector winding, which should be 2.4 ohm at 20°C (68°F).
2. Check for short-circuit to earth on the winding, by connecting the ohmmeter probes between each injector terminal and injector body; meter should read infinity. Renew the injector if the winding is open circuit or short circuit.

### Cold start injector

#### Remove the injectors

1. Depressurise the fuel system.
2. Disconnect the battery.
3. Disconnect the electrical plug from the injector.
4. Release the hose clip and remove the hose from the injector.
5. Remove the two screws (with spring washers) securing the injector to the plenum chamber.
6. Withdraw the cold start injector and remove the gasket.

### Manifold injectors—RH and LH

7. Remove the plenum chamber (see plenum chamber remove and refit).
8. Remove the ram housing (see ram housing remove and refit).

NOTE: To enable any pair of injectors to be released from their locations on either the left- or right-hand side of the intake manifold, it will be necessary to remove the fuel rail from all four injectors in that bank.

WARNING: Before releasing any hoses from the fuel rail, place a cloth around the hose connections to prevent any excess fuel which may be under pressure escaping.

9. Release the four clips securing the injector hoses to the fuel rail.
10. **Left-hand injectors**—release the clip and remove the fuel feed hose from the fuel rail.
    **Right-hand injectors**—release the clip and remove the spill return hose to the pressure regulator at the fuel rail.
11. Remove the electrical plugs from the injectors.
12. Carefully prise the fuel rail out of the injector hoses.

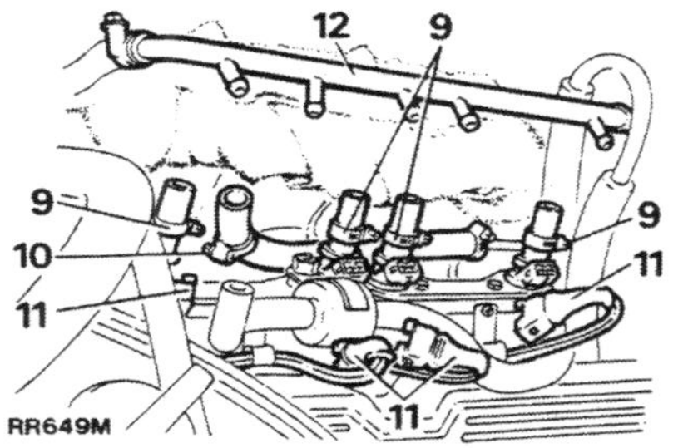

Left-hand injectors

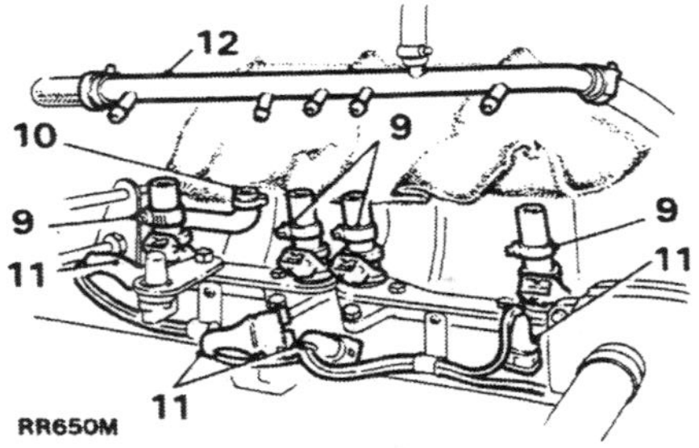

Right-hand injectors

13. Remove the two bolts securing the injector retaining plate.

*Continued*

NOTE: **Left-hand front injectors**—remove the single nut securing the solenoid operated air valve to the injector retaining plate, remove the air valve to give access to the retaining plate securing nut.
**Right-hand rear injectors**—remove the single nut securing the over-run fuel shut off valve to the injector retaining plate, remove the valve to give access to the retaining plate securing nut.

14. Withdraw the pair of injectors complete with retaining plate.
15. Remove the single screw securing the injector harness retaining clip to the injector location plate.
16. Remove the two bolts securing the location plate to the intake manifold.
17. Remove the location plate to give access to the small rubber seating washers located in the intake manifold injector bores.
18. Pull the rubber seats from the bore and discard.

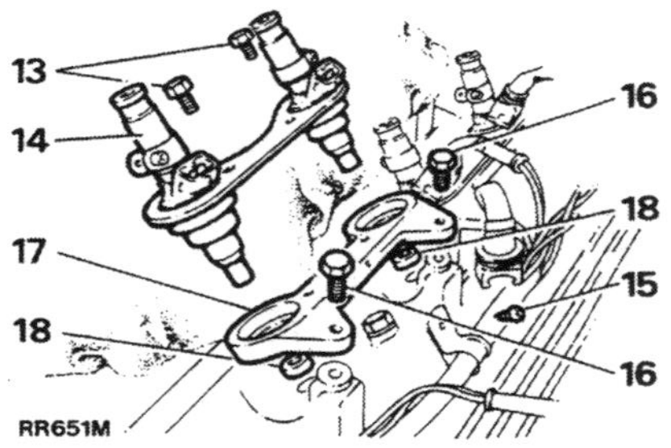

19. Remove the injectors from the retaining plate by releasing the large rubber seating washers from their registers.
20. Withdraw the injector(s) from the retaining plate.

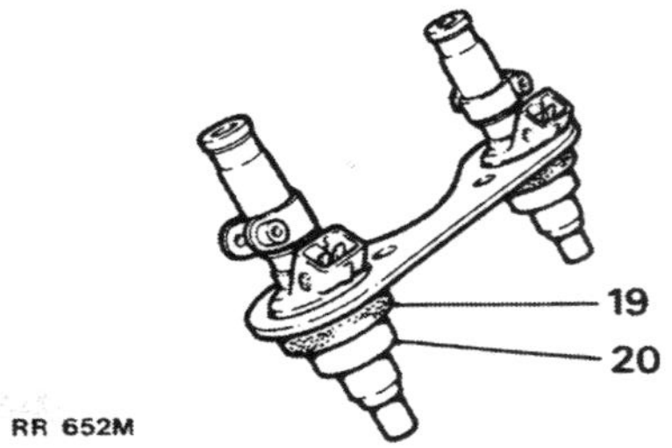

### Refitting the injector(s)

21. Fit **new** seating rubbers in the intake manifold before refitting the injectors.
22. Refit the injectors with the electrical plug connections facing away from the intake manifold.
23. Re-connect the fuel rail to the injector hoses ensuring that the rail is pushed firmly into position, tighten all hose clips securely.
24. Refit the ram housing and plenum chamber.
25. Refit all ancillary items.
26. Refit fuel pump relay and re-connect the battery.
27. Start the engine and visually inspect for fuel leaks around all hose connections.

## FUEL INJECTION SYSTEM

### INTAKE MANIFOLD

**Removal and refit**

**Removing**

1. Depressurise the fuel system.
2. Drain the cooling system.
3. Remove the plenum chamber and ram housing.
4. Remove the extra air valve.
5. Disconnect the electrical multi-plugs from the injectors.
6. Remove the multi-plugs from the thermotime switch and coolant sensor.
7. Disconnect the multi-plug from the cold start injector.
8. Release the four screws and clips securing the injector harness to the left- and right-hand side of the intake manifold.
9. Move the harness away from the intake manifold.
10. Disconnect the top hose from the thermostat housing.
11. Release the two flexible hoses from the rear of the water pump.
12. Disconnect the electrical lead from the water temperature sensor.

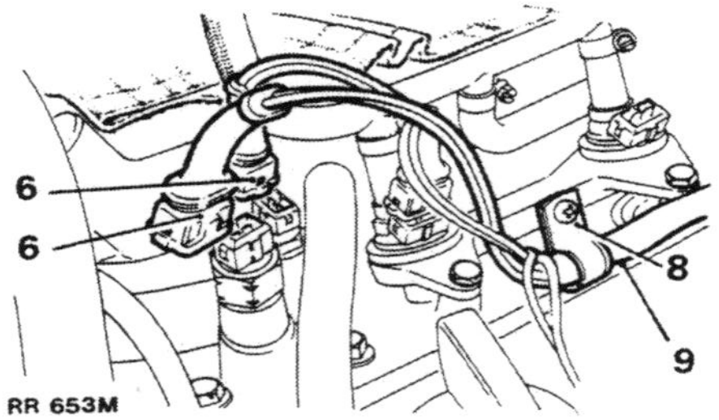

NOTE: If air conditioning is fitted, disconnect the two electrical leads from the sensor in the thermostat housing.

13. Release the two flexible heater hoses from the rear of the intake manifold.

NOTE: To enable the intake manifold rear right-hand securing bolt to be removed it is necessary to remove the fuel pressure regulator, this is removed as follows.

14. Release the clip securing the flexible hose to the rigid fuel supply pipe and remove the hose from the pipe.
15. Remove the single bolt securing the rigid fuel supply pipe to the rear of the intake manifold.
16. Release the nut securing the flexible fuel feed hose to the rigid fuel supply pipe.
17. Release the hose clip and remove the spill return hose at the pressure regulator.
18. Release the hose clip and remove the flexible spill return hose at the side of the pressure regulator.
19. Disconnect the two electrical leads at the vacuum switch.
20. Release the flexible hose from the vacuum switch.
21. Remove the single bolt and spring washer securing the pressure regulator mounting bracket to the rear right-hand side of the intake manifold.
22. Manoeuvre the regulator and mounting bracket, withdraw the assembly from the engine compartment.
23. Release the sparking plug leads from the two plastic retaining clips located at the front of the intake manifold.
24. Remove the twelve bolts and washers securing the intake manifold to the cylinder heads.

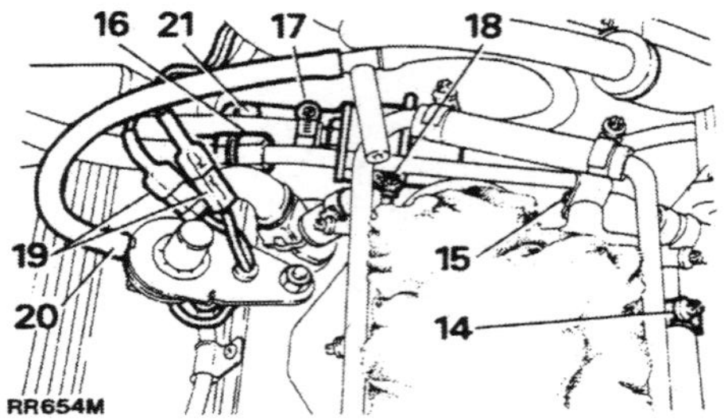

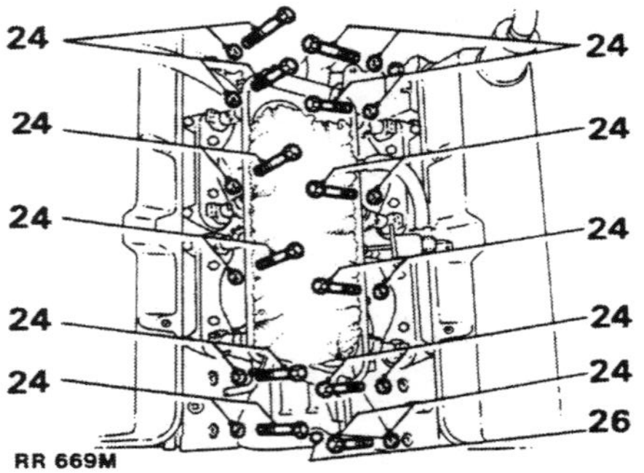

NOTE: The two bolts securing the front of the manifold are of a longer type. Ensure that these two bolts are fitted in the correct positions on re-assembly.

25. Lift off the intake manifold.
26. Remove the two gasket clamps from the top of the cylinder block.
27. Lift off the gasket.
28. Remove the gasket seals.

*Continued*

**Refitting**

29. Using new gasket seals, smear both sides of the seals with silicon grease.
30. Locate the seals in position with their ends engaged in the notches formed between the cylinder heads and block.
31. Apply 'Hylomar' sealant around the outside of the water passage apertures on the cylinder heads, manifold gasket and intake manifold.
32. Fit the manifold gasket with the word 'FRONT' to the front and the open bolt hole to the front right-hand side.
33. Fit the gasket clamps but DO NOT fully tighten the bolts at this stage.
34. Locate the intake manifold onto the cylinder heads, clean the threads of the manifold securing bolts and coat them with thread lubricant sealer 3M EC776.
35. Fit all manifold bolts and tighten them a little at a time, evenly, alternate sides working from the centre outwards.
36. Tighten to correct torque—see 'Data section'.
37. Tighten the gasket clamps to the correct torque—see 'Data section'.
38. Reverse removal procedure items 1–23.
39. Start the engine, check for water and fuel leaks.

### RENEW FUEL LINE FILTER

WARNING: THE SPILLING OF FUEL IS UNAVOIDABLE DURING THIS OPERATION. ENSURE THAT ALL NECESSARY PRECAUTIONS ARE TAKEN TO PREVENT FIRE AND EXPLOSION.

1. Depressurise the fuel system.
2. The fuel line filter is located on right-hand chassis side member forward of the fuel tank filler neck. Access to the filter is gained through the right-hand rear wheel arch.
3. Chock the front road wheels.
4. Raise the rear of the vehicle and place on suitable axle stands.
5. Remove the right-hand rear road wheel.
6. Clamp the inlet and outlet hoses to the fuel filter (denoted by the two arrows).
7. Release the two hose clips and disconnect the hoses. Plug the open ends of the hoses to prevent ingress of dirt.
8. Release the filter clamp bolt and slide the filter out of the clamp.
9. Fit a new filter observing the direction of flow arrows stamped on the filter body.
10. Reverse the removal procedure. Start the engine and inspect all hose connections for leakage.

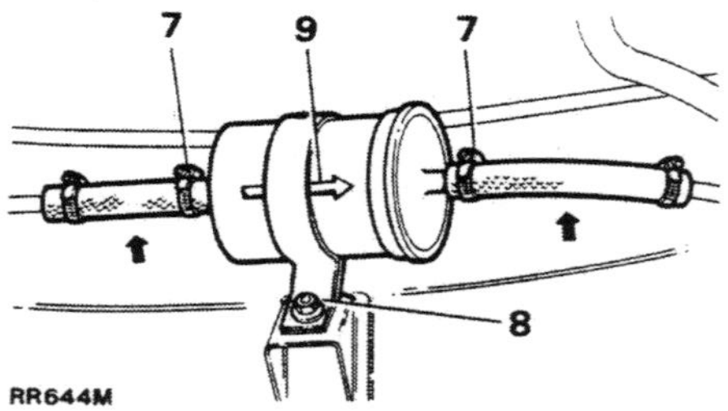

### REMOVE FUEL PUMP

The high pressure fuel pump is located within the fuel tank, and is accessible by removing the tank from the vehicle. The removal of the fuel tank from the vehicle requires the assistance of a second person to support the tank when the fuel feed pipe and electrical terminals are disconnected from the fuel pump.

WARNING: Ensure all the necessary precautions are taken against handling and spillage of fuel.

NOTE: The electrical supply to the fuel pump is taken from the starter motor solenoid and the air flow meter switch via the diode pack and relays. The pump should therefore only operate when the engine is being turned by the starter motor or whilst air flow holds the switch closed.

1. Drive the vehicle onto a suitable hydraulic ramp.
2. Depressurise the fuel system.
3. Disconnect the battery.
4. Remove the drain plug from the bottom of the tank and drain fuel into a suitable container.
5. Remove the inlet pipe from the fuel filter and plug the hose.
6. Disconnect the breather pipe and plug the hose.
7. Release the two hose clips at the top and bottom of the filler tube, manoeuvre the hose up the filler tube to withdraw the hose from the tank filler neck.

NOTE: If a tow bar is fitted the two tie bars from the towing plate to the chassis must be removed to allow the fuel tank to be removed from the vehicle.

*Continued*

# 19 CARBURETTER FUEL SYSTEM

## RENEW THE AIR CLEANER ELEMENT

1. Remove the air cleaner.
2. Release the end plate clips.
3. Withdraw the end plates.
4. Remove the wing nut, washer and retaining plate.
5. Withdraw the air cleaner elements.
6. Fit new air cleaner elements.
7. Fit new sealing washers.
8. Reverse 1 to 4.

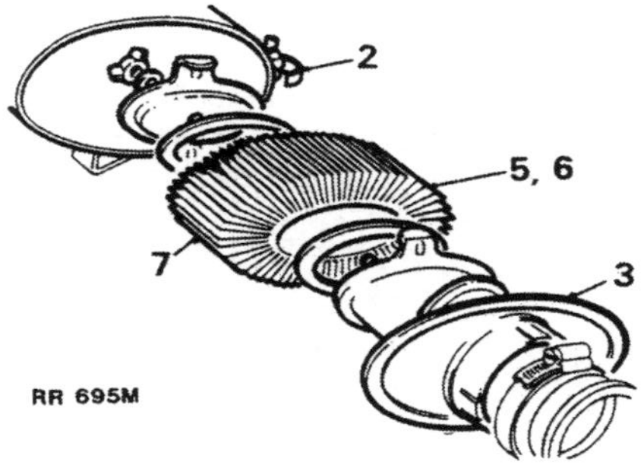

RR 695M

## FUEL SYSTEM—CARBURETTER

### Remove the Air Filter

1. Release the hose clips each side of the air cleaner.
2. Withdraw the air cleaner elbows.

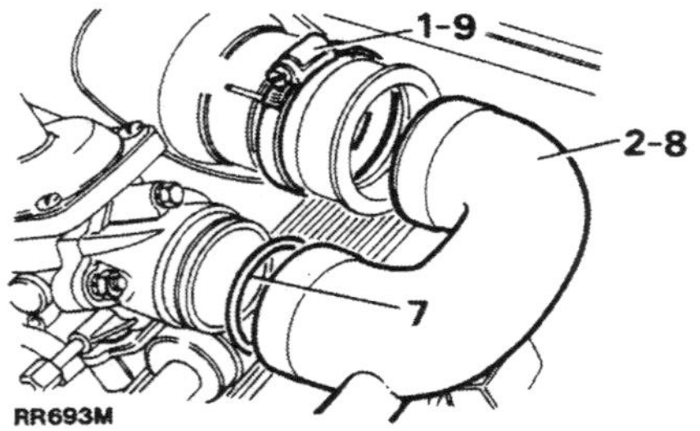

RR693M

3. Detach the choke cable from the clip on the air cleaner.
4. Withdraw the air cleaner from the retaining posts, at the same time disconnecting the hose from the engine breather filter.

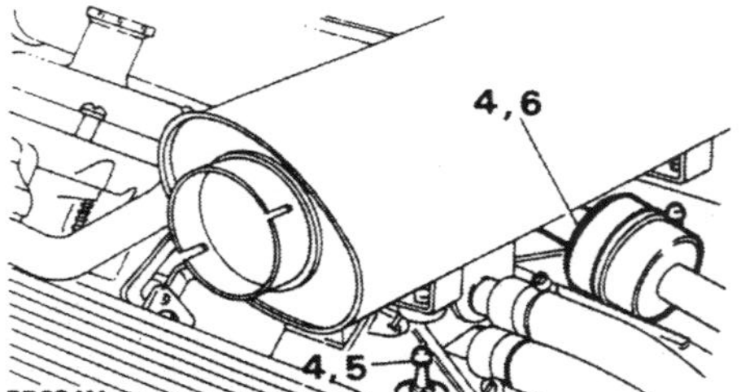

RR694M

### Refitting

5. Fit the air cleaner, locating the rubber mountings over the retaining posts.
6. Connect the engine breather hose at the underside of the air cleaner.
7. Smear the 'O' rings at the carburetter intakes with MS4 grease.
8. Fit the air cleaner elbows.
9. Secure the hose clips.
10. Retain the choke cable in the clip at the front of the air cleaner.

# 19 FUEL INJECTION SYSTEM

8. Remove the four bolts securing the tank to the chassis.
9. With assistance, tilt the left-hand side of the tank downwards as far as the remaining connected fuel pipe will permit, giving access to the fuel feed pipe connection and electrical terminals at the fuel pump.
10. Disconnect the electrical leads from the fuel pump, place the fuel tank on a suitable workbench.

### Remove the fuel pump

NOTE: Plug the ends of the fuel feed pipe and fuel pump outlet aperture to prevent ingress of dirt.

11. Remove the five screws securing the pump to the top of the tank.
12. Withdraw the fuel pump and pump seal from the tank.

### Fit new fuel pump

13. Ensure sealing faces are free from dirt and grease.
14. Fit the fuel pump and a NEW seal to the tank.
15. Refit the fuel tank to the chassis, taking care to relocate the fuel feed pipe grommets between the fuel tank and chassis.
16. Re-connect all fuel pipes.
17. Re-connect the battery.
18. Refill the tank with fuel.
19. Refit fuel pump relay.
20. Start the engine and inspect for fuel leaks around the fuel pump and all hose connections.

## FUEL PIPES—FUEL INJECTION SYSTEM

WARNING: Before disconnecting any part of the fuel system ensure that all necessary precautions are taken against fuel spillage.

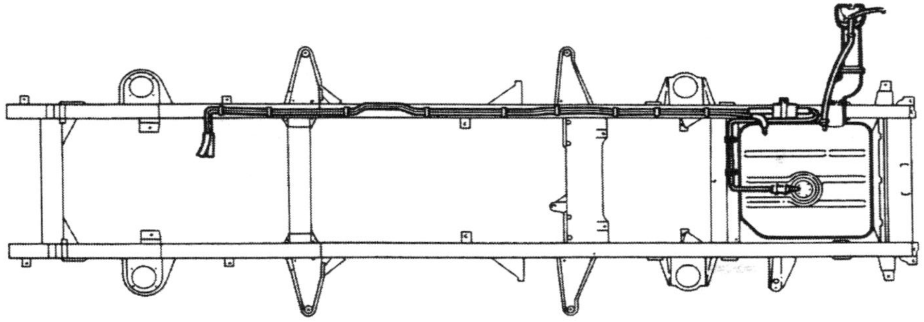

RR646M

# CARBURETTER FUEL SYSTEM | 19

## CARBURETTERS

### Description

Variations in carburetters may be fitted to meet local territory legislation.

### Tamperproofing

These carburetters may be externally identified by a tamperproof sealing tube fitted around the slow running adjustment screw.

The purpose of these carburetters is to more stringently control the air fuel mixture entering the engine combustion chambers and in consequence the exhaust gas emissions leaving the engine.

For this reason the only readily accessible external adjustment on these carburetters is to the throttle settings for fast idle speed and, on some later carburetters, this may require the use of a special tool to adjust the settings.

On all carburetters a deceleration (poppet) valve is incorporated in the throttle (butterfly) disc (A) and consists of a precisely set, spring-loaded plate valve (B). With low manifold induction depression conditions, for instance during over-run with the throttle closed, the valve opens thereby slightly reducing the depression, allowing a correct quantity of fuel and air mixture to sustain engine running which improves the combustion of fuel during these conditions and helps to prevent high-value hydrocarbon emissions.

The deceleration valve is not adjustable (See fault diagnosis).

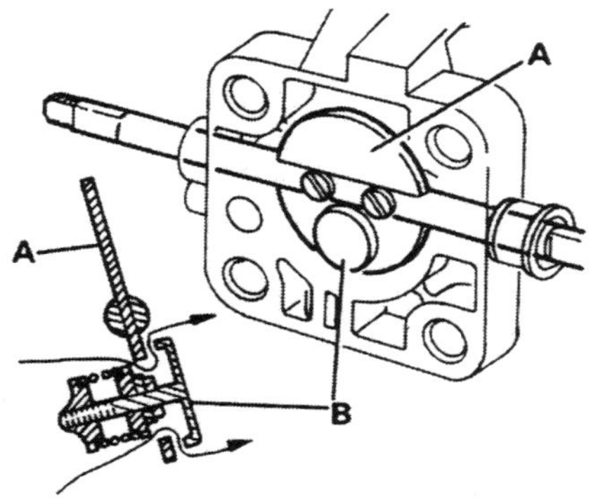

### ENGINE TUNING PROCEDURE—CARBURETTER

Before carrying out 'Engine Tuning' to carburetters it is important that all other engine related setting procedures are undertaken first, i.e. sparking plugs correctly set, hoses and air cleaner correctly fitted and secure. When the engine is running at its normal operating temperature i.e. thermostat open, the following additional checks can be made.

### Check and adjust ignition timing

Equipment required:
Calibrated tachometer
Stroboscopic timing lamp

1. Couple stroboscopic timing lamp and tachometer to engine, following the manufacturers instructions.
2. Disconnect the vacuum pipe from the distributor.
3. Start engine, with no load and not exceeding 3000 rev/min run engine until normal operating temperature is reached. Check that normal idle speed falls within that specified (see 'Engine Tuning Data' for required idle speed dependent upon market).
4. Idle speed should not exceed 750 rev/min and this speed should be achieved by removing a breather hose. NOT by adjusting the carburetter idle screws.
5. If adjustment is necessary, slacken the distributor clamping bolt and rotate the distributor (clockwise to retard or anti-clockwise to advance) until the timing flash coincides with the timing pointer and correct timing mark on the torsional vibration clamper.
6. Re-tighten the distributor clamping bolt securely and re-check the timing setting.
7. Refit vacuum pipe (for vehicle specification see 'Engine Tuning Data' section).

### ADJUSTMENT PROCEDURE—CARBURETTER

To comply with ECE exhaust emission regulations, all carburetters are tamperproofed on the idle adjustment screws. When mixture and idle settings have been finalised, the carburetter must be tamperproofed by fitting a cap to the nylon shroud on the idle adjusting screw.

Should, for any reason, the cap require removal, this can be affected by piercing the cap with a sharp pointed tool and prising out.

The following tools will be required to adjust idle speed, mixture and tamperproof carburetter.

Carburetter adjusting wrench—MS 86
Carburetter jet adjusting tool—MS 80
Tamperproof cap fitting tool—ERC 3786

A numerical code exists for the tamperproofed cap and must be adhered to. Cap fitted by Land Rover Service Departments—ERC 3429.

9. Remove the tamperproof caps from the carburetter idle screws. Using idle screw adjustment tool MS 86 release the locknuts.

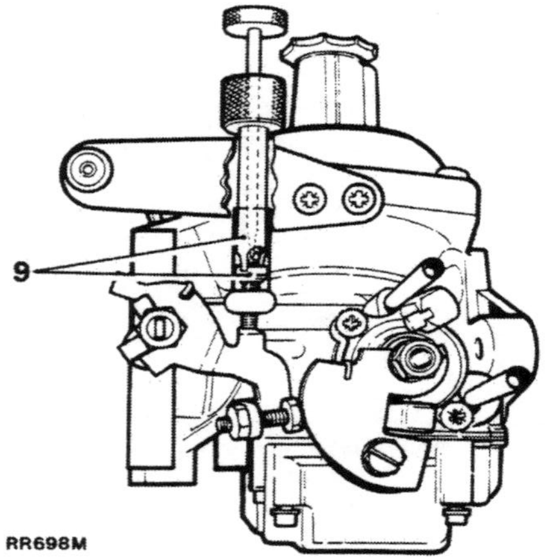

## SLOW RUNNING (IDLE) ADJUSTMENT—Manual and Automatic

### Assuming that mixture levels are correct

1. Check that the throttle control between the pedal and the carburetters is free and has no tendency to stick.
2. Check the throttle cable setting with the throttle pedal in the released position. The throttle linkage must not have commenced movement but commences with the minimum depression of the pedal.
3. Run the engine until it reaches normal operating temperature—thermostat open.
4. Turn ignition off and remove the air cleaner.
5. Slacken the nut which secures the kickdown cable to carburetter inter-connecting link at the right-hand carburetter lever (Automatic only). Slacken the screw securing the throttle cam to the left-hand carburetter lever (Manual only).

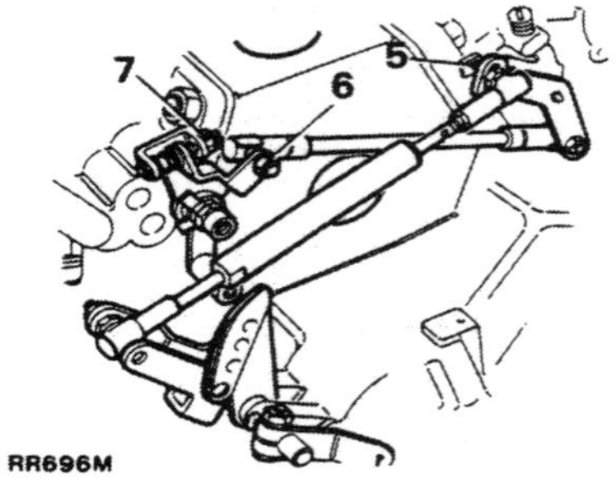

**Automatic Version**

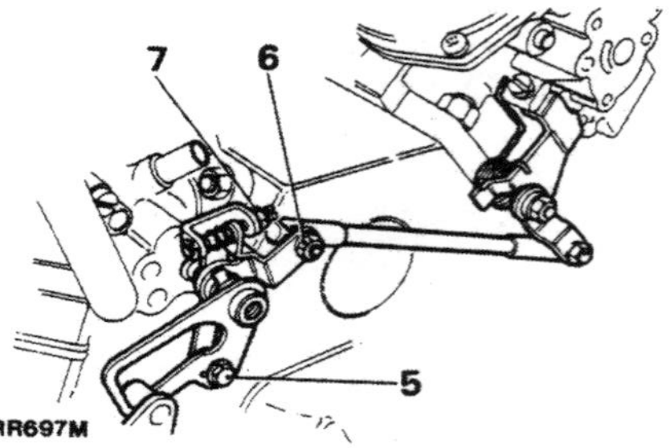

**Manual Version**

6. Disconnect the inter-connecting link between the carburetters, at the left-hand carburetter.
7. Slacken the locknut and release the lost motion adjusting screw on the left-hand carburetter, ensuring that the screw is well clear of the spring loaded pad.
8. Place suitable carburetter balancing equipment across the carburetter intake apertures.

10. Start the engine and check the idle speed using a reliable proprietary tachometer.
11. Check balancer gauge reading.

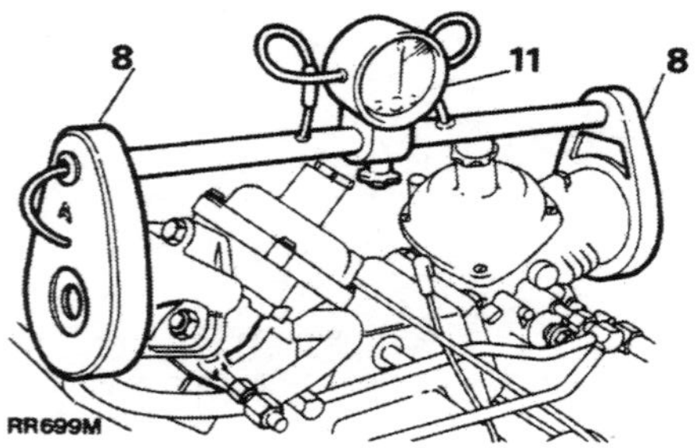

12. If the gauge pointer is in the 'zero' sector no adjustment is required to balance carburetter.
13. If the gauge pointer moves to the right, decrease the air flow through the left-hand carburetter by unscrewing the left-hand idle adjustment screw or increase the air flow through the right-hand carburetter by turning the right-hand idle adjustment screw clockwise. Reverse the procedure if the pointer moves to the left.
14. If the engine idle speed rises too high or drops too low, re-adjust the idle screws on both carburetters ensuring that the pointer on the balancing gauge remains in the 'zero' sector.

*Continued*

# CARBURETTER FUEL SYSTEM

15. When items 13 and 14 are satisfactory, tighten the idle screw locknuts.
16. Re-connect the inter-connecting link to the left-hand carburetter.
17. Hold the throttle lever against the throttle lever stop on the right-hand carburetter and adjust the lost motion screw until contact is made with the spring loaded pad and tighten the locknut.
18. Re-check idle speed and balance. Correct if necessary repeating checks 13 and 14.
19. Check idle CO level.
20. If all checks and adjustments are satisfactory; fit new tamperproof caps—part number ERC 3429 using tool—part number ERC 3786.
21. **Automatic version**—Ensuring that the right-hand carburetter countershaft lever is against the idle stop, tighten the inter-connecting link securing nut.
   **Manual version**—Ensure that the roller in the throttle cam on the left-hand carburetter is seated in the corner of the cam slot, tighten the cam lever securing screw.

## FAST IDLE ADJUSTMENT—LH carburetter only

### Automatic and Manual

Operation of the choke control from 'on' to 'off' should result in a fast idle speed of 1100 rev/min ± 50 when the choke control is approximately 12.7 mm (0.5 in) from the choke off position.

### Check and adjust

1. Slacken choke cable clamping screw at the left-hand carburetter.
2. Pull the choke control knob out and push it into a distance of approximately 12.7 mm (0.5 in) and lock in position.
3. Rotate the fast idle cam allowing the choke cable to slide through the trunnion until the punched mark on the cam flank aligns with the centre of the domed screw on the throttle lever, tighten the choke cable clamping screw.
4. With the cam held in this position release the fast idle adjustment screw locknut.
5. Turn the adjustment screw until contact is made with the face of the fast idle cam, continue adjusting until the specified fast idle speed is achieved. Tighten the locknut.

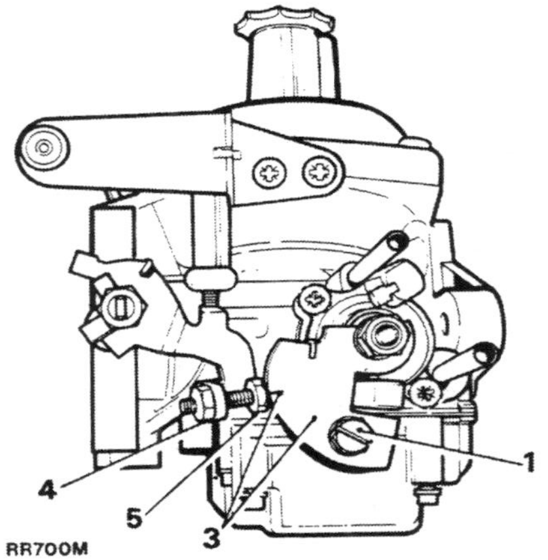

6. Push the choke control knob fully home and check that normal idle speed is regained.

## IDLE MIXTURE ADJUSTMENT

### Automatic and Manual

Service Tools—MS 80 Mixture Adjustment Tool.
Carburetter Balancing Gauge.

Mixture adjustment should be carried out when the vehicle is up to normal operating temperature, i.e. warm air intake valve open.

### Check and adjust

1. Remove the piston damper plug, and using special tool MS 80 adjust the mixture. Locate the outer sleeve of the tool to engage a machined slot to prevent the air valve twisting. Turn the inner tool clockwise to enrich the mixture and anti-clockwise to weaken it. After every adjustment the tool should be removed from the carburetter to allow engine to stabilise. Run engine at 2000 rev/min to aid stabilisation.
2. When the mixture is correctly adjusted, the engine speed will remain constant or may fall slowly a small amount as the air valve is lifted.

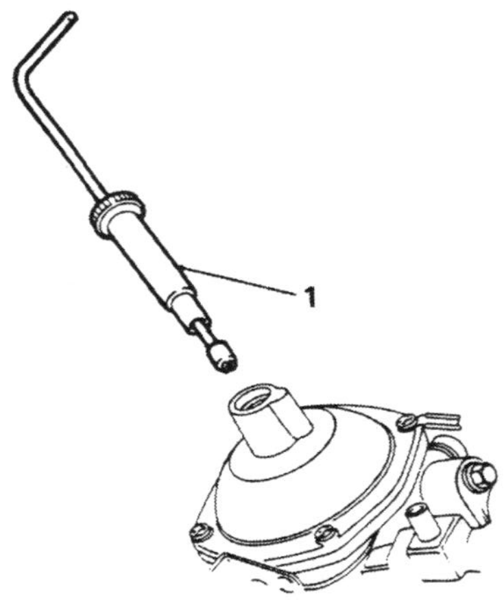

3. Place balancer on the carburetter adaptors, ensuring that there are no air leaks. If the engine stalls or decreases considerably in speed, the mixture is too rich. If the engine speed increases, the mixture is too weak.
4. If necessary, remove balancer and re-adjust the mixture, then refit the tool.

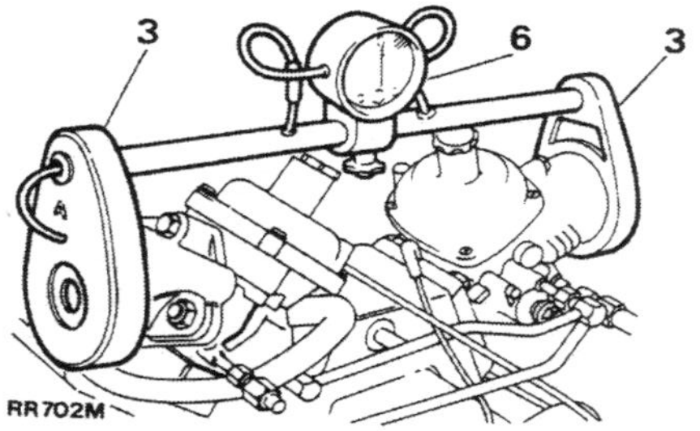

5. Check balancer gauge reading.
6. If the gauge pointer is in the 'zero' sector, no adjustment is required.
7. If the gauge pointer moves to the right, decrease the air flow through the left-hand carburetter by unscrewing the slow running screw or increase the air flow through the right-hand carburetter by turning clockwise the slow running screw. Reverse the procedure if the pointer moves to the left.
8. If the engine idle speed (slow running) rises too high or drops too low during balancing adjust to the correct idle speed, whilst maintaining the gauge pointer in the zero sector.

9. Remove balancer. With the mixture setting and carburetter balance correctly adjusted the difference in engine rev/min with the balancing tool on or off will be negligible, approximately ± 25 rev/min.
10. Check CO levels.

**NOTE: The flow test of the carburetter may require the needle to be set within the limits of ± ½ turn of adjuster from a 'flush' condition. Before any needle component is changed the setting of the needle from 'flush' should be noted and reproduced when refitting. If difficulty is experienced with carburation the needle should be set with the shoulder flush with the face of the piston for investigation.**

When using the adjusting tool a positive stop will be felt when the needle reaches full rich position. In the anti-clockwise direction there is no stop and it is possible to disengage the needle from the adjusting screw if more than two turns are made from the datum position. Should disengagement occur it can be rectified by applying light pressure, in an upwards direction to the shoulder of the needle at the piston face, whilst turning in a clockwise direction.

## CHECK CO LEVEL

Use a proprietary non-dispersive infra-red exhaust gas analyser.

11. Insert the probe of the analyser as far as possible into the exhaust tail pipe, start the engine and allow a one to one and a half minute stabilisation period.
12. Check that the correct idle speed (slow running) is maintained and observe the CO reading against that given in the data section. If necessary re-adjust the mixture setting to achieve the correct CO level.

**NOTE: For local territory emission level requirements see 'Engine Tuning Data' in specification section.**

# CARBURETTER FUEL SYSTEM | 19

14. Float chamber retaining screws (6 off)
15. Butterfly
16. Butterfly retaining screws (2 off)
17. Butterfly (throttle) spindle
18. Throttle spindle lever assembly
19. Cold start assembly (left-hand carburetter only)
20. Slow running screw assembly
21. Tamper-proof sleeve
22. Temperature compensator assembly
23. Temperature compensator rubber seals (large and small)
24. Float chamber joint washer

## KEY TO LH CARBURETTER

1. Top cover and retaining screws
2. Air valve return spring
3. Air valve
4. Diaphragm
5. Metering needle
6. Metering needle retaining screw
7. Diaphragm retaining ring and plate
8. Screw (4 off) retaining diaphragm to air valve
9. Damper assembly
10. Float chamber
11. Float assembly
12. Float assembly spindle
13. Needle valve

---

# CARBURETTER FUEL SYSTEM | 19

## CARBURETTERS—Manual & Automatic

### Removing

1. Remove the air cleaner.
2. Disconnect the choke cable.
3. Disconnect the emission control pipes.
4. Disconnect the distributor vacuum pipe.
5. Disconnect the throttle linkages.
6. Disconnect the main fuel supply pipe.
7. Disconnect the choke fuel supply pipe.
8. Remove the eight retaining nuts and lift off the carburetters.
9. Withdraw the joint washers, insulator and liner.

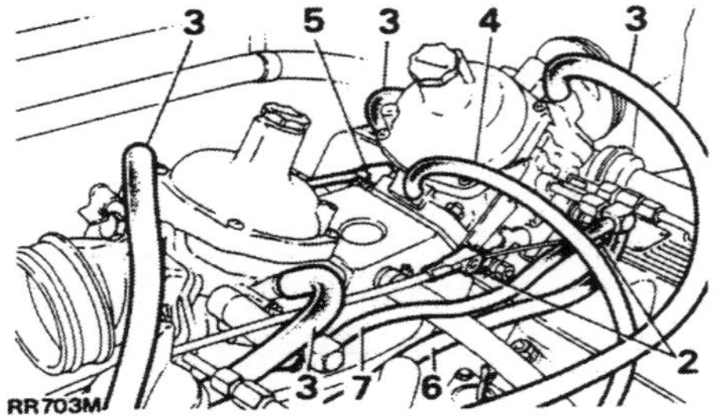

### Refitting

10. Locate a new joint washer on the inlet manifold.
11. Fit the insulator, aligning the arrows.
12. Fit the liner fully into the insulator, engaging the three tabs into the recesses.
13. Locate a joint washer on the insulator.
14. Reverse 1 to 8.
15. Fit the air cleaner.
16. Tune and adjust the carburetters.

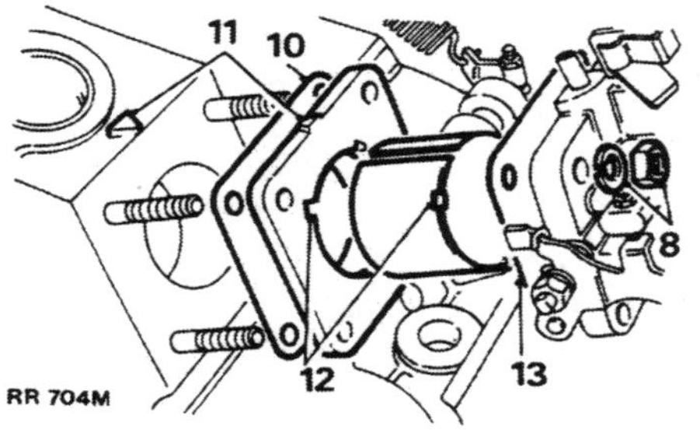

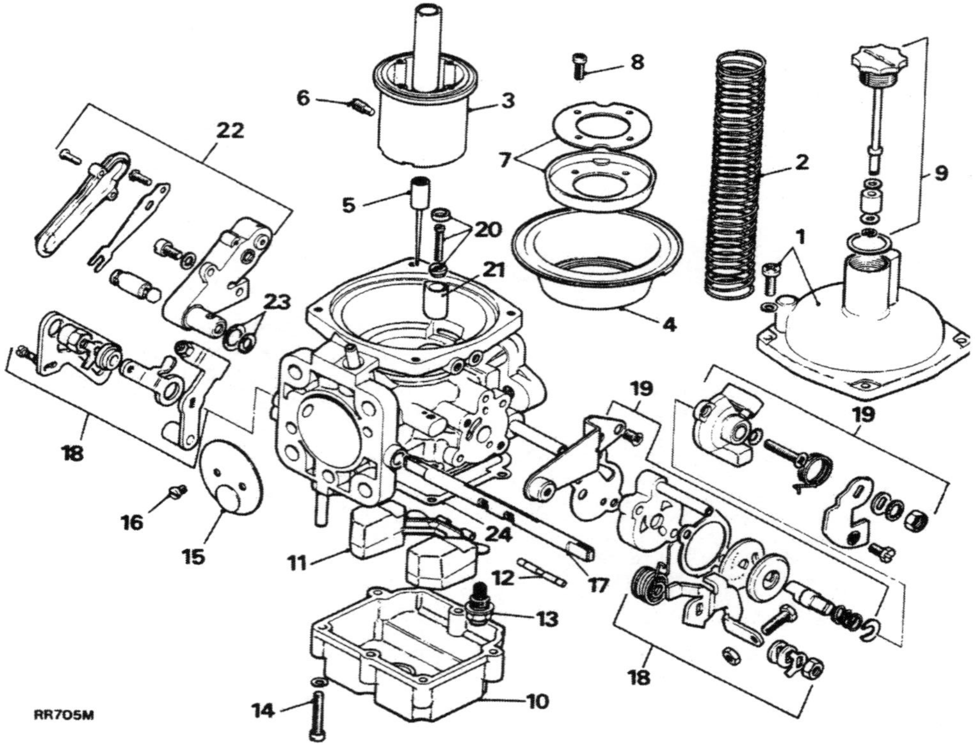

# CARBURETTER FUEL SYSTEM

## CARBURETTER OVERHAUL

### DISMANTLE

#### Remove the piston assembly

1. Remove the carburetters from the engine.
2. Release the four screws and withdraw the top cover and spring.
3. Withdraw the air valve, shaft and diaphragm assembly.
4. Remove the metering needle retained by a locking screw.
5. Release the four screws and separate the diaphragm from the air valve.

#### Remove the float chamber

6. Release the six screws and remove the float chamber and joint washer.
7. Release the float assembly and spindle from the carburetter body.
8. Unscrew the needle valve and washer from carburetter body.

#### Dismantle carburetter body

9. Make location marks, as illustrated, to assist correct assembly, on the throttle butterfly, spindle and carburetter body.
10. Right-hand carburetter. Release the two screws and remove the butterfly and withdraw the spindle.
11. Left-hand carburetter. Remove the left-hand lever assembly. Release the two screws, remove the butterfly and withdraw the spindle.
12. Left-hand carburetter. Release the two retaining screws and shake-proof washers and remove the cold start assembly and joint washers.
13. Dismantle the cold start assembly but DO NOT remove the discs from the spindle.
14. If necessary, dismantle the throttle spindle lever assemblies from both carburetters.

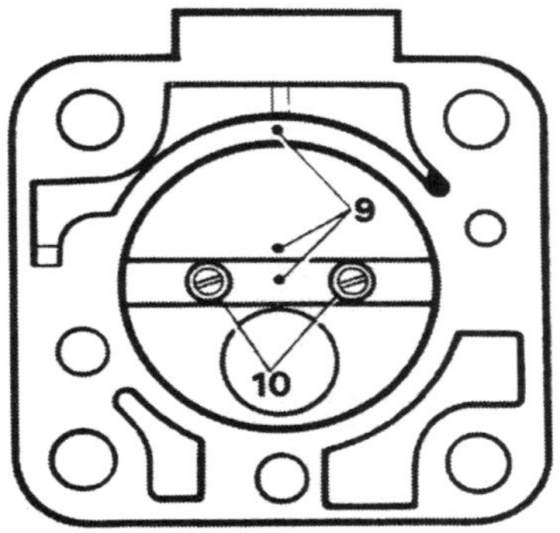

#### Slow running adjustment screws

Do not attempt to remove these screws or break the tamper-proof seals. See Cautionary note under "Tune and Adjust".

#### Remove temperature compensator

15. Release the two screws and withdraw the temperature compensator unit complete.
16. Remove the large and small rubber washers.

### CLEANING AND INSPECTION

#### Cleaning

17. When cleaning fuel passages do not use metal tools (files, scrapers, drills etc.) which could cause dimensional changes in the drillings or jets. Cleaning of all components should be affected using clean fuel and, where necessary, a moisture-free air blast.

#### Joint washers and seals

18. New gaskets and seals should be used throughout carburetter rebuild. A complete set of gaskets is available for replacement purposes. Inspect metering needle; it is machined to very close limits and should be handled with care. Examine for wear, bend and twist; renew if necessary.

#### Diaphragm

19. Examine the faces for deep scores which would lead to leakage taking place when assembled.
20. Examine the diaphragm for deterioration, damage and punctures. Do not use any cleaning chemicals on the diaphragm only clean lint free rag.

#### Float assembly

21. Examine the two plastic floats and check for punctures and damage.
22. Check the spindle and retaining clips for wear.
23. Inspect the needle valve assembly for wear. Renew the valve if there is any tendency for the needle to stick.

#### Cold start assembly

24. Examine all the cold start components for wear and corrosion and the machined faces for scores.

## ASSEMBLE CARBURETTERS

### Cold start—LH carburetter

25. Place the spring on the cold start spindle.
26. Fit the spring retaining clip.
27. Check that the discs slide easily on the spindle.
28. Place the cold start spindle on the starter face.
29. Place the starter cover in position.
30. Fit the return spring over the spindle.
31. Rotate the spindle until the oval port in the end disc is aligned with the oval port in the starter face.
32. Fit the cold start lever.
33. Engage the return spring over the lug on the starter cover and the back of the cold starter lever.
34. Place the cold start gasket onto the carburetter body.
35. Fit the cold start assembly to the carburetter body, and check for ease of operation.

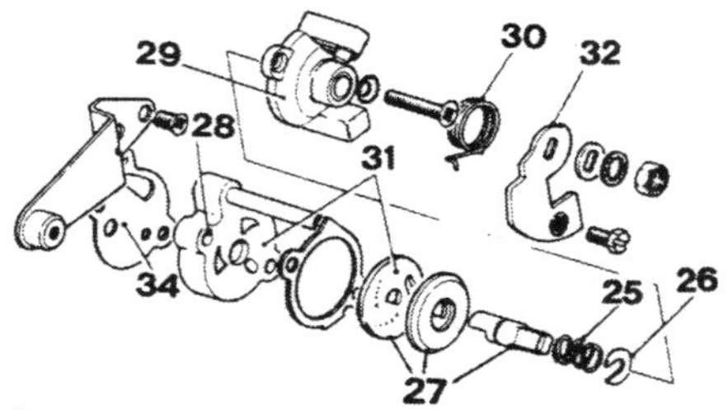

### Throttle spindle, LH carburetter

36. Place the return spring over either end of the spindle.
37. Fit the throttle stop and fast idle lever and secure with spacers, tab washer and nut.
38. Insert the throttle spindle from the cold start side of the carburetter body and fit the throttle return spring on the fast idle adjustment lug. Tension spring half a turn.

NOTE: To enable the throttle butterfly to be centralised, remove the tamperproof cap (if fitted) from the idle adjustment screw, slacken the screw until the lever is free to move without restrictions.

39. Fit the throttle butterfly, maintaining the previously marked alignment. Leave the retaining screws loose. Actuate the throttle several times to centralize the butterfly, then tighten the retaining screws and lock by peening ends.
40. Fit the lost motion lever to the opposite end of the spindle.
41. Fit the throttle lever locating the raised tab between the spring loaded pad and adjustment screw.
42. Place the spacer on the spindle.
43. Place the tab washer on the spindle.
44. Fit the sleeve nut, sleeve end first, and engage the tab washer.
45. Fit the throttle adjusting lever (automatic gearbox carburetter versions are not fitted with this lever).

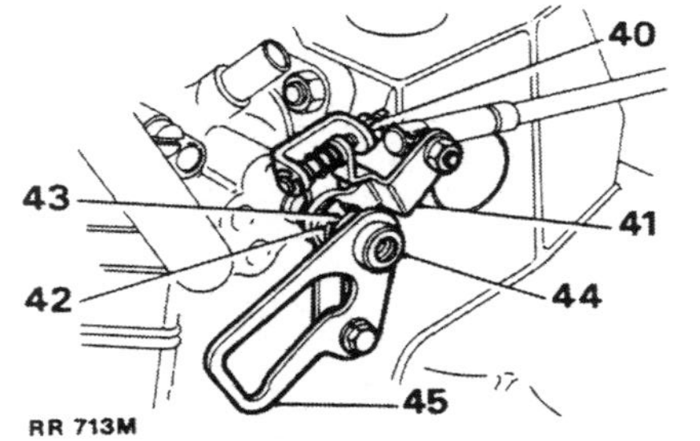

### Throttle spindle—RH carburetter

46. Place the return spring over the threaded end of the spindle.
47. Fit the throttle stop lever.
48. Fit the throttle lever.
49. Fit the spacer and tab washer.
50. Fit the retaining nut.
51. Secure the assembly to the shaft with the tab washer.

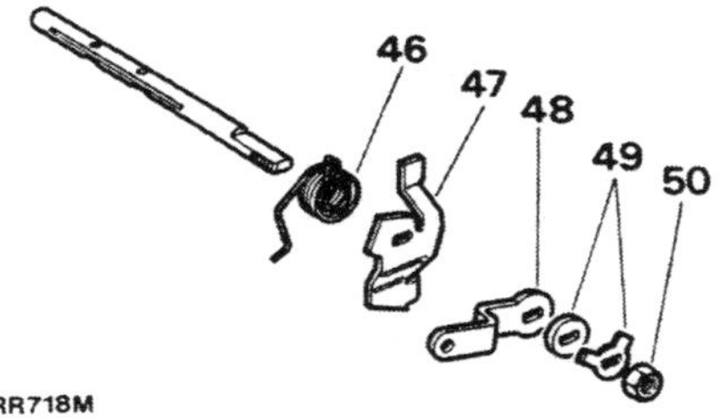

*Continued*

## CARBURETTER FUEL SYSTEM | 19

52. Fit the spindle to the carburetter and assemble the throttle butterfly as described in instruction 39. (See note preceding instruction 39.) Anchor the return spring as illustrated.

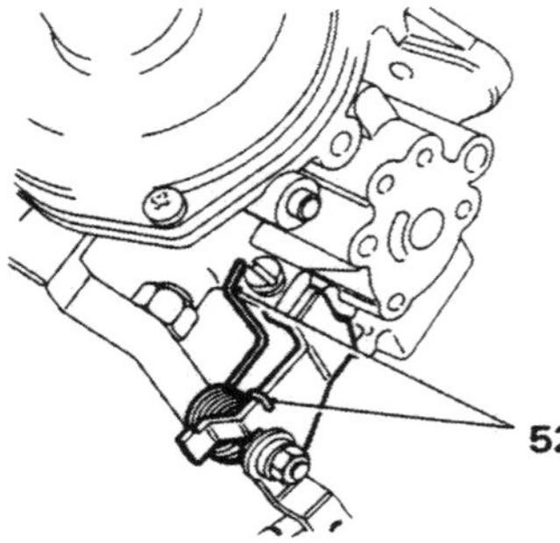

*RR 719M*

53. Fit the kick-down lever. (Automatic gearbox versions only.)

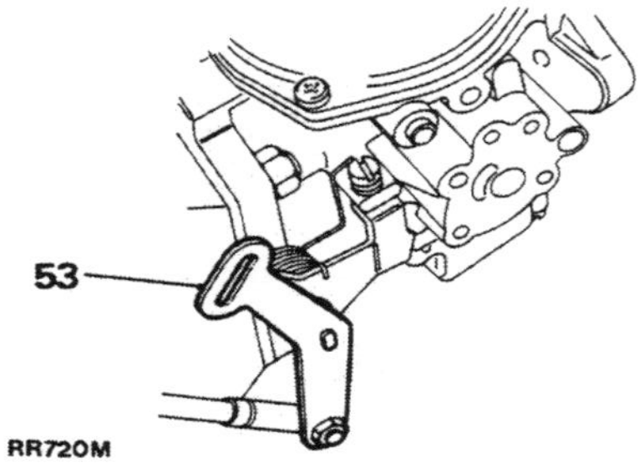

*RR720M*

### Float chamber assembly

54. Fit the needle valve and new washer.
55. Locate the spindle in the float arm and fit the assembly into the retaining clips.

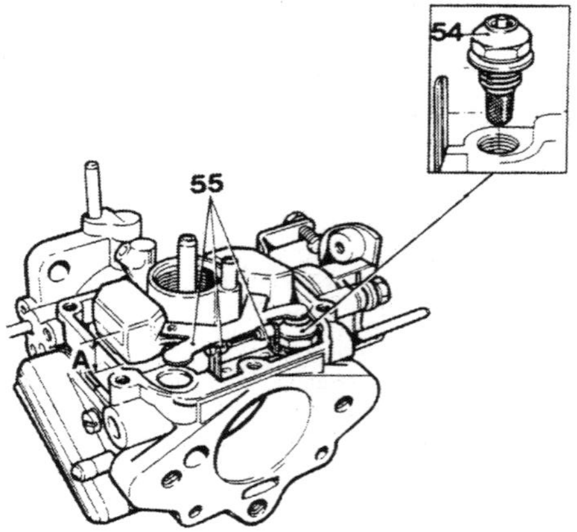

56. Invert the carburetter so that the needle is on its seating and the float tab is contacting the needle. Measure the dimension 'A' between the carburetter gasket face and the highest point on the floats. The correct measurement should be 17 to 18 mm (0.67 to 0.71 in). Adjust by bending the float tab. This dimension must be the same for both floats. The float carrier tab must be maintained at right angles to the needle in the closed position.
57. Fit the float chamber and new gasket and evenly tighten the retaining screws.

*ST1123M*

### Air valve and diaphragm

58. Fit the diaphragm to the air valve with the inner tag locating in the air valve recess.
59. Fit the diaphragm retaining ring and secure with the four screws.
60. Fit the metering needle into the air valve and secure with the locking screw.
61. Insert the air valve and needle into the carburetter and locate the diaphragm outer tag into the recess in the carburetter body.
62. Fit the spring and top cover and secure with the four screws.
63. Fit the damper.

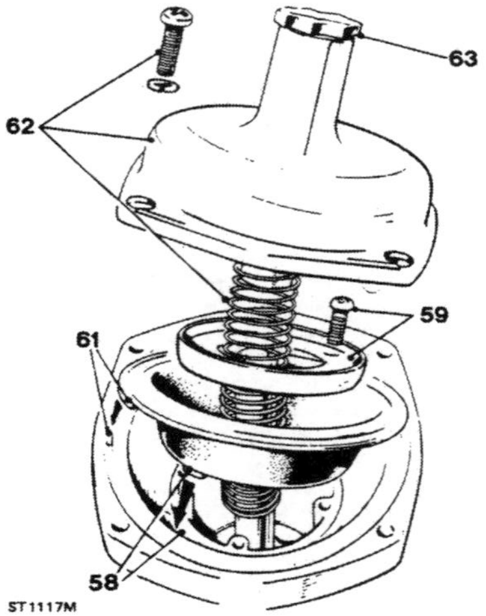

*ST1117M*

### Temperature compensator

64. Clean the carburetter and compensator mating faces.
65. Fit a new inner and outer rubber washer and secure the temperature compensator with the two screws and shake-proof washers.

### Fast idle adjustment—LH carburetter only

66. Slacken the fast idle adjusting screw.
67. Hold the cold start cam lever in the maximum position.
68. Adjust the fast idle adjusting screw against the cam lever until there is 0.61 to 0.66 mm (0.024 to 0.026 in) gap between the top edge of the throttle butterfly and the carburetter barrel wall. Use feeler gauges or a 0.65 mm diameter (No. 72) drill to measure the gap.
69. Secure the locknut on the fast idle adjusting screw without disturbing the adjustment.

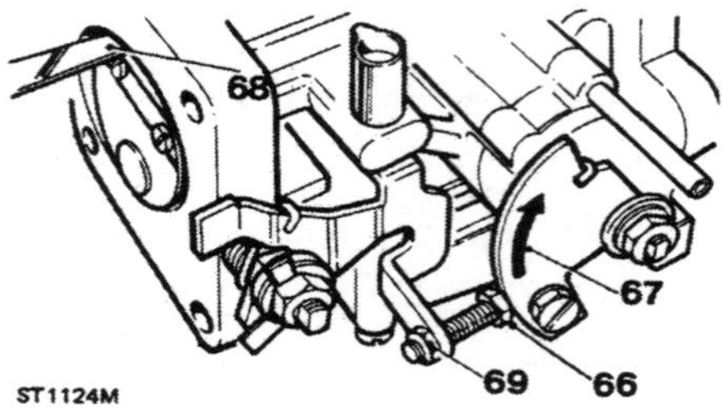

*ST1124M*

---

## CARBURETTER FUEL SYSTEM | 19

### RENEW THE THROTTLE CABLE—manual gearbox

1. Remove the air cleaner.
2. Remove the split pin and clevis pin from the counter shaft lever.
3. Release the two locknuts securing the outer cable to the adjustment bracket.
4. Manoeuvre the inner and outer cable and remove from the adjustment bracket.

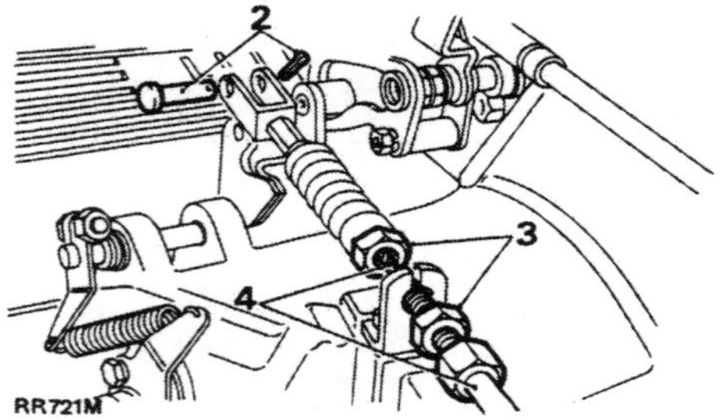

*RR721M*

5. Remove the six screws retaining the lower fascia within the vehicle. Disconnect the dimmer switch multi-plug to enable the fascia panel to be removed to give access to the throttle pedal linkage.
6. Remove the split pin and clevis pin from the top of the throttle pedal.
7. Release the two outer cable locknuts at the throttle cable retaining bracket on the inner bulkhead.
8. Prise out the rubber grommet accessible from the engine compartment.
9. Withdraw the throttle cable complete.

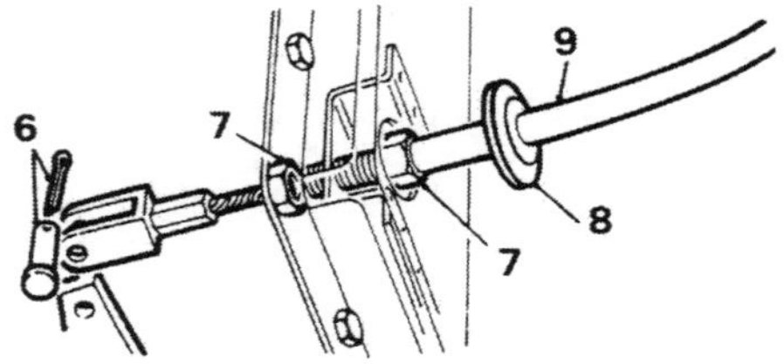

*RR 722M*

### Fit new throttle cable

10. Fit new cable ensuring that new split pins are used.
11. Lightly grease linkage and clevis pins.
12. Adjustment of the cable should be undertaken with the throttle pedal in the fully released position.
13. Adjust the outer cable until the throttle levers operate with the minimum of throttle pedal pressure.
14. Refit the air cleaner and lower fascia panel.

*Continued*

# CARBURETTER FUEL SYSTEM 19

## RENEW CHOKE CONTROL CABLE—carburetter models only

### Manual and Automatic gearbox

1. Release the pinch screw and disconnect the choke cable from the left-hand carburetter fast idle cam.
2. Release the cable from the securing clip in the engine compartment.

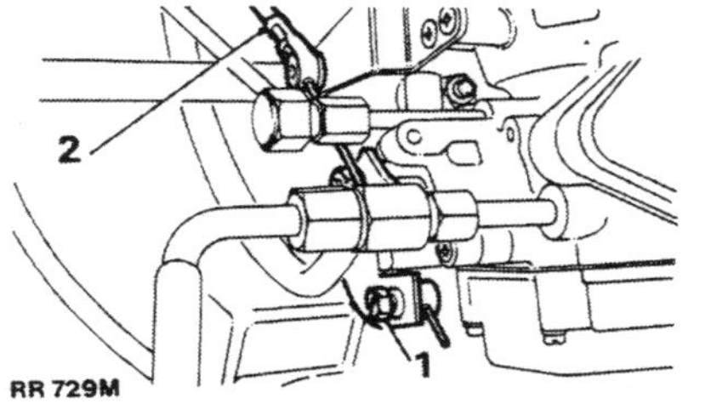

*RR 729M*

3. To gain access to the rear of the choke control knob it is necessary to remove the lower fascia panel.
4. Remove the six lower fascia panel securing screws, lower the panel, disconnect the electrical multi-plug attached to the dimmer control switch and remove the fascia panel.
5. Disconnect the two electrical leads from the choke control switch.
6. Release the locknut and screw, slide the clamp off the switch.
7. Remove the switch from the cable.
8. Release the locknut at the rear of the choke control knob.
9. From within the vehicle, pull the cable assembly through the bulkhead and upper fascia panel.

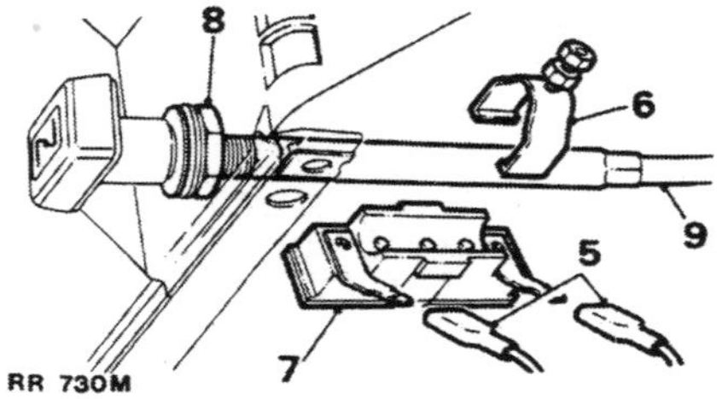

*RR 730M*

### Fit new choke cable

10. Feed the new choke cable through the upper fascia and slide the locknut over the cable.

*Continued*

---

## THROTTLE LINKAGE—Automatic gearbox versions

### Remove and refit

#### Removing

11. Disconnect the throttle cable from the cam lever.
12. Remove the split pin and clevis pin securing the kickdown cable to the countershaft lever.
13. Remove the nut (with spring washer) securing the carburetter inter-connecting link to the countershaft lever.
14. Disconnect the throttle return spring.
15. Release the spring return lever retaining bolt and remove the lever from the countershaft lever spindle.
16. Drift out the small roll pin securing the cam lever to the countershaft lever spindle.

*RR725M*

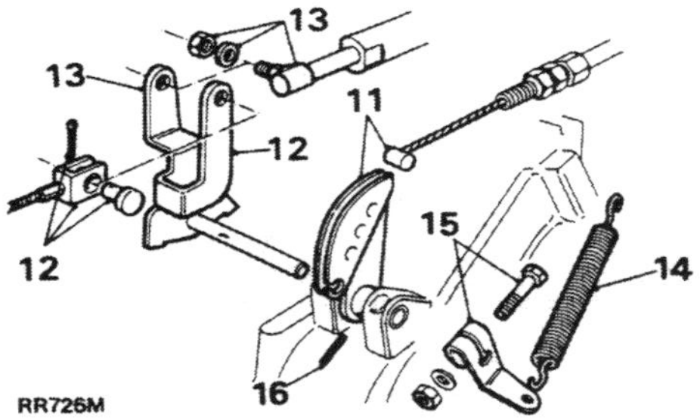

17. Manoeuvre the countershaft lever and withdraw it from its location.

#### Refitting

18. Fit the countershaft lever and cam lever, secure the cam lever to the spindle using a new roll pin.
19. Fit the return spring lever and lightly tighten the securing bolt.
20. Position the return spring lever and countershaft lever so that the holes for the kickdown cable and return spring connection are 172½° ± 2°, and tighten the securing bolt.

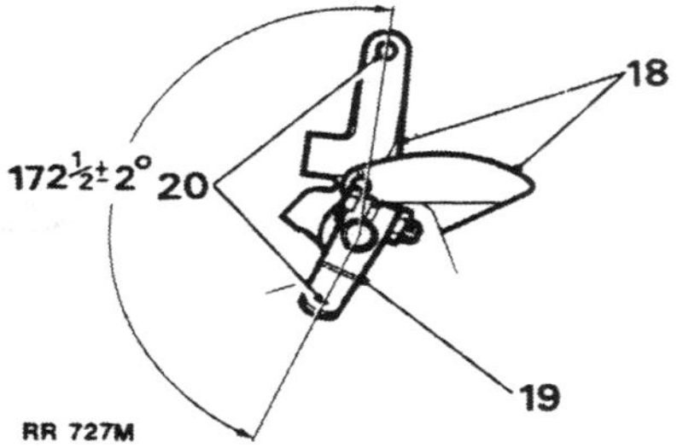

*RR 727M*

21. Reverse the remaining removal instructions 11 to 14.

---

## Renew the throttle cable—Automatic gearbox

15. Remove the six screws securing the lower fascia within the vehicle, lower the fascia and disconnect the multi-plug at the dimmer switch to enable the fascia panel to be removed to give access to the top of the throttle pedal.
16. Remove the split pin and clevis pin from the throttle pedal.
17. Release the two outer cable locknuts at the throttle cable retaining bracket on the inner bulkhead panel.
18. Remove the air cleaner.
19. Prise the rubber grommet from the bulkhead to allow the cable to be manoeuvred into the engine compartment (see illustration RR722M).
20. Release the two outer cable locknuts from the adjustment bracket at the intake manifold.
21. Remove the nipple end of the cable from the cam lever and remove the complete cable assembly.
22. Manoeuvre the inner and outer cable to release it from the adjustment bracket.

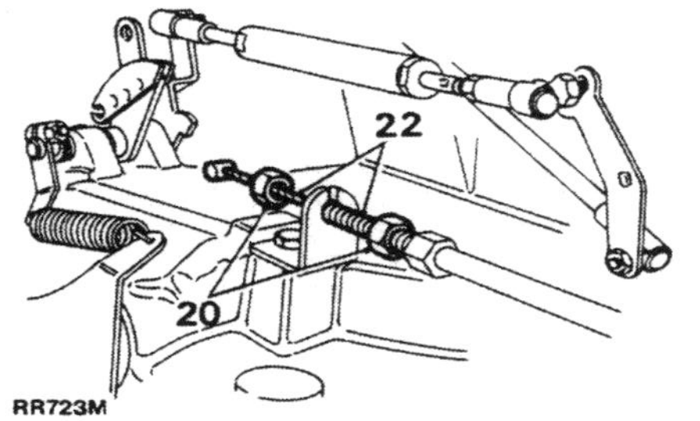

### Fit new throttle cable

23. Fit new cable ensuring that a new split pin is used.
24. Lightly grease clevis pin and linkage at the throttle pedal.
25. Adjustment of the cable should be undertaken with the throttle pedal in the fully released position.
26. Adjust the outer throttle cable until the throttle levers operate with the minimum of throttle pedal pressure.
27. Refit air cleaner.
28. Refit the lower fascia panel.

---

## THROTTLE LINKAGE—Manual gearbox version

### Remove and refit

#### Removing

1. Release the throttle cable from the countershaft lever.
2. Remove the small bolt with spring washer securing the throttle adjustment lever to the left-hand carburetter spindle and remove the lever.
3. Disconnect the throttle return spring.
4. Release the nut and slide the return spring lever off the countershaft spindle.
5. Remove the circlip and plain washer from the countershaft lever.
6. Pull the countershaft spindle out of its location.

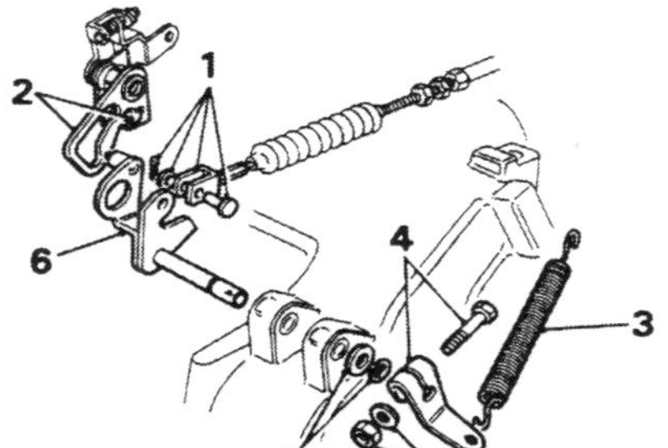

*RR724M*

### Refitting

7. Fit the countershaft assembly and secure in position with the plain washer and circlip.
8. Fit the return spring lever to about the circlip and lightly tighten the securing bolt.
9. Position the lever and countershaft lever so that the holes for the throttle cable and return spring connection are 172½° ± ½°, and tighten the securing bolt.

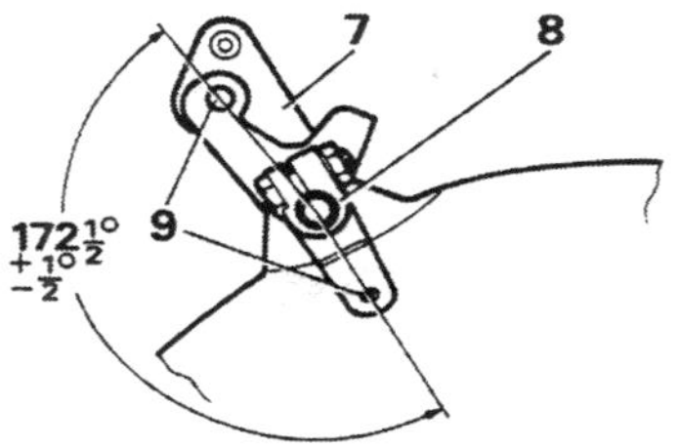

*RR725M*

10. Reverse the remaining removal instructions 1 to 3.

# CARBURETTER FUEL SYSTEM 19

11. Push cable through bulkhead grommet into the engine compartment.

NOTE: Ensure that the grommet is secure in the bulkhead.

12. Tighten the locknut at the rear of the choke cable knob.
13. Feed the cable through the locking screw on the fast idle cam. Do not tighten the clamping screw at this stage.
14. Attach the outer choke cable to the securing clips.
15. Pull the choke control knob out approximately 12.7 mm (0.50 in), rotate the fast idle cam until the stamped mark on the cam flank aligns with the centre of the domed screw on the throttle lever, tighten the choke cable clamping screw.
16. Push the choke control knob in and check that the domed screw is not in contact with the fast idle cam.
17. Fit the choke warning light switch to the outer cable ensuring that the three small pegs on the switch locate with the indentations on the choke cable.
18. Slide the clamp over the switch and secure in position.
19. Re-connect the electrical leads to the switch.
20. Re-connect dimmer switch electrical leads and secure the lower fascia in position.

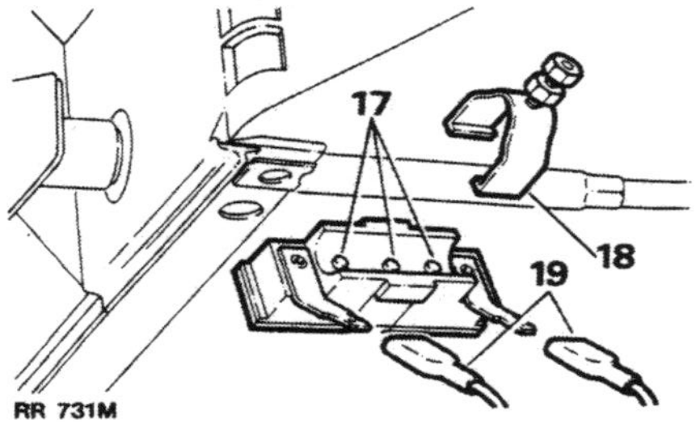

## REMOVE THE THROTTLE PEDAL—all models

1. Release the six securing screws from the lower fascia panel, lower the panel and disconnect the two electrical leads from the dimmer switch, remove the panel from the vehicle.
2. Remove the split pin and clevis pin securing the throttle cable to the throttle pedal.
3. Release the tension from the pedal return spring.
4. Remove the circlip from the pedal pivot pin.
5. Withdraw the pivot pin.
6. Withdraw the throttle pedal.

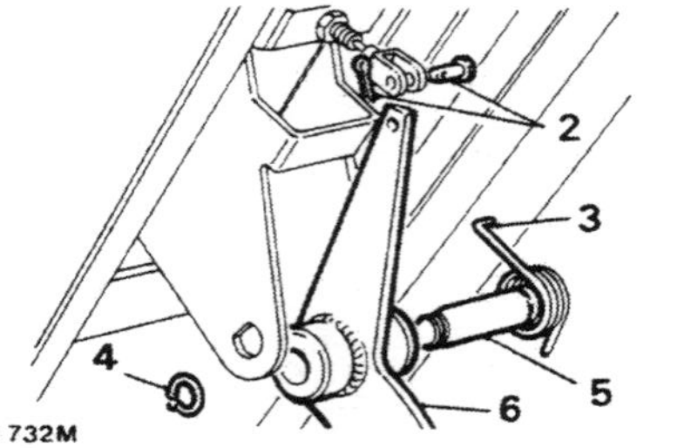

7. Lightly grease the pivot pin and clevis pin before re-assembly.
8. Fit a new split pin to the clevis pin.
9. Reverse the removal procedure.

## Fit throttle pedal

## REMOVE MAIN FUEL FILTER—carburetter versions only

WARNING: Ensure that all the necessary precautions are taken against the spillage of petrol when removing the filter assembly.

1. Release the two pipe unions securing the pipes to the filter. Plug the end of the pipes to prevent ingress of dirt.
2. Remove the two nuts, bolts and washers securing the filter to the inner wing mounting bracket.
3. Remove the filter assembly.

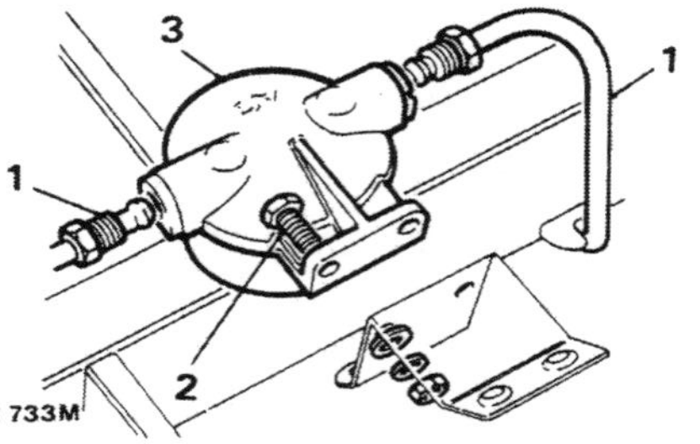

### Refitting

4. Reverse the remove instructions 1 to 3, ensuring there are no fuel leaks around the pipe unions.

## FUEL PIPES

WARNING: Before disconnecting any part of the fuel pipe system ensure that all necessary precautions are taken against fuel spillage.

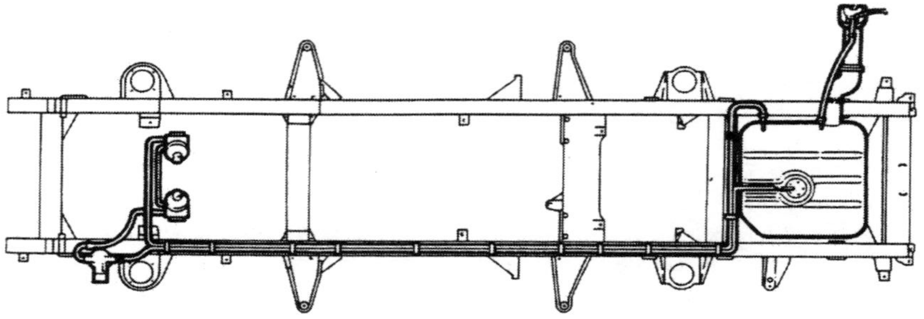

## REMOVE THE FUEL TANK AND FUEL PUMP

NOTE: The removal of the fuel tank from the vehicle requires the assistance of a second person to support the tank when the fuel feed pipe is disconnected from the fuel pump.

The low pressure in-tank fuel pump is located on the top of the fuel tank, access to the pump is gained by removing the fuel tank from the vehicle.

WARNING: Ensure that all the necessary precautions are taken when draining the fuel from the tank.

### Remove the fuel tank

1. Drive the vehicle onto a suitable ramp.
2. Disconnect the battery.
3. Remove the drain plug from the bottom of the fuel tank, drain the fuel into a suitable container. Refit the plug.
4. Disconnect the spill return pipe at the fuel tank, accessible through the rear right-hand wheel arch. Plug the end of the return pipe to prevent ingress of dirt.
5. Disconnect the flexible breather hose from the tank.
6. Release the two hose clips at the top and bottom of the filler tube. Manoeuvre the hose up the filler tube and withdraw the hose off the tank filler neck.

NOTE: If a tow bar is fitted, the two tie-bars from the towing plate to the chassis must be removed allowing the tank to be removed from the vehicle.

7. From beneath the vehicle remove the four bolts, nuts, washers and spring washers securing the tank to the chassis frame.

8. Tilt the left-hand side of the tank downwards as far as the remaining connected fuel pipe will permit, to give access to the fuel feed pipe connection and electrical terminals at the fuel pump.
9. With assistance from someone supporting the tank, disconnect the fuel feed pipe and electrical leads.
10. Carefully lower the tank and place on a suitable workbench.
11. Plug the ends of the fuel feed pipe and fuel pump to prevent ingress of dirt.

### Remove the fuel pump

12. Remove the five screws and washers securing the pump in the tank.
13. Withdraw the fuel pump and sealing washer from the tank.

### Fit the fuel pump

14. Ensure the area around the fuel pump aperture in the tank is free from dirt and grease.
15. Fit the pump and a NEW sealing washer to the tank, secure the pump in position.

### Fit fuel tank

16. With assistance offer up the fuel tank to the chassis, remove the plugs from the fuel pipes and re-connect the hose to the chassis and re-connect the electrical leads.
17. Secure the tank to the chassis and refit all hoses and pipes.
18. Connect the battery, turn ignition on and check for fuel leaks at all hose and pipe connections.

## 19 CARBURETTER FUEL SYSTEM

### V8 ENGINE CARBURETTER
### FAULT DIAGNOSIS

| SYMPTOM | POSSIBLE CAUSE | CURE |
|---|---|---|
| DIFFICULT STARTING WHEN COLD | Insufficient choke action | Check action of cold start unit to ensure that the choke is being applied fully—adjust choke cable. Check position of cold start adjuster—move fully outward. |
| | Fast idle adjustment incorrect | Check and adjust fast idle setting. Check linkage between choke and throttle for distortion. |
| | Float chamber level too low | Check needle valve for sticking—(closed). Check float level setting. Check inlet connection filter for blockage. Check external fuel system in accordance with fuel system fault diagnosis. |
| | Carburetter flooding | Check needle valve for sticking—(open). Float punctured. Fuel pump pressure too high. Float level too high. |
| | No fuel supply to carburetter | Check filters and pump for blockage. Check fuel tank breather and fuel lines for blockage. Remove fuel pump and check operation. Overhaul or fit new pump. |
| | No oil in damper or oil too thin | Check level of oil in damper, and fill to correct level with oil of a viscosity of S.A.E. 20. |
| DIFFICULT STARTING WHEN HOT | Choke sticking 'on' | Check to ensure choke is returning to fully 'off' position; reset as necessary. |
| | Blocked air cleaner | Fit new air cleaner elements. |
| | Float chamber level too high | Check float level setting. Check float arms for distortion. Check needle valve for sticking. Punctured float, fuel pump pressure too high. |
| LACK OF ENGINE POWER | No oil in damper or oil too thin | Check level of oil in damper, and fill to correct level with oil of a viscosity of S.A.E. 20. |
| | Air valve sticking | Check air valve assembly moves freely and returns under spring load—centre jet assembly. Check diaphragm for cracks or porosity. |
| | Water in fuel | If water is present in float chamber, the complete fuel system should be drained, fuel components should be dismantled, inspected for contamination, paying particular attention to filters. |
| ERRATIC SLOW RUNNING OR STALLING ON DECELERATION | Float level too low | Check float chamber level. Check for needle valve sticking. |
| | Incorrect jet setting | Check and reset jet settings in accordance with carburetter overhaul instructions. |
| | Carburetter air leaks | Check throttle spindle and bearings for wear. |
| | Manifold air leaks | Check inlet manifold gasket for leakage. Check inlet manifold for cracks and distortion of mating faces. Check gasket between carburetter and manifold. Check condition of vacuum advance pipe and connections. Check vacuum servo pipes and connections. |
| | Damper oil too thick. No oil in damper | Check and refill to correct level with oil specified. |
| | Air valve sticking | Check air valve assembly moves freely and returns under spring load—centre jet assembly. Check diaphragm for cracks or porosity. |
| EXCESSIVE FUEL CONSUMPTION | Blocked air cleaner | Fit new air cleaner elements. |
| | Damper oil too thick | Replace with correct grade. |
| | Incorrectly adjusted carburetter | Check and reset slow running in accordance with carburetter tune and adjust instructions. |
| | Float level too high | Check and reset float level. |
| | Incorrect needle | Check needle type. |
| | Worn jets and needle | Check and replace as necessary. |
| | Choke sticking 'on' | Check to ensure choke is returning to fully 'off' position, reset as necessary. |
| | Engine fault | See 'Engine Fault' diagnosis. |

16

---

## 26 COOLING SYSTEM

### COOLANT

### Drain and refill

**Draining**

**WARNING: Do not remove the expansion tank filler cap when the engine is hot because the cooling system is pressurised and personal scalding could result.**

1. Remove the expansion tank filler cap by first turning it anti-clockwise a quarter of a turn to allow pressure to escape, then turn it further in the same direction and lift off.

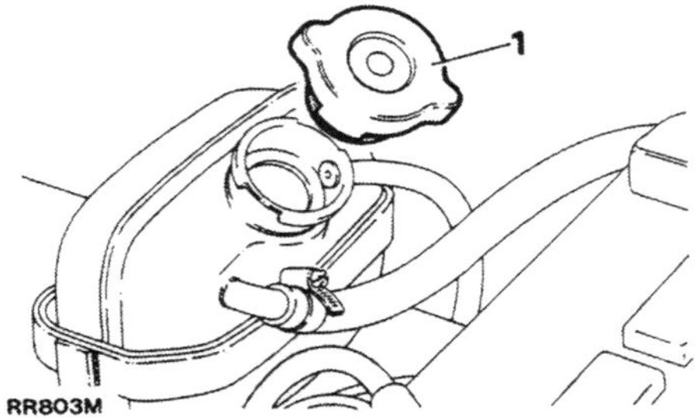

RR803M

2. Remove the radiator filler plug and washer to assist drainage.

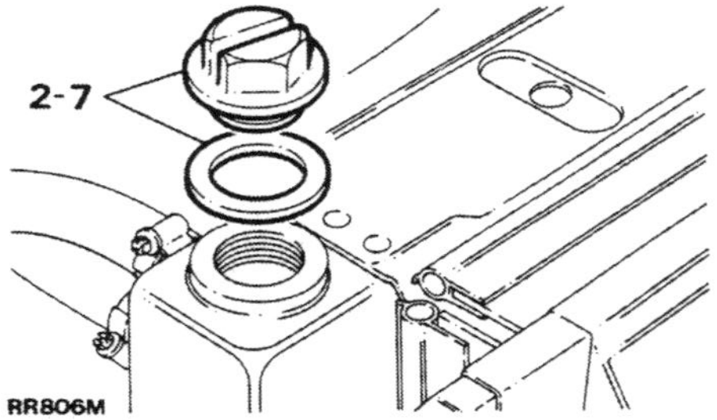

RR806M

3. Disconnect the radiator bottom hose and allow the coolant to drain into a suitable container. Use a clean container if the coolant solution is to be reused. Re-connect the bottom hose after draining and tighten the hose clip.

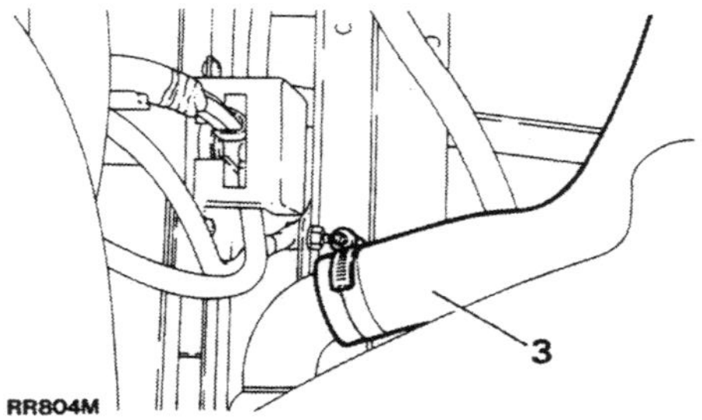

4. Remove the engine drain plugs, one each side of the cylinder block, beneath the exhaust manifolds. Allow the coolant to drain and refit plugs.

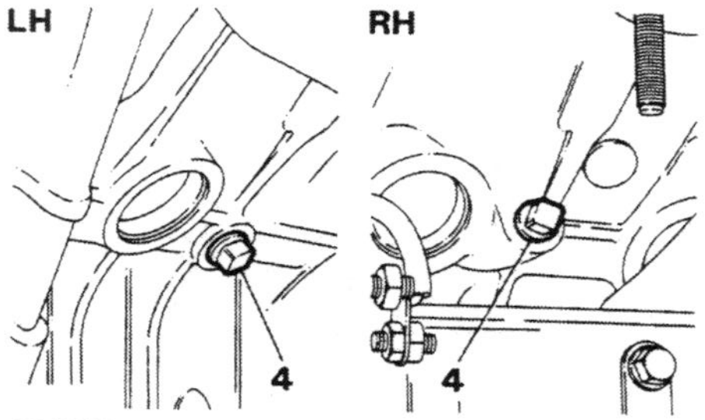

LH  RR805M  RH

See the following coolant requirements before refilling the system.

*continued*

1

# COOLING SYSTEM

## Coolant requirements

### Frost precautions and engine protection

The engine cooling system MUST ALWAYS be filled and topped-up with a solution of water and anti-freeze, winter and summer, or, where frost precautions are not required, water and inhibitor. NEVER use water alone as this may corrode the aluminium alloy.

**CAUTION: Do not use salt water even with an inhibitor, otherwise corrosion will occur. In certain territories where the only available natural water supply has some salt content use only rain or distilled water.**

### Recommended solutions

**Anti-freeze:** Universal Anti-freeze or permanent type ethylene base, without methanol, with a suitable inhibitor for aluminium engines and engine parts.

Use one part of anti-freeze to one part of water.

**Inhibitor:** Marston Lubricants SQ36 inhibitor concentrate.

If frost precautions are not required use a 10% solution of inhibitor, i.e. one part inhibitor to nine parts of water. Inhibitor solution should be drained and flushed out and new inhibitor solution introduced every two years, or sooner where the purity of the water is questionable.

Anti-freeze can remain in the cooling system and will provide adequate protection for two years provided that the specific gravity of the coolant is checked before the onset of the second winter and topped-up with new anti-freeze as required.

Vehicles leaving the factory have the cooling system filled with 50% of anti-freeze mixture. This gives protection against frost down to minus 47°C (minus 53°F). Vehicles so filled can be identified by a label affixed to the windscreen and radiator.

After the second winter the system should be drained and thoroughly flushed. Before adding new anti-freeze examine all joints and renew defective hoses to make sure that the system is leakproof.

See the 'Recommended Lubricants, Fluids and Capacities' section for protection quantities.

### Refilling

5. Pour 4.5 litre (1 gal) of water into the radiator.
6. Add the recommended quantity of anti-freeze or inhibitor.
7. Top-up the radiator with water, refit the radiator filler plug and washer and tighten.
8. Add water to the expansion tank, up to the 'WATER LEVEL' plate.

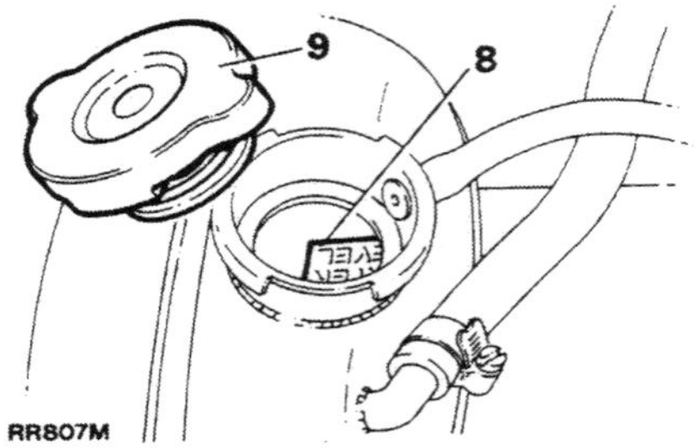

9. Fit the expansion tank filler cap.
10. Run the engine until normal operating temperature is attained, that is, thermostat open.
11. Allow the engine to cool, then check the coolant level and top-up if necessary.

## EXPANSION TANK

### Remove and refit

### Removing

**WARNING: Do not remove the expansion tank filler cap when the engine is hot because the cooling system is pressurised and personal scalding could result.**

1. Remove the expansion tank filler cap by first turning it anti-clockwise a quarter of a turn to allow pressure to escape, then turn it further in the same direction and lift off.

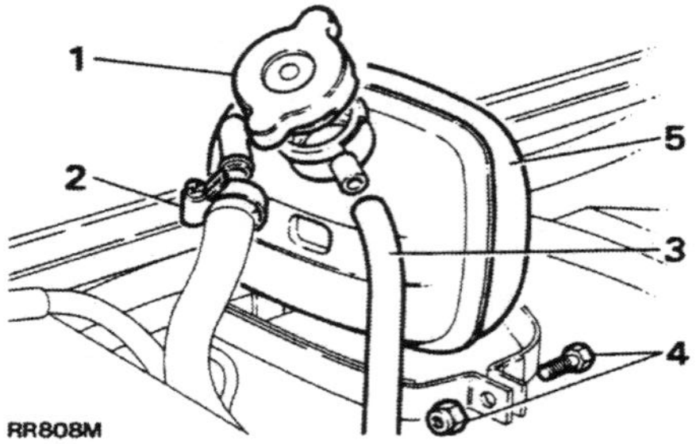

2. Disconnect the hose to the radiator.
3. Disconnect the overflow pipe.
4. Remove the pinch bolt.
5. Lift out the expansion tank.

### Refitting

6. Reverse 2 to 5.
7. Replenish the cooling system.

## FAN BELT

### Check and adjust tension, 1 and 5 to 6

### Remove and refit, 1 to 6

### Removing

1. Slacken the jockey pulley pivot bolt.
2. Pivot the jockey pulley inwards.
3. Lift off the fan belt.

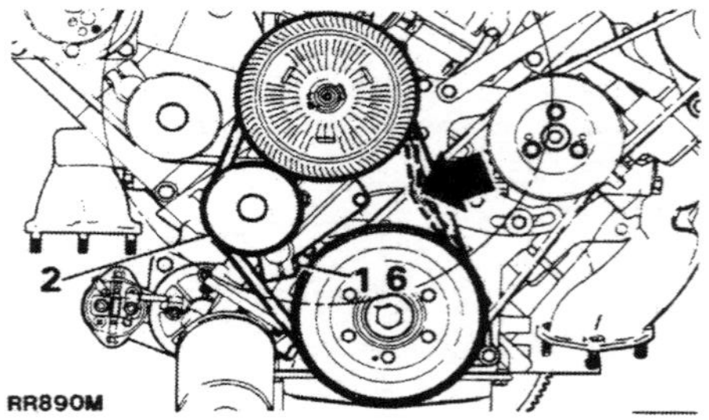

### Refitting

4. Locate the fan belt on the pulleys.
5. Adjust the fan belt by using the jockey pulley to give 11 to 14 mm (0.437 to 0.572 in) free movement when checked midway between the fan and crankshaft pulleys by hand.
6. Tighten the jockey pulley pivot bolt and check the adjustment.

**NOTE: Re-check belt adjustment after approximately 1500 km (1,000 miles) running when a new belt has been fitted.**

## VISCOUS COUPLING, FAN BLADES, PULLEY AND FAN COWL

**Viscous coupling—remove and refit 1 to 5**

**Fan blades—remove and refit 1 to 5 and 8**

**Fan cowl—remove and refit 1 to 4**

**Fan pulley—remove and refit 1 to 7**

### Removing

1. Slacken the nut securing the viscous coupling and fan blade assembly to the water pump spindle.
2. Remove the four fan cowl fixings and lift the cowl out of its lower mountings.

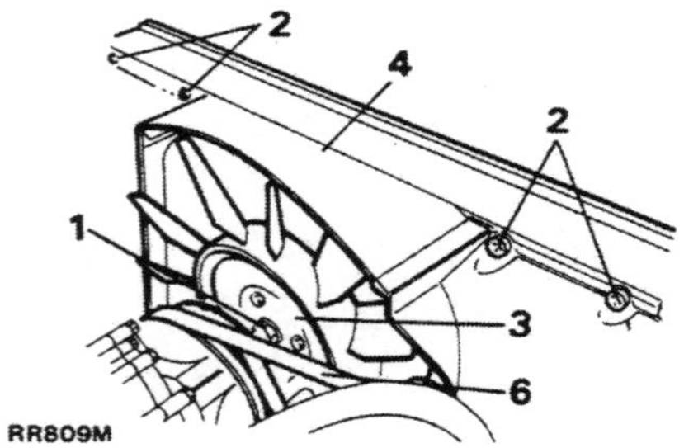

3. Remove the viscous coupling securing nut and withdraw the assembly.
4. Remove the fan cowl, if required.

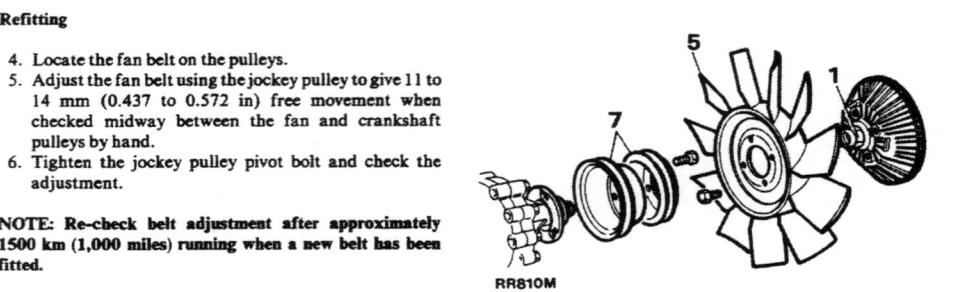

5. Remove the fan blades from the viscous coupling, if required.
6. Remove the fan belt and, if fitted, the compressor belt.
7. Remove the pulley fixings and withdraw the pulley.

*Continued*

# COOLING SYSTEM | 26

8. Vehicles having seven-bladed fans have a detachable boss securing the viscous coupling. Follow 1 to 5 above and remove the bolt to dismantle.

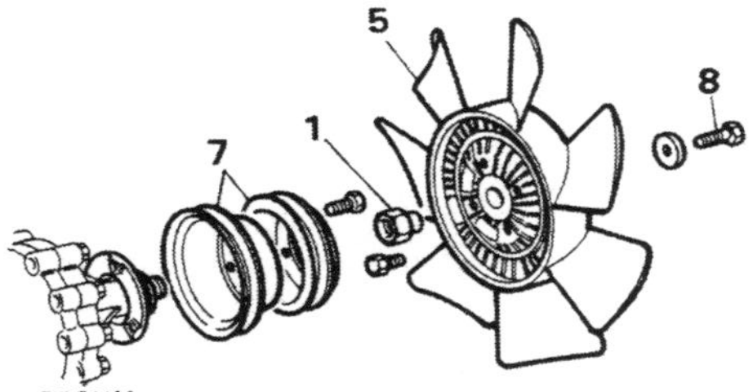

## Refitting

9. Reverse the above procedure, ensuring the fan blades are fitted correctly. Eleven-bladed fans are marked FRONT. Seven-bladed fans are fitted with the deeper dished side towards the viscous coupling. See 'Data' section for correct torque figures.
10. Adjust the fan belt and compressor belt.

## RADIATOR

### Remove and refit

### Removing

1. Drain the cooling system.
2. Remove the fan blades.
3. Remove the fan cowl.
4. Disconnect the top hose from the radiator.
5. Disconnect the hose to the expansion tank. 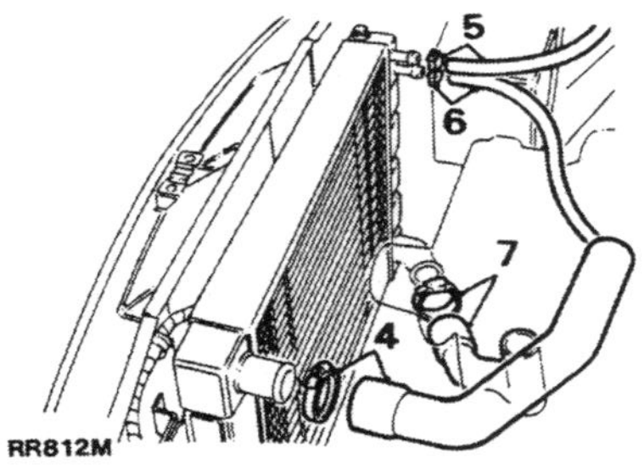
6. Disconnect the hose to the induction manifold.
7. Disconnect the hose from the bottom of the radiator.
8. Remove the fixings from each side of the radiator.
9. Withdraw the radiator from the rubber-mounted spigots.

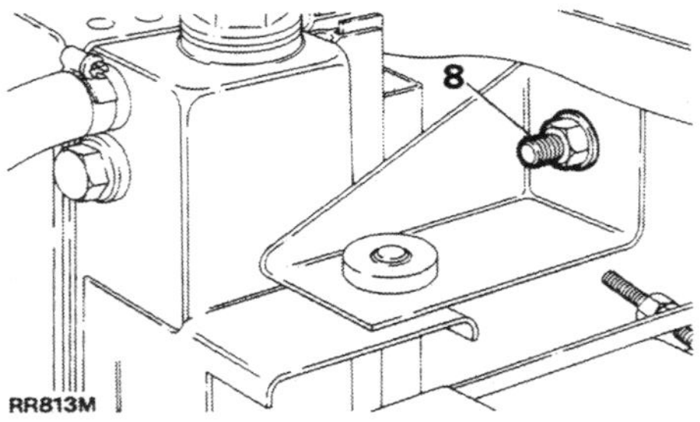

### Refitting

10. Reverse 1 to 9 noting the assembly order of the radiator side fixings and ensuring that the radiator sealing strips are correctly located and secure.

## THERMOSTAT

### Remove and refit

### Test

### Removing

1. Drain the cooling system, sufficient to drain the induction manifold.
2. Disconnect the hose to the radiator.
3. Disconnect the electrical connections to the water temperature switch, if fitted. 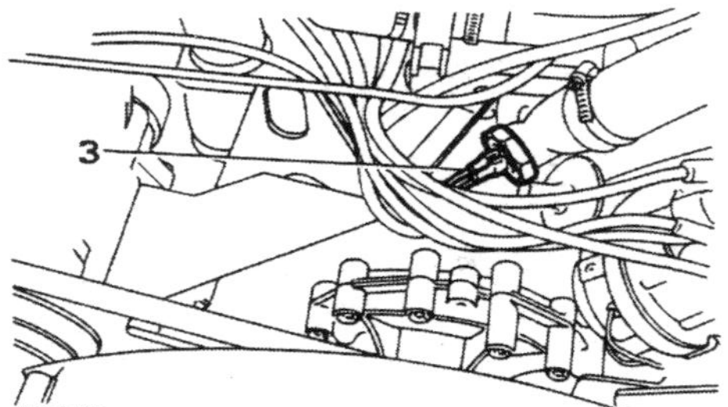
4. Remove the outlet elbow.
5. Withdraw the thermostat.

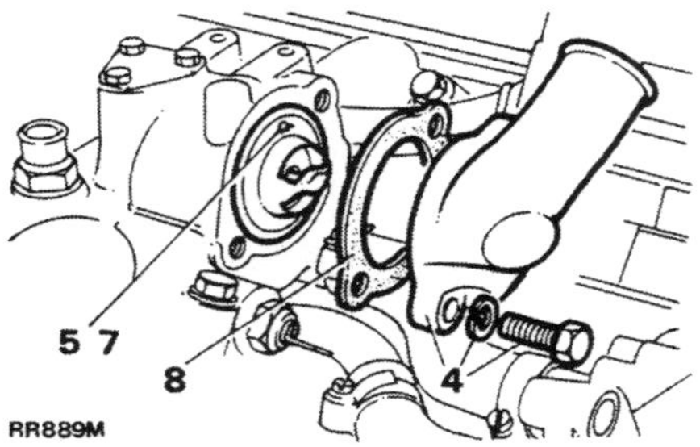

## Testing

6. Note that the rating of the thermostat is 88°C. Place the thermostat in a suitable container half full of water. Heat the water and observe the temperature at which the thermostat opens.

## Refitting

7. Insert the thermostat with the jiggle pin uppermost (12 o'clock).
8. Using a new joint washer, fit the outlet elbow and tighten to the correct torque, see 'Data' section.
9. Reverse 1 to 3.

## WATER PUMP

### Remove and refit

### Removing

1. Drain the engine cooling system.
2. Remove the fan belt and, where fitted, the compressor belt.
3. Remove the fan blades and pulley.
4. Remove the air control valve from the support bracket (carburetter vehicles only).
5. Release the alternator adjusting link and the power-steering pump fixings.
6. Disconnect the inlet hose from the water pump.
7. Remove the water pump.

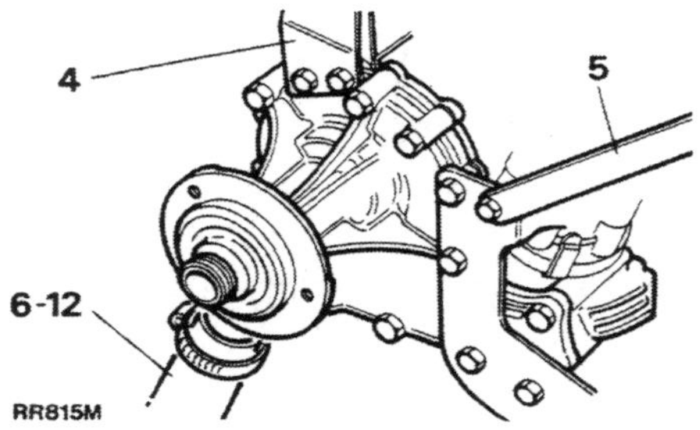

## Refitting

8. Lightly grease a new joint washer and place it in position on the timing cover.
9. Clean the threads of the four long bolts and smear with Loctite 572 thread lubricant-sealant.
10. Locate the water pump in position.
11. Locate the alternator adjusting link and power-steering pump bracket.
12. Leave the alternator adjusting link loose and tighten the remaining water pump housing bolts evenly and to the correct torque, see 'Data' section.
13. Connect the inlet hose to the water pump.
14. Fit the fan pulley.
15. Fit the air-control valve to the support bracket.
16. Fit and adjust the fan belt, power-steering pump belt and, where applicable, the compressor belt.
17. Fit the fan blade assembly.
18. Refill the cooling system.

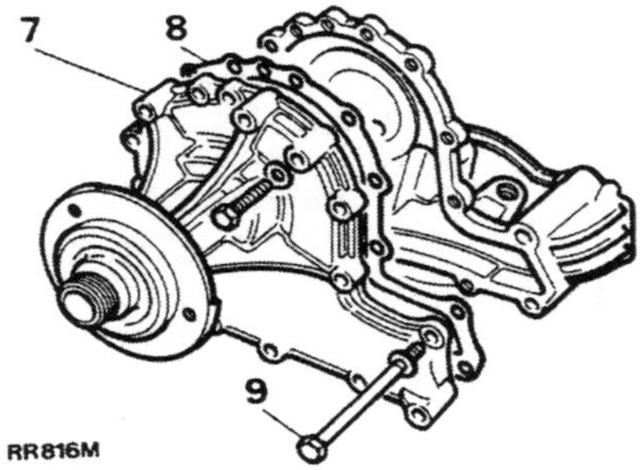

*Continued*

5

# MANIFOLD AND EXHAUST SYSTEM | 30

13. Disconnect the inlet hose to the heater.
14. Disconnect the return hose from the heater.
15. Disconnect the return hose to the radiator.
16. Disconnect the return hose from the top of the induction manifold.

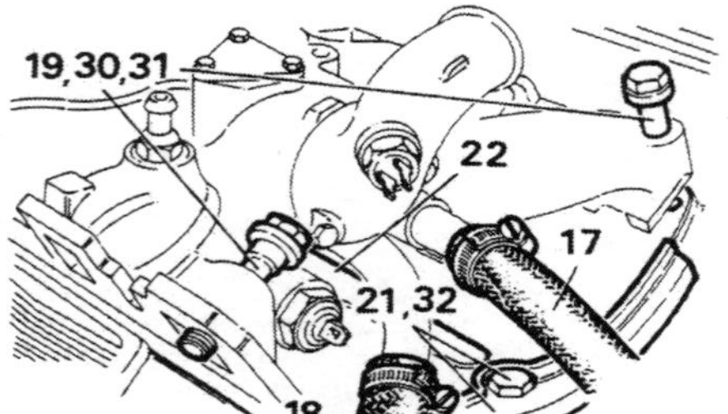

RR883M

17. Disconnect the outlet hose from the manifold.
18. Disconnect the heater return hose from the manifold.
19. Evenly slacken and withdraw twelve bolts and remove the manifold.
20. Wipe away any coolant lying on the manifold gasket.
21. Remove the gasket clamps.
22. Lift off the gasket.
23. Withdraw the gasket seals.

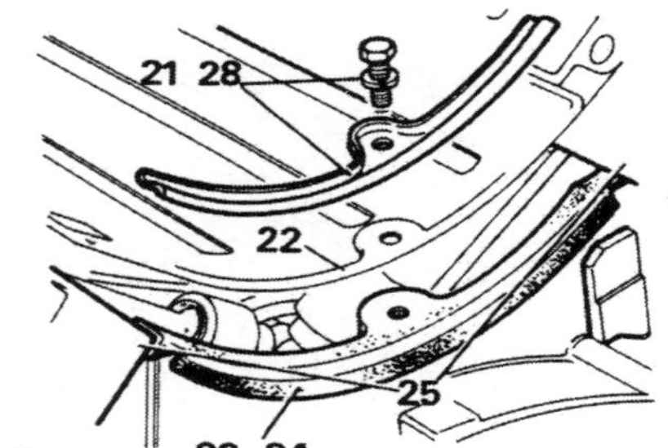

RR884M

## Refitting

24. Using new seals, smear them on both sides with silicon grease.
25. Locate the seals in position with their ends engaged in the notches formed between the cylinder head and block.
26. Apply 'Hylomar' sealing compound SQ32M on the corners of the cylinder head, manifold gasket and manifold, around the water passage joints.
27. Fit the manifold gasket with the word 'FRONT' to the front and the open bolt hole at the front RH side.

*Continued*

## INDUCTION MANIFOLD

### Remove and refit

NOTE: Refer to electronic fuel injection, section 19, for remove and refit of EFI induction manifold.

### Removing

1. Drain the cooling system.
2. Remove the air cleaner.
3. Remove the engine breather filter.
4. Disconnect the throttle cable from the carburetter and manifold.

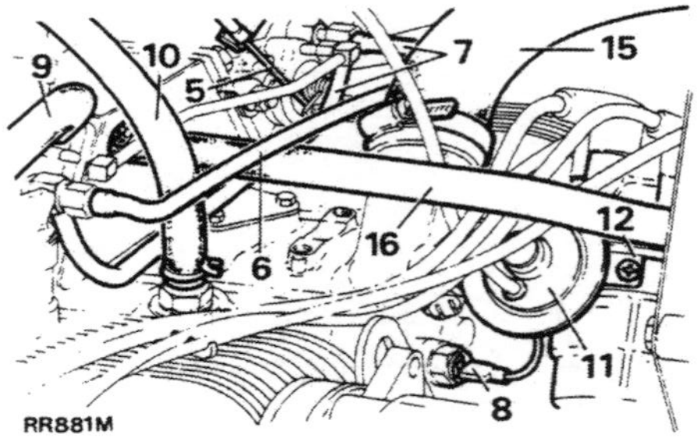

RR881M

5. Disconnect the choke cable from the carburetter.
6. Disconnect the fuel spill return pipe from the RH carburetter.
7. Remove the fuel supply pipe from the carburetters.

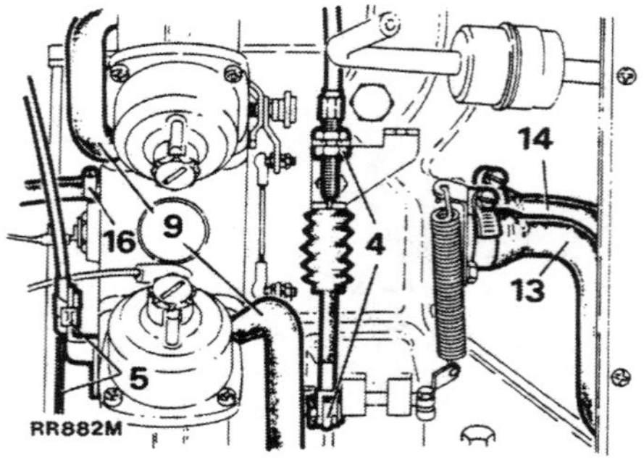

RR882M

8. Disconnect the lead from the water temperature transmitter.
9. Disconnect the flame trap hoses from the carburetters.
10. Disconnect the vacuum pipe for the brake servo.
11. Disconnect the vacuum pipe from the distributor.
12. Release the distributor cap.

# 26 | COOLING SYSTEM

## COOLING SYSTEM FAULT DIAGNOSIS

| SYMPTOM | POSSIBLE CAUSE | CURE |
|---|---|---|
| A—EXTERNAL LEAKAGE | 1. Loose hose clips<br>2. Defective rubber hose<br>3. Damaged radiator seams<br>4. Excessive wear in the water pump<br>5. Loose core plugs<br>6. Damaged gaskets<br>7. Leaks at the heater connections or plugs<br>8. Leak at the water temperature gauge plug | 1. Tighten<br>2. Renew<br>3. Rectify<br>4. Renew<br>5. Renew<br>6. Renew<br>7. Rectify<br>8. Tighten |
| B—INTERNAL LEAKAGE | 1. Defective cylinder head gasket<br>2. Cracked cylinder wall<br>3. Loose cylinder head bolts | 1. Renew. Check engine oil for contamination and refill as necessary<br>2. Renew cylinder block<br>3. Tighten. Check engine for oil contamination and refill as necessary |
| C—WATER LOSS | 1. Boiling<br>2. Internal or external leakage<br>3. Restricted radiator or inoperative thermostat | 1. Ascertain the cause of engine overheating and correct as necessary<br>2. See items A and B<br>3. Flush radiator or renew the thermostat as necessary |
| D—POOR CIRCULATION | 1. Restriction in system<br>2. Insufficient coolant<br>3. Inoperative water pump<br>4. Loose fan belt<br>5. Inoperative thermostat | 1. Check hoses for crimps, reverse-flush the radiator, and clear the system of rust and sludge<br>2. Replenish<br>3. Renew<br>4. Adjust<br>5. Renew |
| E—CORROSION | 1. Excessive impurity in the water<br>2. Infrequent flushing and draining of system<br>3. Incorrect anti-freeze mixtures | 1. Use only soft, clean water together with correct anti-freeze or inhibitor mixture<br>2. The cooling system should be drained and flushed thoroughly at least once a year<br>3. Certain anti-freeze solutions have a corrosive effect on parts of the cooling system. Only recommended solutions should be used |
| F—OVERHEATING | 1. Poor circulation<br>2. Dirty oil and sludge in engine<br>3. Radiator fins choked with chaff, mud, etc.<br>4. Incorrect ignition timing<br>5. Insufficient coolant<br>6. Low oil level<br>7. Tight engine<br>8. Choked or damaged exhaust pipe or silencer<br>9. Dragging brakes<br>10. Overloading vehicle<br>11. Driving in heavy sand or mud<br>12. Engine labouring on gradients<br>13. Low gear work<br>14. Excessive engine idling<br>15. Inaccurate temperature gauge<br>16. Defective thermostat | 1. See item D<br>2. Refill<br>3. Use air pressure from the engine side of the radiator and clean out passages thoroughly<br>4. Check engine electronic equipment<br>5. See item D<br>6. Replenish<br>7. New engines are very tight during the 'running-in' period and moderate speeds should be maintained for the first 1,000 miles (1,500 km)<br>8. Rectify or renew<br>9. Adjust brakes<br>10. In the hands of the operator<br>11. In the hands of the operator<br>12. In the hands of the operator<br>13. In the hands of the operator<br>14. In the hands of the operator<br>15. Renew<br>16. Renew |
| G—OVERCOOLING | 1. Defective thermostat<br>2. Inaccurate temperature gauge | 1. Renew<br>2. Renew |

# MANIFOLD AND EXHAUST SYSTEM

28. Fit the gasket clamps but do not fully tighten the bolts at this stage.
29. Locate the manifold onto the cylinder head.
30. Clean the threads of the manifold securing bolts and then coat them with Thread Lubricant-sealer 3M EC776.
31. Fit all the manifold bolts and tighten them a little at a time, evenly, alternate sides working from the centre to each end. Finally tighten to the correct torque, see Data section.

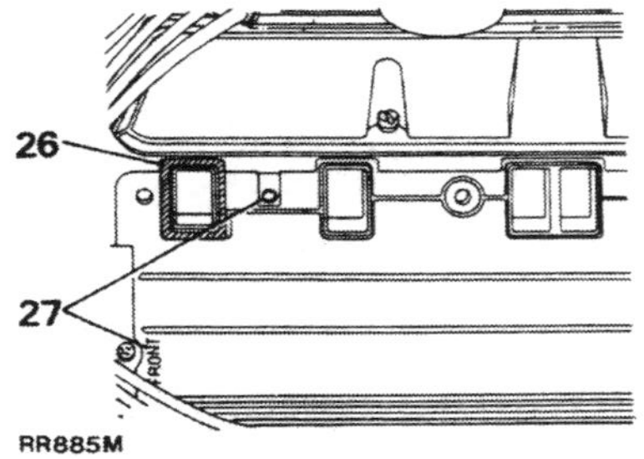

32. Tighten the gasket clamp bolts to the correct torque, see Data section.
33. Reverse 1 to 18.
34. Run the engine and check for water leaks.

## EXHAUST SYSTEM COMPLETE

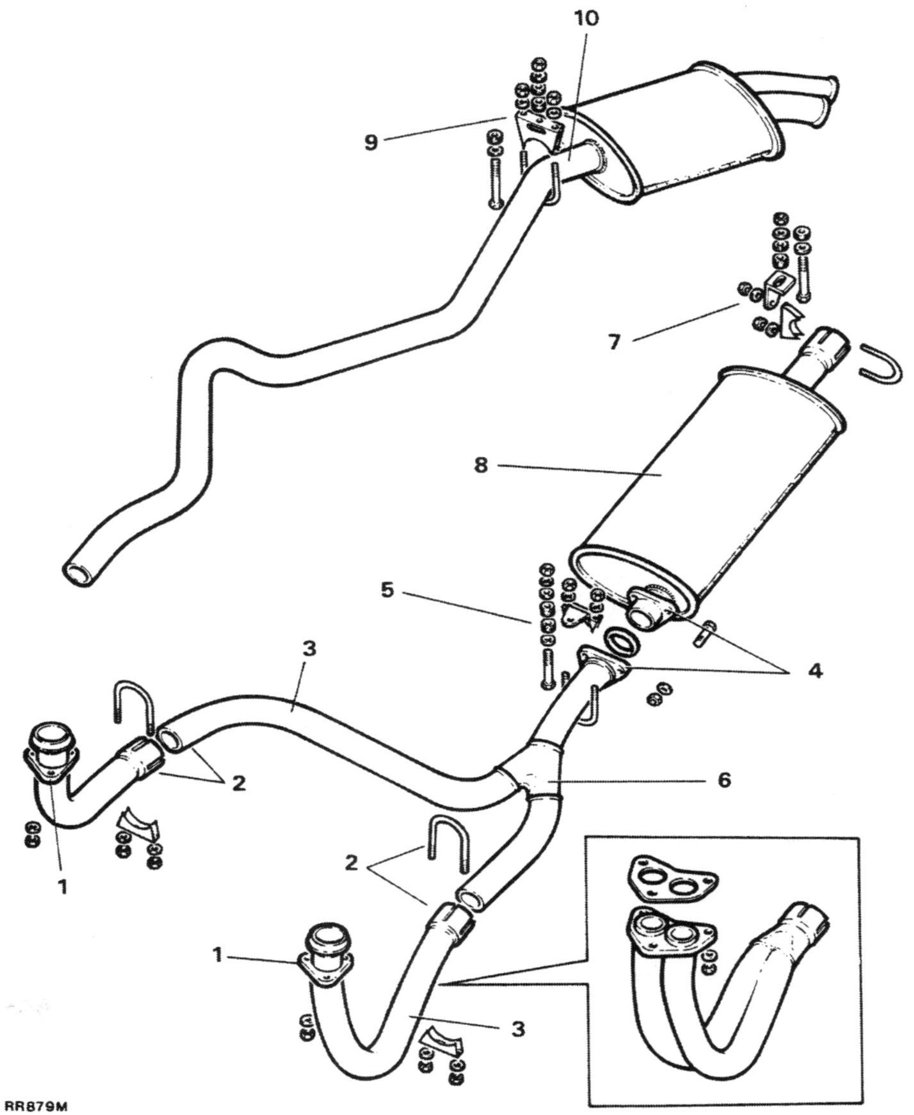

**Remove and refit, 1 to 11**

**Front pipe, left hand 1 to 3 and 12**

**Front pipe, right hand 1 to 3 and 12**

**Silencer 4, 7, 8 and 13**

**Intermediate pipe 1 to 6 and 14**

**Tail pipe 7, 9, 10 and 15**

NOTE: Ensure that no exhaust leaks are evident in either a new or an old exhaust system, as this will affect vehicle performance.

### Removing

Note that a gasket is fitted between manifold and down pipe on EFI models—see inset.

1. Disconnect the front pipe(s) from the manifold(s).
2. Slacken the U-bolts securing the front and intermediate pipes.
3. Withdraw the front pipe(s).
4. Remove three bolts securing the intermediate pipe to the main silencer and withdraw the olive.
5. Remove the U-bolt from the pipe mounting bracket.
6. Withdraw the intermediate pipe.
7. Remove the U-bolt from the pipe mounting bracket.
8. Withdraw the silencer.
9. Remove the U-bolt from the tail pipe mounting bracket.
10. Withdraw the tail pipe.

### Refitting

NOTE: Apply Firegum Putty, Part No 15608 to the joints between the downpipe and intermediate pipe and between the silencer and tailpipe.

11. Complete system, reverse 1 to 10.
12. Front pipe, reverse 1 to 3.
13. Silencer, reverse 4, 7 and 8.
14. Intermediate pipe, reverse 1 to 6.
15. Tail pipe, reverse 7, 9 and 10.

# 30 MANIFOLD AND EXHAUST SYSTEM

## EXHAUST MANIFOLD

### Remove and refit

**Left hand**

**Right hand**

#### Removing

1. Disconnect the front exhaust pipe from the manifold and (where fitted) remove the hot air box.
2. Tap back the bolt locking tabs and remove eight bolts with lock tabs and washers.
3. Remove the manifold.

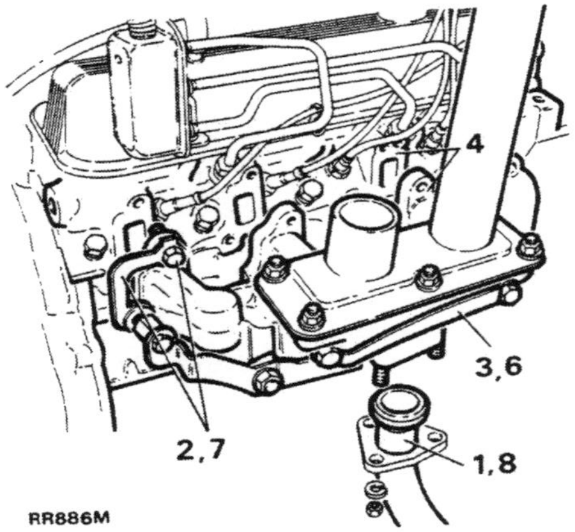

#### Refitting

4. Ensure that the mating surfaces of the cylinder head and exhaust manifold are clean and smooth.
5. Coat the exhaust manifold (cylinder head mating faces) with 'Foliac J 166' or 'Moly Paul' anti-seize compound.
   'Foliac J 166' is manufactured by Rocol Ltd, Rocol House, Swillington, Leeds, England.
   'Moly Paul' is manufactured by K.S. Paul Products Ltd, Nobel Road, London N18.
6. Place the manifold in position on the cylinder head and fit the securing bolts, lockplates and plain washers. The plain washers are fitted between the manifold and lockplates.
7. Evenly tighten the manifold bolts to the correct torque, see Data section, and bend over the lockplate tabs.
8. Reconnect the front exhaust pipe.

# 33 CLUTCH

## HYDRAULIC SYSTEM

### Bleed

#### Procedure

NOTE: **During the following procedure, keep the fluid reservoir topped up to avoid introducing air to the system. Use only the hydraulic fluid recommended in Section 09.**

1. Attach a length of suitable tubing to the slave cylinder bleed screw.
2. Place the free end of the tube in a glass jar containing clutch fluid.
3. Slacken the bleed screw.
4. Pump the clutch pedal, pausing at the end of each stroke, until the fluid issuing from the tubing is free of air with the tube free end below the surface of the fluid in the container.
5. Hold the clutch pedal down. Keeping the free end of the tube below the fluid, tighten the bleed screw.

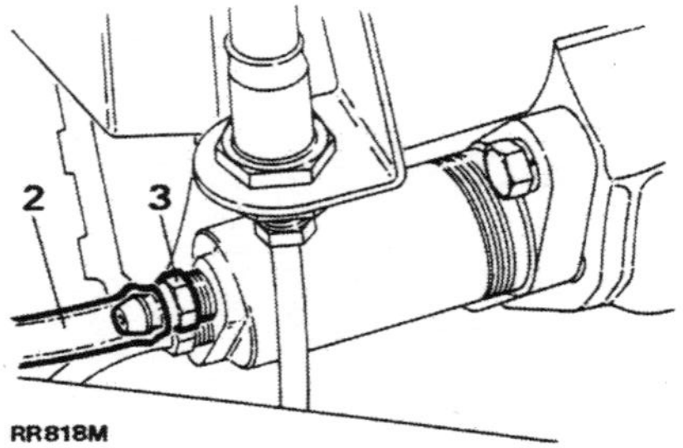

## CLUTCH ASSEMBLY

### Overhaul

#### Clutch pressure plate

Renew the pressure plate if the diaphragm spring fingers are worn or if the pressure plate shows signs of wear, cracks or burning.

#### Clutch driven plate

Renew the driven plate if the centre hub splines are worn or if the lining is contaminated, burned or unevenly worn.

### Remove and refit

Service tool: 18G 79 clutch centralising tool

#### Removing

1. Remove the engine.
2. Mark the clutch cover fitted position relative to the flywheel.
3. Remove the clutch cover securing bolts, working evenly and diagonally.
4. Do not disturb the three bolts located in the apertures in the clutch cover.
5. Remove the clutch assembly.
6. Withdraw the clutch driven plate.

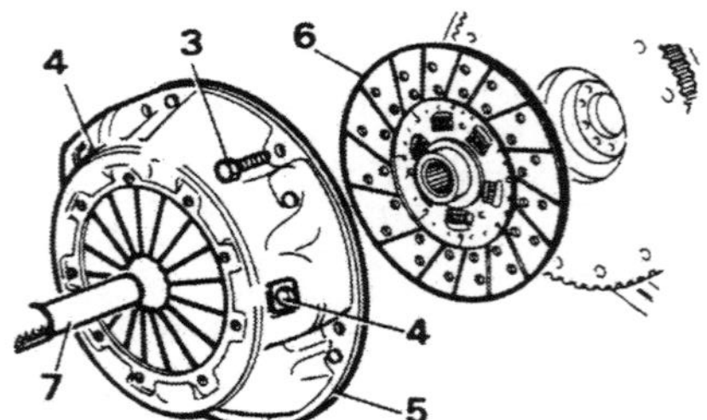

#### Refitting

NOTE: **As a precaution against the clutch plate sticking, lubricate the splines using Rocol MV 3 or Rocol MTS 1000 grease.**

7. Reverse 5 and 6, aligning the assembly marks. Centralising tool 18G 79.
8. Secure the cover fixings evenly, working in a diagonal sequence. Finally tighten to the correct torque, see Data section.
9. Fit the engine.

# CLUTCH 33

## MASTER CYLINDER

### Remove and refit

#### Removing

1. Evacuate the hydraulic fluid from the system.
2. Disconnect the fluid pipe at the master cylinder. Plug the master cylinder fluid port and seal the end of the hydraulic pipe to prevent ingress of foreign matter.
3. Remove the lower fascia panel.
4. Remove the master cylinder fixings at the dash panel.
5. Remove the pivot bolt and sleeves to free the push rod from the clutch pedal.
6. Withdraw the master cylinder.

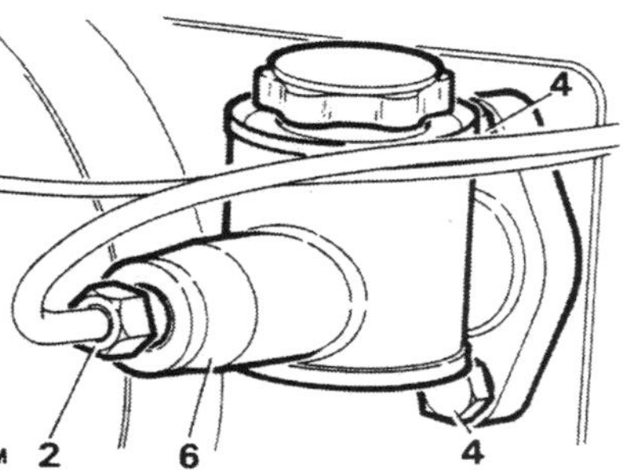

#### Refitting

7. Fit the master cylinder and dash fixings.
8. Fit the push-rod to the pedal. Do not tighten the pivot bolt nut at this stage.
9. Check the brake pedal setting.
10. Back off the lower stop bolt.
11. Align the clutch pedal to the same angle as the brake pedal by turning the pivot bolt and integral cam.
12. Tighten the pivot bolt securing nut.
13. Fully depress the pedal.
14. Adjust the lower stop bolt to touch the pedal then continue a further turn.
15. Fit the fluid pipe to the master cylinder.
16. Bleed and replenish the hydraulic system.
17. Fit the lower fascia panel.

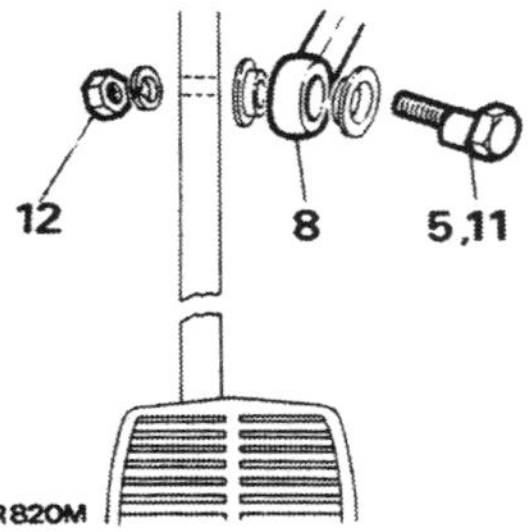

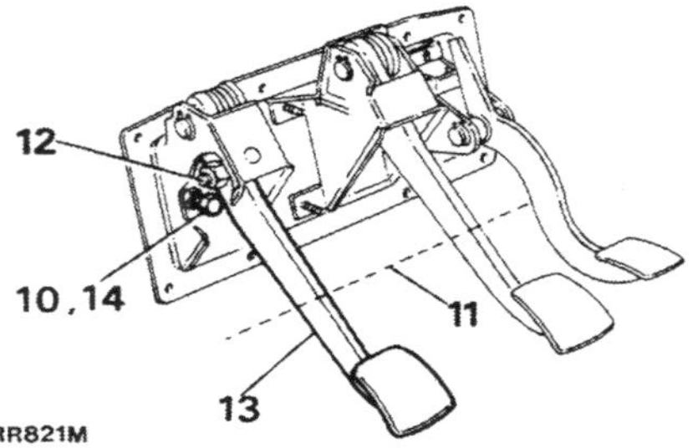

## MASTER CYLINDER

### Overhaul

#### Dismantling

1. Remove the master cylinder.
2. Pull back and remove the rubber sealing boot from the pushrod.
3. Depress the push-rod and extract the circlip.
4. Withdraw the push-rod assembly.
5. Withdraw the piston assembly.
6. Withdraw the retainer and spring.
7. Remove the two piston seals and the piston washer.

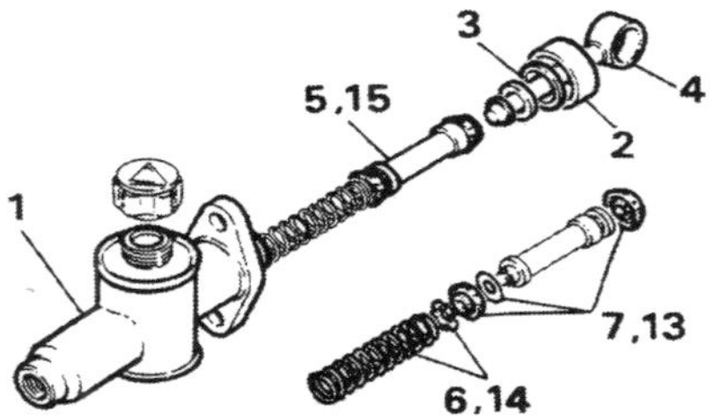

#### Inspecting

8. Clean all the components thoroughly using new hydraulic fluid. Dry, using a lint-free cloth.
9. Examine the cylinder bore and piston, ensure that they are smooth to the touch with no corrosion, score marks or ridges. If there is any doubt, fit new replacements.
10. Replace the seals and rubber boot using new components. These items are all included in the master cylinder overhaul kit.
11. Ensure that the feed and by-pass ports are not obstructed.
12. Ensure the reservoir cap vent is clear.

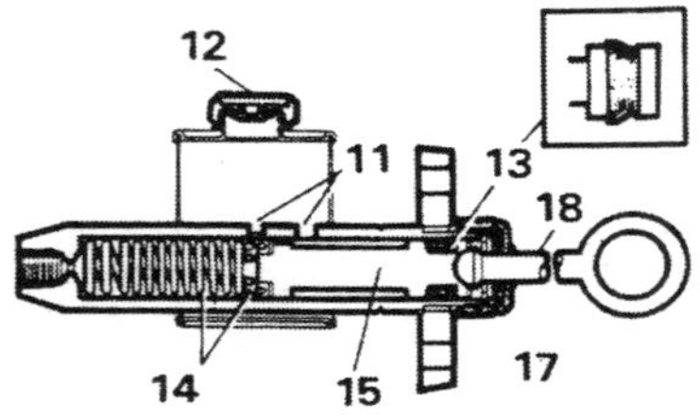

#### Assembling

NOTE: Scrupulous cleanliness is essential, ensure that the hands are free of grease or dirt. Lubricate the cylinder bore and rubber seals with new hydraulic fluid before assembling.

13. Fit a new piston washer and the thinner of the two piston seals, lip last, over the piston nose, up against the drilled piston head. Fit the thicker seal into the piston groove with the lip facing towards the seal at the opposite end.
14. Insert the spring and retainer into the master cylinder bore.
15. Insert the piston and seal assembly, ensuring that the seal lips do not bend back.
16. Reverse 3 and 4, correctly locating the circlip.
17. Gently stretch the new rubber boot over the push-rod, pack with rubber grease, and fit securely into its locating groove.
18. Operate the push-rod several times to ensure free movement of the internal components.
19. Fit the master cylinder.

## RELEASE BEARING ASSEMBLY

### Remove and refit

#### Removing

1. Remove the engine.
2. Remove the clutch slave cylinder.
3. Withdraw the retainer staple.
4. Withdraw the bearing and sleeve. If required, press the bearing off the sleeve. Fit the replacement bearing with the domed face outwards from sleeve.
5. Remove the spring clip and bolt.
6. Withdraw the release lever assembly.

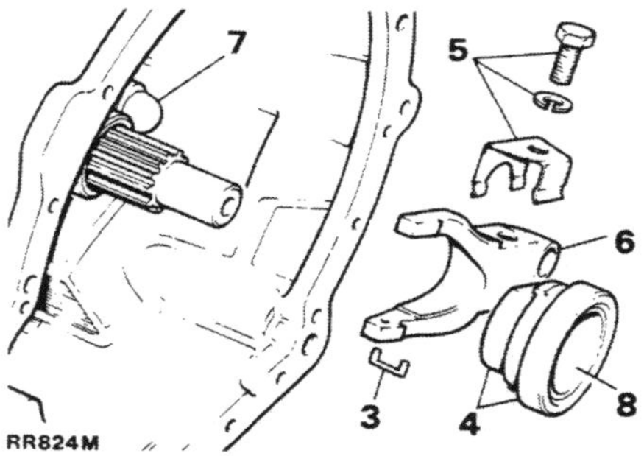

#### Refitting

7. Smear the pivot with grease and fit the release lever and retain with the spring clip and bolt.
8. Smear the release bearing sleeve inner diameter with molybdenum disulphide base grease.
9. Reverse 1 to 4.

# CLUTCH

## CLUTCH PEDAL

### Remove and refit

#### Removing

1. Remove the lower fascia panel.
2. Remove the pedal bracket fixings at the cab dash panel.
3. Disconnect the brake fluid pipes and electrical connection at the brake master cylinder.
4. Disconnect the fluid pipe at the clutch master cylinder.
5. Withdraw the pedal bracket assembly into the engine compartment.

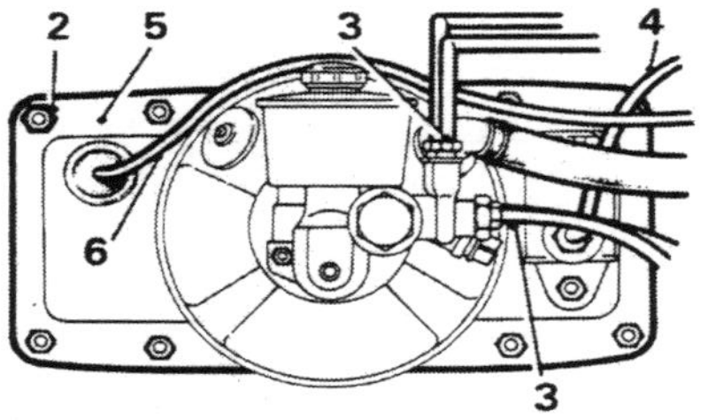

6. Disconnect the accelerator control cable at the pedal.
7. Withdraw the pedal bracket assembly from the vehicle.
8. Remove the pivot bolt nut.
9. Withdraw the pivot bolt and bearing sleeves which retain the master cylinder push-rod.
10. Remove the pedal spindle circlip.
11. Withdraw the spindle.
12. Lift out the return spring.
13. Withdraw the pedal.
14. If required, press out the spindle bushes. Press in replacements and lubricate.

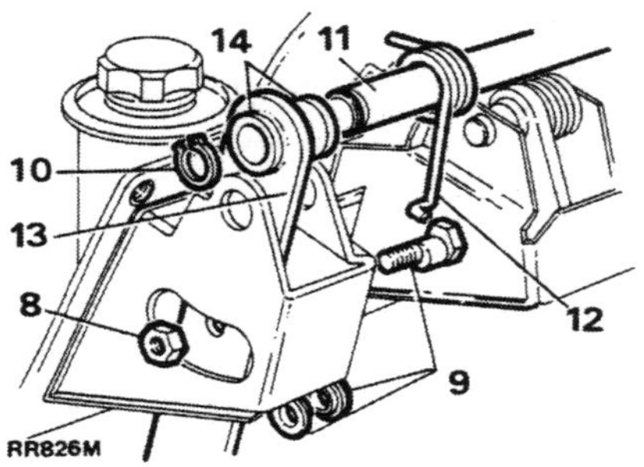

#### Refitting

15. Reverse 9 to 13.
16. Loosely fit the pivot bolt nut.
17. Align the clutch pedal to the same angle as the brake pedal by turning the pivot bolt and integral cam.
18. Tighten the pivot bolt securing nut.

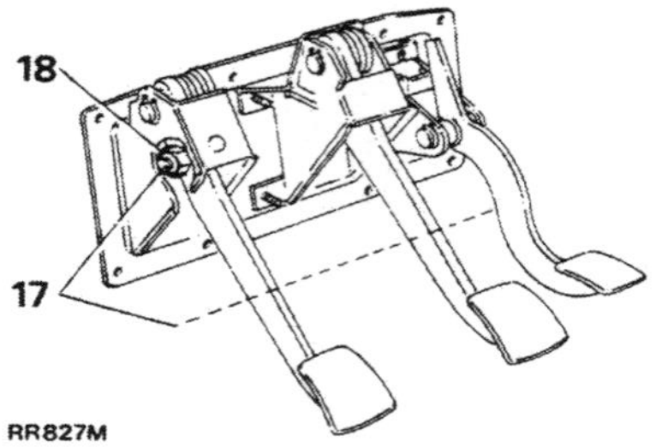

19. Fit the accelerator cable.
20. Offer the pedal bracket assembly and joint washer to the dash panel. Avoid damaging the brake light switch.
21. Reverse 1 to 4.
22. Bleed the brake system.
23. Bleed the clutch system.

## SLAVE CYLINDER

### Remove and refit

#### Removing

1. Evacuate the clutch system fluid at the slave cylinder bleed valve.
2. Disconnect the fluid pipe.
3. Remove the two securing bolts and withdraw the slave cylinder and backing plate.
4. If the dust cover is not withdrawn with the slave cylinder, withdraw it from the bell housing.

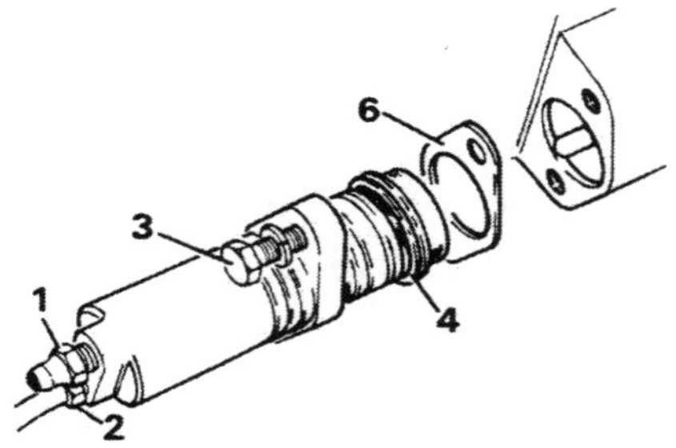

#### Refitting

5. Withdraw the dust cover and backing plate from the slave cylinder.
6. Coat both sides of the backing plate with Hylomar P232M waterproof jointing compound.
7. Locate the backing plate and dust cover in position on the slave cylinder.
8. Fit the slave cylinder, engaging the push-rod through the centre of the dust cover and with the bleed screw uppermost.

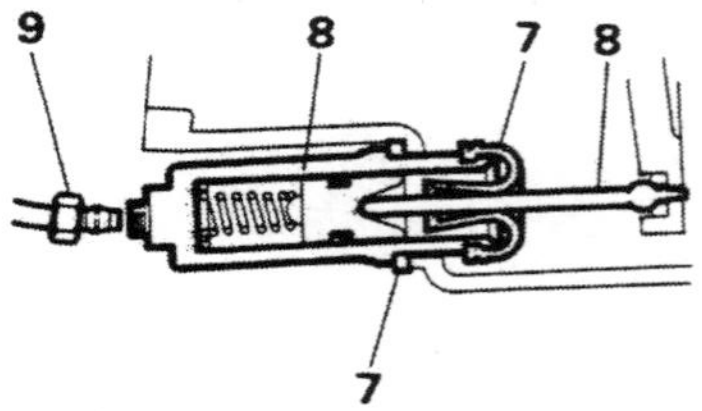

9. Reconnect the fluid pipe.
10. Replenish and bleed the clutch hydraulic system.
11. Check for fluid leaks with the pedal depressed and also with the system at rest.

## SLAVE CYLINDER

### Overhaul

1. Remove the slave cylinder.

#### Dismantling

2. Withdraw the rubber boot.
3. Withdraw the push-rod.
4. Remove the circlip.
5. Extract the piston and seal assembly, applying low pressure air to the fluid inlet if necessary.
6. Withdraw the spring.
7. Remove the bleed valve.

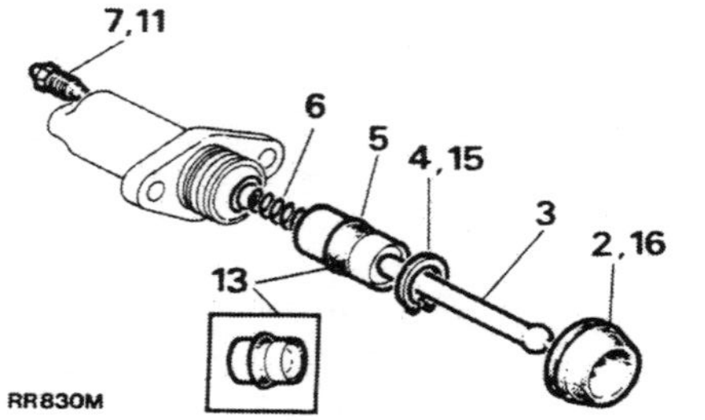

#### Inspecting

8. Clean all components thoroughly using new hydraulic fluid, and dry using lint-free cloth.
9. Examine the cylinder bore and piston which must be free from corrosion, scores and ridges.
10. Replace the seal and rubber boot using the appropriate repair kit.

#### Assembling

NOTE: Scrupulous cleanliness is essential, ensure that hands are free of grease or dirt.

11. Fit the bleed valve. Do not overtighten.
12. Lubricate the seals, piston and bore using new hydraulic fluid.
13. Fit the seal into the piston groove, the lip of the seal towards the fluid inlet end of the cylinder.
14. Enter the piston assembly, spring first, into the cylinder bore. Ensure that the seal lip does not fold back.
15. Secure with the circlip.
16. Fill the rubber boot with rubber grease.
17. Reverse 1 to 3.

# RANGE ROVER WORKSHOP MANUAL 1986 ONWARDS

# Part 3

|  | Section | Page |
|---|---|---|
| General Specification Data | 04 | 88 |
| Torque Wrench Settings | 06 | 89 |
| Manual Gearbox and Transfer Box | 37 | 91 |
| Automatic Gearbox | 44 | 114 |
| Propeller Shafts | 47 | 142 |
| Rear Axles & Final Drive | 51 | 143 |
| Front Axles & Final Drive | 54 | 149 |

**Section Number**

### 04 GENERAL SPECIFICATION DATA

- Manual gearbox and transfer gearbox ratio's — 1
- Automatic gearbox and transfer gearbox ratio's — 1
- Automatic shift speeds — 2
- Propeller shafts — 2
- Rear axle — 2
- Front axle — 2

### 06 TORQUE WRENCH SETTINGS

- Transfer gearbox — 1
- Automatic gearbox — 2
- Manual gearbox — 3
- Front axle — 3
- Rear axle — 4

### 37 LT77 FIVE SPEED GEARBOX

- Remove and refit — 1
- Overhaul — 3

### 37 LT230T TRANSFER GEARBOX

- Renew speedometer drive pinion — 19
- Renew rear output shaft oil seals — 19
- Renew front output shaft oil seals — 19
- Remove and refit — 20
- Overhaul — 26

*Continued*

## Notes

# Special Service Tools

The use of approved special service tools is important. They are essential if service operations are to be carried out efficiently, and safely. The amount of time which they save can be considerable.

Every special tool is designed with the close co-operation of Land Rover Ltd., and no tool is put into production which has not been tested and approved by us. New tools are only introduced where an operation cannot be satisfactorily carried out using existing tools or standard equipment. The user is therefore assured that the tool is necessary and that it will perform accurately, efficiently and safely.

Special tools bulletins will be issued periodically giving details of new tools as they are introduced.

All orders and enquiries from the United Kingdom should be sent direct to V. L. Churchill. Overseas orders should be placed with the local V. L. Churchill distributor, where one exists. Countries where there is no distributor may order direct from V. L. Churchill Limited.

The tools recommended in this Repair Operation Manual are listed in a multi-language, illustrated catalogue obtainable from Messrs. V. L. Churchill at the above address under publication number 2217/2/84 or from Land Rover Ltd., under part number LSM00052TC from the following address: Land Rover Limited, Service Department, Lode Lane, Solihull, West Midlands, England B92 8NW.

Section Number

| 44 | AUTOMATIC GEARBOX | |
|---|---|---|
| | —Fault diagnosis | 1 |
| | —Service tools and data | 9 |

**STAGE I**

| —Inhibitor switch leak elimination and replacement | 17 |
| —Intermediate plate screw plugs leak elimination | 17 |
| —Selector shaft leak elimination | 17 |
| —Oil pan leak elimination and replacement | 18 |
| —Kickdown cable leak elimination | 19 |
| —Extension case leak elimination and replacement | 19 |
| —Oil screen replacement | 20 |
| —Control unit replacement | 21 |
| —Oil inlet sealing rings renewal | 22 |
| —Manual valve operating mechanism | 24 |
| —Governor housing renewal | 24 |
| —Governor hub renewal | 25 |
| —Parking brake mechanism renewal | |

**STAGE II**

| —ZF gearbox, remove and refit | 27 |
| —Eliminating leaks/replacing torque converter | 28 |
| —Eliminating leaks on the pump housing | 29 |
| —Eliminating leaks between gearbox housing and intermediate plate | 30 |
| —Replacing bellhousing | 31 |
| —Replacing pump | 31 |
| —Replacing intermediate plate | 33 |
| —OVERHAUL | 34 |

| 47 | PROPELLER SHAFTS | |
|---|---|---|
| | —Overhaul | 1 |

| 51 | REAR AXLE AND FINAL DRIVE | |
|---|---|---|
| | —Remove and refit axle assembly | 1 |
| | —Overhaul rear differential | 2 |
| | —Rear discs—remove and refit | 8 |
| | —Remove and overhaul rear hubs | 8 |

| 54 | FRONT AXLE AND FINAL DRIVE | |
|---|---|---|
| | —Remove and refit axle assembly | 1 |
| | —Overhaul front differential | 1 |
| | —Front discs—remove and refit | 2 |
| | —Remove and overhaul front hubs | 2 |
| | —Overhaul—stub axle | 5 |
| | Axle shaft | 5 |
| | Constant velocity joint | 6 |
| | Swivel assembly | 7 |

87

# GENERAL SPECIFICATION DATA 04

## TRANSMISSION
**Main gearbox**
Model .................................................. LT77 (manual)
Type .................................................. Five-speed, single helical constant mesh with synchromesh on all forward gears

Ratios:
| | |
|---|---|
| 5th | 0.770 |
| 4th | 1.000 |
| 3rd | 1.397 |
| 2nd | 2.132 |
| 1st | 3.321 |
| Reverse | 3.429 |

**Transfer gearbox**
Model .................................................. LT230T
Type .................................................. Two-speed reduction on main gearbox output. Front and rear drive permanently engaged via a lockable differential

Ratios:
| | |
|---|---|
| High | 1.192 :1 |
| Low | 3.320 :1 |

Overall ratio (including final drive):
| | High Transfer | Low Transfer |
|---|---|---|
| 5th | 3.25 :1 | 9.05 :1 |
| 4th | 4.22 :1 | 11.75 :1 |
| 3rd | 5.89 :1 | 16.41 :1 |
| 2nd | 8.99 :1 | 25.04 :1 |
| 1st | 14.01 :1 | 39.02 :1 |
| Reverse | 14.46 :1 | 40.27 :1 |

## TRANSMISSION
**Main gearbox**
Model .................................................. ZF 4HP 22 (automatic)
Type .................................................. Automatic four-speed and reverse epicyclic gearbox with fluid torque converter and lock-up

Ratio:
| | |
|---|---|
| 4th | 0.728 :1 |
| 3rd | 1.000 :1 |
| 2nd | 1.480 :1 |
| 1st | 2.480 :1 |
| Reverse | 2.086 :1 |

**Transfer gearbox**
Model .................................................. LT230T
Type .................................................. Two-speed reduction on main gearbox output. Front and rear drive permanently engaged via a lockable differential

Ratios:
| | |
|---|---|
| High | 1.192 :1 |
| Low | 3.320 :1 |

Overall ratio (including final drive):
| | High Transfer | Low Transfer |
|---|---|---|
| 4th | 3.08 :1 | 8.553 :1 |
| 3rd | 4.218 :1 | 11.747 :1 |
| 2nd | 6.240 :1 | 17.380 :1 |
| 1st | 10.458 :1 | 29.127 :1 |
| Reverse | 8.797 :1 | 24.501 :1 |

---

## ABBREVIATIONS AND SYMBOLS USED IN THIS MANUAL

|  |  |
|---|---|
| Across flats (bolt size) | AF |
| After bottom dead centre | ABDC |
| After top dead centre | ATDC |
| Alternating current | a.c. |
| Ampere | amp |
| Ampere-hour | amp hr |
| Atmospheres | Atm |
| Before bottom dead centre | BBDC |
| Before top dead centre | BTDC |
| Bottom dead centre | BDC |
| Brake mean effective pressure | BMEP |
| Brake horse power | bhp |
| British Standards | BS |
| Carbon monoxide | CO |
| Centimetre | cm |
| Centigrade (Celsius) | C |
| Cubic centimetre | cm³ |
| Cubic inch | in³ |
| Degree (angle) | deg or ° |
| Degree (temperature) | deg or ° |
| Diameter | dia. |
| Direct current | d.c. |
| Fahrenheit | F |
| Feet | ft |
| Feet per minute | ft/min |
| Fifth | 5th |
| Figure (illustration) | Fig. |
| First | 1st |
| Fourth | 4th |
| Gramme (force) | gf |
| Gramme (mass) | g |
| Gallons | gal |
| Gallons (US) | US gal |
| High compression | h.c. |
| High tension (electrical) | H.T. |
| Hundredweight | cwt |
| Independent front suspension | i.f.s. |
| Internal diameter | i.dia. |
| Inches of mercury | in.Hg |
| Inches | in |
| Kilogramme (force) | kgf |
| Kilogramme (mass) | kg |
| Kilogramme centimetre (torque) | kgf.cm |
| Kilogramme per square centimetre | kg/cm² |
| Kilogramme metres (torque) | kgf.m |
| Kilometres | km |
| Kilometres per hour | km/h |
| Kilovolts | kV |
| King pin inclination | k.p.i. |
| Left-hand steering | LHStg |
| Left-hand thread | LHThd |
| Litres | litre |
| Low compression | l.c. |
| Low tension | l.t. |
| Maximum | max. |
| Metre | m |
| Microfarad | mfd |
| Midget edison screw | MES |
| Millimetre | mm |
| Miles per gallon | mpg |
| Miles per hour | mph |
| Minimum | min |
| minute (angle) | ' |
| Minus (of tolerance) | — |
| Negative (electrical) | — |
| Number | No. |
| Ohms | ohm |
| Ounces (force) | ozf |
| Ounces (mass) | oz |
| Ounce inch (torque) | ozf.in. |
| Outside diameter | o.dia. |
| Paragraphs | para. |
| Part number | Part No. |
| Percentage | % |
| Pints | pt |
| Pints (US) | US pt |
| Plus (tolerance) | + |
| Positive (electrical) | + |
| Pound (force) | lbf |
| Pounds feet (torque) | lbf.ft. |
| Pounds inches (torque) | lbf.in. |
| Pound (mass) | lb |
| Pounds per square inch | lb/in² |
| Radius | r |
| Rate (frequency) | c/min |
| Ratio | : |
| Reference | ref. |
| Revolution per minute | rev/min |
| Right-hand | RH |
| Right-hand steering | RHStg |
| Second (angle) | " |
| Second (numerical order) | 2nd |
| Single carburetter | SC |
| Specific gravity | sp.gr. |
| Square centimetres | cm² |
| Square inches | in² |
| Standard | std. |
| Standard wire gauge | s.w.g. |
| Synchroniser/synchromesh | synchro. |
| Third | 3rd |
| Top dead centre | TDC |
| Twin carburetters | TC |
| United Kingdom | UK |
| Vehicle Identification Number | VIN |
| Volts | V |
| Watts | W |

**SCREW THREADS**
| | |
|---|---|
| American Standard Taper Pipe | NPTF |
| British Association | BA |
| British Standard Fine | BSF |
| British Standard Pipe | BSP |
| British Standard Whitworth | Whit. |
| Unified Coarse | UNC |
| Unified Fine | UNF |

# GENERAL SPECIFICATION DATA

## SHIFT SPEED SPECIFICATION
### Automatic ZF 4HP22 Gearbox

| OPERATION | SELECTOR POSITION | VEHICLE SPEED APPROX. MPH | VEHICLE SPEED APPROX. KPH | ENGINE SPEED APPROX. (RPM) |
|---|---|---|---|---|
| | | **KICK DOWN** | | |
| KD1-2 | D | 35–40 | 56–64 | 4600–5250 |
| KD2-3 | D | 61–67 | 97–107 | 4750–5200 |
| KD4-2 | D | 34–63 | 54–101 | |
| KD4-1 | D | 16–34 | 25–54 | |
| KD3-2 | 3 | 57–64 | 91–102 | |
| | | **FULL THROTTLE** | | |
| FT1-2 | D | 27–33 | 43–53 | 3600–4300 |
| FT2-3 | D | 55–61 | 88–98 | 4350–4800 |
| FT3-4 | D | 75–82 | 120–131 | 3980–4330 |
| FT4-3 | D | 61–68 | 98–109 | |
| FT3-2 | 3 | 40–47 | 64–75 | |
| | | **ZERO THROTTLE** | | |
| ZT4-3 | D | 19–26 | 30–42 | |
| ZT3-2 | D | 12–16 | 19–25 | |
| ZT2-1 | D | 6–7 | 10–12 | |
| | | **PART THROTTLE** | | |
| PT4-3 | D | 47–55 | 75–88 | |
| PT3-2 | D | 29–37 | 46–59 | |
| PT2-1 | D | 10–12 | 16–25 | |
| | | **LIGHT THROTTLE** | | |
| LT1-2 | D | 8–10 | 13–16 | 1180–1220 |
| LT2-3 | D | 18–23 | 29–37 | 1420–1820 |
| LT3-4 | D | 27–31 | 43–50 | 1430–1650 |
| Lock Up | IN | 40–44 | 64–70 | 1480–1620 |
| | OUT | 39–42 | 62–67 | 1430–1560 |

**PROPELLER SHAFTS**
Type ........ Open type 50.8 mm (2 in) diameter
Universal joints ........ 03EHD standard shafts

**REAR AXLE**
Type ........ Spiral bevel, fully floating shafts
Ratio ........ 3.54 :1

**FRONT AXLE**
Type ........ Spiral bevel enclosed constant velocity joint
Angularity of joint on full lock ........ 32°
Ratio ........ 3.54 :1

# TORQUE WRENCH SETTINGS

## TORQUE WRENCH SETTINGS – TRANSFER GEARBOX LT230T

| COMPONENT | DESCRIPTION | QUANTITY | Nm | lbf ft |
|---|---|---|---|---|
| Pinch bolt, operating arm | 6 × 25,0 mm bolt | 1 | 7 to 10 | 5 to 7 |
| Gate plate to grommet plate | 6 × 20,0 mm screw | 4 | 7 to 10 | 5 to 7 |
| End cover | 6 × 20,0 mm screw | 2 | 7 to 10 | 5 to 7 |
| Speedometer cable retainer | 6 mm nut | 1 | 7 to 10 | 5 to 7 |
| Rear output/speedometer housing | 6 × 30,0 mm stud | 1 | See note | |
| Locating plate to gear change housing | 5 mm self lock nut | 2 | 5 to 7 | 4 to 5 |
| Bottom cover to transfer case | 8 × 30,0 mm bolt | 10 | 22 to 28 | 16 to 21 |
| Front output housing to transfer case | 8 × 30,0 mm bolt | 7 | 22 to 28 | 16 to 21 |
| Front output housing to transfer case | 8 × 90,0 mm bolt | 1 | 22 to 28 | 16 to 21 |
| Cross shaft housing to front output housing | 8 × 55,0 mm bolt | 6 | 22 to 28 | 16 to 21 |
| Gear change housing | 8 × 55,0 mm bolt | 2 | 22 to 28 | 16 to 21 |
| Pivot shaft | 8 mm nut | 1 | 22 to 28 | 16 to 21 |
| Connecting rod | 8 mm nut | 2 | 22 to 28 | 16 to 21 |
| Anti-rotation plate intermediate shaft | 8 × 20,0 mm screw | 1 | 22 to 28 | 16 to 21 |
| Front output housing cover | 8 × 25,0 mm screw | 7 | 22 to 28 | 16 to 21 |
| Gear change housing | 8 × 25,0 mm screw | 2 | 22 to 28 | 16 to 21 |
| Bracket to extension housing | 8 × 25,0 mm screw | 2 | 22 to 28 | 16 to 21 |
| Finger housing to front output housing | 8 × 25,0 mm screw | 3 | 22 to 28 | 16 to 21 |
| Mainshaft bearing housing | 8 × 25,0 mm screw | 2 | 22 to 28 | 16 to 21 |
| Brake drum | 8 × 20,0 mm screw | 2 | 22 to 28 | 16 to 21 |
| Gearbox to transfer box | 10 × 40,0 mm bolt | 3 | 40 to 50 | 29 to 37 |
| Gearbox to transfer box | 10 × 45,0 mm bolt | 1 | 40 to 50 | 29 to 37 |
| Bearing housing to transfer gearbox | 10 × 35,0 mm bolt | 6 | 40 to 50 | 29 to 37 |
| Speedometer housing to transfer gearbox | 10 × 30,0 mm screw | 5 | 40 to 50 | 29 to 37 |
| Speedometer housing to transfer gearbox | 10 × 45 mm screw | 1 | 40 to 50 | 29 to 37 |
| Selector finger to cross shaft high/low | 10 mm grub screw | 1 | 22 to 28 | 16 to 21 |
| Selector fork, high/low to shaft | 10 mm grub screw | 1 | 22 to 28 | 16 to 21 |
| Transmission brake | 10 × 25,0 mm bolt | 4 | 65 to 80 | 48 to 59 |
| Intermediate shaft stake nut | 20,0 mm nut | 1 | 130 to 140 | 96 to 104 |
| Gate plate to grommet plate (auto only) | 6 × 20,0 mm screw | 4 | 7 to 10 | 5 to 7 |
| Gate plate to gear change housing (manual versions only) | 6 × 16,0 mm countersunk screw | 2 | 7 to 10 | 5 to 7 |

*Continued*

# TORQUE WRENCH SETTINGS

## TORQUE WRENCH SETTINGS (Continued)

| | | Nm | lbf ft. |
|---|---|---|---|
| Gearbox to transfer case | 10 mm nut | 2 | 2 |
| Gearbox to transfer case | 10 mm studs | 2 | 2 |
| Oil drain plug | 12 × 14 mm hexagon head | 40 to 50 | 29 to 37 |
| Differential case | 10 × 60 mm bolt | See note | |
| Output flanges | 20 mm self locking nut | 25 to 35 | 19 to 26 |
| Differential case rear | 50 mm nut | 55 to 64 | 40 to 47 |
| Link arm and cross shaft | ¼ inch UNF | 146 to 179 | 108 to 132 |
| lever to ball joint | self locking nut | 66 to 80 | 50 to 60 |
| Oil filler/level plug | ¾ inch taper thread | 8 to 12 | 6 to 9 |
| Transfer breather | ⅛ inch B.S.P. | 25 to 35 | 19 to 26 |
| | | 14 to 16 | 10 to 12 |

**NOTE:** Studs to be assembled into casings with sufficient torque to wind them fully home, but this torque must not exceed the maximum figure quoted for the associated nut on final assembly.

## ZF 4HP22 AUTOMATIC GEARBOX

| | Nm | lbf/ft. |
|---|---|---|
| Coupling shaft to mainshaft | 36 to 48 | 26 to 34* |
| Filler tube to sump | 35 to 42 | 25 to 30 |
| Gear change lever to gearbox | 22 to 28 | 16 to 21 |
| Cooler pipe adaptor to gearbox | 36 to 48 | 26 to 34 |
| Securing screws—clutch F. | 10 | 7 |
| Securing screw—parking pawl | 10 | 7 |
| Securing screws—pump | 50 | 37 |
| Intermediate plate plugs (M20) | 40 | 29 |
| Intermediate plate plugs (M14) | 46 | 34 |
| Bell housing mounting bolts | 10 | 7 |
| Governor mounting screws | 23 | 17 |
| Extension housing bolts | 8 | 6 |
| Control unit mounting bolts | 10 | 7 |
| Sump plug | 8 | 6 |
| Mounting screws for sump | 35 to 42 | 25 to 30* |
| Drive plate to converter | 36 to 48 | 26 to 34 |
| Gearbox to engine | 36 to 48 | 26 to 34 |
| Strut (nyloc nut end) | 7 to 10 | 5 to 7 |
| Bottom cover to converter housing | 7 to 10 | 5 to 7 |
| Cover—converter housing | 35 to 46 | 25 to 33* |
| Drive plates to crankshaft adaptor | 77 to 90 | 55 to 65 |
| Adaptor to crankshaft | | |

**NOTE:** * These bolts must have threads coated with Loctite 270 prior to assembly.

## MAIN GEARBOX (FIVE-SPEED)

| | Nm | lbf ft. |
|---|---|---|
| Bottom cover to clutch housing | 7 to 10 | 5 to 7 |
| Oil pump body to extension case | 7 to 10 | 5 to 7 |
| Clip to clutch release lever | 7 to 10 | 5 to 7 |
| Attachment plate to gearcase | 22 to 28 | 16 to 21 |
| Extension case to gearcase | 22 to 28 | 16 to 21 |
| Pivot—clutch lever to bell housing | 22 to 28 | 16 to 21 |
| Guide clutch release sleeve | 22 to 28 | 16 to 21 |
| Slave cylinder to clutch housing | 65 to 80 | 48 to 59 |
| Front cover to gearcase | 22 to 28 | 16 to 21 |
| 5th support bracket | 40 to 47 | 30 to 35 |
| Clutch housing to gearbox | 65 to 80 | 48 to 59 |
| Plug—detent spring | 14 to 16 | 10 to 12 |
| Oil drain plug | 25 to 35 | 19 to 26 |
| Oil filter plug | 22 to 28 | 16 to 21 |
| Breather | 204 to 231 | 150 to 170 |
| Oil level plug | 7 to 10 | 5 to 7 |
| Upper gear lever to lower gear lever | 22 to 28 | 16 to 21 |
| 5th gear retaining nut | 22 to 28 | 16 to 21 |
| Attachment plate to gear change housing | 22 to 28 | 16 to 21 |
| Gear change housing to extension case | 7 to 10 | 5 to 7 |
| Plunger housing to gear change housing | 36 to 45 | 27 to 33 |
| Adjustment plate to gearchange housing | 22 to 28 | 16 to 21 |
| Cover to gear change housing | 22 to 28 | 16 to 21 |
| Bell housing to cylinder block bolts | 22 to 28 | 16 to 21 |
| Pivot plate to bell housing | | |
| Yoke to selector shaft | | |
| Locknut—plus reverse knockout | | |

## FRONT AXLE

| | Nm | lbf ft. |
|---|---|---|
| Hub driving shaft to hub | 41 to 52 | 30 to 38 |
| Brake disc to hub | 65 to 80 | 48 to 59 |
| Stub axle to swivel pin housing | 60 to 70 | 44 to 52* |
| Brake caliper to swivel pin housing | 75 to 88 | 55 to 65 |
| Upper swivel pin to swivel pin housing | 68 to 88 | 50 to 65* |
| Lower swivel pin to swivel pin housing | 68 to 88 | 50 to 65* |
| Oil seal retainer to swivel pin housing | 9 to 12 | 7 to 9 |
| Swivel bearing housing to axle case | 65 to 80 | 48 to 59* |
| Pinion housing to axle case | 36 to 46 | 26 to 34 |
| Crown wheel to differential housing | 55 to 61 | 40 to 45 |
| Differential bearing cap to pinion housing | 80 to 100 | 59 to 74 |
| Differential drive flange to propeller shaft | 41 to 52 | 30 to 38 |
| Mudshield to bracket lower swivel pin | 9 to 12 | 7 to 9 |
| Bevel pinion nut | 95 to 163 | 70 to 120 |
| Draglink to hub arm | 40 | 30 |
| Panhard rod to axle bracket | 88 | 65 |
| Radius arm to axle | 190 | 140 |
| Radius arm to chassis side member | 190 | 140 |

* These bolts to be coated with Loctite 270 prior to assembly.

## LT77 FIVE SPEED GEARBOX — 37

### LT77 FIVE SPEED GEARBOX AND LT230T TRANSFER BOX

1. Install the vehicle on a ramp and disconnect the battery.
2. Remove both main and transfer gear lever knobs.
3. Remove the cubby box liner, disconnect the window lift switches and release the handbrake inner cable from the handbrake lever. Release the console fixings and withdraw the console from the vehicle.
4. Remove the noise insulation pad to gain access to the gaiter. Release the four screws and drill out all the pop-rivets securing the gaiter to the top of the gearbox.
5. Disconnect the reverse light electrical leads and pull the leads through the tunnel aperture.
6. Remove the lower lock nut from the high/low operating rod.
7. Remove the spring clip from the clevis pin securing the differential lock lever to the high/low gear change housing.
8. Remove the two pan head screws securing the gear change housing top cover. Raise the cover to give access to one of the gear change housing to extension case securing bolts located by the reverse plunger assembly, and remove the bolt.
9. Remove the three remaining gear change housing to extension case securing bolts, remove the housing complete with transfer gear housing and gaiter.
10. Remove the nut securing the outer handbrake cable to the top of the tunnel and feed the inner and outer cable through its aperture to the underside of the vehicle.
11. Remove the nuts and bolts securing the fan cowl to the radiator. Move the cowling away from the radiator and hang it loosely around the fan blade assembly.
12. Release the transmission breather pipes, speedometer cable, and starter motor harness from the clips at the rear of the engine.
13. **Fuel injection models**—Disconnect the airflow meter to plenum chamber air intake hose.
14. Raise the ramp.
15. Remove the eight nuts and bolts securing the chassis cross member and using a suitable means of spreading the chassis, remove the cross member.
16. Place a suitable container under the transmission, remove the three drain plugs, allow the oil to drain and refit the plugs. Clean filter on the extension housing plug before refitting.
17. Remove the intermediate exhaust pipe and silencer section as follows:
    (a) Release the connection to the front pipe assembly at the flanges.
    (b) Release the connection to the rear section at the flange immediately behind the silencer.
    (c) Release the "U" bolt retaining the pipe to the bracket attached to the transfer box.
18. Mark the drive flanges for reassembly and disconnect the front propeller shaft from the transfer box.
19. Similarly, disconnect the rear propeller shaft.
20. Disconnect the speedometer cable from the rear of the transfer box.
21. Remove the two bolts and withdraw the clutch slave cylinder from the bell housing.

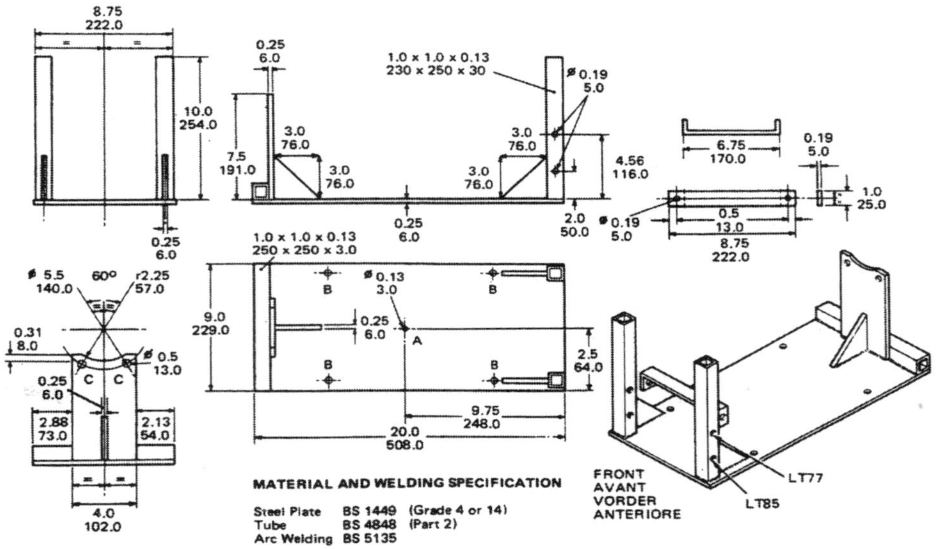

---

## 06 — TORQUE WRENCH SETTINGS

### REAR AXLE

| | Nm | lbf ft |
|---|---|---|
| Axle shaft to hub | 41 to 52 | 30 to 38 |
| Brake disc to hub | 65 to 80 | 48 to 59 |
| Stub axle rear to axle case | 60 to 70 | 44 to 52 |
| Brake caliper to axle case | 75 to 88 | 55 to 65 |
| Pinion housing to axle case | 36 to 46 | 26 to 34 |
| Crown wheel to differential case | 55 to 61 | 40 to 45 |
| Differential bearing cap to pinion housing | 80 to 100 | 59 to 74 |
| Differential drive flange to propeller shaft | 41 to 52 | 30 to 38 |
| Mudshield to axle case | 9 to 12 | 7 to 9 |
| Bevel pinion nut | 95 to 163 | 70 to 120 |
| Lower link to axle | 176 | 130 |
| Pivot bracket ball joint to axle | 176 | 130 |

### CHARTS BELOW GIVE TORQUE SETTINGS FOR ALL SCREWS AND BOLTS USED EXCEPT FOR THOSE THAT ARE SPECIFIED OTHERWISE

| SIZE | METRIC Nm | METRIC lbf ft | SIZE | UNC Nm | UNC lbf ft | UNF Nm | UNF lbf ft |
|---|---|---|---|---|---|---|---|
| M5 | 5–7 | 3,7–5,2 | ¼ | 6,8–9,5 | 5–7 | 8,1–12,2 | 6–9 |
| M6 | 7–10 | 5,2–7,4 | ⁵⁄₁₆ | 20,3–27,1 | 15–20 | 20,3–27,1 | 15–20 |
| M8 | 22–28 | 16,2–20,7 | ⅜ | 35,2–43,4 | 26–32 | 35,2–43,4 | 26–32 |
| M10 | 40–50 | 29,5–36,9 | ⁷⁄₁₆ | 67,8–88,1 | 50–65 | 67,8–88,1 | 50–65 |
| M12 | 80–100 | 59,0–73,8 | ½ | 81,3–101,7 | 60–75 | 81,3–101,7 | 60–75 |
| M14 | 90–120 | 66,4–88,5 | ⅝ | 122,0–149,1 | 90–110 | 122,0–149,1 | 90–110 |
| M16 | 160–200 | 118,0–147,5 | | | | | |

---

**MATERIAL AND WELDING SPECIFICATION**

Steel Plate  BS 1449  (Grade 4 or 14)
Tube  BS 4848
Arc Welding  BS 5135  (Part 2)

ST 538M

FRONT
AVANT
VORDER
ANTERIORE

LT77
LT85

# LT77 FIVE SPEED GEARBOX

22. Manufacture a cradle to the dimensions given in the drawing and attach it to a transmission hoist. To achieve balance of the transmission unit when mounted on the transmission hoist, it is essential that point A is situated over the centre of the lifting hoist ram. Drill fixing holes B to suit hoist table. Secure the transmission unit to the lifting bracket at point C, by means of the lower bolts retaining the transfer gearbox rear cover.
23. Remove the bottom two bolts from the transfer box rear cover and use them to attach to the rear end of the cradle to the transfer box. Ensure that the tube in the centre of the cradle locates over the extension housing drain plug.
24. Raise the hoist just enough to take the weight of the transmission.
25. Remove the three nuts and bolts securing the transfer box LH and RH mounting brackets to the chassis.
26. Remove the nuts retaining the brackets to the mounting rubbers and remove the brackets.
27. Lower the hoist sufficiently to allow the differential lock electrical leads to be disconnected from the transfer-box.
28. Support the engine under the sump with a jack, placing timber between the jack pad and sump.
29. Remove the bolts from the bell housing.
30. Withdraw the transmission whilst ensuring all connections and chassis are released.

### Separating the transfer box from gearbox

31. Remove the transmission assembly from the hoist and cradle and install it safely on a bench.
32. Remove the clip and withdraw the clevis pin securing the cross-over lever at the pivot point. Release the breather pipes from the clips at the pivot point. NOTE: note the position of the breather pipes for re-assembly.
33. Disconnect the connecting link from the differential lock lever.
34. Place a sling round the transfer box and attach to a hoist.
35. Remove the two nuts and four bolts retaining the transfer box to the extension housing and withdraw the transfer box.

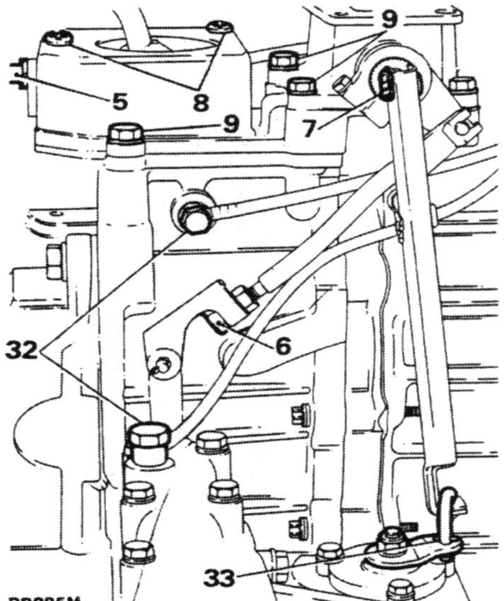

### Assembling transfer box to main gearbox

36. Stand the gearbox vertically on the bell housing face on two pieces of wood to prevent damage to the primary pinnion which protrudes beyond the bell housing face. Lower the transfer gearbox onto the main gearbox, care should be taken to prevent any damage to seals. Secure the transfer gearbox to the main gearbox and tighten all bolts to the specified torque.
37. Refit the differential lock connecting link.
38. Refit the breather pipes and crossover lever.

### Fitting main gearbox and transfer box to engine

39. Fit the cradle to the transmission hoist and the transmission to the cradle as described in instruction 23. Smear Hylomar on bell housing face mating with engine.
40. Locate the gear change housing temporarily on the gearbox and select any gear in the main gearbox to facilitate entry of the primary shaft, after selection remove the housing from the gearbox.
41. Position and raise the hoist to line up with the engine, feed the handbrake cable through the aperture in the tunnel, ensure that any pipes or electrical leads do not become trapped.
42. Secure the engine with seven of the eight bell housing bolts noting that the third bolt up on the left hand side is fitted with a harness clip.
43. Fit the differential lock indicator electrical leads to the transfer box.
44. Check that there are no obstructions and raise the transmission hoist until the gearbox mounting points are in line with the chassis fixing points.
45. Fit the transfer box LH and RH mounting brackets but only partially tighten the securing nuts and bolts.
46. Loosely fit the rubber mounting nuts and lower the transmission onto the mountings. Fully tighten all the securing nuts and bolts.
47. Remove the supporting jack from under the engine sump.
48. Remove the two bolts securing the cradle to the transfer box and remove the cradle and hoist.
49. Refit the two bolts using Loctite 290 on the threads and note that the LH bolt holds a clip for the speedometer cable.
50. Fit the slave cylinder using Hylosil on the gasket and tighten the two bolts evenly to the specified torque.
51. Connect the speedometer cable.
52. Check that the three drain plugs are tight and remove the main gearbox and transfer box filler level plugs. Fill the main gearbox with approximately 1,76 litres (3 pints) of a recommended oil or until it begins to run out of the filler level hole. Fit and tighten the filler plug. Similarly remove the transfer box filler level plug and inject approximately 2,6 litres (4,5 pints) of recommended oil or until it runs out of the filler hole. Apply Hylosil to the threads and fit the plug and wipe away any surplus oil.
53. Line up the marks and fit the front and rear propeller shafts to the transfer box.
54. Fit the exhaust system, and evenly tighten the flange nuts and bolts. Fit the 'U' bolt and secure to the bracket.
55. Fit the bottom cover on the front of the gearbox and tighten the bolts to the specified torque.
56. Expand the chassis and fit the cross-member and secure in position with the eight bolts, lower the ramp.
57. Fit a NEW gasket to the top of the gearbox and refit the gear change housing. Coat the securing bolts with Hylomar PL32 or Loctite 290 and fit the bolts, tighten to the specified torque.
58. Reconnect the high/low operating rod and secure in position with the single lock nut.
59. Reconnect the differential lock lever to the transfer lever selector shaft, fit a NEW spring clip.
60. Refit the gear change housing top cover, apply Hylomar PL32 or loctite 290 to the forward screw tighten screws to the specified torque.
61. Re-connect the electrical leads to the reverse light switch.
62. Secure the gaiter in position using self tapping screws or pop rivets and position the insulation pad.
63. Refit the centre console, reconnect the handbrake cable using a NEW split pin. Secure the outer cable to the top of the tunnel.
64. Connect the electrical plugs to the window lift switches and refit the cubby box liner.
65. Refit the main and transfer gearbox knobs.
66. Refit the radiator cowl and reconnect the battery.
67. Re-clip the transmission breather pipes, speedometer cable and starter motor harness to the rear of the engine.
68. **Fuel Injection models**—Re-connect the plenum chamber to airflow meter air intake hose.

## OVERHAUL LT77 FIVE SPEED GEARBOX

**Service Tools:**

18G705—Bearing remover
18G705-1A—Adaptor for mainshaft
18G705-5—Adaptor for layshaft
18G1400—Remover for synchromesh hub and gear cluster
18G1400-1—Adaptor mainshaft fifth gear
MS47—Hand press
18G47BA—Adaptor for layshaft bearing remover
18G47BAX—Conversion kit
18G284—Impulse extractor
18G284AAH—Adaptor for input shaft pilot bearing track
18G1422—Mainshaft rear oil seal replacer
18G1431—Mainshaft fifth gear and oil seal collar replacer

### Dismantle

1. Place gearbox on a bench with the transfer gearbox removed, ensuring the oil is first drained.
2. Thoroughly clean the exterior of the gearbox case.
3. Remove the clutch release bearing carrier clip.
4. Remove the clutch release bearing.
5. Release the single bolt and remove the spring clip from the clutch release lever. Pull the lever of the clutch release pivot.
6. Remove the clutch release pivot and single bolt retaining the clutch release bearing guide. Withdraw the guide from the input shaft.
7. Remove the six bolts and washers securing the bell housing.

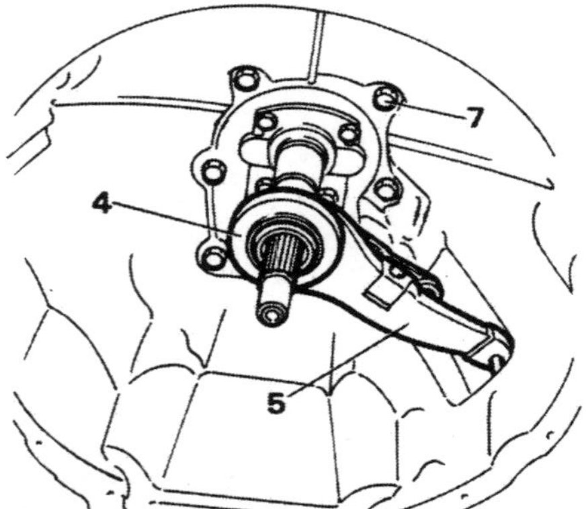

### GEAR HOUSING

8. Carefully ease the bell housing off the dowels and withdraw it from the gearbox.
9. Remove the circlip which retains the mainshaft oil seal collar located at the rear of the gearbox.
10. Using tools 18G705 and 18G705-1A remove the oil seal collar. 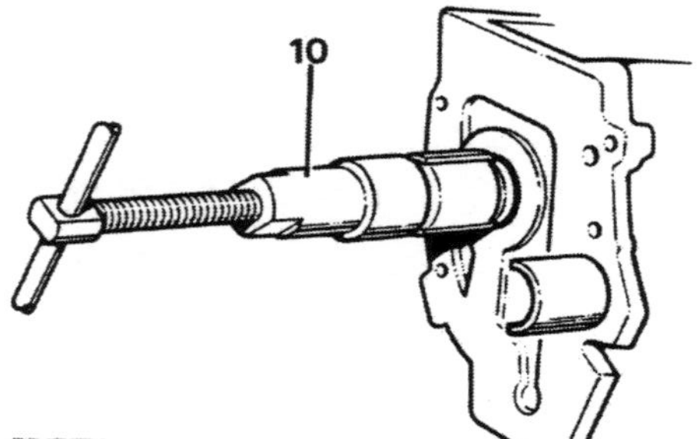

*Continued*

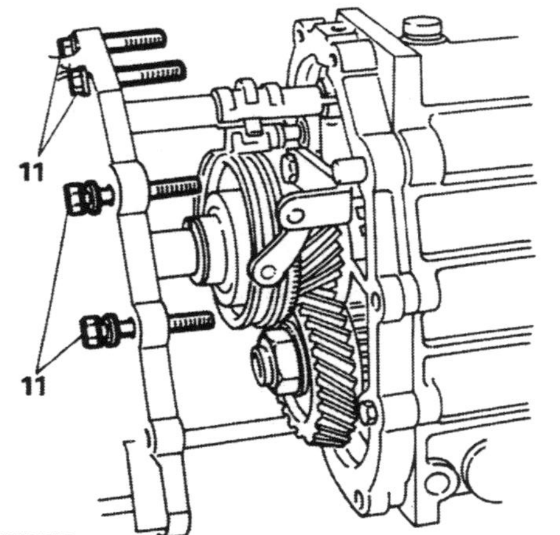

## LT77 FIVE SPEED GEARBOX

11. Remove the ten bolts and spring washers securing the fifth gear extension case to the gearcase; withdraw the extension case and discard the gasket.
12. Fit two dummy bolts (8 × 35 mm) to the casing to retain the centre plate to the main case.
13. Select first or third gear and remove the grub screw securing the selector yoke to the selector shaft, manoeuvre the yoke and withdraw it from the shaft.
14. Remove the oil seal collar 'O' ring from the mainshaft.
15. Withdraw the oil pump drive shaft.
16. Remove the two circlips retaining the 5th selector fork to its bracket and remove the fork and pads.
17. Withdraw the fifth gear selector spool.
18. Remove the two bolts and spring washers and withdraw the fifth gear selector fork and bracket.
19. Engage reverse gear by turning selector rail anti-clockwise and pulling rearwards. Move the fifth speed synchro hub into mesh with the fifth gear. De-stake the retaining nut securing the fifth gear layshaft and remove nut. Select neutral by pushing selector rail inwards and turning clockwise; and return fifth speed synchro hub to its out of mesh position.
20. Release the circlip retaining the fifth gear synchromesh assembly to the mainshaft.
21. Using tools 18G1400-1 and 18G1400 withdraw the selective washer, fifth gear synchromesh hub and cone, fifth gear (driven) and spacer from the mainshaft.
22. Remove the split roller bearing assembly from the mainshaft.
23. Using tools 18G705 and 18G705-1A remove the layshaft fifth gear.
24. Fit suitable guide studs (measuring 8 × 60 mm) to the main gearbox case.
25. Locate the gearbox to a suitable stand.
26. Remove the six bolts and spring washers from the front cover, withdraw the cover and discard the gasket.
27. Remove the input shaft and layshaft selective washers from the gearcase.
28. Remove the two bolts and washers securing the locating boss for the selector shaft front spool, withdraw the locating boss.
29. Withdraw the selector plug, spring and detent ball from the centre plate.
30. Remove the dummy bolts and carefully lift the gearcase, leaving the centre plate and gear assemblies in position. Discard gasket.
31. Insert two slave bolts and nuts to retain the centre plate to the stand; and remove the circlip, pivot pin, reverse lever and slipper pad.
32. Slide the reverse shaft rearwards and lift off the thrust washer, reverse gear and reverse gear spacer.
33. Lift off the layshaft cluster.
34. Remove the input shaft and fourth gear synchromesh cone.
35. Rotate the fifth gear selector shaft clockwise (viewed from above) to align the fifth gear selector pin with the slot in the centre plate.
36. Remove the mainshaft and selector fork assemblies from the centre plate together.
37. Detach the selector fork assembly from the mainshaft gear cluster.
38. Remove the slave bolts from the centre plate and lift the centre plate clear of the stand.

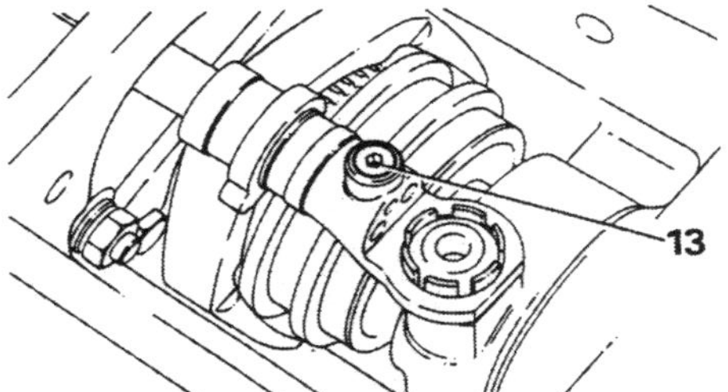

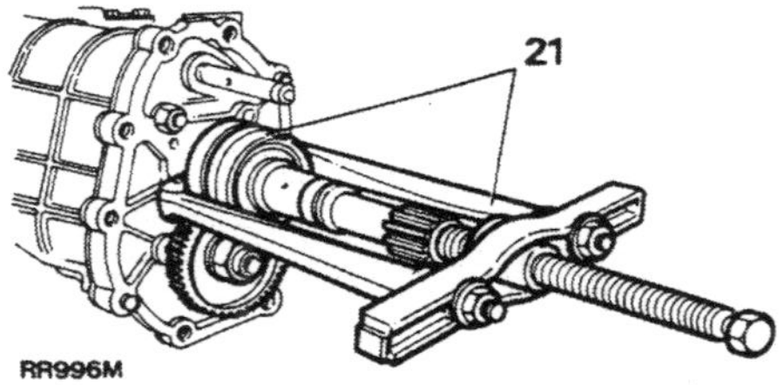

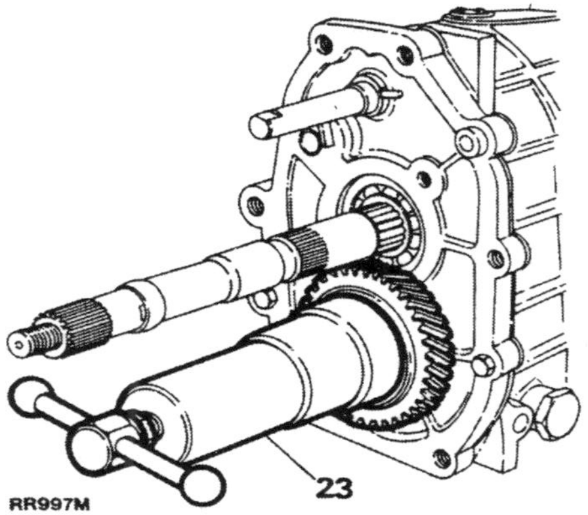

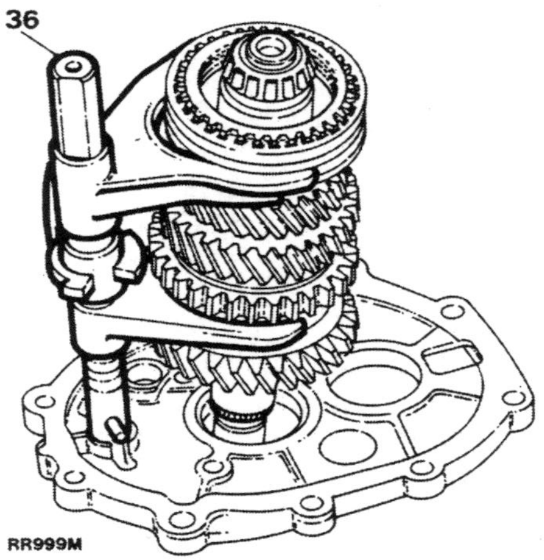

### Front cover

39. Remove and discard the oil seal from the front cover. Do not fit a new oil seal at this stage.

### Layshaft

40. Using press 18G705 and tool 18G705-5 remove the layshaft bearings.

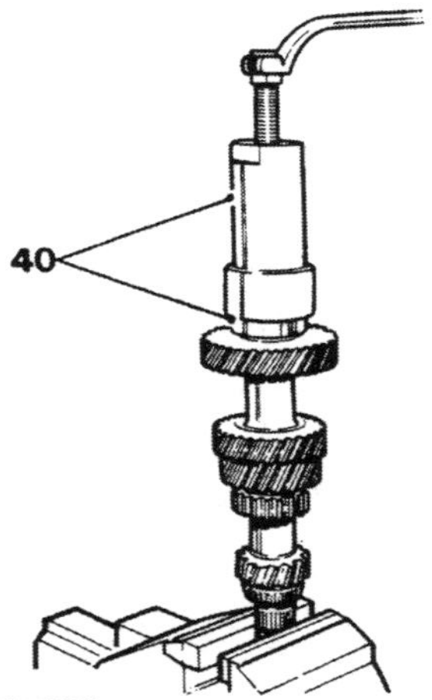

*Continued*

# LT77 FIVE SPEED GEARBOX

## Mainshaft

41. Remove the centre bearing circlip.
42. Using press MS47 and any suitable metal bars, remove the centre bearing, first gear bush, first gear and needle bearings and first gear synchromesh cone.
43. If a difficulty is experienced in removing the first and second gear synchromesh hub, locate underneath the second gear with a suitable tool; and extract the complete synchromesh hub and second gear assemblies using a suitable press.
44. Turn the mainshaft through 180° and using press MS47 and extension, with a support underneath the third speed gear, press the mainshaft through the pilot bearing spacer, third and fourth synchromesh hub, third gear synchromesh cone, third gear and third gear needle roller bearing.

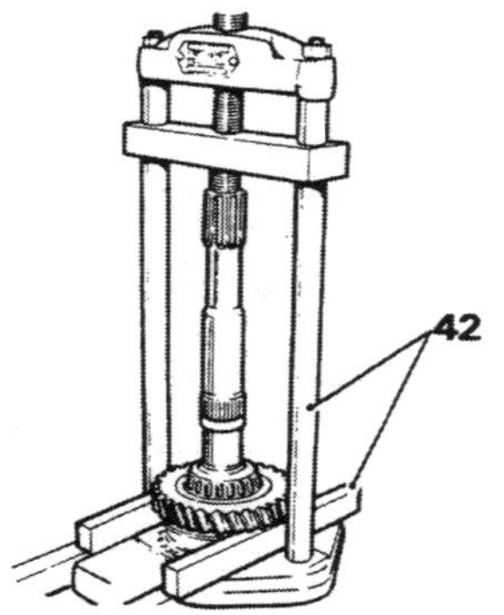

### First and second gear synchromesh assemblies

45. Mark the hub and sleeve to aid reassembly and remove the slipper springs from the front and rear of the first and second gear synchromesh assemblies.
46. Withdraw the slippers and hub from the sleeve.

### Third and fourth gear synchromesh assemblies

47. Mark the hub and sleeve to aid reassembly and remove the slipper springs from the front and rear of the assembly.
48. Withdraw the slippers and hub from the sleeve.

### Extension case

49. Remove the three oil pump housing bolts, spring washers and oil pump gears and housing.
50. Do not withdraw oil pick-up pipe.
51. Remove the plug, washer and filter.
52. Invert casing and extract the oil seal.
53. Press out the ferrobestos bush from the casing.

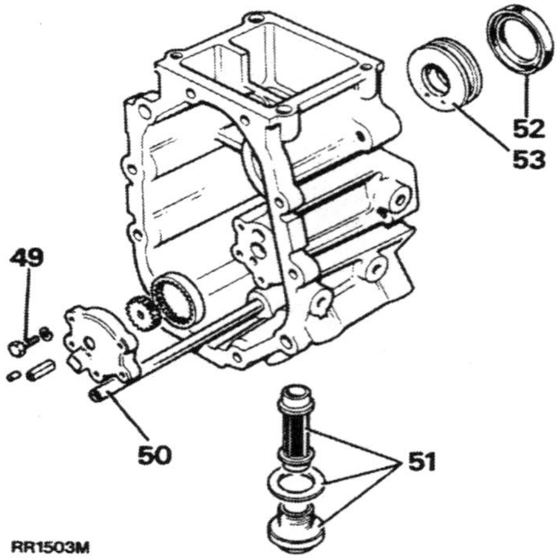

### Input shaft

54. Using tools MS47 and 18G47BA, remove the input shaft bearing.
55. With the aid of tools 18G284AAH and 18G284, extract the pilot bearing track.

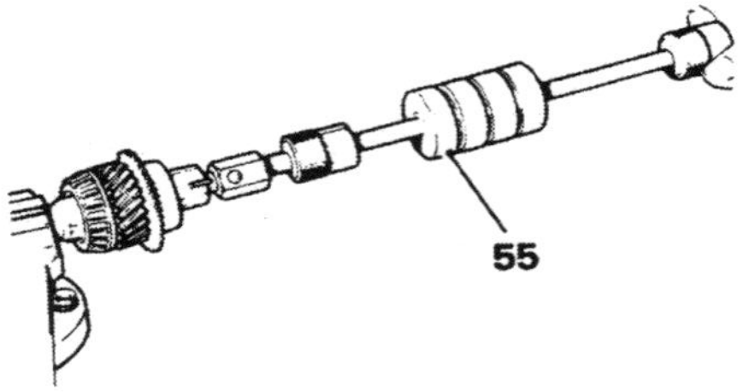

### Reverse idler gear

56. Remove the circlip from the reverse idler gear.
57. Having noted their positions, remove both needle roller bearings and remaining circlip from the gear.

### Fifth gear synchromesh assembly

58. Mark the hub and sleeve to aid reassembly and lever the backing plate off the fifth gear synchromesh assembly.
59. Remove the slipper springs from the front and rear of the assembly.
60. Release the slippers and slide the hub from the sleeve.

### Centre plate

61. Remove the layshaft and mainshaft bearing tracks from the centre plate and reverse pivot post.

### Main gearbox casing

62. Remove the mainshaft and layshaft bearing tracks from the main casing.
63. Remove the plastic oil trough from the front of the casing.

### Selector rail

64. The selector rail is supplied complete with first and second selector fork, pin and fifth speed selector pin. If it is required to replace the first and second selector pin and remove the fifth speed gear selector fork on its own, press out the fifth speed gear selector fork from the selector rail.

### Gear change housing

NOTE: **The upper and lower gear levers are loctited together. If difficulty is experienced in separating these two components, a complete new assembly should be fitted.**

65. Raise the gear lever top cover as far as the gaiter and upper gear lever will permit to gain access to the lower gear lever fixings.
66. Release the single bolt with plain washer securing the reverse plunger to the gear change housing, withdraw the plunger and shims if fitted
67. Remove the two bolts and washers anchoring the bias springs.
68. Release the springs from the register at the gear lever and remove them from the dowels.
69. Remove the remaining two bolts securing the bias adjustment plate to the housing.
70. Remove the lower gear lever and bias plate from the housing.
71. Remove the railko bush from the lower gear lever.
72. Prise the seal from the bottom of the gear change housing.

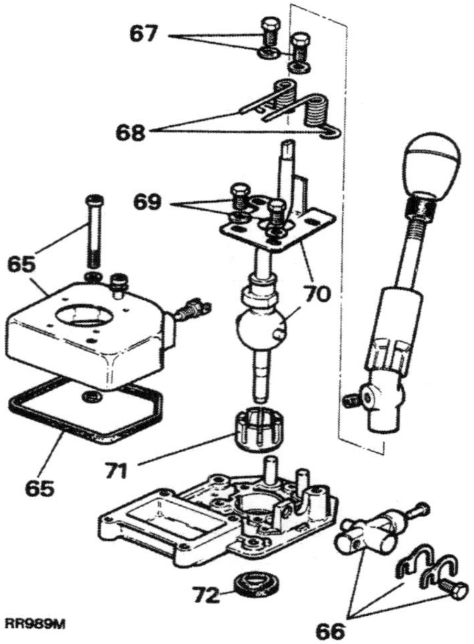

### Reverse gear plunger assembly

NOTE: **The plunger assembly is not a serviceable item. To check that the unit is operating correctly proceed as follows.**

73. Apply a load of between 50 to 60 kg (110 to 130 lb) to the plunger nose. If the plunger is operational within these limits the unit is satisfactory. If the plunger operates outside the limits renew the plunger assembly.

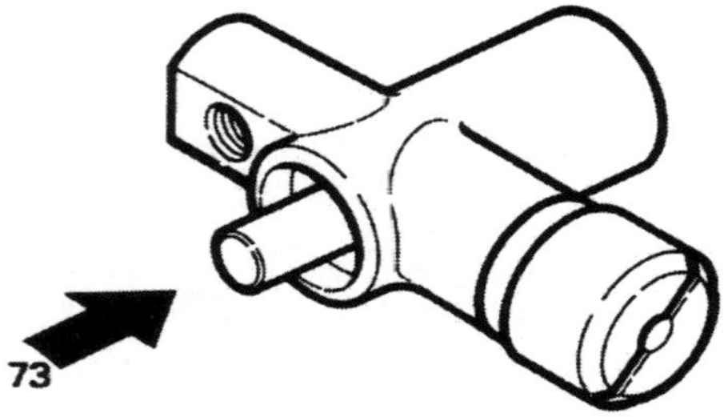

*Continued*

# LT77 FIVE SPEED GEARBOX

1. MAIN GEARCASE
2. CENTRE PLATE
3. EXTENSION CASE
4. BELL HOUSING
5. LAYSHAFT ASSEMBLY
6. REVERSE IDLER ASSEMBLY
7. OIL PUMP ASSEMBLY

## Cleaning and inspection

74. Clean gearcase thoroughly using a suitable solvent. Inspect case for cracks, stripped threads in the various bolt holes, and machined mating surfaces for burrs, nicks or any condition that would render the gearcase unfit for further service. If threads are stripped, install Helicoil, or equivalent inserts.
75. Inspect all gear teeth for chipped or broken teeth, or showing signs of excessive wear. Inspect all spline teeth on the synchromesh assemblies. If there is evidence of chipping or excessive wear, install new parts on reassembly. Check all slippers and slipper springs for wear or breakage. Replace with new parts if necessary.
76. Inspect all circlip grooves for burred edges. If rough or burred, remove condition carefully using a fine file.
77. Ensure all oil outlets are clear of sludge or contamination especially the mainshaft oil ways. Clean with compressed air observing the necessary safety requirements.
78. During the rebuild operation, it is recommended that new roller and needle bearings are fitted.

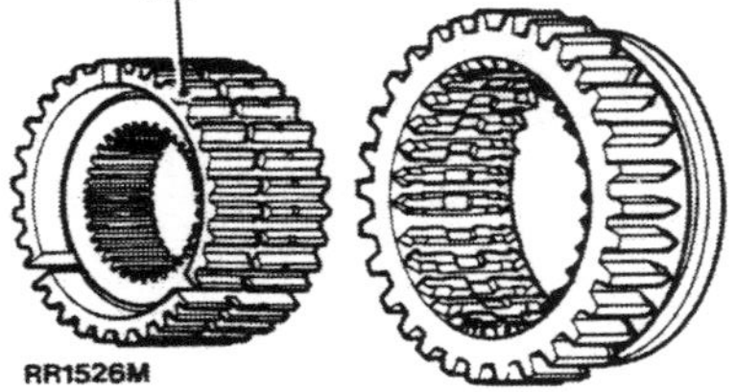

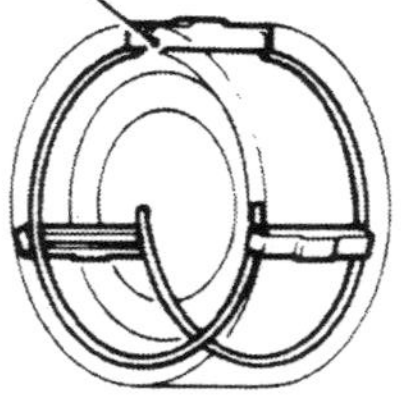

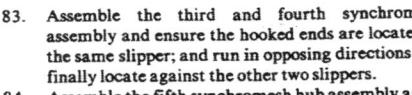

## ASSEMBLY

### Layshaft

79. Using tools MS47 and a suitable tube, fit new bearing cones to the layshaft.

### Synchromesh assemblies

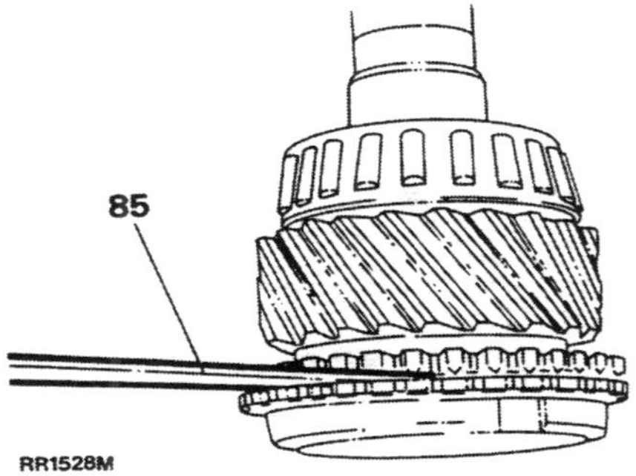

80. With the outer sleeve held, a push-through load applied to the outer face of the synchromesh hub should register 8,2 to 10 kgf m (18 to 22 lbf ft) to overcome the spring detent in either direction.
81. Assemble the first and second synchromesh assembly by locating the shorter splined face towards the second gear.
82. Refit the slippers and locate the slipper springs to each side of the assembly, ensuring that the hooked ends of both slipper springs are located in the same slipper; but running in opposite directions and finishing against the other two slippers.
83. Assemble the third and fourth synchromesh assembly and ensure the hooked ends are located in the same slipper; and run in opposing directions and finally locate against the other two slippers.
84. Assemble the fifth synchromesh hub assembly again ensuring the hooked ends of the springs are located in the same slipper, but running in opposite directions. Fit the backplate onto the rear of the synchromesh hub assembly. Ensure the tag on the backplate locates in the slot on the hub.
85. Check the wear between all the synchromesh cones and gears by pushing the cone against the gear and measuring the gap between the gear and cone. The minimum clearance is 0,64 mm (0.025 in). If this clearance is not met, fit new synchromesh cones.

*Continued*

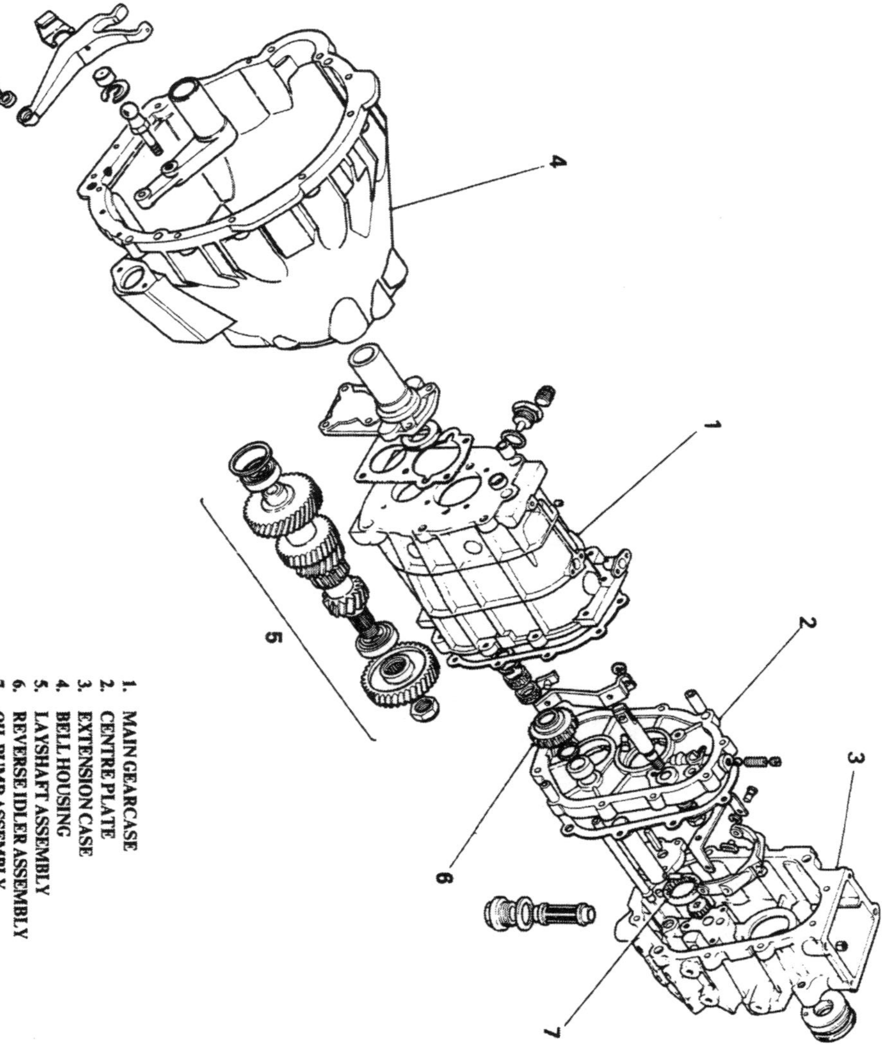

# LT77 FIVE SPEED GEARBOX

11. BIAS PLATE ASSEMBLY
12. UPPER GEAR LEVER ASSEMBLY
13. GEARBOX TOP COVER AND SEAL
14. LOWER GEAR LEVER AND BUSH
15. REVERSE PLUNGER AND SHIMS

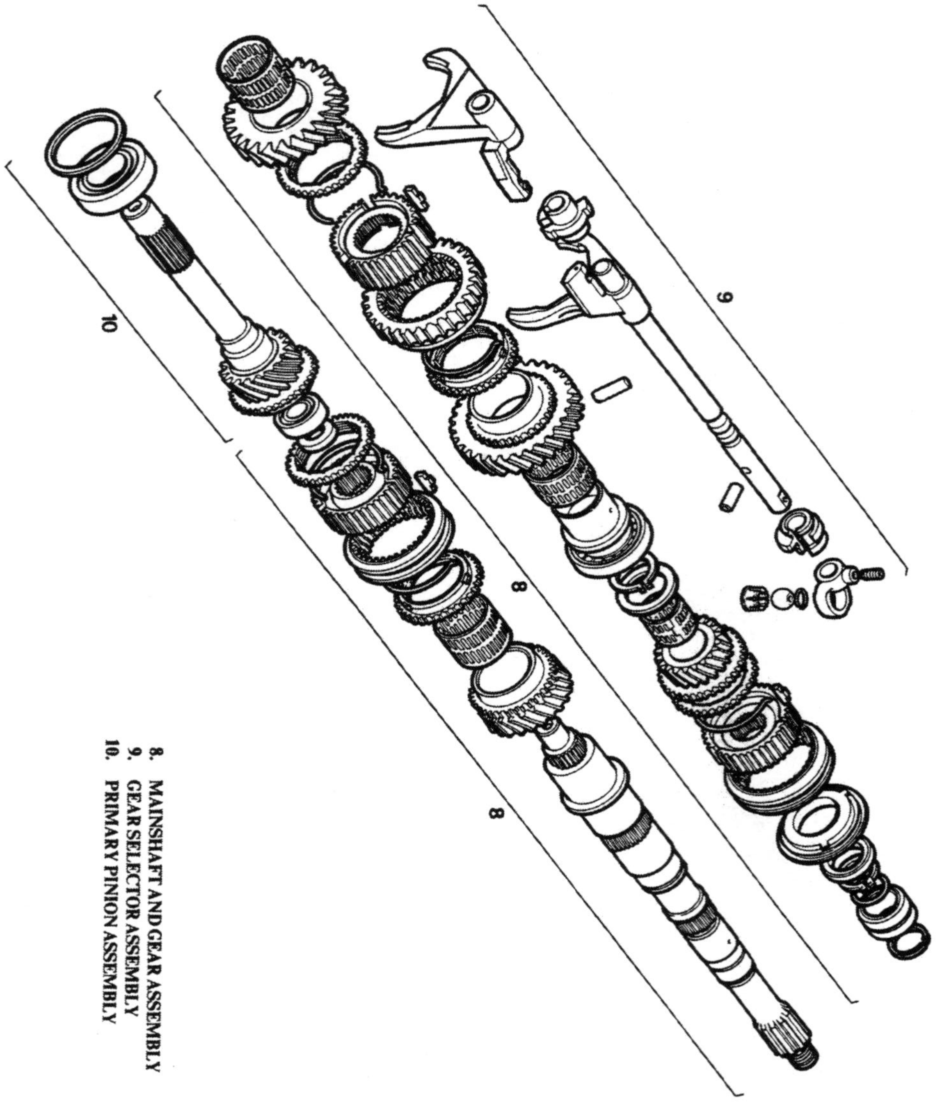

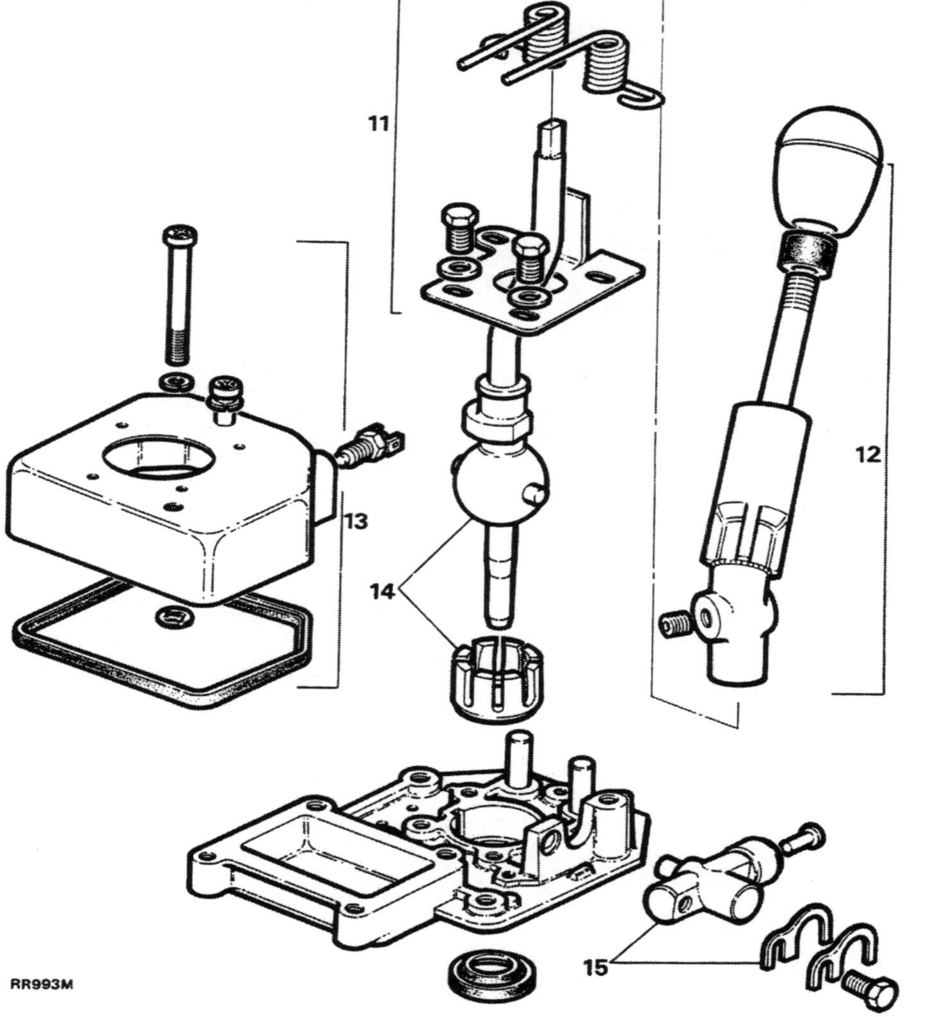

8. MAINSHAFT AND GEAR ASSEMBLY
9. GEAR SELECTOR ASSEMBLY
10. PRIMARY PINION ASSEMBLY

# LT77 FIVE SPEED GEARBOX  37

## Fifth gear end-float

86. Fit the washer and fifth gear to the mainshaft, place a straight edge on the shoulder and using feeler gauges check the end-float between gear and shoulder. End float should not be in excess of 0,25 mm (0.010 in) maximum.
87. If end float is outside limit inspect the washer and gear faces for wear.

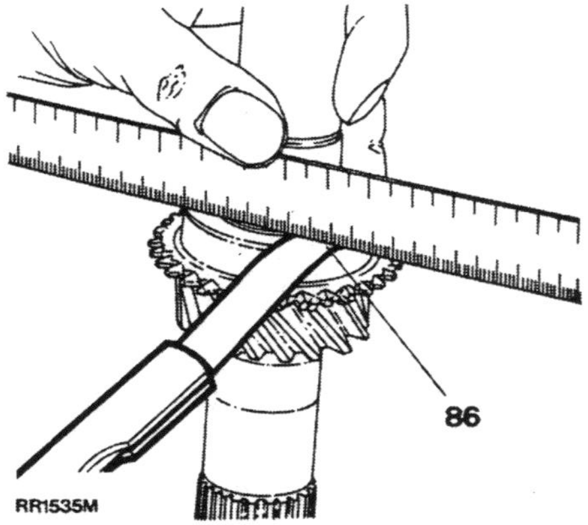

## First gear bush end-float

88. Manufacture a spacer to the dimensions provided in the illustration, this will represent a slave bearing.
89. Lubricate the second gear needle bearing with a light oil and fit the bearing, second gear and synchromesh cone to the mainshaft. It should be noted that the second gear synchromesh cone has larger slipper slots than the other synchromesh cones.
90. Fit the first and second synchromesh hub assembly with the selector fork groove to the rear of the mainshaft.
91. Holding the synchromesh hub firmly in position check the second gear end-float, between the gear and flange on the mainshaft, end-float should not be in excess of 0,25 mm (0.010 in) maximum. If end float is outside the mainshaft limits inspect the flange and gear faces for wear.
92. Fit the first gear bush to the first gear, place the gear and bush on a spacer so that the gear is sitting firmly on the shoulder of the bush.
93. Place a straight-edge across bush and gear and measure the clearance between bush and gear, this should not be in excess of 0,25 mm (0.010 in) maximum.
94. If the clearance is in excess of the maximum permissible limit, renew the bush. If after renewing the bush of a similar type, the end-float is still not within limit, inspect gear faces for wear.
95. Fit the first gear bush and slave bearing spacer and a new circlip, to the mainshaft. When fitting the circlip, care must be taken to ensure it is not opened (stretched) beyond the minimum necessary to pass over the shaft.
96. Press the slave bearing spacer back against the circlip to allow the bush maximum end-float. Measure the clearance between the rear of the first gear bush and front face of the slave bearing spacer with a feeler gauge. The clearance should be within 0,05 mm (0.002 in) maximum. The first gear bush is available with collars of different thickness. Select a bush with a collar to give the required end-float. The bush must be free to rotate easily with the required end-float.
97. Remove the circlip, slave bearing spacer and first gear bush from the mainshaft.
98. First gear bushes are available in the following sizes:

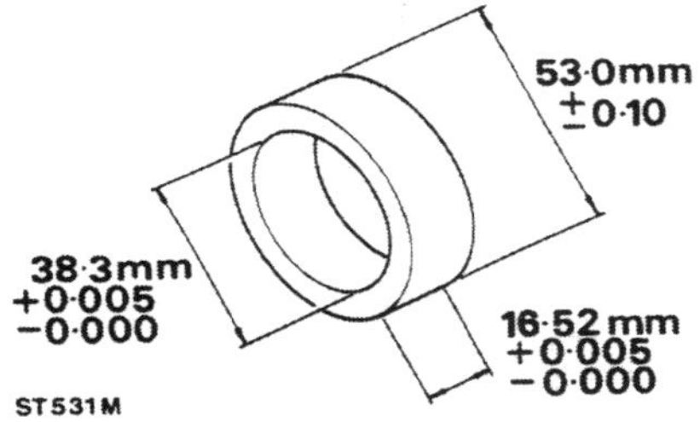

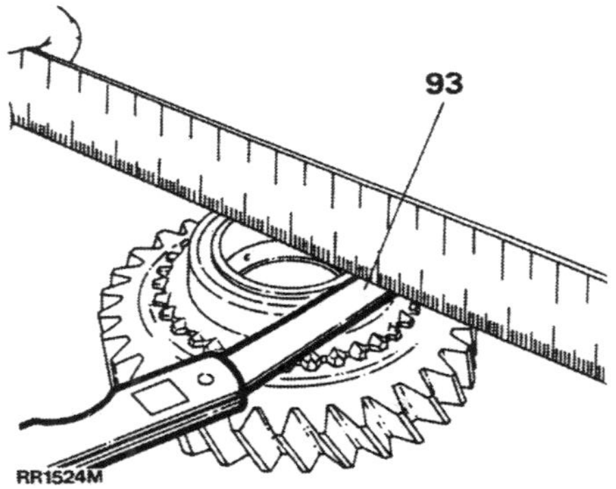

| Part No. | Thickness (mm) |
|---|---|
| FRC5243 | 40,16–40,21 |
| FRC5244 | 40,21–40,26 |
| FRC5245 | 40,26–40,31 |
| FRC5246 | 40,31–40,36 |
| FRC5247 | 40,36–40,41 |

99. Having selected a suitable first gear bush, lubricate the needle bearing and fit to the first gear.
100. Fit the selected bush to the first gear and place first gear synchromesh cone, followed by the first gear assembly to the mainshaft.
101. Using tools MS47, 18G47BA and 18G47BA-X refit the centre bearing and circlip to the mainshaft. After fitting the circlip, ensure the bearing, gear and circlip are in their rearmost position on the mainshaft, this can be achieved by placing two bars under the 1st gear and using press MS47 carefully press the assembly rearwards on the mainshaft.
102. Invert the mainshaft, lubricate the third gear needle roller bearing with light oil, fit to the front end of the mainshaft.
103. Fit the third gear to the mainshaft; and locate the third gear synchromesh cone to the third gear.
104. Fit the third/fourth synchromesh assembly (with the longer boss of the synchromesh hub to the front of the gearbox) to the mainshaft. 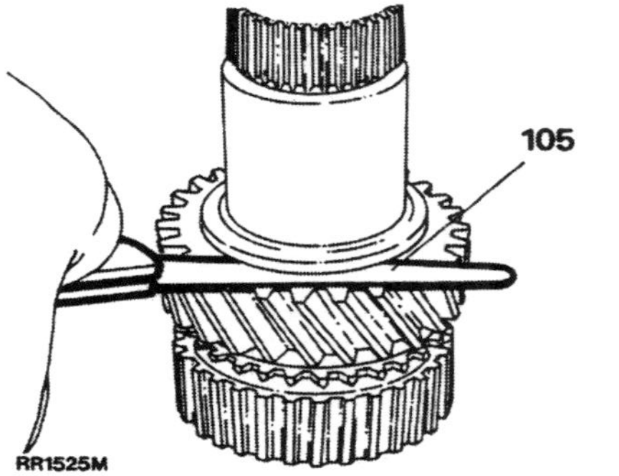
105. Holding the synchromesh assembly firmly in position, check third gear end-float between the gear face and mainshaft flange. The end-float should not be in excess of 0,25 mm (0.010 in) maximum. If end float is outside specified limit check flange and gear faces for wear.
106. Fit the spacer and bearing to the front of the mainshaft.

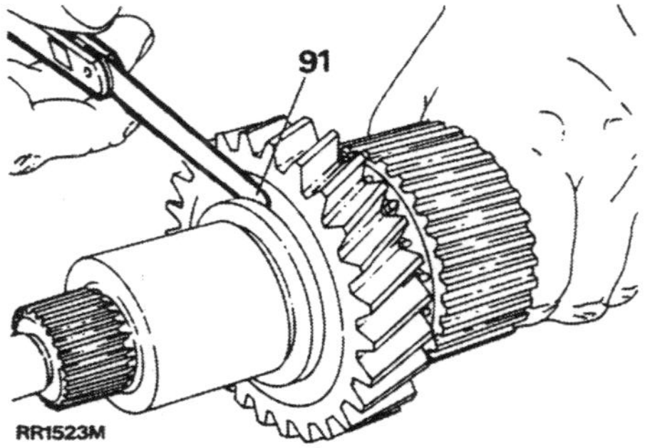

## Input shaft

107. Using tool MS47 and any suitable tube, refit a new pilot bearing track to the input shaft.
108. Fit the input shaft bearing using tools MS47, 18G47BA and 18G47BA-X.

## Reverse gear and shaft

109. Fit a new circlip to the rear of the reverse idler gear, ensuring that the circlip is not stretched beyond the minimum necessary to pass over the shaft.
110. Lubricate with light oil and fit both needle roller bearings. Fit the shorter needle bearing to the rear of the reverse idler gear.
111. Fit a new circlip to the front of the reverse idler gear.

## Extension case

112. Using a suitable press, fit a new ferrobestos bush to the case, ensuring the two drain holes are towards the bottom of the case.

NOTE: If a new extension case is fitted, it is essential that a grub screw is securely fitted in the main oilway located in the rear of the case.

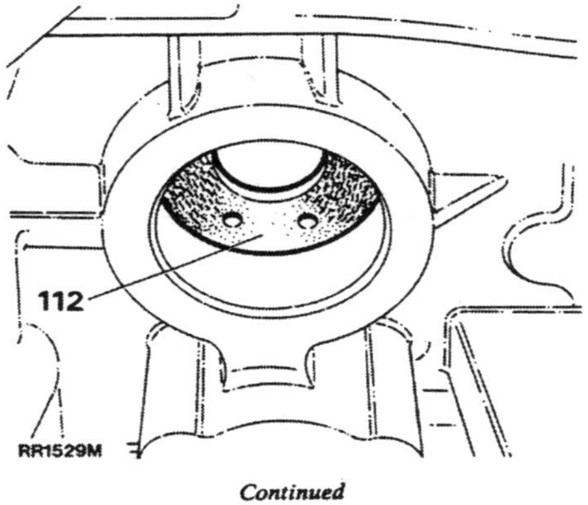

Continued

# LT77 FIVE SPEED GEARBOX

113. With the aid of tool 18G1422, fit a new oil seal to the rear of the extension case. Ensure the seal lips are towards the ferrobestos bush. Lubricate the seal lips with a suitable SAE 140 oil.
114. Assemble and fit the fibre oil pump gears to the oil pump cover, whilst ensuring the centre rotor squared drive faces the layshaft.
115. Fit the three bolts and spring washers to secure the oil pump cover, and tighten to the specified torque.
116. Ensure that the oil pick-up pipe is free of contamination or blockage.
117. Fit a new oil filter, fibre gasket and tighten plug to the specified torque.

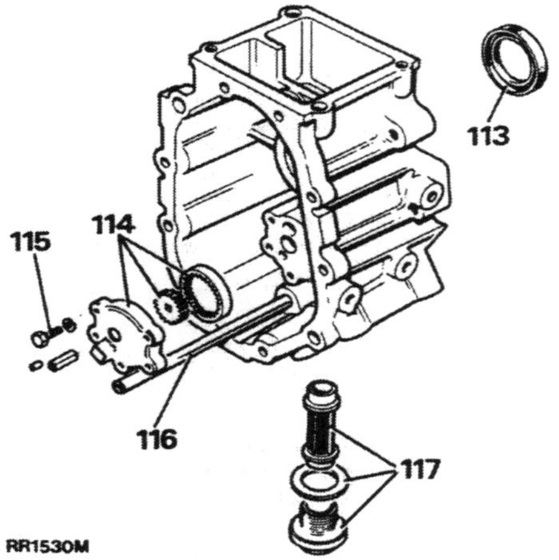

## Centre plate

118. Fit the centre plate to a suitable stand and secure with two slave bolts.
119. Place the new mainshaft and layshaft bearing tracks to the centre plate.
120. Lightly lubricate the selector shaft with a light oil.
121. Take the selector shaft complete with the first and second selector fork, front spool and third and fourth selector fork; engage both selector forks in their respective synchromesh sleeves on the mainshaft, simultaneously engage the selector shaft and mainshaft assemblies in the centre plate, whilst rotating the fifth gear selector pin to align with the slot in the centre plate.
122. Fit the layshaft to the centre plate.

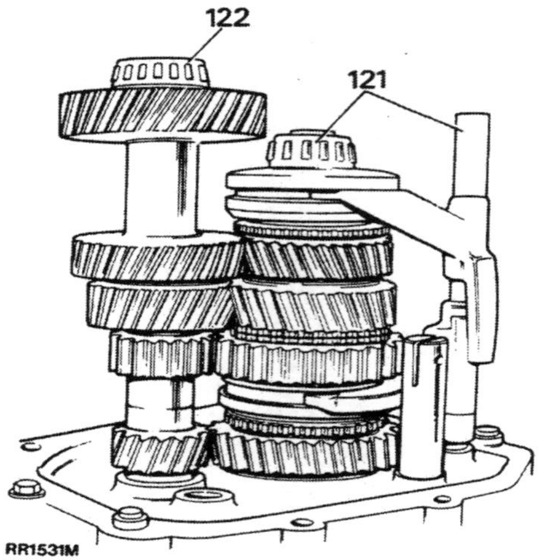

123. Rotate the selector shaft and spool to enable the reverse crossover lever forks to correctly align to the selector pin. Reposition the selector shaft and locate the lever into the slot of the reverse gear pivot shaft. Insert pivot pin and fit a new circlip, ensuring that it is not opened beyond the minimum necessary to pass over the shaft.
124. Fit the slipper pad to the reverse lever. If a new reverse lever pivot shaft has been fitted, it will be necessary to ascertain that its radial location is consistent with the reverse pad slipper engagement/clearance. The radial location is determined during initial assembly.
125. Fit the reverse gear spacer and reverse gear assembly, locating the slipper pad lip to the reverse gear groove. Engage the reverse gear shaft from the underside of the centre plate, ensuring the roll pin is aligned with the slot in the centre plate casing.
126. Prior to assembly lubricate the detent ball and spring with light oil, and fit to the top of centre plate. Smear Hylomar PL32 or Loctite 290 to the plug threads and screw the plug flush with the case. Stake the plug to prevent rotation using a suitable centre punch. Release the slave bolts.
127. Locate the fourth gear synchromesh come to the third/fourth synchromesh assembly.
128. Fit the input shaft to the mainshaft.
129. Fit the reverse gear thrust washer to the reverse gear shaft.
130. Fit a new gasket to the centre plate.

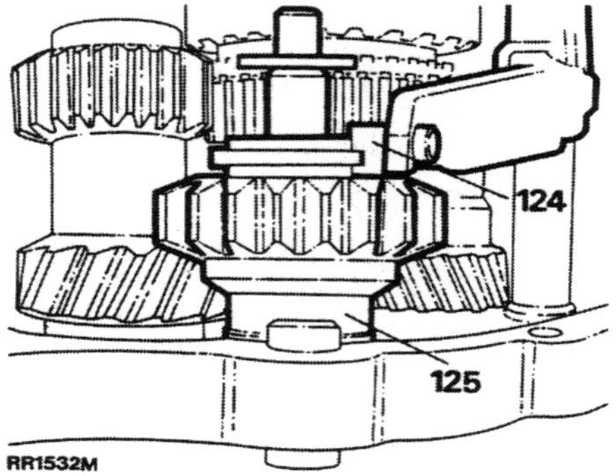

## Main gearbox casing

131. Insert a new plastic oil trough to the back of the main gearbox casing, ensuring the open trough faces the top of the case.
132. Carefully lower the gearcase into position over the gear assemblies. DO NOT USE FORCE. This operation can be assisted by the use of two 8 × 100 mm guide studs. Ensure the centre plate dowels and selector shaft are engaged in their respective locations.
133. Fit the layshaft and input shaft bearing outer tracks.
134. Using 8 × 35 mm slave bolts and plain washers to prevent damaging the rear face of the centre plate, evenly draw the gearcase into position on the plate.
135. Fit the locating shaft front spool to the top of the gearcase using Hylomar PL32 to seal between the spool and gearcase. Smear Loctite 290 or Hylomar PL32 to the bolt threads, tighten bolts and spring washers to the specified torque.
136. Manufacture a layshaft support plate and plain washer to the dimensions provided in the illustration.

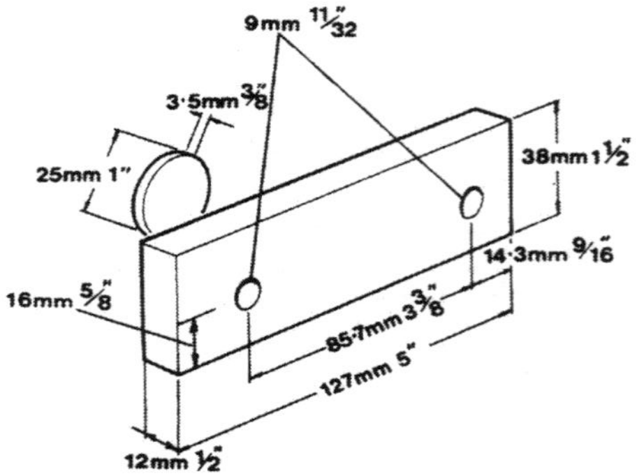

137. The layshaft support plate is fitted using two 8 × 25 mm bolts and washers situated between the support plate with the plain washer situated between the support plate and layshaft. The plate also retains the input shaft bearing outer track.
138. Using a suitable press, fit the fifth gear to the layshaft, fit a new 22 mm stake nut and tighten sufficiently to retain.
139. Locate assembly horizontally in a vice or suitable jig.
140. Fit the fifth speed washer, roller bearing, gear and cone to the mainshaft.
141. Press fit fifth gear synchromesh hub assembly using tool 18G1431. Fit a dummy spacer with an oversize bore to ascertain the correct spacer to provide the specified clearance on the fifth gear. When the correct spacer has been determined, fit the spacer to the mainshaft and secure in position with a new circlip.
142. Measure the clearance between the front spacer and circlip, which should be between 0,055 and 0,005 mm, (0.002–0.0002 in) maximum. Select the appropriate spacer to provide the aforementioned clearance.

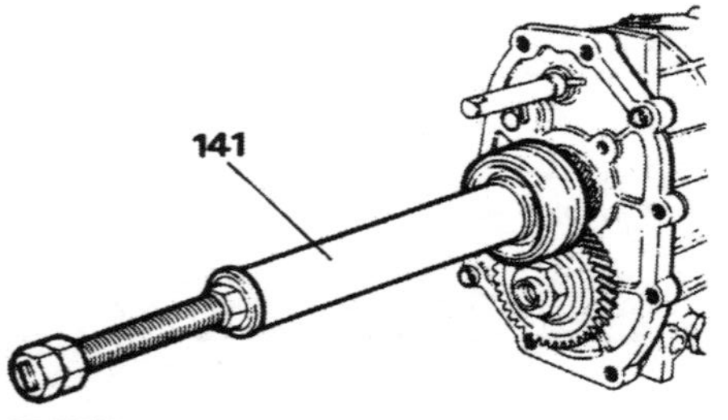

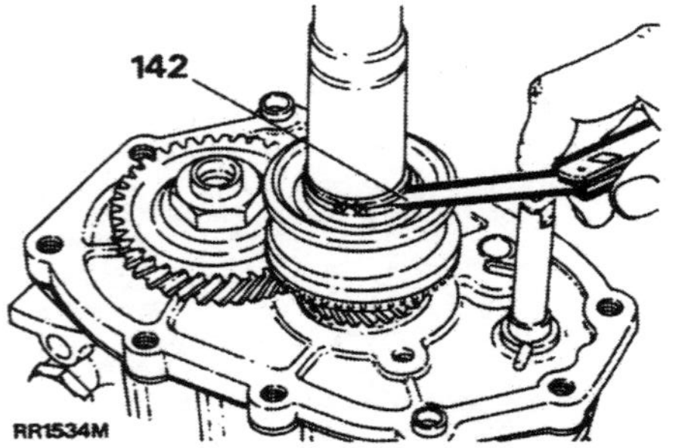

Continued

## LT77 FIVE SPEED GEARBOX

| Part No. | Thickness (mm) | Part No. | Thickness (mm) |
|---|---|---|---|
| FRC5284 | 5,10 | FRC5294 | 5,40 |
| FRC5286 | 5,16 | FRC5296 | 5,46 |
| FRC5288 | 5,22 | FRC5298 | 5,52 |
| FRC5290 | 5,28 | FRC5300 | 5,58 |
| FRC5292 | 5,34 | FRC5302 | 5,64 |

143. Fit the correct selective spacer and new circlip.

### Mainshaft and layshaft end-float

144. Measure and adjust the mainshaft and layshaft end-float as necessary. Remove the layshaft support plate from the front of the gearbox.

| Part No. | Thickness (mm) | Part No. | Thickness (mm) |
|---|---|---|---|
| FRC4327 | 1,51 | FRC4349 | 2,17 |
| FRC4329 | 1,57 | FRC4351 | 2,23 |
| FRC4331 | 1,63 | FRC4353 | 2,29 |
| FRC4333 | 1,69 | FRC4355 | 2,35 |
| FRC4335 | 1,75 | FRC4357 | 2,41 |
| FRC4337 | 1,81 | FRC4359 | 2,47 |
| FRC4339 | 1,87 | FRC4361 | 2,53 |
| FRC4341 | 1,93 | FRC4363 | 2,59 |
| FRC4343 | 1,99 | FRC4365 | 2,65 |
| FRC4345 | 2,05 | FRC4367 | 2,71 |
| FRC4347 | 2,11 | FRC4369 | 2,77 |

145. When ascertaining the mainshaft end-float care must be taken when checking the dial gauge readings to ensure that the end-float only, as distinct from side movement, is recorded. To overcome the difficulty in differentiating between end-float and side movement, wrap approximately ten turns of masking tape around the plain portion of the input shaft below the splines. Ascertain that the rise and fall of the input shaft is not restricted by the tape.

146. Place a mainshaft and layshaft spacer of nominal thickness 1,02 mm on the mainshaft and layshaft bearing tracks, fit the front cover and gasket, tighten bolts and spring washers to the specified torque.

147. Invert the gearbox on the stand. Rotate the mainshaft to correctly seat the bearing.

148. Place a suitable ball bearing in the mainshaft centre and mount the dial gauge on the gearcase with the stylus resting on the ball bearing centre. Zero the gauge.

149. Check the end-float by a 'push-pull' action to the mainshaft. The required mainshaft end-float measurement should be between 0,01 to 0,06 mm (0.0004 to 0.002 in) maximum with no preload.

150. Select the appropriate spacer to give the specified end-float.

151. Rotate the layshaft to correctly seat the bearing.
152. Place a suitable ball bearing in the layshaft centre and mount the dial gauge on the gearcase, with the stylus resting on the ball bearing centre. Zero the gauge.

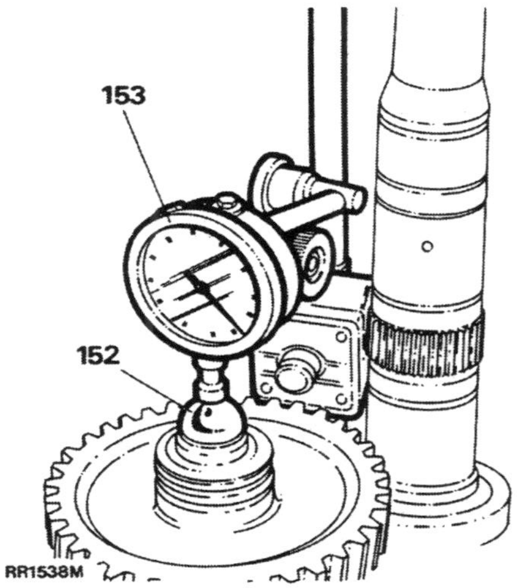

153. With the aid of levers approximately 23 cm long; to prevent component damage, check the end-float by a gentle 'push-lift' action to the layshaft. The required layshaft setting is:
    0,025 mm end-float
    0,025 mm preload.
    Spacer thickness required equals; nominal thickness of spacer, plus end-float obtained. Remove the gauge and ball bearing.

154. Remove the front cover. Having ascertained the mainshaft and layshaft end-float, fit the mainshaft and layshaft spacers of the appropriate thickness to the mainshaft and layshaft bearing tracks. Selective spacers are available in a range of sizes to meet the aforementioned clearance limits.

| Part No. | Thickness (mm) | Part No. | Thickness (mm) |
|---|---|---|---|
| TKC4633 | 1,69 | TKC4649 | 2,17 |
| TKC4635 | 1,75 | TKC4651 | 2,23 |
| TKC4637 | 1,81 | TKC4653 | 2,29 |
| TKC4639 | 1,87 | TKC4655 | 2,35 |
| TKC4641 | 1,93 | TKC4657 | 2,41 |
| TKC4643 | 1,99 | TKC4659 | 2,47 |
| TKC4645 | 2,05 | TKC4661 | 2,53 |
| TKC4647 | 2,11 | TKC4663 | 2,59 |

155. Fit a new oil seal to the front cover, ensuring the seal lips face towards the gearbox. Lubricate the seal lips with SAE 140 gear oil.
156. Mask the splines with masking tape to protect the oil seal, refit the front cover and remove the spline masking tape.
157. Refit the bolts and spring washers having used Hylomar PL32 or Loctite 290 on the bolt threads. Tighten to the specified torque.
158. Remove gearbox from the stand and place suitably supported on the bench.
159. Remove the guide studs fitted to the centre plate.
160. Select reverse gear by turning the selector rail anti-clockwise and pulling rearwards. Move the fifth speed synchromesh hub into mesh with the fifth gear.
161. Tighten the staked nut onto the fifth gear layshaft to the specified torque. Stake the nut with a suitable punch to secure. Select neutral by pushing selector rail inwards and turning clockwise, thereby returning the fifth speed synchromesh hub to its out of mesh position.

### Fifth gear selector fork assembly

162. Fit the fifth selector fork bracket to the centre plate with the bolts and spring washers and tighten to the specified torque.
163. Fit the fifth gear spool to the selector shaft. It should be noted that the longer shoulder of the spool is fitted towards the front of the gearbox.
164. Fit the bronze pads to the 5th gear selector fork (retain with vaseline), engage, the fork with the spool, bracket and synchromesh sleeve, insert the two dowel pins and secure in position with the two circlips.

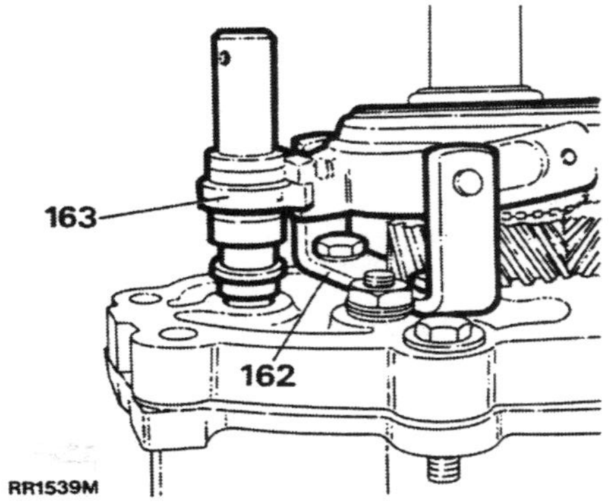

165. Fit the selector yoke to the selector shaft and secure in position with a NEW 10 mm grub screw.

NOTE: The NEW grub screw is encapsulated with loctite during manufacture.

166. Remove the six dummy bolts securing the centre plate to the main casing.
167. Position the gearbox assembly horizontally and fit the oil pump shaft to the pump.

### Extension case

168. Fit a new gasket to the centre plate.
169. Rotate the oil pump shaft to align with the layshaft.
170. Carefully fit the extension case ensuring that the oil pump shaft engages the layshaft.
171. Fit the extension case bolts and spring washers; tighten to specified torque.
172. Cover the mainshaft splines with masking tape and fit a new oil seal collar 'O' ring. Remove the masking tape.
173. Using tool 18G1431 fit a NEW oil seal collar to the mainshaft, ensuring the collar is NOT pushed too far on the shaft, fit only with sufficient clearance to allow the circlip to engage in its groove.

### Bell housing

174. Locate the bell housing to the dowels and fit the two long bolts (12 × 45 mm) with spring and plain washers to the dowel positions. The remaining four bolts (12 × 30 mm) are fitted with spring washers only. Tighten to the specified torque.
175. Slide the release bearing guide over the input shaft and secure in position with the single bolt and clutch release pivot, tighten to the specified torque.

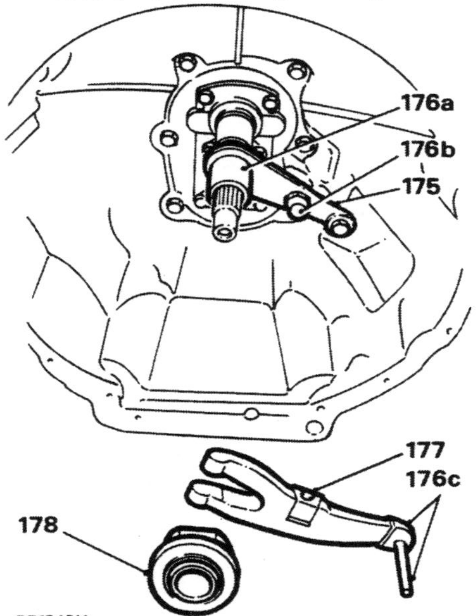

*Continued*

# LT77 FIVE SPEED GEARBOX

176. Prior to assembly, lubricate the following items with a thin film of molybdenum disulphide grease.
    (a) Clutch release lever fulcrum pivot socket.
    (b) The outer diameter of the release bearing guide.
    (c) Ball ends of the clutch operating push rod.
177. Locate the clutch release bearing lever on the pivot ball and secure in position with the spring clip, tighten the clip securing bolt to the specified torque.
178. Slide the clutch release bearing over the bearing guide and locate the NEW nylon clutch release lever clip onto the bearing.

### Gear change housing

179. Fit a new seal to the bottom of the housing, lips of the seal uppermost.
180. Lightly grease the lower gear lever ball with Shell Alvania R3 and fit the railko bush.
181. Fit the assembly to the housing locating the two pegs on the ball in the recesses of the ball seat.
182. Locate the bias adjustment plate, coat two of the retaining bolts with Hylomar PL32 or Loctite 290 and fit them forward of the gear lever. Do not tighten the bolts at this stage.
183. Inspect the bias springs for condition, renew if necessary. Fit the springs onto the posts, coat the remaining two bolts with Hylomar PG32 or Loctite 290 and fit to the housing to secure the springs in position. Do not tighten the bolts at this stage.
184. Carefully lever the free end of the springs around the rear of the gear lever until they are retained by the stop on the adjustment plate.
185. Fit a NEW gasket to the top of the gearbox.
186. Fit the housing to the gearbox and secure in position with two bolts (diagonally opposite).

### Bias adjustment plate setting

187. Select fourth gear and load the gear lever fully to the right hand side of the gearbox.
188. Tighten the four bolts to the specified torque.
189. Select all forward and reverse gears to ensure that gear selection is not inhibited after final tightening of the adjustment plate.

### Reverse plunger assembly setting

190. Fit the assembly to the housing and secure in position with the single bolt and washer.
191. Select first gear. Using feeler gauges, check the gap between the reverse plunger nose and the side of the gear lever. The required setting should be 0,6 to 0,85 mm (0.024 to 0.034 in) clearance. Adjust the gap by adding or removing the shims behind the plunger assembly.

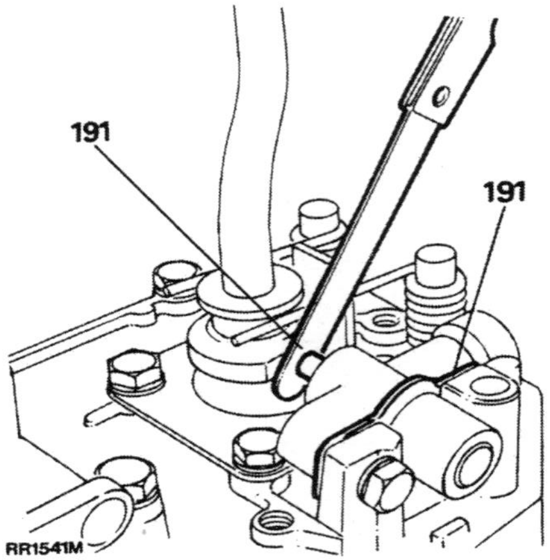

# LT230T TRANSFER GEARBOX

## LT 230T TRANSFER BOX

The following operations can be carried out with the gearbox in the vehicle. For ease of working, the vehicle should be raised on a ramp or placed over a pit.

### RENEW SPEEDOMETER DRIVE PINION

1. Disconnect the battery.
2. Raise the vehicle on a ramp.
3. Remove the speedometer drive clamp and nut and withdraw the cable.
4. Prise out the drive pinion assembly.
5. Push in a new assembly and fit the speedometer cable and secure with the clamp and nut.

### RENEW REAR OUTPUT SHAFT OIL SEAL

**Special tool:** 18G1422

1. Disconnect the battery for safety.
2. Disconnect the rear propshaft from the output flange.
3. Remove the brake drum retaining screws and withdraw the drum.
4. Remove the four back plate bolts that also retain the oil catcher and remove the brake back plate and catcher.

**NOTE: An hexagonal type socket should be used for these bolts.**

5. Remove the output shaft nut, steel washer, felt washer and withdraw the flange.
6. Using the slot provided, lever off the dust cover.
7. Prise out the output shaft oil seal(s).
8. Using special tool 18G1422 fit the double-lipped oil seal, open side inwards, with the seal in contact with the bearing circlip, taking care not to touch the seal lips.
9. Fit the dust cover.
10. Lubricate the surface of the flange which runs in the seal and carefully fit the flange.

**NOTE: To renew the flange bolts first remove the circlip before fitting the flange.**

11. Secure the flange with the nut and washer and tighten to the specified torque.
12. Fit the oil catcher to the back plate using silicone rubber sealant and secure with the two back plate bolts (with plain washers).
13. Fit the brake drum and retain with the two screws.
14. Reconnect the propeller shaft.

### RENEW FRONT OUTPUT SHAFT OIL SEAL

**Special tool:** 18G1422

1. Disconnect the front propeller shaft from the flange and move to one side.
2. Remove the output shaft nut, steel washer, felt washer and withdraw the flange.
3. Remove the oil seal shield.
4. Prise out the oil seal(s).
5. Using special oil seal replacer tool 18G1422 fit the double-lipped oil seal, open side inwards, with the seal in contact with the bearing circlip, taking care not to touch the seal lip.
6. Lubricate the running surface of the flange and fit it together with the oil seal shield.
7. Secure the flange with the nut and washer and tighten to the specified torque.
8. Refit the propeller shaft.

192. Remove the gear change housing from the gearbox.
193. Fit a new fibre washer to the oil drain plug, fit the plug to the gearbox and tighten to the specified torque.
194. Refill the gearbox with the correct quantity and grade of oil as specified in the 'Recommended Lubricants' section.
195. Refit the gearbox oil level plug, and tighten to the specified torque.
196. Refit the gearbox to the vehicle.
197. Refit the gear change housing to the vehicle after the gearbox has been installed in the vehicle.

## 37 LT230T TRANSFER GEARBOX

### REMOVE LT230T TRANSFER GEARBOX

Special tool: 18G 1425—Guide studs (3)
Also, locally manufactured adaptor plate, see below

**Adaptor plate for removing transfer gearbox**

The transfer gearbox should be removed from underneath the vehicle, using a hydraulic hoist. An adaptor plate for locating the transfer gearbox onto the hoist can be manufactured locally to the drawing below. If a similar adaptor plate was made for the LT230R transfer gearbox, it can be modified to suit both the LT230R and LT230T gearboxes by making the modifications shown by the large arrows.

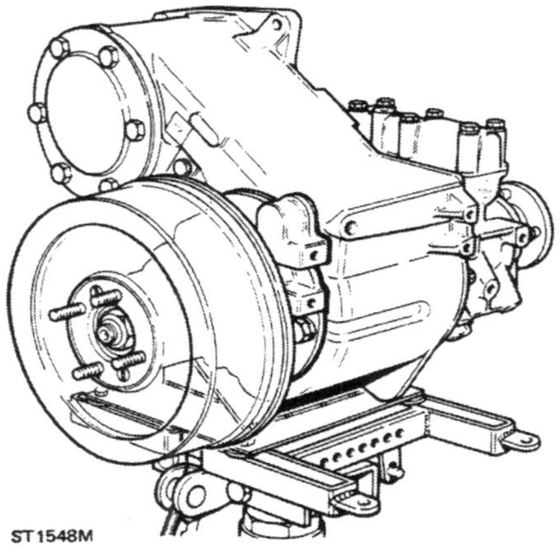

Material: Steel plate BS 1449 Grade 4 or 14.
Holes marked thus * to be drilled to fit hoist being used.

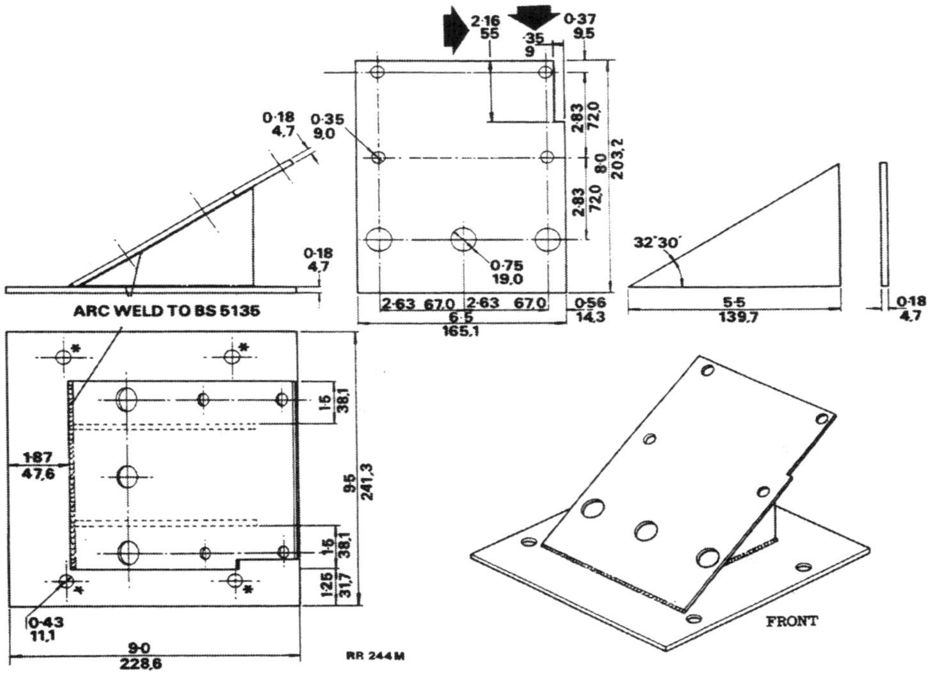

---

## 37 LT230T TRANSFER GEARBOX

**Removing**

1. Install the vehicle on a hydraulic ramp.
2. Open the bonnet.
3. Disconnect the battery.
4. **Fuel Injection models only**—Release the airflow meter to plenum chamber hose.
   **Carburetter models only**—Remove the air intake elbows and withdraw the air cleaner from its location.
5. Remove the four screws securing the cubby box liner to the cubby box and lift out the liner.
6. Carefully prise the window lift, switch panel away from the front of the cubby box.
7. Identify each switch connection for reassembly, disconnect the plugs and remove the switch panel.
8. Remove the main and transfer gearbox knobs.
9. Carefully prise the centre panel out of the floor mounted console and remove it from the vehicle.

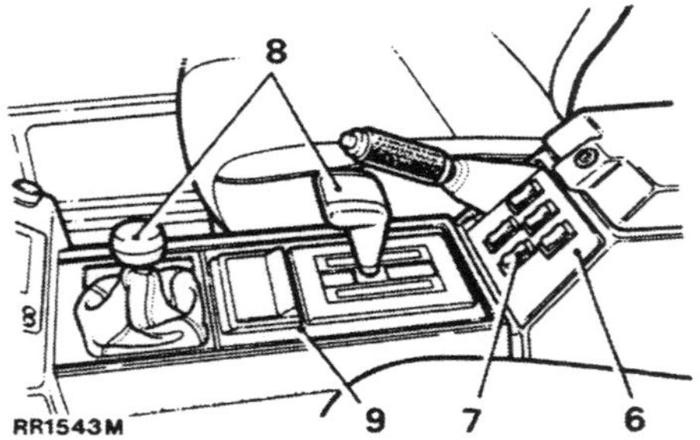

Automatic version illustrated

10. Release the two bolts and two screws securing the console assembly to the gearbox tunnel.
11. Release the handbrake and remove the split pin, clevis pin and washer securing the handbrake cable to the hand brake lever.
12. Carefully manoeuvre the assembly away from the radio housing and remove it from the vehicle.
13. Release the large nut retaining the handbrake outer cable to the top of the gearbox tunnel.
14. Remove the nut and feed the cable through the hole to the underside of the vehicle.

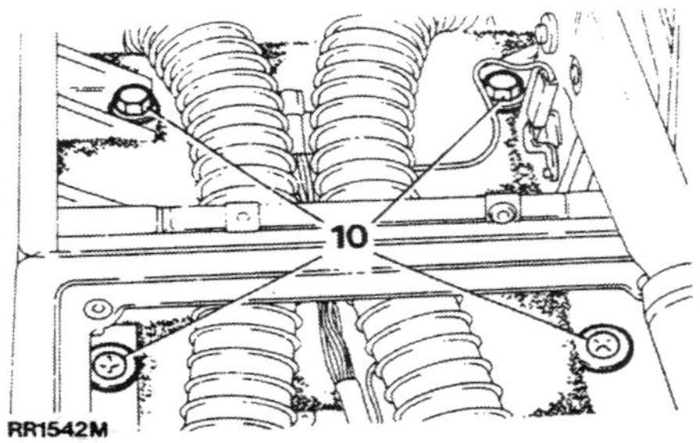

NOTE: The illustration for the following removal instructions is located at top of the following page.

15. Raise the vehicle on the ramp and drain the transfer gearbox.
16. Release the nut and clamp securing the speedometer cable to the rear of the transfer box, withdraw the cable from the speedometer drive pinnion.
17. Release the cable from the clip at the side of the gearbox.
18. Release the four nuts and bolts securing the rear propeller shaft to the rear output flange and tie the shaft to one side.
19. Remove the four nuts and bolts securing the front propeller shaft to the front output flange and tie the shaft to one side.
20. Release the nuts and bolts securing the front down pipe to the front silencer.
21. Release the nut at the rear tailpipe bracket, disconnect the silencer from the down pipe, and tie the rear tail pipe and silencer to one side.
22. Manufacture an adaptor plate in accordance with the drawing, to attach to the gearbox hoist and transfer box to facilitate removal (RR244M).
23. Place four, 30 mm (1.250 in) long spacers between the top of the hoist and the adaptor plate at the securing points and secure the adaptor plate to the hoist.
24. Remove the four central bolts from the transfer box bottom cover, move the hoist into position and secure the adaptor plate to the transfer box.
25. Adjust the hoist to take the weight of transfer box.
26. Remove the tie bar from the transfer gearbox.
27. Remove the rear left-hand side mounting bracket to chassis nuts and bolts.
28. Remove the right-hand side mounting bracket to chassis nuts and bolts.
29. Remove right-hand side mounting bracket to flexible mounting, rubber retaining nut and place bracket aside.
30. Lower the hoist until the rear brake drum clears the rear passenger footwell.
31. Remove the split pin and washers securing the differential lock lever to the connecting rod, and disconnect the lever from the rod.
32. Disconnect the electrical leads from the differential lock switch.
33. Remove the breather pipe from the top of the transfer gearbox.
34. Select low range transfer box gear position.
35. Release the high/low rod lower lock nut and remove the rod from the yoke.

*Continued*

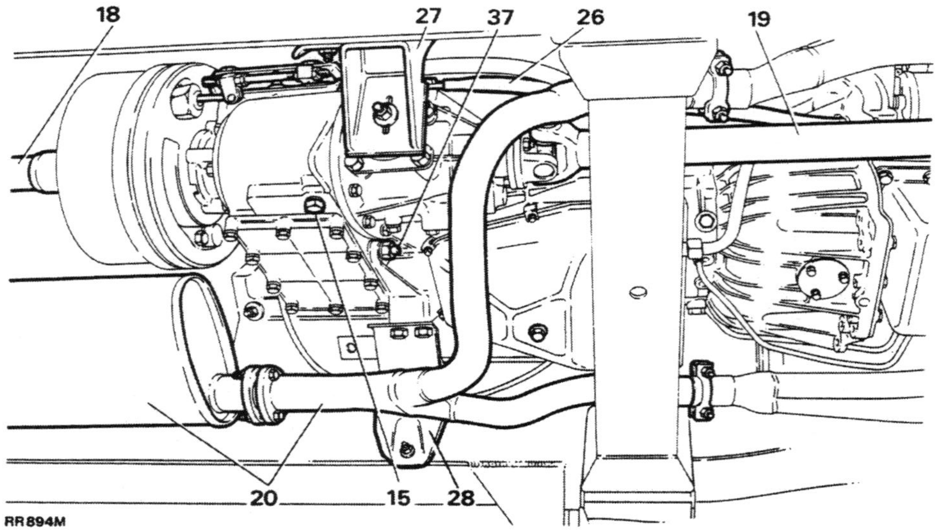

## LT230T TRANSFER GEARBOX

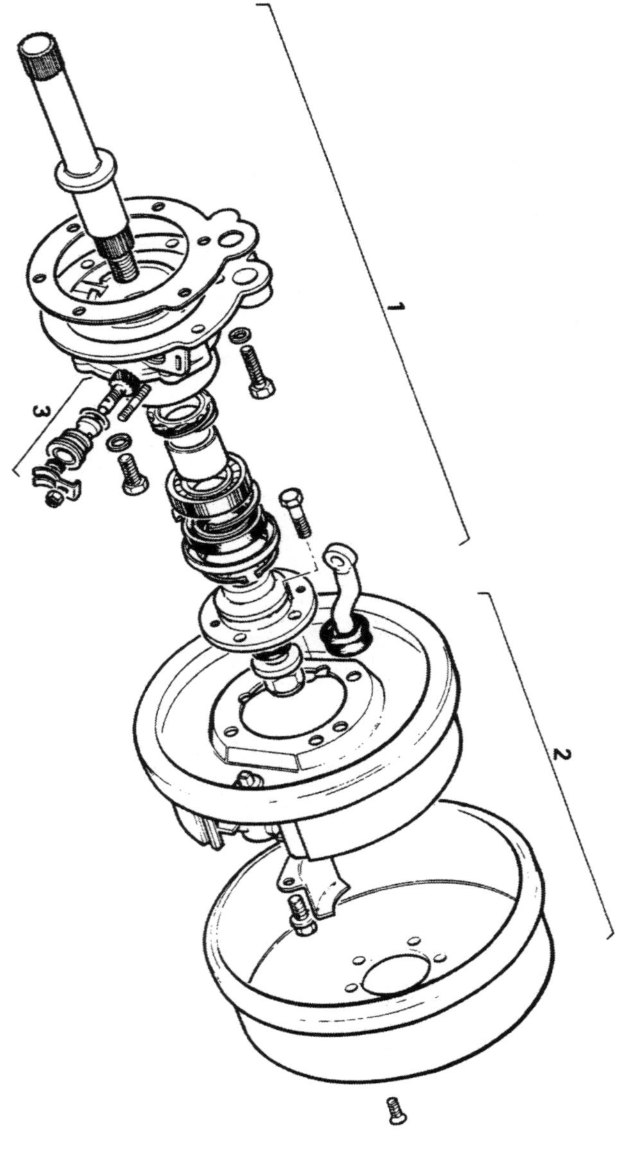

1. Rear output drive assembly.
2. Transmission brake drum assembly.
3. Speedo housing assembly.

36. Place a suitable wooden block between the main gearbox and chassis cross-member, then lower the hoist until the gearbox contacts the wooden block.
37. Remove the upper and lower bolts securing the transfer box to the main gearbox.
38. Fit three guide studs to the gearbox 18G 1425 and manoeuvre the gearbox rearwards to detach it from the main gearbox.

### Refitting

39. Make sure that the joint faces of the transfer box and main gearbox extension case are clean and that the three guide studs, 18G 1425, are fitted to the extension case.
40. Lubricate the oil seal in the joint face of transfer box, secure the transfer box to the adaptor plate on the lifting hoist and raise the hoist until the transfer box can be located over the guide studs.
41. Remove the guide studs and secure the transfer box to the main gearbox extension case. Tighten the nuts and bolts to the correct torque.
42. Complete the refitting procedure by reversing the removal sequence, noting the following important points.
43. After removing the lifting hoist and adaptor plate from the transfer box, clean the threads of the four bolts for the transfer box bottom cover, coat them with Loctite 290, and fit them together with spring washers. Tighten to the correct torque.
44. Refill the transfer box with the correct grade oil to the oil level plug hole.
45. Check, and if necessary top-up the oil level in the main gearbox. Use the correct grade oil.
46. Check the operation of the handbrake and adjust as necessary.

# LT230T TRANSFER GEARBOX

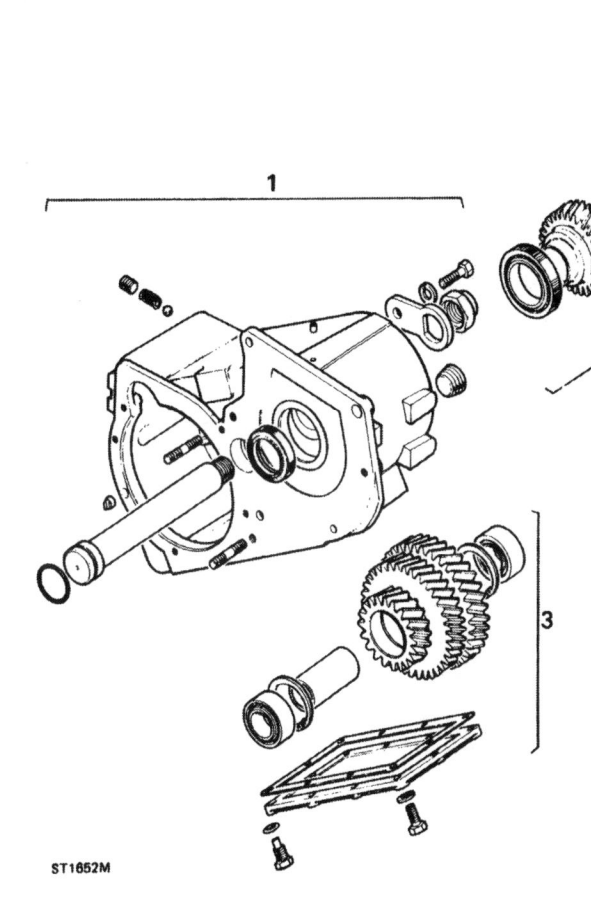

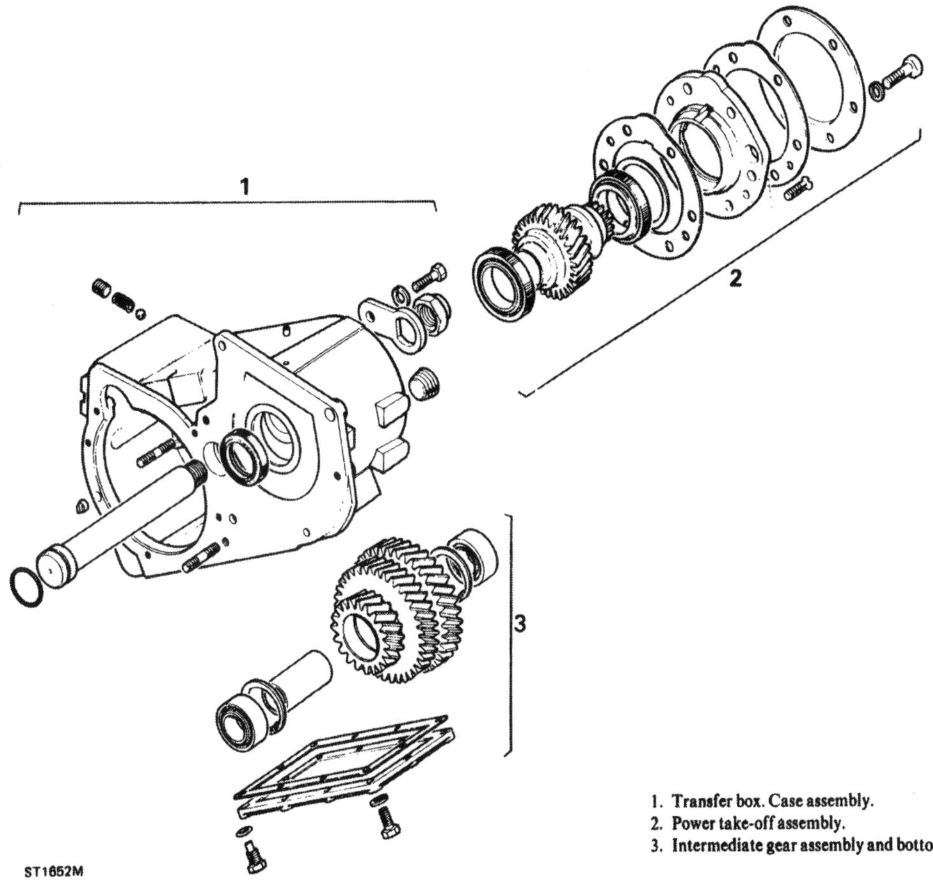

1. Transfer box. Case assembly.
2. Power take-off assembly.
3. Intermediate gear assembly and bottom cover.

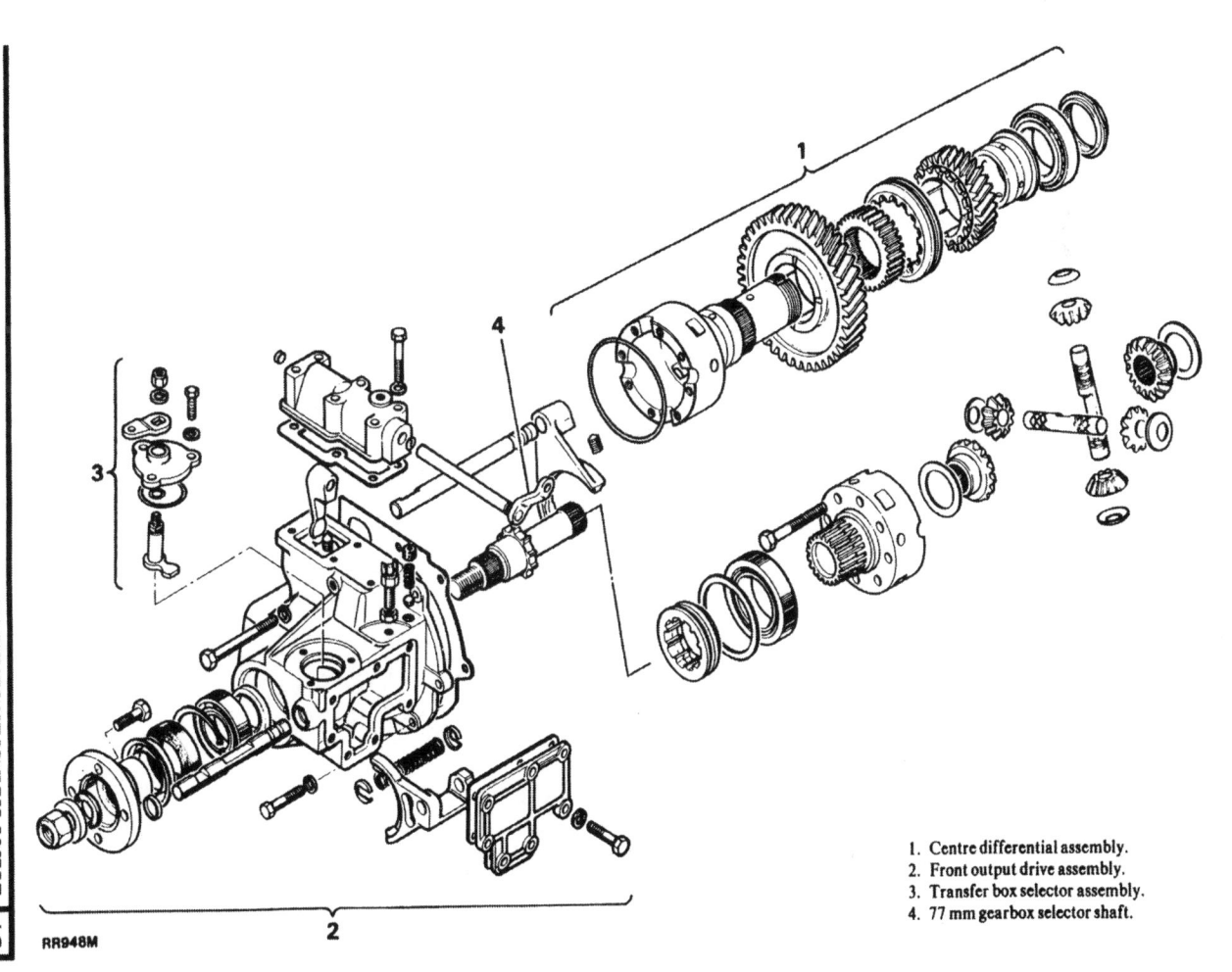

1. Centre differential assembly.
2. Front output drive assembly.
3. Transfer box selector assembly.
4. 77 mm gearbox selector shaft.

# LT230T TRANSFER GEARBOX

## LT230T TRANSFER GEARBOX OVERHAUL

**Service Tools:**
18G47-7—Input gear cluster bearing cones remover/replacer
18G47BB-1—Adaptor centre differential bearing remover
18G47BB-3—Adaptor centre differential bearing remover button
18G257—Circlip pliers
18G1205—Prop flange wrench
18G1271—Oil seal remover
18G1422—Mainshaft rear oil seal replacer
18G1423—Adaptor/socket centre differential stake nut remover/replacer
18G1424—Centre differential bearing replacer
MS47—Hand press
MS550—Bearing and oil seal replacer handle
LST47-1—Adaptor centre differential bearing remover
LST104—Intermediate gear dummy shaft
LST105—Input gear mandrel
LST550-4—Intermediate gear bearing replacer

### TRANSFER BOX DATA

Front bevel gear end-float ............ 0,025 to 0,075 mm (0.001 to 0.003 in)
Rear bevel gear end-float ............ 0,025 to 0,075 mm (0.001 to 0.003 in)
Rear output housing clearance ............ 1,00 mm (0.039 in)
High range gear end-float ............ 0,05 to 0,15 mm (0.002 to 0.006 in)
Front differential bearing pre-load ............ 1.36 to 4.53 kg (3 to 10 lb)
Input bearing pre-load ............ 2.26 to 6.80 kg (5 to 15 lb)
Intermediate shaft bearing pre-load ............ 1.81 to 4.53 kg (4 to 10 lb)

### Transmission brake removal

1. Remove two countersunk screws and withdraw brake drum.
2. Remove four bolts securing the brake back-plate; the two bottom fixings retain the oil catcher.

**NOTE: An hexagonal type socket should be used for these bolts.**

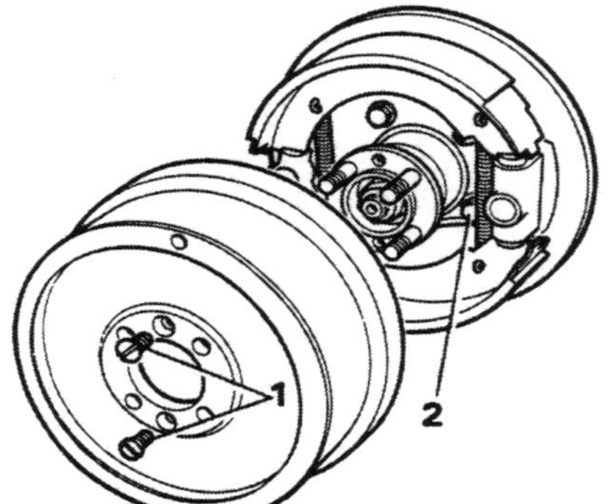

### Bottom cover removal

3. Remove the six bolts and washers retaining the bottom cover.
4. Remove the gasket and bottom cover.

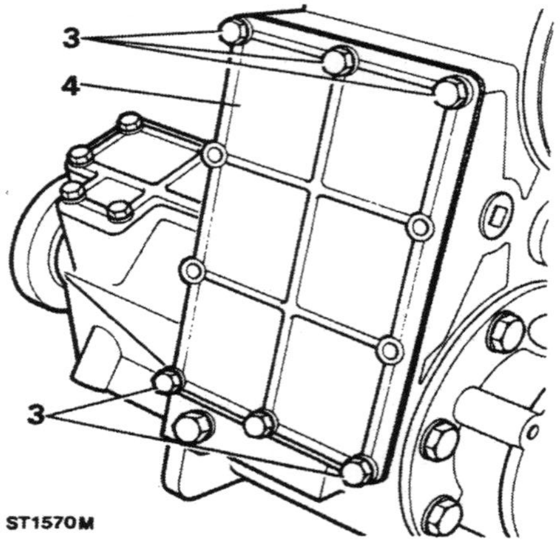

### Intermediate shaft and gear cluster removal

5. Release stake nut from recess in intermediate shaft and remove stake nut and discard.
6. Unscrew the single bolt and remove anti-rotation plate at the rear face of the transfer box.
7. Tap the intermediate gear shaft from the transfer box.
8. Lift out the intermediate gear cluster and bearing assembly.
9. Remove the 'O' rings from the intermediate gear shaft and from inside the transfer box.
10. Remove taper roller bearings and bearing spacer from the intermediate gear cluster assembly.

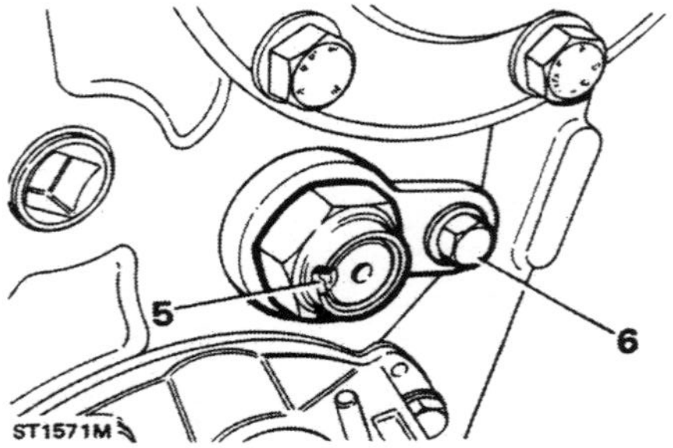

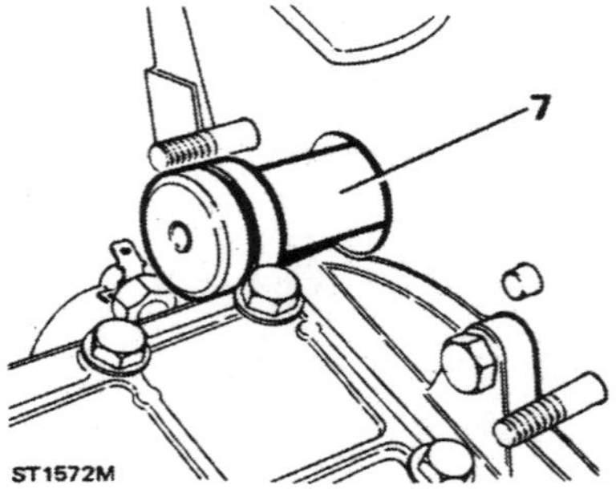

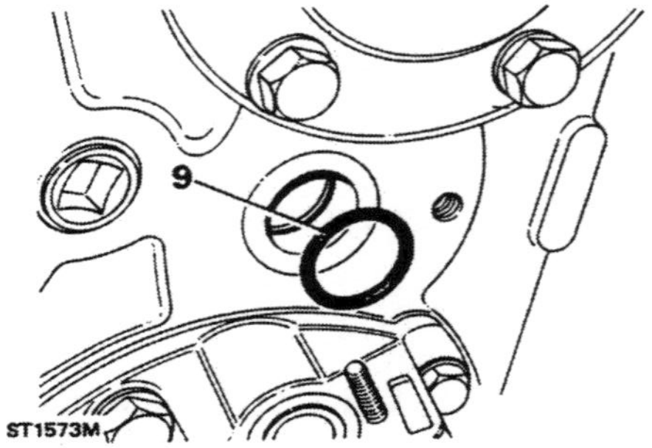

### Power take-off cover removal

11. Remove six bolts and washers retaining the take-off cover and speedo cable clips.
12. Remove the gasket and cover.

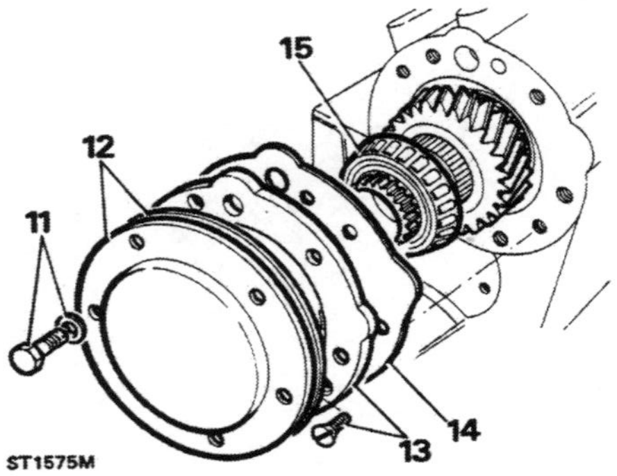

### Input gear removal

13. Remove the two countersunk screws and detach the main shaft bearing housing.
14. Remove the gasket.
15. Withdraw the input gear assembly.
16. Prise out and discard the oil seal at the front of the transfer box casing using service tool 18G 1271.
17. Drift out the input gear front bearing track.

*Continued*

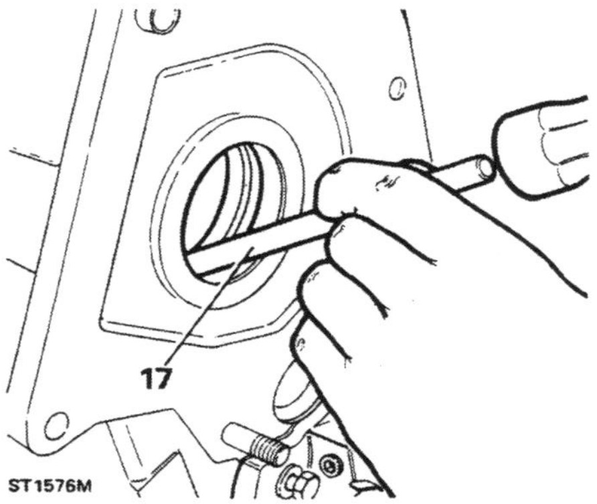

# LT230T TRANSFER GEARBOX

## High/low cross-shaft housing removal

18. Remove the six bolts and washers retaining the cross-shaft housing and earth lead.
19. Remove the gasket and cross-shaft housing.

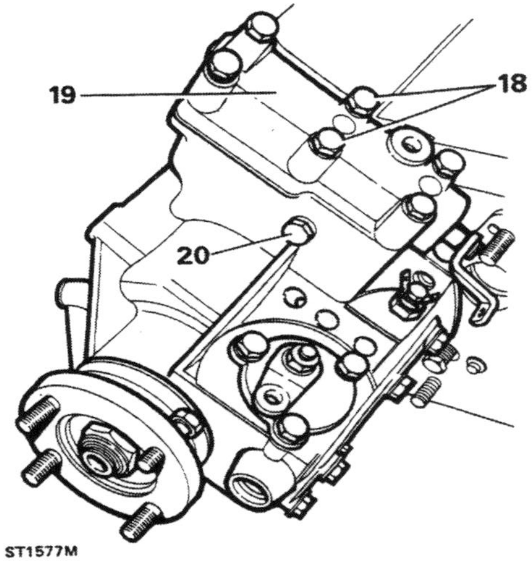

## Front output housing removal

20. Remove the eight bolts and washers and detach the output housing from the transfer box casing, taking care not to mislay the dowel.

## Centre differential removal

21. Remove high/low selector shaft detent plug, spring and retrieve the ball with a suitable magnet.
22. Withdraw the centre differential and selector shaft/fork assembly.

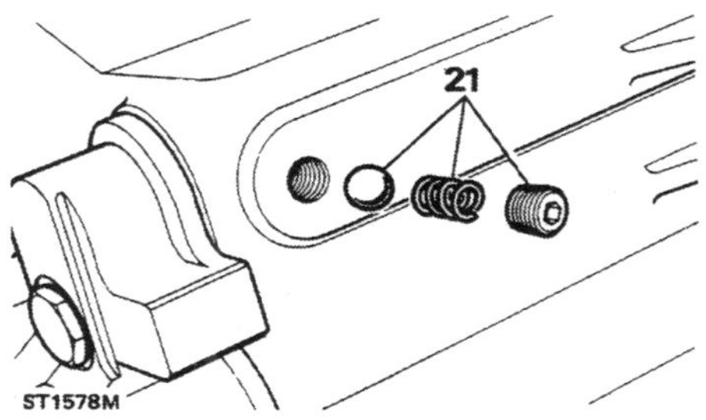

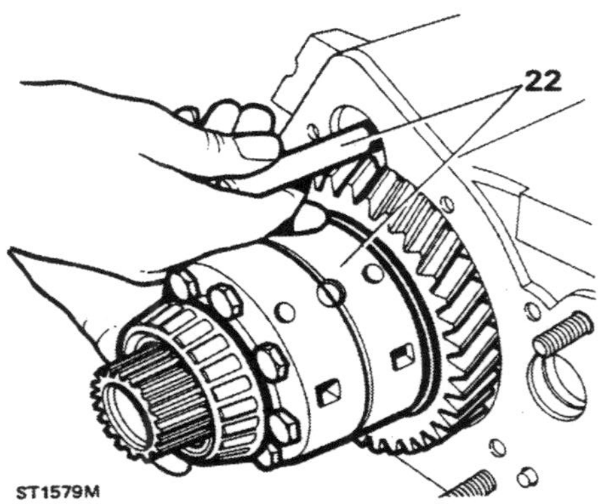

## Rear output housing removal

23. Remove six bolts and washers and detach the rear output housing and shaft assembly from the transfer casing.
24. Remove the gasket.

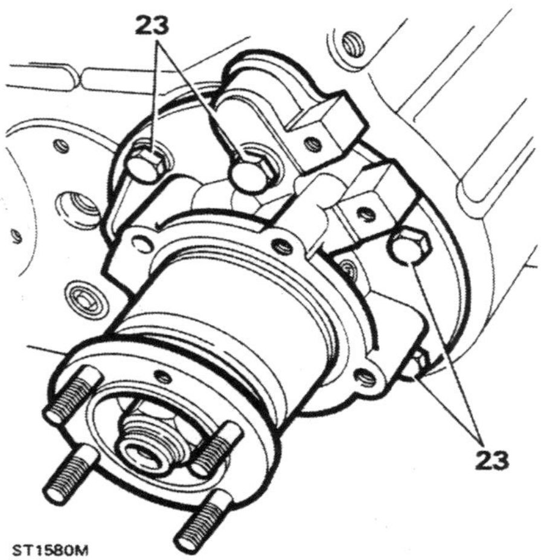

## Transfer case overhaul—dismantling

25. Remove the studs and dowels.
26. Remove the magnetic drain plug and filler/level plug.
27. Drift out differential rear bearing track.
28. Clean all areas of the transfer casing ensuring all traces of 'Loctite' are removed from faces and threads.

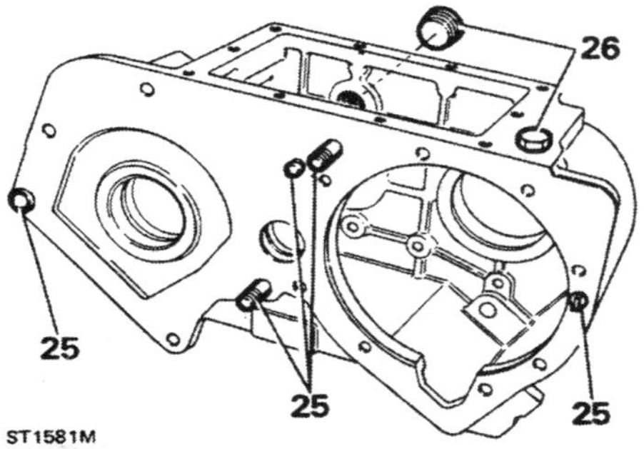

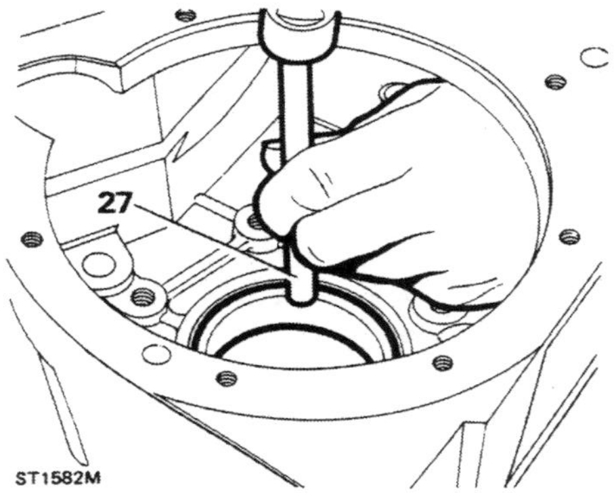

## Transfer case overhaul—reassembling

29. Fit studs and dowels to front face of the transfer casing.

    NOTE: The position of the radial dowel blade is set in line with the circle which is formed by the front output housing fixing holes.

30. Refit magnetic drain plug with new copper washer and tighten to the specified torque, loosely fit the filler/level plug.

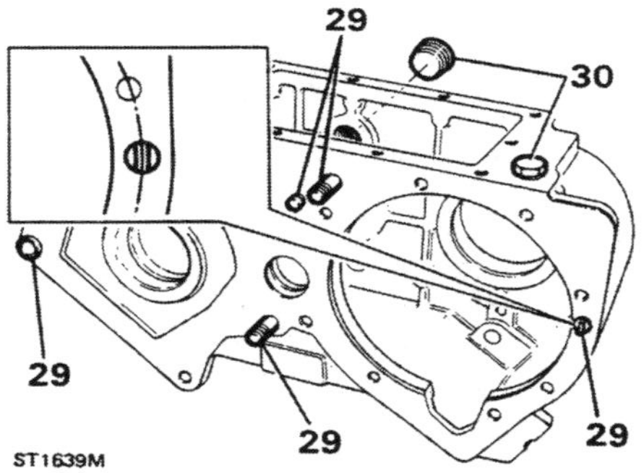

Continued

# LT230T TRANSFER GEARBOX

## Rear output housing overhaul—dismantling

1. Using flange wrench 18G1205 and socket spanner, remove the flange nut, steel and felt washers. Ensure flange bolts are fully engaged in the wrench.
2. Remove output flange with circlips attached. If necessary, use a two-legged puller.

   NOTE: The circlip need only be released if the flange bolts are to be renewed.

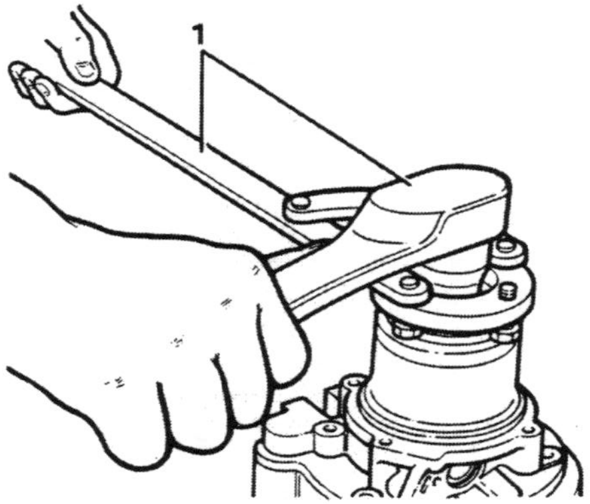

3. Remove speedo-drive housing. This can be prised out with a screwdriver.
4. Remove housing from the vice and drift out the output shaft, by striking the flange end of the shaft.
5. Carefully prise off the oil catch ring using a screwdriver in the slot provided.
6. Prise out and discard the seal from the output housing using tool 18G 1271.
7. Using circlip pliers 18G257, remove the circlip retaining the bearing.
8. Drift out the bearing from the rear of the housing.
9. Remove speedometer gear (driven) from its housing.
10. Remove the 'O' ring and oil seal and discard.
11. Slide off spacer and speedometer drive gear from output shaft.
12. Clean all parts, renew the 'O' ring, oil seals, felt seal and flange nut. Examine all other parts for wear or damage and renew, if necessary.

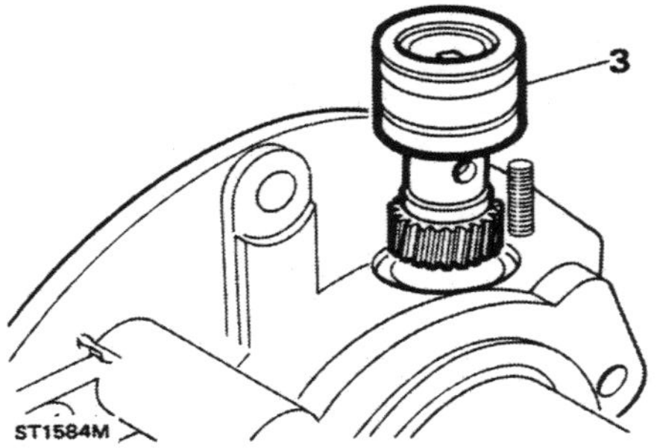

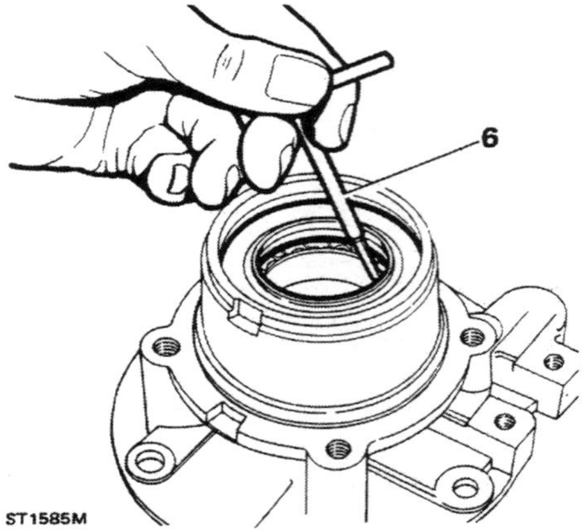

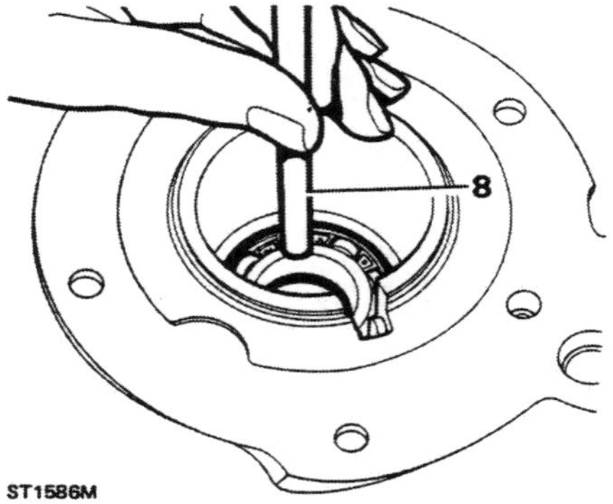

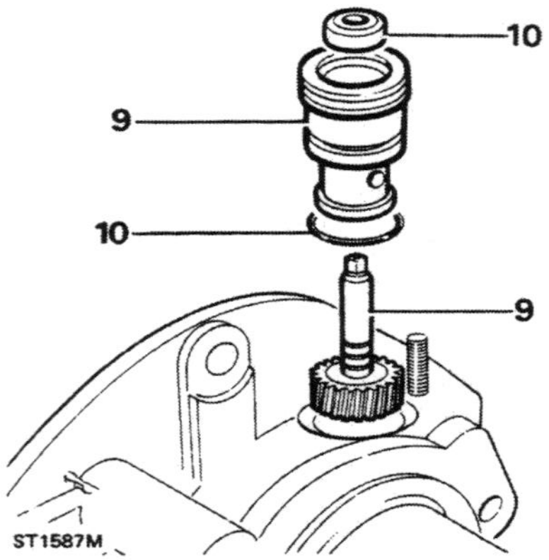

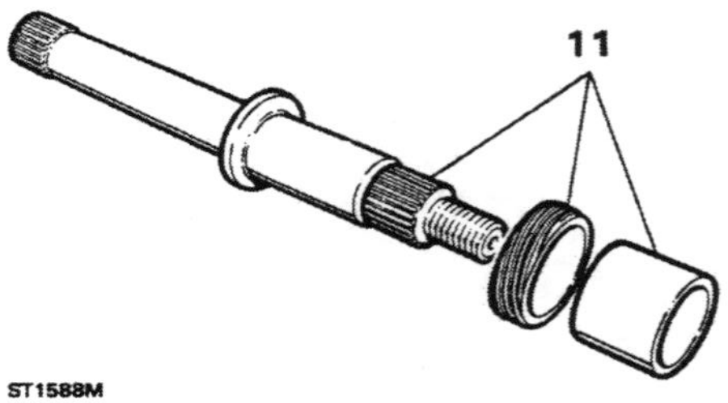

## Reassembling

13. Press output bearing into the housing. Do not use excessive force. To facilitate fitting the bearing, heat the output housing case. (This is not to exceed 100°C.)
14. Retain bearing with circlip, using circlip pliers 18G257.
15. Fit new seal (open side inwards) using tool 18G1422. The seal should just make contact with the bearing circlip.
16. Carefully charge the lips of the seal with clean grease and refit oil catch ring on to output housing.
17. Fit the 'O' ring and oil seal (open side inwards) to speedometer housing.
18. Lubricate the 'O' ring and seal with oil.
19. Locate speedometer gear (driven) in housing and press into position.

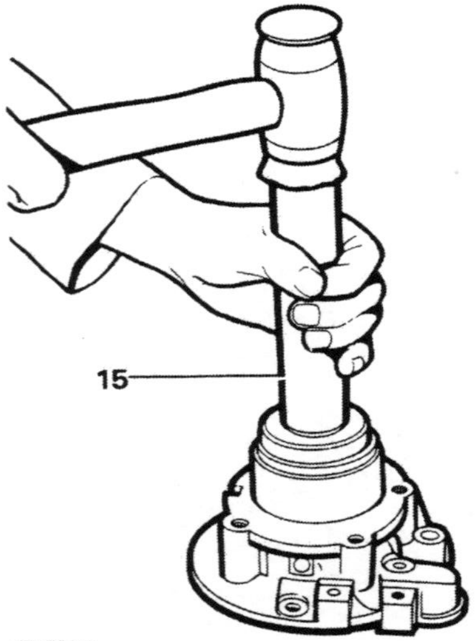

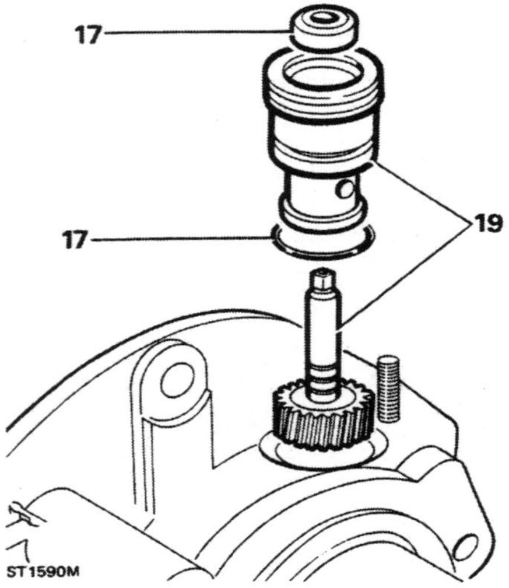

*Continued*

## LT230T TRANSFER GEARBOX

20. Slide drive gear and spacer on to the output shaft.
21. Locate output shaft into the bearing in the housing and drift into position.
22. Locate speedometer gear (driven) housing assembly into the output housing and press in until flush with the housing face.

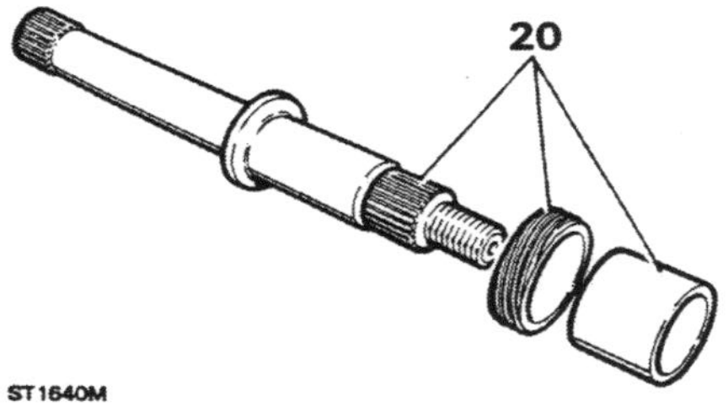

### Centre differential unit overhaul—dismantling

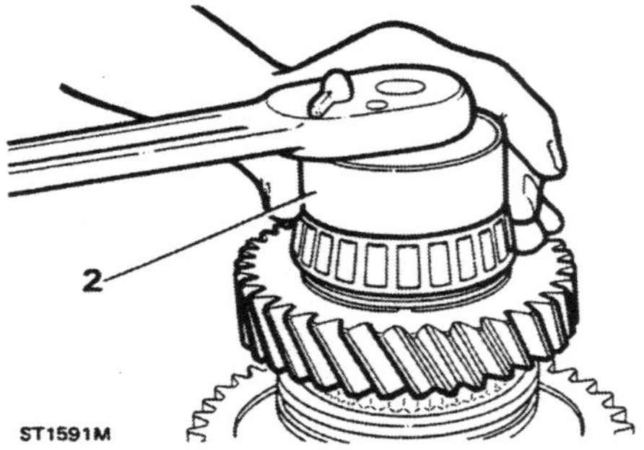

1. Secure centre differential unit to a vice fitted with soft jaws, and release stake nut from recess.
2. Remove stake nut using tool 18G1423 and suitable socket wrench.
3. Remove the differential unit from the vice.
4. Secure hand press MS47 in vice with collars 18G47BB and using button 1847BB/3 remove the rear taper bearing and collars.
5. Remove the high range gear and bush, taking care not to disturb the high/low sleeve.
6. Mark the relationship of the high/low sleeve to the hub and then remove the sleeve.
7. Using a suitable press behind the low range gear carefully remove the high/low hub and low range gear.
8. Substituting collar LST47-1 remove front taper roller bearing.
9. Remove hand press from the vice.
10. Using soft jaws secure the differential unit in the vice by gripping the hub splines.
11. Remove the eight retaining bolts and lift off the front part of the differential unit.
12. Release the retaining ring and remove front upper bevel gear and thrust washer.
13. Remove the pinion gears and dished washers along with the cross shafts.
14. Remove the rear lower bevel gear and thrust washer from the rear part of the differential unit.
15. Remove the rear differential unit from the vice and clean all components; examine for wear or damage and renew if necessary.
16. Clean all components; examine for wear or damage and renew if necessary.
17. Using soft jaws secure the rear differential unit in the vice by gripping the hub splines.
18. Ensure that all differential components are dry to assist in checking end-float.
19. Using a micrometer, measure one of the bevel gear thrust washers and note the thickness.

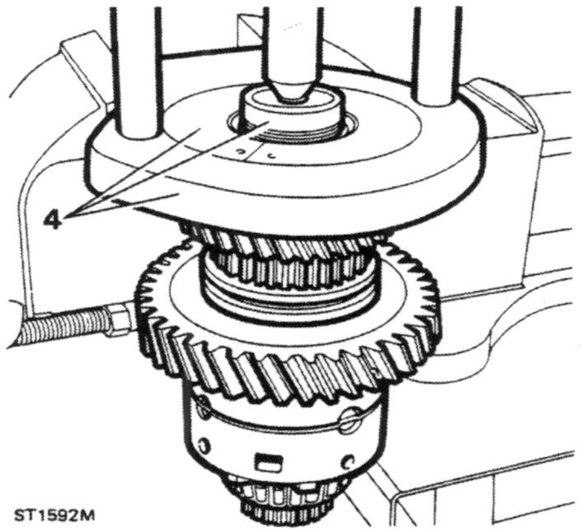

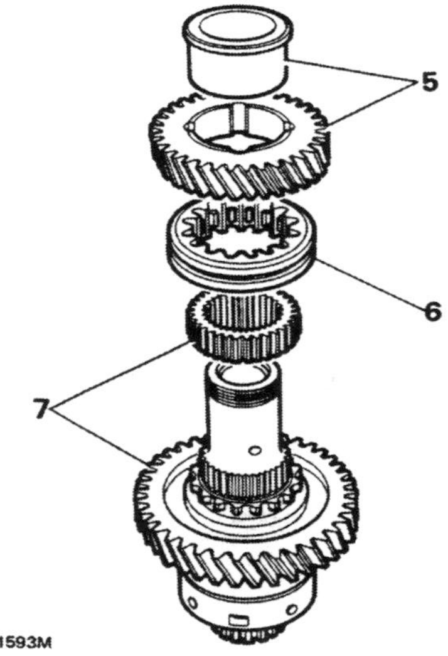

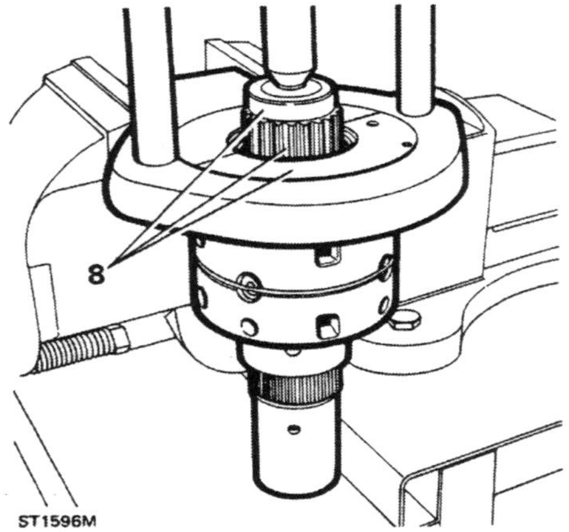

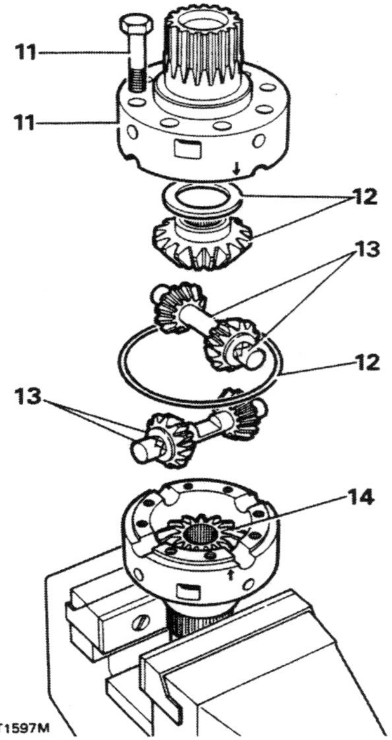

*Continued*

# LT230T TRANSFER GEARBOX

42. Locate the rear differential bearing on to the hub and press it into position using the smaller end of tool 18G1424.
43. Fit the stake nut and tighten to the specified torque using tool 18G1423.
44. Check the end float of the high and low range gears 0,05 to 0,15 mm (0.002 to 0.005 in).

**NOTE: If the clearances vary from those specified in the data, the assembly must be rebuilt using the relevant new parts.**

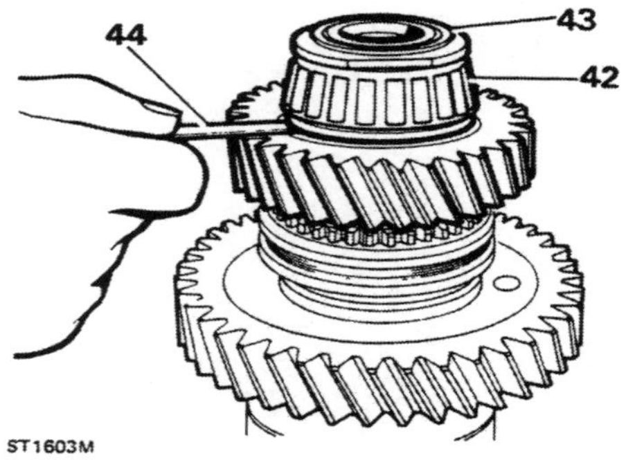

35. Align both units as previously described and secure with the eight bolts to the specified torque.
36. Finally check that the differential gears rotate freely. Locate the front differential bearing onto the front, upper differential shaft and press into position using larger end of tool 18G1424 as shown.
37. Invert the differential unit and secure in the vice.

**NOTE: During the following sequences all parts should be lubricated as they are fitted.**

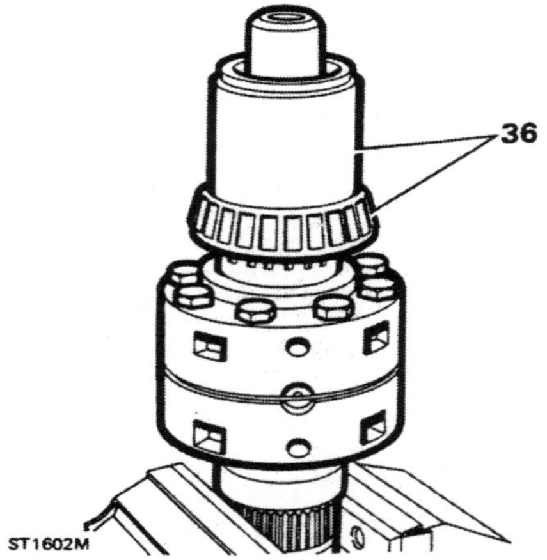

38. Fit the low range gear, with its dog teeth uppermost to the differential assembly.
39. Press the high/low hub on to the differential splines.
40. Slide the high/low selector sleeve on to the high/low hub ensuring that the alignment marks are opposite each other.
41. Fit the bush into the high range gear so that the flange is fitted on the opposite side of the gear to the dog teeth. Slide the bushed gear on to the differential assembly with the dog teeth down.

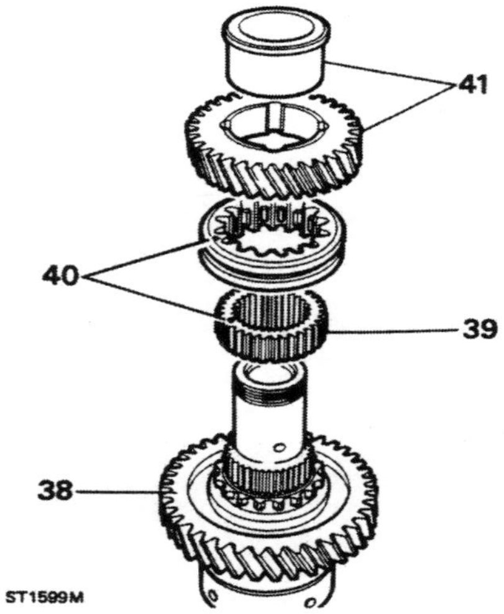

## Reassembling

31. Fit the selected thrust washer and bevel gear into the rear lower differential unit.
32. Assemble both pinion assemblies and dished washers on to their respective shafts and fit the rear differential unit. Secure the assemblies with the retaining ring.
33. Lubricate all the components.
34. Fit the selected thrust washer and bevel gear into the front upper differential unit.

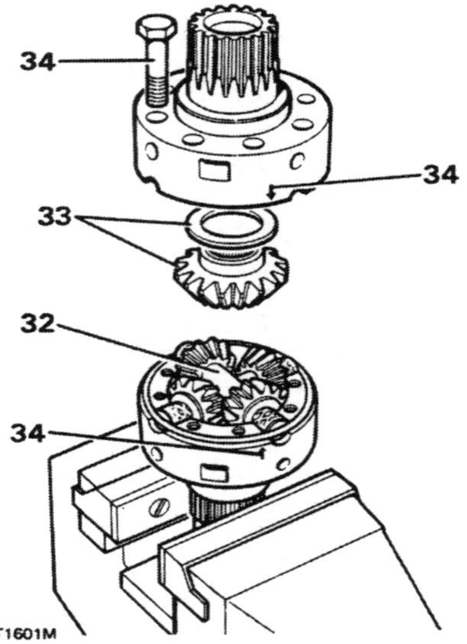

20. Fit the thrust washer and bevel gear to the rear lower differential unit.
21. Assemble both pinion assemblies and dished washers on to their respective shafts and fit to the rear differential unit.
22. Measure the front upper bevel gear thrust washer and note the thickness.
23. Fit the thrust washer and bevel gear to the front unit.
24. Refit the retaining ring and front differential unit, aligning the two engraved arrows marked on both halves of the unit.
25. Fit four bolts equi-spaced and torque to the correct figure.
26. Measure the front bevel gear end-float with feeler gauges through the slots provided in the front differential unit. The end-float must be 0,025 to 0,075 mm (0.001 to 0.003 in) maximum. When measuring use two sets of feeler gauges, one on each side of the front differential unit. This will give a true reading of the end-float.
27. Invert the differential unit and repeat operation 26 for the rear bevel gear end-float.
28. Invert the differential unit and secure in vice and remove the four bolts and lift off the front differential unit.
29. Remove the retaining ring, bevel gear and the washer and both pinion assemblies.
30. Select the correct thrust washers required for final assembly.

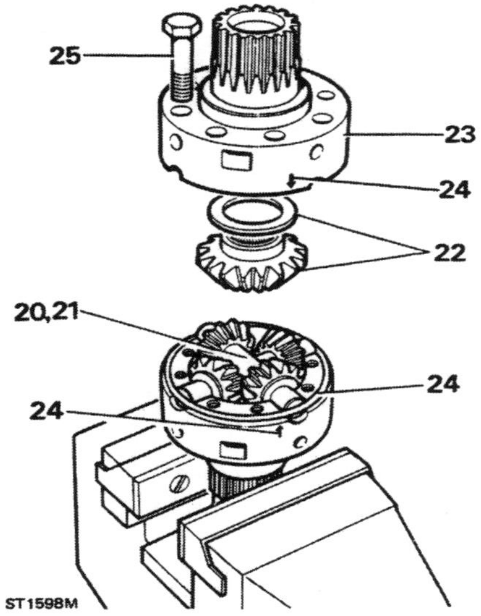

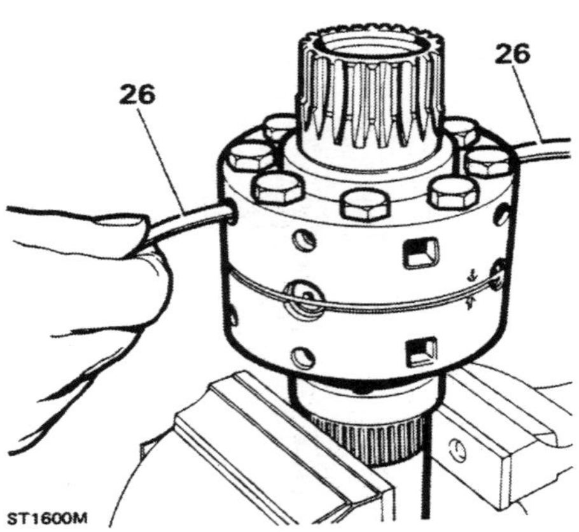

# LT230T TRANSFER GEARBOX

45. Peen the stake nut collar by carefully forming the collar of the nut into the slot as illustrated.

   **CAUTION: A round nose tool must be used for this operation to avoid splitting the collar of the nut.**

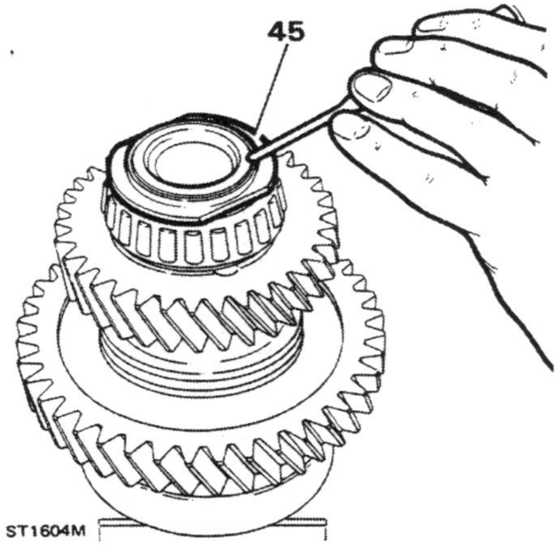

46. Clean and check high/low selector fork assembly for wear and renew if necessary.
47. To renew the selector fork remove the square set screw and slide the fork from the shaft.
48. Fit the new selector fork with its boss towards the three detent grooves. Align the tapped hole in the fork boss with the indent in the shaft nearest to the detent grooves.
49. Apply Loctite 290 to the set screw threads and fit the set screw and tighten to the specified torque.

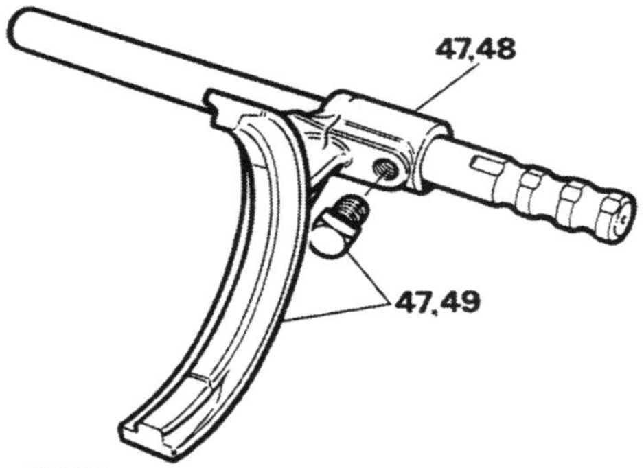

## Centre differential rear bearing track

50. Fit the differential rear bearing track 1,00 mm (0.039 in) below the outer face of casing using a suitable tool as shown.

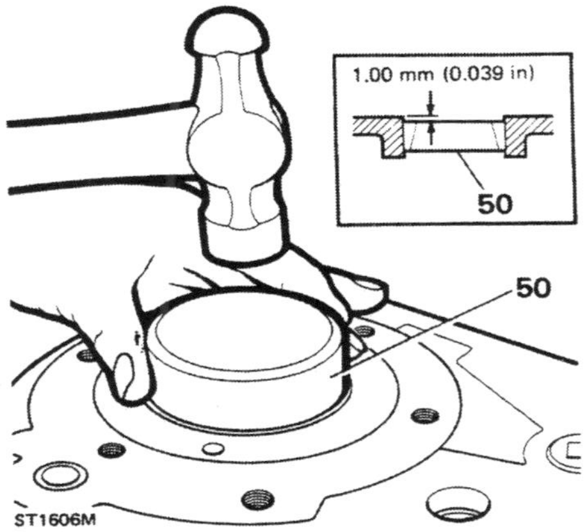

## Centre differential unit refit

1. Fit the selector fork/shaft assembly to the high/low selector sleeve on the differential assembly, with detent groove to the rear of the differential assembly.
2. Locate the differential assembly complete with selector fork into the transfer box casing. It may be necessary to rotate the output shaft to ease fitment, and engage selector shaft into its hole.

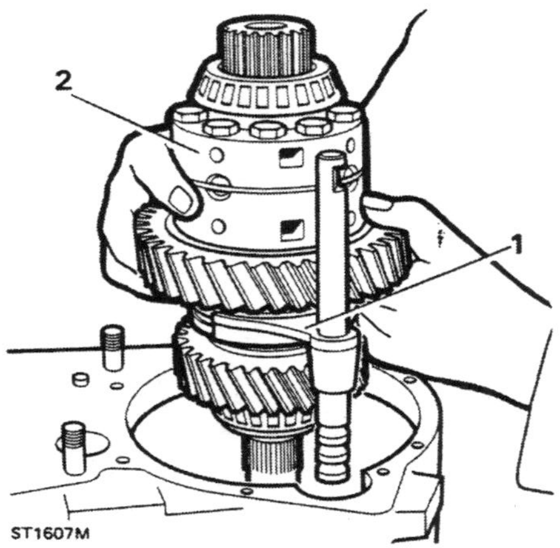

## Front output housing overhaul—dismantling

1. Unscrew seven retaining bolts and washers and remove the differential lock selector side cover and gasket.
2. Unscrew three retaining bolts and washers and lift the differential lock finger housing and actuator assembly from the front output housing.
3. Slacken the locknut and unscrew the differential lock warning light switch.
4. Remove selector shaft detent plug, spring and ball using a suitable magnet.
5. Compress the selector fork spring and remove the two spring retaining caps.

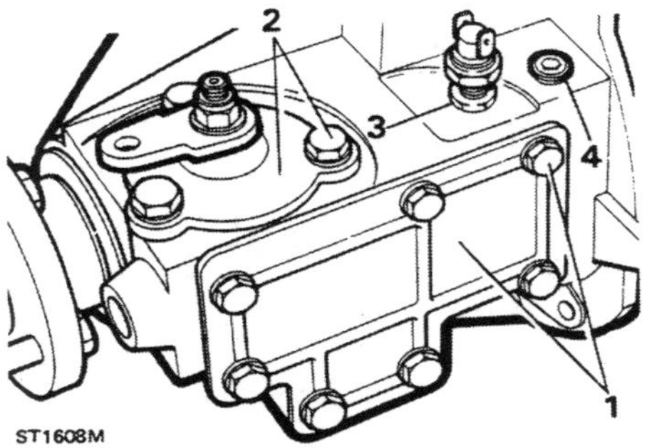

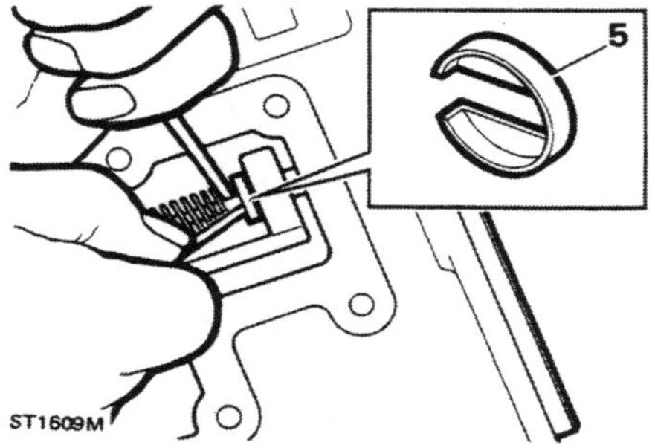

## Rear output housing—refit

1. Grease output housing gasket and position on to the rear face of the transfer box casing.
2. Fit output housing and ensure clearance of 100 mm (0.039 in) between housing face and gasket.
3. Fit the six output housing bolts with Loctite 290 on the threads, with washers and tighten evenly to the correct torque, which will pull the rear bearing into position.

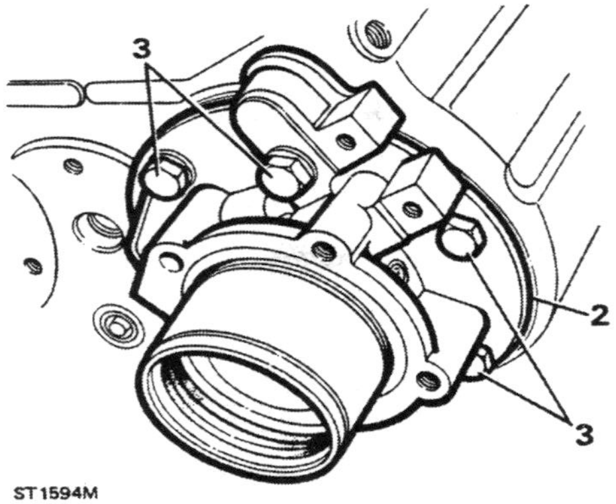

3. Fit selector shaft ball and spring through the side of the transfer box casing.
4. Apply Loctite 290 to detent plug; fit and locate, by screwing gently fully home and then unscrewing two turns.

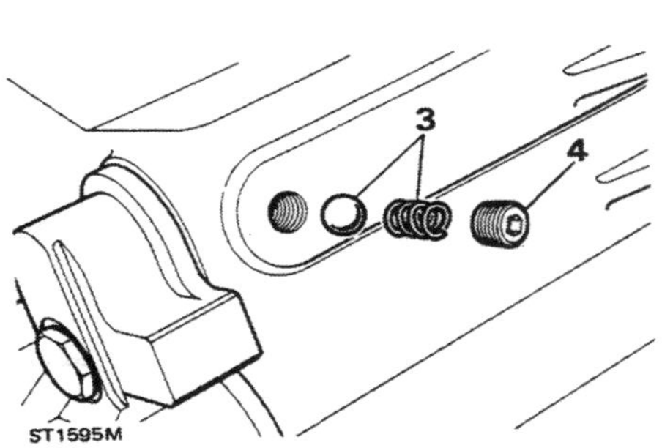

*Continued*

# LT230T TRANSFER GEARBOX

6. Withdraw the selector shaft from the rear of the output housing.
7. Remove the selector fork and spring through the side cover aperture.
8. Remove lock-up sleeve from the rear of the output housing. 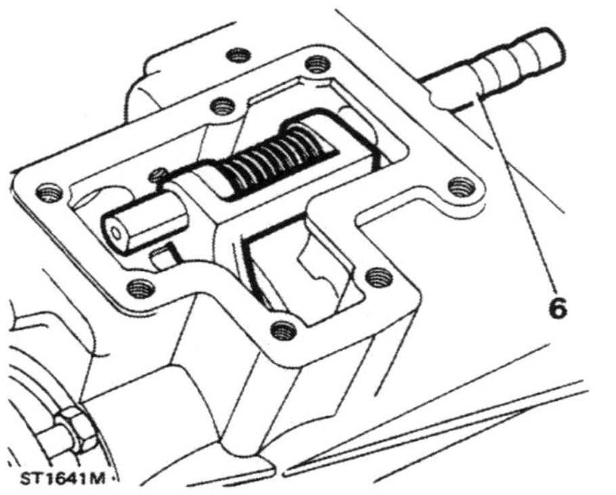
9. Using flange wrench 18G1205 and socket wrench, remove the flange nut, steel and felt washers.

   NOTE: Ensure that flange bolts are fully engaged in the wrench.

10. Remove the output flange with oil seal shield.

    NOTE: These parts need not be separated unless the flange bolts are to be renewed.

11. Drift output shaft rearwards from housing using a soft headed mallet.
12. Slide off the collar from the output shaft. 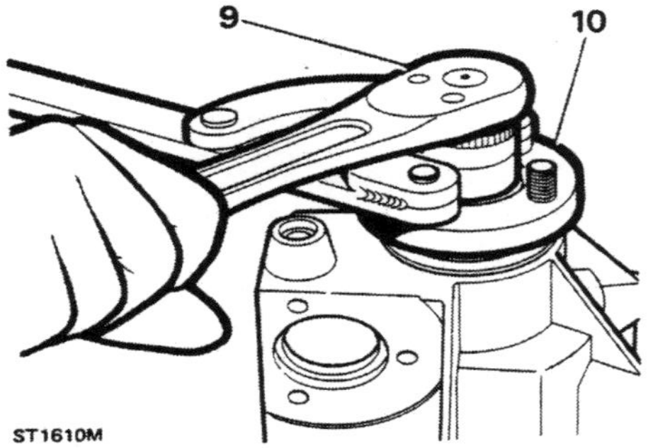
13. Prise out and discard oil seal from output housing using service tool 18G1271.
14. Remove circlip with circlip pliers 18G257. 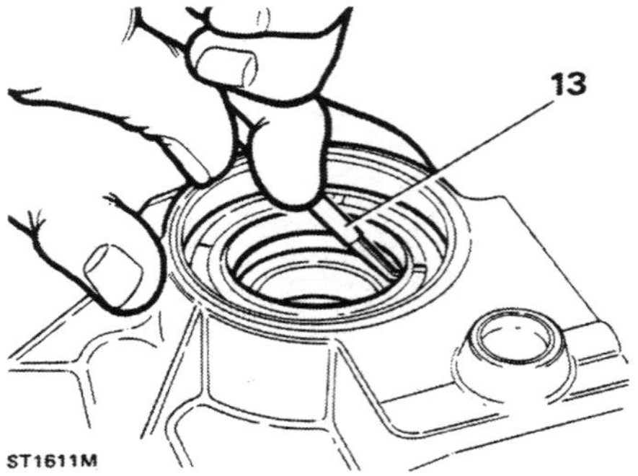
15. Invert housing and drift out bearing from inside the case as shown. 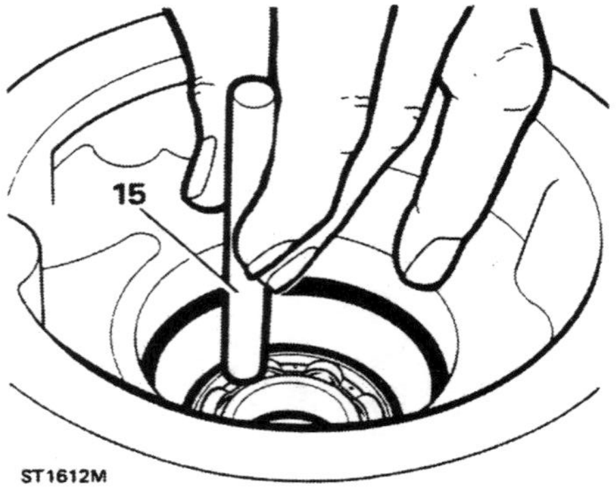
16. Drift out centre differential front taper roller bearing track and shim.
17. Drift out selector shaft cup plug from housing.
18. Clean all components ensuring all traces of 'Loctite' are removed from faces and threads.
19. Examine components for wear or damage and renew if necessary.

    NOTE: Renew oil seal and felt seal and flange nut.

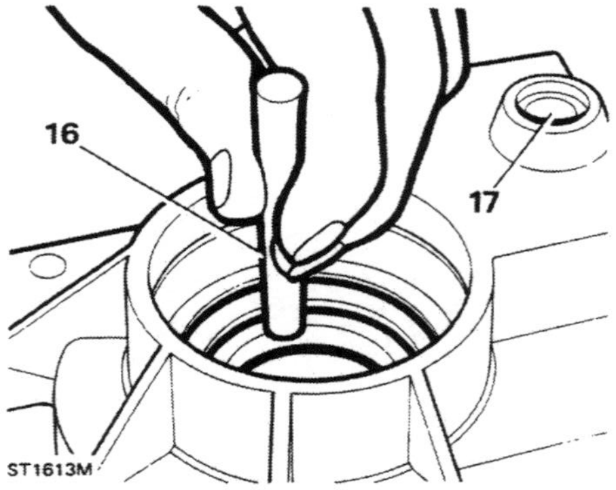

## Reassembling

20. Press the bearing into the housing; do not use excessive force. To facilitate fitting the bearing, heat the front output housing. (This is not to exceed 100°C).
21. Using circlip pliers 18G257, fit the bearing retaining clips.
22. Fit a new oil seal (open side inwards) using replacer tool 18G1422, until the seal just makes contact with the circlip.
23. Carefully charge the lips of the seal with clean grease.
24. Slide collar on to the output shaft, with its chamfered edge towards the dog teeth.
25. Fit the output shaft through the bearing and drift home.

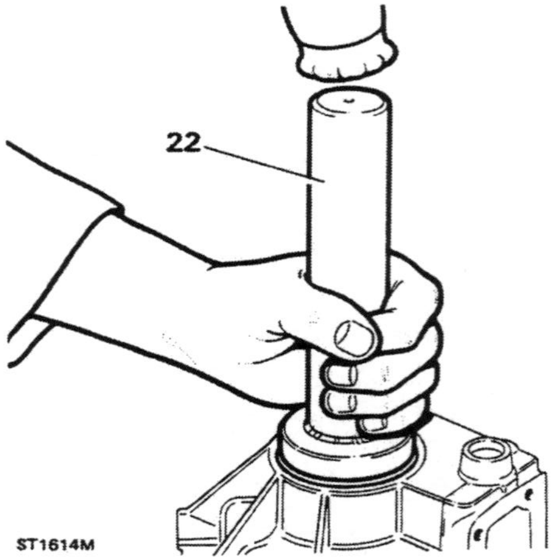

## Adjusting front differential bearing pre-load

26. Measure original differential front bearing track shim.
27. Refit original shim into input housing.
28. Drift differential front bearing track into the housing. 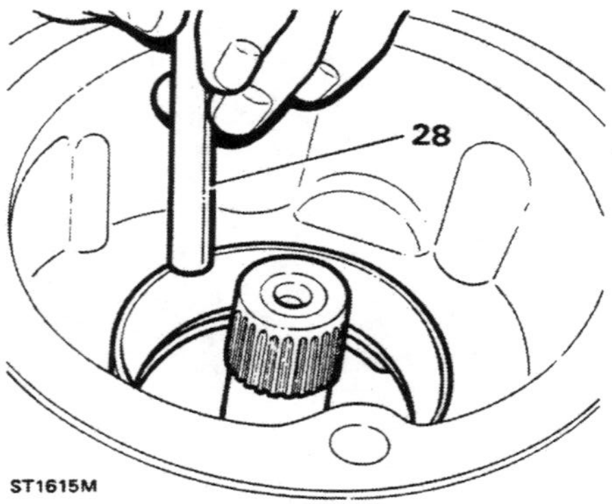
29. Grease and fit new gasket and locate the front output housing on the transfer box casing.
30. Secure housing with the eight retaining bolts and washers, the upper middle bolt being longer than the rest. Do not tighten the bolts at this stage.
31. Engage high or low gear.
32. Check the rolling resistance of the differential using a spring balance and a length of string wound around the exposed splines of the high/low hub.

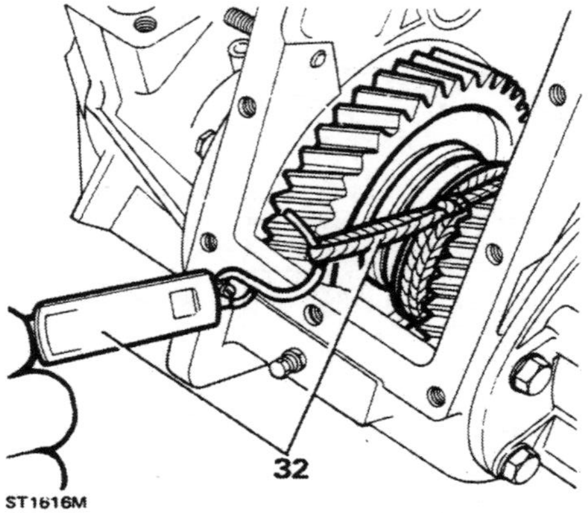

Continued

# LT230T TRANSFER GEARBOX

33. With the correct shim fitted the load to turn should be 1.36 kg to 4.53 kg (3 lb to 10lb). This applies to new or used bearings. (New bearings will register at the top end and used bearings will register at the low end.
34. If the reading is in excess of the above measurements, remove the front output housing assembly from the transfer box casing.
35. Using a suitable extractor, withdraw the centre differential bearing track and change the shim for one of a suitable thickness. (A thinner shim will reduce the rolling resistance).
36. Fit the new shim and drift the differential bearing track back into its housing until fully home.
37. Having obtained the load to turn, prop-up the transfer box casing on the bench with the front face uppermost.
38. Apply Loctite 290 to the threads of the housing retaining bolts and fit the eight bolts and washers into the front output housing and secure to transfer box casing.
39. Fit front output flange, felt washers, steel washers and flange nut.
40. Using flange wrench 18G1205 and torque wrench, pull the output shaft up to the correct position. Check that the oil seal shield does not foul the housing.

NOTE: Ensure that the flange bolts are fully engaged in the wrench.

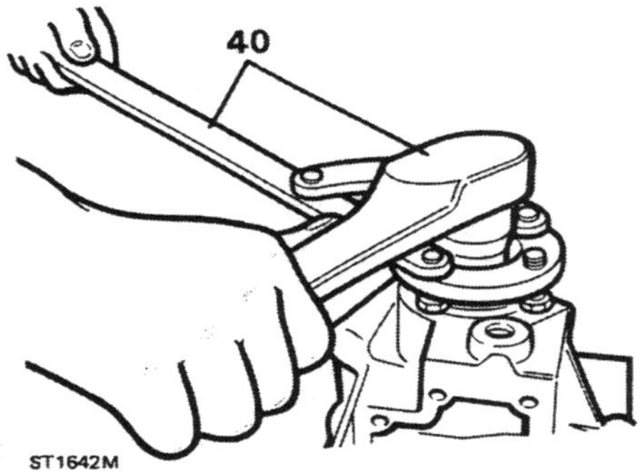

41. Repeat the above operation for the rear output flange.
42. Compress the selector shaft spring and fit to the selector fork.
43. Locate selector fork through front output housing side cover aperture, ensuring that the fork engages in the groove of the lock-up sleeve.
44. Fit selector shaft through the aperture in the front of the output housing and pass it through the selector fork lugs and spring into the rear part of the housing.
45. Rotate the selector shaft until the two flats for the spring retaining caps are at right angles to the side cover plate face.
46. Compress the spring between the fork lugs and slide the retaining caps on to the shaft ensuring the spring is captured with the 'cupped' side of the caps.
47. Drift selector shaft seal cup into position.
48. Fit selector shaft detent ball and spring in the tapped hole on top of the output housing.
49. Apply Loctite 290 to detent plug threads. Screw detent plug gently home and then unscrew two turns.

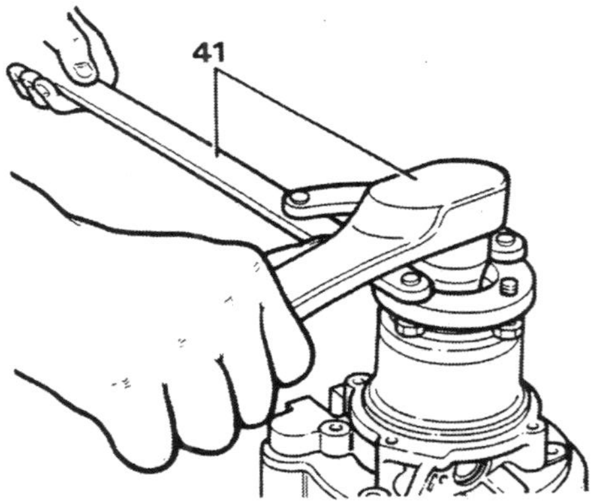

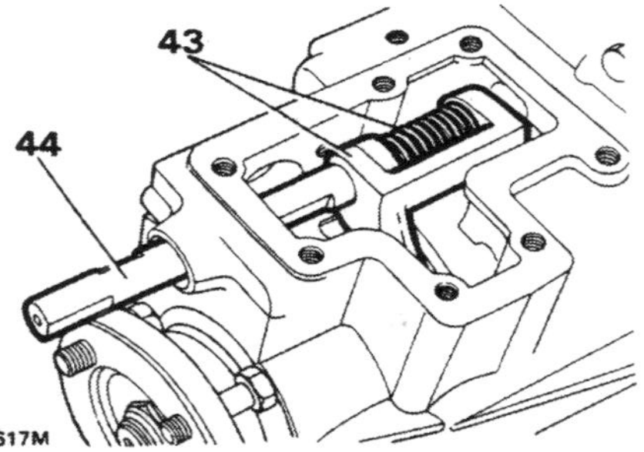

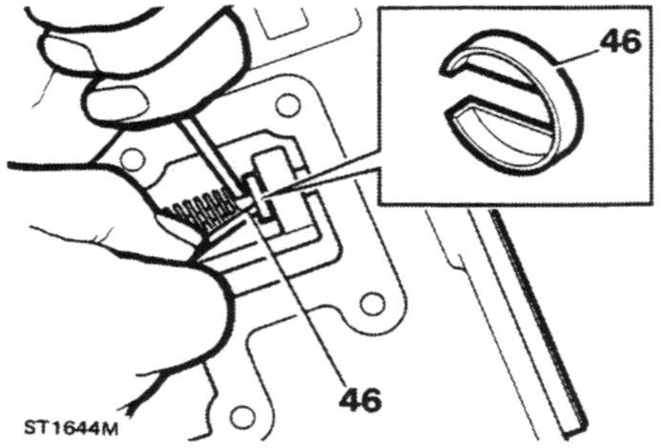

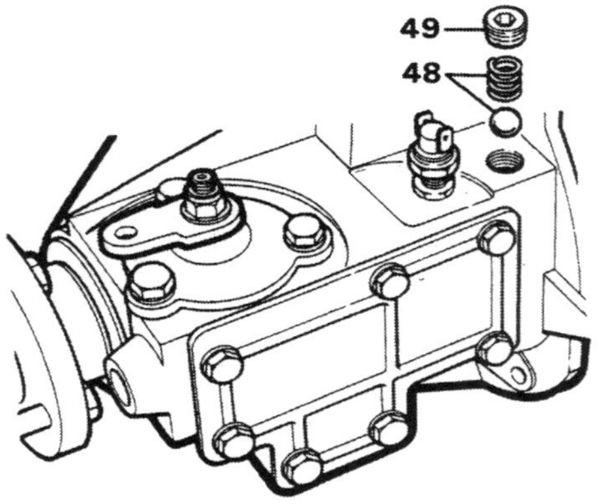

7. Fit the differential lock lever over the pivot shaft so that the lever will face forward to the bend upwards. This lever is then in the correct operating position.
8. Retain the lever with a plain washer and new 'nyloc' nut.

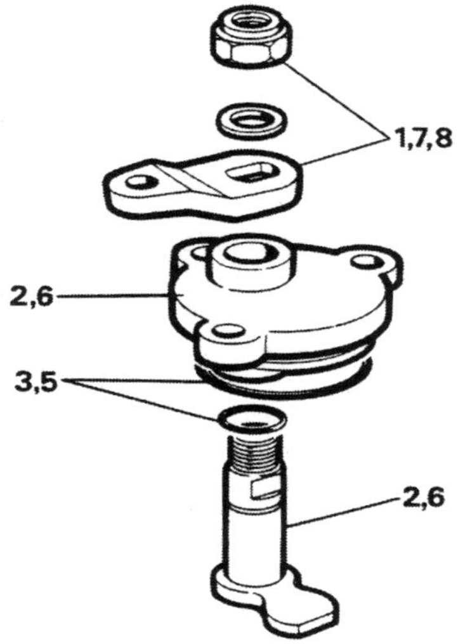

9. Fit the differential lock finger housing into its seating on the front output housing, ensuring that the selector finger is located in the flat of the selector shaft.
10. Apply Loctite 290 to the bolt threads and retain the lock finger housing with the three bolts and washers to the specified torque.

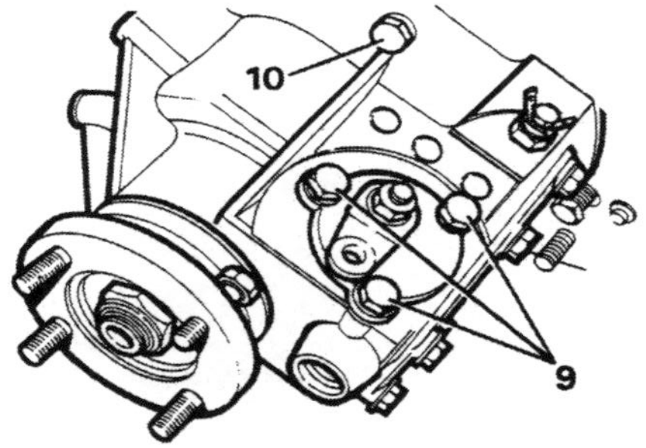

Continued

## Differential lock finger housing overhaul — dismantling

1. Unscrew and discard the 'nyloc' nut and remove the operating lever and washer.
2. Remove the pivot shaft from lock finger housing.
3. Remove the 'O' rings from the pivot shaft and housing and discard.
4. Clean all components; examine for wear or damage and renew if necessary.

### Reassembling

5. Fit new 'O' rings on to pivot shaft and lock finger housing and lubricate with oil.
6. Locate the pivot shaft in the housing.

# LT230T TRANSFER GEARBOX

## High/low cross-shaft housing overhaul

1. Remove the selector finger grub screw and withdraw the cross-shaft from the cross-shaft housing and remove the selector finger.
2. Remove the 'O' ring from the cross-shaft.
3. Drift out selector housing cup plug if necessary.
4. Clean all the components and check for damage or wear, replace if necessary.
5. Apply sealant to a new cup plug and fit so that the cup is just below the chamfer for the cross-shaft bore.
6. Fit new 'O' ring to cross-shaft.
7. Lubricate the shaft and insert into the cross-shaft housing.
8. Fit selector finger ensuring that it aligns with the recess in the cross-shaft.
9. Apply Loctite 290 to the grub screw and secure the selector finger to the cross-shaft and fully tighten to the specified torque.
10. Grease and fit the high/low selector housing gasket on the front output housing.
11. Fit high/low cross-shaft housing, ensuring that the selector finger locates in the slot of the selector shaft, and secure with six bolts and washers to the specified torque.

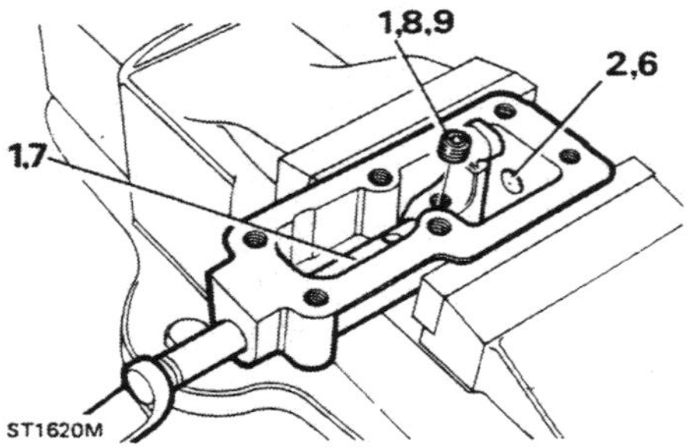

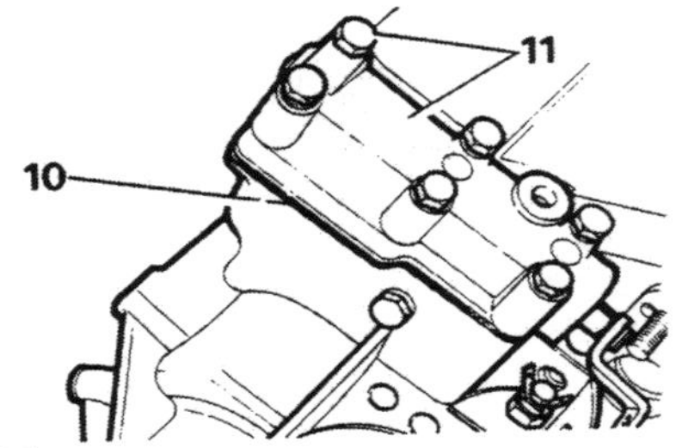

## Input gear overhaul—dismantling

1. Clean the input gear assembly and examine for wear or damage. Remove the bearings only if they are to be renewed.
2. Secure hand press MS47 in the vice and using collars 18G47-7 and button 18G47-BB/3, remove rear taper roller bearing from input gear assembly.
3. Invert input gear assembly in hand press and remove front taper roller bearing.
4. Clean input gear.

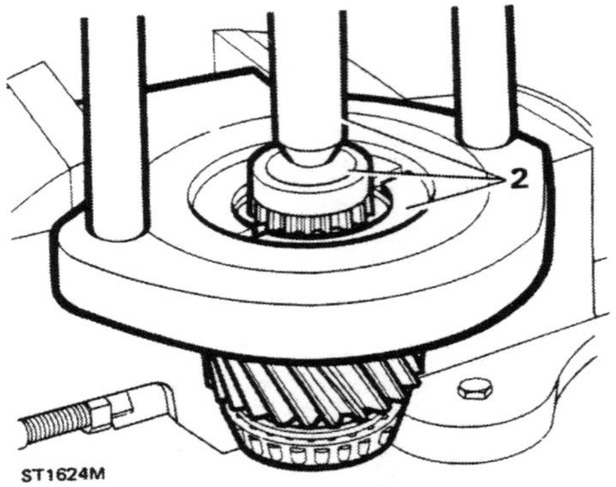

## Reassembling

5. Position rear taper roller bearing on input gear and using hand press MS47 and collars 18G47-7 press the bearing fully home.
6. Invert input gear and fit the front taper roller bearing using the press and collars.

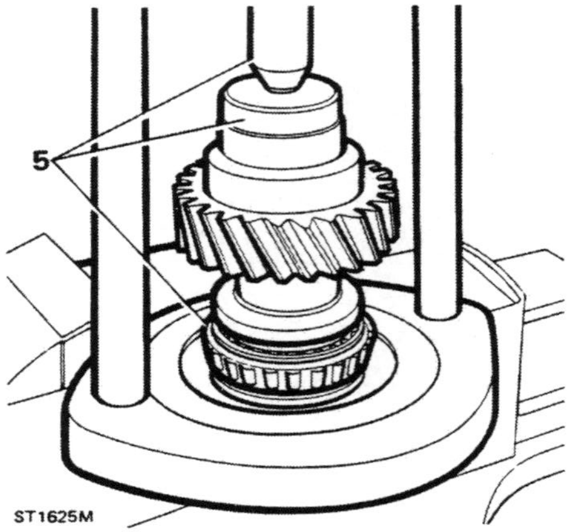

## Checking input gear bearing pre-load

7. Prop up the transfer box casing on the bench with the rear face uppermost.
8. Drift in the front taper bearing track.
9. Reposition transfer box casing so the front face is uppermost and fit oil seal (open side inwards) using replacer tool 18G1422.
10. Lubricate both bearings with clean oil.
11. Fit the input gear assembly into the transfer box casing with the dog teeth uppermost.
12. Secure bearing support plate in the vice. Drift out input bearing track, and remove shim.
13. Clean bearing support plate and shim. Measure original shim and note its thickness.
14. Fit the original shim to the support plate.
15. Locate the bearing track in the support plate and press fully home.
16. Apply grease to the gasket and fit on to the transfer box casing.
17. Fit the bearing support plate on to the transfer box casing and secure with the six bolts, but do not tighten.
18. Fit the service tool LST105 to input gear and engage the spline.
19. Tie a length of string to the split pin and fit it to the service tool as shown.
20. Attach a spring balance to the string and carefully tension the spring until a load to turn the input gear is obtained. A pull of 2,26 kg to 6,80 kg (5 lb to 15 lb) is required.
21. If the reading obtained is outside the above limits, the original shim must be changed.

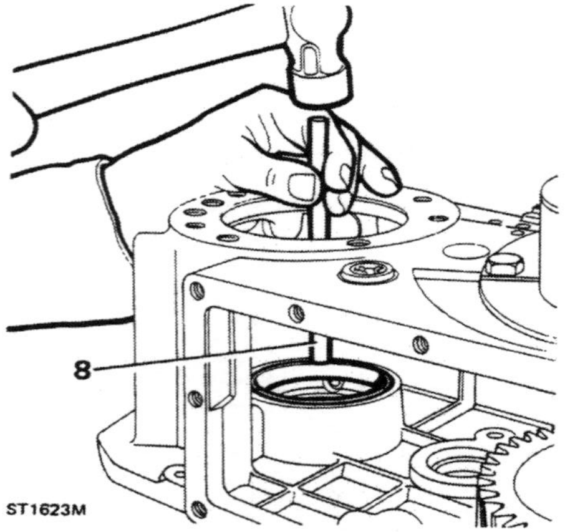

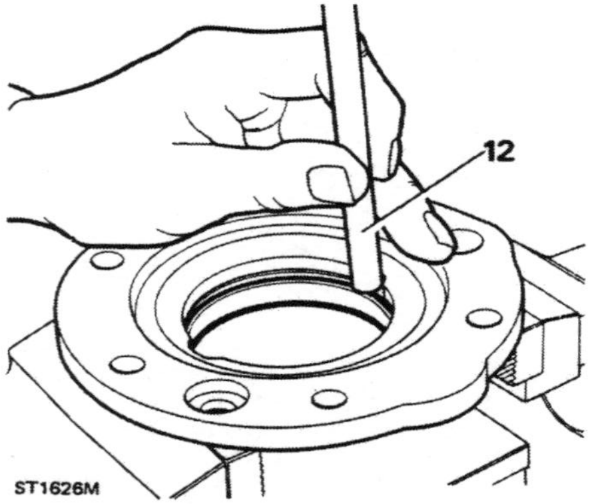

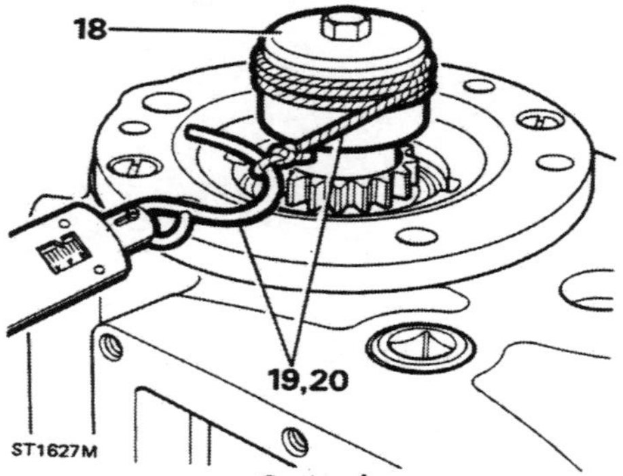

Continued

# LT230T TRANSFER GEARBOX   37

22. Remove the spring balance, string and service tool.
23. Remove the six bolts and the bearing support plate.
24. Drift out the input gear bearing track from the support plate and discard original shim.
25. Select the correct size shim to obtain a load to turn of 2,26 kg to 6,80 kg (5 lb to 15 lb).
26. Fit shim to support plate, locate bearing track and press home.
27. Fit bearing support plate and secure to transfer box casing with the six bolts (do not tighten).
28. Repeat the rolling resistance check as previously described, and note the value obtained.

## Intermediate gear assembly overhaul

1. Drift out intermediate gear bearing tracks.
2. Remove circlips.
3. Clean all intermediate gear components and lock plate. Check for damage or wear and replace as necessary.
4. Fit new circlips into the intermediate gear cluster.
5. Using tools LST550-4 and MS550 fit bearing tracks into the intermediate gear cluster.
6. Fit the 'O' rings to the intermediate shaft and into the intermediate shaft bore at the front of the transfer box casing.

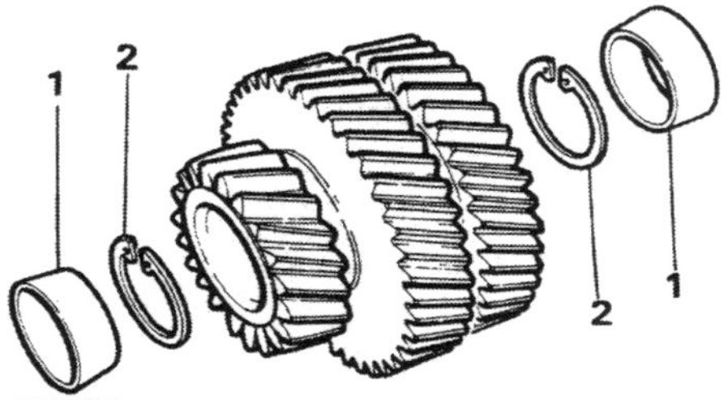

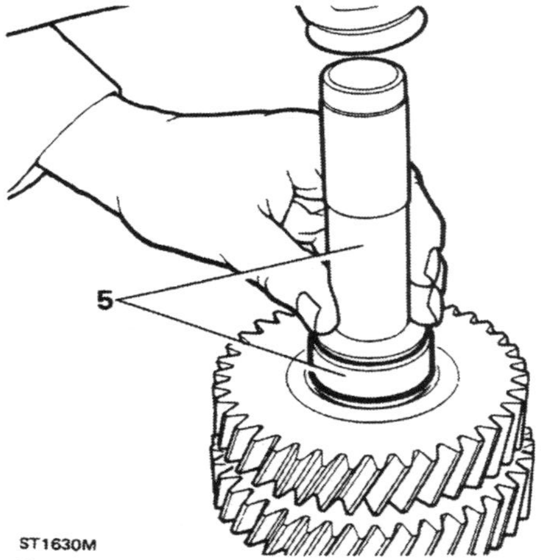

## Intermediate gear reassembly

7. Check for damage to the intermediate shaft thread and if necessary clean up with a fine file or stone.
8. Lubricate the taper roller bearings and intermediate gear shaft.
9. Insert new bearing spacer to gear assembly, followed by the taper roller bearings.
10. Fit dummy shaft LST104 into the intermediate gear cluster.
11. Locate the gear assembly into the transfer box casing from the bottom cover aperture.
12. Insert intermediate shaft from the front of the transfer box casing, pushing the dummy shaft right through as shown and remove. (Making sure that the intermediate gear cluster meshes with the input gear and high range and low range gears.)
13. Turn the intermediate shaft to allow fitting of retaining plate.
14. Fit retaining plate and secure with retaining bolt and washer.
15. Fit the intermediate gear shaft retaining stake nut.

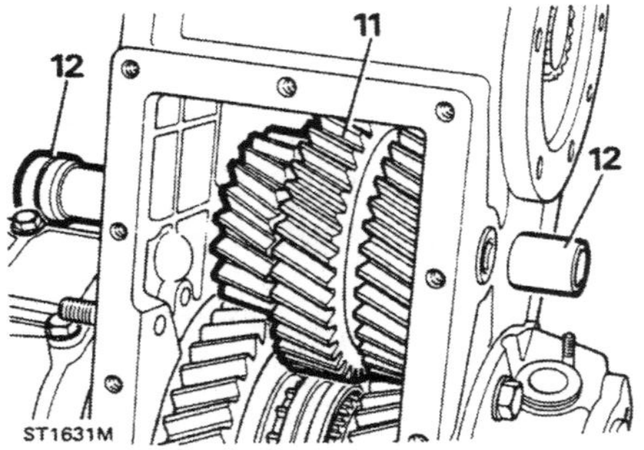

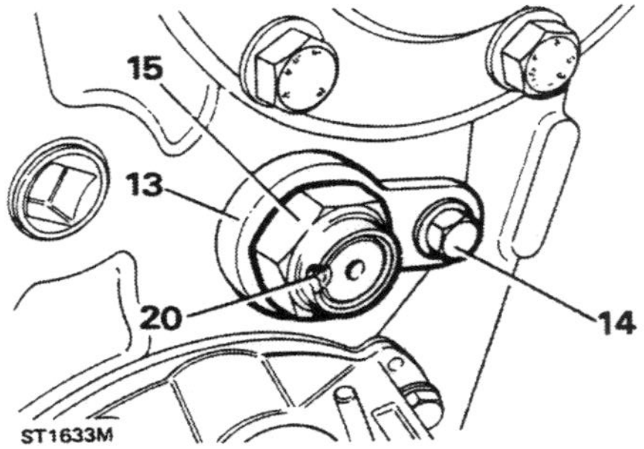

## Adjusting intermediate gear torque-to-turn

16. Select neutral.
17. Fit service tool LST105 to input gear and engage spline.
18. Tie a length of string to a split pin and fit to the service tool as shown. Attach the spring balance to the string.
19. To obtain the correct figures and to collapse the spacer within the intermediate gear cluster, tighten the intermediate shaft nut until the load-to-turn has increased by 3,7 kg (7 lb) ± 1,63 kg (± 3 lb) on that noted when checking input shaft load-to-turn. The torque to tighten the remaining nut will be approximately 203 Nm (150 lb ft).
20. Peel the stake nut by carefully forming the collar of the nut into the intermediate shaft recess, as illustrated.

**CAUTION: A round nose tool must be used for this operation to avoid splitting the collar of the nut.**

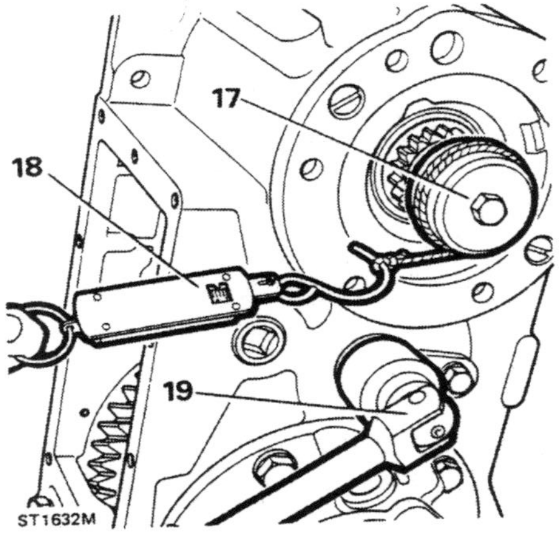

## Power take-off cover — reassemble

21. Clean power take-off cover and gasket face.
22. Fit the two countersunk screws and tighten.
23. Remove the six bolts from the bearing support plate.
24. Apply sealant to the cover plate gasket and fit to the bearing support plate.
25. Apply Loctite 290 to bolt threads and secure the power take-off cover with the six bolts and washers.

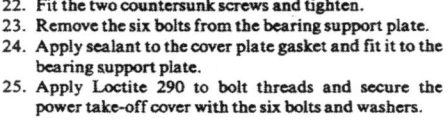

*Continued*

## AUTOMATIC GEARBOX 44

### FAULT DIAGNOSIS: ZF4HP22 Automatic Gearbox

Before refering to the fault diagnosis chart, ensure that the following static checks are firstly carried out:

### INITIAL STATIC CHECKS

| | |
|---|---|
| Check start positions | 'P' & 'N' only |
| Reverse lights | 'R' only |
| Gear engagements | N–D, N–3, N–2, N–1, N–R |
| Full throttle | Engine switched off, check full travel at engine and at pedal |
| Oil level | 'N' selected, engine running at normal running temperature |
| Pressure test | 150 ± 5 lb/in² 2000 rev/min |
| Idle pressure | 100 ± 5 lb/in² (see engine tune data for appropriate idle speeds applicable to model being tested) |

### FIT PRESSURE GAUGE

Service tools:
18G502A—0–300 PSI (0–22 kg/cm²) pressure gauge
18G502K—Flexible hose
LST502–1—Hose adaptor

1. Lift the carpet and sound deadening material from the passenger side of the vehicle, to reveal the plastic grommet in the floor well.
2. Prise the grommet from its location.
3. Plug the end of the pressure gauge hose to prevent ingress of dirt. Feed the hose 18G502K through the hole and manoeuvre it between the floor well and top of the chassis, until it is visible from beneath the vehicle.
4. Drive the vehicle onto a suitable hydraulic ramp.
5. From beneath the vehicle, remove the plug from the hose.
6. Remove the plug from the bottom of the gearbox and fit the adaptor LST502–1 and tighten securely.
7. Fit the hose to the adaptor and tighten securely.
8. Fit the gauge 18G502A to the other end of the hose within the vehicle.
9. Remove the vehicle from the ramp and carry out road test.

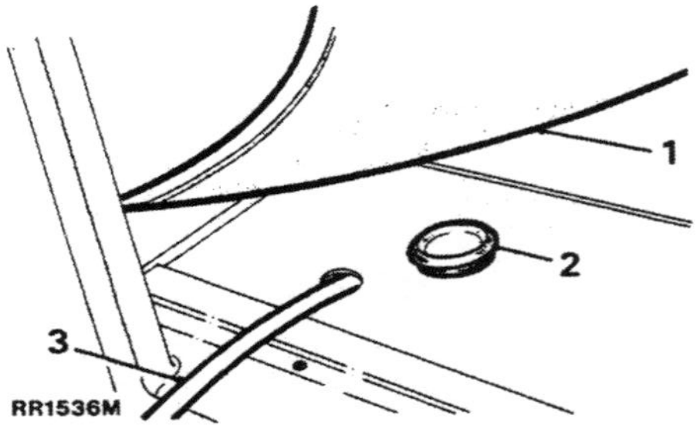

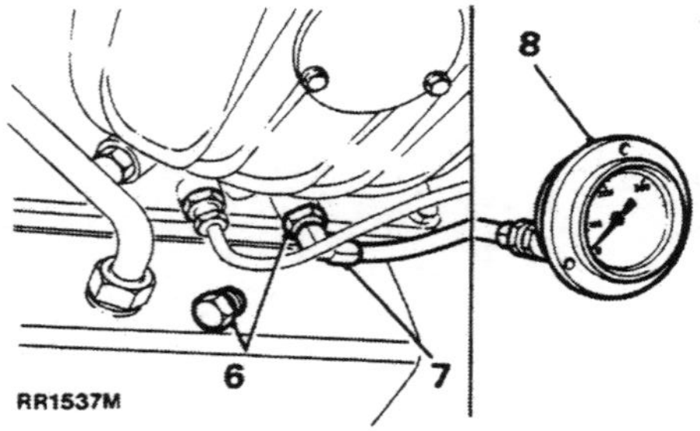

### Remove the pressure gauge

10. Reverse the instructions 1 to 8.

---

## LT230T TRANSFER GEARBOX 37

### Bottom cover—reassemble

26. Clean bottom cover and gasket face.
27. Apply sealant to cover gasket and fit to transfer box casing.
28. Apply Loctite 290 to bolt threads and secure the bottom cover with six bolts and washers.

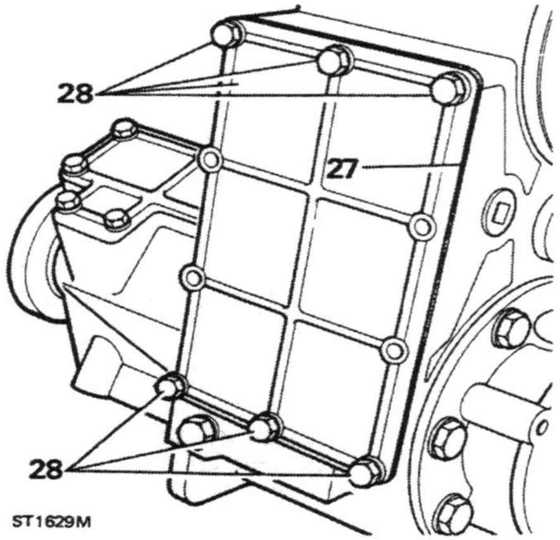

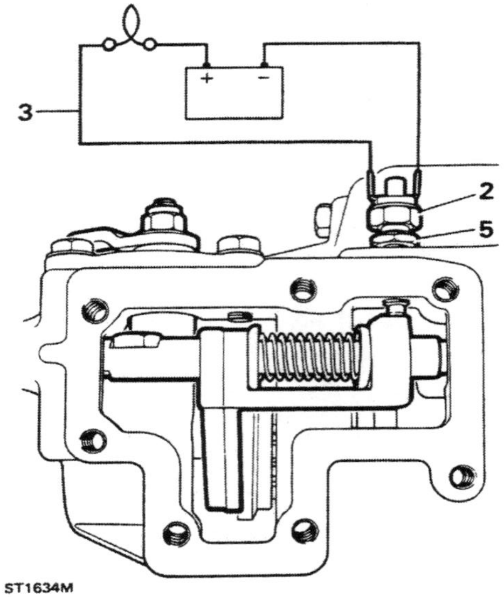

### Transmission brake—reassemble

1. Clean brake backplate and oil catcher and apply sealant to the catcher joint face.
2. Locate brake backplate on the rear output housing with the brake operating lever on the right side of the transfer box casing.
3. Secure the backplate (including the oil catcher) with the four special bolts and tighten using a hexagonal socket to the specified torque.
4. Clean and fit brake drum and secure with two countersunk screws.

### Differential lock switch adjustment

1. Select differential locked position by moving the lock taper towards the right side of the transfer box casing.
2. Apply sealant to the differential lock warning light switch and fit to the top of the front output housing.
3. Connect a test lamp circuit to the differential lock switch.
4. Screw in the lock switch until the bulb is illuminated.
5. Turn in the switch another half a turn and tighten with the locknut against the housing.
6. Disconnect the battery and move the differential lock lever to the left to disengage differential lock.
7. Clean the front output housing side cover.
8. Grease and fit side cover gasket.
9. Apply Loctite 290 to bolt threads, fit side cover and secure with seven bolts and washers.

# AUTOMATIC GEARBOX

## FAULT DIAGNOSIS—ZF4HP22 AUTOMATIC GEARBOX

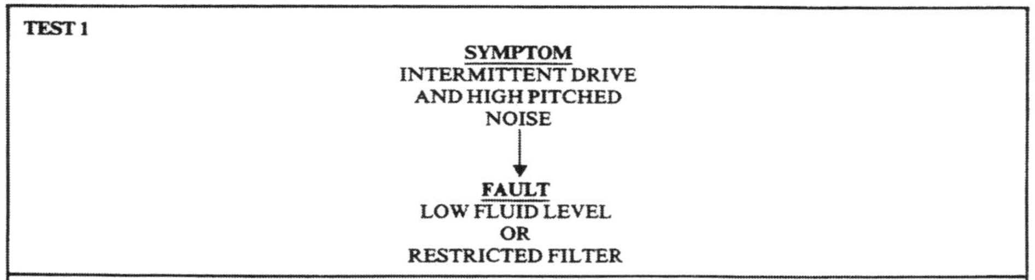

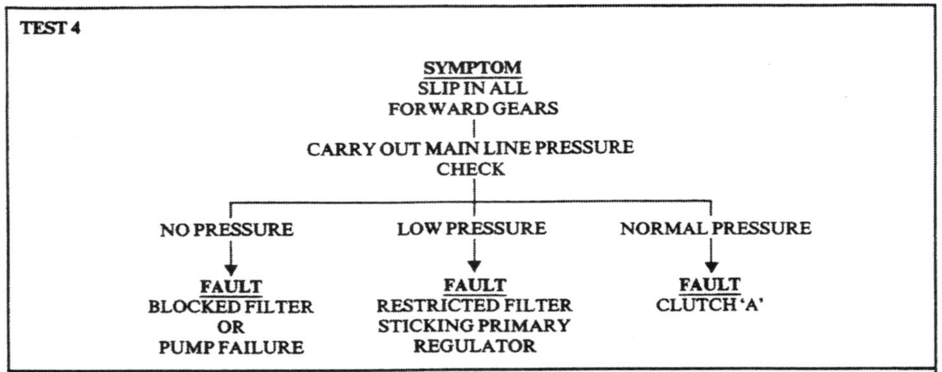

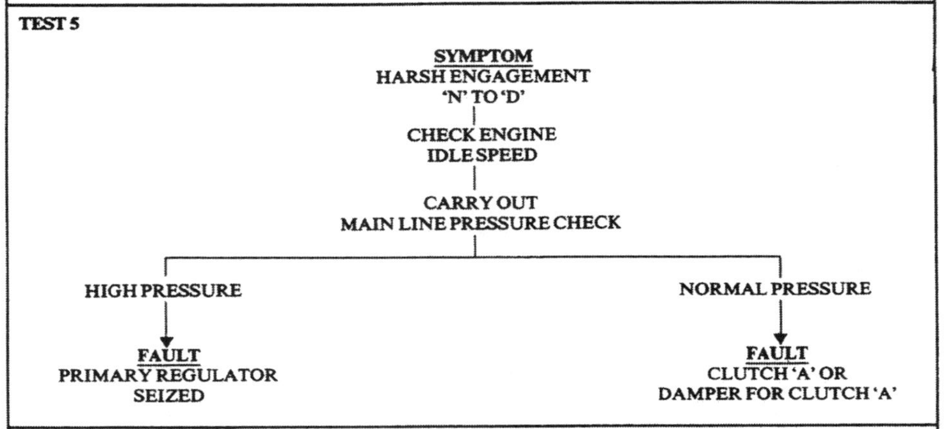

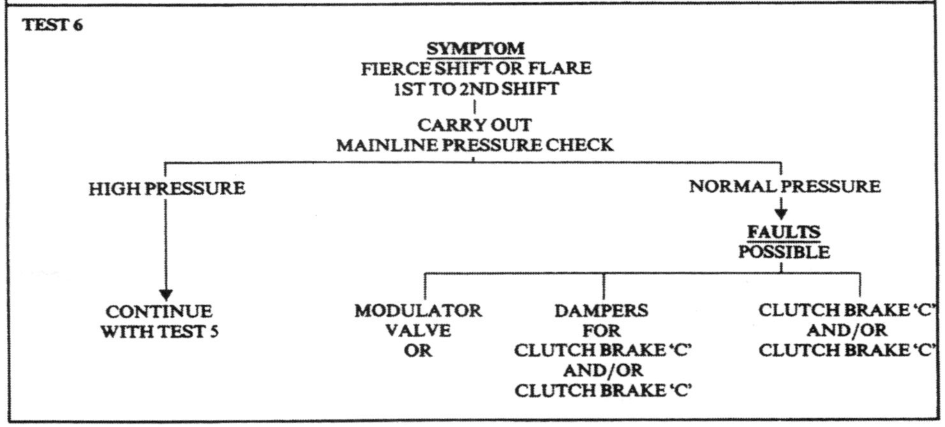

### TEST 1

**SYMPTOM:** INTERMITTENT DRIVE AND HIGH PITCHED NOISE → **FAULT:** LOW FLUID LEVEL OR RESTRICTED FILTER

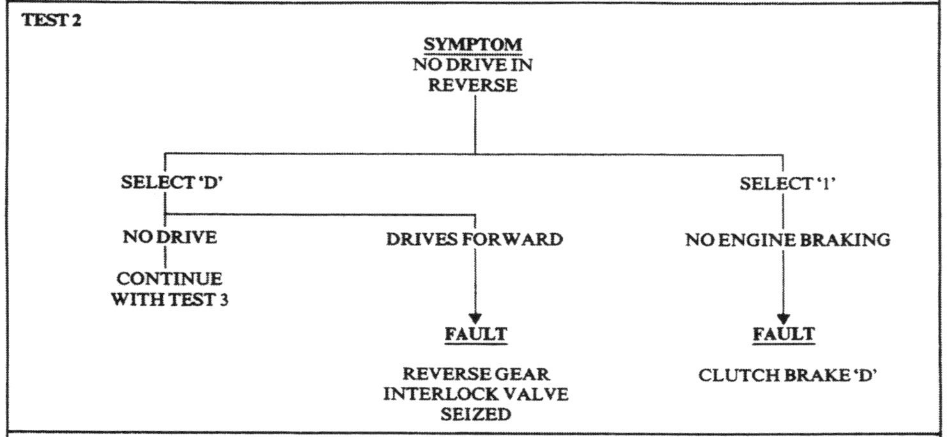

### TEST 2

**SYMPTOM:** NO DRIVE IN REVERSE

- SELECT 'D':
  - NO DRIVE → CONTINUE WITH TEST 3
  - DRIVES FORWARD → **FAULT:** REVERSE GEAR INTERLOCK VALVE SEIZED
- SELECT '1':
  - NO ENGINE BRAKING → **FAULT:** CLUTCH BRAKE 'D'

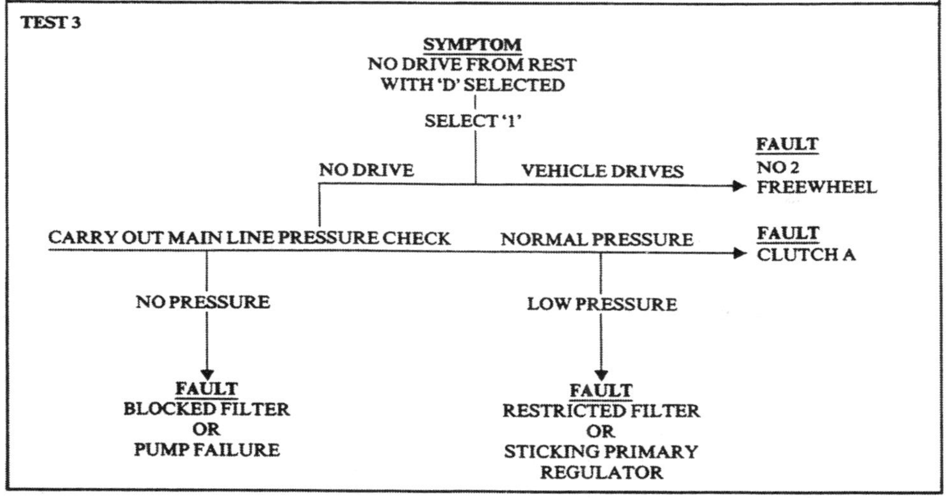

### TEST 3

**SYMPTOM:** NO DRIVE FROM REST WITH 'D' SELECTED

- SELECT '1':
  - VEHICLE DRIVES → **FAULT:** NO 2 FREEWHEEL
  - NORMAL PRESSURE → **FAULT:** CLUTCH A
- NO DRIVE → CARRY OUT MAIN LINE PRESSURE CHECK
  - NO PRESSURE → **FAULT:** BLOCKED FILTER OR PUMP FAILURE
  - LOW PRESSURE → **FAULT:** RESTRICTED FILTER OR STICKING PRIMARY REGULATOR

### TEST 4

**SYMPTOM:** SLIP IN ALL FORWARD GEARS

CARRY OUT MAIN LINE PRESSURE CHECK

- NORMAL PRESSURE → **FAULT:** CLUTCH 'A'
- LOW PRESSURE → **FAULT:** RESTRICTED FILTER OR STICKING PRIMARY REGULATOR
- NO PRESSURE → **FAULT:** BLOCKED FILTER OR PUMP FAILURE

### TEST 5

**SYMPTOM:** HARSH ENGAGEMENT 'N' TO 'D'

CHECK ENGINE IDLE SPEED
CARRY OUT MAIN LINE PRESSURE CHECK

- NORMAL PRESSURE → **FAULT:** CLUTCH 'A' OR DAMPER FOR CLUTCH 'A'
- HIGH PRESSURE → **FAULT:** PRIMARY REGULATOR SEIZED

### TEST 6

**SYMPTOM:** FIERCE SHIFT OR FLARE 1ST TO 2ND SHIFT

CARRY OUT MAINLINE PRESSURE CHECK

- NORMAL PRESSURE → **FAULTS POSSIBLE:** MODULATOR VALVE OR DAMPERS FOR CLUTCH BRAKE 'C' AND/OR CLUTCH BRAKE 'C' OR CLUTCH BRAKE 'C' AND/OR CLUTCH BRAKE 'C'
- HIGH PRESSURE → CONTINUE WITH TEST 5

*Continued*

115

# AUTOMATIC GEARBOX 44

**TEST 7**

SYMPTOM: FIERCE SHIFT OR FLARE 2ND TO 3RD
CARRY OUT MAIN LINE PRESSURE CHECK

- HIGH PRESSURE → CONTINUE WITH TEST 5
- NORMAL PRESSURE → FAULTS POSSIBLE:
  - CLUTCH 'B'
  - MODULATOR VALVE OR CLUTCH 'B' DAMPER

**TEST 8**

SYMPTOM: NO 3RD GEAR

FAULTS POSSIBLE:
- 2-3 SHIFT VALVE
- CLUTCH 'B'

**TEST 9**

SYMPTOM: NO 4TH GEAR 'D' SELECTED

FAULTS POSSIBLE:
- 4TH TO 3RD DOWNSHIFT VALVE SEIZED OR 3RD 4TH UPSHIFT VALVE SEIZED
- CLUTCH BRAKE 'F'

**TEST 10**

SYMPTOM: NONE OR HARSH ENGAGEMENT OF DIRECT DRIVE CLUTCH

NOTE: THE DIRECT DRIVE CLUTCH WILL ONLY ENGAGE IF 4TH GEAR IS ENGAGED AT 40 TO 44 MPH. DISENGAGEMENT OCCURS AT 39 TO 42 MPH.

CARRY OUT ROAD TEST PRESSURE CHECK WITH GAUGE CONNECTED TO TORQUE CONVERTER

- PRESSURE NORMAL → FAULT: FAILED DIRECT DRIVE CLUTCH
  - DIRECT DRIVE CLUTCH AND TORQUE CONVERTER CONTROL VALVE SEIZED
- PRESSURES CORRECT BUT REMAIN HIGH → FAULTS POSSIBLE:
  - HYSTERESIS VALVE SEIZED
  - DIRECT DRIVE CLUTCH CONTROL VALVE SEIZED

**TEST 11**

SYMPTOM: DIRECT DRIVE CLUTCH SHIFT POINT INCORRECT OR AT LOW SPEED

NOTE: DIRECT DRIVE CLUTCH ENGAGEMENT AT LOW SPEED WILL CAUSE VIBRATION IN THE TORQUE CONVERTER.

CARRY OUT MAIN LINE AND TORQUE CONVERTER PRESSURE CHECK

- PRESSURE NORMAL → FAULT: GOVERNOR VALVE STICKING
  - DIRECT DRIVE CLUTCH CONTROL VALVE STICKING
- LOW OR INCORRECT PRESSURE → FAULTS POSSIBLE:
  - HYSTERESIS VALVE STICKING
  - DIRECT DRIVE CLUTCH AND TORQUE CONVERTER CONTROL VALVE STICKING

**TEST 12**

SYMPTOM: DRIVES IN 'D' 1ST BUT IMMEDIATELY UPSHIFTS TO 3RD

FAULT: 2ND TO 3RD SHIFT VALVE SEIZED

*Continued*

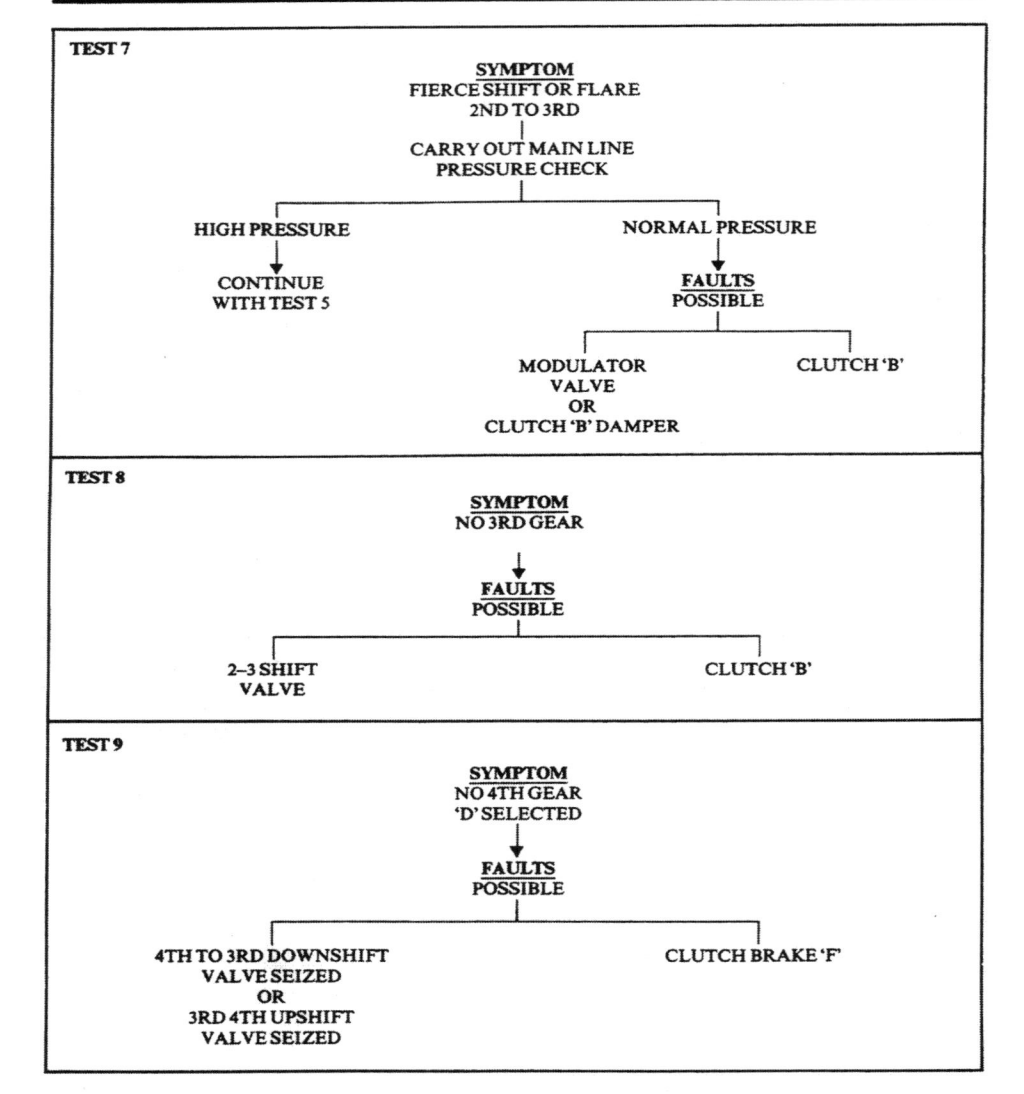

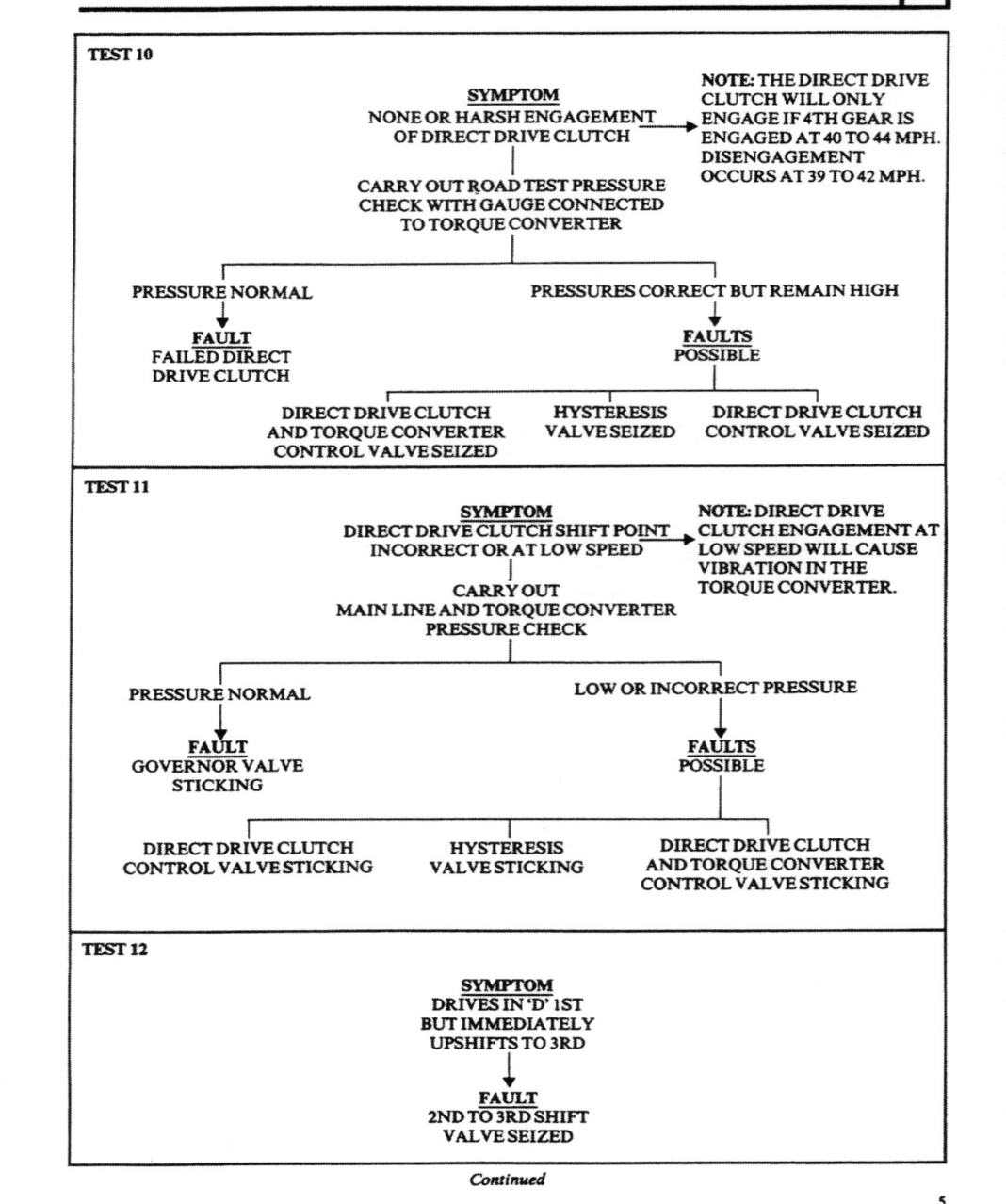

116

# AUTOMATIC GEARBOX

## TEST 13
**SYMPTOM**
WITH 'D' SELECTED VEHICLE STARTS IN 2ND

**FAULTS POSSIBLE**
- 1ST AND 2ND SHIFT VALVE SEIZED
- GOVERNOR SLEEVE STICKING

## TEST 14
**SYMPTOM**
WITH 'D' SELECTED VEHICLE STARTS IN 3RD

**FAULTS POSSIBLE**
- GOVERNOR SLEEVE STICKING
- 1ST TO 2ND AND 2ND TO 3RD SHIFT VALVES SEIZED

## TEST 15
**SYMPTOM**
NO KICKDOWN 4TH TO 3RD

**FAULT**
4TH TO 3RD KICKDOWN VALVE SEIZED

## TEST 16
**SYMPTOM**
UPSHIFTS/DOWNSHIFTS AND KICKDOWN SHIFTS AT INCORRECT ROAD SPEEDS

CHECK THROTTLE KICK DOWN CABLE ADJUSTMENT

CARRY OUT MAIN LINE PRESSURE CHECK

**FAULTS POSSIBLE**
- INCORRECT PRESSURE
  - INCORRECT THROTTLE VALVE ADJUSTMENT
  - PRIMARY REGULATOR STICKING
- NORMAL PRESSURE
  - FAULT GOVERNOR VALVE STICKING

## TEST 17
**SYMPTOM**
NO UPSHIFTS AT LIGHT THROTTLE

**FAULTS POSSIBLE**
- GOVERNOR VALVE STICKING
- SHIFT VALVES STICKING

## TEST 18
**SYMPTOM**
NO ENGINE BRAKING '3' SELECTED 3RD GEAR

**FAULT**
CLUTCH 'C'

## TEST 19
**SYMPTOM**
DELAYED OR NO DOWNSHIFT OCCURS WHEN MAKING A MANUAL SELECTION FROM '3' TO '2'

**FAULTS POSSIBLE**
- GOVERNOR VALVE STICKING
- '2' POSITION INTERLOCK VALVE STICKING
- 2ND AND 3RD UPSHIFT VALVE STICKING

## TEST 20
**SYMPTOM**
AT SPEEDS BELOW 28 MPH WHEN MAKING A MANUAL SELECTION FROM '2' TO '1', 1ST. DOWNSHIFT IS DELAYED OR DOES NOT OCCUR

**FAULTS POSSIBLE**
- GOVERNOR STICKING
- 1ST TO 2ND SHIFT VALVE STICKING
- '1' POSITION INTERLOCK VALVE STICKING

*Continued*

# AUTOMATIC GEARBOX  44

The following repair instructions for the ZF automatic gearbox are divided into three parts. Stage one covers repairs that can be made with the gearbox installed in the vehicle, stage two is with the gearbox removed and stage three a major overhaul procedure.

NOTE: Refer to transfer box section for removal of transfer gearbox.

**Service Tools**

| | | |
|---|---|---|
| LST 108 | Rear oil seal replacer. | |
| LST 109 | Selector linkage setting gauge. | |
| LST 111 | Oil pump rotation sleeve and end float gauge. | |
| LST 112 | Kickdown cable remover. | |
| LST 113 | Control unit inlet oil seals remover/replacer. | |
| LST 114 | Selector shaft oil seal replacer. | |
| TX 27 | Torx bit. | |
| TX 30 | Torx bit. | |
| 18G 1501 | Torque converter remove/refit handles. | |
| LST 115 | B clutch assembly puller hooks. | |
| LST 116 | B clutch 'O' ring and snap ring replacer. | |
| LST 117 | Gear train remover and replacer. | |
| LST 118 | Transmission holding fixture. | |
| LST 1016-1 | Adaptor clutch spring compressor. | |

**Gearbox Data**

Axial end float 0,2 mm to 0,4 mm (0.008 in. to 0.016 in.).
From torque converter boss to torque converter housing face 50 mm (1.96 in.).
Freewheel cage assembly to ring gear, minimum clearance 0,1 mm (0.0039 in.).
Output shaft above cylinder F assembly; dimension 10,00 mm (0.394 in.).
A cylinder protrusion above gearbox front face not more than 8,5 mm (0.33 in.).

In addition to the above service tools, the following items should be manufactured locally to facilitate dismantling and reassembly of the gearbox.

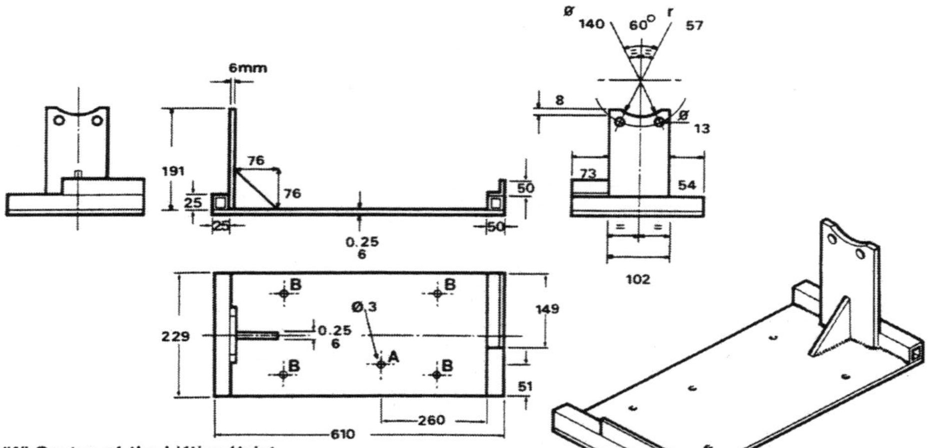

"A" Centre of the Lifting Hoist
"B" Drill fixing holes to suit hoist table

RR739M

**TRANSFER AND AUTOMATIC GEARBOX ADAPTOR PLATE**

---

# AUTOMATIC GEARBOX  44

**TEST 21**

SYMPTOM
'1' SELECTED 1ST GEAR
NO ENGINE BRAKING

↓

FAULT
CLUTCH BAND 3

**TEST 22**

SYMPTOM
'2' SELECTED 2ND GEAR
NO ENGINE BRAKING

↓

FAULT
CLUTCH BAND 1

**TEST 23**

SYMPTOM
VEHICLE DRIVES
FORWARD IN 'N'

↓

FAULT
CLUTCH 'A' SEIZED

# AUTOMATIC GEARBOX

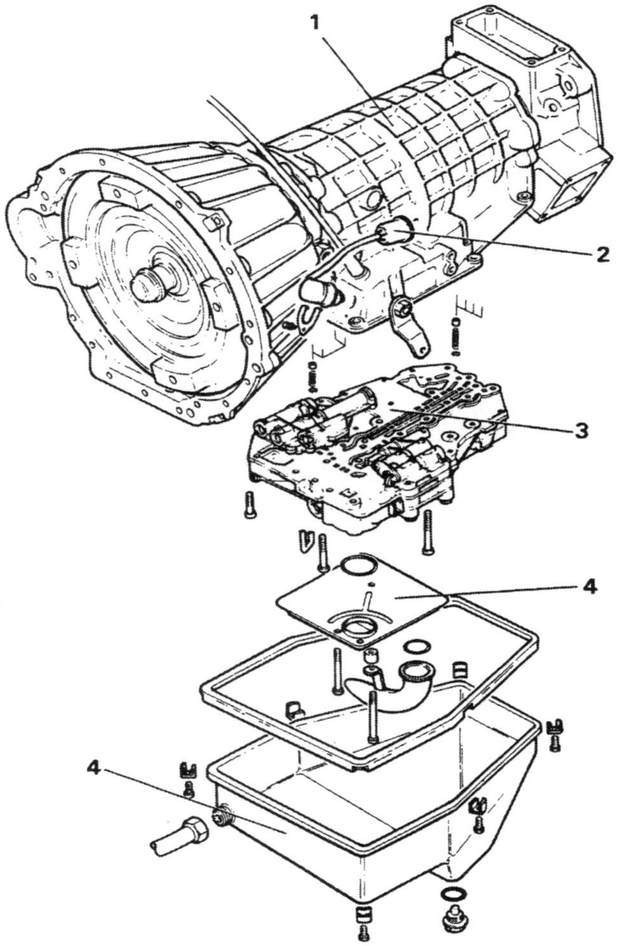

1. GEARBOX ASSEMBLY
2. INHIBITOR SWITCH ASSEMBLY
3. CONTROL UNIT ASSEMBLY
4. FILTER AND SUMP ASSEMBLY

## AUTOMATIC GEARBOX

NOTE: The fixture below can either be manufactured or purchased; fixture number LST 118.

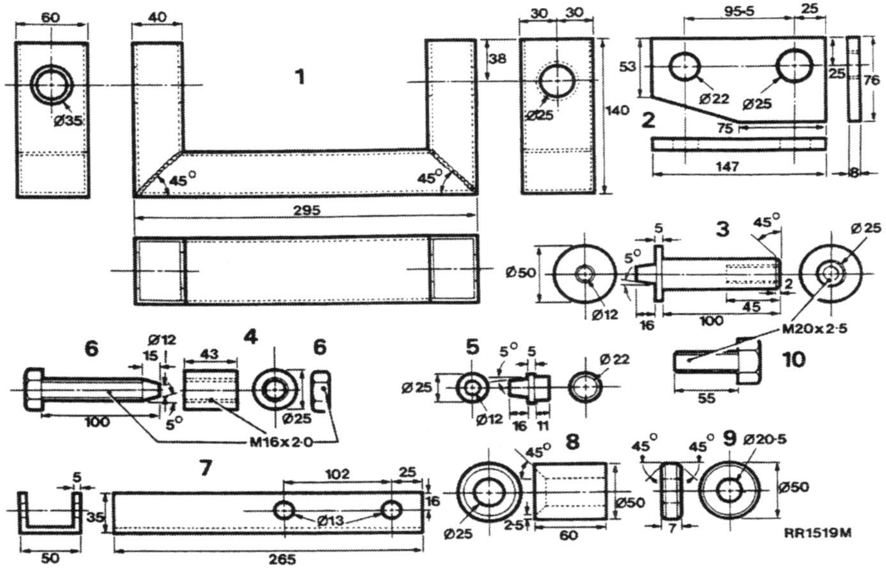

## AUTOMATIC GEARBOX HOLDING FIXTURE

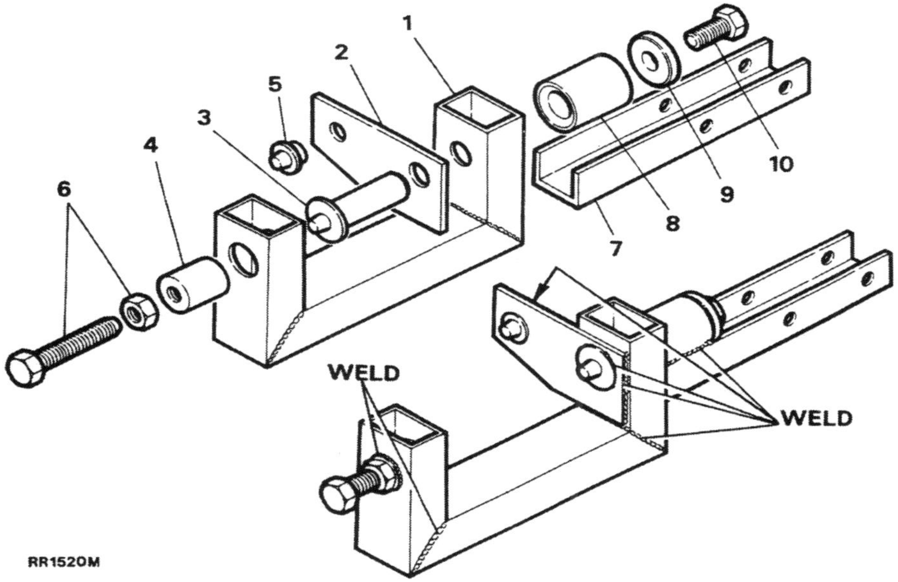

## AUTOMATIC GEARBOX 44

1. A CLUTCH ASSEMBLY
2. B CLUTCH ASSEMBLY

## AUTOMATIC GEARBOX 44

1. TORQUE CONVERTER HOUSING ASSEMBLY
2. GEARBOX PUMP AND CASING ASSEMBLY
3. GOVERNOR AND ADAPTOR HOUSING ASSEMBLY

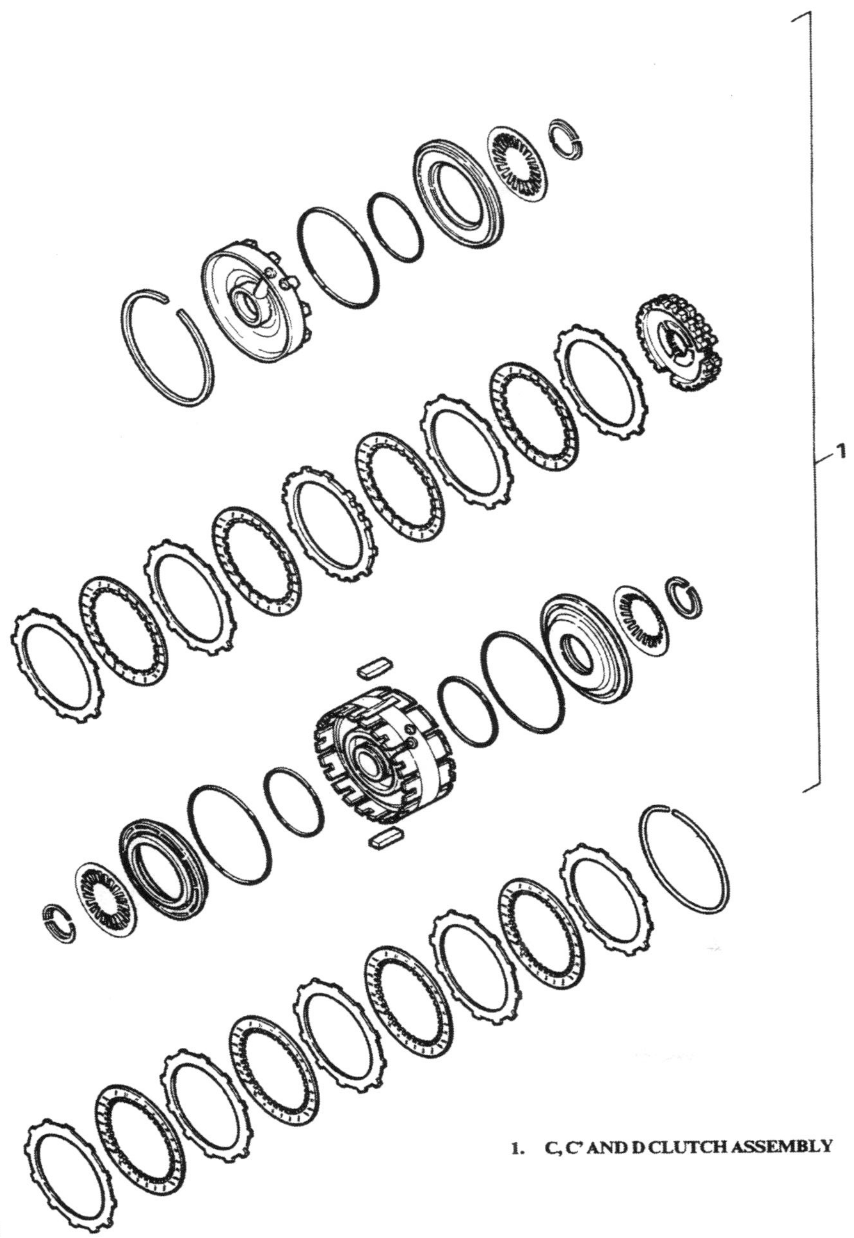

## AUTOMATIC GEARBOX

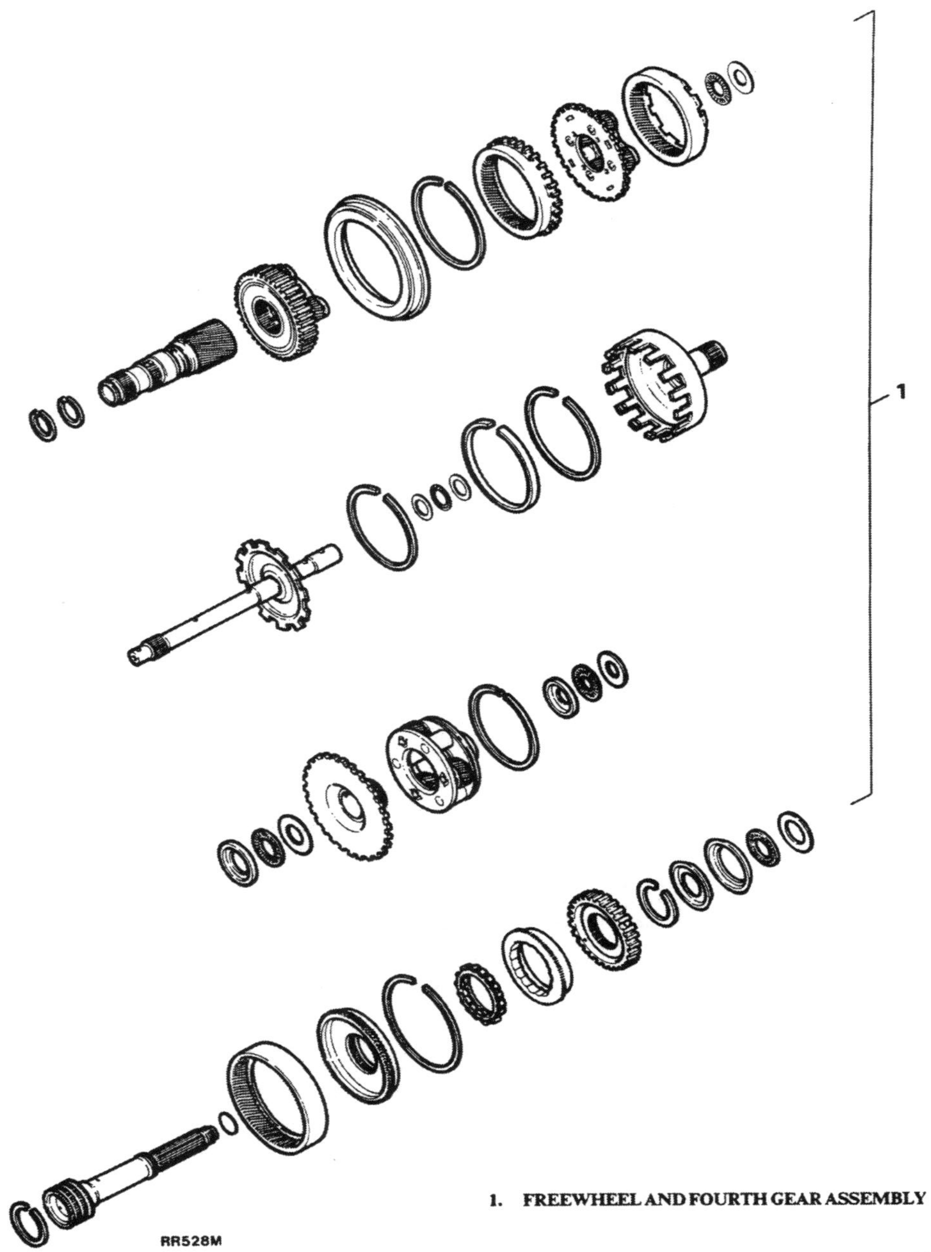

1. FREEWHEEL AND FOURTH GEAR ASSEMBLY

## AUTOMATIC GEARBOX

1. C, C AND D CLUTCH ASSEMBLY

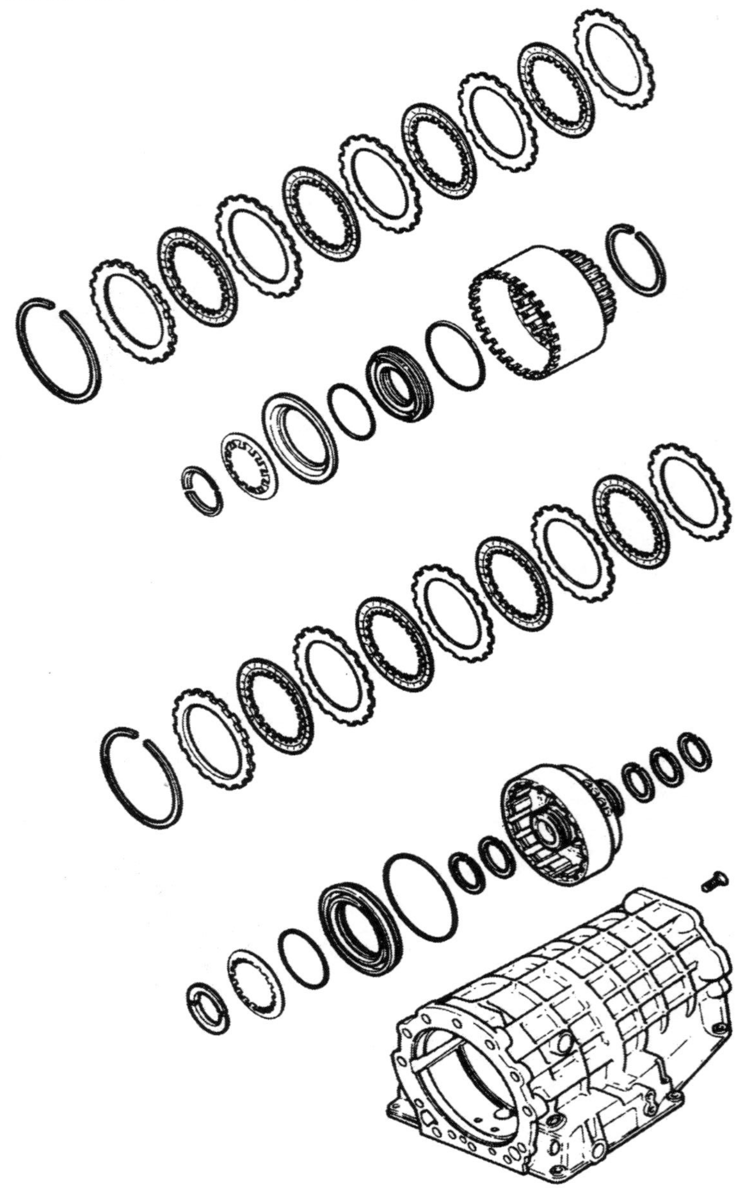

## AUTOMATIC GEARBOX 44

### STAGE I

### Inhibitor switch elimination and replacement

1. Place vehicle on a ramp or over a pit, open the bonnet and disconnect the battery leads.
2. From underneath the vehicle disconnect the inhibitor lead.
3. Undo and remove the bolt and spring washer.
4. Remove the retaining plate.
5. Using a suitable tool remove the inhibitor switch from the casing.
6. Fit a new inhibitor switch, retaining plate if existing one is damaged, spring washer and bolt.
7. Reconnect the inhibitor leads.

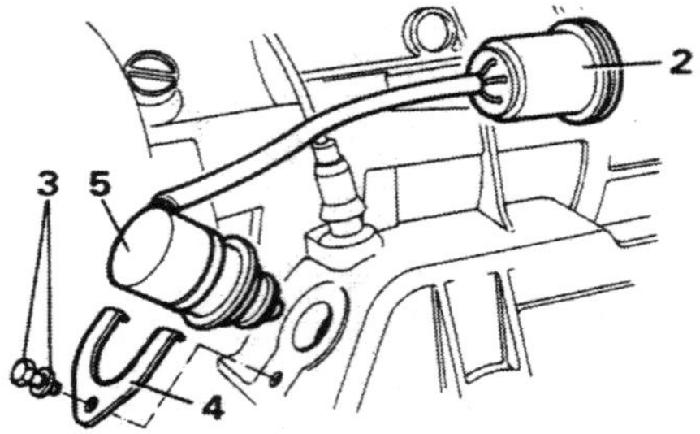

### Intermediate plate screw plugs leak elimination

NOTE: The following procedure is for all four plugs on the plate. Only one of these plugs may be leaking therefore the procedure will only apply to that particular plug or plugs if more than one are faulty.

1. Place the vehicle on a ramp or over a pit, open the bonnet and disconnect the battery leads.
2. From underneath the vehicle, using a suitable spanner remove the two hexagon headed plugs situated in the intermediate plate, catching any oil that may leak from the plate.

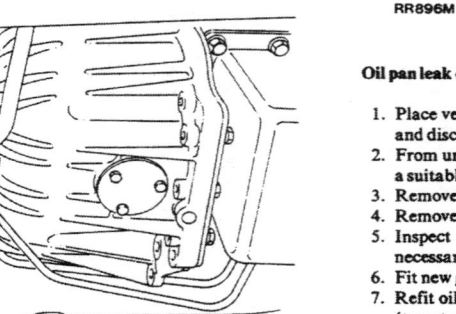

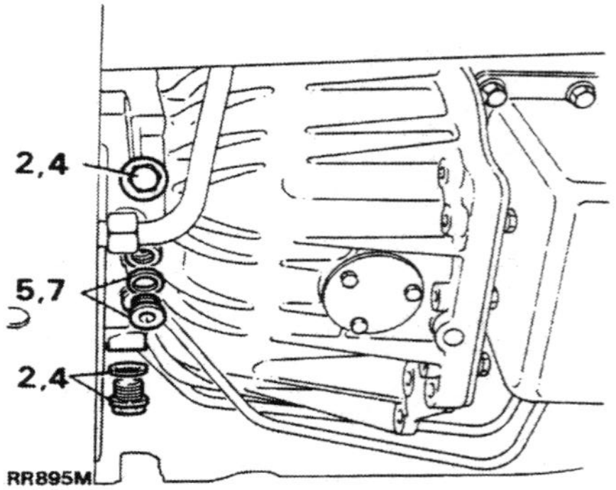

3. Remove and discard the sealing rings.
4. Fit new sealing rings and refit the plugs to the specified torque.
5. Using a suitable hexagon shaped tool, remove the two hexagon socket plugs, catching any oil that may leak from the plate.
6. Remove and discard the sealing rings.
7. Fit new sealing rings and refit the plugs to the specified torque.
8. Connect the battery.
9. Top up the gearbox with the correct oil through the filler level tube located within the engine bay. (See Data section.)
10. Ensuring the vehicle is on level ground with the handbrake applied check oil level while engine is running at idle with neutral selected after selecting each gear.

### Selector shaft leak elimination

1. Place vehicle on a ramp or over a pit, open the bonnet and disconnect the battery leads.
2. From underneath the vehicle remove the nut and gear-change lever.
3. Using a suitable tool remove the oil seal.
4. Fit the new oil seal using the selector shaft oil seal replacer LST 114. For ease of fitment use a light grease or vaseline.
5. Refit gear-change lever, ensure that it is located correctly.
6. Fit and tighten nut to the specified torque.

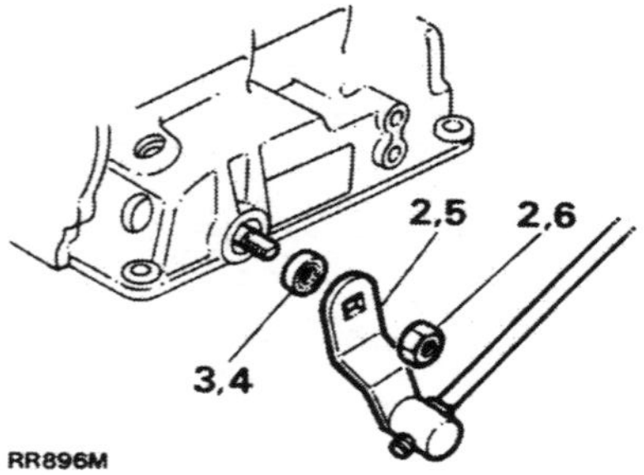

### Oil pan leak elimination and replacement

1. Place vehicle on a ramp or over a pit open the bonnet and disconnect battery leads.
2. From underneath the vehicle drain the gearbox using a suitable container and remove the oil filler level tube.
3. Remove the six retaining plates and bolts.
4. Remove the oil pan and discard the gasket.
5. Inspect oil pan for wear or damage. Replace if necessary.
6. Fit new gasket onto oil pan.
7. Refit oil pan using the six retaining plates and screws (two straight and four corner plates).

1. E AND F CLUTCH ASSEMBLY

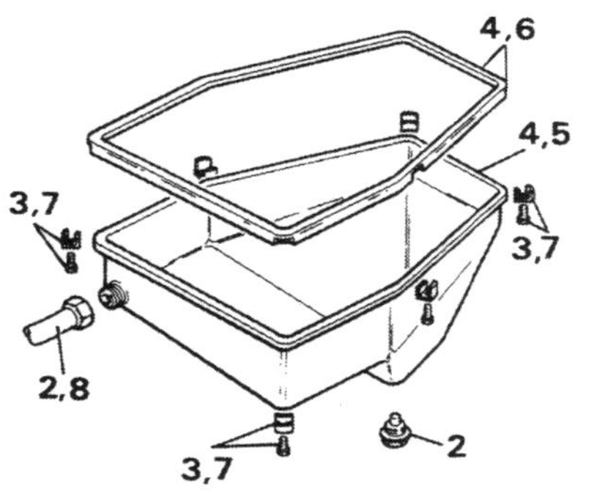

## AUTOMATIC GEARBOX | 44

14. Place the selector linkage setting gauge LST109 in position and gently press the control unit against the tool and tighten all thirteen bolts using TX27 torx bit to the specified torque.

9. Fit the extension case onto the gearbox ensuring the oil seal does not get damaged by the extension shaft.
10. Fit and tighten the nine bolts to the specified torque.
11. From inside the vehicle refit the four bolts which hold the transfer gear selector housing and adaptor bracket.
12. Secure the four bolts to the specified torque.
13. Refit the transfer box as described in section 37.

### Oil Screen Replacement

1. Place the vehicle on a ramp or over a pit, open the bonnet and disconnect the battery leads.
2. From underneath the vehicle drain the gearbox using a suitable container.
3. Discard the oil pan plug seal ring.
4. Remove the filler/level tube from the oil pan.
5. Remove the six retaining plates and bolts.
6. Remove the oil pan and discard the gasket.
7. Using TX27 torx bit undo the three screws which hold the oil screen.

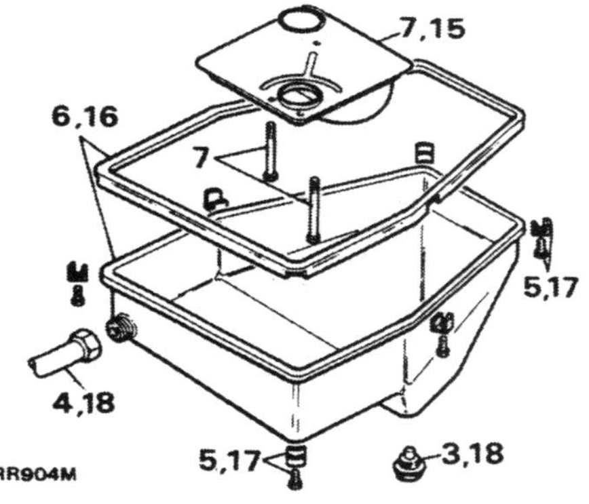

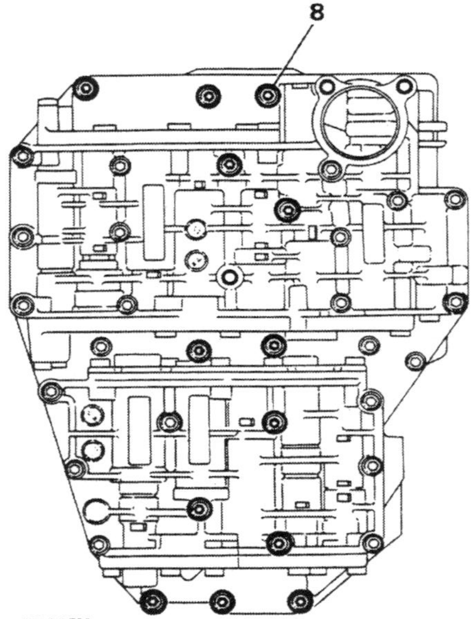

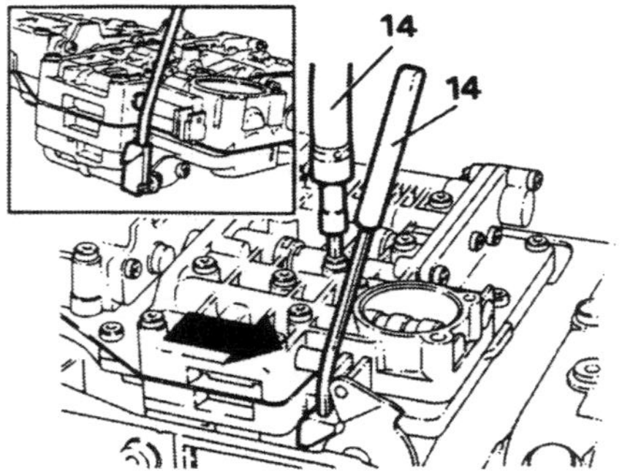

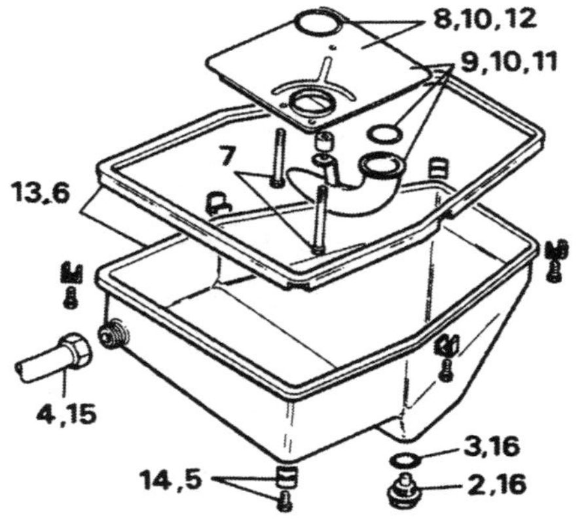

15. Remove setting gauge and fit oil screen using TX27 torx bit to the specified torque.
16. Refit oil pan with new gasket.
17. Refit the six retaining plates and screws (two straight and four corner plates).
18. Reconnect oil filler tube and oil pan, plug with a new seal.
19. Connect the battery leads.
20. Fill the gearbox with the correct oil (see 'Data section').
21. Connect the kickdown cable to the rear of the engine.
22. Adjust the outer cable to achieve a crimp gap of 0.25 to 1.25mm (.006 to .031 in).
23. Hold the outer cable while tightening the locknuts. NOTE: The kickdown cable must be adjusted while the vehicle is running at idle.
24. Ensuring the vehicle is on level ground with the handbrake applied, check oil level while engine is running at idle with neutral selected, after selecting each gear.

### Extension case leak elimination and replacement

1. Remove the transfer box as described in section 37.
2. Using a suitable tool release the four bolts from inside the vehicle holding the transfer gear selector housing and adaptor bracket.
3. From underneath the vehicle using a suitable tool release the nine bolts holding the extension housing.
4. Remove the extension housing and discard the gasket.
5. Place extension housing on the bench and remove the oil seal.
6. Ensure that all the surfaces are clean and the case is free from damage. If damage has been found on the case, replace the case.
7. If the case has to be replaced, fit the two dowels into the case.
8. Fit a new gasket and oil seal using the rear oil seal replacer LST108.

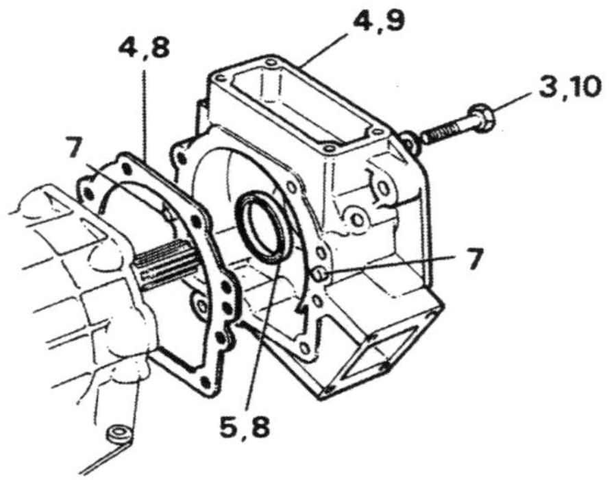

---

## 44 | AUTOMATIC GEARBOX

7. Remove the oil screen, undoing the three bolts using a TX27 Torx bit.
8. Remove the control unit, undoing the thirteen remaining bolts using a TX27 Torx bit.

8. Reconnect oil filler level tube, oil pan plug with a new seal.
9. Connect the battery leads.
10. Fill the gearbox with the correct oil. (See data section.)
11. Ensuring the vehicle is on level ground with the handbrake applied check oil level while engine is running at idle with neutral selected after selecting each gear.
NOTE: If leak persists and old oil pan has been refitted, change the oil pan using the same procedure as above.

### Kickdown cable leak elimination

1. Place the vehicle on a ramp or over a pit, open the bonnet and disconnect the battery leads.
2. Disconnect the kickdown cable from the rear of the engine.
3. From underneath the vehicle, using a suitable container drain the gearbox and discard the gearbox oil pan seal.
4. Remove the oil filler level tube.
5. Remove the six retaining plates and bolts.
6. Remove the oil pan and discard the gasket.

9. Locate the selector cam and remove the nipple holding the kickdown cable from it's seat.
10. Using the kickdown cable remover LST112 remove the cable and it's housing from the casing and discard.
11. Fit new throttle cable with new 'O' ring into the casing.

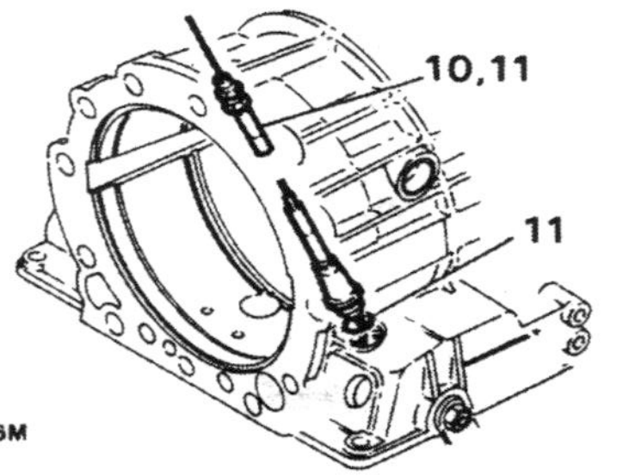

12. Fit the nipple into the cam seat ensuring that the cam free from damage. If damage has been found on the load the cam.
13. Fit the control unit after cleaning the face with a lint free rag, ensuring the selector shaft locates into the gear shift fork and fit the thirteen bolts loosely by hand.

## AUTOMATIC GEARBOX | 44

8. Remove the oil screen and discard the 'O' rings.
9. Separate the oil screen from the suction tube and discard the 'O' ring and oil screen.

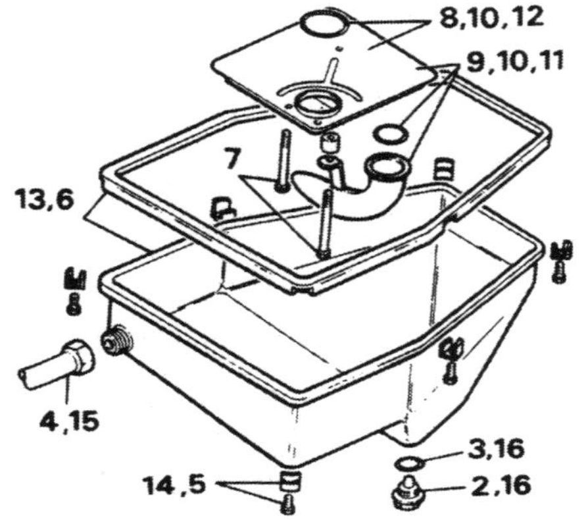

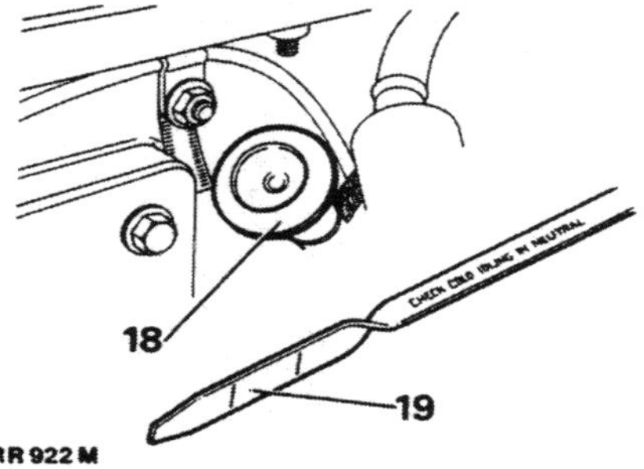

10. Fit two new 'O' rings to the oil screen using a light grease for ease of assembly.
11. Fit the suction tube to the oil screen.
12. Fit the oil screen to the control unit and secure with three bolts using TX27 torx bit tighten to the specified torque.
13. Refit the oil pan using a new gasket.
14. Secure using the six retaining plates and bolts (two straight and four corner plates), tighten to the specified torque.
15. Reconnect the oil level/filler tube.
16. Fit oil pan plug using a new seal.
17. Connect the battery leads.
18. Fill the gearbox with the correct oil through the filler/level tube located within the engine bay. (See Data section.)
19. Ensuring the vehicle is on level ground with the handbrake applied, check oil level while engine is running at idle with neutral selected.

### Control Unit Replacement

1. Place the vehicle on a ramp or over a pit, open the bonnet and disconnect the battery leads.
2. From underneath the vehicle drain the gearbox using a suitable container.
3. Discard the oil pan plug seal ring.
4. Remove the oil filler/level tube from the oil pan.
5. Remove the six retaining plates and bolts.
6. Remove the oil pan and discard the gasket.
7. Using a TX27 torx bit undo the three bolts which hold the oil screen.

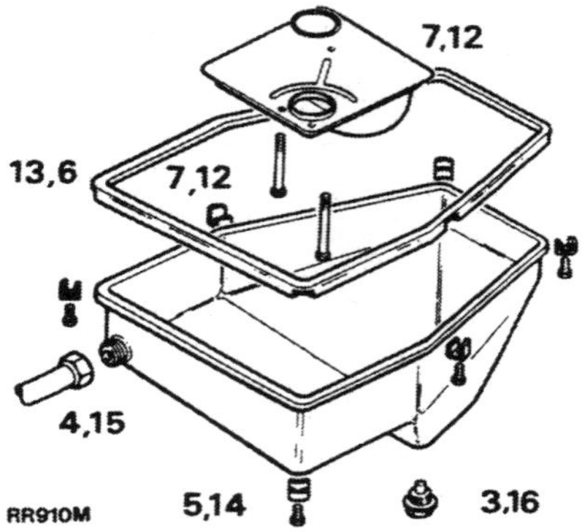

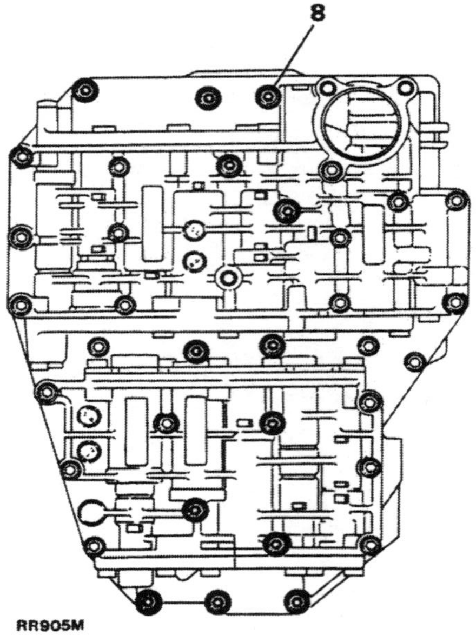

8. Using a TX27 torx bit undo the remaining thirteen bolts retaining the control unit.
9. Clean the surfaces ensuring no damage has occurred to the mounting face of the case, using a lint-free rag.
10. Fit the new control unit ensuring the selector shaft locates into the gear shift fork and fit the thirteen bolts loosely by hand.
11. Place the selector linkage setting gauge LST 109 in position and gently press the control unit against the tool and tighten all thirteen bolts using TX27 torx bit to the specified torque.

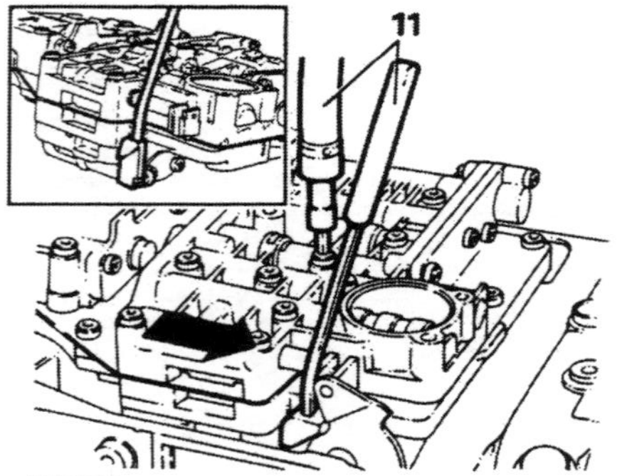

12. Remove the setting gauge and fit the oil screen using TX27 torx bit to the specified torque.
13. Refit the oil pan using a new gasket.
14. Secure with the six retaining plates and bolts (two straight and four corner plates), tighten to the specified torque.
15. Reconnect the oil filler/level tube.
16. Fit oil pan plug using a new seal.
17. Connect the battery leads.
18. Fill the gearbox with the correct oil through the filler/level tube located within the engine bay. (See 'Data' section.)
19. Ensuring the vehicle is on level ground with the handbrake applied, check oil level while engine is running at idle with neutral selected, after selecting each gear.

### Oil Inlet Sealing Rings Renewal

1. Place the vehicle on a ramp or over a pit, open the bonnet and disconnect the battery leads.
2. From underneath the vehicle drain the gearbox using a suitable container.
3. Discard the oil pan plug seal ring.
4. Remove the oil filler/level tube from the oil pan.
5. Remove the six retaining plates and bolts.
6. Remove the oil pan and discard the gasket.
7. Using a TX27 torx bit undo the three bolts which hold the oil screen.

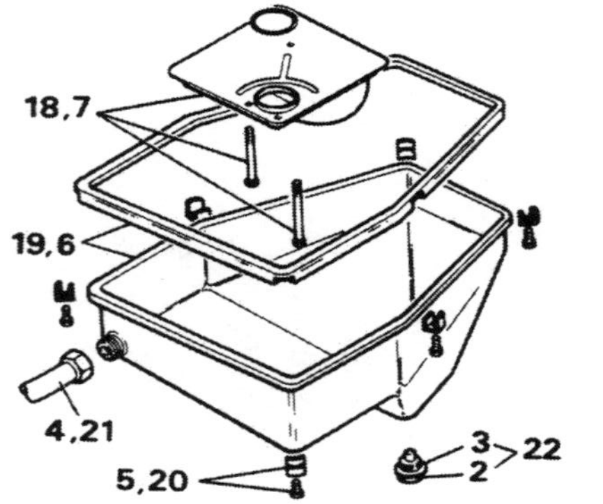

8. Using a TX27 torx bit undo the remaining thirteen bolts retaining the control unit and remove the control unit.
9. Clean the surfaces ensuring no damage has occurred to the mounting face of the case, using a lint-free rag.
10. Using circlip pliers remove the eight oil seals and springs from the gearbox.
11. Using control unit inlet oil seals remover/replacer LST 113 remove the eight oil seals.
12. Clean the orifices and check for damage.

NOTE: If damage has occurred replace the box as described in Stage II.

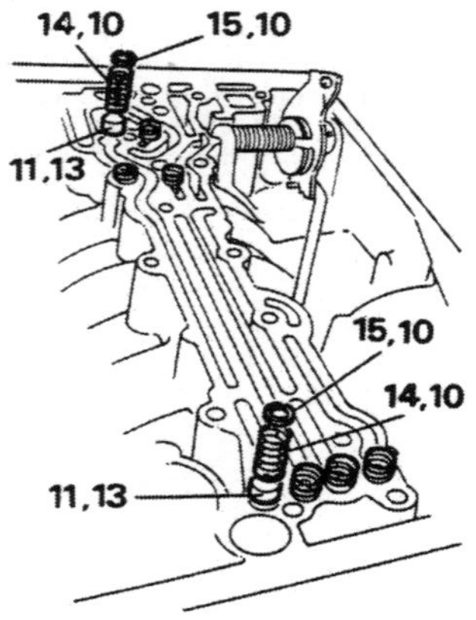

13. Using the control unit inlet oil seal remover/replacer LST 113 fit the new seals ensuring they are seated fully home.

## AUTOMATIC GEARBOX

14. Fit the eight compression springs, the four short ones at the front and the four long ones at the rear of the box.
15. Using the circlip pliers fit the eight circlips which retain the compression springs.
16. Fit the control unit ensuring the selector shaft locates into the gear shift fork and fit the thirteen bolts loosely by hand.
17. Place the selector linkage setting gauge LST 109 in position and gently press the control unit against the tool and tighten all thirteen bolts using TX27 torx bit to the specified torque.
18. Remove the setting gauge and fit the oil screen using TX27 torx bit to the specified torque.
19. Refit the oil pan using a new gasket.
20. Secure with the six retaining plates and bolts (two straight and four corner plates), tighten to the specified torque.
21. Reconnect the oil filler/level tube.
22. Fit oil plug plug using a new seal.
23. Connect the battery leads.
24. Fill the gearbox with the correct oil through the filler/level tube located within the engine bay. (See 'Data' section.)
25. Ensuring the vehicle is on level ground with the handbrake applied, check oil level while engine is running at idle with neutral selected, after selecting each gear.

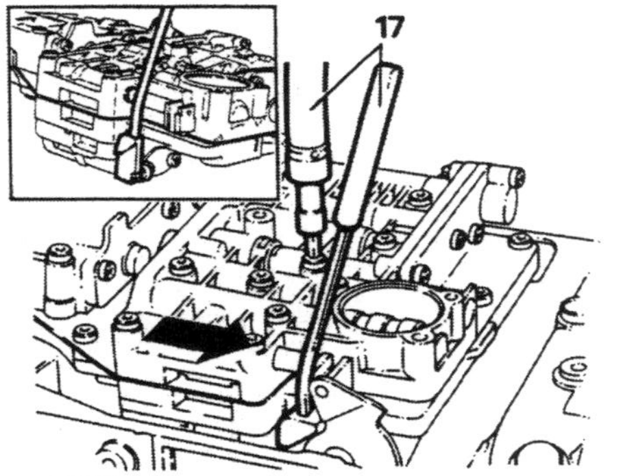

### Manual Valve Operating Mechanism

1. Place the vehicle on a ramp or over a pit, open the bonnet and disconnect the battery leads.
2. From underneath the vehicle, using a suitable container drain the gearbox and discard the gearbox oil pan seal.
3. Remove the oil filler level tube.
4. Remove the six retaining plates and bolts.
5. Remove the oil pan and discard the gasket.
6. Remove the oil screen, undoing the three bolts using a TX27 torx bit.
7. Remove the control unit, undoing the thirteen remaining bolts using a TX27 torx bit.
8. Locate the selector cam and remove the nipple holding the kick-down cable from its seat.
9. Remove the nut and gear change lever.
10. Using a suitable punch drift out the roll pin from the selector shaft and discard it.
11. Using a suitable tool remove the selector shaft from the box, noting the position of the detent plate.
12. Remove the connecting rod complete with detent plate; accelerator cam, spring and using a suitable tool remove the oil seal and discard.
13. Check all parts for wear or damage and replace as necessary.
14. Using selector shaft oil seal replacer LST114, fit the oil seal. For ease of fitment use a light grease or vaseline.
15. Fit connecting rod to detent plate and locate in the box by pushing the selector shaft through from outside of the casing.

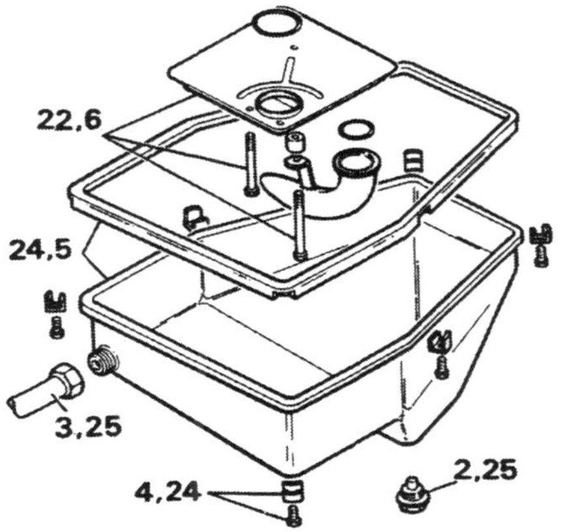

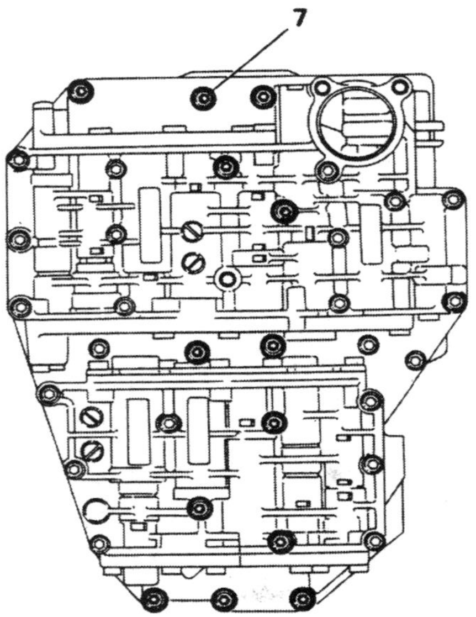

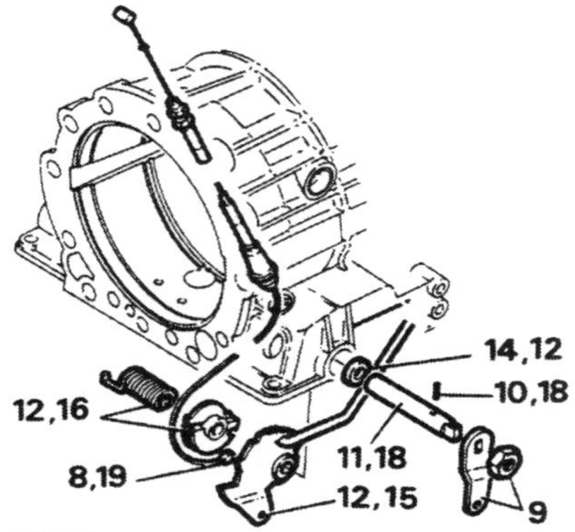

NOTE: The detent plate should go back into the box in the same position as noted earlier.

16. Fit the accelerator cam with the spring.
17. Fit the assembly into the box and secure it by pushing the selector shaft through.
18. Align the hole in the selector shaft with the hole in the detent plate and secure with a new roll pin, using a suitable punch.
19. Fit kickdown cable nipple into the cam seat ensuring that the cam has been turned once before fitment. This will spring load the cam.
20. Fit the control unit ensuring the selector shaft locates into the gear shift fork and fit the thirteen bolts loosely by hand.
21. Place the selector linkage setting gauge LST109 in position and gently press the control unit against the tool and tighten all thirteen bolts using TX27 torx bit to the specified torque.
22. Remove the setting gauge and fit oil screen using TX27 torx bit to the specified torque.
23. Refit the oil pan with a new gasket.
24. Refit the six retaining plates and screws (two straight and four corner plates).
25. Reconnect the oil filler/level tube, oil pan plug with new seal.
26. Connect the battery leads.
27. Fill the gearbox with the correct oil. (See Data section.)
28. Ensuring the vehicle is on level ground with the handbrake applied, check oil level while engine is running at idle with neutral selected, after selecting each gear.

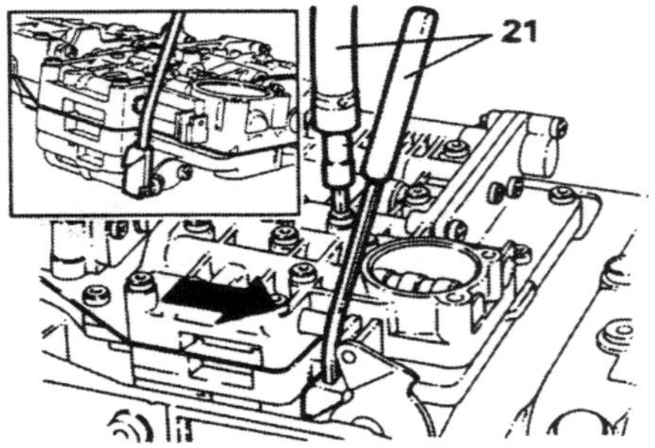

# AUTOMATIC GEARBOX

## Governor Housing Renewal

1. Remove the transfer box as described in section 37.
2. Using a suitable tool release the four bolts from inside the vehicle holding the transfer gear selector housing and adaptor bracket.
3. From underneath the vehicle using a suitable tool release the nine bolts holding the extension housing.
4. Remove the extension housing ensuring that the gasket is not damaged and discard the gasket.
5. Remove the extension shaft and retaining bolt with 'O' ring.
6. Remove the governor assembly with parking wheel.
7. Remove the two screws holding the governor housing using TX27 torx bit.
8. Remove the governor housing complete and discard.
9. Inspect the governor hub and parking wheel for damage, if satisfactory, clean.
10. Fit new governor housing complete to governor hub and parking wheel using TX27 torx bit to the specified torque.
11. Refit the governor assembly with parking wheel onto the output shaft and push the assembly till fully seated.

**NOTE: To avoid damage to 'O' ring use a light grease or vaseline. Ensure the seal rings are snapped together and are seated correctly.**

12. Fit the extension shaft and retaining bolt using a new 'O' ring.
13. Fit new gasket onto rear of gearbox and fit the extension housing, taking care not to damage the seal on assembly.
14. Secure the extension housing using the nine bolts to the specified torque.
15. From inside the vehicle refit the four bolts which retain the transfer gear selector housing and adaptor bracket.
16. Secure the four bolts to the specified torque.
17. Refit the transfer box as described in section 37.

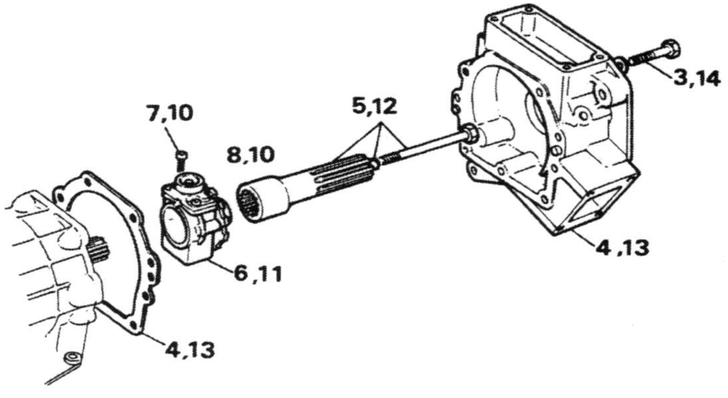

RR586M

## Governor Hub Renewal

1. Remove the transfer box as described in section 37.
2. Using a suitable tool release the four bolts from inside the vehicle holding the transfer gear selector housing and adaptor bracket.
3. From underneath the vehicle using a suitable tool release the nine bolts holding the extension housing.
4. Remove the extension housing ensuring that the seal is not damaged and discard the gasket.
5. Remove the extension shaft and retaining bolt with 'O' ring.
6. Remove the governor assembly with parking wheel.
7. Remove the two screws holding the governor housing using TX27 torx bit.
8. Using a TX27 torx bit unscrew the two bolts and remove the parking wheel and discard governor hub.
9. Remove the security clip and counter-weight.
10. Remove the 'O' ring from off the output shaft and discard.
11. Remove the three seal rings from the 'F' clutch housing shaft.
12. Inspect all parts for damage or wear, replace if necessary.
13. Fit the counterweight and security clip into the new governor hub.
14. Secure governor housing onto governor hub using TX27 torx bit to the specified torque.
15. Fit the parking wheel to the governor hub using TX27 torx bit to the specified torque.
16. Fit three new seal rings onto the F clutch housing shaft and fit 'O' ring onto output shaft.

**NOTE: For ease of fitment of the 'O' ring use a light grease or vaseline.**

17. Fit governor assembly and parking wheel onto the output shaft and push the assembly till fully seated.

**NOTE: To avoid damage to 'O' ring use a light grease or vaseline. Ensure the seal rings are snapped together and are seated correctly.**

18. Fit new gasket onto rear of gearbox and fit the extension housing taking care not to damage the seal or assembly.
19. Fit the extension shaft and retaining bolt using a new 'O' ring.
20. Secure the extension housing using the nine bolts to the specified torque.
21. From inside the vehicle refit the four bolts which retain the transfer gear selector housing and adaptor bracket. Secure the four bolts to the specified torque. Refit the transfer box as described in section 37.

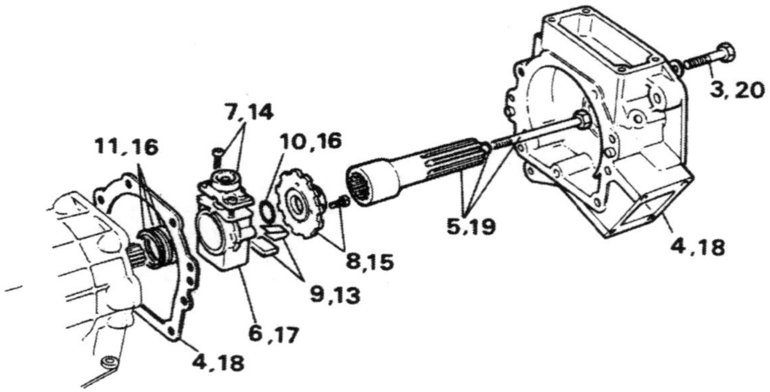

RR919M

## Parking Brake Mechanism Renewal

1. Remove the transfer box as described in section 37.
2. Using a suitable tool, release the four bolts holding the transfer gear selector housing and adaptor bracket.
3. From underneath the vehicle, using a suitable tool, release the nine bolts holding the extension housing.
4. Remove the extension housing ensuring that the seal is not damaged and discard the gasket.
5. Remove the extension shaft and retaining bolt with 'O' ring.
6. Remove the governor assembly with parking wheel.
7. Remove guide plate bolt, using TX27 torx bit.
8. Remove the plate and guide plate from the gearbox case.
9. Remove the pin, pawl and the spring.

**NOTE: Take care when removing park assembly as spring tension will be reduced.**

10. Inspect all parts for wear or damage and replace if necessary.
11. Fit the pin and the leg spring ensuring that the spring is located correctly.
12. Fit the pawl onto the pin and the spring leg into the hole in the pawl. This creates tension in the spring.
13. Fit the plate and guide plate using TX27 torx bit to the specified torque.

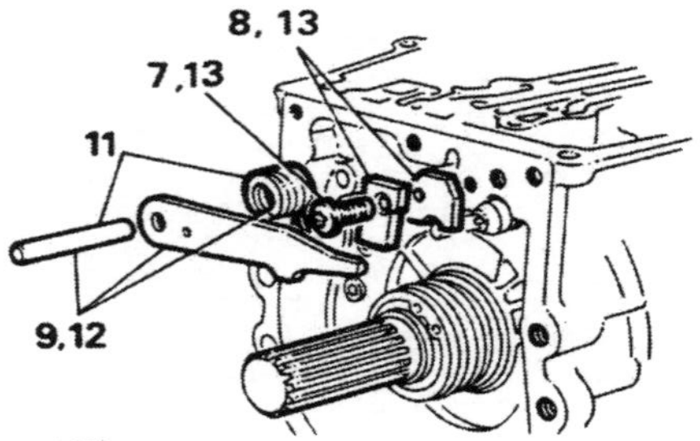

RR584M

# AUTOMATIC GEARBOX 44

14. Refit the governor assembly with parking wheel onto the output shaft and push the assembly till fully seated.

15. Fit new gasket onto rear of gearbox and fit the extension housing, taking care not to damage the seal or assembly.

NOTE: To avoid damage to 'O' ring use a light grease or vaseline. Ensure the seal rings are snapped together and are seated correctly.

16. Fit the extension shaft and retaining bolt using a new 'O' ring.
17. Secure the extension housing using the nine bolts to the specified torque.
18. From inside the vehicle refit the four bolts which retain the transfer gear selector housing and adaptor bracket.
19. Secure the four bolts to the specified torque.
20. Refit the transfer box as described in section 37.

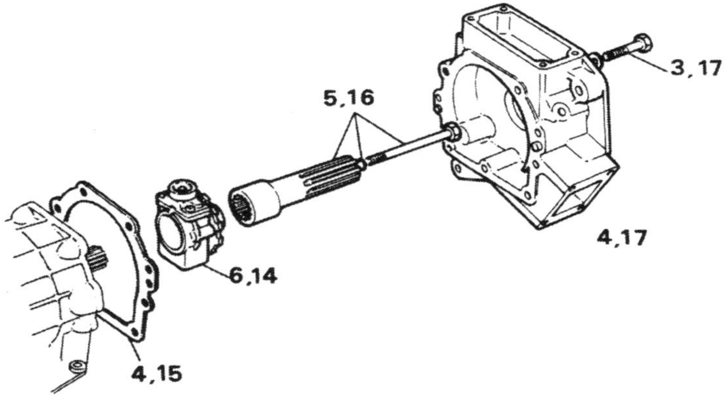

RR920M

## AUTOMATIC GEARBOX 44

### STAGE II

### ZF Gearbox—Remove and Refit

#### Removing

1. Install the vehicle on a hydraulic ramp.
2. Open the bonnet.
3. Disconnect the battery leads.
4. **Fuel Injection models only**—release the airflow meter to plenum chamber hose.
   **Carburetter models only**—Remove the air intake elbows and withdraw the air cleaner from its location.
5. Disconnect the kickdown cable from throttle linkages.
   **Fuel Injection models**: located on the throttle lever bracketry at the rear of the plenum chamber.
   **Carburetter models**: located at the rear of the right-hand carburetter. Remove the transmission dipstick.
6. Remove the fan cowl from the radiator.
7. From inside the vehicle remove the four screws securing the cubby box liner to the cubby box and lift out the liner.
8. Carefully prise the window lift, switch panel away from the front of the cubby box. Identify each switch connection for reassembly, disconnect the plugs and remove the switch panel.
9. Remove the main and transfer gearbox knobs.
10. Carefully prise the centre panel out of the floor mounted console and remove it from the vehicle.

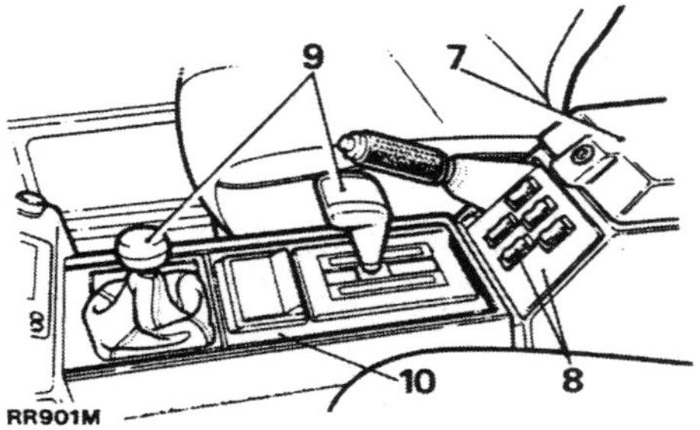

RR1544M

11. Release the two bolts and two screws securing the console assembly to the gearbox tunnel.
12. Release the handbrake and remove the split pin, clevis pin and washer securing the handbrake cable to the handbrake lever.
13. Carefully manoeuvre the console assembly away from the radio housing and remove it from the vehicle.
14. Release the large nut retaining the handbrake outer cable to the top of the gearbox tunnel.
15. Remove the nut and feed the cable through the hole in the tunnel to the underside of the vehicle.
16. Raise the vehicle on the ramp and drain the gearbox.
17. Release the nut and clamp securing the speedometer cable to the rear of the transfer box.
18. Withdraw the cable from the speedometer drive pinnion.

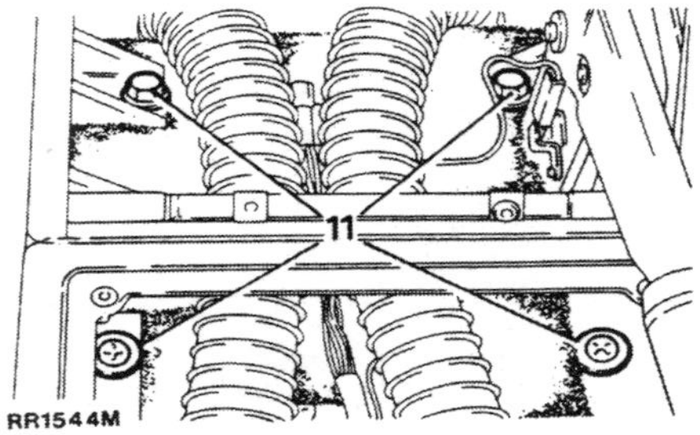

RR901M

19. Release the cable from the clips at the side of the gearbox.
20. Release the four nuts and bolts securing the rear propeller to the rear output flange and tie the shaft to one side.
21. Remove the four nuts and bolts securing the front propeller shaft to the front output flange and tie the shaft to one side.
22. Release the nuts and bolts securing the front downpipe to the front silencer.
23. Release the nut at the rear tailpipe bracket, disconnect the silencer from the downpipe, and tie the rear tail pipe and silencer to one side.
24. Disconnect the oil filler tube from the front of the gearbox oil pan.
25. Disconnect the two oil cooler pipes from the rear of the gearbox bellhousing.

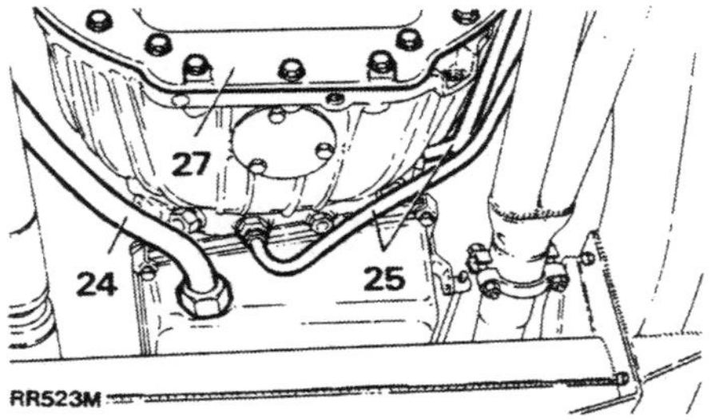

RR52M

26. Remove the bolts securing the cross-member in position, using suitable equipment expand the chassis and withdraw the cross-member.
27. Remove the front cover from the bottom of the torque converter housing and from inside the housing remove the converter drive-bolts.
28. Disconnect the inhibitor switch.
29. Disconnect the selector linkage.
30. Manufacture an adaptor plate in accordance with the drawing (RR739M), to attach to the gearbox hoist and transfer box to facilitate removal.
31. Remove the nuts and bolts holding rear left-hand side, mounting bracket to chassis.
32. Remove the nuts and bolts holding right-hand side mounting bracket to chassis.

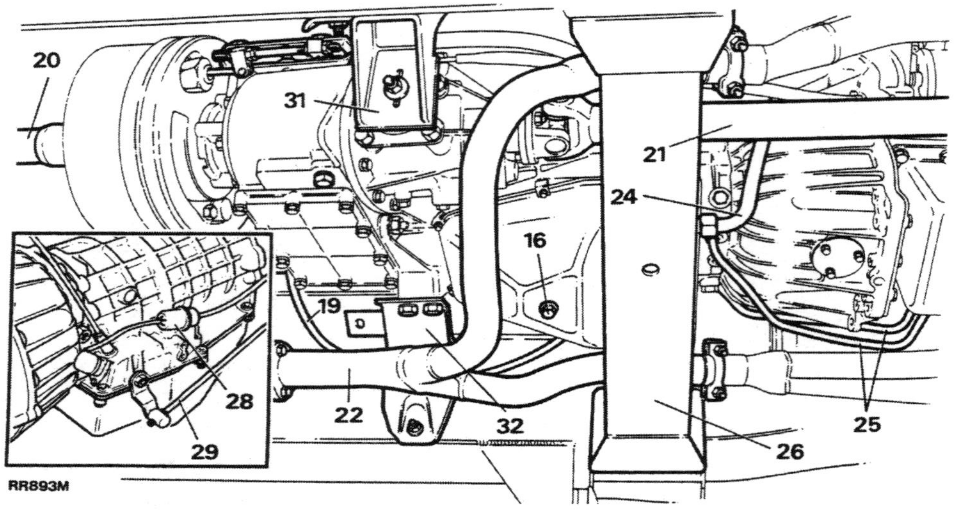

# AUTOMATIC GEARBOX 44

33. Lower the hoist until the rear brake drum clears the rear passenger footwell.
34. Remove the split pin and washers securing the differential lock lever to the connecting rod, and disconnect the lever from the rod.
35. Disconnect the electrical leads from the differential lock switch.
36. Remove the breather pipe from the top of the gearbox.
37. Using a suitable jack support the rear of the engine.
38. Remove the torque converter housing to engine bolts.
39. Carefully withdraw the gearbox and transfer box from the engine taking care not to damage any seats.

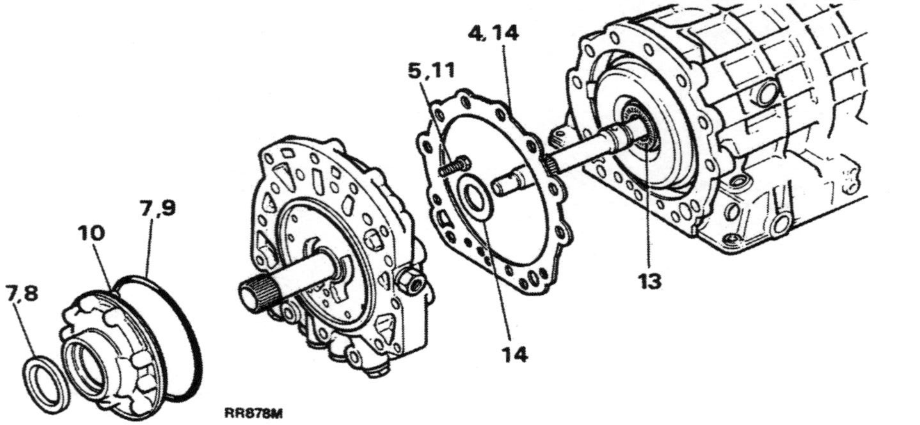

## Refitting

40. Reverse the removal instructions.
41. Refill the gearbox with the correct grade and quantity of oil. See data section.
42. Ensuring the vehicle is on level ground with the handbrake applied, check oil level whilst engine is running at idle with neutral selected, after selecting each gear.

There are several places where leaks can occur at the front of the gearbox. The following are remedies for curing any one of these problems.

## Eliminating leaks/replacing Torque Converter

1. Remove the gearbox/transfer box assembly using the torque converter handles 18G1501, as previously described.
2. Place the gearbox on the bench using the torque converter handles 18G1501, remove the torque converter, taking care not to damage the torque converter/oil pump housing oil seal.
3. Replace with new torque converter using torque converter handles 18G1501, checking that the dimension from the converter fixing bolt boss to the converter housing face is 50 mm (1.96 in). If this dimension is achieved the converter is properly seated in the housing.
4. Refit the gearbox and transfer box assembly as previously described.

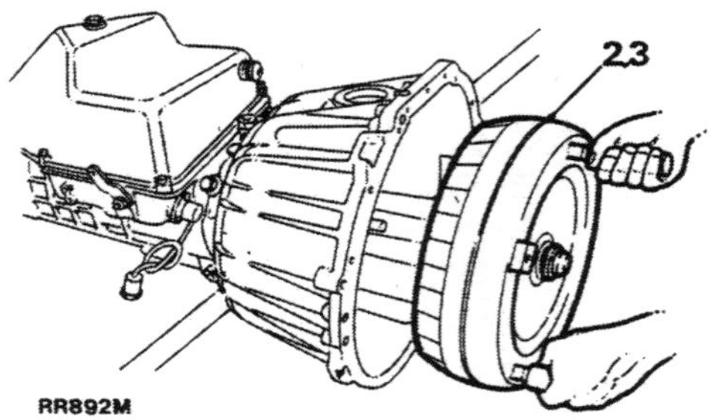

## Eliminating Leaks on the Pump Housing

1. Remove the gearbox/transfer box assembly as previously described.
2. Place the gearbox on the bench and remove the torque converter using torque converter handles 18G1501, taking care not to damage the converter/oil pump housing oil seal.

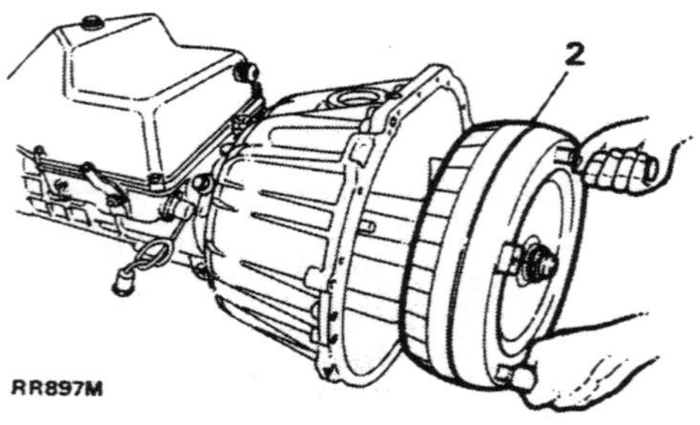

3. Remove the twelve hexagonal bolts (inner ring pattern).

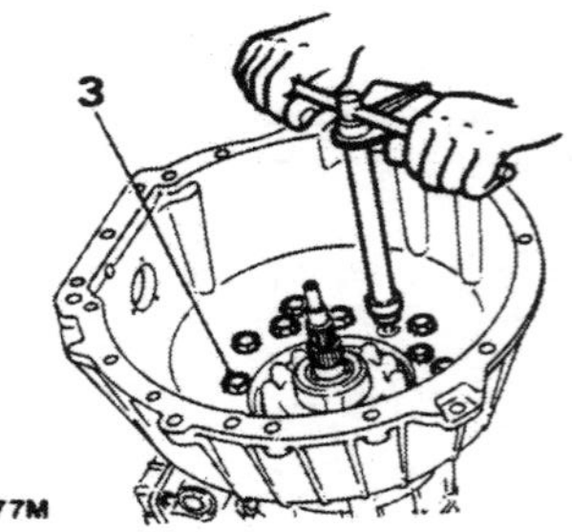

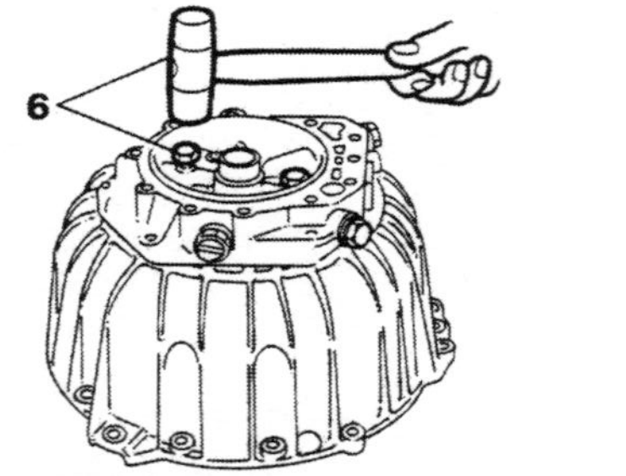

4. Remove bellhousing and pump assembly from gearbox case and discard the gasket.
5. Remove the eight hexagonal bolts on the rear of the pump.
6. Screw in two of the bolts, diagonally opposite each other, tap lightly using a soft headed mallet; this will free the pump assembly from the intermediate plate.
7. Remove the shaft sealing ring and 'O' ring from the pump housing and discard.
8. Using oil seal replacer LST108 fit the shaft seal ring into the pump housing.
9. Fit the 'O' ring onto the circumference of the pump housing.
10. Align the dowel with its hole in the intermediate plate and press the pump housing home.
11. Secure the pump housing to the intermediate plate using the eight hexagonal bolts and tighten to their specified torque.
12. Place the bellhousing and intermediate plate assembly on the bench, front face up. Using the oil pump rotation sleeve LST111, check that the pump gears rotate freely.
13. Before replacing the intermediate plate and bellhousing assembly, check that the thrust washer and axle cage are seated on the A cluch housing.

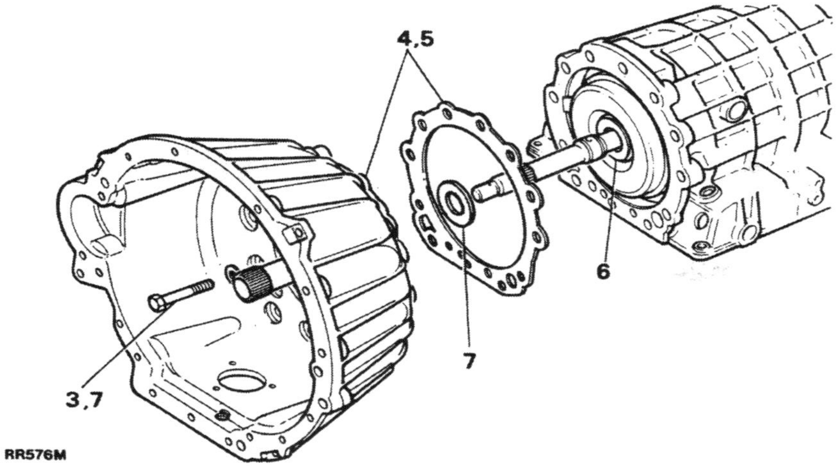

## AUTOMATIC GEARBOX

14. Place the gasket and disc washer onto the bellhousing and intermediate plate assembly using a light grease or vaseline.
15. Fit bellhousing and intermediate plate assembly onto gearcase and secure with the twelve hexagonal bolts tightened to the specified torque.
16. Place the end float gauge LST111 onto the pump housing and check that the axial play is between 0,2–0,4 mm. (0.008 in to 0.016 in). If the end float is excessive or tight, replace existing washer, situated at the rear of the intermediate plate, with a suitable washer to give the required end float as stated above.
17. Refit torque converter into housing using torque converter handles 18G 1501, checking that the dimension from the converter fixing bolt boss to the converter housing face is 50 mm (1.96 in). If this dimension is achieved the converter is properly seated in the housing.
18. Refit the gearbox/transfer box assembly as previously described.

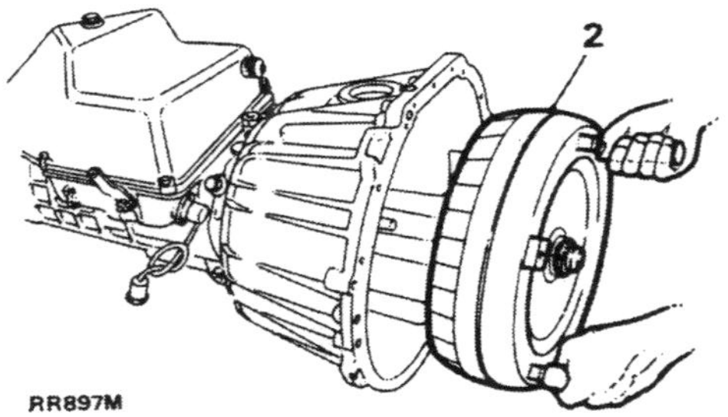

### Eliminating leaks between Gearbox Housing and Intermediate Plate

1. Remove the gearbox/transfer box assembly as previously described.
2. Place the gearbox on the bench and remove the torque converter using torque converter handles 18G 1501, taking care not to damage the converter/oil pump housing oil seal.
3. Remove the 12 hexagonal bolts (inner ring pattern).
4. Remove the bellhousing intermediate plate assembly from gearbox case and discard the gasket.

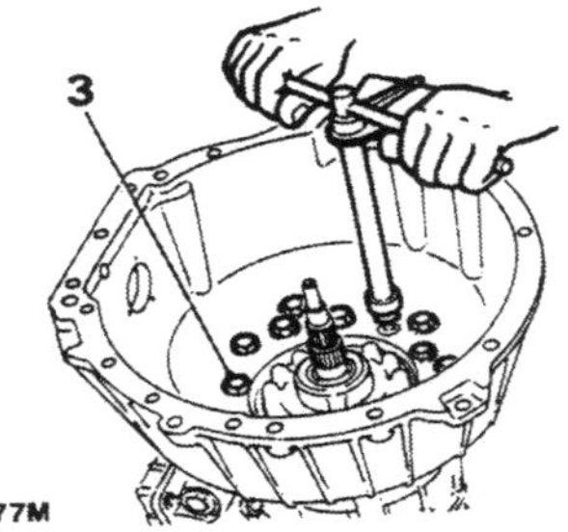

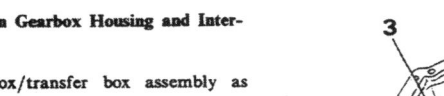

5. Place new gasket onto intermediate plate using a light grease or vaseline.
6. Before replacing the intermediate plate/bellhousing assembly check that the thrust washer and axle cage are seated on the A clutch housing.
7. Fit bellhousing/intermediate plate assembly with disc washer onto gearcase and secure with the twelve hexagonal bolts tightened to the specified torque.

8. Place the end-float gauge LST 111 onto the pump housing and check that the axial play is between 0,2–0,4 mm. (0.008 in to 0.016 in). If the end-float is excessive or tight, replace existing washer, situated at the rear of the intermediate plate, with a suitable washer to give the required end-float as stated above.
9. Refit torque converter into housing using torque converter handles 18G 1501, checking that the dimension from the converter fixing bolt boss to the converter housing case is 50 mm (1.96 in). If this dimension is achieved the converter is properly seated in the housing.
10. Refit the gearbox/transfer box assembly as previously described.

### Replacing Pump

1. Remove the gearbox/transfer box assembly as previously described.
2. Place the gearbox on the bench and remove the torque converter using torque converter handles 18G 1501, taking care not to damage the converter/oil pump housing oil seal.

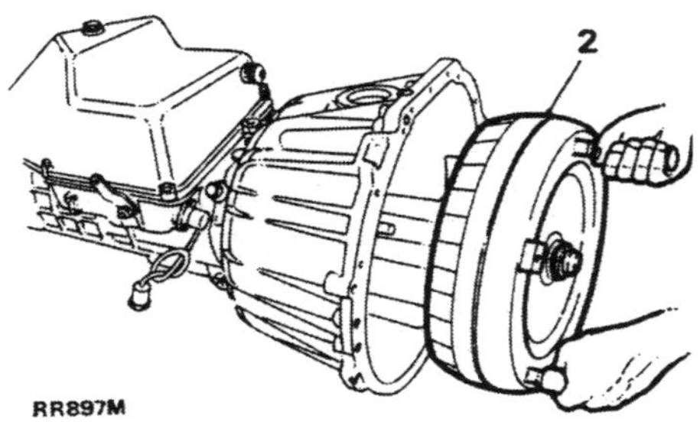

3. Remove the twelve hexagonal bolts (inner ring pattern).

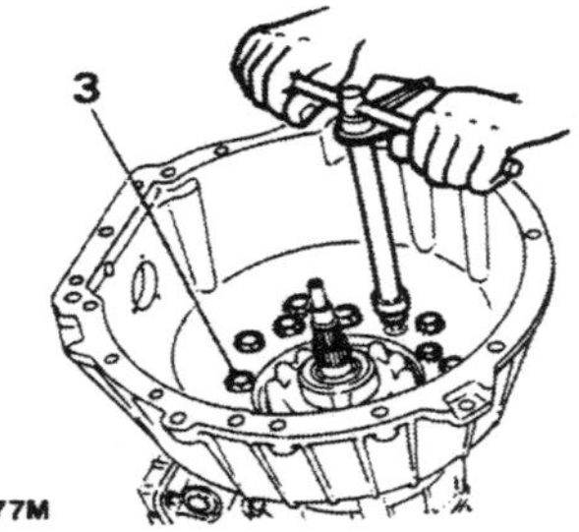

### Replacing Bellhousing

1. Remove the gearbox/transfer box assembly as previously described.
2. Place the gearbox on the bench and using the torque converter handles 18G 1501 remove the torque converter, taking care not to damage the converter/oil pump housing oil seal.

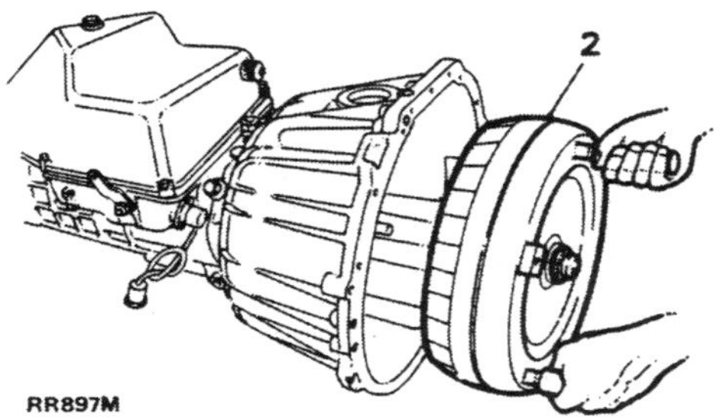

3. Remove the eighteen hexagonal bolts.
4. Remove bellhousing.
5. Fit new bellhousing.
6. Secure bellhousing with the eighteen hexagonal bolts to the specified torque.

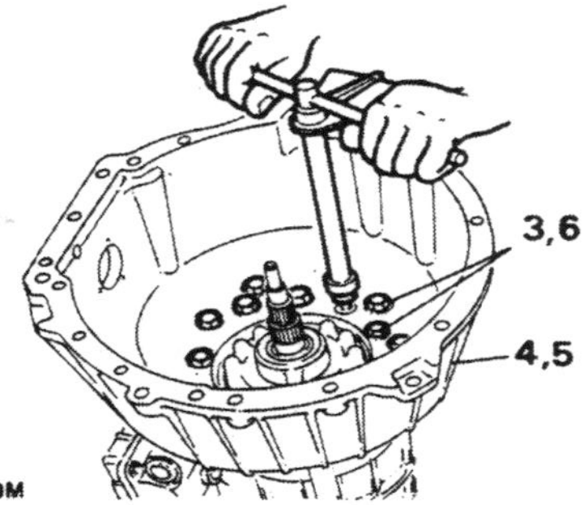

# AUTOMATIC GEARBOX   44

## Replacing Intermediate Plate

1. Remove the gearbox/transfer box assembly as previously described.
2. Place the gearbox on the bench, and remove the torque converter using torque converter handles 18G1501, taking care not to damage the torque converter/oil pump housing oil seal.
3. Remove the twelve hexagonal bolts (inner ring pattern).
4. Remove bellhousing and pump assembly from gearbox case and discard the gasket.
5. Remove the eight hexagonal bolts on the rear of the pump.
6. Screw in two of the bolts, diagonally opposite each other, tap lightly using a soft headed mallet; this will free the pump assembly from the intermediate plate.
7. Remove the 'O' ring from the pump housing and discard.
8. Place the bellhousing and intermediate plate assembly on the bench, front side up.

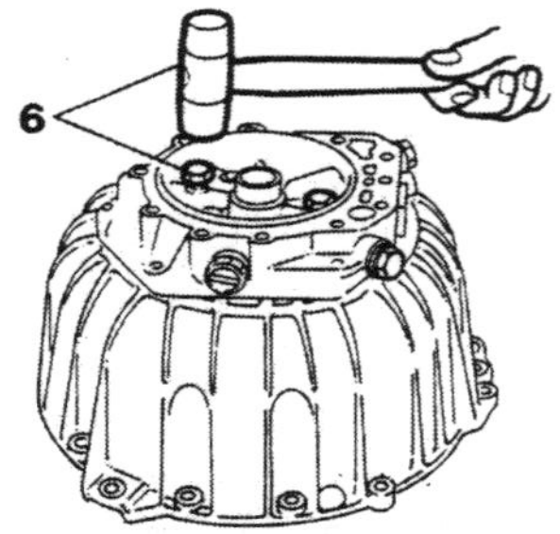

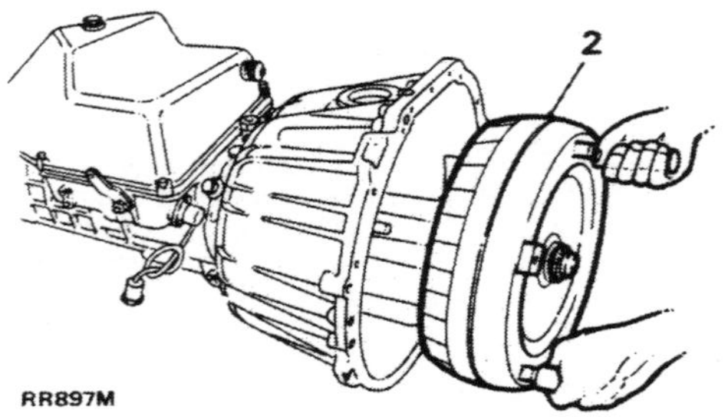

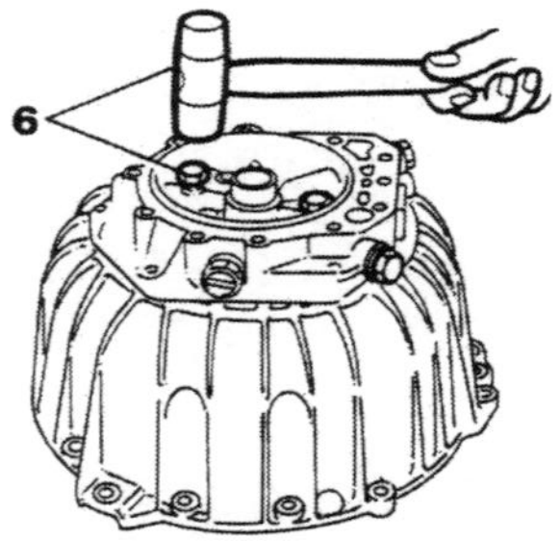

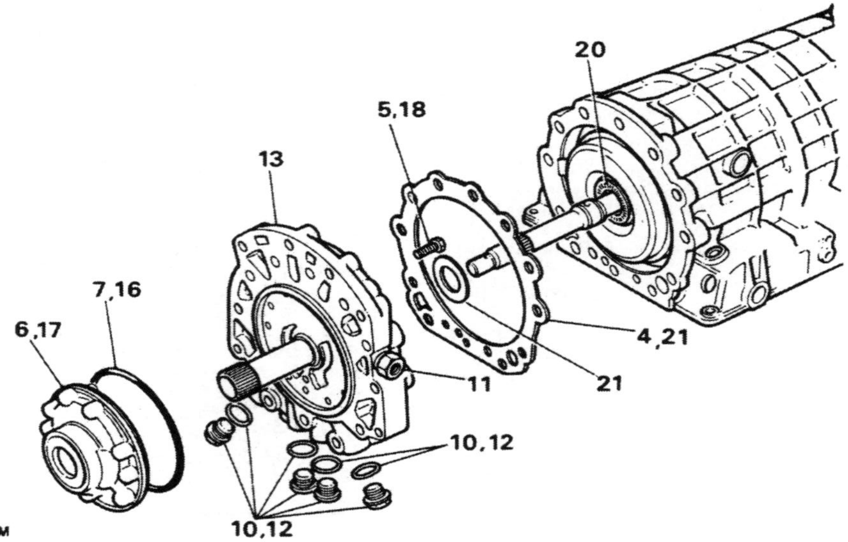

---

## AUTOMATIC GEARBOX   44

4. Remove bellhousing and pump assembly from gearbox case and discard the gasket.
5. Remove the eight hexagonal bolts on the rear of the pump.
6. Screw in two of the bolts, diagonally opposite each other, tap lightly using a soft headed mallet; this will free the pump assembly from the intermediate plate.
7. Fit new pump assembly aligning the dowel with its hole in the intermediate plate and press the pump housing home.
8. Secure the pump housing to the intermediate plate using the eight hexagonal bolts and tighten to their specified torque.
9. Place the bellhousing and intermediate plate assembly on the bench, front face up. Using the oil pump rotation sleeve LST 111, check that the pump gears rotate freely.
10. Before replacing the intermediate plate and bellhousing assembly check that the thrust washer and axle cage are seated on the A clutch housing.
11. Place the gasket and disc washer onto the bellhousing and intermediate plate assembly using a light grease or vaseline.
12. Fit bellhousing and intermediate plate assembly onto gearcase and secure with the twelve hexagonal bolts tightened to the specified torque.
13. Place the end-float gauge LST-111 onto the pump housing and check that the axial play is between 0,2–0,4 mm (0.008 in to 0.016 in). If the end-float is excessive or tight, replace existing washer, situated at the rear of the intermediate plate, with suitable washer to give required end-float as stated above.

**NOTE: If damage is apparent to the bolts they should be replaced.**

14. Refit the torque converter into the housing using torque converter handles 18G 1501, checking that the dimension from the converter fixing bolt boss to the converter housing face is 50 mm (1.96 in). If this dimension is achieved, the converter is properly seated in the housing.
15. Refit the gearbox/transfer box assembly as previously described.

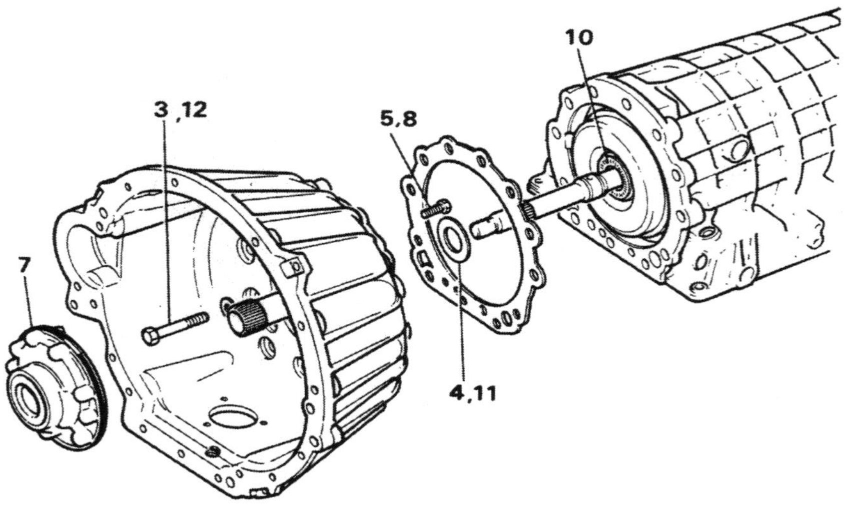

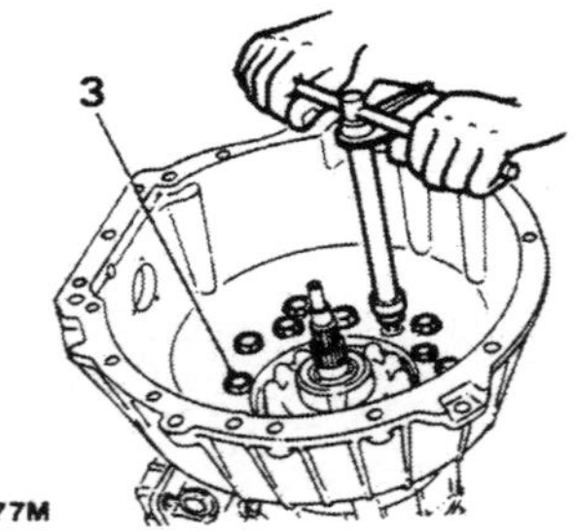

130

# AUTOMATIC GEARBOX  44

9. Remove the six remaining hexagon bolts and remove the bellhousing from the intermediate plate assembly.
10. Remove the four screw plugs and seal rings from the intermediate plate, discard the seal rings.
11. Remove the oil cooler pipe adaptors and fit them into the new intermediate plate.
12. Fit plugs and new seal rings into the new intermediate plate.
13. Fit intermediate plate assembly onto the bellhousing.
14. Secure with six hexagonal bolts (outer ring pattern) and tighten to the specified torque.
15. Place intermediate plate and bellhousing assembly on bench, front face up.
16. Fit the 'O' ring onto the circumference of the pump housing.
17. Align the dowel with its hole in the intermediate plate and press the pump housing home.
18. Secure the pump housing to the intermediate plate using the eight hexagonal bolts and tighten to the specified torque.
19. Place the bellhousing and intermediate plate assembly on the bench, front face up. Using the oil pump rotation sleeve LST111, check that the pump gears rotate freely.
20. Before replacing the intermediate plate and bellhousing assembly check that the thrust washer and axle cage are seated on the A clutch housing.
21. Place the gasket and disc washer onto the bellhousing and intermediate plate assembly using a light grease or vaseline.
22. Fit bellhousing and intermediate plate assembly onto gearcase and secure with the twelve hexagonal bolts tightened to the specified torque.
23. Place the end float gauge LST111 onto the pump housing and check that the axial play is between 0.2–0.4 mm. (0.008–0.016 in.) If end float is excessive or tight, replace existing washer, situated at the rear of the intermediate plate, with suitable washer to give required end float as stated above.
24. Refit the torque converter into the housing using torque converter handles 18G1501, checking that the dimension from the converter fixing bolt boss to the converter housing face is 50 mm (1.96 in.). If this dimension is achieved the converter is properly seated in the housing.
25. Refit the gearbox/transfer box assembly as previously described.

## AUTOMATIC GEARBOX—OVERHAUL

### Remove Torque Converter

NOTE: Refer to Stage II Section for removal of the gearbox from the vehicle.

1. Place gearbox into transmission holding fixture LST118 and tighten.

NOTE: Care must be taken not to over-tighten as casing will distort.

2. Using the torque converter handles 18G1501 remove the converter from the bell housing.

CAUTION: Ensure no damage occurs to the pump bush and seal ring lip when removing the torque converter. The converter is still full of oil even after the gearbox has been drained, so care should be taken when removing the unit.

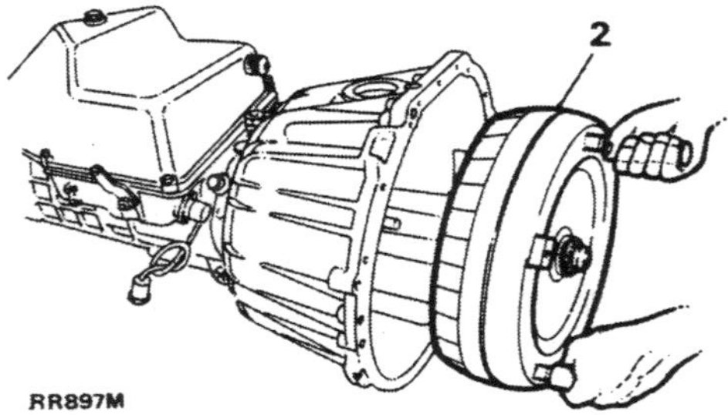

### Remove Valve Body

1. Turn the gearbox upside down in the fixture.
2. Remove the six bolts and retaining plates which hold the oil pan.
3. Remove the oil pan and rubber seal and discard seal.
4. Using torx bit TX27, unscrew the three torx headed bolts which hold the oil screen and remove. Separate the oil screen from the suction tube and discard the two 'O' rings and oil screen.
5. Using torx bit TX27, unscrew the thirteen torx headed bolts which retain the valve block to the gearbox.
6. Using circlip pliers remove the eight circlips.
7. Remove the eight springs (four short springs at the front of the gearbox and four long springs at the rear of the gearbox).
8. Remove the eight sealing rubbers using tool LST113 and discard.
9. Remove the circlip and spring, using tool LST113 remove the restrictor at the rear of the gearbox.

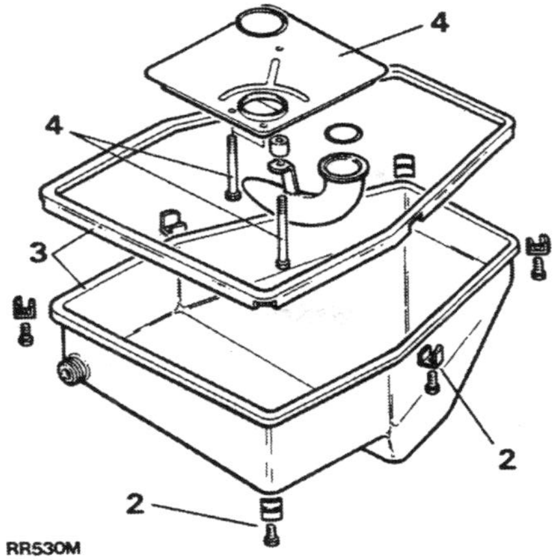

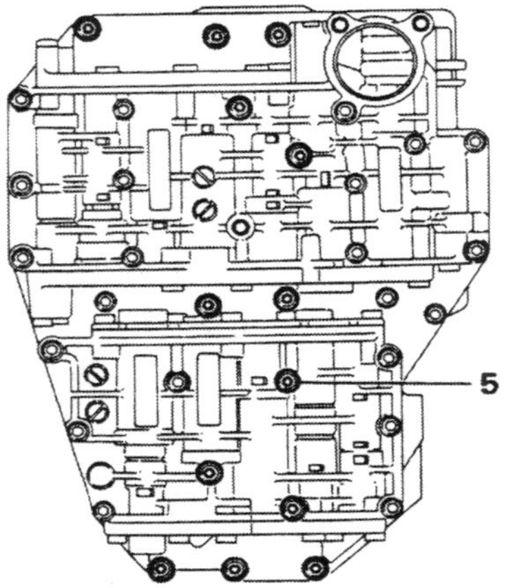

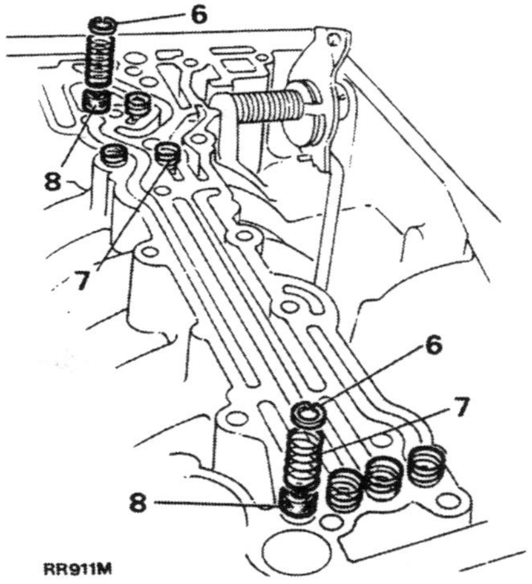

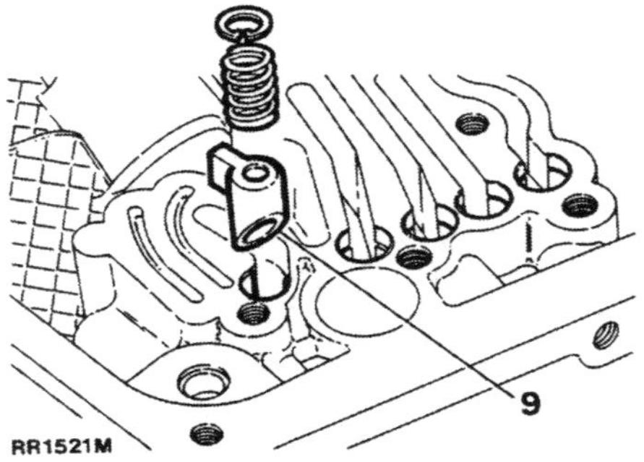

### Remove Parking Pawl and Governor

1. Engage 'Park' position.
2. Using a suitable spanner unscrew the coupling shaft bolt and remove the 'O' ring.
3. Remove the coupling shaft.
4. Remove the nine bolts and washers from the extension housing.
5. Remove the extension housing and gasket from the gearbox and discard the gasket.
6. Engage 'Park' position.
7. Withdraw the parking wheel and governor hub.
8. Unscrew the bolt which retains the guide plate using torx bit TX27.
9. Disengage the spring and remove, also the pin and pawl.

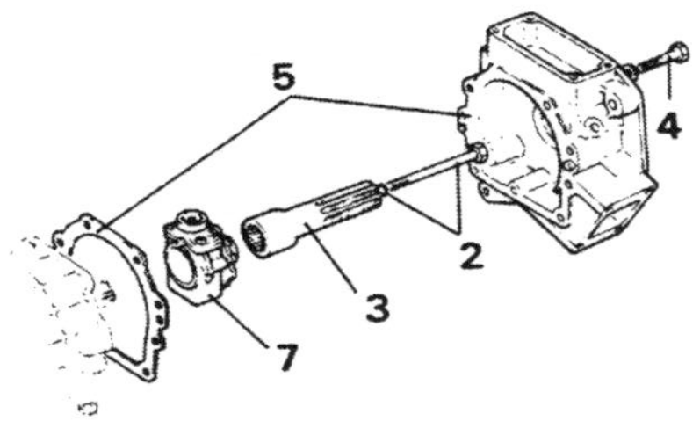

NOTE: When pawl is removed the tension on the spring is reduced.

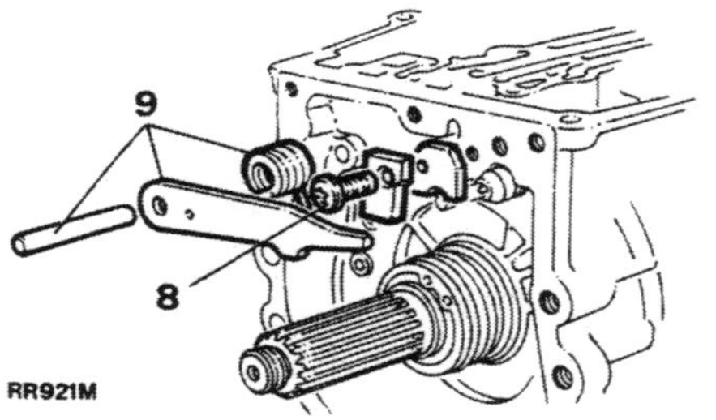

# AUTOMATIC GEARBOX 44

10. Using torx bit TX30 remove the ten torx bolts from rear of casing.

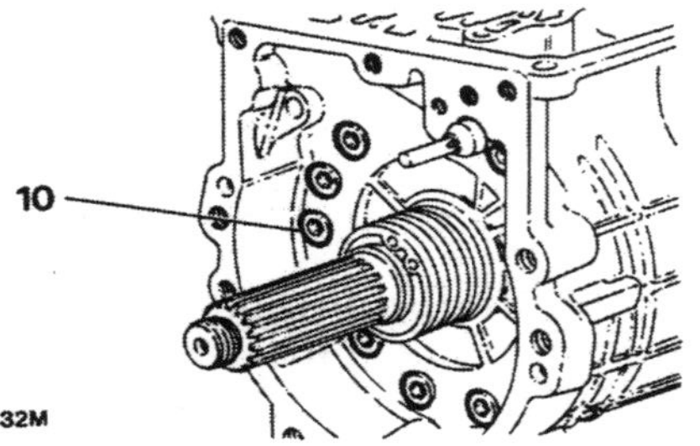

## Remove Inhibitor Switch

1. Using a suitable spanner remove the bolt and spring washer.
2. Remove the retaining plate.
3. Using a suitable tool remove the inhibitor switch from the casing.
4. Discard switch if damaged.

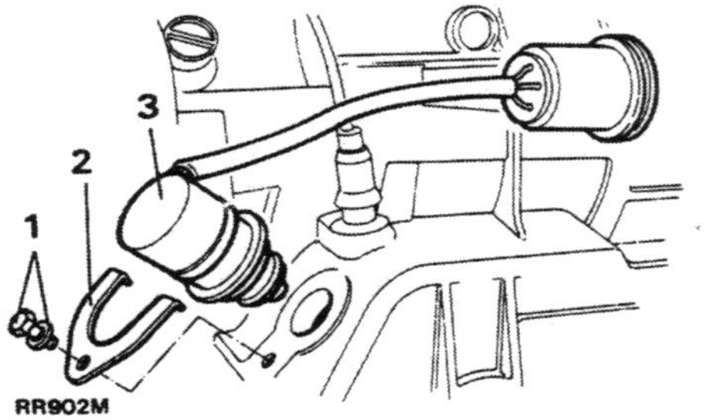

## Remove Bell Housing and Intermediate Plate

1. Using a suitable socket spanner remove the twelve bolts (inside diameter bolt pattern) holding the bell housing.
2. Remove the bell housing and intermediate plate assembly complete, and discard the gasket.
3. Remove the thrust washer, axle bearing and disc washer from the input shaft.

NOTE: Under normal working conditions there is no need to separate the bell housing from the intermediate plate assembly. If damage has occurred to either the bell housing or intermediate plate see appropriate section.

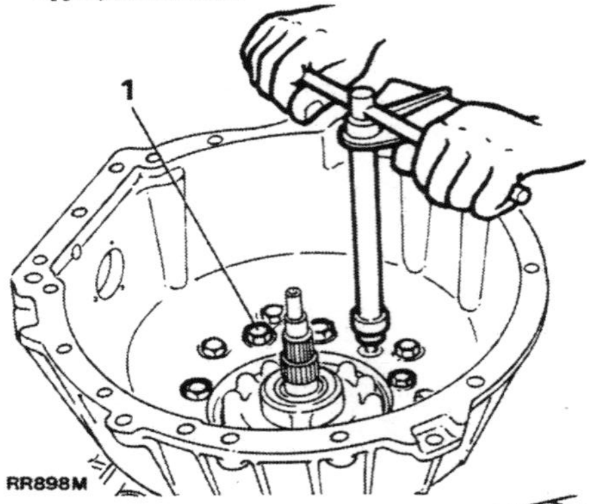

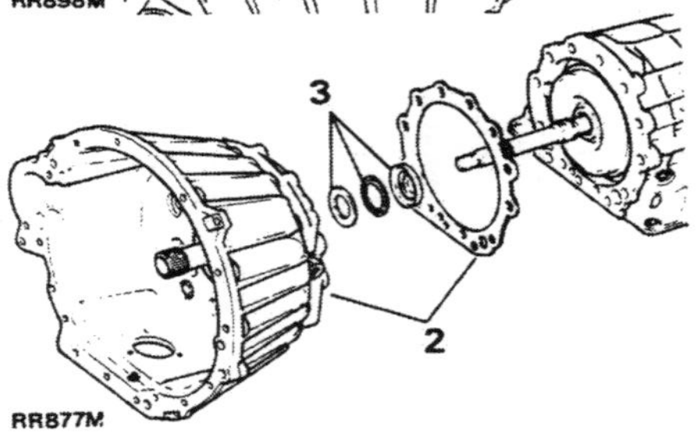

## Remove A Clutch Assembly

1. Turn gearbox with front facing upwards.
2. Remove input shaft and A clutch assembly from gearbox.

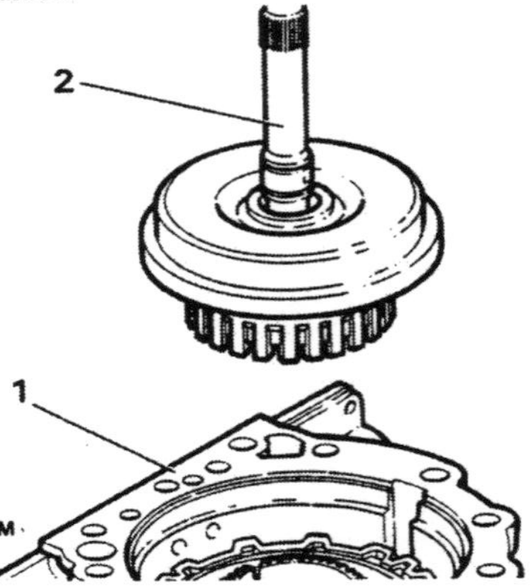

---

# AUTOMATIC GEARBOX 44

3. Remove inner carrier A, disc, axial bearing and thrust washer.

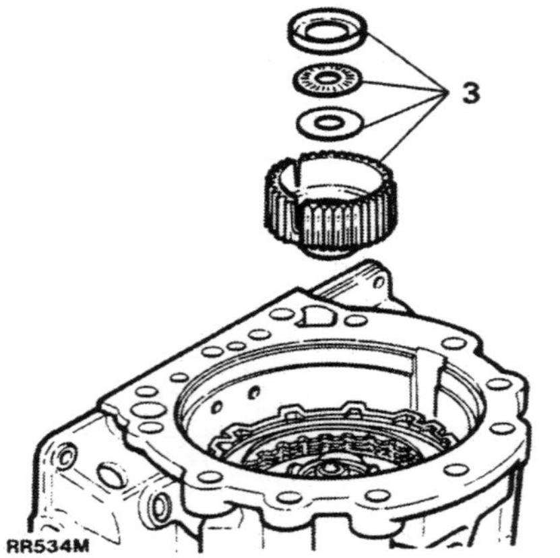

## Remove B Clutch Assembly

1. Using two suitable screwdrivers remove the small snap ring in cylinder B.
2. Using the B clutch assembly puller hooks LST115 remove the B clutch assembly.

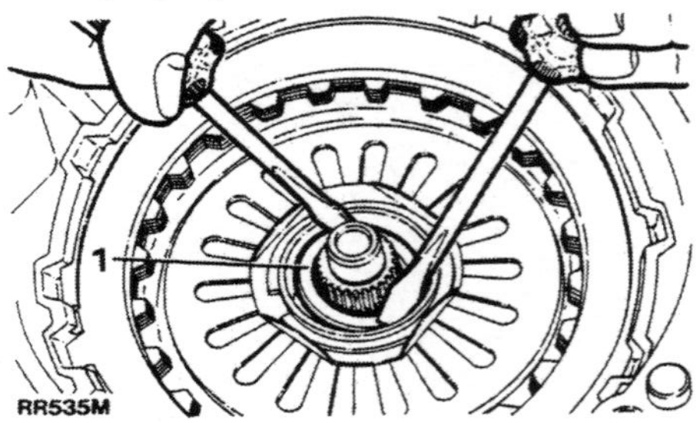

NOTE: To remove assembly, lift up cylinder B until it stops, push assembly back down and lift up again using more weight.

3. Remove support ring and 'O' ring.

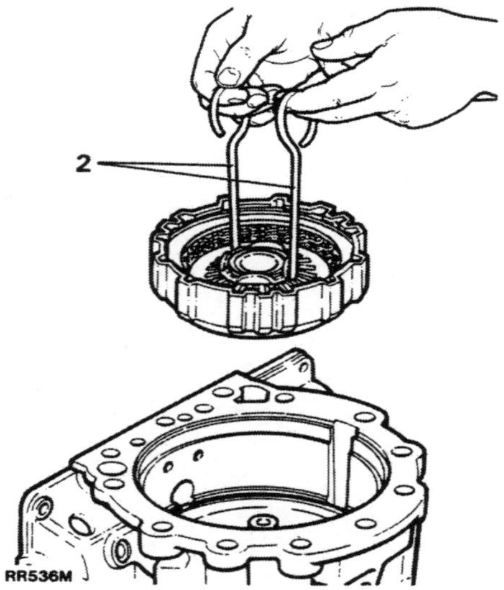

## Remove C, C' and D Clutch Assembly

1. Using a suitable screwdriver remove centreplate snap ring via a hole in the casing.

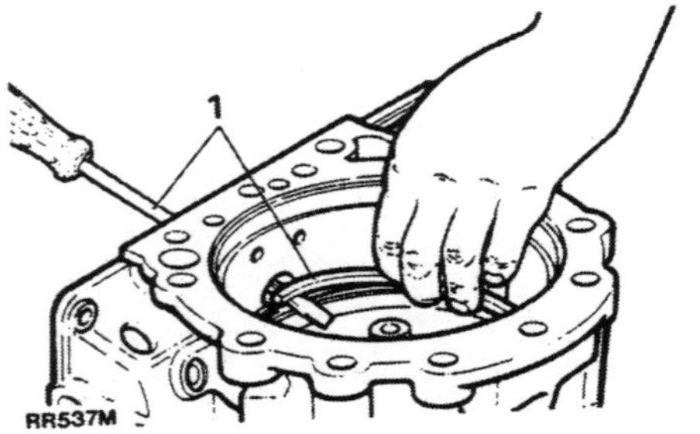

## AUTOMATIC GEARBOX

2. Using tool LST117 attached to intermediate shaft remove C, C' and D clutch assembly.
3. Remove disc, axial bearing and thrust washer.

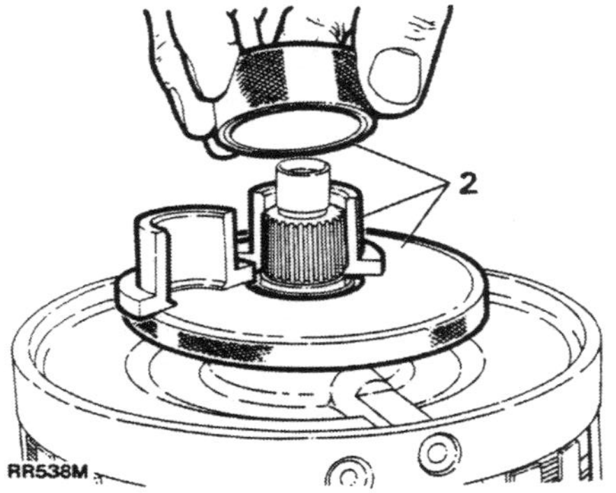

### Remove 4th Gear Assembly

1. Turn gearbox to the horizontal position.
2. Push assembly out from the rear, guiding it from the front of the casing.

### Transmission Gear Selector Assembly

#### Remove and overhaul

1. Remove the kickdown cable from the cam.
2. Using kickdown cable remover LST112, remove the kickdown cable from the casing.
3. Using a punch remove the roll pin from the selector shaft.
4. Using a pair of pliers or grips pull the selector shaft from the casing.
5. Remove the stop washer, connection rod, cam and leg spring.
6. Using a screwdriver prise out the seal ring located in the gearbox casing and discard.
7. Inspect and clean casing ensuring no damage has occurred.

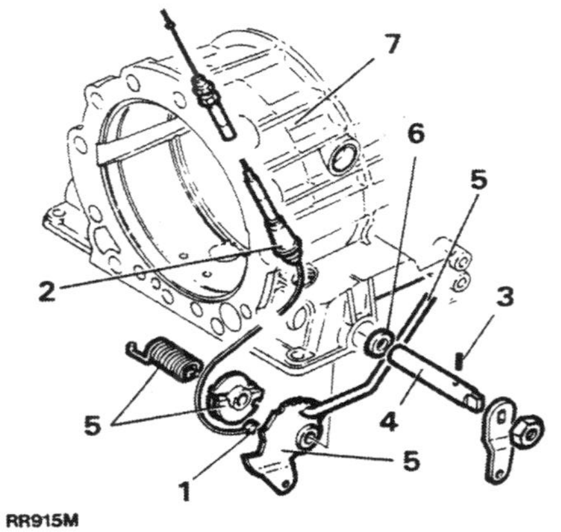

NOTE: At this stage the gearbox is totally stripped.

### 4th Gear Assembly Overhaul

1. Using soft-jawed vice secure the 4th gear assembly by gripping the output shaft.
2. Remove the sungear.
3. Remove the planet gear assembly.

NOTE: Removal of snap-ring on assembly is not necessary unless damage has occurred.

4. Remove the disc washer, axial bearing and thrust washer.
5. Remove assembly from the vice and turn upside down onto the bench.
6. Remove cylinder F from cylinder E.
7. Remove cylinder E from the freewheel 3rd.
8. Remove axial disc, cage and two thrust washers.
9. Using pliers and screwdriver remove the snap-ring on carrier E.

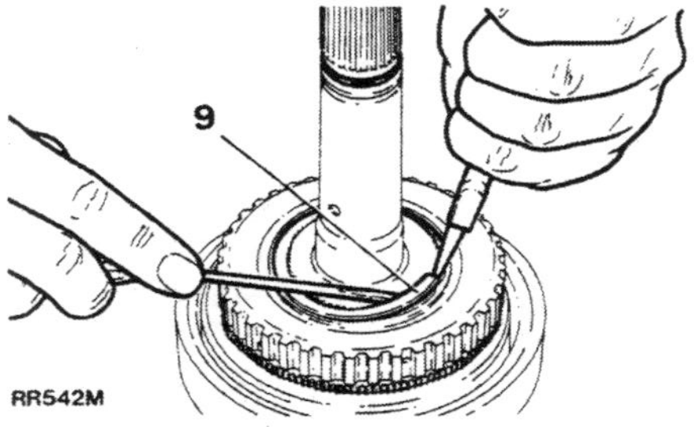

10. Turn the assembly around and remove the output shaft from ring gear by pushing the gear downwards.

NOTE: Do not remove the snap-ring on output shaft.

11. Place ring gear on bench, teeth side down.
12. Remove carrier E from the ring gear assembly.
13. Remove the freewheel cage assembly from the ring gear by using an upward turning motion.
14. Remove the snap-ring retaining the freewheel ring (inner) to the hollow gear.
15. Remove the freewheel ring (inner) from the hollow gear.
16. Remove freewheel cage from freewheel ring (outer).

NOTE: Care should be taken when removing the freewheel ring, which due to the rollers and springs becoming loose may fall out.

17. Remove the snap-ring retaining the clutch plates and steel plates in cylinder F.
18. Remove four clutch plates and five steel plates from cylinder F.

### Assemble

8. Fit new seal ring into gearbox casing using selector shaft oil seal replacer LST114.
9. Fit connection rod with connection rod into the gearbox casing and then feed the selector shaft into the casing.
10. Place the assembly into the gearbox casing and push the selector shaft through until the hole in the shaft aligns with the hole in the stop washer.
11. Fit the leg spring onto the cam.
12. Place the assembly into the gearbox casing and push the selector shaft through until the hole in the shaft aligns with the hole in the stop washer.
13. Using a suitable punch, fit roll pin with the open side facing the rear of the gearbox casing.
14. Fit new kickdown cable assembly into its seat on the gearbox casing.
15. Fit the nipple of the kickdown cable into the cam seat ensuring the cam has been turned once to load the spring.

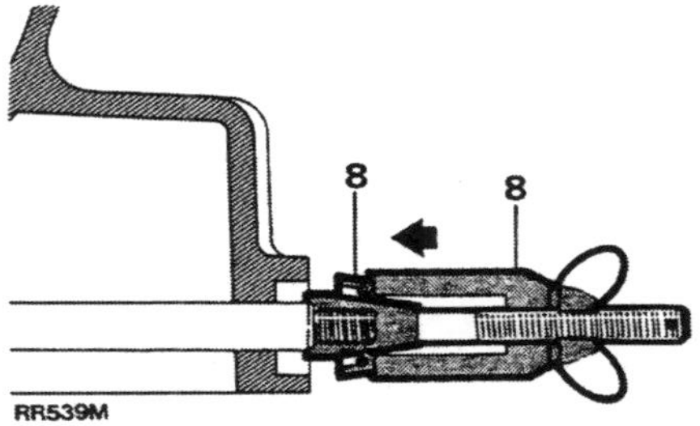

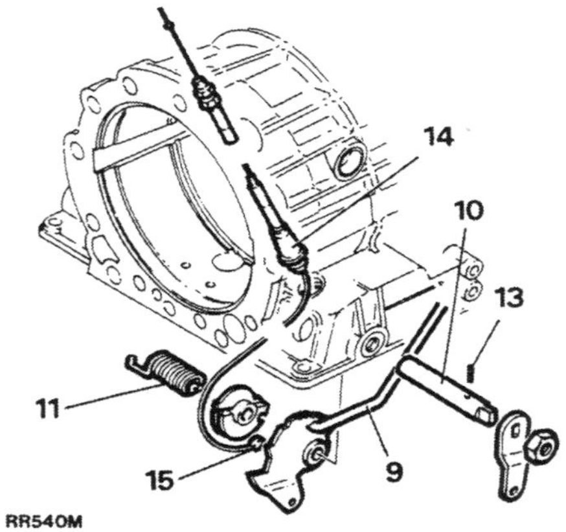

### Fit inhibitor switch

1. Fit new inhibitor switch if existing one was damaged.
2. Replace retaining plate and fix with spring washer and bolt.

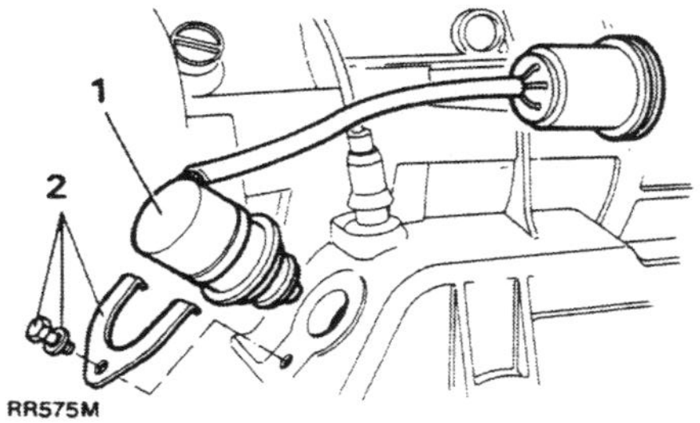

# AUTOMATIC GEARBOX

19. Using clutch spring compressor LST1016-1, press down on the spring plate and remove the split rings.
20. Remove the spring plate.
21. Turn the cylinder upside down, using two small punches placed in the holes (diametrically opposite each other), push down and remove the piston.
22. Remove and discard the two 'O' rings from the piston.

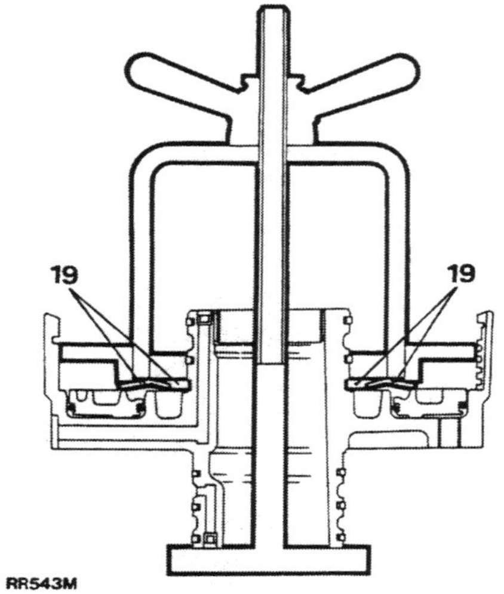

23. Remove the snap ring from cylinder E.
24. Remove the four clutch plates and five steel plates from cylinder E.
25. Using clutch spring compressor LST1016-1, press down on the spring plate and remove the split rings.
26. Remove the pressure plate.
27. Remove piston E by using air pressure directed into the oil feed hole.
28. Remove and discard the two 'O' rings from the piston.

**WARNING: Care should be taken using air pressure when removing the piston.**

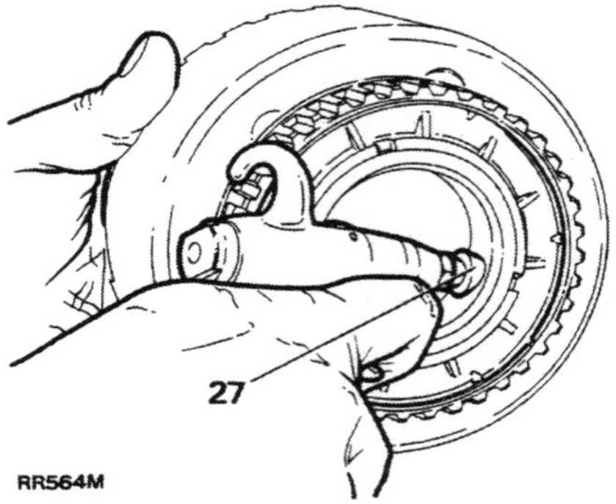

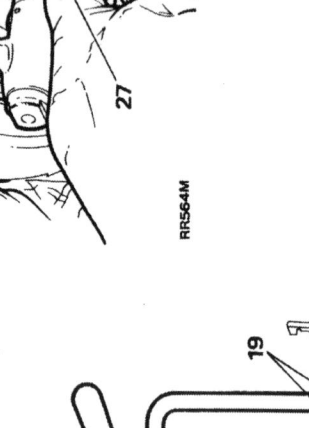

**NOTE: The five sealing rings do not need to be removed on the cylinder unless any damage has occurred to them.**

**NOTE: Do not remove the snap-ring at the bottom of the E cylinder, unless damaged.**

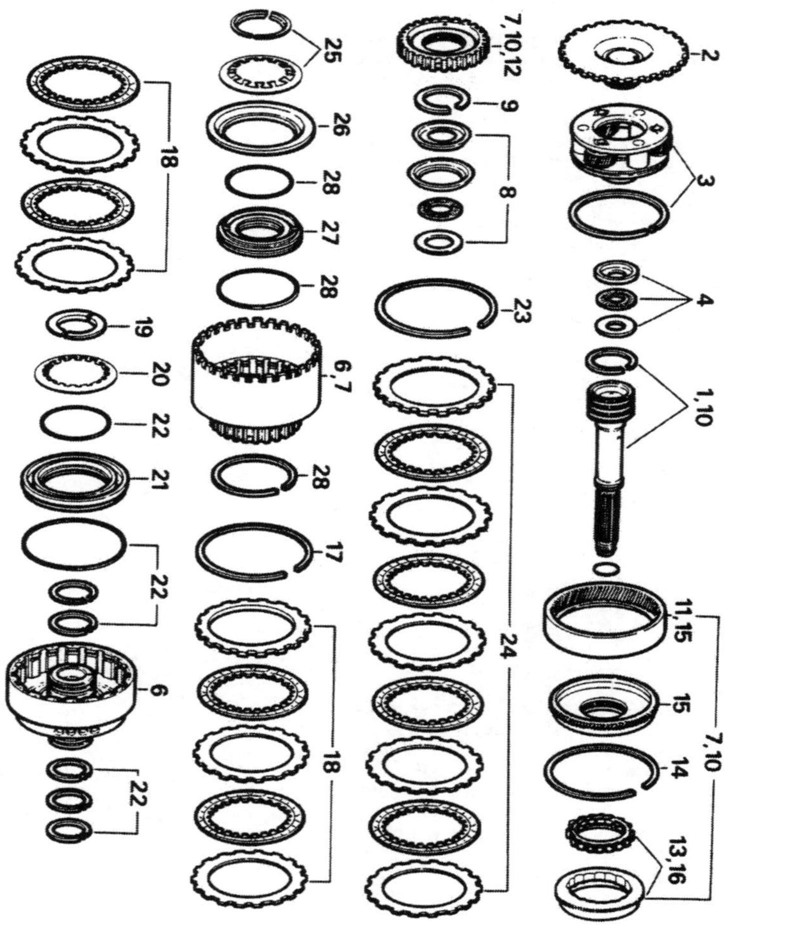

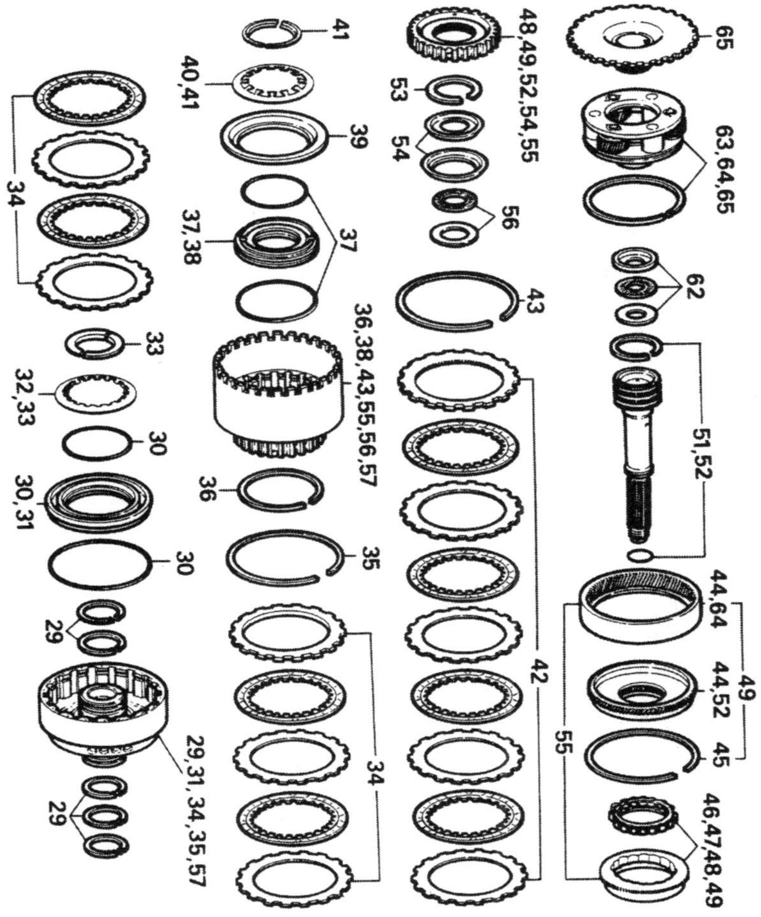

# AUTOMATIC GEARBOX

47. Turn the freewheel cage in the freewheel ring (outer) until rim of the cage has been seated.
48. Fit carrier E to freewheel cage assembly.
49. Fit freewheel cage assembly to ring gear assembly using a clockwise motion.
50. A minimum clearance of 0,1 mm (0.0039 in) should be obtained between the freewheel cage assembly and ring gear.

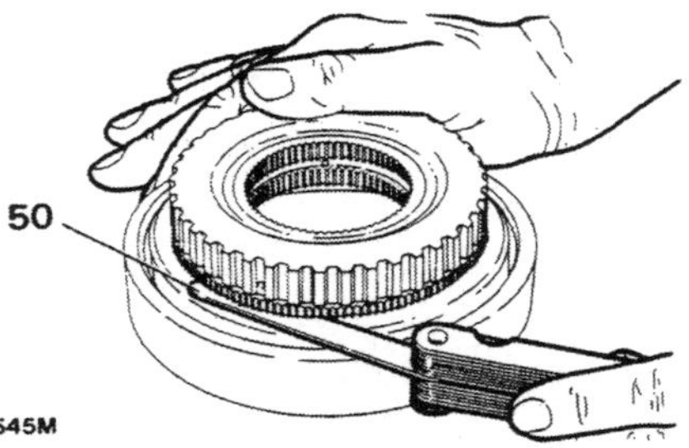

51. Inspect the output shaft for damage to the snap-ring, if any, replace snap-ring; also remove 'O' ring and discard, replace with new 'O' ring.
52. Align inner teeth of carrier E with freewheel ring (inner) teeth and then place freewheel 3rd assembly onto the output shaft.
53. Secure snap-ring into position, retaining the freewheel 3rd.

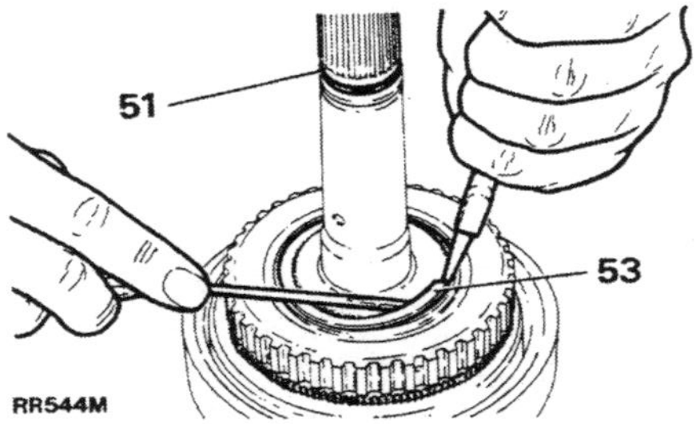

54. Fit the steel thrust washer and then the copper thrust washer onto the freewheel 3rd assembly.
55. Fit cylinder E onto freewheel 3rd assembly using a turning motion, ensuring that the teeth of the end plate line up with the freewheel ring (outer).

NOTE: When correctly assembled, copper thrust washer must be touching cylinder E assembly. The cylinder E assembly will turn in a clockwise direction when holding the output shaft. If the cylinder E assembly is turned in a counter clockwise direction the freewheel will lock up.

## Assemble

29. Fit three seal rings on the outside hub and two seal rings on inside hub of cylinder F if they have been removed.

NOTE: Ensure each seal ring is snapped together.

30. Fit new 'O' rings onto the F piston.

NOTE: For ease of assembly apply petroleum jelly on 'O' rings and stretch the inner 'O' ring to avoid damage on installation.

31. Fit piston F into cylinder F.
32. Fit spring plate using clutch spring compressor LST1016-1.
33. Fit the two halves of the split ring to secure the spring plate in position, then remove the clutch spring compressor.
34. Install the clutch plates and steel plates into the F cylinder starting with a steel plate then clutch plate finishing up with the end plate which is thicker than the normal steel plates.
35. Fit the snap-ring into cylinder F to retain the clutch plate assembly.

NOTE: Do not confuse the steel plates of F clutch with that of the E clutch. The differences are thus: F clutch—steel plates are thicker and the end plate has no inner teeth.

36. Fit new snap-ring at bottom of cylinder E if it has been removed.
37. Fit the two 'O' rings onto the E piston.

NOTE: For ease of assembly apply petroleum jelly.

38. Fit E piston into cylinder E.
39. Fit the pressure plate with 'depression' facing downwards.
40. Fit spring plate using clutch spring compressor LST1016-1.
41. Fit the two halves of the split ring to secure the spring plate in position then remove the clutch spring compressor.
42. Install the clutch plates and steel plates, starting with a steel plate then clutch plate, finishing up with the end plate which is thicker than the normal steel plates.
43. Fit the snap-ring into cylinder E to retain the clutch plate assembly.

NOTE: Do not confuse the steel plates of E clutch with that of the F clutch. The differences are thus: E clutch—steel plates are thinner and the end plate has inner teeth.

44. Fit the freewheel ring (inner) to the hollow gear.
45. Secure using the snap-ring.
46. Fit freewheel cage into the freewheel ring (outer), and press home.

# AUTOMATIC GEARBOX | 44

9. Remove the snap-ring from the hollow gear.
10. Remove the hollow gear from the assembly.
11. Remove the rear planetary set.
12. Remove the thrust washer and axial bearing.
13. Remove the intermediate shaft with the hollow gear complete.
14. Remove the axial bearing and two thrust washers, one from each side of the bearing.
15. Remove the distance ring.

NOTE: The snap-ring in the webshaft need only be removed if damaged.

16. Holding the hollow gear with the rear face uppermost, remove the snap-ring.
17. Disconnect the hollow gear from the intermediate shaft.

## Brakes C, C' and D with planetary sets — overhaul

1. Remove the centre plate assembly.
2. Remove the two brake C' clutches and two steel plates from cylinder C–D.
3. Remove freewheel 2nd complete.
4. Remove the two brake C' clutches and three steel plates from cylinder C–D.
5. Remove the cylinder C–D with brake D assembly.
6. Remove the support ring from the planetary sets assembly.
7. Remove the front planetary set with freewheel assembly.
8. Remove the sunshaft from the assembly.

NOTE: Do not remove the seal ring unless damaged from the sunshaft.

## Refit park mechanism

1. Turn gearbox into a horizontal position.
2. Fit leg spring over pin and place into rear of gearbox.
3. Fit pawl onto pin, to tension spring fit leg of spring into hole of pawl.
4. Fit plate and guide plate and tighten to specified torque using torx bit TX27.

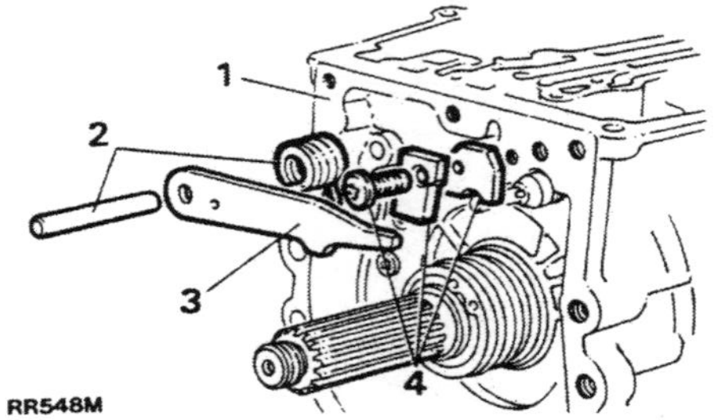

5. Turn gearbox so that the front of the case is uppermost.
6. Fit the disc and axial cage.

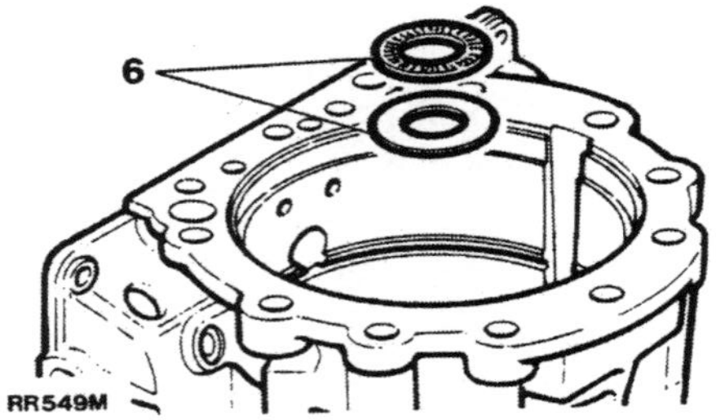

56. Fit the axial cage and axial disc onto the rear of cylinder E.
57. Using a turning motion, fit cylinder F assembly onto cylinder E assembly.
58. When correctly mounted the raised edge of the output shaft will be 10,00 mm (0.393 in) above the top surface of cylinder F assembly.

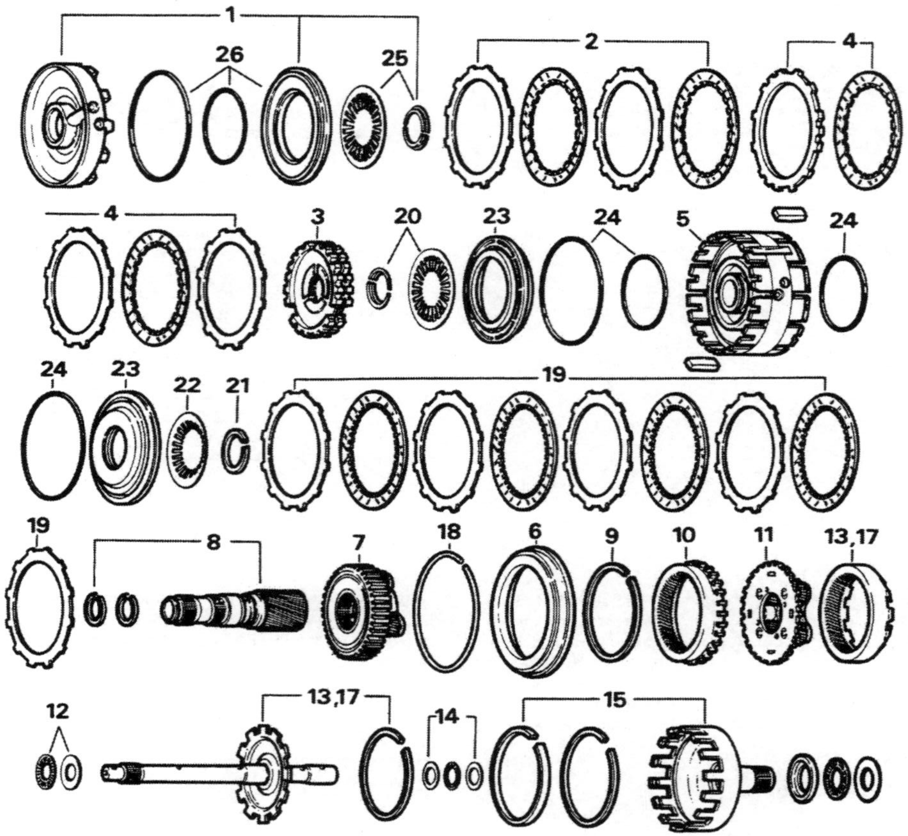

NOTE: Disengagement of end plate and freewheel ring (inner) will occur if end play exceeds 3,00 mm (0.118 in).

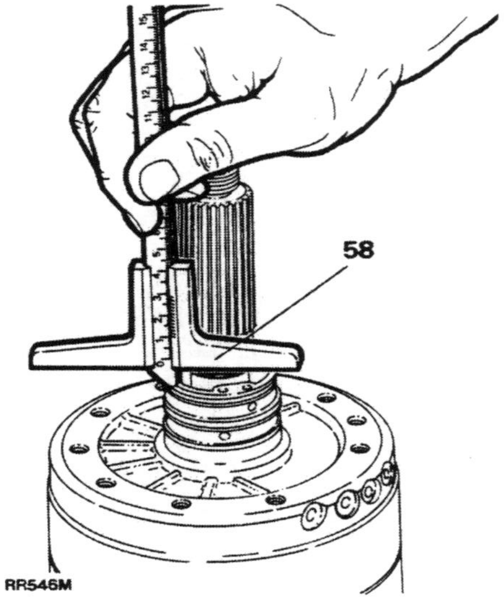

59. Fit the complete 4th gear assembly into the gearbox, ensuring that the oil feed holes in cylinder F line up with the corresponding holes in the gearbox casing.
60. Secure the 4th gear assembly to the gearbox using ten countersunk screws. Tighten screws to the specified torque using torx bit TX30.

NOTE: If screws are not tightened correctly, clutch pressure will be lost in clutch F.

61. Turn the gearbox so that the front of the case is uppermost.
62. Fit the disc washer, axial cage and thrust washer onto the 4th gear assembly.
63. Fit the seal ring onto the planetary case and snap together if ring has been removed.
64. Fit the planetary set into the hollow gear using a turning motion.
65. Fit the sun gear onto the planetary set.

## AUTOMATIC GEARBOX | 44

18. Remove the external snap-ring from the brake D assembly.
19. Remove the four clutch plates and five steel plates from the assembly.
20. Using clutch spring compressor LST1016-1 press down on the spring plate to remove the split rings.
21. Turn cylinder C-D upside down and using the clutch spring compressor, remove the snap-ring with pliers.
22. Remove the spring plate.
23. For ease of removal of both piston C and piston D, use air pressure fed through the oil feed holes.

**WARNING: Care should be taken using air pressure when removing the pistons.**

24. Discard 'O' rings from both piston assemblies.
25. Using the clutch spring compressor remove the spring plate, as previously explained, from the centre plate assembly.
26. To remove the piston use air pressure as previously described and discard 'O' rings.

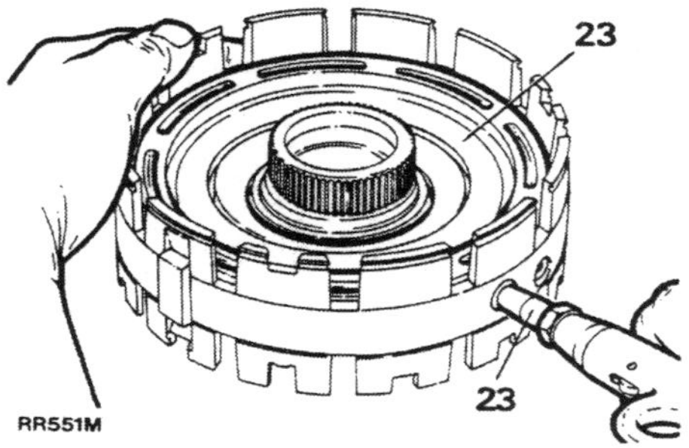

### Assemble

27. Secure the webshaft into a soft-jawed vice.
28. Fit the snap-ring if it has been removed into the lower groove.
29. Fit the distance ring into the webshaft.
30. Place a disc washer and axial cage into the assembly.
31. Assemble together the hollow gear with the intermediate shaft and secure with the snap-ring.
32. Place the other disc washer onto the rear of the intermediate shaft using grease.
33. Fit the intermediate shaft assembly into the webshaft ensuring the disc washer mates up to the axial cage.
34. Using a turning motion fit the rear planetary set into the hollow gear.
35. Fit the front hollow gear into the webshaft assembly and secure with a snap-ring.
36. Insert disc washer and axial cage.
37. Place support ring onto the webshaft assembly.
38. Tap the two fitting pegs down into the slots on the side of the cylinder C-D if they have been removed.
39. Place the two 'O' rings onto piston D.

**NOTE: For ease of assembly, apply petroleum jelly to the 'O' rings.**

40. Fit the D piston into cylinder C-D ensuring that the correct side is selected, that is, the side with the least number of slots in cylinder C-D.
41. Fit the spring plate and using the clutch spring compressor LST1016-1, fit the snap-ring into the groove.
42. Place the two 'O' rings onto piston C.

**NOTE: For ease of assembly, apply petroleum jelly to the 'O' rings.**

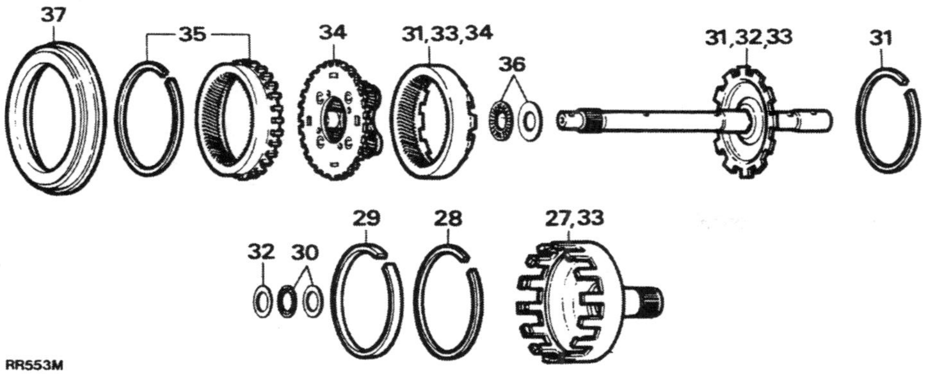

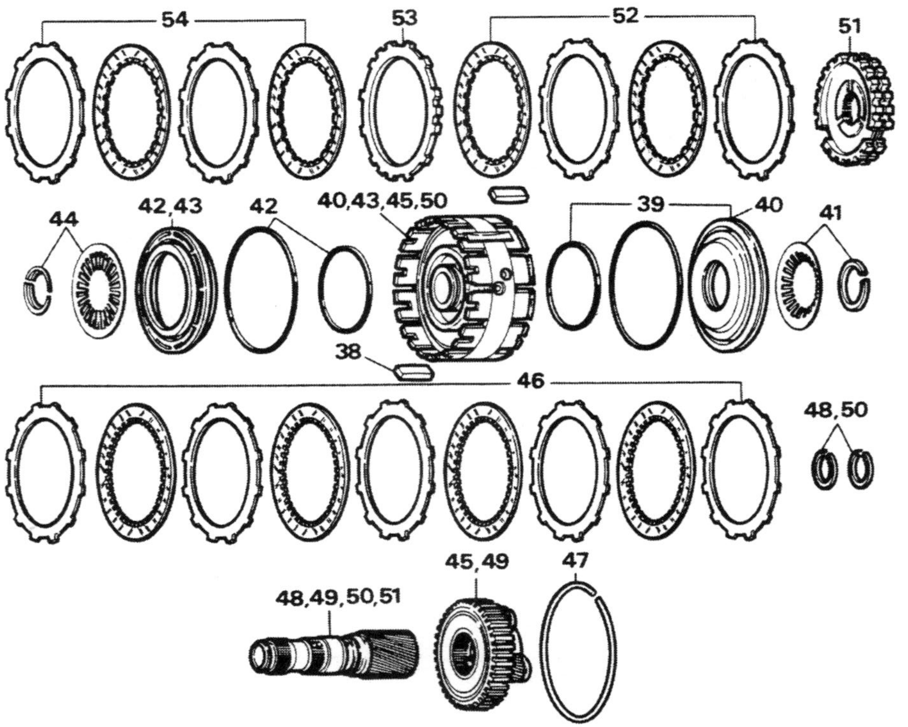

43. Fit the C piston into cylinder C-D.
44. Using the clutch spring compressor LST1016-1, fit the spring plate and the two halves of the split rings.
45. With the clutch D aperture uppermost, fit the planetary set with freewheel 1st gear onto the hub of cylinder C-D.
46. Fit the clutch plates and steel plates starting with a steel plate then a clutch plate, finishing up with the thin end plate.
47. Fit the snap-ring on the outside of the C-D cylinder which secures the D clutch assembly.
48. Fit two seal rings onto the sunshaft and snap together if they have been removed.
49. Fit the sunshaft into the planetary set, splines first.
50. Turning the whole assembly around so the C clutch side is uppermost, fit the assembly into the webshaft assembly.
51. Fit freewheel second onto the sunshaft, before fitting align the upper and lower halves.

**NOTE: To ensure correct fitment of the freewheel second, the top of the assembly is marked with the word 'open'.**

52. Fit the C clutch plates and steel plates starting with a steel plate then a clutch plate into the longer slots of C-D cylinder.
53. Fit end plate which has three groups of three teeth, of which the middle tooth must fit into the short slots in the C-D cylinder.
54. Fit the C' clutch assembly starting with a clutch plate ending with a steel plate.

**NOTE: When fitting these plates ensure teeth on the outside do not go into the 'V' shaped area of the C-D cylinder.**

**If thin steel plates have to be added into the C or C' clutch assembly ensure that these plates are placed on the side nearer to the respective pistons.**

55. Fit the two 'O' rings onto the piston C'.

**NOTE: For ease of assembly use a petroleum jelly.**

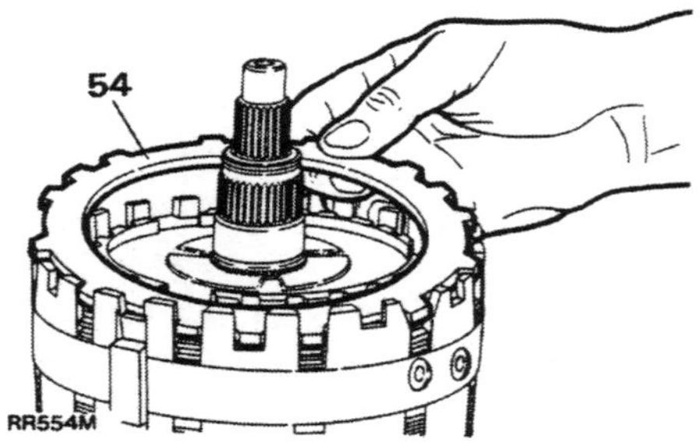

## AUTOMATIC GEARBOX

7. Remove and discard the two 'O' rings on piston B.
8. Remove seal ring on bottom of B cylinder if damage has occurred.

**Assemble**

9. Fit the two 'O' rings onto piston B.

NOTE: For ease of assembly use a petroleum jelly.

10. Install piston B into cylinder B.
11. Place spring plate into cylinder B and using clutch spring compressor LST1016-1, fit retaining washer (lips facing upward) and snap-ring.
12. Fit the clutch and steel plates starting with a steel plate finishing with the steel plate with three sets of three teeth grouped together.
13. Fit snap-ring into the clutch B assembly.
14. Turn upside down and fit seal ring and snap together if removed.
15. Install B clutch assembly into the transmission case, clutch plates facing upwards.
16. Using B clutch 'O' ring and snap-ring replacer LST116, fit 'O' ring, support ring and finally the snap-ring.

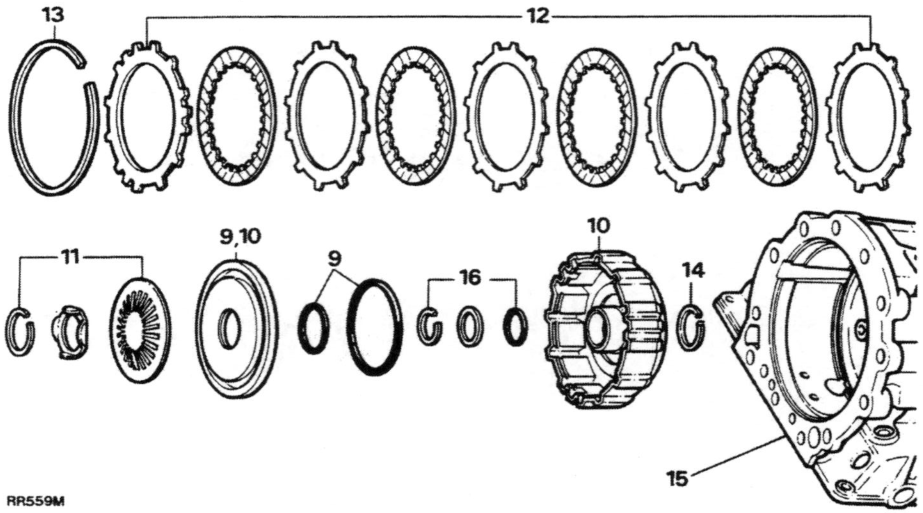

**Clutch B overhaul**

1. Place B clutch assembly with open face upwards.
2. Remove the snap-ring from inside the assembly.
3. Remove the four clutch plates and five steel plates.
4. Using clutch spring compressor LST1016-1, depress spring plate and remove snap-ring and retaining washer.
5. Remove the spring plate.
6. For ease of removal of the piston B, use air pressure fed through the oil feed hole, then turn assembly upside down and tap lightly on the working surface.

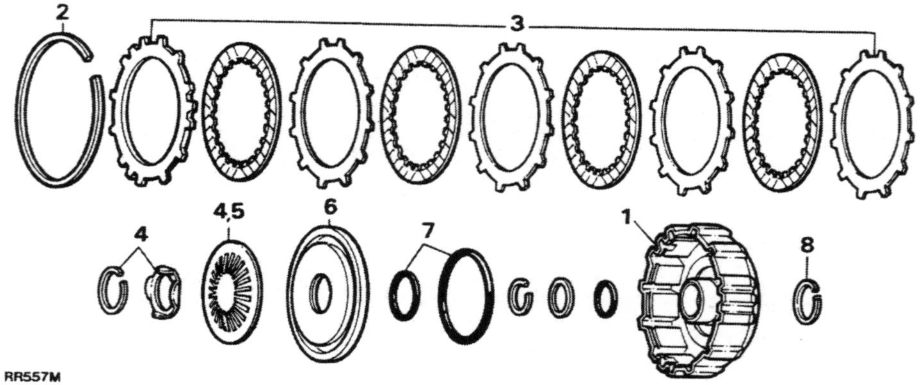

WARNING: Care should be taken using air pressure when removing the piston.

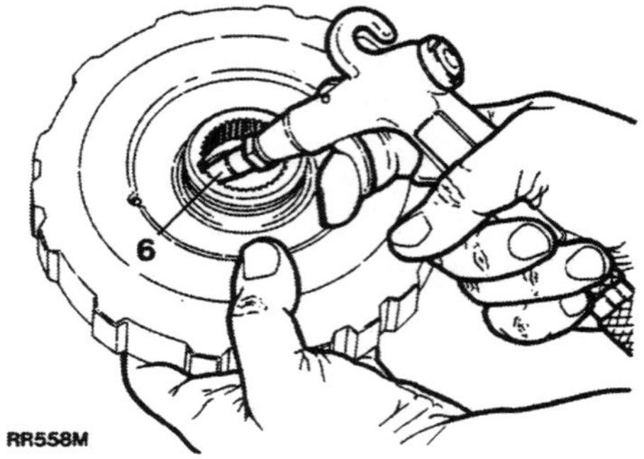

56. Fit C' piston assembly into centre plate.
57. Using clutch spring compressor LST1016-1, fit spring plate and the two halves of the split rings.
58. Place the centre plate onto the C-D cylinder making sure that the 'V' shaped tag on the centre plate fits the 'V' shaped hollow in the C-D cylinder.
59. Remove the C, C' and D clutch assembly from vice and place a greased thrust washer rear face of the webshaft.
60. Fit the whole assembly into the transmission case using the gear train remover/replacer LST117, ensuring that the oil feed holes are aligned with those in the bottom of the casing.
61. Secure the whole assembly with a snap-ring which fits into the groove inside the casing.

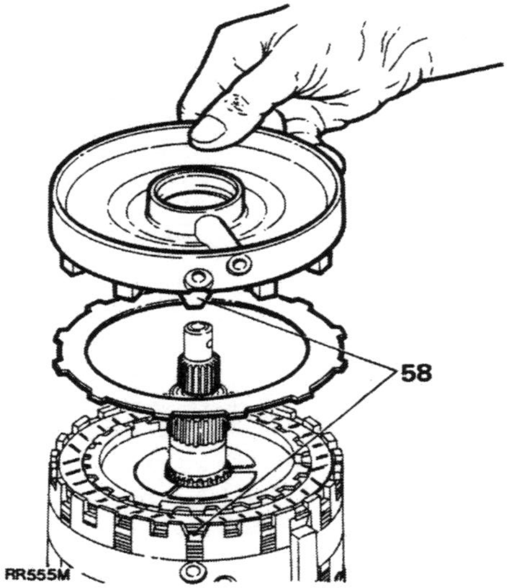

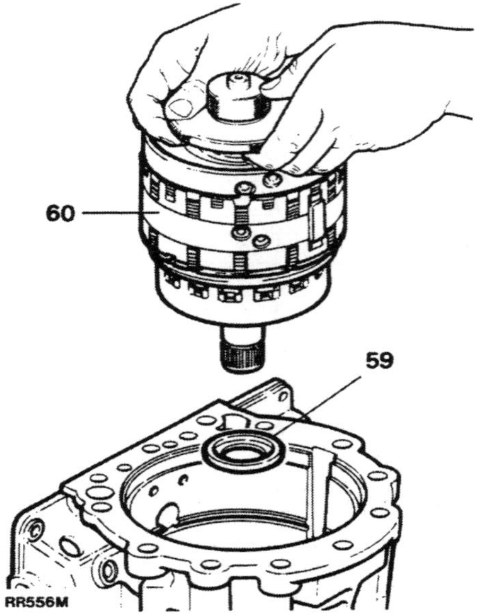

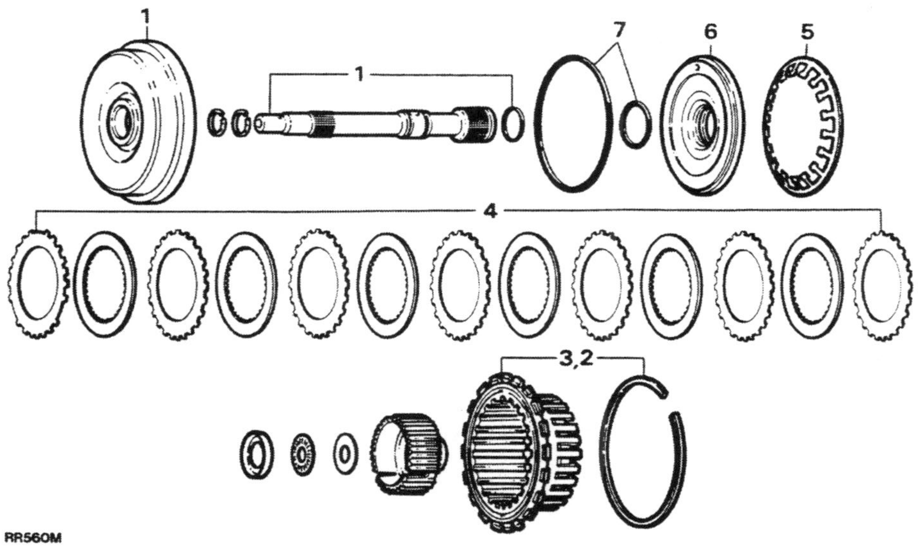

## AUTOMATIC GEARBOX    44

### A clutch assembly overhaul

1. Remove the input shaft by holding the A clutch assembly firmly and pushing the shaft against the working surface. Remove the 'O' ring and discard.
2. Using a suitable press, depress the A–B carrier and remove the snap-ring.
3. Remove carrier A–B.
4. Remove the six clutch plates and seven steel plates.
5. Remove spring plate.
6. For ease of removal of piston A, use air pressure fed through the oil feed hole.

**WARNING: Care should be taken using air pressure when removing the piston.**

7. Remove and discard both 'O' rings on piston A.

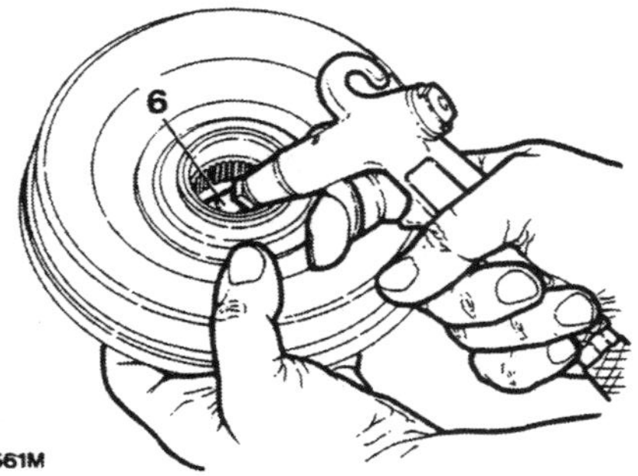

### Assemble

8. Fit the two 'O' rings onto piston A.
9. Place the piston into cylinder A.
10. Fit the spring plate into the cylinder A with the convex side facing the piston.
11. Placing carrier A–B on the bench fit the clutch and steel plates starting with a steel plate and then a clutch plate, finishing with a steel plate.
12. Fit carrier A–B with the clutch assembly onto cylinder A.
13. Using a suitable press, depress the A–B carrier and secure with the snap-ring.
14. Fit the two seal rings, if removed, and an 'O' ring onto the input shaft.
15. Fit the input shaft into the cylinder A assembly and press downwards until shaft meets the stop.
16. Fit the thrust washer onto the input shaft seat.

**NOTE: Use petroleum jelly to retain washer in place.**

17. Install the inner carrier A onto the intermediate shaft within the gearbox.
18. Place the disc washer and axial cage into the inner carrier A.
19. Fit cylinder A assembly into the gearbox using a right to left twisting motion. This will enable the teeth of the clutch plates to mesh into the A–B carrier and inner carrier.

**NOTE: When properly engaged the top of cylinder A should not protrude more than 8,5 mm (0.33 in) above the gearbox front face.**

20. Place the thrust washer and axial cage onto the A cylinder.

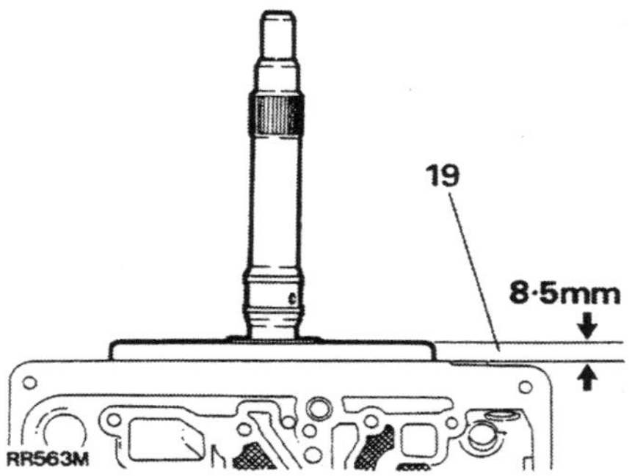

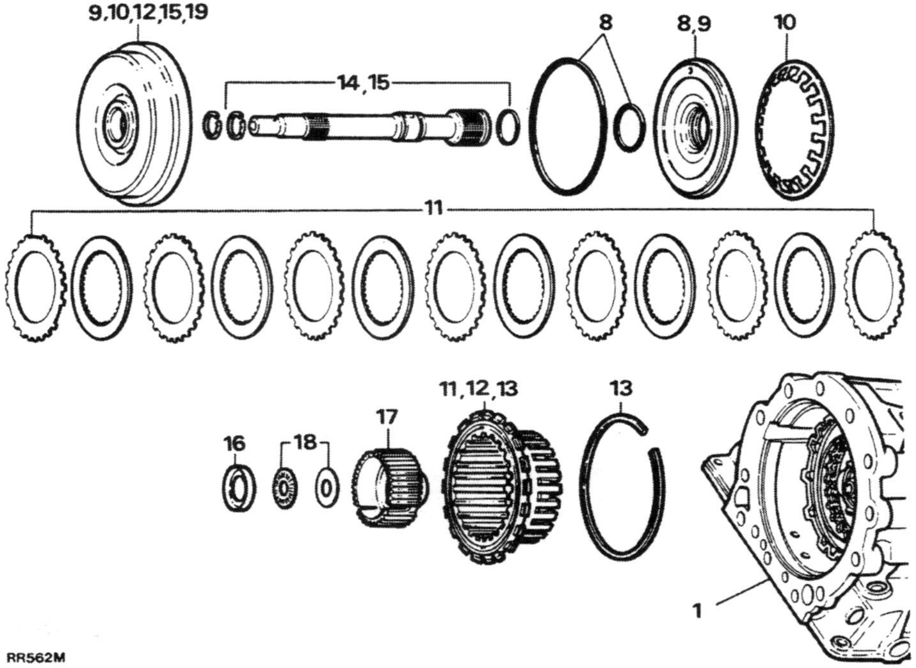

## AUTOMATIC GEARBOX

### Pump, Intermediate Plate and Bell Housing

#### Remove and overhaul

1. Place the bell housing on the bench, open face down.
2. Remove the eight hexagonal bolts on the rear of the pump.
3. Screw in two bolts, diagonally opposite each other, tap lightly using a soft-headed mallet, this will free the pump assembly from the intermediate plate.
4. Remove the six remaining bolts situated on the inside of the bell housing.
5. Separate the bell housing from the intermediate plate.

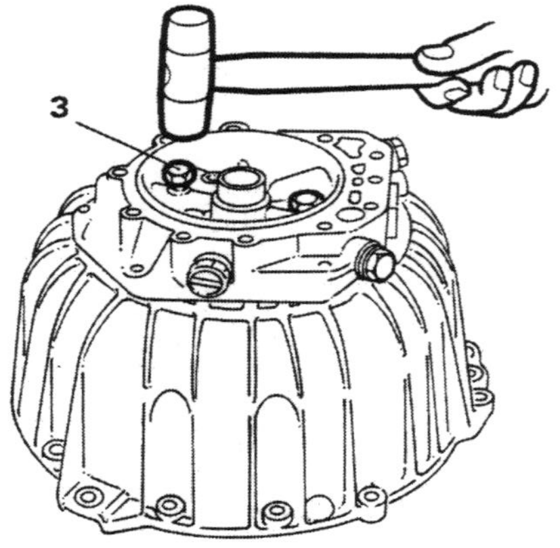

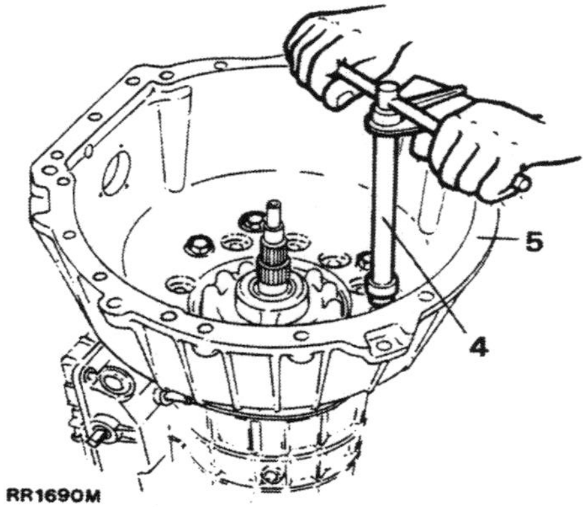

#### Pump assembly

6. Using a suitable tool remove the shaft sealing ring and 'O' ring from the pump housing and discard.
7. Strip, inspect and clean the pump assembly using a lint-free rag.

   **NOTE: If damage has occurred to the assembly, replace the whole pump.**

8. Replace the pump hollow gear and pump gear into pump housing with the marked side of gears facing upwards.
9. Fit the 'O' ring onto the circumference of the pump housing.
10. Using oil seal replacer LST108, fit the shaft seal ring into the pump housing.
11. Fit the alignment pin into the pump housing.

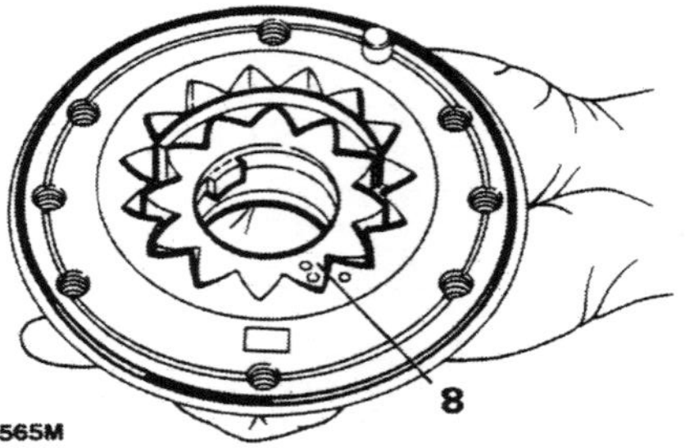

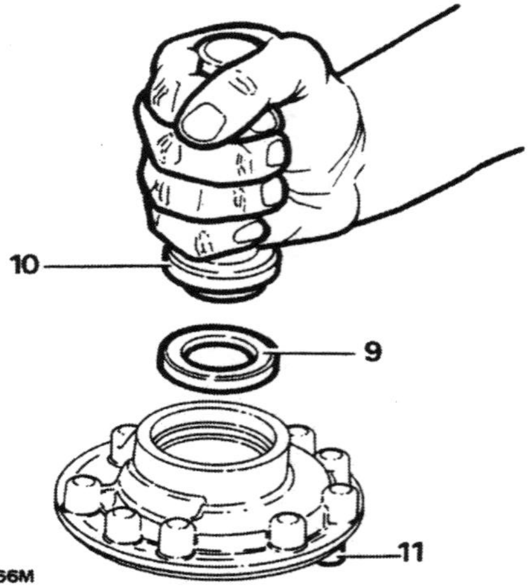

#### Intermediate assembly

12. Remove the four screw plugs and seal rings from the plate and discard the seal ring.
13. Remove the oil cooler pipe adaptors.
14. Inspect and clean the intermediate plate with a lint-free rag.

    **NOTE: If damage is found replace the intermediate plate.**

15. Fit the oil cooler pipe adaptors.
16. Fit the four screw plugs into their correct locations with new seal rings.

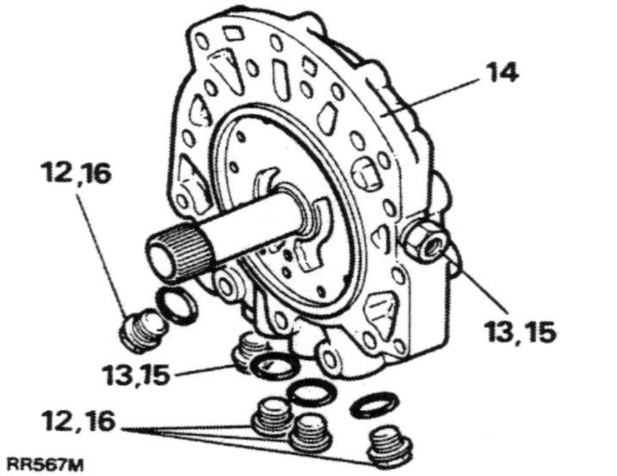

#### Refit bell housing, intermediate plate and pump assembly

17. Inspect and clean the bell housing.

    **NOTE: If damage is found replace the bell housing.**

18. Align the dowel in the pump with its hole in the intermediate plate and press the housing into position.
19. Secure the pump housing to the intermediate plate with the eight hexagonal bolts and tighten to the specified torque.
20. Using the oil pump rotation sleeve LST111 check that the pump gears rotate freely.
21. Place the gasket and disc washer onto the intermediate plate assembly using a petroleum jelly or vaseline.
22. Fit the intermediate plate onto the gearbox.
23. Fit the bell housing onto the intermediate plate assembly.

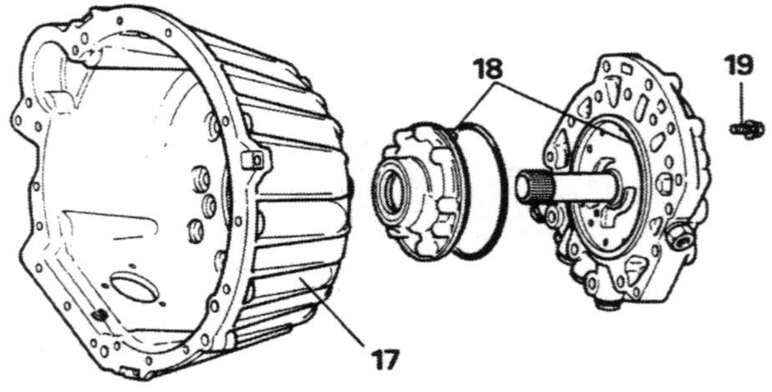

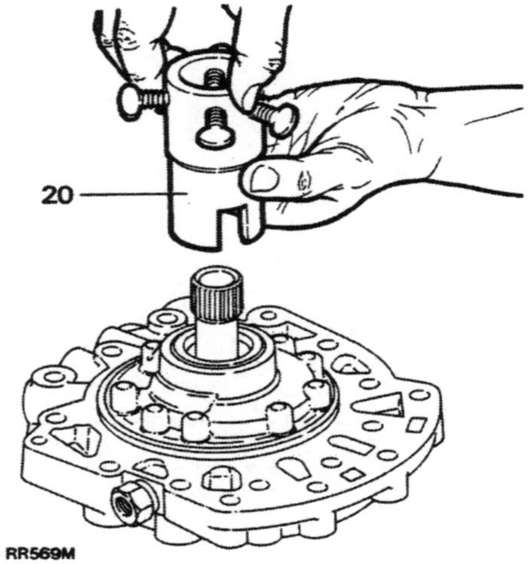

# AUTOMATIC GEARBOX

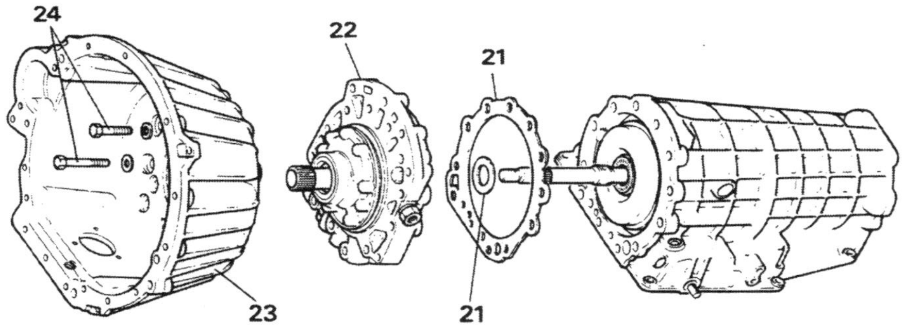

24. Secure with the six short bolts which locate on the outside diameter ring pattern within the bell housing and the twelve long bolts which are located in the inner diameter ring pattern. All the bolts to be tightened to the specified torque.

25. Using the end-float gauge LST111 check the axial clearance 0,2 to 0,4 mm (0.008 to 0.016 in). If the axial clearance is not achieved, remove the bell housing/intermediate plate assembly complete and replace existing disc washer using a thicker or thinner one depending on the reading first taken. Reassemble bell housing/intermediate plate and check the axial clearance once again. Repeat this operation until axial clearance has been achieved.

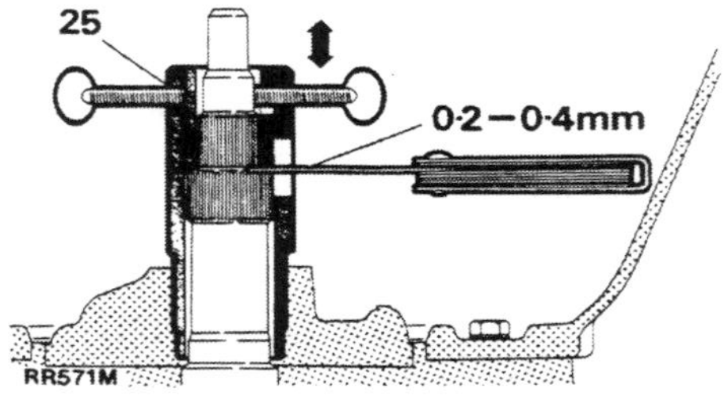

## Check axial clearance

1. Fit LST111 end-float gauge onto the output shaft, making sure the outer shaft engages into the pump.
2. Pressing the output shaft towards the rear of the gearbox and tighten the three screws on the gauge.
3. Measure the clearance and note.
4. Now secure the remaining screw which retains the outer shaft to the inner collar.
5. Pull the whole assembly away from the bell housing, measure the clearance and note.
6. Subtract the first measurement from the second to obtain the axial clearance.

## Extension housing and governor—overhaul

1. Remove the two bolts using torx bit TX27 retaining the parking wheel.
2. Remove the clip and counterweight from inside the governor hub.
3. Remove the two bolts from the top of the governor hub which releases the housing and discard.
4. Release the retaining clip and discard.
5. Remove the pin, spring, piston and weight from the governor housing.
6. Clean and inspect all parts for damage.

16. Fit the governor/parking wheel assembly onto the output shaft and press the assembly until fully seated.

   NOTE: To avoid damage to the 'O' ring use a petroleum jelly. Ensure the seal rings are snapped together and are seated correctly.

17. Fit the extension shaft and retaining bolt using a new 'O' ring.

18. Fit a new seal to the extension housing using the rear oil seal replacer LST108.
19. Fit a new gasket onto the rear of the gearbox and fit the extension housing taking care not to damage the seal on assembly.
20. Secure the extension housing using the nine bolts to the specified torque.

NOTE: Replace any part which may be damaged.

7. Remove the seal ring from the extension housing.
8. Clean and inspect the extension housing for damage.

   NOTE: If the dowels are damaged replace the dowels only. If extension casing is damaged replace the case and dowels.

9. Inspect and clean extension shaft and bolt for damage, replace if necessary.

## Assemble

10. Fit the pin, spring and piston to the governor housing.
11. Fit the weight on the top of the governor housing and secure with a new retaining clip.
12. Fit the counterweight into the governor hub and secure with the clip.
13. Fit the parking wheel and secure with two bolts using torx bit TX27 to the specified torque.
14. Fit the governor housing assembly to the hub and secure with two bolts using torx bit TX27 to the specified torque.
15. Turn gearbox over in holding fixture so that the rear of the box is uppermost.

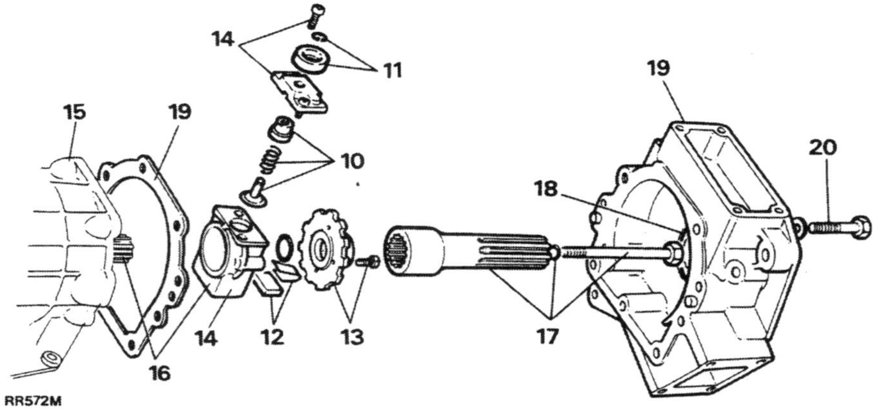

## Refit valve body, oil pan and torque converter

1. Position the gearbox so that the bottom is uppermost.
2. Insert the eight sealing bushes into the oil feed holes using the control unit inlet oil seal remover/replacer LST113.
3. As a test to check the function of the clutch and brake assemblies, insert an air gun into the oil feed holes and exert a pressure of 5 to 6 bar (72.5 to 87 lb/in²).
4. Fit the four short springs into the oil feed holes at the front of the gearbox and four long springs into the oil feed holes at the rear of the gearbox.
5. Fit the eight circlips to retain the springs and sealing bushes.

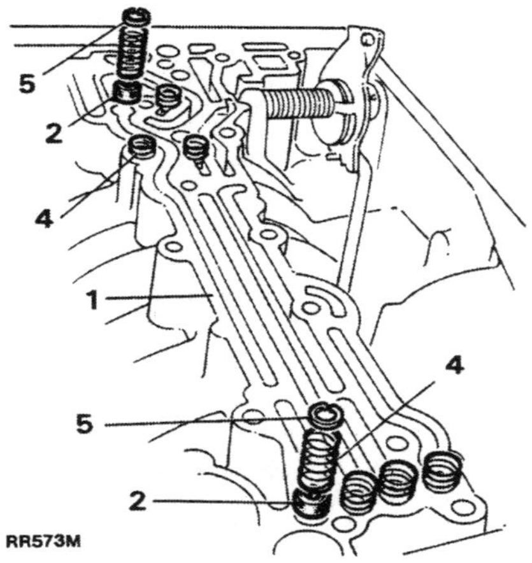

## AUTOMATIC GEARBOX | 44

6. Fit the restrictor, spring and circlip using LST 113 into the hole adjacent to the four rear oil feed holes.

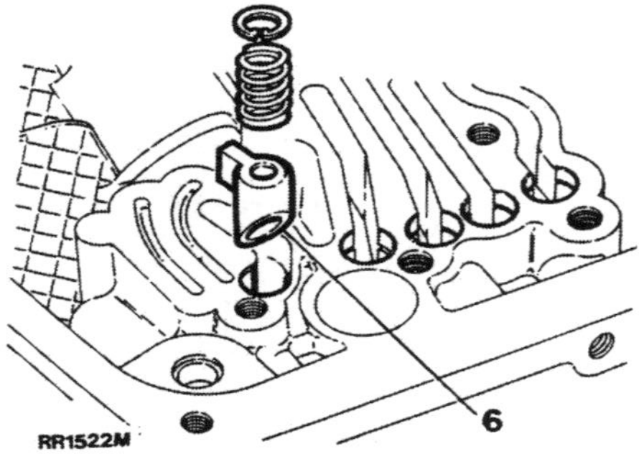

7. Place the control unit ensuring the selector shaft locates into the gear shift fork and fit the thirteen bolts loosely by hand.

8. Place the selector linkage setting gauge LST109 in position and gently press the control unit against the tool and tighten all thirteen bolts using torx bit TX27 to the specified torque.

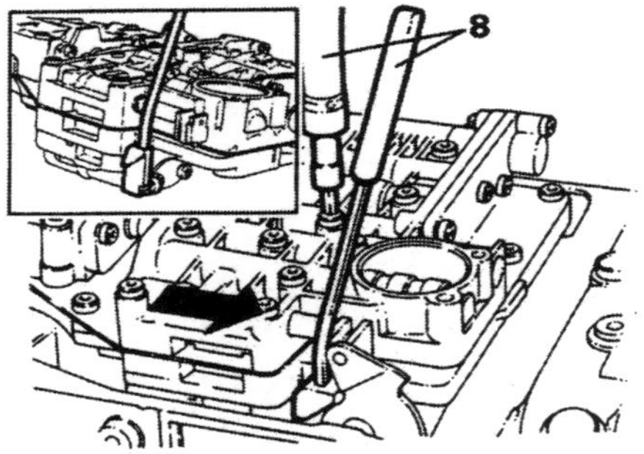

9. Remove the setting gauge and fit the suction and 'O' ring to new oil screen.

10. Fit the new oil screen with new 'O' ring and secure with the three bolts using torx bit TX27 to the specified torque.

11. Fit the oil pan using a new gasket.

12. Secure the oil pan with the six retaining plates (two straight and four corner plates), tighten to the specified torque.

13. Fit oil plug with a new seal.

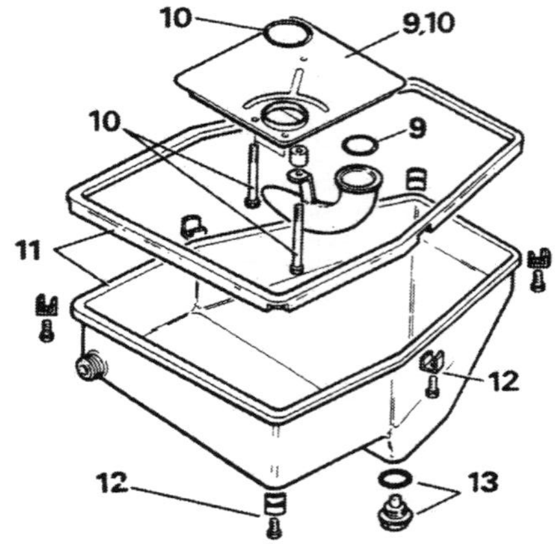

14. Turn the gearbox around until the gearbox is horizontal.

15. Using the torque converter remove/refit handles 18G1501, install the torque converter into the gearbox.

NOTE: Check that the dimension from the converter fixing bolt boss to the converter housing face is 50 mm (1.96 in). If this dimension is achieved the converter is properly seated in the housing.

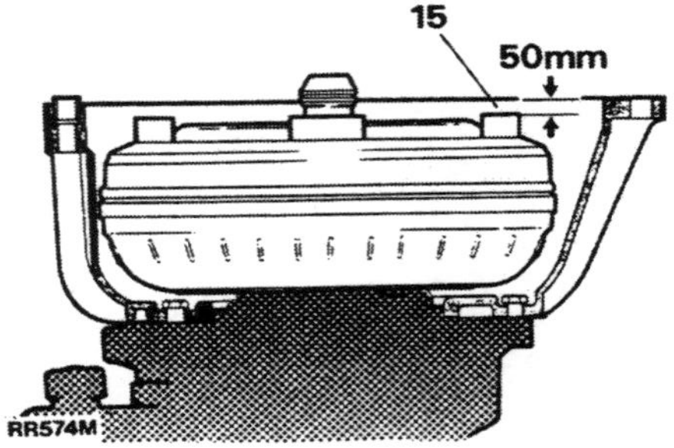

142

## PROPELLER SHAFTS | 47

6. Remove paint, rust, etc. from the vicinity of the universal joint bearing cups and circlips.

NOTE: Before dismantling the propeller shaft joint, mark the position of the spider pin lubricator relative to the journal yoke ears to ensure that the grease nipple boss is reassembled in the correct running position to reduce the possibility of imbalance.

7. Remove the circlips, and grease nipple.

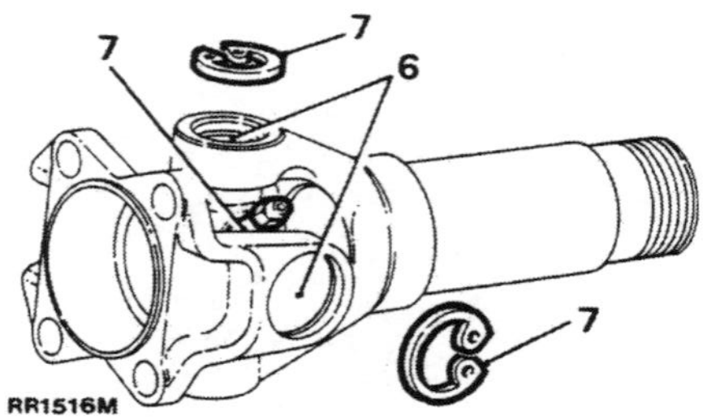

8. Tap the yokes to eject the bearing cups.
9. Withdraw the bearing cups and spider and discard.
10. Repeat instructions 5 to 8 at opposite end of propeller shaft.
11. Thoroughly clean the yokes and bearing cup locations.

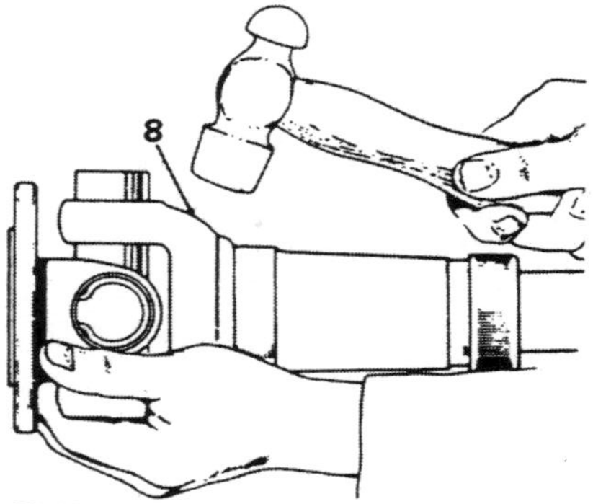

### OVERHAUL PROPELLER SHAFTS

#### Dismantle

1. Place the vehicle on a suitable ramp.
2. Remove the propeller shaft from the vehicle.
3. Note the alignment markings on the sliding member and the propeller shaft.

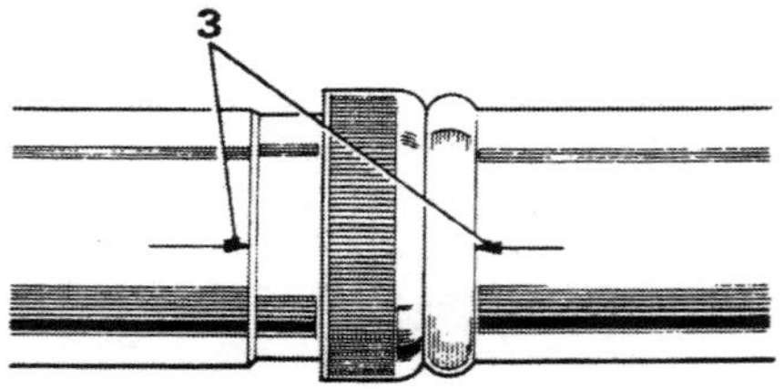

4. Unscrew the dust cap and withdraw the sliding member.

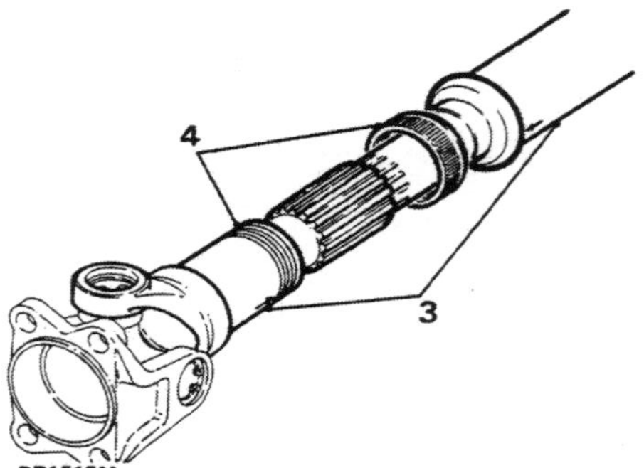

5. Clean and examine the splines for wear. Worn splines or excessive back-lash will necessitate propeller shaft renewal.

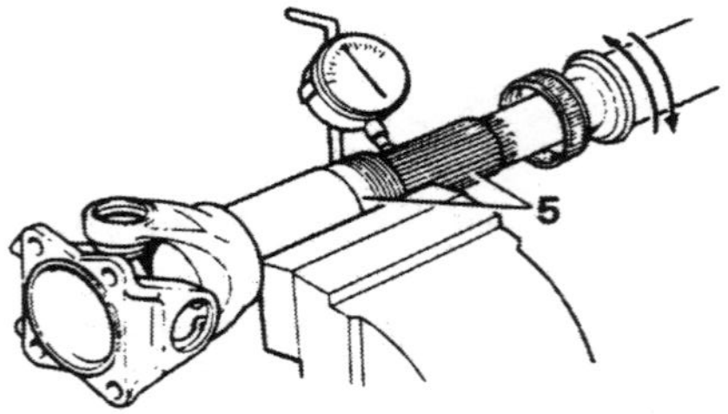

### Assemble

12. Remove the bearing cups from the new spider.
13. Check that all needle rollers are present and are properly positioned in the bearing cups.
14. Ensure bearing cups are one-third full of fresh lubricant. See Lubrication chart.
15. Enter the new spider complete with seals into the yokes of the sliding member flange.

*Continued*

1

## 47 PROPELLER SHAFTS

16. Partially insert one bearing cup into a flange yoke and enter the spider trunnion into the bearing cup taking care not to dislodge the needle rollers.
17. Insert the opposite bearing cup into the flange yoke. Using a vice, carefully press both cups into place taking care to engage the spider trunnion without dislodging the needle rollers.
18. Remove the flange and spider from the vice.
19. Using a flat faced adaptor of slightly smaller diameter than the bearing cups press each cup into its respective yoke until they reach the lower land of the circlip grooves. Do not press the bearing cups below this point or damage may be caused to the cups and seals.
20. Fit the circlips.
21. Engage the spider in the yokes of the sliding member. Fit the bearing cups and circlips as described in instructions 15 to 20.
22. Lubricate the sliding member spines and fit the sliding member to the propeller shaft ensuring that the markings on both the sliding member and propeller shaft align.
23. Fit the tighten the dust cap.
24. Fit the grease nipples to the spider and the sliding member and lubricate.
25. Apply instructions 15 to 20 to the opposite end of the propeller shaft.
26. Fit the grease nipple and lubricate.
27. Fit the propeller shaft to the vehicle.

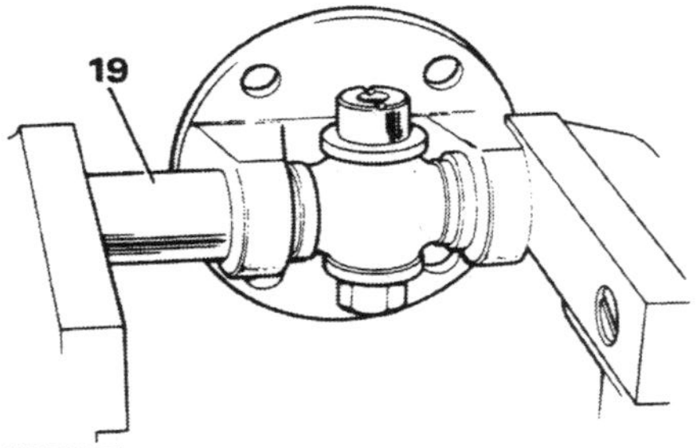

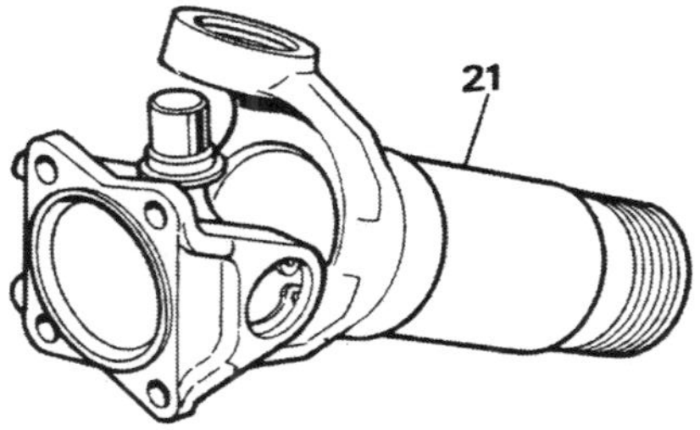

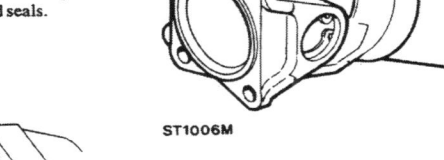

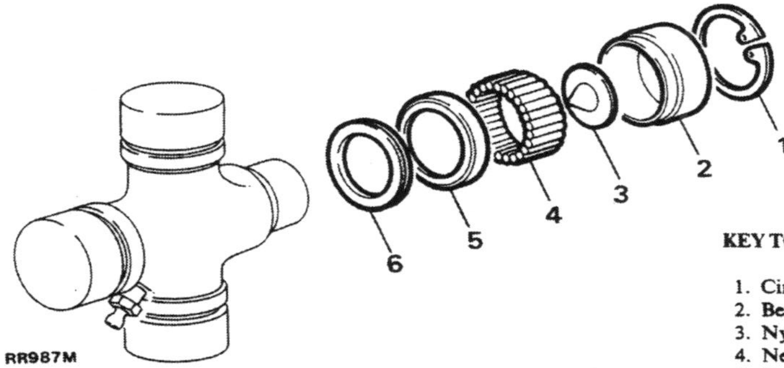

**KEY TO SPIDER ASSEMBLY**

1. Circlip
2. Bearing cup
3. Nylatron washer
4. Needle rollers (27 per cup)
5. Seal retainer
6. Seal

NOTE: When renewing joints use only 03EHD series replacement spider packs.

## REAR AXLE AND FINAL DRIVE 51

8. Mark the differential and propeller shaft drive flanges with identification marks to aid reassembly. Remove the four nuts and bolts, lower the propeller shaft and tie it to one side.
9. Disconnect the pivot bracket ball joint at the axle bracket.
10. Release the bolts and remove the coil spring retaining plates.
11. Lower the axle and remove the road springs.
12. Withdraw the axle assembly.

**Refitting**

13. Position the axle and fit the lower links, and tighten the bolts to the specified torque.
14. Reverse the removal instructions.
15. Tighten the pivot bracket ball joint to axle, to the specified torque.
16. Tighten the propeller shaft to differential drive flange, to the specified torque.
17. Bleed the braking system.

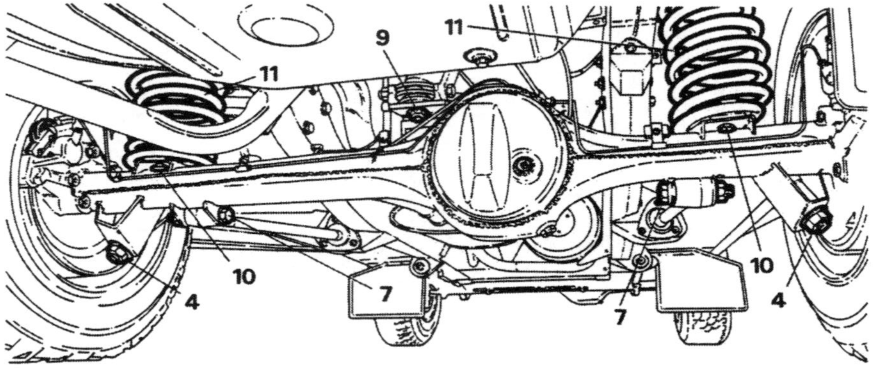

### REAR AXLE

**Remove and refit**

**Removing**

NOTE: The removal and refitting of the axle from the vehicle requires the assistance of two further personnel to steady the axle, when lowering or repositioning the axle.

1. Jack-up the rear of the vehicle and support the chassis.
2. Remove the road wheels.
3. Support the axle weight with a suitable hydraulic jack.
4. Disconnect the shock absorbers.
5. Disconnect the flexible brake hose at the connection under the floor.
6. Disconnect the pad wear multi-plug at the bracket mounted on the underside of the floor. Prise the rubber grommet out of the bracket and feed the plug through the hole.
7. Disconnect the lower links at the axle.

# REAR AXLE AND FINAL DRIVE

## OVERHAUL AXLE DIFFERENTIAL ASSEMBLY

**Service tools:**
18G 1205 flange holder tool;
18G 191 pinion height setting gauge;
18G 191-4 universal setting block;
18G 47-6 pinion head bearing remover/replacer;
LST 106 oil seal replacer;
RO 262757A extractor for pinion bearing caps;
RO 262757-1 replacer—use with RO 262757A;
RO 262757-2 adaptor tail bearing cap replacer;
RO 530105 spanner—differential flange and carrier bearing nuts;
RO 530106 bracket for dial gauge and indicator;
MS 47 press

### DISMANTLE

It is essential that differential components are marked in their original positions and relative to other components so that, if refitted, their initial setting is maintained. Note that the bearing caps must not be interchanged.

1. Remove the differential assembly from the axle.
2. Drift out the roll pin securing the bearing nut locking fingers to the bearing caps. Remove the locking fingers.
3. Slacken the bearing cap bolts and mark the caps for reassembly.
4. Using service tool RO 530105, remove the bearing adjusting nuts.
5. Remove the bearing cap bolts and bearing caps.
6. Lift out of the crown wheel, differential unit and bearings.
7. Remove the split pin securing the pinion flange nut.
8. Remove the pinion flange nut using service tool 18G 1205 to restrain the flange.
9. Withdraw the pinion complete with pinion head bearing and outer bearing shims. Withdraw the shims.
10. Remove the pinion flange oil seal, spacer and bearing. Discard the oil seal.
11. Using service tool RO 262757A, remove the pinion head bearing track and shim and drift out the outer bearing track from the differential housing.
12. Remove the pinion head bearing with service tool MS 47 and adaptor 18G 47-6.
13. Remove the bolts and washers securing the crown wheel to the differential flange. Withdraw the crown wheel.
14. Remove the differential carrier bearings.
15. Remove the circlips securing the differential cross shaft. Extract the cross shaft.
16. Withdraw the differential gears and pinions.
17. Thoroughly clean all components.

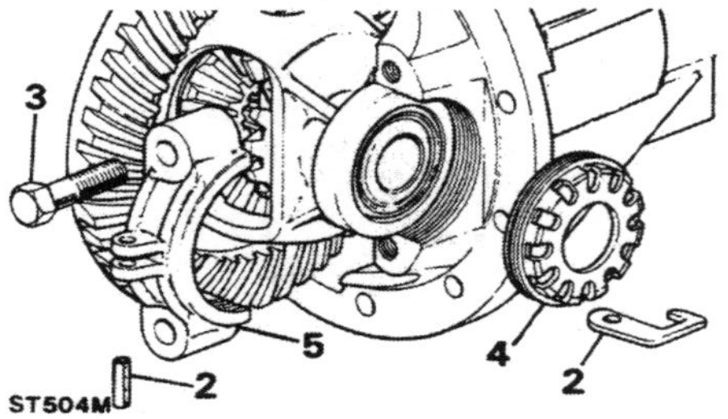

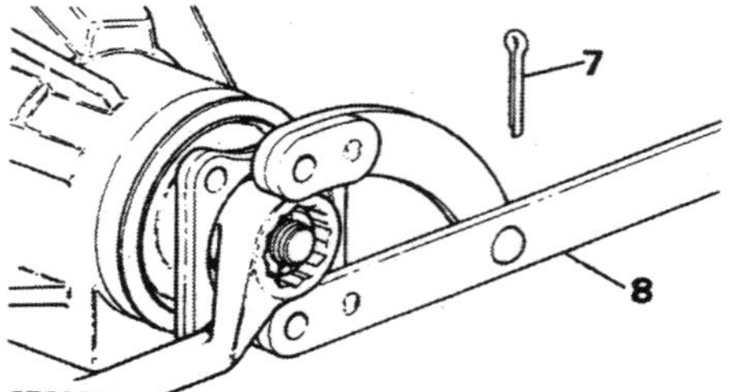

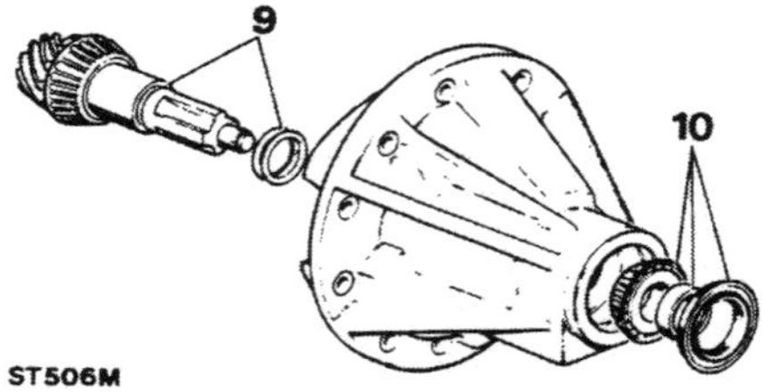

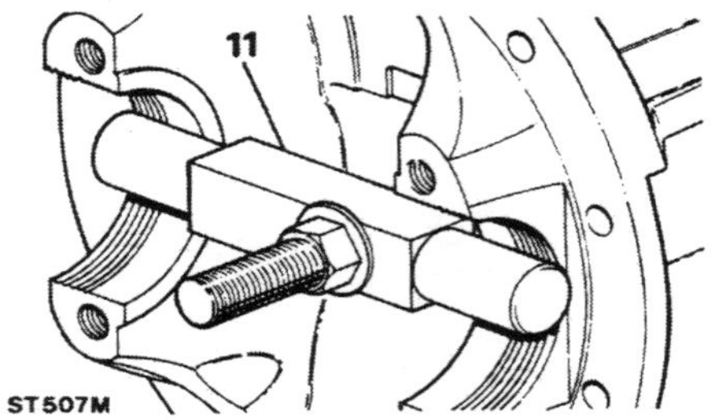

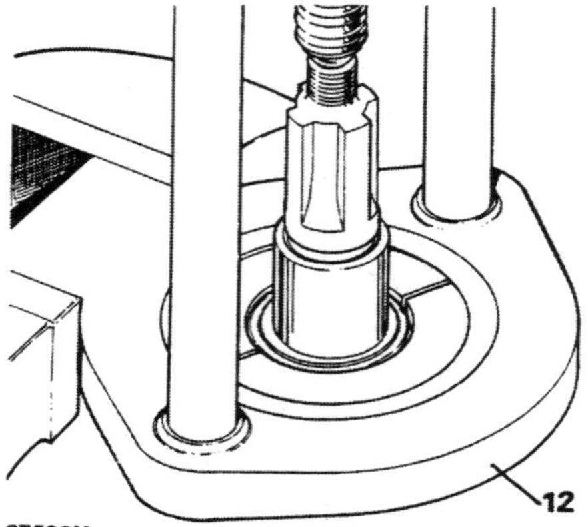

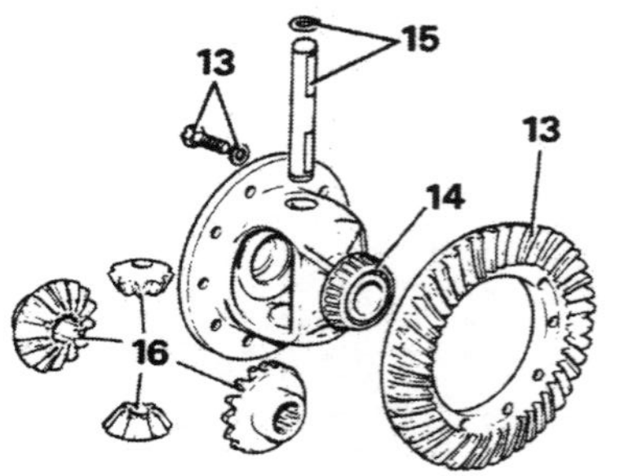

### INSPECTION

18. Check all bearings for wear and/or pitting.
19. Check all gears for wear, scuffing, pitting and damaged teeth.
20. NOTE. The crown wheel and pinion are supplied as a matched set, also the pinion housing and bearing caps.

### ASSEMBLE

#### Differential gears

21. Fit the differential gears to the differential housing.
22. Fit the differential cross shaft and retaining circlips.
23. Check the gear for freedom of rotation and backlash. Nominal backlash should be present. Excessive backlash will necessitate renewal of the gears and/or the differential housing. No provision is made for backlash adjustment.
24. Check that the serial number marked on the pinion end face matches that marked on the crown wheel.
25. Ensuring that the differential housing flange and crown wheel are thoroughly clean fit the crown wheel. Fit the crown wheel bolts and washers and evenly tighten.
26. Fit the carrier bearings using a suitable press or drift and assemble the tracks to the bearings.

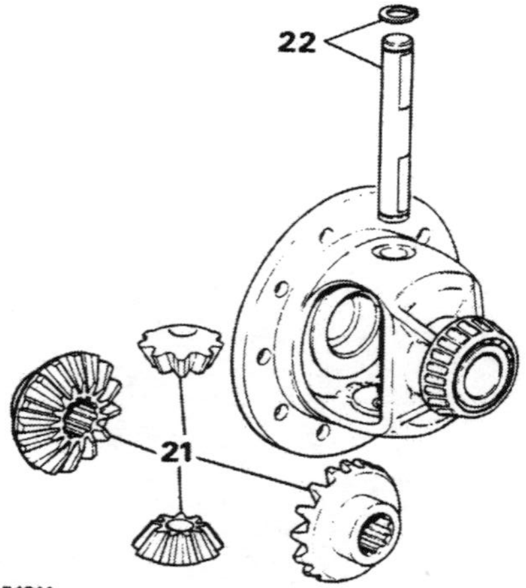

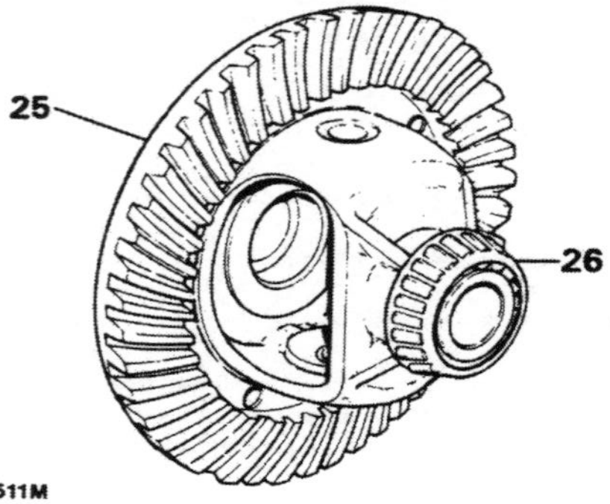

*Continued*

## REAR AXLE AND FINAL DRIVE  **51**

27. Place the differential housing complete with crown wheel and bearings in the pinion housing.
28. Fit the bearing caps and bolts. Do not fully tighten the bolts.
29. Fit the bearing adjusting nuts and adjust to obtain zero end-float.
30. Tighten the bearing cap bolts.
31. Using a dial gauge check the crown wheel for run-out. This should not exceed 0,10 mm (0.004 in). If excessive run-out is recorded remove the crown wheel and examine crown wheel and mounting flange for burrs, grit, etc. Refit the crown wheel and recheck. Run-out, attributable to a buckled or damaged differential housing flange can be corrected only by renewing the differential gear housing.
32. When satisfied that run-out is within the specified limits remove the differential housing from the pinion housing.
33. Remove the crown wheel bolts and refit them using Loctite 'Studlock'. Evenly tighten the bolts to the correct torque.
34. Fit the pinion head bearing track and the original shim to the pinion housing using service tools RO 262757A and RO 262757-1. If the original shim was damaged or mislaid use a new shim of at least 1,27 mm (0.050 in) thickness.
35. Fit the pinion outer bearing track to the pinion housing with service tool RO 262757A and RO 262757-2.

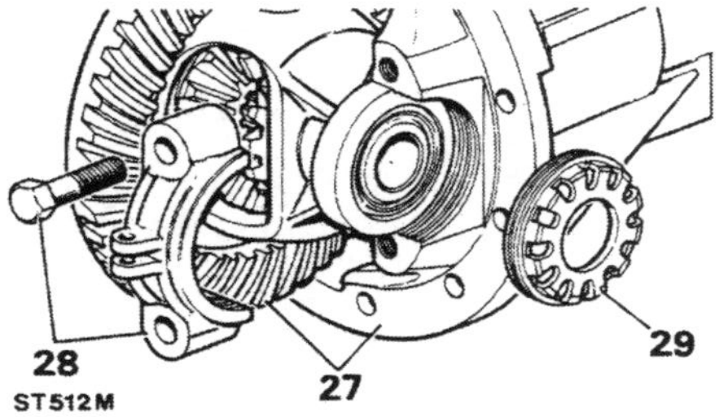

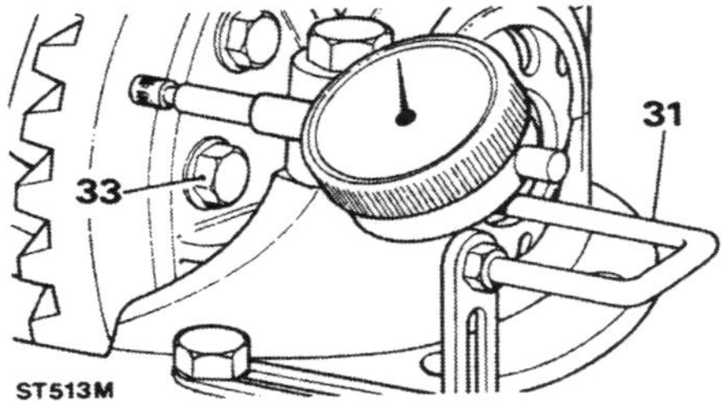

**NOTE: Instructions 34 and 35 are carried out in one operation as illustrated.**

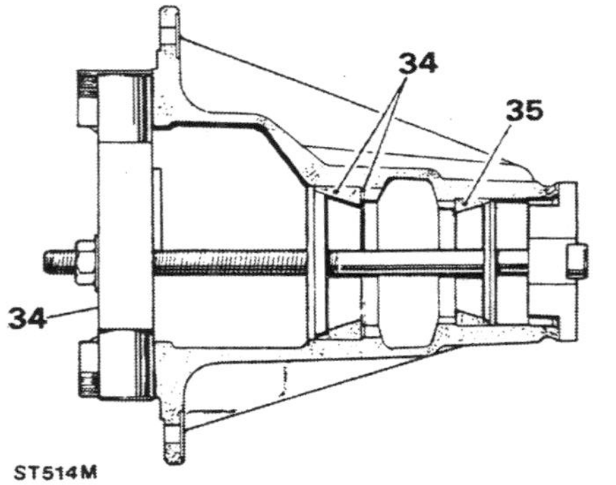

36. Fit the pinion head bearing to the pinion using service tool 18G 47-6.
37. Enter the pinion into its location in the pinion housing. Do not fit the shims for bearing pre-load at this stage.
38. Fit the outer bearing and spacer.
39. Fit the driving flange, washer and nut.

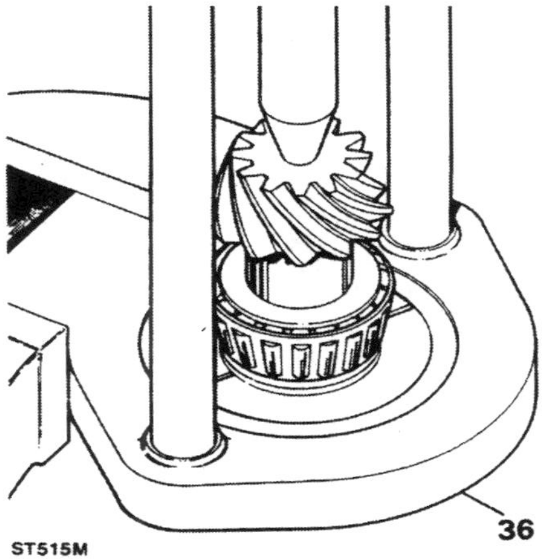

*Continued*

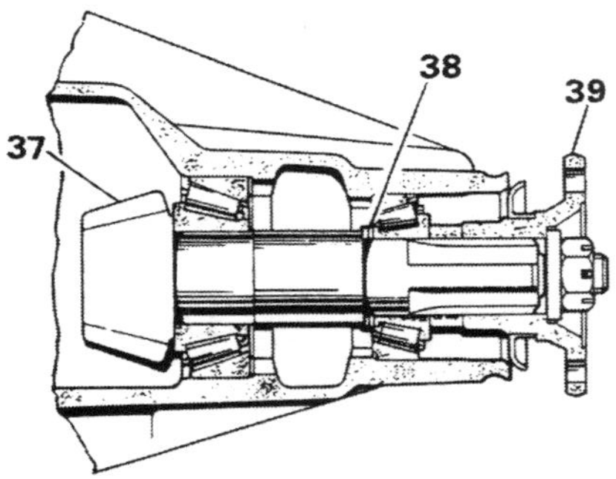

40. Do not fit the oil seal at this stage.
41. Tighten the pinion flange nut slowly until the force required to rotate the pinion is 7,6 to 13 kgf cm (7 to 12 lbf in). This will pre-load the bearings in order to check the pinion height dimension.

### Drive pinion markings

42. The markings on the end face adjacent to the serial number are of no significance during servicing.
43. The figure marked on the end face opposite to the serial number indicates, in thousandths of an inch, the deviation from nominal required to correctly set the pinion. A pinion marked plus (+) must be set below nominal, a minus (−) pinion must be set above nominal. An unmarked pinion must be set at nominal.
44. The nominal setting dimension is represented by the setting gauge block 18G 191-4 which is referenced from the pinion end face to the bottom radius of the differential bearing bore.

### Drive pinion adjustment

45. Ensure that the pinion end face is free of raised burrs around the etched markings.
46. Remove the keep disc from the magnetised base of dial gauge tool 18G 191.
47. Place the dial gauge and setting block on a flat surface and zero the dial gauge stylus on the setting block.

NOTE: The setting block has three setting heights as follows:
39,50 mm Rationalised axle
38,10 mm Pre-Rationalised axle
30,93 mm Salisbury axle

Ensure that the height marked 39,50 mm is used for this differential.

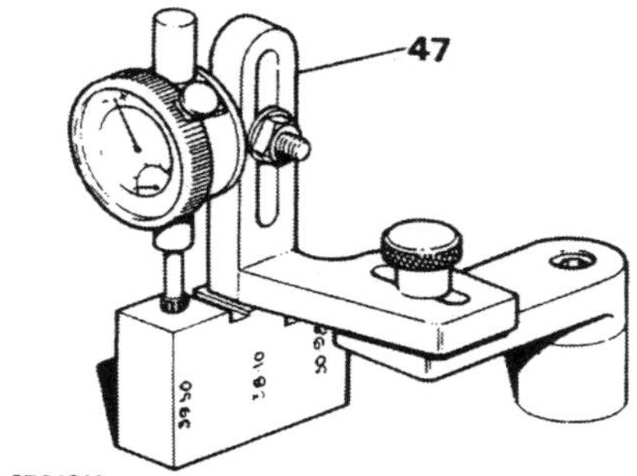

48. Position the dial gauge centrally on the pinion end face with the stylus registering on the lowest point on one differential bearing bore. Note the dial gauge deviation from the zeroed setting.

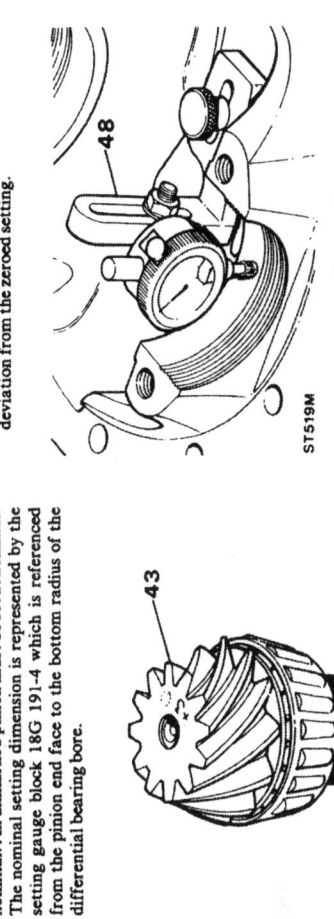

*Continued*

## REAR AXLE AND FINAL DRIVE

49. Repeat on the other bearing bore. Add together the readings than halve the sum to obtain the mean reading. Note whether the stylus has moved up or down from the zeroed setting.

a. Where the stylus has moved down, the amount is equivalent to the thickness of shims that must be removed from under the pinion inner track to bring the pinion down to the nominal position.

b. Where the stylus has moved up, the amount is equivalent to the addition thickness of shims required to bring the pinion up to the nominal position.

50. Before adjusting the shim thickness, check the pinion face marking and if it has a plus (+) figure, subtract that amount in thousandths of an inch from the shim thickness figure obtained in the previous item.

51. Alternatively, if the pinion has a minus (−) figure, add the amount to the shim thickness figure. Adjust the shim thickness under the pinion head bearing track as necessary.

52. Recheck the pinion height setting. If the setting is correct, the mean reading on the dial gauge will agree with the figure marked on the pinion end face. For example, with an end face marking of +3, the dial gauge reading should indicate that the pinion is +0.003 in.

### Bearing pre-load adjustment

53. Remove the pinion flange, pinion, outer bearing and spacer.
54. Slide new shims, of the same thickness as the originals (bearing pre-load) into position on the pinion shaft. If the original shim was damaged or mislaid use a new shim of at least 4,06 mm (0.160 in) thickness.
55. Enter the pinion in its location in the pinion housing and fit the outer bearing and spacer.
56. Fit the driving flange, washer and nut.
57. Do not fit the oil seal at this stage.
58. Tighten the pinion flange nut to the specified torque. The force required to rotate the pinion shaft should be within 7,6 to 13 kgf cm (7 to 12 lbf in) when initial inertia has been overcome. Change the bearing pre-load shim as necessary to obtain this requirement. A thicker shim will reduce pre-load; a thinner shim will increase pre-load.
59. Remove the pinion flange.

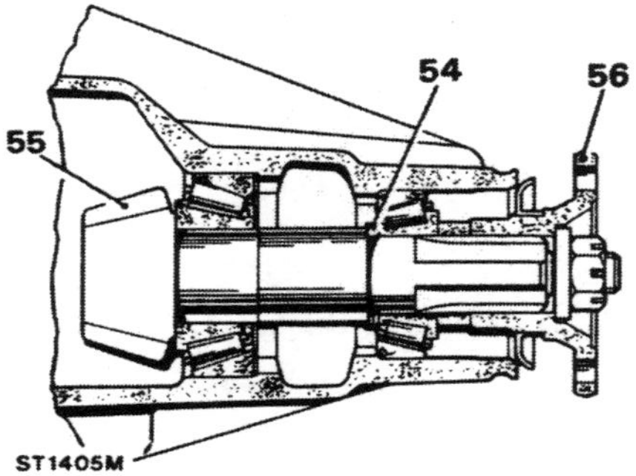

### Fitting Pinion Oil Seal

CAUTION—Before fitting the new seal to the differential, examine the seal to ensure that it is clean, undamaged and that the garter spring is properly located. A small scratch on the seal lips could impair its efficiency.

60. Coat the outer diameter of the new seal width with an all-purpose grease and fit the seal, lip side leading into position on the pinion nose housing and drive the seal into position flush with the end face of the housing using seal replacing tool LST106.

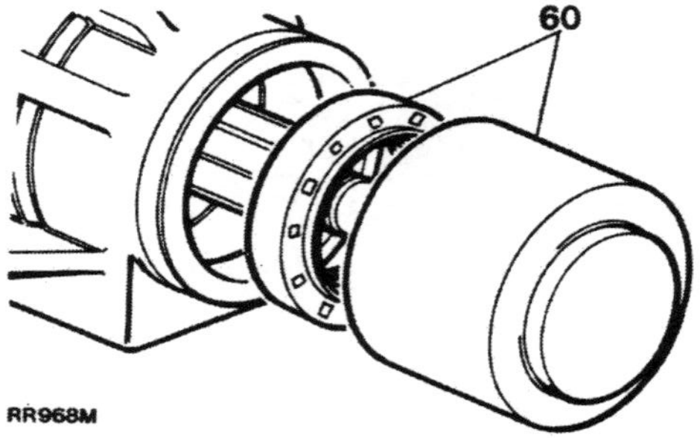

61. Lightly lubricate the seal lips with EP90 oil. Fit the distance piece and flange and secure with washer and nut. Tighten the nut to the specified torque and fit a new split pin.

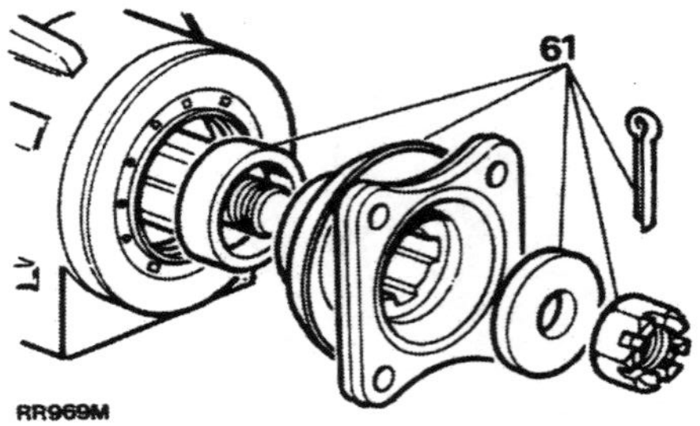

62. Place the differential housing in the pinion housing.
63. Fit the bearing caps and bolts. Do not fully tighten the bolts.
64. Fit the bearing adjusting nuts.
65. Using service tool 530105, slacken the left hand bearing adjustment nut (as illustrated) to produce end float.
66. Tighten the right hand nut until crown wheel/pinion backlash is just removed.
67. Tighten the left hand nut slowly until the crown wheel/pinion backlash is 0,10 to 0,17 mm (0.004 to 0.007 in).
68. Fit the locking fingers and roll pins. If necessary, tighten the adjustment nuts slightly to align the locking finger with a slot.
69. Evenly tighten the bearing cap bolts to the specified torque.
70. Recheck crown wheel/pinion backlash.
71. Lubricate the bearings and gears.

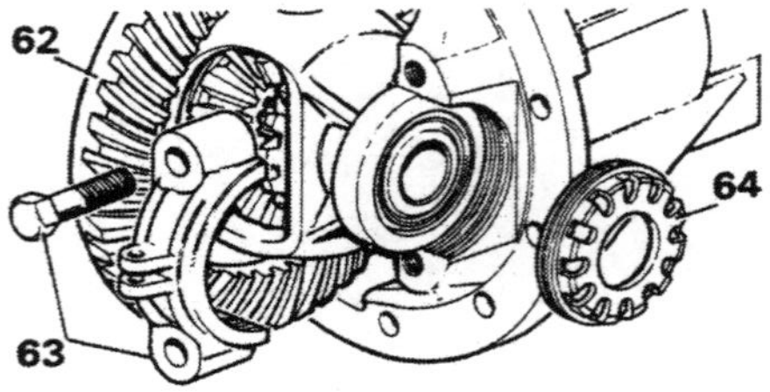

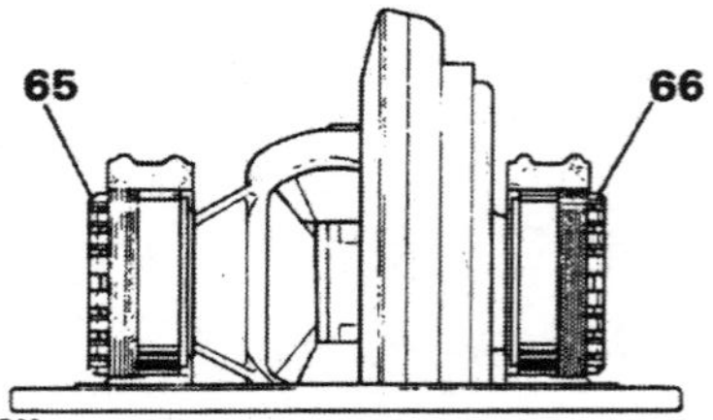

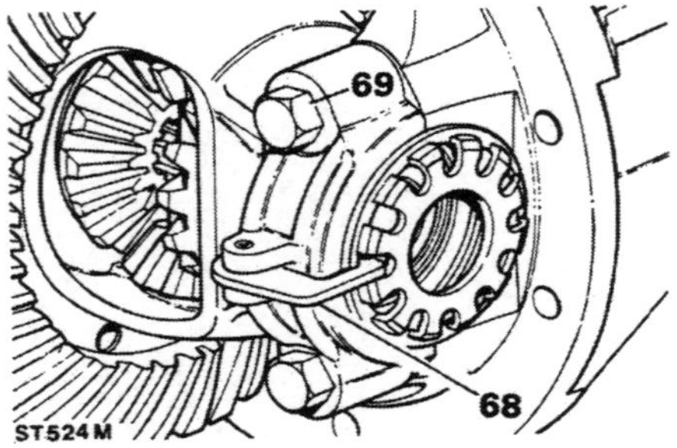

### DATA

Pinion bearing pre-load .................... 7,6 to 13 kgf cm (7 to 12 lbf in)
Crown wheel run-out ....................... 0,10 mm (0.004 in)
Crown wheel/pinion backlash ......... 0,10 to 0,17 mm (0.004 to 0.007 in)

# REAR AXLE AND FINAL DRIVE 51

## REAR DISCS

### Remove and refit

#### Removing

1. Remove the rear hub assembly.
2. Remove the rear disc fixing bolts.
3. Tap off the disc from the rear hub.

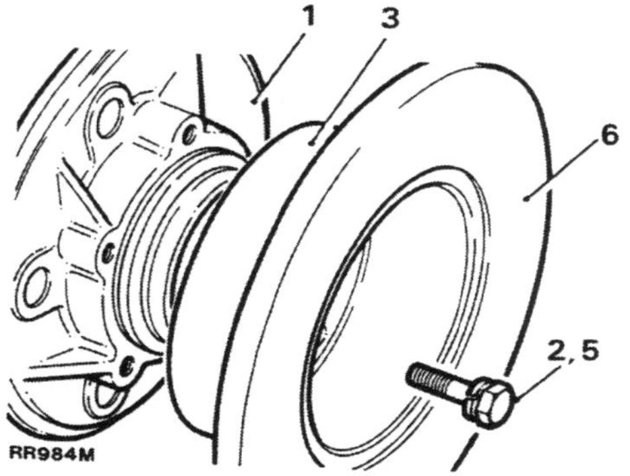

#### Refitting

4. Locate the disc onto the rear hub.
5. Fit the disc fixing bolts. See 'Data Section' for tightening torques.
6. Using a dial test indicator, check the total disc run out, this must not exceed 0,15 mm (0.006 in). If necessary reposition the disc.
7. Fit the rear hub assembly.

#### Disc reclamation

Check the disc thickness marked on the disc boss—this dimension may be reduced to a minimum thickness of 13 mm (0.510 in) front and 12 mm (0.460 in) rear, by machining an equal amount off both sides.

## REMOVE AND OVERHAUL REAR HUBS

Special tools:
Oil Seal replacer LST550-5
Drift for above tool MS550 or 18G134
Hub nut spanner 606435

### Remove

1. Jack-up the vehicle, lower on to axle stands and remove the road wheels.
2. Release the brake pipe from the axle, casing clips and remove the brake caliper retaining bolts and secure the assembly to one side. Take care not to kink the brake pipe.
3. Remove the five bolts securing the axle shaft to the hub and withdraw the shaft.
4. Bend back the lock tab and remove the outer nut using box spanner 606435 and remove the lock washer. Likewise remove the inner nut.
5. Remove the seal track spacer.
6. Withdraw the hub complete with bearing oil seals and brake disc.
7. Remove the inner and outer oil seals.
8. Remove the inner and outer bearing cones.
9. Drift-out the inner and outer bearing cups.
10. Degrease and examine the hub and brake disc and if necessary renew both or whichever part is unserviceable. The disc is attached to the hub with five bolts. Mark the relationship of the hub to the disc if the original parts are to be reassembled.

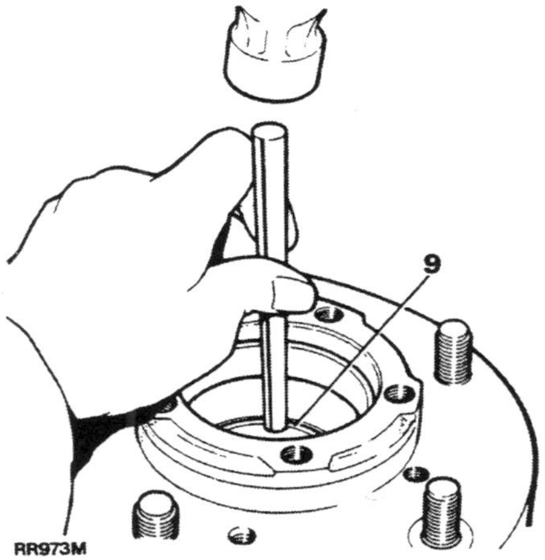

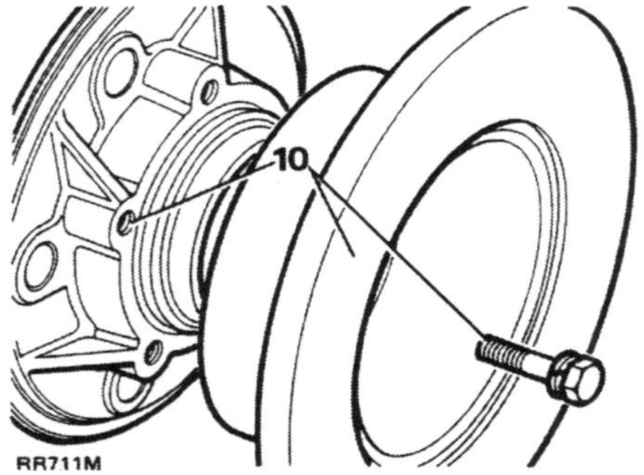

11. Examine the stub axle and in particular check that the inner seal track is smooth and free from blemishes.
12. If necessary remove the six retaining bolts and remove the stub axle complete with the mudshield and joint washer.

### Assemble

13. Using a new joint washer fit the stub axle and mud shield. Coat the threads of the retaining bolts with Loctite 270 and tighten evenly to the correct torque.
14. Fit the new inner and outer bearing cups to the hub.
15. Fit the new inner bearing cone and pack with one of the recommended hub greases.

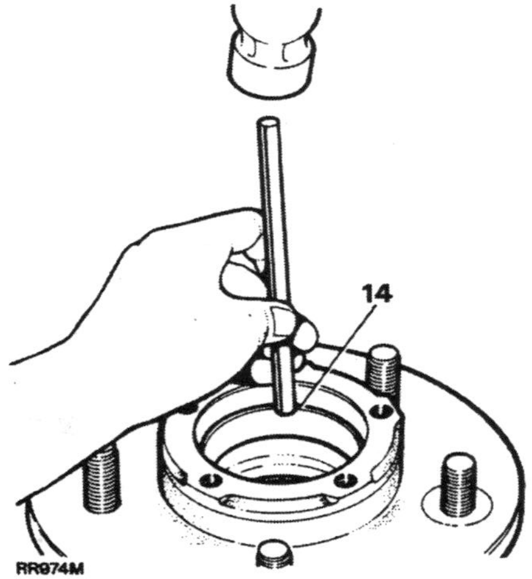

### Fitting new oil seal—inner

16. Clean the hub oil seal housing and ensure that the seal locating surface is smooth and the chamfer on the leading edge is also smooth and free from burrs.
17. Examine the new seal and ensure that it is clean and undamaged and that the garter spring is properly located. Even a small scratch on the seal lip could impair its efficiency.
18. Although the new seal is already pre-greased by the manufacturer, apply one of the recommended hub bearing greases to the outside diameter of the seal before fitting, taking care not to damage the lip.
19. Place the seal, lip side leading, squarely on the hub and using the 76 mm end of seal replacer tool LST 550-5 and drift 550 or 18G 134, drive the seal into position, flush with the end face of the hub.

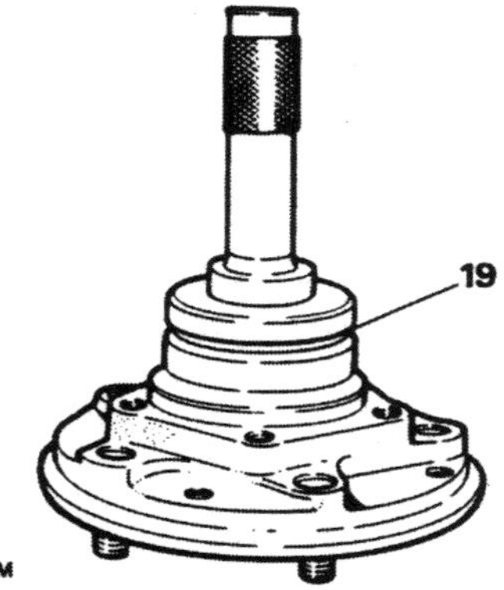

### Fitting outer oil seal

20. Fit the new outer bearing cone and pack with one of the recommended hub greases.

    Carry out instructions 16 to 18.

21. Place the seal, lip side leading, squarely on the hub and using the 72 mm end of seal replacer tool LST 550-5 and drift 550 or 18G 134, drive the seal into position to the depth determined by the tool.
22. Smear the lips of both seals with one of the recommended greases. This is important since a dry seal can be destroyed during the first few revolutions of the hub.

*Continued*

## REAR AXLE AND FINAL DRIVE

### Fitting hub to stub axle

23. Select a new seal track spacer and check that the outer diameter is smooth and free from blemishes and that there are no burrs on the chamfered leading edge.
24. Taking care not to damage the seal lips fit the hub assembly to the stub axle. Do not allow the weight of the hub to rest evenly temporarily on the outer seal otherwise damage and distortion could occur. Therefore hold the hub clear of the stub axle until the seal track spacer is fitted.
25. Carefully fit the seal track spacer, seal lip leading.
26. Fit the hub inner nut and using spanner 606435 tighten the adjusting nut whilst slowly revolving the hub until all end-float is removed, then back-off the nut approximately half a turn.
27. Mount a dial test indicator and bracket on the hub so that the stylus rests in a loaded condition on the nut. Check the end-float which must be 0,013 to 0,10 mm (0.0005 to 0.004 in). Adjust the nut as necessary to achieve this.
28. Fit the lock washer and locknut and tighten against the adjusting nut.
29. Rotate the hub several times to settle the bearings then re-check the end-float.
30. Bend one segment of the lock washer over the adjusting nut and another, diametrically opposite, over the locknut taking care not to damage the outer seal.
31. Using a new joint washer, fit the hub driving shaft and evenly tighten the retaining bolts to the correct torque.
32. Fit the brake caliper and secure with the retaining bolts and tighten to the correct torque. Secure the brake pipes to the axle casing.
33. Fit the road wheels, jack-up the vehicle, remove the axle stands, lower the vehicle to the ground and tighten the road wheel nuts evenly to the correct torque.

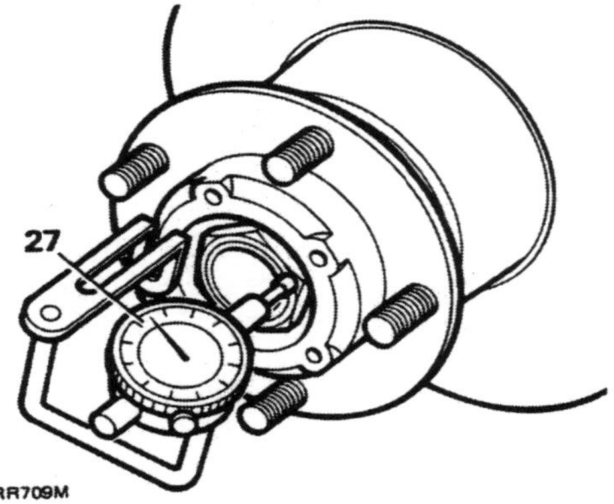

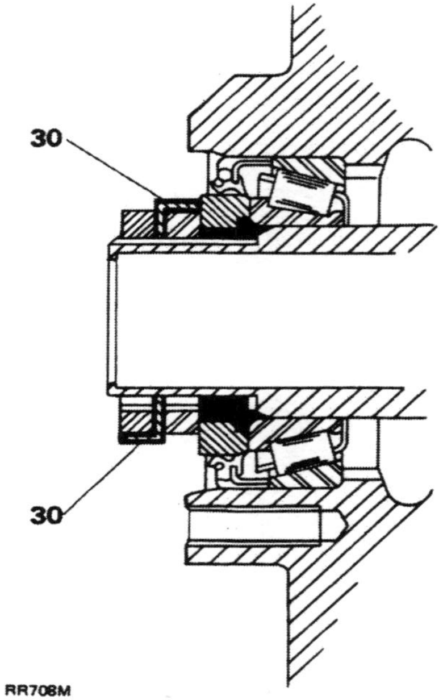

### KEY TO REAR HUB COMPONENTS

1. Rear axle shaft.
2. Joint washer.
3. Locknut.
4. Lock washer.
5. Adjusting nut.
6. Seal track spacer.
7. Outer oil seal.
8. Outer bearing cone.
9. Outer bearing cup.
10. Hub.
11. Inner bearing cup.
12. Inner bearing cone.
13. Inner oil seal.
14. Mudshield.
15. Stub axle.
16. Stub axle joint washer.
17. Brake disc.

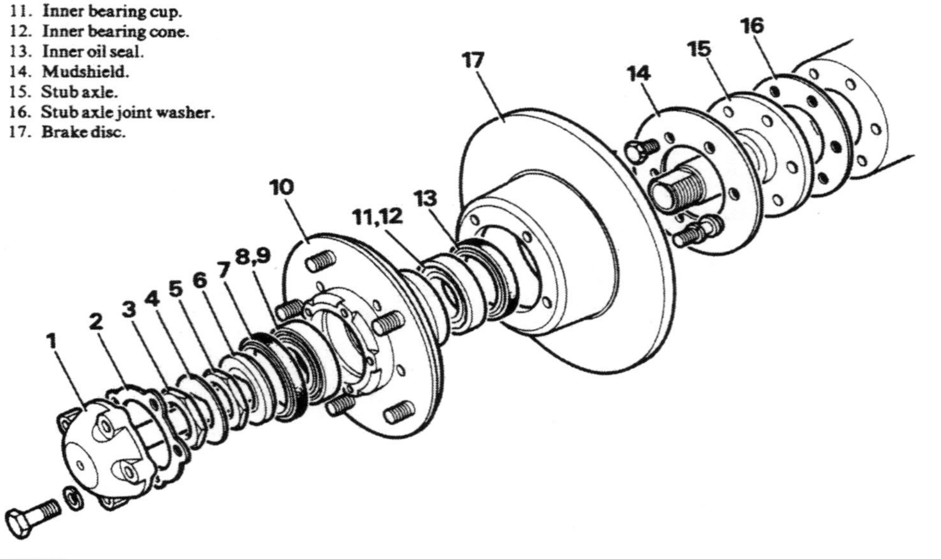

# FRONT AXLE AND FINAL DRIVE | 54

## FRONT AXLE ASSEMBLY

### Remove and refit

#### Removing

**Note:** The removal of the axle from the vehicle will require the assistance of two further personnel to steady the axle when lowering from the vehicle.

1. Jack up the front of the vehicle and support the chassis.
2. Remove the front road wheels.
3. Support the axle weight with a suitable hydraulic jack.
4. Remove the nuts securing the radius arms to the chassis side members.
5. Disconnect the steering damper at the track rod and using a suitable extractor remove the track rod complete from the hub arms.
6. Remove the four nuts and bolts securing the radius arms to the axle brackets.
7. Lower the radius arms and withdraw them from the chassis side members.
8. Remove the two bolts from the top of the swivel pin housings securing the brake pipe brackets. Remove the brackets and refit the bolts to prevent oil leakage.
9. Disconnect the brake pad wear electrical multi-plug at the rear of the caliper (where applicable).
10. Remove the bolts securing the brake calipers and tie the calipers to one side.
11. Remove the nuts and washers securing the shock absorbers to the axle.
12. Using a suitable extractor disconnect the drag link from the hub arm.
13. Remove the two nuts and bolts securing the panhard rod to the axle bracket and lift the rod clear of the axle.
14. Mark the differential and propeller shaft drive flanges with identification marks to aid re-assembly. Remove the four nuts and bolts, tie the propeller shaft to one side.
15. Carefully lower the axle assembly and remove the road springs.
16. Withdraw the axle assembly.

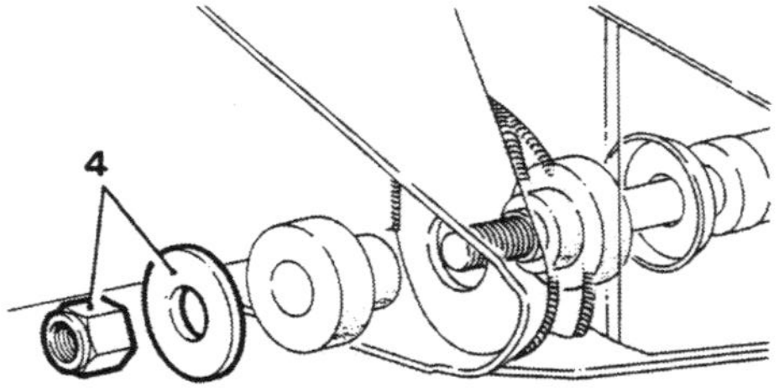

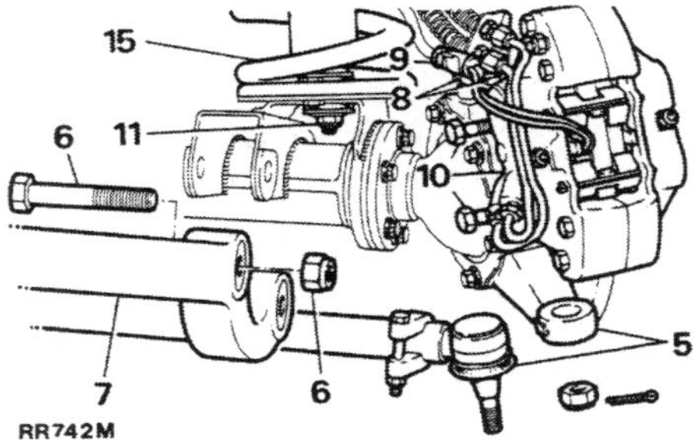

#### Refitting

17. Position the axle under the vehicle, supporting the left hand side of the axle.
18. Reverse the removal instructions.
19. Tighten the propeller shaft to differential bolts to the specified torque.
20. Tighten the panhard rod to axle bracket to the specified torque.
21. Tighten the drag link to hub arm to the specified torque.
22. Tighten the upper swivel pin retaining bolts to the specified torque.
23. Tighten the radius arms to axle bolts to the specified torque.
24. Tighten the radius arms to chassis side member nuts to the correct torque.
25. Tighten the track rod end to the specified torque and fit a NEW split pin.

## FRONT DIFFERENTIAL—OVERHAUL

The front and rear differentials fitted to Range Rover are the same type. When overhauling the front differential refer to the rear differential overhaul procedure in section 51 of this manual.

## FRONT DISCS

### Remove and refit

#### Removing

1. Remove the front hub assembly.
2. Remove the front disc fixing bolts.
3. Tap the disc off the front hub.

*Continued*

## Notes

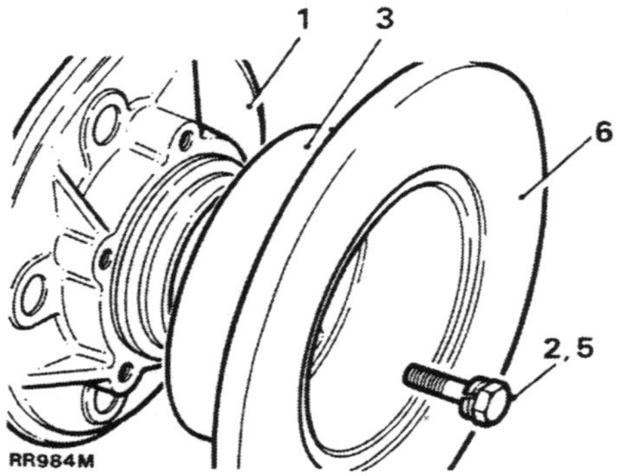

# FRONT AXLE AND FINAL DRIVE 54

## Refitting

4. Locate the disc onto the front hub.
5. Fit the disc fixing bolts. See 'Data section' for tightening torques.
6. Using a dial test indicator, check the total disc run-out, this must not exceed 0,15 mm (0.006 in.). If necessary, reposition the disc.
7. Fit the front hub assembly.

## REMOVE AND OVERHAUL FRONT HUB

**Special tools:**

Oil seal replacer .................. LST 550-5
Drift for above tool............... MS 550 or 18G134
Hub nut spanner .................. 606435

## KEY TO FRONT HUB COMPONENTS

1. Hub driving shaft.
2. Joint washer.
3. Locknut.
4. Lock washer.
5. Adjusting nut.
6. Seal track spacer.
7. Outer oil seal.
8. Outer bearing cone.
9. Outer bearing cup.
10. Hub.
11. Inner bearing cup.
12. Inner bearing cone.
13. Inner oil seal.
14. Mudshield.
15. Stub axle.
16. Stub axle joint washer.
17. Brake disc.

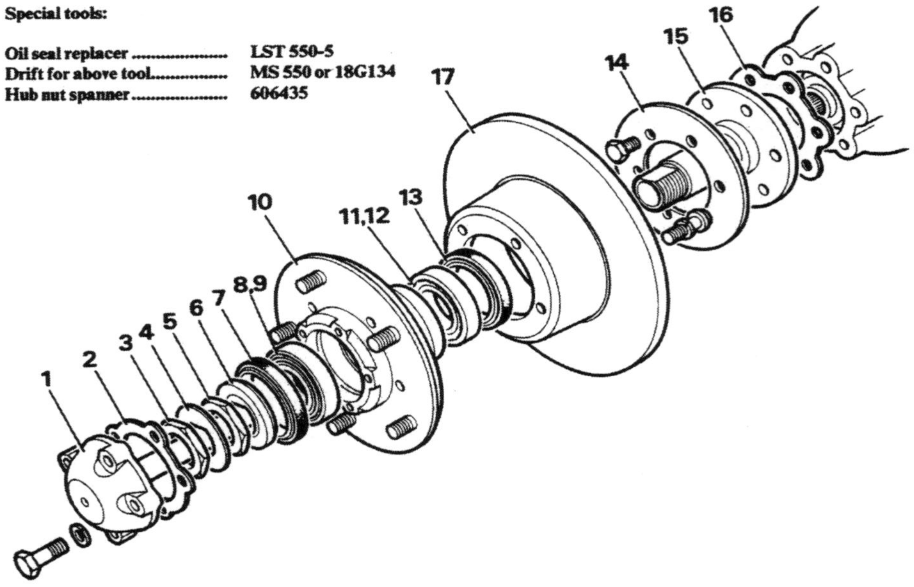

## Remove

1. Jack-up the vehicle, lower on to axle stands and remove the road wheels.
2. Remove the swivel housing top bolt to release the brake hose bracket.
3. Remove the brake caliper retaining bolts and release the assembly from the brake disc and secure to one side.
4. Remove the five bolts retaining the hub driving shaft and withdraw the shaft from the hub.
5. Bend back the lock-tab and remove the outer nut using spanner 606435 and remove the locker. Similarly, remove the inner nut.
6. Remove the seal track spacer.
7. Withdraw the hub complete with bearings, oil seals and brake disc.
8. Remove the inner and outer oil seals.
9. Remove the inner and outer bearing cones.
10. Drift-out the inner and outer bearing cups.

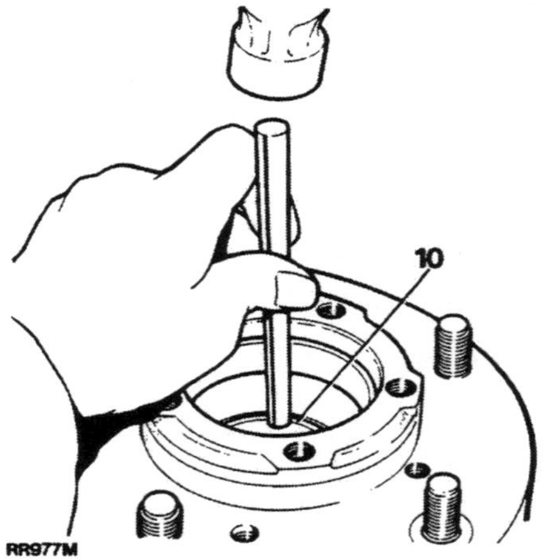

11. Degrease and examine the hub and brake disc and renew both or whichever part is unserviceable. The brake disc is attached to the hub by five bolts. Mark relation of hub to disc before separating.
12. Clean and examine the stub axle and in particular check that the inner seal track diameter is smooth and free from blemishes.
13. If necessary remove the retaining bolts and withdraw the stub axle. Complete with the mudshield and joint washer.

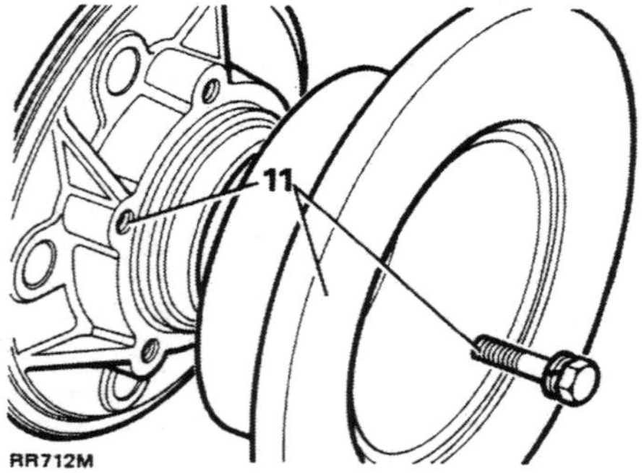

## Assemble

14. Using a new joint washer fit the stub axle and mudshield. Coat the threads of the retaining bolts with Loctite 270 and tighten evenly to the correct torque.
15. Fit the new inner and outer bearing cups to the hub.
16. Fit the new inner bearing cone and pack with one of the recommended hub greases.

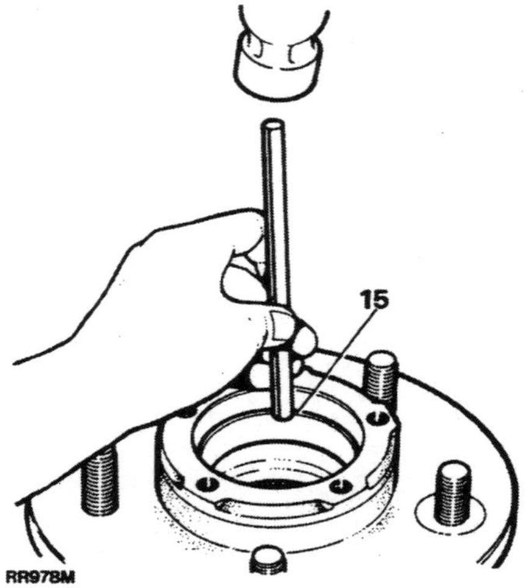

*Continued*

## FRONT AXLE AND FINAL DRIVE

### OVERHAUL STUB AXLE, AXLE SHAFT, CONSTANT VELOCITY JOINT AND SWIVEL ASSEMBLY

Special tool: 18G 284AAH bush extractor

**Remove stub axle, axle shaft and constant velocity joint**

1. Remove the hub complete as described in the operation to overhaul the hub assembly instructions 1 to 14.
2. Drain the swivel pin housing and refit plug.
3. Remove the six bolts retaining the stub axle to the swivel housing.
4. Remove the mud shield.
5. Remove the stub axle and joint washer.

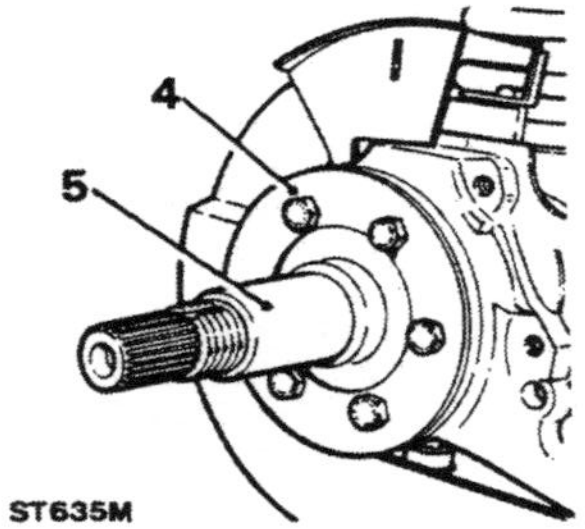

ST635M

6. Pull-out the axle shaft and constant velocity joint from the axle casing.

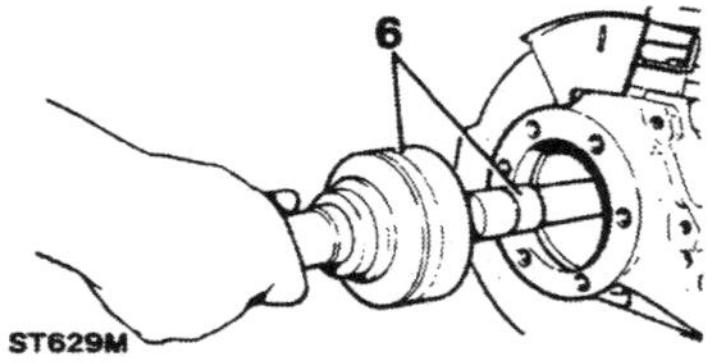

ST629M

*Continued*

---

## FRONT AXLE AND FINAL DRIVE

**Fitting new oil seal — inner**

17. Clean the hub oil seal housing and ensure that the seal locating surface is smooth and the chamfer on the leading edge is also smooth and free from burrs.
18. Examine the new seal and ensure that it is clean and undamaged and that the garter spring is properly located. Even a small scratch on the seal lip could impair its efficiency.
19. Although the new seal is already pre-greased by the manufacturer, apply one of the recommended hub bearing greases to the outside diameter of the seal before fitting.
20. Place the seal, lip side leading, squarely on the hub and using the 76 mm end of seal replacer tool LST 550-5 and drift 550 or 18G 134, drive the seal into position flush with the end face of the hub.

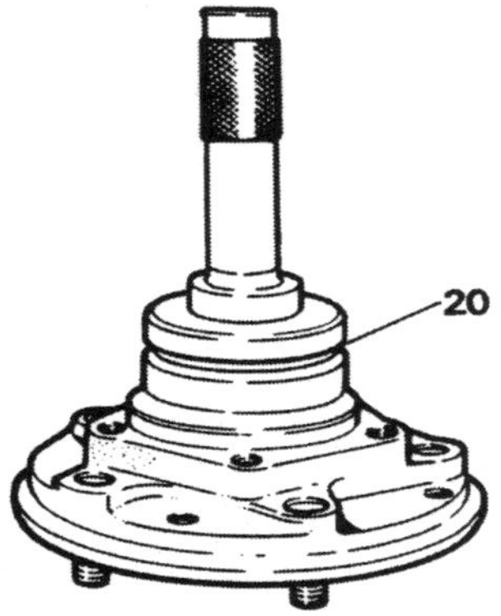

RR979M

**Fitting outer oil seal**

21. Fit the new outer bearing cone and pack with one of the recommended hub greases.
22. Carry out instructions 17 to 19.
23. Place the seal, lip side leading, squarely on the hub and using the 72 mm end of seal replacer tool LST 550-5 and drift 550 or 18G 134, drive the seal into position to the depth determined by the tool.
24. Smear the lips of both seals with one of the recommended greases. This is important since a dry seal can be destroyed during the first few revolutions of the hub.

**Fitting hub to stub axle**

25. Select a new seal track spacer and check that the outer diameter is smooth and free from blemishes and that there are no burrs on the chamfered leading edge.

26. Taking care not to damage the seal lips fit the hub assembly to the stub axle. Do not allow the weight of the hub to rest even temporarily on the outer seal otherwise damage and distortion could occur. Therefore hold the hub clear of the stub axle until the seal track spacer is fitted.
27. Carefully fit the seal track spacer, seal lip leading.
28. Fit the hub inner nut and using spanner 606435 tighten the adjusting nut whilst slowly revolving the hub until all end-float is removed then back-off the nut approximately half a turn.
29. Mount a dial test indicator and bracket on the hub so that the stylus rests in a loaded condition on the nut. Check the end-float which must be 0,013 to 010 mm (0.0005 to 0.004 in). Adjust the nut as necessary to achieve this.

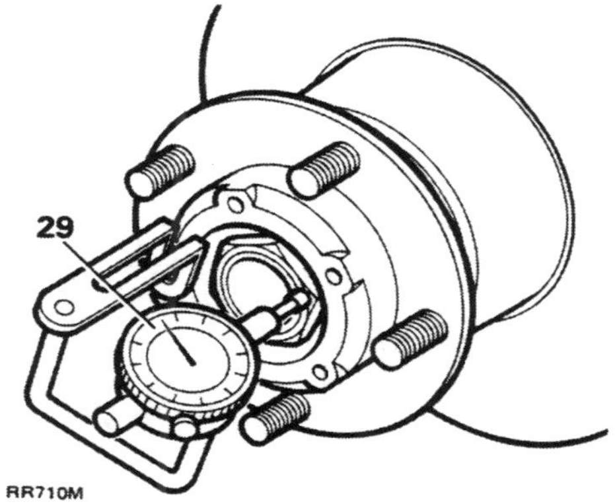

RR710M

30. Fit the locker tab and locknut and tighten against the adjusting nut.
31. Rotate the hub several times to settle the bearings then re-check the end-float.
32. Bend one segment of the locker over the adjusting nut and another, diametrically opposite, over the locknut.
33. Using a new joint washer, fit the hub driving shaft and evenly tighten the retaining bolts to the correct torque.
34. Fit the brake caliper and secure with the retaining bolts and tighten to the correct torque.
35. Fit the swivel housing top bolt and brake hose bracket and tighten to the correct torque.
36. Fit the road wheels, jack-up the vehicle, remove the axle stands, lower the vehicle to the ground and tighten the road wheel nuts evenly to the correct torque.

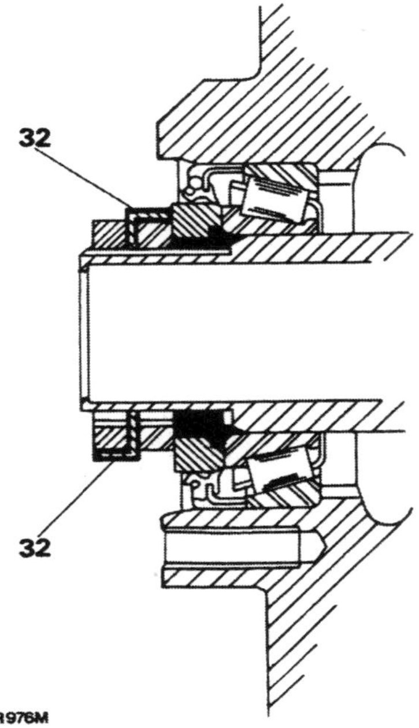

## FRONT AXLE AND FINAL DRIVE

### Remove constant velocity joint from axle shaft

7. Hold the axle shaft firmly in a soft jawed vice.
8. Using a soft mallet drive the constant velocity joint from the shaft.
9. Remove the circlip and collar from the axle shaft.

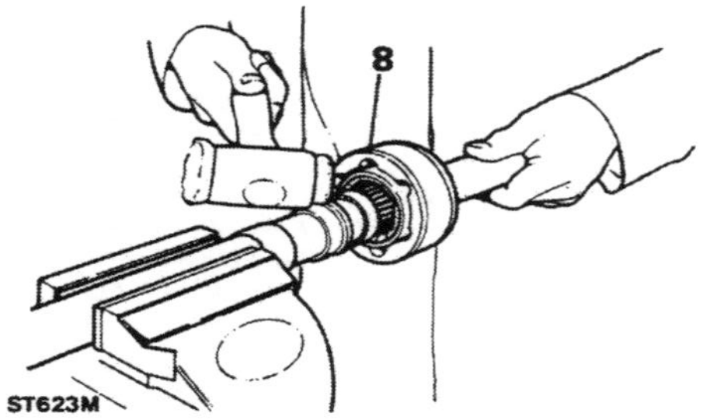

### Dismantle the constant velocity joint

10. Mark the relative positions of the constant velocity joint inner and outer race and the cage for correct reassembly.
11. Tilt and swivel the cage and inner race to remove the balls. 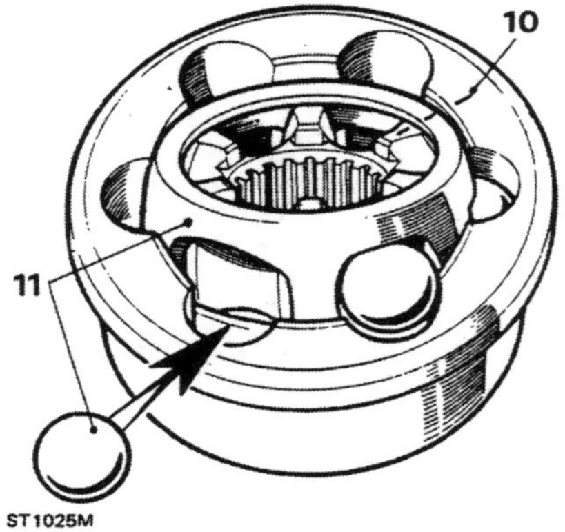
12. Swivel the cage into line with the axis of the joint and turn it until two opposite windows coincide with two lands of the joint housing.
13. Withdraw the cage.
14. Turn the inner track at right angles to the cage with two of the lands opposite the cage openings and withdraw the inner race.
15. The roller bearings must be a light push fit onto the swivel pins. If necessary, drive out the bearing outer tracks and press in replacements.
16. The surface of the bearing housing must be free of corrosion and damage.
17. Examine all components for general condition and examine the inner and outer track, cage, balls and bearing surfaces of the constant velocity joint for damage and excessive wear. Maximum acceptable end-float on the assembled joint 0,64 mm (0.025 in).
18. To assemble the constant velocity joint reverse the dismantling instructions and lubricate with a recommended EP oil.

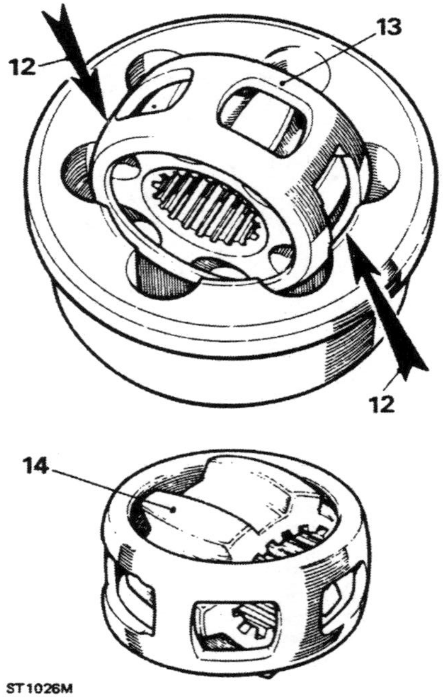

### Fit constant velocity joint to axle

19. Fit the collar and new circlip.
20. Engage the constant velocity joint on the axle shaft splines and using a soft mallet, drive the joint home.

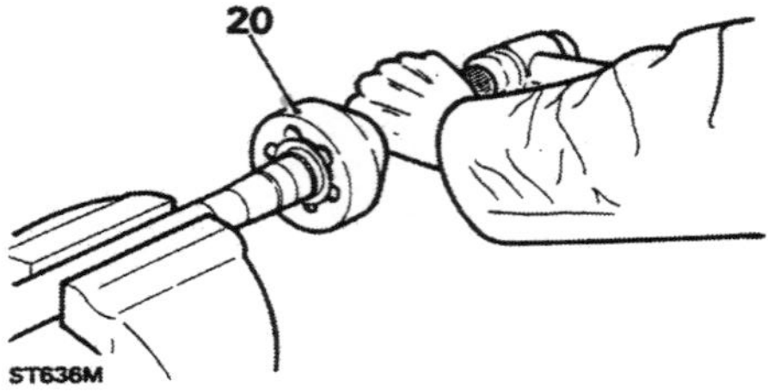

### Renew stub axle intermediate oil seal and bush

21. To remove the bronze bush and oil seal use special tool 18G 284AAH and a slide hammer. Ensure that the fingers of the tool locate behind the oil seal so that the seal and bush are driven-out together. 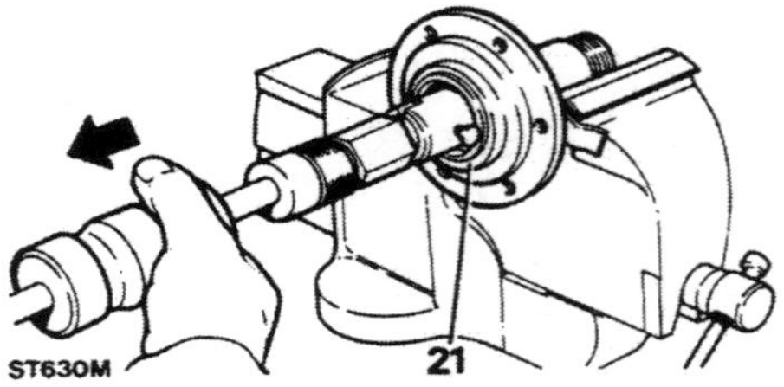
22. With the cavity side leading and using a suitable tube, press-in a new intermediate oil seal and apply grease to the lip.
23. Using a suitable block, press or drive-in the bush up to the shoulder.

### Remove swivel pin housing

24. Remove the brake disc shield secured by one nut and bolt at the bottom front, and one single bolt, behind the shield, in the swivel housing.
25. Disconnect the track-rod end ball joint from the housing.
26. Disconnect the drag-link ball joint.
27. Remove the seven bolts securing the swivel pin housing oil seal and retaining plate and joint washer and release the assembly from the swivel pin housing. Note that whilst the joint washer can be removed at this stage, the oil seal and retaining plate must remain until the swivel pin bearing housing is removed.
28. Remove the two bolts, complete with the brake disc shield bracket, securing the lower swivel pin to the housing.
29. Withdraw the lower swivel pin and joint washer by tapping the small protruding lug.
30. Remove the top swivel pin retaining bolts complete with the brake jump hose bracket.
31. Withdraw the top swivel pin and shims.
32. Remove the swivel pin housing whilst retrieving the lower taper bearing.
33. If the swivel pin housing is to be renewed, remove the drain and level plugs and lock-stop bolt and nut.

### Remove swivel pin bearing housing

34. Remove the seven bolts securing the swivel pin bearing housing to the axle case and remove the housing and joint washer.
35. Remove and discard the swivel pin oil seal and joint washer.

### Overhaul swivel pin bearing housing

36. Prise-out the oil seal from the back of the housing.
37. Drift-out the lower swivel pin bearing track.
38. Drift-out the upper swivel pin bearing track.
39. If worn, pitted or damaged, renew the housing.
40. Press-in the lower swivel pin bearing track.
41. Press-in the upper swivel pin bearing track.
42. With the cavity side trailing press the axle shaft oil seal into the housing and grease.

### Fit swivel pin bearing housing to axle

43. Coat the swivel pin bearing housing to axle casing bolts with Loctite 270.
44. Coat both sides of a joint washer and place in position on the swivel pin bearing housing to axle mating face.
45. Hang the swivel pin bearing housing oil seal, retainer and joint washer over the back of the housing.
46. Fit and secure the swivel pin bearing housing to the axle with the seven bolts tightening evenly to the specified torque.

*Continued*

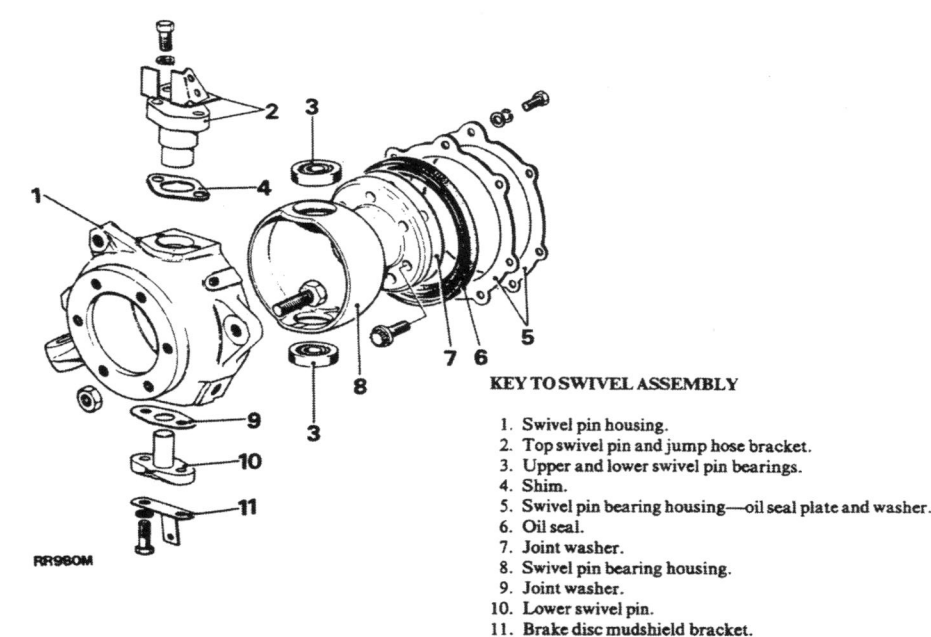

# FRONT AXLE AND FINAL DRIVE

56. Liberally apply—but do not pack—a recommended grease between the lips of the swivel oil seal (2,5 to 4,0 grams).
57. Secure the oil seal and joint washer with the retaining plate and securing bolts tightening evenly to the correct torque.
58. Fit the track-rod and drag link and secure with new split pins.
59. Fit the brake disc shield.
60. Loosely fit the lock stop bolt and nut for later adjustment.

## Fit drive shaft and stub axle

61. Place a new joint washer in position on the swivel pin housing to stub axle mating face.
62. Taking care not to damage the axle shaft oil seals, insert the axle shaft, and when the differential splines are engaged, push the assembly home.
63. Fit the stub axle with the keyway uppermost at 12 o'clock. At this stage it is most important to ensure that the constant velocity joint bearing journal engages fully into the bronze bush in the rear of the stub axle before the stub axle is secured with bolts. Damage to the bush can occur if this precaution is not observed. To ensure proper engagement, grasp the stub axle with one hand and with the other pull the axle shaft into the bush. The shaft and bush are correctly engaged when the end of the axle shaft splines are flush with the end of the stub axle. This condition must be maintained during all ensuing assembly operations.
64. Place the mudshield in position and secure the stub axle to the swivel pin housing with the six bolts using Loctite 270 and evenly tighten to the specified torque.
65. To complete the reassembly, follow instructions 25 to 41 covering front hub overhaul.
66. Check that the swivel pin housing drain plug is tightly fitted and remove the filler and level plug.
67. Inject approximately 0.35 litre (0.6 pint) of recommended EP oil until the oil begins to run out of the level hole. Fit and tighten the plug and wipe away any surplus oil.
68. Set the steering lock-stop bolts to provide a minimum clearance of 40,5 mm (1.59 in) between the tyre wall and radius arm. Tighten the locknut.

## KEY TO SWIVEL ASSEMBLY

1. Swivel pin housing.
2. Top swivel pin and jump hose bracket.
3. Upper and lower swivel pin bearings.
4. Shim.
5. Swivel pin bearing housing—oil seal plate and washer.
6. Oil seal.
7. Joint washer.
8. Swivel pin bearing housing.
9. Joint washer.
10. Lower swivel pin.
11. Brake disc mudshield bracket.

55. To check the top swivel pin pre-load attach a spring balance to the track-rod ball joint bore and pull the balance to determine the effort required to turn the swivel. The resistance, once the initial inertia has been overcome, should be 1.16 to 1.46 kg (2.6 to .2 lb). If necessary, adjust by removing or adding shims to the top swivel pin as required. When the setting has been achieved coat the bolts with Loctite 270 and re-tighten the bolts to the specified torque and bend over the lock tabs of the locking washer.

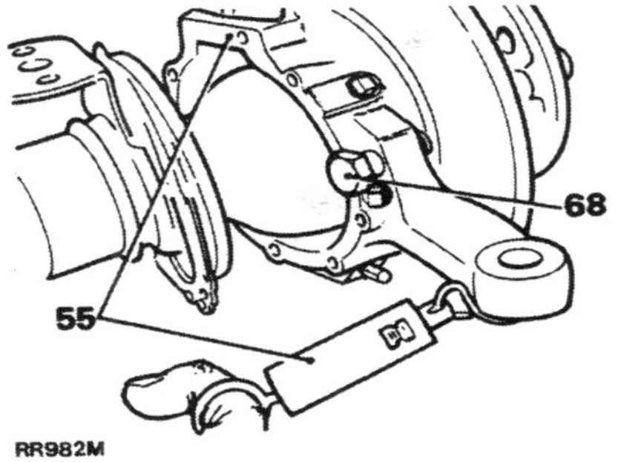

## Fit swivel pin housing

47. Grease and fit the upper swivel pin bearing to the bearing housing.
48. Grease and place the lower swivel pin bearing onto the swivel pin housing.
49. Place the swivel pin housing in position over the swivel in bearing housing.
50. Coat a joint washer both sides with sealing compound and place in position on the lower swivel pin.
51. Fit the lower pin with lip outboard. Do not secure with bolts at this stage.
52. Fit the top swivel pin with existing shims and fit the securing bolts, jump hose bracket and locking washer (do not tighten the bolts at this stage).
53. Coat the threads of the lower swivel pin bolts with Loctite 270 and fit, together with the brake disc shield bracket, tighten the bolts to the specified torque and bend over the lock tabs of the mudshield bracket.
54. Tighten the top swivel pin securing bolts to the specified torque.

# RANGE ROVER WORKSHOP MANUAL 1986 ONWARDS

# Part 4

|  | Section | Page |
|---|---|---|
| General Specification Data | 04 | 156 |
| Torque Wrench Settings | 06 | 157 |
| Steering | 57 | 158 |
| Front Suspension | 60 | 171 |
| Rear Suspension | 64 | 173 |
| Brakes | 70 | 176 |
| Body | 76 | 189 |

Section
Number

## 04 GENERAL SPECIFICATION DATA

—Steering 1
—Suspension 1
—Brakes 2

## 06 TORQUE WRENCH SETTINGS

—Steering 1
—Front suspension 1
—Rear suspension 1
—Brakes 2
—Road wheels 2
—Body 2

## 57 STEERING

—Power steering box—remove and refit 1
—Power steering box—overhaul 2
—Power steering system—bleed 10
—Power steering system—test 11
—Power steering system—adjust 12
—Power steering pump drive belt—remove and refit 12
—Power steering pump drive belt—adjust 13
—Power steering pump—remove and refit 14
—Power steering pump—overhaul 16
—Power steering—fault diagnosis 17
—Coupling shaft and universal joints—remove and refit 18
—Steering column—remove and refit 19
—Steering column assembly—overhaul 20
—Steering column lock and ignition/starter switch—remove and refit 23
—Steering wheel—remove and refit 24
—Drop arm—remove and refit 25
—Drop arm ball joint—overhaul 25
—Track rod and linkage—remove and refit
—Steering damper—remove and refit
—Drag link and drag link ends—remove and refit
—Front wheel alignment—check and adjust
—Steering lock stop—check and adjust

*Continued*

## 60 FRONT SUSPENSION

—Panhard rod—remove and refit 1
—Radius arm—remove and refit 1
—Front road spring—remove and refit 2
—Front shock absorber—remove and refit 2
—Bump stop—remove and refit 3

## 64 REAR SUSPENSION

—Rear road spring—remove and refit 1
—Rear shock absorber—remove and refit 1
—Levelling unit—functional check 2
—Levelling unit—remove and refit 2
—Levelling unit ball joints—remove and refit 3
—Bump stop—remove and refit 4
—Upper suspension link—remove and refit 4
—Lower suspension link—remove and refit 5

## 70 BRAKES

—Description—brake system 1
—Layout of brake pipe system 2
—Bleed the brakes 4
—Brake pressure reducing valve—remove and refit 5
—Master cylinder—overhaul 6
—Pedal assembly—overhaul 10
—Brake servo—overhaul 11
—Front brake caliper—overhaul 17
—Rear brake caliper—overhaul 19
—Transmission brake lever—remove and refit 21
—Transmission brake—overhaul 22
—Transmission brake cable—remove and refit 24

## 76 BODY

—Body repairs—general information 1
—Paintwork—general information 3
—Chassis frame—alignment check 4
—Body—introduction 6
—Welding 6
—Inner body shell—welding charts 7
—Headlining, front and rear—remove and refit 12
—Roof panel—remove and refit 12
—Bonnet—remove and refit 12
—Decker panel—remove and refit 12
—Front wing—remove and refit 13
—Rear corner panel—remove and refit 13
—Rear wing—remove and refit 13
—Rear quarter panel—interior—remove and refit 13
—Rear quarter panel—exterior—remove and refit 14

155

# GENERAL SPECIFICATION 04

## STEERING DATA

### STEERING

**Power steering**

| | |
|---|---|
| Make/type | Adwest Varamatic/linkage |
| Ratio | Variable: straight ahead 17.5:1 |
| Steering wheel turns, lock-to-lock | 3.375 |
| Steering wheel diameter | 431.8 mm (17 in) |
| Front wheel alignment | 1,2 to 2,44 mm (0.046 to 0.093 in) toe out |
| Camber angle | 0° Check with vehicle in static unladen condition, that is, vehicle with water, oil and five gallons of fuel. Rock the vehicle up and down at the front to allow it to take up a static position. |
| Castor angle | 3° |
| Swivel pin inclination | 7° |

## ROADSPRING DATA

| † Specification | Part Number | Colour Code | Rating | Free Length | No. of Coils |
|---|---|---|---|---|---|
| A | 572315 | Blue Stripe | 2375.1 kg/m (133 lb/in) | 391.16 mm (15.4 in) | 7.18 |
| B | NRC4306 | Blue & White Stripe | 2375.1 kg/m (133 lb/in) | 417.57 mm (16.44 in) | 7.55 |
| C | NRC8113 | Pink & Purple Stripe | 3182.1 kg/m (178.2 lb/in) | 418.36 mm (16.47 in) | 8.75 |
| D | NRC8477 | Green & Pink Stripe | 3182.1 kg/m (178.2 lb/in) | 461.67 mm (18.176 in) | 8.75 |
| E | NRC2119 | Green Stripe | 2678.7 kg/m (150 lb/in) | 409.70 mm (16.13 in) | 7.63 |
| F | NRC4305 | Red & Yellow Stripe | 2678.7 kg/m (150 lb/in) | 436.4 mm (17.18 in) | 7.65 |

| Standard Suspension | † Specification Right Side | † Specification Left Side | Heavy Duty Suspension | † Specification Right Side | † Specification Left Side |
|---|---|---|---|---|---|
| RHD Front | A | A | RHD Front | E | E |
| LHD Front | A | B | LHD Front | E | E |
| RHD Rear | C | C | | | F |
| LHD Rear | C | *D | | | |

* Specification 'C' spring fitted to early production vehicles.

## 76 BODY (Continued)

| | |
|---|---|
| —Front door—remove and refit | 14 |
| —Rear passenger door—remove and refit | 15 |
| —Upper tailgate—remove and refit | 16 |
| —Upper tailgate lock—remove and refit | 16 |
| —Lower tailgate—remove and refit | 17 |
| —Lower tailgate lock—remove and refit | 17 |
| —Front door glass and regulator—four-door—remove and refit | 18 |
| —Rear door glass and regulator—remove and refit | 19 |
| —Front door lock, outside and inside door handles—four door—remove and refit | 20 |
| —Front door lock and handle—adjustment | 20 |
| —Rear door lock, outside and inside door handles—remove and refit | 21 |
| —Front door glass and regulator—two door—remove and refit | 22 |
| —Front quarter vent—two door—remove and refit | 23 |
| —Front door lock—two door—remove and refit | 23 |
| —Private lock—two door—remove and refit | 23 |
| —Body side glass—two door—remove and refit | 24 |
| —Windscreen glass—remove and refit | 25 |
| —Tailgate glass—remove and refit | 25 |
| —Front seat—two door—remove and refit | 25 |
| —Front seat—four door—remove and refit | 25 |
| —Front seat base—two and four door—remove and refit | 26 |
| —Cubby box—remove and refit | 27 |
| —Floor mounted console—remove and refit | 27 |
| —Lower fascia panel—remove and refit | 28 |
| —Radio housing | 28 |
| —Centre fascia panel—remove and refit | |
| —Louvre fascia panel—remove and refit | |
| —Fascia top rail—remove and refit | |
| —Front spoiler—remove and refit | |

156

## 06 TORQUE WRENCH SETTINGS

### STEERING

| | Nm | lbf ft |
|---|---|---|
| Clamp bolt nuts | 14 | 10 |
| Ball joint nuts | 40 | 30 |
| Universal joint pinch bolt | 35 | 26 |
| Drop arm nut | 176 | 130 |
| Steering wheel nut | 38 | 28 |
| Sector shaft cover to steering box | 22 to 27 | 16 to 20 |
| Tie bar to steering box | 81 | 60 |
| Steering box to chassis | 81 | 60 |
| Power steering pump mounting | 35 | 26 |
| Flow control valve cap, power steering pump | 40 to 47 | 30 to 35 |
| Power steering pump cover to body screws | 20 to 23 | 15 to 17 |
| Pulley bolt, power steering pump | 14 to 16 | 10 to 12 |
| Union bolt, inlet adaptor, power steering pump | 38 to 41 | 28 to 30 |
| Steering column bracket nuts | 27 | 20 |

### FRONT SUSPENSION

| | Nm | lbf ft |
|---|---|---|
| Drag link to axle | 40 | 30 |
| Securing ring for mounting turret | 13.5 | 10 |
| Radius arm to axle/chassis | 176 | 130 |
| Mounting arm to chassis | 89 | 65 |
| Panhard rod to axle | 89 | 65 |
| Panhard rod to mounting bracket | 89 | 65 |
| Tie bar to Panhard rod | 81 | 60 |

### REAR SUSPENSION

| | Nm | lbf ft |
|---|---|---|
| Levelling unit to rear axle | 176 | 130 |
| Top link to levelling unit | 115 | 85 |
| Top link to mounting bracket | 176 | 130 |
| Upper joint to levelling unit | 34 | 25 |
| Lower joint to levelling unit | 34 | 25 |
| Bottom link to axle | 176 | 130 |
| Top link bracket to rear cross member | 127 | 90 |
| Levelling unit to cross member | 47 | 35 |
| Shock absorber to axle | 52 | 40 |

### BRAKES

Brake pipe connections to:

| | Nm | lbf ft |
|---|---|---|
| —Brake calipers | 12 | 9 |
| —Jump hoses to brackets | 11-13.5 | 8-10 |
| —Jump hose to three-way connection | 12 | 9 |
| —Front caliper jump hoses | 11-13.5 | 8-10 |
| —Rear caliper jump hoses | 11-13.5 | 8-10 |
| PDWA switch | 12 | 9 |
| Master cylinder to servo | 1.4-1.7 | 1-1.3 |
| Brake caliper to swivel pin housing | 13.6-17 | 10-12.5 |
| Handbrake linkage to transfer box | 75-88 | 55-65 |
| Brake disc to hubs | 26-32 | 19-24 |
|  | 65-80 | 48-59 |

| | Nm | lbf in |
|---|---|---|
| Bleed screws | 9-11 | 80-100 |
| Support plate and tube to valve body (servo) | 2-3 | 16-20 |
| Separator shell to valve body (servo) | 13.5-17 | 120-150 |
| Clamp ring to servo | 1-1.6 | 10-15 |

---

## 04 GENERAL SPECIFICATION DATA

### SHOCK ABSORBERS (DAMPERS)
Type .................................................. Telescopic, double-acting non-adjustable
Bore diameter ..................................... 35.47 mm (1.375 in)

### SUSPENSION—All Models
Type .................................................. Coil springs controlled by telescopic dampers front and rear
Front ................................................. Transverse location of axle by Panhard rod, and fore and aft location by two radius arms
Rear .................................................. Fore and aft movement inhibited by two tubular trailing links. Lateral location of axle by a centrally positioned 'A' bracket bolted at the apex to a ball joint mounting. A levelling unit is positioned between the ball joint and upper cross member

### BRAKES

**Foot brake**
Type .................................................. Disc
Operation ........................................... Hydraulic, servo assisted, self-adjusting

**Front brake**
Type .................................................. Outboard discs with four pistons
Disc diameter ..................................... 298.17 mm (11.75 in)
Total pad area .................................... 381 cm² (15 in²)
Pad swept area ................................... 4193.5 cm² (650.7 in²)
Pad material ....................................... DON 230
Pad wear indicator .............................. Inboard pad right-hand caliper

**Rear brake**
Type .................................................. Outboard discs with two pistons
Disc diameter ..................................... 290.0 mm (11.42 in)
Total pad area .................................... 317.34 cm² (49.2 in²)
Pad swept area ................................... 3199.2 cm² (496 in²)
Pad material ....................................... DON 230
Pad wear indicator .............................. Inboard pad left-hand caliper

**Handbrake**
Type .................................................. Transmission drum brake—cable operated
Drum diameter ................................... 254 mm (10 in)
Width ................................................. 70 mm (2.75 in)
Lining material ................................... DON 269

# STEERING 57

## POWER STEERING BOX

**Remove and refit**

Service tools:
Ball joint extractor—18G1063
Drop arm extractor—MS252A

**NOTE:** It is important that whenever any part of the system, including the flexible piping, is removed or disconnected, that the utmost cleanliness is observed.

All ports and hose connections should be suitably sealed off to prevent ingress of dirt, etc. If metallic sediment is found in any part of the system, the complete system should be checked, the cause rectified and the system thoroughly cleaned.

Under no circumstances must the engine be started until the reservoir has been filled. Failure to observe this rule will result in damage to the pump.

Metric pipe fittings are now used with 'O' ring pipe ends on the fittings to the steering box.

Follow normal 'O' ring replacement procedure whenever pipes are disconnected.

Ensure that compatible metric components are used when fitting replacement pipes.

**Removing**

1. Park the vehicle on level ground.
2. Prop open the bonnet.
3. Remove the filler cap from the power steering fluid reservoir. Disconnect the pipes from the pump. Drain and discard the fluid. Replace the filler cap.
4. Disconnect the flexible hoses from the steering box.
5. Blank off all disconnected hose connections to prevent ingress of foreign matter.
6. Jack up and support the chassis front end. Alternatively, raise the vehicle on a ramp.

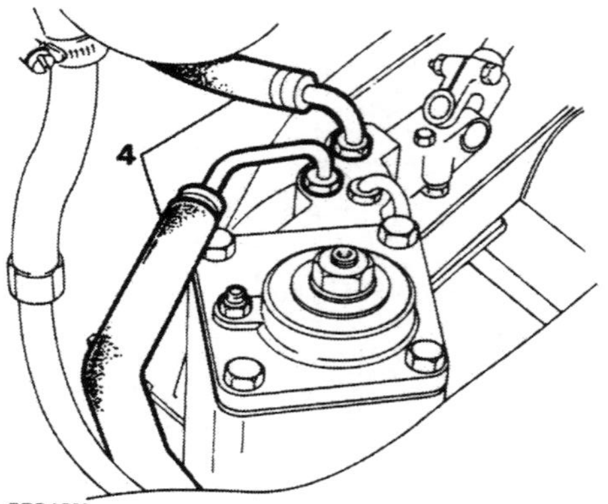

**WARNING: Whichever method is adopted, it is essential that the wheels are chocked and the handbrake applied.**

7. Right-hand drive vehicles only—disconnect the lead from the oil pressure switch.
8. Uncouple the drag link and remove, using ball joint extractor 18G1063.
9. Remove the drop arm, using drop arm extractor MS252A.
10. Remove the pinch bolt attaching the universal joint to the power steering box. 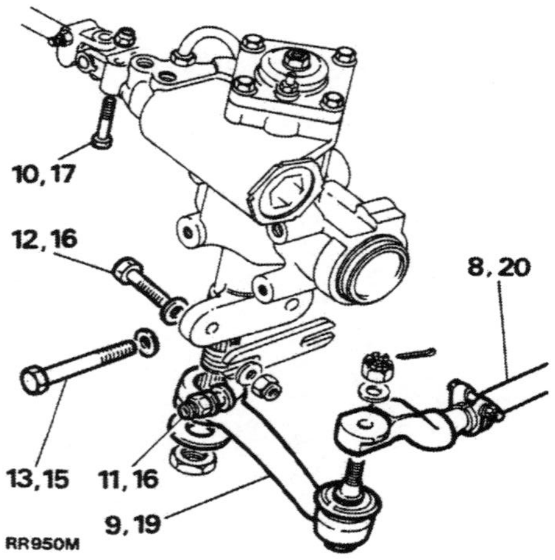
11. Slacken the nut securing the tie bar to the chassis.
12. Remove the bolts securing the tie bar to the steering box and move the tie bar aside.
13. Remove the fixings attaching the power steering box to the chassis side member.
14. Withdraw the power steering box.

*continued*

## 06 TORQUE WRENCH SETTINGS

### ROAD WHEELS

| | Nm | lbf ft |
|---|---|---|
| Wheel nuts | | |
| Alloy wheels | 122–129 | 90–95 |
| Steel wheels | 102–115 | 75–85 |

### BODY

| | Nm | lbf ft |
|---|---|---|
| Rear seat belt inertia reel to body | 68–81 | 50–60 |
| Rear seat belt guide bracket to body | 20–27 | 15–20 |

# STEERING 57

## Refitting

15. Refit the steering box to the chassis side member and tighten the four Nyloc nuts to the correct torque.
16. Refit the tie bar to the steering box, and tighten the tie bar securing nut to the correct torque.
17. Reconnect the pinch bolt, attaching the universal joint to the power steering box, and tighten to the correct torque.
18. Reconnect the lead to the oil pressure switch.
19. Refit the drop arm.
20. Refit the drag link and secure.
21. Lower the vehicle to ground level.
22. Remove the blanking plugs and reconnect the flexible hoses to the steering box.
23. Remove the blanking plug and refit the flexible hose to the power steering pump.
24. Ensure that the steering wheel is correctly aligned when the wheels are in the straight-ahead position.

NOTE: It may be necessary to remove the steering wheel and reposition on the splines to obtain this condition. See steering wheel—remove and refit.

25. Remove the filler cap from the power steering fluid reservoir. Fill the reservoir to the oil level mark with the recommended fluid (see Section 09) and bleed the power steering system. See power steering system—bleed.
26. Check the fluid level and replace the filler cap.
27. Check, and if necessary, adjust the steering box.
28. Test the steering system for leaks, with the engine running, by holding the steering hard on full lock in both directions.

CAUTION: Do not maintain this pressure for more than 30 seconds in any one minute, to avoid causing the oil to overheat and possible damage to the seals.

29. Close the bonnet.
30. Road test the vehicle.

## POWER STEERING BOX OVERHAUL

Service tools:
'C' Spanner—LST120
Worm adjusting socket—LST119
Drop arm extractor—MS252A
Ring expander—606602
Ring compressor—606603
Seal saver, sector shaft—606604
Seal saver, valve and worm—R01015
Torque setting tool—R01016

### Dismantle

1. Remove the steering box from the vehicle, and withdraw the drop arm.

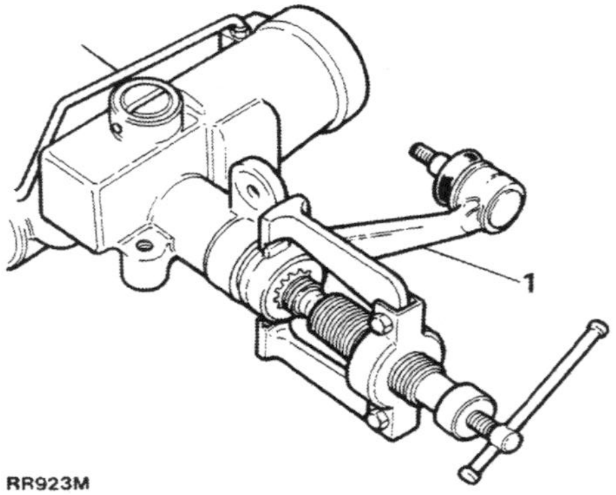

2. Rotate the retainer ring, as necessary, until one end is approximately 12 mm (0.500 in) from the extractor hole.
3. Lift the cover retaining ring from the groove in the cylinder bore, using a suitable pointed drift applied through the hole provided in the cylinder wall.
4. Complete the removal of the retainer ring, using a screwdriver.
5. Turn on left lock (LH steering) until the piston pushes out the end cover (for RH steering models, turn on right lock).

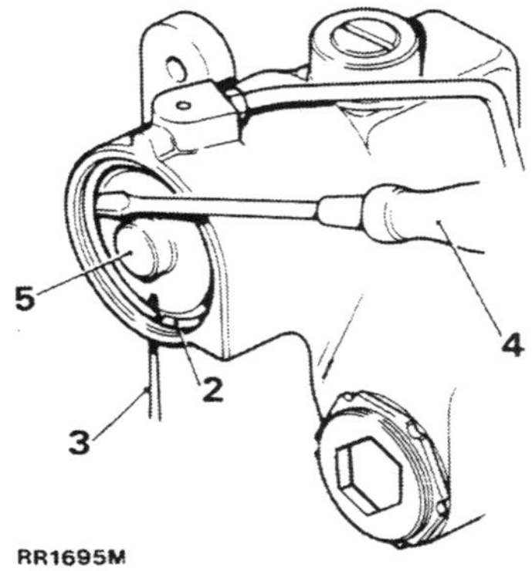

6. Slacken the grub screw retaining the rack pad adjuster.
7. Remove the rack pad adjuster.
8. Remove the sector shaft adjuster locknut.
9. Remove the sector shaft cover fixings.
10. Screw in the sector shaft adjuster until the cover is removed.
11. Slide out the sector shaft.

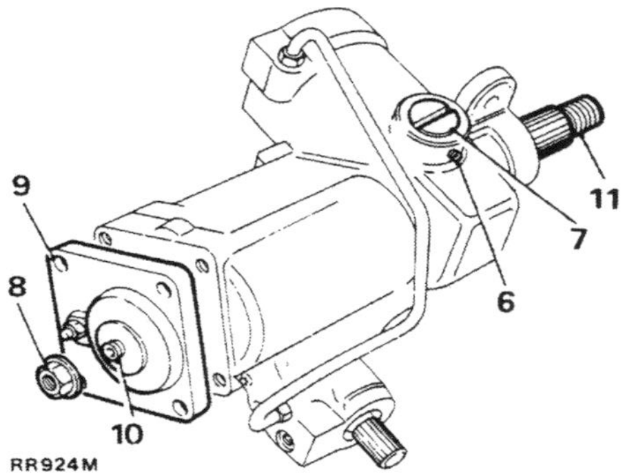

12. Withdraw the piston, using a suitable bolt screwed into the tapped hole in the piston.
13. Remove the worm adjuster locknut using 'C' spanner, LST120.
14. Remove the worm adjuster using socket LST119.

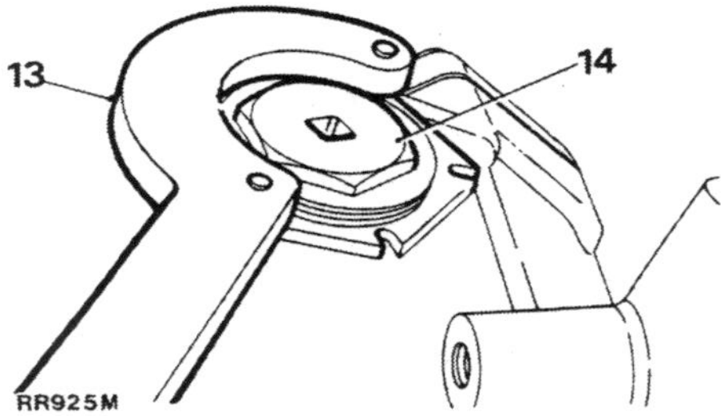

15. Tap the splined end of the spindle shaft to free the bearing.
16. Withdraw the bearing cup and caged ball bearing assembly.

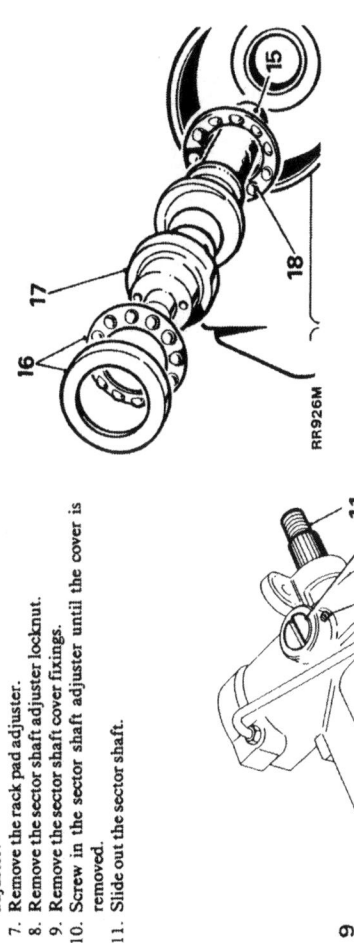

17. Withdraw the valve and worm assembly.
18. Withdraw the inner bearing ball race and shims. Retain the shims.

Continued

# STEERING 57

## Steering box seals

19. Remove the circlip and seals from the sector shaft housing bore.

    NOTE: Do not remove the sector bush unless replacement is required. Refer to instruction 22.

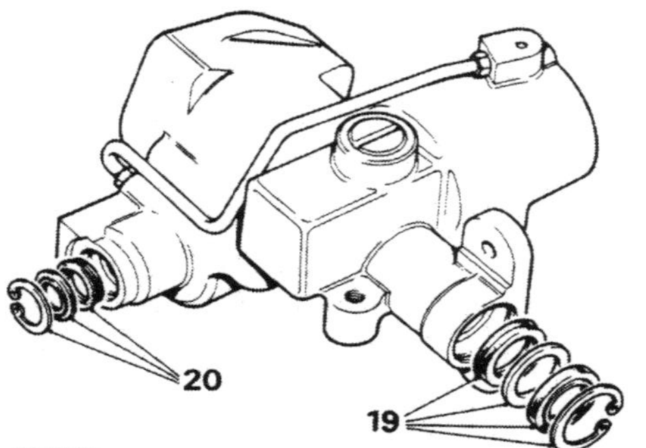

20. Remove the circlip and seals from the input shaft housing bore.

    NOTE: Do not remove the input shaft needle bearing unless replacement is required.

## Inspecting

21. Discard all rubber seals and provide replacements.

    NOTE: A rubber seal is fitted behind the plastic ring on the rack piston. Discard the seal also the plastic ring and provide replacements.

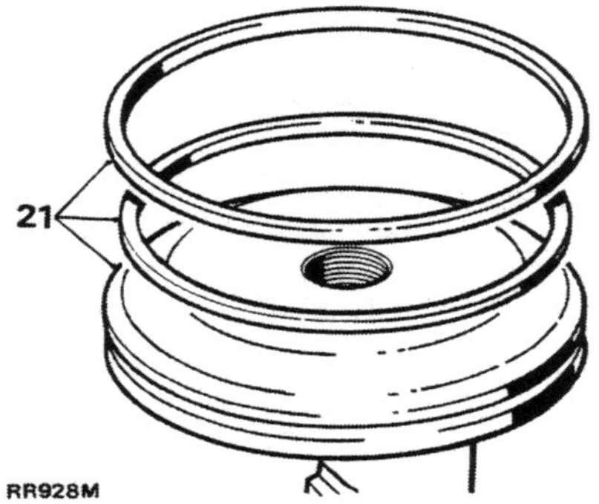

## Steering box casing

22. If necessary, replace the sector shaft bush, using suitable tubing as a drift.
23. Examine the piston bore for traces of scoring and wear.
24. Examine the inlet tube seat for damage. If replacement is necessary this can be undertaken by using a suitable tap.
25. Examine the feed tube for signs of cracking.

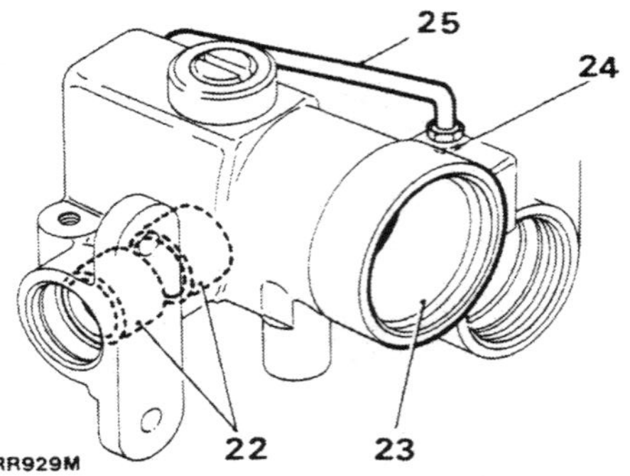

## Sector shaft assembly

26. Check that there is no side play on the rollers.
27. If excessive side play on the roller does exist renew the sector shaft.
28. Check the condition of the adjuster screw threads.
29. Examine the bearing areas on the shaft for excessive wear.
30. Examine the gear teeth for uneven or excessive wear.

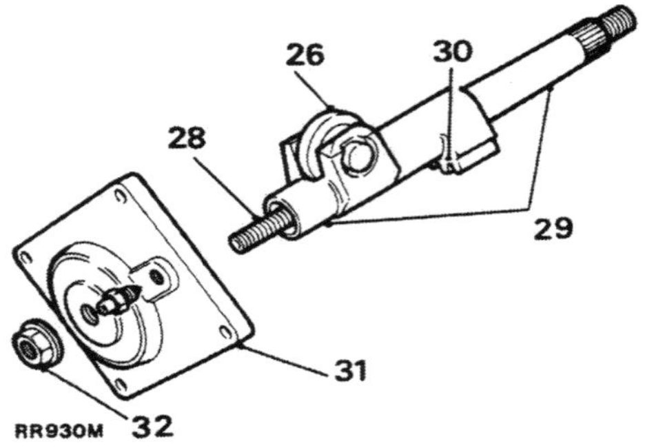

## Sector shaft cover assembly

31. The cover, bush and seat are supplied as a complete assembly for replacement purposes.

## Sector shaft adjuster locknut

32. The locknut functions also as a fluid seal and must be replaced at overhaul.

## Valve and worm assembly

33. Examine the valve rings which must be free from cuts, scratches and grooves. The valve rings should be a loose fit in the valve grooves.
34. Remove the damaged rings ensuring that no damage is done to the seal grooves.
35. If required, fit replacement rings, using the ring expander 606602. Both rings and tool may be warmed if found necessary. Use hot water for this purpose. Then insert into the ring compressor 606603 to cool.

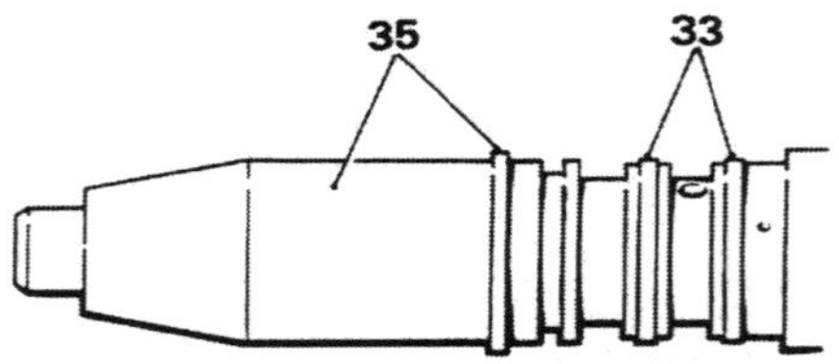

    NOTE: The expander will not pass over rings already fitted. These rings must be discarded to allow access then renewed.

36. Examine the bearing areas for wear. The areas must be smooth and not indented.
37. Examine the worm track which must be smooth and not indented.
38. Check for wear on the torsion bar assembly pins; no free movement should exist between the input shaft and the worm.

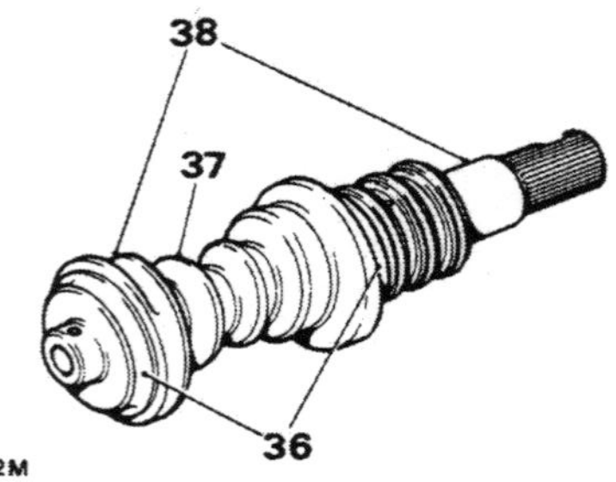

    NOTE: Any sign of wear makes it essential that the complete valve and worm assembly is renewed.

## Ball bearing and cage assemblies

39. Examine the ball races and cups for wear and general condition.
40. If the ball cage has worn against the bearing cup, fit replacements.
41. Bearing balls must be retained by the cage.
42. Bearings and cage repair are carried out by the complete replacement of the bearings and cage assembly.
43. To remove the inner bearing cup and shim washers, jar the steering box on the work bench, or use a suitable extractor.

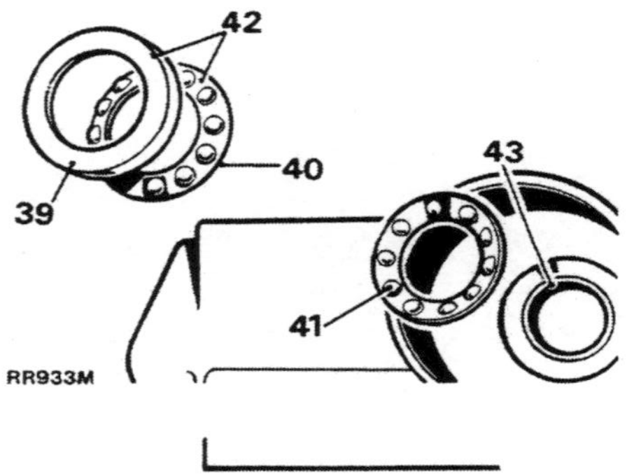

    NOTE: Should difficulty be experienced at this stage, warm the casing and the bearing assembly. Cool the bearing cup using a suitable mandrel and jar the steering box on the bench.

*Continued*

# STEERING  57

### Rack thrust pad and adjuster

44. Examine the thrust pad for scores.
45. Examine the adjuster for wear in the pad seat.
46. Examine the nylon pad for distortion and adjuster grub screw assembly for wear.

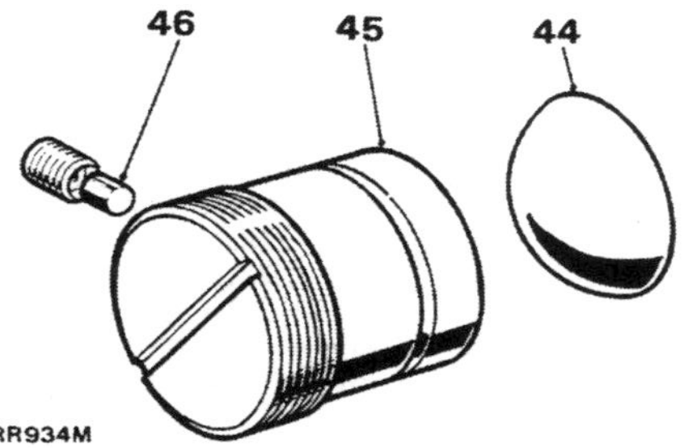

### Rack and piston

47. Examine for excessive wear on the rack teeth.
48. Ensure the thrust pad bearing surface is free of scores and wear.
49. Ensure that the piston outer diameters are free from burrs and damage.
50. Examine the seal and ring groove for scores and damage.
51. Fit a new rubber ring to the piston. Warm the white nylon seal and fit this to the piston. Slide the piston assembly into the cylinder with the rack tube outwards. Allow to cool.

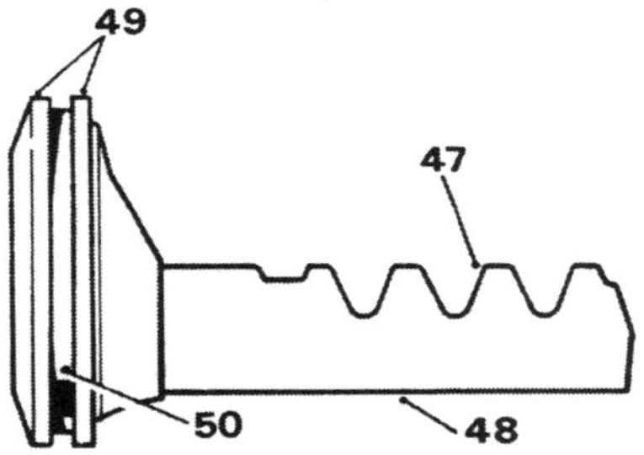

### Input shaft needle bearing

52. If necessary, replace the bearing. The replacement must be fitted squarely in the bore (numbered face of the bearing uppermost). Then, carefully push the bearing in until it is flush with the top of the housing bore. Ideally, the bearing will be just clear of the bottom of the housing bore.

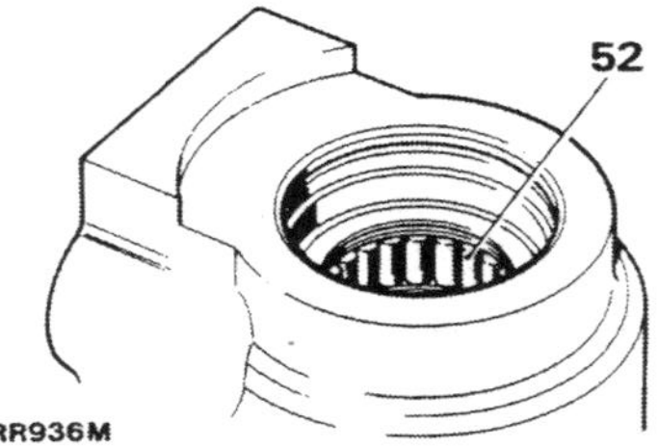

### Reassemble

NOTE: When fitting replacement oil seals, these must be lubricated with recommended fluid. Also ensure that absolute cleanliness is observed during assembly.

### Input shaft oil seal

53. Fit the seal, lipped side first, into the housing. When correctly seated, the seal backing will lie flat on the bore shoulder.
54. Fit the extrusion washer and secure with the circlip.

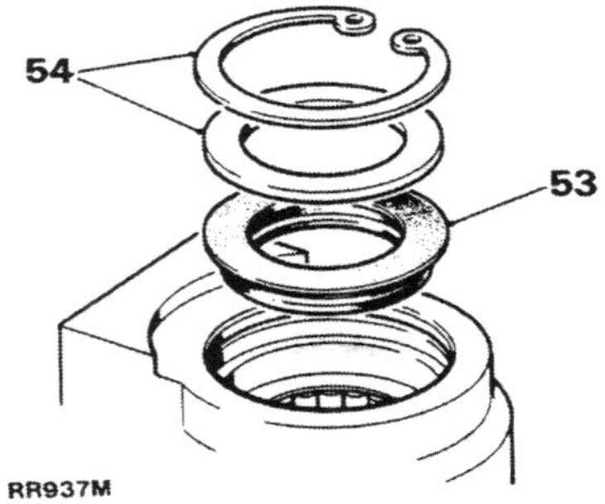

### Sector shaft seal

55. Fit the oil seal, lipped side first.
56. Fit the extrusion washer.
57. Fit the dirt seal, lipped side last.
58. Fit the circlip.

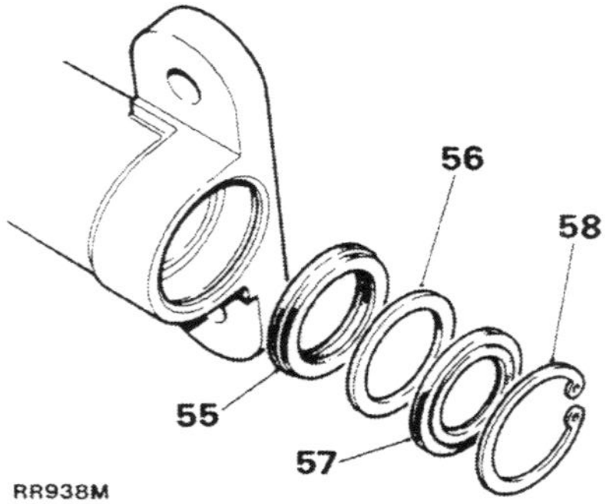

### Fitting the valve and worm assembly

59. If removed, refit the original shim washer(s) and the inner bearing cap. Only vaseline must be used as an aid to assembling the bearings.

NOTE: If the original shims are not available, fit shim(s) of 0.76 mm (0.030 in) nominal thickness.

60. Fit the inner cage and bearings assembly.
61. Fit the valve and worm assembly, using seal saver RO1015 to protect the input shaft seal.
62. Fit the outer cage and bearings assembly.
63. Fit the outer bearing cup.
64. Renew the worm adjuster sealing ring and loosely screw the adjuster into the casing. Fit the locknut, but do not tighten.
65. Turn in the worm adjuster until the end-float at the input is almost eliminated.
66. Measure and record the maximum rolling distance of the valve and worm assembly, using a spring balance and cord coiled around the torque setting tool RO1016.
67. Turn in the worm adjuster to increase the figure recorded in instruction 66 by 1.8 to 2.2 kg (4 to 5 lb) at 1.250 in (31.7 mm) radius to settle the bearings, then back off the worm adjuster until the figure recorded in instruction 66 is increased by 0.9 to 1.3 kg (2 to 3 lb) only, with locknut tight. Use worm adjusting socket LST119 and 'C' spanner LST120.

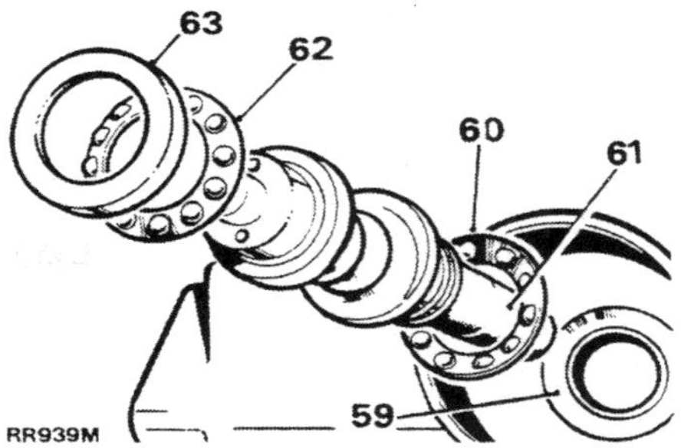

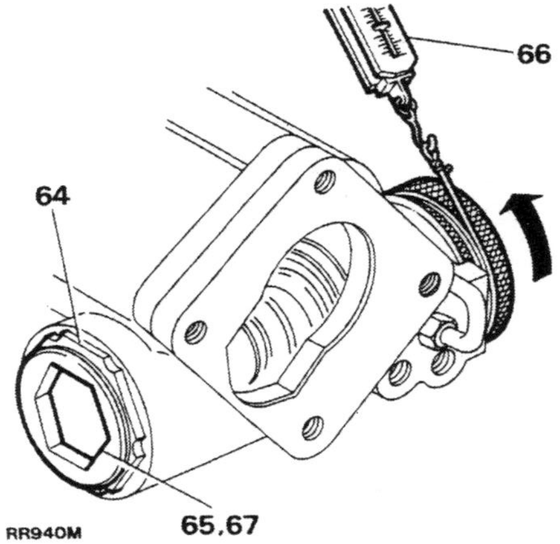

Continued

# STEERING

## Fitting the rack and piston

68. Screw a suitable bolt into the piston head for use as an assembly tool.
69. Fit the piston and rack assembly so that the piston is 63.5 mm (2.5 in) approximately from the outer end of the bore.
70. Feed in the sector shaft using seal saver 606604 aligning the centre gear pitch on the rack with the centre gear tooth on the sector shaft. Push in the sector shaft, and, at the same time rotate the input shaft about a small arc to allow the sector roller to engage the worm.

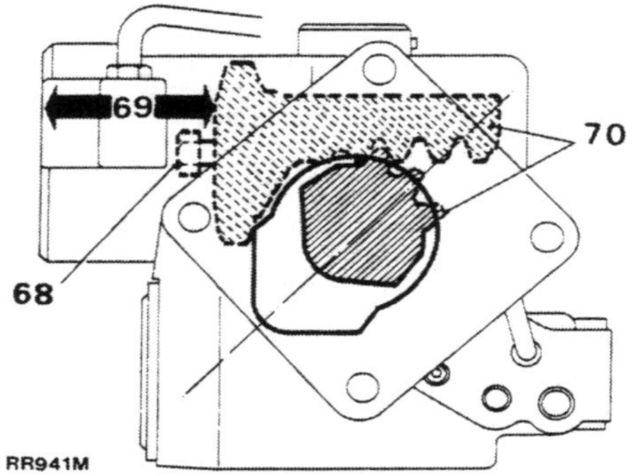

## Fitting the rack adjuster

71. Fit the sealing ring to the rack adjuster.
72. Fit the rack adjuster and thrust pad to engage the rack. Back off a half turn on the adjuster.
73. Loosely fit the nylon pad and adjuster grub screw assembly to engage the rack adjuster.
73. Loosely fit the nylon pad and adjuster grub screw assembly to engage the rack adjuster.

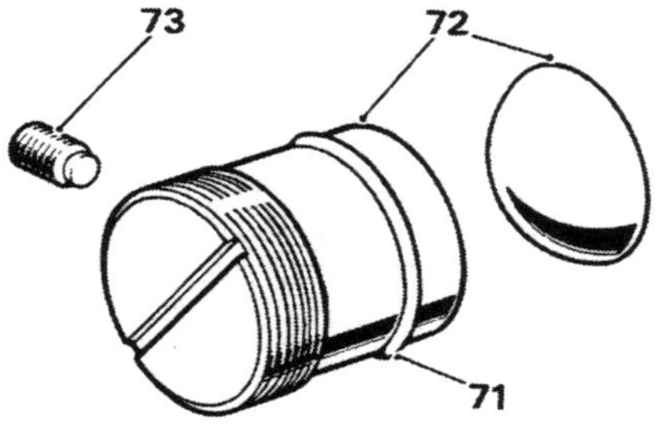

## Fitting the sector shaft cover

74. Fit the sealing ring to the cover.
75. Screw the cover assembly fully on to the sector shaft adjuster screw.
76. Position the cover on to the casing.
77. Tap home the cover. If necessary back off on the sector shaft adjuster screw to allow the cover to joint fully with the casing.

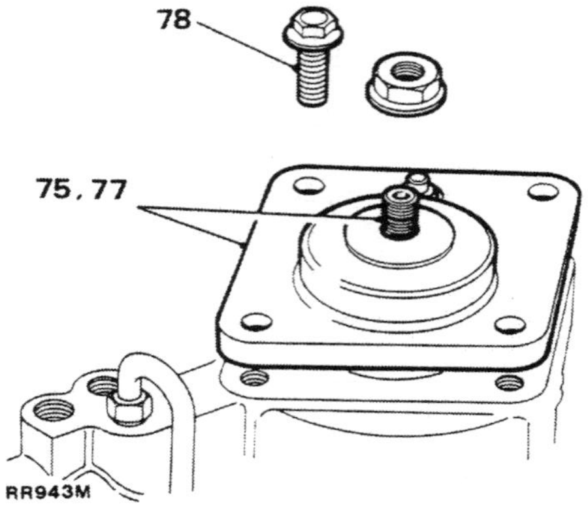

NOTE: Before tightening the fixings, rotate the input shaft about a small arc to ensure that the sector roller is free to move in the valve worm.

78. Fit the cover fixings and tighten to the correct torque.

## Fitting the cylinder cover

79. Fit the square section seal to the cover.
80. Remove the slave bolt and press the cover into the cylinder just sufficient to clear the retainer ring groove. 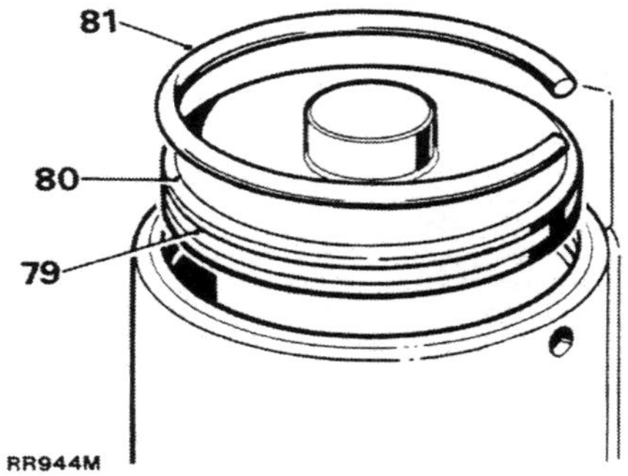
81. Fit the retainer ring to the groove with one end of the ring positioned 12 mm (0.5 in) approximately from the extractor hole.

## Adjusting the sector shaft

82. Set the worm on centre by rotating the input shaft half the total number of turns from either lock.
83. Rotate the sector shaft adjusting screw anti-clockwise to obtain backlash between the input shaft and the sector shaft.
84. Rotate the sector shaft adjusting screw clockwise until the backlash is just eliminated.
85. Measure and record the maximum rolling resistance at the input shaft, using a spring balance, cord and torque tool RO1016.
86. Hold still the sector shaft adjuster screw and loosely fit a new locknut.
87. Turn in the sector shaft adjuster screw until the figure recorded in instruction 88 is increased by 0.9 to 1.3 kg (2 to 3 lb) with the locknut tightened.

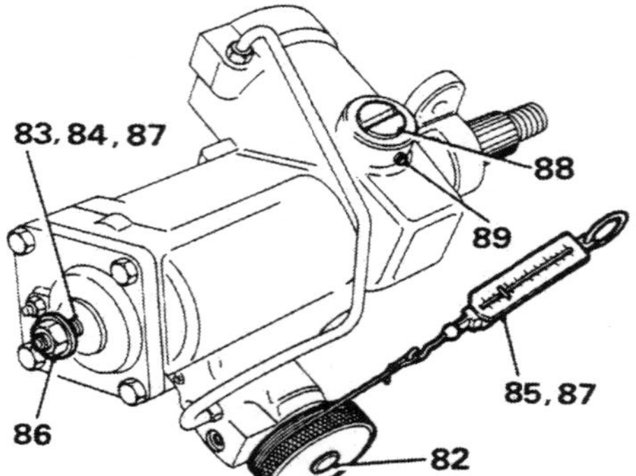

## Adjusting the rack adjuster

88. Turn in the rack adjuster to increase the figure recorded in 90 by 0.9 to 1.3 kg (2 to 3 lb). **The final figure may be less than that but must not exceed 7.25 kg (16 lb)**.
89. Lock the rack adjuster in position with the grub screw.

## Torque peak check

With the input shaft rotated from lock-to-lock, the rolling resistance torque figures should be greatest across the centre position (1½ turns approximately from full lock) and equally disposed about the centre position.

The condition depends on the value of shimming fitted between the valve and worm assembly inner bearing cup and the casing. The original shim washer value will give the correct torque peak position unless major components have been replaced.

NOTE: **During the following 'Procedure', the stated positioning and direction of the input shaft applies for both LH and RH boxes. However, the procedure for shim adjustment where necessary, differs between LH and RH steering boxes and is described under the applicable LH and RH stg. headings.**

## Procedure

90. With the input coupling shaft toward the operator, turn the shaft fully anti-clockwise.
91. Check the torque figures obtained from lock-to-lock using a spring balance cord and torque tool RO1016.

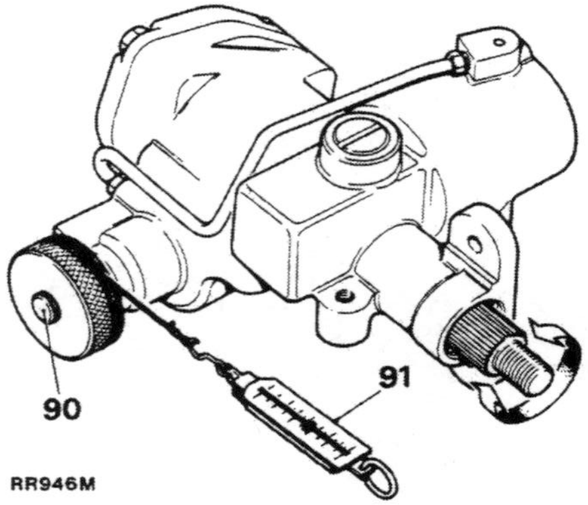

## Adjustments

92. Note where where the greatest figures are recorded relative to the steering position. If the greatest figures are not recorded across the centre of travel (i.e. steering straight-ahead position), adjust as follows:

   **LH steering models.** If the torque peak occurs before the centre position, add to the shim washer value; if the torque peak occurs after the centre position, subtract from the shim washer value.
   
   **RH steering models.** If the torque peak occurs before the centre position, subtract from the shim washer value; if the torque occurs after the centre position, add to the shim washer value.

Shim washers are available as follows:
0.03 mm, 0.07 mm, 0.12 mm and 0.24 mm (0.0015 in, 0.003 in, 0.005 in and 0.010 in).

NOTE: **Adjustment of 0.07 mm (0.003 in) to the shim value will move the torque peak area by ¼ turn approximately on the shaft.**

## STEERING

93. Fit the drop arm to the steering box using a new tab washer. Tighten the nut to the correct torque and bend over tab.
94. Refit the steering box to the vehicle.
95. Replenish the system with the correct grade of fluid. Refer to Recommended Lubricants and Power Steering System—bleed.
96. Test the system for leaks, with the engine running, by holding the steering hard on full lock in both directions.

NOTE: Do not maintain this pressure for more than 30 seconds in any one minute to avoid overheating the fluid and possibly damaging the seals.

97. Road test the vehicle.

### POWER STEERING SYSTEM

#### Bleed

1. Fill the steering fluid reservoir to the mark on the side of the reservoir with one of the recommended fluids.
2. Start and run the engine until it attains normal operating temperature.
3. Check and correct the reservoir fluid level.

NOTE: During the carrying out of items 4, 5 and 6, ensure that the steering reservoir is kept full. Do not increase the engine speed or move the steering wheel.

4. Run the engine at idle speed, slacken the bleed screw. When fluid seepage past the bleed screw is observed, retighten the screw.
5. Ensure that the fluid level is in alignment with the mark on the reservoir.
6. Wipe off all fluid released during bleeding.
7. Check all hose joints, pump and steering box for fluid leaks under pressure by holding the steering hard on full lock in both directions.

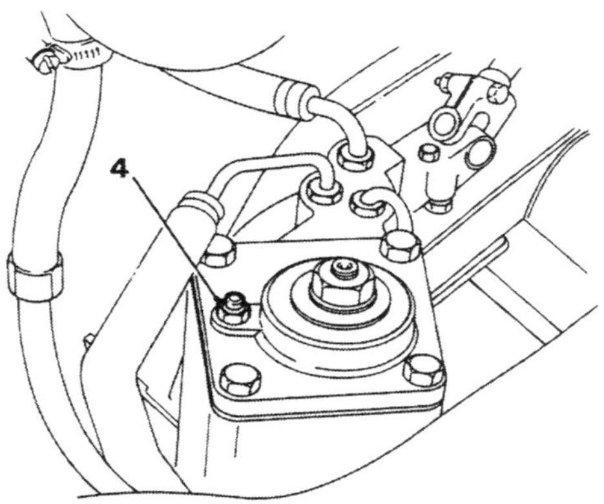

CAUTION: Do not maintain this pressure for more than 30 seconds in any one minute, to avoid causing the oil to overheat and possible damage to the seals. The steering should be smooth lock-to-lock in both directions, that is, no heavy or light spots when changing direction when the vehicle is stationary.

8. Carry out a short road test. If necessary, repeat the complete foregoing procedure.

### POWER STEERING SYSTEM

#### Test

If there is a lack of power assistance for the steering the pressure of the hydraulic pump should be checked first before renewing any components of the system. The fault diagnosis chart should also be used to assist in tracing faults in the power steering system.

#### Procedure

1. The hydraulic pressure test gauge is used for testing the power steering system. This gauge is calibrated to read up to 140 kgf/cm² (2000 lbf/in²) and the normal pressure which may be expected in the power steering system is 77 kgf/cm² (1100 lbf/in²).
2. Under certain fault conditions of the hydraulic pump it is possible to obtain pressures up to 105 kgf/cm² (1500 lbf/in²). Therefore, it is important to realise that the pressure upon the gauge is in direct proportion to the pressure being exerted upon the steering wheel. When testing, apply pressure to the steering wheel very gradually while carefully observing the pressure gauge.
3. Check, and if necessary replenish, the fluid reservoir.
4. Examine the power steering units and connections for leaks. All leaks must be rectified before attempting to test the system.
5. Check the steering pump drive belt for condition and tension, rectify as necessary.
6. Assemble the test equipment and fit to the vehicle.
7. Open the tap in the adaptor.
8. Bleed the system but exercise extreme care when carrying out this operation so as not to overload the pressure gauge.

9. With the system in good condition, the pressures should be as follows:
   (a) Steering wheel held hard on full lock and engine running at 1,000 rev/min, the pressure should be 70 to 77 kgf/cm² (1000 to 1100 lbf/in²).
   (b) With the engine idling and the steering wheel held hard on full lock, the pressure should be 28 kgf/cm² (400 lbf/in²) minimum.

These checks should be carried out first on one lock, then on the other.

CAUTION: Under no circumstances must the steering wheel be held on full lock for more than one minute, otherwise there will be a tendency for the oil to overheat and possible damage to the seals may result.

10. Release the steering wheel and allow the engine to idle. The pressure should be 7 kgf/cm² (100 lbf/in²).
11. If the pressures recorded during the foregoing test are outside the specified range, or pressure imbalance is recorded, a fault exists in the system. To determine if the fault is in the steering box or the pump, close the adaptor tap for a period not exceeding five seconds.
12. If the gauge fails to register the specified pressure, the pump is inefficient and the pump relief valve should be examined and renewed as necessary.
13. Repeat the foregoing test after renewing the relief valve and bleeding the system. If the pump still fails to achieve the specified pressures, the pump should be overhauled or a new unit fitted.
14. If pump delivery is satisfactory and low pressure or marked imbalance exists, the fault must be in the steering box valve and worm assembly.

### ADJUST POWER STEERING BOX

NOTE: The condition of adjustment which must be checked is one of minimum backlash without overtightness when the wheels are in the straight-ahead position.

1. Jack up the front of the vehicle until the wheels are clear of the ground.

WARNING: Wheels must be chocked in all circumstances.

2. Gently rock the steering wheel about the straight-ahead position to obtain the 'feel' of the backlash present. This backlash must not be more than 9.5 mm (0.375 in).
3. Continue the rocking action whilst an assistant slowly tightens the steering box adjuster screw after slackening the locknut until the rim movement is reduced to 9.5 mm (0.375 in) maximum.
4. Tighten the locknut, then turn the steering wheel from lock to lock and check that no excessive tightness exists at any point.
5. Lower the vehicle to ground level and remove the wheel chocks.
6. Road test the vehicle.

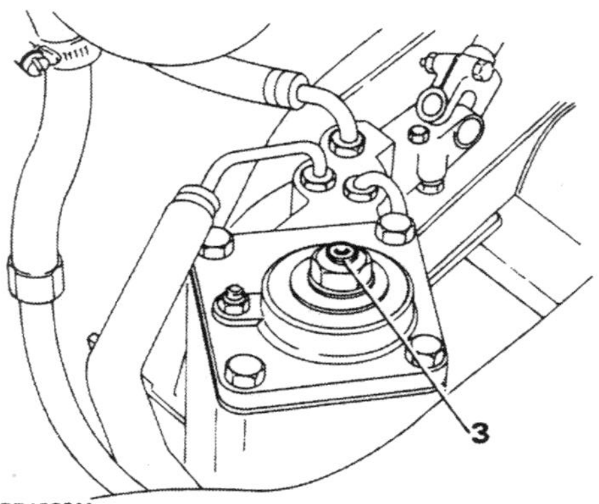

# 57 STEERING

## POWER STEERING FLUID RESERVOIR

### Remove and refit

#### Removing

1. Prop open the bonnet.
2. Remove the reservoir filler cap.
3. Disconnect the return hose from the steering box. Drain the fluid completely from the reservoir.

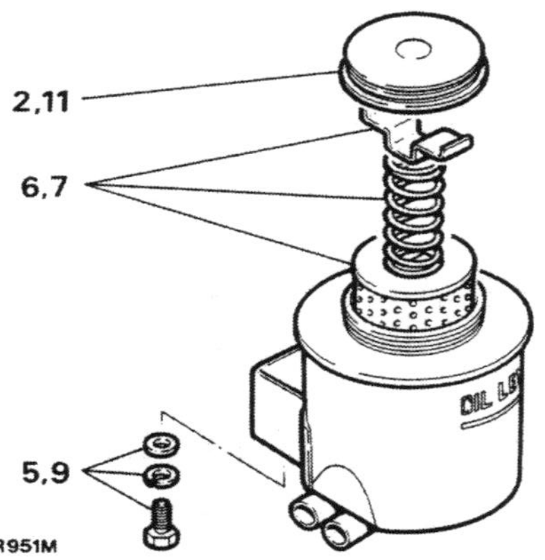

**CAUTION: it is most important that this fluid is not re-used.**

4. Refit the return hose to the steering box.
5. Remove the fixings attaching the reservoir to the wing valance. Disconnect the flexible hoses and withdraw the reservoir.

**NOTE: If the reservoir is not to be refitted immediately, the hoses must be sealed to prevent the ingress of foreign matter.**

6. Depress the spring retainer and withdraw the filter element.

#### Refitting

7. Fit a new filter element.
8. Refit and tighten the fixings attaching the reservoir to the wing valance.
9. Reconnect the flexible hoses to the reservoir. Tighten the clips.
10. Fill the reservoir to the prescribed level with one of the recommended fluids (Section 09) and bleed the power steering system. See power steering system—bleed.
11. Fit the reservoir filler cap.
12. Close the bonnet.

## POWER STEERING PUMP DRIVE BELT

### Adjust

#### Procedure

1. Prop open the bonnet.
2. Check, by thumb pressure, the belt tension between the crankshaft and the pump pulley. There should be a free movement of between 11 to 14 mm (0.437 to 0.562 in). If adjustment is necessary, proceed as follows:
3. Slacken the nut securing the pump mounting bracket to the cylinder head.
4. Slacken the bolt securing the pump lower bracket to the slotted adjustment link.
5. Pivot the pump as necessary and adjust until the correct belt tension is obtained.
6. Maintaining the tension, tighten the pump adjusting bolt and the top pivot nut.

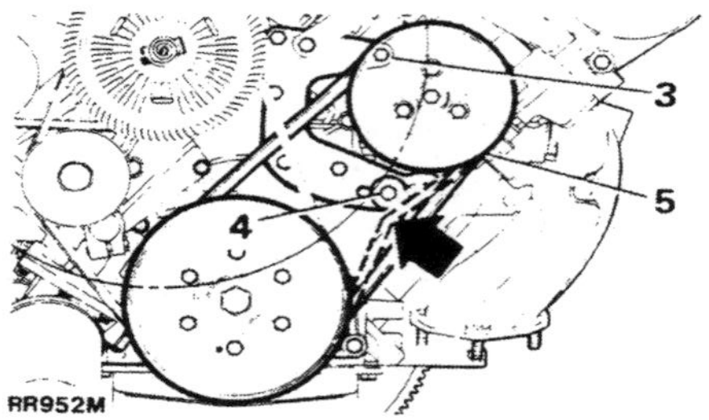

**NOTE: Check the alternator drive belt tension after adjusting the power steering pump belt.**

7. Close the bonnet.

## POWER STEERING PUMP DRIVE BELT

### Remove and refit

#### Removing or preparing for the fitting of a new belt

1. Prop open the bonnet.
2. Slacken the jockey pulley bolt and remove the fan belt.
3. Slacken the alternator mountings and remove the drive belt.
4. Slacken the power steering pump mountings.
5. Pivot the pump and remove the drive belt.

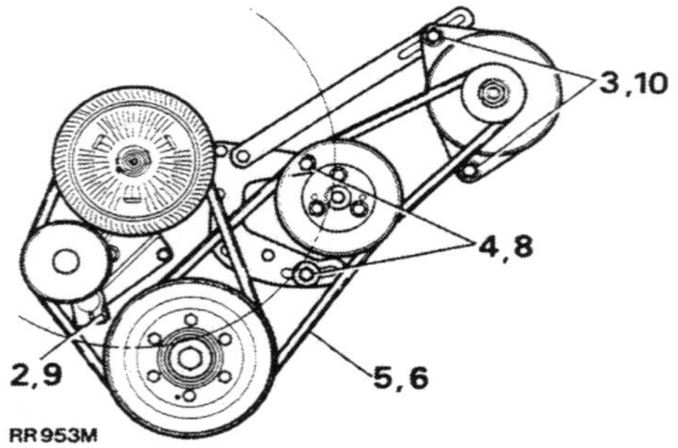

# STEERING 57

5. Replace the reservoir filler cap.
6. Disconnect the outlet hose from the pump. Blank off the orifices to prevent ingress of foreign matter.
7. Slacken and remove the adjuster bolt below the pulley.
8. Slacken and remove the nut on the front mounting bracket.
9. Remove four bolts and withdraw the front mounting bracket complete with the alternator adjusting link.
10. Slide off the power steering pump.

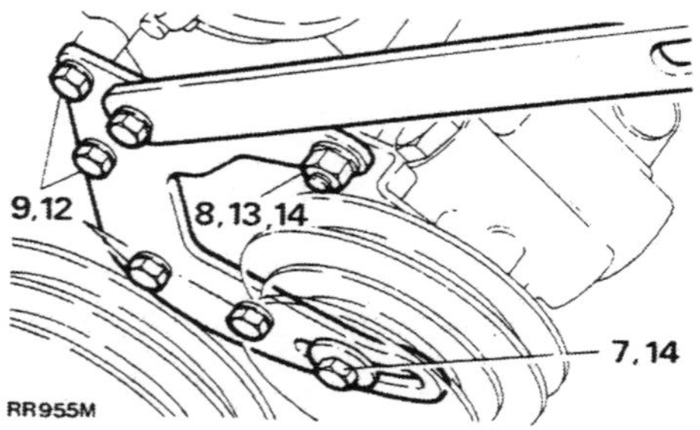

### Refitting

6. Locate the driving belt over the crankshaft and pump pulleys.
7. Adjust the position of the pump to give a driving belt tension of 11 to 14 mm (0.437 to 0.562 in) movement when checked by thumb pressure midway between the crankshaft and pump pulleys.
8. Maintaining the tension, tighten the pump adjusting bolt and the top pivot nut.
9. Refit the fan belt and adjust the tension to give 11 to 14 mm (0.437 to 0.562 in) movement when checked by thumb pressure midway between the crankshaft and water pump pulleys.
10. Refit the alternator drive belt and adjust to give 11 to 14 mm (0.437 to 0.562 in) movement when checked midway between the power steering pump and alternator pulleys.

**NOTE: Whenever a new belt is fitted, it is most important that its adjustment is rechecked after approximately 1.500 km (1,000 miles) running.**

11. Close the bonnet.

## POWER STEERING PUMP

### Remove and refit

#### Removing

1. Prop open the bonnet.
2. Remove the alternator.
3. Remove the filler cap from the power steering fluid reservoir.
4. Disconnect the inlet hose from the pump, retaining the copper crimp washer and rubber gasket. Drain the fluid into a suitable container and blank off the orifices against the ingress of dirt.

**CAUTION: Under no circumstances must the fluid be re-used.**

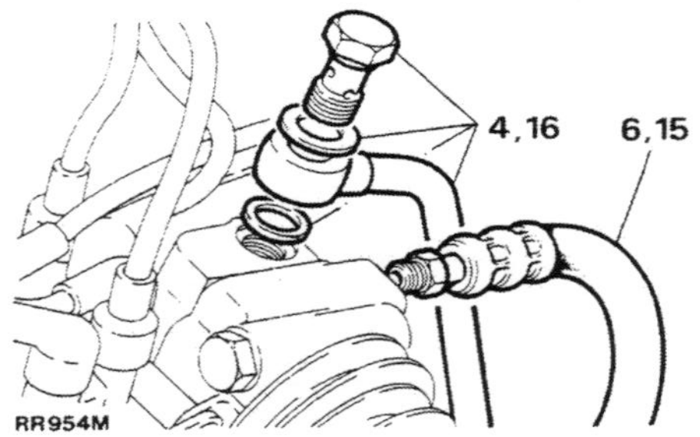

### Refitting

11. Offer up the power steering pump, locating the driving belt over the pulley.
12. Refit the pump mounting bracket and tighten the four bolts.
13. Refit the nut to the front mounting bracket.
14. Adjust the position of the pump to give a driving belt tension of 11 to 14 mm (0.437 to 0.562 in) movement when checked by thumb pressure midway between the crankshaft and pump pulleys. Fit and tighten the pump adjuster bolt below the pulley, also tighten the nut at the front mounting bracket.
15. Remove the blank and reconnect the outlet hose to the pump.

*Continued*

13

# STEERING

16. Replace the square sectioned rubber gasket to the groove around the inlet port and replace the inlet hose to the pump. Tighten the union bolt to the correct torque.
17. Fit the alternator, fit and adjust the drive belt to the correct tension.
18. Remove the filler cap from the steering fluid reservoir.
19. Fill the steering fluid reservoir to the level mark with one of the recommended fluids (Section 09).
20. Replace the filler cap.
21. Bleed the power steering system.
22. Test the steering system for leaks with the engine running, by holding the steering hard on full lock in both directions.

**CAUTION: Do not maintain this pressure for more than 30 seconds in any one minute, to avoid causing the oil to overheat and possible damage to the seals.**

23. Close the bonnet.
24. Road test the vehicle.

## POWER STEERING PUMP OVERHAUL—Series 30

### Dismantle

1. Remove the steering pump from the vehicle.
2. Clean the exterior of the pump and drain off any oil.
3. Remove the bolt, spring washer and large plain washer securing the pulley to the pump shaft.
4. Using a suitable puller, withdraw the pulley. Do not attempt to hammer the shaft from the pulley, or lever the pulley from the shaft, as this may cause internal damage.
5. Withdraw the square key from the shaft.
6. Remove the four bolts and spring washers securing the bearing retainer plate and front mounting plate to the pump body. Remove the plates.
7. Remove the three bolts and spring washers securing the rear mounting plate to the pump body and remove the plate.
8. Clamp the pump body in a vice, ensuring that the jaws are protected.
9. Remove the blank from the inlet port.

**NOTE: The tubular steel venturi flow director under the inlet adaptor is pressed into the cover and should not be removed.**

10. Remove the six Allen screws securing the cover to the pump body. Separate the cover from the body vertically to prevent the parts falling out.
11. Remove the pump from the vice.
12. Remove the 'O' ring seals from the grooves in the pump body.
13. Carefully tilt the pump body, and remove the six rollers.
14. Draw the carrier off the shaft, and remove the drive pin.
15. Remove the shaft from the body.
16. Remove the cam and the cam lock peg from the pump body.

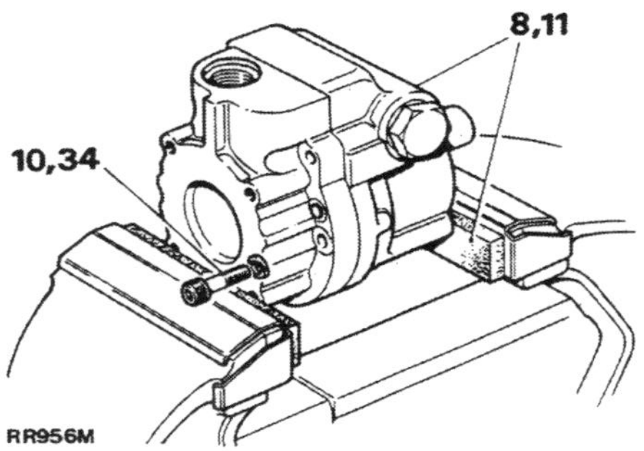

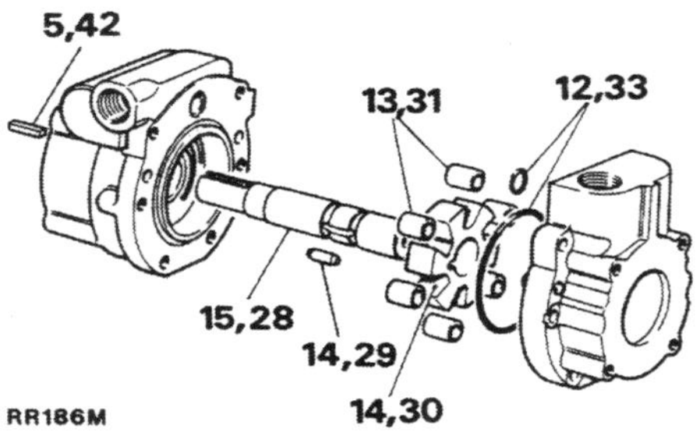

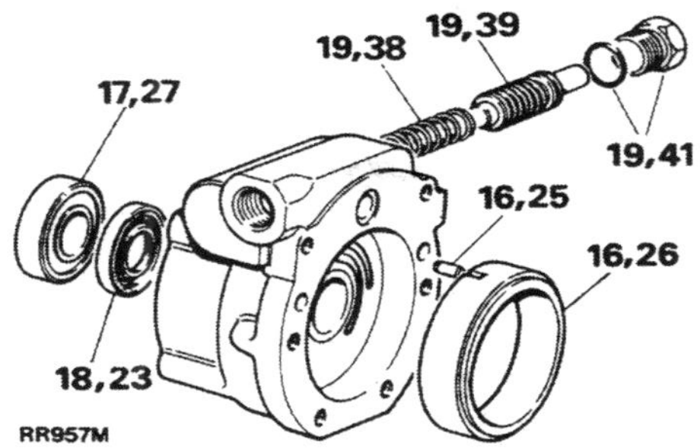

### Inspection

20. Wash all parts in a suitable solvent, air dry, or wipe clean with a lint-free cloth if air is not available.
21. Check the pump body and cover for wear. Renew either part, if faces or bushes are worn.
22. Check the pump shaft around the drive pin slot. Remove any burrs.

**NOTE: Ensure that the aluminium restrictor in the output port is thoroughly cleaned but not dislodged.**

### Reassemble

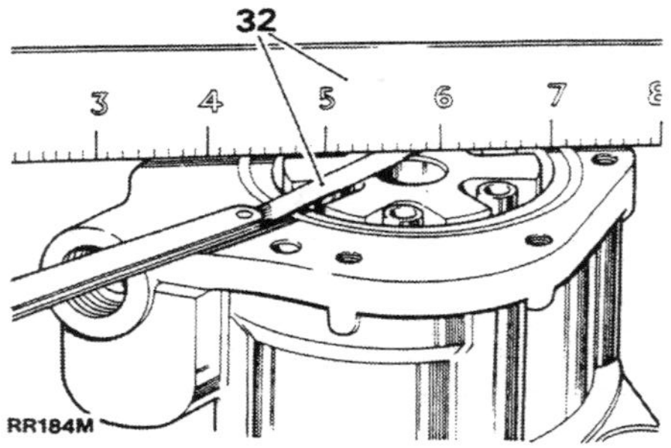

23. Carefully examine a new shaft seal to ensure that it is clean and undamaged. Smear the sealing lips with grease and apply a fine smear of 'Wellseal' to the pump body where the outside diameter of the oil seal locates (applies to metal cased seals). Place the seal square to the housing recess with the lip towards the inside of the housing.
24. Press the seal into position approximately 0.80 mm ($\frac{1}{32}$ in) below the seal housing face, ensuring that it does not tilt.
25. Replace the cam lock peg into the location in the body.
26. Renew the cam if worn or damaged. Refit the cam, ensuring that it seats correctly in the body and that the slot locates over the locking peg.
27. Fit a new sealed bearing onto the pump shaft.
28. Insert the shaft and bearing assembly into the seal side of the body.
29. Refit the carrier drive pin in the shaft.
30. Replace the carrier and replace in position, ensuring that the greater angle on the carrier teeth is in the leading position as illustrated.
31. Inspect the rollers, paying particular attention to the finish on the end. Renew the rollers if scored, damaged or oval. Refit the rollers to the carrier.
32. Using a straight edge across the cam surface, and a feeler gauge, check the end clearance of the carrier and rollers in the pump body. If the end clearance is more than 0.05 mm (0.002 in) renew the carrier and rollers.
33. Smear a fine trace of Loctite 275 to the pump body in a 'figure of 8' outside the 'O' rings and inside the bolt holes. Install new 'O' rings to the body of the pump.
34. Refit the cover on the pump body and secure with six Allen screws and spring washers.
35. Tighten the Allen screws, in diagonal sequence, checking that the shaft rotates freely and does not bind. Finally, tighten to the correct torque.
36. Refit the rear mounting plate to the pump body and secure with three bolts and spring washers.
37. Refit the front mounting plate and the bearing retainer plate to the pump body and secure with four bolts and spring washers.
38. Refit the flow control valve spring in the bore. The spring tension should be 8 to 9 lbf (11 to 12 Nm) at 21 mm (0.820 in). If not, renew the spring.
39. Replace the valve in the bore, inserting the spring so that the exposed ball end enters last. Ensure that the valve is not sticking.
40. Renew the 'O' ring on the valve cap and assemble in the pump. Tighten the cap to the correct torque.
41. Refit the pulley key.
42. Refit the pulley to the shaft and secure with the special washer, spring and washer and bolt. Tighten the bolt to the correct torque.
43. Refit the steering pump to the vehicle.

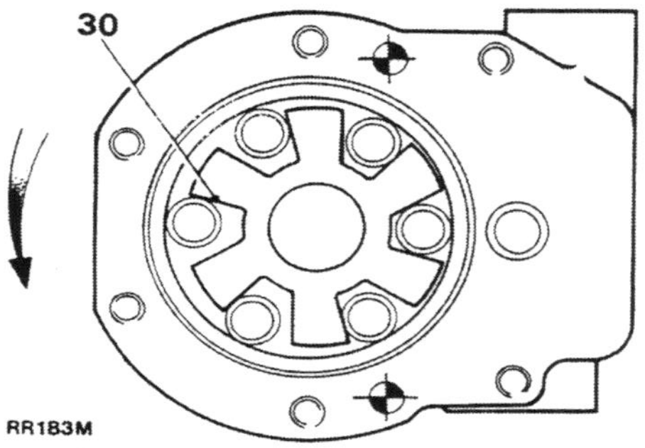

# STEERING

## STEERING COLUMN

### Remove and refit

Service tool: 18G1014 Extractor for steering wheel

**IMPORTANT:** The steering column is of a 'safety' type and incorporates shear pins. Therefore do not impart shock loads to the steering column at any time.

### Removing

1. Remove the steering wheel using extractor 18G1014.
2. Remove the lower fascia panel, driver's side.
3. Disconnect the electrical leads from the steering column switches.
4. Remove the top pinch bolt, universal joint to steering column.
5. Remove the fixings, steering column to toe board.
6. Remove the fixings, steering column to dash bracket.
7. Withdraw the steering column assembly.

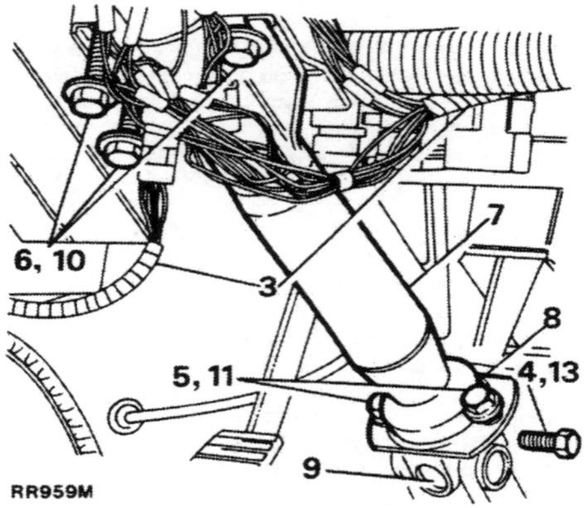

### Refitting

8. Position the sealing gasket on the end of the column assembly.
9. Feed the steering shaft through the tow board and engage the drive splines at the coupling shaft.
10. Loosely fit the column upper fixings.
11. Tighten upper and lower fixings.
12. Tighten upper and lower fixings.
13. Fit universal joint pinch bolt, and tighten to the correct torque.
14. Reverse 1 to 3.

## COUPLING SHAFT AND UNIVERSAL JOINTS

### Remove and refit

### Removing

1. Note the position of the steering wheel spokes.
2. Remove one pinch bolt from the top universal joint to the steering column.
3. Remove two pinch bolts from the lower universal joint.
4. Withdraw the shaft and universal joints.
5. Withdraw the lower universal joint from the shaft.

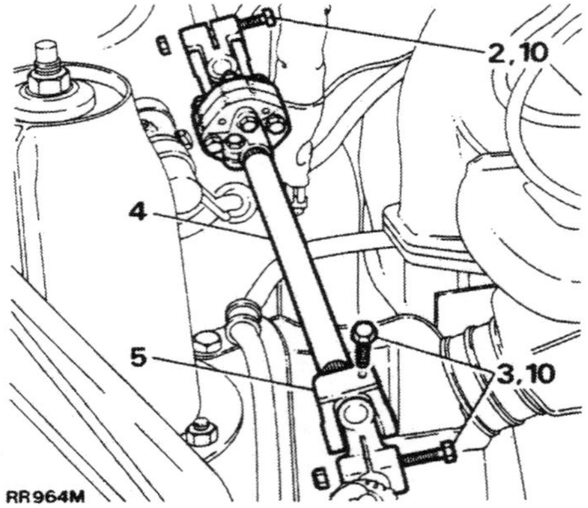

**NOTE:** Do not dismantle the upper coupling joint. The steering shaft, rubber coupling and top universal joint is only available as an assembly.

### Refitting

6. Position the lower universal joint on the shaft.
7. Offer the shaft assembly to the steering column, aligning the pinch bolt hole with the flat on the column.
8. Position the steering wheel as noted in 1.
9. Fit the lower universal joint to the steering box shaft.
10. Fit the pinch bolts, and tighten to the correct torque.

---

## POWER STEERING

### FAULT DIAGNOSIS

| SYMPTOM | CAUSE | TEST ACTION | CURE |
|---|---|---|---|
| INSUFFICIENT POWER ASSISTANCE WHEN PARKING | (1) Lack of fluid | Check hydraulic fluid tank level | If low, fill and bleed the system |
| | (2) Engine idling speed too low | Try assistance at fast idle | If necessary, reset idle speed |
| | (3) Driving belt slipping | Check belt tension | Adjust the driving belt |
| | (4) Defective hydraulic pump and/or pressure relief valve | (a) Fit pressure gauge between high pressure hose and steering pump, with steering held hard on full lock, see Note 1 below, and 'Power steering system test' | If pressure is outside limits (high or low) after checking items 1 and 3, see Note 2 below. |
| | | (b) Release steering wheel and allow engine to idle. See 'Power steering system test' | If pressure is greater, check steering box for freedom and self-centring action |
| POOR HANDLING WHEN VEHICLE IS IN MOTION | Lack of castor action | | It is most important that this screw is correctly adjusted. See instructions governing adjustment |
| | Steering too light and/or over-sensitive | Check for loose torsion bar fixings on steering box valve and worm assembly | Fit new valve and worm assembly |
| HYDRAULIC FLUID LEAKS | Damaged pipework, loose connecting unions, etc. | Check by visual inspection; leaks from the high pressure pipe lines are best found while holding the steering on full lock with engine running at fast idle speed (see Note 1 below) | Tighten or renew as necessary |
| | | Check 'O' rings on pipework | Renew as necessary |
| | | (Leaks from the steering box tend to show up under low pressure conditions, that is, engine idling and no pressure on steering wheel) | |
| EXCESSIVE NOISE | (1) If the high pressure hose is allowed to come into contact with the body shell, or any component not insulated by the body mounting, noise will be transmitted to the car interior | Check the loose runs of the hoses | Alter hose route or insulate as necessary |
| | (2) Noise from hydraulic pump | Check oil level and bleed system | If no cure, change hydraulic pump |

Note 1: Never hold the steering wheel on full lock for more than 30 seconds in any one minute, to avoid causing the oil to overheat and possible damage to the seals.

Note 2: High pressure — In general it may be assumed that excessive pressure is due to a faulty relief valve in the hydraulic pump.
Low pressure — Insufficient pressure may be caused by one of the following:

1. Low fluid level in reservoir } Most usual cause of insufficient pressure
2. Pump belt slip
3. Leaks in the power steering system
4. Faulty relief valve in the hydraulic pump
5. Faulty steering box valve and worm assembly
6. Leak at piston sealing in steering box
7. Worn components in either steering box or hydraulic pump

# STEERING 57

## STEERING COLUMN ASSEMBLY

### Overhaul

1. Remove the steering column assembly.

### Dismantling

2. Remove the lighting switch from the lower shroud.
3. Remove the direction indicator switch from the column assembly.
4. Remove the wiper/washer switch from the column assembly.

### Removing the steering column lock assembly, 5 to 8

5. Drill a hole in each sheared bolt to accept an 'easy-out' extractor.
6. Remove the sheared bolts.
7. Detach the end cap.
8. Withdraw the column lock assembly.
9. Lift off the cam for the self-cancelling switch from the column.

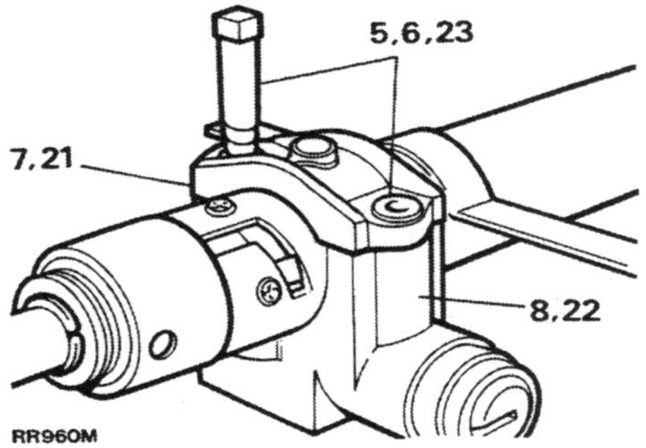

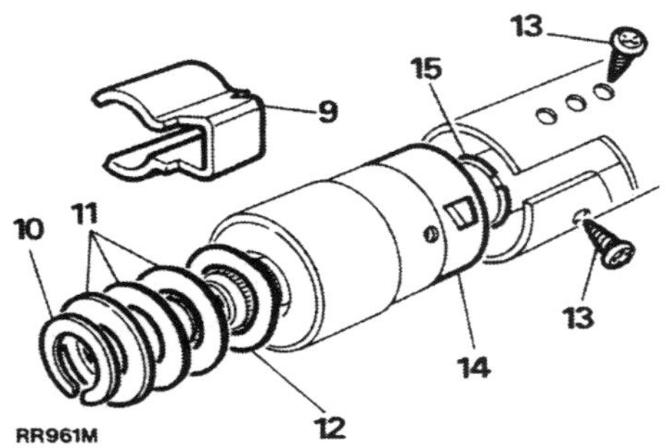

### Removing the top bearing assembly, 10 to 15

10. Remove the circlip.
11. Withdraw the thrust washer and shim washer(s).
12. Withdraw the wave washer.
13. Remove the fixings, top bearing assembly to steering column.
14. Withdraw the top bearing assembly.
15. Remove the retaining ring from groove in the steering column.
16. If required, renew the mesh cover, using heat fusing to join together the replacement cover edges.

**IMPORTANT: The steering column is now dismantled as far as is permitted. A replacement steering column comprises outer column, inner column and lower bearing assembly.**

### Assembling

17. Reverse 10 to 15.

### Checking the top bearing end load, items 18 to 20

18. Hold down the thrust washer and shims fully against the wave washer spring load.
19. Measure the clearance between the thrust washer and the circlip. The clearance must be 0.12 mm (.005 in). 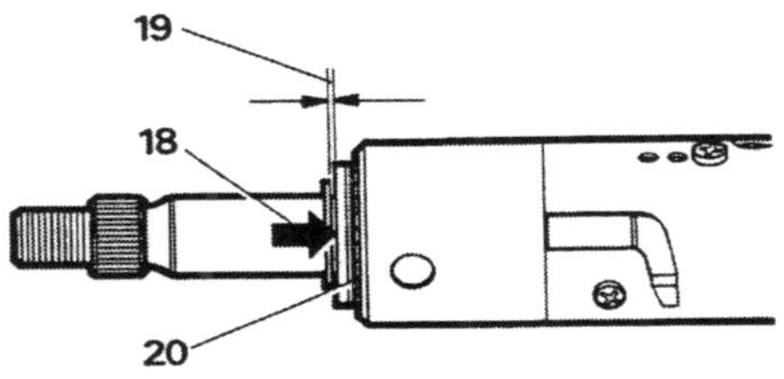
20. Adjust the clearance as necessary by fitting replacement shim washer(s) which are available in the range 0.127 to 0.762 mm (.005 to .030 in) in 0.127 (.005) stages.

### Fitting the steering column lock

21. Position the steering lock cap on the outer column, locating the spigot in the hole provided.
22. Offer the lock assembly to the column.
23. Fit the shear bolts to retain the cap and lock assembly.
24. Tighten the bolts sufficient to shear the heads.
25. Reverse 1 to 4.

---

8. Remove the sheared bolts.
9. Detach the end cap.
10. Withdraw the column lock assembly.
11. Remove the ignition/starter switch.

### Refitting

12. Position the steering lock cap on the outer column, locating the spigot in the hole provided.
13. Offer the lock to the column.
14. Fit the shear bolts to retain the cap and lock.
15. Tighten the bolts sufficient to shear off the heads.
16. Fit the ignition/starter switch.
17. Reverse 1 to 6.

## STEERING COLUMN LOCK AND IGNITION/STARTER SWITCH

### Remove and refit

**IMPORTANT: The steering column is of a 'safety' type and incorporates shear pins. Therefore do not impart shock loads to the steering column at any time.**

Ignition/starter switch only remove and refit is described in Electrical Section 86.

### Removing

1. Remove the lower fascia panel, driver's side.
2. Remove the top shroud.
3. Remove the lighting switch from the lower shroud.
4. Remove the steering column fixings at the toe board.

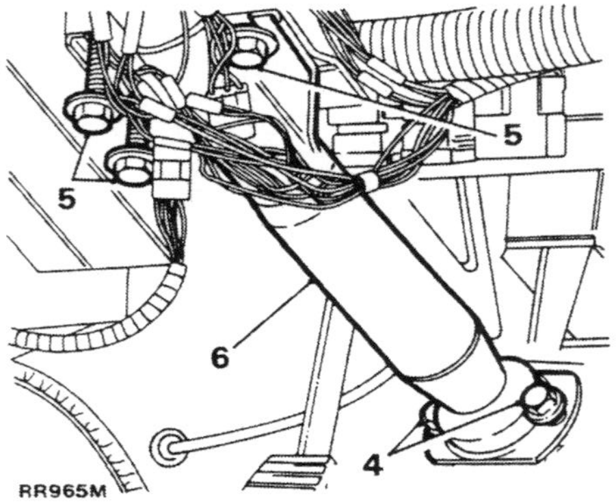

5. Remove the fixings, steering column to dash bracket.
6. Lower the steering column to gain access to the column lock fixings and remove the insulating cover.
7. Drill a hole in each sheared bolt to accept an 'Easiout' extractor.

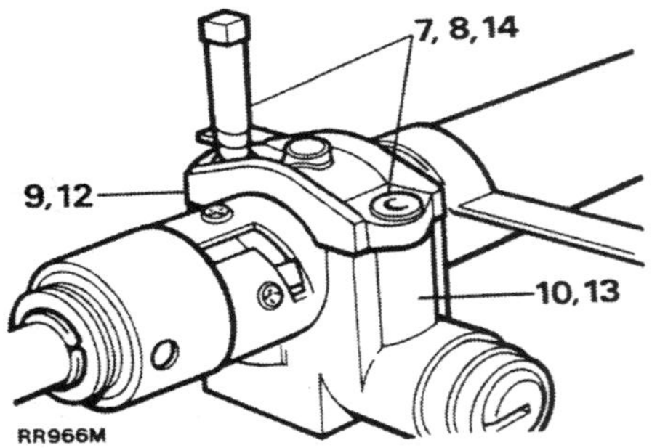

---

## STEERING WHEEL

### Remove and refit

Service tools: 18G 1014—Steering wheel remover
18G 1014-2—Adaptor pins

**IMPORTANT: The steering column is of a 'safety' type and incorporates shear pins. Therefore do not impart shock loads to the steering column during removing and refitting the steering wheel or at any time.**

### Removing

1. Remove the securing screw and withdraw the trim pad.
2. Remove the steering wheel nut and washer.
3. Extract the steering wheel using service tool 18G 1014. Ensure the adaptor pins are inserted to their full length and tighten the cap screws.

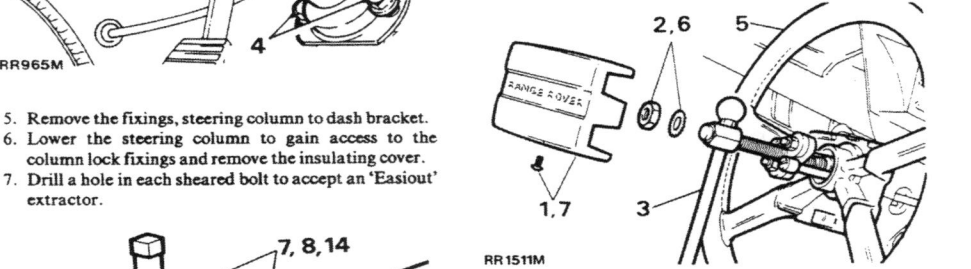

### Refitting

4. Position the road wheels 'straight ahead'.
5. Position the steering wheel on the column splines, in the straight ahead position.
6. Fit the nut and washer, tighten to the correct torque.
7. Fit the trim pad and tighten the securing screw.

# STEERING 57

## DROP ARM

### Remove and refit

Service tools: 18G1063 Ball joint extractor
MS252A Drop arm extractor

### Removing

1. Disconnect the drag link from the drop arm ball joint, using extractor 18G1063.

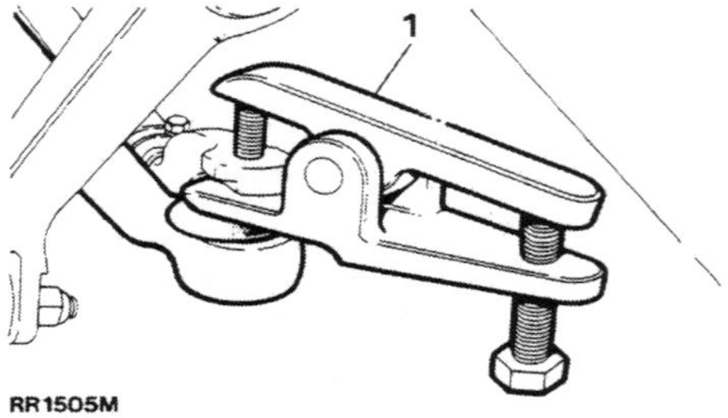

2. Remove the drop arm from the steering box rocker shaft, using extractor MS252A.

NOTE: The drop arm ball joint is integral with the drop arm.

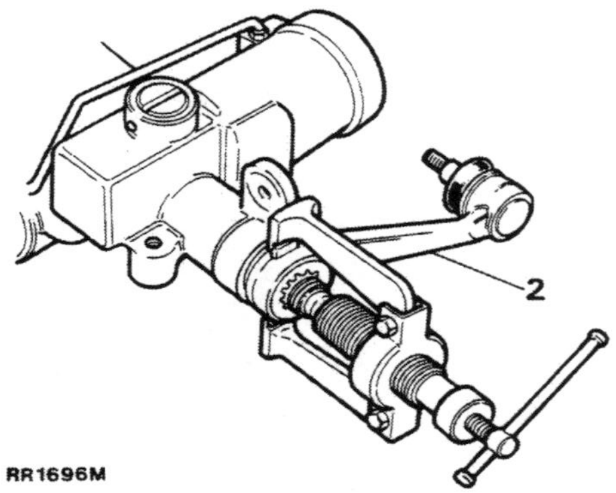

### Refitting

3. Set the steering box in the midway lock-to-lock position.
4. Fit the drop arm in position, aligning the dead splines.
5. Fit the drop arm fixings and tighten to the correct torque.
6. Fit the drag link and tighten to the correct torque.

## DROP ARM BALL JOINT

### Overhaul

The drop arm ball joint can be overhauled and there is a repair kit available which consists of the following items.

| | |
|---|---|
| Ball pin | Ball lower socket |
| Retainer | Spring |
| Spring rings | 'O' ring |
| Dust cover | Cover-plate |
| Ball top socket | Circlip |

### Dismantle

1. Remove the drop arm from the vehicle and clean the exterior.
2. Remove the spring rings and prise off the dust cover.
3. In the interests of safety, position the ball joint under a press to relieve the spring tension and support the housing both sides of the ball pin, as illustrated. Apply pressure to the cover plate and remove the circlip and slowly release the pressure.

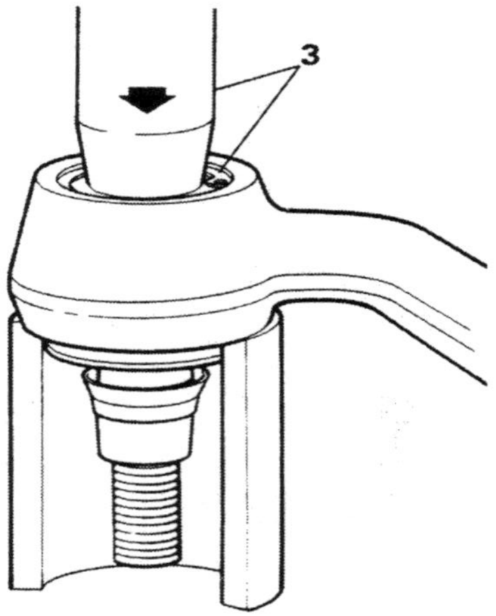

WARNING: Personal injury could result if the circlip is removed without pressure being applied and maintained to the cover plate.

4. Remove the spring, top socket and 'O' ring.

5. Since the ball pin cannot be removed with the retainer in position, tap the threaded end of the ball pin to release the retainer and to remove the pin from the housing.

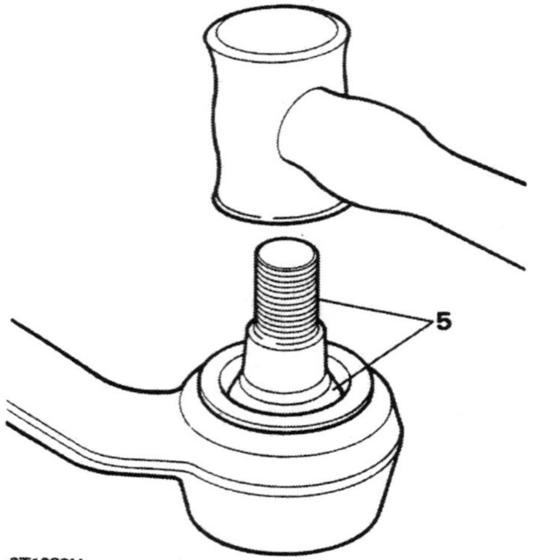

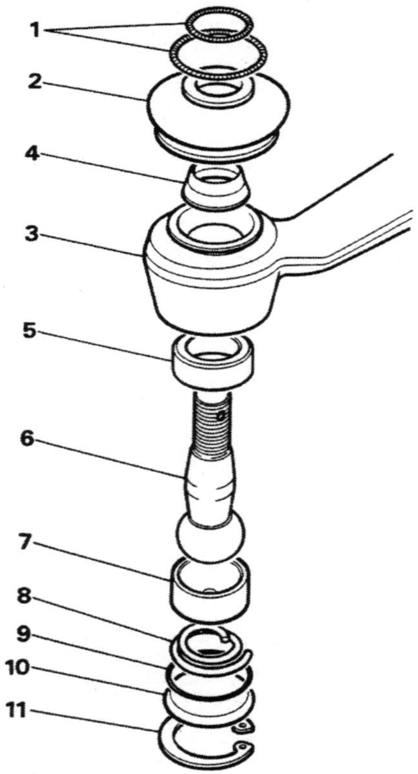

### KEY TO BALL JOINT

1. Spring rings
2. Dust cover
3. Ball housing
4. Retainer
5. Bottom socket
6. Ball pin
7. Top socket
8. Spring
9. 'O' ring
10. Cover-plate
11. Circlip

*Continued*

# STEERING 57

6. Using a sharp-edged punch or chisel, drive the ball lower socket from the housing. Should difficulty be experienced, apply gentle heat to the housing and then continue to drive the socket from the housing.

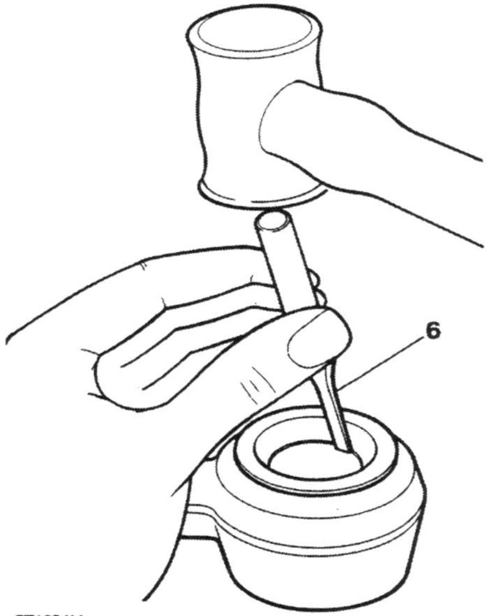

7. Clean the housing and remove any burrs.

## Assemble

8. Press in the lower socket squarely up to the shoulder.
9. Dip the ball in Duckhams LB10 grease, or equivalent and fit to the housing and pack with grease.
10. Fit the top socket.
11. Fit the spring, small diameter towards the ball.
12. Fit the 'O' ring and using the same method as for removing the circlip, compress the cover plate and secure with the circlip. Ensure that the circlip is fully seated in the machined groove.
13. Press the retainer onto the ball pin so that the top edge is level with the edge of the taper.

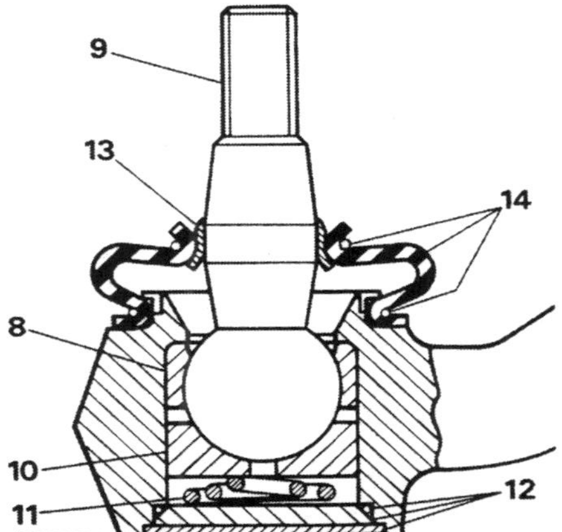

## TRACK ROD AND LINKAGE

### Remove and refit

Service tool: 18G 1063—Ball joint extractor

### TRACK ROD

### Removing

1. Jack-up and support chassis.
2. Disconnect the steering damper at the track rod.
3. Disconnect the track rod at the ball joints, using extractor 18G 1063.
4. Withdraw the track rod complete.

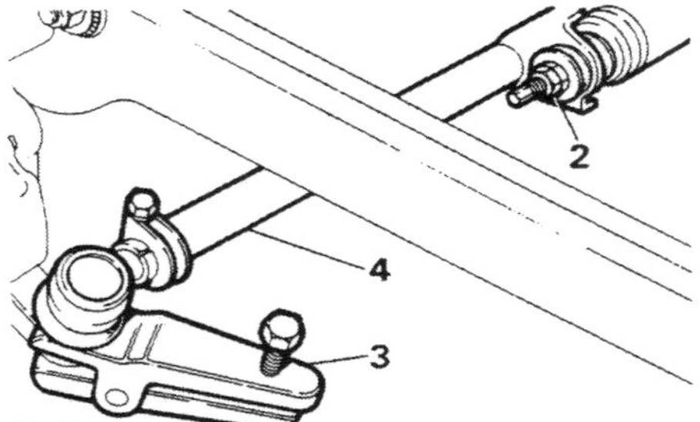

### LINKAGE

### Removing

5. Slacken the clamp bolts.
6. Unscrew the ball joints.
7. Unscrew the track rod adjuster, left-hand thread.

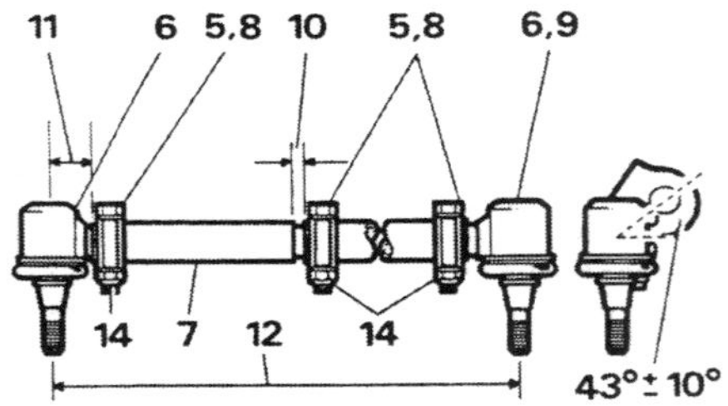

## Refitting

8. Fit the replacement parts. Do not tighten the clamp pinch bolts at this stage.
9. Screw in a ball joint to the full extent of the threads.
10. Set the adjuster dimensionally to the track rod as illustrated, to 8.9 mm (.350 in).
11. Set the adjuster end ball joint dimensionally, as illustrated to 28.57 mm (1.125 in).
12. The track rod effective length of 1230.0 mm (48.4 in) is subject to adjustment during the subsequent wheel alignment check.

## TRACK ROD

### Refitting

13. Fit the track rod and tighten the ball joint nuts to the correct torque.
14. Check the front wheel alignment.
15. Reverse 1 and 2.

**CAUTION: A track rod that is damaged or bent must be renewed. No attempt should be made to repair or straighten it.**

## STEERING DAMPER

### Remove and refit

### Removing

1. Remove the fixings at the differential case bracket.
2. Remove the fixings at the track rod bracket. 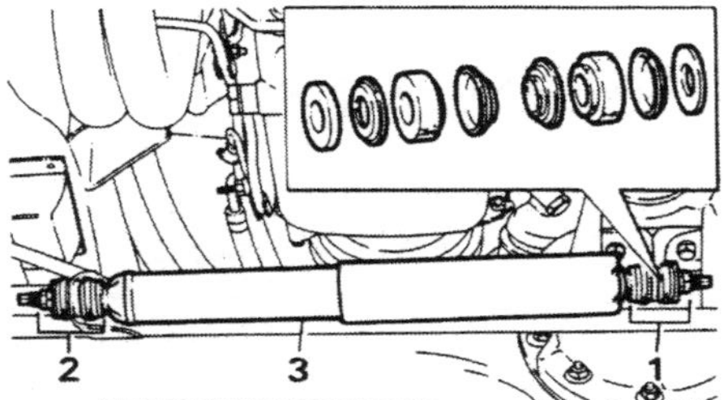
3. Withdraw the steering damper.

### Refitting

4. Reverse 1 to 3.

14. Fit the dust cover and retain with the two spring rings.
15. Fit the drop arm to the steering box using a new lock washer. Tighten the retaining nut to the correct torque and bend over the lock washer.
16. Assemble the ball pin to the drag link, see instructions for fitting drag link and track rod, and tighten the castle nut to the correct torque and secure with a new split pin.

## STEERING

### DRAG LINK AND DRAG LINK ENDS

**Remove and refit**

Service tool 18G 1063 — Extractor for ball joint

**Removing**

1. Jack-up and support chassis.
2. Remove the passenger's side road wheel.
3. Disconnect the drag link ball joint at the swivel housing arm, using extractor 18G 1063.
4. Disconnect the drag link end at the drop arm ball joint, using extractor 18G 1063.
5. Withdraw the drag link.

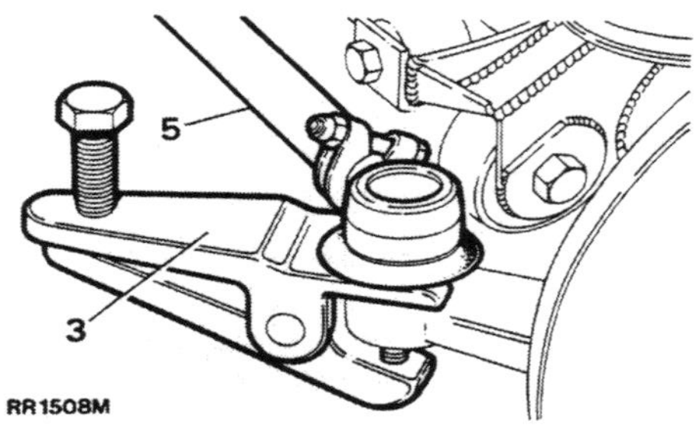

**DRAG LINK ENDS**

**Removing**

6. Slacken the clamp bolts.
7. Unscrew the ball joint.
8. Unscrew the cranked end.

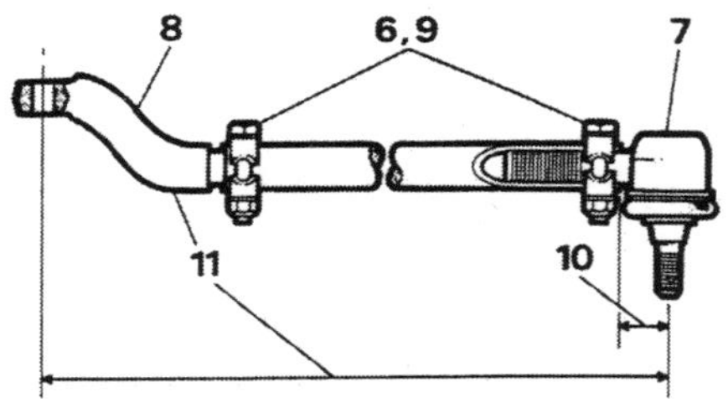

**Refitting**

9. Fit the replacement ends. Do not tighten the clamp bolts at this stage.
10. Set the ball joint dimensionally to the drag link, as illustrated, to 28.57 mm (1.125 in).
11. Adjust the cranked end to obtain the nominal overall length of 919.0 mm (36.2 in). The length is finally adjusted during refitting.

**DRAG LINK**

**Refitting**

12. Fit the drag link. Tighten the ball-joint nuts to the correct torque.
13. Check, and if necessary, set the steering lock stops.
14. Turn the steering and ensure that full travel is obtained between the lock stops. Adjust the drag link length to suit.
15. Using a mallet, tap the ball joints in the direction indicated so that both ball pins are in the same angular plane.
16. Tighten the clamp bolts to the correct torque. 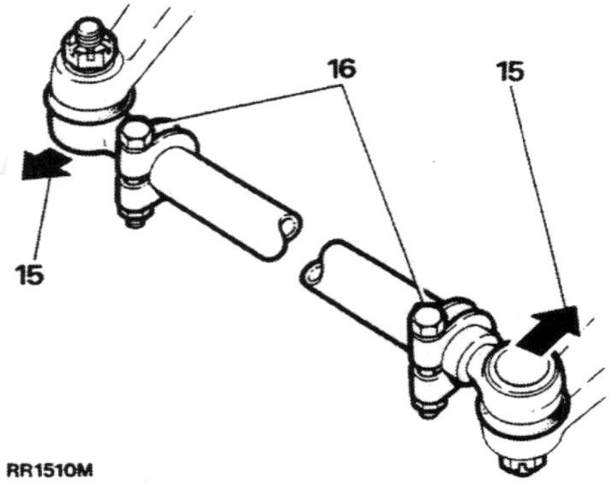
17. Reverse 1 and 2.

CAUTION: A drag link that is damaged or bent must be renewed. No attempt should be made to repair or straighten it.

### STEERING LOCK STOPS

**Check and adjust**

**Checking**

1. Measure the stop bolt protrusion as illustrated. This must be 40.5 mm (1.59 in).

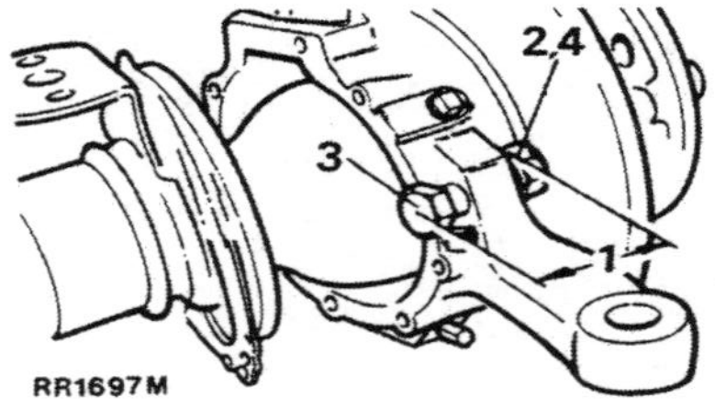

**Adjusting**

2. Slacken the stop bolt locknut.
3. Turn the stop bolt in or out as required.
4. Tighten the locknuts.
5. Check the wheels position at full lock.

### FRONT WHEEL ALIGNMENT

**Check and adjust**

**Checking**

Toe-out dimensions

NOTE: No adjustment is provided for castor, camber or swivel pin inclination.

1. Set the vehicle on level ground with the road wheels in the straight-ahead position.
2. Push the vehicle back then forwards for a short distance to settle the linkage.
3. Measure the toe-out at the horizontal centre-line of the wheels.
4. Toe-out must be 1.2 to 2.4 mm (.04 to .09 in).
5. Check-tighten the clamp bolts fixings to the correct torque.

**Adjusting**

6. Slacken the adjuster sleeve clamp bolts.
7. Rotate the adjuster to lengthen or shorten the track rod.
8. Check the toe-out setting as in instructions 1 to 4. When the toe-out is correct, lightly tap the steering linkage ball joints, in the directions illustrated, to the maximum of their travel to ensure full unrestricted working travel.
9. Finally, tighten the clamp bolts to the correct torque.

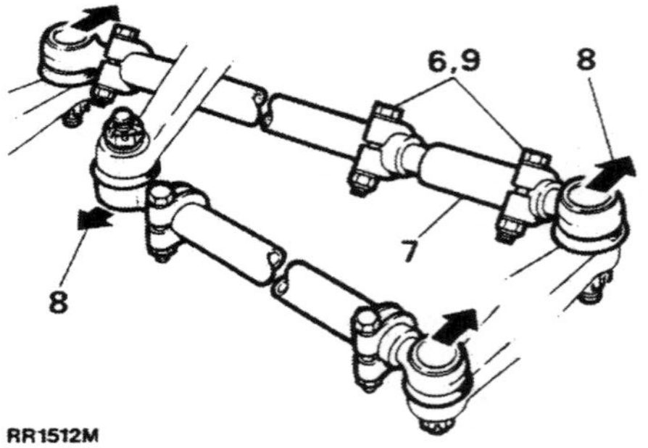

# FRONT SUSPENSION | 60

## PANHARD ROD

**Remove and refit**

### Removing

1. Remove the fixings at the mounting arm.
2. Remove the fixings at the axle bracket.
3. Withdraw the Panhard rod.
4. If required press out the bushes.
5. Fit replacement bushes central in the rod.

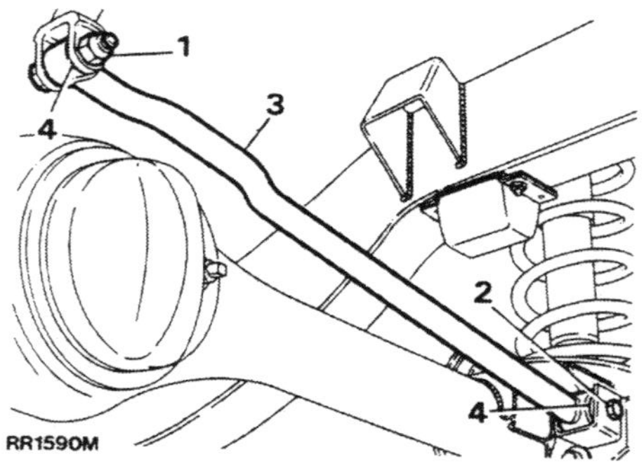

### Refitting

6. Reverse 1 to 4. Tighten the fixings to the correct torque.

## RADIUS ARM

**Remove and refit**

**Service tool:**
18G 1063 Extractor for ball joint

### Removing

1. Remove the road wheel.
2. Support the chassis.
3. Support the axle weight.
4. Remove the fixings, radius arm to chassis side member.
5. Disconnect the track rod at the ball joint. Extractor 18G 1063.
6. Remove the fixings, radius arm to axle.
7. Lower the radius arm front end to clear the axle and withdraw.
8. If required, press out the bush assemblies.

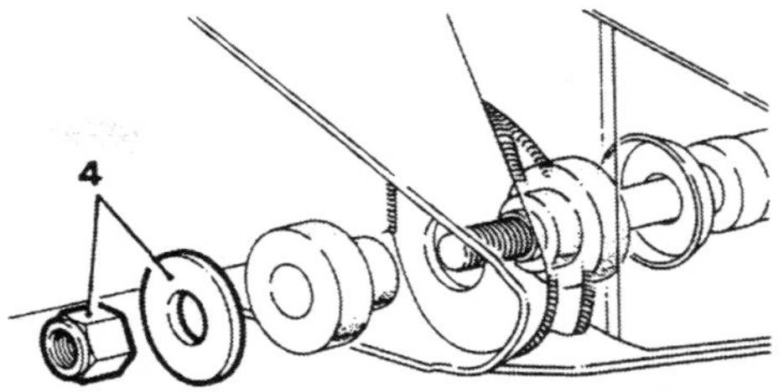

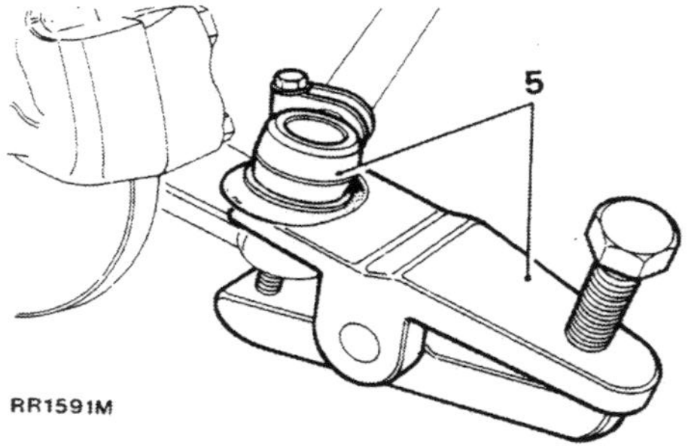

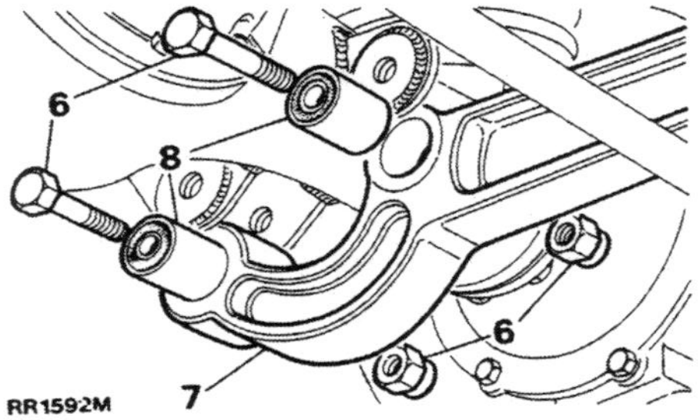

### Refitting

9. Fit the replacement bushes centrally in the arm.
10. Reverse 1 to 7. Tighten the fixings to the correct torque.

# Notes

# FRONT SUSPENSION

## FRONT ROAD SPRING

### Remove and refit

#### Removing

1. Remove the front shock absorber.

   **WARNING: During the following procedure avoid over stretching the brake hoses. If necessary, turn back the hose connector locknuts to allow the hoses to follow the axle.**

2. Lower the axle sufficient to free the road spring.
3. Withdraw the road spring.

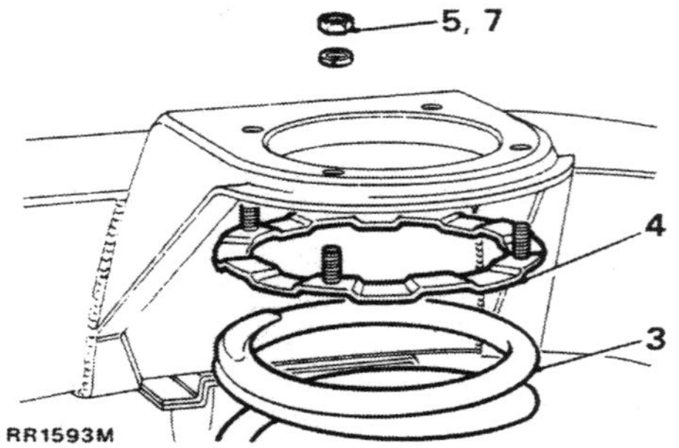

4. Withdraw the shock absorber bracket securing ring.

#### Refitting

5. Reverse 4. Retain in position with a slave nut.
6. Reverse 2 and 3.
7. Remove the nut retaining the securing ring.
8. Fit the front shock absorber.

## FRONT SHOCK ABSORBER

### Remove and refit

#### Removing

1. Remove the road wheel.
2. Support the axle weight.
3. Remove the shock absorber lower fixing.

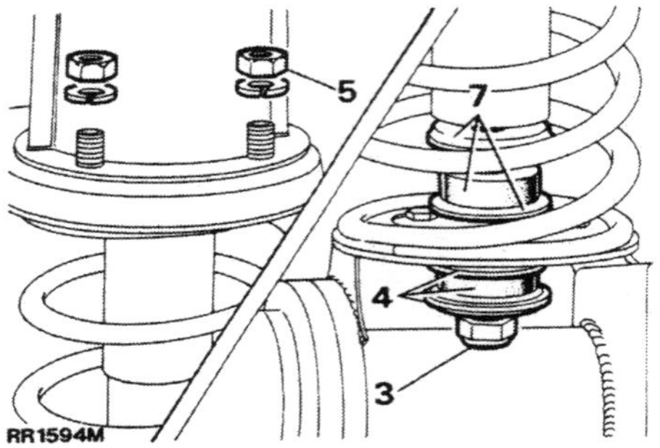

4. Withdraw the cupwasher, rubber bush and seating washer.
5. Remove the shock absorber bracket fixings.
6. Withdraw the shock absorber and bracket complete.

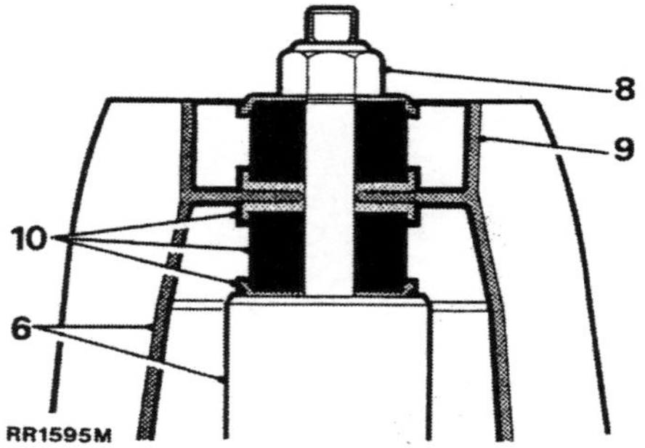

7. Withdraw the seating washer, rubber bush and cupwasher.
8. Remove the fixings, shock absorber to mounting bracket.
9. Withdraw the mounting bracket.
10. Lift off the top seating washer, rubber bush and cupwasher.

#### Refitting

11. Reverse 1 to 10.

## BUMP STOP

### Remove and refit

#### Removing

1. Remove the fixings.
2. Withdraw the bump stop assembly.

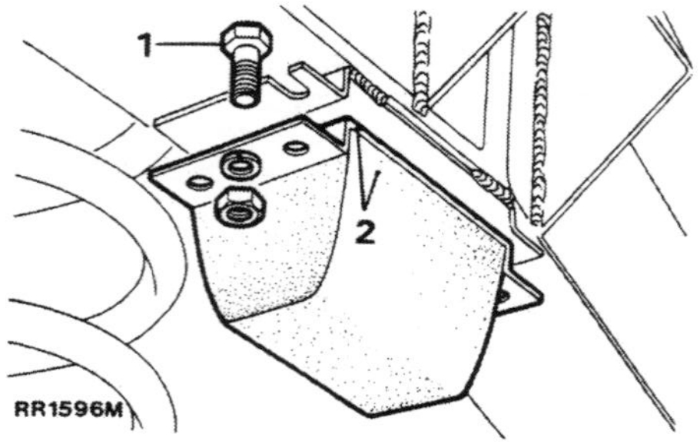

#### Refitting

3. Position the fixing bolts in the slots in the chassis brackets.
4. Fit the bump stop assembly.

## REAR SUSPENSION [64]

### REAR ROAD SPRING

**Remove and refit**

**Service tool:**
RO1006 Extractor for ball joint.

**Removing**

1. Raise rear of vehicle and support chassis.
2. Remove road wheels.
3. Support the axle weight.
4. Disconnect the shock absorbers at one end.
5. Remove fixings, pivot bracket ball joint to axle.
6. Extract the ball joint pin from the tapered housing. Extractor RO1006.
7. Lower the axle further, sufficient to free the road spring from the upper seat.

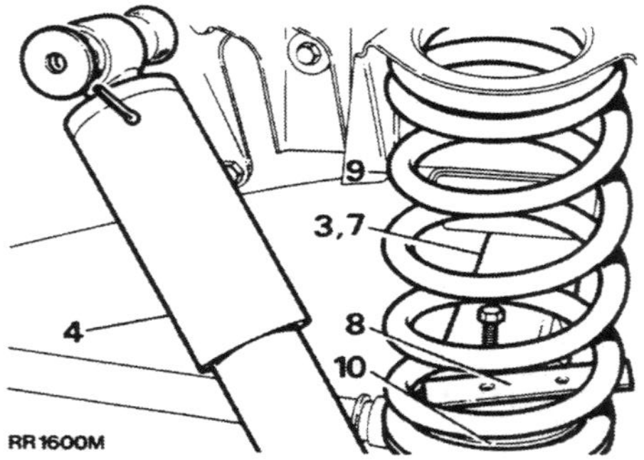

**WARNING: Avoid lowering the axle further than necessary otherwise the rear brake flexible hose will become stretched.**

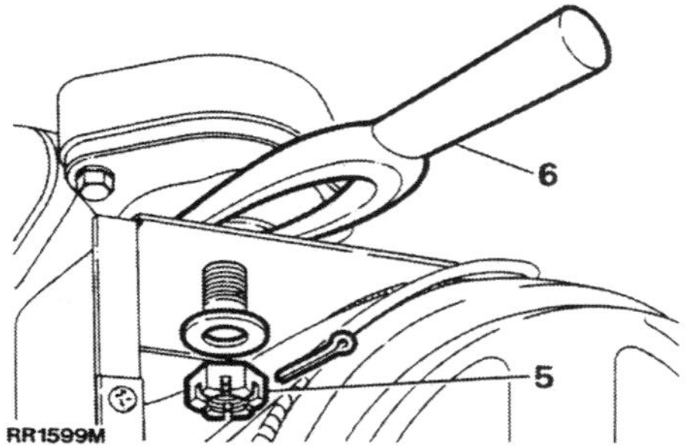

8. Remove the spring retainer plate.
9. Withdraw the road spring.
10. Lift off the spring seat.

**Refitting**

11. Reverse 1 to 10. Tighten the pivot bracket ball joint nut to the correct torque.

### REAR SHOCK ABSORBER

**Remove and refit**

**Removing**

1. Remove road wheel.
2. Support the rear axle.
3. Remove the fixings and withdraw the shock absorber from the axle bracket.
4. Remove upper fixings.
5. Withdraw the shock absorber.
6. If required, remove the mounting bracket at the chassis side member.
7. If required, lift out the mounting rubbers at the upper end.

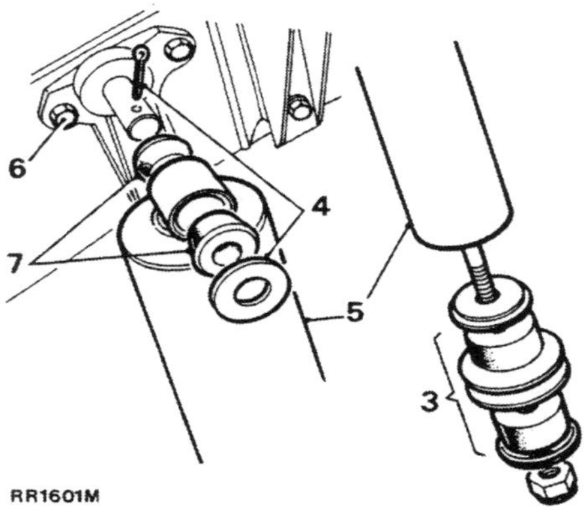

**Refitting**

8. Reverse items 7 and 8 as applicable.
9. Reverse items 1 to 6.

## Notes

## REAR SUSPENSION

### LEVELLING UNIT

#### Functional check

A Boge Hydromat levelling unit is located in the centre of the rear axle.

When the vehicle is unladen the levelling unit has little effect. The unit is self-energising and hence the vehicle has to be driven before the unit becomes effective, the time taken for this to happen being dependent upon the vehicle load, the speed at which it is driven and the roughness of the terrain being crossed.

If the vehicle is overloaded the unit will fail to level fully and more frequent bump stop contact will be noticed.

Should the vehicle be left for a lengthy period, e.g. overnight, in a laden condition, it may settle. This is due to normal internal fluid movement in the unit and is not detrimental to the unit performance.

Before carrying out the checks below, verify that the vehicle is being operated within the specified maximum loading capabilities. If the levelling unit is then believed to be at fault, the procedure below should be followed.

1. Check the levelling unit for excessive oil leakage and if present the unit must be changed. Slight oil seepage is permissible.
2. Remove excessive mud deposits from underneath the vehicle and any heavy items from inside the vehicle that are not part of the original equipment.
3. Measure the clearance between the rear axle bump pad and the bump stop rubber at the front outer corner on both sides of the vehicle. The average clearance should be in excess of 67 mm (2.8 in). If it is less than this figure remove the rear springs and check their free length against the 'Road Spring Data'. Replace any spring whose free length is more than 20 mm (0.787 in) shorter than the figure given. If after replacing a spring the average bump clearance is still less than 67 mm (2.8 in), replace the levelling unit. 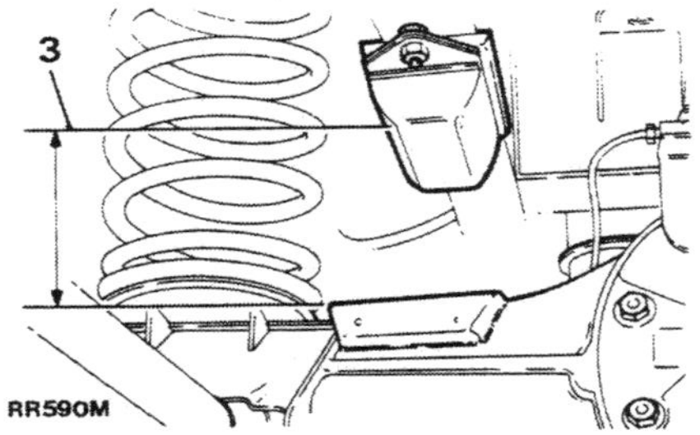
4. With the rear seat upright, load 450 kg (992 lb) into the rear of the vehicle, distributing the load evenly over the floor area. Check the bump stop clearance, with the driving seat occupied.
5. Drive the vehicle for approximately 5 km (3 miles) over undulating roads or graded tracks. Bring the vehicle to rest by light brake application so as not to disturb the vehicle loading. With the driving seat occupied, check the bump stop clearance again.
6. If the change in clearance is less than 20 mm (0.787 in) the levelling unit must be replaced.

### LEVELLING UNIT

#### Remove and refit

#### Removing

WARNING: The levelling unit contains pressurised gas and must not be dismantled nor the casing screws removed. Repair is by replacement of complete unit only.

1. Raise and support the chassis rear end.
2. Support the axle weight.
3. Disconnect the suspension upper links at the pivot bracket. 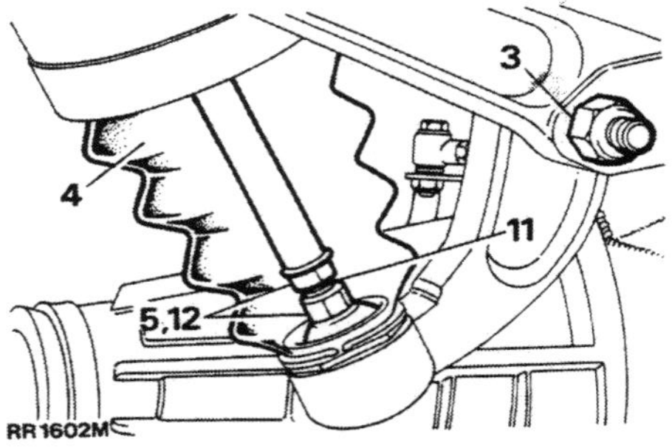
4. Ease up the lower gaiter.
5. Unscrew the lower ball joint at the levelling unit push rod, using thin jawed spanners.
6. Remove the top bracket fixings at the cross member.
7. Withdraw the levelling unit and top bracket complete.
8. Ease back the upper gaiter.
9. Unscrew the upper ball joint at the levelling unit, using thin jawed spanners.
10. Withdraw the upper and lower gaiters and their retaining spring rings.

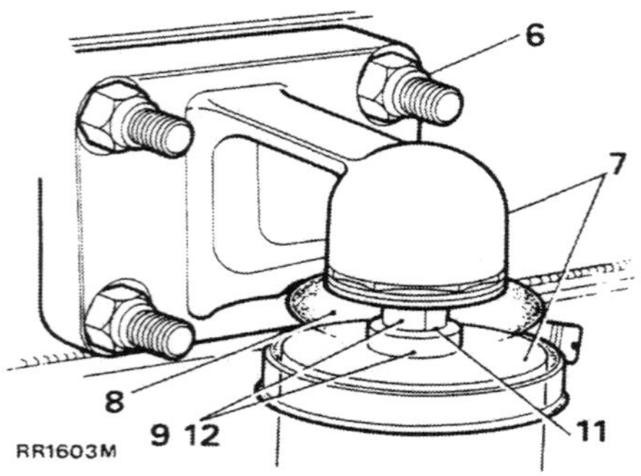

#### Refitting

11. Smear 'Loctite' grade CVX or suitable equivalent sealant on to ball pin threads.
12. Reverse items 1 to 10. Do not fully tighten the fixings until all items are in their fitted position. Finally tighten to the correct torque.

### LEVELLING UNIT BALL JOINTS

#### Remove and refit

Service tools: RO 1006 Extractor for axle bracket ball joint

#### Removing

1. Remove the levelling unit.
2. Remove the split pin and nut at the rear axle bracket.
3. Extract the ball pin from the axle bracket. Extractor RO 1006. 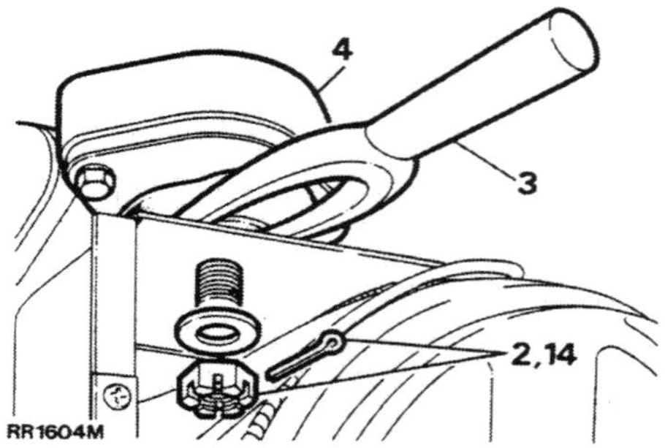
4. Withdraw the pivot bracket complete with ball joints.
5. Unscrew the ball joint assembly for the levelling unit.
6. Remove the ball joint assembly for the axle bracket. 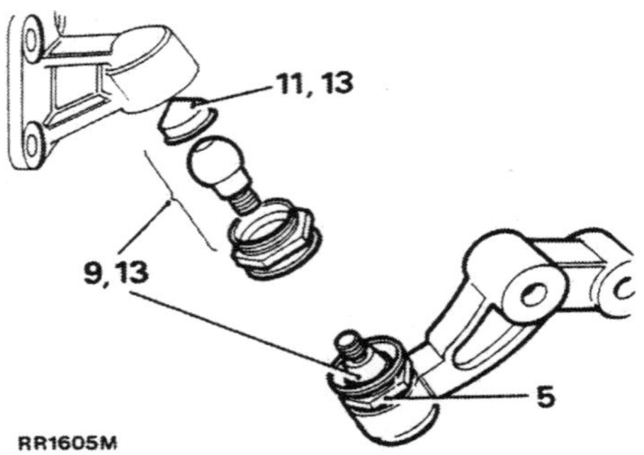
7. Replacement ball joints are supplied as complete assemblies, less fixings, and are pre-packed with grease.
8. The ball joint for the axle bracket must not be dismantled.
9. The ball joints for the levelling unit may be dismantled and cleaned if required.
10. Pack the ball joint with Dextragrease GP or an equivalent grease when assembling.
11. Ensure that the ball seating is square in its housing before refitting.

#### Refitting

12. Press the knurled ball joint into the pivot bracket. 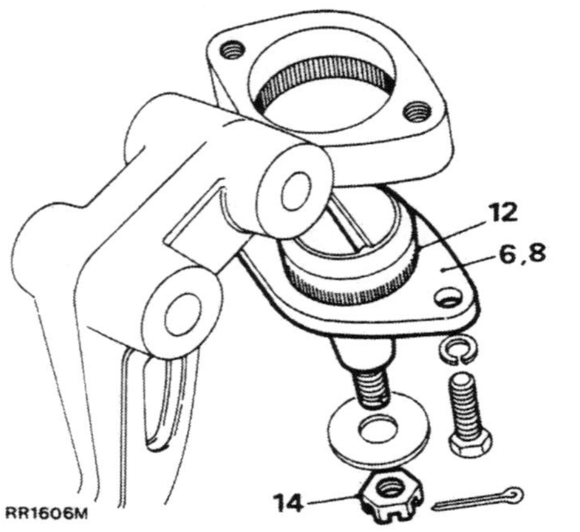
13. Screw the ball joints for the levelling unit into the mounting brackets. If the ball joints do not screw in easily and fully, remove and refit the assemblies ensuring that the plastic seats do not foul in the housings. Tighten to the correct torque.
14. Fit the pivot bracket complete with ball joints to the rear axle. Tighten to the correct torque.
15. Fit the levelling unit.

# REAR SUSPENSION

## BUMP STOP

### Remove and refit

### Removing

1. Remove the fixings.
2. Withdraw the bump stop assembly.

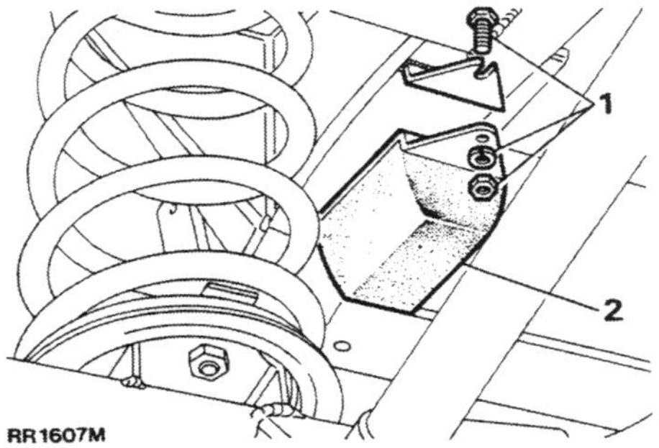

### Refitting

3. Position the fixing bolts in the slots in the chassis brackets.
4. Fit the bump stop assembly, position the shoulder on the carrier to suit the chassis configuration.

## UPPER SUSPENSION LINK

### Remove and refit 1 to 6 and 9

### BUSH

### Remove and refit 7 and 8

### Removing

1. Jack up under the chassis until the rear axle is freely suspended.
2. Remove the fixings, upper link bracket to frame.

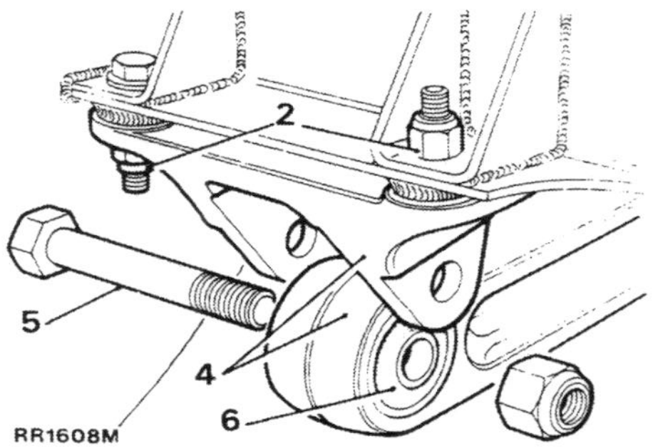

3. Remove the fixings, upper links to pivot bracket.
4. Withdraw the upper link complete with frame bracket.
5. Remove the fixing bolt.
6. Separate link and bush assembly from bracket.

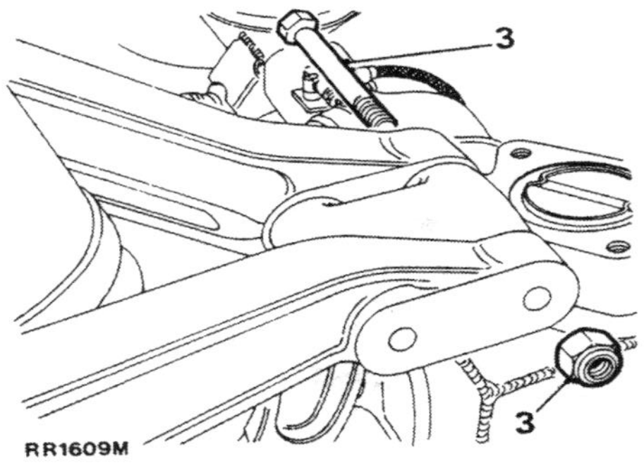

### Replacing the bush

7. Press out the bush assembly.
8. Fit the replacement bush assembly central in the housing.

### Refitting

9. Reverse 1 to 6. Do not fully tighten the fixings until all components are in position.

## LOWER SUSPENSION LINK

### Remove and refit 1 to 7, 10 to 12

### BUSH

### Remove and refit 8 and 9

### Removing

1. Jack up the rear end or use a ramp for accessibility.
2. LH side only. Remove the shock absorber lower fixings and withdraw the shock absorber from the axle bracket.
3. Remove the link rear fixings.

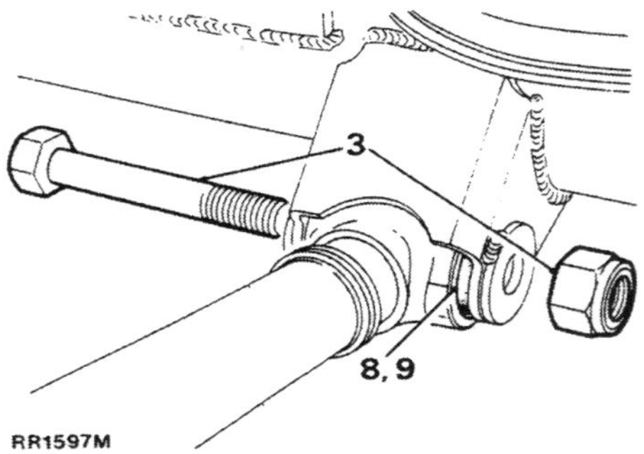

4. Remove the mounting bracket fixings at the side member bracket.
5. Withdraw lower link complete with mounting bracket.
6. Remove the locknut.
7. Withdraw the mounting bracket from the lower link.

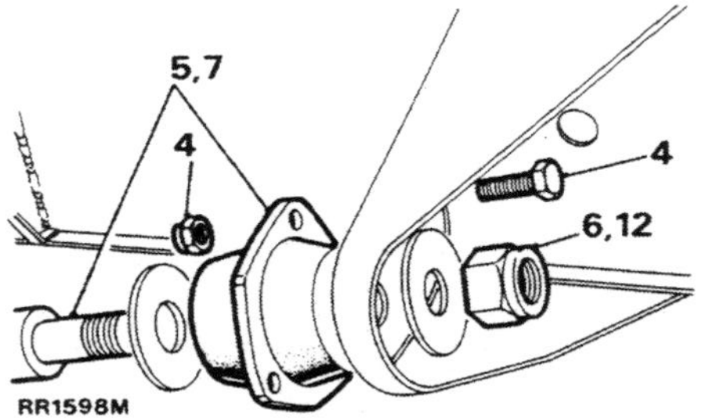

### Replacing the bush

8. Press out the bush assembly from the rear end.
9. Fit the replacement bush assembly central in the housing.

### Refitting

10. Reverse items 6 and 7. Do not tighten the locknut at this stage.
11. Reverse items 3 to 5.
12. Lower the vehicle, remove the jack and allow the axle to take up its static laden position.
13. Tighten the locknut to the correct torque.

# BRAKES

## BRAKE SYSTEM

### Description

The hydraulic braking system fitted to the Range Rover is of the dual line type, incorporating primary and secondary hydraulic circuits.

**NOTE:** References made to primary and secondary do not imply main service brakes or emergency brakes but denote hydraulic line identification.

■ PRIMARY HYDRAULIC CIRCUIT — EMERGENCY BRAKE

□ SECONDARY HYDRAULIC CIRCUIT — SERVICE BRAKE

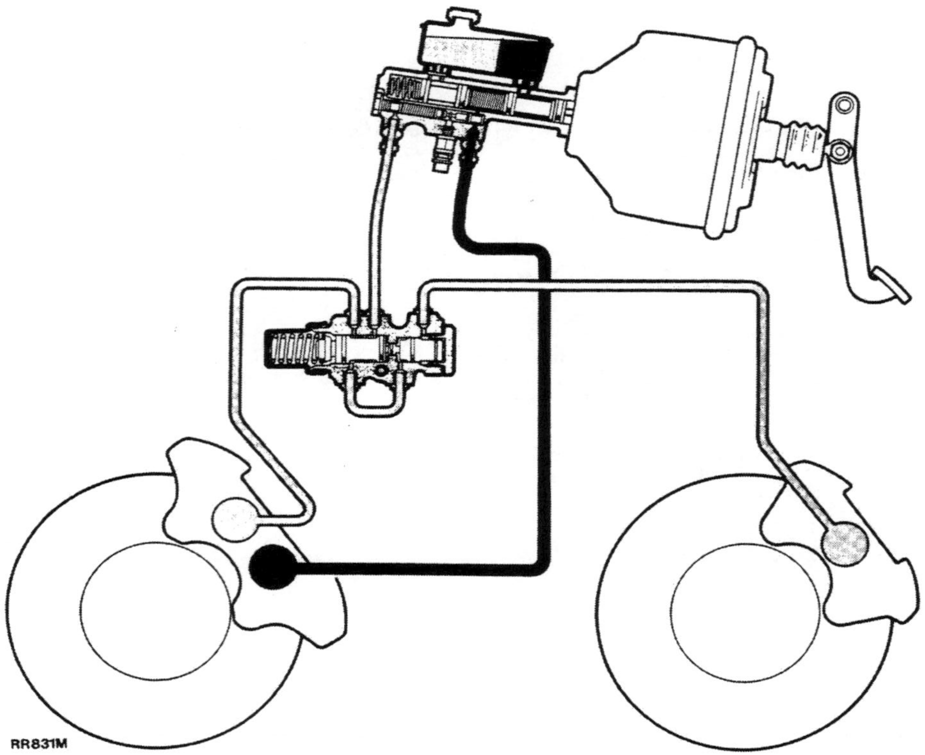

The brake pedal is connected to a vacuum-assisted mechanical servo which in turn operates a tandem master cylinder. The front disc brake calipers each house four pistons, the lower pistons are fed by the primary hydraulic circuit, the upper pistons by the secondary hydraulic circuit. The rear disc brake calipers each house two pistons and these are fed by the secondary hydraulic circuit via a pressure reducing valve.

A brake failure switch incorporated in the master cylinder will illuminate a panel warning light if a failure occurs in either the primary or secondary hydraulic circuits.

*Continued*

## Notes

# BRAKES

The brake fluid reservoir is divided, the front section (section closest to the servo) feeds the primary circuit and the rear section feeds the secondary circuit. Under normal operating conditions both the primary and secondary hydraulic circuits operate simultaneously on brake pedal application. In the event of a failure in the primary circuit the secondary circuit will still function and operate front and rear calipers. Alternatively, if the secondary circuit fails, the primary circuit will still function and operate the lower pistons in the front calipers.

If the servo should fail, both hydraulic circuits will still function but would require greater pedal pressure.

The hand-operated transmission brake is completely independent of the hydraulic circuits.

Brake pad wear sensors are incorporated into the front right-hand side, inboard brake pad and rear left-hand side, inboard pad. The sensors will illuminate a brake pad wear warning light in the instrument binnacle, when pad thickness has been reduced to approximately 3 mm (0.118 in).

## BRAKE SYSTEM LAYOUT – RIGHT-HAND STEERING

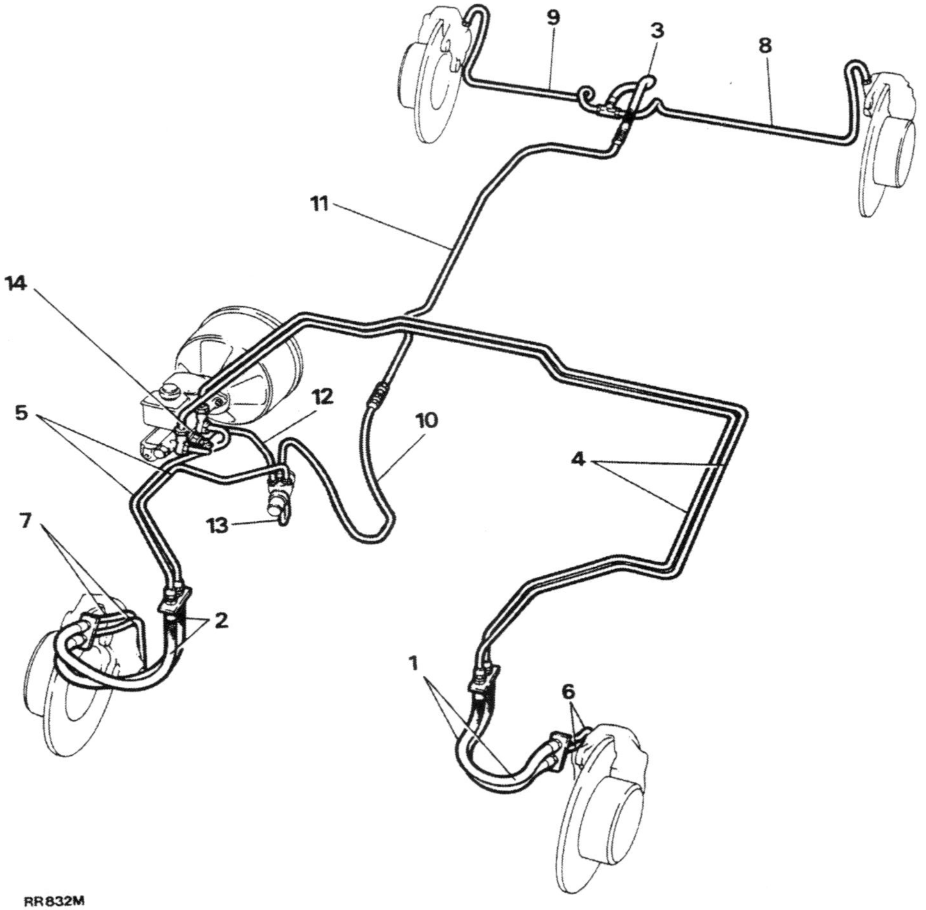

### HOSES

1. Front left-hand flexible hoses
2. Front right-hand flexible hoses
3. Intermediate flexible hose

### PIPES

4. Feed to front left-hand hose connector
5. Feed to front right-hand hose connector
6. Feed to front left-hand caliper
7. Feed to front right-hand caliper
8. Feed to rear left-hand caliper
9. Feed to rear right-hand caliper
10. Feed to two-way connector
11. Feed to intermediate hose
12. Feed to pressure reducing valve
13. Transfer pipe – pressure reducing valve
14. Brake failure warning switch

## BRAKES

### Bleed

The hydraulic system comprises two completely independent sections. The rear calipers and the upper pistons in the front calipers form the secondary section, while the lower pistons in the front calipers form the primary section. The following procedure covers bleeding the complete system, but it is permissible to bleed one section only if disconnections are limited to that section.

Bleeding will be assisted if the engine is run or a vacuum supply is connected to the servo.

**WARNING: IF THE ENGINE IS RUNNING DURING THE BRAKE BLEEDING PROCESS ENSURE THAT NEUTRAL GEAR IS SELECTED AND THAT THE HANDBRAKE IS APPLIED.**

When bleeding any part of the secondary section, almost full brake pedal travel is available. When bleeding the primary section only, brake pedal travel will be restricted to approximately half.

Before commencing to bleed the system it is necessary to slacken off the brake failure warning switch to prevent the shuttle valve restricting the hydraulic fluid flow.

1. Disconnect the leads from the switch.
2. Unscrew the switch and insert the 'C' washer between the switch and master cylinder before depressing the brake pedal.
3. After completion of bleeding, screw in the switch and tighten to the correct torque. See 'Data section'.

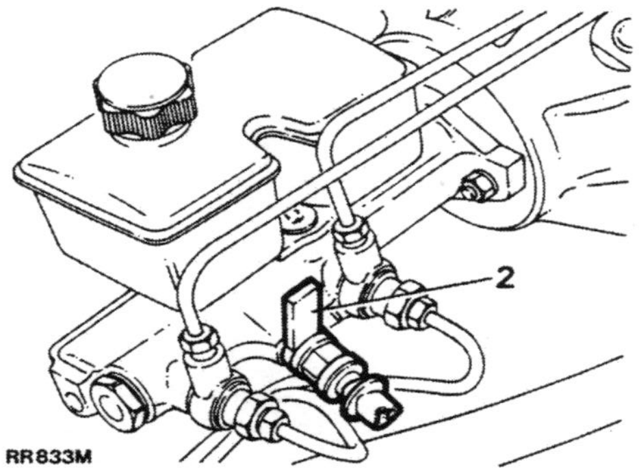

**NOTE: When bleeding the system commence with the caliper furthest from the master cylinder and bleed from the screw on the same side as the fluid inlet pipes, then close the screw and bleed from the screw on the opposite side of the same caliper. Tighten the bleed screws to the correct torque. See 'Data section'.**

### Bleeding

1. Fill the fluid reservoir with the correct fluid, see 'Data section'.

   **NOTE: The correct fluid level must be maintained throughout the procedure of bleeding.**

2. Connect a bleed tube to the bleed screw on the rear caliper furthest from the master cylinder.
3. Submerge the free end of the bleed tube in a container of clean brake fluid.
4. Slacken the bleed screw.
5. Operate the brake pedal fully and allow to return.

   **NOTE: Allow at least five seconds to elapse with the foot right off the pedal to ensure that the pistons fully return before operating the pedal again.**

6. Repeat 5 until fluid clear of air bubbles appears in the container, then keeping the pedal fully depressed, tighten the bleed screw.
7. Remove the bleed tube and replace the dust cap on the bleed screw.
8. Repeat 1 to 7 for the other rear caliper.
9. Remove the front wheel on the side furthest from the master cylinder.
10. Connect a bleed tube to the primary bleed screw on the front caliper furthest from the master cylinder.
11. Connect a bleed tube to the secondary bleed screw on the same side of the caliper as the primary screw.
12. Repeat 3 to 7 for the front caliper, bleeding from the two screws simultaneously.
13. Connect a bleed tube to the other screw on the front caliper furthest from the master cylinder.
14. Repeat 3 to 7 for the second secondary screw on the front caliper.
15. Refit the front wheel.
16. Repeat 9 to 15 for the front caliper nearest the master cylinder.
17. Remove the 'C' washer and tighten the PDWA switch to the correct torque. See 'Data section'.

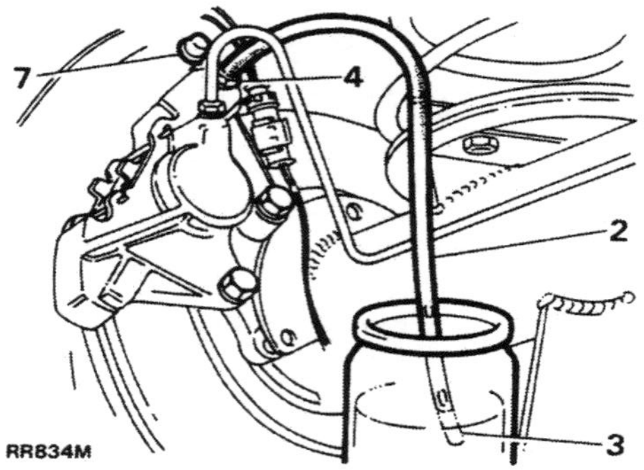

### BRAKE PRESSURE REDUCING VALVE

**Remove and refit**

**Removing**

1. Remove all dust, grime, etc. from the vicinity of the pressure reducing valve fluid pipe unions.
2. Disconnect the inlet and outlet fluid pipes from the pressure reducing valve. Plug the pipes and reducing valve ports to prevent the ingress of foreign matter.
3. Remove the nut, spring washer, bolt, plain washer and distance piece securing the reducing valve to the vehicle.
4. Withdraw the pressure reducing valve from the engine compartment.

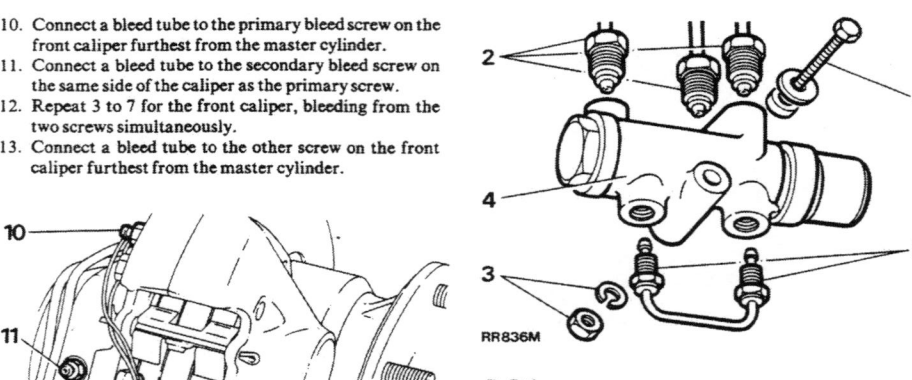

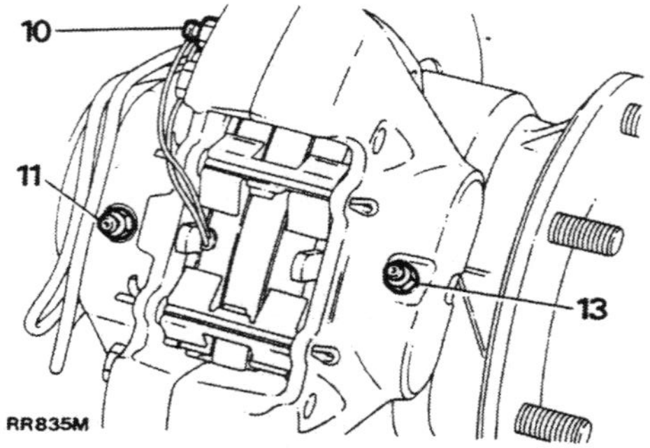

**Refitting**

5. Reverse the removal instructions.
6. Bleed the brake systems.

**NOTE: The pressure reducing valve is not a serviceable item, in the event of a failure or damage, a new unit must be fitted.**

# BRAKES 70

## KEY TO MASTER CYLINDER

1. Securing screws
2. Reservoir
3. Reservoir seals
4. End plug
5. Shuttle valve
6. Secondary piston stop pin
7. Nuts and washers securing cylinder to servo
8. Primary piston
9. Circlip
10. Secondary piston
11. PDWA switch
12. 'C' washer

## MASTER CYLINDER—OVERHAUL

**CAUTION: Brake fluid is corrosive, if any fluid comes into contact with body paintwork, immediately wipe clean with a soft cloth.**

1. Disconnect the brake pipes from the master cylinder and plug the outlet ports.
2. Disconnect the electrical plug from the PDWA switch.
3. Remove the two nuts and spring washers securing the cylinder to the servo unit.
4. Remove the reservoir filler cap and drain off the surplus fluid.
5. Remove the two screws securing the reservoir to the master cylinder.
6. Lift the reservoir off the master cylinder.
7. Prise the two reservoir sealing rubbers from the master cylinder.
8. Secure the master cylinder in a vice and push the primary piston down the bore and withdraw the secondary piston stop pin.
9. Press down the primary piston and remove the circlip.
10. Withdraw the primary piston assembly.
11. Apply a high pressure air line to the secondary outlet port to expel the secondary piston assembly.
12. Remove the PDWA switch from the side of the master cylinder.
13. Remove the large end plug and copper washer retaining the shuttle valve in the master cylinder.
14. Apply a high pressure air line to the primary outlet port to expel the shuttle valve from its bore.

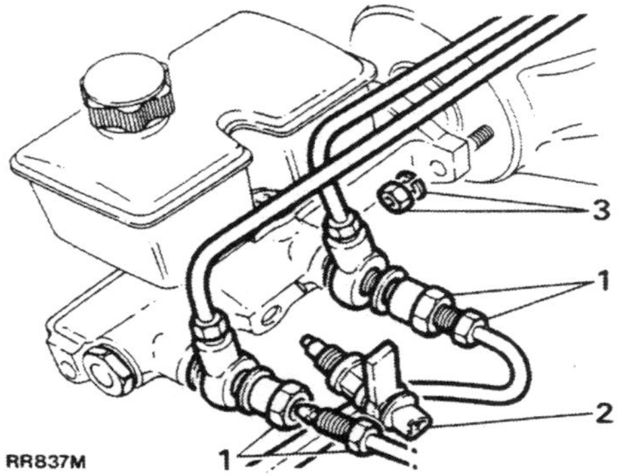

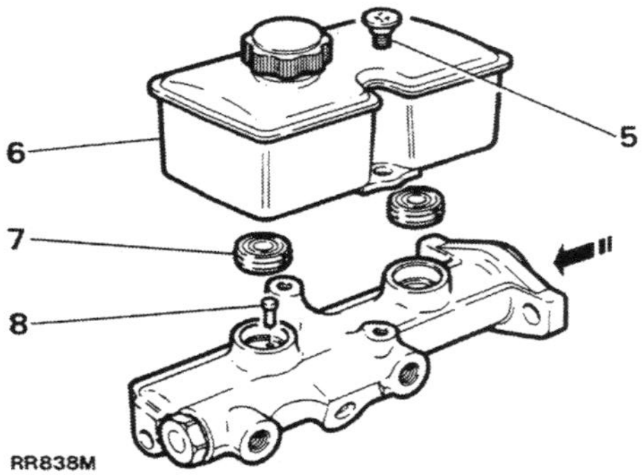

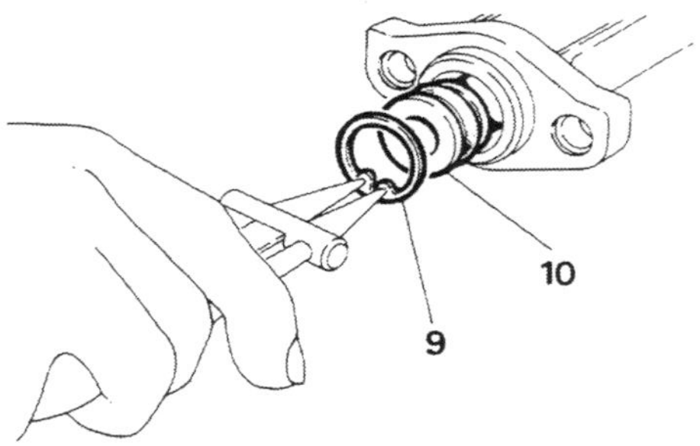

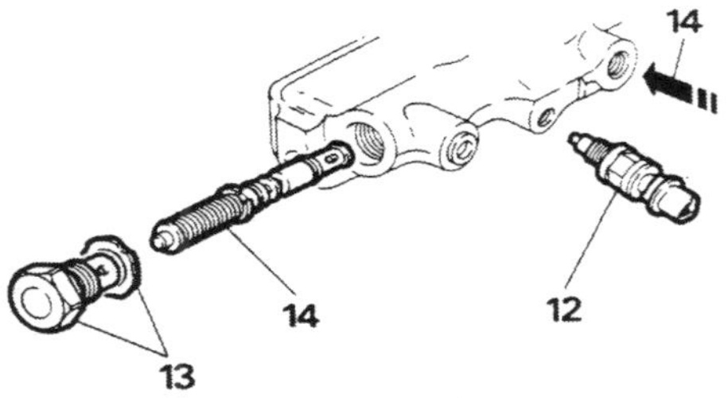

*Continued*

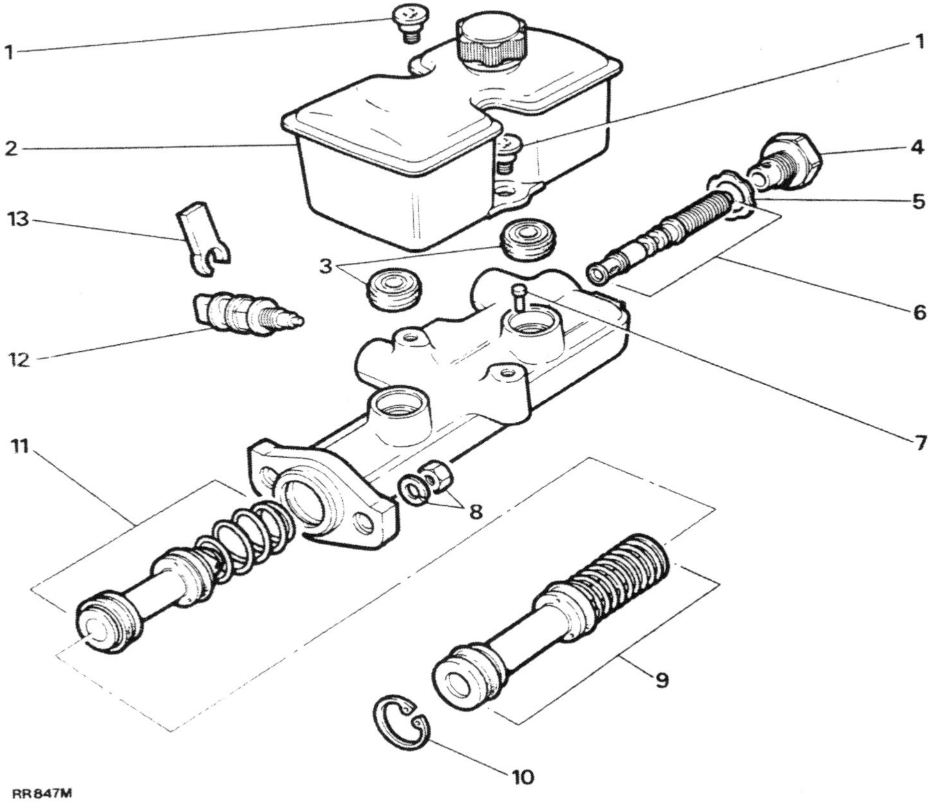

179

# BRAKES 70

## RENEWING PRIMARY PISTON SEALS

15. Remove the retaining screw from the primary piston and remove the following items:
    - (A) Spring retainer.
    - (B) Piston spring.
    - (C) Circlip.
    - (D) Seal retainer.
    - (E) Recuperating seal.
    - (F) Washer.
16. Taking care not to damage the piston, prise off the outer seal.

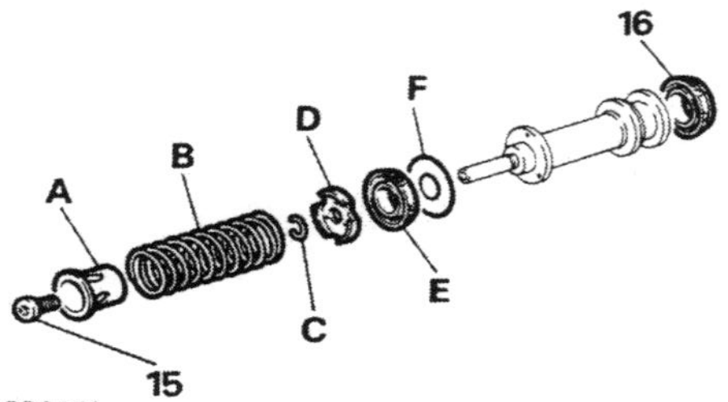

RR841M

## RENEWING SECONDARY PISTON SEALS

19. Remove the following items from the secondary piston:
    - (A) Spring.
    - (B) Seal retainer.
    - (C) Recuperating seal.
    - (D) Washer.
20. Taking care not to damage the piston prise off the outer seal.

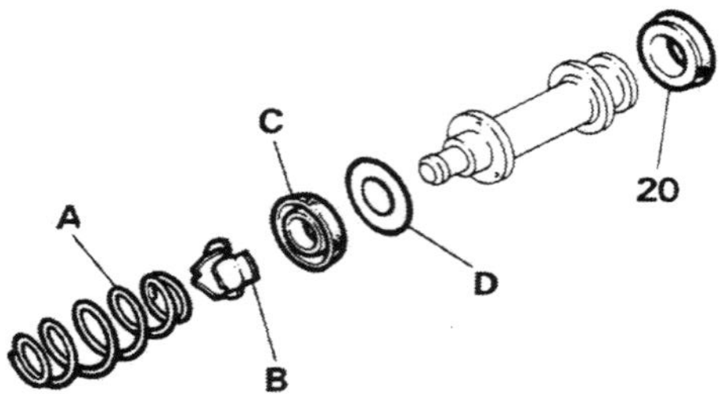

RR843M

17. Fit a new outer seal using the same procedure as for the primary piston outer seal by squeezing the seal between the finger and thumb into an ellipse and press the raised part of the seal over the flange using the fingers of the other hand.
18. Fit a new recuperating seal and assemble the parts in the reverse order of removal. Compress the spring and secure the assembly with the retaining screw. Tighten the screw securely.

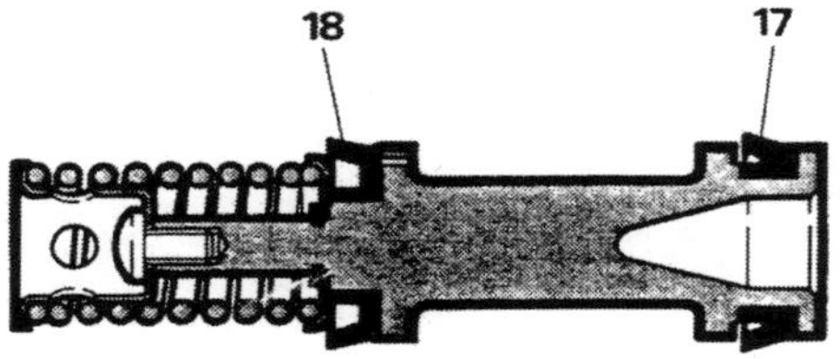

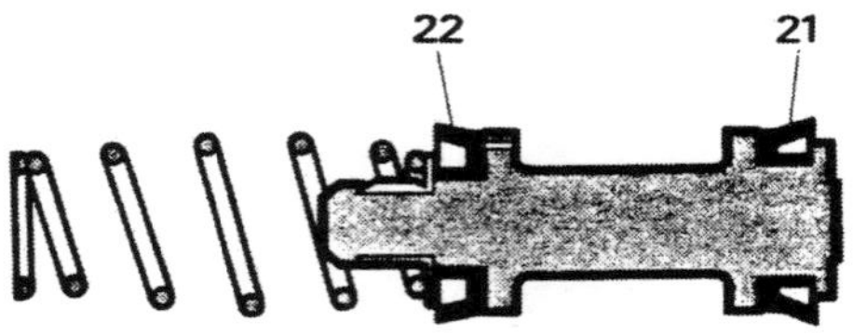

RR842M

## RENEW SHUTTLE VALVE 'O' RINGS

23. Carefully remove the two 'O' rings from the valve taking care not to damage the piston.
24. Fit new 'O' rings.

NOTE: The 'O' rings should not be rolled along the piston, but should be stretched slightly and eased down the piston and into the grooves.

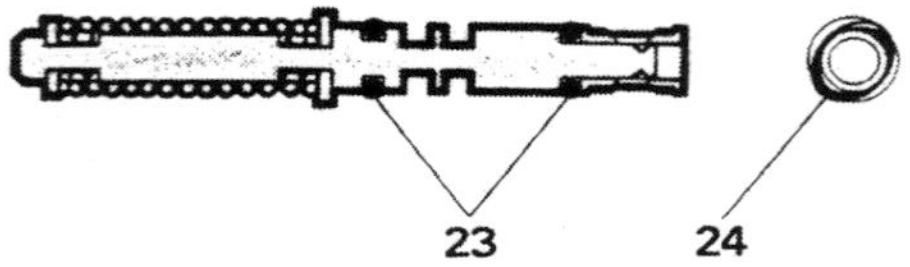

RR845M

## ASSEMBLING MASTER CYLINDER

It is important that the following instructions are carried out precisely, otherwise damage could be caused to the new seals when inserting the plungers into the cylinder bore. Generous amounts of new brake fluid should be used to lubricate the parts during assembly. Never use old fluid or any other form of cleaning and lubricating material. Cleanliness throughout is essential.

25. Clamp the cylinder in a vice and lubricate the secondary piston seals and cylinder bore. Offer the piston assembly to the cylinder until the recuperation seal is resting centrally in the mouth of the bore. Gently introduce the piston with a circular rocking motion, as illustrated. Whilst ensuring that the seal does not become trapped, ease the seal into the bore and slowly push the piston down in one continuous movement.

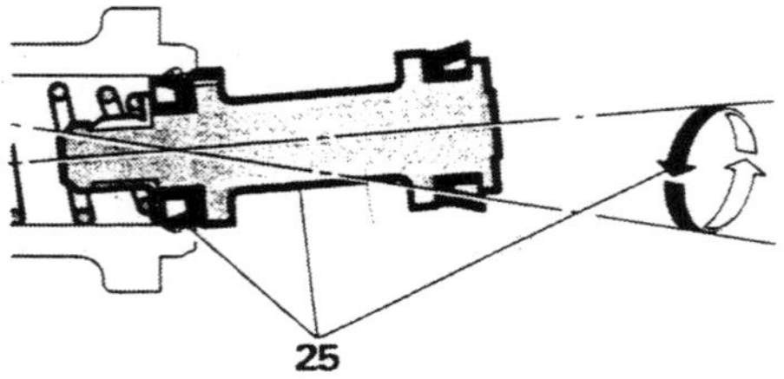

RR846M

26. Slowly press the piston down the bore and fit the secondary piston stop-pin.
27. Fit the primary plunger assembly using the same method as for the secondary plunger. Press the plunger down and secure the assembly with the circlip.
28. Lubricate the 'O' rings and fit the shuttle valve. Fit the end plug using a new copper sealing washer and tighten the plug securely.
29. Fit the plastic 'C' washer to the end of the PDWA switch and screw the switch into the master cylinder.
30. Fit new seals to the bottom of the reservoir.
31. Press the reservoir into the top of the master cylinder and secure in position with the two retaining screws.
32. Fit the master cylinder to the servo and secure with the two nuts and spring washers and tighten to the correct torque. See 'Data section'.
33. Bleed the brakes. After final bleed remove the 'C' washer from the PDWA switch and tighten the switch to the correct torque. See 'Data section'.

# BRAKES 70

## PEDAL ASSEMBLY—OVERHAUL

### Remove the pedal assembly

1. Remove the lower fascia panel.
2. Disconnect the servo operating rod from the brake pedal.
3. Remove the four nuts and spring washers securing the brake pedal and servo assemblies to the engine compartment closure panel.
4. Withdraw the brake pedal assembly sufficient to disconnect the electrical leads from the stop light switch.
5. Lift the pedal assembly clear.

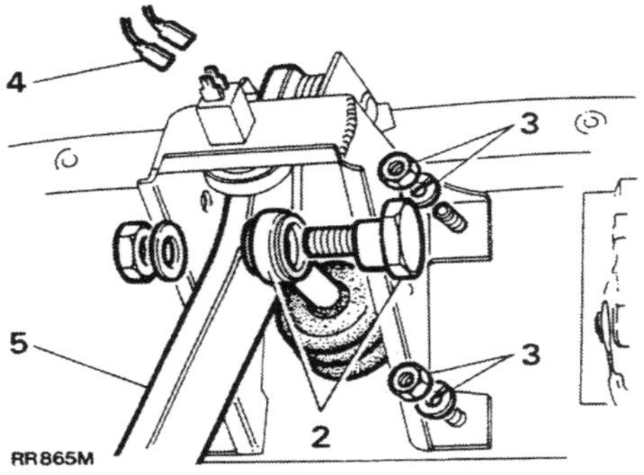

### DISMANTLING

6. Disconnect the pedal return spring.
7. Remove the circlip from the 'D' shaped end of the pedal shaft.
8. Withdraw the pedal shaft.
9. Withdraw the pedal from the box.
10. Remove the bushes from the pedal pivot tube.

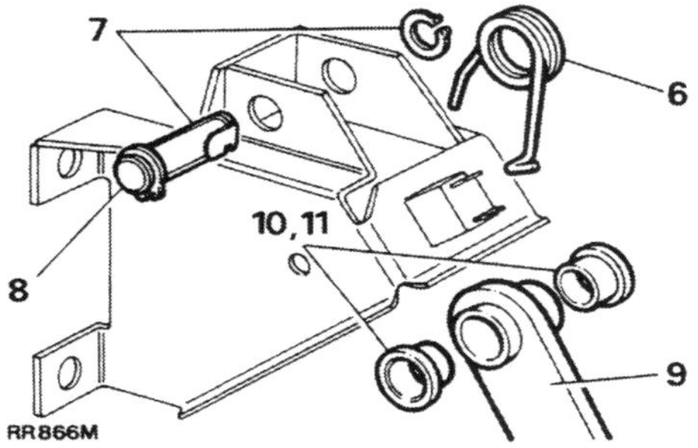

### Assembling

11. Press the new bushes into the pedal pivot tube. If necessary, ream the bushes to 15.87 mm plus 0.05 mm (.625 in plus .002 in).
12. Lightly oil the bushes and pedal shaft.

### Refit the pedal assembly

13. Refit the assembly to the engine compartment closure panel, tighten the four nuts securely.
14. Connect the electrical leads to the stop light switch.
15. Connect the servo operating rod to the brake pedal with the pivot bolt eccentric, in the forward position. Do not fully tighten the pivot bolt nut.
16. Turn the pivot bolt to bring the brake pedal back until it just contacts the rubber buffer, then secure the pivot bolt nut.
17. Refit the lower fascia.

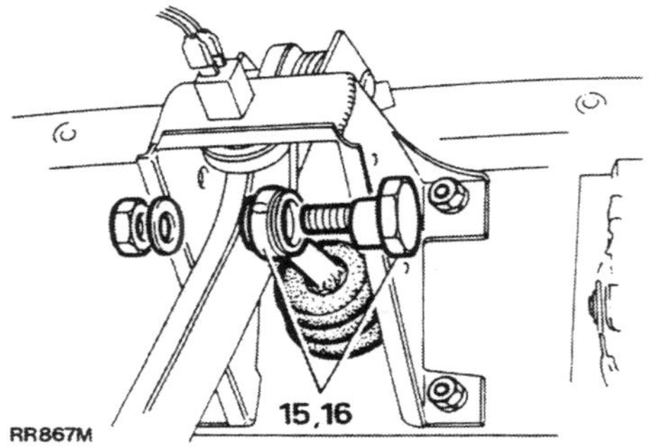

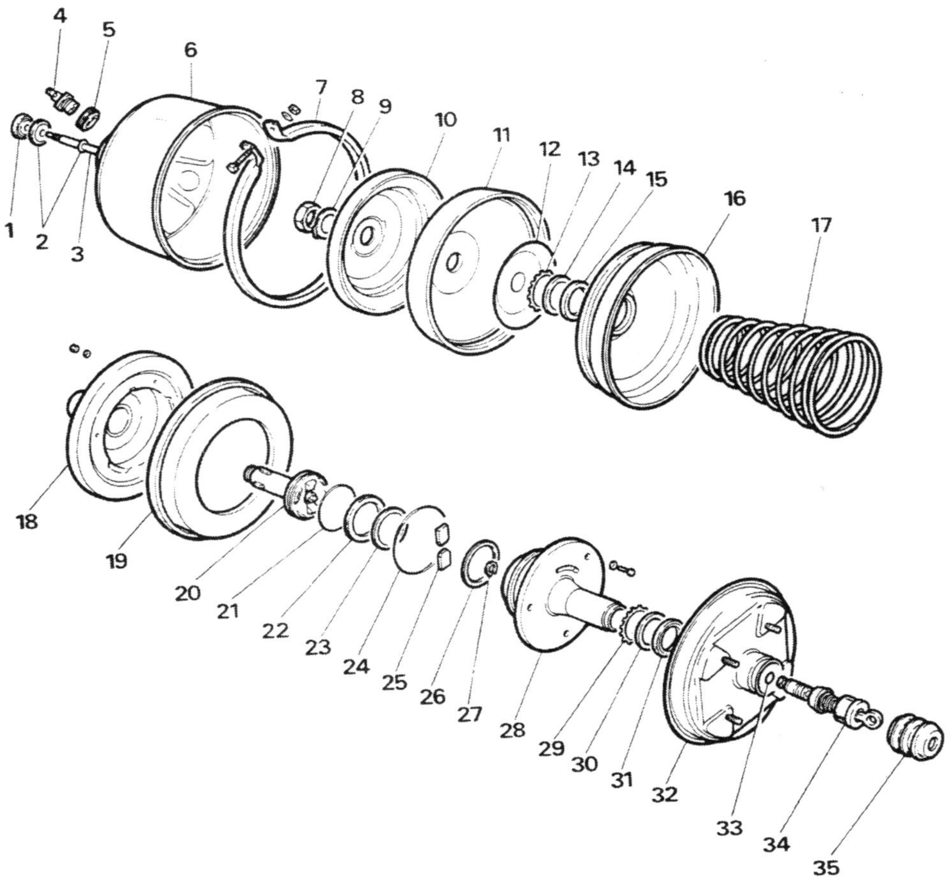

### KEY TO BRAKE SERVO

1. Push rod seal
2. Retainer
3. Push rod
4. Non-return valve
5. Seal
6. Servo shell
7. Clamp ring
8. Locknut
9. Lock washer
10. Support plate
11. Diaphragm
12. Backing washer
13. Bearing retainer
14. Nylon bearing
15. Seal
16. Separator shell
17. Spring
18. Diaphragm support and tube
19. Diaphragm
20. Reaction piston
21. Rubber 'O' ring
22. Sponge seal
23. Backing ring
24. Rubber 'O' ring
25. Levers
26. Bearing ring
27. Circlip
28. Valve body
29. Bearing retainer
30. Nylon bearing
31. Seal
32. Servo cover
33. Rubber 'O' ring
34. Valve/push rod assembly
35. Rubber boot

# BRAKES

## BRAKE SERVO—Overhaul

### REMOVING THE SERVO

1. Remove the master cylinder from the servo.
2. Disconnect the servo vacuum hose from the non-return valve.
3. Remove the lower fascia within the vehicle to gain access to the brake pedal linkage and servo retaining nuts located on the rear of the engine compartment closure panel.
4. Disconnect the brake pedal linkage.
5. Remove the four nuts with spring washers securing the servo to the closure panel.
6. Remove the servo from the engine compartment and place on a suitable clean bench.
7. Thoroughly clean the exterior of the servo.

**CAUTION: CARE MUST BE TAKEN TO ENSURE THAT ALL COMPONENTS ARE SCRUPULOUSLY CLEAN BEFORE UNDERTAKING THE OVERHAUL PROCEDURE. ANY DIRT OR GRIT WITHIN THE SERVO COULD RESULT IN DAMAGE TO THE DIAPHRAGMS.**

### DISMANTLING THE SERVO

8. Before dismantling the servo lightly stamp identification marks on the servo shell, clamp ring and cover plate to aid the assembly procedure.
9. Remove the nut, bolt and plain washer securing the clamping to the servo.
10. Remove the rubber boot from the valve/push rod assembly.
11. Slide the servo cover off the valve body.
12. Using two screwdrivers inserted between the separator shell outer lip and servo shell, carefully lever the diaphragm assembly out of the shell.
13. Bend back the tabs of the locking washer.

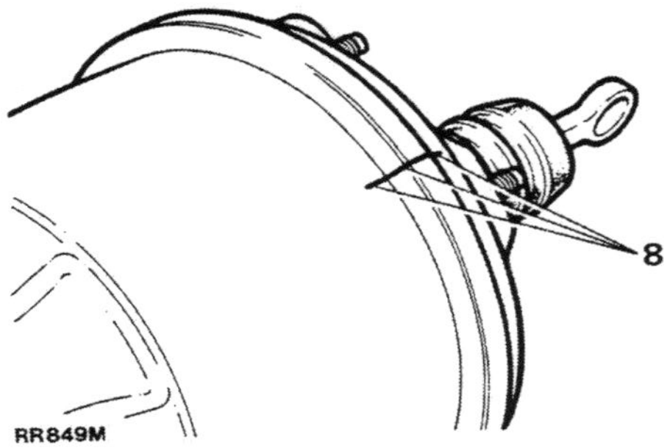

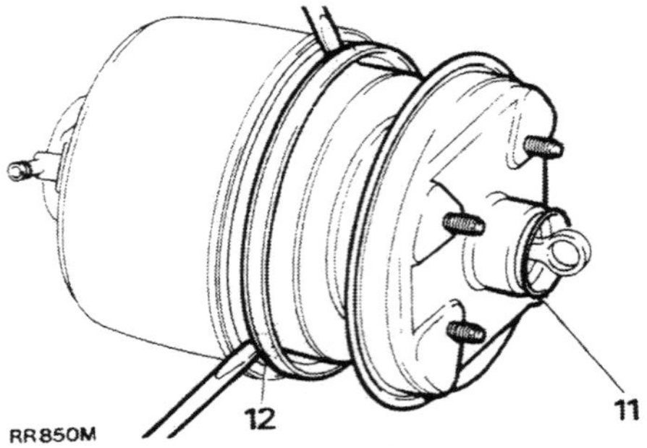

**WARNING: Assistance may be required, to hold the diaphragm assembly together whilst the locknut is released, to prevent the unit separating due to internal spring pressure.**

14. Remove the locknut and separate the assembly.
15. Remove the support plate, diaphragm and backing washer from the separator shell.
16. Prise the serrated bearing retainer out of the separator shell.
17. Remove the nylon bearing and rubber seal from the separator shell. Note their position for reassembly.
18. Remove the three screws, locknuts plain and fibre washers securing the diaphragm to the valve body.
19. Remove the diaphragm support and tube from the front of the diaphragm.
20. Remove the diaphragm from the valve body.
21. Pull the piston out of the valve body.
22. Remove the 'O' rings from the valve body and piston.
23. Remove the sponge seal and backing ring from the valve body.
24. Lift out the two levers and remove the bearing ring.
25. Release the circlip securing the valve/push rod assembly in the valve body.
26. Prise the end cap from the opposite end of the valve body and withdraw the valve/push rod assembly complete.
27. Remove the small 'O' ring from the assembly.

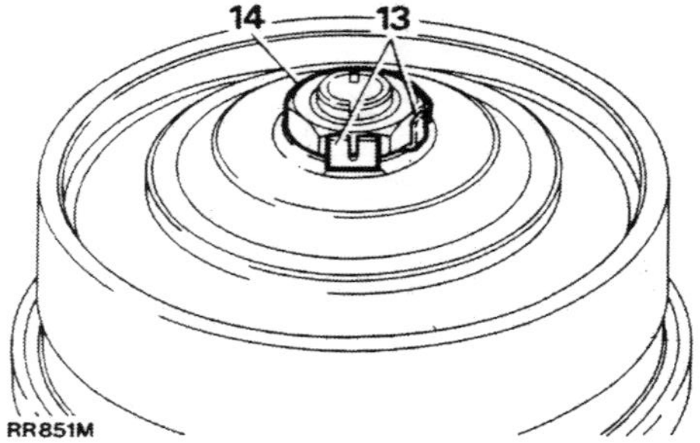

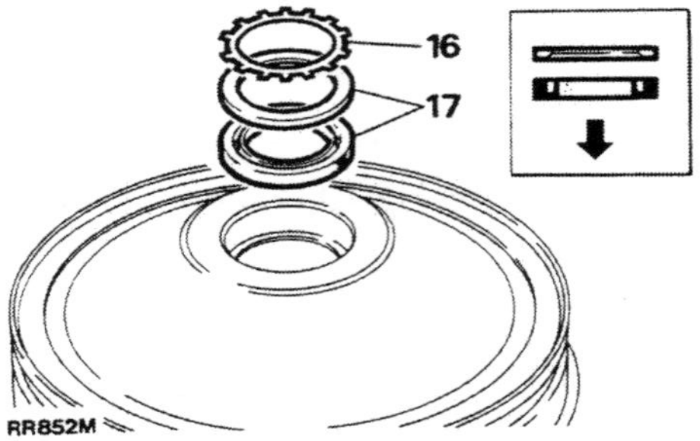

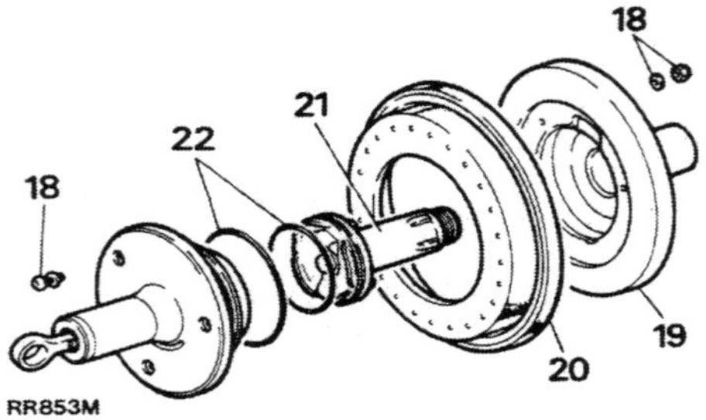

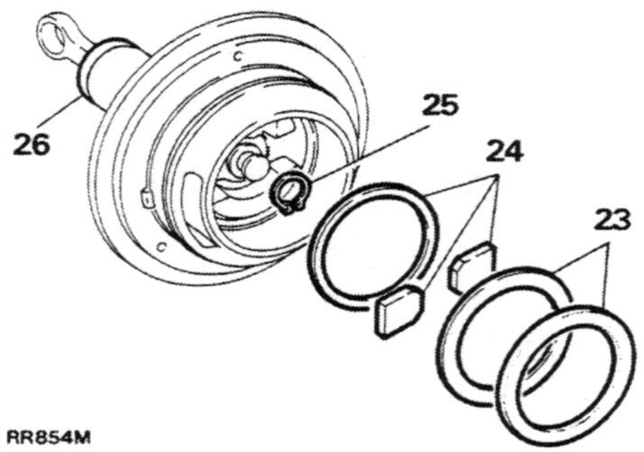

**NOTE: The valve/push rod assembly cannot be further dismantled, if the seals and assembly are in a poor condition, renew the complete assembly.**

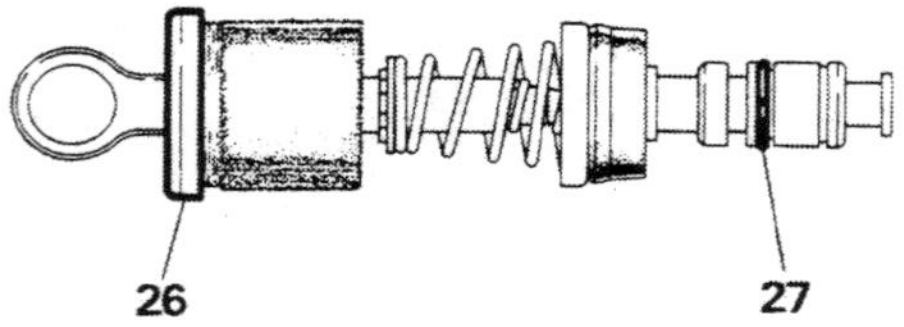

28. Remove the push rod, seal, retainer and washer from the servo shell.

**NOTE: If the push rod is to be renewed the operating length must be set after the servo has been assembled, see assembling procedure instruction number 61.**

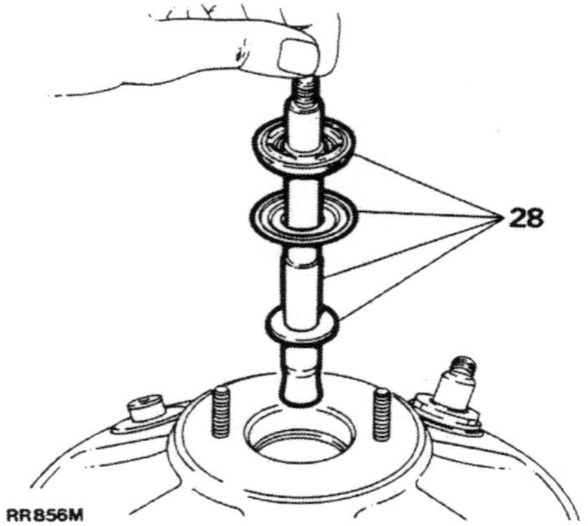

*Continued*

# BRAKES

## ASSEMBLE THE SERVO

Carefully inspect all parts for wear and damage. Scrupulous cleanliness of all parts of the servo is essential.

29. If a new valve/push rod assembly is being fitted it will be necessary to compress the spring located in the centre of the valve, to enable the circlip to be inserted into the groove, thus tensioning the seal.

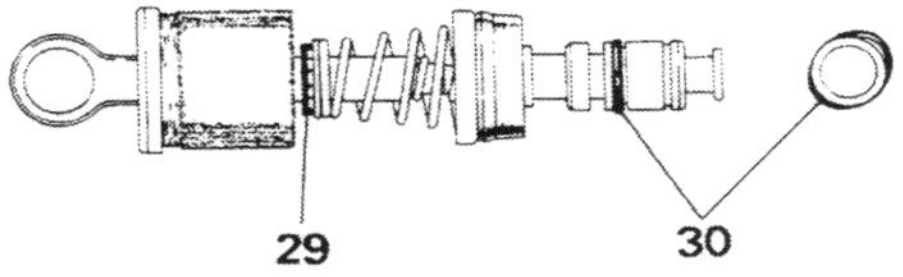

RR857M

30. Fit a new 'O' ring to the end of the valve. DO NOT ROLL THE 'O' RING INTO THE GROOVE. Carefully stretch the seal and ease it down over the valve and into the groove.

31. Lubricate the seals with Lockheed disc brake lubricant and push the assembly fully into the bore of the valve body.

32. Clamp the eye of the assembly in a bench vice. Press the valve body down to expose the circlip groove at the end of the valve/push rod assembly and fit the circlip.

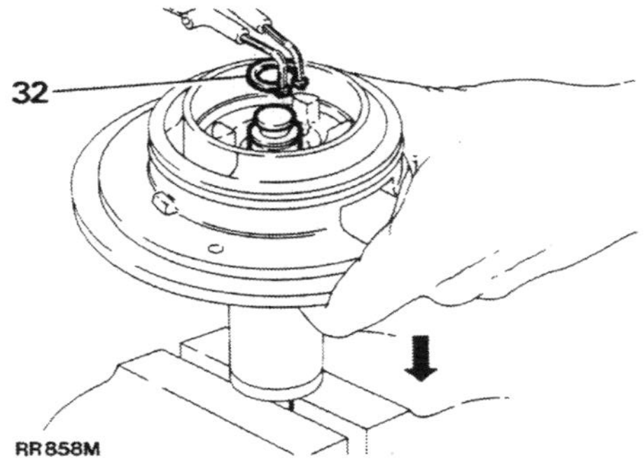

RR858M

33. Fit the bearing ring (thicker of the two rings), levers, backing ring and sponge seal to the valve body,

NOTE: Ensure that the chamfered edges of the levers are fitted firmly into the groove at the end of the push rod assembly.

34. Feed the sponge filter and felt pad into the opposite end of the valve body and press the end cap into position.

35. Lightly coat the new valve body 'O' ring with disc brake lubricant and ease the seal into the groove.

36. Lightly coat the new reaction piston 'O' ring with disc brake lubricant and ease the seal into the groove.

37. Push the piston into the valve body ensuring that the two projections on the piston are located over the levers in the valve body, push the piston firmly into position.

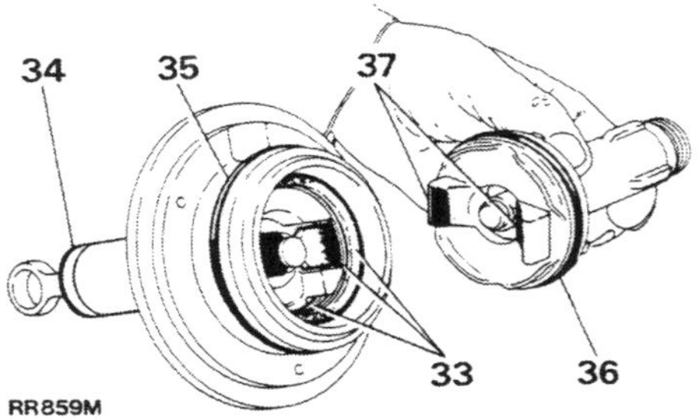

RR859M

38. Place the new diaphragm onto the valve body, locating the inner diaphragm bore onto the shoulder just outside the three bolt holes.

NOTE: DO NOT LUBRICATE THE DIAPHRAGM.

39. Fit the diaphragm support plate and tube to the valve body ensuring that the indent in the support plate lines up with the corresponding projection on the valve body.

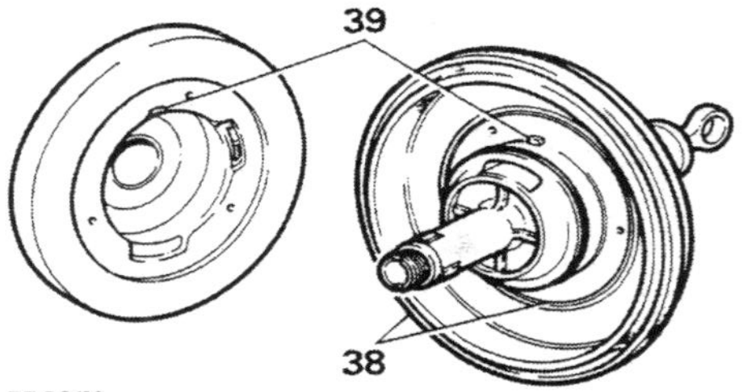

RR981M

40. Fit the three securing screws washers and nuts, entering the bolts from the valve body side.

NOTE: The fibre washer is inserted behind the screw head.

**Tighten to the correct torque see 'Data Section'.**

41. Coat the new nylon bearing and seal with Lockheed disc brake lubricant and insert the seal into the separator shell, with the open edge of the seal facing upwards.

42. Fit the nylon bearing washer with the chamfered inner edge downwards.

43. Using a suitable service tool, example MS 550 press in the new retaining ring, convex side first, until the retainer sits firmly on the nylon bearing.

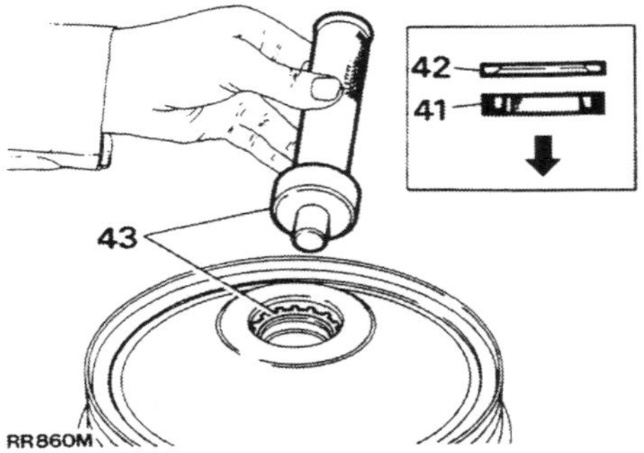

RR860M

44. Fit the backing washer to the concave side of the separator shell diaphragm.

45. Fit the diaphragm to the separator shell, concave side first, locate the sealing lip over the edge of the separator.

46. Place the support plate onto the face of the diaphragm.

47. Mount the valve body into a bench vice.

**CAUTION: DO NOT OVERTIGHTEN THE VICE. Coat the support plate tube with disc brake lubricant.**

48. Place the spring onto the valve body, largest diameter to sit on support plate and tube.

49. Holding the separator shell assembly together place it on to the top of the spring.

50. Compress the complete assembly until the thread on the end of the reaction piston protrudes beyond the separator shell assembly.

51. Fit a new tab washer and screw on the large nut.

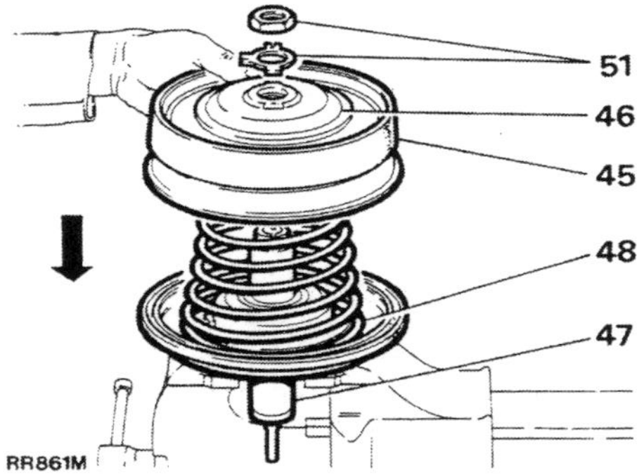

RR861M

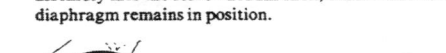

52. Remove the complete assembly from the vice. Tighten the locknut to the correct torque. See 'Data Section'.

53. Bend up the tabs on the lock washer.

54. Locate the outer edge of the valve body diaphragm into the indent around the separator shell and push the assembly into the servo vacuum shell, ensure that the diaphragm remains in position.

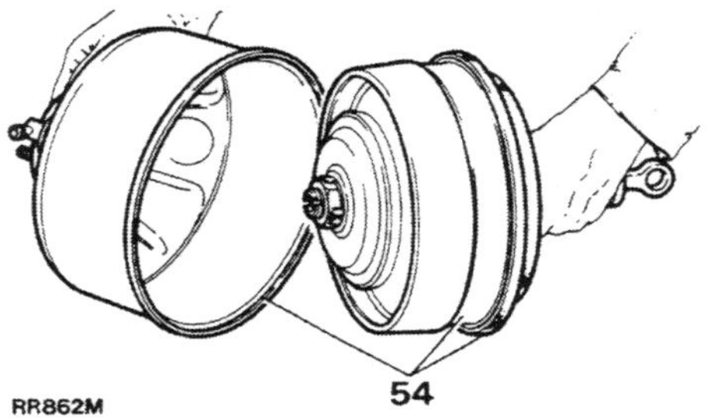

RR862M

55. Coat the new nylon bearing and seal for the servo end cover with disc brake lubricant and press the seal into the cover.

**NOTE: THE OPEN FACE OF THE SEAL DOWNWARDS.**

56. Fit the new bearing retainer using a suitable service tool example: MS 550. Ensure the retainer sits firmly on top of the nylon bearing.

57. Lightly coat the valve body with disc brake lubricant ease the end cover onto the assembly and align the identification marks. Locate the beaded edge of the diaphragm between the end cover and servo shell.

58. Fit the clamp ring, bolt and nut, rotate the ring until the identification mark lines up with those on the servo unit. Tighten the clamp ring to the correct torque. See 'Data section'.

59. Fit a new rubber boot over the valve/push rod assembly.

*Continued*

# BRAKES

60. Fit a new nylon washer, retainer and seal to the push rod, fit the push rod to the servo.

NOTE: Flat face of the seal to the bottom of the seal recess in the servo shell.

(a). Nylon washer.
(b). Retainer.
(c). Seal.

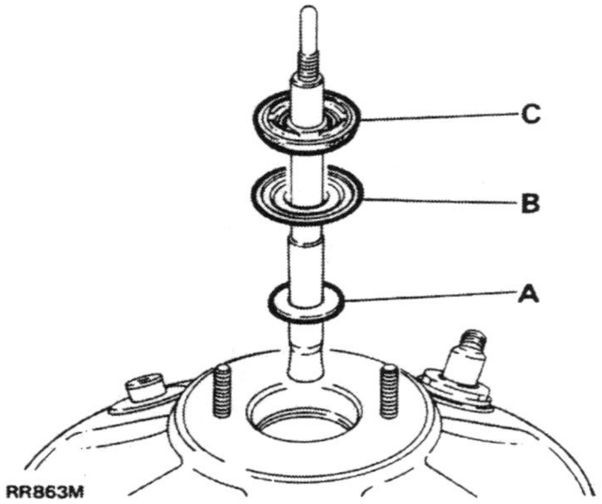

61. Run two nuts down the studs at the master cylinder joint face of the servo. Set the top of the nuts to a dimension of 22.1–21.9 mm (0.871–0.861 inch).
62. Lay a straight edge across the top of the two nuts. Check the height of the push rod to the bottom of the straight edge.
63. If the push rod is out of the limits specified adjust as follows.
64. Remove the push rod from the servo, clamp the small domed end of the rod in a bench vice and detach the main stem.
65. Remove the small spacer. Increase or decrease the size of the spacer accordingly until the correct dimension is attained.

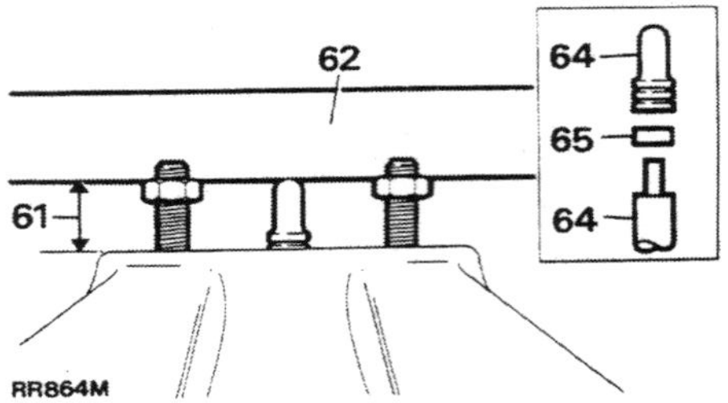

NOTE: If a new push rod is fitted, the operating length must be checked, if adjustment to the rod is required, adjust as follows.

66. Coat the push rod seal with Lockheed disc brake lubricant and fit to the servo.
67. Carefully prise the non-return valve and seal out of the servo shell.
68. Fit new seal.
69. Inspect the non-return valve for condition, renew if necessary.
70. Fit non-return valve.
71. Install the servo into the vehicle and tighten all bolts to the correct torque see 'Data section'.
72. Reconnect the servo operating rod to the brake pedal with the pivot bolt eccentric in the forward position, do not fully tighten the pivot bolt nut.
73. Turn the pivot bolt to bring the brake pedal back until it just contacts the rubber then secure the pivot bolt nut.

## FRONT BRAKE CALIPER ASSEMBLY

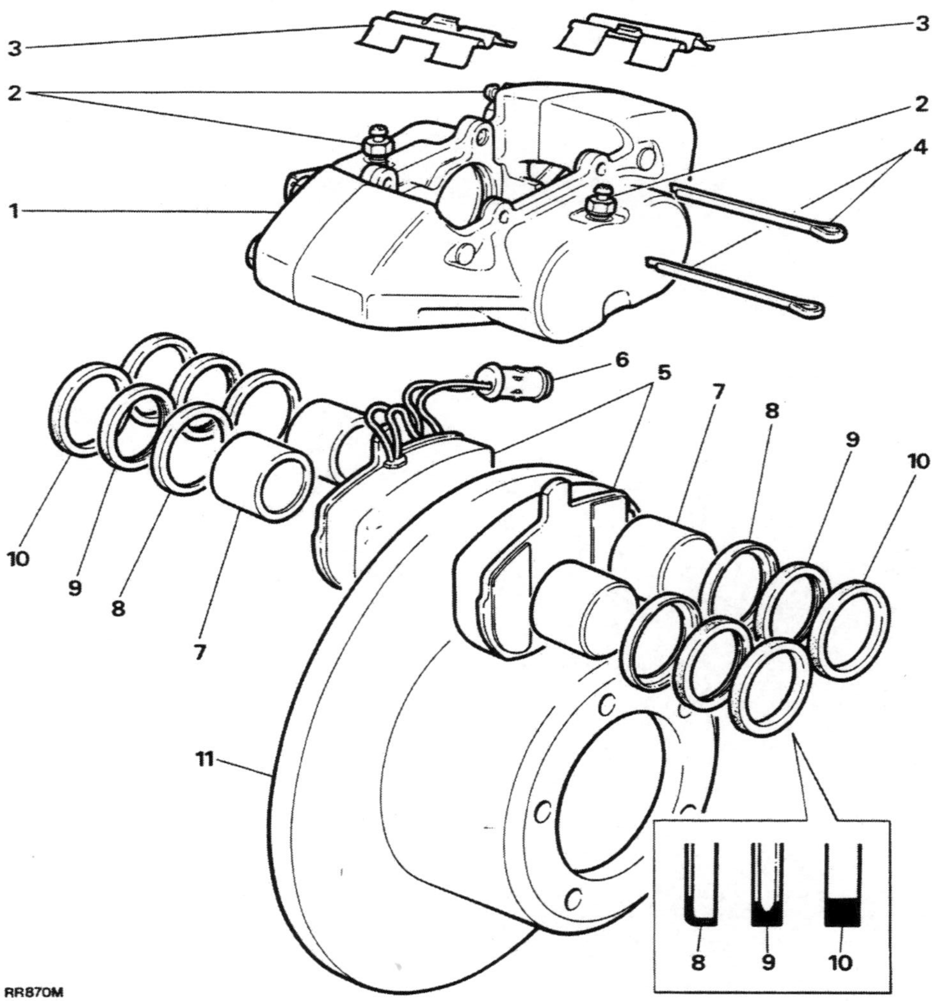

### KEY TO CALIPER

1. Caliper
2. Bleedscrews
3. Pad retaining springs
4. Retaining pins
5. Friction pads
6. Pad wear indicator plug
7. Piston
8. Wiper seal retainer
9. Wiper seal
10. Fluid seal
11. Brake disc

# BRAKES 70

## REMOVE AND OVERHAUL FRONT BRAKE CALIPERS

**Service tool:**
18G 672—Piston clamp

NOTE: Pad wear warning indicators are incorporated into the front right-hand inboard pad and the rear left hand inboard pad.

### Remove caliper

1. Slacken the front wheel retaining nuts, jack up the front of the vehicle and lower onto code stands and remove the wheels.
2. Expose the two flexible brake hoses by moving the coiled protective covering.
3. Clamp both hoses to prevent loss of brake fluid, and disconnect the hoses from the caliper.
4. Disconnect the pad wear warning indicator (front right-hand side only).
5. Remove the retaining pins and springs, withdraw the pads. If the same pads are to be refitted, identify them for assembly to their original locations.
6. Remove the two bolts and withdraw the caliper from the disc.

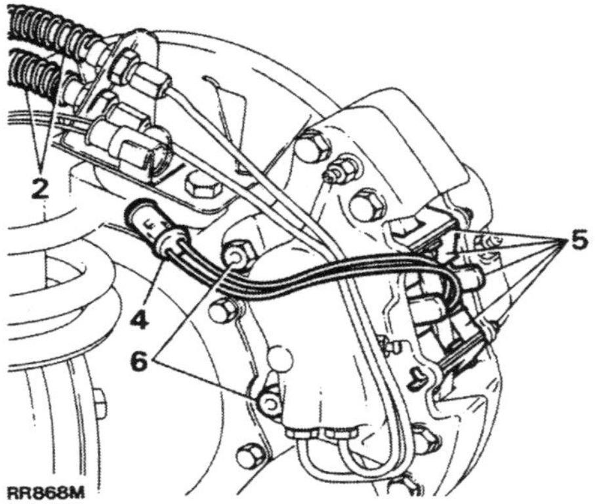

### DISMANTLE AND OVERHAUL

#### Do not separate the caliper halves

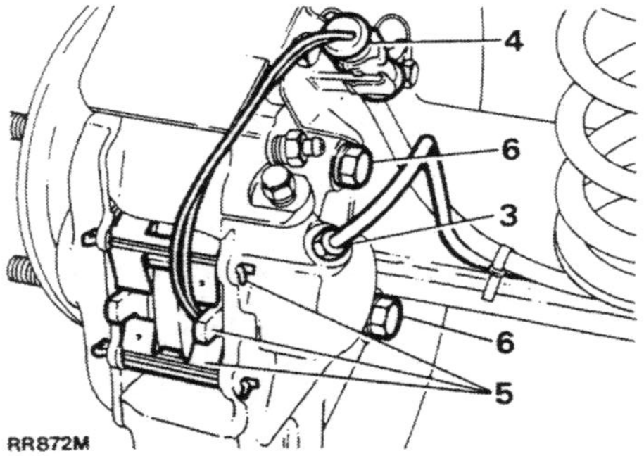

7. Clean the outer surfaces of the caliper with methylated spirit.
8. Using special tool 18G 672, clamp the pistons in the mounting half of the caliper and gently, keeping fingers clear, and with CAUTION, apply air pressure to the fluid inlet port to expel the rim half pistons. Since it is unlikely that both pistons will expel at the same time, regulate the rate with a suitable piece of timber between the appropriate piston and caliper.

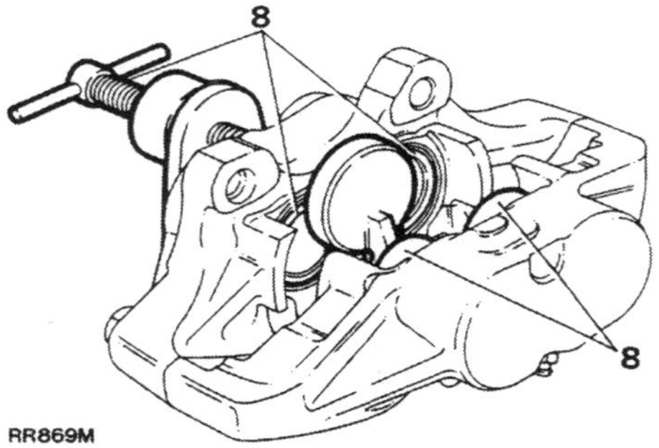

18

2. Clamp the brake flexible hose to the caliper and tighten to the correct torque, see data section.
3. Remove the clamps from the hoses.
4. Lightly smear the back and edges of the pads with disc brake lubricant carefully avoiding the friction material.
5. Insert the pads and retaining springs and secure with new pins and splay the ends. Note the correct position of the retaining springs.

**NOTE: Ensure that the friction pad with the wear indicator is fitted to the inboard side of the front right-hand caliper.**

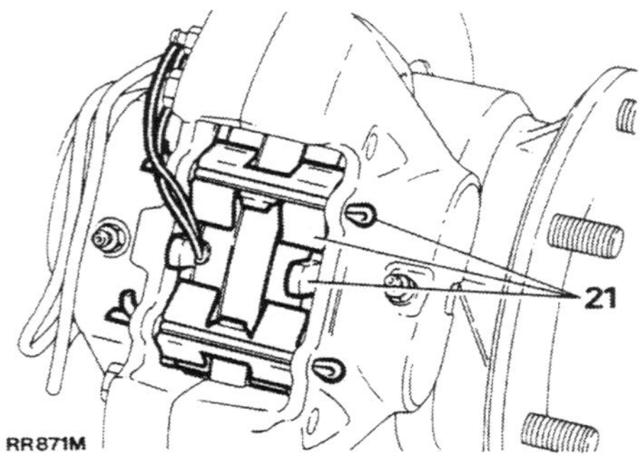

18. Connect the brake flexible hose to the caliper and tighten to the correct torque, see data section.
19. Remove the clamps from the hoses.
20. Lightly smear the back and edges of the pads with disc brake lubricant carefully avoiding the friction material.
21. Insert the pads and retaining springs and secure with new pins and splay the ends. Note the correct position of the retaining springs.

9. Finally, remove the pistons keeping them identified with their respective bores.
10. Remove the wiper seal retainer by inserting a blunt screwdriver between the retainer and the seal and prise the retainer carefully from the mouth of the bore.
11. Taking care not to damage the seal grooves, extract the wiper seal and fluid seal.
12. Clean the bores, pistons and particularly the seal grooves with clean brake fluid or methylated spirit only. If the caliper or pistons are corroded or if their condition is not perfect the parts must be renewed.

### Assemble rim-half pistons

13. Coat a new fluid seal with Lockheed disc brake lubricant. Ease the seal into the groove in the bore using only the fingers and ensure that it is properly seated. The fluid seal and the groove are not the same in section so that when the seal is seated it feels proud to the touch at the edge furthest away from the mouth of the bore.
14. Smear the appropriate piston with disc brake lubricant and insert it squarely into the bore by hand only. Do not tilt the piston during insertion and leave approximately 8 mm projecting from the bore.
15. Coat a new wiper seal with disc brake lubricant and fit it to a new seal retainer. Slide the assembly, seal first, over the protruding piston and into the bore recess. Remove the piston clamp from the mounting half and use the clamp to press home the seal retainer and piston.

### Mounting rim-half pistons

16. Clamp the rim-half pistons and carry out the same procedure as for removing and fitting the rim-half pistons and seals, instructions 8 to 15.

### Fit calipers and pads to vehicle

17. Fit the caliper to the axle and secure with the two bolts tightening evenly to the correct torque, see data section.

22. Reconnect the pad wear indicator electrical plug.
23. Bleed both the primary and secondary brake systems at the front calipers.
24. When the foregoing instructions have been completed on both calipers, depress the brake pedal firmly several times to locate the friction pads.
25. Fit the road wheels, remove the axle stands and finally tighten the road wheel nuts.
26. Road test the vehicle, remembering that if new friction pads have been fitted they are not 'bedded-in' and may take several hundred miles before the brakes are at maximum efficiency.

## REMOVE AND OVERHAUL REAR BRAKE CALIPERS

**Service tool:**
18G 672—Piston clamp

### Remove caliper

1. Slacken the rear road wheel nuts and jack up the rear of the vehicle, and lower onto axle stands and remove the wheels.

### Dismantle and overhaul

#### Do not separate the caliper halves

7. Clean the outer surfaces of the caliper with methylated spirit.
8. Expel the pistons from their bores by applying air pressure to the fluid inlet port. Since it is unlikely that both pistons will expel at the same time, regulate the rate with a suitable piece of timber inserted between the two pistons.

*Continued*

19

# BRAKES

9. Finally, remove the pistons keeping them identified with their respective bores.
10. Remove the wiper seal retainer by inserting a blunt screwdriver between the retainer and the seal and prise the retainer carefully from the mouth of the bore.
11. Taking care not to damage the seal grooves, extract the wiper seal and fluid seal.
12. Clean the bores, pistons and particularly the seal grooves with clean brake fluid or methylated spirit only. If the caliper or pistons are corroded or if their condition is not perfect the parts must be renewed.

## REAR BRAKE CALIPER ASSEMBLY

**LH Rear Caliper Illustrated**

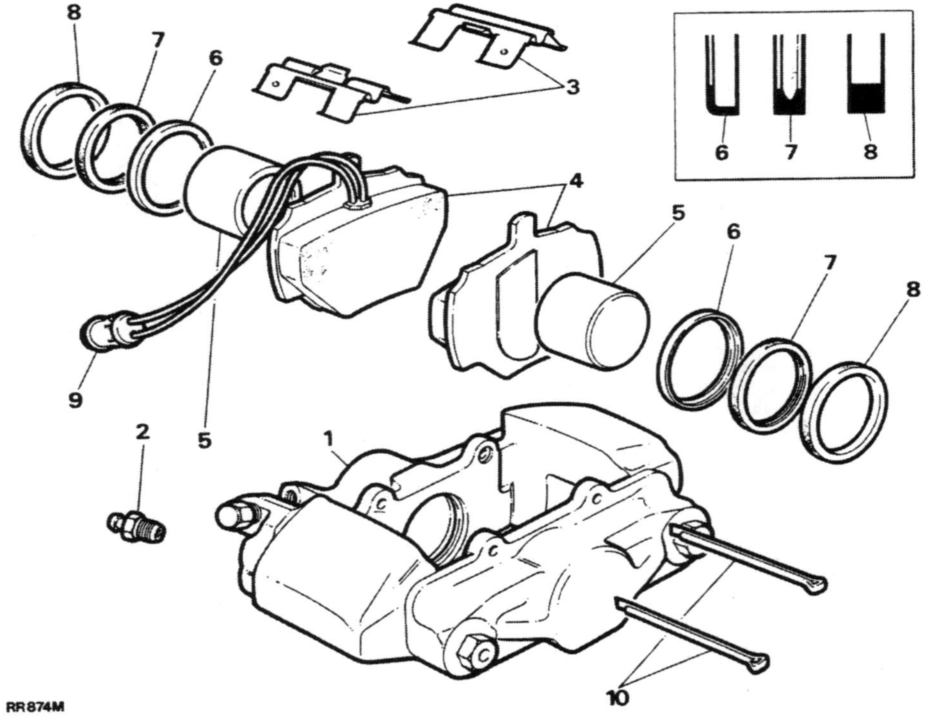

### KEY TO CALIPER

1. Caliper.
2. Bleed screw.
3. Pad retaining springs.
4. Friction pads.
5. Piston.
6. Wiper seal retainer.
7. Wiper seal.
8. Fluid seal.
9. Pad wear indicator plug.
10. Retaining pins.

### Assemble rim-half piston

13. Coat a new fluid seal with Lockheed disc brake lubricant. Ease the seal into the groove in the bore using only the fingers and ensure that it is properly seated. The fluid seal and the groove are not the same in section so that when the seal is seated it feels proud to the touch at the edge furthest away from the mouth of the bore.

*Continued*

14. Smear the appropriate piston with disc brake lubricant and insert it squarely into the bore by hand only. Do not tilt the piston during insertion and leave approximately 8 mm projecting from the bore.
15. Coat a new wiper seal with disc brake lubricant and fit it to a new seal retainer. Slide the assembly, seal first, over the protruding piston and into the bore recess.
16. Using special tool 18G 672—piston clamp, press home the seal retainer and piston.

### Mounting rim-half piston

17. Carry out the same procedure as for removing and fitting the rim-half piston and seals, instructions 8 to 16.

### Fit calipers and pads to vehicle

18. Fit the caliper to the axle and secure with the two bolts tightening evenly to the correct torque, see 'Data section'.
19. Connect the brake pipes to the calipers and remove the clamp from the flexible brake hose above the rear axles, see 'Data section' for brake pipe to caliper tightening torque.
20. Lightly smear the back and edges of the pads with disc brake lubricant carefully avoiding the friction material.
21. Insert the pads and retaining springs and secure with new pins and splay the ends. Note the correct position of the retaining springs.

**NOTE: Ensure that the friction pad with the wear indicator is fitted to the inboard side of the rear left-hand brake caliper.**

22. Reconnect the pad wear indicator electrical multi-plug. 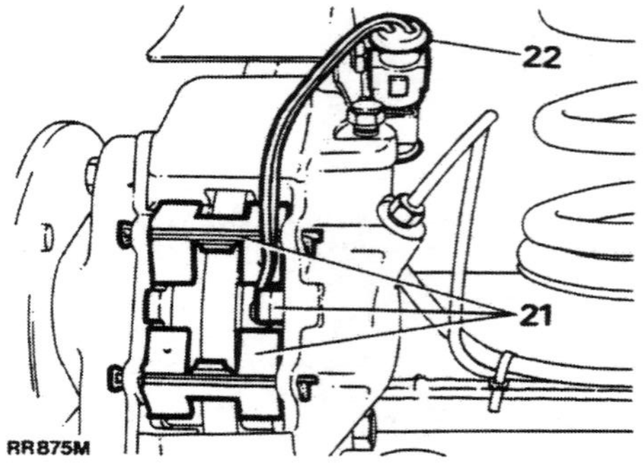
23. Bleed the secondary brake system at the rear calipers, starting at the caliper furthest away from the master cylinder.
24. When the foregoing instructions have been completed on both calipers, depress the brake pedal firmly several times to locate the friction pads.
25. Fit the road wheels, remove the axle stands and finally tighten the road wheel nuts, see data.
26. Road test the vehicle, remembering that if new friction pads have been fitted they are not 'bedded-in' and may take several hundred miles before the brakes are at maximum efficiency.

## TRANSMISSION BRAKE LEVER (HANDBRAKE)

**Remove and refit**

### Removing

1. Disconnect the battery.
2. Chock the road wheels and release the handbrake.
3. Open the cubby box lid and remove the four screws securing the cubby box liner to the outer surround and lift out the liner.
4. Remove the two electrical leads at the rear of the cubby box cigar lighter and release the two heater hoses from their retaining clips.
5. Remove the five electrical multi-plugs from the rear of the window lift switch panel, noting their position for re-assembly.
6. Remove the main gear selector knob and transfer gearbox knob.
7. Carefully prise the centre panel out of the gearbox tunnel mounted console.

**NOTE: On Automatic gearbox models it is necessary to remove the two selector panel illumination bulbs from the graphics panel.**

8. Remove the four bolts and plain washers, two are located immediately behind the handbrake lever accessible from inside the cubby box. The remaining two bolts are located forward of the handbrake lever accessible from the floor mounted console aperture.

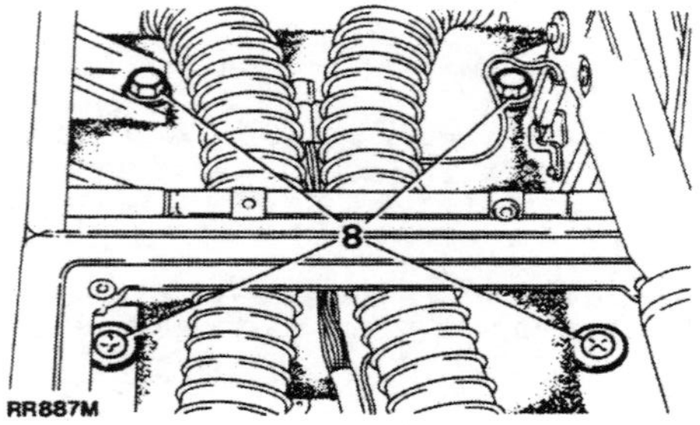

*Continued*

# BRAKES

9. Ease the cubby box and console assembly rearwards to release the location tab from the radio housing and lift the assembly off the gearbox tunnel.
10. Release the split pin and remove the clevis pin, plain and thackeray washer securing the handbrake cable to the lever.
11. Disconnect the electrical lead from the hand brake warning switch.
12. Release the handbrake cable outer retaining nut.
13. Remove the remaining single bolt with plain washer securing the front of the handbrake mounting bracket.
14. Withdraw the handbrake assembly and remove the cable outer.

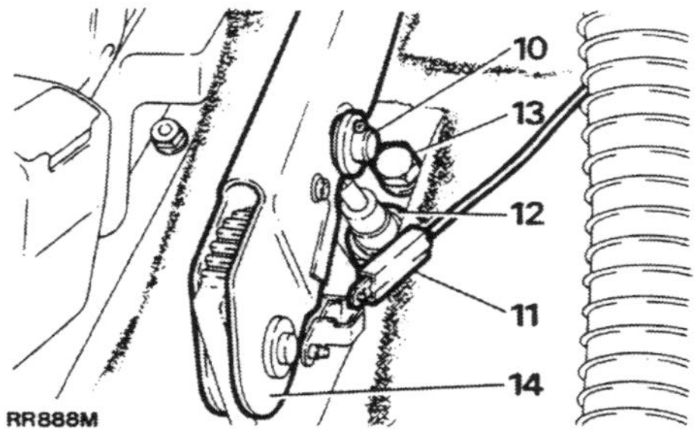

## Refitting

15. Attach the cable outer to the handbrake mounting bracket and tighten the retaining nut securely.
16. Fit the assembly to the gearbox tunnel, lightly secure in position with the front retaining bolt.
17. Fit the inner cable to the handbrake assembly, secure in position with a new split pin.
18. Fit the cubby box console assembly, securing in position with the two screws, two bolts with plain washers.
19. Tighten the two handbrake securing bolts.
20. Reverse the remaining removal instructions, ensuring that the electrical wiring in the cubby box assembly is arranged to prevent it becoming trapped between any mating faces.

## OVERHAUL TRANSMISSION BRAKE

**WARNING: Do not use an air line to remove dust from the brake assembly. Asbestos dust from the brake linings can be a serious health risk, if inhaled.**

### DISMANTLING

1. Disconnect the battery and chock the road wheels for safety.
2. Disconnect the propeller shaft from the output flange.
3. Remove the two screws and withdraw the brake drum. Skim if excessively scored or oval.
4. Remove the split pin and clevis pin connecting the drawlink to the actuating lever.
5. Remove the brake shoes complete with pull-off springs. Note position of springs in relation to the shoes.
6. Remove the four bolts securing back plate to transfer box and withdraw the back plate complete with oil catcher.

### Remove and overhaul expander assembly

7. Remove the rubber dust cover.
8. Remove the expander and draw link.
9. Remove the retainer spring plate.
10. Remove the locking plate.
11. Remove the packing plate and withdraw the expander assembly from the back plate.
12. Remove the two plungers and rollers.

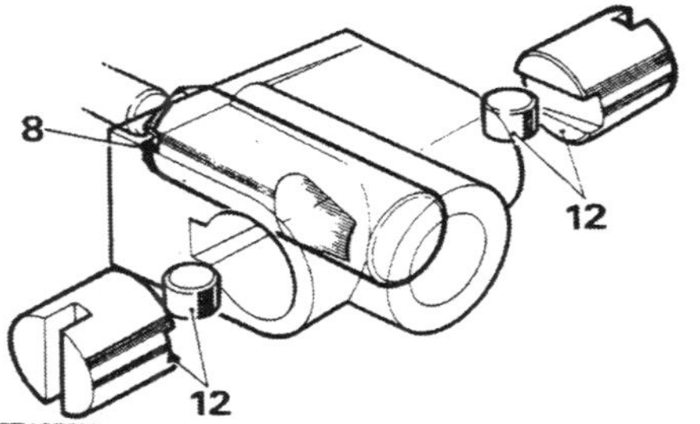

13. Clean all parts in Girling cleaning fluid and allow to dry. Examine the components for wear and discard if unsatisfactory.

### Assemble expander assembly

14. Grease and fit the expander and drawlink.
15. Grease and fit the plungers and rollers noting that the highest end of the ramp on the plungers is fitted towards the back plate. Secure the assembly with a rubber band to prevent the plungers falling out and place to one side for assembly to back plate.

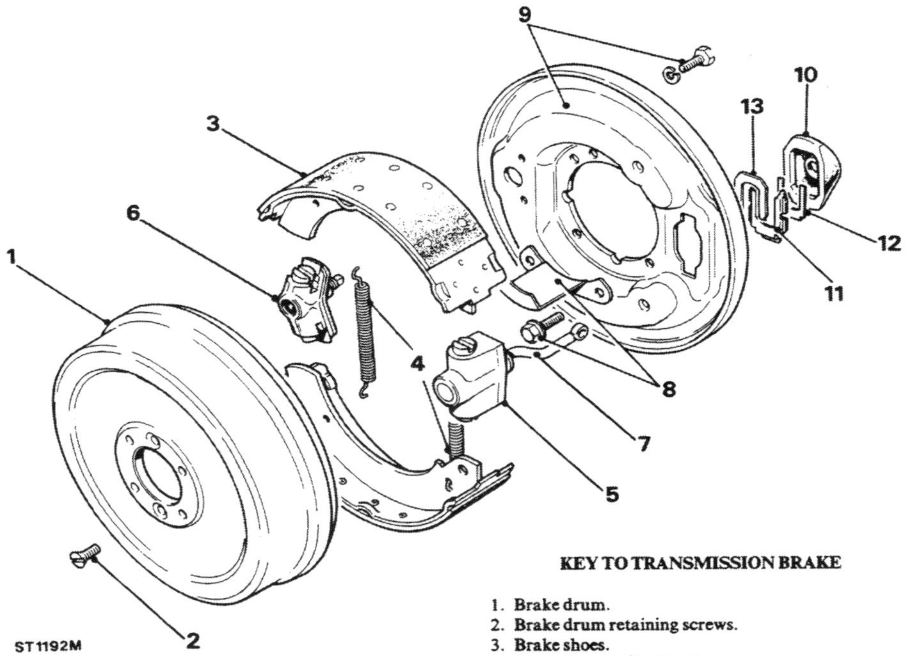

## KEY TO TRANSMISSION BRAKE

1. Brake drum.
2. Brake drum retaining screws.
3. Brake shoes.
4. Brake shoes pull-off springs.
5. Expander assembly.
6. Adjuster assembly.
7. Draw link.
8. Oil catcher.
9. Back plate and retaining bolts.
10. Dust cover.
11. Locking plate.
12. Packing plate.
13. Spring plate.

### Remove and overhaul adjuster assembly

16. Remove the two bolts and withdraw the adjuster assembly from the back plate.
17. Remove the plungers.
18. Screw the adjuster cone inwards to remove from the housing.
19. Clean the parts in Girling cleaning fluid and discard any unsatisfactory components.

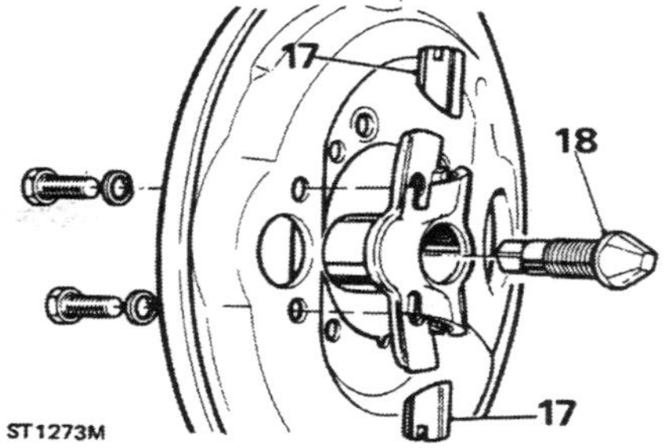

### Assemble adjuster assembly

20. Grease and screw in the adjuster cone.
21. Grease and fit the adjuster plungers and align the chamfered ends with the adjuster cone. Note that the two plungers are identical and can be fitted to either bore. Secure the assembly with a rubber band to prevent the plungers falling out.

*Continued*

# BRAKES

## ASSEMBLE

NOTE: If the brake linings are oil-soaked check and if necessary renew the output shaft oil seal.

22. Position the expander assembly on the inside of the back plate and secure with the following plates at the rear of the back plate.
23. Packing piece.
24. Locking plate.
25. Retainer spring.
26. Fit the rubber dust cover.
27. Fit the adjuster assembly to the back plate with the two bolts but do not fully tighten at this stage.
28. Fit the back plate and mud shield to the transfer box with the four bolts and tighten to the correct torque.
29. Fit new pull-off springs to relined brake shoes and fit to the back plate. Note that the fully lined end of the lower shoe must be toward the expander assembly and the fully lined end of the upper shoe towards the adjuster assembly.
30. Fit the brake drum and secure with the two screws.
31. Connect the expander drawlink to the actuating lever with a new clevis pin, washer and split pin.
32. Turn the adjuster cone fully in and tighten the two retaining bolts left slack in instruction 27.
33. Slacken off the adjuster two 'clicks' and firmly apply the hand lever to centralise the shoes. The drum should then rotate freely.
34. Adjust the handbrake cable to give the pawl two 'clicks' free movement on the ratchet before the third 'click' fully expands the shoes against the drum.
35. Connect the propeller shaft and evenly tighten the retaining nuts to the correct torque.
36. Remove chocks from wheels and connect the battery.

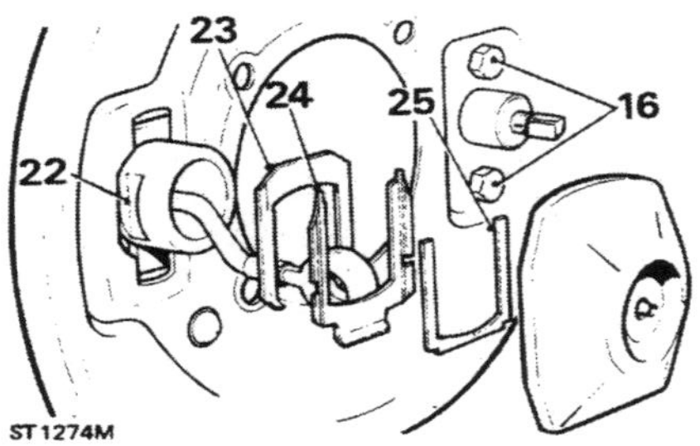

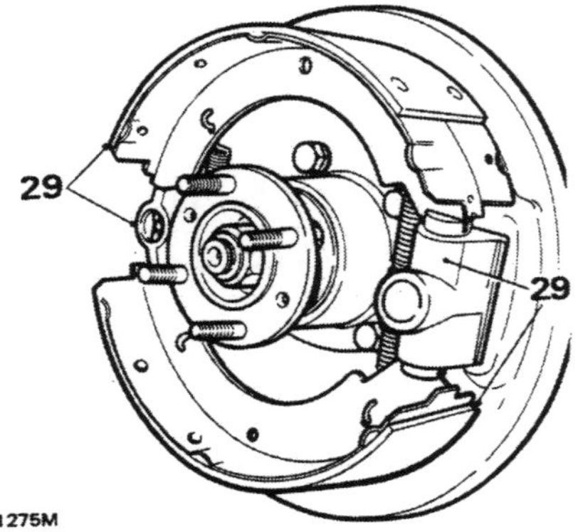

## TRANSMISSION BRAKE CABLE (HANDBRAKE)

### Remove and refit

#### Removing

1. Set the vehicle on level ground and chock the wheels.
2. Release the handbrake.

#### From inside the vehicle

3. Remove the two screws securing the liner in the cubby box.
4. Lift out the liner to gain access to the bottom of the handbrake pivot bracket.
5. Remove the split pin and clevis pin from the handbrake lever.
6. Release the nut securing the handbrake outer cable to the top of the tunnel. Slide the nut up the cable and push the inner and outer cable through the floor panel to the underside of the vehicle.

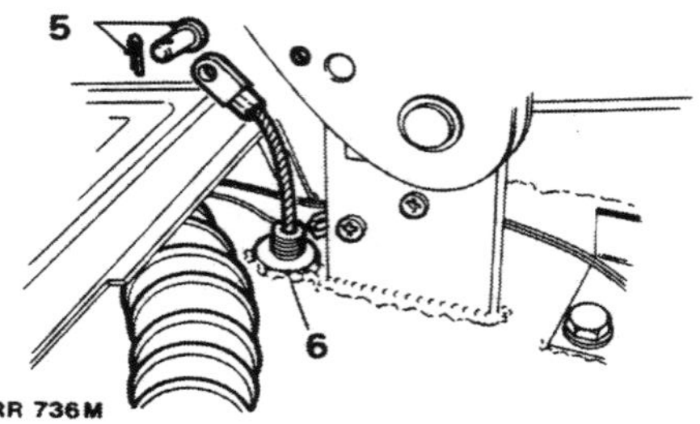

#### From underneath the vehicle

7. Remove the split pin, plain and thackery washer and clevis pin securing the adjustment link to the brake drum actuating lever.
8. Remove the four bolts with spring washers securing the handbrake linkage assembly to the side of the transfer gearbox.
9. Withdraw the linkage from beneath the vehicle complete, with handbrake cable attached.
10. Remove the split pin, plain and thackery washer securing the cable to the handbrake linkage.
11. Withdraw the clevis pin.
12. Release the locknuts securing the handbrake outer cable to the retaining bracket.
13. Remove the cable from the bracket.

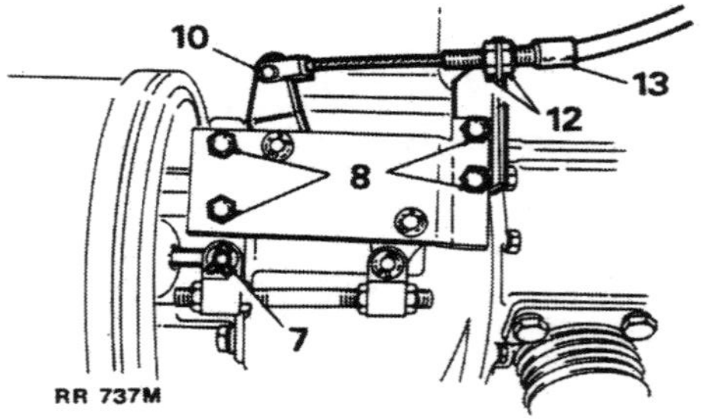

### Fit new cable

14. Feed the handbrake cable through the floor aperture and secure the outer cable in position with the retaining nut.
15. Secure the cable to the handbrake lever, using a new split pin.
16. Fit the other end of the cable to the handbrake linkage using a new split pin.
17. Secure the outer cable to the retaining bracket and lightly tighten the two locknuts.

NOTE: To ensure that adjustment is not exhausted on the adjustment link, initial handbrake setting should be taken up at the outer cable locknuts.

18. Apply the handbrake, the brake shoes should make contact with the drum on the first notch of the handbrake quadrant. Adjust the locknuts on the outer cable as necessary. Release the handbrake.
19. Refit the handbrake linkage assembly to the side of the transfer box, tighten the bolts to the specified torque, see 'data' section.
20. Rotate the brake drum adjuster clockwise until the shoes are fully expanded against the drum.
21. Release the four locknuts at the adjustment link, rotate the link until the clevis pin holes in the adjustment link and brake drum actuating lever line up.
22. Refit the clevis pin, washers and split pin.
23. Apply the handbrake, and slacken the brake drum adjuster until the handbrake lever fully operates the brake shoes on the third or fourth handbrake quadrant notch.

## Body repairs, general information

1. The Range Rover body consists of a steel frame to which alloy outer panels are attached. The radiator grille, front deck panel, front wings, side door outer panels, body side outer panels, roof, rear floor and upper rear quarter panels are made from a special light magnesium-aluminium alloy known as 'Birmabright'.

2. 'Birmabright' was developed for aircraft use, and it is much stronger and tougher than pure aluminium. It melts at a slightly lower temperature than pure aluminium and will not rust nor corrode under any normal circumstances. It is work-hardening, and so becomes hard and brittle when hammered, but it is easily annealed. Exposed to the atmosphere, a hard oxide skin forms on the surface of it.

## Panel beating 'Birmabright'

3. 'Birmabright' panels and wings can be beaten out after accidental damage in the same way as sheet steel. However, under protracted hammering the material will harden, and then it must be annealed to prevent the possibility of cracking. This is quite easily done by the application of heat, followed by slow air-cooling, but as the melting point is low, heat must be applied slowly and carefully. A rough but very useful temperature control is to apply oil to the cleaned surface to be annealed. Play the welding torch on the underside of the cleaned surface and watch for the oil to clear, which it will do quite quickly, leaving the surface clean and unmarked. Then allow to cool naturally in the air, when the area so treated will again be soft and workable. Do not quench with oil or water. Another method is to clean the surface to be annealed and then rub it with a piece of soap. Apply heat beneath the area, as described above, and watch for the soap stain to clear. Then allow to cool, as for the oil method. When applying the heat for annealing, always hold the torch some little distance from the metal, and move it about, so as to avoid any risk of melting it locally.

## Gas welding 'Birmabright'

4. A small jet must be used, one or two sizes smaller than would be used for welding sheet steel of comparable thickness. For instance, use a No. 2 nozzle for welding 18 swg (0.048 in.) sheet, and a No. 3 for 16 swg (0.064 in.) sheet.

5. The flame should be smooth, quiet and neutral, have a brilliant inner core with a well defined rounded end. The hottest point of the flame is close to the jet, and the flame should have a blue to orange envelope becoming nearly colourless at the end.

6. A slightly reducing flame may also be used, that is, there may be a slight excess of acetylene. Such a flame will have a brilliant inner core with a feathery white flame and a blue to orange envelope.

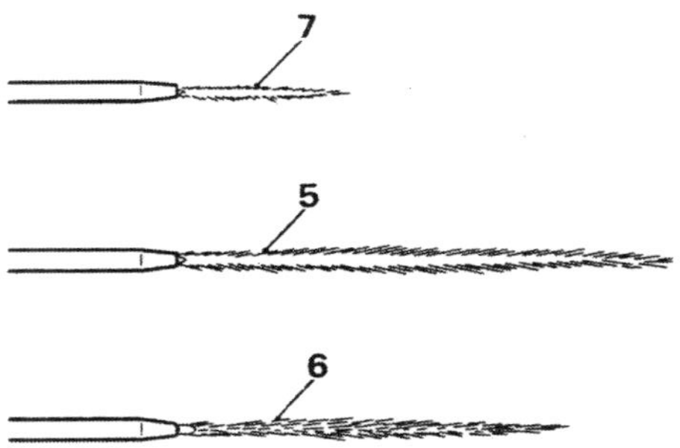

7. Do not use an oxydising flame, which has a short pointed inner core bluish white with a bluish envelope.

8. Use only 5 per cent magnesium/aluminium welding rod (5 Mg/A). Sifalumin No. 27 (MG.5 Alloy) (Use Sifbronze Special flux with this rod) or a thin strip cut of parent metal—that is to say, a strip cut from an old and otherwise useless 'Birmabright' panel or sheet. Do not use too wide or thick a strip, or trouble may be experienced in making it melt before the material which is being welded.

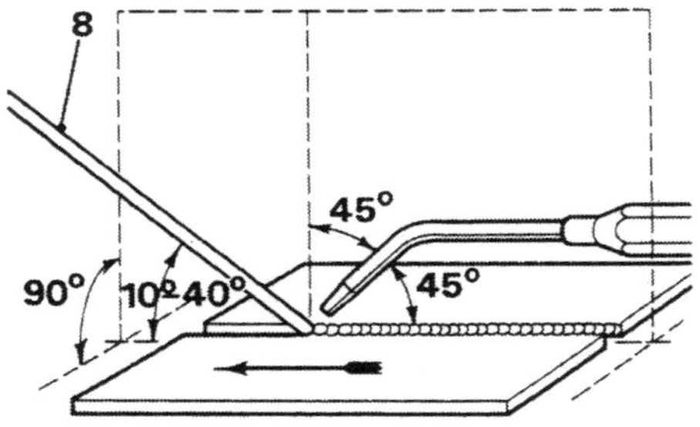

9. Clean off all grease and paint, dry thoroughly and then clean the edges to be welded, and an area at least half an inch on either side of the weld, with a stiff wire scratch-brush or wire wool. Cleanliness is essential. Also clean the welding rod or strip with wire wool. A special acid flux must be used, and we recommend 'Hari-Kari' which is obtainable from:
The Midland Welding Supply Co. Ltd.,
105 Lakey Lane,
Birmingham 28, England.
or
Sifbronze Special Flux, which is obtainable from:
Suffolk Iron Foundry (1920) Ltd.,
Sifbronze Works,
Stowmarket, England.

11. A small quantity of 'Hari-Kari' may be made into a paste with water, following the directions on the tin, and the paste must be applied to both surfaces to be welded, and also to the rod. In the case of Sifbronze Special Flux use in powder form as directed. Remember that aluminium and its alloys do not show 'red-hot' before melting, and so there is nothing about the appearance of the metal to indicate that it has reached welding temperature. A little experience will enable the operator to gauge this point, but a useful guide is to sprinkle a little sawdust over the work; this will sparkle and char when the right temperature is approached; a piece of dry wood rubbed over the hot metal will sparkle at the point of contact.

12. As the flux used is highly acid, it is essential to wash it off thoroughly immediately after a weld is completed. The hottest possible water should be used, with wire wool or a scratch-brush. Very hot soapy water is good, because of the alkaline nature of the soap, which will tend to 'kill' the acid.

13. It is strongly recommended that a few welds are made on scrap metal before the actual repair is undertaken if the operator is not already experienced in welding aluminium and its alloys.

14. The heat of welding will have softened the metal in the area of the repair, and it may be hardened again by peening with a light hammer. Many light blows are preferable to fewer heavy ones. Use a 'dolly' or anvil behind the work to avoid denting and deformation, and to make the hammering more effective. Filing of surplus metal from the weld will also help to harden the work again.

### Welding tears and patching

15. If a tear extends to the edge of a panel, start the weld from the end away from the edge and also at this point drill a small hole to prevent the crack spreading, then work towards the edge.

16. When welding a long tear, or making a long welded joint, tack the edges to be welded at intervals of from 2 in to 4 in (50 to 100 mm) with spots. This is done by melting the metal at the starting end and fusing into it a small amount of the filler rod, repeating the process at the suggested intervals. After this, weld continuously along the joint from right to left, increasing the speed of the weld as the material heats up.

17. After the work has cooled, wash off all traces of flux as described above, and file off any excess of build-up metal.

18. When patching, cut the patch to the correct shape for the hole to be filled, but of such sizes as to leave a gap of $\frac{1}{32}$ in (0.80 mm) between it and the panel, and then weld as described above. Never apply an 'overlay' patch.

### Electric welding

19. **CAUTION. The battery earth lead must be disconnected before commencing electric welding, otherwise the alternator will be damaged.**

20. At the Rover factory the 'Argon-Arc' process is used, and this is very satisfactory, since all atmospheric oxygen is excluded from the weld by the Argon gas shield. For all body repair work normally undertaken by a Distributor's or Dealer's service department, the gas welding method is sufficient and quite satisfactory.

### Spot welding

21. Spot welding is largely used in the manufacture of Range-Rover bodies, but this is a process which can only be carried out satisfactorily by the use of the proper apparatus. Aluminium and its alloys are very good conductors of heat and electricity, and thus it is most important to maintain the right conditions for successful spot welding. The correct current density must be maintained, and so must the 'dwell' of the electrodes. Special spot welding machines have been developed, but they are expensive, and though the actual work can be carried out by comparatively unskilled labour, supervision and machine maintenance must be in the hands of properly qualified persons.

### Riveting

22. Where both sides of the metal are accessible and it is possible to use an anvil or 'dolly', solid aluminium rivets may be used, with a suitable punch or 'pop' to ensure clean rounded head on the work. For riveting blind holes, 'pop-rivets' must be used. These are inserted and closed by special 'Lazy-Tong' 'pop-rivet' pliers.

### Painting 'Birmabright'

23. Refer to the procedure detailed in Paintwork.

## PAINTWORK

### General information

### Body panels

1. Range Rover body panels are manufactured from a special aluminium-alloy known as 'Birmabright' and the following paintwork procedure should be followed on these panels.

### Painting 'Birmabright'

2. The area to be painted must be flatted to remove the hard oxide skin which forms on the surface of the alloy when exposed to the atmosphere. Degrease and dry the area, then apply a suitable etch-primer. Unless an etch-primer is used, paint is liable to come away as it cannot 'key' into the hard oxide of an untreated alloy surface and the use of ICI Etching Primer P565-5002 is recommended. It is quick and easy to apply, and it prolongs the life of the paint film by ensuring excellent adhesion.

### Application

3. The activated Etching Primer has a limited pot-life of about eight hours at normal temperatures and should not be used after this time, as it may have inferior adhesion and corrosion resistance. Any Etching Primer which has been mixed for more than eight hours must be thrown away, and not returned to the can.

4. Apply Etching Primer as soon as possible after cleaning, and paint as soon as the pre-treatment is completed. Undue delay may cause the surface to be contaminated again and thus nullify the treatment. Do not leave pre-treated work overnight before it is painted.

5. Etching Primer, when followed by a suitable paint system, gives a film which is very resistant to moisture, but the Etching Primer itself is water sensitive. It should therefore be coated with paint as soon as possible when it is dry.

6. Activate the Etching Primer by mixing it with an equal volume of Activator P273-5021 and allow to stand for 10 minutes.

7. Adjust the spraying viscosity of the mixture if necessary to 22-25 sec. BSB4 Cup by adding small quantities of Thinner 851-565; never add more Activator.

8. Apply by spray to a clean, dry surface in a thin uniform coat, rather than a thick heavy one which may impair adhesion.

9. Air dry for at least 15 minutes before applying undercoat by spray or for two hours before brush application. If required, these times can be shortened by force drying, this also gives increased hardness to the film.

10. Subsequent painting follows normal paintshop practice.

11. When wet flatting the subsequent paint layers take care not to rub through to the Etching Primer. If this does occur allow to dry out thoroughly, dry flat the area and spot in with Etching Primer.

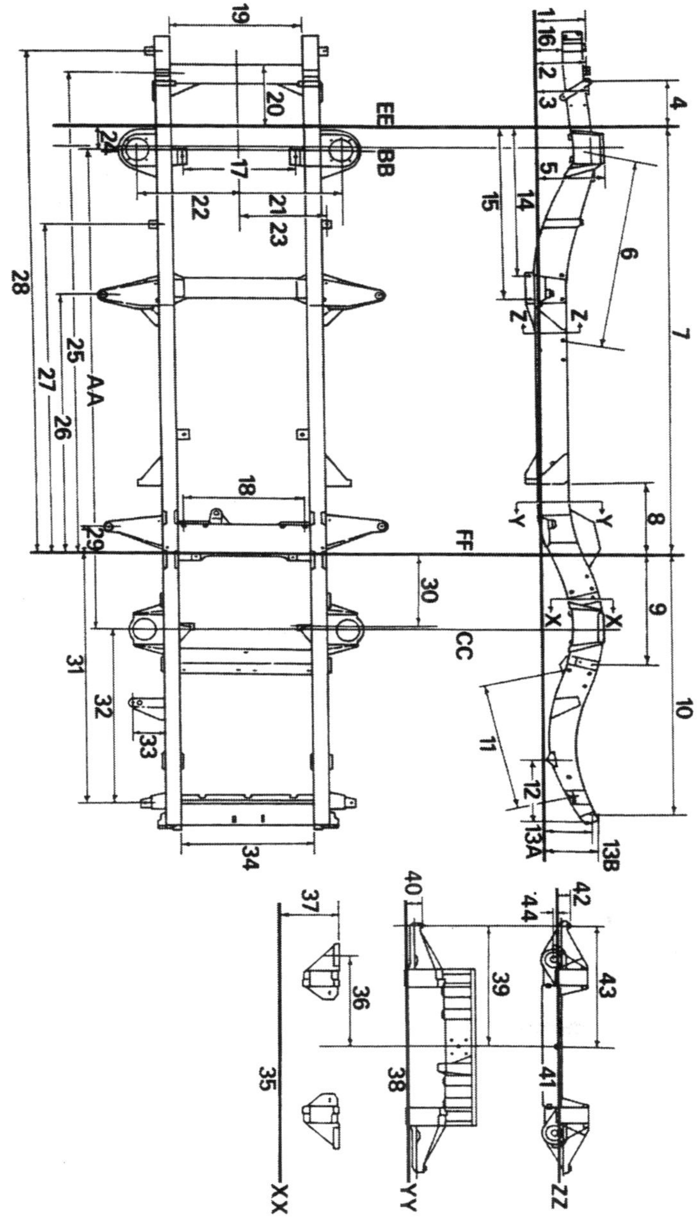

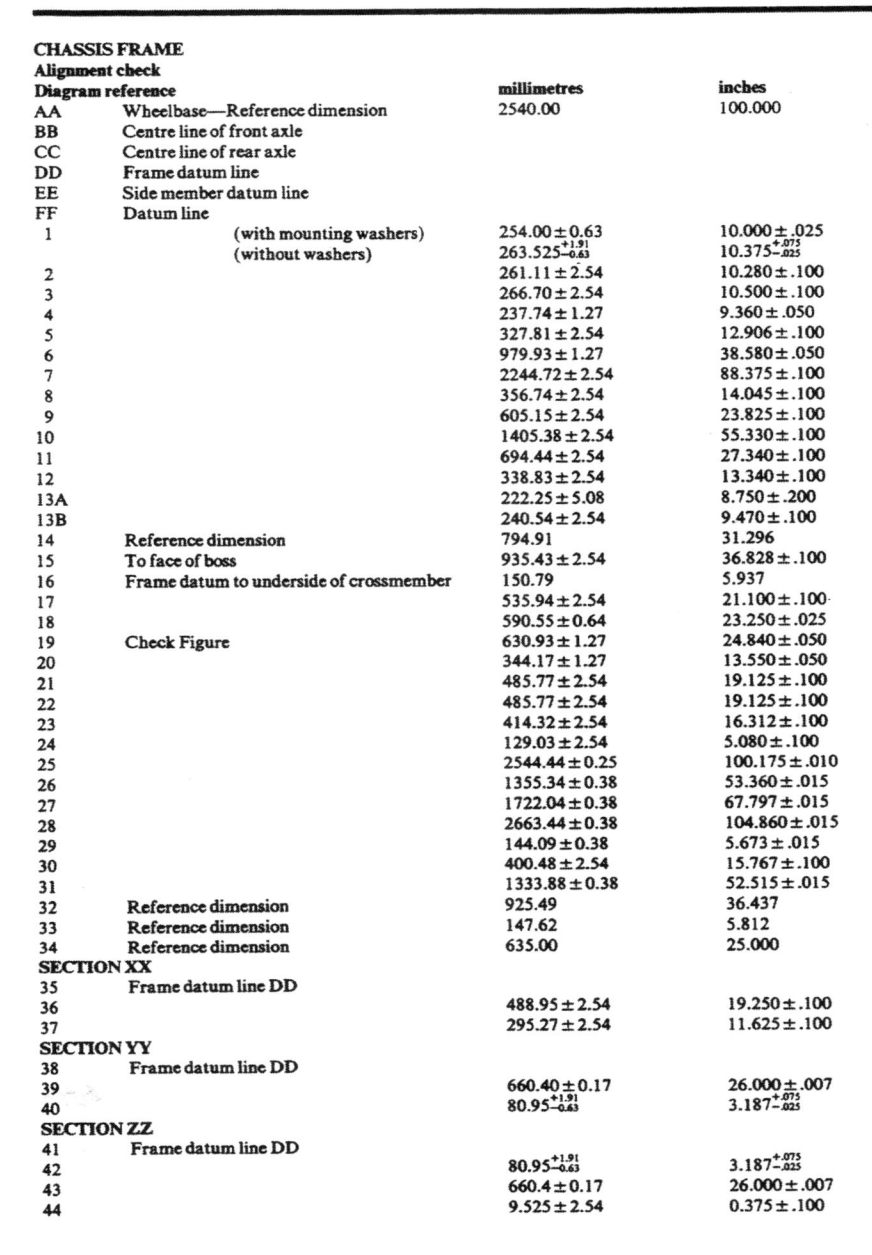

# BODY 76

## CHASSIS FRAME
### Alignment check

| Diagram reference | | millimetres | inches |
|---|---|---|---|
| AA | Wheelbase—Reference dimension | 2540.00 | 100.000 |
| BB | Centre line of front axle | | |
| CC | Centre line of rear axle | | |
| DD | Frame datum line | | |
| EE | Side member datum line | | |
| FF | Datum line | | |
| 1 | (with mounting washers) | 254.00 ± 0.63 | 10.000 ± .025 |
|  | (without washers) | 263.525$^{+1.91}_{-0.63}$ | 10.375$^{+.075}_{-.025}$ |
| 2 | | 261.11 ± 2.54 | 10.280 ± .100 |
| 3 | | 266.70 ± 2.54 | 10.500 ± .100 |
| 4 | | 237.74 ± 1.27 | 9.360 ± .050 |
| 5 | | 327.81 ± 2.54 | 12.906 ± .100 |
| 6 | | 979.93 ± 1.27 | 38.580 ± .050 |
| 7 | | 2244.72 ± 2.54 | 88.375 ± .100 |
| 8 | | 356.74 ± 2.54 | 14.045 ± .100 |
| 9 | | 605.15 ± 2.54 | 23.825 ± .100 |
| 10 | | 1405.38 ± 2.54 | 55.330 ± .100 |
| 11 | | 694.44 ± 2.54 | 27.340 ± .100 |
| 12 | | 338.83 ± 2.54 | 13.340 ± .100 |
| 13A | | 222.25 ± 5.08 | 8.750 ± .200 |
| 13B | | 240.54 ± 2.54 | 9.470 ± .100 |
| 14 | | 794.91 | 31.296 |
| 15 | Reference dimension | 935.43 ± 2.54 | 36.828 ± .100 |
| 16 | To face of boss | 150.79 | 5.937 |
| 17 | Frame datum to underside of crossmember | 535.94 ± 2.54 | 21.100 ± .100 |
| 18 | | 590.55 ± 0.64 | 23.250 ± .025 |
| 19 | Check Figure | 630.93 ± 1.27 | 24.840 ± .050 |
| 20 | | 344.17 ± 1.27 | 13.550 ± .050 |
| 21 | | 485.77 ± 2.54 | 19.125 ± .100 |
| 22 | | 485.77 ± 2.54 | 19.125 ± .100 |
| 23 | | 414.32 ± 2.54 | 16.312 ± .100 |
| 24 | | 129.03 ± 2.54 | 5.080 ± .100 |
| 25 | | 2544.44 ± 0.25 | 100.175 ± .010 |
| 26 | | 1355.34 ± 0.38 | 53.360 ± .015 |
| 27 | | 1722.04 ± 0.38 | 67.797 ± .015 |
| 28 | | 2663.44 ± 0.38 | 104.860 ± .015 |
| 29 | | 144.09 ± 0.38 | 5.673 ± .015 |
| 30 | | 400.48 ± 2.54 | 15.767 ± .100 |
| 31 | | 1333.88 ± 0.38 | 52.515 ± .015 |
| 32 | | 925.49 | 36.437 |
| 33 | Reference dimension | 147.62 | 5.812 |
| 34 | Reference dimension | 635.00 | 25.000 |
| **SECTION XX** | | | |
| 35 | Frame datum line DD | 488.95 ± 2.54 | 19.250 ± .100 |
| 36 | | 295.27 ± 2.54 | 11.625 ± .100 |
| 37 | | | |
| **SECTION YY** | | | |
| 38 | Frame datum line DD | 660.40 ± 0.17 | 26.000 ± .007 |
| 39 | | 80.95$^{+1.91}_{-0.63}$ | 3.187$^{+.075}_{-.025}$ |
| 40 | | | |
| **SECTION ZZ** | | | |
| 41 | Frame datum line DD | 80.95$^{+1.91}_{-0.63}$ | 3.187$^{+.075}_{-.025}$ |
| 42 | | 660.4 ± 0.17 | 26.000 ± .007 |
| 43 | | 9.525 ± 2.54 | 0.375 ± .100 |
| 44 | | | |

RR1545M

# BODY

## BODY

### Introduction:

The information which follows is concerned solely with the 'Monocoque' assembly of the inner body shell on Range Rover models.

Body repairs often require the removal of mechanical and electrical units and associated wiring. Where necessary, reference should be made to the relevant section of the Repair Manual for removal and refitting instructions.

The inner body shell is of 'Monocoque' construction and to gain access to the repair area, it may be necessary to remove exterior body panels, all exterior body panels are bolted to the inner body shell to facilitate easier panel removal and renewal or repair.

It is expected that a repairer will select the best and most economic repair method possible, making use of the facilities available. The instructions given are intended to assist a skilled body repairer by expanding approved procedures for panel replacement with the objective of restoring the car to a safe running condition and effecting a repair which is visually acceptable.

### WELDING

The following charts and illustrations show the locations and types of weld for securing the body side assembly, tailgate frame assembly and the front valance and wheel arch assembly. Before undertaking any spot weld joints to the inner body, it is advisable to make a test joint using offcuts of the damaged components, and to use this test piece to perform a weld integrity test.

Spot welding is satisfactory if the joints do not pull apart. If the weld pulls a hole or tears the metal the weld is satisfactory. It is defective if the weld joint pulls apart or if there are signs of burning, porosity or cracking evident.

### PREPARATION

Thoroughly clean all areas to be welded, remove any sealants and corrosion protectives from around original panels. Align and clamp all new panels in position and check relationship to one another.

## INNER BODY SHELL ASSEMBLY

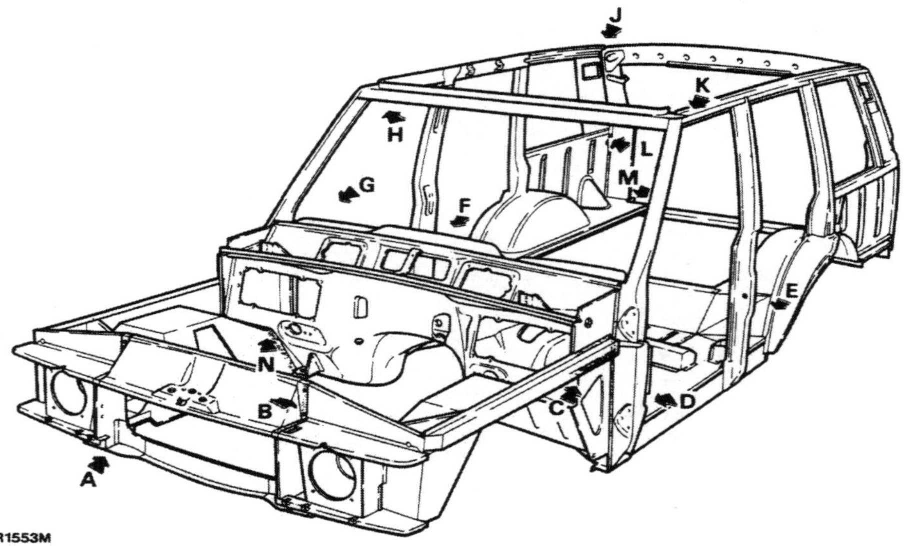

RR1553M

| LOCATION | FACTORY JOINT (minimum number of spot welds quoted) |
|---|---|
| A. Front cross member to valance and wheel arch assembly | 6 spot welds, 20 mm pitch |
| B. Bonnet locking platform to valance and wheel arch assembly | 10 spot welds, 25 mm pitch |
| C. Valance and wheel arch assembly to dash and tunnel assembly | 16 spot welds, 25 mm pitch |
| D. Body side complete to dash and tunnel assembly | 10 spot welds, 65 mm pitch |

*Continued*

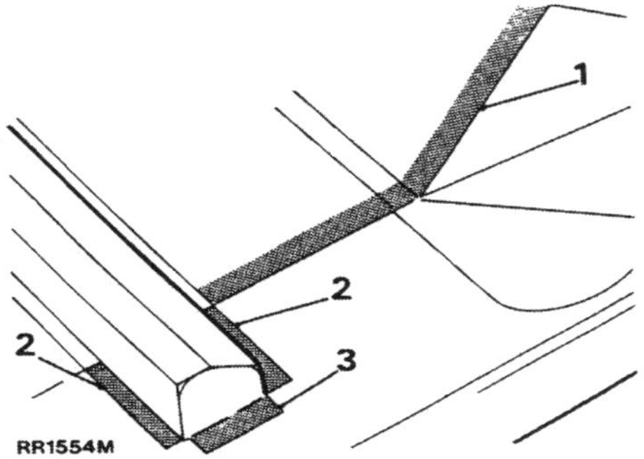

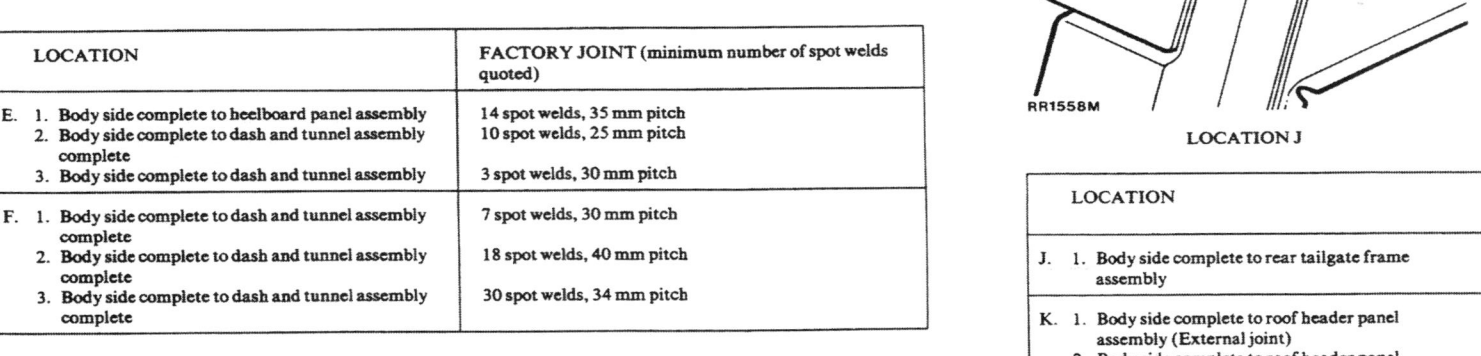

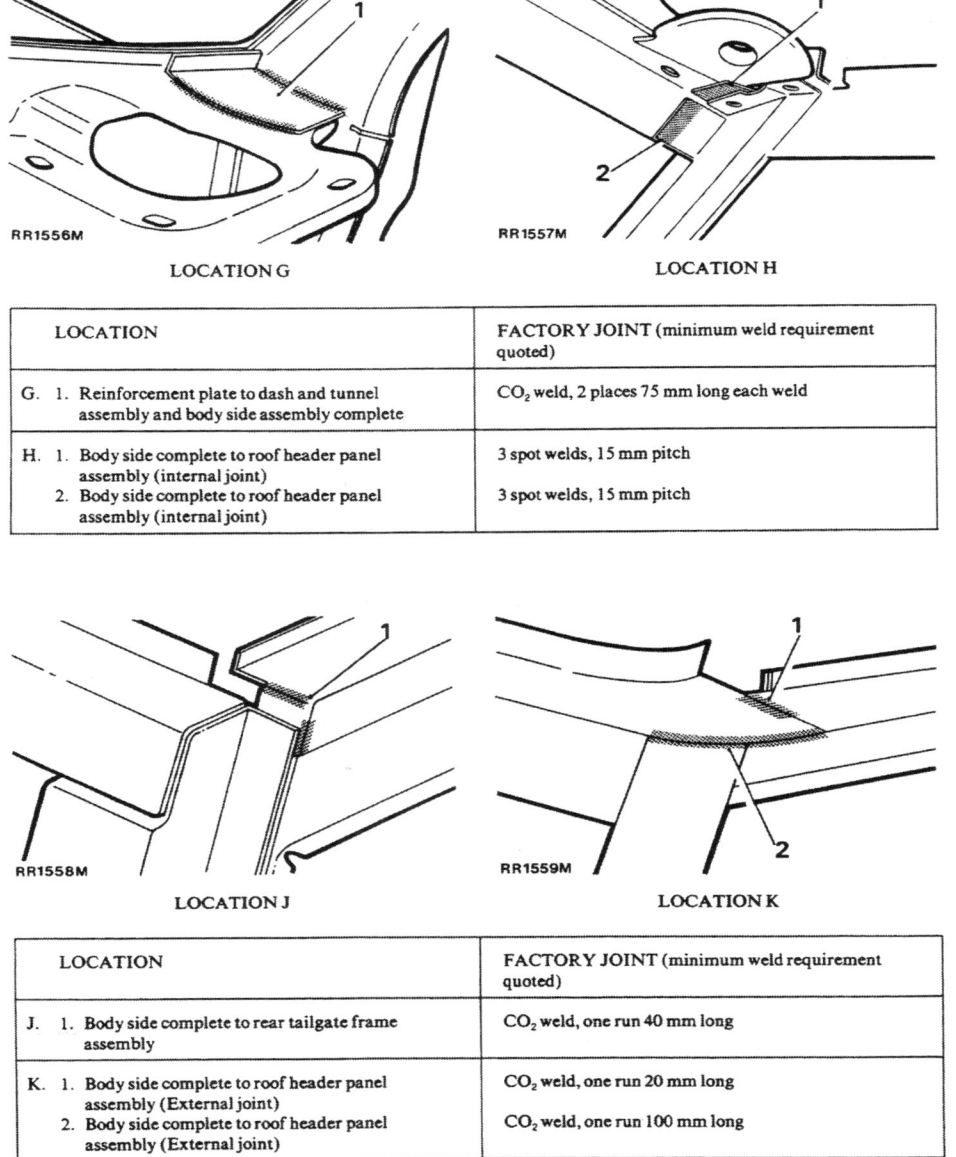

# BODY 76

### LOCATION G

### LOCATION H

### LOCATION J

### LOCATION K

| LOCATION | FACTORY JOINT (minimum weld requirement quoted) |
|---|---|
| G. 1. Reinforcement plate to dash and tunnel assembly and body side assembly complete | $CO_2$ weld, 2 places 75 mm long each weld |
| H. 1. Body side complete to roof header panel assembly (internal joint) | 3 spot welds, 15 mm pitch |
| 2. Body side complete to roof header panel assembly (internal joint) | 3 spot welds, 15 mm pitch |

| LOCATION | FACTORY JOINT (minimum weld requirement quoted) |
|---|---|
| J. 1. Body side complete to rear tailgate frame assembly | $CO_2$ weld, one run 40 mm long |
| K. 1. Body side complete to roof header panel assembly (External joint) | $CO_2$ weld, one run 20 mm long |
| 2. Body side complete to roof header panel assembly (External joint) | $CO_2$ weld, one run 100 mm long |

### LOCATION E

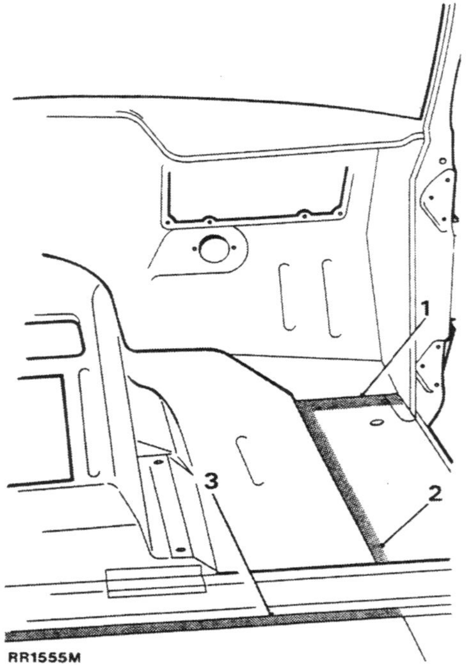

### LOCATION F

| LOCATION | FACTORY JOINT (minimum number of spot welds quoted) |
|---|---|
| E. 1. Body side complete to heelboard panel assembly | 14 spot welds, 35 mm pitch |
| 2. Body side complete to dash and tunnel assembly complete | 10 spot welds, 25 mm pitch |
| 3. Body side complete to dash and tunnel assembly | 3 spot welds, 30 mm pitch |
| F. 1. Body side complete to dash and tunnel assembly complete | 7 spot welds, 30 mm pitch |
| 2. Body side complete to dash and tunnel assembly complete | 18 spot welds, 40 mm pitch |
| 3. Body side complete to dash and tunnel assembly complete | 30 spot welds, 34 mm pitch |

193

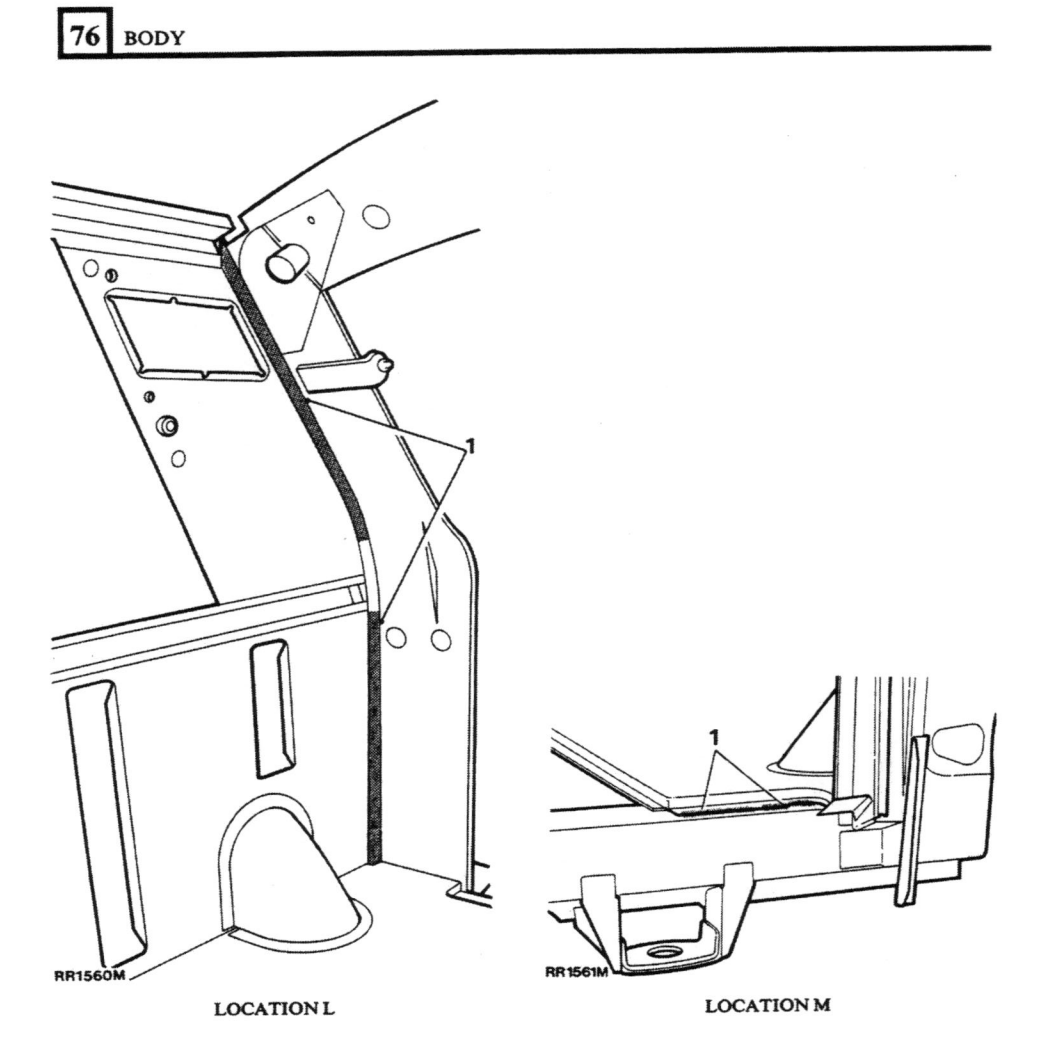

# BODY 76

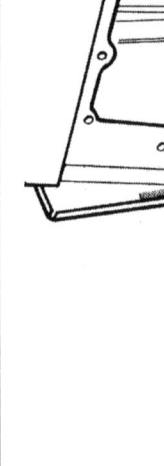

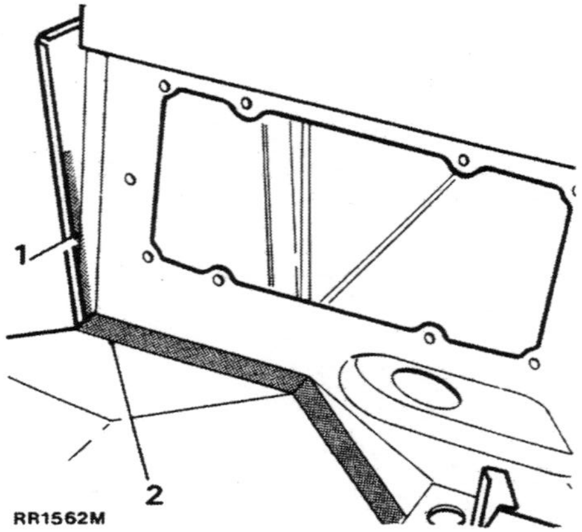

## LOCATION N

| LOCATION | FACTORY JOINT (minimum number of spot welds quoted) |
|---|---|
| N. 1. Valance and wheel arch assembly to dash and tunnel assembly | 4 spot welds, 45 mm pitch |
| 2. Valance and wheel arch assembly to dash and tunnel assembly | 15 spot welds, 25 mm pitch |

## LOCATION M

## LOCATION L

| LOCATION | FACTORY JOINT (minimum weld requirement quoted) |
|---|---|
| L. 1. Body side complete to rear tailgate side member | 32 spot welds, 30 mm pitch |
| M. 1. Body side complete to rear tailgate bottom cross member | $CO_2$ weld, 2 runs 40 mm long |

# BODY

## FRONT AND REAR HEADLINING AND ROOF PANEL

### Remove and refit

### Removing

1. Remove the two roof lamp assemblies.
2. Remove the rear view mirror and mounting bracket.
3. Remove the two sun visors and centre retaining bracket.
4. Remove the rear passenger grab rails.
5. Prise out the four plastic retaining clips securing the front and rear sections of the headlining to the roof panel.
6. Lower the front headlining and remove it from the vehicle.
7. Remove the two plastic retaining clips securing the end of the rear headlining, located adjacent to the upper tailgate hinges.
8. Lower the headlining and disconnect the electrical leads from the rear radio speakers, and remove the lining from the vehicle.
9. Remove the screws (with washers) from around the inner edge of the roof panel.
10. Lift the roof panel from the body and remove any previous sealing compounds from around the edge of the roof panel and body.

### Refitting

11. Reverse the removal instructions 1 to 9.
12. Apply a suitable waterproof sealant to roof and body mating faces.

## BONNET — DECKER PANEL — FRONT WING

### Remove and refit

### Removing

1. Open the bonnet and disconnect the under-bonnet lamp electrical leads.
2. Disconnect the windscreen washer tube at the 'T' joint.
3. Release the four bolts securing the bonnet to the hinges, and lift the bonnet clear of the vehicle.
4. Remove the four cross-head screws securing the hinges to the bulkhead mounting brackets.
5. Remove the wiper arms and two nuts securing the wheel boxes to the decker panel and remove the two exterior sealing rubbers.
6. Remove the nine cross-head screws securing the front of the decker panel.
7. Remove the four bolts (with spring and plain washers) retaining the decker panel to the top of the wing, retrieve the nylon spacing washers from between decker panel and wing.
8. Remove the four cross-head screws retaining the panel to the 'A' post mounting brackets located above the front door hinges.

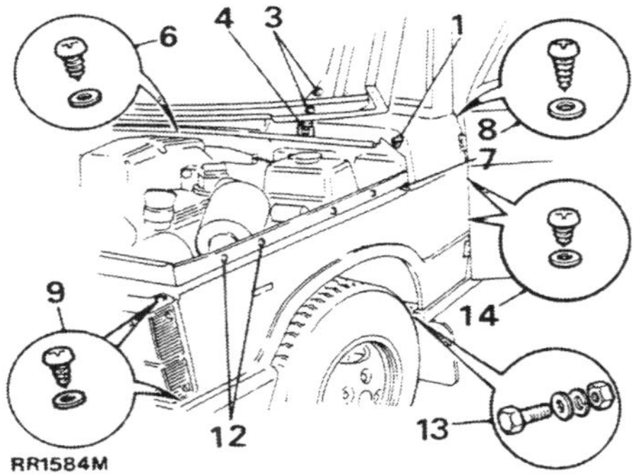

NOTE: If a radio aerial is fitted, remove it from the panel to enable the panel to be lifted clear of the body.

9. Remove the two cross-head screws from the top of the front side light assembly, manoeuvre the assembly out of the two bottom location holes and disconnect the electrical plug at the rear of the lamp. Remove the two screws (with plain washers) from the bottom of the side light aperture.
10. Disconnect the side flasher lamp electrical plug, located inside the engine compartment and feed the plug and grommet through the hole in the inner wing to the underside of the exterior wing.
11. Remove the three fixings securing the bumper wrap around end cap and remove the moulding from the bumper.

NOTE: If a front spoiler is fitted, remove the single bolt which secures the end of the spoiler to the front of the wheel arch, located forward of the road wheel at the bottom of the wing.

12. Remove the five nuts and bolts (with plain and spring washers) securing the top edge of the wing to the wheel arch and valance assembly.
13. Remove the two bolts (with plain washers) securing the sill finishing strip to the bottom of the wing.
14. Remove the two cross-head screws securing the wing to the mounting bracket attached to the 'A' post located in between the front door hinges.

9. From inside the tailgate area remove all the cross-head screws retaining the top of the wing.
10. Remove the rear wing and cover panel complete.
11. Remove the seven bolts (with plain and spring washers) securing the wing to the corner panel and separate the two panels.

### Refitting

12. Apply a suitable waterproof sealant to the wing and cover panel mating faces, fit the bolts and ensure both panels align before final tightening.
13. Fit the assembly to the vehicle ensuring the door shut face to wing and corner panel to lower tailgate are in alignment before the final tightening of screws and fitting of pop-rivets.
14. Coat the underside of panels with a suitable underseal.
15. Reverse the remaining removal instructions.

## BODY REAR QUARTER PANEL — INTERIOR

### Remove and refit

### Removing

1. Remove the rear passenger grab rails.
2. Remove the rear interior light assembly.
3. Remove the four plastic retaining clips securing the front and rear headlining sections to the centre of the roof panel.
4. Remove the two rear plastic retaining clips located adjacent to the upper tailgate fixings from the rear headlining section.
5. Lower the headlining and disconnect the electrical leads from the two radio speakers.
6. Remove the rear headlining section from the vehicle.

NOTE: If rear seat belts are fitted it will be necessary to remove them to enable the interior panel to be detached, these are removed as follows.

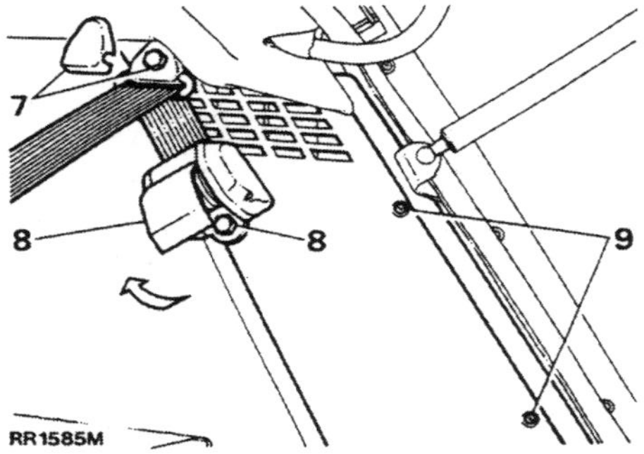

### Refitting

15. Apply a suitable underseal to the inner face of the wing.
16. Ensure that before final tightening of the wing securing bolts the wing aligns with the edge of the front door.
17. Locate the decker panel under the lip of the windscreen sealing rubber.
18. Fit the bonnet ensuring that before final tightening of the securing bolts, the bonnet aligns with the decker panel, wing and front grille.
19. Reverse the remaining removal instructions.

## BODY REAR CORNER PANEL AND WING — TWO AND FOUR DOOR MODELS

### Remove and refit

### Removing

1. Remove the fuel tank filler cap (right hand rear wing only).
2. Remove the fuel tank filler neck (right hand rear wing only).
3. Remove the rear stowage area parcel shelf, shelf side panels and spare wheel.
4. Remove the two nuts (with washers) securing the bumper wrap around end cap and slide the cap off the bumper.
5. Release the rear tail light cluster from the rear corner panel and disconnect the electrical plug.
6. Drill out all the pop-rivets securing the front of the wing (two door model only) and the corner panel (two and four door).
7. Remove the two nuts and bolts securing the front of the wing to the 'D' post located beneath the wheel arch (four door only).
8. Release the single nut and bolt retaining the mud-flap to the bottom of the wing.
Two door models only: Remove the single bolt from the bottom of the wing forward of the road wheel.

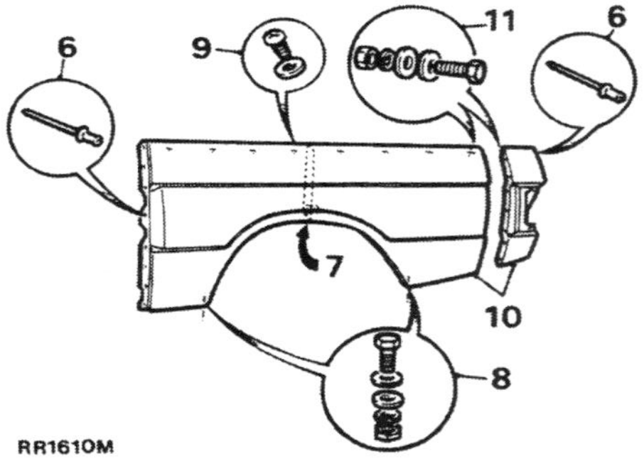

*Continued*

# BODY 76

## REAR PASSENGER DOOR

### Remove and refit

### Removing

1. Disconnect the battery negative lead.
2. Pull the convolute grommet out of the face of the 'B' post.
3. Withdraw the electrical leads from the 'B' post until the electrical plugs are exposed and disconnect the plugs.
4. Remove the two bolts (with spring and plain washers) securing the door check strap.
5. Open the door slightly, support the door and remove the six bolts (with spring washers) securing the hinges to the 'B' post.
6. Lift the door clear.

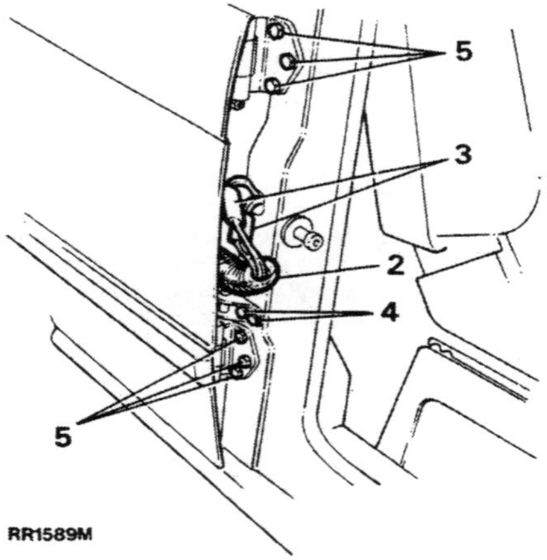

### Refitting

7. Reverse the removal procedure.
8. Adjustment to the rear doors is made by releasing the six hinge securing bolts (hinge to 'B' post) and moving the door either rearwards, forward, up or down in the door aperture.
9. Adjustment to the door striker is identical to front door adjustment.

14. The door lock striker can be adjusted by slackening the striker and moving it in the appropriate direction or adding and subtracting spacing washers between the striker and 'B' post.

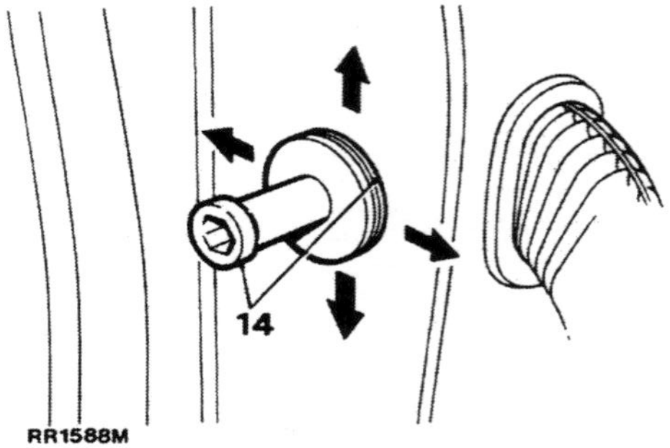

---

7. Detach the plastic cover from the upper seat belt guide bracket and remove the single bolt.
8. Remove the clip-on black plastic cover from the seat belt inertia reel. Remove the two retaining bolts and place the reel to one side.
9. Release the two small cross-head screws and remove the panel from the vehicle.

### Refitting

10. When refitting the seat belt assembly ensure the belts are not twisted and that the securing bolts are tightened to the correct torque.
11. Reverse the removal procedure.

## BODY REAR QUARTER PANEL — EXTERIOR

### Remove and refit

### Removing

1. Remove the rear passenger grab rails.
2. Remove the rear interior light assembly.
3. Remove the rear headlining section.
4. Remove the rear seat belts (if fitted).
5. Remove the two securing screws and detach the interior quarter panel.
6. Remove the three nuts (with plain washers) securing the outer quarter panel to the body side. The nuts are accessible through the large holes located adjacent to the rear window.
7. Remove the four screws securing the quarter panel to the inside edge of the tailgate aperture.
8. Withdraw the panel.

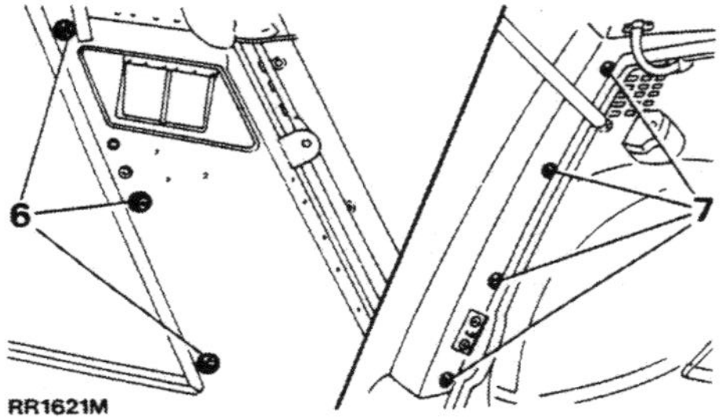

### Refitting

9. Reverse the removal instructions.
10. Using a suitable tool carefully ease the rear side window sealing rubber over the front edge of the exterior quarter panel.

## FRONT DOOR

### Remove and refit

### Removing

1. Disconnect the battery negative lead.
2. Open the appropriate door to be removed.
3. Remove the two screws and detach the side trim panel from the front footwell.
4. Carefully pull the door wiring harness from above the fascia until the electrical plugs are exposed.
5. Disconnect the electrical plugs and feed them through the aperture in the 'A' post.
6. Drive out the roll pin from the door check strap.
7. With assistance support the door and remove the screws, securing door to hinges.
8. Lift the door clear.

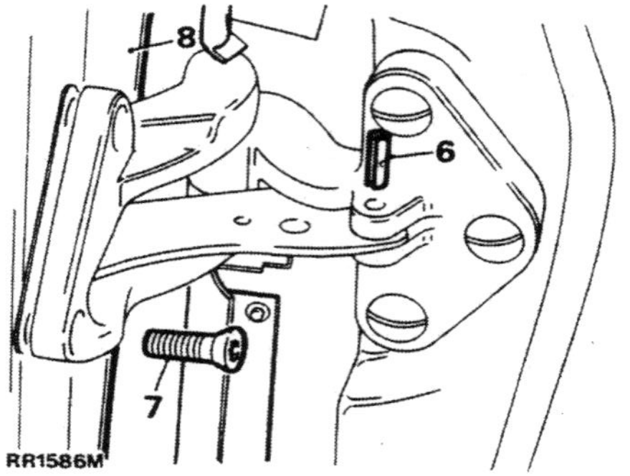

### Refitting

9. Refit the door and feed the electrical cables through the 'A' post.
10. Fully open the door and reconnect the electrical leads, take up the slack in the leads and clip them securely to the footwell side panel, refit the convolute grommet to the 'A' post aperture.
11. Check the location of the door and the operation of the door lock. If necessary, adjust the door and striker plate.
12. By the addition of shims between the hinge and door or hinge and 'A' post to take the door forward or rearward in the aperture.
13. By slackening the six screws securing the hinges to the door, the door can be adjusted up and down or in and out of the aperture.

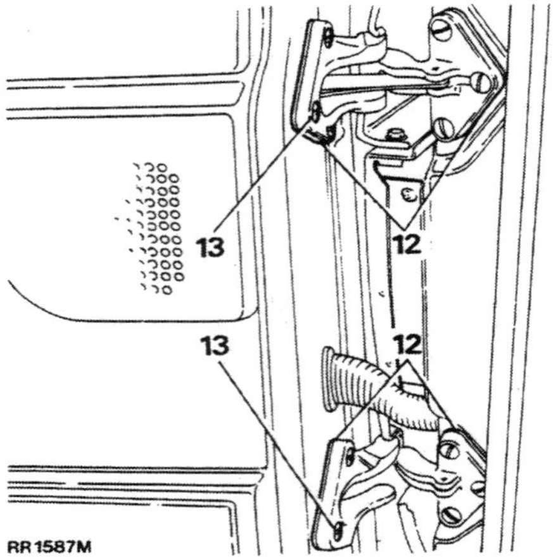

## BODY

## UPPER TAILGATE

### Remove and refit

### Removing

1. Disconnect the battery negative lead.
2. Remove the tailgate wiper arm
3. Open the tailgate and remove the four screws securing the two rear screen electrical lead shrouds located at either end of the tailgate.
4. Manoeuvre the shroud away from the screen and out of the headlining to reveal the electrical connections, disconnect the leads.
5. Prise the two stays off the tailgate.
6. Remove the four hinge to tailgate screws. 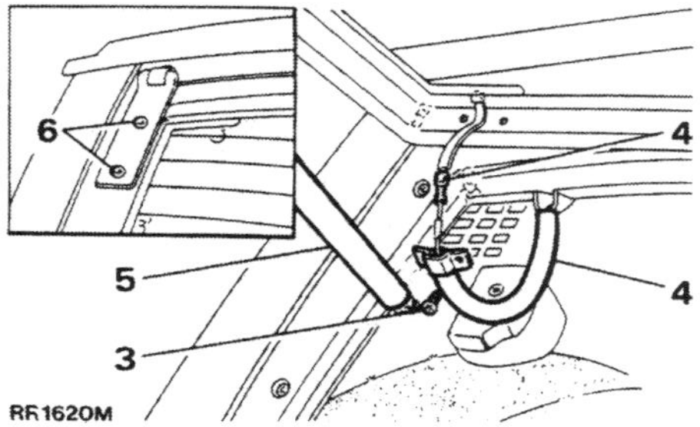
7. Lift the upper tailgate clear.

### Refitting

8. Reverse the removal instructions.

## UPPER TAILGATE — LOCK

### Remove and refit

### Removing

1. Remove the two cross-head screws securing the upper tailgate release handle.
2. Ease the sealing rubber away to reveal the single screw securing the centre of the release handle, remove the screw and detach the handle from the lock mechanism complete with key barrel.

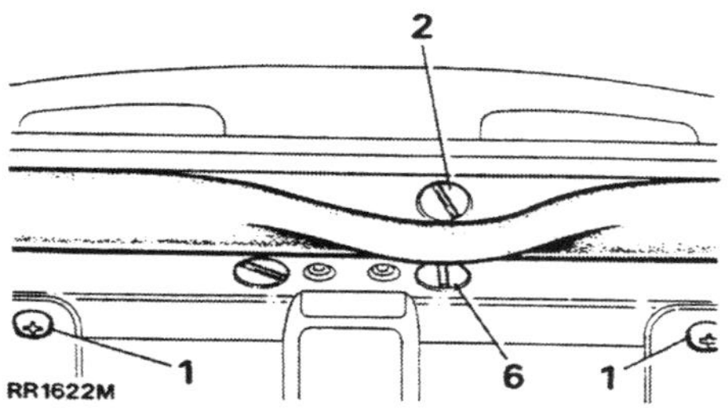

   NOTE: At this stage the barrel can be removed from the handle by releasing the two small cross-head screws at the joint-face. Remove the retaining plate and release the spring and key barrel from its bore.

3. Release the eight screws and remove the operating rod covers located either side of the centre lock mechanism.
4. Remove the four screws and two nuts securing the lock catches to the sides of the tailgate.
5. Release the two small locknuts on the operating rods and rotate the hexagonal connecting rod until the side catches can be removed.
6. Release the two screws securing the centre lock mechanism and remove the unit complete with operating rods.

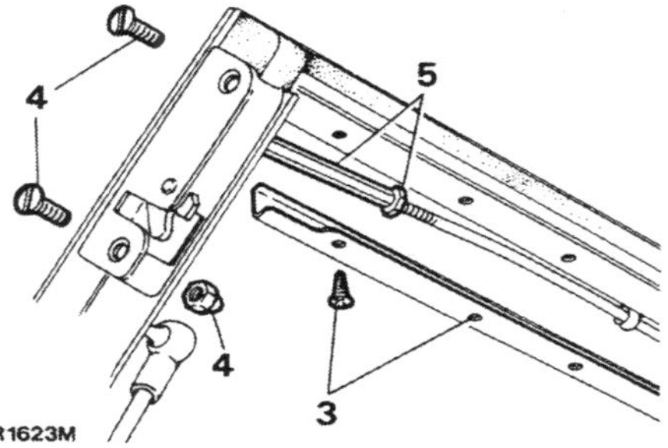

### Refitting

7. Reverse the removal procedure.
8. Adjustment of the side catches may be required after assembly, this is achieved by releasing the locknuts on the operating rods, and rotating the hexagonal link clockwise or anti-clockwise to shorten or extend the length of the operating rods.

## LOWER TAILGATE

### Remove and refit

### Removing

1. Disconnect the electrical leads from the rear number plate lamp.
2. Remove the four cross-head screws securing the sealing rubber around the tailgate hinge bolts.
3. Remove the fixings, tailgate to hinges.
4. Disconnect the check straps.
5. Withdraw the tailgate.

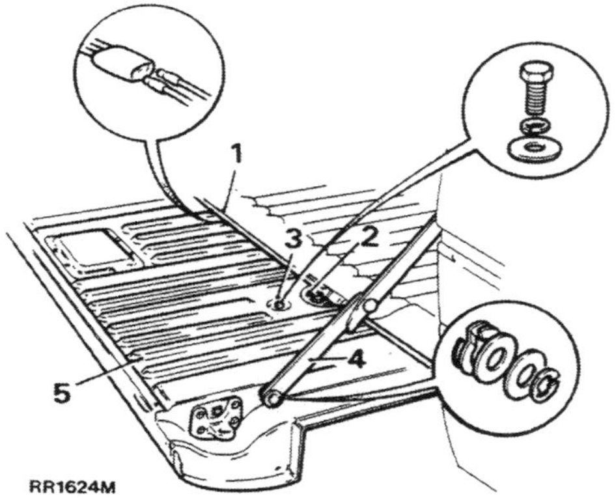

### Refitting

6. Reverse 1 to 4.

7. Close the lower tailgate and check the operation of the lock, the bolts should engage automatically and release when the handle is moved fully right against spring pressure.

### Adjusting the lock 8 to 12

8. Slacken the locknuts at the lock end of the adjuster.
9. Slacken the locknuts at the bolt end of the adjuster, noting that they have a LEFT HAND thread.
10. Turn the adjuster as required to move the bolt in or out.
11. Secure the adjuster locknuts.
12. The eye brackets at each side of the tailgate can also be adjusted to align with their locating dowels, by slackening the fixings, slightly repositioning the brackets and retightening the fixings.

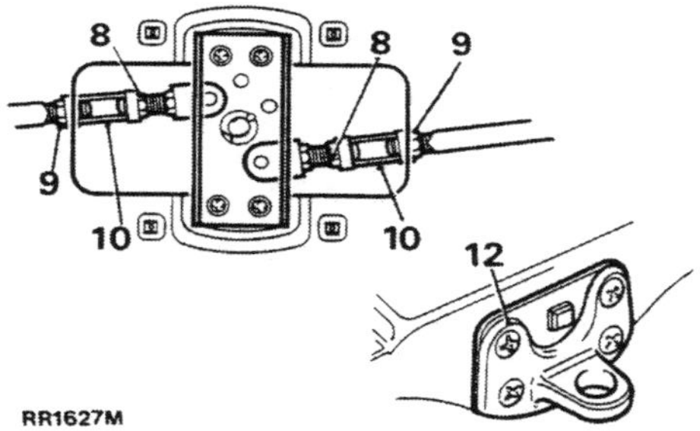

## LOWER TAILGATE LOCK

### Adjust 1 and 8 to 13

### Remove and refit 1 to 7 and 13

### Removing

1. Remove the lock cover plate.
2. Remove the fixings from the lock mounting plate.
3. Disconnect either one of the bolt arms from the door lock.
4. Withdraw the lock complete with the fixed bolt arm.
5. Withdraw the loose bolt arm.

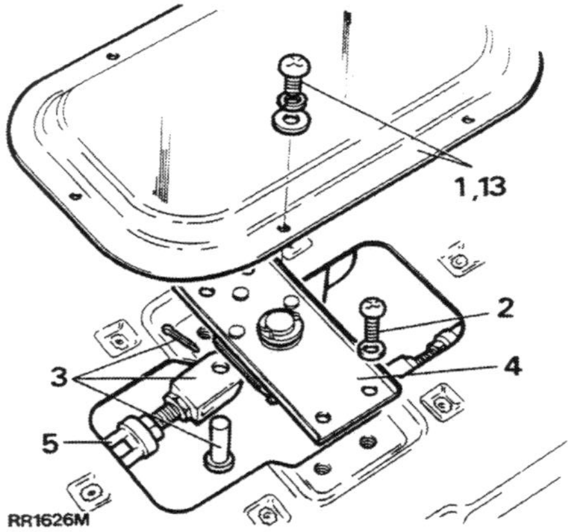

...

13. Fit the lock cover plate.

# BODY 76

## FRONT DOOR GLASS AND REGULATOR—Four door models

### Remove and refit

**Removing**

1. Ensure the window is in its fully closed position and secure it with adhesive tape to prevent the window dropping down.
2. Disconnect the battery.
3. Detach the armrest/door-pull finisher to reveal the two securing screws.
4. Remove the two securing screws (with plain washers) to enable the armrest/door-pull to be detached from the inner door panel.
5. Remove the interior door handle finisher button to reveal the screw retaining the handle surround.
6. Remove the screw and detach the handle surround from the inner door panel.
7. Detach the inner door trim pad by inserting a screwdriver between the trim pad and inner door panel, gently prising out the nine plastic securing clips from their respective holes in the inner door panel.
8. Disconnect the two speaker connections inside the door and remove the door trim pad complete with speaker.
9. Remove the plastic weather sheet.
10. Remove the window lift motor (refer to electrical section).
11. Remove the four window regulator retaining bolts with shakeproof washers from the inner door panel.
12. Remove the two screws with shakeproof washers retaining the lower window lift channel and slide the channel off the stud.
13. Disengage the lifting arm stud from the upper lifting channel, manoeuvre the window regulator and remove it from the lower centre aperture in the inner door panel.
14. Carefully disengage the lifting arm stud from the glass lifting channel and remove the window regulator from the lower aperture in the inner door panel.

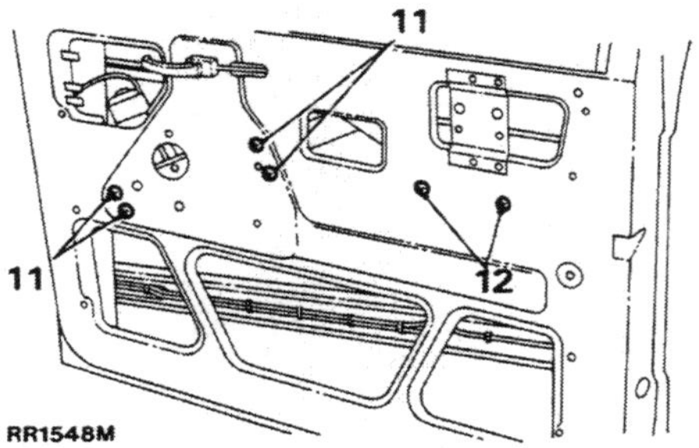

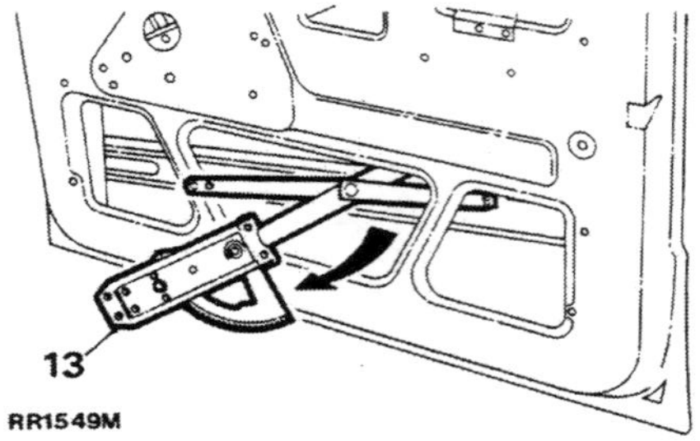

**Refitting**

26. Reverse the removal instructions, items 1 to 25.

**NOTE: When refitting the door glass frame, ensure it is repositioned to suit the door aperture before fully tightening the door frame securing bolts.**

## REAR DOOR GLASS AND REGULATOR

### Remove and refit

**Removing**

1. Ensure the window is in its fully closed position and secure it with adhesive tape over the top of the door to prevent the window dropping down.
2. Disconnect the battery.
3. Remove the armrest/door-pull finisher to reveal the two securing screws.
4. Remove the two securing screws (with plain washers) and detach the armrest/door-pull from the inner door panel.
5. Disconnect the window lift switch multi-plug at the rear of the armrest/door-pull.
6. Remove the interior handle finisher button to reveal the screw retaining the handle surround.
7. Remove the screw and detach the handle surround from the door trim pad.
8. Remove the door trim pad by inserting a screwdriver between the trim pad and the inner door panel, gently prising out the six plastic clips from their respective holes around the edges of the trim.
9. Remove the sill locking button.
10. Remove the inner door capping from its keyhole location.
11. Remove the plastic weather sheet.
12. Remove the window lift motor (refer to electrical section).
13. Remove the four window regulator securing screws (with shakeproof washer).
14. Remove the inner door capping off its keyhole location.
15. Remove the sill locking knob.
16. Remove the exterior driving mirror (refer to exterior driving mirrors removal and refit in electrical section).
17. Remove the waist rail seal from the top of the door panel.
18. Remove the two bolts (shakeproof and plain washers) from the hinge face of the door which secure the front door frame.
19. Remove the single bolt (spring and plain washer) from inside the door which secures the door rear glass frame.
20. Remove the bolt (spring and plain washer) from the recessed hole in the front of the inner door panel under the exterior driving mirror mounting plate.
21. Remove the single screw (spring and plain washer) from inside the door securing the bottom front glass channel.
22. Remove the single screw (spring and plain washer) from inside the door securing the bottom rear glass channel.
23. Lift the door glass frame complete with glass out of the door panel and remove to a suitable bench.
24. Remove the adhesive tape securing the glass to the frame.
25. Slide the glass out of the door frame channel.
15. Remove the waist rail seal from the top of the door panel.
16. Remove the single bolt (spring and plain washers) from inside the door which secures the bottom of the short rear glass run channel.
17. Remove the two bolts (spring and plain washers) from the hinge face of the door which secure the front door frame.
18. Remove the two bolts (spring and plain washers) from the shut face of the door which secure the rear door frame.
19. Lift out the door frame with the glass in position and remove to a suitable workbench.
20. Remove the tape and slide the glass out of the door frame channel.

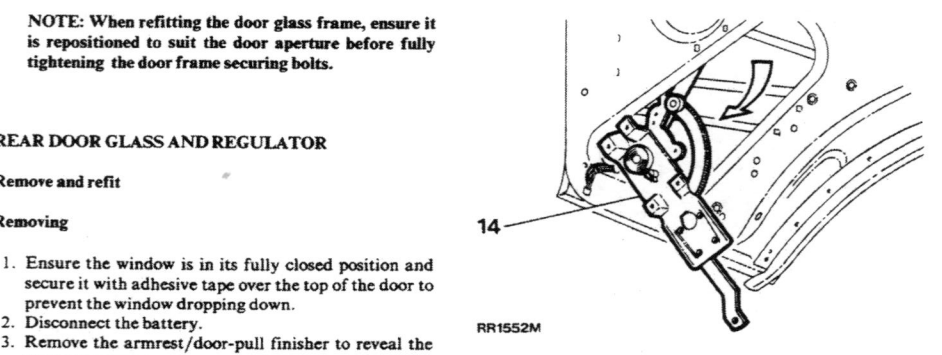

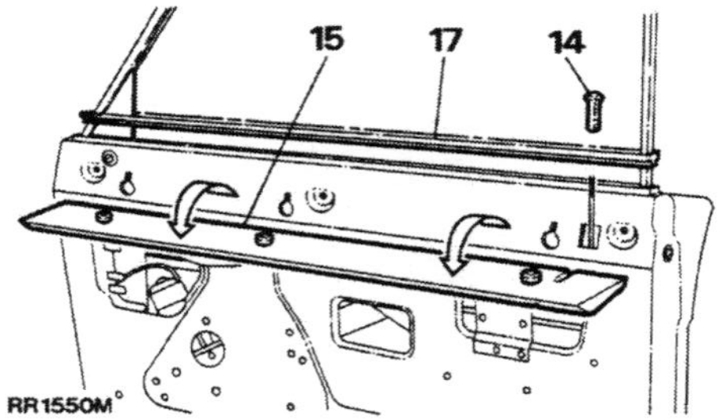

**Refitting**

21. Reverse the removal procedure items 1 to 20.

**NOTE: When refitting the door glass frame securing bolts, ensure that the door frame is adjusted to suit the door aperture before fully securing the frame to the door.**

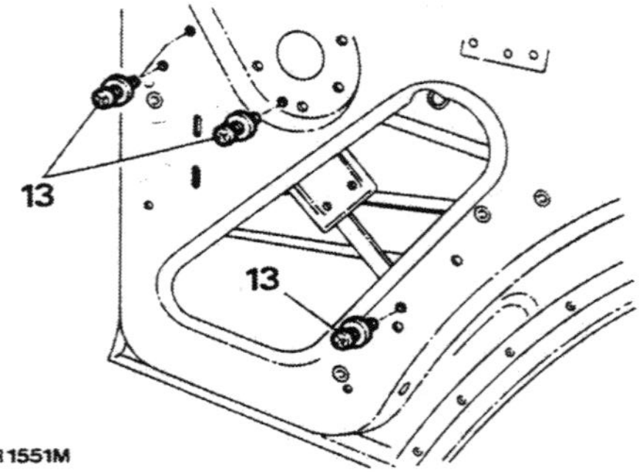

## BODY

### FRONT DOOR LOCK, OUTSIDE AND INSIDE DOOR RELEASE HANDLES — FOUR DOOR MODELS

**Remove and refit**

**Removing**

1. Remove the window lift motor (refer to electrical section).
2. Remove door glass and regulator (refer to door glass and regulator remove and refit).
3. Remove door actuator units (refer to electrical section).
4. Disconnect the control rod from the private key operated lock by releasing the metal clip at the bottom of the rod.
5. Disconnect the control rod from the outside door release handle by pulling it out of the plastic ferrule.
6. Disconnect the control rod connector between the inside door release handle and the door lock by releasing the metal clip and pulling one of the control rods out of the plastic connecting block. This is accessible through the small centre cut-out in the door panel. (The control rod also passes through a guide bracket on the inside of the inner door panel.)
7. Release the door lock by removing the two countersunk screws from the door shut face and the single screw (with shakeproof washer) on the inner door panel.
8. Withdraw the lock through the lower rear cut-out on the inner door panel.

**NOTE: If necessary the following items can also be removed.**

9. Remove the two nuts (with shakeproof washers) and retaining bracket securing the **outside door release handle** to the outer door panel, accessible through the upper rear cut-out on the inner door panel.
10. Carefully detach the outside door release handle from the outer door panel.
11. Remove the two screws securing the **inside door release handle** to the inner door panel.
12. Withdraw the handle from its location with half the connecting rod still attached.
13. Unclip the spring tensioned end of the connecting rod from the door release handle.

**Refitting**

14. Reverse the removal procedure items 1 to 13.

**NOTE: When refitting the door glass frame, ensure that it is positioned to suit the door aperture before fully tightening the door frame securing bolts.**

### ADJUSTMENT — FRONT DOOR LOCK AND HANDLE ASSEMBLY

**Inside door release handle to lock**

Refit the inside door release handle surround before any adjustment is made, allowing the handle to be set from the correct operating position. Rotate the spring tensioned nyloc nut at the opposite end of the interior handle connecting rod, clockwise or anti-clockwise to shorten or extend the operating length.

**Outside door release handle to lock**

Disconnect the small connecting rod at the rear of the outer door release handle by releasing the small metal clip, rotate the rod clockwise or anti-clockwise to shorten or extend the operating length.

**NOTE: Door release movement should be effective before the total handle movement is exhausted to provide a small over-throw movement.**

### REAR DOOR LOCK, OUTSIDE AND INSIDE DOOR RELEASE HANDLES

**Remove and refit**

**Removing**

1. Ensure the window is in its fully closed position.
2. Remove all the interior door trim (refer to door trim and regulator remove and refit, items 2 to 11).
3. Disconnect the control rod from the inside door release handle by pulling the rod out of its location at the door lock.
4. Disconnect the sill locking control rod from the door lock by releasing the metal clip.
5. Disconnect the control rod from the outside door release handle by pulling it out of the plastic ferrule.
6. Release the door lock by removing the two countersunk screws from the door shut face and the single screw (with shakeproof washer) on the inside of the door. Retrieve any spacing washers which may be fitted between the inner door panel and the lock.
7. Withdraw the lock through the upper rear aperture in the inner door panel.

**NOTE: If necessary the following items can also be removed.**

8. Remove the two nuts (with shakeproof washers) and retaining bracket securing the **outside door release handle** to the outer door panel, accessible through the upper rear cut-out or the inner door panel.
9. Carefully detach the outside door release handle from the outer door panel.
10. Remove the two screws (with plain washers) securing the **inside door release handle** to the inner door panel.

11. Withdraw the handle from its location with the connecting rod still attached.
12. Unclip the spring tensioned end of the connecting rod from the door release handle.

**Still locking quadrants**

13. Use a small screwdriver, or 3.175 mm diameter (⅛ in) rod, to press the plastic locking pins through the respective square inserts in the inner door panel, until they can be retrieved from inside the door.
14. Release the quadrants from the inner door panel and unhook the respective connecting rods.
15. Withdraw the quadrant from inside the door.

**NOTE: When refitting the quadrants the plastic locking pins are entered into the square insert from outside and pressed in flush.**

**Refitting**

16. Reverse the removal procedure 1 to 15.

### ADJUSTMENT — REAR DOOR LOCK AND HANDLE ASSEMBLY

**Outside door release handle to lock**

Disconnect the short cranked connecting rod at the rear of the outer door release handle, rotate the rod clockwise or anti-clockwise to shorten or extend the operating length.

**NOTE: Door release should be effective before the total handle movement is exhausted to provide a small over-throw movement.**

### FRONT DOOR GLASS AND REGULATOR — TWO DOOR MODEL

**Remove and refit**

**Removing**

1. Remove the armrest.
2. Remove both interior door handles.
3. Remove both door-pull handles.
4. Remove the window regulator handle.
5. Prise the upper trim panel from the door.
6. Prise the lower trim panel from the door.

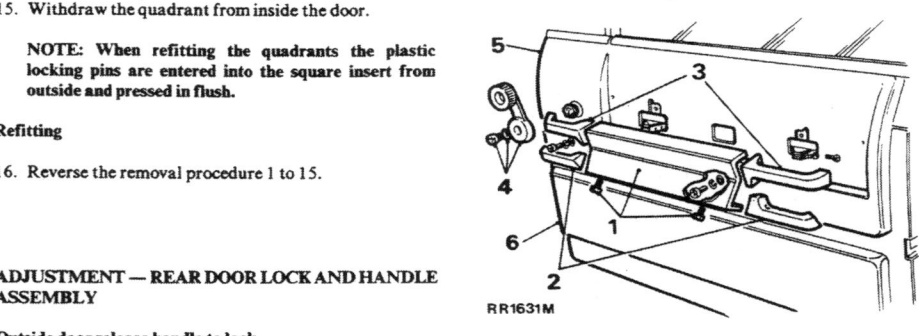

RR1631M

7. Withdraw the seals from the top edge of the door.
8. Remove the door glass frame.
9. Withdraw the glass.

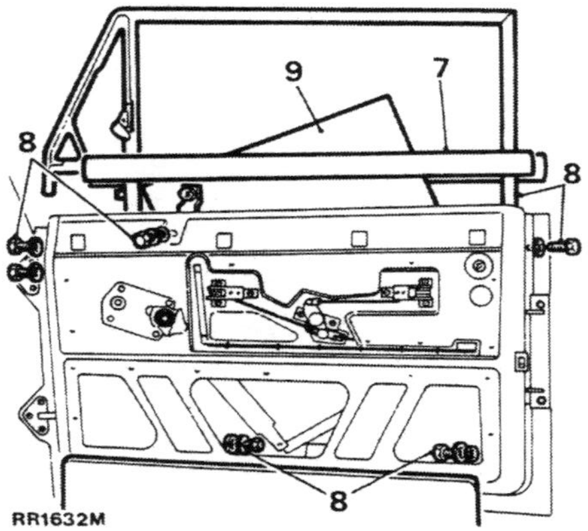

RR1632M

*Continued*

# BODY

10. Remove the four bolts securing the window regulator.
11. Withdraw the regulator out of the centre lower aperture.

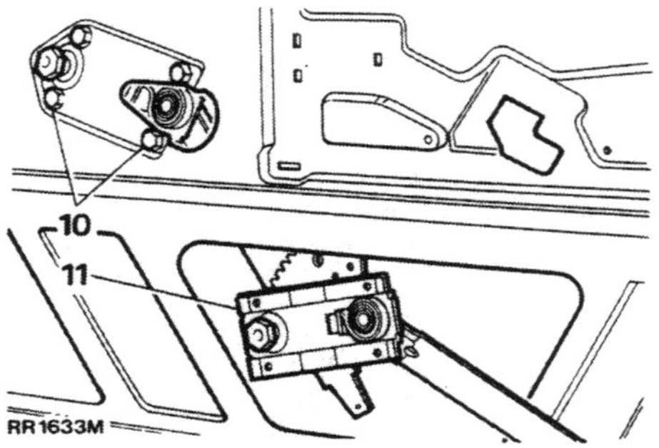

## FRONT QUARTER VENT—TWO DOOR MODELS

### Remove and refit

### Removing

1. Remove the interior door handles.
2. Detach the upper and lower trim panels.
3. Remove all the door glass frame securing bolts (see Door glass and regulator removal instructions).
4. Remove the seals from the top edge of the door.
5. Pull the frame out of the door until access is gained to the nut, spring and washers at the bottom of the vent.
6. Remove the fixings and withdraw the quarter vent.

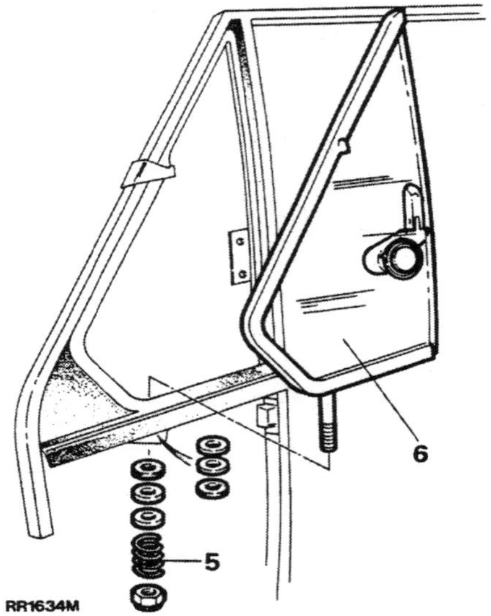

### Refitting

7. Reverse the removal instructions.

### Refitting

12. Fit the door frame and glass, do not fit the securing bolts at this stage.
13. Push the glass down the frame until the regulator runner is located in the glass channel.
14. Position the regulator and frame in the door panel and fit all retaining bolts.
15. Reverse the remaining removal instructions.

---

## FRONT DOOR LOCK—TWO DOOR MODEL

### Remove and refit

### Removing

1. Remove the front door glass and frame (see door glass and regulator removal instructions).
2. Release the two nuts securing the exterior door handle, remove the handle from the door.
3. Disconnect the operating rod from between the interior locking lever and the lock.
4. Disconnect the operating rod at the private key lock.
5. Disconnect the operating rod from the interior door handle relay, at the lock end.
6. Remove the three screws securing the lock to the door panel, and withdraw the lock.

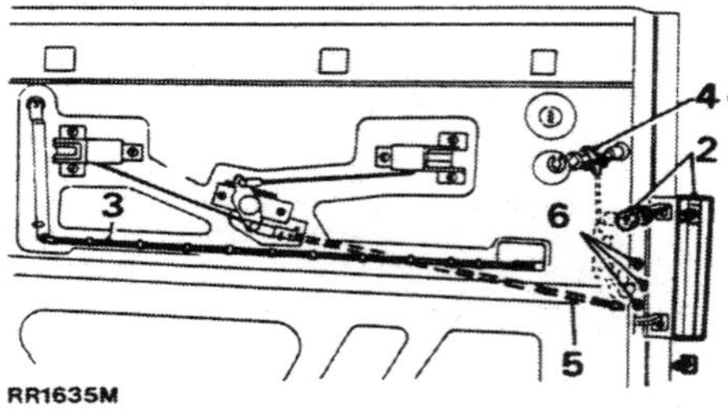

### Refitting

7. Reverse the removal instructions.

## PRIVATE LOCK, FRONT DOOR—TWO DOOR MODEL

1. Fully close the side door window.
2. Remove the interior door handles.
3. Carefully prise the upper and lower trim panels away from the door.
4. Disconnect the operating rod from the lever.
5. Using a bent piece of metal pull the spring clip off the private lock.
6. Manoeuvre the lock and remove it from the door.

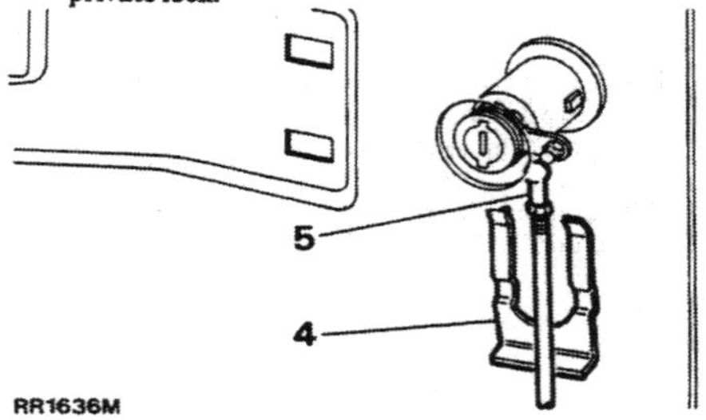

### Refitting

7. Reverse the removal instructions.

## BODY SIDE GLASS—TWO DOOR MODEL

Front—remove and refit 3 to 7

Rear—remove and refit 1, 2 and 6

### Removing

1. Remove the headlining rear section.
2. Remove the expander strip from the channel in the rubber moulding around the glass.
3. Lift the tongue of the spring clips from each of the front glass runners.
4. Slide both runners clear of the glass.
5. Lift out the front glass.
6. Lift out the rear glass.

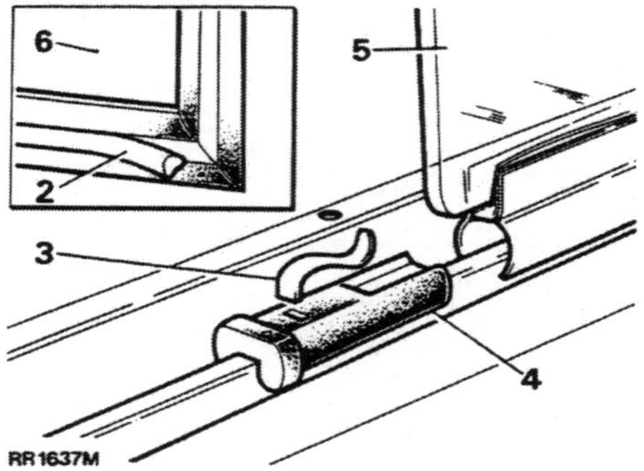

### Refitting

7. Reverse 1 to 6.

## WINDSCREEN GLASS

### Remove and refit

### Removing

1. Remove the windscreen wiper arms.
2. Remove the expander strip from the channel in the rubber moulding around the glass.
3. Ease the bottom edge of the windscreen glass from the rubber moulding.
4. Lift the windscreen glass clear.

*Continued*

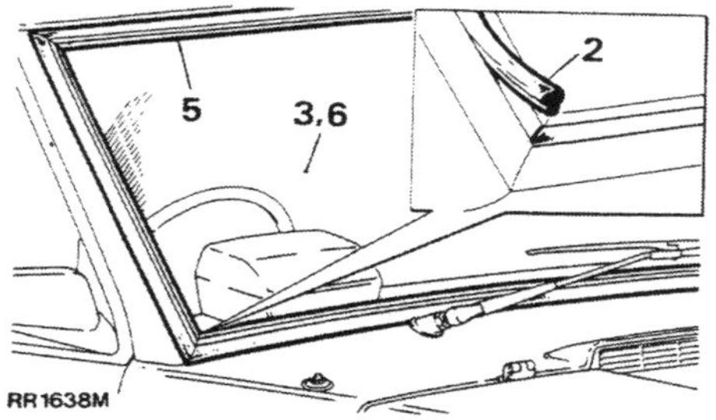

# BODY

## FRONT SEAT—TWO DOOR MODEL

**Remove and refit**

**Removing**

1. Remove the nut and bolt from the front of each seat slide.
2. Remove the six screws and the outside seat base cowling.

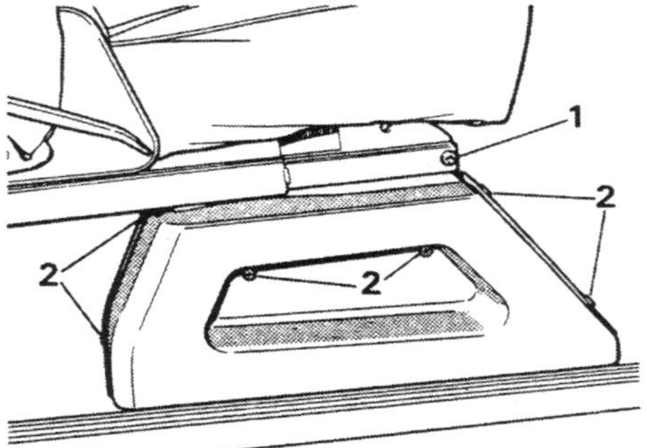

3. Remove the two screws and the plate from the outside rear of the seat base.
4. Remove the upper bolt from the rear retention bar and push the bar downward.
5. Release the retaining spring below the seat squab.

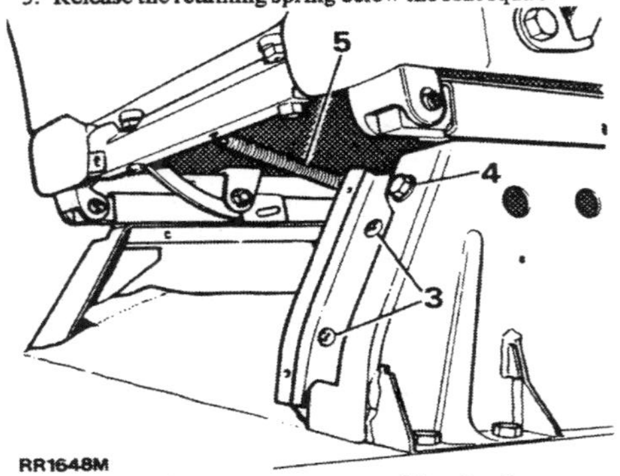

RR1647M

6. Working from the rear seat slide the front seat rearward and manoeuvre over the rear kick plate.
7. Withdraw the front seat.

NOTE: It may be found necessary to remove the 'B' post trim if fouling is experienced.

**Refitting**

8. Reverse instructions 1 to 7.
9. Refit the 'B' post trim, if removed.

## TAILGATE GLASS

**Remove and refit**

**Removing**

1. Remove the upper tailgate.
2. Remove the lock.
3. Remove the lift handle and trim.
4. The upper tailgate glass and frame are serviced as one unit.

**Refitting**

5. Reverse instructions 1 to 3.

RR1638M

**Refitting**

5. Smear soft soap around the windscreen glass location channel in the rubber moulding.
6. Locate the bottom edge of the windscreen glass into the rubber moulding.
7. Use a suitable tool tapered to a thin end, to prise the rubber moulding over the windscreen glass all the way round.
8. Using a suitable tool insert the expander rubber into the channel around the windscreen moulding.

---

## CUBBY BOX AND FLOOR MOUNTED CONSOLE ASSEMBLY

**Remove and refit**

**Removing**

1. Disconnect the battery negative lead.
2. Remove the four cross-head screws securing the cubby box liner to the cubby box and withdraw the liner.
3. Disconnect the electrical leads to the rear passenger cigar lighter.
4. Disconnect the electrical multi-plugs from the rear of the window lift switches, labelling each plug for identification on re-assembly.
5. Prise out the window lift panel complete with switches.

NOTE: To enable the cubby box/console assembly to be removed, disconnect the handbrake cable from the handbrake lever, to allow the lever to be raised to its uppermost position.

6. Remove the main and transfer gear selector knobs.
7. Carefully prise the inset panel around the main gear selector away from the outer surround.

NOTE: Automatic transmission models—Disconnect the graphics illumination bulbs from the selector panel.

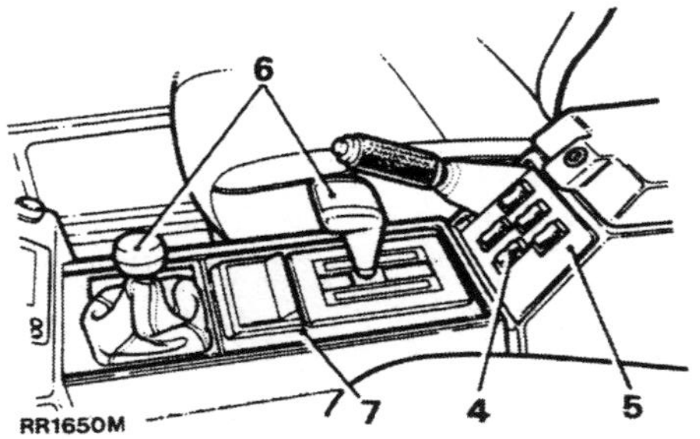

RR1650M

8. Remove the two bolts (with washers) and two screws (with washers) securing floor-mounted console to the gearbox tunnel.
9. Ease the assembly rearwards to detach the small location tab at the front of the console from the bottom of the radio housing.
10. Manoeuvre the assembly from the handbrake and gear levers and remove it from the vehicle.

*Continued*

---

## FRONT SEAT—FOUR DOOR MODEL

**Remove and refit**

**Removing**

1. Remove the four cross-head screws securing the outside seat base cowling.
2. Slide the seat forward and remove the four Allen type screws located inside the seat slide channel at the rear of the seat.
3. Move the seat rearwards and remove the two Allen type screws with nuts located inside the runner channel at the front of the seat.

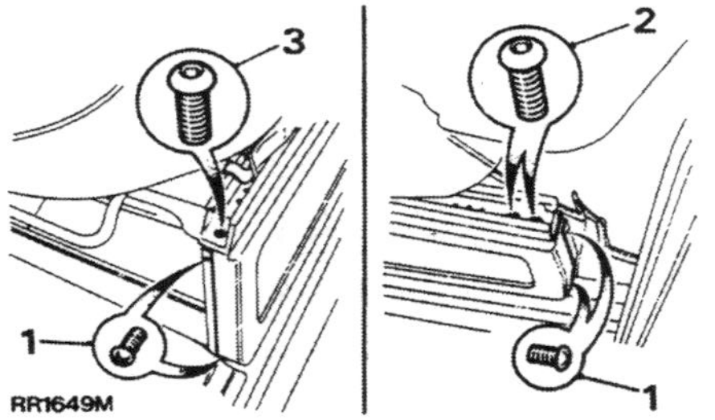

RR1649M

4. Remove the seat from the vehicle.

**Refitting**

5. Reverse the removal procedure.

## FRONT SEAT BASE—TWO AND FOUR DOOR MODELS

**Remove and refit**

**Removing**

1. Remove the seat.
2. Release the front and rear footwell carpets to give access to the seat belt bolts.
3. Remove the twelve bolts (with washers) securing the seat base to the floor.
4. Disengage the eyebolts, seat base to chassis.

NOTE: Four door Fuel Injection models only. Disconnect the multi-plug from the electronic control unit located under the right-hand front seat.

5. Withdraw the seat base.

**Refitting**

6. Reverse the removal procedure.

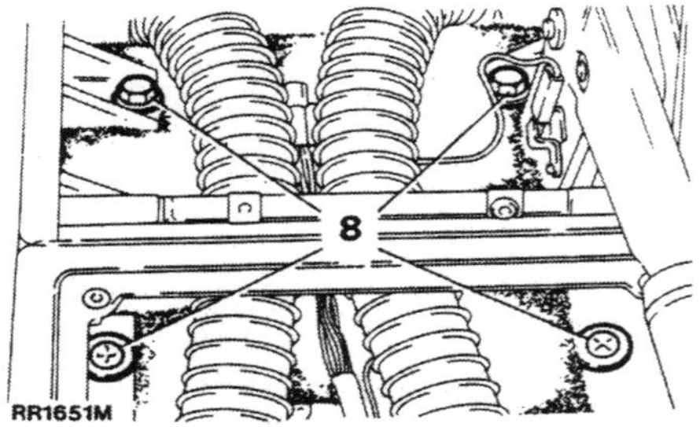

## BODY

### LOUVRE FASCIA PANEL

**Remove and refit**

**Removing**

1. Disconnect the battery negative lead.
2. Carefully prise the clock out of the fascia, disconnect the electrical leads and remove the bulb and holder from its holder.
3. Remove the single screw securing the end of the fascia panel located adjacent to the passenger courtesy light switch.
4. **AIR CONDITIONED MODELS**—Remove the four louvres to give access to the eight securing screws located at the top of the louvre apertures.
   **NON-AIR CONDITIONED MODELS**—Remove the five screws securing the parcel shelf, withdraw the shelf to give access to the four nuts and bolts (with plain washers) located behind the louvre fascia panel retaining the panel to the top rail. Prise out the two centre louvres and release the four screws located within the louvre apertures.
5. **AIR CONDITIONED MODELS**—Remove the remaining screw above the air-conditioning control panel.
6. Remove the three screws securing the bottom of the panel to the centre fascia panel (both models). Remove the remaining three screws above the blower motor trim panel (air-conditioned models only).
7. Ease the fascia panel forward and disconnect the electrical leads from the rear of the air conditioning switch (air conditioned models only).
8. Withdraw the panel from the vehicle.

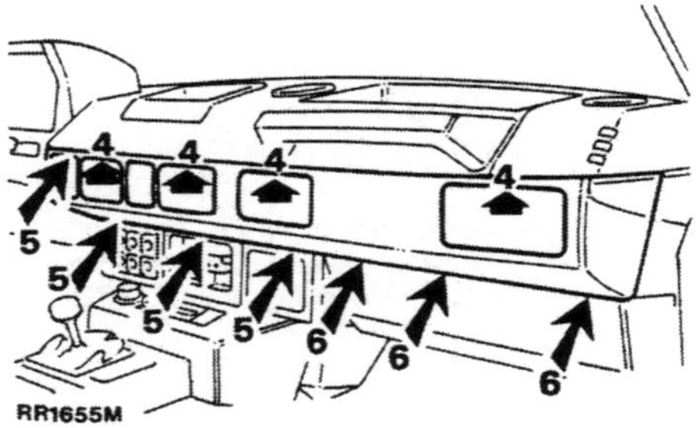

**Refitting**

9. Reverse the removal procedure.

### CENTRE FASCIA PANEL

**Remove and refit**

**Removing**

1. Disconnect the battery negative lead.
2. Remove the lower fascia panel to give access to the single screw securing the side of the centre fascia and release the screw (driver's side).
3. Remove the cubby box and floor-mounted console assembly.
4. Remove the radio and radio housing.
5. Detach the fuse box cover and remove the three screws securing the main and auxiliary fuse box body to the fascia.
6. Pull the four heater control knobs off their levers.
7. Remove the two screws at the top of the heater graphics panel. Pull the panel away from the fascia and remove the illumination bulbs from their locations.
8. Withdraw the graphics panel.
9. Prise the auxiliary switch panel away from the fascia.
10. Identify each switch multi-plug to aid re-assembly, and disconnect them from the switches.
11. Release the three screws securing the top of the centre fascia to the louvre fascia panel.
12. Remove the two screws securing the side of the centre fascia unit (front passenger side).
13. Manoeuvre the centre fascia away from the louvre fascia and remove it from the vehicle.

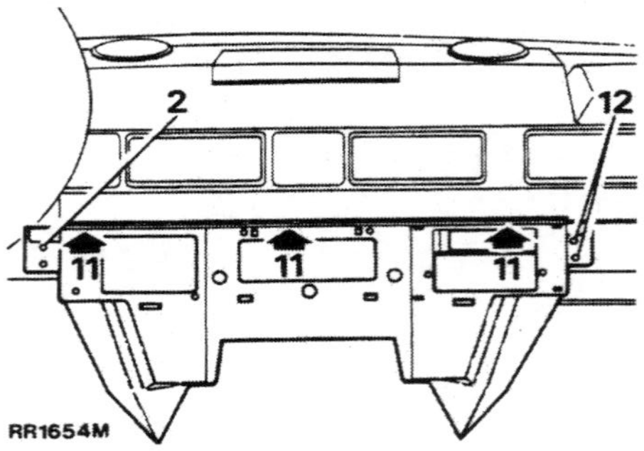

**Refitting**

14. Reverse the removal procedure ensuring that all electrical multi-plugs are fitted correctly and secure in their respective switches.
15. Arrange all electrical wiring and harnesses so that they do not become trapped between any mating faces.

---

7. Lower the fascia and disconnect the electrical plug at the rear of the rheostat switch.
8. Remove the panel from the vehicle.

**Refitting**

9. Reverse the removal procedure.

### RADIO HOUSING

**Remove and refit**

**Removing**

1. Disconnect the battery negative lead.
2. Remove the cubby box and floor-mounted console assembly.
3. Remove the radio from the housing (see Radio—remove and refit in electrical section).
4. Remove the single screw securing the housing to the gearbox tunnel.
5. Tilt the front of the housing upwards and manoeuvre it away from the centre fascia unit as far as the cigar lighter electrical leads will permit.
6. Disconnect the leads from the rear of the cigar lighter and remove radio housing complete with lighter unit.

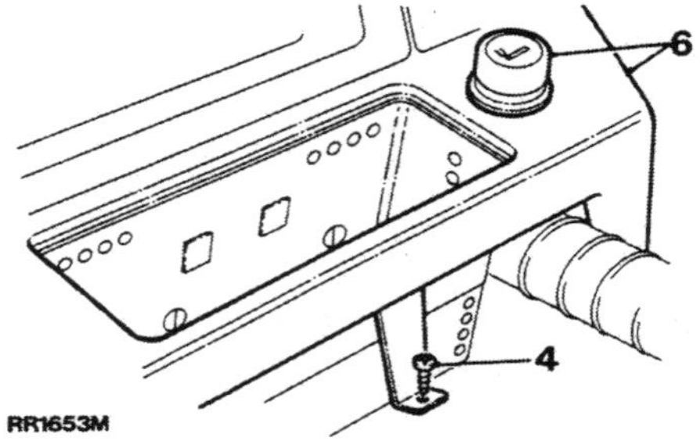

**Refitting**

7. Reverse the removal procedure ensuring the radio electrical leads do not become trapped when refitting the radio to the housing.

---

11. Reverse the removal procedure ensuring that all electrical plugs are fitted securely and correctly. Arrange the electrical wiring beneath the console to ensure it does not become trapped between any joint faces.

### LOWER FASCIA PANEL

**Remove and refit**

**Removing**

1. Disconnect the battery negative lead.
2. Prise the louvre from the fascia (Air conditioning models only).
3. Remove the two cross-head screws from the bottom of the panel located above the pedals.
4. Remove the single cross-head screw from the side of the panel adjacent to the front door courtesy light switch.
5. Remove the two cross-head screws from the top of the louvre aperture.
6. Remove the single cross-head screw above the rheostat switch.

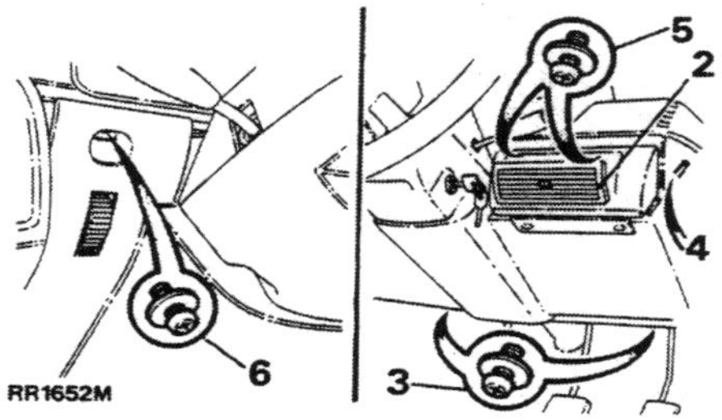

## 76 BODY

### FASCIA TOP RAIL

**Remove and refit**

**Removing**

1. Disconnect the battery negative lead and remove the steering wheel.
2. Remove the lower fascia panel.
3. Remove the cubby box and floor mounted console assembly.
4. Remove the radio and radio housing.
5. Detach the heater control graphics panel, auxiliary switch panel and fuse box from the centre fascia panel.
6. Remove the centre fascia panel.
7. Remove the louvre fascia panel.
8. Remove the instrument binnacle (see instrument binnacle Remove and refit in Electrical section).
9. Remove the rubber mat from the passenger map tray.
10. Remove the five screws, nuts and washers securing the grab rail and fascia top rail to the inner bulkhead.

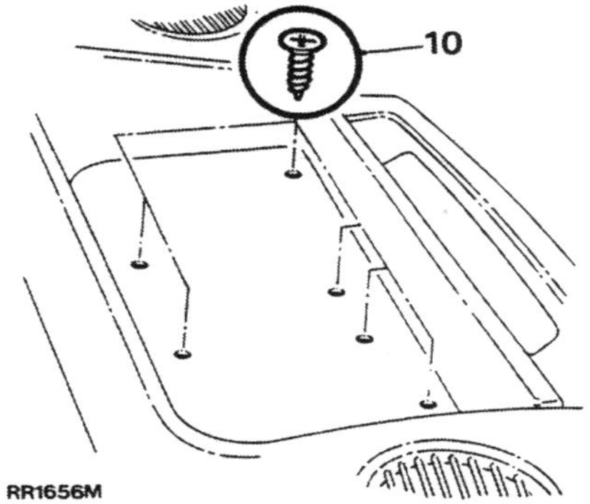

RR1656M

11. Disconnect side and centre air vent hoses from the heater unit.
12. Withdraw the fascia top rail.

**Refitting**

13. Reverse the removal procedure ensuring that the fascia top rail locates on three brackets below the windscreen inner sealing rubber.
14. Ease the sealing rubber over the top of the fascia top rail.

### FRONT SPOILER—option

A front spoiler is fitted as a standard item on Range Rover Vogue and as an option on all other Range Rover models. The spoiler, if fitted, will reduce the vehicle approach angle by approximately 10°. Where the vehicle is expected to perform on rough or hilly terrain, it is advisable to remove the spoiler to prevent any damage which may occur due to ground contact.
The spoiler may be removed as follows.

1. Disconnect the electrical connections at the rear of both auxiliary driving lamps, accessible through the front wheel arches.
2. Remove the two screws (with spring washers) securing the centre of the spoiler.
3. Remove the four nuts (with spring washers) located behind the front bumper above the driving lamp pockets, accessible from beneath the vehicle.
4. Remove the two bolts, nuts and washers securing the outer edges of the spoiler to the corners of the front wings, located forward of the front road wheels.

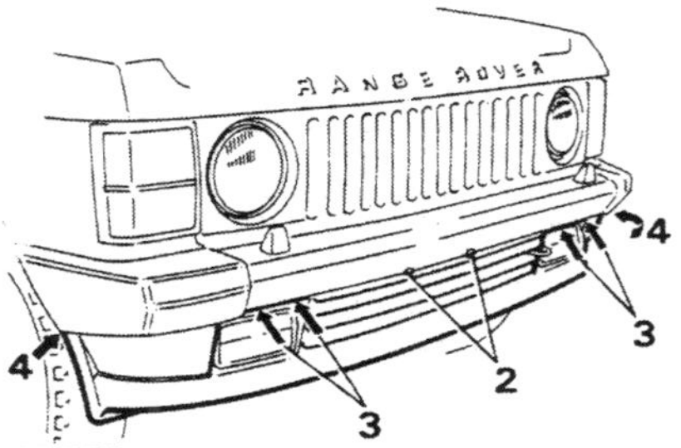

RR738M

5. Remove the spoiler complete with driving lamps.

**Refitting**

6. Reverse the removal instructions.

## Notes

203

28

# RANGE ROVER WORKSHOP MANUAL 1986 ONWARDS

# Part 5

|  | Section | Page |
|---|---|---|
| General Specification Data | 04 | 206 |
| Torque Wrench Settings | 06 | 207 |
| Heating & Ventilation | 80 | 208 |
| Air Conditioning | 82 | 209 |
| Wipers & Washers | 84 | 219 |
| Electrical | 86 | 223 |

Section
Number

| 04 | GENERAL SPECIFICATION DATA | |
|---|---|---|
| | —Air conditioning | 1 |
| | —Electrical | 1 |
| | —Wipers and washers | 2 |

| 06 | TORQUE WRENCH SETTINGS | |
|---|---|---|
| | —Air conditioning | 1 |
| | —Electrical | 1 |

| 80 | HEATING AND VENTILATION | |
|---|---|---|
| | —Heater controls | 1 |
| | —Heater unit | 1 |
| | —Heater fan motor | 2 |
| | —Fan motor resistance unit | 2 |
| | —Heater radiator | 2 |

| 82 | AIR CONDITIONING | |
|---|---|---|
| | —Description | 1 |
| | —General service information | 1 |
| | —Periodic maintenance | 3 |
| | —Service valves | 3 |
| | —Electrical supply—switches and fuses | 5 |
| | —Circuit diagram | 6 |
| | —Fault diagnosis | 7 |
| | —Charging and testing equipment | 10 |
| | —Depressurising | 10 |
| | —Evacuating | 11 |
| | —Sweeping | 11 |
| | —Charging | 11 |
| | —Leak test | 12 |
| | —Pressure test | 12 |
| | —System test | 13 |
| | —Compressor—remove and refit | 13 |
| | —Compressor—drive belt adjustment | 13 |
| | —Compressor—oil level check | 13 |
| | —Condenser—remove and refit | 13 |

*Continued*

Section
Number

| 82 | AIR CONDITIONING (continued) | |
|---|---|---|
| | —Condenser—fans and motors—remove and refit | 15 |
| | —Receiver drier—remove and refit | 16 |
| | —Evaporator—remove and refit | 17 |
| | —Evaporator expansion valve—remove and refit | 17 |
| | —Evaporator—blower assembly—remove and refit | 19 |
| | —Thermostat—remove and refit | 19 |
| | —Fan control switch—remove and refit | 20 |

| 84 | WIPERS AND WASHERS | |
|---|---|---|
| | —Description | 1 |
| | —Washer reservoir—remove and refit | 1 |
| | —Headlamp power wash pump—remove and refit | 1 |
| | —Windscreen washer pump—remove and refit | 1 |
| | —Tailgate glass washer pump—remove and refit | 2 |
| | —Windscreen washer jets—remove and refit | 2 |
| | —Windscreen washer tubes—remove and refit | 2 |
| | —Wiper arms—remove and refit | 3 |
| | —Wiper blades—remove and refit | 4 |
| | —Windscreen wiper motor—remove and refit | 4 |
| | —Wiper motor—check | 5 |
| | —Tailgate glass washer jet—remove and refit | 6 |
| | —Tailgate glass tube—remove and refit—overhaul | 8 |
| | —Tailgate glass wiper motor—remove and refit | 8 |
| | —Headlamp power wash jets—remove and refit | 8 |
| | —Headlamp power wash hose—remove and refit | 8 |

| 86 | ELECTRICAL | |
|---|---|---|
| | —Location of electrical equipment | 1 |
| | —Fault diagnosis | 2 |
| | —Description | 4 |
| | —Alternator—remove and refit | 4 |
| | —overhaul | 6 |
| | —Battery—remove and refit | 9 |
| | —Horns—remove and refit | 9 |
| | —Spark plugs—clean and adjust | 10 |
| | —Electronic ignition—description | 11 |
| | —Distributor—remove and refit | 12 |
| | —overhaul electrical components | 13 |
| | —Coil—remove and refit | 14 |
| | —Amplifier—remove and refit | 14 |
| | —Ignition timing | 15 |
| | —Constant energy ignition system—checks | 18 |
| | —Headlamp assembly—remove and refit | 18 |
| | —Headlamp bulb—remove and refit | 18 |
| | —Auxiliary driving lamps—remove and refit | 18 |
| | —bulb replacement | 18 |

*Continued*

205

# GENERAL SPECIFICATION DATA | 04

**AIR CONDITIONING**
System .................................................. A.R.A.
Compressor ............................................ Sanden SD510

**ELECTRICAL**
System .................................................. 12 volt, negative earth
Battery
Make/type ............................................. Chloride maintenance free 9-plate—210/85/90
Chloride maintenance free 14-plate—380/120/90

**ALTERNATOR**
Manufacturer ......................................... Lucas
Type ..................................................... 133/65
Polarity ................................................. Negative earth
Brush length
New ................................................... 20 mm (0.78 in)
Worn, minimum free protrusion from brush box .. 10 mm (0.39 in)
Brush spring pressure flush with brush box face .. 136 to 279 g (5 to 10 oz)
Rectifier pack output rectification .......... 6 diodes (3 live side and 3 earth side)
Field winding supply rectification .......... 3 diodes
Stator windings .................................... 3 phase—delta connected
Field winding rotor poles ...................... 12
Maximum speed ................................... 15,000 rev/min
Shaft thread ........................................ M15—1.5—6 g
Winding resistance at 20°C ................... 3.2 ohms
Control ................................................. Dual—battery sensed with machine sensed safety control
Regulator—type ................................... 15 TR
voltage ............................................. 13.6 to 14.4 volts
Nominal output
Condition ........................................... Hot
Alternator speed ................................ 6000 rev/min
Control voltage .................................. 14 volt
Amp .................................................. 65 amp

| REPLACEMENT BULBS | | TYPE | |
|---|---|---|---|
| Headlamps | 12V | 60/55W | (Halogen) |
| Headlamps—France (amber) | 12V | 60/55W | (Halogen) |
| Auxiliary driving lamps | 12V | 55W H3 | (Halogen) |
| Sidelamps | 12V | 4W | bayonet fitting |
| Stop/tail lamps | 12V | 5/21W | bayonet fitting |
| Reverse lamps | 12V | 21W | bayonet fitting |
| Rear fog guard lamps | 12V | 21W | bayonet fitting |
| Direction indicator lamps | 12V | 4W | bayonet fitting |
| Side repeater lamps | 12V | 4W | bayonet fitting |
| Number plate lamps | 12V | 1.2W | bulb/holder unit |
| Instrument panel lamps and warning lamps | 12V | 2W | capless |
| Ignition warning lamp (Instrument panel) | 12V | 10W | 'Festoon' |
| Interior roof lamps | 12V | 2W | bayonet fitting |
| Clock illumination | 12V | 1.2W | capless |
| Cigar lighter illumination | 12V | 10/5W | capless |
| Door edge/puddle lamps | 12V | 1.2W | capless |
| Auxiliary switch panel illumination (green) | 12V | 1.2W | capless |
| Heated rear screen warning lamp (amber) | 12V | 1.2W | capless |
| Hazard warning lamp (red) | 24V | 5W | capless |
| Automatic graphics illumination | 12V | 1.2W | capless |
| Air conditioning graphics illumination | | | |

# Notes

## 04 GENERAL SPECIFICATION DATA

### DISTRIBUTOR
| | |
|---|---|
| Make/type | Lucas 35DM8 |
| Firing angles | 0°—45°—90° etc. ± 1° |
| Application | 12V Negative earth |
| Pick-up air gap adjustment | 0.20 mm–0.35 mm (0.008 in–0.014 in) |
| (Pick-up limb/reluctor tooth) | |
| Pick-up winding resistance | 2k–5k ohms |

### FUSES
| | |
|---|---|
| Type | Autofuse (Blade type) |
| | Blow ratings to suit individual circuits |

### HORNS
| | |
|---|---|
| Make/type | Klarnix (Mixo) TR99 |

### IGNITION MODULE
| | |
|---|---|
| Make/type | Lucas 2CE 12-volt electronic |

### SPARK PLUGS
| | |
|---|---|
| Make/type | Champion N9YC |
| Gap | 0.70 to 0.80 mm (0.030 to 0.032 in) |

### STARTER MOTOR
| | |
|---|---|
| Make/type | Lucas 3M100 pre-engaged |
| British spring tension | 1020 gms (36 ozs) |
| Minimum brush length | 9.5 mm (.375 in) |

### TAILGATE WIPER MOTOR
| | |
|---|---|
| Make/type | Lucas 14W PM |
| Armature end float | 0.05–0.25 mm (0.002 in–0.010 in) |
| Minimum brush length | 4.8 mm (0.187 in) |
| Running current (rack disconnected) | 1.5 amps |
| Wiper speed after 60 seconds | 46–52 rev/min |

### WINDSCREEN WIPER MOTOR
| | |
|---|---|
| Make/type | Lucas 28W 2-speed |
| Running current (rack disconnected) | 1.5 amps at 39–45 rev/min (normal speed) |
| Rotary link speed | 60–73 rev/min (high speed) |

## 06 TORQUE WRENCH SETTINGS

### TORQUE WRENCH SETTINGS

#### AIR CONDITIONING
| | Nm | lbf ft |
|---|---|---|
| Compressor hose | 34–40 | 24–29 |
| Receiver drier hose | 14–21 | 10–15 |
| Receiver drier switch | 21–25 | 15–19 |
| Compressor oil filler plug | 8–12 | 6–9 |

#### ELECTRICAL
| | Nm | lbf ft |
|---|---|---|
| Alternator mounting bracket to cylinder head | 34 | 25 |
| Alternator to mounting bracket | 24 | 17 |
| Alternator to adjusting link | 24 | 17 |
| Alternator shaft nut | 27.2 to 47.5 | 20 to 35 |
| Alternator through bolts | 4.5 to 6.2 | 3.3 to 4.6 |
| Alternator rectifier bolts | 3.4 to 3.96 | 2.5 to 2.9 |
| Distributor clampbolt | 19 to 22 | 14 to 16 |
| Distributor pick-up bearing plate support pillars | 1.0 to 1.2 | 9 to 11 lbf in |
| Distributor pick-up barrel nuts | 1.1 to 1.5 | 10 to 12 lbf in |
| Spark plug | 13.8 to 16.2 | 10 to 12 |
| Starter motor to engine bolts | 40.6 to 47.4 | 30 to 35 |
| Starter motor through bolts | 10.8 | 8 |
| Solenoid fixing stud nut | 6 | 4.5 |
| Solenoid upper terminal nut | 4 | 3 |
| Reverse light switch | 20 to 27 | 15 to 20 |
| Wiper motor yoke retaining bolts | 1.35–1.8 | 12–16 |

Charts below give torque settings for all screws and bolts used—except for those that are specified.

| SIZE | NM | LB/FT |
|---|---|---|
| M5 | 5–7 | 3.7–5.2 |
| M6 | 7–10 | 5.2–7.4 |
| M8 | 22–28 | 16.2–20.7 |
| M10 | 40–50 | 29.5–36.9 |
| M12 | 80–100 | 59.0–73.8 |
| M14 | 90–120 | 66.4–88.5 |
| M16 | 160–200 | 118.0–147.5 |

| SIZE | UNC | | UNF | |
|---|---|---|---|---|
| | LB/FT | NM | LB/FT | NM |
| ¼ | 5–7 | 6.8–9.5 | 6–9 | 8.1–12.2 |
| ⁵⁄₁₆ | 15–20 | 20.3–27.1 | 15–20 | 20.3–27.1 |
| ⅜ | 26–32 | 35.2–43.4 | 26–32 | 35.2–43.4 |
| ⁷⁄₁₆ | 50–65 | 67.8–88.1 | 50–65 | 67.8–88.1 |
| ½ | 60–75 | 81.3–101.7 | 60–75 | 81.3–101.7 |
| ⅝ | 90–110 | 122.0–149.1 | 90–110 | 122.0–149.1 |

# HEATING AND VENTILATION  80

## HEATER CONTROLS

### Remove and refit

#### Removing

1. Remove the transmission lever surround.
2. Remove the radio mounting console.
3. Remove the centre console unit.
4. Disconnect the electrical leads from the control switch.
5. Disconnect the relay rod from the 'SCREEN-CAR' lever.
6. Disconnect the relay rod from the 'VENT' lever.
7. Disconnect the control cables from each side of the heater unit.
8. Remove the four securing nuts and washers and withdraw the heater controls assembly.

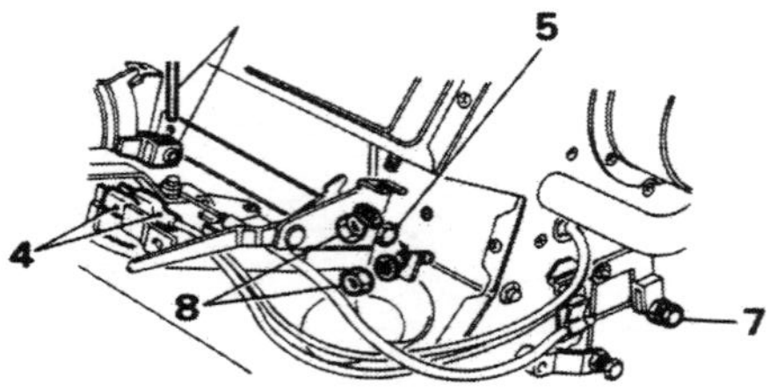

#### Refitting

9. Reverse 5 to 8.
10. Check that the control levers give full movement of the flaps. If necessary, adjust at the relay rod or cable end fixings.
11. Connect the electrical leads to the control switch, the Black/White lead connects to the front terminal.
12. Reverse 1 to 3.

## HEATER UNIT

### Remove and refit

#### Removing

1. Drain the cooling system.
2. Remove the air cleaner (carburetter vehicles).
3. Disconnect the water inlet hose from the heater.
4. Disconnect the water outlet hose from the heater.
5. Remove the lower fascia panel.
6. Remove the transmission lever surround.
7. Remove the centre console unit.
8. Disconnect the six hoses from the heater unit.
9. Disconnect the electrical leads from the heater unit.
10. Remove the four heater mounting bolts.
11. Remove the heater unit.

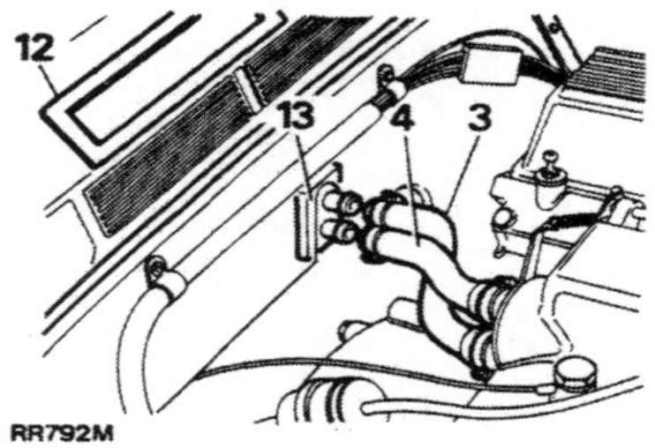

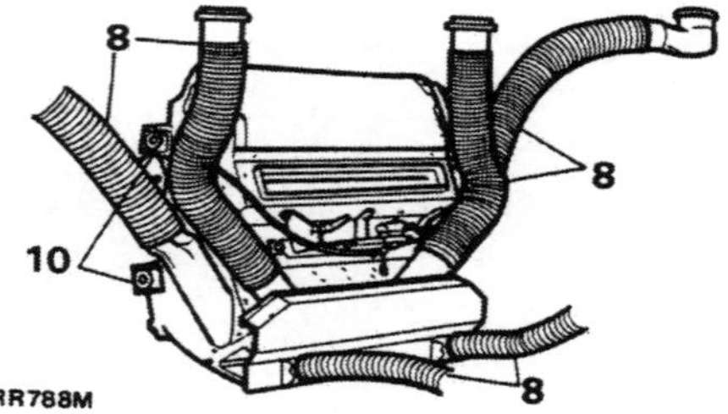

#### Refitting

12. Check that the seal for the fresh air intake is in place on the back of the heater unit.
13. Check that the seal for the heater radiator is in place on the radiator pipes.
14. Reverse 1 to 11.

*Continued*

# Notes

## HEATING AND VENTILATION

### HEATER FAN MOTOR

**Remove and refit**

**Removing**
1. Remove the heater unit.
2. Disconnect the electrical leads.
3. Disconnect the air cooling hose.
4. Remove five screws and withdraw the fan and motor assembly.
5. Slacken two grub screws securing the fan to the motor spindle and remove the fan.
6. Drill out three pop-rivets and remove the fan motor.

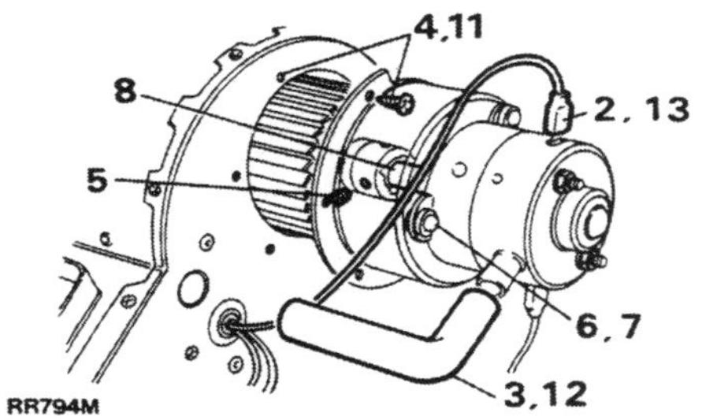

**Refitting**
7. Locate the fan motor and secure using suitable pop rivets.
8. Engage the fan motor spindle into the boss on the fan and secure the grub screws.
9. Locate the fan motor in position, engaging the fan spindle into the bearing.
10. Hold the fan motor firmly in position and check that the fan rotates freely. If necessary, adjust the position of the fan on the motor spindle.
11. Align the air cooling hose connection and refit the five screws securing the fan motor assembly.
12. Connect the air cooling hose.
13. Connect the electrical leads.
14. Refit the heater unit.

### FAN MOTOR RESISTANCE UNIT

**Remove and refit**

**Removing**
1. Remove the heater unit.
2. Disconnect the electrical leads from the fan motor.
3. Disconnect the air cooling hose.
4. Remove the fan motor and fan assembly.
5. Disconnect the electrical leads from the control switch.
6. Drill out the two pop-rivets securing the resistance unit.
7. Withdraw the resistance unit complete with leads.

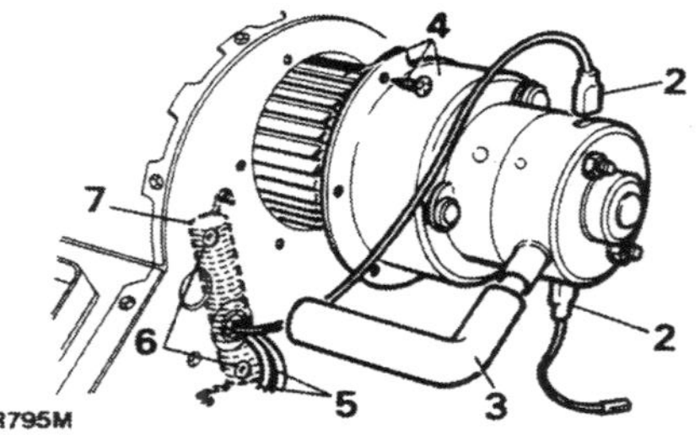

**Refitting**
8. Reverse 1 to 7. When connecting the electrical leads, connect the Black/White resistance unit lead to the front terminal on the control switch.

### HEATER RADIATOR

**Remove and refit**

**Removing**
1. Remove the heater unit.
2. Remove the fixings from the cam bracket for the fresh air flap, and move the cam and bracket assembly aside.
3. Remove the lock-washers from the four flap spindles.
4. Remove all the drive screws from the left hand side cover.
5. Withdraw the left hand side cover complete with the fresh air flap.
6. Withdraw the heater radiator complete with seals.
7. Withdraw the seals from the heater radiator.

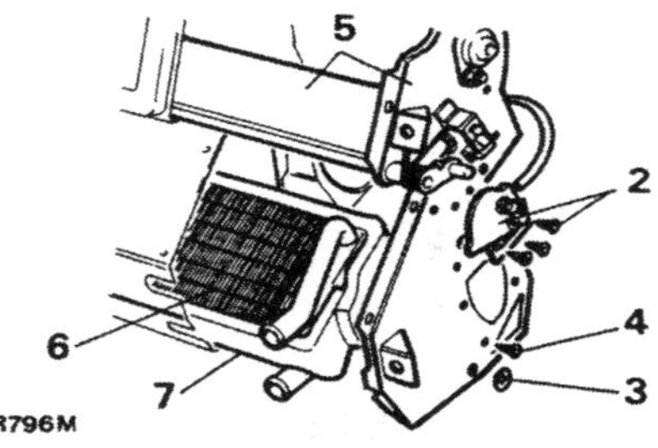

**Refitting**
8. Apply Bostik sealing compound around the flange of the left hand side cover.
9. Reverse 1 to 7.

---

## AIR CONDITIONING

### AIR CONDITIONING — A.R.A. SYSTEM

**Description**

The A.R.A. air conditioning system comprises four units:
1. An engine-mounted compressor.
2. A condenser mounted in front of the radiator.
3. A receiver/drier unit located in the engine compartment.
4. An evaporator unit mounted behind the fascia.

The four units are interconnected by hoses carrying refrigerant, and the evaporator is linked into the vehicle ventilation system.

**WARNING: Under no circumstances should refrigerant pipes be disconnected without first depressurising the system.**

**Cold refrigerant circuit**

The function of the refrigeration circuit is to cool the evaporator.

*1 Compressor*

The compressor draws vaporized refrigerant from the evaporator. It is compressed, and thus heated, and passed on to the condenser as a hot, high pressure vapour.

*2 Condenser*

The condenser is mounted directly in front of the vehicle radiator. It consists of a refrigerant coil mounted in a series of thin cooling fins to provide the maximum heat transfer in a minimum amount of space. Airflow across the condenser is induced by vehicle movement and is assisted by two electric condenser fans.

The refrigerant enters the inlet at the top of the condenser as a heat laden high pressure vapour. As this vapour passes down through the condenser coils, heat will follow its natural tendency and flow from the hot refrigerant vapour into the cooler air flowing across the condenser coils and fins.

When the refrigerant vapour reaches the temperature and pressure that will induce a change of state a large quantity of latent heat will be transferred to the outside air. The refrigerant will change from a high pressure HOT VAPOUR to a high pressure WARM LIQUID.

*3 Receiver drier*

This unit filters, removes moisture, and acts as a reservoir for the liquid. To prevent icing inside the system, extreme precautions are taken during servicing to exclude moisture. The receiver drier should be considered as a second stage insurance to prevent the serious consequences of ice obstructing the flow. A sight glass provided in the unit top enables a visual check to be made of the high pressure liquid flow.

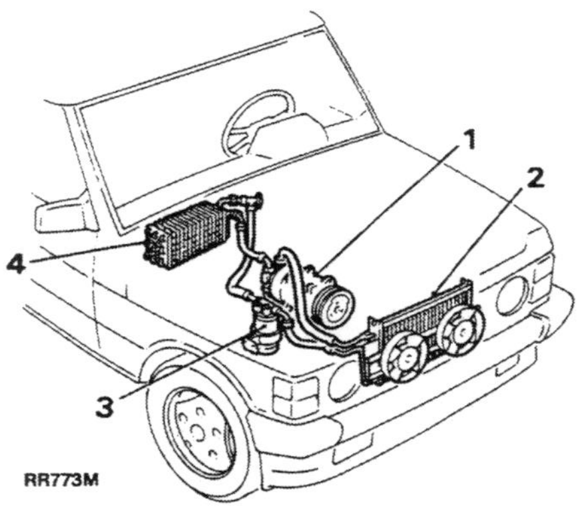

*4 Expansion valve and evaporator*

High pressure liquid refrigerant is delivered to the expansion valve. A severe pressure drop occurs across the valve and as the refrigerant enters the evaporator space at a temperature of approximately −6°C it boils and vaporizes. As this change of state occurs, a large amount of latent heat is absorbed. The evaporator is therefore cooled and as a result heat is extracted from the air flowing across the evaporator. The air flow is controlled by two evaporator fans regulated by the air conditioner fan control.

*Second cycle*

Vaporized refrigerant is then drawn from the evaporator by the compressor and a second cycle commences.

### GENERAL SERVICE INFORMATION

**Introduction**

Before any component of the air conditioning system is removed, the system must be depressurised. When the component is replaced, the system must be evacuated to remove all traces of old refrigerant and moisture. Then the system must be recharged with new refrigerant.

Any service operation that requires the loosening of a refrigerant line connection should be performed only by qualified service personnel. Refrigerant and/or oil will escape whenever a hose or pipe is disconnected.

All work involving the handling of refrigerant requires special equipment, a knowledge of its proper use and attention to safety measures.

*Continued*

# AIR CONDITIONING

## Servicing equipment

The following equipment is required for full servicing of the air conditioning system.

Charging trolley
Leak detector
Tachometer
Safety goggles
Refrigerant charging line gaskets
Thermometer: 20°C to −60°C (68°F to −76°F)
Valve core removers
Compressor dipstick

## SERVICING MATERIALS

Refrigerant: Refrigerant 12, which includes Freon 12 or Arcton 12.

**CAUTION: Methychloride refrigerants must not be used.**

Nominal charge weight:
RHD vehicles: 1.25 kg (44 oz)
LHD vehicles: 1.08 kg (38 oz)
Compressor oil: See Recommended Lubricants.

## PRECAUTIONS IN HANDLING REFRIGERANT

Refrigerant 12 is transparent and colourless in both the gaseous and liquid state. It has a boiling point of −29.8°C (−21.7°F) at atmospheric pressure and at all normal pressures and temperatures it becomes a vapour. The vapour is heavier than air, non-flammable, and non-explosive. It is non-poisonous except when in contact with an open flame, and non-corrosive until it comes in contact with water.

The following precautions in handling Refrigerant 12 should be observed at all times.
DO NOT — leave refrigerant drum without its heavy cap fitted.
— carry refrigerant drum inside a vehicle.
— subject refrigerant drums to high temperature.
— weld or steam clean near an air conditioning system.
— expose eyes to liquid refrigerant, ALWAYS wear goggles.
— discharge refrigerant vapour into an area with an exposed flame or into an engine intake. Heavy concentrations of refrigerant in contact with naked flame produces a toxic gas.
— allow liquid refrigerant to contact bright metal, it will tarnish metal and chrome surfaces, and combined with moisture can seriously corrode all metal surfaces.

## PRECAUTIONS IN HANDLING REFRIGERANT LINES

**WARNING: Always wear safety goggles when opening refrigerant connections.**

(a) When disconnecting any pipe or flexible connection the system must be discharged of all pressure. Proceed cautiously, regardless of gauge readings. Open connections slowly, keeping hands and face well clear, so that no injury occurs if there is liquid in the line. If pressure is noticed, allow it to bleed off slowly.

(b) Lines, flexible end connections and components must be capped immediately they are opened to prevent the entrance of moisture and dirt.

(c) Any dirt or grease on fittings must be wiped off with a clean alcohol dampened cloth. Do not use chlorinated solvents such as trichloroethylene. If dirt, grease or moisture cannot be removed from inside the pipes, they must be replaced with new pipes.

(d) All replacement components and flexible end connections are sealed, and should only be opened immediately prior to making the connection.

(e) Ensure the components are at room temperature before uncapping, to prevent condensation of moisture from the air that enters.

(f) Components must not remain uncapped for longer than fifteen minutes. In the event of delay, the caps must be replaced.

(g) Receiver/driers must never be left uncapped as they contain Silica Gel crystals which will absorb moisture from the atmosphere. A receiver/drier left uncapped must be replaced, and not used.

(h) The compressor shaft must not be rotated until the system is entirely assembled and contains a charge of refrigerant.

(j) A new compressor contains an initial charge of 135 ml (4.6 UK fluid oz) of oil when received, part of which is distributed throughout the system when it has been run. The compressor contains a holding charge of gas when received which should be retained until the hoses are connected.

(k) The receiver/drier should be the last component connected to the system to ensure optimum dehydration and maximum moisture protection of the system.

(l) All precautions must be taken to prevent damage to fittings and connections. Slight damage could cause a leak with the high pressures used in the system.

(m) Always use two spanners of the correct size, one on each hexagon, when releasing and tightening refrigeration unions.

(n) Joints and 'O' rings should be coated with refrigeration oil to aid correct seating. Fittings which are not lubricated with refrigerant oil are almost certain to leak.

(o) All lines must be free of kinks. The efficiency of the system is reduced by a single kink or restriction.

(p) Flexible hoses should not be bent to a radius less than ten times the diameter of the hoses.

(q) Flexible connections should not be within 50 mm (2 in) of the exhaust manifold.

(r) Completed assemblies must be checked for refrigeration lines touching metal panels. Any direct contact of lines and panels transmits noise and must be eliminated.

## PERIODIC MAINTENANCE

Routine servicing, apart from visual checks, is not necessary. The visual inspections are as follows:

### Condenser

With a hose pipe or air line, clean the face of the condenser to remove flies, leaves, etc. Check the pipe connections for signs of oil leakage.

### Compressor

Check hose connections for signs of oil leakage. Check flexible hoses for swelling. Examine the compressor belt for tightness and condition. Checking the compressor oil level and topping-up is only necessary after charging the system or in the event of a malfunction of the system.

### Receiver/Drier

Examine the sight glass for bubbles with the system operating. Check connections for leakage.

### Evaporator

Examine the refrigeration connections at the unit. If the system should develop a fault, or if erratic operation is noticed, refer to the fault diagnosis chart.

## SERVICE VALVES

There are two types of service valves in operation: 'Stem' and 'Schrader'.

### Stem type

Stem type service valves allow for the isolation of the compressor from other parts of the system. When these valves are used in conjunction with the liquid line quick-disconnect fittings, the three major assemblies of the system can be removed from the vehicle with a minimal loss of refrigerant. In addition, it is possible to remove major assemblies for repair of components which are not part of the refrigeration system, or provide access to parts of the vehicle which are obstructed by the air conditioning system, without fully discharging the system. A thorough understanding of the stem type service valve is necessary before undertaking servicing or repair involving the air conditioning system.

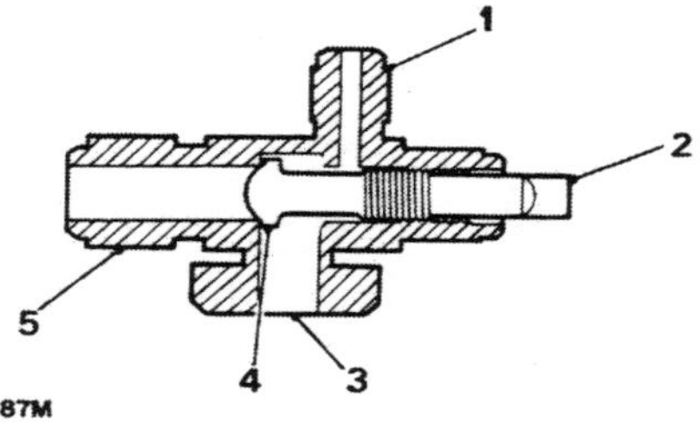

ST1387M

Stem type service valve
1. Service port.
2. Valve stem.
3. Compressor port.
4. Valve seat.
5. Hose connector.

**NOTE: A special wrench should be used to adjust the valve to prevent damage to the stem.**

The stem type service valve has three positions, the operation of which is explained as follows.

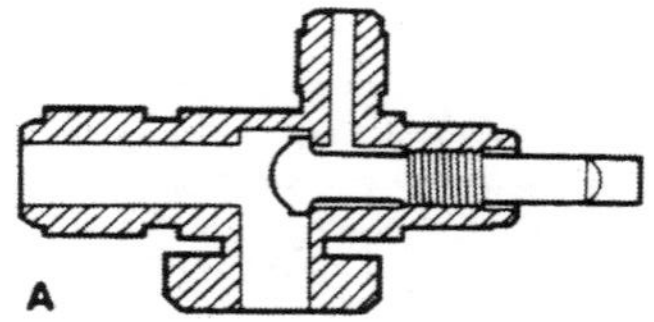

RR1734M

A. ON: FULLY ANTICLOCKWISE — Normal operating position, and the position which is used for connecting and disconnecting the manifold gauge set, is the 'on' position. The stem is turned fully anticlockwise. This seals the service gauge port from receiving any refrigerant flow.

*Continued*

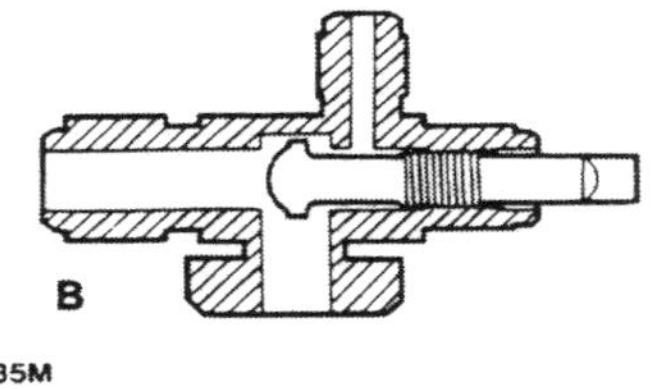

# AIR CONDITIONING

B. MID (Test) POSITION — After the service gauge manifold has been installed (the valve stem is in the on position), turn the valve stem the required number of turns clockwise. This will put the valve stem seat midway in the service valve and allow full system operation while permitting refrigerant pressure to reach the gauges.

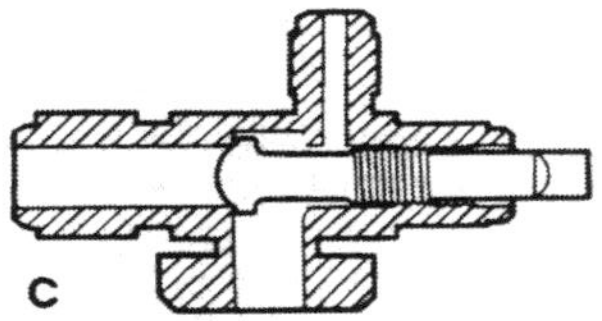

C. OFF: FULLY CLOCKWISE — With the service valve stem turned fully clockwise, the valve will block passage of refrigerant flow through the system. As illustrated, the refrigerant flow to or from the compressor (depending on whether it is high side or low side) is blocked.

**WARNING: NEVER operate the air conditioning system with the service valves in the OFF POSITION, it will cause severe damage to the compressor.**

### Schrader type

These are secured to the head of the compressor, and the suction and discharge flexible end connections are secured to them by unions.

The service valves are identified as suction or low pressure, and discharge or high pressure. Whilst they are identical in operation they are not interchangeable, as the connections are of different sizes.

As the name suggests, these valves are for service purposes, providing connections to external pressure/vacuum gauges for test purposes. In combination with charging and testing equipment they are used to charge the system with refrigerant.

### Schrader service valve

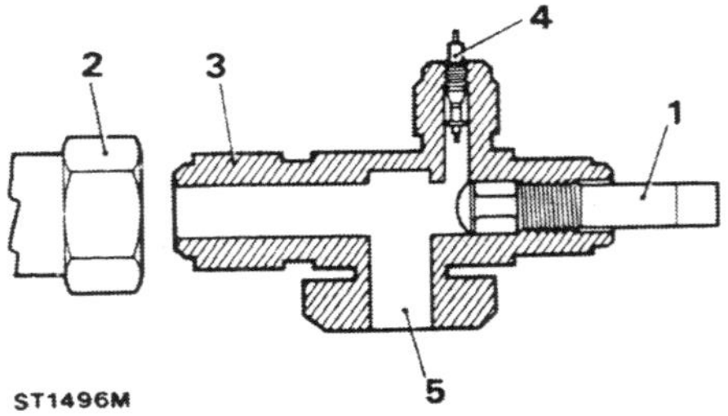

1. Valve stem.
2. Hose connection.
3. Service valve.
4. Schrader valve core.
5. Compressor port.

**NOTE: A special wrench should be used to adjust the valve to prevent damage to the stem.**

The Schrader type service valve has two positions, the operation of which is explained as follows.

A. ON: FULLY ANTICLOCKWISE — Normal operating position, and the position which is used for connecting and disconnecting the manifold gauge set, is the 'on' position. The stem is turned fully anticlockwise. This seals the service gauge port from receiving any refrigerant flow.

B. OFF: FULLY CLOCKWISE — With the service valve stem turned fully clockwise, the valve will block passage of refrigerant flow through the system. As illustrated, the refrigerant flow to or from the compressor (depending on whether it is high side or low side) is blocked.

**WARNING: NEVER operate the air conditioning system with the service valves in the OFF POSITION, it will cause severe damage to the compressor.**

### Valve core remover

Where Schrader valve depressors are not fitted to the testing equipment lines, valve core removers can be used.

### Valve core removal

The use of valve core removers will facilitate servicing operations and should be used as follows:
1. Close all valves on the charging trolley.
2. Remove the service valve cap and seals from the valve core remover.
3. Withdraw the plunger as far as possible and connect the core remover to the service valve.
4. Connect the hose to the core remover.
5. Depress the plunger until it contacts the valve core. Unscrew the valve until it is free. Withdraw the plunger to its full extent.

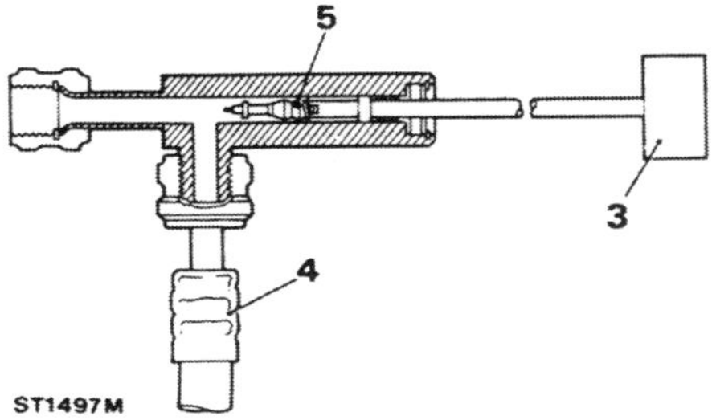

Service valve caps must be replaced when service operations are completed. Failure to replace caps could result in refrigerant loss and system failure.

### Electrical supply switches and fuses

The four main components of the air conditioning system draw current from their own separate relays. The air conditioning system is mastered from the starter relay and is switched 'off' during engine cranking. Each component in turn is energised and controlled by a series of relays and switches of various types as indicated by the circuit diagram.

The relays are mounted on the engine compartment closure panel. Both condenser fans operate together when the air conditioning circuit is switched 'on' or when the ignition is switched 'on' and the coolant temperature is high, on automatic vehicles only.

Electronic Fuel Injection vehicles have a solenoid operated air valve, which is activated when the compressor clutch is energized. This increases engine rev/min when the air conditioning system is operative at idle speed.

The four Autofuse type fuses are located in the fuse box mounted on the lower fascia panel. They are numbered A1 to A4. It is essential that a fuse of the same value is used when fitting a replacement.

*Continued*

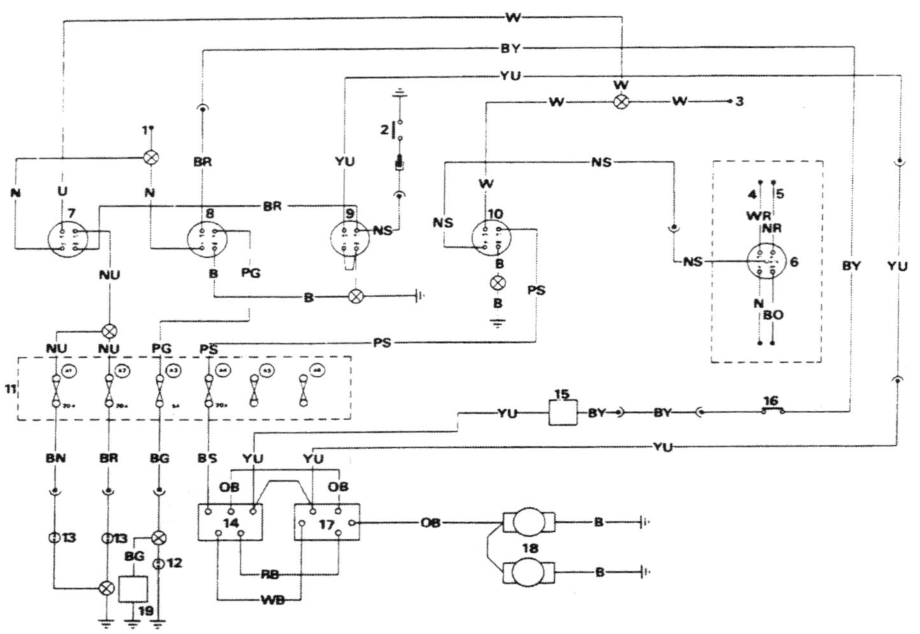

# AIR CONDITIONING

## Electrical System Fault Diagnosis

| FAULT | CAUSE | REMEDY |
|---|---|---|
| A. MOTOR INOPERATIVE OR SLOW RUNNING | 1 Incorrect voltage.<br>2 Open or defective fuse or relay.<br>3 Loose wire connection, including ground.<br>4 Switch open or defective.<br>5 Tight, worn, or burnt motor bearings.<br>6 Open rotor windings.<br>7 Worn motor brushes.<br>8 Shaft binding—blade misaligned.<br>9 Defective resistor board. | 1 Check voltage.<br>2 Check and replace as necessary.<br>3 Check system wires; tighten all connections.<br>4 Replace switch.<br>5 Replace motor.<br>6 Replace motor.<br>7 Replace motor.<br>8 Check alignment. Repair or replace as necessary.<br>9 Rectify or replace. |
| B. CLUTCH INOPERATIVE | 1 Incorrect voltage.<br>2 Open or defective fuse or relay.<br>3 Defective thermostat control or pressure switch.<br>4 Shorted or open field coil.<br>5 Bearing seized (clutch will not disengage).<br>6 Refrigeration circuit problem causing heavy load and excessive drive torque. | 1 Check voltage.<br>2 Check and replace as necessary.<br>3 Replace thermostat or pressure switch.<br>4 Replace coil.<br>5 Replace bearing.<br>6 Check and rectify. |
| C. CLUTCH NOISY | 1 Incorrect alignment.<br>2 Loose belt.<br>3 Compressor not mounted securely.<br>4 Bearing in clutch-pulley assembly not pressed in properly.<br>5 Low voltage to clutch.<br>6 Clutch will not spin freely.<br>7 Oil on clutch face.<br>8 Slipping clutch.<br>9 Overloaded or locked compressor.<br>10 Icing. | 1 Check alignment; repair as necessary.<br>2 Adjust to proper tension.<br>3 Repair as necessary.<br>4 Remove clutch and replace bearing.<br>5 Check connections and voltage.<br>6 Refer to B5 above.<br>7 Check compressor seals for leaks.<br>8 Refer to C5 above.<br>9 Repair or replace compressor.<br>10 Check for suction line frosting. Replace expansion valve if necessary. Replace receiver/drier if necessary. |
| D. CONDENSER AND/OR EVAPORATOR VIBRATION | 1 Motor and/or blades improperly mounted.<br>2 Blade corrosion or foreign matter build-up.<br>3 Excessive wear of motor bearings. | 1 Check mountings, adjust as necessary.<br>2 Clean blades with solvent or other non-inflammable cleaner.<br>3 Replace motor. |

## KEY TO CIRCUIT DIAGRAM

1. Terminal post.
2. Engine water temperature switch (Automatic only).
3. Ignition feed.
4. Crank feed.
5. Starter solenoid.
6. Starter relay.
7. Fan relay.
8. Compressor clutch relay.
9. Air conditioning controlled fan relay.
10. Air conditioning relay (Ignition controlled).
11. Auxiliary fuse box.
12. Compressor clutch.
13. Fans.
14. Control switch.
15. Thermostat.
16. High pressure switch.
17. Resistor.
18. Blower motors.
19. Solenoid operated air valve.

### Circuit symbols

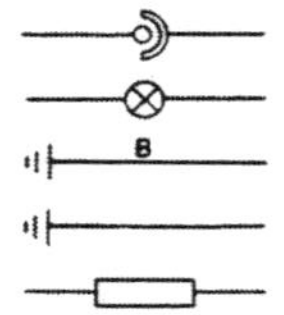

Plug and socket
Clinch connector
Earth connection via cables
Earth connection via bolts
Snap connectors

### Cable colour code

| B | Black | G | Green | R | Red |
| U | Blue | O | Orange | W | White |
| N | Brown | P | Purple | Y | Yellow |

The last letter of a colour code denotes the tracer.

RR613M

## AIR CONDITIONING

### Refrigeration System Fault Diagnosis

For any refrigeration system to function properly all components must be in good working order. The unit cooling cycle and the relationship between air discharge temperature and ambient temperature and the pressures at the compressor can help to determine proper operation of the system.

The length of any cooling cycle is determined by such factors as ambient temperature and humidity, thermostat setting, compressor speed and air leakage into the cooled area, etc. With these factors constant, any sudden increase in the length of the cooling cycle would be indicative of abnormal operation of the air conditioner.

The low and high side pressures at the compressor will vary with changing ambient temperature, humidity, cab temperature and altitude.

The following conditions should be checked after operating the system for several minutes:

1. All high side lines and components should be warm to the touch.
2. All low side lines should be cool to the touch.
3. Inlet and outlet temperatures at the receiver/drier should be at the same temperature (warm). Any very noticeable temperature difference indicates a blocked receiver/drier.
4. Heavy frost on the inlet to the expansion valve may indicate a defective valve or moisture in the system.
5. With ambient humidity between 30% and 60%, compressor pressures and evaporator air discharge temperature should fall within the general limits given in the table below.

| Type of Weather | Evaporator Air Temp °F (°C) | Low Side Pressure lb/in² (kg/cm²) | High Side Pressure lb/in² (kg/cm²) |
|---|---|---|---|
| Cool Day 70–80°F (21–27°C) | 35–45°F (1.7–7.2°C) | 15–20 (1.1–1.4) | 160–200 (11.2–14) |
| Warm Day 80–90°F (27–32°C) | 40–50°F (4.4–10°C) | 20–25 (1.4–1.8) | 190–240 (13.4–16.9) |
| Hot Day Over 90°F (Over 32°C) | 45–60°F (7.2–15.6°C) | 25–30 (1.8–2.1) | 220–270 (15.5–19) |

NOTE:
1. Low and high side pressures are guides not specific limits.
2. Evap. air temperatures will be lower on dry days, higher on humid days.

## AIR CONDITIONING

| FAULT | CAUSE | REMEDY |
|---|---|---|
| A. HIGH HEAD PRESSURE | 1 Overcharge of refrigerant. | 1 Purge with bleed hose until bubbles start to appear in sight glass; then, add sufficient refrigerant gas to clear sight glass. |
| | 2 Air in system. | 2 Slowly blow charge to atmosphere. Install new drier, evacuate and charge system. |
| | 3 Condenser air passage clogged with dirt or other foreign matter. | 3 Clean condenser of debris. |
| | 4 Condenser fan motor defective. | 4 Replace motor. |
| B. LOW HEAD PRESSURE | 1 Undercharge of refrigerant; evident by bubbles in sight glass while system is operating. | 1 Evacuate and recharge the system. Check for leakage. |
| | 2 Split compressor gasket or leaking valves. | 2 Replace gasket and/or reed valve; Install new drier, evacuate and charge the system. |
| | 3 Defective compressor. | 3 Repair or replace compressor. |
| C. HIGH SUCTION PRESSURE | 1 Slack compressor belt. | 1 Adjust belt tension. |
| | 2 Refrigerant flooding through evaporator into suction line; evident by ice on suction line and suction service valve. | 2 Check thermobulb. Bulb should be securely clamped to clean horizontal section of copper suction pipe. |
| | 3 Expansion valve stuck open. | 3 Replace expansion valve. |
| | 4 Compressor suction valve strainer restricted. | 4 Remove and clean or replace strainer. |
| | 5 Leaking compressor valves, valve gaskets and/or service valves. | 5 Replace valves and/or gaskets. Install new drier, evacuate and charge the system. |
| | 6 Receiver/drier stopped; evident by temperature difference between input and output lines. | 6 Install new drier, evacuate and charge the system. |
| D. LOW SUCTION | 1 Expansion valve thermobulb not operating. | 1 Warm thermobulb with hand. Suction should rise rapidly to 20 lb. or more. If not, replace expansion valve. |
| | 2 Expansion valve sticking closed. | 2 Check inlet side screen. Clean if clogged. Refer to C-2 and C-3. |
| | 3 Moisture freezing in expansion valve orifice. Valve outlet tube will frost while inlet hose tube will have little or no frost. System operates periodically. | 3 Install new drier, evacuate and charge the system. |
| | 4 Dust, paper scraps, or other debris restricting evaporator blower grille. | 4 Clean grilles as required. |
| | 5 Defective evaporator blower motor, wiring, or blower switch. | 5 Refer to Fault Diagnosis Chart for Electrical System. |
| E. NOISY EXPANSION VALVE (steady hissing) | 1 Low refrigerant charge; evident by bubbles in sight glass. | 1 Leak test. Repair or replace components as required. |
| F. INSUFFICIENT COOLING | 1 Expansion valve not operating properly. | 1 Refer to C-2, C-3, D-1 and E. |
| | 2 Low refrigerant charge—evident by bubbles in sight glass. | 2 Refer to B-1 and E. |
| | 3 Compressor not pumping. | 3 Refer to B-2 and B-3. |
| G. COMPRESSOR BELT SLIPPING | 1 Belt tension. | 1 Adjust belt tension. |
| | 2 Incorrect alignment of pulleys or worn belt not riding properly. | 2 Refer to A-1 through A-4 and C-6. |
| | 3 Excessive head pressure. | 3 Repair as needed. |
| | 4 Nicked or broken pulley. | 4 Replace as needed. |
| | 5 Seized compressor. | 5 Replace compressor. |
| H. ENGINE NOISE AND/OR VIBRATION | 1 Loose or missing mounting bolts. | 1 Repair as necessary. |
| | 2 Broken mounting bracket. | 2 Repair as necessary. |
| | 3 Loose flywheel or clutch retaining bolt. | 3 Repair as necessary. |
| | 4 Rough idler pulley bearing. | 4 Replace bearing. |
| | 5 Bent, loose, or improperly mounted engine drive pulley. | 5 Repair as necessary. |
| | 6 Defective compressor bearing. | 6 Replace bearing. |
| | 7 Insecure mountings of accessories: generator, power steering, air filter, etc. | 7 Repair as necessary. |
| | 8 Excessive head pressure. | 8 Refer to A-1, A-2, A-3, A-4 and C-6. |
| | 9 Incorrect compressor oil level. | 9 Refer to Compressor Oil Level Check. |

# AIR CONDITIONING

## CHARGING AND TESTING EQUIPMENT

This is standard equipment for the servicing of automotive air conditioning systems, and is used for all testing, trouble shooting, evacuating and charging operations.

Various designs of charging and testing equipment are available depending upon the manufacturer chosen by the user. As slight variations do occur it is recommended that the operator adheres to the appropriate manufacturers' instructions for the equipment in use.

**WARNING: The air conditioning system is charged with a high pressure, potentially toxic refrigerant. Repairs or servicing MUST only be carried out by an operator familiar with both the vehicle system and the charging and testing equipment.**

**All operations must be carried out in a well-ventilated area away from open flame and heat sources.**

**Always wear safety goggles when opening refrigerant connections.**

Refrigerant 12 evaporates so rapidly at normal atmospheric pressures and temperatures that it tends to freeze anything it contacts. Extreme care must be taken to prevent any liquid refrigerant from contacting the skin and especially the eyes. Should any liquid refrigerant get into the eyes, use a few drops of sterile mineral oil to wash them out and then wash the eyes with a weak solution of boric acid. Seek medical attention immediately even though the initial irritation has ceased after first-aid treatment. Always wear safety goggles when opening refrigerant connections.

**WARNING: Open connections slowly, keeping the hands and face well clear, so that no injury occurs if there is liquid in the line. If pressure is noticed allow it to bleed off slowly.**

### Depressurising

1. Connect the equipment according to the manufacturer's instructions.
2. Run the blue, low pressure hose to an open-topped container of approximately one litre capacity. Attach the hose to the container so that it will not blow out of the container. The purpose of the container is to collect any oil carried by the refrigerant.
3. When depressurising adjust the refrigerant flow to ensure that oil is not blown out of the container.
4. Measure the amount of oil discharged from the system. Add an equal amount of new oil to the system during the charging operation. Discard the old oil.

**NOTE: If it is necessary to disconnect the compressor hoses, the compressor should be sealed by fully closing the relevant service valve (turn fully clockwise). It is essential to ensure that both service valves are open before operating the compressor. Similarly any other component of the refrigeration system should be capped immediately when disconnected.**

### Evacuate

Whenever the system has been opened to the atmosphere it is necessary that the system be evacuated to remove all air and moisture. It is also an essential preliminary operation to charging the system with Refrigerant 12. The evacuate operation also provides a check for leaks due to faulty connections.

### Evacuating

1. Depressurise the system as previously described, and connect the charging and testing equipment referring to the manufacturer's instructions.
2. Slowly open the vacuum control valve. If the vacuum is applied to the system too quickly, the residual oil may be drawn out.
3. In evacuating the system it is necessary to lower the pressure so that the boiling point of water in the system is lower than the surrounding air temperature. At an ambient temperature of 23.8°C (75°F), it is necessary to lower the system pressure to 29.5 in Hg vacuum to bring the boiling point of water to 22°C (72°F). Atmospheric pressure (and vacuum gauge readings) decrease at altitude increases by approximately 25 mm (1 in) Hg per 300 m (100 ft). The following chart provides a guide to the various gauge readings at differing altitudes, for the same 10 mm (0.4 in) Hg absolute pressure.

| Altitude, ft | Vacuum Reading in Hg |
|---|---|
| 0 | 29.5 |
| 1,000 | 28.5 |
| 2,000 | 27.4 |
| 3,000 | 26.4 |
| 4,000 | 25.4 |
| 5,000 | 24.5 |
| 6,000 | 23.5 |
| 7,000 | 22.6 |
| 8,000 | 21.8 |
| 9,000 | 20.9 |
| 10,000 | 20.1 |

4. The low side gauge should indicate a vacuum of 660 mm (26 in) Hg within five minutes.
5. If 660 mm (26 in) Hg of vacuum is not achieved within five minutes, it signifies either the system has a leak or the vacuum pump is defective. Initially check the vacuum pump, if the pump proves to be functioning properly then investigate for a leak in the air conditioning system.
6. Stop the vacuum pump and allow the vacuum to hold for fifteen minutes, then check that there is no pressure rise (a loss of vacuum) evident on the compound gauge. Any pressure rise denotes a leak which must be rectified before proceeding further. Refer to the heading titled 'Leak Detection' later in this section.
   With the system satisfactorily evacuated, the system is ready for charging with refrigerant.

### Sweep

**NOTE: This operation is in addition to evacuating, and is to remove moisture from systems that have been open to atmosphere for a long period, or that are known to contain excessive moisture.**

### Sweeping

1. Fit a new liquid receiver drier, as detailed under the heading 'Receiver Drier'.
2. Connect the charging and testing equipment and follow the equipment manufacturer's instructions for sweeping.
3. Maintain the vacuum for twenty minutes. The air conditioning system is now ready for charging with refrigerant.

### Charge

**CAUTION: Do not charge liquid refrigerant into the compressor. Liquid cannot be compressed; and if liquid refrigerant enters the compressor inlet valve, severe damage is possible; in addition, the oil charge may be absorbed into the refrigerant, causing damage when the compressor is operated.**

### Charging

1. Ensure that the air conditioning system is evacuated as previously described.
2. Follow the equipment manufacturer's instructions for charging the system with refrigerant.
3. Ensure that the full charge of refrigerant (RHD 1.25 kg (44 oz), LHD 1.08 kg (38 oz) is drawn into the system.
4. After completing the procedure check the air conditioning system is operating satisfactorily by carrying out a pressure test, as described later in the Section.

**CAUTION: Do not overcharge the air conditioning system as this will cause excessive head pressure.**

### Leak test

The following instructions refer to an electronic type refrigerant leak detector which is the safest, most sensitive and widely used.

1. Place the vehicle in a well ventilated area but free from draughts, as leakage from the system could be dissipated without detection.
2. Follow the instructions issued by the manufacturer of the particular leak detector being used. Certain detectors have visual and audible indicators.
3. Commence searching for leaks by passing the detector probe around all joints and components, particularly on the underside, as the refrigerant gas is heavier than air.
4. Insert the probe into an air outlet of the evaporator. Switch the air conditioning blower on and off at intervals of ten seconds. Any leaking refrigerant will be gathered in by the blower and detected.
5. Insert the probe between the magnetic clutch and compressor to check the shaft seal for leaks.
6. Check all service valve connections, valve plate, head and base plate joints and back seal plate.
7. Check the condenser for leaks at the pipe connections.
8. If any leaks are found, the system must be depressurised before attempting rectification. If repairs by brazing are necessary, the component must be removed from the vehicle and all traces of refrigerant expelled before heat is applied.
9. After repairs check the system for leaks and evacuate prior to charging.

*Continued*

# AIR CONDITIONING

## Pressure test

1. Fit the charging and test equipment as previously described.
2. Start the engine.
3. Run the engine at 1,000 to 1,200 rev/min with the heat control set to cold (blue) zone, air distribution control to the central position, recirculation control to recirculation and the air conditioning control 'ON' with fan speed to maximum.
4. Note the ambient air temperature control in the immediate test area in front of the vehicle, and check the high pressure gauge readings—discharge side—against table 1.

### Table 1

| Ambient Temperature | | Compound Gauge Readings | | High Pressure Gauge Readings | |
|---|---|---|---|---|---|
| °C | °F | kgf/cm² | lbf/in² | kgf/cm² | lbf/in² |
| 16 | 60 | 1.05-1.4 | 15-20 | 7.0-10.2 | 100-150 |
| 26.7 | 80 | 1.4-1.75 | 20-25 | 9.8-13.3 | 140-190 |
| 38 | 100 | 1.75-2.1 | 25-30 | 11.6-15.8 | 180-225 |
| 43.5 | 110 | 2.1-2.45 | 30-35 | 15.1-17.5 | 215-250 |

The pressure gauge readings will vary within the range quoted with the rate of flow of air over the condenser, the higher readings resulting from a low air flow. It is recommended that a fan is used for additional air flow over the condenser if the system is to be operated for a long time. Always use a fan if temperatures are over 26.7°C (80°F) so that a consistent analysis can be made of readings.

5. If the pressure readings are outside the limits quoted, refer to the fault diagnosis chart at the beginning of this section.
6. Stop the engine.
7. Close both service ports (turn fully anti-clockwise) and close all valves on the charging and test equipment. Disconnect the charging lines from the compressor. Refit the blanking caps to the compressor valve stems, port connections and charging lines.
8. Close the bonnet.

## System test

1. Place the vehicle in a ventilated, shaded area free from excessive draught, with the doors and windows open.
2. Check that the surface of the condenser is not restricted with dirt, leaves, flies, etc. Do not neglect to check the surface between the condenser and the radiator. Clean as necessary.
3. Switch on the ignition and the air conditioner air flow control. Check that the blower is operating efficiently at low, medium and high speeds. Switch off the blower and the ignition.
4. Check that the evaporator condensate drains are open and clear.
5. Check the tension of the compressor driving belt, and adjust if necessary.
6. Inspect all connections for the presence of refrigerant oil. If oil is evident, check for leaks, and rectify as necessary.

**NOTE: The compressor oil is soluble in Refrigerant 12 and is deposited when the refrigerant evaporates from a leak.**

7. Start the engine.
8. Set the temperature control switch to maximum cooling and switch the air conditioner blower control on and off several times, checking that the magnetic clutch on the compressor engages and releases each time.
9. With the temperature control at maximum cooling, warm up the engine and fast idle at 1,000 rev/min. Check the sight glass for bubbles or foam. The sight glass should be generally clear after five minutes running, occasional bubbles being acceptable. Continuous bubbles may appear in a serviceable system on a cool day, or if there is insufficient air flow over the condenser at a high ambient temperature.
10. Repeat at 1,800 rev/min.
11. Gradually increase the engine speed to the high range, and check the sight glass at intervals.
12. Check for frosting on the service valves and evaporator fins.
13. Check the high pressure pipes and connections by hand for varying temperature. Low temperature indicates a restriction or blockage at that point.
14. Switch off the air conditioning blower and stop the engine.
15. If the air conditioning equipment is still not satisfactory, proceed with the pressure test as previously described in this section.

---

# AIR CONDITIONING

## COMPRESSOR DRIVE BELT

### Adjust

1. Slacken the jockey wheel securing bolt.
2. Adjust the position of the jockey wheel until the correct tension is obtained. The belt must be tight with 4 to 6 mm (0.19 to 0.25 in) total deflection when checked by hand midway between the pulleys on the longest run.
3. Tighten the securing bolt and recheck the tension.

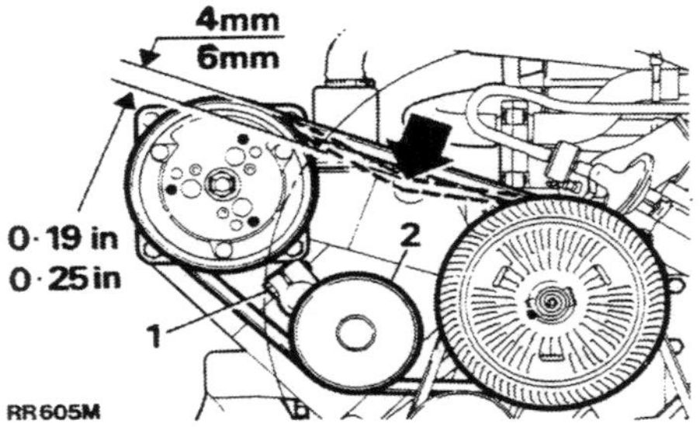

## COMPRESSOR

### Remove and refit

### Removing

1. Place the vehicle in a ventilated area away from open flames and heat sources.
2. Stop the engine and secure the bonnet in an open position.
3. Depressurise the air conditioning system.
4. Using goggles to protect the eyes, and wearing gloves, disconnect the suction and discharge unions from the back of the compressor. Cap the flexible end connections and service valves immediately.
5. Disconnect the lead to the compressor magnetic clutch at the connector.
6. Slacken the jockey pulley securing bolt and release the driving belt.
7. Remove the two compressor mounting bolts and lift compressor clear.

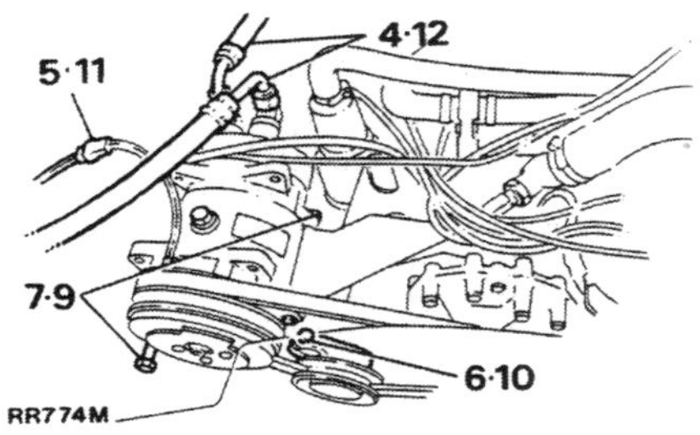

## COMPRESSOR OIL LEVEL

### Check

**NOTE:** The compressor oil level should be checked whenever any components, including the compressor are removed and refitted, or when a pipe or hose has been removed and reconnected or, if a refrigerant leak is suspected. All compressors are factory charged with 135 ± 15 ml (4.6 ± 0.5 UK fl oz) of oil. When the air conditioning equipment is operated some of the oil circulates throughout the system with the refrigerant, the amount varying with engine speed. When the system is switched off the level of oil remains in the pipe lines and components, so the level of oil in the compressor is reduced, by approximately 30 ml (1 UK fl oz). The compressor oil level must finally be checked after the system has been fully charged with refrigerant and operated to obtain a refrigerated temperature of the car interior. This ensures the correct oil balance throughout the system.

### Refitting

8. If a new compressor is being fitted, drain the oil from the new compressor. Drain and measure the oil from the old compressor. Measure new oil equal to the amount drained from the old compressor. Add 30 ml (1 UK fluid oz) of new oil to this amount and refill the new compressor.
9. Locate the compressor in position, fit and tighten the mounting bolts.
10. Fit the compressor driving belt and adjust as described under 'Compressor drive belt—adjust'.
11. Connect the lead to the compressor magnetic clutch at the connector.
12. Refit the suction and discharge flexible end connectors to the service valves, lubricating the flares and threads of the unions with compressor oil.
13. Evacuate the air conditioning system, maintaining the vacuum for ten minutes.
14. Charge the air conditioning system.

*Continued*

## 82 AIR CONDITIONING

The compressor is not fitted with an oil level dipstick, and a suitable dipstick must be made locally from 3 mm (0.125 in) diameter soft wire in accordance with the accompanying illustration. After shaping, mark the end of the dipstick with six graduations 3 mm (0.125 in) apart.

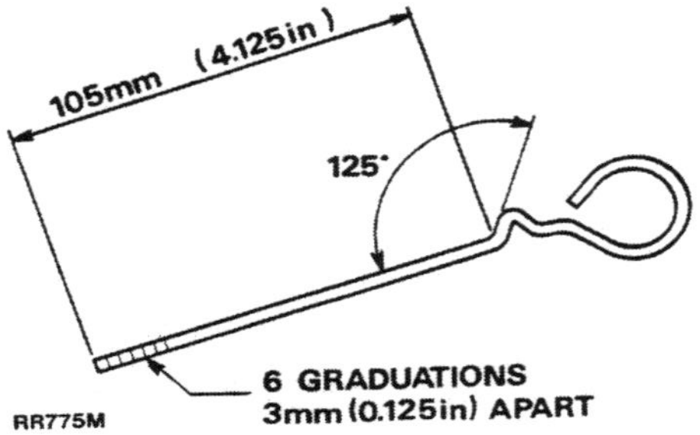

**6 GRADUATIONS 3mm (0.125in) APART**
**125°**
**105mm (4.125in)**
RR775M

### Procedure

1. Open the bonnet.
2. Fit the charging and testing equipment.
3. Start the engine and turn the temperature control to maximum cooling position, and the air flow control to HIGH speed. Operate the system for five minutes at 1,200–1,500 rev/min.

**NOTE: It is important to open the valve slowly during the following item to avoid a sudden pressure reduction in the compressor crankcase that could cause a large amount of oil to leave the compressor.**

4. Reduce the engine speed to idling, and SLOWLY open the suction side valve on the test equipment until the compound gauge reads 0 or a little below.
5. Stop the engine at this point and quickly open the suction valve and discharge valve.
6. Loosen the oil filler plug and unscrew it slowly by five turns to bleed off crankcase pressure.

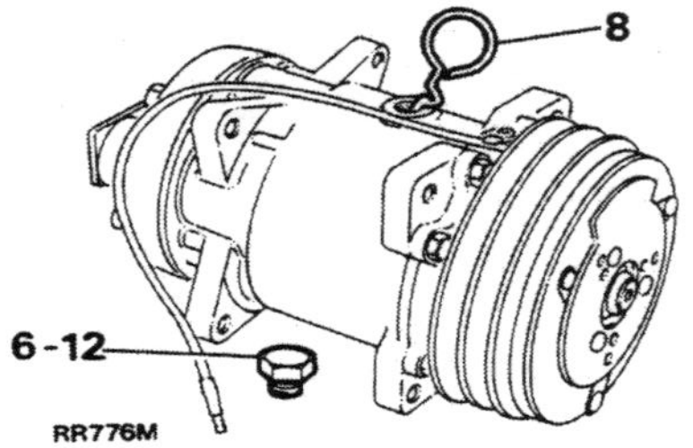

7. Remove the oil filler plug. Looking through the oil filler orifice, centralise the internal components by slowly turning the compressor clutch plate. This will enable the dipstick to be inserted to its full depth.
8. Wipe the dipstick and insert to its stop position, ensuring the angle of the dipstick is flush with the surface of the filler orifice.
9. Withdraw the dipstick and count the number of graduations to determine the depth of oil.
10. The acceptable level is two to four graduations, if required add or remove oil until the mid-range figure is obtained. Use only the correct compression oil. See Recommended Lubricants, Section 09.
11. Lubricate a new 'O' ring with compressor oil, fit it over the threads of the level plug without twisting, and install the level plug loosely.
12. Evacuate the air from the compressor using the vacuum pump on the charging and testing equipment, following the equipment manufacturer's instructions. Tighten the filler plug to the correct torque, see Section 09.
13. Close fully the suction and discharge valves.

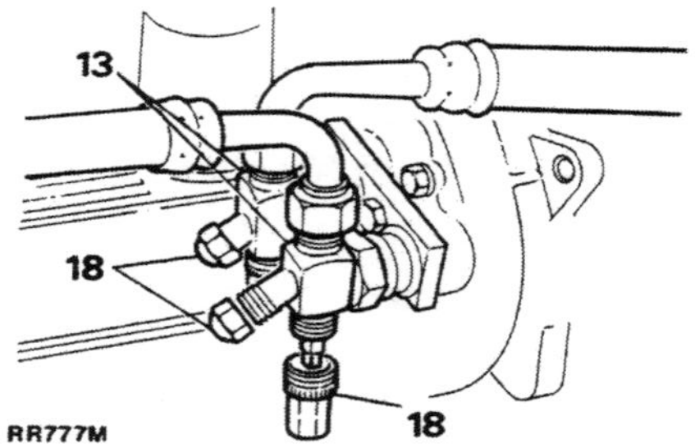

RR777M
**18**
**13**
**18**

14. Start and run the engine at 1,200 rev/min and check for leak at the compressor level plug. Do not overtighten to correct a leak. In the event of a leak isolate the compressor as previously described in items 4 to 6, and check the 'O' ring seats for dirt, etc.
15. Stop the engine.
16. Close all valves on the charging and testing equipment.
17. Disconnect the charging lines from the compressor.
18. Refit the blanking caps to the compressor valve stems and gauge connections, and to the charging lines.
19. Close the bonnet.

**6–12**
**8**
RR776M

---

## AIR CONDITIONING 82

### CONDENSER

#### Remove and refit

#### Removing

1. Open the bonnet and disconnect the battery.
2. Depressurise the air conditioning system.
3. Remove six screws and withdraw the front grille panel.
4. Remove the radiator.
5. Disconnect the two fan motor wiring connections.

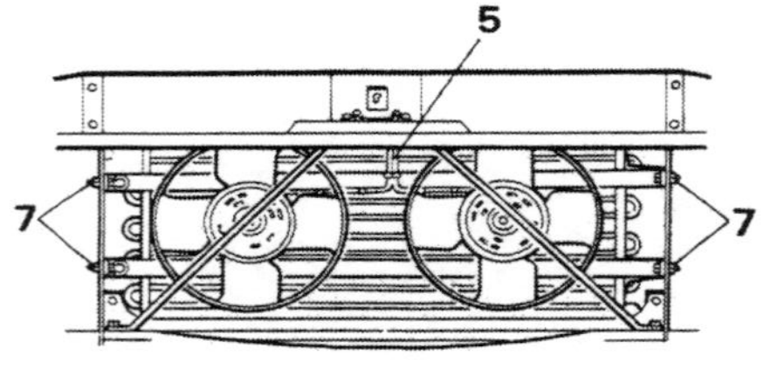

RR778M

6. Using two spanners on each union, carefully disconnect the pipes at the condenser end. Fit blanks to the exposed ends of the pipes.
7. Remove four bolts securing the condenser and remove condenser complete with fan motor assemblies.

**CAUTION: Before carrying out instruction 6 protect the eyes with safety goggles and wear protective gloves.**

#### Refitting

8. Reverse instructions 3 to 7 above.
9. Add 30 ml (1 UK fl oz) of the correct oil to the compressor to compensate for oil loss if a new condenser is to be fitted.
10. Evacuate the air conditioning system.
11. Charge the system.
12. Carry out a leak test on the disturbed joints.
13. Check the air conditioning operation by carrying out a System Test.

### CONDENSER FANS AND MOTORS

#### Remove and refit

#### Removing

1. Open the bonnet and disconnect the battery.
2. Remove the six screws and withdraw the grille panel.
3. (Automatic vehicles only) remove the transmission oil cooler by disconnecting the wiring, detaching the two oil pipes and two mounting bolts each side.
4. Disconnect the two fan motor wiring connectors.
5. Release the wiring securing clips.

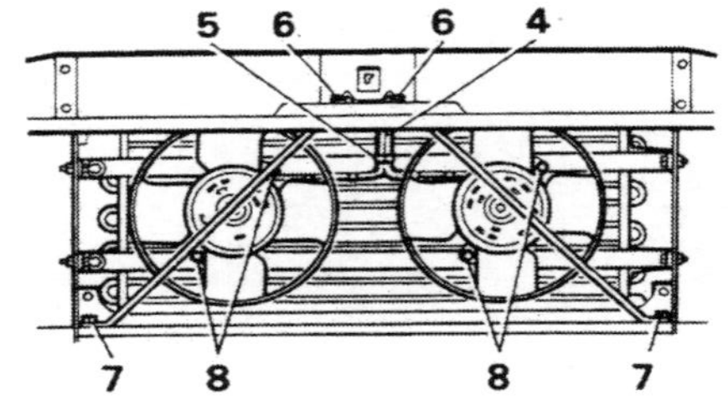

RR779M

6. Slacken the two upper bolts securing the left- and right-hand bonnet striker support stays.
7. Remove the lower bolts securing the lower ends of the stays and pivot both stays forward.
8. Remove the two nuts and washers securing each motor and withdraw the fan motor assemblies from the vehicle.
9. Remove the blanking caps from the fan centres.

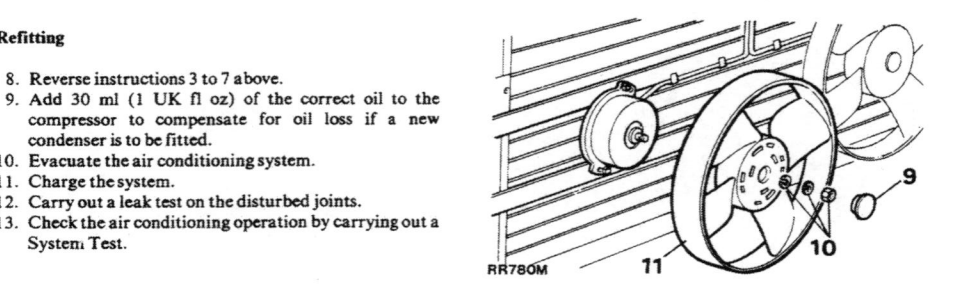

RR780M

10. Remove the securing nut and washers.
11. Withdraw the fan blades from the motor spindle.

#### Refitting

12. Reverse 1–11 above, ensuring that the wire is correctly clipped and no fouling of the fan blades occurs.
13. Automatic only: check transmission fluid level.

# AIR CONDITIONING

## RECEIVER DRIER

### Remove and refit

**CAUTION: Immediate blanking of the receiver drier is important. Exposed life of the unit is only 15 minutes.**

### Removing

1. Connect the gauge set.
2. Discharge the complete system.
3. Protect the eyes with safety goggles and wear gloves during operations 4 and 5.
4. Disconnect the electrical leads at the snap connectors and carefully unscrew the pressure switches from the receiver drier. Blank the exposed connections immediately.
5. Carefully disconnect the two hose connections. Use a second spanner to support the hose adaptor. Blank the exposed connections immediately.
6. Remove one bolt, nut and washers securing the mounting bracket to the wing valance.
7. Remove the clamp bolts, washers and nuts.
8. Withdraw the receiver drier from the mounting bracket.

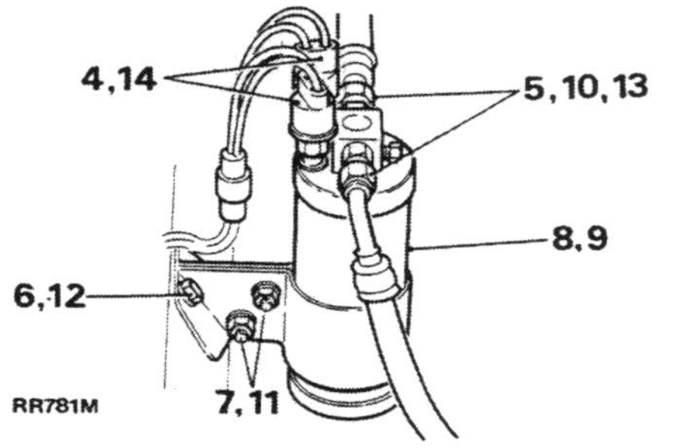

### Refitting

9. Insert the receiver drier into the mounting bracket with the inlet and outlet connections correct to the refrigerant circuit as shown.
10. Connect the two hose connections finger tight. Use refrigerant compressor oil on all mating surfaces to assist leakage prevention.
11. Fit the clamp bolts, washers and nuts.
12. Secure the mounting bracket to the wing valance.
13. Tighten the two hose connections to the correct torque. Use a second spanner to support the hose adaptor.
14. Carefully refit the pressure switches to the receiver drier. Use refrigerant compressor oil on all mating surfaces to assist leakage prevention and tighten the switches to the correct torque. Reconnect the electrical leads.
15. To compensate for oil loss, add 15 ml (½ UK fl oz) of the correct oil to the compressor.
16. Evacuate the complete system.
17. Charge the complete system.
18. Perform a leak test on any disturbed joints.
19. Carry out a functional check.
20. Disconnect the gauge set.

## DASHBOARD UNIT—ARA

### EVAPORATOR

### Remove and refit

**Expansion valve**
Remove and refit 1 to 22 and 34 to 56

**Hose—compressor to evaporator**
Remove and refit 1 to 5 and 50 to 56

**Hose—receiver drier to evaporator**
Remove and refit 1 to 24 and 32 to 56

**Blower assembly**
Remove and refit 1 to 5 and 50 to 56

### Removing

1. Open the bonnet and connect the gauge set.
2. Depressurise the system.
3. Isolate the battery.
4. Protect the eyes with safety goggles and wear gloves during instruction 5.
5. Disconnect the evaporator hoses from the compressor and the receiver drier. Use a second spanner to support the hose adaptors and blank all the exposed connections immediately.
6. Disconnect the dash unit electrical harness at the underbonnet relays and release the cable from the clips.
7. Working inside the vehicle, withdraw the lower fascia panel and remove the screws securing the lower edge of the centre console.
8. Remove the six screws securing the lower edge of the louvre panel to the console and evaporator case.
9. Remove the heater control panel and knobs from the centre console.
10. Remove the centre console.
11. Carefully prise out the four air vents.
12. Remove the screws securing the evaporator plenum and louvre panel to the dash top panel.
13. Withdraw the thermostat sensor from the evaporator.
14. Withdraw the louvre panel clear of the dash.
15. Depress the left end of the plenum and remove the air hoses from the upper panel.

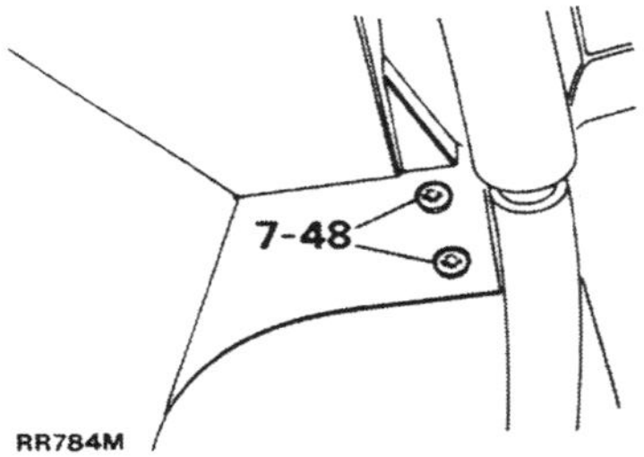

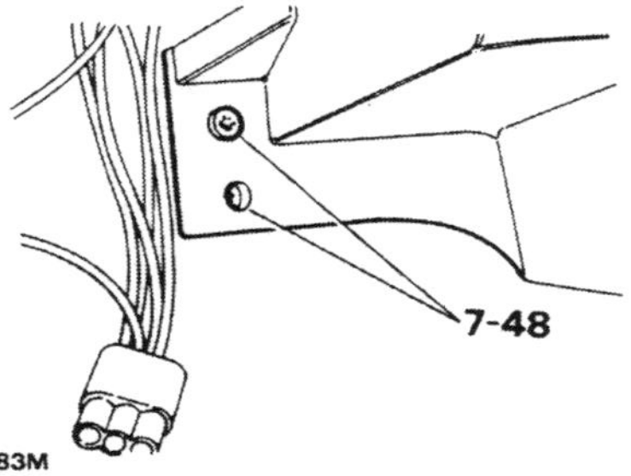

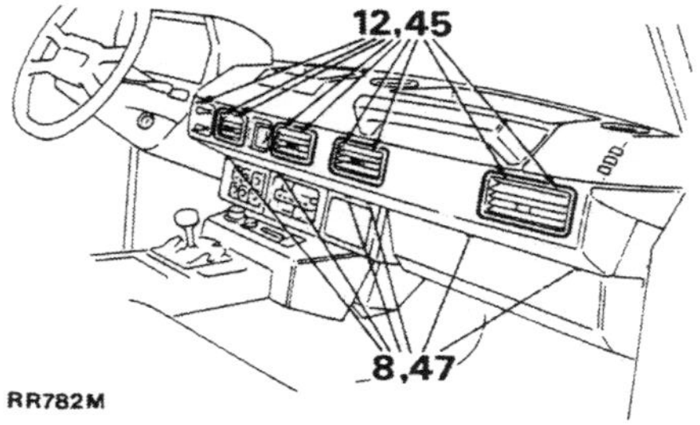

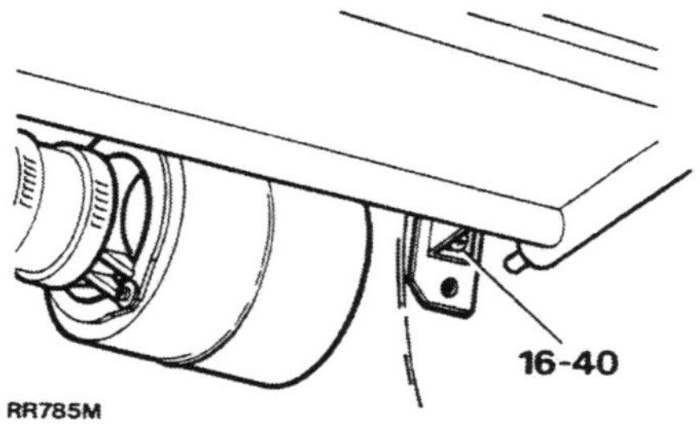

*Continued*

# AIR CONDITIONING

16. Remove the screws securing the lower right evaporator bracket.
17. Support the evaporator case and remove the two nuts securing the case and reinforcing strip to the upper mounting bracket.
18. Carefully withdraw the refrigerant hoses, electrical harness and evaporator condensate tubes through the bulkhead and remove the rear left-hand air hose from the plenum. Remove the evaporator and plenum assembly from the vehicle.

## Dismantling

19. Remove the insulation from the evaporator and expansion valve hose connections.
20. Disconnect the hoses from the expansion valve and evaporator. Use a second spanner to support the hose adaptors and blank all the exposed connections immediately.
21. Unclip the sensor coil from the evaporator outlet pipe.
22. Carefully unscrew the expansion valve from the evaporator. Blank the exposed connections immediately.
23. Unplug the electrical harness at the connector on the left-hand blower casing.
24. Remove the eight securing screws and detach the blower units from the evaporator case.
25. Remove the screws securing the upper evaporator/plenum casing to the evaporator and lower casing.
26. Remove the heater seal and lift off the upper casing.
27. Remove the insulation pad and the two screws securing the evaporator to the lower casing.
28. Withdraw the evaporator from the casing.

## Assembling

29. Secure the evaporator to the lower casing.
30. Fit the insulation pad and run the electrical leads under the evaporator outlet pipe.
31. Secure the casings together with the screws and refit the heater seal.
32. Refit and secure the blower units to the evaporator casing.
33. Connect the electrical lead to the connector on the left-hand blower casing.
34. Assemble the expansion valve to the evaporator with the inlet facing downwards. Use refrigerant compressor oil on all mating surfaces to assist leakage prevention. Tighten the connection to the correct torque.
35. Clip the sensor coil to the evaporator outlet pipe.
36. Connect the hoses to the evaporator and expansion valve. Use new 'O' rings and refrigerant compressor oil on all mating surfaces to assist leakage prevention. Tighten the connections to the correct torque.
37. Wrap all exposed metal at the hose connections with 'prestite' tape.

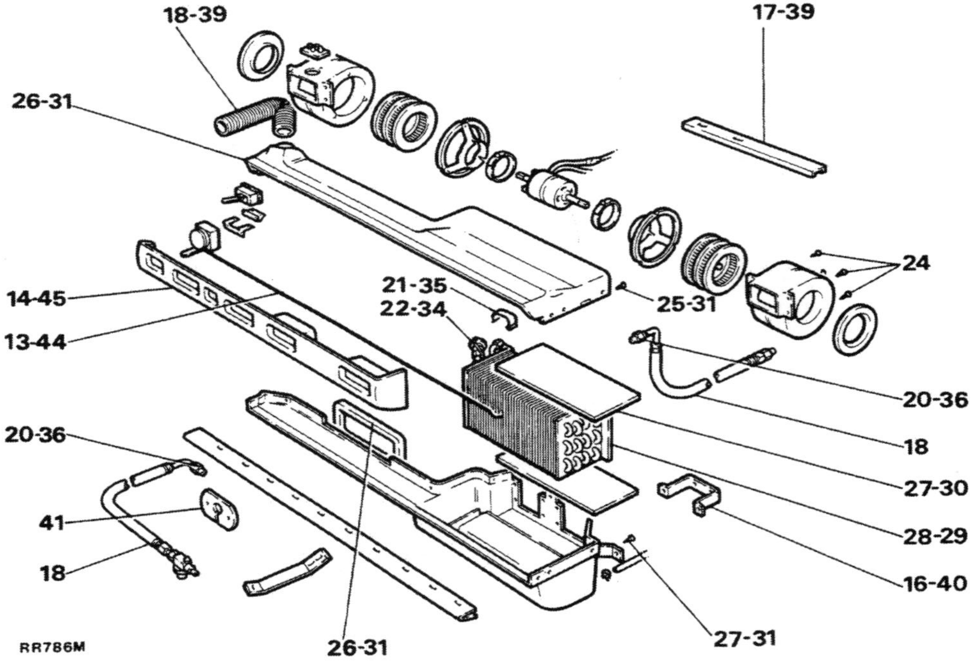

## Refitting

38. Place the evaporator assembly on the floor of the vehicle and route the electrical harness and the refrigerant hoses through the bulkhead.
39. Lift the unit into the mounting position and connect the rear left-hand air hose. Fit the reinforcement strip and secure the casing to the upper bracket with two nuts.
40. Secure the lower right mounting bracket to the vehicle.
41. Feed the hoses, electrical harness and evaporator condensate tubes through the bulkhead. Ensure that the apertures and grommets are adequately sealed against ingress of dust and moisture.
42. Depress the left end of the plenum and connect the two upper air hoses.
43. Position the left-hand of the plenum so that the opening is centered over the fresh air outlet of the heater.
44. Carefully push the thermostat pipe into the evaporator fins.
45. Refit the louvre panel and secure the plenum casing and grille panel to the dash top panel with screws.
46. Refit the centre console.
47. Refit the six screws securing the louvre panel to the centre console and evaporator case.
48. Secure the lower edge of the console and refit the lower fascia panel.
49. Working under the bonnet, connect the dash harness to the two relays.
50. Connect the two refrigerant hoses to the compressor and receiver drier. Use refrigerant compressor oil on all mating surfaces to assist leakage prevention. Tighten the connections to the correct torque.
51. To compensate for oil loss, add 45 ml (1.5 UK fl oz) of the correct oil to the compressor.
52. Evacuate the system.
53. Charge the complete system.
54. Perform a leak test on any accessible disturbed joints.
55. Perform a functional check.
56. Disconnect the gauge set.

## CONTROL PANEL

### Thermostat
Remove and refit—1 to 6 and 10 to 14

### Fan control switch
Remove and refit—1 to 9 and 14

### Removing

1. Disconnect the battery.
2. Pull the finger tip control knobs off the control levers.
3. Carefully prise the four air vents out of the louvre panel.
4. Remove the nine screws securing the louvre panel to the upper panel.
5. Remove the six screws securing the louvre panel to the lower panel.
6. Withdraw the panel only as far as the electrical leads will permit.
7. Disconnect the multi-plug from the rear of the fan control switch.
8. Remove the two securing screws.
9. Withdraw the fan control switch.
10. Withdraw the thermostat sensor from the fins of the evaporator.
11. Disconnect the electrical leads from the rear of the thermostat.
12. Remove the two securing screws.
13. Withdraw the thermostat.

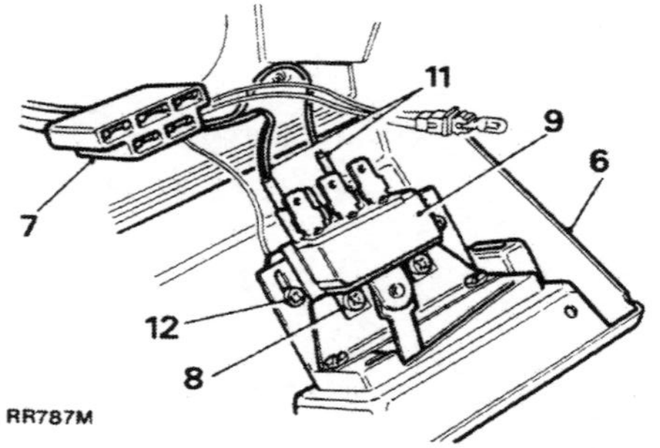

### Refitting

14. Reverse the removal procedure.

# WIPERS AND WASHERS

## WIPERS AND WASHERS

### Description

**Headlamp power wash**

A headlamp power wash is fitted as standard equipment on Vogue models and as an optional extra on other models. The power wash is actuated when the windscreen washer control is used with the headlamps illuminated.

**Washer reservoir**

A combined underbonnet washer reservoir supplies the windscreen, tailgate glass and headlamp power wash. Three separate supply pipes are used with an electric pump for each facility.

NOTE: See Electrical Section for wiring diagram.

### WASHER RESERVOIR

**Remove and refit**

Removing

1. Slacken the hose-clip and remove the power wash hose.
2. Disconnect the tubing from the washer pumps.
3. Disconnect the electrical leads from the pumps. 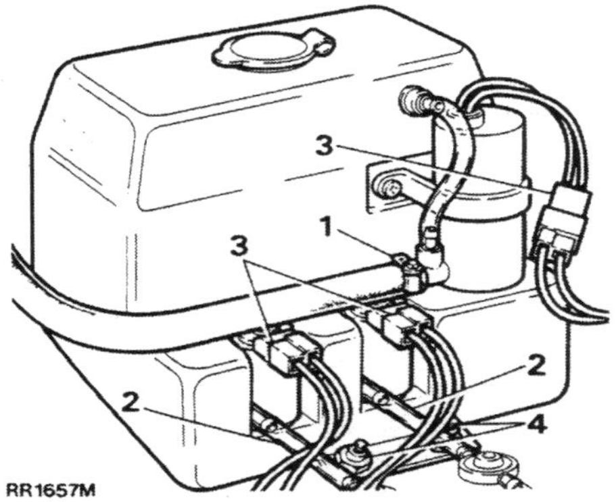
4. Remove the three securing bolts from below and withdraw the reservoir.

Refitting

5. Reverse instructions 1 to 4.

## POWER WASH PUMP

**Remove and refit**

Removing

1. Disconnect the hose from the power wash pump.
2. Disconnect the electrical leads from the pump.
3. Remove two screws and withdraw the power wash pump.

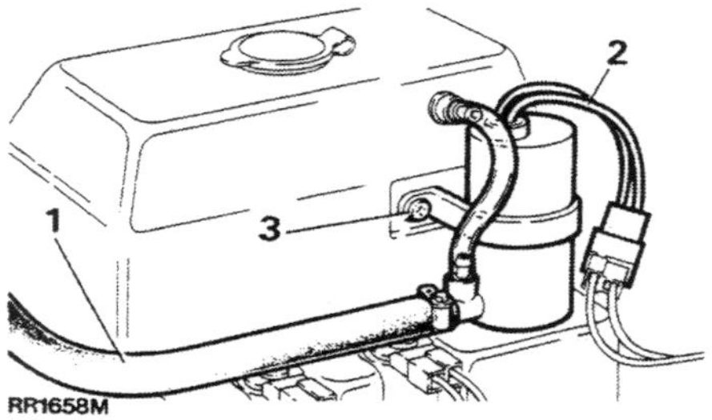

Refitting

4. Reverse instructions 1 to 3.

## WINDSCREEN WASHER PUMP

## TAILGATE GLASS WASHER PUMP

**Remove and refit**

Removing

1. Disconnect the washer tubing from the pump.
2. Disconnect the electrical leads from the pump.

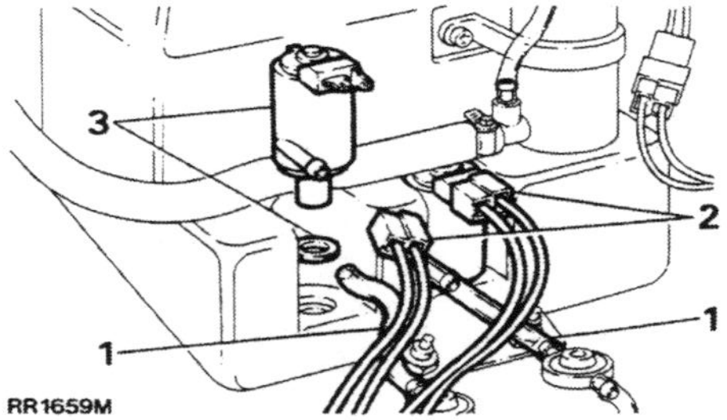

*Continued*

## Notes

## 84 | WIPERS AND WASHERS

3. Pull the top of the pump away from its location in the reservoir and remove the pump. Note the position of the sealing gasket in the reservoir.

**Refitting**

4. Reverse instructions 1 to 3.

### WINDSCREEN WASHER JETS

**Remove and refit**

**Removing**

1. Disconnect tubes from the jets.
2. Remove the locknut and washer securing jet to bonnet.
3. Remove jet.

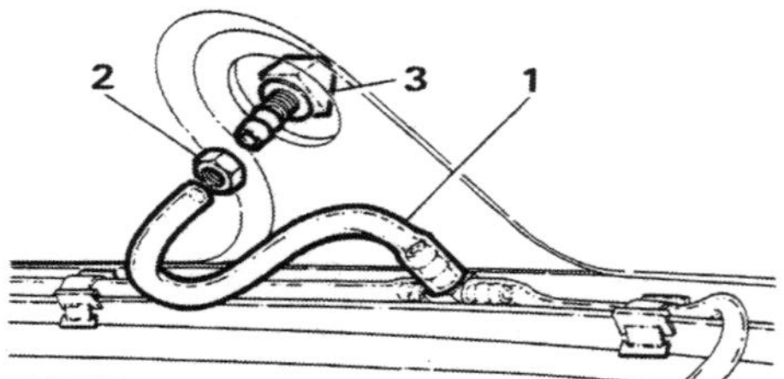

**Refitting**

4. Reverse instructions 1 to 3.

### WINDSCREEN WASHER TUBES

**Remove and refit**

**Removing**

1. Disconnect tubing from reservoir pump.
2. Disconnect tubing from washer jets.
3. Disconnect tubing from three-way tee piece.
4. Release tubing from edge clips.

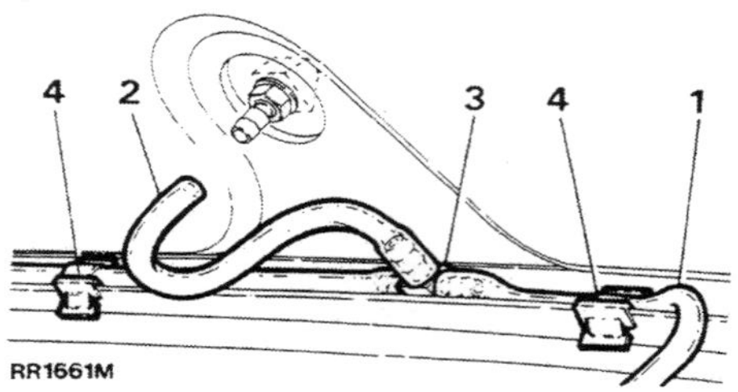

## WIPERS AND WASHERS | 84

4. Remove bonnet.
5. Remove front decker panel.
6. Remove the spring clips securing the primary links to the wheelbox spindle links.
7. Remove the spring clips securing the primary links to the motor crank.
8. Remove the primary links.
9. Remove the lower grommet from the wheelboxes.

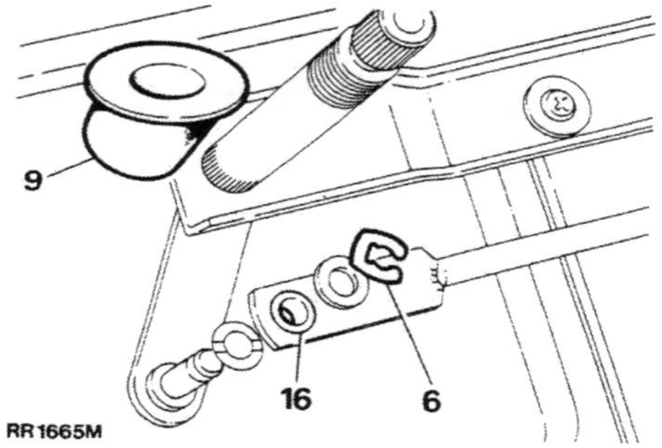

### WIPER BLADES

**Remove and refit**

**Removing**

1. Pull the wiper arm away from the glass.
2. Lift the spring clip and withdraw the blade from the arm.

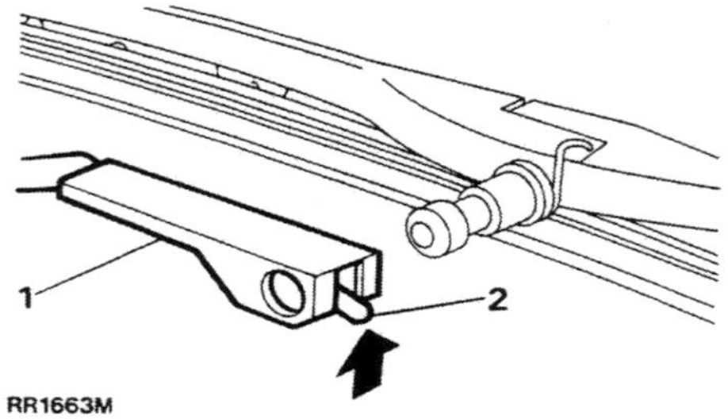

**Refitting**

3. Reverse instructions 1 and 2.

### WINDSCREEN WIPER MOTOR, LINKAGE AND WHEEL BOXES

**Remove and refit**

**Removing**

1. Remove wiper arms.
2. Remove locknuts from wheel boxes.
3. Remove grommet from wheel boxes.

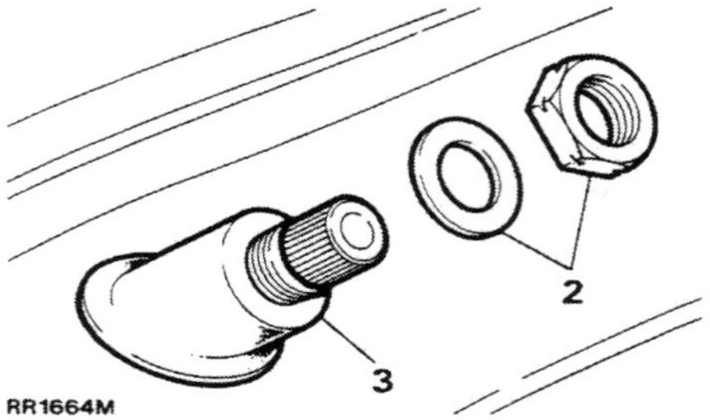

10. Remove the screws securing the motor and linkage assembly to the bulkhead.
11. Gently ease the unit out of its mounting aperture and disconnect the electrical leads at the plug and socket.
12. Withdraw the unit.

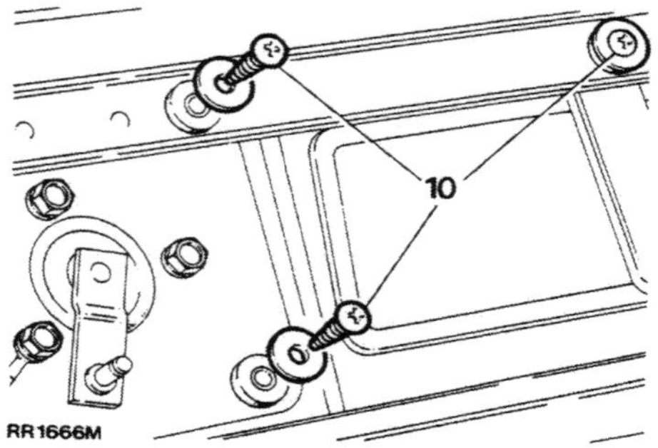

### WIPER ARMS

**Remove and refit**

**Removing**

1. Hold back the small spring clip which retains the wiper arm on the spindle boss, by means of a suitable tool.
2. Gently prise off the wiper arm from the spindle boss.

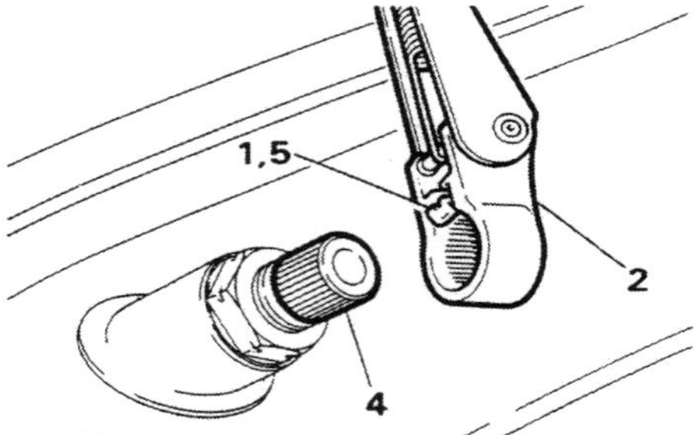

**Refitting**

3. Allow the motor to move to the 'park' position.
4. Push the arm on to the boss, locating it on the splines so that the wiper blades are just clear of the screen rail.
5. Ensure that the spring retaining clip is located in the retaining groove on the spindle boss.

NOTE: A non-return valve is fitted between the washer motors and jets. Ensure that this valve is correctly fitted when replacing the washer tubes.

*Continued*

220

# WIPERS AND WASHERS

13. Remove the three bolts securing the motor to the mounting plate.
14. Separate the motor from the mounting plate by pulling the motor crank through the grommet.

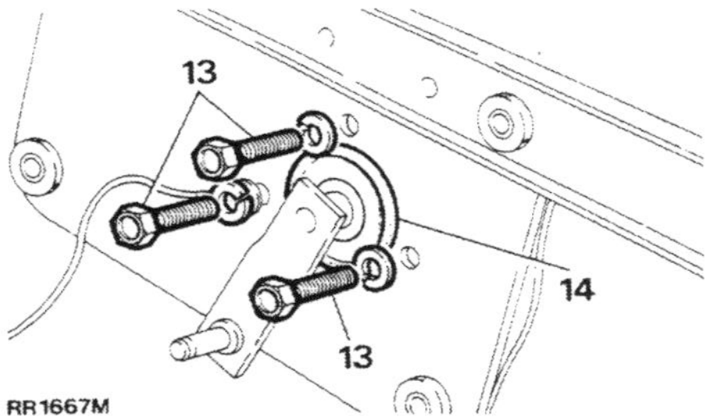

**Refitting**

15. Reverse the removal procedure.
16. When replacing the primary links ensure that they are mounted with the bushes on the inside, that is, towards the wiper motor. The shorter primary link is mounted on the driver's side.
17. Replace the front decker panel.
18. Replace the bonnet.
19. Replace the wiper arms.

## WIPER MOTOR CHECKING

NOTE: These instructions apply to both Windscreen and Tailgate glass wiper motors.

If unsatisfactory operation of the wiping equipment is experienced, a systematic check to determine the origin of the fault should be carried out as follows:

1. Check the blades for signs of excessive friction. Frictional blades will greatly reduce the wiping speed of the motor and cause increased current draw which may damage the armature. Check by substitution.
2. Check the motor light running current and speed with the motor coupling link disconnected from the wiper spindle transmission linkage. Connect a first grade moving coil ammeter in series with the motor supply and measure the current consumption when the motor is switched on. Check the operating speed by timing the speed of rotation of the motor coupling link. The results should compare with the figures given in the technical data.
3. If the motor does not run satisfactorily or takes higher than normal current, then a fault is apparent and should be investigated.
4. If the current consumption and speed of the motor are satisfactory, then a check should be carried out for proper functioning of the transmission linkage and wiper arm spindles.

NOTE: Service parts are not available for the windscreen wiper motor with the exception of the rotary connecting link.
**The complete unit must be renewed in the event of a failure.**

## TAILGATE GLASS WASHER JET

**Remove and refit**

**Removing**

1. Lower or remove the headlining rear section to gain access to the jet securing nut and washer tube connection.
2. Disconnect the washer tube, and remove the grommet. Drain away any water in the tube to avoid damaging the trim.
3. Hold the base of the jet and remove the nut and washer.
4. Withdraw the washer jet from outside the vehicle.

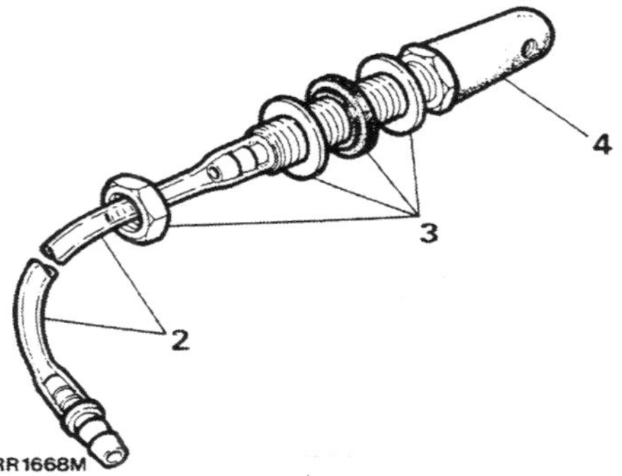

**Refitting**

5. Reverse instructions 1 to 4 as appropriate.

NOTE: If the vehicle wiring connections are disconnected and an alternative supply source is applied it is essential that the correct polarity is observed.
Failure to observe this will cause the motor to rotate in the reverse direction, which may result in the limit switch contacts being damaged.

## TAILGATE GLASS WASHER TUBE

**Remove and refit**

**Removing**

1. Remove the rear headlining.
2. Disconnect the washer tube from the pump and washer jet.
3. Secure a length of strong cord to the rear end of the tube. On withdrawing the tube, the cord will follow the route of the tube through the body trim.
4. Remove the tube by feeding it in stages towards the engine compartment.

**Refitting**

5. Reverse instructions 1 to 4 securing the cord to the tube and using it to draw the tube through the body trim.

## TAILGATE GLASS WIPER MOTOR

**Remove and refit**

Wiper wheelbox remove and refit instructions 1 to 8 and 11.
Wiper rack remove and refit instructions 1 to 10 and 11.

**Removing**

1. Lower or remove the headlining rear section to gain access to the wiper motor assembly.
2. Remove the wiper arm and blade.
3. Remove one nut and spacer to release the wiper wheelbox from the body.

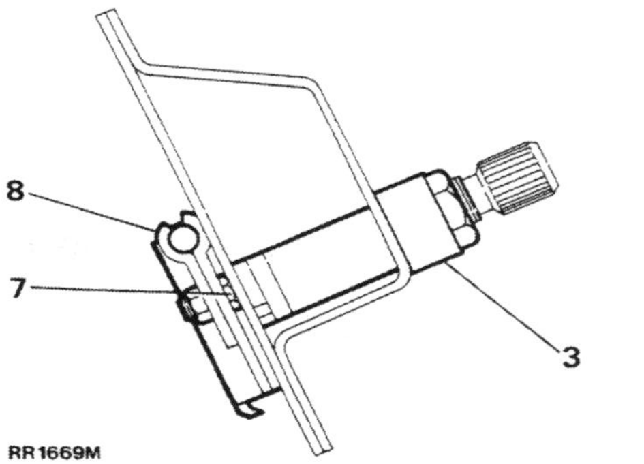

4. Disconnect the electrical leads from the wiper motor, including the earth lead.
5. Support the wiper motor and remove the bolt and bracket.
6. Withdraw the wiper motor from the body, complete with rack tube and wheelbox.

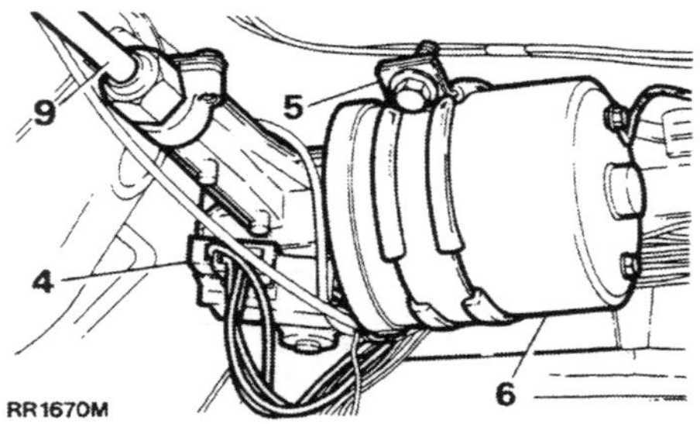

7. Slacken two bolts and nuts on the wheelbox to release it from the rack.
8. Slide the wheelbox and end rack tube off the rack.
9. Unscrew the rack tube nut and slide the tube off the rack.
10. Remove the wiper motor gearbox cover and withdraw the rack.

**Refitting**

11. Reverse instructions 1 to 10, ensuring that the wheelbox is correctly located on the rack before tightening the two wheelbox bolts and nuts.

*Continued*

# 84 | WIPERS AND WASHERS

## TAILGATE GLASS WIPER MOTOR

### Overhaul

### Dismantle

1. Remove the wiper motor and drive from the vehicle.
2. Remove the wiper motor gearbox cover.
3. Remove the circlip and plain washer securing the connecting rod.
4. Withdraw the connecting rod.
5. Withdraw the flat washer.
6. Remove the circlip and washer securing the shaft and gear.
7. Clean any burrs from the gear shaft and withdraw the gear. Do not detach the crankpin mounting plate.
8. Withdraw the dished washer.
9. Note the alignment marks on the yoke and gearbox. To ensure correct rotation of the motor, the mark on the yoke must be adjacent to the arrow-head marked on the gearbox case.
10. Remove the yoke securing bolts.
11. Withdraw the yoke and armature.
12. Remove the brush gear assembly.
13. Remove the limit switch.

### INSPECTION AND TEST

14. Check the brushes for excessive wear, if they are worn to 4.8mm (0.190 in) in length, fit a new brush gear assembly.
15. Using a push type gauge, check that the brush spring pressure is 170gf (6ozf) when the bottom of the brush is level with the bottom of the slot in the brush box. Fit a new brush gear assembly if the springs are not satisfactory.
16. Check the armature windings for open or short circuit using armature testing equipment.
17. Test the armature insulation using a 110V, 15W test lamp. Lighting of the lamp indicates faulty insulation. Renew the armature if faulty.
18. Examine the gear wheel for damage or excessive wear.

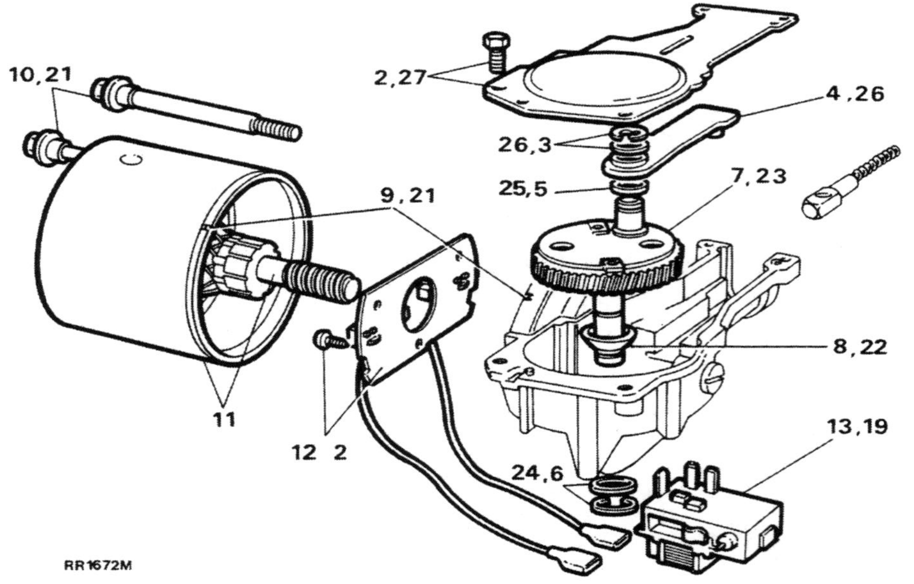

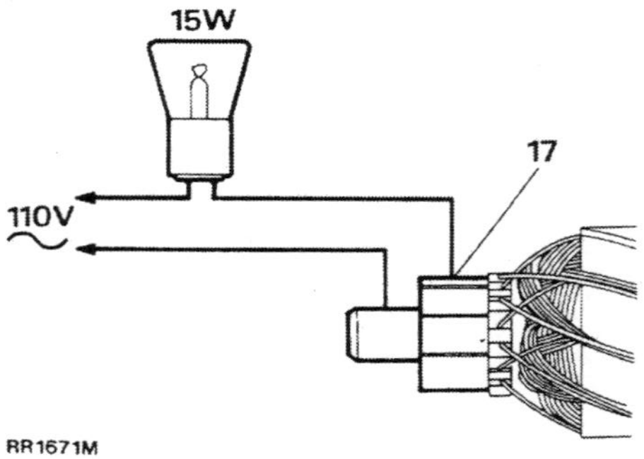

### ASSEMBLE

Use Ragosine Listate Grease to lubricate the gear wheel teeth, armature shaft worm gear, connecting rod and pin, cable rack and wheelbox gear wheels.
Use Shell Turbo 41 oil sparingly to lubricate the bearing bushes, armature shaft bearing journals, gear wheel shaft and wheelbox spindles. Thoroughly soak the felt washer in the yoke bearing with oil.

19. Fit the limit switch.
20. Fit the brush gear assembly.
21. Fit the armature and yoke to gearbox using alignment marks, secure with the yoke retaining bolts tightening to the correct torque. If a replacement armature is being fitted slacken the thrust screw to provide end-float for fitting the yoke.
22. Fit the dished washer beneath the gear wheel with the concave side towards the gear wheel.
23. Fit the gear wheel to the gearbox.
24. Secure the gear wheel shaft with the plain washer and circlip.
25. Fit the larger flat washer over the crankpin.
26. Fit the connecting rod and secure with the smaller plain washer and circlip.
27. Fit the gearbox cover and secure with the retaining screws.
28. Connect the electrical leads between the wiper motor and limit switch.
29. To adjust the armature shaft end-float, hold the yoke vertically with the adjuster screw uppermost. Carefully screw-in the adjuster until resistance is felt, then back-off one quarter turn.

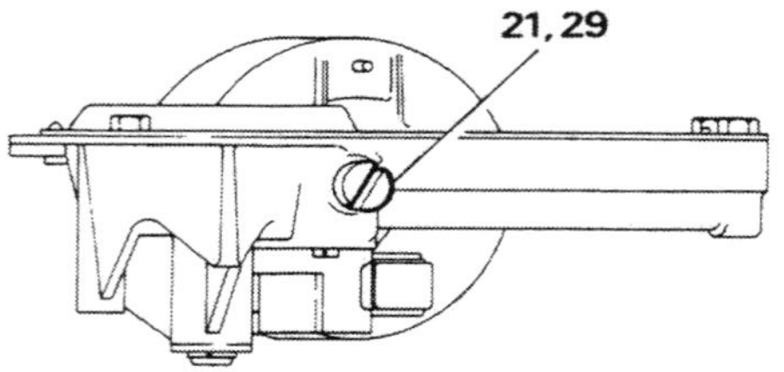

*Continued*

## ELECTRICAL 86

**LOCATION OF ELECTRICAL EQUIPMENT**

1. Battery
2. Air conditioning compressor (if fitted)
3. Horns
4. Oil pressure switch
5. Water temperature switch
6. Electronic distributor
7. Alternator
8. Starter motor
9. Fuel shut-off relay (fuel injection models only)
10. Power resistor (fuel injection models only)
11. Coil and amplifier assembly
12. Relays
13. Wiper motor—front screen
14. Choke warning light switch (carburetter models only)
15. Relays
16. Heater
17. Window lift motor (front right-hand door)
18. Door lock actuator (front right-hand door)
19. Electronic control unit (fuel injection models only)
20. Relays (fuel injection models only)
21. Handbrake warning light switch
22. Window lift motor (front left-hand door)
23. Door lock actuator (front left-hand door)
24. Window lift motor (rear left-hand door)
25. Door lock actuator (rear left-hand door)
26. Electrical in-tank fuel pump
27. Window lift motor (rear right-hand door)
28. Door lock actuator (rear right-hand door)
29. Wiper motor—rear screen
30. Fuel cut-off relay (carburetter models only)

NOTE: Right-hand drive vehicle illustrated: Certain electrical components may change position on left-hand drive vehicles, but can be found in a similar location on the opposite side of the vehicle.

For full information on fuel injection related items—see fuel injection section of manual.

To identify individual relays (items 12 and 15) see relays in electrical section of manual.

---

## 84 WIPERS AND WASHERS

### HEADLAMP POWER WASH JET

**Remove and refit**

**Removing**

1. Disconnect the hose from the power wash jet.
2. Remove the nut securing the jet to the bumper.
3. Remove the power wash jet.

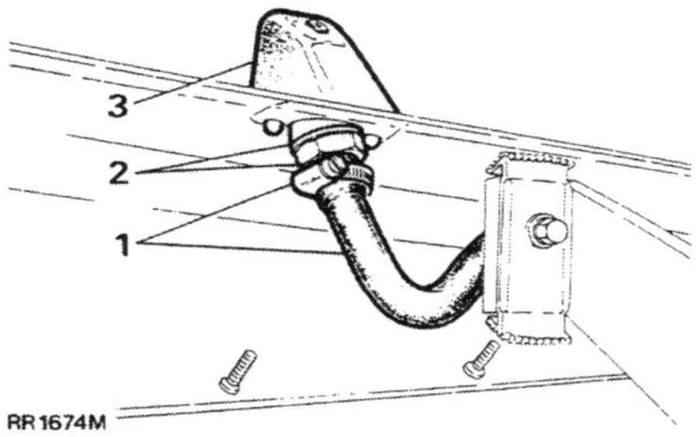

**Refitting**

4. Reverse instructions 1 to 3.

### POWER WASH HOSE

**Remove and refit**

**Removing**

1. Disconnect the hoses from the jets and the washer pump.
2. Release the clips locating the hose.
3. Remove the power wash hose.

**Refitting**

4. Reverse instructions 1 to 3 ensuring the pipe run is correct.

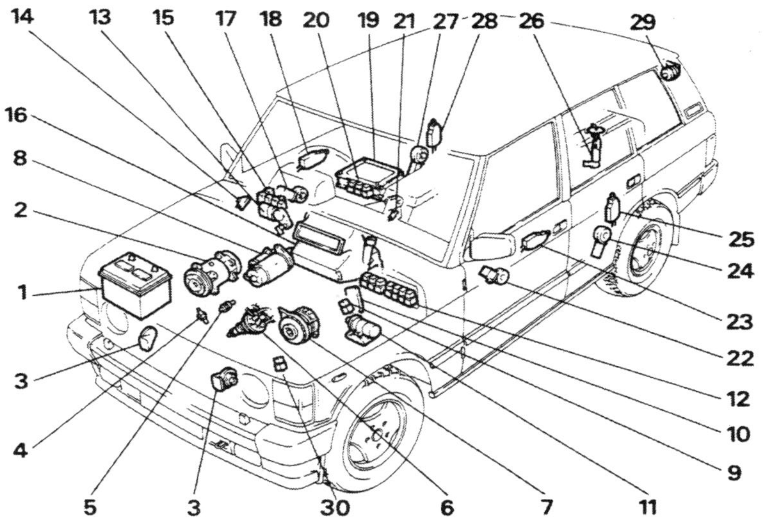

# ELECTRICAL 86

## FAULT DIAGNOSIS

| SYMPTOM | POSSIBLE CAUSE | CURE |
|---|---|---|
| A— Battery in low state of charge | 1. Broken or loose connection in alternator circuit | 1. Examine the charging and field circuit wiring. Tighten any loose connections and renew any broken leads. Examine the battery connection |
| | 2. Current voltage regulator not functioning correctly | 2. Adjust or renew |
| | 3. Slip rings greasy or dirty | 3. Clean |
| | 4. Brushes worn, not fitted correctly or wrong type | 4. Renew |
| B— Battery overcharging, leading to burnt-out bulbs and frequent need for topping-up | 1. Current voltage regulator not functioning correctly | 1. Renew |
| C— Lamps giving insufficient illumination | 1. Battery discharged | 1. Charge the battery from an independent supply or by a long period of daylight running |
| | 2. Bulbs discoloured through prolonged use | 2. Renew |
| D— Lamps light when switched on but gradually fade out | 1. Battery discharged | 1. Charge the battery from an independent supply or by a long period of daylight running |
| E— Lights flicker | 1. Loose connection | 1. Tighten |
| F— Failure of lights | 1. Battery discharged | 1. Charge the battery from an independent supply or by a long period of daylight running |
| | 2. Loose or broken connection | 2. Locate and rectify |
| G— Starter motor lacks power or fails to turn engine | 1. Stiff engine | 1. Locate cause and remedy |
| | 2. Battery discharged | 2. Charge the battery either by a long period of daytime running or from independent electrical supply |
| | 3. Broken or loose connection in starter circuit | 3. Check and tighten all battery, starter and starter switch connections and check the cables connecting these units for damage |
| | 4. Greasy or dirty slip rings | 4. Clean |
| | 5. Brushes worn, not fitted correctly or wrong type | 5. Renew |
| | 6. Brushes sticking in holders or incorrectly tensioned | 6. Rectify |
| | 7. Starter pinion jammed in mesh with flywheel | 7. Remove starter motor and investigate |
| H— Starter noisy | 1. Starter pinion or flywheel teeth chipped or damaged | 1. Renew |
| | 2. Starter motor loose on engine | 2. Rectify, checking pinion and the flywheel for damage |
| | 3. Armature shaft bearing | 3. Renew |
| J— Starter operates but does not crank the engine | 1. Pinion of starter does not engage with the flywheel | 1. Check operation of starter solenoid. If correct, remove starter motor and investigate |
| K— Starter pinion will not disengage from the flywheel when the engine is running | 1. Starter pinion jammed in mesh with the flywheel | 1. Remove starter motor and investigate |
| L— Engine will not start | 1. The starter will not turn the engine due to a discharged battery | 1. The battery should be recharged by running the car for a long period during daylight or from an independent electrical supply |
| | 2. Sparking plugs faulty, dirty or incorrect plug gaps | 2. Rectify or renew |
| | 3. Defective coil or distributor | 3. Remove the lead from the centre distributor terminal and hold it approximately 6 mm (¼ in) from some metal part of the engine while the engine is being turned over. If the sparks jump the gap regularly, the coil and distributor are functioning correctly. Renew a defective coil or distributor |
| | 4. A fault in the low tension wiring circuit | 4. Examine all the ignition cables and check that the bottom terminals are secure and not corroded |
| | 5. Faulty amplifier | 5. Check or renew |
| | 6. Air gap out of adjustment | 6. Adjust |
| | 7. Controls not set correctly or trouble other than ignition | 7. See Starting Procedure in the Owner's Instruction Manual |
| M— Engine misfires | 1. Distributor incorrectly set | 1. Adjust |
| | 2. Faulty coil or reluctor | 2. Renew |
| | 3. Faulty sparking plugs | 3. Rectify |
| | 4. Faulty carburetter | 4. Check and rectify |

*Continued*

| SYMPTOM | POSSIBLE CAUSE | CURE |
|---|---|---|
| N— Frequent recharging of the battery necessary | 1. Alternator inoperative | 1. Check the brushes, cables and connections or renew the alternator |
| | 2. Loose or corroded connections | 2. Examine all connections, especially the battery terminals and earthing straps |
| | 3. Slipping fan belt | 3. Adjust |
| | 4. Voltage control out of adjustment | 4. Renew |
| | 5. Excessive use of the starter motor | 5. In the hands of the operator |
| | 6. Vehicle operation confined largely to night driving | 6. In the hands of the operator |
| | 7. Abnormal accessory load | 7. Superfluous electrical fittings such as extra lamps, etc. |
| | 8. Internal discharge of the battery | 8. Renew |
| P— Alternator not charging correctly | 1. Slipping fan belt | 1. Adjust |
| | 2. Voltage control not operating correctly | 2. Rectify or renew |
| | 3. Greasy, charred or glazed slip rings | 3. Clean |
| | 4. Brushes worn, sticking or oily | 4. Rectify or renew |
| | 5. Shorted, open or burn-out field coils | 5. Renew |
| Q— Alternator noisy | 1. Worn, damaged or defective bearings | 1. Renew |
| | 2. Cracked or damaged pulley | 2. Rectify |
| | 3. Alternator out of alignment | 3. Rectify |
| | 4. Excessive brush noise | 4. Check for rough or dirty slip rings, badly seating brushes, incorrect brush tension, loose brushes and loose field magnets. Rectify or renew |
| R— Defective distributor (refer to distributor overhaul and test procedure) | 1. Air gap incorrectly set | 1. Adjust |
| | 2. Distributor cap cracked | 2. Renew |
| | 3. Faulty pick-up or reluctor | 3. Renew |
| | 4. Excessive wear in distributor shaft bushes, etc. | 4. Renew |
| | 5. Rotor arm and flash shield cracked or showing signs of tracking | 5. Renew |
| S— Mixture control warning light fails to appear when engine reaches running temperature | 1. Mixture control already pushed in | 1. In the hands of the operator |
| | 2. Broken connection in warning light circuit | 2. Rectify |
| | 3. Blown bulb | 3. Renew |
| | 4. Faulty thermostat switch (at cylinder head) | 4. Renew |
| | 5. Faulty manual switch (at mixture control) | 5. Renew |
| | 6. Broken operating mechanism at manual switch | 6. Rectify |
| T— Mixture control warning light remains on with engine at running temperature | 1. Mixture control out | 1. Push control right in |
| | 2. Faulty manual switch | 2. Renew |
| | 3. Broken operating mechanism at manual switch | 3. Rectify |
| U— Poor performance of horns | 1. Low voltage due to discharged battery | 1. Recharge |
| | 2. Bad connections in wiring | 2. Carefully inspect all connections and horn push |
| | 3. Loose fixing bolt | 3. Rectify |
| | 4. A faulty horn | 4. Renew |
| V— Central door locking does not operate (on all four doors) | 1. Battery discharged | 1. Recharge |
| | 2. Control unit in driver's door lock actuator faulty | 2. Renew |
| | 3. Loose or broken connection in driver's door | 3. Locate and rectify |
| | 4. Blown fuse | 4. Renew |
| W— Central door locking does not operate (on one door only) | 1. Loose or broken connection | 1. Locate and rectify |
| | 2. Lock actuator failure | 2. Renew |
| | 3. Faulty lock | 3. Rectify |
| | 4. Mechanical linkages disconnected | 4. Locate and rectify |
| X— Window lift will not operate | 1. Motor failure | 1. Renew |
| | 2. Loose or broken connection | 2. Locate and rectify |
| | 3. Faulty switch | 3. Renew |
| | 4. Mechanical linkage faulty | 4. Rectify |
| Y— Exterior mirrors fail to operate | 1. Loose or broken connection | 1. Locate and rectify |
| | 2. Faulty switch | 2. Renew |
| | 3. Mirror motor failure | 3. Renew |

# ELECTRICAL EQUIPMENT

## DESCRIPTION

The electrical system is Negative earth, and it is most important to ensure correct polarity of the electrical connections at all times. Any incorrect connections made when reconnecting cables may cause irreparable damage to the semiconductor devices used in the alternator and regulator. Incorrect polarity would also seriously damage any transistorised equipment such as radio and tachometer etc.

Before carrying out any repairs or maintenance to an electrical component, always disconnect the battery.

## ALTERNATOR

The V-drive fan belt used with alternators is not the same as that used with d.c. machines. Only use the correct Rover replacement fan belt. Occasionally check that the engine and alternator pulleys are accurately aligned.

It is essential that good electrical connections are maintained at all times. Of particular importance are those in the charging circuit (including those at the battery) which should be occasionally inspected to see that they are clean and tight. In this way any significant increase in circuit resistance can be prevented.

Do not disconnect battery cables while the engine is running or damage to the semi-conductor devices may occur. It is also inadvisable to break or make any connections in the alternator charging and control circuits while the engine is running.

The Model 15TR electronic voltage regulator employs micro-circuit techniques resulting in improved performance under difficult service conditions. The whole assembly is encapsulated in silicone rubber and housed in an aluminium heat sink, ensuring complete protection against the adverse effects of temperature, dust, and moisture etc.

The regulating voltage is set during manufacture to give the required regulating voltage range of 14.2 ± 0.2 volts, and no adjustment is necessary. The only maintenance needed is the occasional check on terminal connections and wiping with a clean dry cloth.

The alternator system provides for direct connection of a charge (ignition) indicator warning light, and eliminates the need for a field switching relay or warning light control unit. As the warning lamp is connected in the charging circuit, lamp failure will cause loss of charge. Lamp should be checked regularly and a spare carried.

When using rapid charge equipment to re-charge the battery, the battery must be disconnected from the vehicle.

## ALTERNATOR

### Remove and refit

#### Removing

1. Disconnect battery earth lead.
2. Disconnect leads from alternator.
3. Slacken alternator fixings, pivot alternator inwards and remove fan belt.
4. Remove alternator.

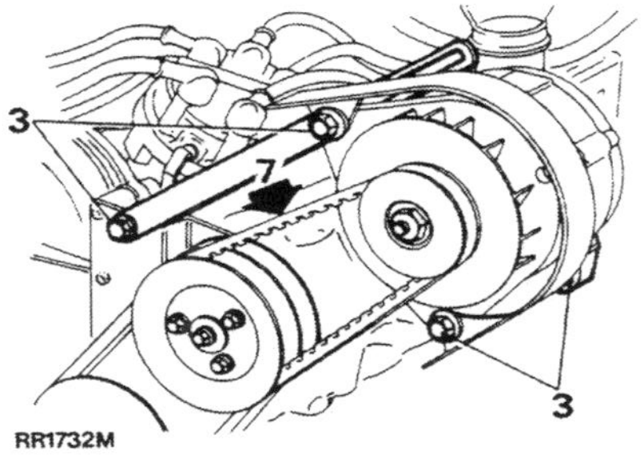

#### Refitting

5. Attach the alternator adjustment bracket and attach the alternator to the bracket.

**NOTE: The fan guard is attached to the front fixing and the adjustment bracket bolt.**

6. Slacken the alternator adjustment bracket and attach the alternator to the bracket.
7. Fit the fan belt and adjust the belt tension.
8. Connect the wiring plug to the alternator.
9. Connect the battery.

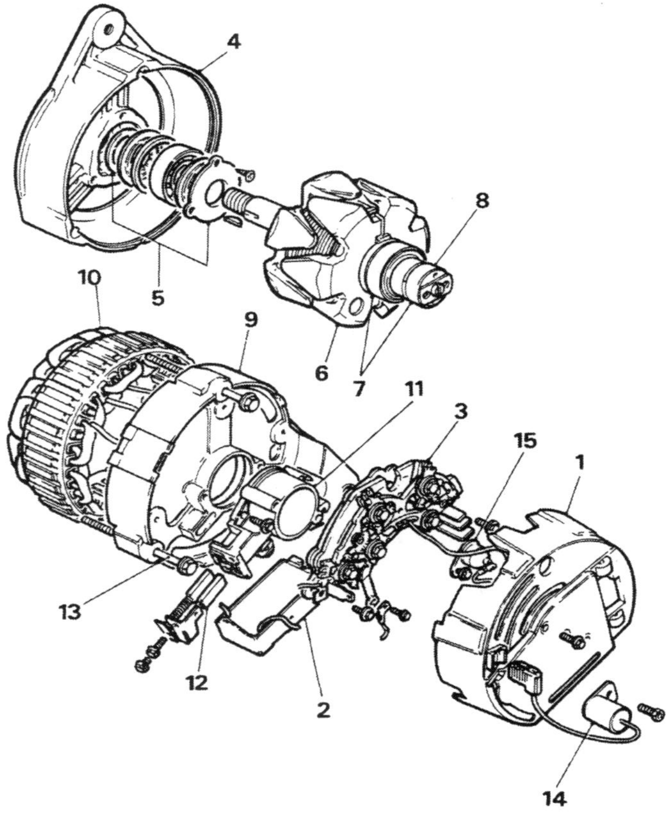

**Alternator type 133/65**

1. Cover
2. Regulator
3. Rectifier
4. Drive end bracket
5. Bearing assembly
6. Rotor
7. Slip ring end bearing
8. Slip rings
9. Slip ring end bracket
10. Stator
11. Brush box
12. Brushes
13. Through bolt
14. Suppressor
15. Surge protection diode

# ELECTRICAL

## ALTERNATOR — Lucas — 133/65

### Overhaul

**Including Test (Bench)**

**NOTE: Alternator charging circuit** — The ignition warning light is connected in series with the alternator field circuit. Bulb failure would prevent the alternator charging, except at very high engine speeds, therefore, the bulb should be checked before suspecting an alternator failure.

### Precautions

Battery polarity is NEGATIVE EARTH, which must be maintained at all times.

A separate control unit is fitted; instead a voltage regulator of micro-circuit construction is incorporated on the slip ring end bracket, inside the alternator cover.

Battery voltage is applied to the alternator output cable even when the ignition is switched off, the battery must be disconnected before commencing any work on the alternator. The battery must also be disconnected when repairs to the body structure are being done by arc welding.

### Surge protection device

Some protection of the alternator is provided by a surge protection device, to absorb high transient voltages.

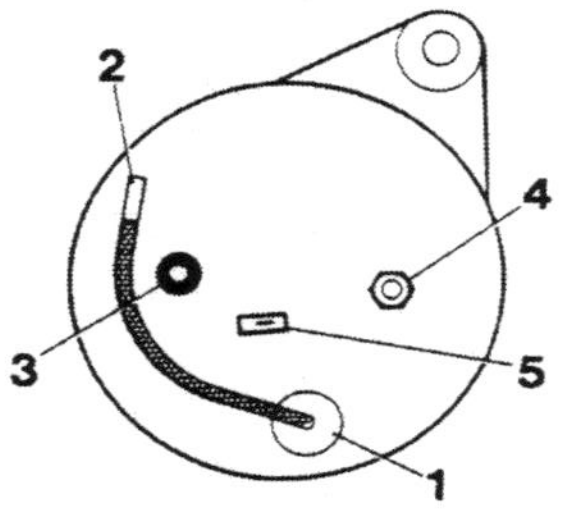

1. Suppression capacitor
2. Positive suppression terminal
3. IND terminal
4. + output terminal
5. Sensing terminal

### Testing in position
**Surge protection device**

If the alternator output falls to zero, the fault may be caused by the surge protection device failing safe, short-circuit, or a fault in the alternator circuit. Check the surge protection device as follows:

1. Check that the fan belt is correctly tensioned.
2. Withdraw the connectors from the alternator.
3. Disconnect the suppressor and remove the alternator cover.
4. Remove the surge protection device.
5. Refit the alternator cover, and connectors. Ensure that all circuit connections are clean and tight.
6. Start and run the engine. If the alternator output is now normal, fit a new surge protection device.

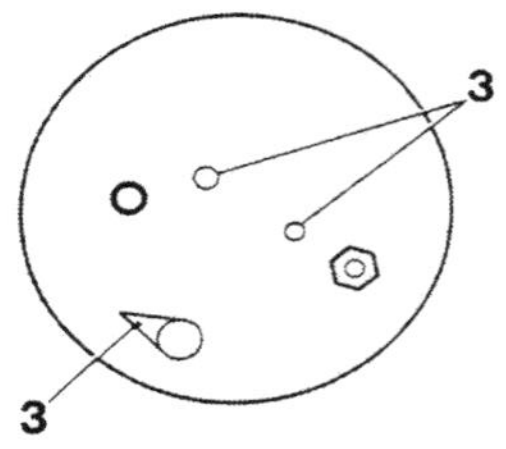

### Output test

7. Check that the fan belt is correctly tensioned and that all charging circuit connections are secure.
8. Run the engine at fast idle until normal operating temperature is attained.
9. Stop the engine.
10. Withdraw the connectors from the alternator.
11. Disconnect the suppressor and remove the alternator cover.
12. Connect the regulator case to the alternator frame.
13. Connect a 0–60 ammeter between the alternator and the battery.
14. Connect a 0–20 voltmeter across the battery terminals.
15. Connect a 15 ohm 35 amp variable resistor across the battery terminals.

**CAUTION: Do not leave the variable resistor connected across the battery terminals for longer than is necessary to carry out the following test, items 16 and 17.**

16. Start the engine and run at 750 rev/min. The warning light bulb should be extinguished.
17. Increase the engine speed to 3000 rev/min, and adjust the variable resistance until the voltmeter reads 13.6 volts. The ammeter reading should then be approximately 60 amps. Any appreciable deviation from this figure will necessitate removing and dismantling the alternator. If the output test is satisfactory, proceed with the regulator test.

### Regulator test

18. Disconnect the variable resistor and remove the connection between the regulator and the alternator frame.
19. With the remainder of the circuit connected as for the alternator output test, start the engine and run at 3000 rev/min, until the ammeter shows an output current of less than 10 amperes.
20. The voltmeter should now give a reading of 13.6 to 14.4 volts. Any appreciable deviation from this (regulating) voltage indicates a faulty regulator which must be replaced.
21. If the foregoing output and regulator tests show the alternator and regulator to be performing satisfactorily, disconnect the test circuit, reconnect the alternator terminal connector and proceed with the charging circuit resistance test.

### Charging circuit resistance test

22. Connect a low range voltmeter between either of the alternator terminals marked + and the positive terminal of the battery.

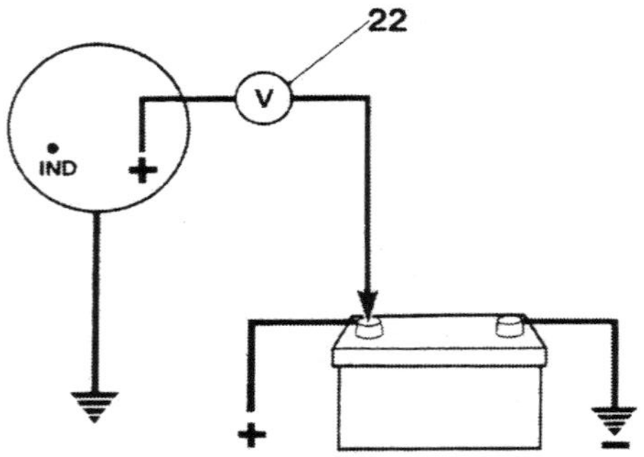

23. Switch on the headlamps and start the engine. Set the throttle to run at approximately 3000 rev/min. Note the voltmeter reading.

24. Transfer the voltmeter connections to the frame of the alternator and the negative terminal of the battery, and again note the voltmeter reading.

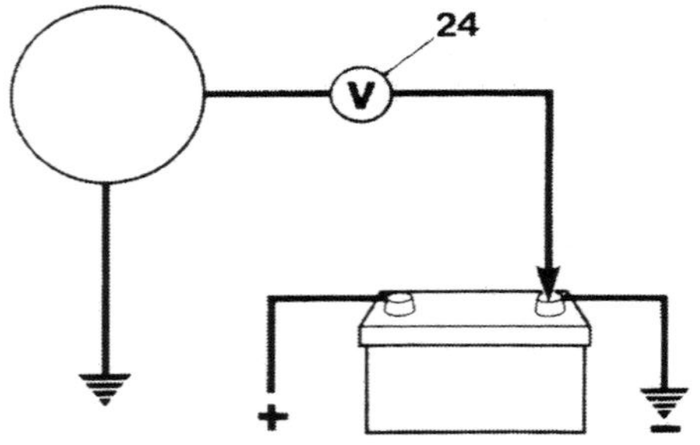

25. If the reading exceeds 0.5 volt on the positive side or 0.25 volt on the negative side, there is a high resistance in the charging circuit which must be traced and remedied.

### Testing — alternator removed

26. Withdraw the connectors from the alternator.
27. Remove the alternator.
28. Disconnect the suppressor and remove the alternator cover.
29. Detach the surge protection device.
30. Disconnect the lead and remove the rectifier assembly.
31. Note the arrangement of the brush box connections and remove the screws securing the regulator to the brush box and withdraw. This screw also retains the inner brush mounting plate in position.
32. Remove the screw retaining the outer brush box in position and withdraw both brushes.
33. Check brushes for wear by measuring length of brush protruding beyond brush box moulding. If length is 10 mm (0.4 in) or less, brush must be renewed.
34. Check that brushes move freely in holders. If brush is sticking, clean with petrol moistened cloth or polish sides of brush with fine file.
35. Check brush spring pressure using push-type spring gauge. Gauge should register 136 to 279 g (5 to 10 oz) when brush is pulled back until face is flush with housing. If reading is outside these limits, renew brush assembly.
36. Remove the two screws securing the brush box to the slip ring end bracket and lift off the brush box assembly.

*continued*

# ELECTRICAL 86

37. Securely clamp alternator in a vice and release the stator winding cable ends from the rectifier by applying a hot soldering iron to the terminal tags of the rectifier. Prise out the cable ends when the solder melts.

**CAUTION: When soldering or unsoldering connections to diodes take care not to overheat the diodes or bend the pins. During soldering operations, diode pins should be gripped lightly with a pair of long nosed pliers which will act as a thermal shunt.**

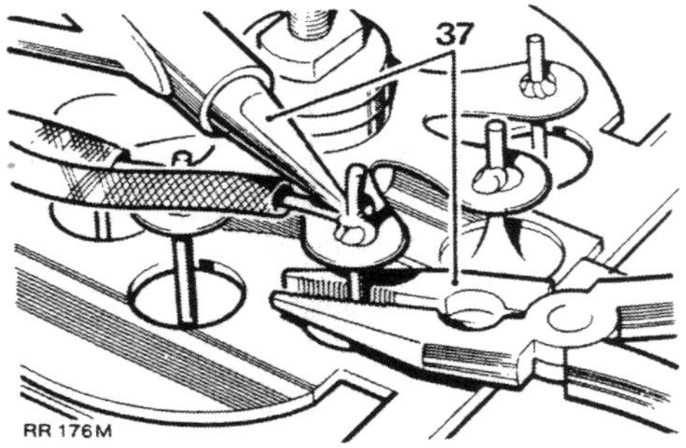

38. Remove the two remaining screws securing the rectifier assembly to the slip ring end bracket and lift off the rectifier assembly. Further dismantling of the rectifier is not required.
39. Remove the slip ring end bracket bolts and lift off the bracket.
40. Connect a 12 volt battery and a 36 watt test lamp to two of the stator connections. Repeat the test replacing one of the two stator connections with the third. If test lamp fails to light in either test, stator must be renewed.
41. Using a 110 volt a.c. supply and a 15 watt test lamp, test for insulation between any one of the three stator connections and stator laminations. If test lamp lights, stator must be renewed.
42. Clean surfaces of slip rings using petrol moistened cloth.
43. Inspect slip ring surfaces for signs of burning; remove burn marks using very fine sandpaper. On no account should emery cloth or similar abrasives be used, or any attempt made to machine the slip rings.
44. Note the position of the stator output leads in relation to the alternator fixing lugs, and lift the stator from the drive end bracket.
45. Connect an ohmmeter or a 12 volt battery and an ammeter to the slip rings. An ohmmeter reading of 3.2 ohms or an ammeter reading of 4 amps should be recorded.
46. Using a 110 volt a.c. supply and a 15 watt test lamp, test for insulation between one of the slip rings and one of the rotor poles. If the test lamp lights, the rotor must be renewed.
47. To separate the drive end bracket and rotor, remove the shaft nut, washers, woodruff key and spacers from the shaft.
48. Remove bearing retaining plate by removing the three screws. Using a press, drive the rotor shaft from the drive end bracket.
49. If necessary, to remove the slip rings or the slip ring end bearing on the rotor shaft, unsolder the outer slip ring connection and gently prise the slip ring off the shaft, repeat the procedure for the inner slip ring connection. Using a suitable extraction tool, withdraw the slip ring bearing from the shaft.

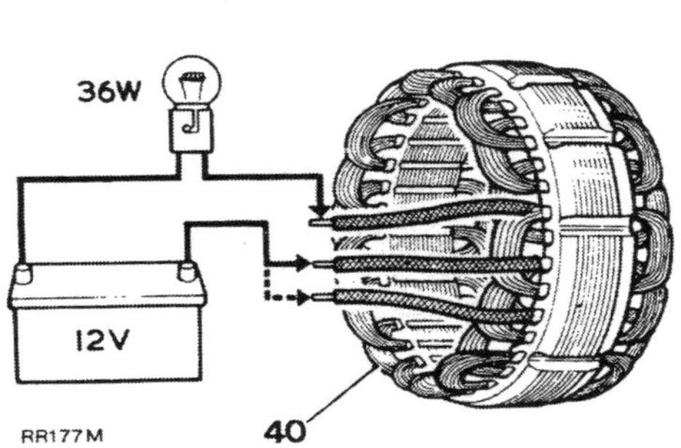

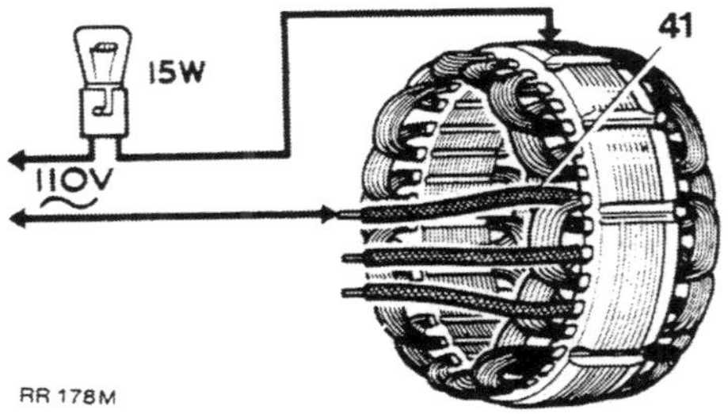

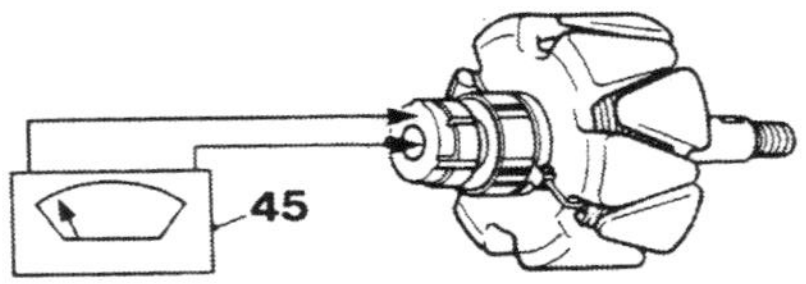

*Continued*

## Reassembling

50. Reverse the dismantling procedure, noting the following points.
    (a) Use Shell Alvania 'RA' to lubricate bearings.
    (b) When refitting slip ring end bearing, ensure it is fitted with open side facing rotor.
    (c) Use Fry's H.T.3 solder on slip ring field connections.
    (d) When refitting rotor to drive end bracket, support inner track of bearing. Do not use drive end bracket to support bearing when fitting rotor.
    (e) Tighten through-bolts evenly.
    (f) Fit brushes into housings before fitting brush moulding.
    (g) Tighten shaft nut to the correct torque, see Torque Wrench Settings.
    (h) Refit regulator pack to brush moulding.
51. Reconnect the leads between the regulator; brush box and rectifier, as illustrated.
52. Refit the alternator.

## HORNS

### Remove and refit

#### Removing

1. Disconnect the battery.
2. Remove radiator grille.
3. Remove the headlamp surround.
4. Remove the nut and serrated washer securing the horn in position.
5. Disconnect the electrical leads.
6. Withdraw the horns.

**NOTE: Twin horns are fitted. An identification letter is stamped on the front outer rim of the horn: 'H' — high note, 'L' — low note.**

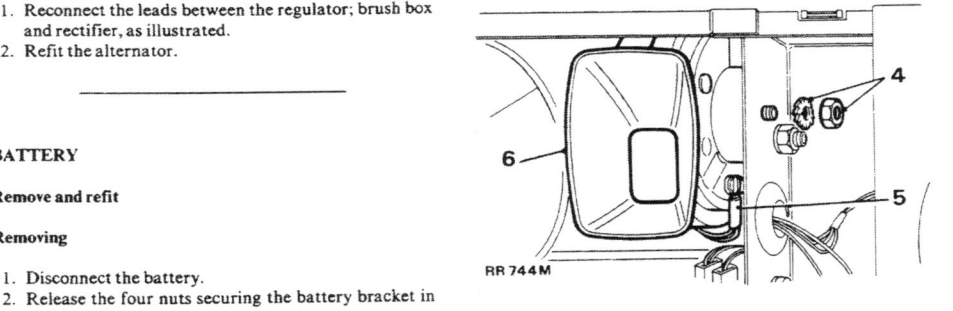

#### Refitting

7. Reverse removal procedure.

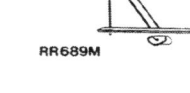

## BATTERY

### Remove and refit

#### Removing

1. Disconnect the battery.
2. Release the four nuts securing the battery bracket in position.
3. Remove the bracket from the studs.
4. Remove the battery.

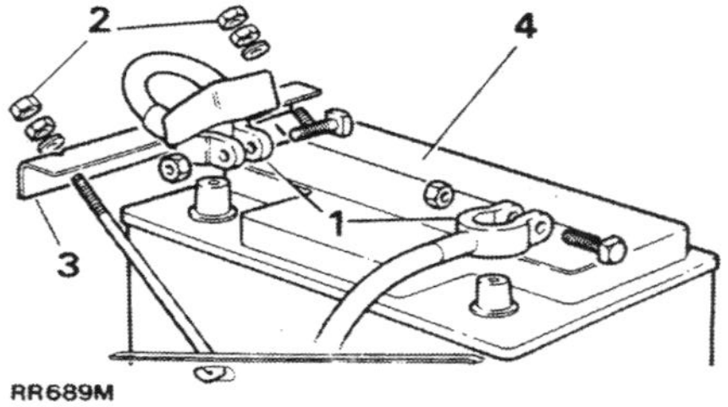

#### Refitting

5. Reverse the removal procedure.

**NOTE: Smear the battery clamps and terminals with petroleum jelly before refitting.**

# ELECTRICAL 86

## ELECTRONIC IGNITION

A Lucas model 35DM8 distributor is employed. This has a conventional advance/retard vacuum unit and centrifugal automatic advance mechanism.

A pick-up module, in conjunction with a rotating timing reluctor inside the distributor body, generates timing signals. These are applied to an electronic ignition amplifier unit fitted under the ignition coil mounted on top of the left front wing valance.

## MAINTENANCE

80,000 km (48,000 miles).
Remove the distributor cap and rotor arm and wipe inside with a nap-free cloth.

**DO NOT DISTURB the clear plastic insulating cover which protects the magnetic pick-up module.**

### Service Parts

1. Cap
2. H.T. brush and spring
3. Rotor arm
4. Insulation cover (Flash shield)
5. Pick-up and base plate assembly
6. Vacuum unit
7. 'O' ring oil seal

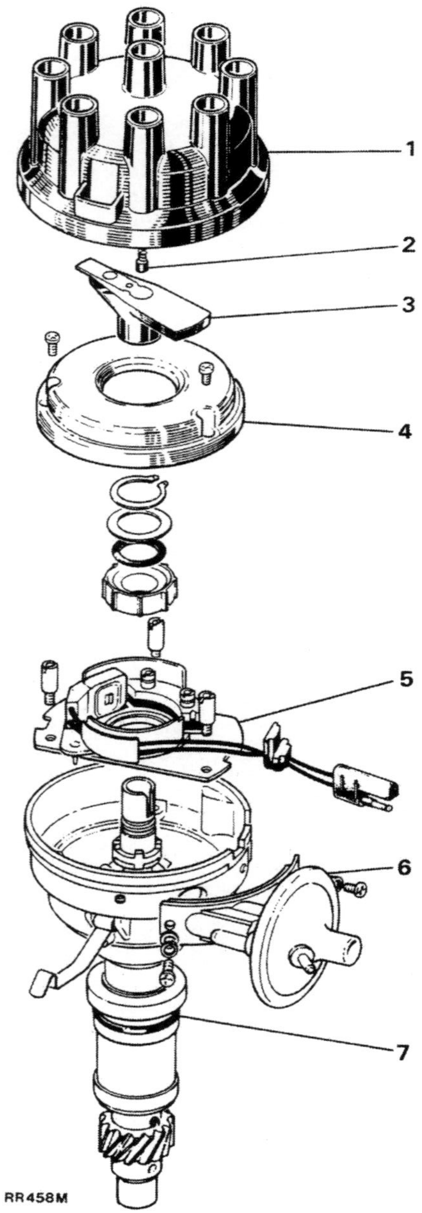

# ELECTRICAL 86

## SPARK PLUGS

### Remove, clean, adjust and refit

#### Removing

1. Withdraw leads by gripping end shrouds. DO NOT pull leads alone.

   **NOTE: Remove the hot air pipe for access to the RH plugs as necessary.**

2. Remove sparking plugs and washers.

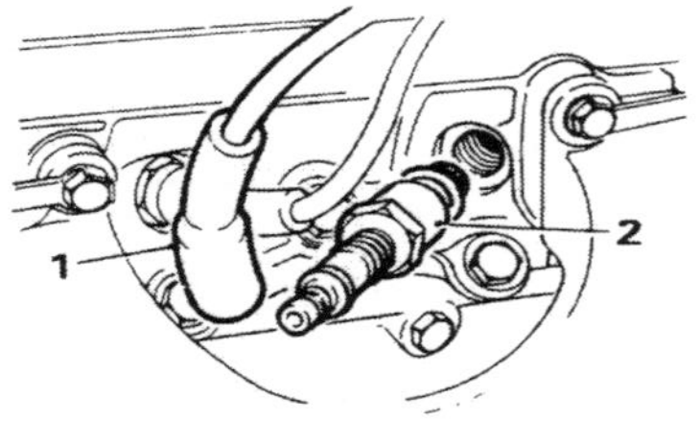

#### Cleaning

3. Fit plug in plug cleaning machine.
4. Wobble plug with circular motion while operating abrasive blast for a maximum of four seconds.

   **CAUTION: Excessive abrasive blasting will erode insulator nose.**

5. Change to air blast only and continue to wobble plug for a minimum of thirty seconds to remove abrasive grit from plug cavity.
6. Wire brush plug threads, open gap slightly.
7. Using point file, square off electrode surfaces.
8. Set electrode gap: 0.80 mm (0.030 in).

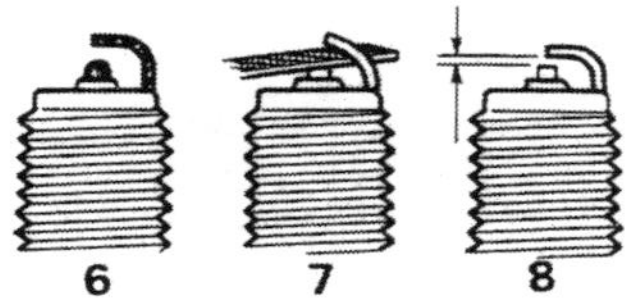

9. Test plugs in accordance with cleaning machine manufacturers instructions. If satisfactory, refit plugs in engine.

10. Examine high tension leads, including coil to distributor lead, for insulation cracking or corrosion at end contacts. Fit new leads as necessary.
11. In addition to correct firing order, high tension leads must also be fitted in correct relation to each other to avoid cross firing.

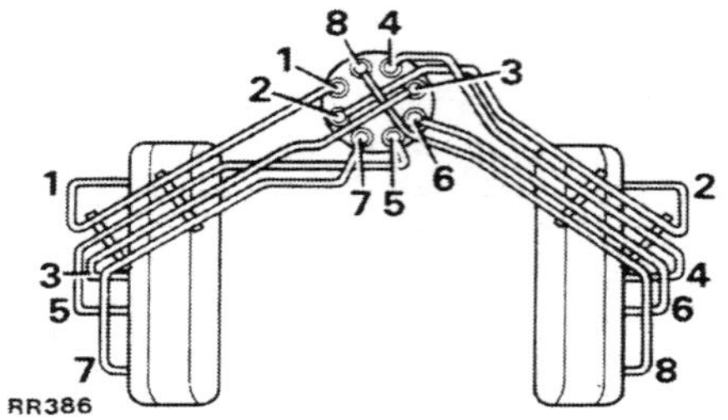

12. Leads at distributor cap must be connected as illustrated—Figures 1 to 8 inclusive—indicate plug lead numbers. RH—Right-hand side of engine when viewed from rear. LH—Left-hand side of engine when viewed from rear.

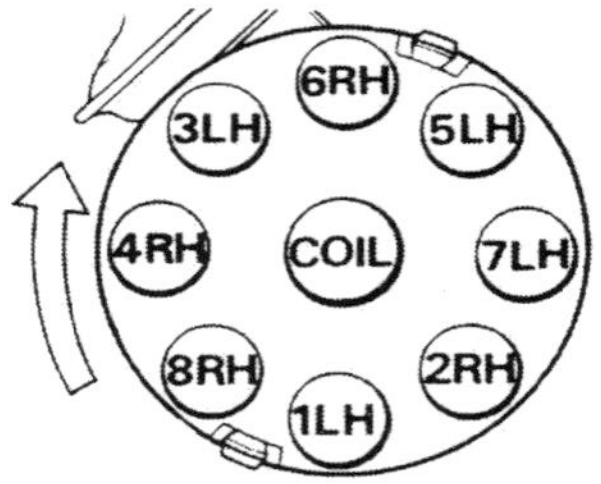

13. When pushing leads on plugs ensure ferrules within shrouds are firmly seated on plugs. A guide is that shroud ends are within 6 mm (0.250 in) of metal body of plugs.

    **NOTE: If new plugs are necessary refer to technical data section.**

228

# ELECTRICAL 86

## DISTRIBUTOR

### Remove and refit

#### Removing

1. Disconnect battery.
2. Disconnect vacuum pipe(s).
3. Remove distributor cap.
4. Disconnect low tension lead from coil.
5. Mark distributor body in relation to centre line of rotor arm.
6. Add alignment marks to distributor and front cover.
7. Release the distributor clamp and remove the distributor.

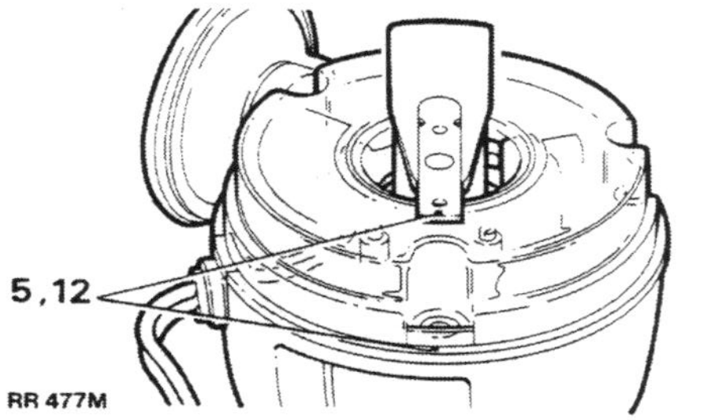

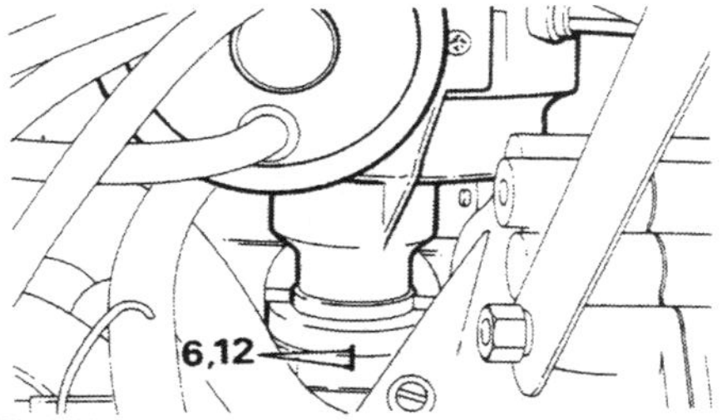

NOTE: Marking distributor enables refitting in exact original position, but if engine is turned while distributor is removed, complete ignition timing procedure must be followed.

RR477M
RR476M

### Refitting

NOTE: If a new distributor is being fitted, mark body in same relative position as distributor removed.

8. Leads for distributor cap should be connected as illustrated.
   Figures 1 to 8 inclusive indicate plug lead numbers.
   RH—Right-hand side of engine, when viewed from the rear.
   LH—Left-hand side of engine, when viewed from the rear.

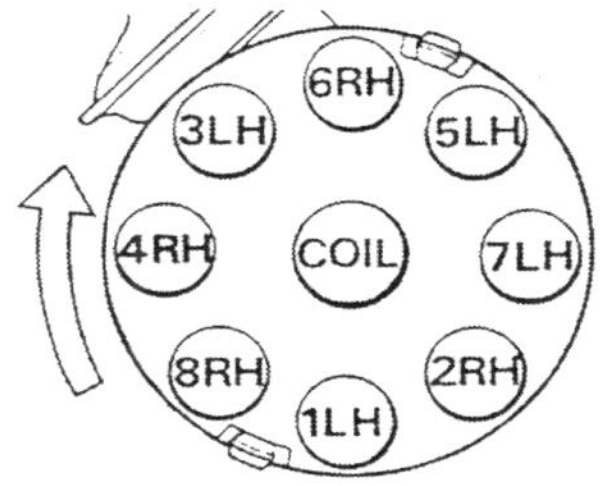

RR616M

9. If engine has not been turned whilst distributor has been removed, proceed as follows (items 10 to 17).
10. Fit new 'O' ring seal to distributor housing.
11. Turn distributor drive until centre line of rotor arm is 30° anti-clockwise from mark made on top edge of distributor body.
12. Fit distributor in accordance with alignment markings.

NOTE: It may be necessary to align oil pump drive shaft to enable distributor drive shaft to engage in slot.

13. Fit clamp and bolt. Secure distributor in exact original position.
14. Connect vacuum pipe to distributor and low tension lead to coil.
15. Fit distributor cap.
16. Reconnect battery.
17. Using suitable electronic equipment, set the ignition timing as follows.
18. If, with distributor removed, engine has been turned it will be necessary to carry out the following procedure.
19. Set engine—No. 1 piston to static ignition timing figure (see data section) on compression stroke.
20. Turn distributor drive until rotor arm is approximately 30° anti-clockwise from number one sparking plug lead position on cap.
21. Fit distributor to engine.
22. Check that centre line of rotor arm is now in line with number one sparking plug lead on cap. Reposition distributor if necessary.
23. If distributor does not seat correctly in front cover, oil pump drive is not engaged. Engage by lightly pressing down distributor while turning engine.
24. Fit clamp and bolt leaving both loose at this stage.
25. Set the ignition timing statically to within 2°–3° of T.D.C.
26. Connect the vacuum pipe to the distributor.
27. Fit low tension lead to coil.
28. Fit distributor cap.
29. Reconnect the battery.
30. Using suitable electronic equipment set the ignition timing.

## DISTRIBUTOR—LUCAS 35DM8

### Overhaul

### DISTRIBUTOR CAP

1. Unclip and remove cap.
2. Renew cap if known to be faulty.
3. Clean cap with a nap-free cloth.

### ROTOR ARM

4. Pull rotor arm from keyed shaft.
5. Renew rotor arm if known to be faulty.

### INSULATION COVER (Flash shield)

6. Remove cover, secured by three screws.
7. Renew cover if known to be faulty.

### VACUUM UNIT

8. Remove two screws from vacuum unit securing bracket, disengage vacuum unit connecting rod from pick-up base plate connecting peg, and withdraw vacuum unit from distributor body.

### PICK-UP AND BASE PLATE ASSEMBLY

9. Use circlip pliers to remove the circlip retaining the reluctor on rotor shaft.
10. Remove the flat washer and then the 'O' ring recessed in the top of the reluctor.
11. Insert the blade of a small screwdriver beneath the reluctor and prise it partially along the shaft, sufficient to enable it to be gripped between the fingers and withdrawn from the shaft.

NOTE: Coupling ring fitted beneath reluctor.

12. Remove pick-up and base plate assembly, secured by three support pillars.

NOTE: Do not disturb the two barrel nuts securing the pick-up module, otherwise the air gap will need re-adjustment.

13. Renew pick-up and base plate assembly if module is known to be faulty, otherwise check pick-up winding resistance (2k–5k ohm).

## RE-ASSEMBLY

14. This is mainly a reversal of the dismantling procedure, noting the following points:

## LUBRICATION

### Apply clean engine oil:

a. Three drops to felt pad reservoir in rotor shaft.

### Apply Chevron SR1 (or equivalent) grease.

b. Auto advance mechanism.
c. Pick-up plate centre bearing.
d. Pre tilt spring and its rubbing area (pick-up and base plate assy).
e. Vacuum unit connecting peg (pick-up and base plate assy).
f. The connecting peg hole in vacuum unit connecting rod.

### Apply Rocol MHT (or equivalent) grease:

g. Vacuum unit connecting rod seal (located in vacuum unit where connecting rod protrudes).

NOTE: Applicable only to double acting vacuum units.

## FITTING PICK-UP AND BASE PLATE ASSEMBLY

15. Pick-up leads must be prevented from fouling the rotating reluctor. Both leads should be located in plastic carrier as illustrated. Check during re-assembly.

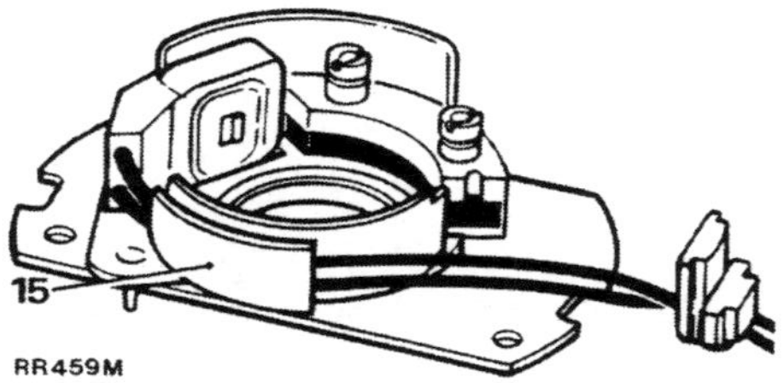

RR459M

## REFITTING RELUCTOR

16. Slide reluctor as far as it will go on rotor shaft, then rotate reluctor until it engages with the coupling ring beneath the pick-up base plate. The distributor shaft, coupling ring and reluctor are 'keyed' and rotate together.

*Continued*

## ELECTRICAL 86

### PICK-UP AIR GAP ADJUSTMENT

17. The air gap between the pick-up limb and reluctor teeth must be set within the specified limits, using a non-ferrous feeler gauge.

**NOTE: When the original pick-up and base plate assembly has been refitted the air gap should not normally require resetting as it is pre-set at the factory. When renewing the assembly the air gap will require adjusting to within the specified limits. Refer to 'Engine Tuning Data' for ignition timing data.**

### IGNITION COIL

**Remove and refit**

Removing

1. Disconnect the battery.
2. Disconnect the electrical leads from the coil.
3. Remove the two retaining nuts and washers securing the coil to the amplifier.
4. Lift the coil off the amplifier.

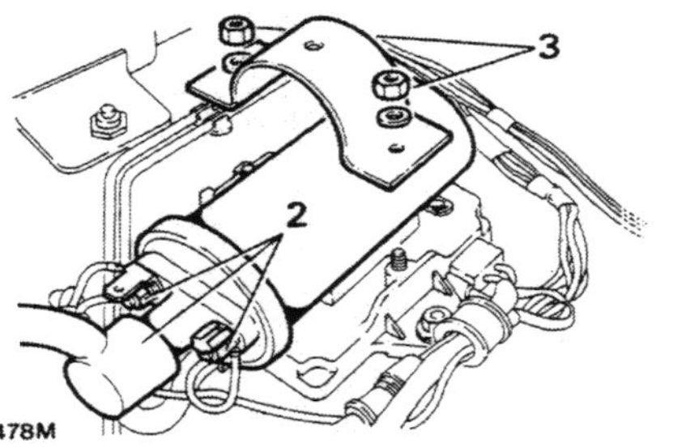

Refitting

5. Reverse the removal procedure.

RR478M

### AMPLIFIER

**Remove and Refit**

Removing

1. Disconnect the battery.
2. Disconnect the electrical leads from the amplifier and coil.
3. Remove the two retaining nuts with washers securing the coil to the amplifier.
4. Remove the two bolts, nuts, spring washers and plain washers securing the amplifier to the valance.
5. Remove the amplifier from the valance.

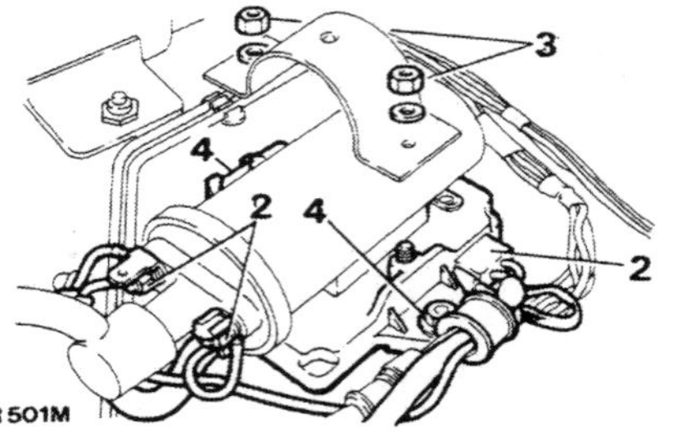

RR501M

Refitting

6. Reverse the removal procedure, ensuring that all electrical leads are correctly reconnected.

**NOTE: The amplifier is not serviceable, in the event of a fault a new amplifier must be fitted.**

### IGNITION TIMING

1. It is essential that the following procedures are adhered to. Inaccurate timing can lead to serious engine damage and additionally create failure to comply with the emission regulations applying to the country of destination. If the engine is being checked in the vehicle and is fitted with an air conditioning unit the compressor must be isolated.

2. On initial engine build, or if the distributor has been disturbed for any reason, the ignition timing must be set statically to within 2°–3° of T.D.C. (This sequence is to give only an approximation in order that the engine may be started) ON NO ACCOUNT MUST THE ENGINE BE STARTED BEFORE THIS OPERATION IS CARRIED OUT.

**Equipment required**

Calibrated Tachometer
Stroboscopic lamp

3. Couple stroboscopic timing lamp and tachometer to engine following the manufacturers instructions.
4. Disconnect the vacuum pipes from the distributor.
5. Start engine, with no load and not exceeding 3,000 rpm, run engine until normal operating temperature is reached. (Thermostat open). Check that the normal idling speed falls within the tolerance specified in the data section.
6. Idle speed for timing purposes must not exceed 750 rpm, and this speed should be achieved by removing a breather hose NOT BY ADJUSTING IDLE SETTING SCREWS.
7. With the distributor clamping bolt slackened turn distributor until the timing flash coincides with the timing pointer and the correct timing mark on the rim of the torsional vibration damper as shown in the engine tuning section.
8. Retighten the distributor clamping bolt securely. Recheck timing in the event that retightening has disturbed the distributor position.
9. Refit vacuum pipes.
10. Disconnect stroboscopic timing lamp and tachometer from engine.

### LUCAS CONSTANT ENERGY IGNITION SYSTEM 35DM8 PRELIMINARY CHECKS

Inspect battery cables and connections to ensure they are clean and tight. Check battery state of charge if in doubt as to its condition.

Inspect all LT connections to ensure that the HT leads are correctly positioned and not shorting to earth against any engine components. The wiring harness and individual cables should be firmly fastened to prevent chafing.

### PICK-UP MODULE AIR GAP SETTINGS

Air gap settings vary according to vehicle application.

**NOTE: The gap is set initially at the factory and will only require adjusting if tampered with or when the pick-up module is replaced.**

**Test Notes**

(i) The ignition must be switched on for all checks.
(ii) Key to symbols used in the charts for Tests 2.

✓ Correct Reading   H High Reading   L Low Reading

(iii) Use feeler gauges manufactured from a non-magnetic material when setting air gaps.

### TEST 1:

**Check HT Sparking**

Remove coil/distributor HT lead from distributor cover and hold approximately 6 mm (0.25 in) from the engine block. Switch the ignition 'on' and operate the starter. If regular sparking occurs, proceed to Test 6. If no sparking proceed to Test 2.

Test 1

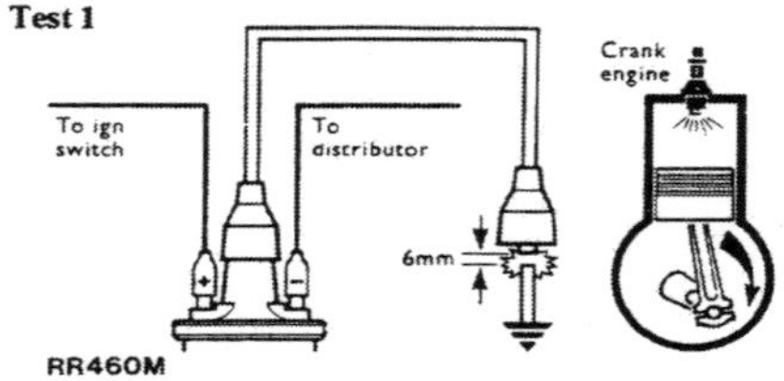

RR460M

### TEST 2:

**Amplifier Static Checks**

Switch the ignition 'on'

(a) Connect voltmeter to points in the circuit indicated by the arrow heads and make a note of the voltage readings.

**NOTE: Only move the voltmeter POSITIVE lead during tests 2, 3 and 4.**

(b) Compare voltages obtained with the specified values listed below:

**EXPECTED READINGS**

1. More than 11.5 volts
2. 1 volt max below volts at point 1 in test circuit
3. 1 volt max below volts at point 1 in test circuit
4. 0 volt—0.1 volt

Test 2

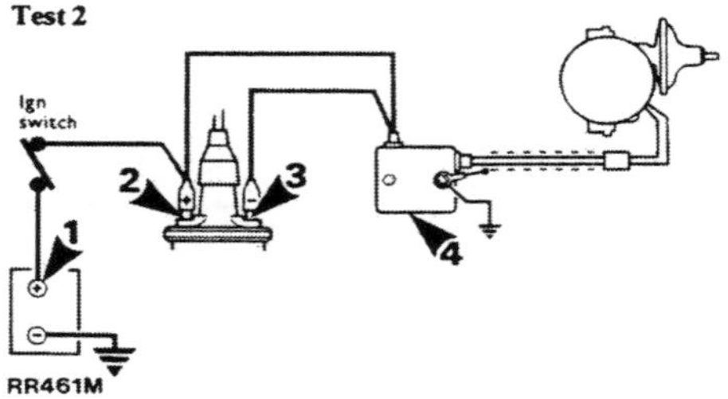

RR461M

*Continued*

# ELECTRICAL 86

(c) If all readings are correct proceed to Test 3.
(d) Check incorrect reading(s) with chart to identify area of possible faults, i.e. faults listed under heading 'Suspect'.

| 1 | 2 | 3 | 4 | SUSPECT |
|---|---|---|---|---------|
| L | | | | DISCHARGED BATTERY |
| ✓ | L | L | | IGN. SWITCH AND/OR WIRING |
| ✓ | ✓ | L | ✓ | COIL OR AMPLIFIER |
| ✓ | ✓ | ✓ | H | AMPLIFIER EARTH |

## TEST 3:

### Check Amplifier Switching

Disconnect the high tension lead between the coil and distributor.

Connect the voltmeter between battery positive (+ve) terminal and H.T. coil negative (−ve) terminal, the voltmeter should register zero volts.

Switch the ignition 'on' then crank the engine. The voltmeter reading should increase just above zero, in which case proceed with Test 5.

If there is no increase in voltage during cranking proceed to Test 4.

### Test 3

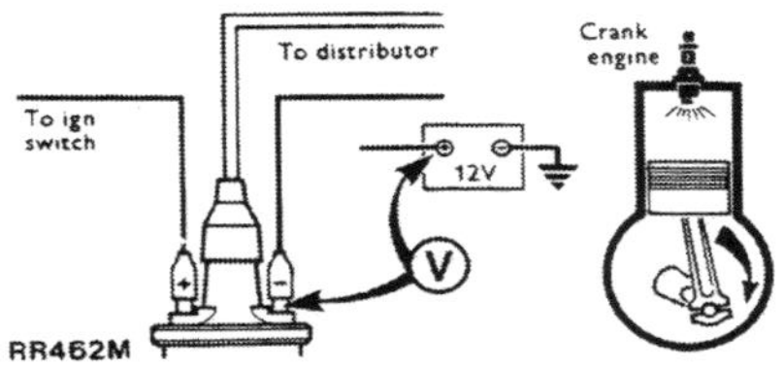

RR462M

## TEST 4:

### Pick-Up Coil Resistance
### Applications with Separate Amplifier

Disconnect the pick-up leads at the harness connector. Connect the ohmmeter leads to the two pick-up leads in the plug.

The ohmmeter should register between 2k and 5k ohm if pick-up is satisfactory. Change the amplifier if ohmmeter reading is correct. If the engine still does not start carry out Test 5.

Change the pick-up if ohmmeter reading is incorrect. If the engine still does not start proceed to Test 5.

### Test 4

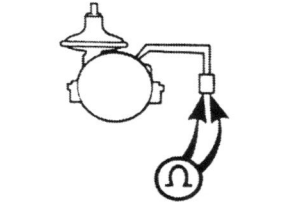

RR463M

## TEST 5:

### Check H.T. Sparking

Remove existing coil/distributor H.T. lead and fit test H.T. lead to coil chimney. Hold free end about 6 mm (0.25 in) from the engine block and crank the engine.

H.T. sparking good, repeat test with original H.T. lead, if then no sparking, change H.T. lead. If sparking is good but engine will not start, proceed to Test 6.

If no sparking, replace coil.

If engine will not start carry out Test 6.

### Test 5

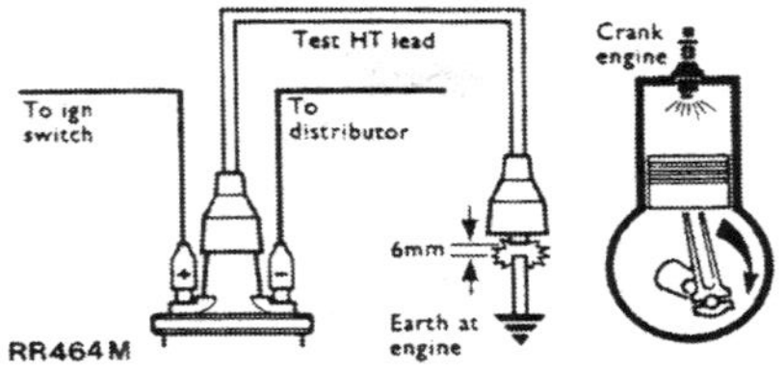

RR464M

Continued

## TEST 6:

### Check Rotor Arm

Remove distributor cover. Disconnect coil H.T. lead from cover and hold about 3 mm (0.13 in) above rotor arm electrode and crank the engine. There should be no H.T. sparking between rotor and H.T. lead. If satisfactory carry out Test 7.

H.T. sparking, replace rotor arm.

If engine will not start carry out Test 7.

## TEST 7:

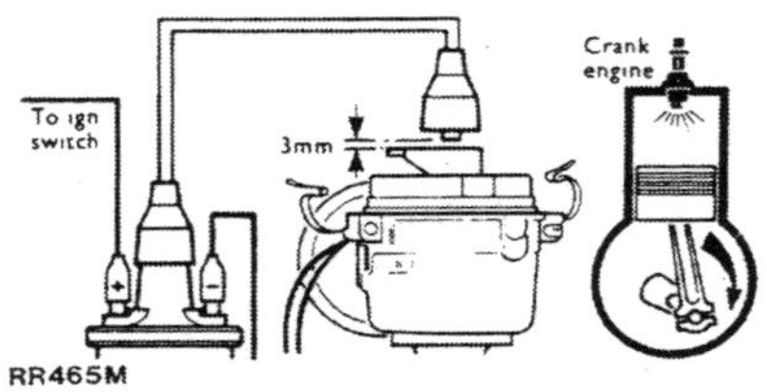

### Visual and H.T. Cable Checks

Examine:
1. Distributor Cover.
2. Coil Top.
3. H.T. Cable Insulation.
4. H.T. Cable Continuity.
5. Sparking Plugs.

NOTE:
1. Reluctor.
2. Rotor and Flash Shield.

Should be:
Clean, dry, no tracking marks.
Clean, dry, no tracking marks.
Must not be cracked, chafed or perished.
Must not be open circuit.
Clean, dry, and set to correct gap.

Must not foul pick-up or leads.
Must not be cracked or show signs of tracking marks.

### Test 6

RR465M

# ELECTRICAL 86

## HEADLAMP ASSEMBLY

### Remove and refit

### Removing

1. Disconnect the battery.
2. Remove screws and washers securing the headlamp frame to the body.
3. Ease the headlamp assembly forward and disconnect.
4. Remove the two adjusting screws and one clamp to separate lamp unit from the frame.
5. Remove the rubber seal.
6. Separate the lamp unit from rim by loosening the three retaining screws.

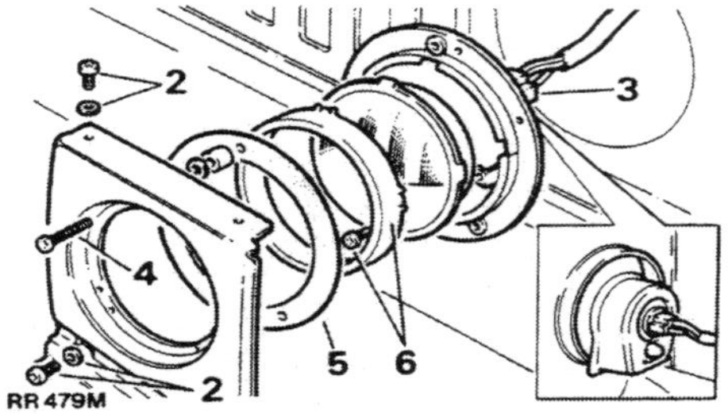

### Refitting

7. Reverse removal procedure.

## HEADLAMP BULB

### Remove and refit (Halogen)

### Removing

1. Prop open the bonnet. Two large clearance holes are provided, one on each side of the front valance, to give access to the respective bulb holders in the headlamp reflectors.

NOTE: To obtain access to the right-hand clearance hole it will be necessary to remove the battery from the vehicle.

2. Disconnect the multi-plug lead.
3. Remove the rubber dust cover.
4. Release the bulb retaining spring clip.
5. Remove the faulty bulb.

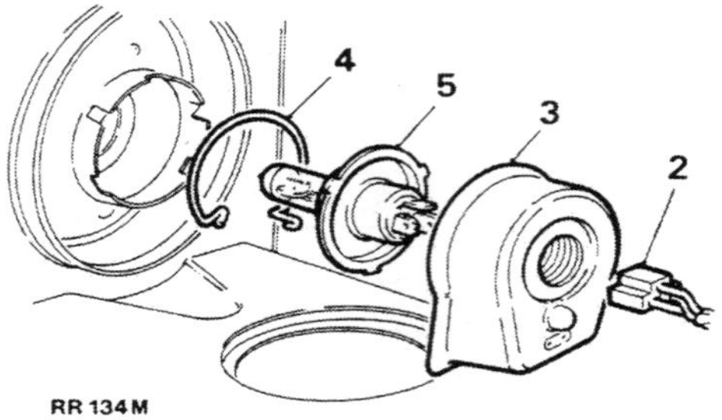

RR 479M

RR 134M

### Refitting

6. Fit the correct 'Halogen' type. The bulb holder is keyed to facilitate fitting.

**IMPORTANT: Do not touch the quartz envelope of the bulb with the fingers. If contact is accidentally made wipe gently with methylated spirits.**

7. Reverse remaining removal procedure.

## AUXILIARY DRIVING LAMP—RH AND LH

### Remove and refit

### Bulb replacement

1. Disconnect the battery.
2. The auxiliary driving lamp securing nut is located beneath the front wing adjacent to the front body fixing. Access to the lamp is gained through the front wheel arch.
3. Disconnect the electrical plug.
4. Remove the single nut and washer.

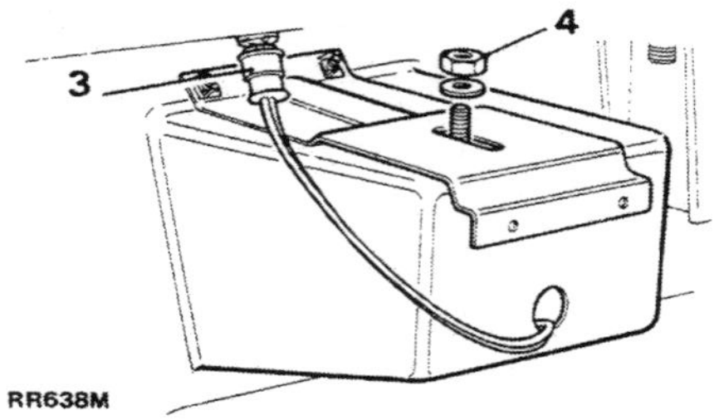

RR638M

5. From the front of the vehicle, manoeuvre the lamp and remove it from the spoiler aperture.

6. Remove the two screws securing the cover to the rear of the lamp.
7. Withdraw the cover.
8. Disconnect the lucar connector.
9. Release the spring clip securing the bulb to the lamp unit.
10. Remove the bulb.

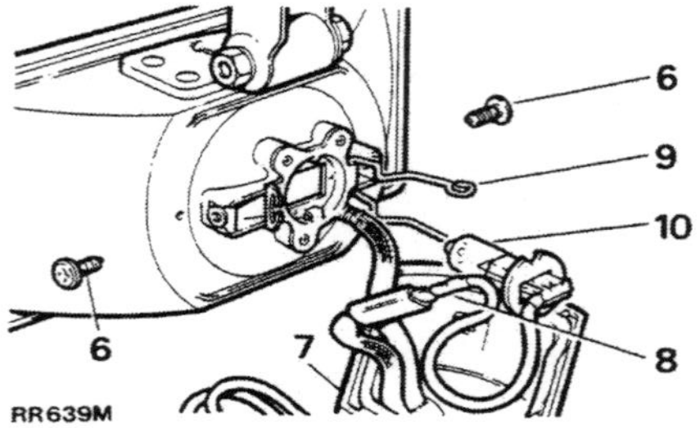

RR639M

### Refitting

11. Reverse the removal procedure.
12. Fit a new bulb ensuring that the two notches on the bulb body locate with the registers on the lamp unit.

## SIDE LIGHT AND FLASHER LAMP ASSEMBLY — RH AND LH AND BULB

### Remove and refit

### Removing

1. Disconnect the battery.
2. Remove the four screws securing the lamp lens.
3. Release the lamp lens from the lamp body.
4. Remove the foam rubber seal.
5. Remove the two bayonet fitting bulbs.
6. Remove the two screws securing the lamp body.
7. Ease the lamp body forward to reveal the electrical connection.

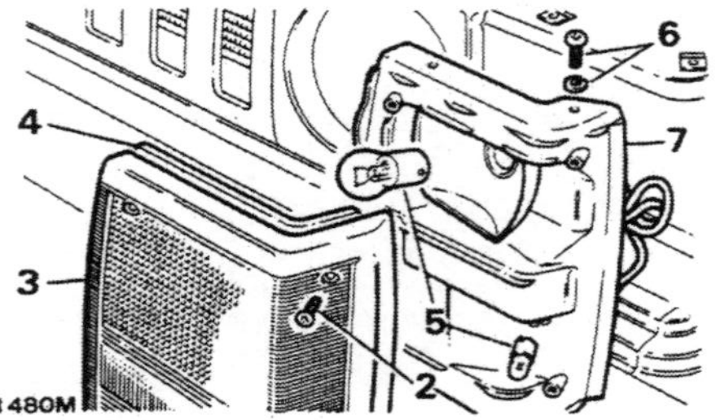

RR 480M

8. Disconnect the electrical plug at the rear of the lamp body.
9. Remove the lamp body.

# ELECTRICAL 86

## REFLECTORS

### Remove and refit

### Removing

1. Remove the four screws securing reflector.
2. Remove reflector.
3. Remove rubber seal.

NOTE: To remove the rubber seal completely it is necessary to remove the tail light lens.

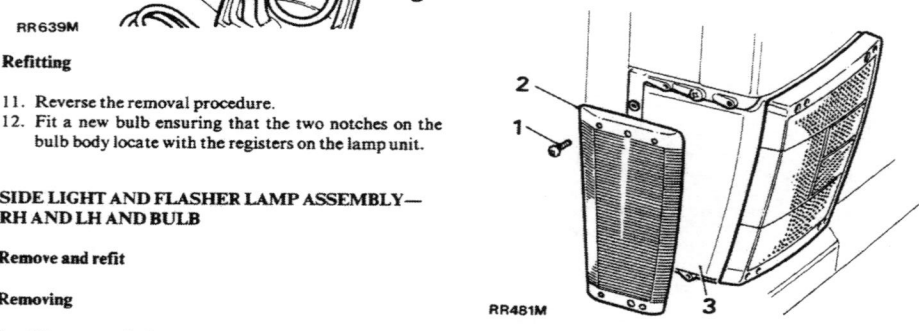

RR481M

### Refitting

4. Reverse the removal procedure.

### Refitting

10. Reverse the removal procedure.

# ELECTRICAL 86

## TAIL, STOP, REVERSE, FOG GUARD AND FLASHER LAMP ASSEMBLY—RH AND LH

### Remove and refit

#### Removing

1. Disconnect the battery.
2. Remove the four lens retaining screws.
3. Remove lens.
4. Remove sealing rubber.

NOTE: To remove the sealing rubber complete it is necessary to remove the side reflector lens.

5. Remove the bulbs.
6. Remove the four screws securing the lamp unit to the body.
7. Remove the two through-screws from the reflector side, which also secure the lamp unit to the body.
8. Ease the lamp unit forward and disconnect leads at moulded connectors.

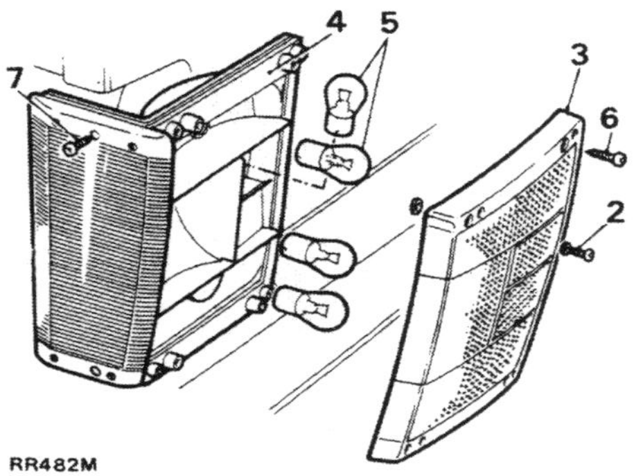

#### Refitting

9. Reverse the removal procedure.

RR482M

## UNDER BONNET LAMP ASSEMBLY

### Remove and refit

#### Removing

1. Disconnect the battery.
2. Remove the two securing screws.
3. Remove the lamp glass.
4. Pull the five-watt 'wedge' type bulb from the bulb holder.
5. Disconnect the electrical leads located below the bonnet lamp switch attached to the inner wing.
6. Pull the rubber grommet off the leads and pull the lamp and leads up through the bonnet stiffener channel.

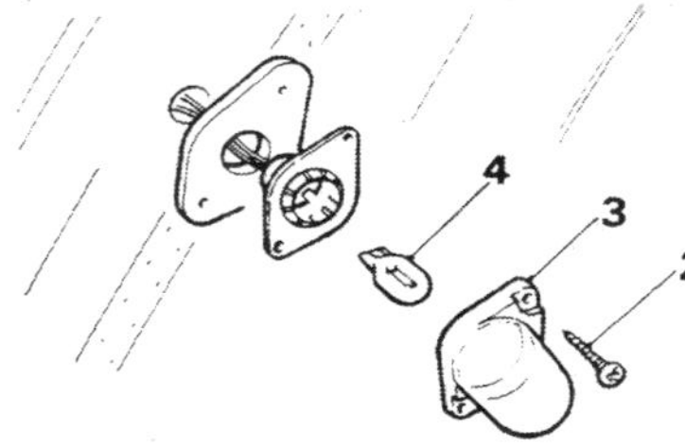

RR483M

#### Refitting

7. Reverse operations 1 to 6.

NOTE: A piece of bent wire will be needed to pull the electrical leads out of the channel exit hole when fitting a new lamp assembly.

## SIDE REPEATER LAMPS

### Remove and refit

#### Removing

1. Disconnect the battery.
2. Remove the single screw retaining the lamp lens.
3. Remove the lamp lens and rubber seal.
4. Remove the four-watt bayonet fitting bulb.
5. Remove the two nuts and spring washers from the rear of the lamp body accessible from behind the front wing.
6. Remove the earth wire from the rear of the lamp.
7. Disconnect the twin snap connector from within the engine compartment located directly behind the lamp.
8. Remove the lamp body from the outer wing.

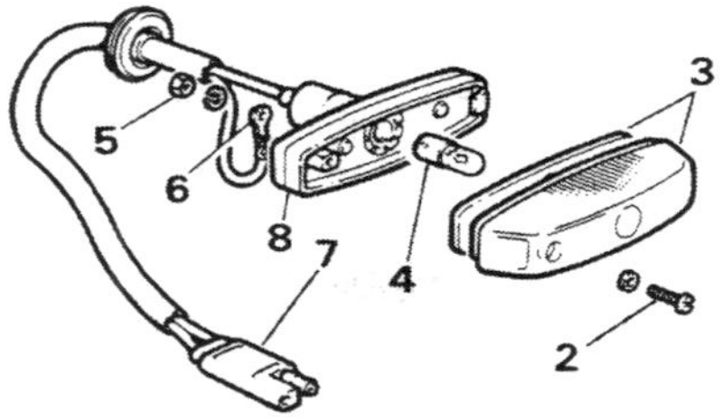

RR457M

#### Refitting

9. Reverse the removal procedure.

20

---

# ELECTRICAL 86

6. Disconnect the door edge lamp and puddle lamp two pin electrical plugs within the door, accessible through the lower centre and outer apertures of the inner door panel.
7. Release the door edge lamp electrical leads from the retaining clips.
8. Prise the lamps out of the door and withdraw the electrical leads.

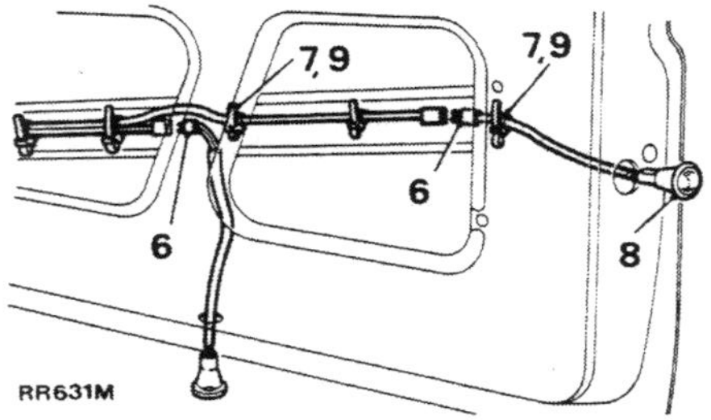

RR631M

#### Refitting

9. Reverse the removal procedure.

NOTE: Ensure the door edge lamp wiring harness is securely clipped to the lower stiffener plate within the door to prevent damage occurring to the electrical leads when the door glass is in its lowest position.

## DOOR EDGE LAMPS/PUDDLE LAMPS

### Bulb replacement

1. Disconnect battery.
2. Carefully prise out the lamp lens.
3. Withdraw the lamp body from the door ONLY as far as the electrical leads will permit.
4. Pull the bulb from the holder.

## HEATER/VENTILATION CONTROL PANEL

### Bulb replacement

The heater/ventilation control panel is illuminated by four 12-volt 1.2-watt 'wedge' type (capless) bulbs. In the event of a bulb failure a replacement bulb can be fitted as follows:

1. Pull the four finger tip knobs off the control levers.
2. Remove the two screws at the top of the panel.
3. Carefully ease the panel away from the centre console only as far as the electrical leads will permit.
4. Pull the appropriate bulb holder out of the rear of the panel.
5. Pull the bulb from the holder.
6. Renew the bulb and push the bulb holder firmly back into its location at the rear of the panel.

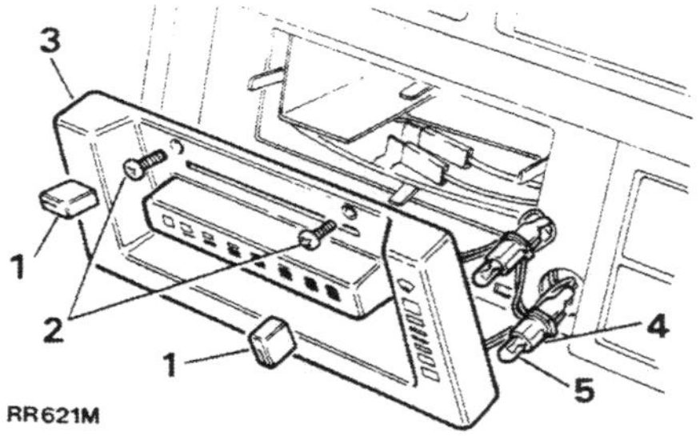

RR621M

#### Refitting

7. Ensuring that the electrical leads do not become trapped between the panel console and operating levers, refit the panel.

## DOOR EDGE LAMPS/PUDDLE LAMPS

Incorporated into the front door assemblies are door edge lamps and puddle lamps, these are located on the shut face and bottom of the door. The lamps are activated by the courtesy light switches when either front door is opened and will immediately switch off when both doors are closed.

### Remove and Refit

#### Removing

1. Ensure the side door glass is fully closed.
2. Disconnect the battery.
3. Remove the interior door handle and arm rest/door pull from the door.
4. Carefully release the interior door trim pad from the inner door panel.
5. Peel back the lower half of the plastic weather sheet.

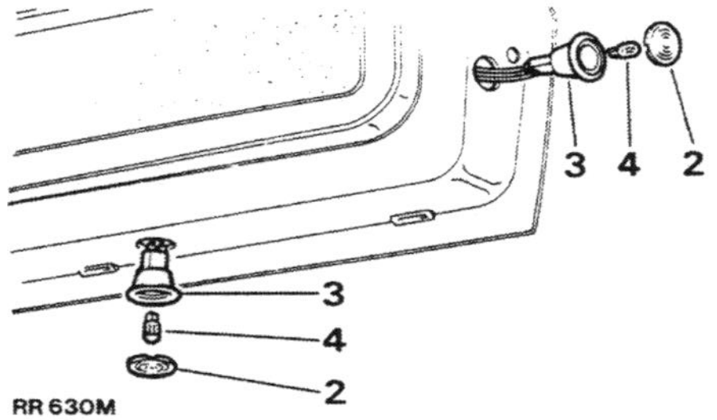

RR630M

5. Renew the bulb and refit the lamp lens.
6. Push the lamp into the door.
The correct bulb type is a 12-volt 10/5-watt capless.

21

# ELECTRICAL 86

## AUTOMATIC GEAR SELECTOR—PANEL ILLUMINATION

### Bulb replacement

1. Disconnect the battery.
2. Unclip the cover from the top of the gear selector knob.
3. Remove the circlip retaining the detent button.
4. Withdraw the detent button.
5. Remove the lower circlip above the gear selector knob securing nut.
6. Remove the securing nut.
7. Withdraw the serrated washer.
8. Slide the selector knob off the shaft.

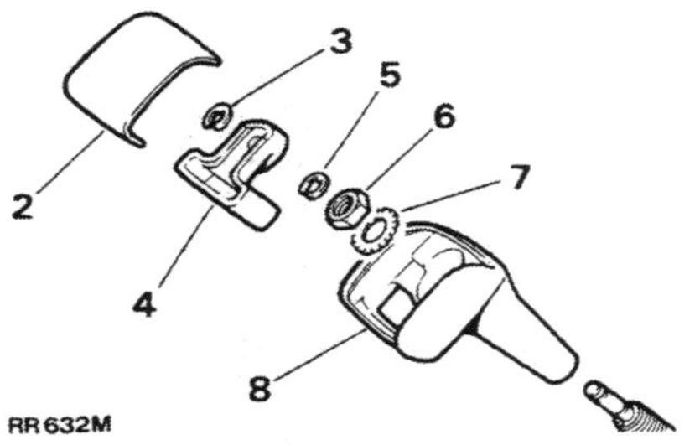

RR632M

9. Carefully prise the inset panel out of the floor mounted console, complete with selector illumination panel and ash tray.
10. The two illumination bulbs are located on the reverse side of the illumination panel.
11. Pull the appropriate bulb holder from its location.
12. If necessary, to facilitate easier removal of the bulb holders, remove the four screws securing the illumination panel to the outer surround panel.

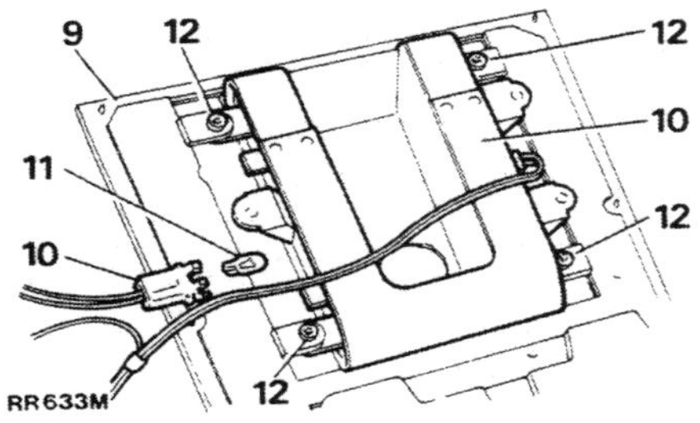

RR633M

13. Pull the bulb from the holder. The correct bulb type is a 24-volt 5-watt 'wedge' base (capless).

### Refitting

14. Reverse the removal procedure ensuring that the electrical leads beneath the floor mounted console do NOT become trapped between mating surfaces.
15. To prevent damage to the gear selector knob (Automatic only) on reassembly do NOT overtighten the retaining nut, see data section for correct torque.

## AIR CONDITIONING CONTROL PANEL ILLUMINATION—Bulb replacement

1. Disconnect the battery.
2. Pull the finger tip control knobs off the control levers.
3. Carefully prise the four air vents out of the fascia panel.
4. Remove the fifteen screws securing the panel to the upper and lower panels.
5. Withdraw the panel only as far as the electrical leads will permit.
6. Pull the bulb holder from the rear right-hand side of the control panel.
7. Pull the bulb from the holder.

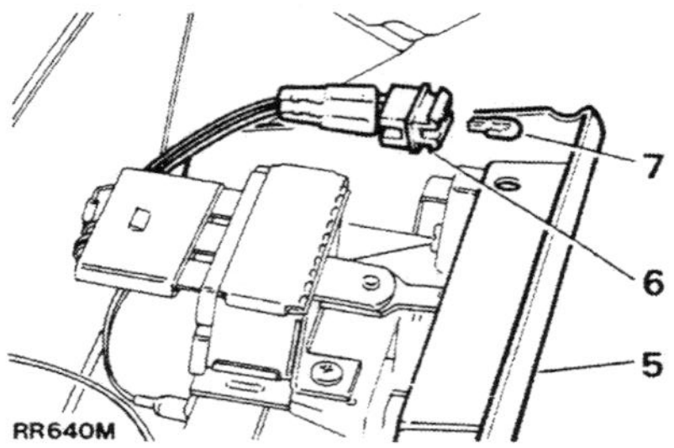

RR640M

### Refitting

8. Reverse the removal procedure. The correct bulb type is a 12-volt 1.2-watt 'wedge' base (capless).

## NUMBER PLATE LAMP ASSEMBLY AND BULB REPLACEMENT

### Remove and refit

### Removing

1. Disconnect the battery.
2. Remove the two self-tapping screws and fibre washers.
3. Detach the lens surround and lamp lens.
4. Remove the bulb.

*Continued*

## INTERIOR ROOF LAMPS CIRCUIT DELAY

### Remove and refit

The roof lamp circuit incorporates a delay function which is designed to allow the lamps to remain on for 12 to 18 seconds after either of the front doors are closed.

NOTE: Switching on the ignition (with both doors closed) will immediately over-ride this feature, switching the interior lamps off.

### Removing

1. Disconnect the battery.
2. Remove the six screws securing the lower fascia panel.
3. Lower the fascia panel to gain access to the red delay unit attached to the steering column support bracket.
4. Remove the delay unit by pushing the unit up off its retaining bracket, to clear the steering column support bracket.
5. Pull the red multi-plug off the delay unit.

### Refitting

6. Reverse the removal operations.

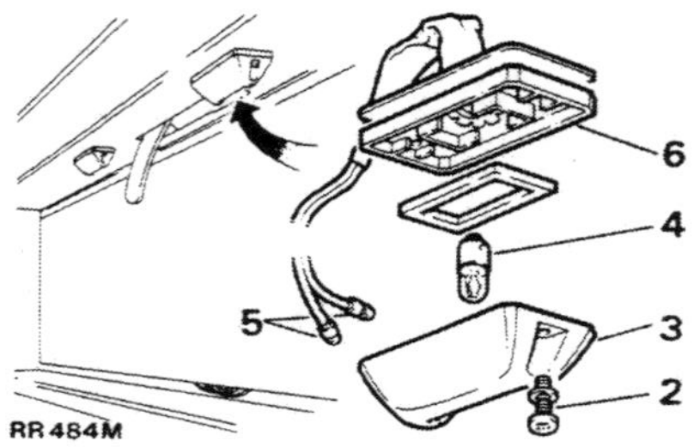

RR484M

### Refitting

8. Reverse the removal procedure.

## INTERIOR ROOF LAMPS

### Remove and refit

The interior roof lamps are operated automatically via the side door and tailgate courtesy switches or by an independent switch located on the auxiliary switch panel.

### Removing

1. Disconnect the battery.
2. Remove the lens from the courtesy lamp by pressing upward and turning it anti-clockwise.
3. Withdraw bulb from spring clip holder.
4. Remove screws securing lamp base to roof panel.
5. Lower the lamp to reveal the cable snap connections.
6. Disconnect the electrical connections.

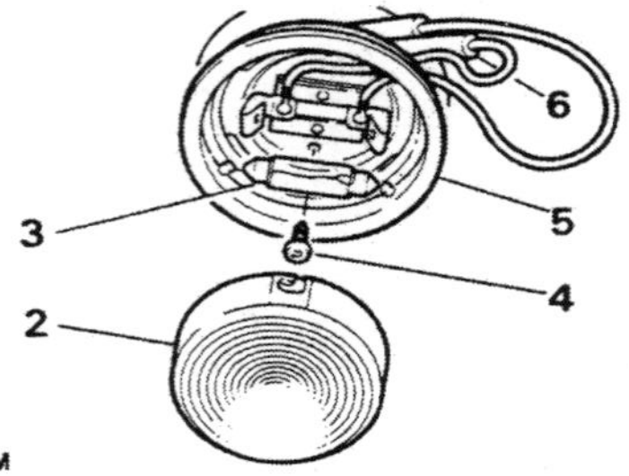

RR485M

### Refitting

7. Reverse the removal procedure.

# ELECTRICAL 86

## STARTER MOTOR

### Remove and refit

#### Removing

1. Place car on a suitable ramp.
2. Disconnect the battery.
3. Disconnect the leads from the solenoid and starter motor and remove the exhaust heat shield.
4. Remove the two bolts securing the starter motor to the flywheel housing.
5. Remove starter motor from underneath the vehicle.
6. Remove the two bolts securing the solenoid to the starter motor.
7. Withdraw the solenoid from the solenoid housing.

#### Refitting

8. Reverse the removal procedure.

RR 489M

## STARTER SOLENOID

### Remove and refit

#### Removing

1. Place car on a suitable ramp.
2. Disconnect the battery.
3. Disconnect the leads from the solenoid and starter motor and remove the exhaust heat shield.
4. Remove the two bolts securing the starter motor to the flywheel housing.
5. Remove the starter motor from underneath the vehicle.
6. Remove the two bolts securing the solenoid to the starter motor. 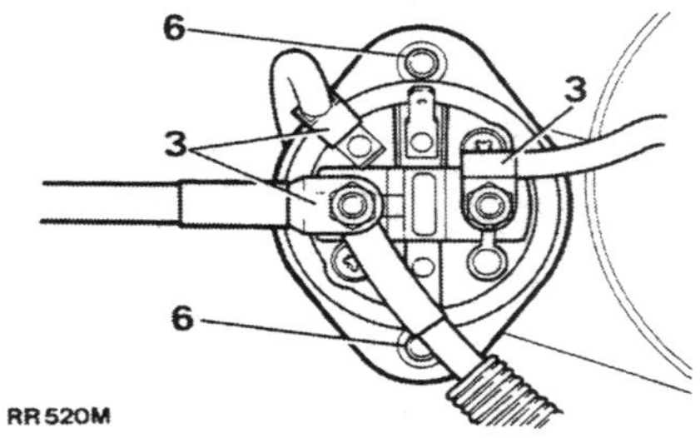
7. Withdraw the solenoid from the solenoid housing.

#### Refitting

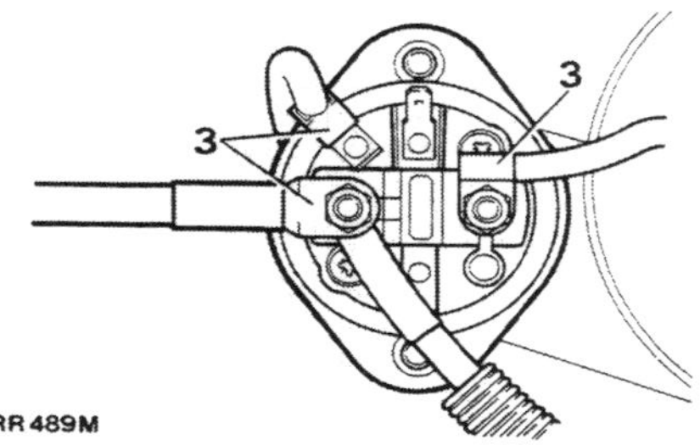

6. Reverse the removal procedure.

RR 520M

---

## STARTER MOTOR—Lucas 3M100PE

### Overhaul

#### Dismantling

1. Remove the starter motor.
2. Remove the connecting link between the starter and the solenoid terminal 'STA'.
3. Remove the solenoid from the drive end bracket.
4. Grasp the solenoid plunger and lift the front end to release it from the top of the drive engagement lever.
5. Remove the end cap seal.
6. Using an engineer's chisel, cut through a number of the retaining ring claws until the grip on the armature shaft is sufficiently relieved to allow the retaining ring to be removed.
7. Remove the two through bolts.
8. Partially withdraw the commutator end cover and disengage the two field coil brushes from the brush box.
9. Remove the commutator end cover.
10. Withdraw the yoke and field coil assembly.
11. Remove the retaining ring from the drive engagement lever pivot-pin, using the method previously described.
12. Withdraw the pivot pin.
13. Withdraw the armature.
14. Using a suitable tube, remove the collar and jump ring from the armature shaft.
15. Slide the thrust collar and the roller clutch drive and lever assembly off the shaft.
16. Remove the intermediate bracket and seals.

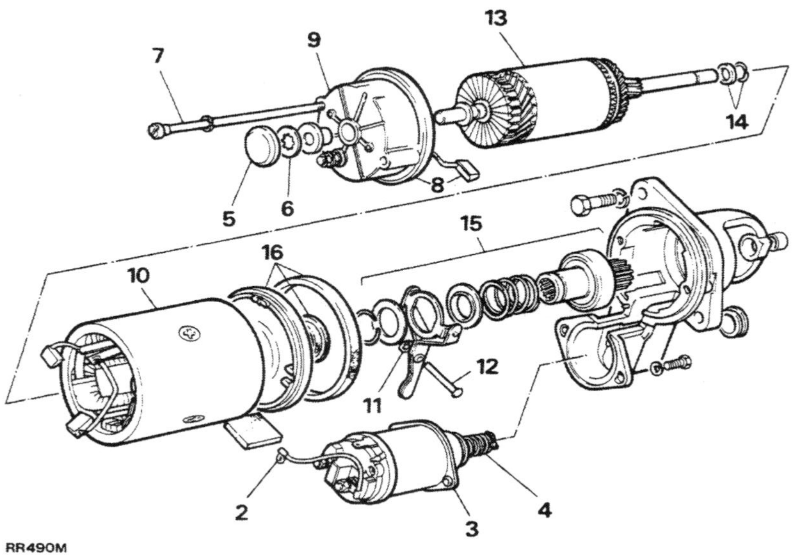

RR490M

*Continued*

# ELECTRICAL  86

## Inspecting

### Clutch

17. Check that the clutch gives instantaneous take-up of the drive in one direction and rotates easily and smoothly in the other direction.
    Ensure that the clutch is free to move round and along the shaft splines without any tendency to bind.

    **NOTE: The roller clutch drive is sealed in a rolled steel cover and cannot be dismantled.**

18. Lubricate all clutch moving parts with Shell SB 2628 grease for cold and temperate climates or Shell Retinax 'A' for hot climates.

### Brushes

19. Check that the brushes move freely in the brush box moulding. Rectify sticking brushes by wiping with a petrol moistened cloth.

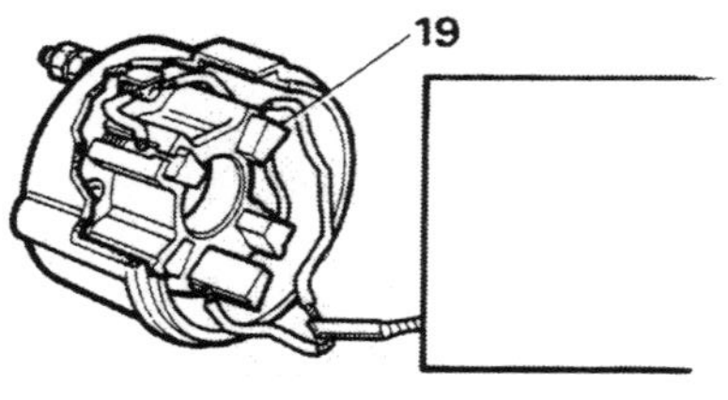

*RR 491M*

20. Fit new brushes if they are damaged or worn to approximately 9.5 mm (0.375 in).
21. Using a push-type spring gauge, check the brush spring pressure. With new brushes pushed in until the top of the brush protrudes about 1.5 mm (0.065 in) from the brush box moulding, the spring pressure reading should be 1.0 kgf (36 ozf).

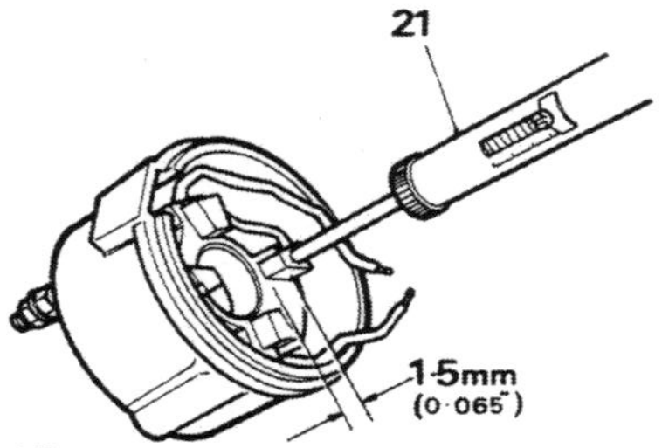

*RR 492M*

22. Check the insulation of the brush springs by connecting a 110-volt a.c. 15-watt test lamp between a clean part of the commutator end cover and each of the springs in turn. The lamp should not light.

### Armature

23. Check the commutator. If cleaning only is necessary, use a flat surface of very fine glass paper, and then wipe the commutator surface with a petrol moistened cloth.
24. If necessary, the commutator may be machined providing a finished surface can be obtained without reducing the thickness of the commutator copper below 3.5 mm (0.140 in), otherwise a new armature must be fitted. Do not undercut the insulation slots.
25. Check the armature insulation by connecting 110-volt a.c. 15-watt test lamp between any one of the commutator segments and the shaft. The lamp should not light, if it does light fit a new armature.

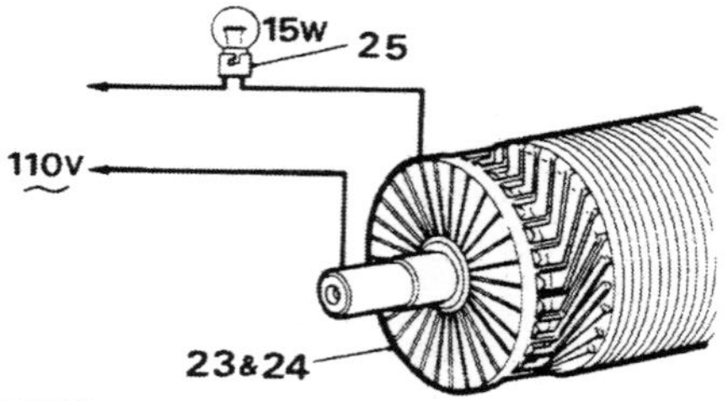

*RR 493M*

### Field coil insulation

26. Disconnect the end of the field winding where it is riveted to the yoke, by filing away the riveted over end of the connecting-eyelet securing rivet, sufficient to enable the rivet to be tapped out of the yoke.

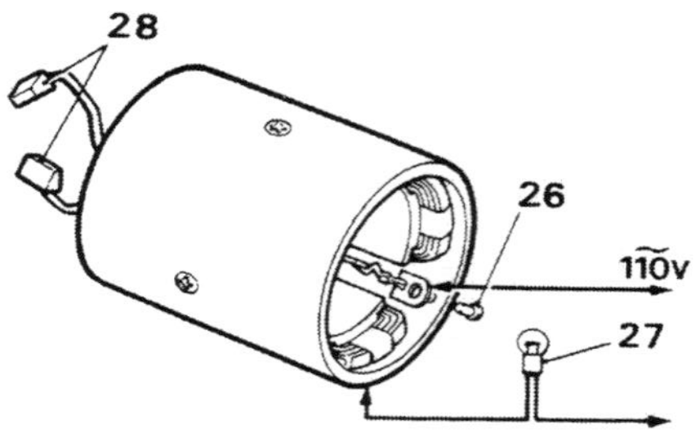

*RR 494M*

27. Connect a 110-volt a.c. 15-watt test lamp between the disconnected end of the winding and a clean part of the yoke.
28. Ensure that the brushes or bare parts of their flexibles are not touching the yoke during the test.
29. The lamp should not light, if it does light, fit a new field coil assembly.
30. Resecure the end of the field winding to the yoke.

### Field coil continuity

31. Connect a 12-volt battery operated test lamp between each of the brushes in turn and a clean part of the yoke.
32. The lamp should light, if it does not light, fit a new field coil assembly.

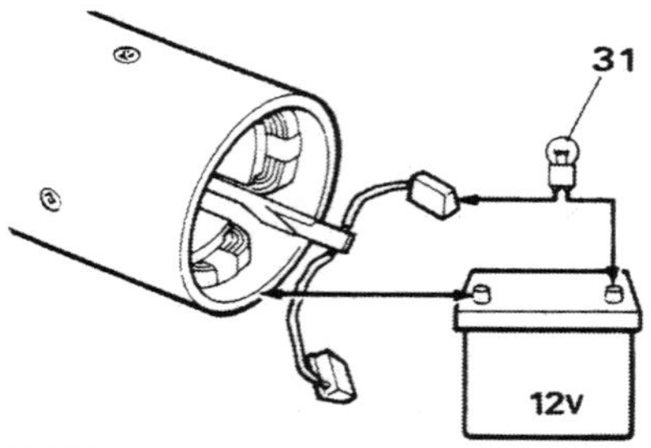

*RR 495M*

### Solenoid

33. Disconnect all cables from the solenoid terminals and connectors.
34. Connect a 12-volt battery and a 12-volt 60-watt test lamp between the solenoid main terminals. The lamp should not light, if it does light, fit new solenoid contacts or a new solenoid complete.
35. Leave the test lamp connected and, using the same 12-volt battery supply, energise the solenoid by connecting 12-volt between the small solenoid operating 'Lucar' terminal blade and a good earth point on the solenoid body.

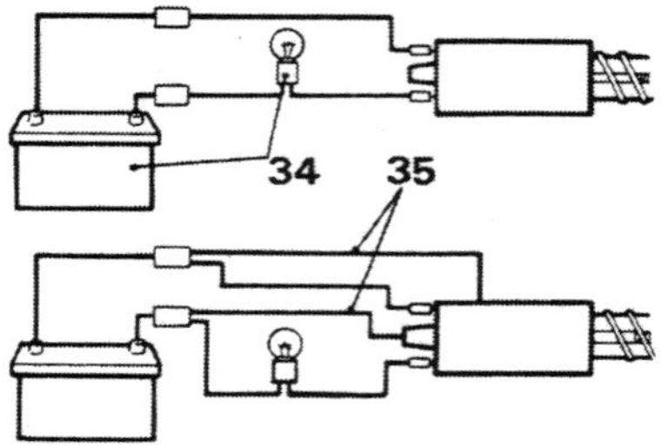

36. The solenoid should be heard to operate and the test lamp should light with full brilliance, otherwise fit new solenoid contacts or a new solenoid complete.

## Reassembling

37. Reverse 1 to 15, including the following:
38. Fit the commutator end cover before refitting the solenoid to facilitate assembly of the block shaped grommet which, when assembled, is compressed between the yoke, solenoid and fixing bracket.
39. Ensure that the internal thrust washer is fitted to the commutator end of the armature shaft.
40. Tighten the through bolts, solenoid fixing and upper terminal nuts to the correct torque—see Torque Wrench Settings.
41. Set the armature end float by driving the retaining ring on the armature shaft into a position that provides a maximum of 0.25 mm (0.010 in) clearance between the retaining ring and the bearing bush shoulder.

*RR 496M*

## ELECTRICAL 86

### FUSE BOX

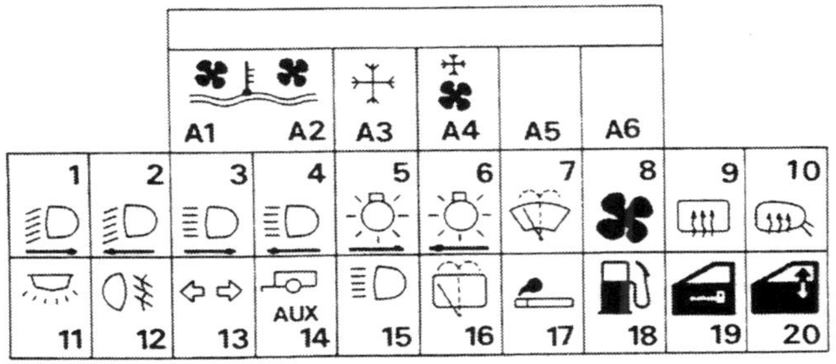

| | A1 | A2 | A3 | A4 | A5 | A6 |
|---|---|---|---|---|---|---|
| | ❋ | ❋ | ✣ | ✣ | | ☘ |
| 1 | 2 | 3 | 4 | 5 | 6 | 7 | 8 |
| 11 | 12 | 13 | 14 | 15 | 16 | 17 | 18 | 19 | 20 |

RR622M

**AUXILIARY FUSE PANEL** —applicable to Air conditioning only

| Fuse No. | Colour Code | Fuse Value | Circuit Served | Ignition Key Controlled |
|---|---|---|---|---|
| A1 | Yellow | 20 amp | Air Conditioning fan—option | IGN |
| A2 | Yellow | 20 amp | Air Conditioning fan—option | IGN |
| A3 | Tan | 5 amp | Air Conditioning Compressor clutch—option | IGN |
| A4 | Yellow | 20 amp | Air Conditioning Blower Motor—option | IGN |
| A5 | — | — | Spare | — |
| A6 | — | — | Spare | — |

**MAIN FUSE PANEL**

| Fuse No. | Colour Code | Fuse Value | Circuit Served | Ignition Key Controlled |
|---|---|---|---|---|
| 1 | Brown | 7.5 amp | RH headlamp dipped beam | — |
| 2 | Brown | 7.5 amp | LH headlamp dipped beam | — |
| 3 | Brown | 7.5 amp | RH headlamp main beam | — |
| 4 | Brown | 7.5 amp | LH headlamp main beam | — |
| 5 | Tan | 5 amp | RH side and panel lamps | — |
| 6 | Tan | 5 amp | LH side lamps | — |
| 7 | Blue | 15 amp | Front wash/wipe motor | Aux |
| 8 | Yellow | 20 amp | Heater motor | Ign |
| 9 | Blue | 15 amp | Heated rear screen | Ign |
| 10 | Violet | 3 amp | Electric mirror heating element—option | — |
| 11 | Blue | 15 amp | Interior lights, under bonnet illumination, clock, headlamp flash, horns | — |
| 12 | Red | 10 amp | Rear fog guard (from dipped headlamps) | — |
| 13 | Blue | 15 amp | Directional indicators, stop lights, reverse lights, electric mirror motors | Ign |
| 14 | Blue | 15 amp | Auxiliary circuit to trailer | — |
| 15 | Blue | 15 amp | Auxiliary driving lamps | — |
| 16 | Red | 10 amp | Rear wash/wipe motor | Aux |
| 17 | Yellow | 20 amp | Cigar lighters (front and rear) | Ign |
| 18 | Red | 10 amp | Fuel pump | Ign |
| 19 | Red | 10 amp | Central door locking—option | — |
| 20 | White | 25 amp | Electric window lifts—option | Aux |

NOTE: Radio/Cassette combination—option. An in-line type 7 amp fuse is incorporated into the power input lead of the unit.

---

## ELECTRICAL 86

### RELAYS—Identification

Incorporated in the vehicle electrical circuits are several relays, some of which are located behind the lower fascia panel attached to the steering column support bracket. The remaining relays are located in the engine compartment attached to the closure panel, these relays are accessible having removed the black protective cover.

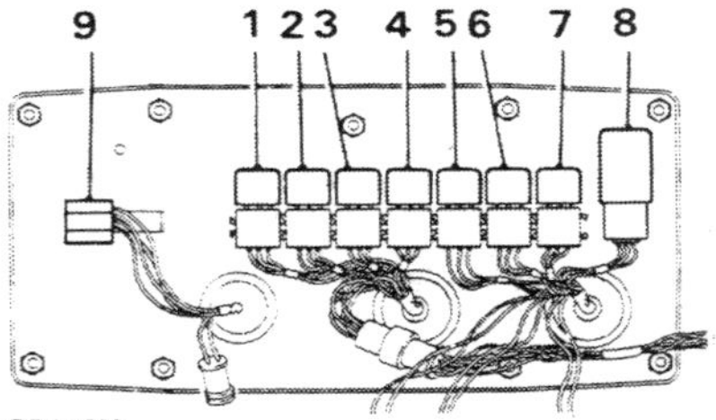

RR618M

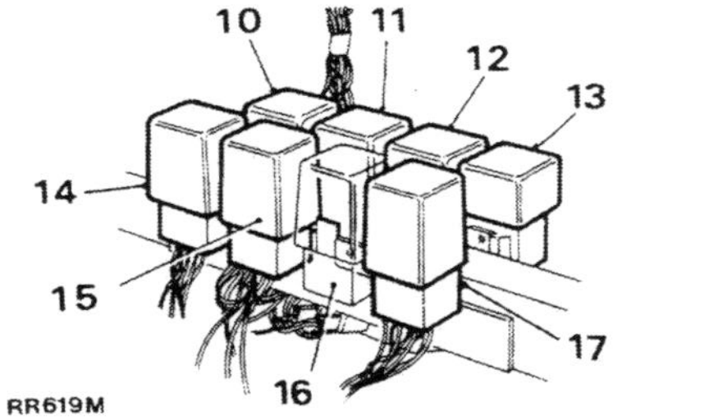

RR619M

Viewed from inside the vehicle with the lower fascia removed.

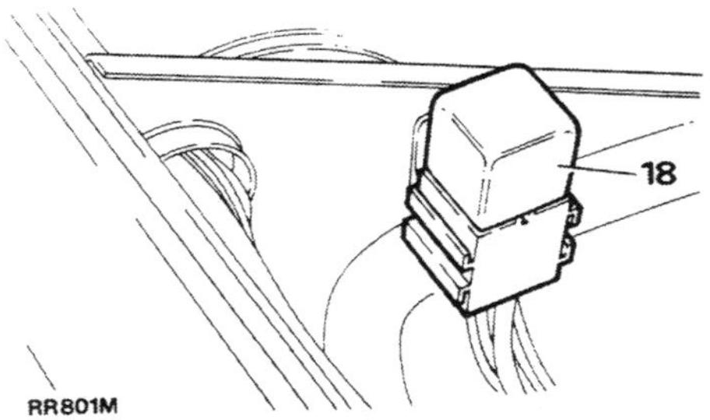

RR801M

Fuel cut-off relay (carburetter models only) located in the engine compartment attached to the left hand valance stiffener bracket.

Closure panel viewed from the engine compartment. (RH drive vehicle illustrated—LH drive is a mirror image of RH drive).

**NOTE: Refer to fuel injection section of manual (Book 2) for full information E.F.I. related to relays.**

| | RELAY | CIRCUIT DIAGRAM ITEM NUMBER | | COLOUR |
|---|---|---|---|---|
| 1. | Air conditioning controlled fan relay | 9. | On air conditioning circuit diagram | Natural |
| 2. | Air conditioning relay (ignition controlled) | 10. | Air conditioning circuit diagram | Natural |
| 3. | Compressor clutch relay | 8. | On air conditioning circuit diagram | Natural |
| 4. | Fan relay | 7. | On air conditioning circuit diagram | Natural |
| 5. | Starter solenoid relay | 122. | On main circuit diagram | Natural |
| 6. | Heated rear window relay | 126. | On main circuit diagram | Natural |
| 7. | Brake failure warning check relay | 103. | On main circuit diagram | Natural |
| 8. | Headlamp wash timer unit | 73. | On main circuit diagram | Natural |
| 9. | Ignition pick-up point | — | | — |
| 10. | Hazard and flasher relay | 96. | On main circuit diagram (only fitted to vehicles with air conditioning or split charge facility) | Blue |
| 11. | Voltage sensitive switch | 5. | On split charge circuit diagram | Yellow |
| 12. | Interior lamp delay | 129. | On main circuit diagram | Red |
| 13. | Auxiliary driving lamps relay | 38. | On main circuit diagram | Natural |
| 14. | Rear wiper delay | 43. | On main circuit diagram | Black |
| 15. | Rear wiper relay | 136. | On main circuit diagram | Black |
| 16. | Speed warning unit (Saudi only) | 138. | On main circuit diagram | Green |
| 17. | Front wiper delay | 78. | On main circuit diagram | Black |
| 18. | Fuel cut off relay (carburetter models) | 121. | On main circuit diagram | Natural |

---

### FUSE BOX—Main and Auxiliary

**Remove and refit**

**Remaining**

1. Disconnect the battery.
2. Remove the clip-on fuse box cover
3. Remove the fuses from the main and auxiliary fuse boxes.

NOTE: The fuses in the six-way auxiliary fuse box will only be present when air-conditioning is fitted to the vehicle.

4. Remove the single screw securing the top auxiliary fuse box to the fuse box surround.
5. Unclip the opposite end of the fuse box.
6. Remove the two screws securing the main fuse box to the lower centre fascia panel.
7. Withdraw the auxiliary fuse box surround.
8. Manoeuvre the main and auxiliary fuse box to enable them to be withdrawn through the fuse box aperture.
9. Remove the leads from the fuse boxes, by inserting a small screwdriver into each back of the lucar connections, withdraw the leads from the rear of the fuse box.

**Refitting**

10. Reverse the removal instructions ensuring that all leads are refitted to the correct fuse socket (refer to main circuit diagram).

NOTE: When refitting the leads to the fuse box, the retaining tabs on the back of the lucar connectors must be in their raised position to prevent the leads being pushed out of the rear of the fuse box when the fuse is refitted.

# ELECTRICAL

**RELAYS**—(Mounted on the engine compartment closure panel).

## Remove and refit

### Removing

1. Lift the bonnet.
2. Disconnect the battery.
3. Remove the bolt securing the relay protective cover, located on the front of the engine compartment closure panel.
4. Remove the cover.
5. Pull the appropriate relay off its multi-plug.

### Refitting

6. Reverse the removal procedure.

**RELAYS**—(Mounted on the steering column support bracket).

## Remove and Refit

### Removing

1. Disconnect the battery.
2. Remove the six screws securing the lower fascia panel.
3. Lower the fascia panel, disconnect the electric leads from the dimming control switch and remove the fascia panel.
4. Locate the appropriate relay on the relay mounting bracket, carefully pull the relay off the multi-plug.

### Refitting

5. Reverse the removal procedure.

## AUXILIARY SWITCH PANEL

The auxiliary switch panel contains four 'push-push' type switches (five when auxiliary driving lamps are fitted) which incorporate integral symbols for identification. (The sixth switch aperture is fitted with a blank cover, which is removable, to facilitate the fitting of an extra switch if required).
The symbols are illuminated by two bulbs which become operational when the vehicle lights are on.

The heated rear screen and hazard warning switches (2 and 5) are also provided with individual warning lights, illuminated when the switches are operated.

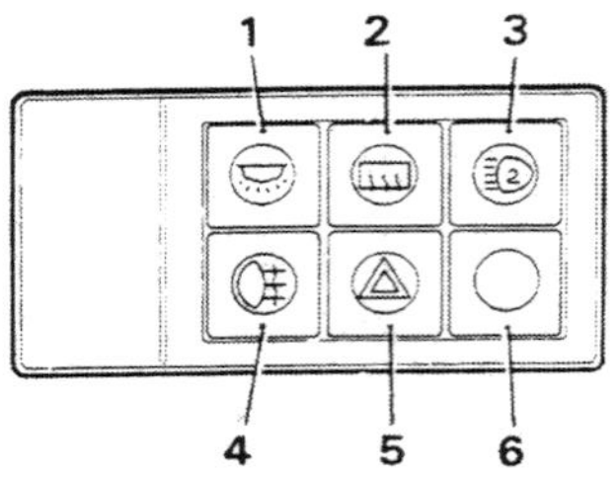

RR627M

1. Interior roof and tailgate lights.
2. Heated rear screen.
3. Auxiliary driving lamps—option.
4. Rear fog guard lamps.
5. Hazard warning.
6. Blank.

## AUXILIARY SWITCH PANEL

### Remove and Refit

#### Removing

1. Disconnect the battery.
2. Carefully prise the auxiliary switch panel surround away from the centre console.
3. Withdraw the switch panel as far as the electrical leads will permit.
4. Unclip the multi-plugs at the rear of the switches by depressing the retaining lugs.
5. Pull the plugs from the switches.
6. Remove the switch assembly complete.

**NOTE: If necessary each individual switch can now be removed as follows.**

7. Depress the small retaining lugs on the top and bottom of the switch and push the switch(es) through the front of the switch surround.

#### Refitting

8. Reverse the removal procedure.

**NOTE: To aid identification and location of multi-plug to switch a coloured plastic tab is attached to each switch body which corresponds with an appropriate coloured multi-plug.
The switches if removed, should always be refitted in their original position.**

*Continued*

---

## Auxiliary switch panel warning lights—bulb replacement (switches 2 and 5).

### To replace either bulb

1. Disconnect the battery.
2. Carefully prise the switch panel surround away from the centre console.
3. Unclip the multi-plug from the rear of the appropriate switch and disconnect the plug.
4. The warning light bulb is located in the multi-plug and is removed by pulling the bulb from its location.
5. Renew the bulb and refit the multi-plug.

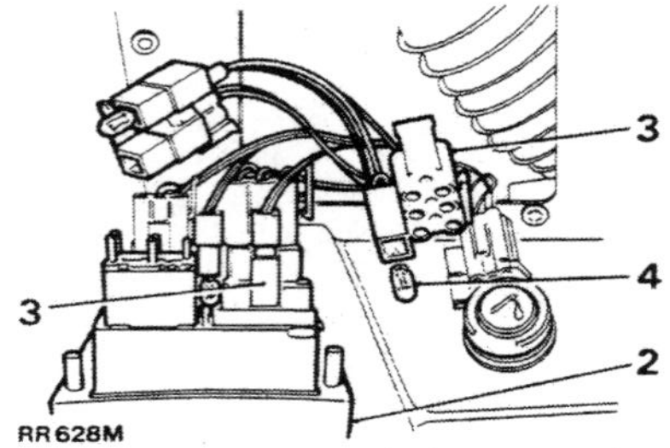

6. Press the auxiliary switch panel back into the centre console.
Red bulb—Hazard warning/Amber bulb—Heated rear screen. The correct bulb type is a 12-volt 1.2-watt 'wedge' base (capless).

### Auxiliary switch panel illumination

The auxiliary panel green illumination bulbs are located in the rear fog and hazard warning multi-plugs, each bulb is positioned in the centre of a group of four switches.

1. Disconnect the battery.
2. Carefully prise the switch panel surround away from the centre console to give access to the multi-plugs at the rear of the switches.
3. Unclip and pull the multi-plug from the rear of the appropriate switch.
4. Pull the green illumination bulb from its location.

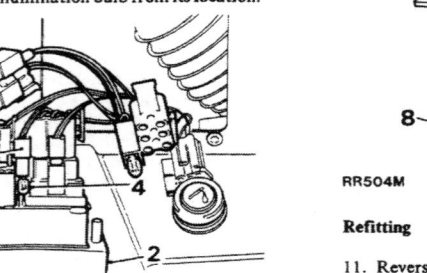

RR628M

5. Renew the bulb and refit the multi-plug.
6. Press the auxiliary panel surround back into the centre console.

The correct bulb type is a 12-volt 1.2-watt 'wedge' base (capless).

## IGNITION STARTER SWITCH

### Remove and refit

#### Removing

1. Disconnect the battery.
2. Remove the steering wheel centre cover.
3. Remove the lower fascia panel.
4. Remove the four screws securing bottom shroud to top shroud.
5. Remove the single screw securing top shroud to switch housing bracket.
6. Remove top shroud and lower the bottom shroud.

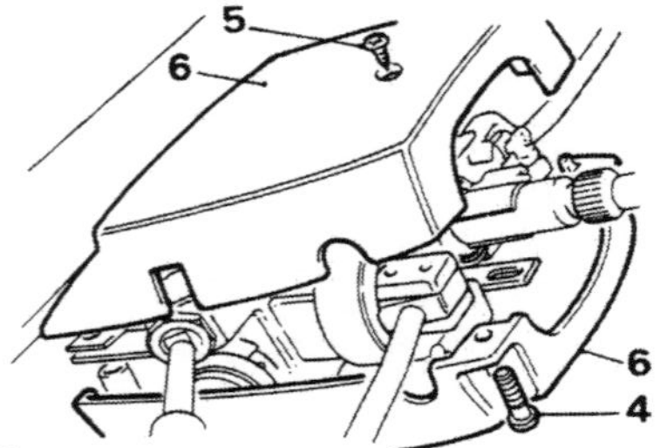

RR503M

7. Disconnect the ignition switch cable at the multi-plug.
8. Remove the rubber cover protecting the switch.
9. Remove the single screw securing the ignition/starter switch to the housing.
10. Withdraw the switch.

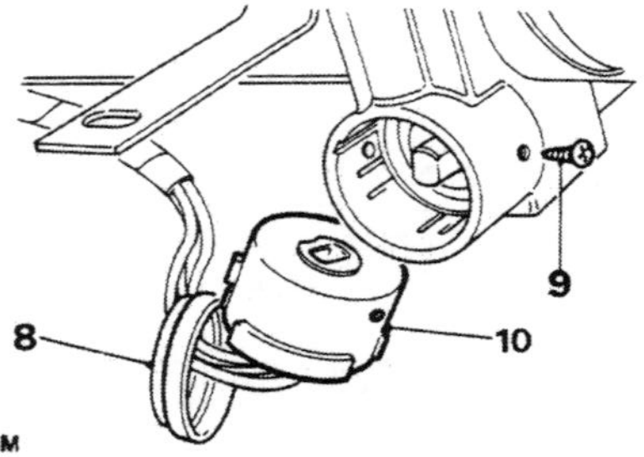

RR504M

### Refitting

11. Reverse the removal procedure.

## ELECTRICAL 86

### STEERING COLUMN CONTROLS

The steering column switch layout has been standardised for left- and right-hand drive vehicles and is as follows:—

### LEFT-HAND CONTROLS

Lower switch—Main lighting switch.
Upper switch—Main and dipped beam, direction indicators and horn.

### RIGHT-HAND CONTROLS

Lower switch—Rear screen programmed/wash/wipe.
Upper switch—Windscreen programmed wash/wipe.

### MAIN LIGHTING SWITCH

**Remove and refit**

Removing

1. Disconnect the battery.
2. Remove the steering wheel centre cover.
3. Remove the lower fascia panel.
4. Remove the four screws securing bottom shroud to top shroud.
5. Remove the single screw securing top shroud to switch housing bracket.
6. Remove top shroud, and lower the bottom shroud.
7. Disconnect cables at snap connectors.
8. Loosen the switch retaining lock-nut.
9. Slide switch unit away from its bracket.

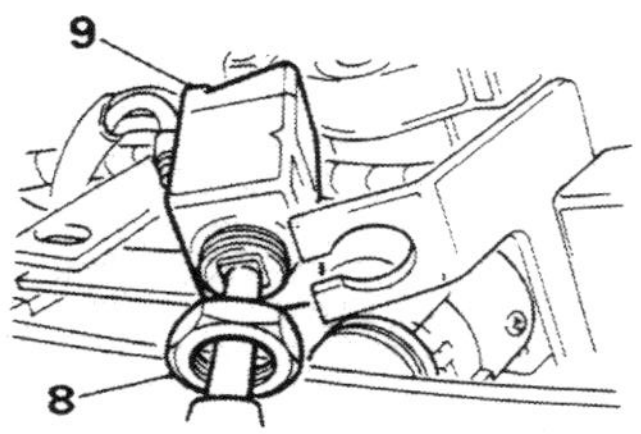

RR505M

Refitting

10. Reverse the removal procedure.

### WINDSCREEN PROGRAMMED WASH WIPE SWITCH

### MAIN AND DIPPED BEAM, DIRECTION INDICATORS AND HORN SWITCH

**Remove and refit**

Removing

1. Disconnect the battery.
2. Remove the steering wheel centre cover.
3. Remove the lower fascia panel.
4. Remove the four screws securing the bottom shroud to the top shroud.
5. Remove the single screw securing the top shroud to the switch housing bracket.
6. Remove the top shroud and lower the bottom shroud.
7. Disconnect the electrical leads at the multi-plugs.
8. Remove the two screws securing the windscreen wash/wipe switch to the column switch bracket.
9. Remove the switch to give access to the screws securing the main and dipped beam switch.
10. Release the two screws securing the upper switch to the switch bracket.
11. Slide the switch and bracket off the steering column.

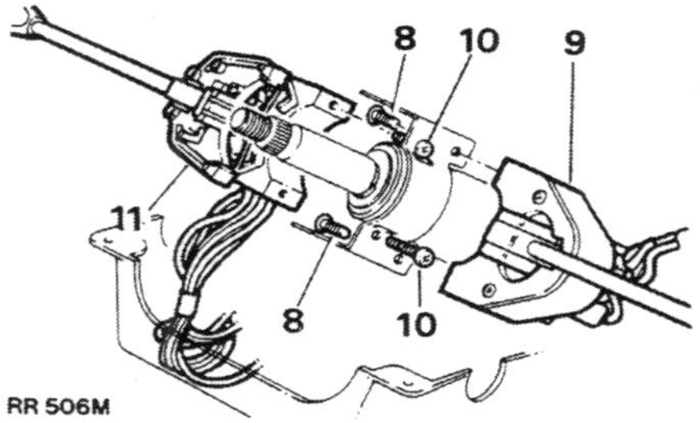

RR 506M

Refitting

12. Reverse the removal procedure.

### REAR TAILGATE SWITCH

**Remove and refit**

Removing

1. Disconnect the battery.
2. Remove the single screw securing the switch to the tailgate aperture.
3. Withdraw the switch.
4. Disconnect the electrical lead.

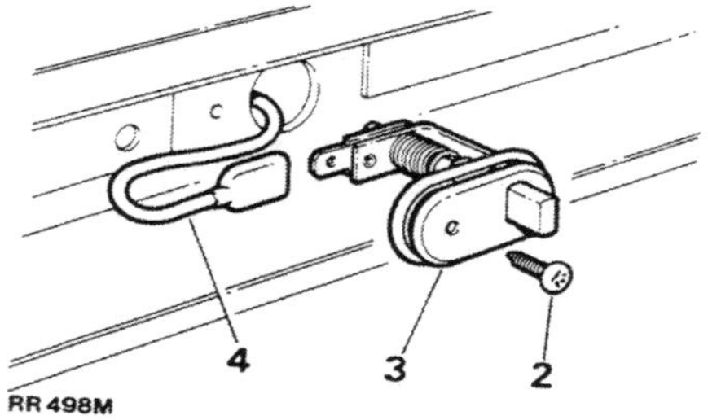

RR 498M

Refitting

5. Reverse the removal procedure.

### REAR SCREEN PROGRAMMED WASH WIPE SWITCH

**Remove and refit**

Removing

1. Disconnect the battery.
2. Remove the steering wheel centre cover.
3. Remove the lower fascia panel.
4. Remove the top and bottom shroud (refer to main and dipped beam switch removal and refit-items 4 to 7).
5. Disconnect the electrical leads at the multi-plug.
6. Remove the four small bolts securing the switch to the switch mounting bracket.
7. Remove the switch from the bracket.

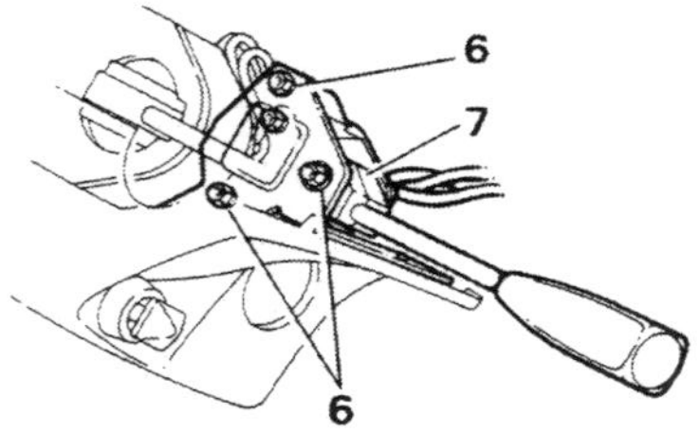

RR 507M

Refitting

8. Reverse the removal procedure.

### DOOR PILLAR SWITCH

**Remove and refit**

Removing

1. Disconnect the battery.
2. Remove screw securing switch to door pillar.
3. Withdraw switch.
4. Disconnect electrical lead from connector blade.

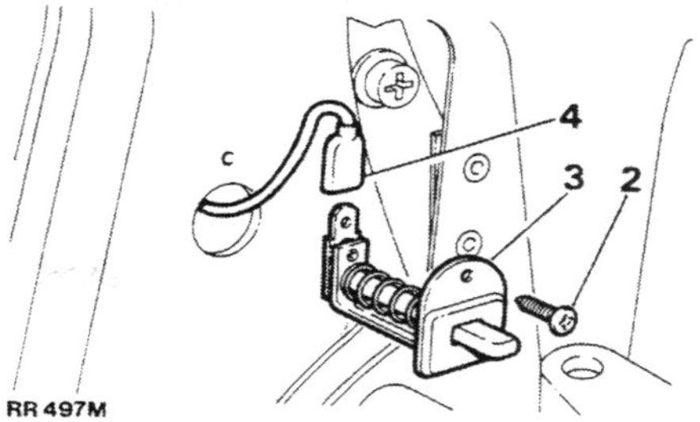

RR 497M

Refitting

5. Reverse removal procedure.

### UNDER BONNET ILLUMINATION SWITCH

**Remove and refit**

Removing

1. Disconnect the battery.
2. Remove the single screw securing the switch to the decker panel.
3. Withdraw the switch.
4. Disconnect the electrical lead.

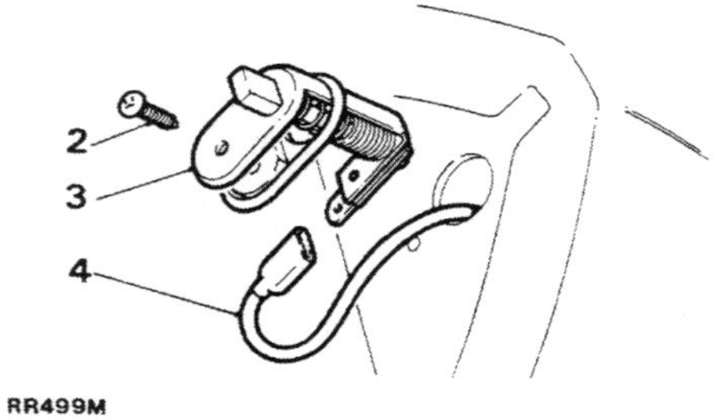

RR499M

Refitting

5. Reverse the removal procedure.

# ELECTRICAL 86

## CIGAR LIGHTER—radio housing

### Remove and refit

### Removing

1. Disconnect the battery.
2. Remove the High/Low range gear knob.
3. Remove the main gearbox knob.
   **Manual gearbox versions**—unscrew the knob.
   **Automatic gearbox versions**—See Automatic gear selector panel illumination.
4. Remove the cubby box liner and release the handbrake cable from the handbrake lever, prise the inset panel out of the floor mounted console.
   **Automatic gearbox versions**—Pull the two illumination bulbs from the selector panel.
5. Release the cubby box from its four floor mounted fixings.
6. Raise the front of the cubby box and console assembly and ease the unit away from the radio housing.

**NOTE: It is necessary to remove the radio from its housing to facilitate easy removal of the radio console. The radio is removed as follows.**

7. Carefully prise the outer surround away from the radio.
8. Slide the retaining clip to the left.
9. Lift the radio out of the housing.
10. Disconnect the electrical connections and aerial plug from the rear of the radio.
11. Remove the single screw securing the housing to the top of the gearbox tunnel.
12. Pull the housing away from the lower fascia.
13. Disconnect the electrical leads at the rear of the cigar lighter.
14. Remove the push in switch from the lighter outer body.
15. Depress the outer plastic surround where denoted by the arrows and push the outer body through the surround.
16. Manoeuvre the plastic surround and remove it from the radio housing.

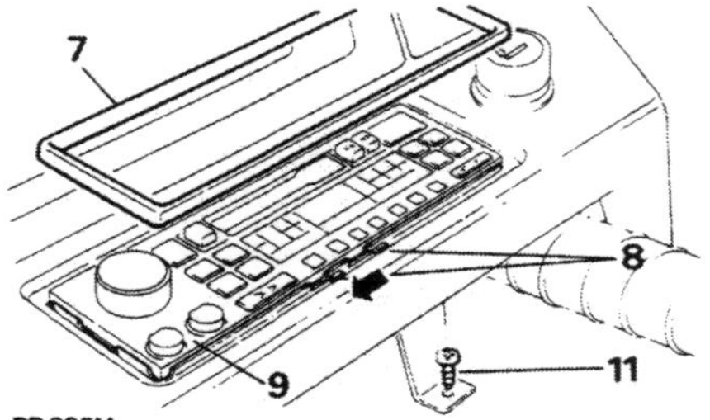

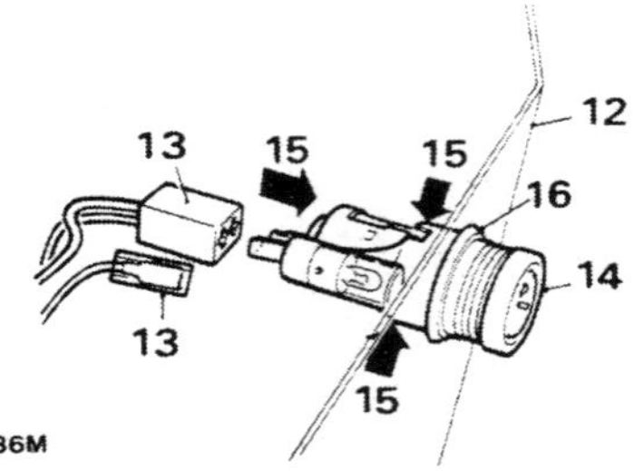

## CIGAR LIGHTER ILLUMINATION—Bulb replacement

17. Remove the bulb holder from the plastic surround.
18. Pull the bulb from the holder. The correct bulb type is a 12v 1.2 watt wedge base (capless).

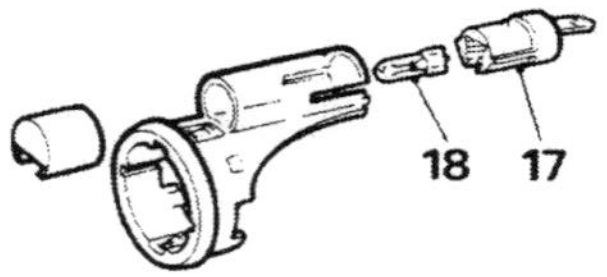

### Refitting

19. Reverse the removal procedure.

## CIGAR LIGHTER—Cubby Box

The rear cigar lighter is located in the bottom of the cubby box, access to the rear of the lighter is gained through heater/air vent duct below the rear ashtray. Follow instructions 13 to 16 of CIGAR LIGHTER—radio housing to remove the lighter from the cubby box.

---

## CIGAR LIGHTER

### Removing

1. Drive the vehicle onto a suitable ramp.
2. Disconnect the battery.
3. Disconnect the multi-plug.
4. Release the clamp bolt and remove the clamp.
5. Withdraw the switch from its location.

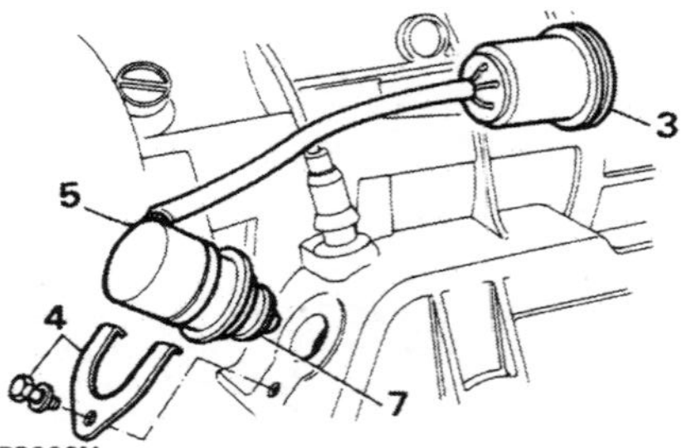

### Refitting

6. Reverse the removal instructions.
7. Fit a NEW 'O' ring to the switch.

## REVERSE LIGHT SWITCH—MANUAL GEARBOX (5 SPEED)

### Remove and refit

### Removing

The reverse light switch is located at the rear of the gear selector housing and is accessible from underneath the vehicle.

1. Drive vehicle onto a ramp.
2. Disconnect the battery.
3. From underneath the vehicle, disconnect the electrical leads.
4. Release the locknut securing the switch in position.
5. Unscrew the reverse light switch from the gear selector housing.

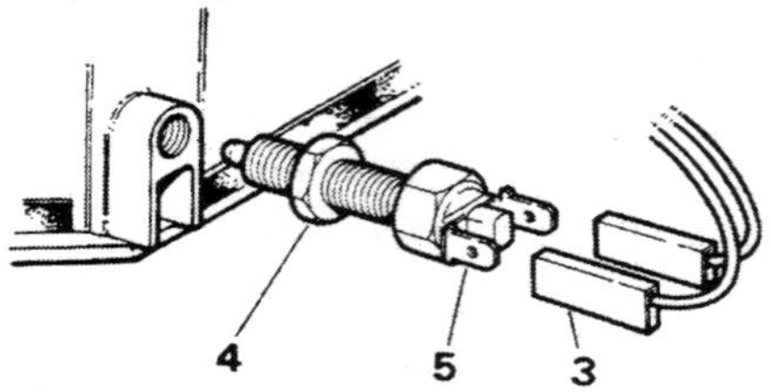

### Refitting

**NOTE: The reverse light switch will require re-setting on reassembly.**

6. Select reverse gear.
7. Connect a 12-volt supply to either of the switch terminals.
8. Connect a test lamp to the remaining terminal.
9. Screw the switch into the housing until the test lamp is illuminated. Rotate the switch a further half-turn.
10. Tighten the locknut to secure the switch in position.
11. Re-connect the switch electrical leads.
12. Re-connect the battery.
13. Drive the vehicle off the ramp.

## REVERSE LIGHT SWITCH—START INHIBITOR SWITCH

### Automatic gearbox

### Remove and refit

The reverse light switch is an integral part of the start inhibitor switch and is located on the left-hand side of the gearbox above the front of the gearbox sump and is accessible from beneath the vehicle.

*Continued*

## OIL PRESSURE WARNING SWITCH

### Remove and refit

### Removing

1. Disconnect the battery.
2. Disconnect the electrical lead from the switch.
3. Unscrew the switch unit.
4. Remove switch and sealing washer.

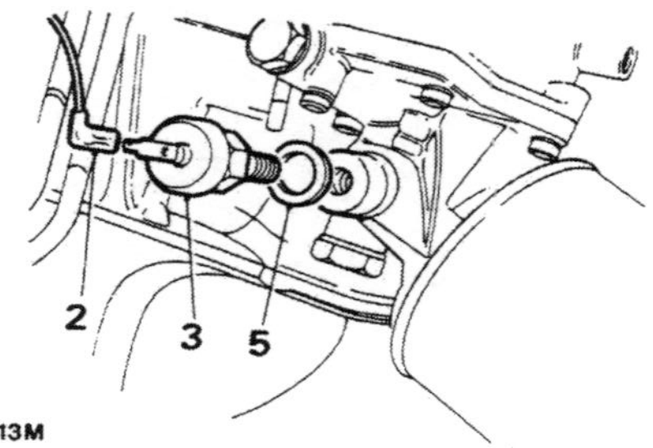

### Refitting

5. Reverse the removal procedure, using a NEW sealing washer.

# ELECTRICAL 86

## COOLANT TEMPERATURE TRANSMITTER —Carburetter Models only

### Remove and refit

### Removing

1. Disconnect the battery.
2. Disconnect the electrical lead from the transmitter and the air cleaner hose.
3. Remove the transmitter from the inlet manifold.

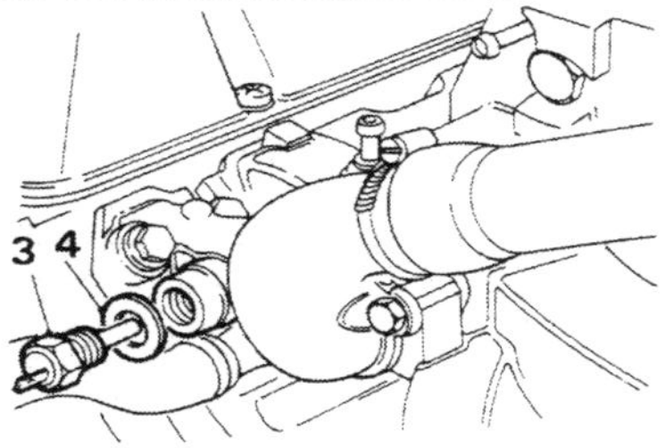

### Refitting

4. Reverse the removal procedure, using a NEW joint washer.

## STOP LIGHT SWITCH

### Remove and refit

### Removing

1. Disconnect the battery.
2. Remove the lower fascia panel.
3. Depress the foot brake.
4. Remove the rubber protector from switch (where fitted).
5. Remove the hexagon nut.
6. Withdraw the switch.
7. Disconnect the electrical leads.

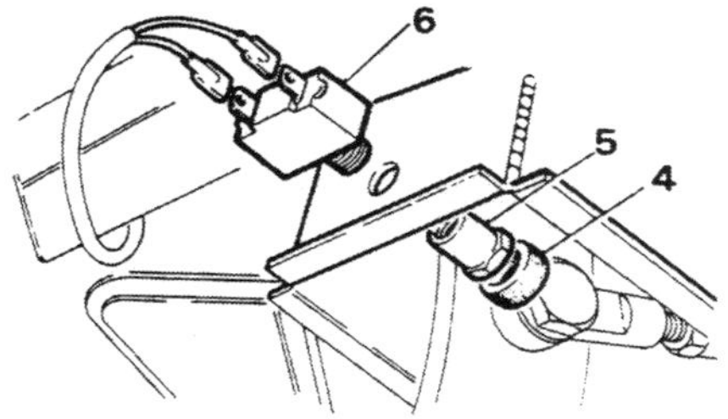

### Refitting

8. Reverse the removal procedure.

## HANDBRAKE WARNING SWITCH

### Remove and refit

### Removing

1. Disconnect the battery.
2. Apply the handbrake.
3. To gain access to the warning switch located on the side of the handbrake mounting bracket, it is necessary to remove the cubby box liner.
4. Remove the four screws securing the cubby box liner and lift out the liner.
5. Carefully pull the rear warm air flow hose away from the side of the handbrake mounting bracket to give access to the two screws securing the switch in position.
6. Remove the two screws.
7. Manoeuvre the switch around the front of the handbrake mounting bracket and disconnect the electrical lead.
8. Withdraw the switch.

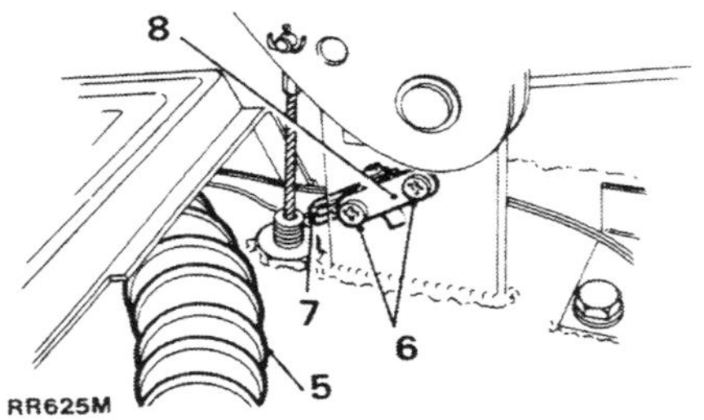

### Refitting

9. Reverse the removal procedure.

## CHOKE WARNING LIGHT SWITCH—Carburetter Versions only

### Remove and refit

### Removing

1. Disconnect the battery.
2. Remove the six screws securing the lower fascia panel.
3. Remove the lower fascia panel to give access to the rear of the upper fascia.
4. Remove the two electrical leads from the switch.
5. Remove the screw and clip securing the switch to the choke cable and slide the clip off the switch.
6. Remove the switch.

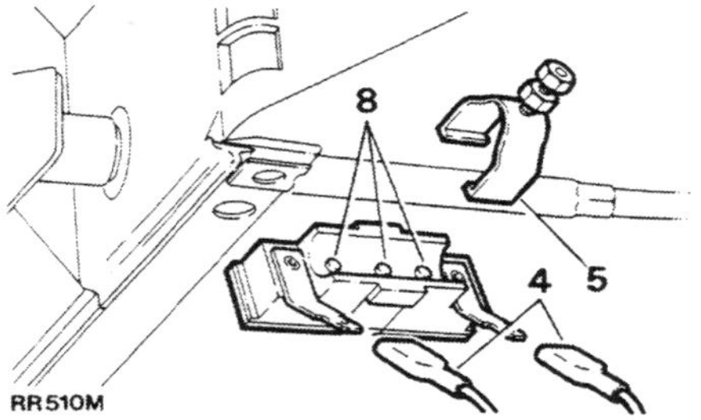

### Refitting

7. Reverse the removal procedure.
8. Ensure that the three pegs on the switch locate in the corresponding holes on the choke outer cable.

## EXTERIOR DRIVING MIRRORS—4 DOOR ONLY

1. The mirror housing is hinged vertically and should be set in one of the two fixed angle positions provided to suit the respective left or right side mirror location.
2. Additionally, for safety and convenience, the mirror housing is designed to fold completely forwards or rearwards against the vehicle body.

NOTE: Flat mirrors are fitted to Australian vehicles.

### Setting the mirror—manual version

3. The glass angle is finely adjusted by moving it vertically or horizontally as required.

### Setting the mirror—electrically operated

4. Fine adjustment is controlled by an electric motor inside the mirror housing. This is operated by individual finger-tip operated controls fitted on either side of the steering column lower cover. To adjust, move the head of the appropriate control to the left, right, up or down as required. The mirror selected will respond accordingly.

5. The mirror also incorporates a demist facility, activated by operation of the rear window demist switch.

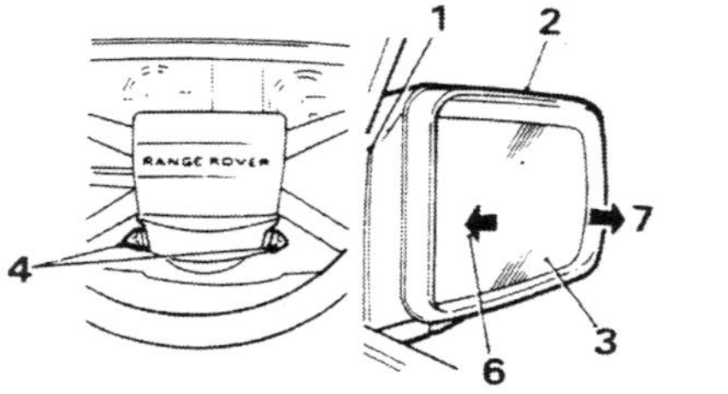

### Renewing the mirror glass—manual and electric versions

6. Press the inner (wider) end of the glass inwards to its full extent.
7. Insert the fingers under the outer (narrower) end of the glass, and pull outwards until the glass is released from its four retaining clips.
8. On electrical versions disconnect the two demister leads attached to back of the glass unit.
9. To replace the glass, locate the inner (wider) end of the glass in the mirror housing first.
10. Carefully press the outer (narrower) end of the glass inwards until it is safely held by its four retaining clips.
11. Reset the fine adjustment as required.

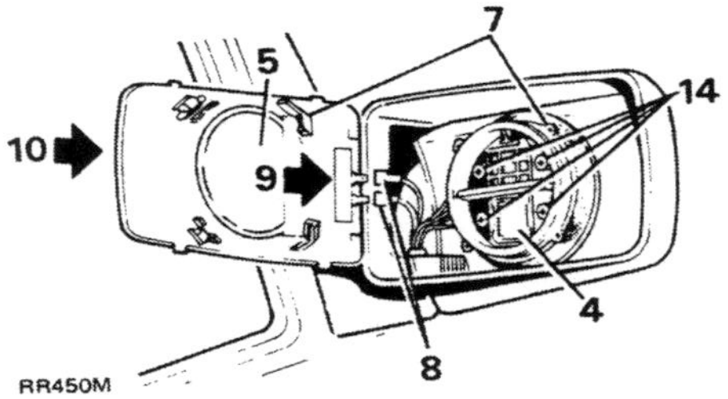

## EXTERIOR DRIVING MIRRORS

## ELECTRIC MOTORS

### Remove and Refit

### Removing

12. Disconnect the battery.
13. Remove the mirror glass, as described in items 6 to 8.
14. Remove the four self-tapping screws securing the motor assembly to the mirror body.

*Continued*

# ELECTRICAL

15. Manoeuvre the motor assembly to reveal the electrical connections on the rear of the motor.
16. Pull the leads from the rear of the motor assembly.

**Refitting**

17. Reverse operations 12 to 16, ensuring that the electrical leads are correctly refitted (see electric mirror, circuit diagram).

## EXTERIOR DRIVING MIRRORS—FINGERTIP CONTROLLED SWITCHES

**Remove and Refit**

Service tool: 18G 1014 Extractor for steering wheel.
18G 1014-2 Adaptor pins.

**Removing**

18. Disconnect the battery.
19. Release the screw retaining the centre cover to the steering wheel and remove the cover.
20. Remove the retaining nut and washer securing the steering wheel.
21. Remove the steering wheel using the correct service tool.
22. Remove the lower fascia panel by releasing the six securing screws.
23. Remove the four screws securing the bottom shroud to the top shroud.
24. Remove the single screw securing the top shroud to the switch housing bracket.
25. Remove the top shroud.
26. Release the light switch locknut and remove the switch from the mounting bracket.
27. Manoeuvre the bottom shroud to clear the ignition switch/steering lock assembly.
28. Retrieve the small spacing collar located on the forward left hand side of the bottom shroud.
    Note: It is important that the spacing collar is refitted on assembly.
29. Pull the multi-plug from the rear of the fingertip controlled mirror switch.
30. Carefully prise off the fingertip button at the operating end of the switch.
31. Unscrew the black plastic retaining collar securing the switch to the bottom shroud.
32. Remove the switch from the shroud.

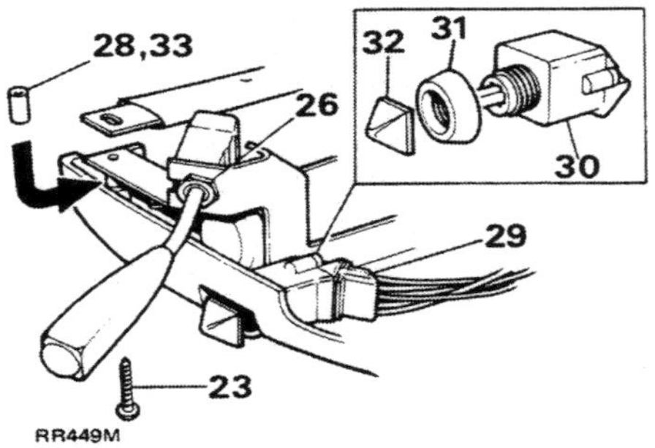

**Refitting**

33. Reverse operations 18 to 32, ensuring that the black spacing collar is refitted.

NOTE: To prevent damage to the electrical wiring within the top and bottom shroud, the leads should be arranged carefully to avoid contact with mating faces on re-assembly.

## EXTERIOR DRIVING MIRRORS

### COMPLETE ASSEMBLY

**Remove and Refit**

**Removing**

34. Disconnect the battery.
35. Carefully prise off the interior finisher plate to reveal the three securing screws and electric wiring.
36. Disconnect the two electrical plugs (one two pin, one three pin).
37. Supporting the exterior mirror assembly remove the three securing screws (with plain and spring washer).
38. Pull the inner mounting plate away from the inner door frame complete with the two retaining clips.
39. Detach the mirror assembly from the outer door frame.
40. Remove the sealing rubber.

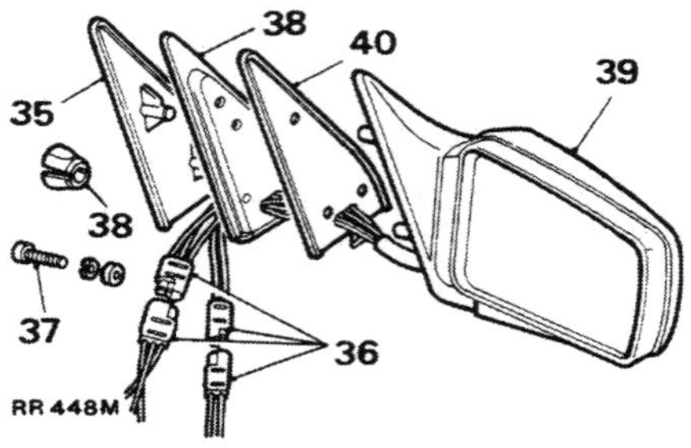

**Refitting**

41. Reverse the operations 34 to 40.

NOTE: To prevent damage to the electrical wiring do not push the leads down inside the door casing.

## INSTRUMENT ILLUMINATION
### ELECTRONIC DIMMING CONTROL

The electronic dimming control switch is located on the lower fascia panel adjacent to the steering column. Rotate the control upwards to fully illuminate the instruments and downwards to reduce intensity.
The dimming control unit also controls the clock, heater and cigar lighter illumination.

**Remove and refit**

**Removing**

1. Disconnect the battery.
2. Remove the lower fascia panel by releasing the six securing screws.
3. Disconnect the dimming control multi-plug.
4. Remove the two screws securing the dimmer control switch to the under-side of the lower fascia panel.

**Refitting**

Reverse operations 1 to 4.

## INSTRUMENT BINNACLE WARNING LIGHT SYMBOLS

- Direction indicator — left turn (green)
- Direction indicator — right turn (green)
- Park brake on — Australia only (red)
- Headlamp main beam on (blue)
- Trailer connected — flashes with direction indicators (green)
- Rear fog guard lamps on (amber)
- Ignition on (red)
- Automatic gearbox oil temperature high (red)
- Engine oil pressure, low (red)
- Cold start, engaged (amber)
- Fuel indicator, low (amber)
- Differential lock engaged (amber)
- Transmission handbrake on — except Australia (red)
- Brake padwear (amber)
- Brake fluid pressure, failure (red)

NOTE: The ignition and engine oil pressure symbols will be automatically illuminated when the ignition is switched on and extinguished when the engine is running. The brake fluid pressure symbol will also be illuminated while the ignition key is being held over to actuate the starter, confirming that the warning circuit is functioning correctly.

*Continued*

## ELECTRICAL 86

**Instrument Pack**

1. Fuel gauge
2. Temperature gauge
3. Voltage stabiliser
4. Ignition warning bulb (with separate red holder unit)
5. Panel/warning lights bulb/holder
6. Printed circuit input tags (for harness connection)
7. Speedometer drive unit
8. Printed circuit
9. Tachometer
10. Warning lights panel
11. Instrument case (front)
12. Curved lens
13. Wire connecting clips
14. Binnacle housing
15. Speedometer

*Continued*

## RENEWAL OF PANEL AND WARNING LIGHT BULBS

1. Disconnect the battery.
2. Unclip the back of the cowl from the instrument binnacle to give access to the panel and warning light bulbs in the back of the instrument case.
3. Remove the appropriate bulb holder unit by rotating it anti-clockwise and withdrawing it.

NOTE: The No charge ignition warning light, identified by its red coloured bulb holder, is of a higher wattage and is the only bulb which can be pulled from its holder and replaced independently.

4. Fit a new bulb holder unit and rotate clockwise to lock in position. The correct bulb type is a 1.2 watt bulb/holder unit, except the ignition bulb which is 2 watt wedge base type.
5. Refit the cowl and reconnect the battery.

NOTE: If difficulty is experienced in changing bulbs, due to the limited space available the instrument binnacle fixings should be removed to enable the binnacle to be raised above the fascia as far as other connections permit. See 'Instrument Binnacle removal' below for details of binnacle mounting bracket fixing.

## INSTRUMENT BINNACLE

### Remove and refit

**Remove**

1. Disconnect the battery.
2. Remove the lower fascia by releasing the six retaining screws.
3. Remove the four nuts (with spring and plain washers) from under the top fascia rail which secure the instrument binnacle to the vehicle.
4. Unclip the binnacle cowl, from the rear, to provide access to the two-part speedometer cable.
5. Disconnect the two-part speedometer cable from the speedometer drive on the back of the instrument case. Alternatively, from under the top fascia rail, release the cable connector ring at the intermediate clamped connection. This is located some 470 mm. (18.5 in.) from the speedometer drive. This connection is provided to facilitate separate renewal of either the upper or lower part of the cable in service.
6. Disconnect the two multi-plugs from the printed circuit connectors.
7. Lift the instrument binnacle from the top fascia rail and transfer it to the workbench.

NOTE: On LHD vehicles, where an over-speed buzzer is fitted, the intermediate speedometer cable connections are threaded and retain a sensor unit between the two parts of the cable.

A lead from the sensor unit is connected to a black two-pin socket (black and white leads) below the binnacle, above the steering column area. The adjacent buzzer will be audible at approximately 120 kph. (75 mph.).

### Removing Instrument Pack

9. Having removed the instrument binnacle from the vehicle, detach the binnacle mounting bracket. This is secured to the instrument case by two screws and to the bottom of the binnacle bezel by two smaller screws.
10. Remove the two screws retaining the top of the bezel to the front housing and detach the bezel.
11. Separate the instrument case from the binnacle housing by releasing the two wire clips.
12. Detach the curved lens from the binnacle housing by releasing the wire clip at the top.

### Refitting Instrument Pack to Binnacle

13. Reverse removal instructions 9 to 12.

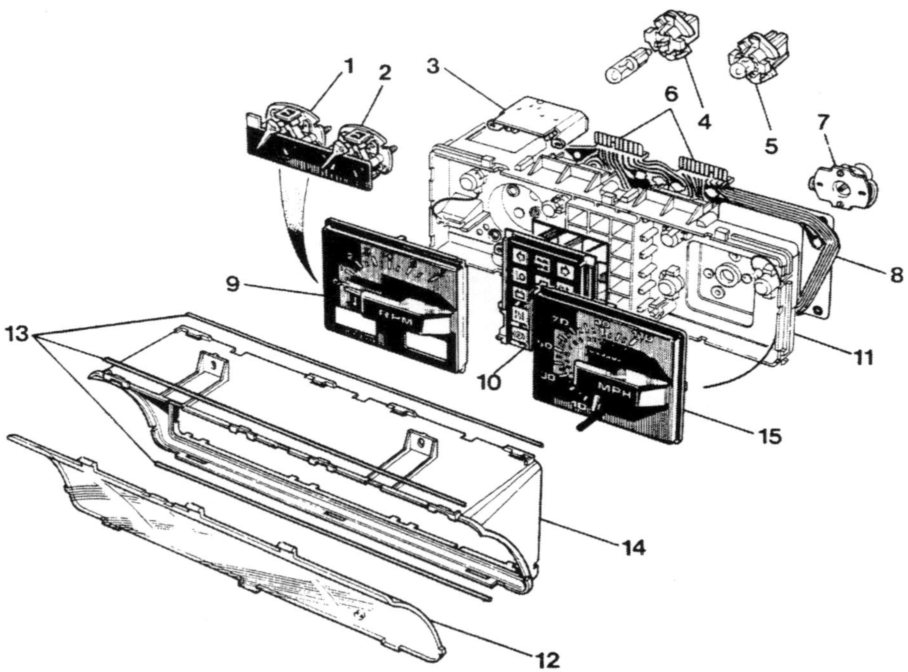

### Refitting

8. Reverse the removal instructions 1 to 7.

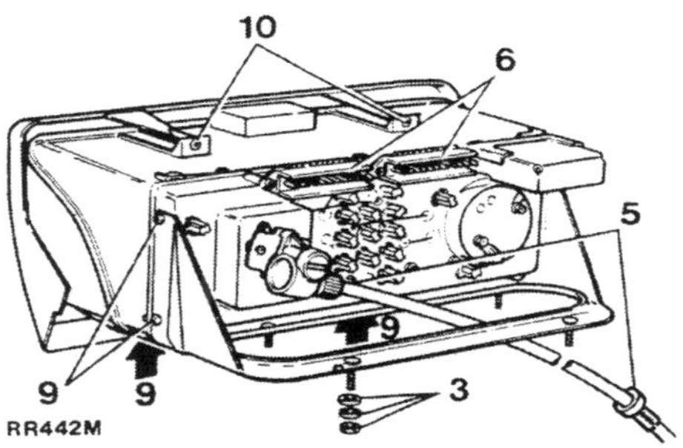

41

243

## ELECTRICAL 86

### Renewing panel and warning lamp bulbs

14. Remove the appropriate bulb holder unit from the back of the instrument case by rotating the bulb holder anti-clockwise and withdrawing it.

   NOTE: The No charge ignition warning light, identified by its red coloured bulb holder is of a higher wattage and is the only bulb which can be separated from its holder and replaced independently.

15. Fit a new bulb holder unit to the printed circuit and rotate clockwise to lock in position.
   The correct bulb type is a 1.2 watt bulb/holder, except the ignition bulb which is 2 watt wedge base type.

### Instrument case (back)

1. Locating pegs
2. Panel light bulbs
3. Speedometer securing screws
4. Speedometer drive securing screws
5. Harness connectors
6. Warning light bulbs
7. No charge warning light bulb (red holder)
8. Temperature gauge securing nuts
9. Fuel gauge securing nuts
10. Tachometer securing nuts
11. Multi-function unit
12. Printed circuit
13. Pull-up resistor—high temperature gearbox oil
14. Park/hand brake
15. Brake failure
16. Brake pad wear

### PRINTED CIRCUIT HARNESS CONNECTIONS

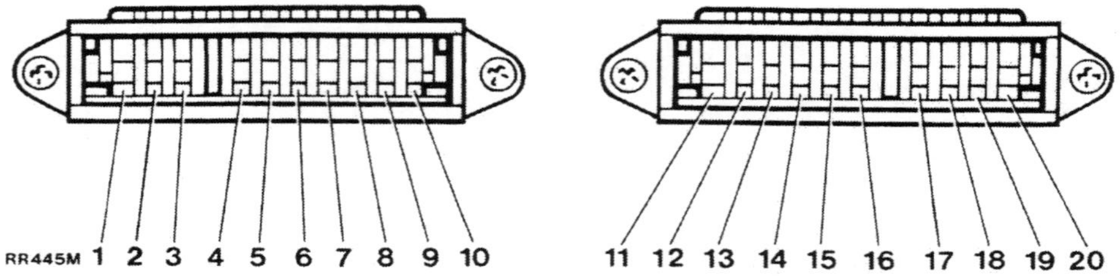

(Sequence of connections viewed from back of instrument case.)

### CIRCUIT SERVED

1. Tacho signal
2. Ignition switch 12v+
3. Ignition warning light
4. Trailer warning light
5. Main beam warning light
6. Earth
7. Direction indicators left hand
8. Rear fog warning light
9. Direction indicators right hand
10. Oil pressure warning light
11. High oil temperature warning light (Auto gearbox)
12. Cold start warning light
13. Differential lock warning light
14. Brake failure warning light
15. Brake pad wear warning light
16. Panel illumination warning light
17. Brake pad wear warning light
18. Park brake warning light
19. Fuel tank gauge
20. Coolant temperature gauge

21. Additional wired circuit on RHD vehicles for alternative Australian park brake warning light symbol. Connected to a black two-pin socket (white & black/pink leads) located under the binnacle, above the steering column area.

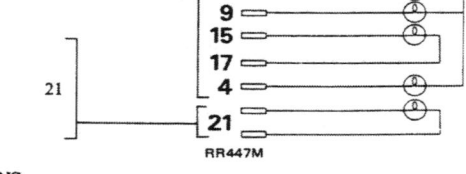

### MULTI-FUNCTION UNIT CONNECTIONS

A — 12v+ supply
B — Input to high oil temp warning light circuit
C — Tachometer
D — Tacho drive
E — Spare
F — 10v+ stabilised
G — Input to low fuel warning light circuit
H — Tacho signal
I — Low fuel warning light
J — 12v+ protected
K — High oil temperature warning light
L — Earth

Continued

### Removing printed circuit

16. Remove the two tachometer nuts (with washers) to release the printed circuit connecting tags.
17. Remove the four nuts (with washers) securing the fuel and temperature gauges to release the printed circuit from the fixing studs.
18. Release the two screws retaining the multi-function unit and lift off to release the printed circuit connecting tag.
19. Remove the two harness connectors, retained by four screws, to release the printed circuit tags.
20. Carefully ease the printed circuit from its four locating pegs.

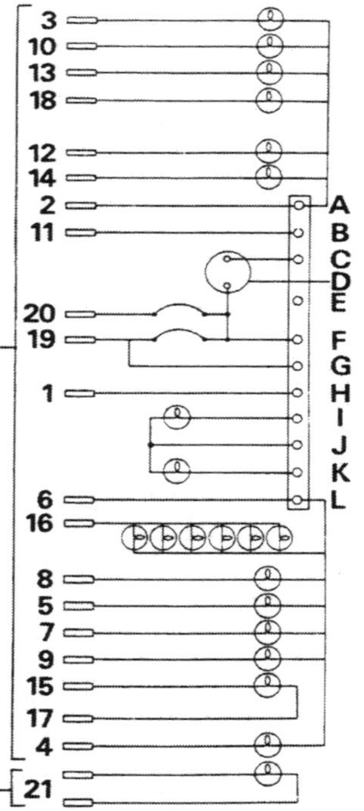

Continued

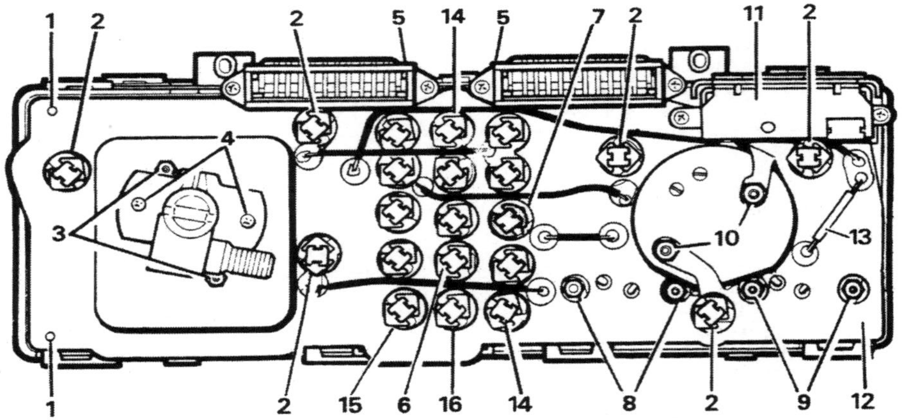

# ELECTRICAL 86

### Refitting the Printed Circuit

21. Reverse the removal procedure, items 16 to 20.
22. Ensure that the fuel and temperature gauge mounting studs are correctly located before pressing the printed circuit on to its four locating pegs.

### Removing Tachometer

23. Carefully prise the needle shroud from the tachometer and disconnect the fibre optic element underneath the shroud.
24. Remove the two nuts (with washers) at the back of the instrument case which retains the tachometer and release the printed circuit tags.
25. Slacken the four nuts retaining the fuel and temperature gauges and carefully manoeuvre the tachometer from the front of the instrument case.

### Refitting the Tachometer

26. Reverse the removal procedure, items 23 to 25.

### Removing Fuel and Temperature Gauge Unit

27. Carefully prise the needle shroud from the tachometer and disconnect the fibre optic element underneath the shroud.
28. Remove the two nuts (with washers) retaining the tachometer and release the printed circuit tags.
29. Remove the four nuts (with washers) retaining the fuel and temperature gauges and carefully manoeuvre the fuel and temperature gauge unit from the front of the instrument case.

### Refitting the Fuel and Temperature Gauges

30. Locate the fuel and temperature gauge unit in the instrument panel but **do not fit the washers and nuts at this stage**.
31. Feed the fibre optic element through the aperture in the tachometer then locate the tachometer in the instrument panel.
32. Position the printed circuit tags over the two tachometer studs, fit the washers and fit and tighten the retaining nuts.
33. Fit the washers to the four fuel and temperature gauge studs and fit and tighten the retaining nuts.

### Removing the Speedometer and Speedometer Drive Unit

34. Carefully prise the needle shroud from the speedometer and disconnect the fibre optic element underneath the shroud.
35. Remove the two hexagonal headed screws (with washers) at the back of the instrument case which retain the speedometer.
36. Carefully remove the speedometer from the front of the instrument case.
37. To release the speedometer drive unit, remove the two self-tapping screws securing it to the back of the instrument case.

### Refitting the Speedometer and Speedometer Drive Unit

38. Reverse the removal procedure items 34 to 37 ensuring that the rubber gasket is fitted behind the speedometer drive unit.

---

## CLOCK

### Remove and refit

### Removing

1. Disconnect the battery.
2. Carefully prise the clock out of the fascia panel to reveal the electrical connections.
3. Disconnect the two electrical leads.
4. Remove the illumination lead complete with holder and bulb.

**NOTE: The clock is illuminated by a 2-watt bayonet type bulb.**

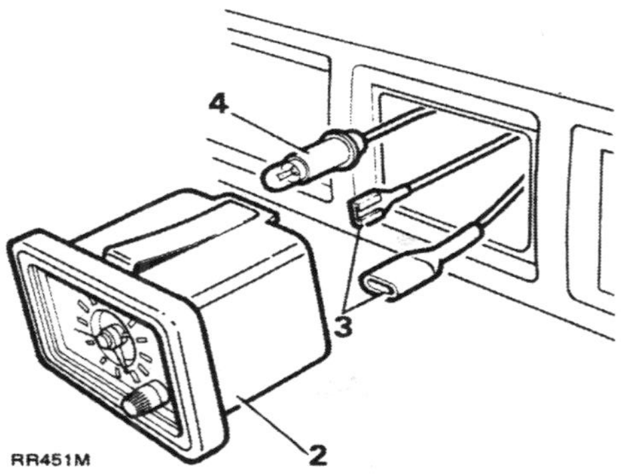

### Refitting

5. Reverse the removal procedure.

---

# ELECTRICAL 86

## SPEEDOMETER CABLE ASSEMBLY

The speedometer cable is a two part assembly, consisting of an upper cable, connected to the rear of the binnacle and a lower cable connected to the speedometer drive housing at the rear of the transfer gearbox. The two cables are joined by a connector ring behind the lower fascia, this connection is provided to facilitate separate renewal of either the upper or lower part of the cable in service.

To remove the upper cable refer to instrument binnacle removal.

### Lower Speedometer cable

### Remove and refit

### Removing

1. Disconnect the battery.
2. Remove the lower fascia panel by releasing the six retaining screws.
3. Release the cable connector ring between the upper and lower cables and withdraw the cable and grommet from the bulkhead.
4. Remove the single nyloc nut and clamp securing the cable to the speedometer drive housing at the rear of the transfer gearbox.
5. Release the cable from the two retaining clips.

**NOTE: On left-hand-drive vehicles with automatic gearbox the speedometer cable is secured by a further two clips located above the cross-member attached to the chassis side-member.**

6. Withdraw the cable from the speedometer drive housing.

### Refitting

7. Reverse the removal procedure.

---

## WINDOW LIFT SWITCHES

### Remove and refit

### Removing

1. Disconnect the battery.
2. Carefully prise the window lift switch surround away from the front of the cubby box.
3. Disconnect the multi-plug at the rear of the switch(es).
4. Apply pressure to the rear of the switch to push it through the surround.

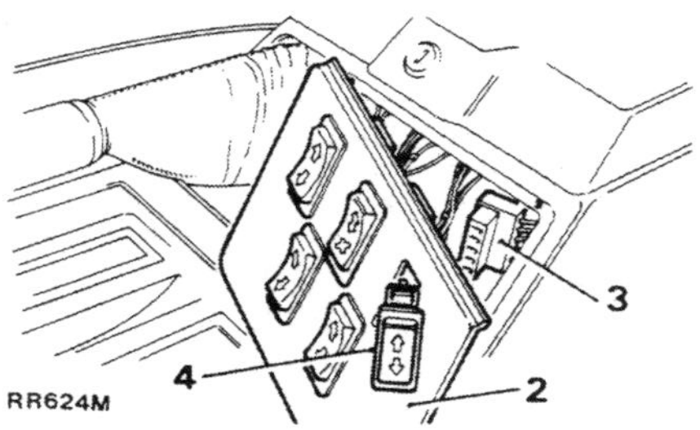

---

## WINDOW LIFT MOTOR — Rear doors

### Remove and refit

### Removing

15. Ensure the side door glass is in its fully closed position and secure it with adhesive tape.
16. Disconnect the battery.
17. Remove the arm-rest/door-pull finisher to reveal the two securing screws.
18. Remove the two screws (with plain washers) and detach the arm-rest/door-pull from the inner door panel. To enable the arm-rest/door-pull to be removed from the door, the window operating switch multi-plug must be disconnected from the rear of the switch.

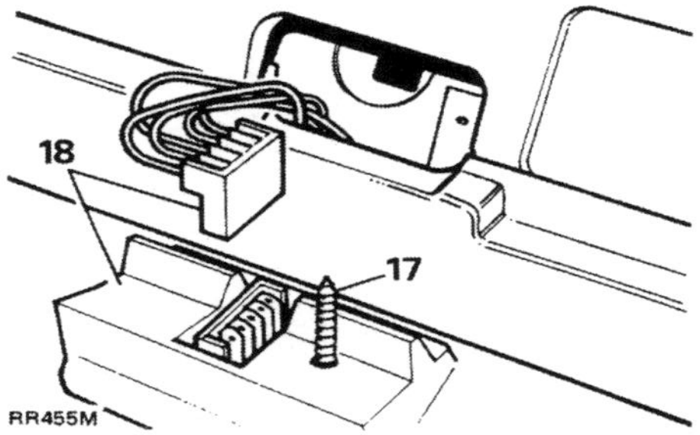

**NOTE: At this stage the window operating switch can be removed by applying a little pressure to the rear of the switch to push it through the door-pull handle.**

19. Remove the interior handle finisher button to reveal the screws retaining the handle surround.

*Continued*

# ELECTRICAL | 86

20. Remove the screw and detach the handle surround from the door trim pad.
21. Remove the door trim pad by inserting a screwdriver between the trim pad and inner door panel, gently prising out the six plastic securing clips from their respective holes in the inner door panel.
22. Displace the bottom half of the plastic weather sheet to reveal the window lift motor.
23. Release the lift motor wiring harness from the retaining clips.
24. Disconnect the lift motor harness snap connections from the main door harness.
25. Supporting the lift motor release the three bolts securing the motor to the inner door panel.
26. Withdraw the lift motor from the lower aperture in the inner door panel.

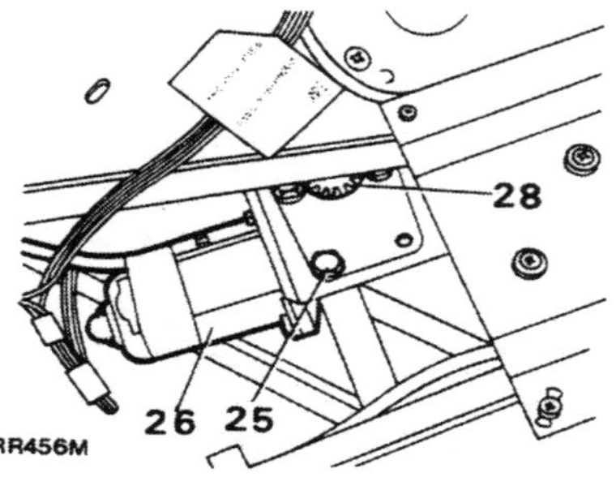

### Refitting

27. Reverse operations 15 to 26.
28. Ensure the lift motor drive gear is engaged and correctly aligned with the window lift linkage before fitting the securing bolts.

## WINDOW LIFT MOTOR—Front doors

### Remove and refit

### Removing

1. Ensure that the side door glass is in its fully closed position and secure it with adhesive tape.
2. Disconnect the battery.
3. Detach the arm-rest/door-pull by removing the two securing screws.
4. Remove the two screws (with plain washers) to enable the arm-rest/door-pull to be detached from the inner door panel.
5. Remove the interior door handle finisher button to reveal the screw retaining the handle surround.

6. Remove the screw and detach the handle surround from the inner door trim pad.
7. Detach the inner door trim pad by inserting a screwdriver between the trim pad and inner door panel gently prising out the nine plastic securing clips from their respective holes in the inner door panel.
8. Disconnect the two radio speaker connections behind the trim pad, remove the trim pad complete with speaker.

NOTE: **At this stage the speaker can be removed by releasing the four nuts (with plain washers) located on the back of the trim pad.**

9. Peel back the front top corner of the plastic weather sheet to reveal the window lift motor.
10. Release the window lift motor wiring harness from the three retaining clips to allow the harness to be pulled out of the aperture at the front of the inner door panel.
11. Disconnect the window lift motor multi-plug from the main door harness.
12. Supporting the motor, remove the three securing bolts.
13. Withdraw the motor through the top front aperture of the door.

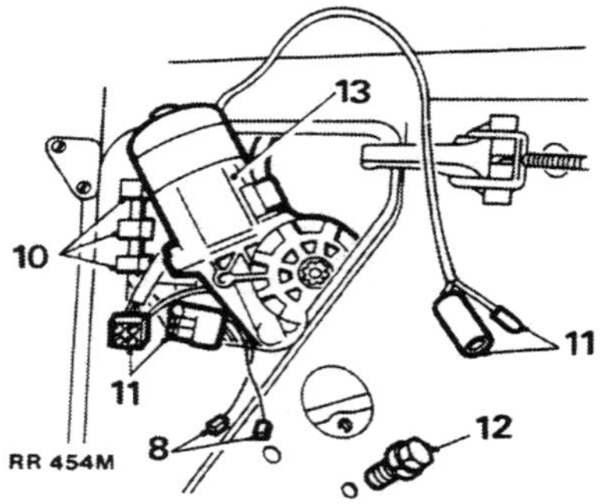

### Refitting

14. Reverse operations 1 to 13.

NOTE: **Ensure that the drive gear is engaged and correctly aligned with the window lift linkage before fitting the securing bolts.**

## ELECTRICAL | 86

8. Remove the four screws (with plain washers) securing the lock actuator mounting plate to the inner door panel.
9. Release the clip retaining the electrical cable.
10. Manoeuvre the actuator assembly to detach the operating rod 'eye' from the hooked end of the actuator link on the door lock.
11. Withdraw the actuator assembly from the door until the electrical cable is pulled out of its channel sufficiently to expose the connectors which can then be detached.
12. Remove the actuator assembly from the door.
13. The actuator unit may be changed if necessary by removing the two rubber mounted screws which secure it to the mounting plate.

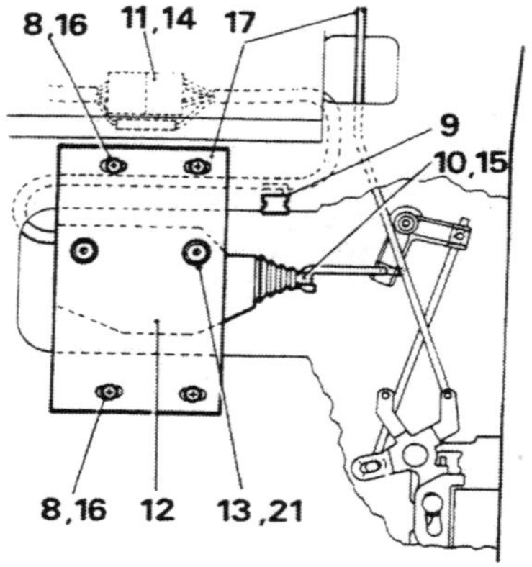

## ELECTRICALLY OPERATED CENTRAL DOOR LOCKING SYSTEM

An electrically operated central door locking system is fitted as an option on four door models.

Locking or unlocking the drivers door from outside by key operation, or from inside by sill knob automatically locks or unlocks all four doors.

Front and rear passenger doors can be independently locked or unlocked from inside the vehicle by sill knob operation but can be overridden by further operation of the drivers door locking control.

On rear doors only a child safety lock is provided which can be mechanically pre-set to render the interior door handles inoperative.

Failure of an actuator will not affect the locking of the remaining three doors. The door with the inoperative actuator can still be locked or unlocked manually.

NOTE: **The door lock actuator units contain non-serviceable parts. If a fault should occur replace the unit concerned with a new one.**

Before carrying out any maintenance work disconnect the battery.

## FRONT DOOR ACTUATOR UNITS

### Remove and refit

### Removing

1. Ensure the window is in its fully closed position.
2. Remove the arm-rest/door-pull finisher button to reveal the two retaining screws.
3. Remove the interior door handle finisher button to reveal the screw retaining the handle surround.
4. Release the screw and remove the handle surround from the interior door trim pad.

NOTE: **On models fitted with manually operated side door windows it is necessary to remove the window regulator handle to enable the trim pad to be removed.**

5. Release the door trim pad by inserting a screwdriver between the trim pad and the inner door panel, gently prising out the nine plastic clips from their respective holes around the edges of the trim pad.
6. Disconnect the two speaker connections inside the door and remove the door trim pad complete with speaker.
7. Peel back the top of the plastic weather sheet at the rear of the inner door panel to expose the lock actuator unit.

### Refitting

14. Locate the actuator assembly in the inner door panel and fit the electrical cable connectors. The cable, and connectors, are pulled back into the channel from the font end and the cable clip refitted.
15. Manoeuvre the actuator assembly to engage the operating rod 'eye' on the hooked actuator link.
16. Loosely fit the actuator mounting plate to the inner door panel with the four screws, setting the mounting plate in the centre of the slotted holes.
17. Ensure that manual operation of the sill locking control is not restricted by the operation of the actuator operating rod and vice versa, resetting the mounting plate as necessary.
18. Reconnect the vehicle battery.
19. Check that electrical operation of the door lock occurs when the sill locking control is moved through half of its total movement. Reset the mounting plate if necessary and tighten the four screws.

NOTE: **The above adjustment ensures that the full tolerance on the switching operation is utilised.**

# ELECTRICAL

## REAR DOOR ACTUATOR UNITS

### Remove and refit

Instructions as for front doors with the following exceptions:

20. No radio speaker is involved.
21. The electrical cable and plug is retained to the inner door panel by two spring clips and is immediately accessible through the large aperture in the door.
22. Instruction 19 does not apply to rear actuator units which are not fitted with switches.

NOTE: If necessary the lock actuator may be detached from its mounting plate to facilitate the removal of the lock actuator from the connector rod inside the door panel.

## FUEL TANK GAUGE UNIT

### Remove and refit

Service tool–18G 1001 Locking spanner

### Removing

1. Disconnect the battery.

NOTE: Ensure the fuel level is below the level of the gauge unit.

2. Chock the front wheels, raise the rear wheels clear of the ground and support the vehicle on stands.
3. Remove the left side rear wheel to provide easy access to the gauge unit which is fitted in the side of the fuel tank.
4. Disconnect the electrical leads from the gauge unit.
5. Using tool 18G 1001 release the tank unit locking ring.
6. Remove the gauge unit and sealing washer.

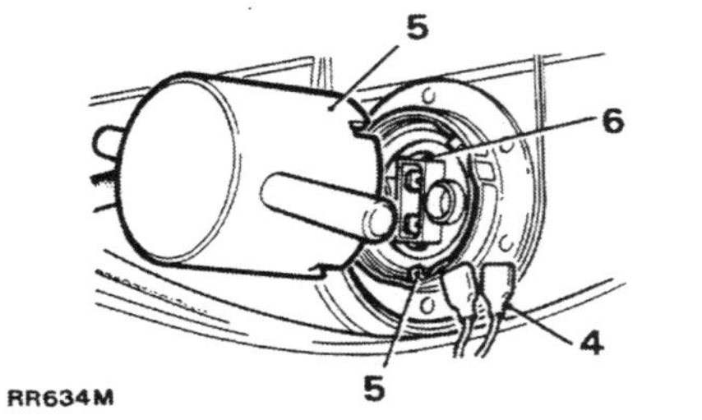

### Refitting

7. Fit a NEW sealing washer and locate the gauge unit in the tank ensuring that the notch in the periphery of the gauge unit locates with the register in the gauge aperture of the tank.
8. Refit the electrical leads.
   Green/Black lead to top terminal.
   Black lead to lower terminal.
9. Reverse the removal procedure.

## SPLIT CHARGING FACILITY—OPTION

The circuit provides an additional source of electrical supply allowing separate charging and discharging of an additional battery for auxiliary equipment without affecting the charge state of the vehicle's main battery.

A terminal bracket, heavy duty relay and cables are fitted on the left hand front wing valance adjacent to the power steering reservoir.

The additional battery, leads and fixing clamps are not included in the option.

The split charging system is controlled by a voltage sensitive switch which energises the relay when the ignition voltage exceeds a pre-set level, thus supplying current to the positive terminal on the terminal bracket. Conversely, if the ignition voltage falls below the pre-set level the split charging circuit will cut out.

NOTE: The split charge facility is not available on fuel injection models.

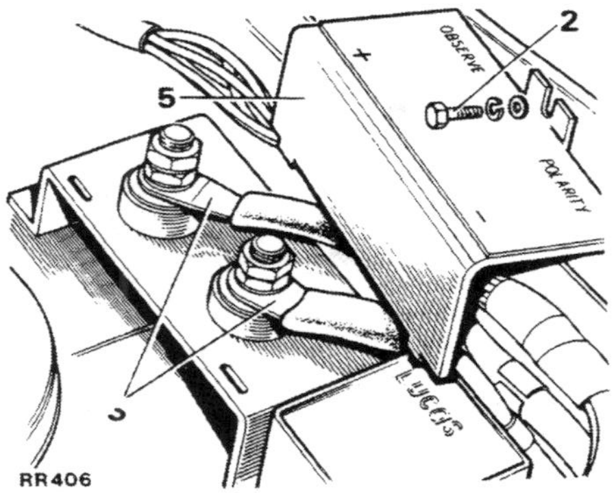

### To operate split charging system

1. Install the additional battery.
2. Remove the fixing bolt securing the terminal bracket cover.
3. Ensure that the positive and negative leads are correctly connected to the terminals and the battery in accordance with the markings on the terminal bracket cover.
4. Start the engine. As soon as the alternator is charging the vehicle battery and the voltage exceeds the pre-set level the split charge function will operate.
5. After charging replace terminal bracket cover.

## TRAILER SOCKET—OPTION

Incorporated in the vehicle electrical circuit is a facility for fitting a seven pin trailer lighting socket.

The pick-up point is located behind the left-hand rear tail light cluster and is accessible by removing the tail light assembly.

The pick-up point consists of a seven pin pre-wired plug, a separate auxiliary fused line feed and reverse light lead.

1. Disconnect the battery.
2. Remove the rear tail light assembly and disconnect the electrical plug.
3. Remove the protective cap from the trailer pick-up point plug.
4. Feed a seven core cable (fitted with a pre-wired plug to one end—suitable for connection to pick-up point) down between the inner and outer body panels through the rear light aperture.
5. Feed the cable alongside the existing rear lighting harness.
6. Pull the cable through the aperture between the chassis side member and fuel tank.
7. Fit two retaining clips to the cable and secure it to the rear end cross member.
8. Connect the electrical leads to the vehicle trailer socket. (Refer to current trailer wiring regulations).
9. Secure trailer socket to the tow bar.
10. If it is necessary to provide a line feed and reverse light feed, provision is made for this by the presence of two extra leads in the rear light aperture. Means of identification are as follows:

    Fused auxiliary line feed—**Pink lead**
    Reverse light feed—**Green/Brown lead**

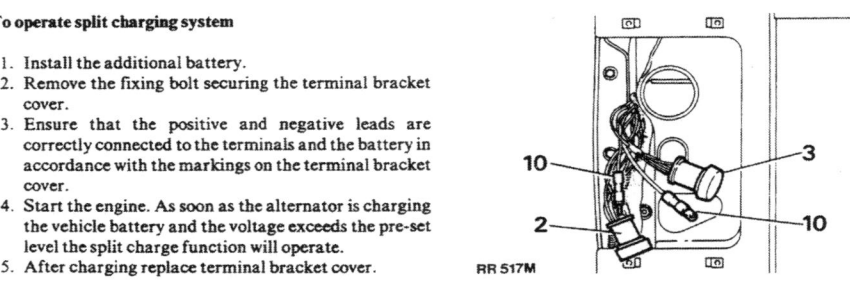

11. Refit rear tail light.
12. Reconnect the battery.

# ELECTRICAL

## ELECTRICAL EQUIPMENT — CIRCUIT DIAGRAMS

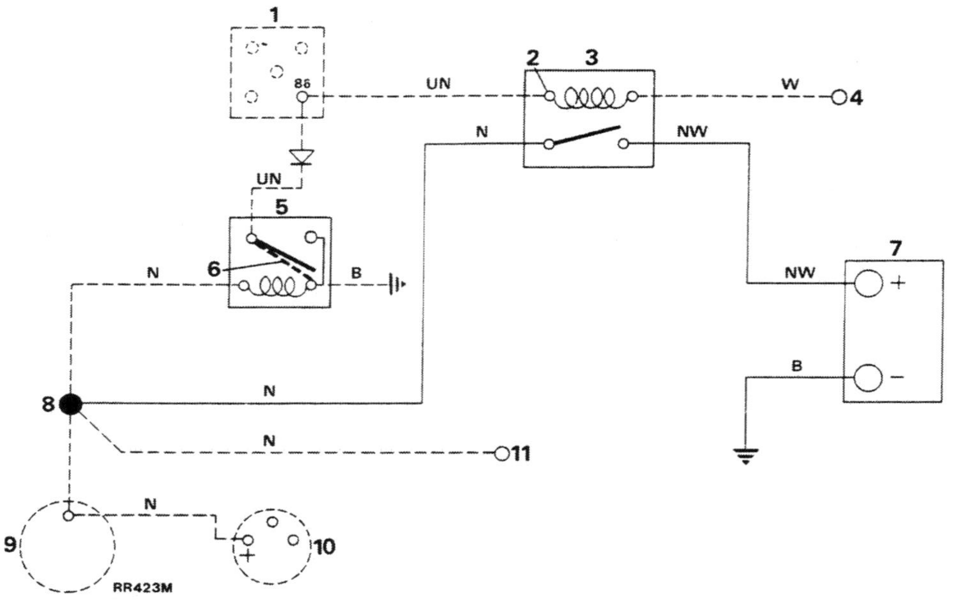

## OPTIONAL ELECTRICAL EQUIPMENT — RANGE ROVER 2- AND 4-DOOR MODELS

### Split Charge Circuit Diagram

1. Heated rear window relay
2. Pick-up point for split charge relay (items 106 and 125 on main circuit diagram)
3. Split charge relay
4. Fuse box
5. Voltage sensitive switch
6. Link wire (removed from plug when voltage sensitive switch is fitted)
7. Terminal box auxiliary battery
8. Terminal post
9. Starter motor
10. Alternator
11. Vehicle battery

NOTE: Chain dotted lines indicate existing parts.

## OPTIONAL ELECTRICAL EQUIPMENT — RANGE ROVER 4-DOOR MODELS

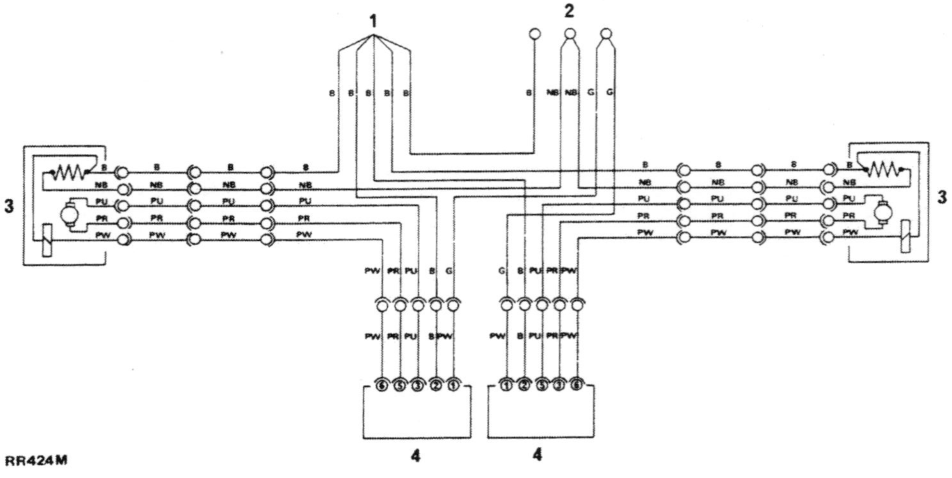

### Electric Mirrors Circuit Diagram

1. Clinch
2. Main cable connections (item 75 on main circuit diagram)
3. Mirrors
4. Mirror switches

## Key to cable colours

B Black  G Green  K Pink  L Light  N Brown  O Orange  P Purple  R Red  S Slate  U Blue  W White  Y Yellow

The last letter of colour code denotes the trace colour

Connector via plug   Snap Connectors   Permanent on-line connections   Earth Connections via cable   Earth Connections via bolts

# Notes

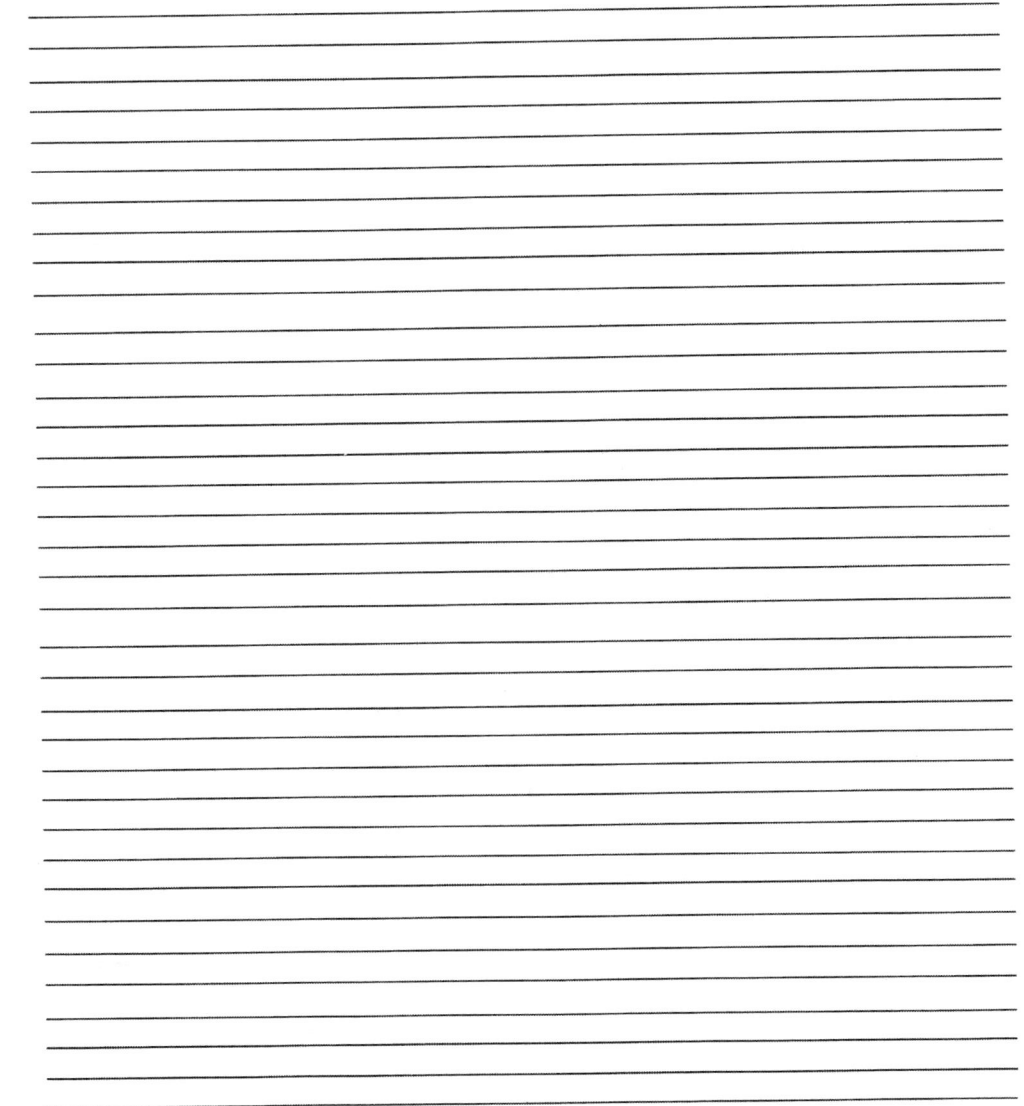

## ELECTRICAL

**OPTIONAL ELECTRICAL EQUIPMENT—RANGE ROVER 4-DOOR MODELS**

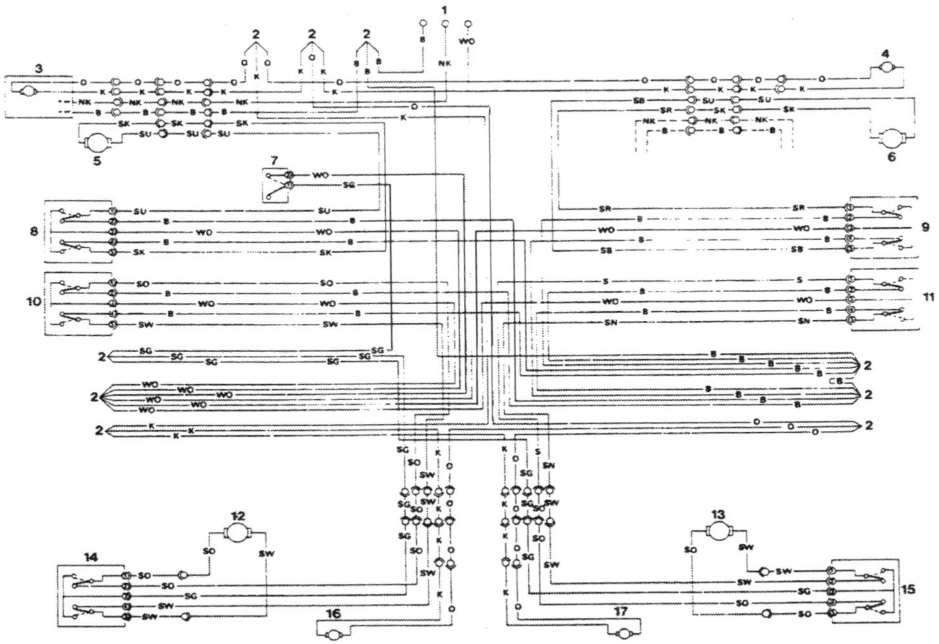

### Window Lifts and Door Locks Circuit Diagram

1. Main cable connections (item 107 on main circuit diagram)
2. Clinches
3. Switch unit central door locking (drivers door)
4. Lock unit central door locking (front passenger door)
5. Window lift motor left-hand front
6. Window lift motor right-hand front
7. Isolator switch
8. Window lift switch left-hand front
9. Window lift switch right-hand front
10. Window lift switch left-hand rear
11. Window lift switch right-hand rear
12. Window life motor left-hand rear
13. Window lift motor right-hand rear
14. Window lift switch left-hand rear door
15. Window lift switch right-hand rear door
16. Lock unit central door locking left-hand rear
17. Lock unit central door locking right-hand rear

## CIRCUIT DIAGRAM LEGEND—1986 onwards

1. Front interior lamp
2. Rear interior lamp
3. LH front door switch
4. RH front door switch
5. Tailgate switch
6. LH rear door switch
7. RH rear door switch
8. RH stop lamp
9. LH stop lamp
10. LH front indicator lamp
11. LH rear indicator lamp
12. LH side repeater lamp
13. RH front indicator lamp
14. RH rear indicator lamp
15. RH side repeater lamp
16. RH auxiliary driving lamp
17. LH auxiliary driving lamp
18. Auxiliary driving lamp switch
19. RH headlamp dip
20. LH headlamp dip
21. RH headlamp main
22. LH headlamp main
23. RH rear fog lamp
24. LH fog lamp
25. RH number plate lamp
26. RH side lamp
27. RH tail lamp
28. LH number plate lamp
29. LH side lamp
30. LH tail lamp
31. Radio illumination
32. Switch illumination
33. Switch illumination
34. Automatic selector illumination
35. Automatic selector illumination
36. LH door lamps
37. RH door lamps
38. Interior lamp delay
39. Automatic transfer oil temperature switch
40. Diode
41. Interior lamp switch
42. Stop lamp switch
43. Auxiliary lamps delay
44. Rheostat
45. Front cigar lighter illumination
46. Clock illumination
47. Heater illumination
48. Heater illumination
49. Heater illumination
50. Heater illumination
51. LH horn
52. RH horn

53. Tachometer
54. Instrument illumination (6 bulbs)
55. Trailer warning
56. RH indicator warning light
57. LH indicator warning light
58. Rear fog warning light
59. Head lamp warning light
60. High transfer oil temperature warning light
61. Low fuel warning light
62. Multi-function unit in binnacle
63. Fuel indicator gauge
64. Cold start warning light—(carburetter versions only)
65. Differential lock warning light
66. Ignition warning light
67. Brake failure warning light
68. Brake pad wear warning light
69. Oil pressure warning light
70. Park brake warning light
71. Park brake warning light (Australia)
72. Water temperature gauge
73. Headlamp wash timer (option)
74. Headlamp wash pump (option)
75. Head electric mirrors (option)
76. Trailer socket (option)
77. Front screen wash
78. Front wiper delay
79. Wiper motor
80. Steering column switches
81. Cold start warning lamp switch (carburetter only)
82. Differential lock switch
83. Brake failure switch
84. Diode
85. Front brake pad wear
86. Rear brake pad wear
87. Diode
88. Oil pressure switch
89. Park brake switch
90. Pick-up point-park brake warning light (Australia)
91. Water temperature transducer
92. Light switch
93. Rear fog lamp switch
94. Main fuse box
95. Heater motor and switch unit
96. Flasher unit

97. Hazard switch
98. Hazard warning lamp
99. Reverse lamp switch
100. Heated rear screen
101. Starter solenoid
102. Alternator
103. Brake failure warning lamp check relay
104. Fuel tank unit
105. Air conditioning (option)
106. Split charge relay (option)
107. Electric windows and central door locking (option)
108. Under bonnet illumination switch
109. Reverse lamps
110. Fuel pump
111. Pick-up point for petrol ignition wiring
112. Terminal post
113. Battery
114. LH rear speaker (option)
115. RH rear speaker (option)
116. LH front speaker
117. RH front speaker
118. Radio (option)
119. Radio fuse
120. Radio choke
121. Fuel shut-off relay (carburetter models only)
122. Starter solenoid relay
123. Ignition start switch
124. Start inhibitor switch (Automatic)
125. Split charge relay (option)
126. Heated rear window relay
127. Diode
128. Heated rear window switch
129. Voltage switch (option)
130. Heated rear window warning lamp
131. Bonnet lamp
132. Cigar lighter (dash)
133. Cigar lighter (cubby box)
134. Clock
135. Rear screen wash motor
136. Rear wiper delay
137. Rear wash wipe switch
138. Rear wiper relay
139. Rear wiper motor
140. Constant energy unit
141. Distributor
142. Ignition pick-up point (petrol injection only)
143. Not used

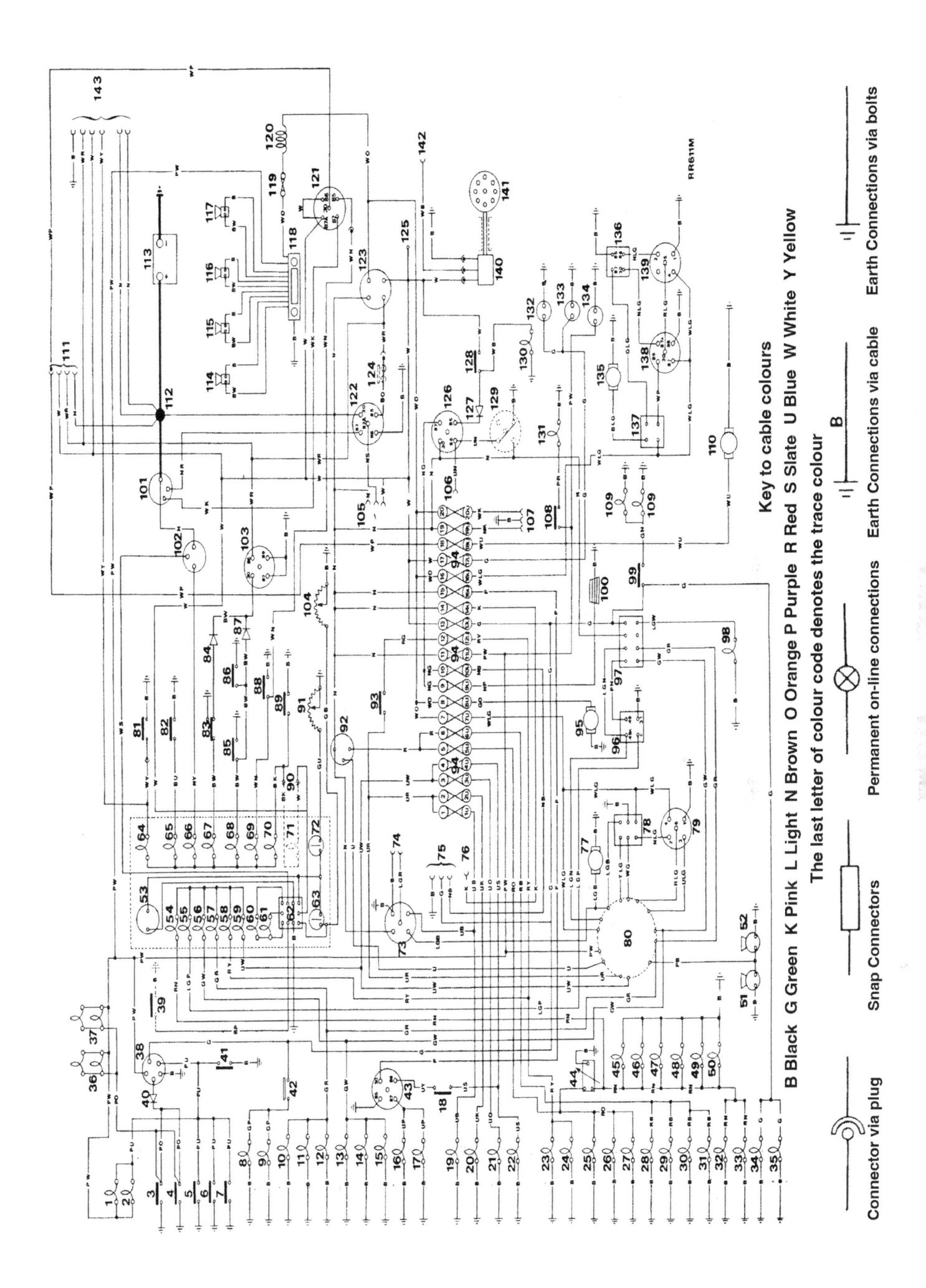

# Range Rover
# 1987
# Workshop Manual Supplement
# Publication Number LSM180WS1 (edition 2)

|  | Section | Page |
|---|---|---|
| Introduction | 01 | 254 |
| General Specification Data | 04 | 255 |
| Engine Tuning Data | 05 | 256 |
| Lubricants and Fluids | 09 | 258 |
| Maintenance | 10 | 260 |
| Fuel System | 19 | 262 |
| Clutch | 33 | 268 |
| Steering | 57 | 269 |
| Brakes | 70 | 273 |
| Body | 76 | 275 |
| Heating & Ventilation | 80 | 278 |
| Air Conditioning | 82 | 279 |
| Wipers & Washers | 84 | 280 |
| Electrical | 86 | 281 |

Land Rover
Lode Lane,
Solihull
West Midlands
B92 8NW
England

Section
Number

## 01 INTRODUCTION
   1

## 04 GENERAL SPECIFICATION DATA

| | |
|---|---|
| Steering pump | 1 |
| Coil | 1 |
| Distributor | 1 |
| Ignition module | 1 |
| Starter motor (Petrol models) | 1 |
| Starter motor (Diesel models) | 1 |
| Tailgate wiper motor | 1 |
| Replacement bulbs | 2 |

## 05 ENGINE TUNE DATA

| | |
|---|---|
| General data | 1 |
| Fuel injection models | 2 |
| Carburetter models (United Kingdom and Europe) | 3 |
| Carburetter models (Rest of World and Gulf States) | 4 |

## 09 LUBRICANTS AND FLUIDS
   1-4

## 10 MAINTENANCE

| | |
|---|---|
| Steering box fluid | 1 |
| Carburetter dampers | 1 |
| Drive belts-adjust | 2 |
| Handbrake cable-adjust | 4 |

## 19 FUEL SYSTEM

| | |
|---|---|
| SU Carburetter | 1 |
|   Overhaul | 6 |
|   Tune and adjust | 10 |
| Fuel pipe layout | |
|   Fuel injection models | 10 |
|   Carburetter models | 11 |
|   Diesel models | 12 |

## 33 CLUTCH
Hydraulic damper (Diesel models only)    1

Section
Number

## 57 STEERING

**Torque Wrench Settings**

| | |
|---|---|
| Steering wheel - remove and refit | 1 |
| Power steering fluid reservoir - remove and refit | 1 |
| Power steering pump (Petrol models only) | |
|   Adjust | 2 |
|   Drive belt - remove and refit | 3 |
|   Pump - remove and refit | 4 |
| Power steering pump (Diesel models only) | |
|   Adjust - remove and refit | 6 |

## 70 BRAKES

| | |
|---|---|
| Description | 1 |
| Hand brake cable - remove and refit - adjust | 3 |
| Brake pressure reducing valve - remove and refit | 4 |

## 76 BODY

| | |
|---|---|
| Front door stowage bins - remove and refit | 1 |
| Lower tailgate release mechanism - remove and refit | 1 |
| Lower tailgate striker - adjust | 2 |
| Assisted bonnet lift - remove and refit | 2 |
| Radiator grille - remove and refit | 4 |
| Fuel filler flap - remove and refit | 4 |
| Asymmetric split rear seat - remove and refit | 5 |
| Locking mechanism - remove and refit | 6 |
| Adjustment | 6 |

## 80 HEATING AND VENTILATION

| | |
|---|---|
| Heater and air conditioning controls - remove and refit | 1 |
| Recirculating/fresh air solenoid switch - remove and refit | 2 |
| Vacuum unit (recirculating/fresh air flap) - remove and refit | 2 |

## 82 HEATER AND AIR CONDITIONING
Circuit diagram    1

## 84 WIPERS

| | |
|---|---|
| Tailgate wiper arm - remove and refit | 1 |
| Tailgate wiper motor - remove and refit | 1 |

# INTRODUCTION 01

## INTRODUCTION

A number of major and minor detail changes are introduced on Range Rover Petrol and Diesel models for the 1987 model year.

Specification for individual vehicles may vary, but all models will include some of the new features and options summarised below.

- General Specification Data (petrol models only).
- Engine Tuning Data (petrol models only).
- Revised drive belt arrangement (petrol models only).
- SU carburetters and linkage
- Steering wheel.
- Power steering pump and fluid reservoir.
- Revised handbrake linkage.
- Front door stowage bins.
- Lower tailgate release mechanism.
- Assisted bonnet lift.
- Radiator grille.
- Fuel filler flap
- Asymmetric split rear seats
- Revised front lamps.
- Revised fuse box.
- Distributor (petrol models only).
- Starter motor.
- Rear wiper motor.
- Rear upper tailgate screen incorporating radio antenna.
- Column control switches.
- Revised instrument binnacle to include new warning symbols for low level coolant and screen wash fluid.
- Exterior driving mirror controls.
- Central locking on fuel filler flap.
- Revised heater and controls.

Service and repair information for the new equipment is included in this supplement, which should be used in conjunction with the existing Workshop Manual LSM 180WM and Diesel Model Supplement LSM 227WS.

Section
Number

## 86 ELECTRICAL

| | |
|---|---|
| Location of electrical equipment | 1 |
| Distributor - Lucas 35 DLM8 | 2 |
| Remove and refit | 3 |
| Overhaul | 4 |
| Ignition coil - remove and refit | 6 |
| Ignition timing - adjust | 6 |
| Lucas constant energy ignition system 35 DLM8 | 7 |
| Preliminary checks | 7 |
| Starter motor Lucas M78R (Petrol models only) | 10 |
| Overhaul and test | 10 |
| Starter motor Bosch 544 (Diesel models only) | 13 |
| Remove, test and refit | 13 |
| Headlamp assembly/sealed beam unit - remove and refit | 18 |
| Headlamp alignment | 18 |
| Sidelight and flasher lamp assembly - right hand and left hand | 19 |
| Remove and refit - bulb replacement | 19 |
| Number plate assembly | 20 |
| Remove and refit - bulb replacement | 20 |
| Direction indicator side repeater lamp | 21 |
| Remove and refit - bulb replacement | 22 |
| Auxiliary switch panel | 23 |
| Relays - identification | 23 |
| Fuse box - identification | 23 |
| Steering column controls - remove and refit | 23 |
| Main lighting switch and rear screen programmed wash/wipe switch | |
| Remove and refit | 23 |
| Windscreen programmed wash/wipe switch and main and dipped beam | |
| Direction indicators and horn switch - remove and refit | 24 |
| Hazard warning switch - bulb replacement | 24 |
| Column switch illumination - bulb replacement | 25 |
| Exterior driving mirror controls | 25 |
| Adjusting | 26 |
| Remove and refit | 26 |
| Instrument binnacle | 27 |
| Warning light symbols | 28 |
| Instrument case (back) | 28 |
| Printed harness connections | 29 |
| Circuit served | 30 |
| Central locking - fuel filler flap actuator unit | |
| Remove and refit | 31 |
| Differential lock warning lamp assembly | 31 |
| Remove and refit - bulb replacement | |
| Radio antenna amplifier - remove and refit | 32 |
| Headlamp dim/dip lighting (United Kingdom only) | 32 |
| Remove and refit control unit | |
| Coolant level sensor - remove and refit | 33 |
| Vacuum sensor - brake servo (Diesel models only) | 33 |
| Remove and refit | 34 |
| Brake check unit - remove and refit | 34 |
| Window lifts and door locks - circuit diagram | 35 |
| Electric mirrors - Circuit diagram | 36 |
| Main circuit diagram - Right hand steering | 38 |
| Main circuit diagram - Left hand steering | |

254

# GENERAL SPECIFICATION DATA | 04

## GENERAL SPECIFICATION DATA

### STEERING PUMP

| | |
|---|---|
| Make/type | Hobourn-Eaton series 200 |
| Operating pressure - straight ahead position - at idle | 7 kgf/cm$^2$ (100 lbf/in$^2$) maximum |
| Full lock (left or right) at idle | 28 kgf/cm$^2$ (400 lbf/in$^2$) minimum |
| Full lock (left or right) 1000 rev/min | 70-77 kgf/cm$^2$ (1000-1110 lbf/in$^2$) |

### COIL

| | |
|---|---|
| Make/type | Lucas 32C5 |

### DISTRIBUTOR

| | |
|---|---|
| Make/type | Lucas 35 DLM8 |
| Rotation | Clockwise |
| Air gap | 0.20 - 0.35mm (0.008 - 0.014) |

### IGNITION MODULE

| | |
|---|---|
| Make/type | Lucas 9EM amplifier module, distributor mounted |

### STARTER MOTOR (Petrol models)

| | |
|---|---|
| Make/type | Lucas M78R pre-engaged |
| Minimum brush length | 3.5mm (0.138 in) |
| Minimum commutator diameter | 28.8mm (1.13 in) |

### STARTER MOTOR (Diesel models)

| | |
|---|---|
| Make/type | Bosch 544 pre-engaged |
| Minimum brush length | 15.5mm (0.610 in) |
| Minimum commutator diameter | 39.5mm (1.55 in) |

### TAILGATE WIPER MOTOR

| | |
|---|---|
| Make/type | IMOS (non-serviceable) |
| Running current, wet screen at 20°C ambient | 1.0 to 2.8 amps |
| Wiper speed, wet screen at 20°C ambient | 37 to 43 cycles per minute |

## Notes

## 05 ENGINE TUNING DATA

### ENGINE TUNING DATA

| | |
|---|---|
| Type | V8 Cylinder |
| Firing order | 1.8.4.3.6.5.7.2 |
| Cylinder Numbers - Left bank | 1.3.5.7. |
| Right bank | 2.4.6.8. |
| No. 1 Cylinder location | Pulley end of left bank |
| Timing marks | On crankshaft vibration damper |
| Spark plugs | |
| Make/type | Champion N9YC |
| Gap | 0.85 - 0.95mm (0.033 - 0.038 in) |
| Distributor | |
| Make/type | Lucas 35DLM8 Electronic |
| Air gap | 0.20 - 0.35mm (0.008 - 0.014 in) |

## 04 GENERAL SPECIFICATION DATA

### REPLACEMENT BULBS

| | TYPE | | |
|---|---|---|---|
| Headlamps | 12V | 60/55W | (Halogen) sealed beam |
| Headlamps - France amber | 12V | 60/55W | (Halogen) |
| Auxiliary driving lamps | 12V | 55W H3 | (Halogen) |
| Sidelamps (Exterior lights) | 12V | 5W | bayonet fitting |
| Stop/tail lamps | 12V | 5/21W | bayonet fitting |
| Reverse lamps | 12V | 21W | bayonet fitting |
| Rear fog guard lamps | 12V | 21W | bayonet fitting |
| Direction indicator lamps | 12V | 21W | bayonet fitting |
| Number plate lamps | 12V | 5W | capless |
| Instrument panel lamps and warning lamps | 12V | 1.2W | bulb/holder unit |
| Ignition warning lamp (Instrument panel) | 12V | 2W | capless |
| Interior roof lamps | 12V | 10W | 'Festoon' |
| Clock illumination | 12V | 2W | bayonet fitting |
| Cigar lighter illumination | 12V | 1.2W | capless |
| Door edge/puddle lamps (Interior lights) | 12V | 5W | capless |
| Auxiliary switch panel illumination (green) | 12V | 1.2W | capless |
| Heated rear screen warning lamp (amber) | 12V | 1.2W | capless |
| Hazard warning lamp | 12V | 1.2W | capless |
| Automatic graphics illumination | 24V | 5W | capless |
| Heater/air conditioning graphics illumination | 12V | 1.2W | capless |
| Differential lock warning lamp | 12V | 2W | bayonet fitting |
| Column switch illumination | 12V | 1.2W | capless |
| Rear fog warning lamp (amber) | 12V | 1.2W | capless |

256

# ENGINE TUNING DATA

## FUEL INJECTION MODELS

Compression ratio .................... 9.35:1 and 8.13:1
Fuel injection system ............... Lucas 'L' system

### Valve timing

|  | Inlet | Exhaust |
|---|---|---|
| Opens | 24° BTDC | 62° BBDC |
| Closes | 52° ABDC | 14° ATDC |
| Duration | 256° | 256° |
| Valve peak | 104° ATDC | 114° BTDC |

Idle speed ............................. 700 - 800 rev/min

### Ignition timing

Ignition timing at ..................... 600 rev/min *

### Ignition timing

Dynamic or static .................... TDC ± 1°
Exhaust gas CO content at idle .... 0.5 - 1.0% max

### Distributor

Make/type ............................ Lucas 35 DLM8 electronic
Rotation ............................... Clockwise
Air gap ................................ 0.20mm - 0.35mm (0.008 - 0.014 in)
Serial number ........................ 42649

### Centrifugal advance

Distributor decelerating speeds * ... -1600 distributor advance
                                       -1100
                                       -600
No centrifugal advance below ....... 100 rev/min

Fuel .................................... 94 min octane - 9.35:1
                                       90 min octane - 8.13:1

*Vacuum pipe disconnected

11° to 14°
9° to 11°
1° 30' to 4°

---

# ENGINE TUNING DATA

## UK AND EUROPE

### CARBURETTER MODELS

Compression ratio .................... 9.35:1

### Valve timing

|  | Inlet | Exhaust |
|---|---|---|
| Opens | 36° BTDC | 74° BBDC |
| Closes | 64° ABDC | 26° ATDC |
| Duration | 280° | 280° |
| Valve peak | 99° ATDC | 119° BTDC |

### Carburetters

Type .................................. 2 x SU HIF44
Specification number ............... FZX 2006
Needle ................................ BGD
Idle speed (engine hot) ............ 700 - 800 rev/min
Fast idle speed (engine hot) ...... 1050 - 1150 rev/min
Mixture setting
-CO at idle .......................... 0.5% - 2.5% pulsair connected

### Ignition

Distributor make/type ............... Lucas 35 DLM8 electronic
Direction of rotation ................ Clockwise
Air gap ................................ 0.20mm - 0.33mm (0.008 - 0.014 in)
Distributor serial number .......... 42650

### Centrifugal advance

Decelerating check with vacuum pipe disconnected

Distributor decelerating speeds ..... -2800 Distributor advance     5° 30' to 9°
                                     -1450                          6° to 8°
                                     -1000                          2° 30' to 4° 30'

No advance below ..................... 250 rev/min

### Ignition timing

Dynamic or static .................... 6° ± 1° BTDC at 750 rev/min max (vacuum pipe disconnected)

Fuel .................................... 94 min Octane

Delay valve ........................... Orange (manual)
                                       Blue (automatic)

# LUBRICANTS AND FLUIDS 09

## Recommended Lubricants and fluids

Use only the recommended grades of oil set out below.

These recommendations apply to temperate climates where operational temperatures may vary between -10°C (14°F) and 35°C (95°F).

| COMPONENTS | BP | CASTROL | DUCKHAM | ESSO | MOBIL | PETROFINA | SHELL | TEXACO |
|---|---|---|---|---|---|---|---|---|
| Petrol engine sump Dash pots (Carburetter models only) Oil can | BP Visco 2000 (15W/40) or BP Visco Nova (10W/30) | Castrol GTX (15W/50) or Castrolite (10W/40) | Duckhams 15W/50 Hypergrade Motor Oil | Esso Superlube (15W/40) | Mobil Super 10W/40 or Mobil 1 Rally Formula | Fina Supergrade Motor Oil 15W/40 or 10W/40 | Shell Super Motor Oil 15W/40 or 10W/40 | Havoline Motor Oil 15W/40 or Eurotex HD (10W/30) |
| Diesel engine sump | BP Vanellus C3 Extra (15W/40) | Castrol Turbomax (15W/40) | | Mobil Delvac 1400 Super (15W/40) | | | Shell Myrina (15W/40) | |
| | The following list of oils are for emergency use only if the above oils are not available. They can be used for topping up without detriment, but if used for engine oil changing, they are limited to a maximum oil 5,000 km (3,000 miles) between oil and filter changes. Use only oils to MIL-L-2104D or CCMC D2 or API Service levels CD or SE/CD - 15W/40 | | | | | | | |
| Automatic gearbox | BP Vanellus C3 Multigrade 15W/40 | Castrol Deusol RX Super | Duckhams Hypergrade (15W/40) | Esso Essolube XD-3(15W/40) | Mobil Delvac Super (15W/40) | Fina Dilano HPD (15W/40) | Shell Rimula X (15W/40) | Texaco URSA Super Plus (15W/40) |
| | BP Autran DX2D | Castrol TQ Dexron IID | Duckhams Fleetmatic CD or Duckhams D-Matic | Esso ATF Dexron IID | Mobil ATF 220D | Fina Dexron IID | Shell ATF Dexron IID | Texamatic Fluid 9226 |
| Manual gearbox | BP Autran G | Castrol TQF | Duckhams Q-Matic | Esso ATF Type G | Mobil ATF 210 | Fina Purfimatic 33C | Shell Donax TF | Texamatic Type G |
| Front and Rear differential Swivel pin housings and LT230T Transfer gearbox | BP Gear Oil SAE 90EP | Castrol Hypoy SAE 90EP | Duckhams Hypoid 90 | Esso Gear Oil GX 85W/90 | Mobil Mobilube HD90 | Fina Pontonic MP SAE 80W/90 | Shell Spirax 90EP | Texaco Multigear Lubricant EP 85W/90 |
| Propeller shaft Front and Rear | BP Energrease L2 | Castrol LM Grease | Duckhams LB 10 | Esso Multi-purpose Grease H | Mobil Grease MP | Fina Marson HTL 2 | Shell Retinax A | Marfak All Purpose Grease |
| Power steering box and fluid Reservoir | BP Autran DX2D * | Castrol TQ Dexron IID * | Duckhams Fleetmatic CD or Duckhams D-Matic * | Esso ATF Dexron IID * | Mobil ATF 220D * | Fina Dexron II * | Shell ATF Dexron IID * | Texamatic Fluid 9226 * |
| Brake and clutch reservoirs | Brake fluids having a minimum boiling point of 260°C (500°F) and complying with FMVSS 116 DOT3 or DOT4 | | | | | | | |
| Lubrication nipples (hubs, ball joints etc.) | BP Energrease L2 | Castrol LM Grease | Duckhams LB 10 | Esso Multi-purpose Grease H | Mobil Grease MP | Fina Marson HTL 2 | Shell Retinax A | Marfak All Purpose Grease |
| Ball joint assembly Top Link | Dextragrease Super GP | | | | | | | |
| Seat slides Door lock striker | BP Energrease L2 | Castrol LM Grease | Duckhams LB 10 | Esso Multi-purpose Grease H | Mobil Grease MP | Fina Marson HTL 2 | Shell Retinax A | Marfak All purpose grease |

NLGI-2 Multi-purpose Lithium-based Grease

* Or fluids listed for manual gearbox

---

# ENGINE TUNING DATA 05

## R.O.W./GULF STATES

### CARBURETTER MODELS

Compression ratio .................................. 8.13:1

### Valve timing

| | Inlet | Exhaust |
|---|---|---|
| Opens | 30° BTDC | 68° BBDC |
| Closes | 75° ABDC | 37° ATDC |
| Duration | 285° | 285° |
| Valve peak | 106° ATDC | 112° BTDC |

### Carburetters

Type .................................. 2 x HIF44
Specification number .................................. FZX 2005
Needle .................................. BGC
Idle speed (engine hot) .................................. 700 - 800 rev/min
Fast idle speed (engine hot) .................................. 1050 - 1150 rev/min
Mixture setting - CO at idle .................................. 0.5% to 2.5%

### Ignition

Distributor make/type .................................. Lucas 35DLM8 electronic
Direction of rotation .................................. Clockwise
Air gap .................................. 0.20mm - 0.35mm (0.008 - 0.014 in)
Distributor serial number .................................. 42652

### Centrifugal advance

Decelerating check with vacuum pipe disconnected

Distributor decelerating speeds
- -2300 Distributor advance 10° 30' to 12° 30'
- -1800 8° to 10°
- -1200 3° 30' to 5° 30'

No advance below .................................. 450 rev/min

### Ignition timing

Dynamic or stati .................................. 6° ± 1° BTDC at 750 rev/min max (vacuum pipe disconnected)

Fuel .................................. 90 min octane

# LUBRICANTS AND FLUIDS

| | |
|---|---|
| Bonnet plinth | Graphite Lock Grease Type 'R' |
| Door locks (anti-burst) Inertia reels | **DO NOT LUBRICATE.** These components are 'life' lubricated at the manufacturing stage. |
| Battery lugs Earthing surfaces Where paint has been removed | Petroleum jelly **NOTE:** Do not use silicon grease |
| Fuel | **Petrol engines** - 9.35:1 97 octane (4 star rating in UK)<br>- 8.13:1 90 octane (2 star rating in UK) with standard ignition timing<br>**Diesel engines** - 22:1 Diesel fuel oil, distillate, diesel fuel, automotive gas oil or Derv fuel to British standard 2869, 1967 Class A1 |
| Windscreen washers | Universal screen washer fluid |
| Engine cooling | For all Petrol and Diesel Models use an ethylene glycol based anti-freeze (containing no methanol) with non-phosphate corrosion inhibitors suitable for use in aluminium engines to ensure the protection of the cooling system against frost and corrosion in all seasons.<br>Use one part anti-freeze to one part water for protection down to -36°C (-33°F) **IMPORTANT:** Coolant solution must not fall below proportions one part anti-freeze to three parts water, i.e. minimum 25% anti-freeze in coolant otherwise damage to engine is liable to occur.<br>When anti-freeze is not required, the cooling system must be flushed out with clean water and refilled with a solution of one part Marstons SQ36 inhibitor to nine parts water i.e. minimum 10% inhibitor in coolant. |
| Air conditioning system Refrigerant | **METHYLCHLORIDE REFRIGERANTS MUST NOT BE USED** Use only with refrigerant 12. This includes 'Freon 12' and 'Arcton 12' |
| Compressor oil | Shell Clavus 68, BP Energol LPT68, Sunisco 4GS Texaco Capella E Wax Free 68 |

## LUBRICANTS AND FLUIDS

### RECOMMENDED LUBRICANTS AND FLUIDS - ALL CLIMATES AND CONDITIONS

| COMPONENTS | SERVICE CLASSIFICATION | | AMBIENT TEMPERATURE °C |
|---|---|---|---|
| | Specification | SAE Classification | -30 -20 -10 0 10 20 30 40 50 |
| **Petrol models** Engine sump Dash pots (carburetter models only) Oil can | Oils must meet BLS.22.OL.07 or | 5W/30 5W/40 5W/50 | |
| | CCMC3 or API service levels SF | 10W/30 10W/40 10W/50 | |
| | Oils must meet BLS.22.OL.02 | 15W/40 15W/50 | |
| | or CCMC G1 or G2 | 20W/40 20W/50 | |
| | or API service levels SE or SF | 25W/40 25W/50 | |
| Diesel models engine sump | SHPD oils meeting CCMC D3 | 10W/30 15W/40 | |
| | * **Emergency only:** Oils meeting MIL-L-2104D or CCMCD2 or API CD | | |
| Main Gearbox Automatic | ATF Dexron IID | | |
| Main Gearbox manual | ATF M2C33 (F or G) | | |
| Transfer gearbox Final drive units Swivel pin housings | API GL4 or GL5 MIL-L-2105 or MIL-L-2105B | 90 EP 80W EP | |
| Power steering | ATF M2C 33 (F or G) or ATF Dexron IID | | |

* Oils for emergency use only if the above oils are not available. They can be used for topping up without detriment, but if used for engine oil changing, they are limited to a maximum of 5,000 km (3,000 miles) between oil and filter changes.

## MAINTENANCE

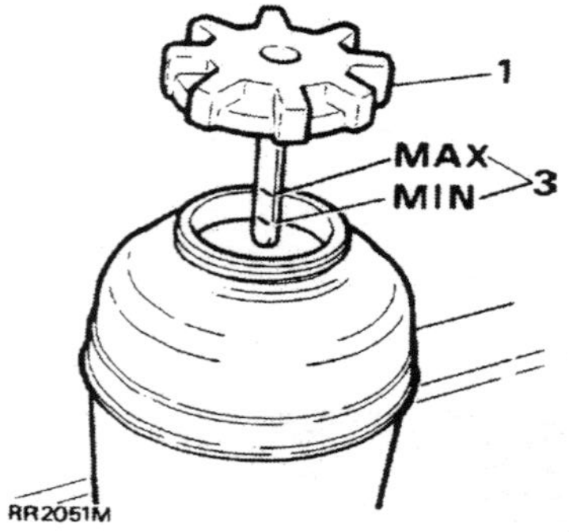

### CARBURETTER DAMPER- Topping up

1. Unscrew the cap from the top of the carburetter suction chamber and withdraw the cap and plunger.
2. Top-up with clean engine oil to bring the level to the top of the hollow piston rod.
3. Screw the cap firmly into the carburetter.

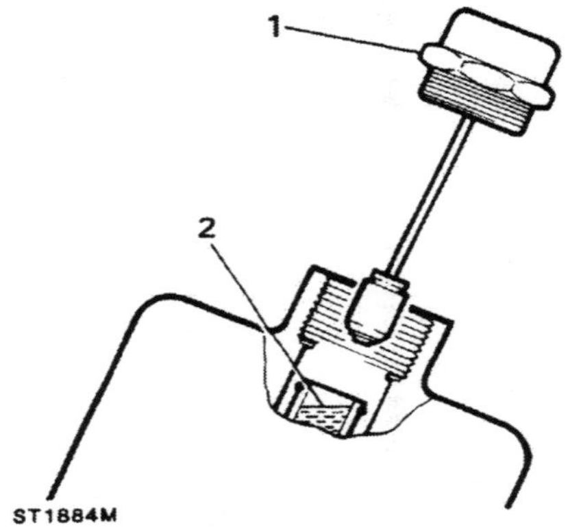

### MAINTENANCE

### TOP UP STEERING BOX FLUID

1. Clean and remove the reservoir cap.
2. Wipe the reservoir dipstick clean and refit the cap.
3. Remove the cap again and check that the fluid level registers between the **MAX** and **MIN** level markings on the dipstick.
4. Top up as necessary and refit the cap.

### MAINTENANCE SCHEDULES

All Range Rovers fitted with a manual gearbox must have the main and transfer gearbox oils changed at 20,000 km (12,000 mile) service intervals.

## LUBRICANTS AND FLUIDS

| | |
|---|---|
| Lubrication nipples (hubs, ball joints propeller shafts, etc) | NLGI-2 Multi-purpose lithium based grease |
| Brake and clutch | Universal Brake Fluids or other Brake Fluids having a minimum boiling point of 260°C (500°F) and complying with FMVSS 116 DOT3 or DOT4 |
| Windscreen | Universal screen washer fluid |
| Engine cooling system | For all Petrol and Diesel models use an ethylene glycol based anti-freeze (containing no methanol) with non-phosphate corrosion inhibitors suitable for use in aluminium engines to ensure the protection of the cooling system against frost and corrosion in all seasons. Use one part anti-freeze to one part water for protection down to -36°C (-33°F) **IMPORTANT:** Coolant solution must not fall below proportions of one part anti-freeze to three parts water i.e. minimum 25% anti-freeze in coolant otherwise damage to engine is liable to occur. When anti-freeze is not required, the cooling system must be flushed out with clean water and refilled with a solution of one part Marstons SQ36 inhibitor to nine parts water, i.e. minimum 10% inhibitor in coolant. |
| Air conditioning Refrigerant | **METHYCHLORIDE REFRIGERANTS MUST NOT BE USED** Use only with refrigerant 12. This includes 'Freon 12' and 'Arcton 12' |
| Compressor oil | Shell Clavus 68, BP Energol LPT 68, Sunisco 4GS, Texaco Capella E Wax Free 68. |

260

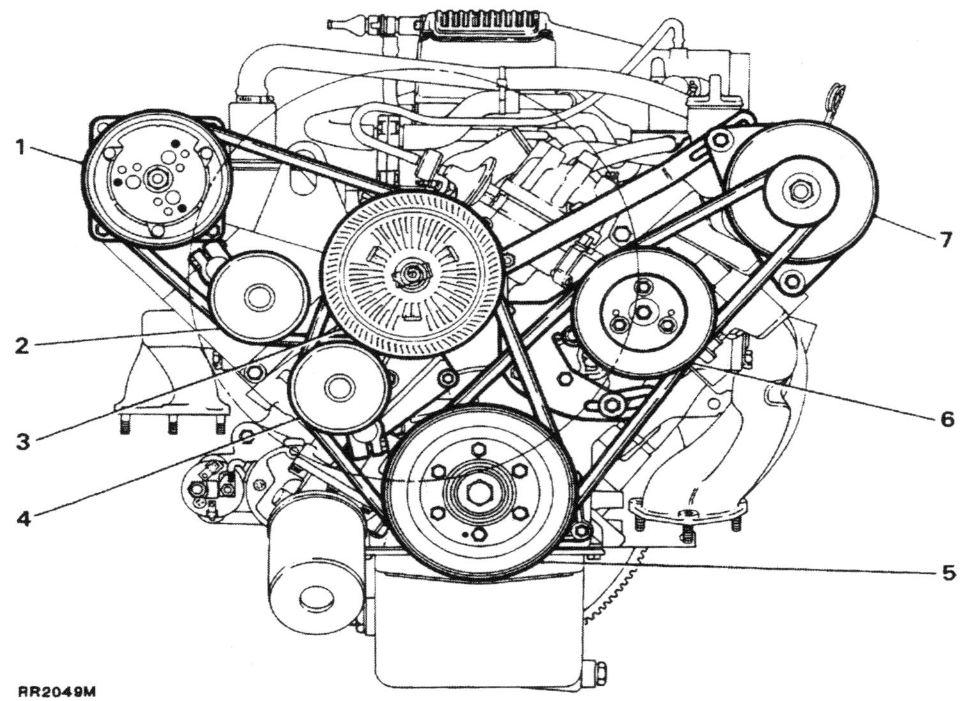

# MAINTENANCE

1. Air conditioning compressor.
2. Jockey wheel.
3. Viscous fan-water pump unit.
4. Jockey wheel.
5. Crankshaft.
6. Power steering pump.
7. Alternator.

**WARNING: DISCONNECT THE BATTERY NEGATIVE TERMINAL BEFORE ADJUSTING DRIVE BELTS TO AVOID THE POSSIBILITY OF THE VEHICLE BEING STARTED.**

## MAINTENANCE

ILLUSTRATION A

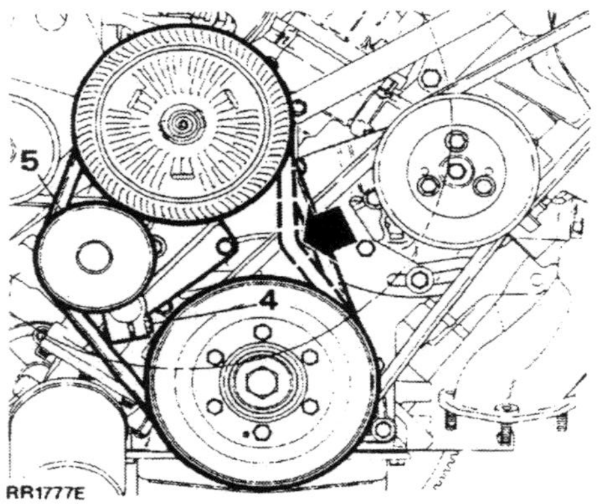

ILLUSTRATION B

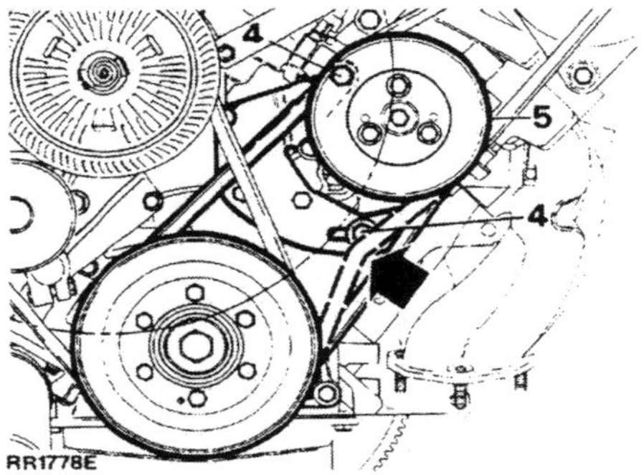

ILLUSTRATION C

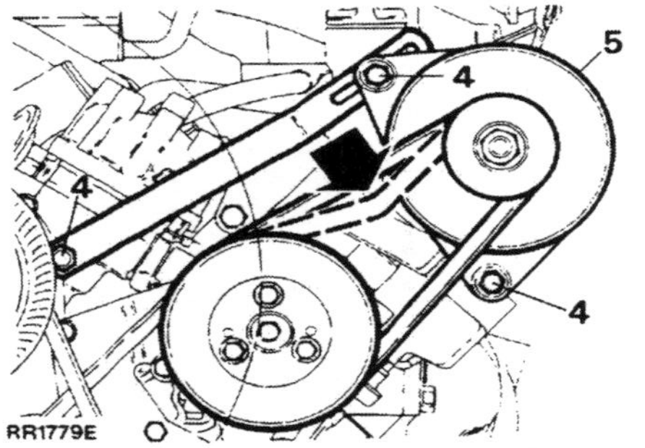

### DRIVE BELTS-adjust or renew (Petrol models only)

#### COMPRESSOR DRIVE BELT

The belt must be tight with not more than 4 to 6mm (0.19 to 0.25 in) total deflection when checked by hand midway between the pulleys on the longest run.

Where a belt has stretched beyond the limits, a noisy whine or knock will often be evident during operating, if necessary adjust as follows:

1. Slacken the jockey wheel securing bolt.
2. Adjust the position of the jockey wheel until the correct tension is obtained.
3. Tighten the securing bolt and re-check the belt tension.

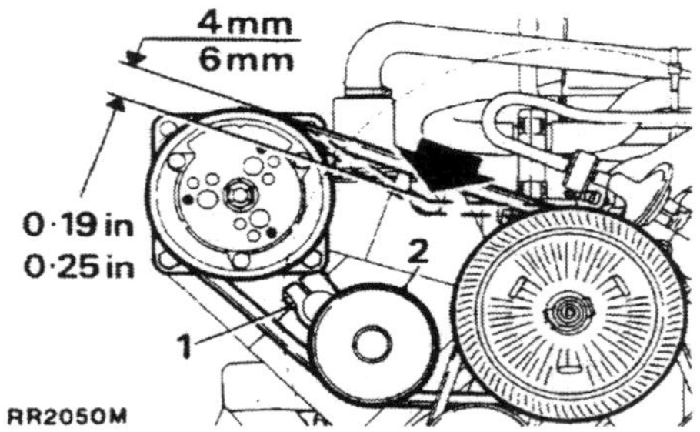

### Check driving belts, adjust or renew as necessary

1. Examine the following belts for wear and condition and renew if necessary.

   (A) Crankshaft-Jockey Pulley-Water pump
   (B) Crankshaft-Steering Pump
   (C) Steering Pump-Alternator

# MAINTENANCE

2. Each belt should be sufficiently tight to drive the appropriate auxiliary unit without undue load on the bearings.
3. Slacken the bolts securing the unit to its mounting bracket.
4. Slacken the appropriate pivot bolt or jockey wheel and the fixing at the adjustment link where applicable.
5. Pivot the unit inwards or outwards as necessary and adjust until the correct belt tension is obtained.
6. Belt tension should be approximately 11 to 14mm (0.437 to 0.562 in) at the points denoted by the bold arrows.
7. Tighten all unit adjusting bolts. Check adjustment again, when a new belt is fitted, after approximately 1,500 km (1,000 miles) running.

## ADJUST HANDBRAKE CABLE

The handbrake lever acts on a transmission brake drum at the rear of the transfer box.

1. Set the vehicle on level ground and select 'P' in main gearbox. Disconnect the battery negative lead.
2. Fully release the handbrake.
3. From underneath the vehicle slacken the two locknuts securing the handbrake outer cable to the mounting bracket, to enable the brake drum to be adjusted without putting any tension on the handbrake outer cable.

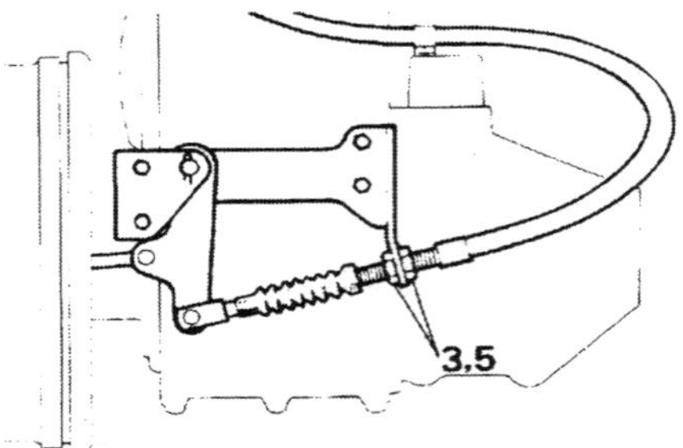

RR189E

4. Rotate the adjuster on the brake drum back plate clockwise, until the brake shoes are fully expanded against the drum.

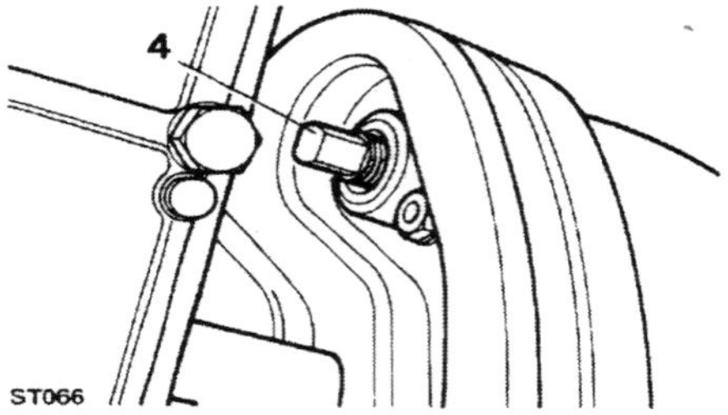

ST066

5. Rotate the two outer cable locknuts until contact is made with the mounting bracket, tighten the two nuts consecutively to prevent any movement occurring on the outer cable.
6. Slacken the adjuster on the back of the brake drum until the handbrake lever becomes fully operational on the second or third notch of the handbrake ratchet.
7. Lightly grease the handbrake linkage with a general purpose grease.

**CAUTION: DO NOT over-adjust the handbrake, the drum must be free to rotate when the handbrake is released, otherwise serious damage will result.**

# FUEL SYSTEM

## CARBURETTER OVERHAUL–S.U. HIF 44

**Right hand**
(Horizontal integral float chamber)

### DISMANTLE

1. Remove the carburetters from the engine and clean the exteriors with a suitable solvent.
2. Remove the two nuts and spring washers and withdraw the air intake adaptor and joint washer.
3. Unscrew and remove the piston damper assembly and drain the oil.
4. Remove the three screws and lift-off the suction chamber complete with piston and spring.
5. Remove the spring clip from the top of the piston rod and withdraw the piston and spring.
6. Unscrew the metering needle guide locking screw. Attempt to remove assy with fingers. If difficulty is experienced then holding the needle as close to the piston as possible in a soft jawed vice with a sharp pull, withdraw the needle, guide and spring assembly.
7. Remove the four screws and withdraw the float chamber cover plate and sealing ring.
8. Remove jet adjusting lever retaining screw and spring.
9. Withdraw the jet complete with the bi-metal lever and separate the lever from the jet.
10. Unscrew and remove the float pivot spindle and plain washer, and remove the float.
11. Lift-out the needle valve.
12. Unscrew and remove the needle valve and filter.
13. Unscrew and remove the jet bearing nut.
14. Invert the carburetter body to allow the jet bearing to fall out. If the bearing sticks, carefully tap it out from the bridge side.
15. Remove the piston guide peg.
16. Remove the suction chamber-to-body sealing ring.
17. Unscrew and remove the mixture adjusting screw and seal. Use thin nosed pliers to finally withdraw the screw.
18. Bend-back the cam lever nut lock tabs and remove the nut and lock washer.
19. Remove the cam lever and spring.
20. Remove the end seal cover and seal.
21. Remove the two screws and withdraw the cold start valve body and seal together with the cold valve spindle. Also collect the paper joint washer.
22. Note the position of the throttle levers and return spring.
23. Bend-back the lock washer tabs and remove the throttle lever nut.
24. Remove the lock washer, bush washer and throttle actuating lever.
25. Release the throttle return spring and remove the throttle adjusting lever from the throttle butterfly spindle and remove the return spring.
26. Hold the butterfly closed and mark the relationship of the butterfly to the carburetter flange.
27. Remove the butterfly two retaining screws and withdraw the butterfly from the spindle.
28. Withdraw the throttle butterfly spindle from the carburetter body together with the two seals.
29. Clean all components with petrol or de-natured alcohol ready for inspection. Do not use abrasives for the removal of stains or deposits.

### INSPECTION

30. Examine the throttle spindle and bearings for excessive axial clearance.
31. Check the float needle and seating for wear and the float for punctures and renew if necessary.
32. Check the condition of all rubber seals, 'O' rings and joint washers and renew if necessary. The float cover plate seal must be renewed.
33. Examine the carburetter body for cracks and damage.
34. Ensure that the inside of the suction chamber is clean and fit the piston into the chamber without the spring. Hold the assembly horizontally and spin the piston. The piston should spin freely in the suction chamber without any tendency to stick.
35. Inspect the metering needle for wear, scores and distortion. Check also that it has the correct designation number - see Engine Tuning Data, Section 05.

# FUEL SYSTEM

36. Examine the bi-metal jet lever for cracks.
37. Check all springs for cracks and distortion.

## ASSEMBLE

### Fit throttle butterfly

38. Fit the throttle spindle to the carburetter body and insert the throttle disc into the spindle in its original position. Secure the disc with new screws and ensure that before tightening the throttle disc is correctly positioned and closes properly. Splay the split ends of the screws to prevent turning.
39. Fit new seals to both ends of the throttle spindle ensuring that they are fitted the correct way round.

### Fit cold start assembly

40. Fit a new 'O' ring to the valve body and assemble the valve spindle to the valve body.
41. Fit a new paper joint washer to the valve noting that the half-moon cut-out in the washer is clearance for the top retaining screw.
42. Fit the starter assembly to the carburetter body and secure with the two screws.
43. Fit the end seal and cover.

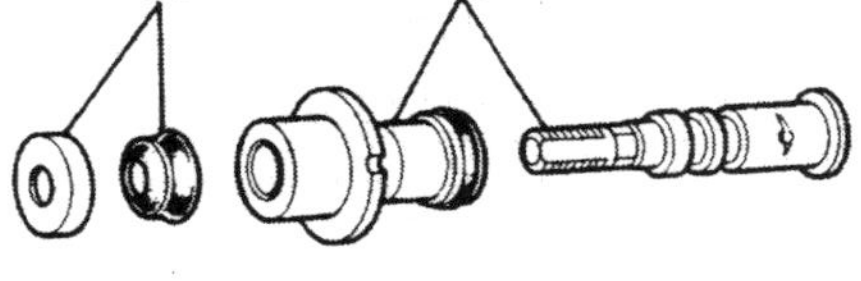

ST1873M

44. Fit the return spring.
45. Fit the cam lever and tension the spring. Fit a new lock washer and secure with the nut and bend the tabs over a convenient flat.
46. Adjust the coils of the spring, if necessary, to prevent coil binding.

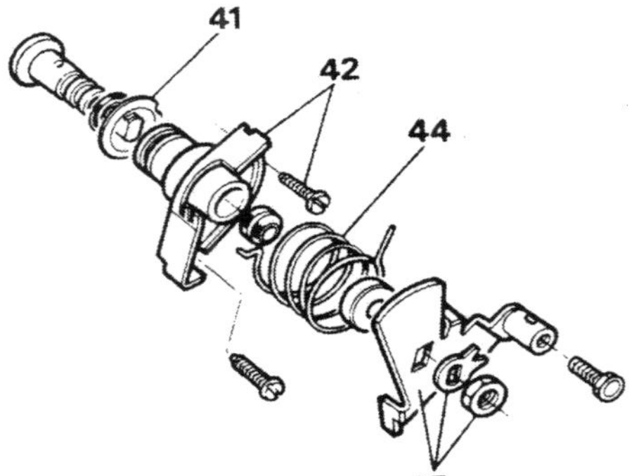

ST1874M

## KEY TO S.U. CARBURETTER COMPONENTS

1. Piston damper.
2. Spring clip.
3. Suction chamber.
4. Piston.
5. Piston spring.
6. Suction chamber retaining screws -3 off.
7. Needle retaining screw.
8. Needle bias spring.
9. Needle guide.
10. Needle.
11. Suction chamber sealing ring.
12. Throttle adjusting screw and seal.
13. Piston key and retaining screw.
14. Mixture adjusting screw and seal.
15. Carburetter body.
16. Throttle butterfly and retaining screws.
17. Throttle spindle.
18. Throttle spindle seals - 2 off.
19. Float chamber.
20. Float chamber cover and retaining screws.
21. Float chamber cover seal.
22. Jet assembly.
23. Jet bearing.
24. Jet bearing nut.
25. Bi-metal jet lever.
26. Jet retaining screw and spring.
27. Float needle.
28. Float needle seat.
29. Float needle seat filter.
30. Float.
31. Float pivot spindle.
32. Cold start and cam lever assembly.
33. Throttle adjusting lever and lost motion assembly.
34. Throttle actuating lever.
35. Bush washer.
36. Throttle lever assembly retaining nut and lock washer.

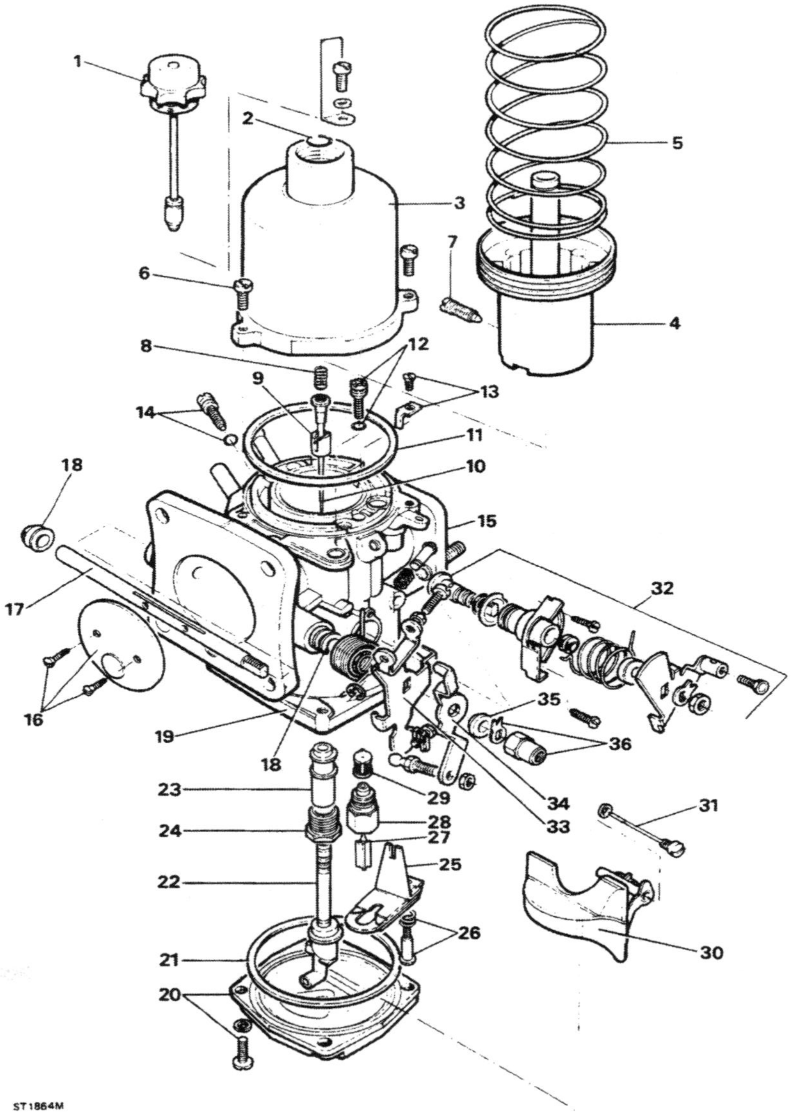

# FUEL SYSTEM

## Fit throttle lever assembly

47. Fit the return spring so that the longest leg rests against the throttle adjusting screw housing.
48. Fit the throttle adjusting lever and lost motion assembly and tension the return spring.
49. Fit the throttle actuating lever.
50. Fit the bush washer and lock washer.
51. Fit and tighten the special nut and bend the lock tabs over a convenient flat.

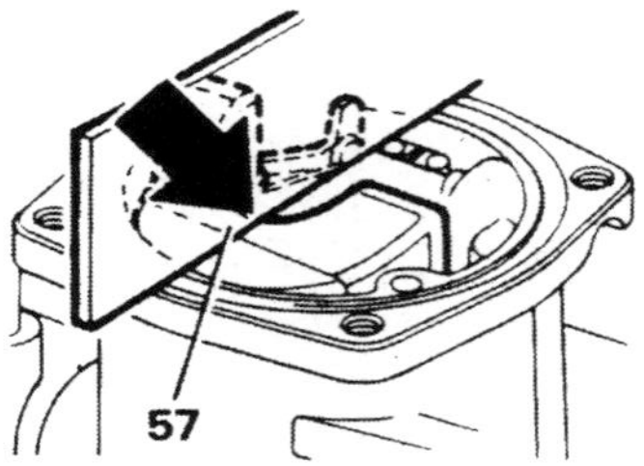

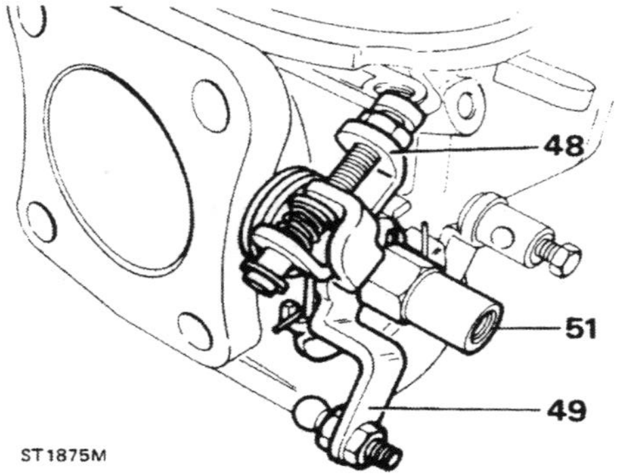

## Fit jet and float assembly

52. Fit the jet bearing, long end towards the float.
53. Fit the jet bearing nut.
54. Clean or renew the filter and fit the float needle seat.
55. Fit the needle valve, spring loaded pin uppermost.
56. Fit the float and secure with the pivot pin.
57. Hold the carburetter in the inverted position so that the needle valve is closed by the weight of the float only. Check using a straight edge that the point on the float, arrowed on the illustration, is 0.5 to 1.5mm (0.020 to 0.059 in) below the level of the float chamber face, dimension 'A'.
58. Adjust the float position by carefully bending the brass pad until the correct dimension is achieved. After adjustment, check that the float pivots freely about the spindle.
59. Assemble the jet to the bi-metal jet lever and ensure that the jet head moves freely in the cut-out.
60. Fit the jet and bi-metal jet lever to the carburetter and secure with the spring loaded jet retaining screw.
61. Fit the mixture adjusting screw.

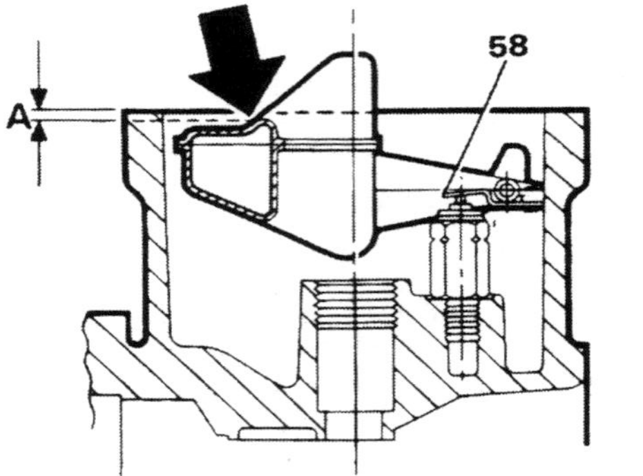

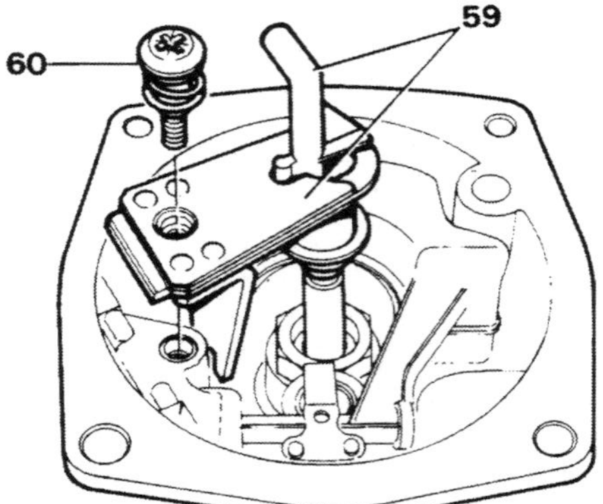

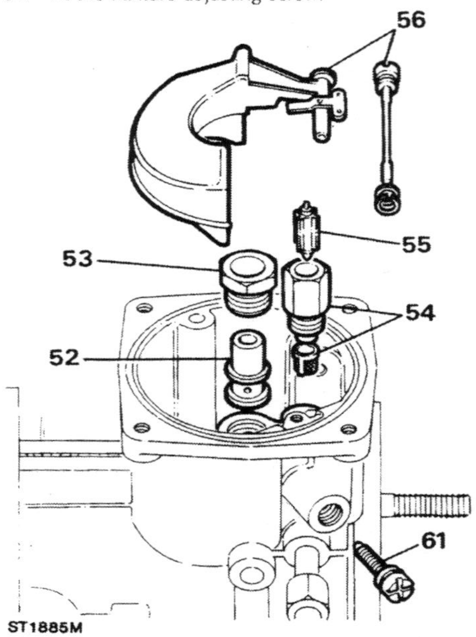

## Fit piston and suction chamber

64. Fit the needle, spring and guide assembly to the piston ensuring that the etched arrow head on the needle locating guide is aligned between the piston transfer holes, as illustrated.
65. Secure and ensure that when the screw is tightened the guide is flush with the piston and that the screw locates in the guide slot.
66. Fit the piston key to the carburetter body using a new screw. Tighten the screw and splay the end.
67. Fit a new suction chamber sealing ring to the groove in the carburetter body.
68. To prevent the piston spring being 'wound-up' during assembly, temporarily fit the piston and suction chamber less the spring to the body, and pencil mark the relationship of the chamber to the body. Remove the suction chamber and fit the spring to the piston. Hold the suction chamber above the spring and piston, align the pencil marks and lower the chamber over the spring and piston, taking care not to rotate the suction chamber. Secure the chamber to the body with the three screws, tightening evenly and check that the piston moves freely.
69. Hold the piston at the top of its stroke and fit the spring clip.
70. Fit the piston damper.
71. Using a new joint washer, fit the air intake adaptor and secure with the two nuts and spring washers.

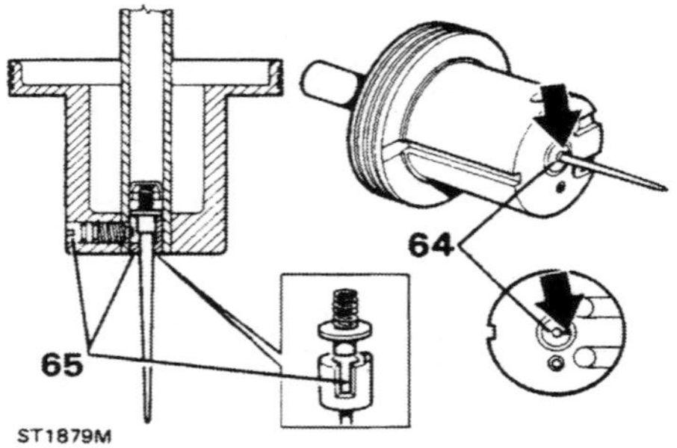

62. Adjust until the jet is flush with the carburetter bridge, then re-adjust jet three and one-half turns clockwise.

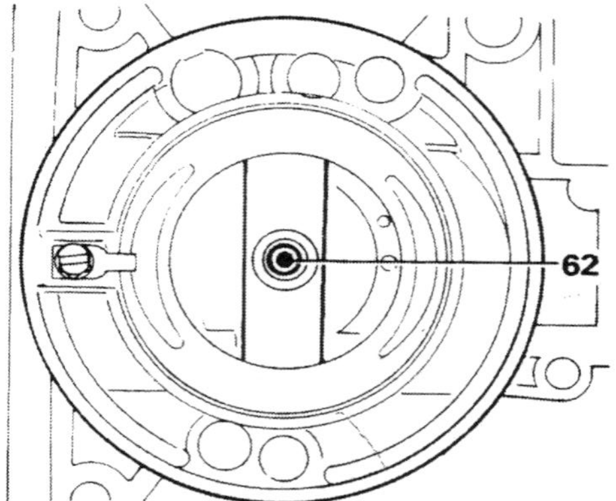

63. Using a new sealing ring, fit the float chamber cover, noting that it can only be fitted one way. Secure with the four screws and spring washers and evenly tighten.

# FUEL SYSTEM

72. Fit the carburetters to the inlet manifold, ensuring that the joint washers, deflector and insulator are fitted in the sequence illustrated. The insulator must be fitted with the arrow head uppermost and pointing inwards towards the manifold. Secure with the four nuts and spring washers and tighten evenly to the correct torque.

   A. Joint washer.
   B. Deflector-teeth pointing inwards.
   C. Joint washer.
   D. Insulator.
   E. Joint washer.

73. Connect the linkages, tune and adjust the carburetters.

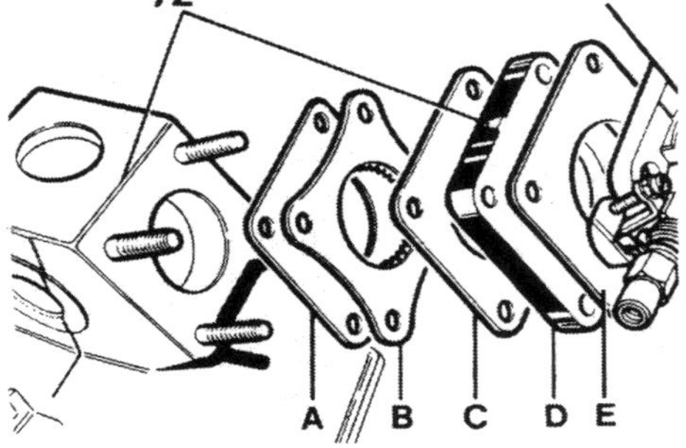

## TUNE AND ADJUST - SU HIF 44 CARBURETTERS

**Special tools:**

Carburetter balancer 605330 or B89 Non - dispersive infra - red exhaust gas analyser.

### General Requirements Prior To Tuning Carburetters.

Accurate engine speed is essential during carburetter tuning, therefore the distributor pick up air gap and ignition timing must be checked together with the vacuum advance system.

Whenever possible the ambient air temperature of the tuning environment should be between 15° to 26° C. (60° to 80° F). When checking engine speed, use an independent and accurate tachometer.

Idling adjustments should be carried out on a fully warmed up engine, that is, at least 5 minutes after the thermostat has opened. This should be followed by a run of one minute duration at an engine speed of approximately 2,500 rev/min before further adjustments or checks are carried out. This cycle may be repeated as often as required. It is important that the above cycle is adhered to, otherwise overheating may result and settings may be incorrect. The piston dampers must always be kept topped-up with the correct grade of oil.

Before any attempt is made to check settings a thorough check should be carried out to ensure that the throttle linkage between the pedal and carburetters is free and has no tendency to stick. Ensure that the choke control lever is is pushed fully down.

**NOTE: References to left and right hand are as from the drivers seat.**

### TAMPER - PROOFING

To comply with E.C.E regulations the idle speed and mixture adjusting screws must be tamper - proofed following any adjustments. A red blanking plug; **Part number - JZX 1258** must be fitted into the mixture screw recess and a red cap; **Part number - JZX 1197** fitted over the idle adjustment screw (throttle adjustment screw).

## TUNE AND ADJUST

The following instructions apply to both carburetters unless otherwise stated.

### CARBURETTER BALANCE

Using **balancer 605330**

1. Disconnect the inter connecting link between the two carburetters, fit the balancer to the carburetter intakes and ensure that there are no air leaks, if necessary, zero the gauge with the adjustment screw.

2. Start the engine, and if necessary allow it to reach normal operating temperature. If the needle moves to the right, decrease the air flow through the left hand carburetter by unscrewing the idle screw. Alternatively, increase the air flow through the right hand carburetter by screwing down the idle screw. Reverse the procedure if the pointer moves to the left. When the desired setting has been achieved reconnect the inter connecting link between the two carburetters.

Using **balancer B89**

3. Disconnect the inter-connecting throttle link between the two carburetters.

4. Back-off the idle adjusting screw on each carburetter, clear of the throttle lever.

5. Turn each throttle adjusting screw so that it just touches the throttle lever, then turn the screws by equal amounts to achieve an approximate idle speed of 700 to 800 rev/min. Press the balancer firmly over the carburetter intake. Press or withdraw the control on the side of the balancer to adjust the meter needle reading to approximately half scale, and note the reading.

7. Without altering the position of the balancer control, place the balancer on the second carburetter intake and adjust the idle screw as necessary to achieve the same reading.

8. Alternately, adjust and check the balance of both carburetters until an idle speed of 700 to 800 rev/min is obtained.

9. Reconnect the throttle inter connecting link, and again check the idle speed and balance.

### MIXTURE SETTING

10. Ensure that the engine is at normal operating temperature. Remove the air cleaner, air intake elbows and mixture adjustment screw blanking plugs.

11. Mark the relationship of the suction chamber to the carburetter body, remove the retaining screws and lift off the suction chamber complete with pistons.

12. To achieve a datum setting for the mixture screw, turn it anti - clockwise until the jet is level with the carburetter bridge. Check by placing a straight edge across the bridge and adjust as necessary so that the jet just touches the straight edge.

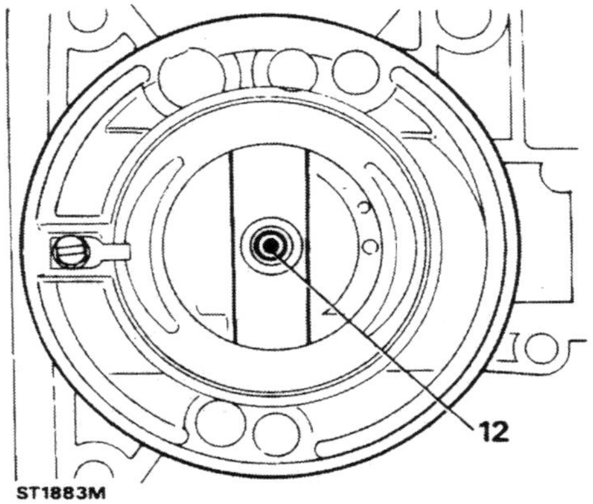

13. Refit the suction chamber and piston, evenly tighten the retaining screws. Check that the piston moves freely without sticking. Top - up the piston damper.

14. Turn the mixture adjustment screw three and one-half turns.

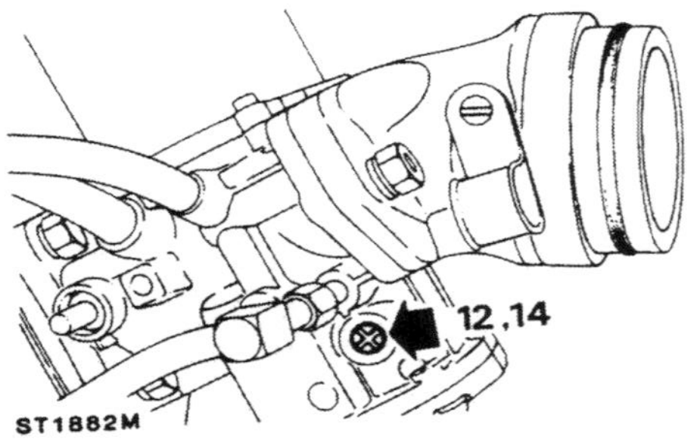

## FUEL SYSTEM

**MANUAL GEARBOX MODELS**

**AUTOMATIC GEARBOX MODELS**

15. Insert the probe of an infra - red exhaust gas analyser as far as possible up the exhaust pipe, start the engine and allow a one and one half minute stabilisation period.
16. Adjust the mixture screw on both carburetters by equal amounts, rich or weak to achieve a CO reading of 0.5 to 2.5%.
17. If after approximately two minutes the CO level is not satisfactory run the engine at 2000 rev/min for one minute to stabilise the equipment, continue the setting procedure until a stable CO reading of 0.5 to 2.5% at an idle speed of 700 - 800 rev/min is obtained.

**IDLE SPEED AND LINKAGE ADJUSTMENT**

18. Check that the engine is at normal operating temperature.
19. **Manual gearbox models**- Slacken the nut, at the left hand carburetter securing the inter-connecting link ball to the throttle cam lever. **Automatic gear box models**- Slacken the lower nut, securing the inter - connecting link ball to the throttle lever at the right hand carburetter.
20. Disconnect the inter - connecting link between the carburetters at the left hand carburetter.
21. At the right hand carburetter, release the lock nut and slacken off the lost motion adjustment screw, until it is well clear of the spring loaded pad.
22. If necessary adjust the idle screw to maintain the correct idle speed. Check the CO level and carburetter balance, adjust if required.
23. Re - connect the inter - connecting link to the left-hand carburetter.
24. Hold the right hand throttle lever against the idle screw stop and adjust the inter - connecting screw until contact is made with the spring loaded pad, tighten the lock nut.
25. Check the idle speed and balance. Adjust the lost motion screw to restore balance if necessary.
26. **Manual gearbox models**-Ensuring that the roller is firmly seated in the lower corner of the cam lever, tighten the nut which secures the inter - connecting link ball to the cam lever.
    **Automatic gearbox models**-Ensuring that the kick down cable linkage is firmly on its idle stop, tighten the inter - connecting link ball securing nut at the right hand carburetter.

**FAST IDLE ADJUSTMENT**

27. Pull out the cold start control (choke) until the scribed line on the left hand fast idle cam is in-line with the centre of the fast idle screw head.
28. Check that the scribed line on the right hand fast idle cam is similarly in-line with the fast idle screw head. If there is mis-alignment, slacken the fast idle cam link rod screw at the right hand carburetter and move the cam until the scribed line coincides with the centre of the screw head. Tighten the cam rod screw.
29. Turn each fast idle screw clockwise until just clear of the cam.
30. Turn the fast idle screw of the leading (left-hand) carburetter down (clockwise) until a slight change in engine speed is noted.
31. Similarly turn the fast idle screw of the second carburetter (right-hand) down until a further slight change of engine speed is noted.
32. Adjust the fast idle screws of both carburetters by equal amounts to achieve a fast idle speed of 1100 to 1150 rev/min.
33. Push the cold start (choke) fully home then pull it out again to its full extent and re-check the fast idle speed.
34. Fit the appropriate blanking plug and cap to the mixture screw recess and idle adjusting screw.
35. Fit the carburetter air intake elbows and air cleaner.

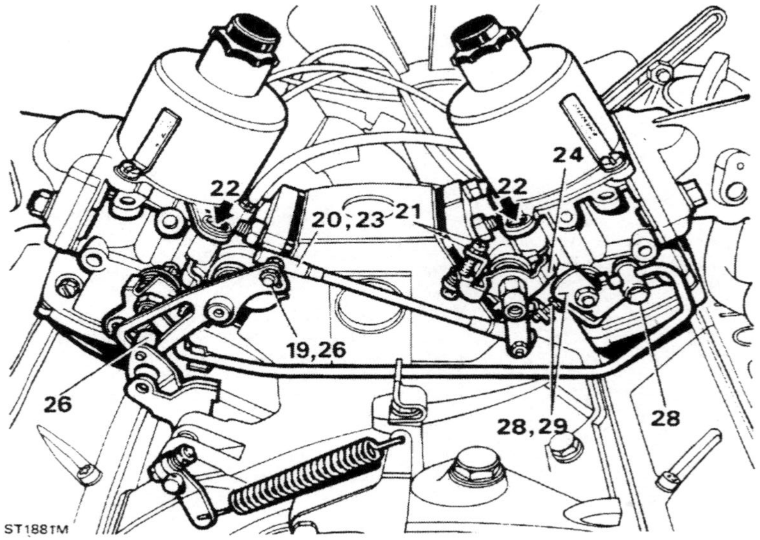

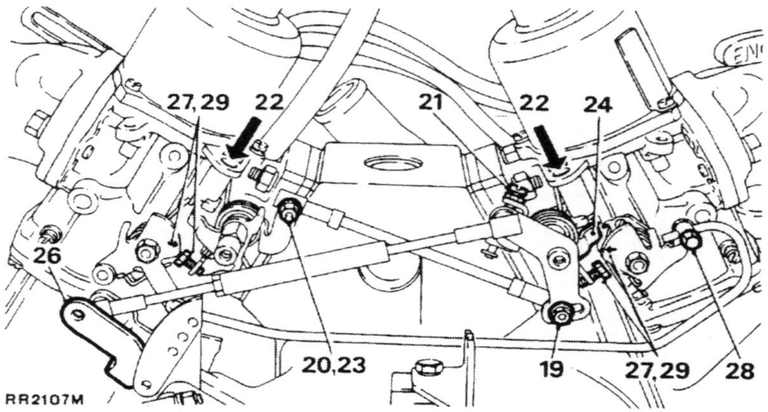

# FUEL SYSTEM

## FUEL PIPE LAYOUT

Fuel injection, Carburetter and Diesel models.

**WARNING: The spillage of fuel during the disconnection of any of the pipes in the fuel system is unavoidable, ensure that all necessary precautions are taken to prevent fire or explosion.**

## FUEL INJECTION MODELS

NOTE: Depressurise the fuel system before attempting to disconnect any pipes within the fuel system.

### KEY

1. Fuel feed hose to fuel rail.
2. Spill return hose to fuel tank.
3. Rigid fuel feed pipe.
4. Rigid spill return pipe.
5. Fuel filter.
6. Rigid fuel feed pipe to filter.
7. Breather hose.
8. In-tank fuel pump.
9. Breather valve and pipes (To evaporative loss system - when system fitted).
10. Fuel filler neck.
11. Fuel tank.

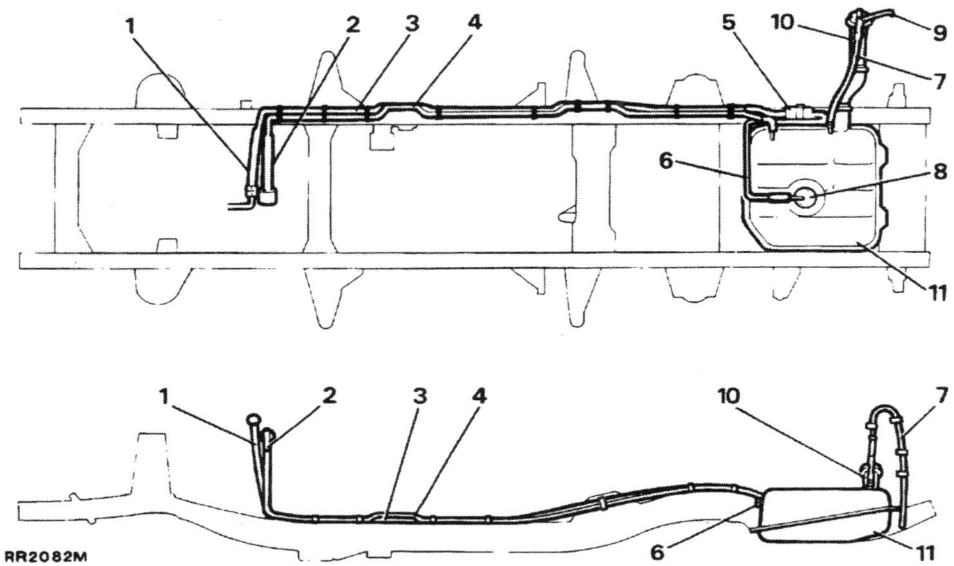

## CARBURETTER MODELS

### KEY

1. Fuel feed pipe.
2. Spill return pipe.
3. Fuel feed balance pipe.
4. Fuel filter.
5. Breather hose.
6. Breather valve and pipes (To evaporative loss system - when system fitted).
7. In-tank fuel pump.
8. Fuel filler neck.
9. Fuel tank.

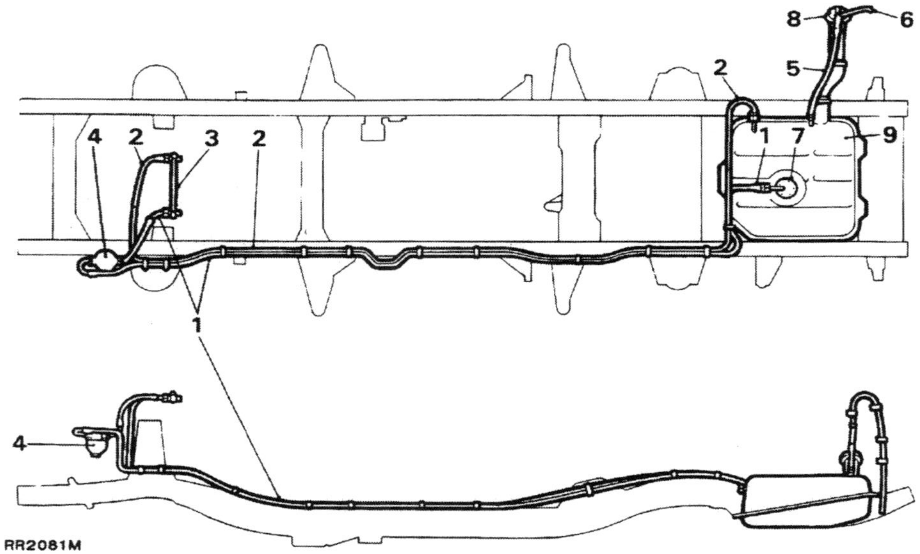

# CLUTCH 33

## HYDRAULIC DAMPER
(Diesel models only)

### Remove and refit

**Removing**

1. Raise the vehicle on a suitable hydraulic ramp.
2. Thoroughly clean the area around the slave cylinder and damper.
3. Using a recognised hose clamp, clamp the flexible hose to prevent excessive loss of hydraulic fluid.
4. Remove the flexible hose and rigid fluid pipe from the damper.
5. Remove the two bolts securing the damper to the mounting bracket.
6. Withdraw the damper.

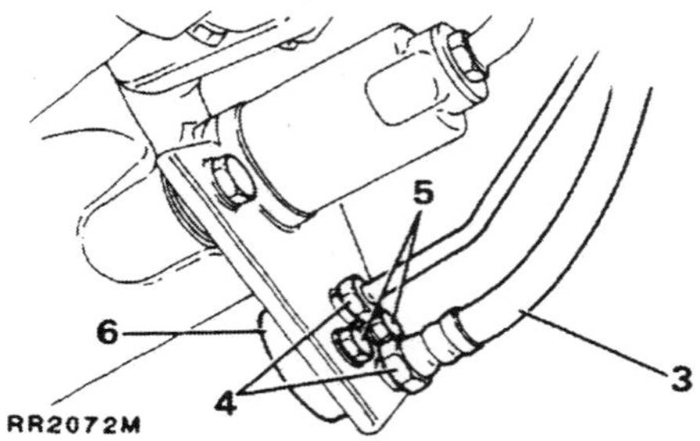

**Refitting**

7. Reverse the removal procedure.
8. Bleed the clutch hydraulic system. (Refer to main workshop manual).
9. Inspect for fluid leaks around hose and pipe connections.

# FUEL SYSTEM

## DIESEL MODELS

### KEY

1. Fuel feed pipe to distributor pump.
2. Fuel feed pipe - lift pump to filter.
3. Fuel filter.
4. Lift pump.
5. Spill return pipe.
6. Fuel feed - sedimenter to lift pump.
7. Sedimenter.
8. Fuel feed pipe - fuel tank to sedimenter.
9. Assembly - fuel pick-up pipes.
10. Fuel tank.
11. Breather hose.
12. Breather valve and pipes.
13. Fuel filler neck.

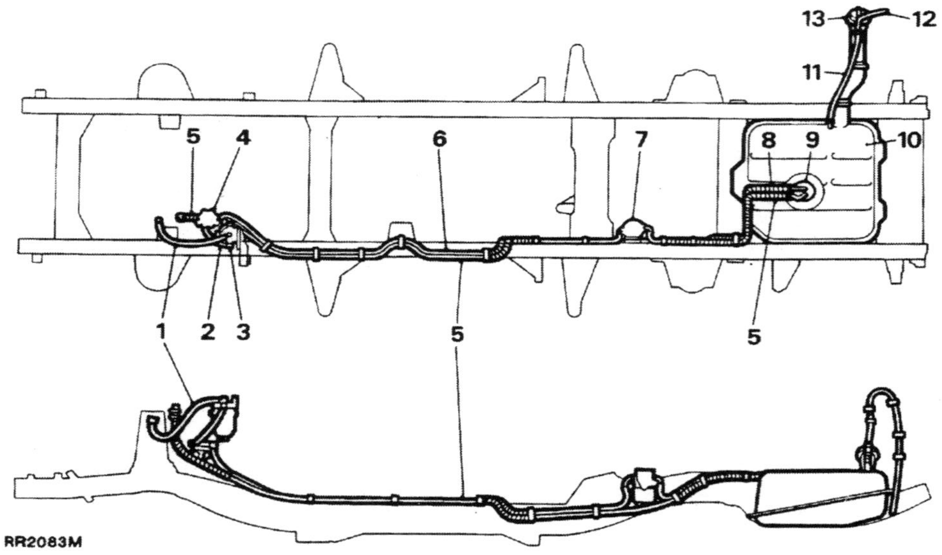

# STEERING

## TORQUE WRENCH SETTINGS

| STEERING | Nm | lbf ft | lbf in |
|---|---|---|---|
| Ball joint nuts | 40 | 30 | - |
| Clamp bolt nuts | 14 | 10 | - |
| Steering column bracket nuts | 27 | 20 | - |
| Steering wheel nut | 38 | 28 | - |
| Universal joint pinch bolt | 35 | 26 | - |
| PAS box |  |  |  |
| - Drop arm nut | 176 | 130 | - |
| - Sector shaft cover to steering box | 22-27 | 16-20 | - |
| - Steering box to chassis | 81 | 60 | - |
| - Steering box fluid pipes 14mm thread | 15 | 11 | - |
| - Steering box fluid pipes 16mm thread | 20 | 15 | - |
| - Tie bar to steering box | 81 | 60 | - |
| PAS pump |  |  |  |
| - High pressure fluid pipe | 20 | 15 | - |
| - Power steering pump mounting | 35 | 26 | - |
| - Pulley bolts, power steering pump | 8-12 | 6-9 | - |
| - Hose clamp | 3 | - | 27 |
| PAS reservoir |  |  |  |
| - Hose clamp | 3 | - | 27 |

# STEERING 57

7. Place the steering wheel on the column splines, fit the nut and washer and tighten to the specified torque.
8. Refit the steering wheel centre cover.
9. Re-connect the battery.

## POWER STEERING FLUID RESERVOIR

### Remove and refit

### Removing

1. Place a drain tray beneath the power steering box.
2. Prop open the bonnet.
3. Remove the reservoir filler cap.
4. Disconnect the fluid return hose from the power steering box. Drain the fluid completely from the reservoir and reconnect the hose.

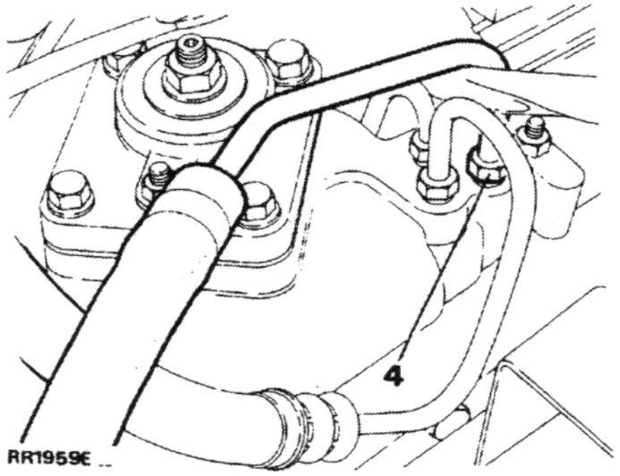

**CAUTION:** Power steering fluid is corrosive. Should any fluid seep onto paintwork, chassis or any other component immediately wipe clean. It is most important that any power steering fluid drained from the power steering system is not re-used.

5. Carefully dispose of unwanted fluid.
6. Release the pinch bolt and remove the reservoir from the bracket.

## STEERING WHEEL

### Remove and refit

### Removing

Service Tools:
18G 1014 Steering wheel remover
18G 1014-2 Adaptor pins

**NOTE:** The steering column is of a 'safety' type and incorporates shear pins. Therefore do not impart shock loads to the steering column during removing and refitting the steering wheel or at any time.

1. Disconnect the battery negative lead.
2. Ensure the road are in the straight ahead position to enable the steering wheel to be fitted in its correct location on re-assembly.
3. Carefully ease the centre trim pad off the steering wheel.
4. Restraining the steering wheel remove retaining nut and serrated washer.
5. Extract the steering wheel using service tool 18G 1014. Ensure the extractor pins are inserted in the threads up to the shoulder of the pins.

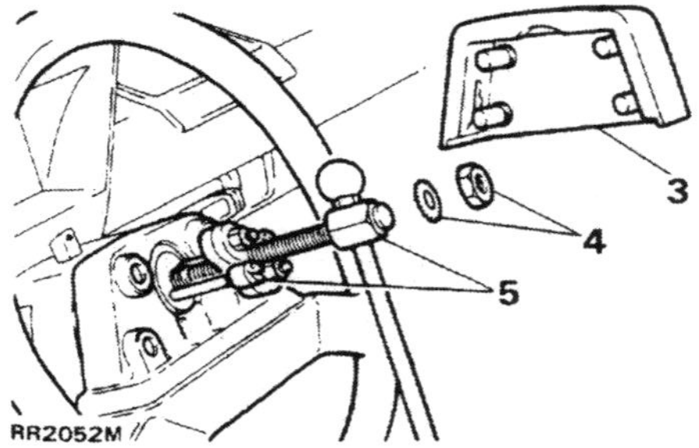

### Refitting

6. Ensure the road wheels are in the straight ahead position.

**CAUTION:** Do not apply shock loads to the steering wheel to ensure that it is firmly fitted in position.

# STEERING

7. Release the hose clips and remove the flexible hoses, with draw the the reservoir from the engine compartment.

NOTE: If the reservoir is not to be refitted immediately, the hoses must be sealed to prevent the ingress of foreign matter.

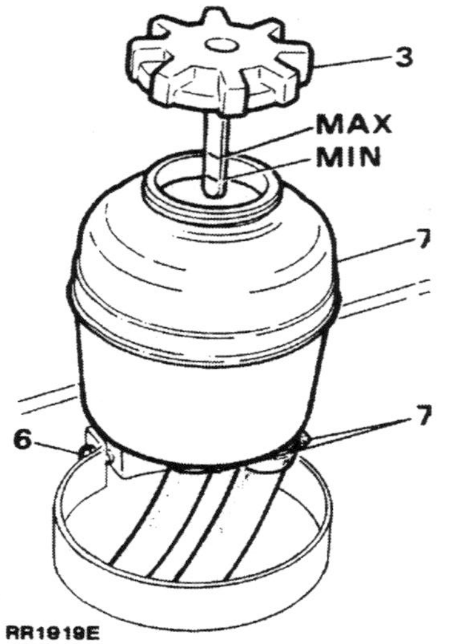

NOTE: The reservoir contains an integral filter which is not serviceable, however, in normal use the reservoir unit should last the life of the vehicle.

Should the power steering system malfunction and under inspection it is found that the steering fluid has been contaminated by foreign matter the fluid reservoir MUST be renewed.

### Refitting

8. Reconnect the flexible hoses to the reservoir.
9. Tighten the hose clips securely.
9. Fit the reservoir to the bracket and tighten the pinch bolt securely.
10. Fill the reservoir to the prescribed level on the dipstick with the recommended fluid. Bleed the power steering system. (See recommended fluids and bleed procedure in main workshop manual).
11. Fit the reservoir filler cap.
12. Close the bonnet.

## POWER STEERING PUMP DRIVE BELT
(Petrol models only)

### Adjust

### Procedure

1. Prop open the bonnet and disconnect the battery negative lead.
2. Check, by thumb pressure, the belt tension between the crankshaft and the pump pulley. There should be a free movement of between 11 to 14mm (0.437 to 0.562 in).

CAUTION: To prevent damage occurring to the steering pump DO NOT use a lever on the outer casing of the pump when tensioning the belt.

3. Slacken the two nuts at the side of the pump to enable the pump to be pivoted.
4. Slacken the bolt securing the pump lower bracket to the slotted adjustment link.
5. Pivot the pump (in the direction of the bold arrow) as necessary and adjust until the correct belt tension is obtained.

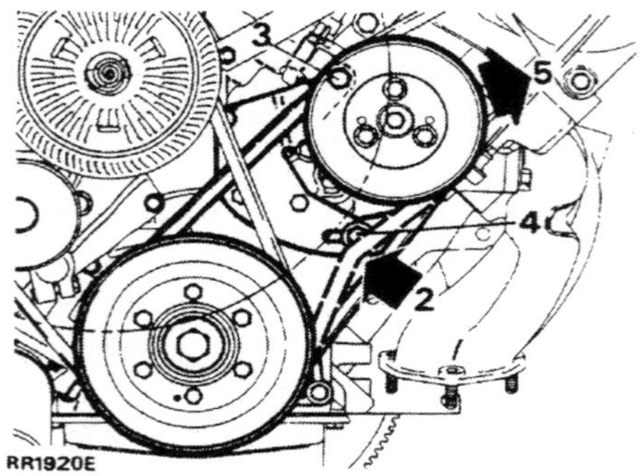

6. Maintaining the tension, tighten the pump adjusting bolt and the top pivot nuts.

NOTE: Check the alternator drive belt tension after adjusting the power steering pump belt.

7. Reconnect the battery negative lead and close the bonnet.

## POWER STEERING PUMP DRIVE BELT
(Petrol models only)

### Remove and refit

Removing or preparing for the fitting of a new belt.

1. Prop open the bonnet and disconnect the battery negative lead.
2. Slacken the jockey pulley bolt and remove the fan belt.
3. Slacken the alternator mountings and remove the drive belt.
4. Slacken the power steering pump mountings.
5. Pivot the pump and remove the drive belt.

### Refitting

6. Locate the driving belt over the crankshaft and pump pulleys.
7. Adjust the position of the pump to give a driving belt tension of 11 to 14mm (0.437 to 0.562 in) movement when checked by thumb pressure midway between the crankshaft and pump pulleys.

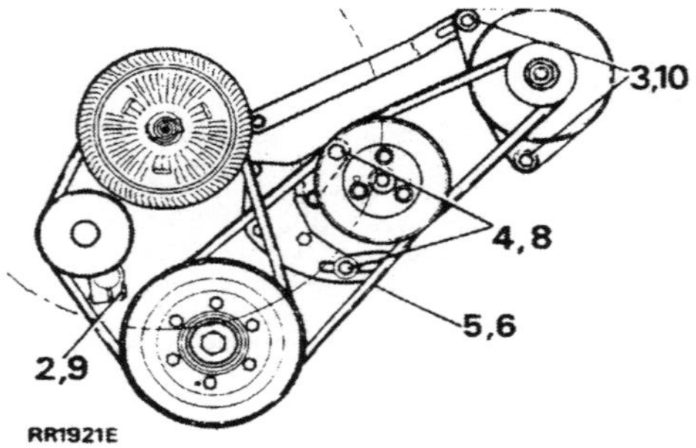

CAUTION: To prevent damage occurring to the steering pump DO NOT use a lever on the outer casing of the pump when tensioning the belt.

8. Maintaining the tension, tighten the pump adjusting bolt and the top pivot nut.
9. Refit the fan belt and adjust the tension to give 11 to 14mm (0.437 to 0.562 in) movement when checked by thumb pressure midway between the crankshaft and wear pump pulleys.
10. Refit the alternator drive belt and adjust to give 11 to 14mm (0.437 to 0.562 in) movement when checked midway between the power steering pump and alternator pulleys.

NOTE: Whenever a new belt is fitted, it is important that its adjustment is rechecked after approximately 1,500 km (1,000 miles) running.

11. Reconnect the battery negative lead and close the bonnet.

# STEERING

## POWER STEERING PUMP
(Petrol models only)

**NOTE: The power steering pump is not a serviceable item. In the event of failure or damage a new pump must be fitted.**

### Remove and refit

#### Removing

1. Disconnect the battery negative lead.
2. Slacken the alternator pivot bolts and adjustment link bolts, pivot the alternator inwards and remove the drive belt.
3. Slacken the water pump drive belt jockey pulley and remove the drive belt.
4. Remove the left hand bank spark plug leads and detach the distributor cap, place the leads and cap to one side.
5. Disconnect the electrical plug from the distributor amplifier unit.
6. Slacken the two nuts securing the power steering pump pivot bracket.
7. Release the three bolts securing the pulley to the steering pump, do not remove them at this stage.
8. Release the bottom adjustment bolt below the steering pump and pivot the pump inwards towards the water pump to enable the drive belt to be removed.
9. Remove the three bolts with plain washers retaining the pulley to the pump and withdraw the pulley.

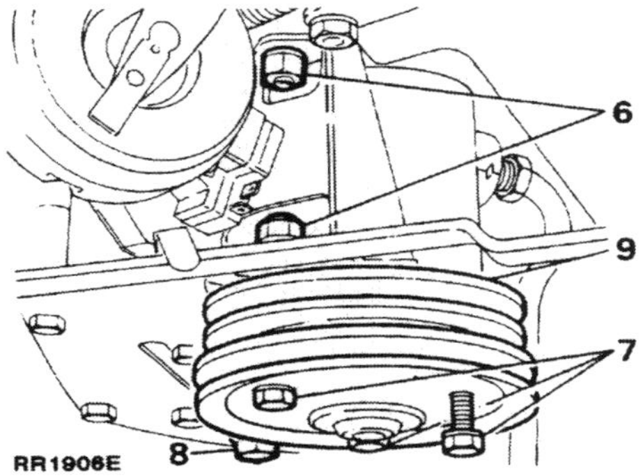

**NOTE: Place a drain tray under the vehicle to catch any power steering fluid which will seep from the pump when the fluid pipe is disconnected. Wipe away any fluid which comes into contact with paintwork or other components.**

**CAUTION: Under no circumstances must the fluid be re-used.**

10. Disconnect the fluid pipe from the side of the pump, plug the pipe and pump apertures to prevent ingress of dirt.
11. Remove the three bolts securing the pump to the pivot bracket, manoeuvre the pump out of the bracket and withdraw it from the engine compartment as far as the remaining connected fluid hose will permit.
12. Release the clip securing the hose to the pump, remove the hose and plug both apertures to prevent ingress of dirt.

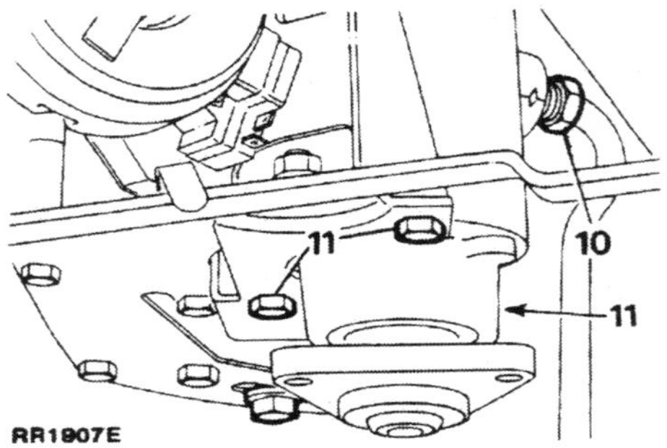

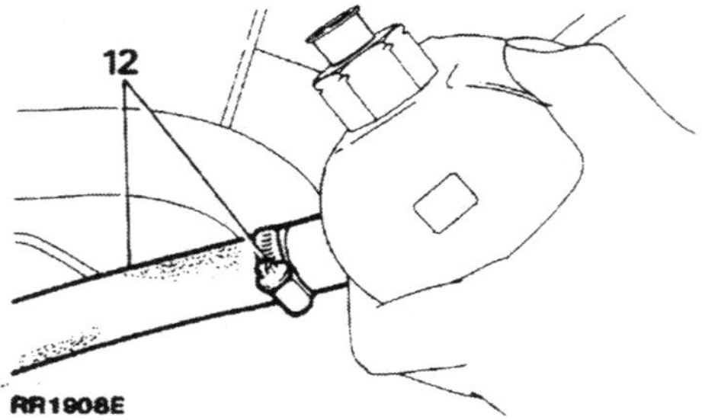

#### Refitting

13. Remove the plug from the fluid hose and secure the hose to a **NEW** pump. Tighten the hose clip securely.
14. Manoeuvre the pump into the pivot bracket and secure in position with the three retaining bolts. Tighten the bolts to the specified torque.
15. Remove the plugs from the fluid pipe and steering pump apertures and fit the pipe. Tighten the pipe securely.
16. Fit the pulley to the steering pump drive flange, coat the three bolts with Loctite and fit to the steering pump, do not fully tighten the bolts at this stage.
17. Refit the crankshaft to steering pump drive belt, pivot the steering pump outwards to tension the belt, tighten the pivot bolts securely. Check that the belt deflects approximately 11 to 14 mm (0.437 to 0.562) when checked by thumb pressure midway between the crankshaft and pump pulleys.
18. Tighten the three steering pump pulley retaining bolts to the specified torque.
19. Reverse the remaining removal instructions.
20. Bleed the power steering system.
21. Test the power steering system for leaks with the engine running, holding the steering on full lock in both directions.

**CAUTION: Do not maintain this pressure for more than 30 seconds in any one minute, to avoid causing the oil to overheat and possible damage to the seals.**

22. Close the bonnet.
23. Road test the vehicle.

# STEERING 57

## POWER STEERING PUMP
(Diesel models only)

NOTE: The power steering pump is not a serviceable component, in the event of failure or damage a new pump must be fitted.

### Remove and refit

#### Removing

1. Access to the steering pump is gained from beneath the vehicle.
2. Install the vehicle on a suitable hydraulic ramp, apply the handbrake and disconnect the battery negative terminal.
3. Release the pump pivot and adjustment bolts, detach the drive belt.
4. Place a drain tray beneath the vehicle to catch any fluid which may seep from the pump. Release the hose clip and remove the bottom hose from the pump, quickly seal both hose and pump aperture to prevent excessive loss of fluid and ingress of foreign matter.
5. Remove the pivot and adjustment bolts, manoeuvre the pump until access is gained to the fluid pipe at the rear of the pump.
6. Remove the bolt and release the banjo fitting, seal both hose and pump aperture to prevent ingress of foreign matter.
7. Withdraw the pump from the vehicle.

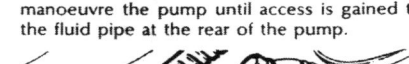

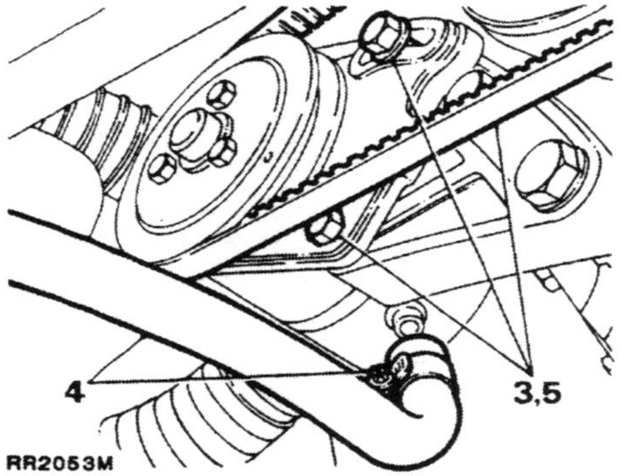

#### Refitting

8. Reverse the removal instructions ensuring that the hoses are fitted securely.
9. Tension the drive belt until deflection is between 7 to 12 mm (.275 to .472 inch) at mid point on the longest run of the belt, tighten the adjustment bolts.

CAUTION: To prevent damage occurring to the steering pump DO NOT use a lever on the outer casing of the pump when tensioning the belt.

10. Replenish the steering fluid reservoir with the recommended fluid. (Refer to main workshop manual).
11. Bleed the power steering system. (Refer to main workshop manual).
12. Test the power steering system for leaks with the engine running, holding the steering on full lock in both directions.

CAUTION: Do not maintain this pressure for more than 30 seconds in any one minute, to avoid causing the fluid to overheat, which could result in possible damage to the seals.

RR2053M

## POWER STEERING
### FAULT DIAGNOSIS

| SYMPTOM | CAUSE | TEST ACTION | CURE |
|---|---|---|---|
| INSUFFICIENT POWER ASSISTANCE WHEN PARKING | (1) Lack of fluid | Check hydraulic fluid reservoir level | If low, fill and bleed the system |
| | (2) Driving belt | Check belt tension | Adjust the driving belt |
| | (3) Defective hydraulic pump | (a) Fit pressure gauge between high pressure hose and steering pump with steering held hard on full lock, see Note 1 and 'Power Steering System Test' | If pressure is outside limits (high or low) after checking items 1 and 2, see Note 2 |
| | | (b) Release steering wheel and allow engine to idle. See 'Power Steering System Test') | If pressure is greater, check box for freedom and self-centering action |
| POOR HANDLING WHEN VEHICLE IS IN MOTION | Lack of castor action | This is caused by over-tightening the rocker shaft backlash adjusting screw on top of steering box. | It is most important that this screw is correctly adjusted. See instructions governing adjustment |
| HYDRAULIC FLUID LEAKS | Damaged pipework, loose connecting unions etc. | Check by visual inspection; leaks from the high pressure lines are best found while holding the steering on full lock with engine running at fast idle speed (See Note 1). | Tighten or renew as necessary |
| | | Check 'O' rings on pipework | Renew as necessary |

NOTE: Leaks from the steering box tend to show up under low pressure conditions, that is, engine idling and no pressure on steering wheel.

# STEERING

| SYMPTOM | CAUSE | TEST ACTION | CURE |
|---|---|---|---|
| EXCESSIVE NOISE | (1) If the high pressure hose is allowed to come into contact with the body shell, or any component not insulated by the body mounting, noise will be transmitted to the car interior | Check the loose runs of the hoses | Alter hose route or insulate as necessary |
|  | (2) Noise from hydraulic pump | Check oil level and bleed system | If no cure, change hydraulic pump |

Note 1. Never hold the steering wheel on full lock for more than 30 seconds in any one minute, to avoid causing the oil to overheat and possible damage to the seals.

Note 2. High pressure. In general it may be assumed that excessive pressure is due to a fault in the hydraulic pump.
Low pressure- Insufficient pressure may be caused by one of the following:

1. Low fluid level in reservoir ) Most usual cause of
2. Pump belt slip                ) insufficient pressure
3. Leaks in the power steering system
4. Hydraulic pump not delivering correct pressure
5. Fault in steering box valve and worm assembly
6. Leak at piston sealing in steering box
7. Worn components in either steering box or hydraulic pump

### Steering pump

Make/type .......... Hoboum series 200
Operating pressure - straight ahead position - at idle ...... 7 kgf/cm² (100 p.s.i.) maximum
Full lock (left or right) at idle ...... 28 kgf/cm² (400 p.s.i.) minimum
Full lock (left or right) 1000 rev/min ...... 70-77 kgf/cm² (1000-1100 p.s.i.)

---

# BRAKES

If the servo should fail, both hydraulic circuits will still function but would require greater pedal pressure.

The handbrake is completely independent of the hydraulic circuits.

Brake pad wear sensors are incorporated into the front and rear right hand side, inboard brake pads. The sensors will illuminate a brake pad wear warning light in the instrument binnacle, when pad thickness has been reduced to approximately 3mm (0.118 in).

**CAUTION: THOROUGHLY CLEAN ALL BRAKE CALIPERS, PIPES AND FITTINGS BEFORE COMMENCING WORK ON ANY PART OF THE BRAKE SYSTEM. FAILURE TO DO SO COULD CAUSE FOREIGN MATTER TO ENTER THE SYSTEM AND CAUSE DAMAGE TO SEALS, AND PISTONS WHICH WILL SERIOUSLY IMPAIR THE BRAKE SYSTEM EFFICIENCY.**

To ensure the brake system efficiency is not impaired the following warnings must be adhered to:-

**WARNING:**

DO NOT use brake fluid previously bled from the system.

DO NOT use old or stored brake fluid.

ENSURE that only new fluid is used and that it is taken from a sealed container.

DO NOT flush the brake system with any fluid other than the recommended brake fluid.

The brake system should be drained and flushed at the recommended service intervals.

## BRAKE SYSTEM

### Description

The hydraulic braking system fitted to the Range Rover is of the dual line type, incorporating primary and secondary hydraulic circuits.

**NOTE: References made to primary and secondary do not imply main service brakes or emergency brakes but denote hydraulic line identification.**

The brake pedal is connected to a vacuum-assisted mechanical servo which in turn operates a tandem master cylinder. The front disc brake calipers each house four pistons, the upper pistons are fed by the primary hydraulic circuit, the lower pistons by the secondary hydraulic circuit. The rear disc brake calipers each house two pistons and these are fed by the secondary hydraulic circuit via a pressure reducing valve.

A brake failure switch incorporated in the master cylinder will illuminate a panel warning light if a failure occurs in either the primary or secondary hydraulic circuits.

The brake fluid reservoir is divided, the front section (section closest to the servo) feeds the primary circuit and the rear section feeds the secondary circuit. Under normal operating conditions both the primary and secondary hydraulic circuits operate simultaneously on brake pedal application. In the event of a failure in the primary circuit the secondary circuit will still function and operate front and rear calipers.

Alternatively, if the secondary circuit fails, the primary circuit will still function and operate the upper pistons in the front calipers.

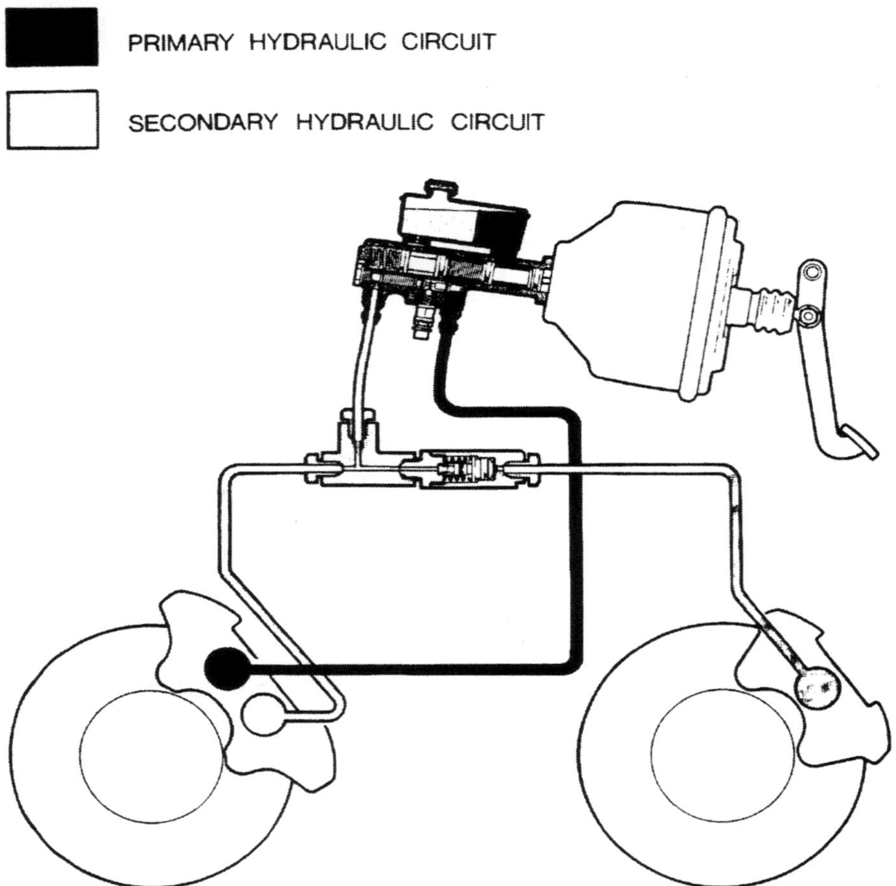

# BRAKES 70

PRIMARY HYDRAULIC CIRCUIT

SECONDARY HYDRAULIC CIRCUIT

RR1944E

**WARNING:** Some components on your vehicle, such as gaskets and friction surfaces (brake linings, clutch discs or automatic transmission brake bands), may contain asbestos. Inhaling asbestos dust is dangerous to your health and the following essential precautions must be observed:-

- Work out of doors or in a well ventilated area and wear a protective mask.

- Dust found on the vehicle or produced during work on the vehicle should be removed by extraction and not by blowing.

- Dust waste should be dampened, placed in a sealed container and marked to ensure safe disposal.

- If any cutting, drilling etc., is attempted on materials containing asbestos the item should be dampened and only hands tools or low speed power tools used.

# BRAKES 70

9. Release the outer cable from the 'P' clip located on top of the transfer gearbox, and withdraw the cable assembly from the vehicle.

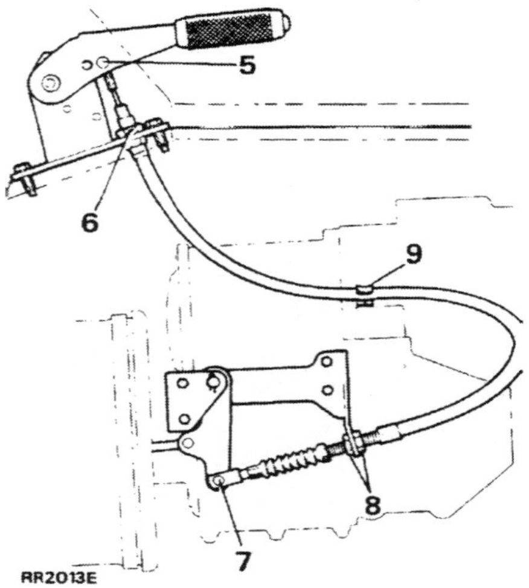

## HANDBRAKE CABLE

### Remove and refit

### Removing

1. Set the vehicle on level ground and chock the road wheels and select 'P' in the main gearbox.
2. Disconnect the battery negative terminal and release the handbrake.

### From inside the vehicle

3. Remove the four screws securing the liner in the cubby box.
4. Lift out the liner to gain access to the bottom of the handbrake pivot bracket.
5. Remove the split pin and clevis pin from the handbrake lever.
6. Release the nut securing the handbrake outer cable to the top of the handbrake mounting bracket. Slide the nut up the cable and push the inner and outer cable through the floor panel to the underside of the vehicle.

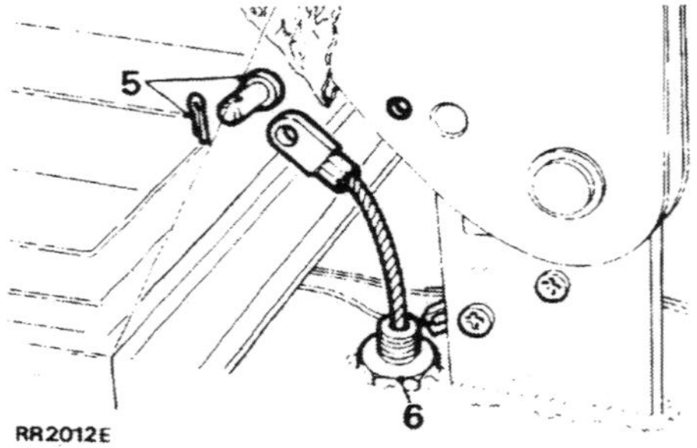

### From underneath the vehicle

7. Remove the split pin, plain washer and clevis pin securing the adjustment link to the brake drum actuating lever.
8. Release the locknuts securing the handbrake outer cable to the retaining bracket.

RR2012E

### Fit new cable

10. Feed the handbrake cable assembly through the floor aperture and secure the outer cable in position with the retaining nut.
11. Secure the cable to the handbrake lever, using a new split pin.
12. Secure the outer cable into the 'P' clip.
13. Position the outer cable into the retaining bracket bolted to the side of the transfer gearbox and loosely secure in position with the two outer cable lock nuts.
14. Reconnect the outer cable to the brake drum actuating lever. Fit the clevis pin, plain washer and new split pin.
15. Rotate the brake drum adjuster clockwise until the brake shoes are fully expanded against the brake drum.
16. Tighten the two brake cable outer lock nuts to secure the cable to its mounting bracket.

RR2013E

# BODY 76

## FRONT DOOR STOWAGE BINS

### Remove and refit

#### Removing

1. Remove the seven fixings securing the stowage bin to the inner door trim pad.
2. Withdraw the stowage bin.

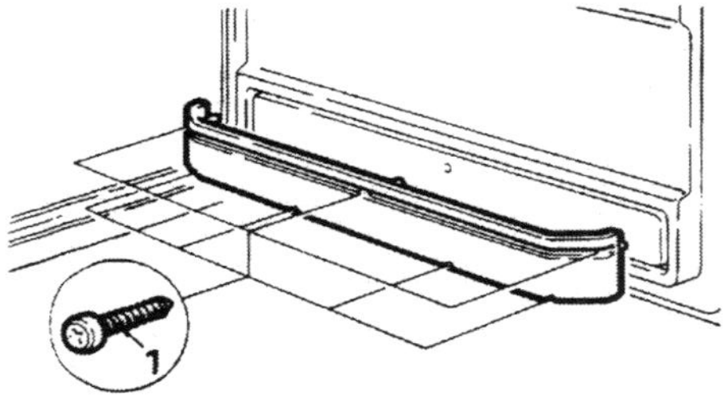

#### Refitting

3. Reverse the removal procedure.

## LOWER TAILGATE RELEASE MECHANISM

### Remove and refit

#### Removing

1. Open and raise the upper tailgate.
2. Release and lower, the lower tailgate.
3. Lift the trim panel off the tailgate inner panel.
4. Remove the screws securing the lock cover plate.
5. Remove the cover plate complete with handle release mechanism.
6. Remove the two screws and detach the handle release actuator lever.
7. Remove the two nyloc nuts and detach the handle release retaining bracket.
8. Withdraw the handle release mechanism from the cover plate.
9. Release the spring clips securing the operating rods to the internal tailgate release mechanism.
10. Remove the screws securing the exterior locks at either side of the tailgate.
11. Withdraw the exterior locks with operating rods.

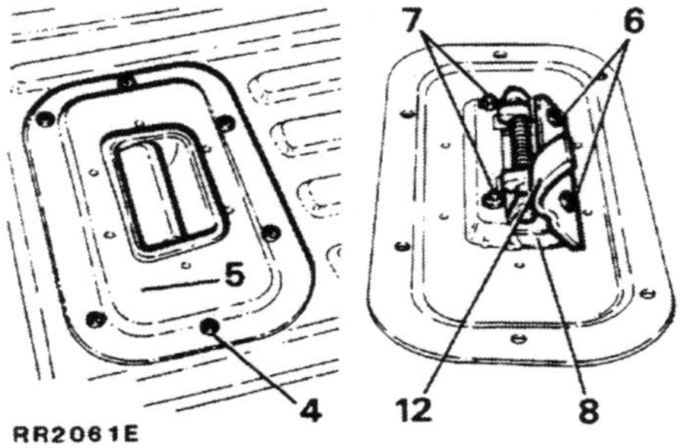

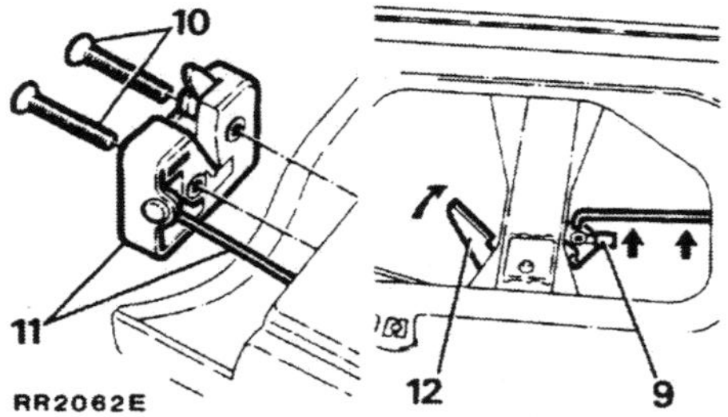

#### Refitting

12. Reverse the removal procedure, lightly grease the handle release actuator lever and internal tailgate operating lever.

---

# 70 BRAKES

17. Apply the handbrake, and slacken the brake drum adjuster until the handbrake lever fully operates the brake shoes on the second or third notch of the handbrake ratchet.
18. Refit the cubby box liner.

## BRAKE PRESSURE REDUCING VALVE

### Remove and refit

#### Removing

1. Remove all dust, grime, etc., from the vicinity of the pressure reducing valve fluid pipe unions.
2. Disconnect the outlet fluid pipe from the pressure reducing valve. Plug the pipe and reducing valve port to prevent the ingress of foreign matter.
3. Remove the valve from the three-way connector and plug both apertures.
4. Withdraw the pressure reducing valve from the engine compartment.

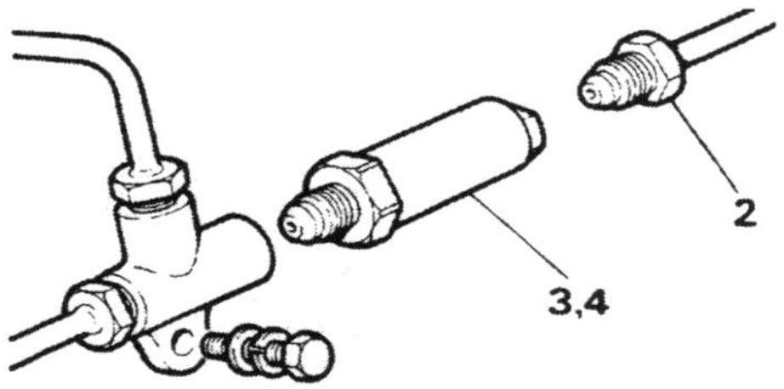

#### Refitting

5. Reverse the removal instructions.
6. Bleed the brake systems.

**NOTE: The pressure reducing valve is not a serviceable item, in the event of failure or damage, a new unit must be fitted.**

275

# BODY

## LOWER TAILGATE STRIKER ADJUSTMENT

### Adjust

1. Open and raise upper tailgate.
2. Open and lower, lower tailgate.
3. Release the striker and move in the appropriate direction, either add or subtract spacing washers between the striker and tailgate aperture.
4. Adjustment is correct when outer profile of tailgate panel aligns with both rear body corner panels.

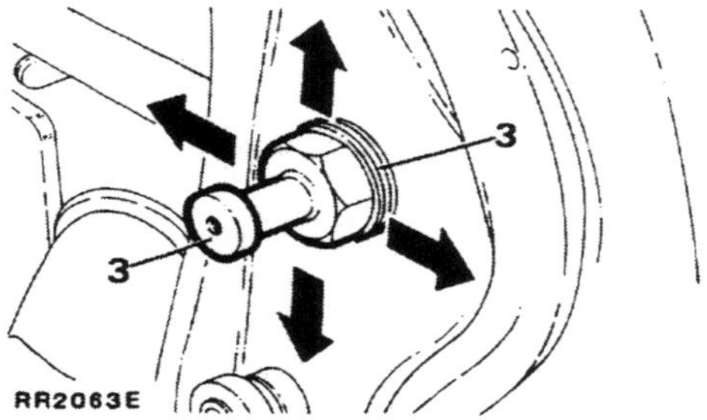

## ASSISTED BONNET LIFT

### Remove and refit

### Removing

**CAUTION: The assisted lift mechanism of the bonnet eases the bonnet open and lift procedure. When the bonnet is fully open, secure the bonnet stay in position. The assisted lift mechanism alone WILL NOT retain the bonnet in its upright position.**

1. Carefully prise the wiper arms off the spindle bosses, noting their position for re-assembly.
2. Raise the bonnet and disconnect the battery negative terminal.
3. Disconnect the electrical lead to the bonnet illumination lamp.
4. Disconnect the screen washer fluid feed pipe at the 'T' joint, remove the feed pipe from the bonnet retaining clip.

NOTE: **The removal of the bonnet will require the assistance of a second person.**

5. Release the four bolts (with plain washers) securing the bonnet to the hinges.

NOTE: **Petrol Models Only:- An engine earth braid is attached to the forward bolt of the left hand hinge.**

6. With assistance lift the bonnet clear of the hinges and store safely to one side, cover bonnet to protect paintwork.
7. Remove the two wiper box wheel nuts and rubber spacers.
8. Remove the four fixings securing the decker panel to the front wings, the front two fixings are accessible from the front of the decker panel, access to the rear two fixings is gained by opening the front doors. Note the nylon spacing washers at each bolt.
9. Remove the nine cross-head screws from the front of the decker panel water channel.
10. Place extension tubes over each bonnet hinge, with assistance lower the hinges, manoeuvre the decker panel off the wiper arm spigot bosses and along the tubes until the panel is clear of the vehicle. Place the panel to one side and cover to protect paintwork.

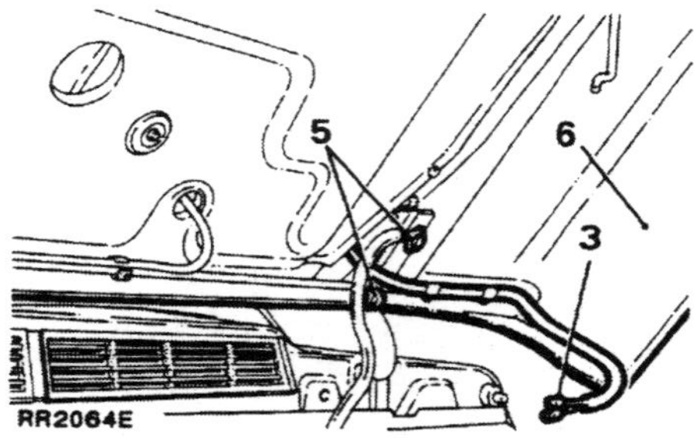

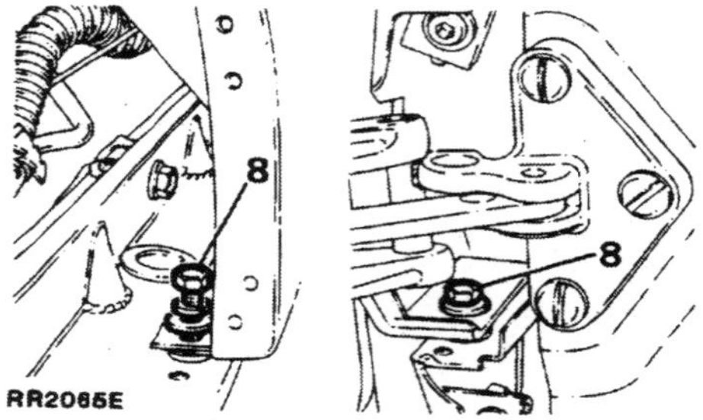

**WARNING: Gradually let the torsion bar spring tension return the hinges to their upright position to prevent the possibility of personal injury or damage to the vehicle.**

11. Place an extension tube over the hinge, lower the hinge until the stop bracket can be removed, with draw the bracket and gradually allow the hinge to return to its upright position.
12. Release the torsion bar from the retaining clip.
13. Manoeuvre the torsion bar until it can be released from the hinge.
14. Release the torsion bar from the retaining bracket.
15. Remove the two bolts (with plain washers) securing the hinge to its mounting bracket.
16. Withdraw the hinge.

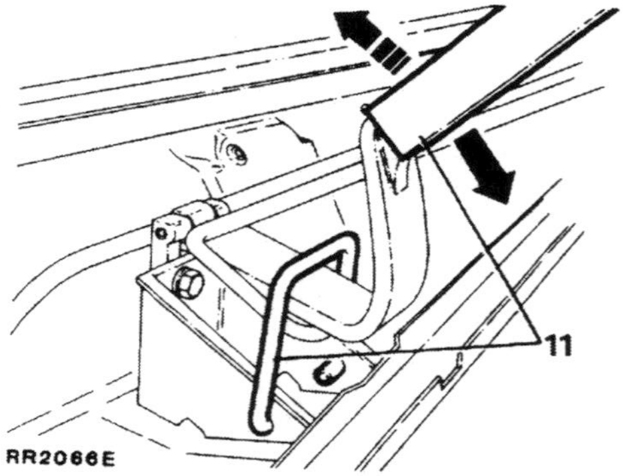

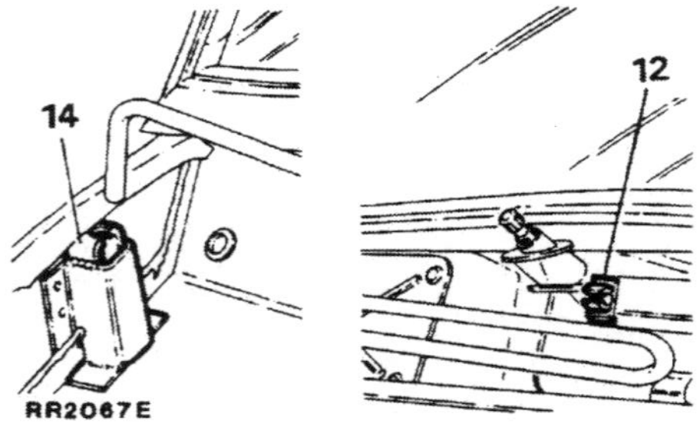

### Refitting

17. Fit the hinge and securely tighten the two retaining bolts (with plain washers).
18. Fit the torsion bar ensuring that it is securely located in the retaining clip and bracket.
19. Reverse the remaining removal instructions.

NOTE: **Petrol Models Only:- Fit the earth braid under the forward bolt of the left hand hinge.**

20. Using a soft blunt implement ease the bottom lip of the windscreen seal up and over onto the face of the decker panel.

# BODY

## RADIATOR GRILLE

### Remove and refit

#### Removing

1. Raise the bonnet and secure the bonnet support.
2. Depress the four upper retaining lugs and ease the grille forward.
3. Lift the grille upwards and withdraw it from the vehicle.

#### Refitting

4. Fit the radiator grille ensuring that the bottom lugs of the grille locate in their respective slots below the headlamp units.
5. Ease the grille rearwards and locate the upper retaining lugs.

## FUEL FILLER FLAP

### Remove and refit

### Adjust

#### Removing

**NOTE: If the vehicle has central locking ensure that the locking system has been released before attempting to open the fuel filler flap.**

1. Open the fuel filler flap.
2. Release the two screws (with plain washers).
3. Withdraw the flap.

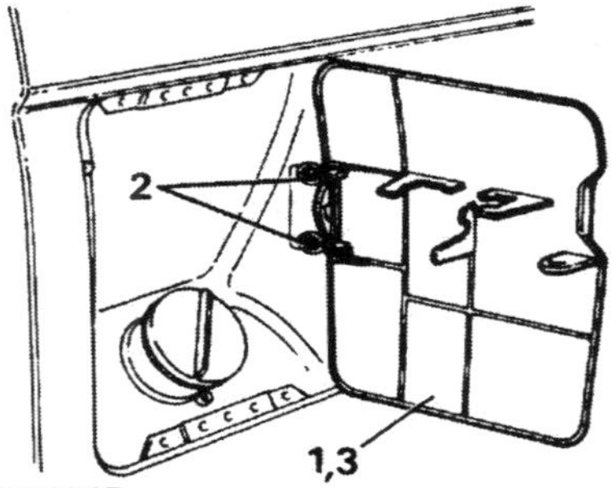

#### Refitting

4. Fit the flap, but do not fully tighten the screws at this stage.
5. Close the flap and check that the outer profile of the flap aligns with the rear wing, adjust by easing the flap in or out of the aperture.
6. Open the flap and securely tighten the screws.

## ASYMMETRIC SPLIT REAR SEAT

### Remove and refit

#### Removing

1. Lift the finger button to release the seat back rest securing catch and fold the seat forward.
2. Whilst folding the seat forward feed the rear seat lower part of the seat belt through the aperture between the seat back and squab.

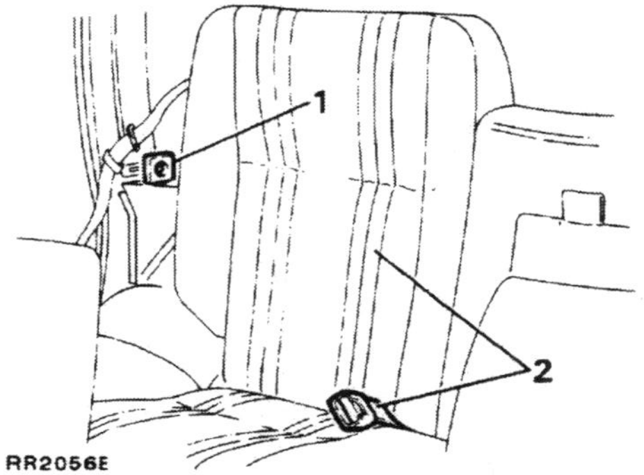

3. Fold the seat fully forward and remove the four rear pivot bracket bolts.
4. Fold the seat back and lift the rear footwell carpet from just below the front of the seat to gain access to the four front fixings securing the pivot brackets, remove the bolts, withdraw the seat assembly from the vehicle.

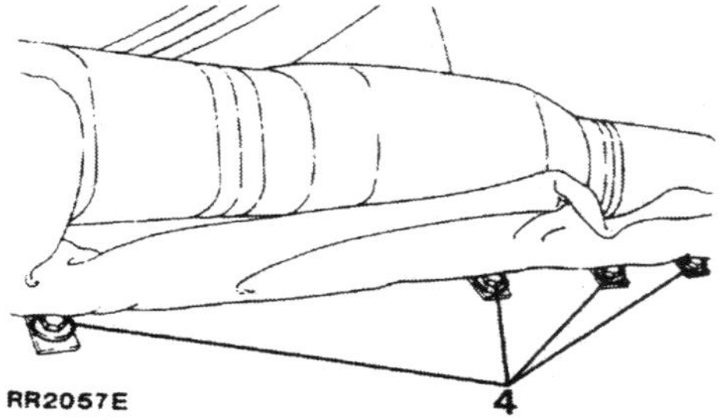

### Refitting

5. Reverse the removal procedure ensuring that all fixings are securely tightened.
6. Adjust the seat mechanism plate if necessary.

# 76 BODY

## ASYMMETRIC SPLIT REAR SEAT-LOCKING MECHANISM

### Remove and refit

#### Removing

1. Lift the finger button and fold the seat back rest forward.
2. Remove the spring clip securing the finger button operating rod to the lock mechanism lever, accessible through the lock mounting bracket, withdraw the finger button.

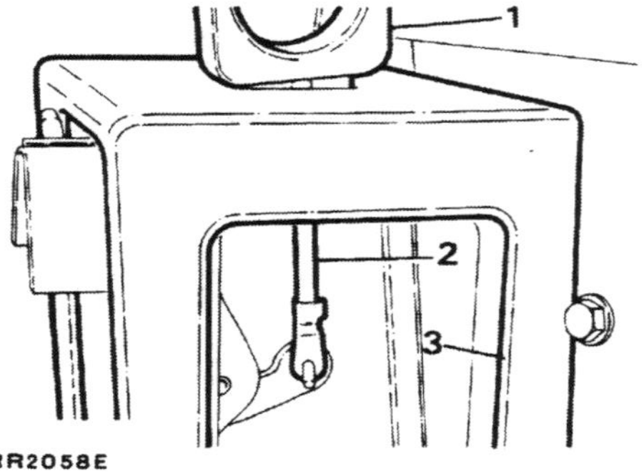

RR2058E

3. Remove the trim covering.
4. Remove the three screws from the face of the lock catch plate.
5. Manoeuvre the catch plate off the lock mechanism.
6. Retrieve the lock mechanism from the mounting bracket aperture.

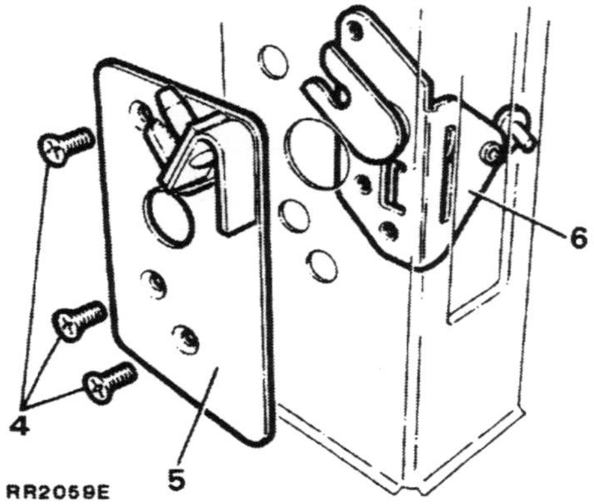

RR2059E

### Refitting

7. Lightly grease the internal lock mechanism.
8. Reverse the removal instructions.
9. Adjust the lock catch plate to align with the seat striker.

## ASYMMETRIC SPLIT REAR-SEAT LOCKING MECHANISM ADJUSTMENT

### Adjust

1. Release the finger button and fold the seat back rest forward.
2. Slacken the three screws securing the catch plate to the lock mechanism.
3. Manoeuvre the lock assembly either horizontally, vertically or diagonally until the catch plate aligns with the striker at the side of the seat.
4. Securely tighten the retaining screws.

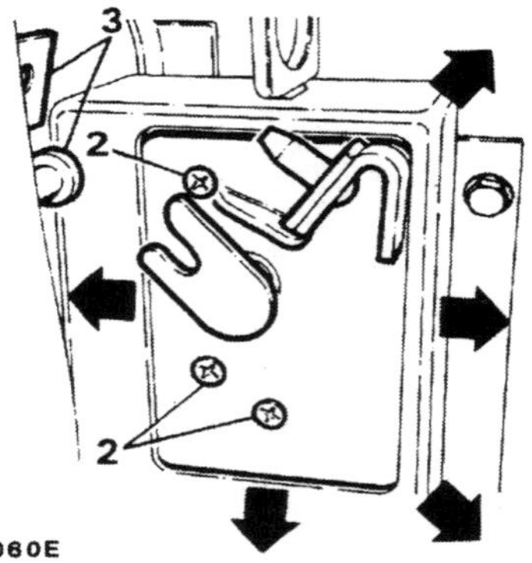

RR2060E

# HEATING AND VENTILATION 80

## HEATING AND VENTILATION

The heater/ventilation and air conditioning controls are integrated in the heater control panel.
The fresh air/recirculating flap in the heater is controlled by a solenoid operated vacuum unit.

## HEATER AND AIR CONDITIONING CONTROLS

### Remove and refit

#### Removing

1. Disconnect the battery negative lead.
2. Remove the transmission lever surround and radio housing.
3. Remove the lower fascia panel.
4. Remove the centre fascia unit.
5. Disconnect the electrical leads from the thermostat, fan speed and recirculate/fresh air switches.
6. Disconnect the relay rod from the 'SCREEN-CAR' lever.
7. Disconnect the relay rod from the 'VENT' lever.
8. Disconnect the control cable from the side of the heater unit.

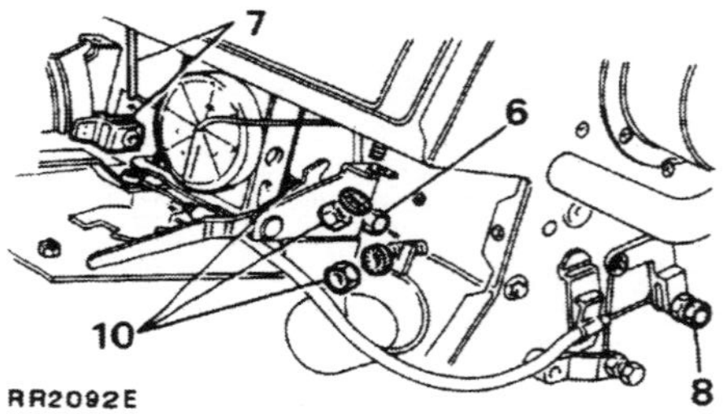

RR2092E

9. Remove the thermostat sensor from the fins of the evaporator.
10. Remove the four securing nuts and washers, and two screws behind the thermostat, and withdraw the heater controls assembly.

11. **Thermostat:** remove the two securing screws. Remove cable from control lever and lift thermostat over lever to remove.
12. **Fan speed switch:** drill out securing rivets, or remove fixing screws, and withdraw the switch.
13. **Air conditioning / fresh air / recirculating switch:** drill out securing rivets, or remove fixing screws and withdraw the switch.

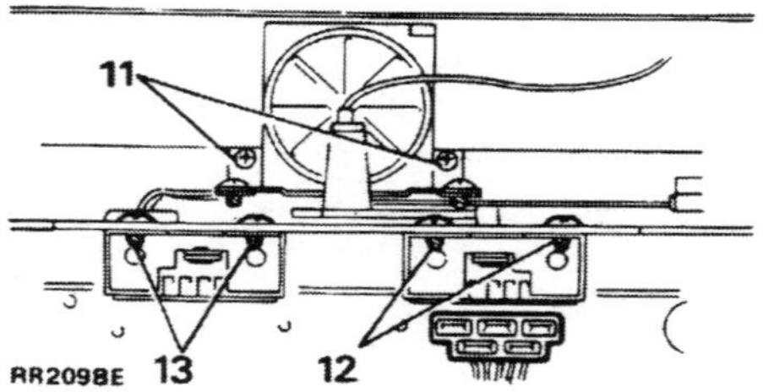

RR2098E

### Refitting

14. Reverse instructions 5 to 13.
15. Check that the control levers give full movement to the flaps. If necessary, adjust at the relay rod or cable end fixings.
16. Reverse instructions 1 to 4.

# AIR CONDITIONING 82

## HEATER AND AIR CONDITIONING- circuit diagram

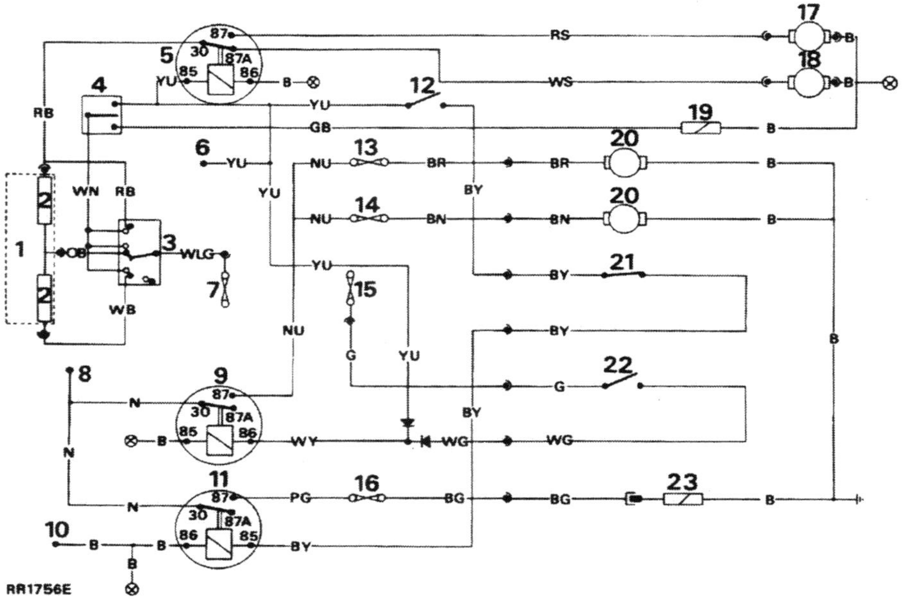

1. Heater unit.
2. Resistors.
3. Fan speed switch
4. Air conditioning/re- circulating/fresh air-switch.
5. Air conditioning/heater relay.
6. Cable connection to ECU.
7. Fuse 8-main fuse panel.
8. Pick up point main cable connection.
9. Fan relay.
10. Ground-via main cable.
11. Compressor clutch relay.
12. Thermostat.
13. Fuse A1-auxiliary fuse panel.
14. Fuse A2-auxiliary fuse panel.
15. Fuse 13-main fuse panel.
16. Fuse A3-auxiliary fuse panel.
17. Air conditioning motor - dashboard unit.
18. Heater motor.
19. Heater recirculating solenoid.
20. Condenser fan motors.
21. High pressure switch
22. Engine water temperature sensor.
23. Compressor clutch.

### Cable colour code

B  Black
U  Blue
N  Brown
G  Green
L  Light
O  Orange
P  Purple
R  Red
S  Slate
W  White
Y  Yellow

The last letter of a colour code denotes the tracer.

### AIR CONDITIONING/HEATER CONTROLS

The air conditioning controls are now incorporated in the heater control panel. See Section 80 -Heating and Ventilation.

---

# 80 HEATING AND VENTILATION

## RECIRCULATING/FRESH AIR SOLENOID SWITCH

### VACUUM UNIT (recirculating/fresh air flap)

**Remove and refit**

1. Disconnect the battery negative lead.
2. Remove the transmission lever surround.
3. Remove the radio mounting console.
4. Remove the centre fascia unit and the lower fascia panel.
5. Disconnect the electrical leads to the solenoid.
6. Disconnect the two vacuum pipes.
7. Remove the solenoid fixings and withdraw the solenoid.
8. Vacuum unit: remove the vacuum pipe from the vacuum unit.
9. Remove the actuating rod securing clip.
10. Remove two fixing nuts and washers and withdraw the vacuum unit.

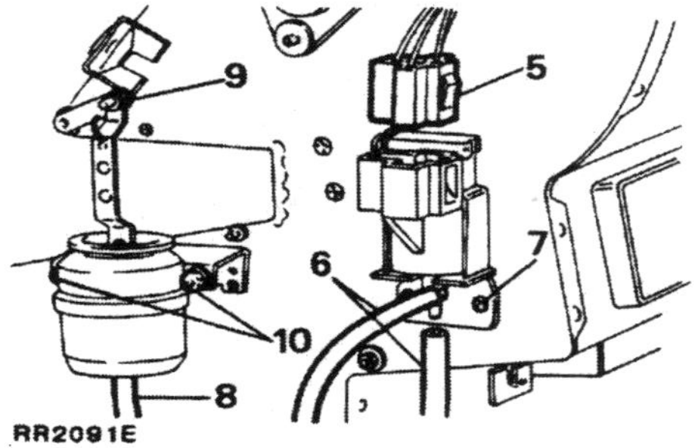

**Refitting**

11. Reverse the removal procedure.

# WIPERS AND WASHERS | 84

## TAILGATE WIPER ARM

### Remove and refit

#### Removing

1. Lift the wiper arm end cap to gain access to the wiper motor spindle.
2. Remove the wiper arm securing nut.
3. Withdraw the wiper arm from the spindle.

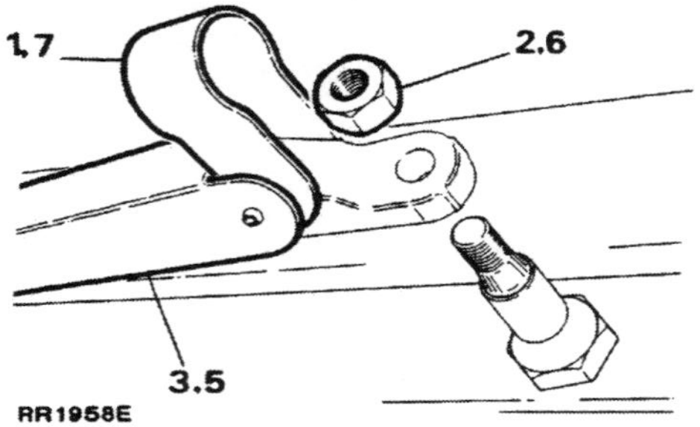

#### Refitting

4. Allow the motor to move to the 'park' position.
5. Fit the wiper arm to the spindle, locating it on the splines so that the wiper blade is just clear of the screen surround.
6. Fit and tighten the securing nut.
7. Push the end cap back into position.
8. Check the correct operation of the wiper.

## TAILGATE WIPER MOTOR

The tailgate wiper motor contains non-serviceable parts, repair is by replacement only.

### Remove and refit

#### Removing

1. Disconnect the battery negative lead.
2. Lower or remove the headlining rear section to gain access to the wiper motor assembly.
3. Remove the wiper arm and blade.
4. Slacken the nut securing the wiper motor to the body. DO NOT remove at this stage.
5. Disconnect the electrical leads at the multi-plug.
6. Remove the two bolts securing the wiper motor to the inner body.
7. Supporting the wiper motor, remove the nut slackened at instruction 4, complete with protective cover, washer and seal. Simultaneously withdraw the wiper motor from the body.

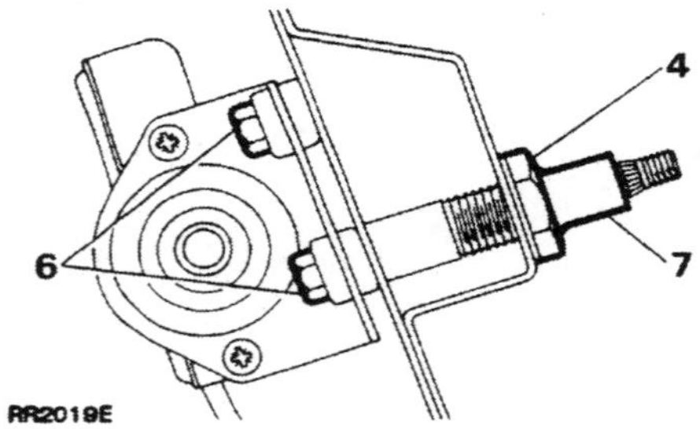

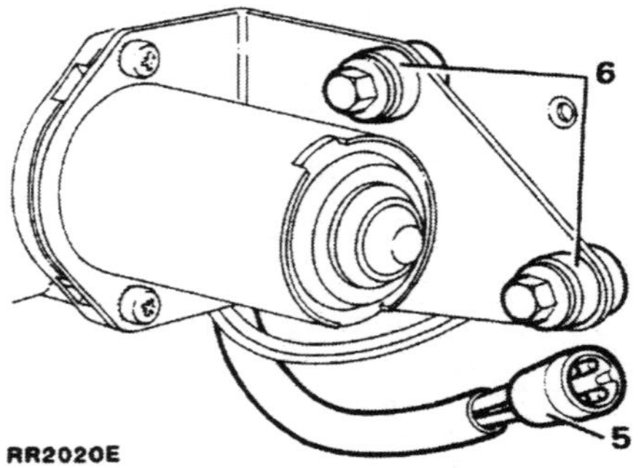

#### Refitting

8. Reverse instructions 1 to 7, ensuring that the spacer is correctly positioned before fitting the motor.

## Notes

# ELECTRICAL 86

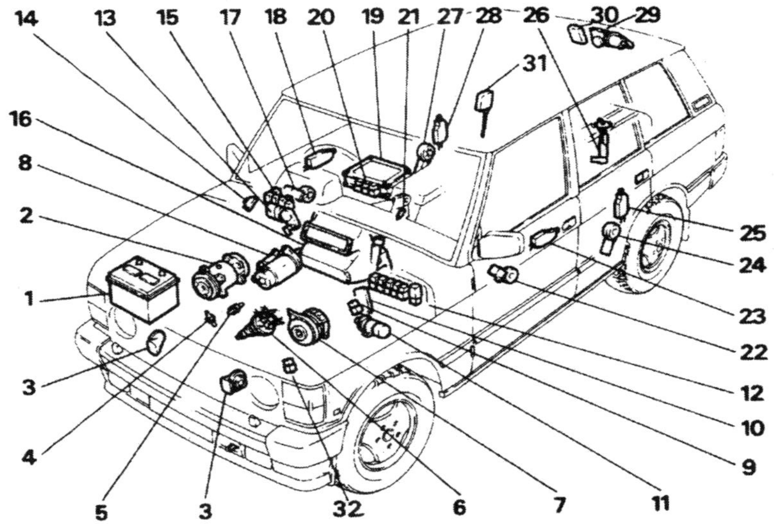

## LOCATION OF ELECTRICAL EQUIPMENT

1. Battery
2. Air conditioning compressor (if fitted)
3. Horns
4. Oil pressure switch
5. Water temperature switch
6. Electronic distributor
7. Alternator
8. Starter motor
9. Fuel shut-off relay (fuel injection models only)
10. Power resistor (fuel injection models only)
11. Coil and amplifier assembly
12. Relays
13. Wiper motor-front screen
14. Choke warning light switch (carburetter models only)
15. Relays
16. Heater
17. Window lift motor (front right-hand door)
18. Door lock actuator (front right-hand door)
19. Electronic control unit (fuel injection models only)
20. Relays (fuel injection models only)
21. Handbrake warning light switch
22. Window lift motor (front left-hand door)
23. Door lock actuator (front left-hand door)
24. Window lift motor (rear left-hand door)
25. Door lock actuator (rear left-hand door)
26. Electrical in-tank fuel pump
27. Window lift motor (rear right-hand door)
28. Door lock actuator (rear right-hand door)
29. Wiper motor-rear screen
30. Radio aerial amplifier
31. Fuel filler lock actuator
32. Fuel cut-off relay (carburetter models only)

**NOTE: Right-hand drive vehicle illustrated: Certain electrical components may change position on left-hand drive vehicles, but can be found in a similar location on the opposite side of the vehicle.**

For full information on fuel injection related items - see Main Workshop Manual.

To identify individual relays (Items 12 and 15) see relays in Electrical Section of Manual.

## Notes

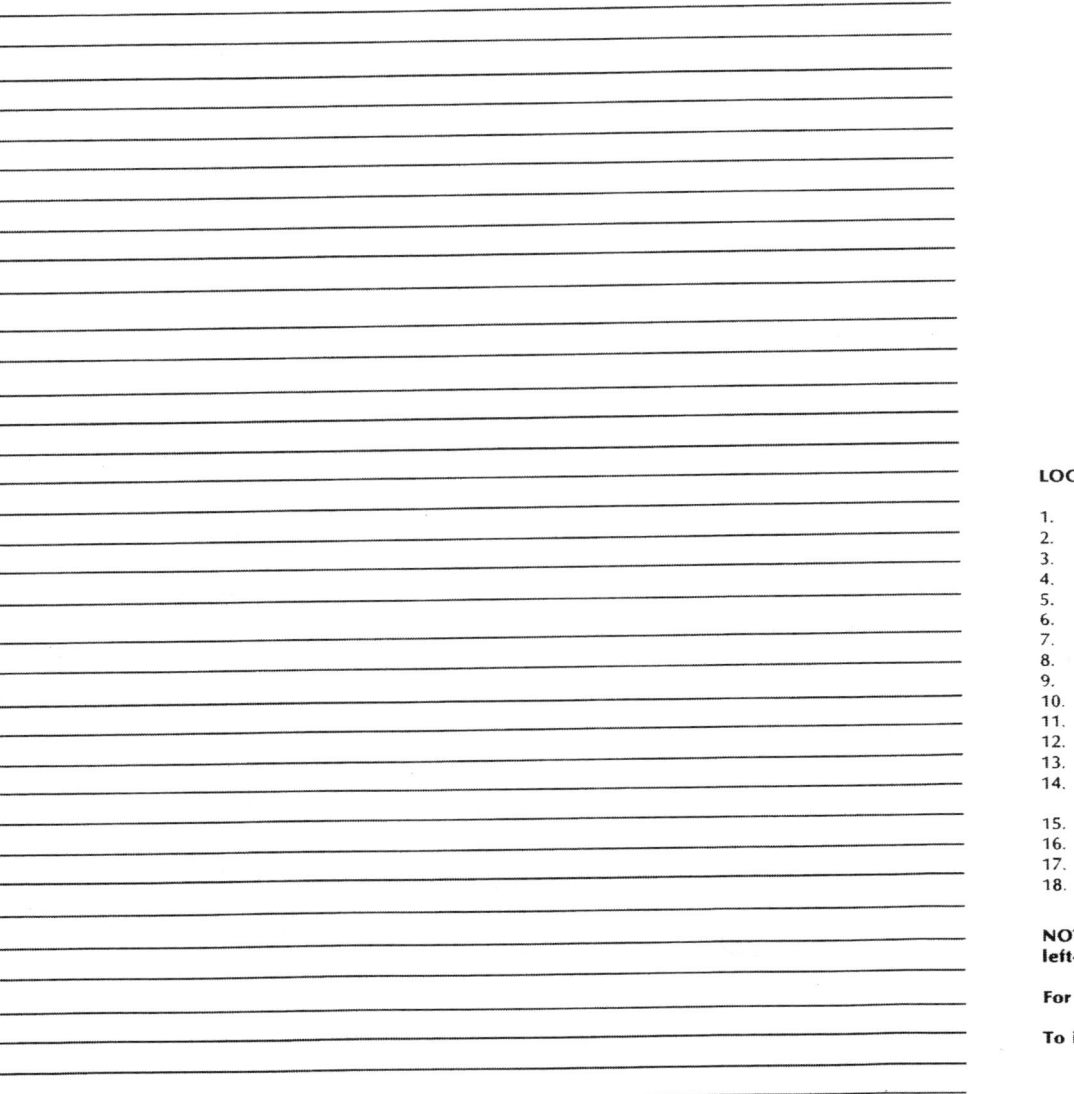

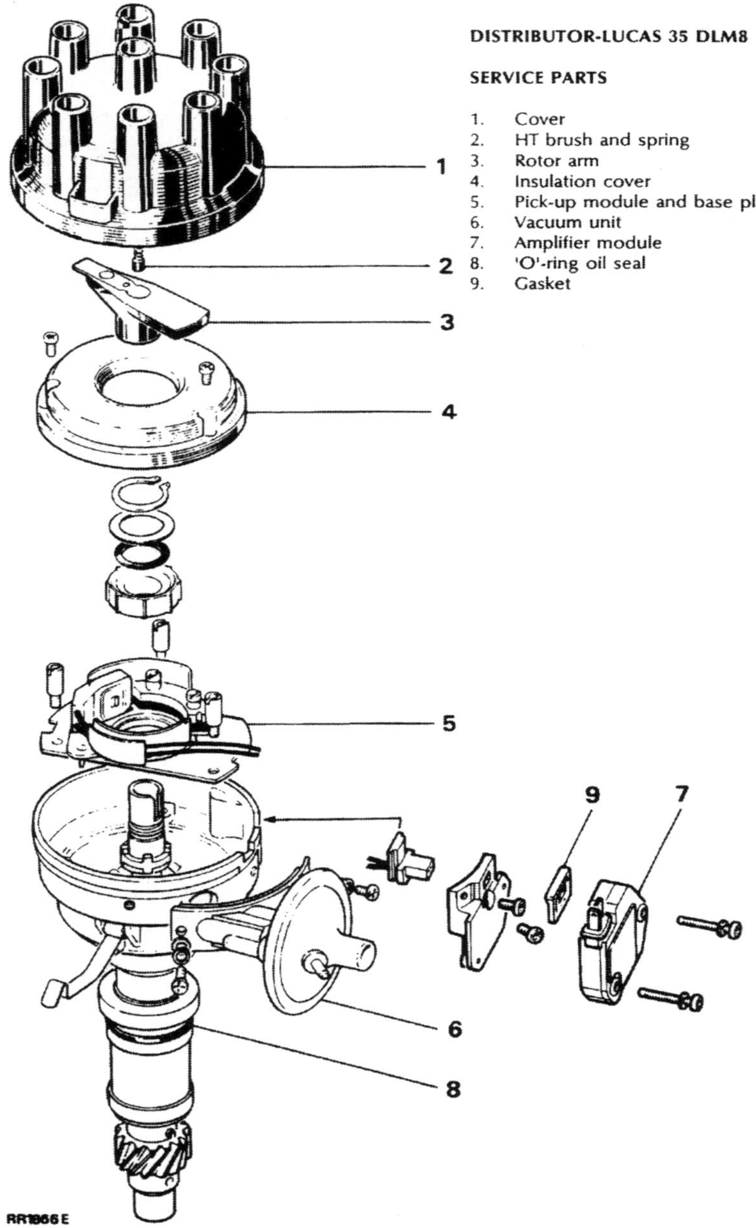

# ELECTRICAL 86

## DISTRIBUTOR-LUCAS 35 DLM8

### SERVICE PARTS

1. Cover
2. HT brush and spring
3. Rotor arm
4. Insulation cover
5. Pick-up module and base plate assembly
6. Vacuum unit
7. Amplifier module
8. 'O'-ring oil seal
9. Gasket

## ELECTRONIC IGNITION

A Lucas 35DLM8 distributor is employed. This has a conventional advance unit and centrifugal automatic advance mechanism.

A pick-up module, in conjunction with a rotating timing reluctor inside the distributor body, generates timing signals. These are applied to an electronic ignition amplifier module mounted on the side of the distributor body.

NOTE: The pick-up air gap is factory set. Do not adjust the gap unless the pick-up is being changed or the base plate has been moved. Use a non-ferrous feeler gauge to set the air gap.

## DISTRIBUTOR

### Remove and refit

### Removing

1. Disconnect the battery negative lead.
2. Disconnect the vacuum pipe.
3. Remove the distributor cap.
4. Disconnect low tension lead from the coil.
5. Mark distributor body in relation to centre line of rotor arm.

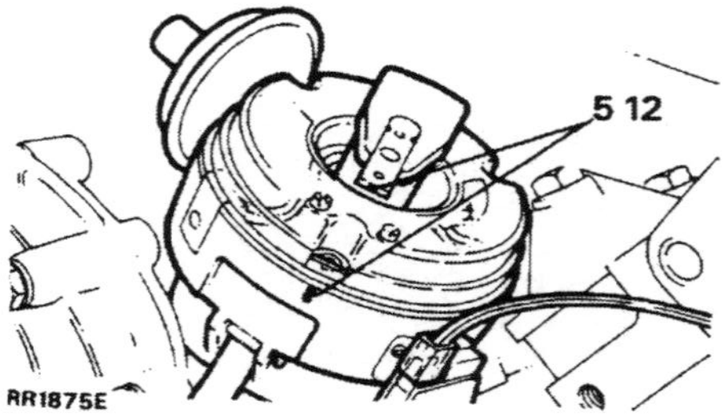

6. Add alignment marks to distributor and front cover.

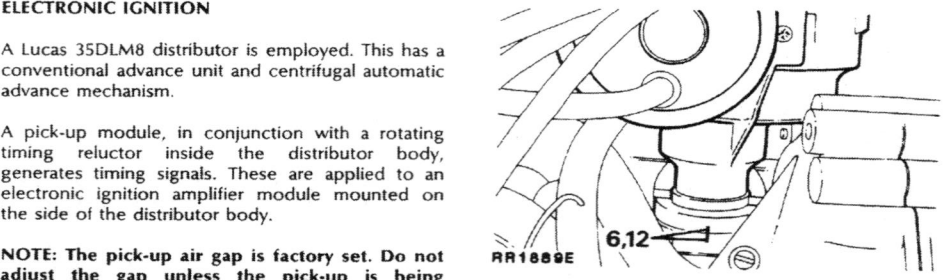

NOTE: Marking the distributor enables refitting in exact original position, but if engine is turned while distributor is removed, complete ignition timing procedure must be followed.

7. Release the distributor clamp and remove the distributor.

### Refitting

NOTE: If a new distributor is being fitted, mark body in same relative position as distributor removed.

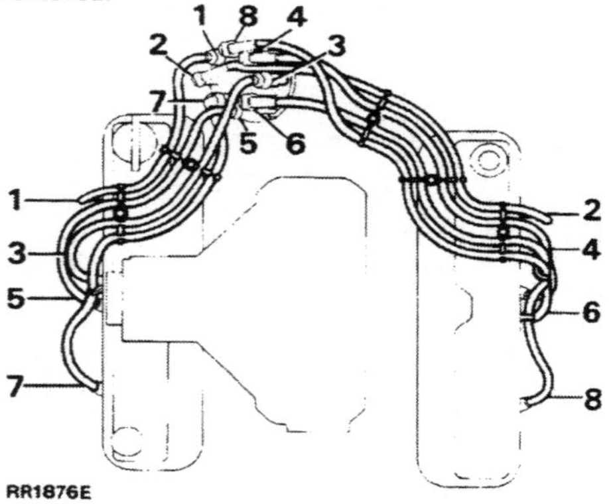

8. Leads for distributor cap should be connected as illustrated. Figures 1 to 8 inclusive indicate plug lead numbers. RH-Right hand side of engine, when viewed from the rear. LH-Left hand side of engine, when viewed from the rear.

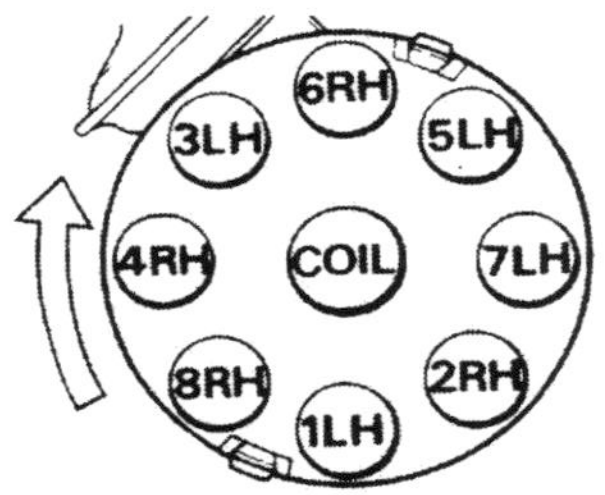

# ELECTRICAL

RR516M

9. If engine has not been turned whilst distributor has been removed, proceed as follows (items 10 to 17). Alternatively proceed to instruction 18.
10. Fit new 'O' ring seal to distributor housing.
11. Turn distributor drive until centre line of rotor arm is 30o anti-clockwise from mark made on top edge of distributor body.
12. Fit distributor in accordance with alignment markings.

**NOTE: It may be necessary to align oil pump drive shaft to enable distributor drive shaft to engage in slot.**

13. Fit clamp and bolt. Secure distributor in exact original position.
14. Connect vacuum pipe to distributor and low tension lead to coil.
15. Fit distributor cap.
16. Reconnect battery.
17. Using suitable electronic equipment, set the ignition timing, see **IGNITION TIMING-Adjust**.
18. If, with distributor removed, engine has been turned it will be necessary to carry out the following procedure.
19. Set engine-No. 1 piston to static ignition timing figure (see Engine Tuning Data- Section 05) on compression stroke.
20. Turn distributor drive until rotor arm is approximately 30o anti-clockwise from number one sparking plug lead position on cap.
21. Fit distributor to engine.
22. Check that centre line of rotor arm is now in line with number one sparking plug lead on cap. Reposition distributor if necessary.
23. If distributor does not seat correctly in front cover, oil pump drive is not engaged. Engage by lightly pressing down distributor while turning engine.
24. Fit clamp and bolt leaving both loose at this stage.
25. Set the ignition timing statically to within 2°-3° of T.D.C.
26. Connect the vacuum pipe to the distributor.
27. Fit low tension lead to coil.
28. Fit distributor cap.
29. Reconnect the battery.
30. Using suitable electronic equipment set the ignition timing, see **IGNITION TIMING-Adjust**.

## DISTRIBUTOR-LUCAS 35DLM8

### Overhaul

### DISTRIBUTOR COVER

1. Unclip and remove the cover.
2. Renew the cover if known to be faulty.
3. Clean the cover and HT brush with a nap free cloth.

### ROTOR ARM

4. Pull rotor arm from shaft.
5. Renew rotor arm if known to be faulty.

### INSULATION COVER (Flash shield)

6. Remove cover, secured by three screws.
7. Renew cover if known to be faulty.

### VACUUM UNIT

8. Remove two screws from vacuum unit securing bracket, disengage vacuum unit connecting rod from pick-up base plate connecting peg, and withdraw vacuum unit from distributor body.

### AMPLIFIER MODULE

9. Remove two screws and withdraw the module.
10. Remove the gasket.
11. Remove two screws securing the cast heatsink and remove the heatsink.

**WARNING: The amplifier module is a sealed unit containing Beryllia. This substance is extremely dangerous if handled. Do not attempt to open or crush the module.**

### PICK-UP AND BASE PLATE ASSEMBLY

12. Use circlip pliers to remove the circlip retaining the reluctor on rotor shaft.
13. Remove the flat washer and then the 'O' ring recessed in the top of the reluctor.
14. Gently withdraw the reluctor from the shaft, taking care not to damage the teeth.

**NOTE: Coupling ring fitted beneath reluctor.**

15. Remove three support pillars and cable grommet. Lift out the pick-up and base plate assembly.

**NOTE: Do not disturb the two barrel nuts securing the pick-up module, otherwise the air gap will need re-adjustment.**

16. Renew pick-up and base plate assembly if module is known to be faulty, otherwise check pick-up winding resistance (2k-5k ohm).

### RE-ASSEMBLY

17. This is mainly a reversal of the dismantling procedure, noting the following points:

### LUBRICATION

Apply clean engine oil:

a. Three drops to felt pad reservoir in rotor shaft.

Apply Omnilube 2 (or equivalent) grease.

b. Auto advance mechanism.
c. Pick-up plate centre bearing.
d. Pre tilt spring and its rubbing area (pick-up and base plate assembly).
e. Vacuum unit connecting peg (pick-up and base plate assembly).
f. The connecting peg hole in vacuum unit connecting rod.

### FITTING PICK-UP AND BASE PLATE ASSEMBLY

18. Pick-up leads must be prevented from fouling the rotating reluctor. Both leads should be located in plastic carrier as illustrated. Check during re-assembly.

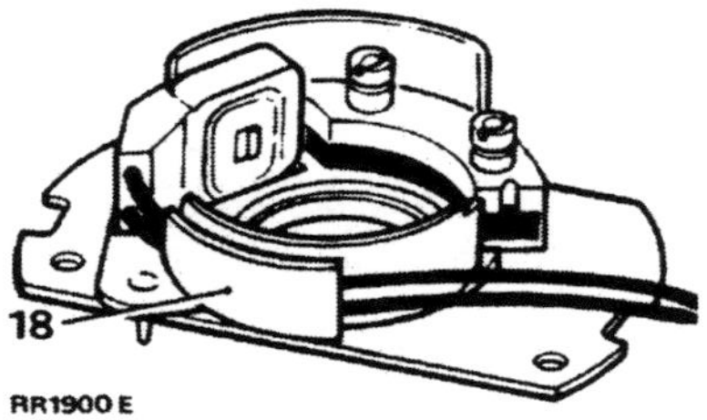

RR1900 E

### REFITTING RELUCTOR

19. Slide reluctor as far as it will go on rotor shaft, then rotate reluctor until it engages with the coupling ring beneath the pick-up base plate. The distributor shaft, coupling ring and reluctor are 'keyed' and rotate together.

## ELECTRICAL

### PICK-UP AIR GAP ADJUSTMENT

20. The air gap between the pick-up limb and reluctor teeth must be set within the specified limits, using a non-ferrous feeler gauge.

NOTE: When the original pick-up and base plate assembly has been refitted the air gap should be checked, and adjusted if necessary.

When renewing the assembly the air gap will require adjusting to within the specified limits.

Refer to 'Engine Tuning Data'

### AMPLIFIER MODULE

21. Before fitting the module, apply MS4 Silicone grease or equivalent heat-conducting compound to the amplifier module backplate, the seating face on distributor body and both faces of the heatsink casting.

### IGNITION COIL

**Remove and refit**

**Removing**

1. Disconnect the battery negative terminal.
2. Disconnect the High Tension and Low Tension electrical leads from the ignition coil.
3. Remove the two bolts securing the coil to the valance.

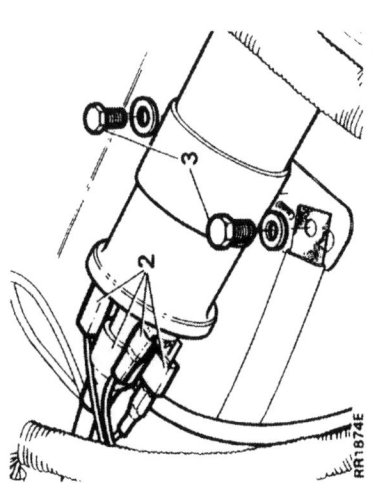

NOTE: An earth braid is located under one of the bolts.

4. Remove the coil from the engine compartment.

**Refitting**

5. Reverse the removal instructions.

NOTE: Ensure that the bolting location for the earth braid is free from paint and grease. Coat the area around the bolt with petroleum jelly.

### IGNITION TIMING

**Adjust**

1. It is essential that the following procedures are adhered to. Inaccurate timing can lead to serious engine damage and additionally create failure to comply with emission regulations. If the engine is being checked in the vehicle and is fitted with an air conditioning unit the compressor must be isolated.

2. On initial engine build, or if the distributor has been disturbed for any reason, the ignition timing must be set statically to within 2°-3° of T.D.C.

(This sequence is to give only an approximation in order that the engine may be started) ON NO ACCOUNT MUST THE ENGINE BE STARTED BEFORE THIS OPERATION IS CARRIED OUT.

**Equipment required**

Calibrated Tachometer
Stroboscopic lamp

3. Couple stroboscopic timing lamp and tachometer to engine following manufacturer's instructions.

4. Disconnect the vacuum pipe from the distributor.
5. Start engine, with no load and not exceeding 3,000 rev/min run engine until normal operating temperature is reached. (Thermostat open). Check that the normal idling speed falls within the tolerance specified in the data section.
6. Idle speed for timing purposes must not exceed 600 rev/min.
7. With the distributor clamping bolt slackened turn distributor until the timing flash coincides with the timing pointer and the correct timing mark on the rim of the torsional vibration damper as shown in the engine tuning section.
8. Retighten the distributor clamping bolt securely. Recheck timing in the event that retightening has disturbed the distributor position.
9. Refit vacuum pipe.
10. Disconnect stroboscopic timing lamp and tachometer from engine.

### LUCAS CONSTANT ENERGY IGNITION SYSTEM 35DLM8-PRELIMINARY CHECKS

Inspect battery cables and connections to ensure they are clean and tight. Check battery state of charge if in doubt as to its condition.

Inspect all LT connections to ensure that they are clean and tight. Check the HT leads are correctly positioned and not shorting to earth against any engine components. The wiring harness and individual cables should be firmly fastened to prevent chafing.

### PICK-UP MODULE AIR GAP SETTINGS

Air gap settings vary according to vehicle application.

NOTE: The gap is set initially at the factory and will only require adjusting if tampered with or when the pick-up module is replaced.

### Test Notes

(i) The ignition must be switched on for all checks.
(ii) Key to symbols used in the charts for Tests 2.

| C | Correct Reading |
| H | High Reading |
| L | Low Reading |

(iii) Use feeler gauges manufactured from a non-magnetic material when setting air gaps.

### TEST 1:

**Check HT Sparking**

Remove coil/distributor HT lead from distributor cover and hold approximately 6mm (0.25 in) from the engine block. Switch the ignition 'On' and operate the starter.

If regular sparking occurs, proceed to Test 6. If no sparking proceed to Test 2.

**Test 1**

### TEST 2:

**Amplifier Static Checks**

Switch the ignition 'On'

(a) Connect voltmeter to points in the circuit indicated by the arrow heads and make a note of the voltage readings.

NOTE: Only move the voltmeter POSITIVE lead during tests 2,3, and 4.

## ELECTRICAL 86

(b) Compare voltages obtained with the specified values listed below:

**EXPECTED READINGS**

1. More than 11.5 volts.
2. 1 volt max below volts at point 1 in test circuit.
3. 1 volt max below volts at point 1 in test circuit.
4. 0 volt-0.1 volt.

**Test 2**

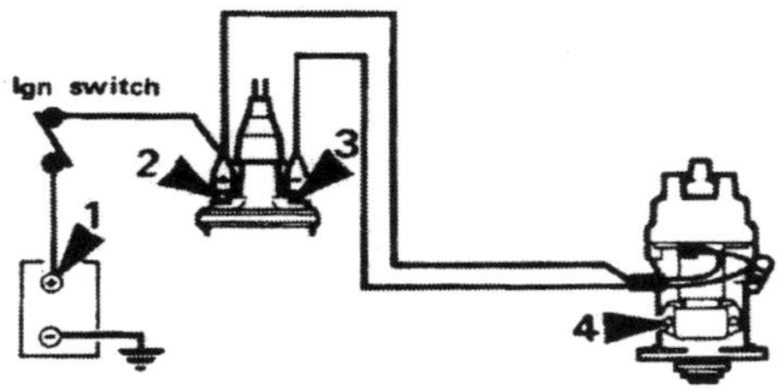

(c) If all readings are correct proceed to Test 3.
(d) Check incorrect reading(s) with chart to identify area of possible faults, i.e. faults listed under heading 'Suspect'.

| 1 | 2 | 3 | 4 | SUSPECT |
|---|---|---|---|---|
| L | C | C | C | DISCHARGED BATTERY |
| C | L | L | C | IGN. SWITCH AND/OR WIRING |
| C | C | L | C | COIL OR AMPLIFIER |
| C | C | C | H | AMPLIFIER EARTH |

### TEST 3:

**Check Amplifier Switching**

Disconnect the high tension lead between the coil and distributor.

Connect the voltmeter between positive (+ ve) terminal and H.T. coil negative (- ve) terminal, the voltmeter should register zero volts.

Switch the ignition 'On' then crank the engine. The voltmeter reading should increase when cranking, in which case proceed with Test 5.

If there is no increase in voltage during cranking proceed to Test 4.

**Test 3**

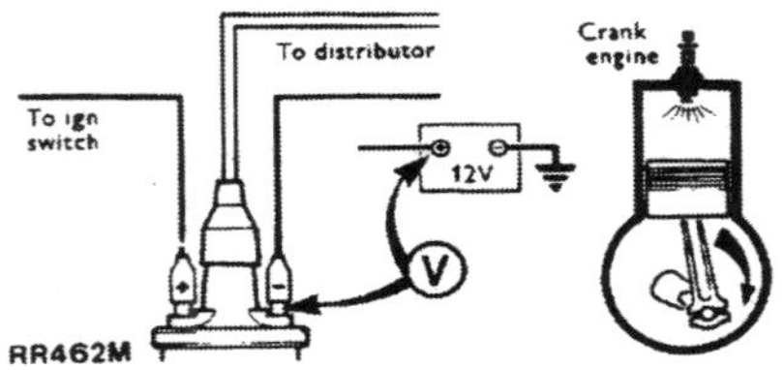

### TEST 4:

**Pick-Up Coil Resistance Applications with Separate Amplifier**

Disconnect the pick-up leads at the harness connector. Connect the ohmmeter leads to the two pick-up leads in the plug.

The ohmmeter should register between 2k and 5k ohm if pick-up is satisfactory. Change the amplifier if ohmmeter reading is correct. If the engine still does not start carry out Test 5.

**Test 4**

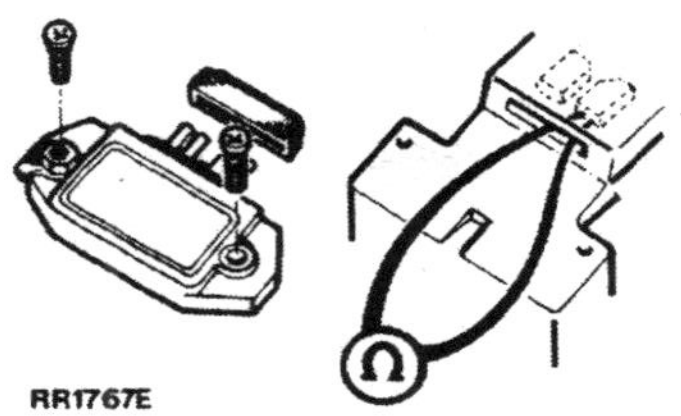

---

## ELECTRICAL 86

Change the pick-up if ohmmeter reading is incorrect. If the engine still does not start proceed to Test 5.

### TEST 5:

**Coil H.T. Sparking**

Remove existing coil/distributor H.T. lead and fit test H.T. lead to coil chimney. Hold free end about 6mm (0.25 in) from the engine block and crank the engine.

H.T. sparking good, repeat test with original H.T. lead, if then no sparking, change H.T. lead. If sparking is good but engine wil not start, proceed to Test 6.

If no sparking, replace coil. If engine will not start carry out Test 6.

**Test 5**

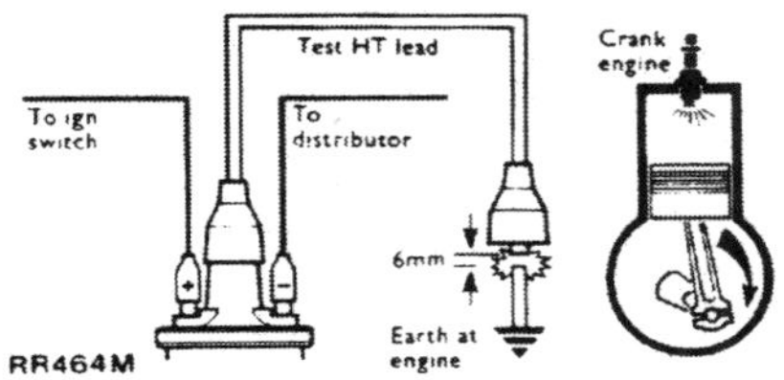

### TEST 6:

**Check Rotor Arm**

Remove distributor cover. Disconnect coil H.T. lead from cover and hold about 3mm (0.13 in) above rotor arm electrode and crank the engine. There should be no H.T. sparking between rotor and H.T. lead. If satisfactory carry out Test 7.

H.T. sparking, replace rotor arm.

If engine will not start carry out Test 7.

**Test 6**

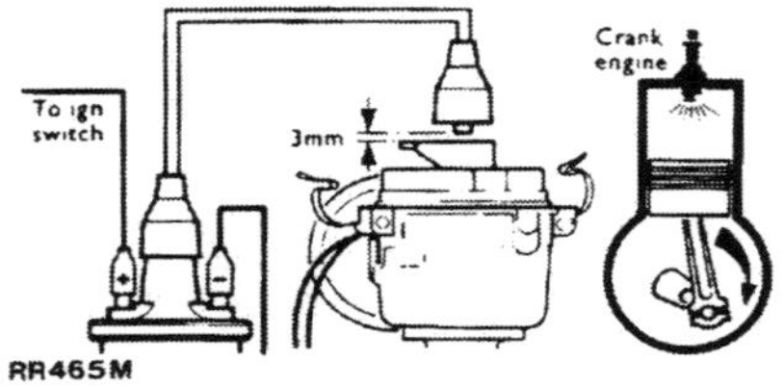

### TEST 7:

**Visual and H.T. Cable Checks**

| Examine: | Should be: |
|---|---|
| 1. Distributor Cover | Clean, dry, no tracking marks |
| 2. Coil Top | Clean, dry, no tracking marks. |
| 3. H.T. Cable Insulation | Must not be cracked, chafed or perished |
| 4. H.T. Cable Continuity | Must not be open circuit |
| 5. Sparking Plugs | Clean, dry, and set to correct gap |

**NOTE:**

1. Reluctor Rotor and Insulation Cover — Must not foul pick-up or leads Must not be cracked or show signs of tracking marks

# ELECTRICAL 86

## STARTER MOTOR-Lucas M78R
(Petrol models)

### Overhaul

**Dismantling**

1. Remove the starter motor.
2. Remove the braid between the starter and the solenoid terminal.
3. Remove the solenoid fixing screws.
4. Withdraw the solenoid body.
5. Lift and remove the solenoid plunger.
6. Remove two nuts and two screws from the commutator end bracket.
7. Remove the commutator end bracket.
8. Remove the grommet from the yoke.
9. Lift the brushbox assembly clear of the armature.
10. Remove the brush springs.
11. Unclip and remove the earth brushes.
12. Remove the insulating plate.
13. Withdraw the brushes and bus bar.
14. Remove the armature from the yoke.
15. Remove the yoke.
16. Remove the intermediate bracket.
17. Loosen and remove the through bolts from the drive end bracket.
18. Remove the sun and planet gears.
19. Push out the drive shaft sprocket assembly from the drive end bracket.
20. Carefully tap the thrust collar from over the jump ring back towards the drive.
21. Prise the jump ring from its locating groove.
22. Remove the drive assembly from the drive shaft.

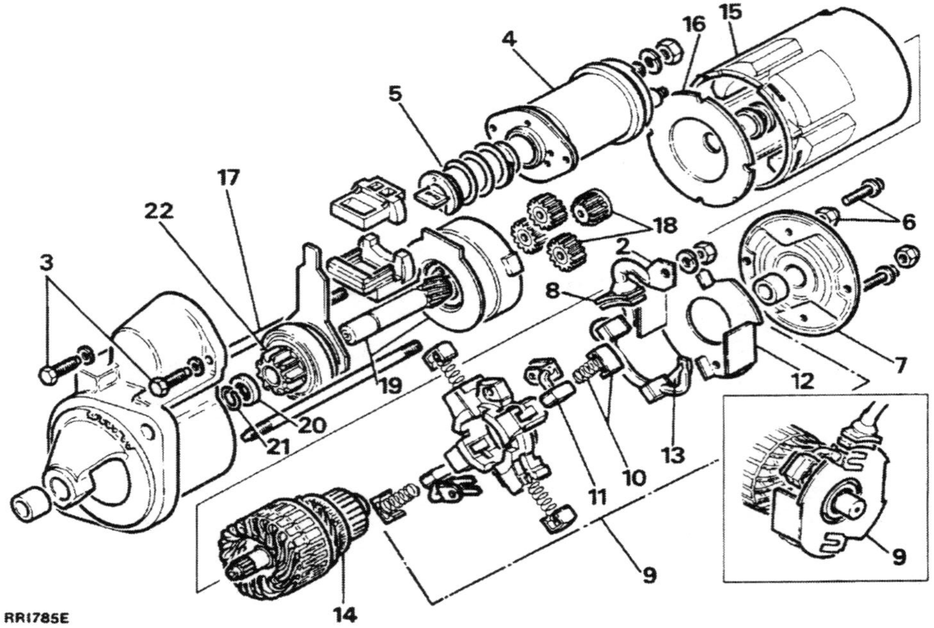

### Inspecting

**Solenoid**

23. Check the continuity and resistance value of windings by connecting an ohmmeter as shown.

(a) Resistance value should be: 1.074 ± 0.035 ohms

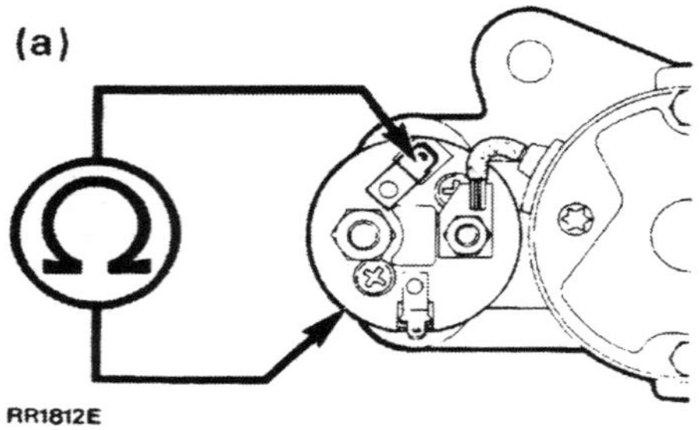

(b) Resistance value should be: 0.298 ± 0.015 ohms

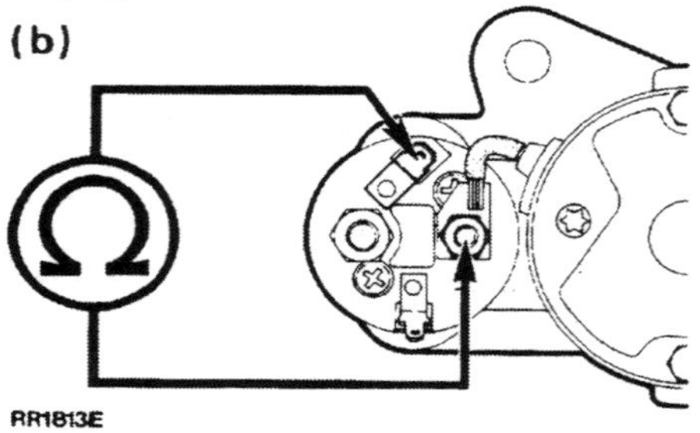

If test results are unsatisfactory replace the solenoid.
If results are correct proceed to 24.

24. Check the contacts by connecting an ohmmeter as shown. Solenoid plunger removed, ohmmeter should read infinity.

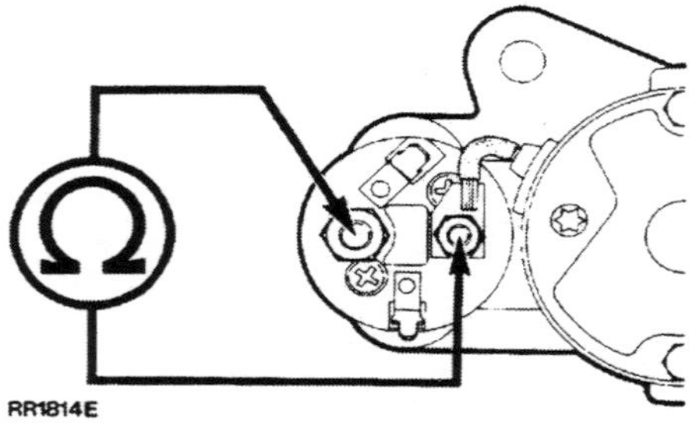

Solenoid plunger operated by hand, ohmmeter should read zero. If test results are unsatisfactory, replace the solenoid. If results are correct proceed to 25.

25. Check operation of spring for freedom of movement.

**Brush gear**

26. Check brush springs and ensure that the brushes move freely in their holders. Clean the brushes with a petrol moistened cloth, if required.

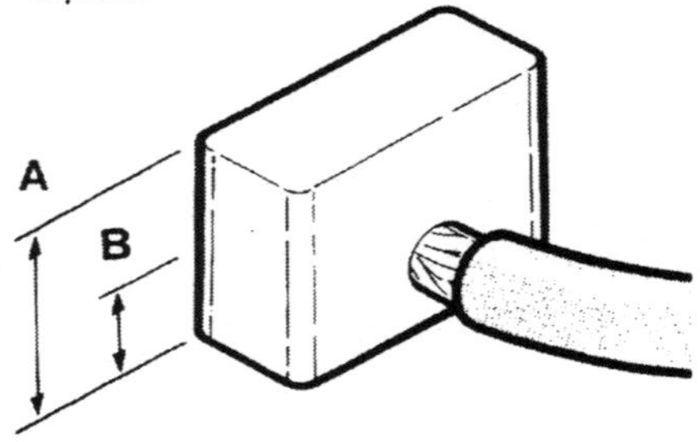

Brush length new, Dimension A is 9mm (0.354 in). Minimum brush length, Dimension B is 3.5mm (0.138 in).

## ELECTRICAL 86

### Armature

27. Check the armature insulation using suitable test equipment. Connect the tester between any one commutator segment and the shaft. The method illustrated uses a 110V, 15W test lamp. If the lamp illuminates the armature is faulty, and a replacement component is required.

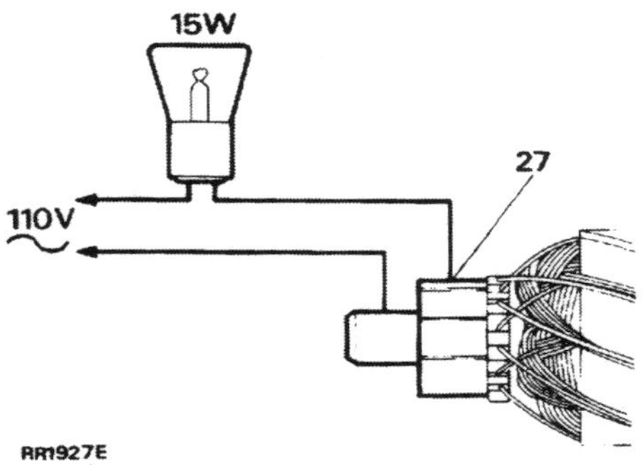

28. If necessary, the commutator may be machined, providing a finished surface can be obtained without reducing the diameter below 28.8mm (1.13 in), otherwise a new commutator must be fitted. Finish the surface with fine emery cloth. Do not undercut the insulation slots.

### Drive assembly

29. Test the roller clutch. The pinion should rotate in one direction only, independent of the clutch body. Replace the unit if unsatisfactory or if teeth are damaged or worn.

### Bearings

30. Renew the bearing bushes if there is evidence of armature fouling magnets or if there is perceptible side play between the shaft and bush.

31. Drive end/intermediate end bracket: press out the bush using a suitable press and mandrel.

32. Press the new bush in, ensuring that on the drive end bracket, the bush is flush with the casting.

33. Commutator end bracket: thread a 9/16" Whitworth or suitable similar tap firmly into the bush. Extract the bush with the tap using a power press in reverse.

**NOTE: Soak new bushes in engine oil for thirty minutes before fitting.**

### Reassembly

34. Reverse the instructions 1 to 22. Smear the teeth and operating collar of the roller clutch with Shell Retinax 'A' grease. Smear the pivot lever of the drive assembly with Mobil 22 grease. Smear the drive shaft sun and planet gears with Rocol BRB1200 grease.

35. Tighten all the fixings to the correct torque—see Torque Wrench Settings.

---

## ELECTRICAL 86

9. Manoeuvre the starter motor and withdraw it from the flywheel housing, place the unit on a suitable workbench.

### DISMANTLE

### SOLENOID

### Remove and test

#### Remove

10. Disconnect the link lead, solenoid to starter motor.
11. Remove the three fixings and withdraw the body of the solenoid from the solenoid housing.
12. Remove the operating fork pivot bolt, withdraw the solenoid plunger from its housings.
13. Inspect the rubber boot on the solenoid for condition, if in poor condition, renew as necessary.
14. Refit the plunger to the solenoid body.

#### Test

15. Connect a 12 volt battery supply and a 12 volt 60 watt test lamp between the solenoid main terminals. The lamp should not light, if it does light, fit a new solenoid.
16. Leave the test lamp connected and using the same 12 volt battery supply between the small solenoid operating 'spade' terminal and a good earth point on the body.

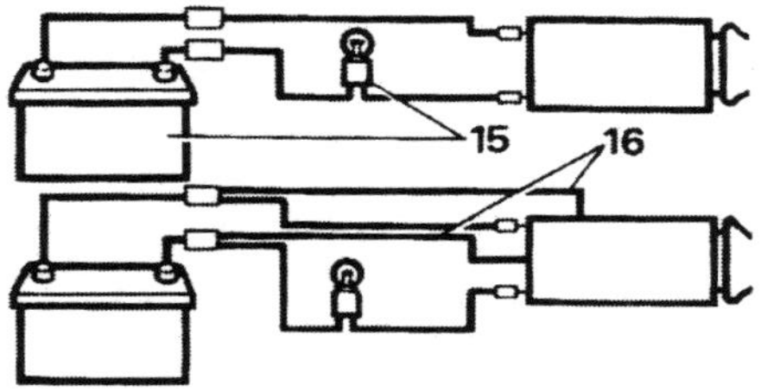

17. The plunger should pull back in the solenoid body and the test lamp should illuminate, otherwise fit a new solenoid.

---

## STARTER MOTOR - Bosch 544 (Diesel models)

The solenoid, brushes and drive gear are the only components that can effectively be replaced on the Bosch 544 starter motor.

Should a major fault occur within the internal windings a new starter motor is to be fitted.

The starter motor is a fully water-proof unit, when dismantling care should be taken to note the location of 'O' rings and seals to ensure that they are refitted on assembly.

As a secondary precaution to ensure the complete water tightness of the starter motor, coat all threads of bolts and joint faces with a Silicon based sealing compound.

### Remove, test and refit

#### Removing

1. Install the vehicle on an hydraulic ramp.
2. Prop open the bonnet and disconnect the two battery negative terminals.
3. Remove the single nut and bolt securing the heat shield to its mounting bracket, adjacent to the starter motor to flywheel housing lower fixing.
4. Remove the two bolts securing the heat shield and mounting bracket to the starter motor end cover.
5. Withdraw the heat shield.
6. Release the two bolts securing the starter motor mounting bracket to the side of the cylinder block, do not fully remove the bolts but allow the bracket to remain as loose as possible
7. Remove the electrical leads from the solenoid.
8. Supporting the starter motor remove the upper and lower fixings, note that an earth braid and heat shield bracket is located under the lower nut.

# ELECTRICAL 86

## BRUSHES

### Remove and test

**Remove**

18. Remove the two screws securing the end cap to the starter motor end cover, note the two small 'O' rings on the screw threads.
19. Remove the 'C' clip and shims from the end of the armature shaft.
20. Remove the two nuts (with plain washers) securing the starter motor end cover, note the position of the two 'O' rings behind the nuts.
21. Withdraw the end cover.
22. Depress the field winding brush springs and open out the four retaining lugs to release the spring and brush from its holder.
23. Withdraw the brush carrier assembly from the commutator, noting that the two negative brushes are still attached to the carrier. Care should be taken to ensure that the two negative brush springs do not jump out of the carrier.
24. 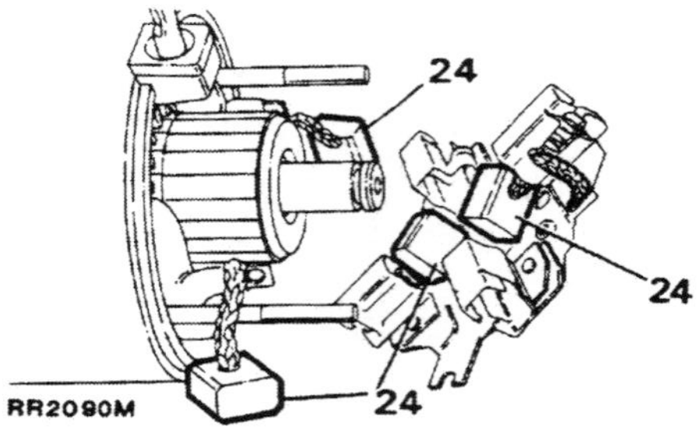 
25. Withdraw the yoke of the starter motor from the drive end bearing housing. Note the position of 'O' rings for reassembly.

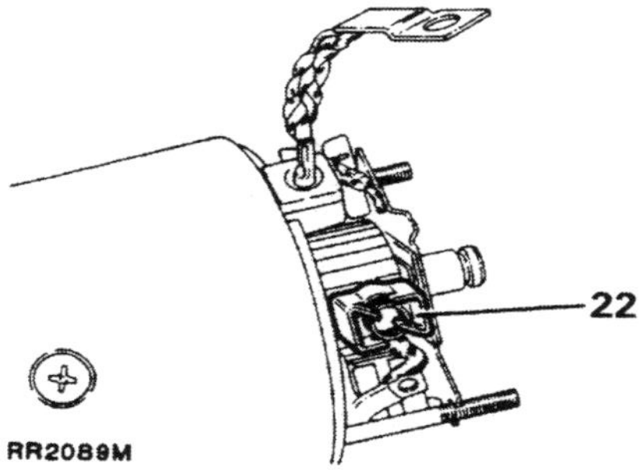

**Test**

26. Check for continuity between link lead and each brush in turn, if no continuity renew starter motor.
27. Check field winding insulation between each brush in turn to a clean paint free part of the yoke, there should be no continuity, if continuity exists renew starter motor.

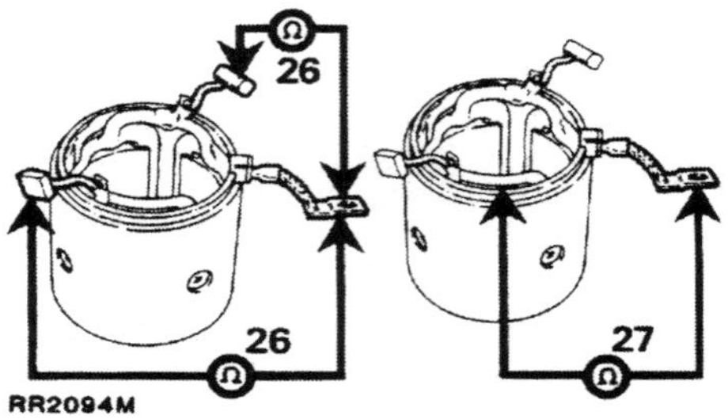

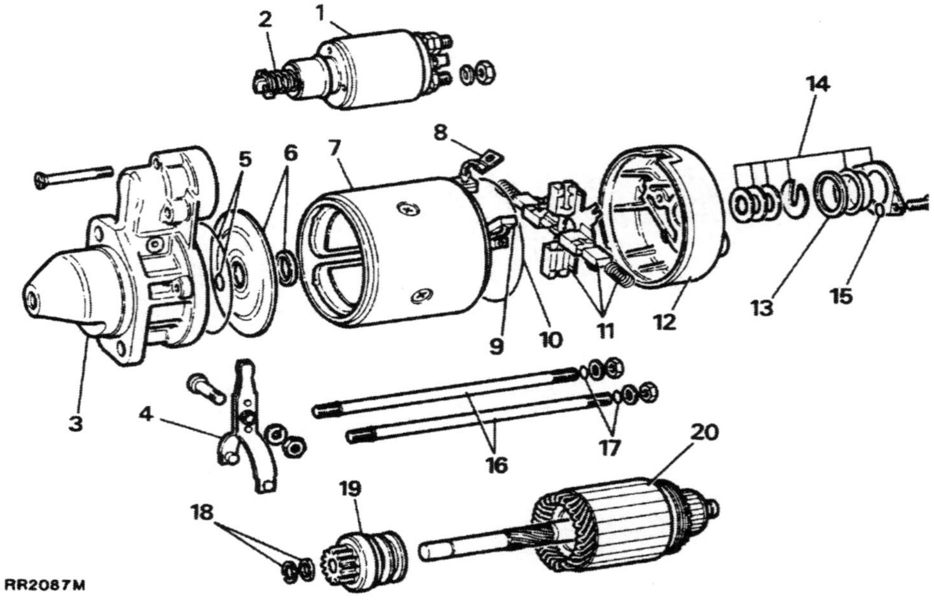

## KEY TO STARTER MOTOR

1. Solenoid
2. Solenoid plunger and spring
3. Drive end - bearing housing
4. Operating fork
5. 'O' Rings - bearing housing (3 off)
6. Seal and end plate
7. Yoke
8. Link lead
9. Field winding brushes
10. 'O' Ring
11. Armature brushes and carrier assembly
12. End cover
13. 'O' Ring
14. Shims, 'C' clip and end cap
15. 'O' Rings (2 off)
16. Through studs
17. 'O' Rings (2 off)
18. Circlip and retainer
19. Drive gear assembly
20. Armature

## ELECTRICAL 86

28. Check the insulation between the brush carrier and brush holder, if continuity exists renew the carrier assembly.

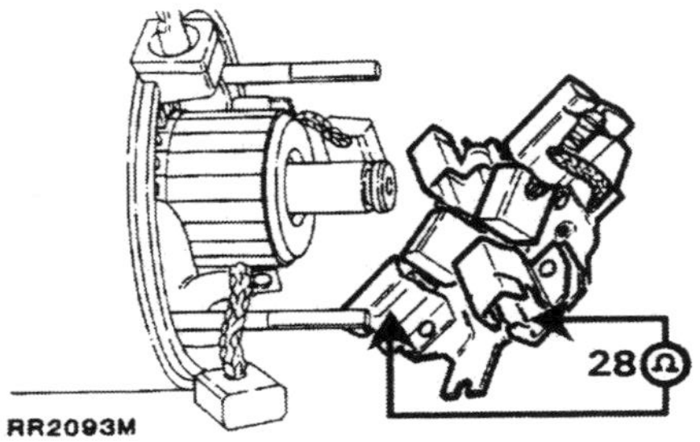

RR2093M

### ARMATURE

**Remove and test**

**Remove**

29. Remove the two long through studs to enable the armature to be withdrawn.
30. Ensure the drive gear is as far forward in the bearing housing as is possible.
31. Withdraw the armature until the drive gear operating fork is visible between the armature end plate and bearing housing.
32. Manoeuvre the armature assembly until the operating fork can be removed from the bearing housing and withdraw the armature. Note the position of the 'O' rings for reassembly.
33. Check the commutator, if cleaning only is necessary use a piece of very fine glass paper on a flat surface to remove any brush score marks, follow by wiping the commutator surface with a petrol moistened cloth.
34. If the commutator is heavily worn or has excessively deep score marks it may be necessary to machine the commutator surface, providing that, when all marks have been removed the finished diameter is not below 39.5mm (1.55 inch).
35. Having machined the commutator each segment must now be undercut to reduce the insulation to a depth of 0.5mm (0.019 inch) below the surfae of the commutator.

**Test**

36. Check to see if there is any continuity between the commutator and armature core, if continuity exists it indicates that the insulation between the windings and core have broken down, renew starter motor.

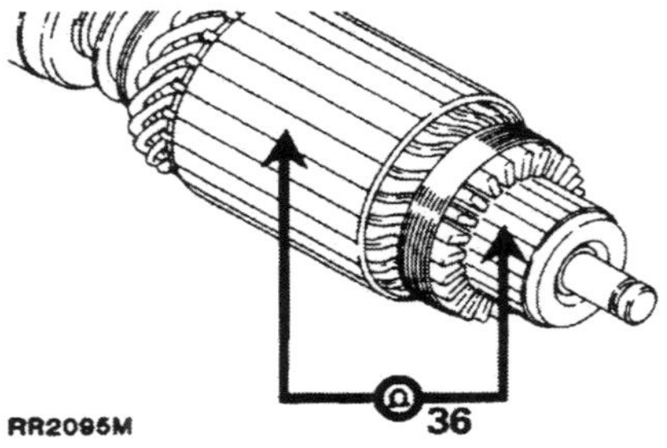

RR2095M

37. Check for continuity between each adjacent pair of commutator segments, if there is no continuity this will indicate that the armature windings have broken down, renew starter motor.

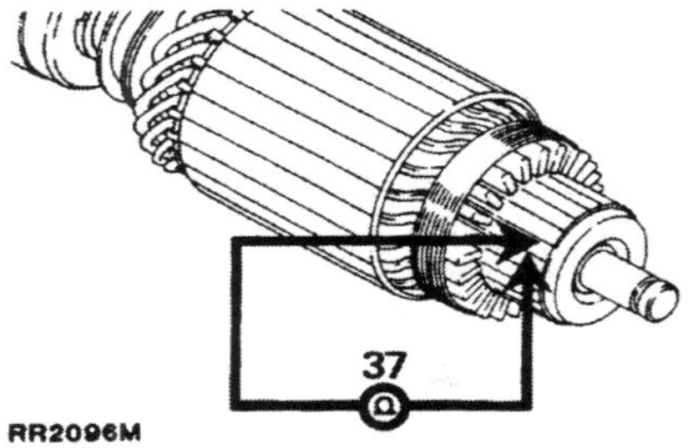

RR2096M

### DRIVE GEAR

**Remove**

38. Using a suitable piece of tube over the armature shaft, carefully tap the circlip retainer down the shaft to expose the circlip.
39. Remove the circlip, inspect the groove to ensure that sharp edges have not been raised during the removal of the circlip, if necessary, with a smooth file remove any sharp edges before removing the drive gear assembly.

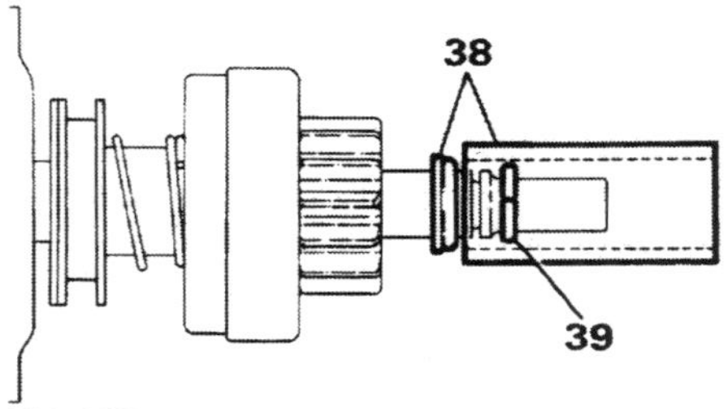

RR2097M

### ASSEMBLE

40. Coat the spiral gear on the armature shaft and the groove in the collar of the drive gear with Shell Retinax 'A' prior to assembly.
41. Ensure that all 'O' rings are in good condition and that they are fitted in their correct locations.
42. Coat all joint faces and screw threads with a Silicon based sealing compound to prevent ingress of water.
43. Reverse the dismantling instructions ensuring that all srews and nuts are securely tightened.

44. Refit the starter motor to the flywheel housing.
45. Reverse instructions 1 to 8 ensuring that the fixing bolts are tightened to the correct torque.
46. Securely tighten the nut which secures the electrical feed lead to the solenoid.
47. Lightly smear petroleum jelly around both of the terminals.

# ELECTRICAL 86

## HEADLAMP ASSEMBLY/SEALED BEAM UNIT

### Remove and refit

#### Removing

1. Disconnect the battery negative lead.
2. Remove the radiator grill- see Body Section 76.
3. Remove three crosshead screws and the headlamp retaining rim.
4. **DO NOT** disturb the two adjusting screws.
5. Withdraw the sealed beam unit and disconnect the wiring plug from the rear of the unit.
6. Remove three securing screws, prise away the grommet and withdraw the headlamp bowl.

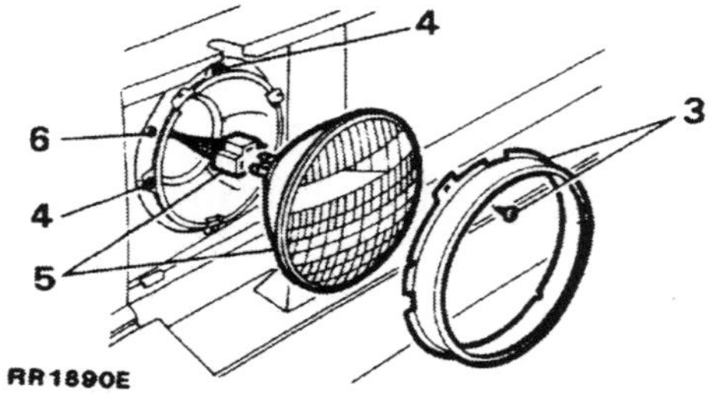

#### Refitting

7. Reverse removal procedure.

RR1890E

## HEADLAMP ALIGNMENT

Headlamp beam setting should only be carried out by qualified personnel using suitable beam setting equipment.

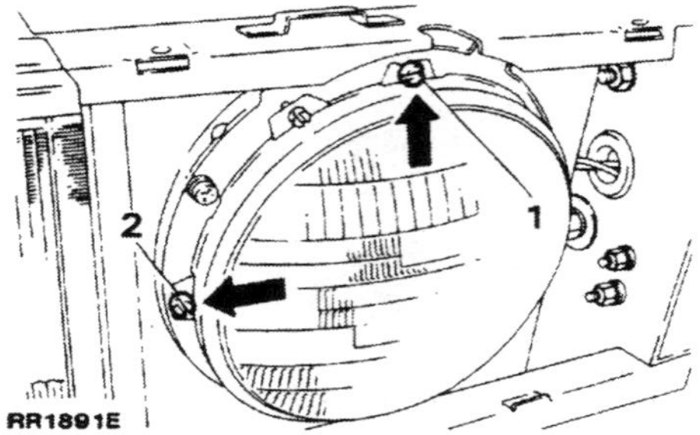

### Adjusting

1. Turn the top adjusting screw anti-clockwise to lower the beam, clockwise to raise the beam.
2. Turn the side adjusting screw anti-clockwise to move the beam to the left, clockwise to move the beam to the right.

RR1891E

---

# ELECTRICAL 86

## NUMBER PLATE LAMP ASSEMBLY AND BULB REPLACEMENT

### Remove and refit

#### Removing

1. Disconnect the battery negative lead.
2. Remove the two self-tapping screws and washers.
3. Detach the lamp assembly.
4. Disconnect the bulb holder and remove the bulb.

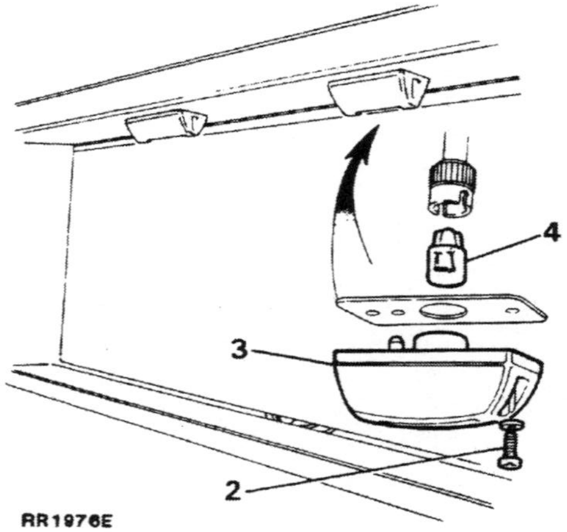

RR1976E

**NOTE: Carefully pull the electrical leads out of the bottom of the lower tailgate panel to reveal the snap connectors.**

5. Disconnect the electrical connections located at the bottom of the lower tailgate.
6. Remove the bulb holder.
7. Carefully pull the electrical leads up through the inside of the lower tailgate panels.

#### Refitting

8. Reverse the removal procedure. The correct bulb 'type' is a 12-volt, 5 watt wedge base (capless).

## SIDELIGHT AND FLASHER LAMP ASSEMBLY - RH AND LH AND BULB

### Remove and refit

#### Removing

1. Open the bonnet and disconnect the battery negative lead.
2. Remove the two screws and plain washers securing the lamp assembly.
3. Lift the assembly away sufficiently to gain access to the rear of the lamp.
4. Remove the waterproof cover.
5. Depress the two retaining clips and withdraw the bulb holder.
6. Remove the required bulb. The direction indicator bulb is located in the upper section of the bulb holder, the side lamp bulb in the lower.
7. Disconnect the multi-plug to remove the complete assembly.

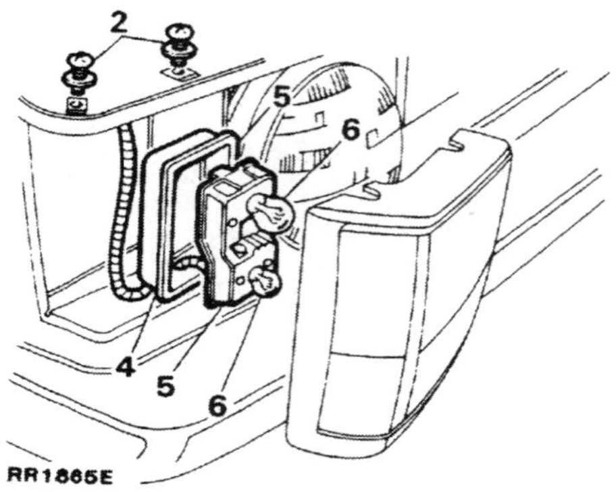

#### Refitting

8. Reverse the removal procedure, ensuring the waterproof cover is located correctly.

RR1865E

## ELECTRICAL 86

### DIRECTION INDICATOR SIDE REPEATER LAMP BULB REPLACEMENT

**Remove and refit**

1. Working at the back of the lamp, through the wheel arch, slightly twist the bulb holder anti-clockwise to release it from the lamp assembly.
2. Remove the bulb.

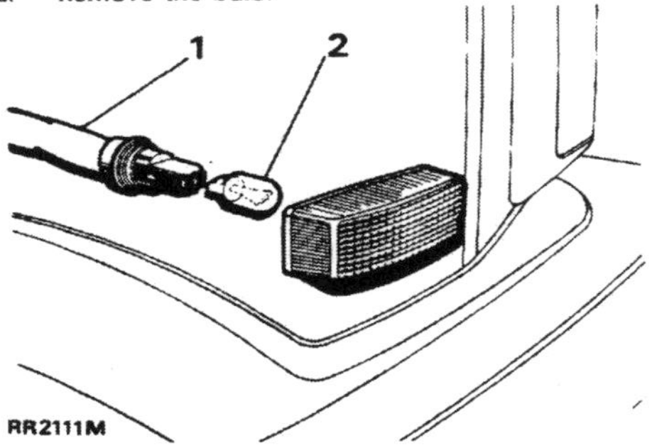

RR2111M

**Refitting**

3. Reverse the removal instructions. The correct bulb 'type' is a 12-volt, 5 watt wedge base (capless).

### AUXILIARY SWITCH PANEL

The auxiliary switch panel contains four 'push-push' type switches which incorporate integral symbols for identification.
(The unused apertures are fitted with blank covers, which are removable, to facilitate the fitting of extra switches if required).
The symbols are illuminated by two bulbs which become operational when the vehicle lights are on.

The rear fog guard lamp and heated rear screen switches (1 and 5) are also provided with individual warning lights, illuminated when the switches are operated.

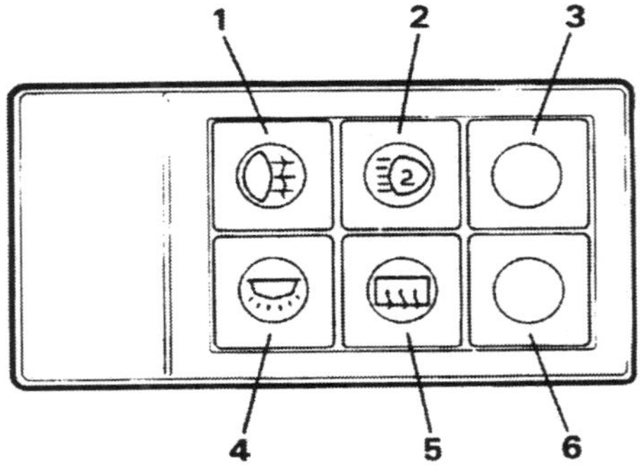

RR2102 M

1. Rear fog guard lamps.
2. Auxiliary driving lamps.
3. Blank.
4. Interior and tailgate lamps.
5. Heated rear screen.
6. Blank.

291

## ELECTRICAL 86

### RELAYS-Identification

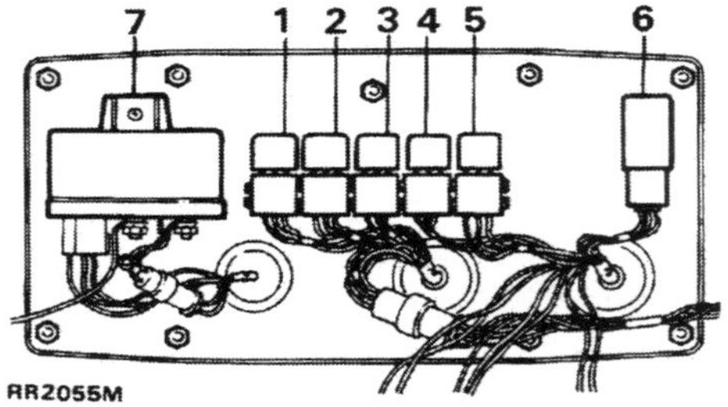

RR2055M

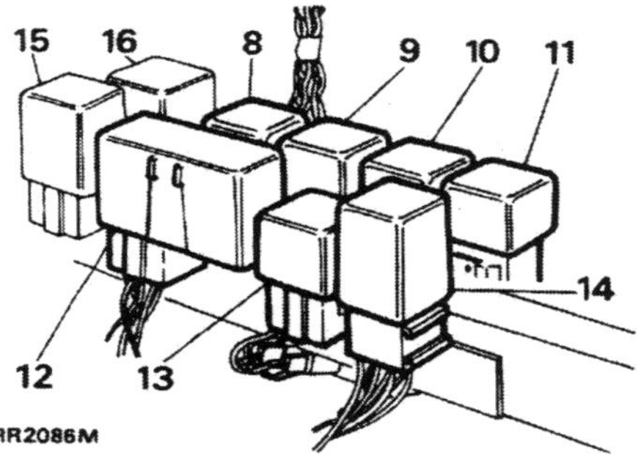

RR2086M

Steering column mounted relays viewed with the lower fascia panel removed.

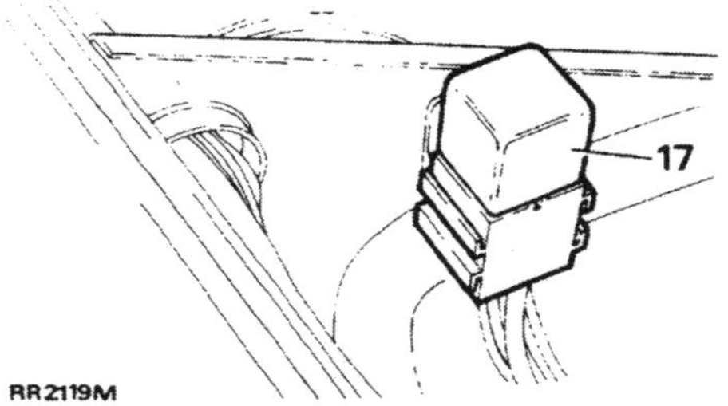

Closure panel viewed from the engine compartment, with protective cover removed.

RR2119M

Fuel cut-off relay (carburetter models only) located in the engine compartment attached to the left-hand valance stiffener brackets.

| Relay | Right-hand Drive Circuit Diagram Item Number | Left-hand Drive Circuit Diagram Item Number | Colour |
|---|---|---|---|
| 1. Air conditioning/heater | Left and Right Hand Drive | 5. Air conditioning | Natural |
| 2. Condenser fan | | 9. | Natural |
| 3. Compressor clutch | | 11. circuit diagram | Natural |
| 4. Starter solenoid relay | 6. | 6. Main circuit diagram | Natural |
| 5. Heated rear window | 66. | 64. Main circuit diagram | Black |
| 6. Headlamp wash timer unit | 19. | 17. Main circuit diagram | Black |
| 7. Glow plug timer (Diesel) | 143. | 150. Main circuit diagram | Blue |
| 8. Flasher/Hazard unit | 75. | 73. Main circuit diagram | Yellow |
| 9. Voltage sensitive switch | 72. | 70. Main circuit diagram | Red |
| 10. Interior lamp delay | 101. | 99. Main circuit diagram | Natural |
| 11. Auxiliary lamp relay | 88. | 86. Main circuit diagram | Black |
| 12. Rear wiper delay | 132. | 139. Main circuit diagram | Black |
| 13. Ignition load relay | 1. | 1. Main circuit diagram | Red |
| 14. Front wiper delay | 15. | 14. Main circuit diagram | Green |
| 15. Overspeed monitor) Saudi | - | 132. Main circuit diagram | Black |
| 16. Buzzer unit ) only | - | 133. Main circuit diagram | Natural |
| 17. Fuel cut-off relay (carburetter models only) | 117. | 115. Main circuit diagram | |

# ELECTRICAL 86

## FUSE BOX

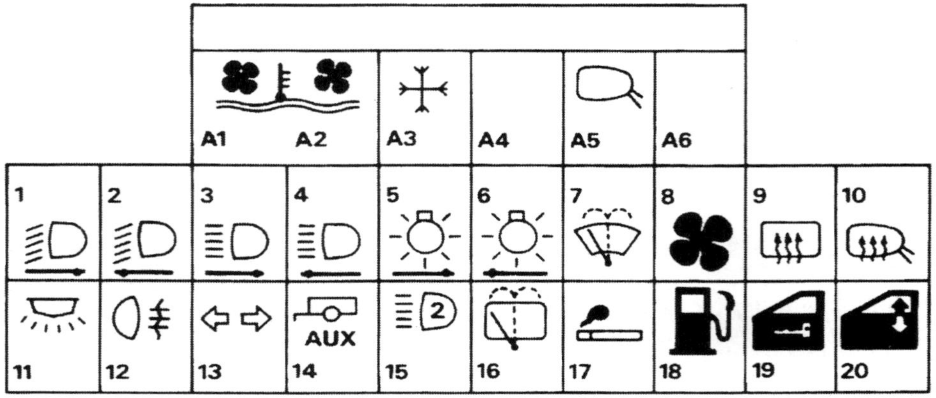

RR2101M

### MAIN FUSE PANEL

| FUSE NO. | COLOUR CODE | FUSE VALUE | CIRCUIT SERVED | IGNITION KEY CONTROLLED |
|---|---|---|---|---|
| 1 | Brown | 7.5 amp | RH headlamp dipped beam and power wash | |
| 2 | Brown | 7.5 amp | LH headlamp dipped beam | |
| 3 | Brown | 7.5 amp | RH headlamp main beam | |
| 4 | Brown | 7.5 amp | LH headlamp main beam | |
| 5 | Tan | 5 amp | RH parking lights and instrument illumination | |
| 6 | Tan | 5 amp | LH parking lights and radio illumination | |
| 7 | Blue | 15 amp | Front wash/wiper motors | AUX |
| 8 | Yellow | 20 amp | Heating/air conditioning motor | AUX |
| 9 | White | 25 amp | Heated rear screen | IGN |
| 10 | Violet | 3 amp | Mirror heaters | IGN |
| 11 | Blue | 15 amp | Headlamp flash,door, underbonnet and internal lamps, radio, clock and horns | |
| 12 | Red | 10 amp | R/LH rear fog lamps | |
| 13 | Blue | 15 amp | Low coolant monitor, stop and reverse lamps, direction indicators, instruments and screen wash fluid monitor | IGN |
| 14 | Blue | 15 amp | Auxiliary feed trailer | AUX |
| 15 | Blue | 15 amp | Auxiliary driving lamps | IGN |
| 16 | Red | 10 amp | Rear wash/wipe motor | AUX |
| 17 | Yellow | 20 amp | Cigar lighters (front and rear),automatic gear selector illumination | IGN |
| 18 | Red | 10 amp | Fuel pump | IGN |
| 19 | Red | 10 amp | Central locking-option | |
| 20 | White | 25 amp | Window lifts-option | AUX |

NOTE: Radio/Cassette combination. An in-line type 5 amp fuse is incorporated in the power input lead of the unit.

### AUXILIARY FUSE PANEL-(A)

| A1 | Yellow | 20 amp | Air conditioning fan-option | IGN |
| A2 | Yellow | 20 amp | Air conditioning fan-option | IGN |
| A3 | Tan | 5 amp | Air conditioning compressor clutch-option | IGN |
| A4 | | | Spare | |
| A5 | Violet | 3 amp | Electric mirror motors-option | IGN |
| A6 | | | Spare | |

---

5. Right hand shroud-remove three securing screws and remove the shroud over the windscreen wash wipe switch.
6. To facilitate reassembly remove the screw securing the two halves of the shroud together from one side only.

**Refitting**

7. If both sides of the shroud have been removed ensure that the plate on the steering column is correctly located in the slot in the shroud.
8. Reverse the removal procedure.

## MAIN LIGHTING SWITCH

## REAR SCREEN PROGRAMMED WASH WIPE SWITCH

**Remove and refit**

**Removing**

1. Remove the steering column shroud from the required side.
2. Disconnect cables at snap connectors.
3. Push the two spring clips locating the switch inwards and remove the switch from its mounting.

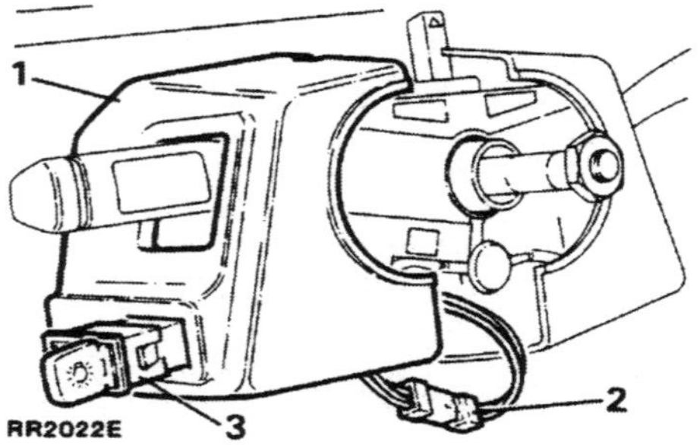

**Refitting**

4. Reverse the removal procedure.

RR2022E

## STEERING COLUMN CONTROLS

The steering column switch layout is as follows:

### LEFT HAND CONTROLS

Lower switch-Main lighting switch
Upper switch-Main and dipped beam, direction indicators and horn.

### RIGHT HAND CONTROLS

Lower switch - Rear screen programmed wash/wipe.
Upper switch - Windscreen programmed wash/wipe.

## STEERING COLUMN SHROUD

Certain operations within the electrical section necessitate removal of the steering column shroud. Unless removal of both sides of the shroud assembly is required, remove ONLY the side necessary for access.

**Remove and refit**

**Removing**

1. Disconnect the battery negative lead.
2. Remove the lower fascia panel.
3. Disconnect the electrical connections to either the master lighting switch or the rear screen wash wipe switch. (Disconnect both if removing the complete shroud).
4. Left hand shroud-remove three securing screws and remove the shroud over the indicator/main beam switch.

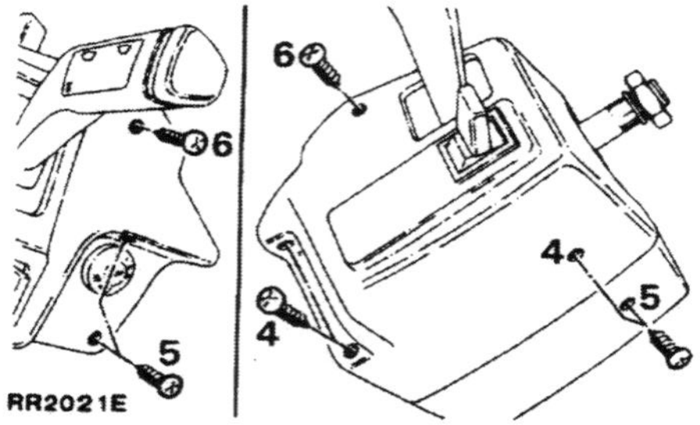

RR2021E

# ELECTRICAL

## WINDSCREEN PROGRAMMED WASH WIPE SWITCH

## MAIN AND DIPPED BEAM, DIRECTION INDICATORS AND HORN SWITCH

### Remove and refit

#### Removing

1. Remove the steering column shroud from the required side.
2. Release the appropriate retaining clip and pull the fibre optic guide from the housing.
3. Depress the retainers at the top and bottom of the switch and pull combined switch assembly away from the steering column switch housing.
4. Lighting, indicator and horn switch: release the two harness multi-plugs from the back of the switch and remove the switch assembly.

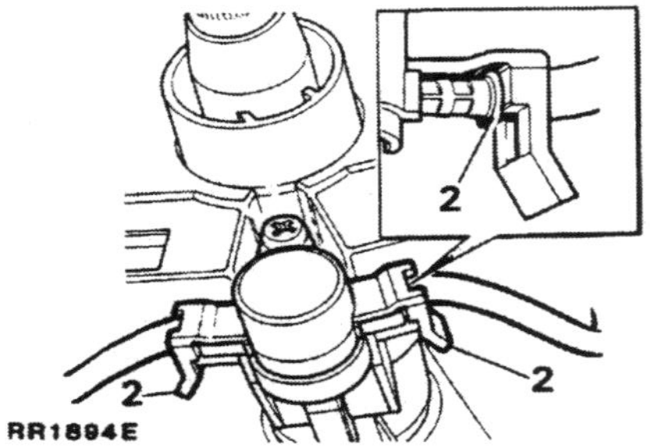

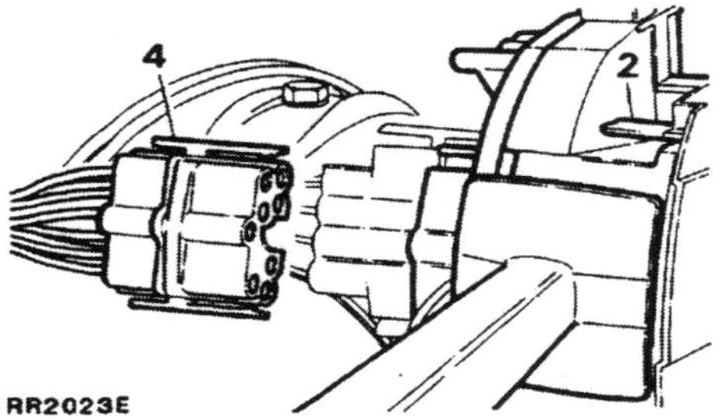

NOTE: Wiper and washer switch: release the harness multi-plug from the back of the switch and remove the switch assembly

#### Refitting

5. Reverse the removal procedure.

## HAZARD WARNING SWITCH BULB REPLACEMENT

### Remove and refit

#### Removing

1. Disconnect the battery negative lead.
2. Pull the hazard switch cover upwards and remove it to gain access to the bulb.
3. Remove the bulb by pulling it upwards. A piece of rubber tubing or adhesive tape attached to the bulb may facilitate removal and refitting.

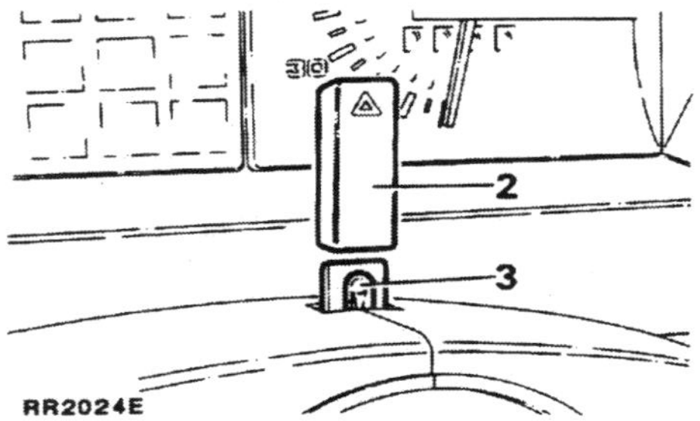

#### Refitting

4. Locate the bulb in its holder and reverse instructions 1 and 3.
The correct bulb is a 12V, 1.2 watt 'wedge' base (capless).

## EXTERIOR DRIVING MIRRORS CONTROLS

### Adjusting

1. Fine adjustment is controlled by an electric motor inside the mirror housing. This is operated by two controls fitted in the fascia panel. To adjust, select left or right hand mirror using the rocker switch.

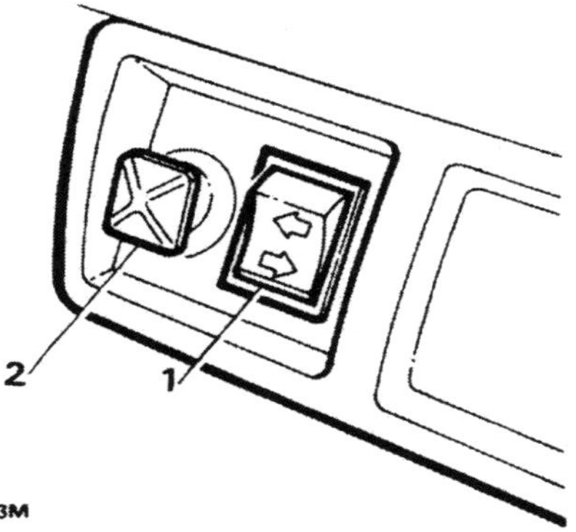

2. Move the head of the finger tip control to the left, right, up or down as required.

## COLUMN SWITCH ILLUMINATION BULB REPLACEMENT

### Remove and refit

#### Removing

1. Disconnect the battery negative lead.
2. Remove the left hand side steering column shroud.
3. Working behind the column switch housing twist the bulb holder through 90° and withdraw.
4. Remove the bulb.

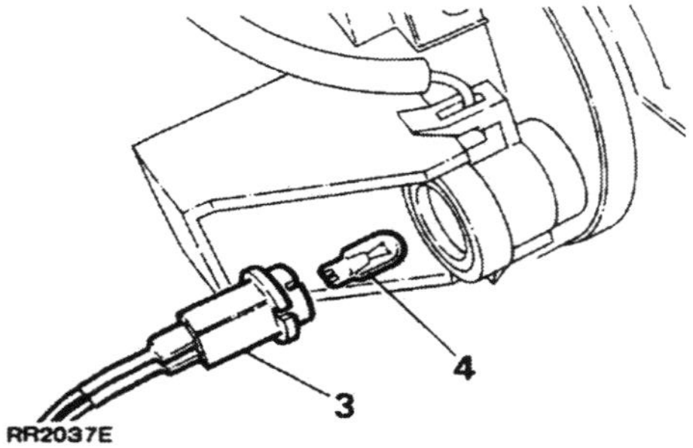

#### Refitting

5. Reverse the removal procedure. The correct bulb type is a 12-volt, 1.2-watt 'wedge' base (capless).

## ELECTRICAL 86

### EXTERIOR DRIVING MIRRORS

### CONTROL SWITCHES

#### Remove and refit

1. Disconnect the battery negative lead.
2. Carefully prise the four air vents out of the fascia panel.
3. Remove the six screws securing the fascia to the lower panel.
4. Remove the nine screws securing the fascia to the upper panel.
5. Withdraw the panel only as far as the electrical leads will permit.
6. Pull the multi-plug from the rear of the fingertip controlled mirror switch.
7. Carefully prise off the fingertip button at the operating end of the switch.
8. Unscrew the black plastic retaining collar securing the switch.
9. Remove the switch from the panel.
10. Disconnect the multi-plug at the rear of the selector switch.
11. Apply pressure to the switch from the rear and remove it from the panel.

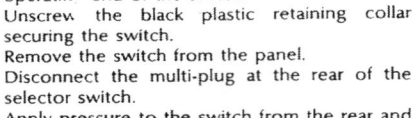

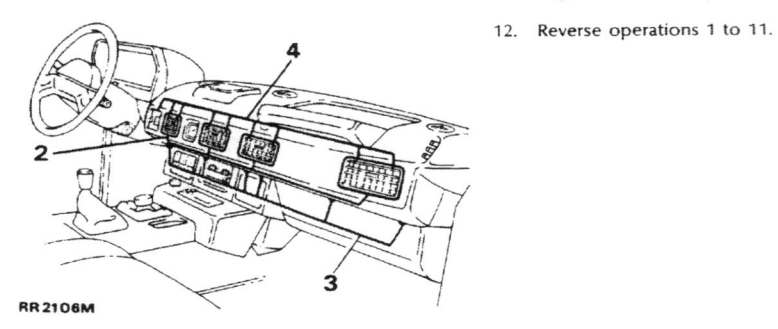

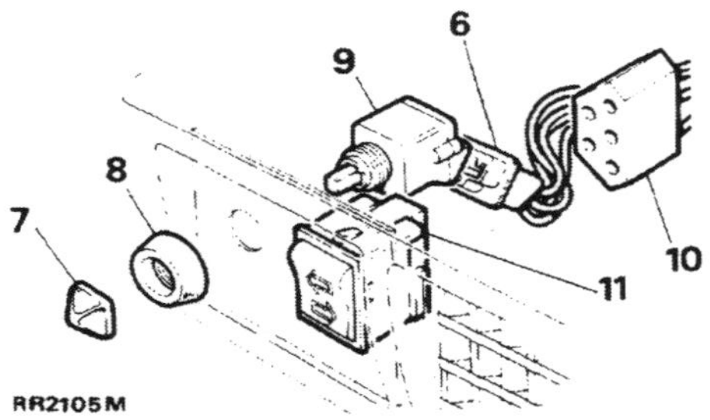

#### Refitting

12. Reverse operations 1 to 11.

RR2106M

RR2105M

---

## ELECTRICAL 86

### INSTRUMENT BINNACLE WARNING LIGHT SYMBOLS

Trailer connected-flashes with direction indicators (green)

Direction indicator-left turn /right turn (green)

Seat belt (red) NOTE: The seat belt warning symbol appears on all binnacles but will only be illuminated when Territory regulations require a seat belt warning system to be fitted.

Headlamp main beam on (blue)

Engine oil pressure low (red)

Cold start control (Petrol models) engaged
Heater plugs operating (Diesel models) (amber)

Ignition on/low charge (red)

Low coolant (red)

Automatic gearbox oil temperature high (red)

Low fuel indicator (amber)

Low wash fluid (amber)

Transmission handbrake on (red)

Brake pad wear (amber)

Brake fluid pressure failure/low fluid level (red)

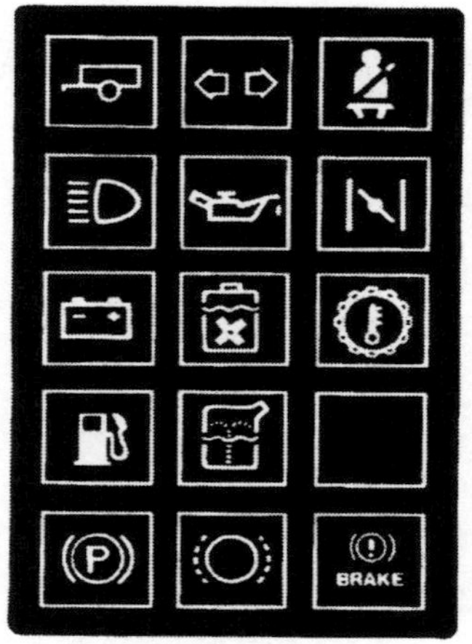

RR2113M

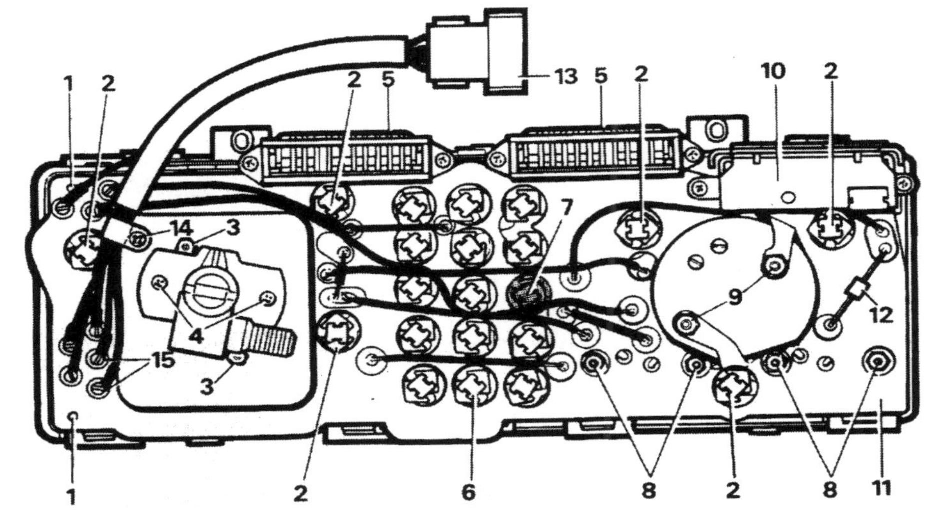

# ELECTRICAL 86

## CIRCUIT SERVED

| | |
|---|---|
| Tacho signal | 1 |
| Ignition switch 12V+ | 2 |
| Low coolant input | 3 |
| Earth-VE | 4 |
| Ignition warning light | 5 |
| Low oil and low pressure warning light | 6 |
| Main beam warning light | 7 |
| Zero volts from dimmer | 8 |
| Trailer warning light | 9 |
| Direction indicators warning light | 10 |
| Seat belts warning light | 11 |
| Cold start warning light | 12 |
| Temperature warning light (automatic gearbox) | 13 |
| Low wash fluid warning light | 14 |
| Not used | 15 |
| 12V+ from dimmer | 16 |
| Brake fail warning light | 17 |
| Panel illumination bulbs (6 off) | 18 |
| Low fuel warning light | 19 |
| Low coolant warning light | 20 |

NOTE: The following 21 to 25 are connected at the single multi-plug located behind the binnacle

| | |
|---|---|
| Brake pad wear warning light | 21 |
| Not used | 22 |
| Fuel tank unit and fuel gauge | 23 |
| Temperature gauge | 24 |
| Not used | 25 |

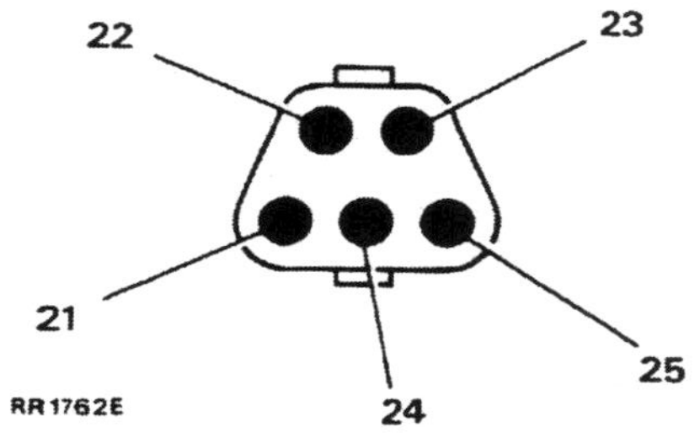

Sequence of connections at the single multi-plug.

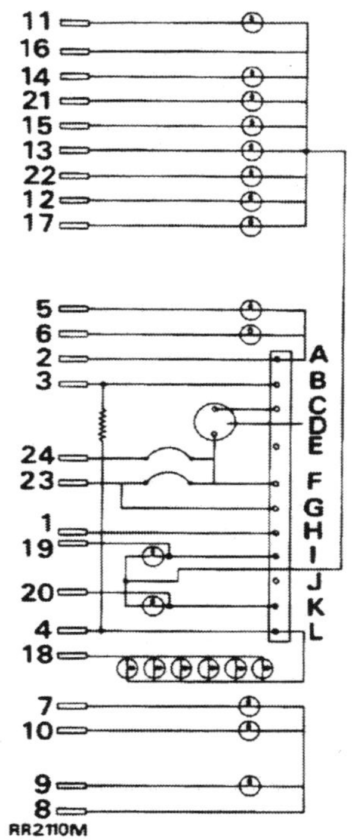

## MULTI-FUNCTION UNIT

A. 12V+ supply
B. Input to low coolant circuit
C. Tachometer drive
D. Tachometer
E. Spare
F. 10V+ stabilised
G. Input to fuel tank unit - stabilised
H. Tachometer signal
I. Low fuel warning light
J. Spare
K. Low coolant warning light
L. Earth

## Instrument case (back)

1. Locating pegs
2. Panel light bulbs
3. Speedometer securing screw
4. Speedometer drive securing screws
5. Harness connectors
6. Warning light bulbs (14)
7. No charge warning light bulb (red holder)
8. Temperature and fuel gauge unit securing nuts
9. Tachometer securing nuts
10. Multi-function unit
11. Printed circuit
12. Pull-up resistor-high temperature gearbox oil
13. Single multi-plug
14. Single multi-plug securing screw
15. Single multi-plug wiring connecting screws

## PRINTED CIRCUIT HARNESS CONNECTIONS

Sequence of connections viewed from back of instrument case.

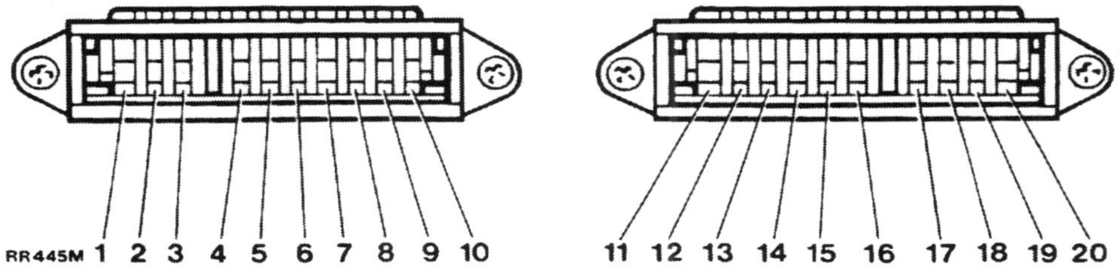

295

# ELECTRICAL 86

## RADIO ANTENNA AMPLIFIER

The radio aerial is incorporated in the rear screen heater element. The signal to the radio is electronically boosted by the amplifier mounted at the rear of the vehicle adjacent to the tailgate screen wiper motor.

### Remove and refit

### Removing

1. Disconnect the battery negative lead.
2. Lower or remove the rear headlining.
3. Remove the electrical leads and the antenna lead to the radio.

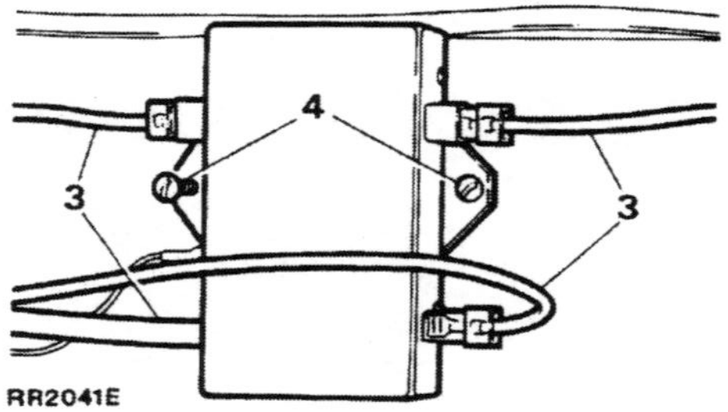

RR2041E

### Refitting

4. Reverse the removal procedure.

## DIFFERENTIAL LOCK WARNING LAMP ASSEMBLY/BULB REPLACEMENT

### Remove and refit

1. Carefully prise the warning lamp out of the radio console.
2. Remove the two wiring connectors and withdraw the lamp assembly, if required.
3. Squeeze the sides of the lamp body to enable the lens surround to be slid back along the body.
4. Remove the amber lens.
5. Remove the bayonet fitting bulb.

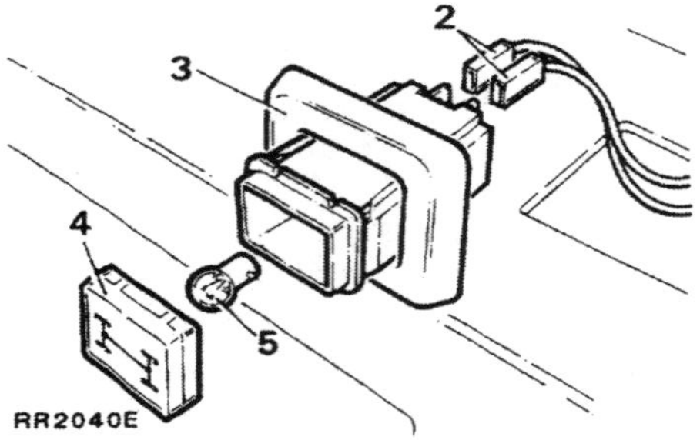

RR2040E

### Refitting

6. Reverse the removal procedure.

The correct bulb type is a 12-volt, 2-watt bayonet fitting.

# 86 ELECTRICAL

## CENTRAL LOCKING-FUEL FILLER FLAP ACTUATOR UNIT

The fuel filler flap is locked when the vehicle central locking is actuated. Repair is by replacement of the actuator unit.

### Remove and refit

### Removing

1. Remove six screws and withdraw the closure panel, situated in the tool stowage area.
2. Ensure that the actuator is in the unlocked position and the fuel filler flap is open.
3. Release two screws and manoeuvre the actuator unit clear of its mounting.

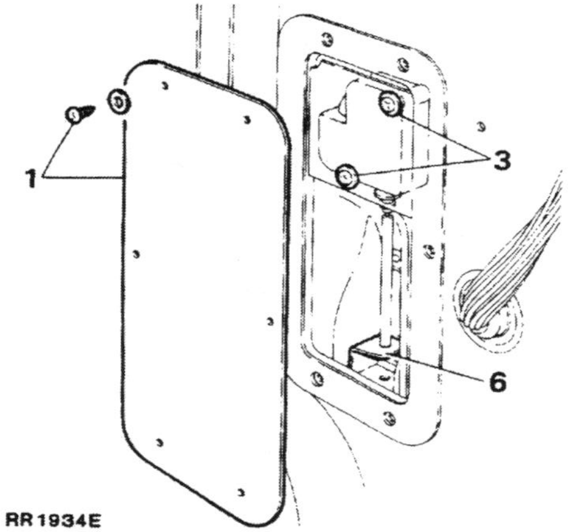

RR1935E

4. Disconnect the wiring plug.
5. Withdraw the actuator.

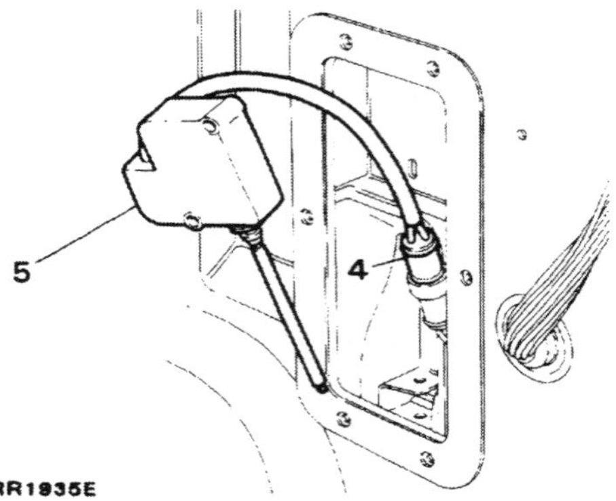

### Refitting

6. Reverse the removal procedure. The actuator mounting holes in the body are elongated. Adjust the position of the actuator to ensure that the rod will pass through the guide brackets without fouling.
7. Check the operation of the central locking system.

RR1934E

296

# ELECTRICAL

## HEADLAMP DIM-DIP LIGHTING (U.K. only)

Vehicles manufactured after October 1st 1986 must comply with new lighting regulations, which prohibit such vehicles being driven on side lights only. This is achieved by the addition of an electronic control unit in the lighting system, so that when the side lights are switched on with the engine running, low voltage current is supplied to the headlamp dipped beam circuit, giving dim-dipped lighting. When the headlamps are switched on, full voltage is automatically restored, giving normal headlamp lighting.

## ELECTRONIC CONTROL UNIT (DIM-DIP LIGHTING)

### Remove and refit

1. Disconnect the battery negative lead.
2. Remove the lower fascia panel.
3. Remove two securing screws and retain the distance piece.
4. Disconnect the wiring connector and withdraw the electronic control unit.

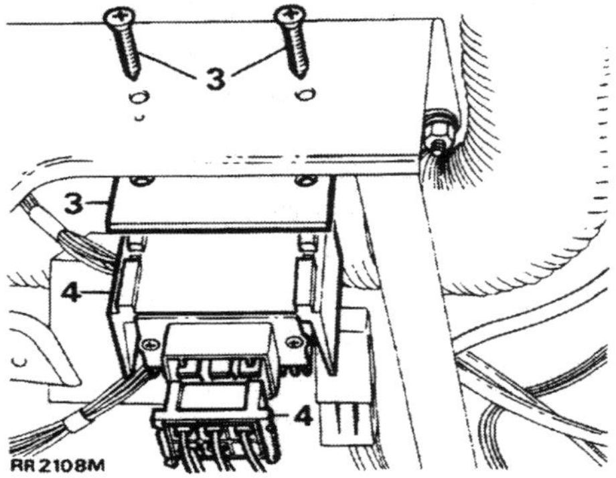

### Refitting

5. Reverse the removal procedure.

## COOLANT LEVEL SENSOR

### Remove and refit

### Removing

**WARNING: Do not remove the expansion tank filler cap when the engine is hot beause the cooling system is pressurised and personal scalding could result.**

1. Disconnect the multi-plug from the sensor.
2. Remove the expansion tank filler cap by first turning it anti-clockwise a quarter of a turn to allow pressure to escape. Then turn it in the same direction and lift off.
3. Release the retaining nut and withdraw the sensor from the expansion tank and disconnect the electrical leads.

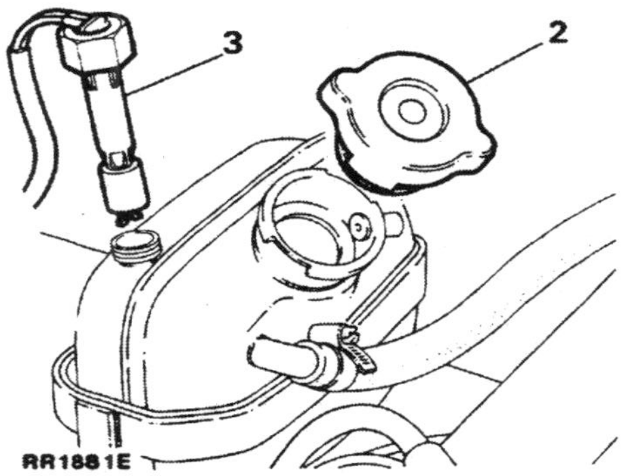

### Refitting

4. Reverse the removal instructions.
5. Start engine and run until normal running temperature is attained, thermostat open. Check for coolant leaks around the sensor.

## VACUUM SENSOR - BRAKE SERVO (Diesel models only)

The sensor will illuminate the brake fail warning light indicating lack of vacuum within the servo. When the engine is started the warning light will remain illuminated until the vacuum is restored.

### Remove and refit

1. Disconnect the battery negative lead.
2. Remove the two wiring connectors from the sensor.
3. Slacken the sensor and remove from the servo.

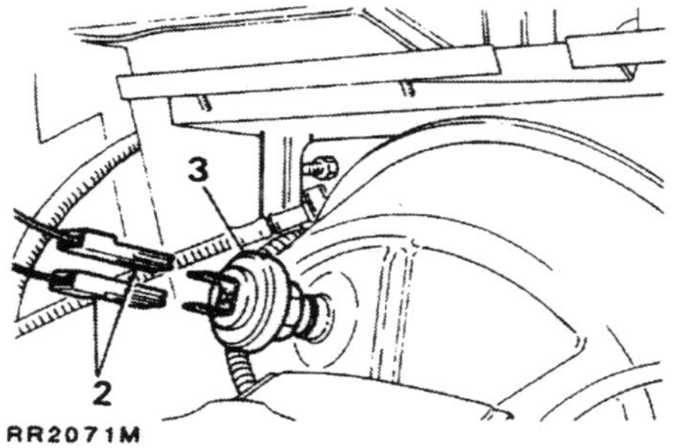

### Refitting

4. Reverse the removal procedure.

## BRAKE CHECK UNIT

### Remove and refit

### Removing

1. Disconnect the battery negative terminal.
2. Release the six screws securing the lower fascia panel below the steering column.
3. Lower the fascia panel and disconnect the multi-plug from the rheostat switch.
4. Pull the brake check unit from the spring clip on the underside of the fascia panel and disconnect the three multi-plugs from the unit.
5. Remove the brake check unit from the vehicle.

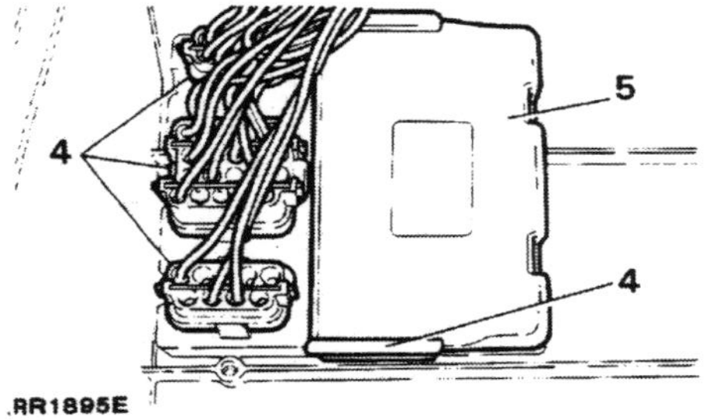

### Refitting

6. Reverse the removal procedure ensuring that the multi-plugs and unit are securely pushed into position.

# ELECTRICAL 86

## ELECTRIC MIRRORS - Circuit diagram

1. Clinch
2. Main cable connections
3. Fuse 10-main fuse panel
4. Mirror motors
5. Change over switch
6. Mirror control switch
7. Earth-via main cable

## CABLE COLOUR CODE

| | |
|---|---|
| B | Black |
| U | Blue |
| N | Brown |
| G | Green |
| O | Orange |
| P | Purple |
| R | Red |
| S | Slate |
| W | White |
| Y | Yellow |

The last letter of a colour code denotes the tracer.

## ELECTRICAL EQUIPMENT - CIRCUIT DIAGRAMS

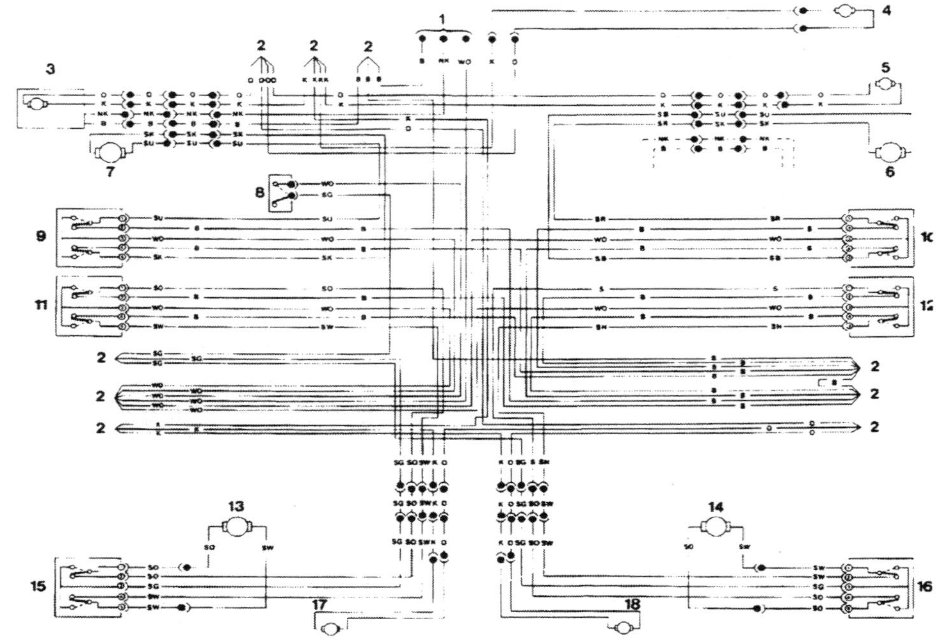

### WINDOW LIFTS AND DOOR LOCKS - Circuit diagram

1. Main cable connections
2. Clinches
3. Switch unit-central door locking (driver's door)
4. Fuel flap actuator
5. Lock unit-central door locking (front passenger door)
6. Window lift motor L/H front
7. Window lift motor R/H front
8. Isolator switch
9. Window lift switch L/H front
10. Window lift switch R/H front
11. Window lift switch L/H rear
12. Window lift switch R/H rear
13. Window lift motor L/H rear
14. Window lift motor R/H rear
15. Window lift switch L/H rear door
16. Window lift switch R/H rear door
17. Lock unit central door locking L/H rear door
18. Lock unit central door locking R/H rear door

## CABLE COLOUR CODE

| | | | | | |
|---|---|---|---|---|---|
| B | Black | G | Green | S | Slate |
| U | Blue | L | Light | W | White |
| N | Brown | O | Orange | Y | Yellow |
| K | Pink | | | | |
| P | Purple | | | | |
| R | Red | | | | |

The last letter of a colour code denotes the tracer.

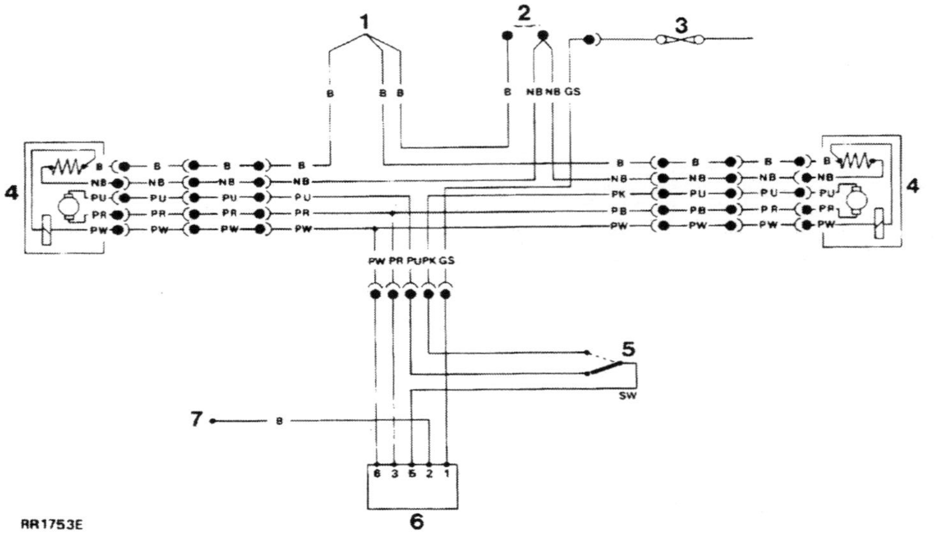

# ELECTRICAL 86

## MAIN CIRCUIT DIAGRAM
### Right Hand Steering - RR2343M & RR2344M

1. Ignition load relay
2. Battery
3. Terminal post
4. Starter solenoid
5. Starter motor
6. Starter relay
7. Starter inhibit switch (Automatic)
8. Ignition switch
9. Tachometer
10. Voltage transformer(dim dip)
11. Ignition warning lamp
12. Alternator
13. Fuse 7
14. Front wipe/wash switch
15. Front wipe delay unit
16. Front wiper motor
17. Front wash switch
18. Front wash pump
19. Headlamp wash timer unit (option)
20. Headlamp wash pump (option)
21. Main lighting switch
22. Fuse 6
23. Fuse 5
24. LH side lamp
25. LH tail lamp
26. Number plate lamp(2 off)
27. Main beam dip/flash switch
28. Radio illumination
29. RH side lamp
30. RH tail lamp
31. Rheostat
32. Fuse 3
33. Fuse 4
34. Fuse 1
35. Fuse 2
36. Rear fog switch
37. Fuse 12
38. Switch illumination (2 off)
39. Cigar lighter illumination (2 off)
40. Heater illumination (4 off)
41. Clock illumination
42. Automatic gear selector illumination (2 off)
43. Instrument illumination (6 off)
44. Rear fog warning lamp
45. LH rear fog
46. RH rear fog
47. LH dip beam
48. RH dip beam
49. LH main beam
50. RH main beam
51. Main beam warning lamp
52. Fuel gauge
53. Fuel gauge sender unit
54. Water temperature gauge
55. Water temperature sender unit
56. Fuse 11
57. Horn switch
58. RH horn
59. LH horn
60. Under bonnet illumination switch
61. Under bonnet light
62. Clock
63. Fuse 19
64. Fuse 20
65. Pick-up point central locking/window lift (option)
66. Heated rear window relay
67. Fuse 9
68. Radio aerial amplifier
69. Heated rear screen
70. Heated rear screen switch
71. Heated rear screen warning lamp
72. Voltage sensitive switch
73. Fuse 13
74. Hazard switch
75. Flasher unit
76. Direction indicator switch
77. Hazard/indicator warning lamp
78. LH rear indicator lamp
79. LH front indicator lamp
80. LH side repeater lamp
81. RH side repeater lamp
82. RH front indicator lamp
83. RH rear indicator lamp
84. Trailer warning lamp
85. Fuse 15
86. Stop lamp switch
87. Reverse lamp switch
88. Auxiliary lamp relay (option)
89. LH stop lamp
90. RH stop lamp

## MAIN CIRCUIT DIAGRAM
### Right Hand Steering - RR2343M & RR2344M

91. LH reverse lamp
92. RH reverse lamp
93. LH auxiliary lamp (option)
94. RH auxiliary lamp (option)
95. Auxiliary lamp switch (option)
96. Fuse 17
97. Dash cigar lighter
98. Cubby box cigar lighter
99. LH interior lamp
100. RH interior lamp
101. Interior lamp delay unit
102. LH door edge lamp
103. RH door edge lamp
104. LH puddle lamp
105. RH puddle lamp
106. Interior lamp switch
107. LH rear door switch
108. RH rear door switch
109. Tailgate switch
110. LH front door switch
111. RH front door switch
112. Differential lock warning lamp
113. Differential lock switch
114. Oil pressure warning lamp
115. Oil pressure switch
116. Fuse 18
117. Fuel cut off relay (carburetter models)
118. Fuel pump(petrol models)
119. Ignition coil
120. Capacitor
121. Distributor
122. EFi Harness plug
123. Fuel shut off solenoid (Diesel)
124. Radio choke
125. Radio fuse
126. Radio and four speakers
127. Ignition pick up points
128. Automatic transmission oil temperature warning lamp
129. Automatic transmission oil temperature switch
130. Fuse 16
131. Rear wash wipe switch
132. Rear wipe delay unit
133. Rear wiper motor
134. Rear screen wash pump
135. Low screen wash fluid level warning lamp
136. Low screen wash switch
137. Low coolant switch
138. Multi-function unit in binnacle
139. Low coolant level warning lamp
140. Low fuel level warning lamp
141. Cold start/diesel glow plug warning lamp
142. Cold start switch - carburetter
143. Glow plug timer (diesel)
144. Glow plugs (diesel)
145. Handbrake warning lamp
146. Brake fail warning lamp
147. Handbrake warning switch
148. Brake fail warning switch
149. Brake pad wear warning lamp
150. Brake pad wear sensors
151. Brake check relay
152. Split charge relay (option)
153. Split charge terminal post (option)
154. Heater/air conditioning connections
155. Fuse 8
156. Coil negative (engine RPM input to ECU.)

## CABLE COLOUR CODE

B — Black
U — Blue
N — Brown
G — Green
L — Light
O — Orange
K — Pink
P — Purple
R — Red
S — Slate
W — White
Y — Yellow

ELECTRICAL | 86

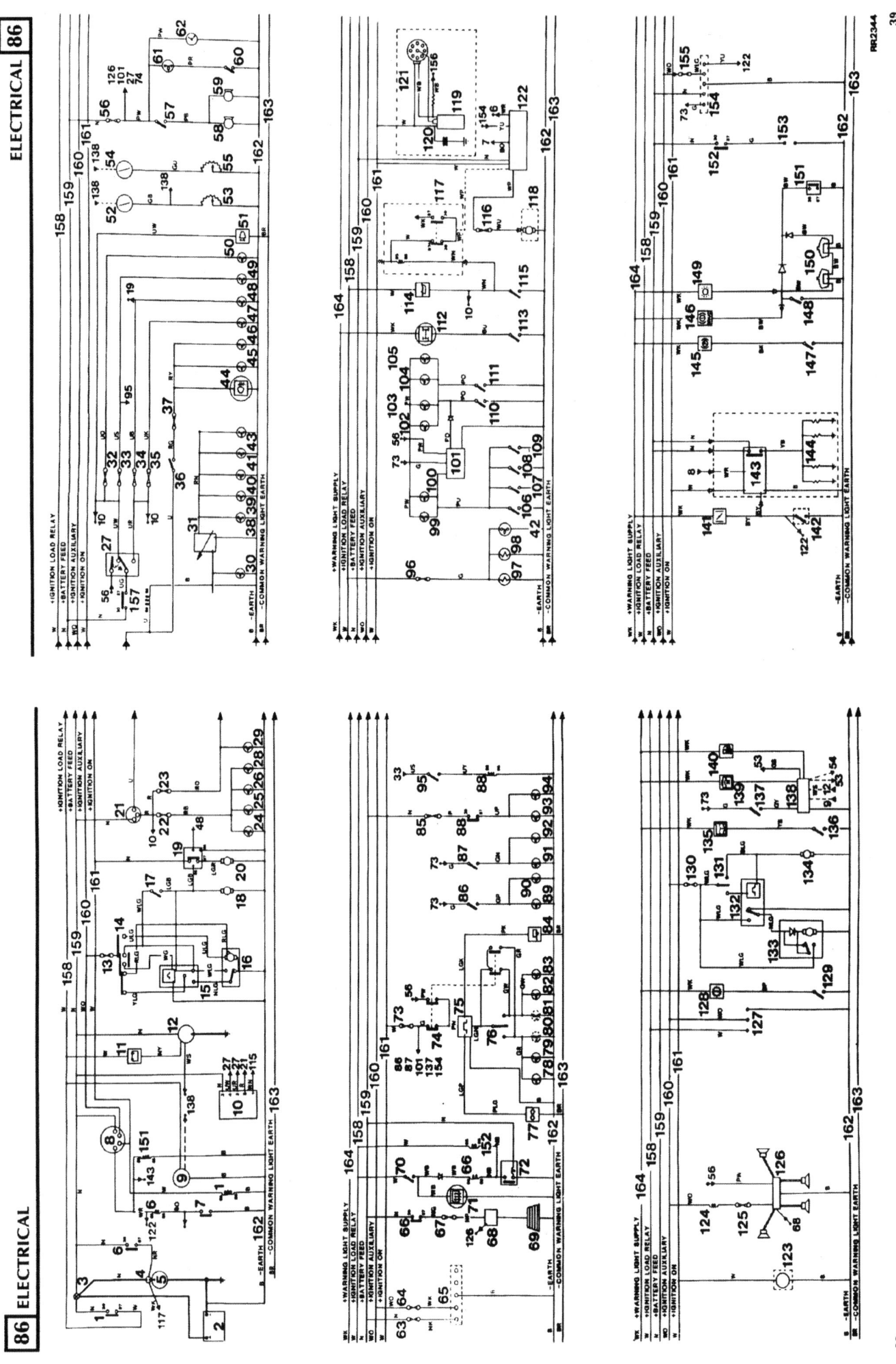

# ELECTRICAL

## MAIN CIRCUIT DIAGRAM
### Left-hand Steering - RR2345M & RR2346M

1. Ignition load relay
2. Battery
3. Terminal post
4. Starter solenoid
5. Starter motor
6. Starter relay
7. Starter inhibit switch (automatic)
8. Ignition switch
9. Tachometer
10. Ignition warning lamp
11. Alternator
12. Fuse 7
13. Front wipe/wash switch
14. Front wipe delay unit
15. Front wiper motor
16. Front wash pump
17. Headlamp wash timer unit (option)
18. Headlamp wash pump (option)
19. Main lighting switch
20. Fuse 6
21. Fuse 5
22. LH side lamp
23. LH tail lamp
24. Number plate lamp (2 off)
25. Main beam dip/flash switch
26. Radio illumination
27. RH side lamp
28. RH tail lamp
29. Rheostat
30. Fuse 3
31. Fuse 4
32. Fuse 1
33. Fuse 2
34. Rear fog switch
35. Fuse 12
36. Switch illumination (2 off)
37. Cigar lighter illumination (2 off)
38. Heater illumination (4 off)
39. Clock illumination
40. Automatic gear selector illumination (2 off)
41. Instrument illumination (6 off)
42. Rear fog warning lamp
43. LH rear fog
44. RH rear fog
45. LH dip beam
46. RH dip beam
47. LH main beam
48. RH main beam
49. Main beam warning lamp
50. Fuel gauge
51. Fuel gauge sender unit
52. Water temperature gauge
53. Water temperature sender unit
54. Fuse 11
55. Horn switch
56. RH horn
57. LH horn
58. Under bonnet illumination switch
59. Under bonnet light
60. Clock
61. Fuse 19
62. Fuse 20
63. Pick-up point central locking/window lift
64. Heated rear window relay
65. Radio aerial amplifier
66. Heated rear screen
67. Heated rear screen switch
68. Heated rear screen warning lamp
69. Voltage sensitive switch
70. Fuse 13
71. Hazard switch
72. Flasher unit
73. Direction indicator switch
74. Hazard/indicator warning lamp
75. LH rear indicator lamp
76. LH front indicator lamp
77. LH side repeater lamp
78. RH side repeater lamp
79. RH front indicator lamp
80. RH rear indicator lamp
81. Trailer warning lamp
82. Fuse 15
83. Stop lamp switch
84. Reverse lamp switch
85. Auxiliary lamp relay
86. LH stop lamp
87. RH stop lamp
88. LH reverse lamp
89. RH reverse lamp

## MAIN CIRCUIT DIAGRAM
### Left-hand Steering - RR2345M & RR2346M

91. LH auxiliary lamp (option)
92. RH auxiliary lamp (option)
93. Auxiliary lamp switch (option)
94. Fuse 17
95. Dash cigar lighter
96. Cubby box cigar lighter
97. LH interior lamp
98. RH interior lamp
99. Interior lamp delay unit
100. LH door edge lamp
101. RH door edge lamp
102. LH puddle lamp
103. RH puddle lamp
104. Interior lamp switch
105. LH rear door switch
106. RH rear door switch
107. Tailgate switch
108. LH front door switch
109. RH front door switch
110. Differential lock warning lamp
111. Differential lock switch
112. Oil pressure warning lamp
113. Oil pressure switch
114. Fuse 18
115. Fuel shut-off relay - carburetter
116. Fuel pump - petrol models
117. Ignition coil
118. Capacitor
119. Distributor
120. EFI Harness plug
121. Fuel shut-off solenoid-Diesel
122. Radio choke
123. Radio fuse
124. Radio
125. Four speakers
126. Seat belt warning lamp
127. Speed transducer, Saudi only
128. Resistor
129. Audible warning unit
130. Transfer box neutral switch
131. Seat buckle switch
132. Overspeed monitor (Saudi only)
133. Overspeed buzzer (Saudi only)
134. Ignition pick up points
135. Automatic transmission oil temperature warning lamp
136. Automatic transmission oil temperature switch
137. Fuse 16
138. Rear wash wipe switch
139. Rear wipe delay unit
140. Rear wiper motor
141. Rear screen wash pump
142. Low screen wash fluid level warning lamp
143. Low screen wash switch
144. Low coolant switch
145. Multi-function unit in binnacle
146. Low coolant level warning lamp
147. Low fuel level warning lamp
148. Cold start/Diesel glow plug warning lamp
149. Choke switch - carburetter
150. Glowplug timer/Diesel
151. Glowplugs/Diesel
152. Handbrake warning lamp
153. Handbrake warning switch
154. Brake fail warning lamp
154a. Brake fail warning lamp
155. Brake pad wear warning lamp
156. Brake pad wear sensors
157. Brake check unit
158. Split charge relay (option)
159. Split charge terminal post
160. Heater/air conditioning connections
161. Fuse 8
162. Coil negative (engine RPM input to ECU)

## CABLE COLOUR CODE

B — Black
U — Blue
N — Brown
G — Green
L — Light
O — Orange
K — Pink
P — Purple
R — Red
S — Slate
W — White
Y — Yellow

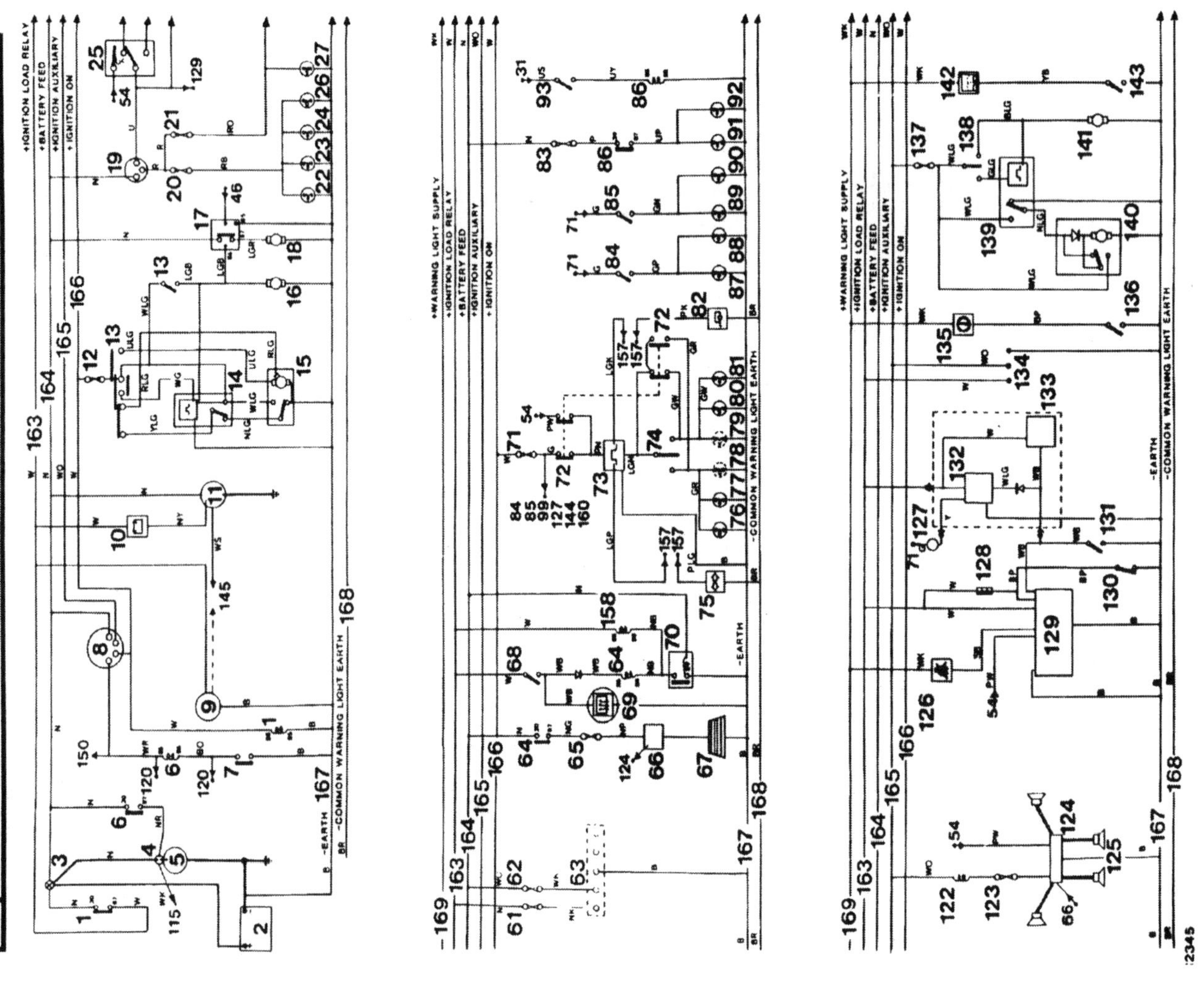

+IGNITION LOAD RELAY
+BATTERY FEED
+IGNITION AUXILIARY
+IGNITION ON
+WARNING LIGHT SUPPLY
-EARTH
-COMMON WARNING LIGHT EARTH


42



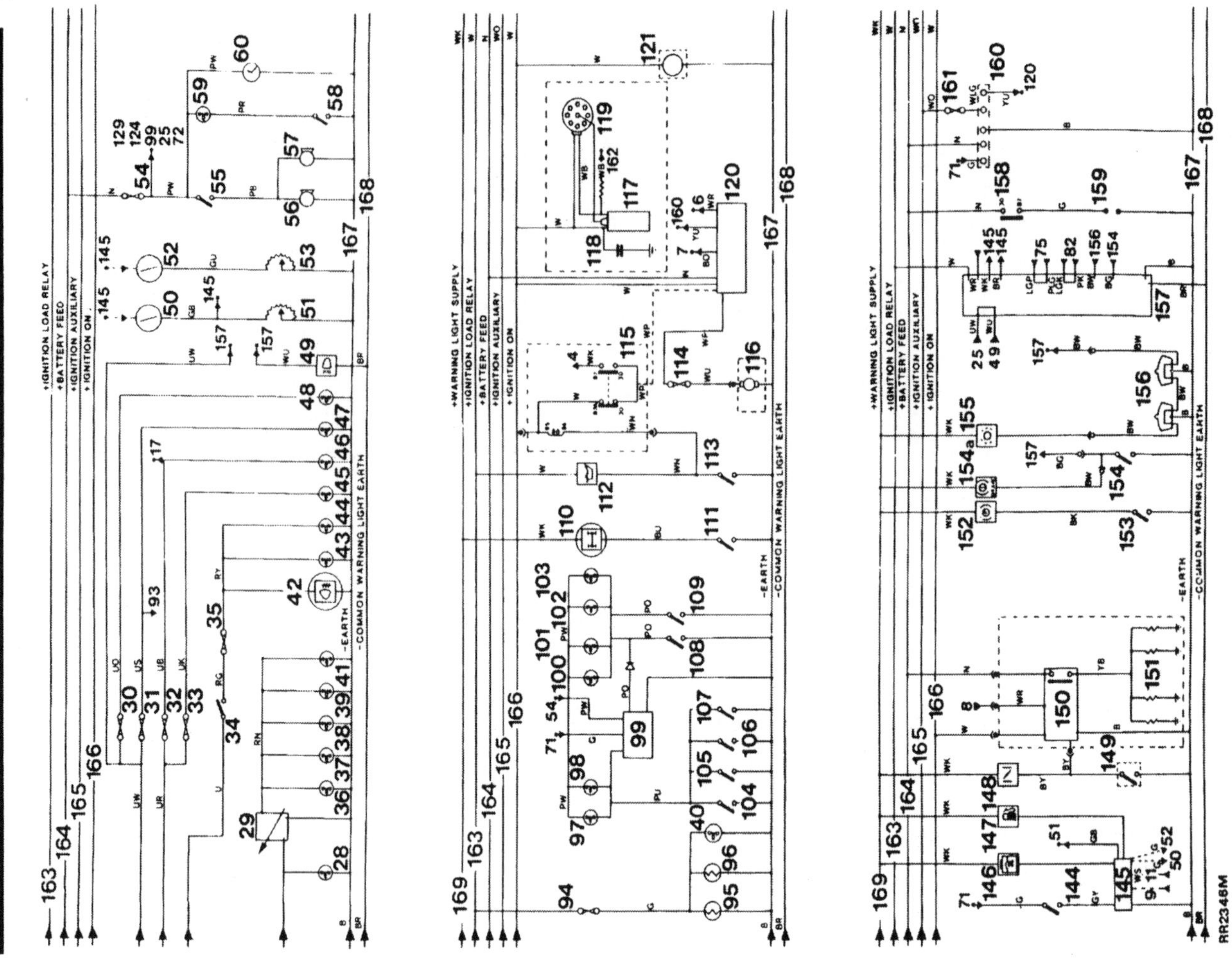

+IGNITION LOAD RELAY
+BATTERY FEED
+IGNITION AUXILIARY
+IGNITION ON
+WARNING LIGHT SUPPLY
-EARTH
-COMMON WARNING LIGHT EARTH
RR2346M


43

**Range Rover
1988
Catalytic Exhaust System
Workshop Manual Supplement
Publication Number LSM180WS2**

|  | Section | Page |
|---|---|---|
| Introduction | 01 | 306 |
| General Specification Data | 04 | 307 |
| Engine Tuning Data | 05 | 309 |
| Torque Wrench Settings | 06 | 310 |
| Lubricants and Fluids | 09 | 311 |
| Maintenance | 10 | 312 |
| Emission Control | 17 | 319 |
| Fuel System | 19 | 322 |
| Manifold & Exhaust System | 30 | 342 |
| Propeller Shafts | 47 | 343 |
| Electrical | 86 | 344 |

**Land Rover
Lode Lane,
Solihull
West Midlands
B92 8NW
England**

Section
Number

## 01 INTRODUCTION ............................................. 1

## 04 GENERAL SPECIFICATION DATA

- Electrical .................................................. 4
- Fuel system ............................................... 1
- Transmission .............................................. 2

## 05 ENGINE TUNING DATA

- Engine tuning data ....................................... 1

## 06 TORQUE VALUES

- Engine .................................................... 1

## 09 RECOMMENDED LUBRICANTS, FLUIDS AND CAPACITIES

- Capacities ................................................. 2
- Fuel requirements ......................................... 2
- Lubrication ................................................ 2
- Recommended lubricants ................................... 1

## 10 MAINTENANCE

- Charcoal canister ......................................... 9
- Cooling system ........................................... 12
- Distributor ............................................... 11
- Drive belt adjustment .................................... 13
- Filters .................................................... 7
- Ignition .................................................. 10
- Lubrication ............................................... 6
- Maintenance schedule ..................................... 1

## 17 EMISSION CONTROL

- Adsorption canister - remove and refit .................... 3
- Catalytic converters ...................................... 3
- Description ............................................... 1
- Evaporative emission control system ....................... 4
- Fuel expansion tank - remove and refit .................... 6
- Lambda (oxygen) sensor - remove and refit ................. 6
- Vacuum delay valve - test - remove and refit .............. 5

Continued

Notes

304

Section
Number

## 19 FUEL INJECTION SYSTEM

- Air cleaner - remove and refit — 25
- Air flow sensor - remove and refit — 25
- By-pass air valve (stepper motor) - remove and refit — 27
- Circuit diagram — 1
- Component location — 2
- Continuity test procedure — 7
- Continuity test using a multi meter — 8
- Coolant temperature thermistor - remove and refit — 30
- Depressurize fuel system — 34
- Description — 3
- Electronic control unit (ECU) - remove and refit — 28
- Electronic fuel injection relays - remove and refit — 28
- Engine setting procedure — 6
- Engine tuning procedure — 24
- Fuel line filter - remove and refit — 37
- Fuel pipes - layout — 40
- Fuel pressure regulator - remove and refit — 34
- Fuel pump - remove and refit — 39
- Fuel rail - injectors, LH and RH - remove and refit — 35
- Fuel tank - remove and refit — 38
- Fuel temperature thermistor - remove and refit — 29
- Intake manifold - remove and refit — 36
- Plenum chamber - remove and refit — 32
- Resetting throttle levers — 30
- Ram housing - remove and refit — 33
- Speed transducer - remove and refit — 27
- Throttle cable - renew — 31
- Throttle pedal - remove and refit — 31
- Throttle potentiometer - remove, refit and setting — 26

## 30 MANIFOLD AND EXHAUST SYSTEM

- Exhaust manifold - remove and refit — 2
- Exhaust system complete — 1
- Exhaust system - remove and refit — 2

## 47 PROPELLER SHAFTS

- Propeller shafts - fitting — 1

Section
Number

## 86 ELECTRICAL

- Alternator drive belt - adjust — 8
- Alternator - overhaul and test — 10
- Alternator - remove and refit — 8
- Coil - remove and refit — 18
- Constant energy ignition system - checks — 19
- Description - Electrical equipment — 7
- Distributor - overhaul electrical components — 16
- Distributor - remove and refit — 15
- Electronic ignition - description — 15
- Electric mirrors - circuit diagram — 28
- Fault diagnosis — 2
- Headlamp assembly/bulb replacement - remove and refit — 22
- Ignition timing - adjust — 18
- Instrument binnacle - warning light symbols — 23
- Location of electrical equipment — 1
- Main circuit diagram — 30
- Multi - function unit connections — 27
- Printed circuit harness connections — 26
- Relays - identification — 24
- Relays - remove and refit — 25
- Window lifts and doorlocks - cicuit diagram — 28

305

# RANGE ROVER

## INTRODUCTION 01

## INTRODUCTION

This Workshop Manual Supplement covers the introduction of Lucas 'Hot Wire' air flow sensor with microprocessor controlled fuel injection system and catalytic exhaust system.

**NOTE: This supplement is designed to assist skilled operators in the effective maintenance and repair of Range Rover vehicles, and should be used in conjunction with the MAIN WORKSHOP MANUAL and SUPPLEMENT LSM180WS1.**

## Notes

# RANGE ROVER

## GENERAL SPECIFICATION DATA

### FUEL SYSTEM

Fuel system type ..................... Lucas hot wire system electronically controlled
Fuel pump-make/type ............. AC Delco-high pressure (electrical) immersed in the fuel tank
Fuel pump delivery pressure ..... 2.4-2.6 kgf/cm² (34-37 p.s.i.)
Fuel filter .............................. Bosch in-line filter 'canister' type

**Airflow Sensor**

Make and type ....................... Lucas 'Hot Wire' 3AM

**Injectors**

Make and type ....................... Lucas 8NJ

**Electronic Control Unit**

Make and type ....................... Lucas 13CU

**Fuel pressure regulator**

Make and type ....................... Lucas 8RV

**Fuel temperature sensor**

Make and type ....................... Lucas 6TT

**Coolant temperature sensor**

Make and type ....................... Lucas 3TT

**Bypass Airvalve (Stepper motor)**

Make and type ....................... Lucas 2ACM

**Throttle potentiometer**

Make and type ....................... Lucas 215SA

**Lambda sensor**

Make and type ....................... Lucas 3LS

**Continued**

# GENERAL SPECIFICATION DATA | 04

## TRANSMISSION

**Transfer gearbox-LT230**
Type .................................. Two speed reduction on main gearbox output. Front and rear drive permanently engaged via a lockable differential.

**Transfer gearbox ratios**
High .................................. 1.222:1
Low ................................... 3.320:1

**Automatic gearbox**
Model ................................ ZF4HP22
Type .................................. Four speed and reverse epicyclic with fluid torque converter and lock up.

4th ..................................... 0.728:1
3rd ..................................... 1.000:1
2nd .................................... 1.480:1
1st ..................................... 2.480:1
Reverse ............................. 2.086:1

**Overall ratio (final drive):**

| | High transfer | Low transfer |
|---|---|---|
| 4th | 3.15:1 | 8.55:1 |
| 3rd | 4.32:1 | 11.75:1 |
| 2nd | 6.40:1 | 17.38:1 |
| 1st | 10.72:1 | 29.13:1 |
| Reverse | 9.02:1 | 24.50:1 |

**Propeller shafts**
Type .................................. Solid bar 28.6 mm (1.125 in) diameter
Front .................................
Rear .................................. 51 mm (2 in) diameter
Universal joints ................. Open type O3EHD

---

# GENERAL SPECIFICATION DATA | 04

## SHIFT SPEED SPECIFICATION
### Automatic ZF4HP22 Gearbox

| OPERATION | SELECTOR POSITION | VEHICLE SPEED APPROX. MPH | VEHICLE SPEED APPROX. KPH | ENGINE SPEED APPROX. (RPM) |
|---|---|---|---|---|
| **KICKDOWN** | | | | |
| KD4-3 | D | 78-95 | 125-153 | |
| KD3-2 | 3 (D) | 56-61 | 90-98 | |
| KD2-1 | 2 (D,3) | 27-34 | 43-55 | |
| KD3-4 | D | NOT APPLICABLE | | 4750-5200 |
| KD2-3 | D (3) | 59-64 | 95-103 | 4600-5250 |
| KD1-2 | D (3,2) | 34-39 | 55-63 | |
| **FULL THROTTLE** | | | | |
| FT4-3 | D | 60-66 | 97-106 | |
| FT3-2 | 3 (D) | 39-45 | 63-72 | 3980-4330 |
| FT3-4 | D (3) | 73-79 | 117-127 | 4350-4800 |
| FT2-3 | D (3) | 54-59 | 87-95 | 3950-4650 |
| FT1-2 | D (3,2) | 29-34 | 47-55 | |
| **ZERO THROTTLE** | | | | |
| ZT4-3 | D (3) | 19-25 | 31-40 | |
| ZT3-2 | D (3,2) | 12-15 | 19-24 | |
| ZT2-1 | | 6-7 | 10-11 | |
| **PART THROTTLE** | | | | |
| PT4-3 | D | 46-53 | 74-85 | |
| PT3-2 | D (3) | 29-36 | 47-58 | |
| PT2-1 | D (3,2) | 10-12 | 16-19 | |
| **LIGHT THROTTLE** | | | | |
| LT3-4 | D | 26-30 | 42-48 | 1430-1650 |
| LT2-3 | D (3) | 18-22 | 29-35 | 1420-1820 |
| LT1-2 | D (3,2) | 9-10 | 14-16 | 1180-1220 |
| **TORQUE CONVERTER** | | | | |
| Lock Up (IN) | D | 50-53 | 80-85 | 1875-2000 |
| Unlock (OUT) | D | 48-51 | 77-82 | 1825-1930 |

NOTE: The speeds given in the above chart are approximate and only intended as a guide. Maximum shift changes should take place within these tolerance parameters.

# RANGE ROVER

## 04 GENERAL SPECIFICATION DATA

### ELECTRICAL

**Battery**
Make/type ..................................... Chloride maintenance free 14-plate-380/120/90

**Alternator**
Manufacturer ................................. Lucas
Type ............................................... 133/80
Polarity .......................................... Negative ground
Brush length
  New, minimum free protrusion
  from brush box ............................ 20 mm (0.78 in)
  Worn, minimum free protrusion
  from brush box ............................ 10 mm (0.39 in)
Brush spring pressure flush with brush box face ... 136 to 279 g (5 to 10 oz)
Rectifier pack output rectification ...... 6 diodes (3 positive side and 3 ground side)
Field winding supply rectification ...... 3 diodes
Stator windings .............................. 3 phase-delta connected
Field winding rotor poles ................ 12
Maximum speed ............................ 16,000 rev/min
Winding resistance at 20°C ............ 2.6 ohms
Control ........................................... Field voltage sensed regulation
Regulator-type ................................ 15 TR
  voltage ......................................... 13.6 to 14.4 volts
Nominal output
  Condition ..................................... Hot
  Alternator speed ........................... 6000 rev/min
  Control voltage ............................. 14 volt
  Amp ............................................. 80 amp

### REPLACEMENT BULBS

| | | | TYPE |
|---|---|---|---|
| Headlamps | ) | 12V 60/55W | (Halogen) |
| Auxiliary driving lamps | ) | 12V 55W H3 | (Halogen) |
| Sidelamps | ) | 12V 5W | bayonet |
| Tail lamps | ) Exterior | 12V 5/21W | bayonet |
| Reverse lamps | ) lights | 12V 21W | bayonet |
| Stop lamps | ) | 12V 21W | bayonet |
| Direction indicator lamps | ) | 12V 4W | bayonet |
| Rear side marker lamps | ) | 12V 5W | capless |
| Number plate lamps | ) | | |
| Instrument panel lamps and warning lamps | ) | 12V 1.2W | bulb/holder unit |
| Ignition warning lamp (Instrument panel) | ) | 12V 2W | capless |
| Interior roof lamps | ) | 12V 10W | 'Festoon' |
| Clock illumination | ) | 12V 2W | bayonet |
| Cigar lighter illumination | ) | 12V 1.2W | capless |
| Door edge/puddle lamps | ) Interior | 12V 5W | capless |
| Auxiliary switch panel | ) lights | | |
| illumination (green) | ) | 12V 1.2W | capless |
| Heated rear screen warning lamp (amber) | ) | 12V 1.2W | capless |
| Hazard warning lamp | ) | 24V 5W | capless |
| Automatic graphics illumination | ) | 12V 1.2W | capless |
| Heater/air conditioning graphics illumination | ) | 12V 2W | bayonet |
| Differential lock warning lamp | ) | 12V 1.2W | capless |
| Column switch illumination | ) | | |

---

# RANGE ROVER

## 05 ENGINE TUNING DATA

### ENGINE TUNING DATA

Type .............................................. V8
Firing order ................................... 1-8-4-3-6-5-7-2
Cylinder Numbers
  Left bank ..................................... 1-3-5-7
  Right bank ................................... 2-4-6-8
No 1 Cylinder location .................. Pulley end of left bank
Timing marks ................................ On crankshaft vibration damper
Spark plugs
  Make/type ................................... Champion RN12YC
  Gap ............................................. 0.85-0.95mm (0.033-0.038 in)
Coil
  Make/type ................................... Lucas 32C5
Compression ratio ........................ 8.13:1
Fuel injection system ................... Lucas Hot-wire air flow sensor system electronically controlled

**Valve Timing**

| | Inlet | Exhaust |
|---|---|---|
| Opens | 24° BTDC | 62° BBDC |
| Closes | 52° ABDC | 14° ATDC |
| Duration | 256° | 256° |
| Valve peak | 104° ATDC | 114° BTDC |

Idle speed .................................... 665 to 735 rev/min
Ignition Timing at ......................... 800 rev/min max
Ignition Timing
  Dynamic ...................................... 6° ± 1° B.T.D.C.

**Distributor**
Make/type ..................................... Lucas 35DLM8 electronic
Rotation ........................................ Clockwise
Air gap .......................................... 0.20-0.35mm (0.008-0.014 in)
Despatch number .......................... 42620

**Centrifugal Advance**
Decelerating check-vacuum hose disconnected
Distributor rpm decelerating speeds        Distributor advance
2300 ............................................. 8° to 11°
1600 ............................................. 8° 54' to 11°
600 ............................................... 1° 18' to 3° 18'
No centrifugal advance below ........ 150 rev/min

**Fuel**
Unleaded ....................................... 95 octane

309

# RANGE ROVER

## TORQUE VALUES 06

## TORQUE WRENCH SETTINGS

Lubricants/sealants have been specified in certain applications for assembly purposes.

* These bolts must have threads coated with Loctite 572 prior to assembly. For this purpose it is necessary to use an approved dispenser to apply the sealant/lubricant to the first three threads of the bolts.
** These bolts must have threads coated in lubricant EXP16A (Marston Lubricants) prior to assembly.
*** These bolts must have threads coated in sealant Loctite 270 prior to assembly.

It is essential that all bolts are securely tightened and it is imperative that the correct torques values are adhered to.

### ENGINE

| | Nm | ft lb |
|---|---|---|
| Alternator mounting bracket to cylinder head | 35 - 43 | 26 - 32 |
| Alternator to mounting bracket | 22 - 28 | 16 - 21 |
| Alternator to adjusting link | 22 - 28 | 16 - 21 |
| Chainwheel to camshaft | 54 - 61 | 40 - 45 |
| Connecting rod nut | 47 - 54 | 35 - 40 |
| Cylinder head: | | |
|   Outer row | 54 - 61 | 40 - 45 * |
|   Centre row | 88 - 95 | 65 - 70 * |
|   Inner row | 88 - 95 | 65 - 70 * |
| Distributor clamp nut | 19 - 22 | 14 - 16 |
| Exhaust manifold to cylinder heads | 19 - 22 | 14 - 16 |
| Fan to viscous unit | 26 - 32 | 19 - 24 |
| Flexible drive plate to crankshaft adaptor plate | 35 - 46 | 26 - 34 *** |
| Adaptor plate to crankshaft | 77 - 90 | 57 - 66 |
| Intake manifold to cylinder heads | 34 - 41 | 25 - 30 |
| Lifting eye to cylinder heads | 35 - 43 | 26 - 32 |
| Main bearing cap bolts | 68 - 75 | 50 - 55 |
| Main bearing cap rear bolts | 88 - 95 | 65 - 70 ** |
| Manifold gasket clamp bolt | 14 - 20 | 10 - 15 |
| Oil pump cover to timing cover | 11 - 14 | 8 - 10 |
| Oil plug | 24 - 30 | 18 - 22 |
| Oil relief valve plug | 40 - 47 | 30 - 35 |
| Oil sump drain plug | 40 - 47 | 30 - 35 |
| Oil sump to cylinder block | 7 - 11 | 5 - 8 |
| Oil sump rear to cylinder block | 17 - 20 | 13 - 15 |
| Rocker cover to cylinder head | 7 - 10 | 5 - 7 |
| Rocker shaft bracket to cylinder head | 34 - 40 | 25 - 30 |
| Spark plug | 19 - 22 | 14 - 16 |
| Starter motor attachment | 41 - 47 | 30 - 35 |
| Damper to crankshaft | 257 - 285 | 190 - 210 |
| Timing cover to cylinder block | 24 - 30 | 18 - 22 * |
| Viscous unit to water pump hub | 40 - 50 | 30 - 37 |
| Water pump pulley to water pump hub | 8 - 12 | 6 - 9 * |
| Water pump timing cover to cylinder block | 24 - 30 | 18 - 22 |
| Water jacket to plenum chamber | 11 - 14 | 8 - 10 * |
| Plenum chamber to ram housing | 22 - 28 | 16 - 21 |
| Ram housing to intake manifold | 20 - 27 | 15 - 20 |
| Thermostat housing to intake manifold | 24 - 30 | 18 - 22 |

Continued

1

# LUBRICANTS, FLUIDS AND CAPACITIES 09

## RECOMMENDED LUBRICANTS AND FLUIDS

| COMPONENT | SPECIFICATION | Viscosity |
|---|---|---|
| Engine | Use oils to API service levels SE or SF or SE/CC or SF/CC | 5W/20<br>5W/30<br>5W/40<br>10W/30<br>10W/40<br>10W/50<br>15W/40<br>15W/50<br>20W/40<br>20W/50 |
| Automatic gearbox | ATF Dexron IID | |
| Transfer gearbox<br>Final drive units<br>Swivel pin housings | APIGL4 or GL5<br>MIL-L-2105 or<br>MIL-L-2105B | Viscosity<br>90 EP<br>80W EP |
| Power steering | ATF Dexron IID | |
| Brake reservoir | Brake fluid must have a minimum boiling point of 260°C (500°F) and comply with FMVSS/116/DOT 4 | |
| Lubrication nipples (hubs, ball joints, etc.) | NLGI-2 multipurpose lithium based grease | |

AMBIENT TEMPERATURE

°C: -30  -20  -10  0  +10  +20  +30  +40  +50
°F: -22  -4  +14  +32  +50  +68  +86  +104  +122

Continued

## 06 TORQUE VALUES

Charts below give torque values for all screws and bolts used except for those that are specified otherwise.

| SIZE | METRIC |  | SIZE | UNC |  | UNF |  |
|---|---|---|---|---|---|---|---|
|  | Nm | ft lb |  | Nm | ft lb | Nm | ft lb |
| M5 | 5-7 | 3.7-5.2 | 1/4 | 6.8-9.5 | 5-7 | 8.1-12.2 | 6-9 |
| M6 | 7-10 | 5.2-7.4 | 5/16 | 20.3-27.1 | 15-20 | 20.3-27.1 | 15-20 |
| M8 | 22-28 | 16.2-20.7 | 3/8 | 35.3-43.4 | 26-32 | 35.3-43.4 | 26-32 |
| M10 | 40-50 | 29.5-36.9 | 7/16 | 67.8-88.1 | 50-65 | 67.8-88.1 | 50-65 |
| M12 | 80-100 | 59.0-73.8 | 1/2 | 81.3-101.7 | 60-75 | 81.3-101.7 | 60-75 |
| M14 | 90-120 | 66.4-88.5 | 5/8 | 122.0-149.1 | 90-110 | 122.0-149.1 | 90-110 |
| M16 | 160-200 | 118.0-147.5 | | | | | |

311

# MAINTENANCE 10

## RANGE ROVER MAINTENANCE - FIRST 1,600 KM SERVICE ONLY

Check front wheel alignment

Check PCV system for leaks and hoses for security and condition (Positive crankcase ventilation).

Check operation of throttle mechanical linkage and transmission cable

Check ignition wiring and HT leads for security, and fraying

Check/adjust all drive belts

Check operation of all instruments fuel and temperature gauges, warning indicators, lamps, horns and audio unit

Check operation of front and rear screen and headlamp wash/wipers

Check operation of rear view mirrors and for security, cracks and crazing

Check condition and security of seats, seat belt mountings, belts and buckles

Check exhaust system for leaks, security and damage

Check tyres comply with Manufacturers specification

Check tyres for cuts, lumps, bulges, uneven wear, tread depth and road wheels for damage

Check and adjust tyre pressures including spare

Check and tighten road wheel retaining nuts

Continued

---

Renew engine oil

Renew transfer box oil

Renew front axle oil

Renew rear axle oil

Renew steering swivel housing oil

Check/top up power steering fluid

Check/top up automatic transmission fluid

Check for oil/fluid leaks from:- Suspension, dampers and self levelling unit, engine and transmission units, front and rear axles

Check brake pipes/unions for security, chafing, leaks and corrosion

Check power steering system for leaks, hydraulic pipes/unions for security, chafing and corrosion

Check fuel system for leaks

Check cooling and heater systems for leaks, hoses for security and condition

Check security and operation of park brake

Check foot brake operation

Check condition and security of steering unit, joints and gaiters

Check/adjust steering box

---

# 09 LUBRICANTS, FLUIDS AND CAPACITIES

| Capacities (approx.)* | Litres | Imperial unit | US unit |
|---|---|---|---|
| Engine sump and filter from dry | 5.68 | 10 pints | 12.0 pints |
| Gearbox from dry-automatic ZF | 9.1 | 16 pints | 20 pints |
| Transfer gearbox from dry | 2.5 | 4.4 pints | 5.3 pints |
| Front axle from dry | 1.7 | 3.0 pints | 3.6 pints |
| Front axle swivel pin housing (each) | 0.35 | 0.6 pints | 0.7 pints |
| Rear axle from dry | 1.7 | 3.0 pints | 3.6 pints |
| Power steering box and reservoir | 2.9 | 5.0 pints | 6.0 pints |
| Cooling system | 11.4 | 20 pint | 24 pints |
| Fuel tank | 76.4 | 16.8 gallon | 20 gallons |

**NOTE:** * All levels must be checked by dipstick or level plugs as applicable.

When draining oil from the ZF automatic gearbox, oil will remain in the torque converter, refill to high level on dipstick only.

## LUBRICATION PRACTICE

The engine is filled with special oil to protect it during the running-in-period. The engine must be drained after 1600 km (1,000 miles) and refilled with an appropriate lubricant.

Use a high quality oil of the correct viscosity range and service classification in the engine during maintenance and when topping up. The use of oil not to the correct specification can lead to high oil and fuel consumption and ultimately to damaged components.

Oil to the correct specification contains additives which disperse the corrosive acids formed by combustion and prevent the formation of sludge which can block the oilways. Additional oil additives should not be used. Always adhere to the recommended servicing intervals.

**WARNING: Many liquids and other substances used in motor vehicles are poisonous and should under no circumstances be consumed and should be kept away from open wounds. These substances among others include anti-freeze, brake fluid, fuel, windscreen washer additives, lubricants and various adhesives.**

## FUEL REQUIREMENTS

The engine is designed to use only unleaded fuel. Unleaded fuel must be used for the emission control system to operate properly. Its use will also reduce spark plug fouling, exhaust system corrosion and engine oil deterioration.

Using fuel that contains lead will result in damage to the emission control system. The effectiveness of the catalysts in the catalytic converters will be seriously impaired if leaded fuel is used. The vehicle is equipped with an electronic fuel injection system, which includes two oxygen sensors. Leaded fuel will damage the sensors, and will deteriorate the emission control system.

Regulations require that pumps delivering unleaded fuel be labelled **UNLEADED**. Only these pumps have nozzles which fit the filler neck of the vehicle fuel tank.

## FUEL WITH A RATING OF AT LEAST 95 OCTANE SHOULD BE USED.

Using unleaded fuel with an octane rating lower than stated above can cause persistent, heavy 'spark knock' ('spark knock' is a metallic rapping noise). If severe, this can lead to engine damage. If a heavy spark knock is detected even when using fuel of the recommended octane rating, check the ignition timing system.

**CAUTION: Do not use oxygenated fuels such as blends of methanol/ gasoline or ethanol/gasoline (e.g. 'Gasohol'). Take care not to spill fuel during refuelling.**

312

# MAINTENANCE

## RANGE ROVER EMISSION MAINTENANCE

| MAINTENANCE INTERVALS KM X 1000 | 10 | 20 | 30 | 40 | 50 | 60 | 70 | 80 | 90 | 100 | 110 | 120 | 130 | 140 | 150 | 160 |
|---|---|---|---|---|---|---|---|---|---|---|---|---|---|---|---|---|
| Renew PCV intake filter check and clean PCV system (Positive crankcase ventilation) | | | | | | | | | | | | | | | | • |
| Renew engine oil | • | • | • | • | • | • | • | • | • | • | • | • | • | • | • | • |
| Renew engine oil filter | • | • | • | • | • | • | • | • | • | • | • | • | • | • | • | • |
| Renew fuel filter | | | | | | • | | | | | | • | | | | |
| Renew air cleaner element/check/clean dump valve | | | | • | | | | • | | | | • | | | | • |
| Renew charcoal canister | | | | | | | | | | | | | | | | • |
| Renew spark plugs | | | | • | | | | • | | | | • | | | | • |
| Renew catalytic converters | | | | | | | | | | | | | | | | • |
| Renew oxygen sensors | | | | | | | | • | | | | | | | | • |
| Check exhaust system for leaks, security and damage | • | • | • | • | • | • | • | • | • | • | • | • | • | • | • | • |
| Check condition of driving belts-adjust if required | • | • | • | • | • | • | • | • | • | • | • | • | • | • | • | • |
| Check/top up cooling system | • | • | • | • | • | • | • | • | • | • | • | • | • | • | • | • |
| Check ignition wiring and HT leads for fraying, chafing and deterioration | | | | | | | | | | | | | | | | • |
| Clean distributor cap and rotor arm, check for cracks and tracking. Lubricate rotor spindle with rotor arm removed | | | | | | | | | | | | • | | | | |
| Check/adjust ignition timing | | | | | | | | | | | | • | | | | |
| Check engine idle speed | | | | | | | | • | | | | • | | | | • |
| Check fuel evaporative loss control system for leaks | | | | | | | | | | | | • | | | | |
| Check fuel filler cap seal for leaks | | | | | | | | • | | | | | | | | |
| Check fuel pipes, filler hoses and connections for leaks and security | | | | | | | | | | | | • | | | | |
| Check engine emission control system hoses, tubes and vacuum lines for security and condition | | | | | | | | • | | | | | | | | • |
| Check operation of electronic control unit/systems | | | | | | | | | | | | | | | | • |
| Check operation of auxiliary emission control devices | | | | | | | | | | | | | | | | • |

The Emission Maintenance above is necessary to keep the vehicle in compliance with the Swiss Emission regulations up to 80,000 Km or 5 years. The manufacturer of Range Rover recommends that the sequence of maintenance operations be carried out throughout the life of the vehicle in order to maintain vehicle performance and reliability.

## RANGE ROVER MAINTENANCE

| MAINTENANCE INTERVALS KM X 1000 | 10 | 20 | 30 | 40 | 50 | 60 | 70 | 80 | 90 | 100 | 110 | 120 | 130 | 140 | 150 | 160 |
|---|---|---|---|---|---|---|---|---|---|---|---|---|---|---|---|---|
| Check/top up transfer box oil | • | • | • | • | • | • | • | • | • | • | • | • | • | • | • | • |
| Renew transfer box oil | | | | | | | | | • | | | | | | | |
| Check/top up front axle oil | • | • | • | • | • | • | • | • | • | • | • | • | • | • | • | • |
| Renew front axle oil | | | | • | | | | | | | | • | | | | |
| Check/top up rear axle oil | • | • | • | • | • | • | • | • | • | • | • | • | • | • | • | • |
| Renew rear axle oil | | | | • | | | | | | | | • | | | | |
| Check/top up steering swivel housing oil | • | • | • | • | • | • | • | • | • | • | • | • | • | • | • | • |
| Renew steering swivel housing oil | | | | | | | | | | | | • | | | | |
| Renew automatic transmission fluid | | | | | | | | | | | | • | | | | |
| Renew automatic transmission fluid filter | | | | | | | | | | | | • | | | | |
| Check/top up power steering fluid | • | • | • | • | • | • | • | • | • | • | • | • | • | • | • | • |
| Check/top up automatic transmission fluid | • | • | • | • | • | • | • | • | • | • | • | • | • | • | • | • |
| Check/top up brake fluid | • | • | • | • | • | • | • | • | • | • | • | • | • | • | • | • |
| Check for oil/fluid leaks from: Suspension, dampers and self levelling unit | • | • | • | • | • | • | • | • | • | • | • | • | • | • | • | • |
| Check foot brake operation | • | • | • | • | • | • | • | • | • | • | • | • | • | • | • | • |
| Check condition and security of steering unit, joints and gaiters | | • | | • | | • | | • | | • | | • | | • | | • |
| Check/adjust steering box | | • | | • | | • | | • | | • | | • | | • | | • |
| Check front wheel alignment | | • | | | | • | | | | • | | | | • | | |
| Check PCV system for leaks and hoses for security and condition (Positive crankcase ventilation) | | | | • | | | | • | | | | • | | | | • |
| Check and clean PCV system and breathers (Positive crankcase ventilation) | | | | | | | | • | | | | | | | | • |
| Check/clean air cleaner dump valve and element | | | | • | | | | • | | | | • | | | | • |
| Check operation of throttle mechanical linkage and transmission cable | | • | | • | | • | | • | | • | | • | | • | | • |
| Check battery condition | | • | | • | | • | | • | | • | | • | | • | | • |
| Clean and grease battery terminals | | | | • | | | | • | | | | • | | | | • |
| Check ignition wiring and HT leads for security, fraying, chafing, deterioration | | | | • | | | | • | | | | • | | | | • |
| Check distributor cap and rotor arm, check for cracks and tracking | | | | • | | | | • | | | | • | | | | • |
| Lubricate distributor rotor spindle with rotor arm removed | | | | • | | | | • | | | | • | | | | • |
| Clean/adjust spark plugs | | | | | | • | | | | | | • | | | | |
| Check/adjust ignition timing | | | | | | • | | | | | | • | | | | |

Continued

# MAINTENANCE

## RANGE ROVER MAINTENANCE
### Continued

| MAINTENANCE INTERVALS KM X 1000 | 10 | 20 | 30 | 40 | 50 | 60 | 70 | 80 | 90 | 100 | 110 | 120 | 130 | 140 | 150 | 160 |
|---|---|---|---|---|---|---|---|---|---|---|---|---|---|---|---|---|
| Check condition and security of seats, seat belt mountings, belts and buckles | • | • | • | • | • | • | • | • | • | • | • | • | • | • | • | • |
| Check tightness of propeller shaft coupling bolts |  | • |  | • |  | • |  | • |  | • |  | • |  | • |  | • |
| Check exhaust system for leaks, security and damage | • | • | • | • | • | • | • | • | • | • | • | • | • | • | • | • |
| Check tyres comply with Manufacturer's specification | • | • | • | • | • | • | • | • | • | • | • | • | • | • | • | • |
| Check tyres for cuts, lumps, bulges, uneven wear, tread depth and road wheels for damage |  | • |  | • |  | • |  | • |  | • |  | • |  | • |  | • |
| Check/adjust tyre pressures including spare | • | • | • | • | • | • | • | • | • | • | • | • | • | • | • | • |
| Check/adjust headlamp and auxiliary lamp alignment | • |  |  |  |  | • |  |  |  |  |  | • |  |  |  |  |
| Check operation of all instruments, fuel and temperature gauges, warning indicators, lamps, horns and audio unit | • | • | • | • | • | • | • | • | • | • | • | • | • | • | • | • |
| Check operation of front and rear screen and headlamp wash/wipers | • | • | • | • | • | • | • | • | • | • | • | • | • | • | • | • |
| Clear sun roof drain tubes | • |  | • |  | • |  | • |  | • |  | • |  | • |  | • |  |
| Check operation of all doors, hood, tailgate locks and window controls | • | • | • | • | • | • | • | • | • | • | • | • | • | • | • | • |
| Lubricate all locks (not steering lock), hinges and door check mechanisms | • | • | • | • | • | • | • | • | • | • | • | • | • | • | • | • |
| Check operation of heater and air conditioning systems |  | • |  | • |  | • |  | • |  | • |  | • |  | • |  | • |
| Check brake pipes/unions for security, chafing, leaks and corrosion | • | • | • | • | • | • | • | • | • | • | • | • | • | • | • | • |
| Check power steering system for leaks, hydraulic pipes/unions for security, chafing and corrosion | • | • | • | • | • | • | • | • | • | • | • | • | • | • | • | • |
| Check fuel pipes for leaks | • | • | • | • | • | • | • | • | • | • | • | • | • | • | • | • |
| Check cooling and heater systems for leaks, hoses for security and condition |  | • |  | • |  | • |  | • |  | • |  | • |  | • |  | • |
| Check/top up cooling system | • | • | • | • | • | • | • | • | • | • | • | • | • | • | • | • |
| Inspect brake pads for wear, calipers for leaks and discs for condition | • | • | • | • | • | • | • | • | • | • | • | • | • | • | • | • |
| Check security and operation of hand brake | • | • | • | • | • | • | • | • | • | • | • | • | • | • | • | • |
| Lubricate hand brake mechanical linkage | • | • | • | • | • | • | • | • | • | • | • | • | • | • | • | • |
| Lubricate propeller shaft universal joints | • | • | • | • | • | • | • | • | • | • | • | • | • | • | • | • |
| Lubricate propeller shaft sealed sliding joints |  | • |  | • |  | • |  | • |  | • |  | • |  | • |  | • |

## MAINTENANCE

### It is recommended that:

At 30,000 Km intervals or every 18 months, whichever is the sooner, the hydraulic brake fluid should be completely renewed.

At 60,000 Km intervals or every 3 years, whichever is the sooner, all hydraulic brake fluid, seals and flexible hoses should be renewed, all working surfaces of the master cylinder, wheel cylinders and caliper cylinders should be examined and renewed where necessary.

At 60,000 Km intervals remove all suspension dampers, test for correct operation, refit or renew as necessary.

At two yearly intervals or at the onset of the second winter the cooling system should be drained, flushed and refilled with the required water and anti-freeze solution.

The battery electrolyte level should be checked and topped up if required three times per year in high ambient temperatures, and once per year in moderate ambient temperatures.

**NOTE:** Climatic and operating conditions affect maintenance intervals to a large extent; in many cases, therefore, the determination of such intervals must be left to the good judgement of the owner or to advice from a Range Rover Authorised Dealer, but the recommendations will serve as a firm basis for maintenance work.

Vehicles operating under arduous conditions will require more frequent servicing, therefore, at a minimum, the maintenence intervals should be reduced by half.

## MAINTENANCE

### LUBRICATION

This first part of the maintenance section covers renewal of lubricating oils for the major units of the vehicle and other components that require lubrication, as detailed in the 'Maintenance Schedules'. Refer to Section 09 for Capacities and Recommended Lubricants.

Vehicles operating under severe conditions of dust, sand, mud and water should have the oils changed and lubrication carried out at more frequent intervals than that recommended in the maintenance schedules.

Draining of used oil should take place after a run when the oil is warm. Always clean the drain and filler-level plugs before removing. In the interests of safety disconnect the vehicle battery to prevent the engine being started and the vehicle moved inadvertently, while oil changing is taking place.

Allow as much time as possible for the oil to drain completely except where blown sand or dirt can enter the drain holes. In these conditions clean and refit the drain plugs immediately the main bulk of oil has drained.

Where possible, always refill with oil of the make and specification recommended in the lubrication charts and from sealed containers.

### RENEW ENGINE OIL AND FILTER

#### DRAIN THE OIL

1. Before changing the oil ensure that the vehicle is level on either hoist or ground.
2. Run the engine to warm the oil; switch off the ignition and disconnect the battery for safety.
3. Place an oil tray under the drain plug.
4. Remove the drain plug in the bottom of the sump at the left-hand side. Allow oil to drain away completely. Fit new copper washer and replace the plug, tighten to the correct torque value.

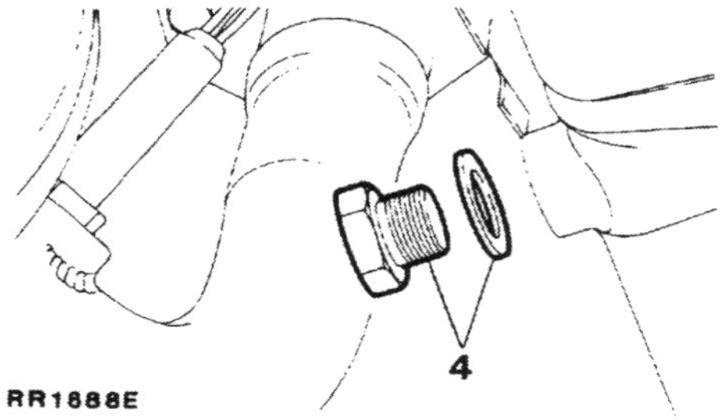

#### Fit new oil filter

5. Place an oil tray under the engine.
6. Unscrew the filter counter -clockwise, using a strap wrench as necessary.
7. Clean the oil pump mating face and coat the rubber washer of the new filter with clean engine oil, screw the filter on clockwise until the rubber sealing ring touches the machined face, tighten a further half turn by hand only. **DO NOT** overtighten.

#### Refill sump with oil

8. Check that the drain plug is tight.
9. Clean the outside of the oil filler cap, remove it from the extension filler neck and clean the inside.
10. Pour in the correct quantity of new oil of the correct grade from a sealed container to the high mark on the dipstick and firmly replace the filler cap. **DO NOT FILL ABOVE 'HIGH' MARK.** Reconnect the battery.
11. Run the engine and check for leaks from the filter. Stop the engine, allow the oil to run back into the sump for a few minutes, then check the oil level again and top up if necessary.

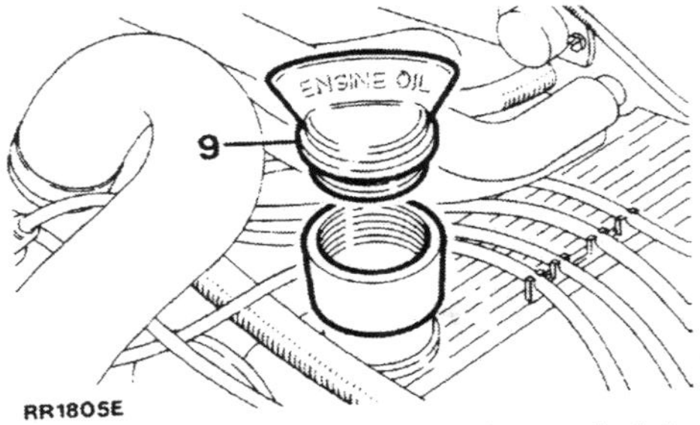

### FILTERS

#### CLEAN POSITIVE CRANKCASE VENTILATION BREATHER FILTER

1. Release the hose clamp and pull the hose off the canister.
2. Unscrew the canister and remove it from the rocker cover.
3. Remove the large 'O' ring from the threaded end of the canister.
4. Visually inspect the condition of the wire screen within the canister, if in poor condition fit a new assembly, if in an acceptable condition clean the screen as follows:
5. Immerse the canister in a small amount of solvent and allow time for the solvent to dissolve and loosen any engine fume debris within the canister.
6. Remove canister from solvent bath and allow to dry out in still air.

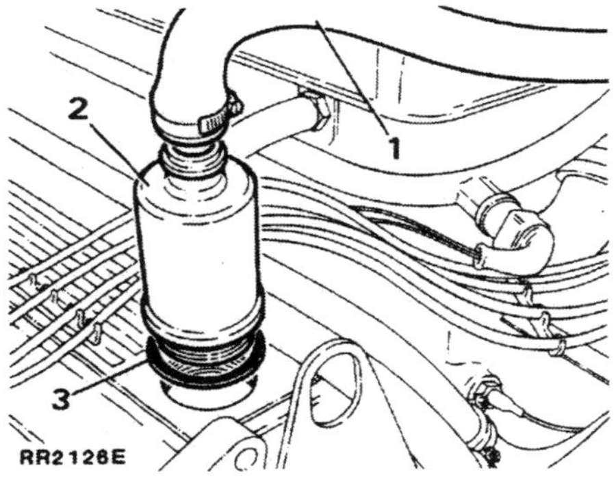

**WARNING: Do not use a compressed air line to remove any remaining solvent or particles of debris within the canister as this could cause fire or personal injury.**

#### Refitting the breather/filter

7. Fit a new rubber 'O' ring.
8. Screw the flame trap canister into the rocker cover, hand tight only.
9. Refit hose and tighten hose clamp securely.

### CLEAN PLENUM CHAMBER VENTILATION PASSAGEWAY

The cleaning of the plenum chamber ventilation passageway can be carried out without removing the plenum chamber from the ram housing.

**CAUTION: Care must be taken to prevent debris from the passageway passing beyond the throttle valve disc.**

**WARNING: Safety glasses must be worn when performing this operation. Ensure that debris is not blown into the atmosphere which could be harmful to other personnel within the vicinity.**

1. Disconnect the battery negative terminal.
2. Release the hose clamp and remove the hose from the plenum chamber.
3. Remove the crankcase ventilation hose from the side of the plenum chamber.
4. Clean out the passageway into the throttle bore.
5. Remove the small 'T' piece between the crankcase ventilation hoses and check that it is free from blockages, clean as necessary.
6. Refit the 'T' piece and hoses, tighten the hose clamps securely.

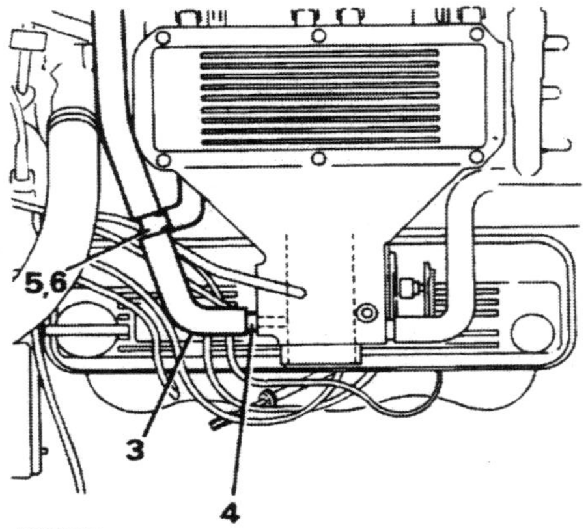

# MAINTENANCE

## AIR CLEANER ELEMENT

### Remove and refit

#### Removing

1. Release the two clips securing the air cleaner to the airflow sensor.
2. Release the two nuts and bolts securing the air cleaner to the left hand valance mounting bracket.
3. Detach the airflow sensor from the air cleaner, and lay carefully to one side.
4. Detach the air cleaner from the centre mounting bracket and withdraw from the engine compartment.
5. Remove the large 'O' ring from the outlet tube of the air cleaner, inspect for condition, fit a new 'O' ring if in poor condition.
6. Unclip the three catches securing the inlet tube to the air cleaner canister and remove the inlet tube.
7. Remove the nut and end plate securing the air cleaner element in position.
8. Withdraw the air cleaner element and discard.
9. Inspect the dump valve for condition and check that it is clear of obstructions.

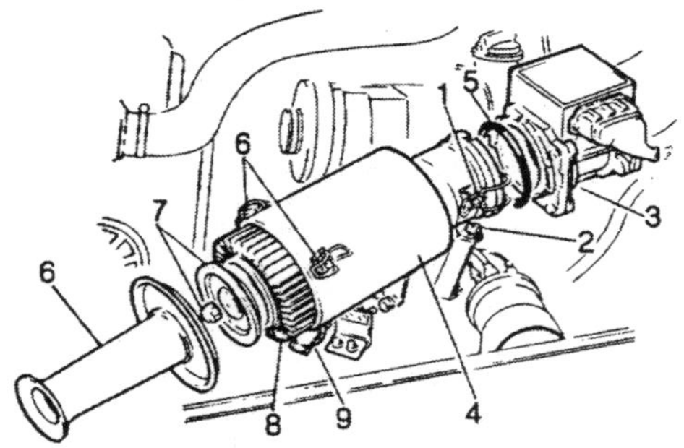

#### Refitting

10. Fit new element and secure in position.
11. Refit the inlet tube to the air cleaner canister.
12. Refit the air cleaner to the mounting bracket and tighten the two nuts and bolts.
13. Clip the air flow sensor to the air cleaner.

## FUEL LINE FILTER

**WARNING: The spilling of fuel is unavoidable during this operation. Ensure that all necessary precautions are taken to prevent fire and explosion.**

1. Depressurise the fuel system. (Refer to Fuel Injection System - Section 19)
2. The fuel line filter is located on the right hand chassis side member forward of the fuel tank filler neck. Access to the filter is gained through the right hand rear wheel arch.
3. Thoroughly clean the immediate area around the hose connections to prevent ingress of foreign matter into the fuel system.
4. Loosen the two hose clamps nearest the filter to enable the hoses to be removed from the filter canister. Plug the end of the hoses to prevent ingress of dirt.
5. Release the securing bolt and bracket and remove the filter from the chassis side member.

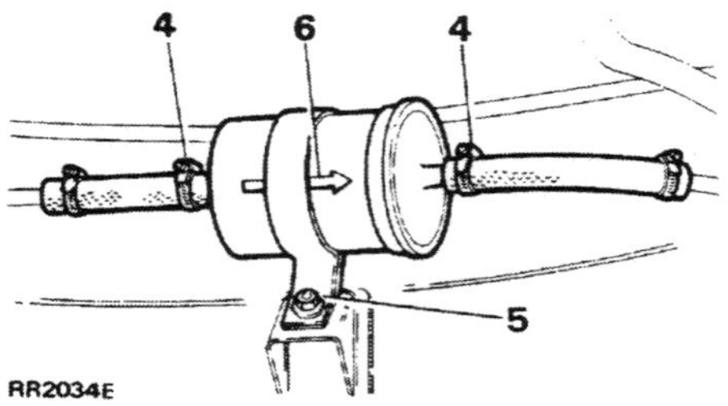

### Fit new filter

6. Fit a new filter observing the direction of flow arrows stamped on the canister.
7. Start the engine and inspect for fuel leaks around the hose connections.

## CHARCOAL CANISTER

1. Disconnect from the canister:
   (i) Canister line to fuel tank.
   (ii) Canister purge line.
2. Loosen the clamp pinch bolt
3. Remove the canister.

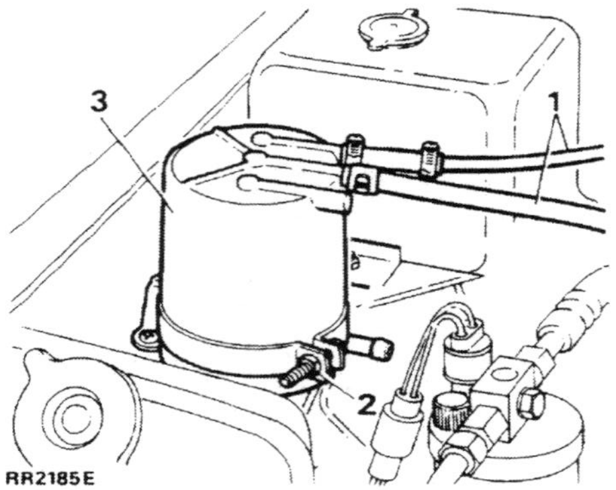

### Fit new canister

4. Secure the canister in the clamp.
5. Reverse instructions 1 and 2 above.

**WARNING: The use of compressed air to clean a charcoal canister or to clear a blockage in the evaporative system is highly dangerous. An explosive gas present in a fully saturated canister may be ignited by the heat generated when compressed air passes through the canister.**

## POSITIVE CRANKCASE VENTILATION INTAKE FILTER

1. Pry the filter holder upwards to release it from the rocker cover.
2. Discard the sponge filter.

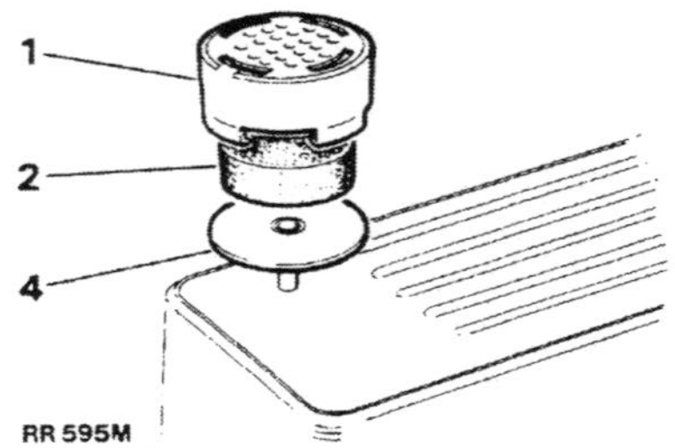

### Fit new filter

3. Insert a new filter into the plastic body.
4. Push the filter holder onto the rocker cover until it clips firmly into place.

### CHECK

Check ignition wiring and high tension leads for fraying, chafing and deterioration.

### CHECK

Check/adjust ignition timing. (See Engine Tuning Data/Procedure) using suitable electronic equipment.

### FUEL SYSTEM

Check all hose connections for leaks and hose deterioration, fit new hoses or tighten hose clamps as necessary.

# MAINTENANCE

## IGNITION TIMING

### Adjust

1. It is essential that the following procedures are adhered to. Inaccurate timing can lead to serious engine damage and additionally create failure to comply with emission regulations. If the engine is being checked in the vehicle, the air conditioning compressor must be disengaged.
2. On initial engine build, or if the distributor has been disturbed for any reason, the ignition timing must be set statically to 6° ± 1° B.T.D.C.
(This sequence is to give only an approximation in order that the engine may be started) **ON NO ACCOUNT MUST THE ENGINE BE STARTED BEFORE THIS OPERATION IS CARRIED OUT.**

### Equipment required

Calibrated Tachometer
Stroboscopic lamp

3. Couple stroboscopic timing lamp and tachometer to engine following the manufacturer's instructions.
4. Disconnect the vacuum hose from the distributor.
5. Start engine, with no load and not exceeding 3,000 rev/min run engine until normal operating temperature is reached. (Thermostat open). Check that the normal idling speed falls within the tolerance specified in the data section.
6. Idle speed for timing purposes must not exceed 800 rev/min.
7. With the distributor clamping bolt loosened turn distributor until the timing flash coincides with the timing pointer and the correct timing mark on the rim of the torsional vibration damper as shown in the engine tuning section.
8. Retighten the distributor clamping bolt securely. Recheck timing in the event that retightening has disturbed the distributor position.
9. Refit vacuum hose.
10. Disconnect stroboscopic timing lamp and tachometer from engine.

---

## IGNITION

### spark plugs

1. Take great care when fitting spark plugs not to cross-thread the plug, otherwise costly damage to the cylinder head will result.
2. Check or replace the spark plugs as applicable.
3. It is important that only the correct type of spark plugs are used for replacements.
4. Incorrect grades of plugs may lead to piston overheating and engine failure.

### To remove spark plugs proceed as follows:

5. Disconnect the battery negative lead and remove the leads from the spark plugs.
6. Remove the plugs and washers.

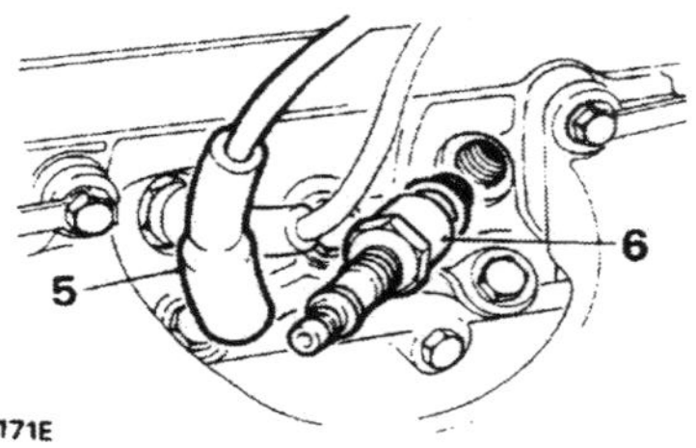

7. Set the electrode gap to the recommended clearance.
8. When pushing the leads onto the plugs, ensure that the shrouds are firmly seated on the plugs.

### Fitting H.T. leads

9. Ensure that replacement H.T. leads are refitted in their spacing cleats in accordance with the correct layout illustrated.
Failure to observe this instruction may result in cross-firing between two closely fitted leads which are consecutive in the firing order.

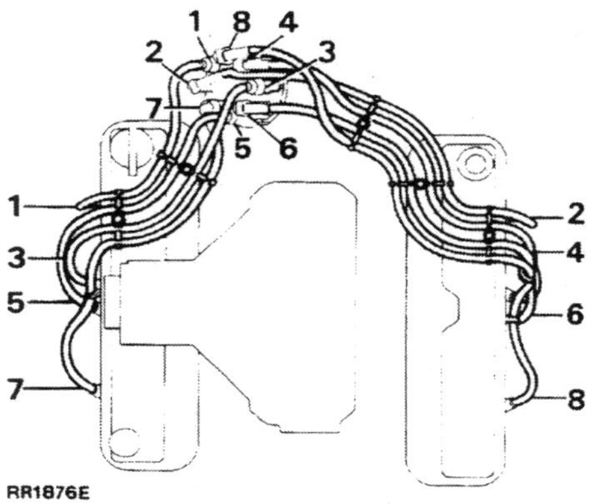

### DISTRIBUTOR-LUCAS 35DLM8

The electronic ignition employs a Lucas 35DLM8 distributor.
The internal operating parts of the distributor are pre-set at the factory and should not normally require resetting.
Adjustments should only be made, if the unit is known to be faulty or damaged. Maintenance of the distributor consists of the following items.

1. Clean outer surfaces of distributor cap to remove dirt, grease etc.
2. Unclip the cap, check cap for signs of cracking.
3. Wipe inside cap with a lint free cloth.
4. Check rotor arm, cap and flash shield for signs of tracking.
5. Apply a spot of clean engine oil into the rotor spindle before fitting the rotor arm.
**DO NOT DISTURB the clear plastic insulating cover (flash shield) which protects the magnetic pick-up module.**

## MAINTENANCE 10

### CHECK/TOP UP COOLING SYSTEM

1. To prevent corrosion of the aluminium alloy engine parts it is imperative that the cooling system is filled with a solution of water and phosphate free anti-freeze, winter or summer. Never fill or top up with plain water.

**WARNING: Do not remove the filler cap when engine is hot because the cooling system is pressurised and personal scalding could result.**

2. When removing the filler cap, first turn it anti-clockwise a quarter of a turn and allow all pressure to escape, before turning further in the same direction to lift it off.

3. With a cold engine, the correct coolant level should be up to the top of the sensor float which is approximately 1.00 inch (25 mm) below the filler neck.

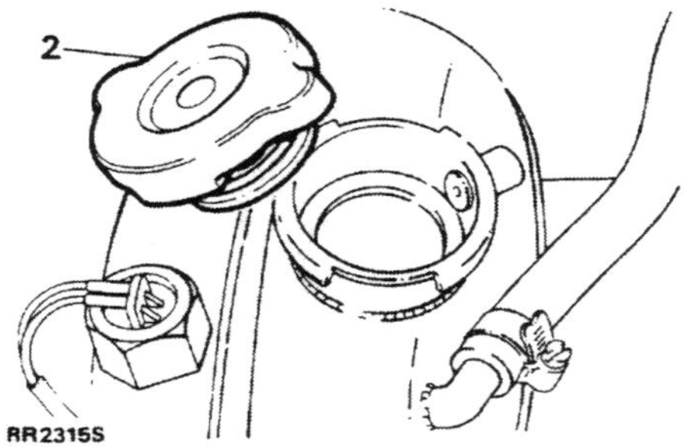

4. When replacing the filler cap it is important that it is tightened down fully, not just to the first stop. Failure to tighten the filler cap properly may result in water loss, with possible damage to the engine through over-heating. Use soft water whenever possible, if local water supply is hard, rainwater should be used.

**Check cooling/heater systems for leaks and hoses for security and condition.**

Cooling system hoses should be changed at the first signs of deterioration.

The cooling system should be drained and flushed at 2 year intervals or at the onset of the second winter. Refer to Coolant Requirements in the main workshop manual.

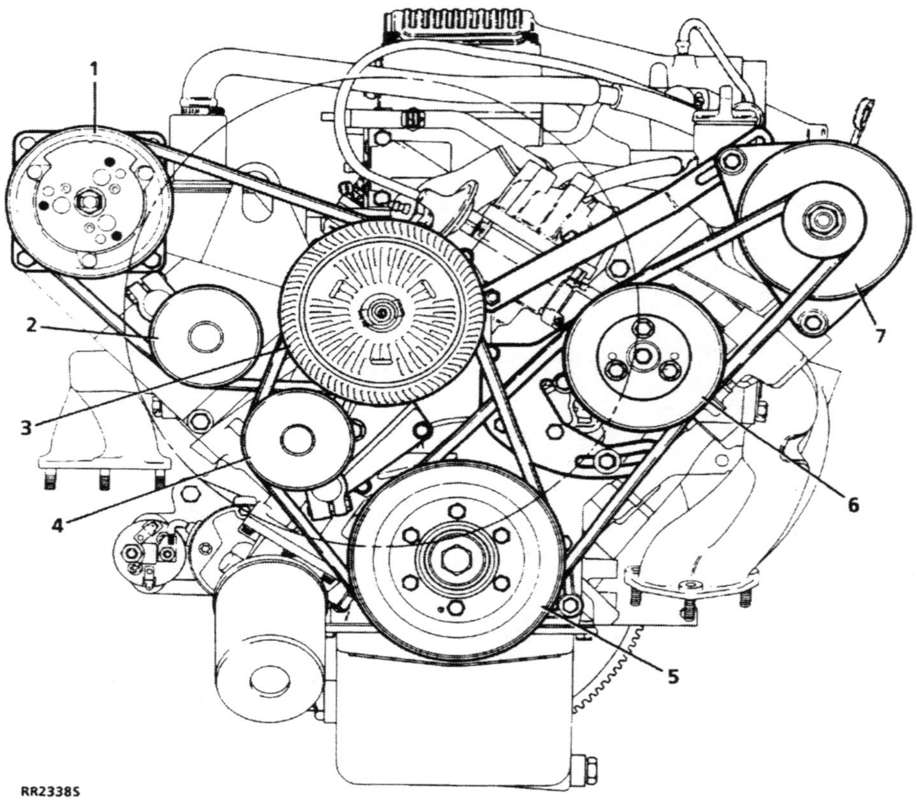

---

**RR2338S**

1. Air conditioning compressor.
2. Idler pulley
3. Viscous fan-water pump unit.
4. Idler pulley
5. Crankshaft.
6. Power steering pump.
7. Alternator

**WARNING: DISCONNECT THE BATTERY NEGATIVE TERMINAL BEFORE ADJUSTING DRIVE BELTS TO AVOID THE POSSIBILITY OF THE VEHICLE BEING STARTED.**

# MAINTENANCE

## DRIVE BELTS-adjust or fit new belts

### COMPRESSOR DRIVE BELT

The belt must be tight with not more than 4 to 6mm (0.19 to 0.25 in) total deflection when checked by hand midway between the pulleys on the longest run.

Where a belt has stretched beyond the limits, a noisy whine or knock will often be evident during operating, if necessary adjust as follows:

1. Loosen the idler pulley securing bolt.
2. Adjust the position of the idler pulley until the correct tension is obtained.
3. Tighten the securing bolt and recheck the belt tension.

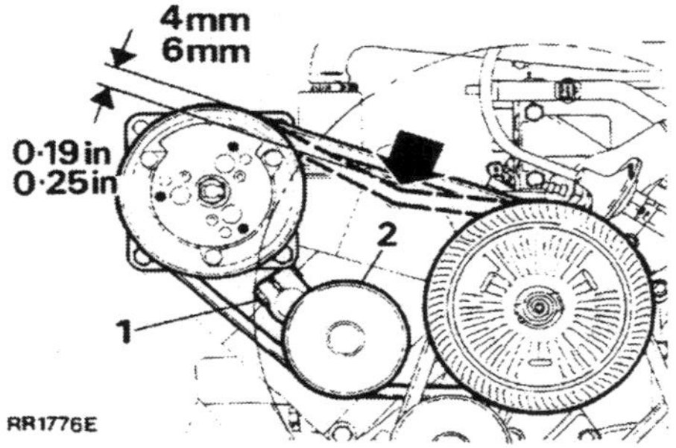

### Check driving belts, adjust or fit new belts as necessary.

1. Examine the following belts for wear and condition and fit new belts if necessary:
   (A) Crankshaft-Idler pulley-Water Pump
   (B) Crankshaft-Steering Pump
   (C) Steering Pump-Alternator

**ILLUSTRATION A**

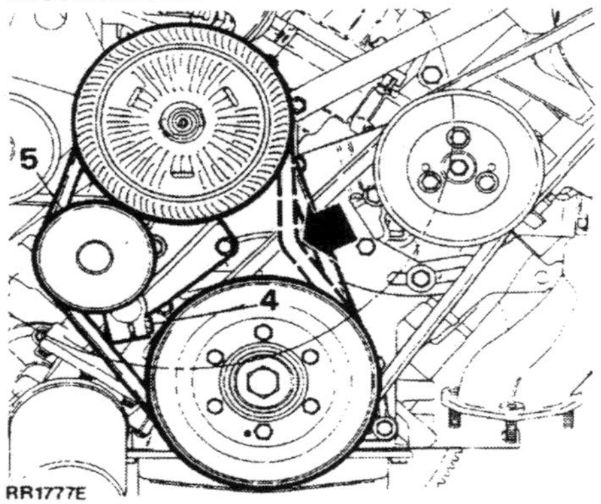

**ILLUSTRATION B**

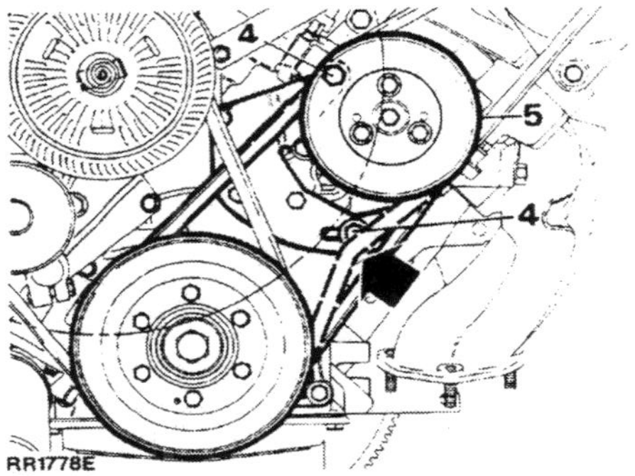

**ILLUSTRATION C**

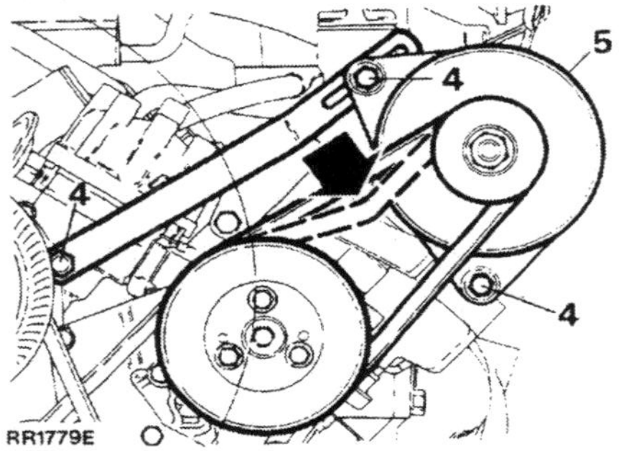

2. Each belt should be sufficiently tight to drive the appropriate auxiliary without undue load on the bearings.
3. Loosen the bolts securing the unit to its mounting bracket.
4. Loosen the appropriate pivot bolt or idler pulley and the fixing at the adjustment link where applicable.
5. Pivot the unit inwards or outwards as necessary and adjust until the correct belt tension is obtained.
6. Belt deflection should be approximately 4 to 6mm (0.19 to 0.25 in) at the points denoted by the bold arrows.
7. Tighten all unit adjusting bolts.
8. When fitting a new drive belt, tension the belt as described above. Reconnect the battery, start and run the engine at fast idle for 3 to 5 minutes, after which time the belt must be re-checked and retensioned if necessary.

---

# EMISSION CONTROL

## EMISSIONS

### EMISSION AND EVAPORATIVE CONTROL

Range Rovers are fitted during manufacture with various items of emission and evaporative control equipment to ensure that they meet stringent exhaust emission regulations.

Unauthorised replacement or modification of the emission or evaporative control equipment could seriously impair the efficiency of the emission system.

### EVAPORATIVE EMISSION CONTROL SYSTEM

This system prevents fuel vapour from reaching the atmosphere. The system consists of a fuel expansion tank located between the inner right hand body side and rear right hand wing, and an adsorption canister located in the engine compartment attached to the front right hand valance.

When the fuel expands in the fuel tank due to temperature increase it is vented into the bottom of the expansion tank, any liquid fuel can be siphoned back into the main tank. Fuel vapour is directed as the fuel cools through the outlet pipe at the top of the expansion tank to the adsorption (charcoal) canister by means of a pipe running along the underside of the vehicle.

A restrictor located in the purge line at the plenum chamber controls purge line flow.

The adsorption canister containing activated charcoal is used to store fuel vapour from the fuel tank. Filter pads are fitted above and below the charcoal to prevent ingress of foreign matter of charcoal into the purge line. Emissions from the fuel tank enter the top of the canister and the purging air enters at the bottom. The canister is purged of its vapours by the vacuum generated within the plenum chamber, the vapour being drawn into the plenum chamber and burnt with the in-going mixture.

Continued

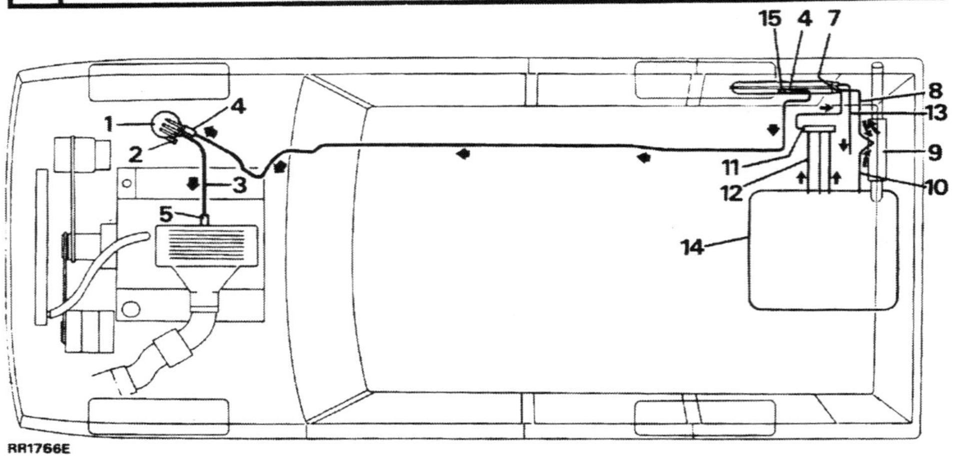

# EMISSION CONTROL

## KEY TO DIAGRAM

1. Charcoal canister
2. Air inlet to canister
3. Purge line to plenum chamber
4. Connector hoses with restrictors
5. Restrictor in purge line
6. Fuel expansion tank
7. Fuel vapour pipe from manifold
8. Breather hose with anti-surge valve
9. Fuel tank filler neck
10. Filler neck breather hose
11. Manifold
12. Fuel vapour pipes from fuel tank (3 off)
13. Pressure relief valve and hose
14. Fuel tank
15. Float/rollover valve
16. Grommet

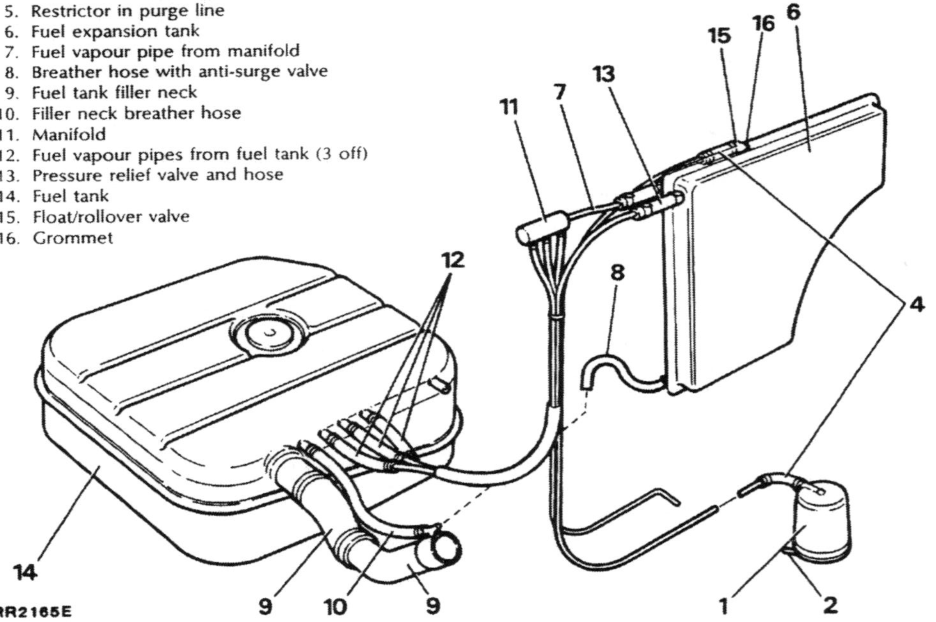

## CATALYTIC CONVERTORS

Three catalytic convertors are fitted into the exhaust system to reduce carbon monoxide, oxides of nitrogen and hydrocarbon emissions. The two down pipes from the exhaust manifolds each house an oxygen sensor located forward of the catalytic convertors.

The active constituents of the catalytic device are platinum and rhodium. In order for the device to function correctly, it is necessary to control very closely the oxygen concentration in the exhaust gas entering the catalyst. This is achieved by the use of a fuel control system which continuously monitors the oxygen content of the exhaust gas by means of the oxygen sensor and adjusts the mixture level to obtain the required oxygen content.

**Unleaded** fuel must be used on catalyst equipped vehicles, and labels to indicate this are displayed on the instrument panel and inside the fuel filler flap. The filler neck is designed to accommodate unleaded fuel pump nozzles only.

The emission control system fitted to this engine is designed to keep emissions within legislated limits, providing the engine is correctly maintained and is in sound mechanical condition.

## ADSORPTION (CHARCOAL) CANISTER

### Remove and Refit

### Removing

1. Disconnect from the canister:-
   (i) Canister line to expansion tank
   (ii) Canister purge line
2. Loosen the clamp nut screw.
3. Remove the canister.
4. Remove the short hose from the inlet vapour pipe and check that the restrictor is free from blockages.

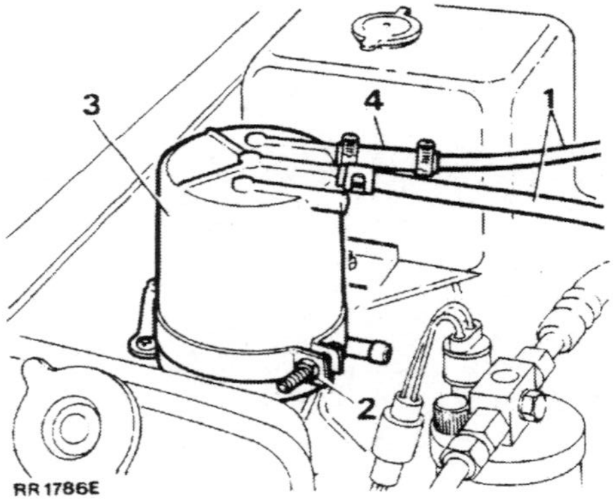

### Refitting

5. Secure the canister in the clamp.
6. Reverse instructions 1 and 2 above.

**WARNING:** The use of compressed air to clean an adsorption canister or clear a blockage in the evaporative system is very dangerous. An explosive gas present in a fully saturated canister may be ignited by the heat generated when compressed air passes through the canister.

# EMISSION CONTROL

## RANGE ROVER

## FUEL EXPANSION TANK

### Remove and Refit

The fuel expansion tank is located between the right hand rear wing and inner body side assembly, access to the tank is gained by removing the rear wing and body corner panel.

**WARNING: Ensure all necessary precautions are taken against the spillage of fuel when disconnecting the expansion tank hoses.**

### Removing

1. Depressurise the fuel system. (see Depressurising procedure in Fuel Injection System-section 19 page 34)
2. Disconnect negative battery terminal.
3. Remove the rear lamp cluster.
4. Remove the wrap around bumper end cap.
5. Remove the rear wing and corner panel assembly.
6. Release the three hose clamps and remove the three hoses from the expansion tank.
7. Release the hose clamp and remove the hose from the float valve located on top of the expansion tank.
8. Remove the three bolts retaining the bottom of the expansion tank.
9. Lift the trim covering the vehicle tool kit at the right hand side of the rear stowage area to gain access to the two expansion tank securing bolts located below the rear side glass and remove the two bolts.
10. Remove the tank from the vehicle.
11. Remove the short hose connection from the top vapour hose and check that the restrictor in the hose is free from blockages.
12. While the tank is still removed from the vehicle check the operation of the float/rollover valve as follows:

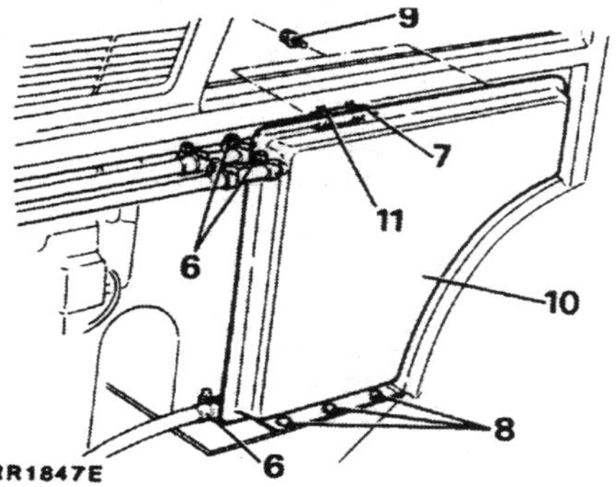

    A. Seal the top two outlet pipes.
    B. Apply air pressure at 2 p.s.i to the bottom pipe. With the tank in its upright position air flow will pass through the valve. Rotate the tank 90° onto its side air flow should not pass through the valve.
    C. Disconnect the air supply to the tank. With the bottom pipe sealed fill the tank with mineral spirit, hold the tank in its upright position, the float valve should shut off and prevent fluid passing through the valve.
    D. If the valve does not operate accordingly with the above instructions; replace the float valve.

    **NOTE: DO NOT remove the float valve unless faulty. If a new valve is fitted, always fit a new grommet.**

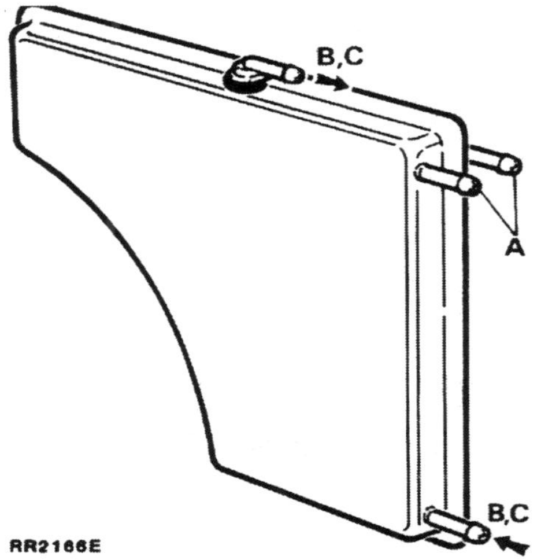

13. Before refitting the tank remove the breather hose attached to the top of the filler neck, identify filler neck to breather hose end to aid reassembly. Vigourously shake the hose and listen for valve ball movement; no sound from the valve-replace the hose assembly.

### Refitting

14. Refit the expansion tank ensuring that all hose and pipe connections are secure and that all hose clamps are securely tightened.

## VACUUM DELAY VALVE

The coloured side of the vacuum delay valve should always be fitted to the hose from the distributor.

### Remove and Refit

### Removing

1. Pull the two flexible hoses from the delay unit.
2. Remove the unit from its retaining clip and withdraw it from the engine compartment.

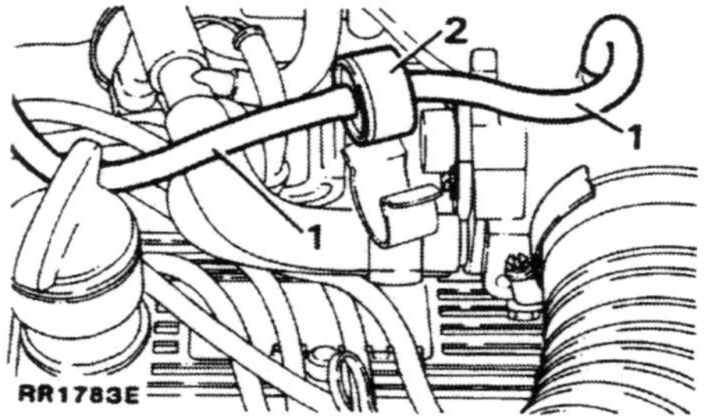

### Refitting

3. Reverse the removal instructions ensuring that the coloured side of the valve is fitted to the longer hose from the distributor.

### Test: Check Valve Air Flow

1. Attach a 10.00 ± .250 cu.inch vacuum tank to the coloured side of the valve.
2. Expose the black side to atmospheric pressure.
3. **Expected result:** The time required for the vacuum to drop from minus 20 inch Hg to minus 2 inch Hg will be 0.5 seconds maximum.
4. Vacuum recovery air flow: Attach a 22.75 ± .5 cu.inch vacuum tank to the black side of the valve.
5. Expose the coloured side to atmospheric pressure.
6. **Expected result:** The time required for the vacuum to drop from 16 inch Hg to 8 inch Hg will be 240 to 360 seconds.

### Test: External Leakage Check

1. Seal the coloured side of the valve and attach a short flexible pipe to the other end.
2. Submerge the valve in water and orally blow through the valve.
3. If any external leakage is noticeable, fit a new valve.

If the delay valve does not comply with any of the test results, replace the unit.

Continued

# FUEL INJECTION SYSTEM |19|

**FUEL INJECTION - Circuit Diagram**

1. 40 way connector to Electronic Control Unit (ECU).
2. Lambda sensor (left side - bank A).
3. Lambda sensor (right side - bank B).
4. By-pass air valve (stepper motor) (fast idle).
5. Lambda sensor screened ground.
6. Fuse 18 - main fuse panel.
7. Pick-up point E.F.I. warning symbol (instrument binnacle).
8. Fuel pump.
9. Ignition switch.
10. Speed transducer (road speed input).
11. Neutral switch (automatic gearbox) (load input).
12. Pick-up point air conditioning circuit (load input).
13. Battery.
14. Diagnostic plug.
15. In-line resistor.
16. Coil/v-e (engine RPM input).
17. Coolant temperature thermistor (sensor) (input).
18. Fuel temperature thermistor (sensor) (input).
19. Throttle potentiometer.
20. Air flow sensor.
21. Fuel pump relay.
22. Main relay.
23. Injectors-1 to 8.

= = Denotes screened ground.

**Cable colour code**

| B | Black | G | Green | R | Red | S | Grey |
|---|---|---|---|---|---|---|---|
| U | Blue | O | Orange | W | White | K | Pink |
| N | Brown | P | Purple | Y | Yellow | LG | Light green |

The last letter of a colour code denotes the tracer.

---

# EMISSION CONTROL |17|

## LAMBDA (OXYGEN) SENSOR

### Remove and Refit

The removal of the sensors from the exhaust system must only be carried out when the engine is cold.

### Removing

1. Disconnect the battery negative lead.
2. Disconnect the electrical plugs from the sensors.
3. Unscrew and remove the sensors from the two exhaust downpipes.

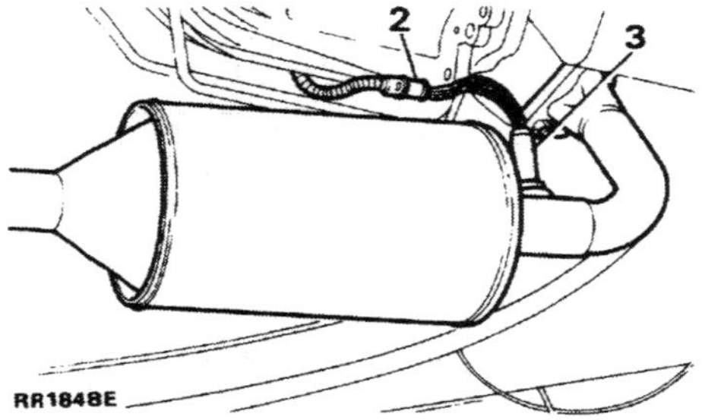

### Refitting

4. Coat the threads of the sensors with anti-seize compound.

**CAUTION: To ensure that the efficiency of the sensor is not impaired, DO NOT allow anti-seize compound to come into contact with the sensor nose.**

5. Reverse the removal procedure.

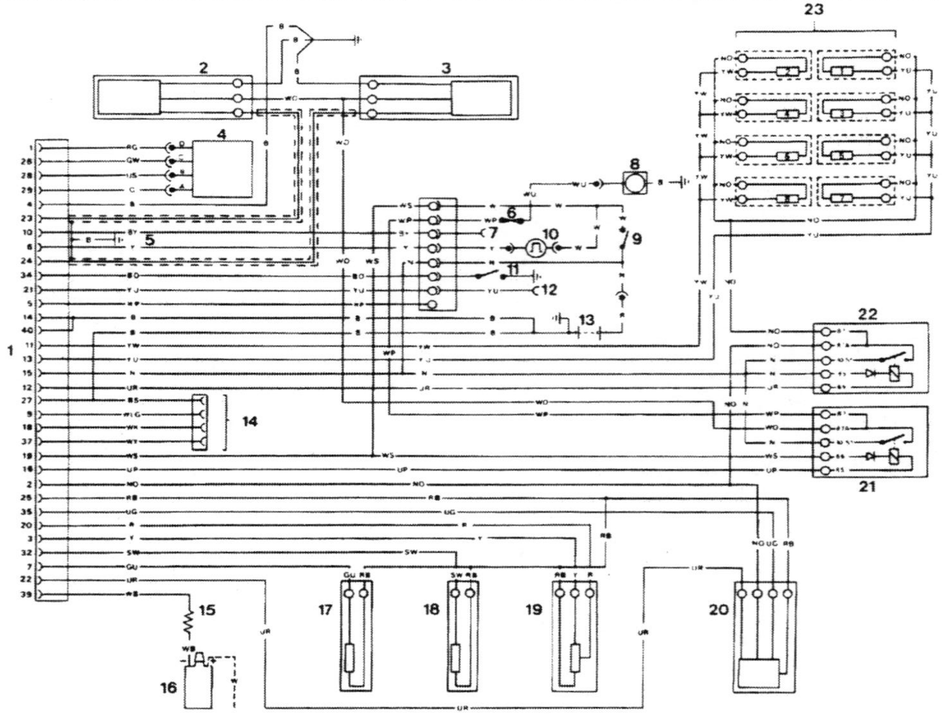

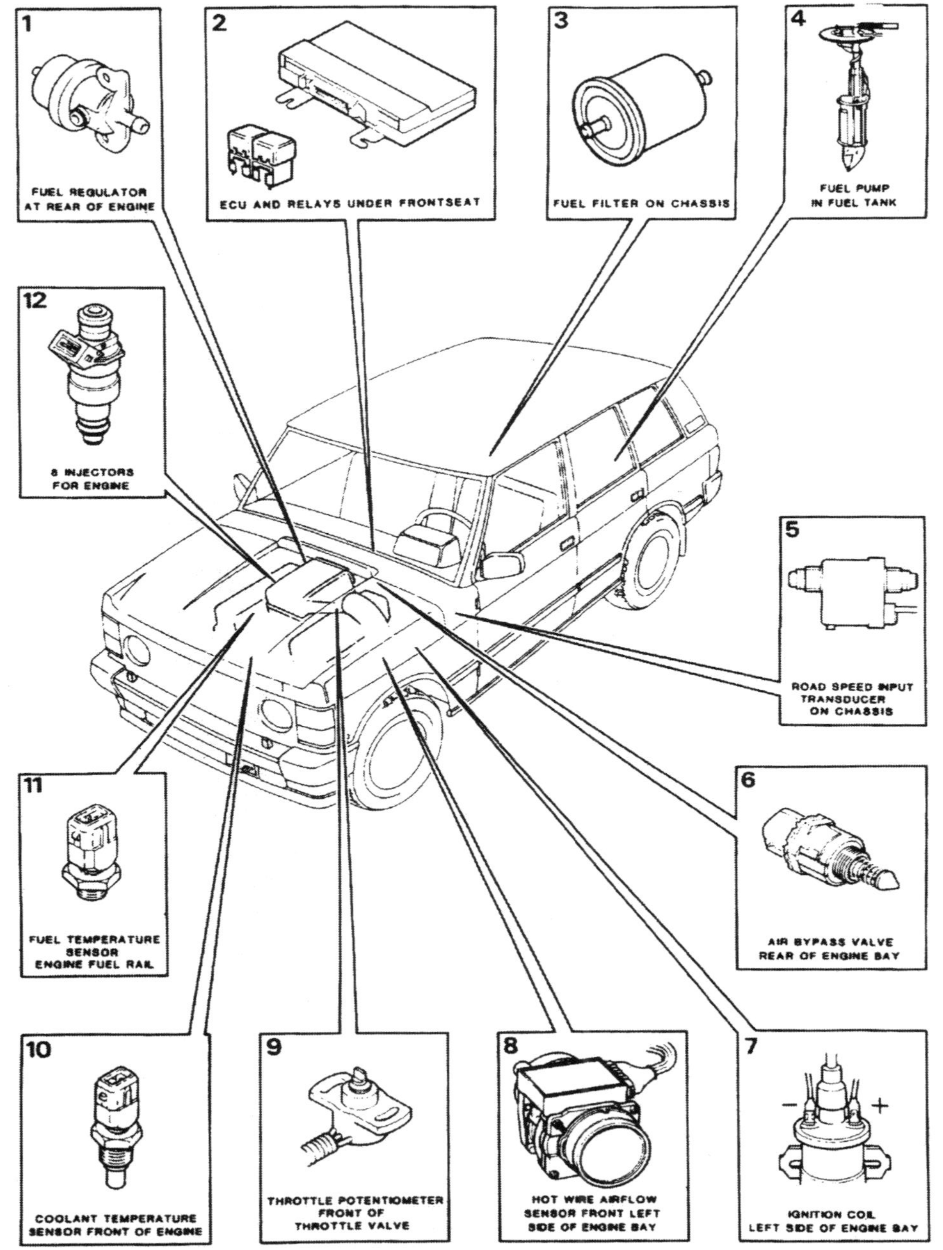

# FUEL INJECTION SYSTEM

## RANGE ROVER

### INTRODUCTION

The Electronic Fuel Injection system provides a reliable and efficient microprocessor controlled fuel management system.

The function of the system is to supply the exact amount of fuel directly into the inlet manifold according to the prevailing engine operating conditions.

To monitor these conditions, various sensors are fitted to the engine to measure engine parameters. Data from the sensors is received by the Electronic Control Unit (E.C.U.), the E.C.U. will then determine the exact amount of fuel required at any condition.

The E.C.U. having received data from the sensors produces pulses, the length of which will determine the simultaneous open time of each bank of injectors in turn, which will govern the amount of fuel injected.

### DESCRIPTION

### ELECTRONIC CONTROL UNIT-ECU

The Electronic Fuel Injection system is controlled by the E.C.U. which is located under the front right hand seat. The control unit is a microprocessor with integrated circuits and components mounted on printed circuit boards. The E.C.U. is connected to the main harness by a 40 pin plug.

### INJECTORS

The eight fuel injectors are fitted between the pressurised fuel rail and inlet manifold. Each injector comprises a solenoid operated needle valve with a movable plunger rigidly attached to the nozzle valve. When the solenoid is energised the plunger is attracted off its seat and allows pressurised fuel into the intake manifold.

### ENGINE COOLANT TEMPERATURE THERMISTOR (SENSOR)

The coolant thermistor (sensor) is located by the front left hand branch of the intake manifold. The thermistor provides engine coolant information to the E.C.U. The E.C.U. on receiving the signal from the thermistor will lengthen slightly the time that the injectors are open, and reducing this time as the engine reaches normal operating temperature.

### FUEL TEMPERATURE THERMISTOR (SENSOR)

The fuel temperature thermistor (sensor) is located in the fuel rail forward of the ram housing. The thermistor sends fuel temperature data to the E.C.U, the E.C.U. on receiving the data will adjust the injector open time accordingly to produce good hot starting in high ambient temperatures.

### BYPASS AIR VALVE (STEPPER MOTOR)

The bypass valve is screwed into a housing attached to the rear of the plenum chamber, between the plenum chamber and bulkhead. The bypass valve has two windings which enable the motor to be energised in both directions thus opening or closing the air valve as required by the E.C.U.
The bypass valve will open and allow extra air into the plenum chamber to maintain engine idle speed when the engine is under increased (Electrical and Mechanical) loads.
The bypass valve will control engine idle speed when the vehicle is stationary.

### LAMBDA SENSORS (0₂ SENSORS)

The two Lambda sensors are located forward of the catalysts mounted in the exhaust downpipes.
The sensors monitor the oxygen content of the exhaust gases and provide feedback information of the air/fuel ratio to the E.C.U. Each sensor is heated by an electrical element to improve its response time when the ignition is switched on.

Continued

# FUEL INJECTION SYSTEM

## FUEL PRESSURE REGULATOR

The fuel pressure regulator is mounted in the fuel rail at the rear of the plenum chamber. The regulator is a mechanical device controlled by plenum chamber vacuum, it ensures that fuel rail pressure is maintained at a constant pressure difference of 2.5 bar above that of the manifold. When pressure exceeds the regulator setting excess fuel is returned to the fuel tank.

## FUEL PUMP

The electric fuel pump is located in the fuel tank, and is a self priming 'wet' pump, the motor is immersed in the fuel within the tank.

## AIR FLOW SENSOR

The hot-wire air flow sensor is mounted on a bracket attached to the left hand valance, rigidly connected to the air cleaner and by hose to the plenum chamber inlet neck.

The air flow sensor consists of a cast alloy body through which air flows. A proportion of this air flows through a bypass in which two wire elements are situated: one is a sensing wire and the other is a compensating wire. Under the control of an electronic module which is mounted on the air flow sensor body, a small current is passed through the sensing wire to produce a heating effect. The compensating wire is also connected to the module but is not heated, but reacts to the temperature of the air taken in, as engine intake air passes over the wires a cooling effect takes place.

The electronic module monitors the reaction of the wires in proportion to the air stream and provides output signals in proportion to the air mass flow rate which are compatible with the requirements of the E.C.U.

## THROTTLE POTENTIOMETER

The throttle potentiometer is mounted on the side of the plenum chamber inlet neck and is directly coupled to the throttle valve shaft.

The potentiometer is a resistive device supplied with a voltage from the E.C.U. Movement of the throttle pedal causes the throttle valve to open, thus rotating the wiper arm within the potentiometer which in turn varies the resistance in proportion to the valve position. The E.C.U. lengthens the injector open time when it detects a change in output voltage (rising) from the potentiometer.

In addition the E.C.U. will weaken the mixture when it detects the potentiometer output voltage is decreasing under deceleration and will shorten the length of time the injectors are open.

When the throttle is fully open, the E.C.U. will detect the corresponding throttle potentiometer voltage and will apply full load enrichment. This is a fixed percentage and is independent of temperature. Full load enrichment is also achieved by adjusting the length of the injector open time.

When the throttle is closed, overrun fuel cut off or idle speed control may be facilitated dependant on other inputs to the E.C.U.

## ROAD SPEED TRANSDUCER

The road speed transducer is fitted between the upper and lower speedometer cables. It is mounted on a bracket located on the left hand chassis side member adjacent to the rear engine mounting. The transducer provides road speed data to the ECU. The ECU in turn detects vehicle movement from the road speed input and ensures that idle speed control mode is disengaged. Should the speed transducer fail in service the ECU idle speed control would become erratic.

## RELAYS

The two electronic fuel injection relays are located under the front right hand seat mounted forward of the E.C.U. The main relay is energised via the E.C.U when the ignition is switched on and supplies current to the fuel injection system. The fuel pump relay is energised by the E.C.U. which in turn operates the fuel pump to pressurise the fuel system.

## E.F.I. WARNING SYMBOL (Instrument binnacle)

An E.F.I. warning symbol incorporated into the instrument binnacle will illuminate when the E.C.U. detects that it cannot maintain correct air/fuel ratio due to a fault in one of the following fuel injection system components.

Air flow sensor.
Lambda sensor.
Water temperature thermistor. (sensor)
Throttle potentiometer.

If the symbol illuminates when the engine is idling or the vehicle is being driven it indicates a failure of one of the four functions, the vehicle should be driven with care, and the cause rectified, refer to test procedure for the particular functions. Should one of the functions fail, the vehicle can still be driven due to a limp home feature incorporated into the fuel injection system.

# FUEL INJECTION SYSTEM

## FUEL INJECTION SYSTEM

**CAUTION:** The fuel system incorporates fine metering components that would be affected by any dirt in the system; therefore it is essential that working conditions are scrupulously clean. If it is necessary to disconnect any part of the fuel injection system, the system MUST be depressurised. All openings left open after the removal of any component from the fuel system, MUST be sealed off to prevent ingress of dirt.

### ENGINE SETTING PROCEDURE

If a major overhaul has been undertaken on the fuel injection/engine system, the following check and adjustments must be carried out before attempting to start the engine.

A. **Throttle potentiometer setting** - see 'Throttle switch - potentiometer setting procedure.
B. **Spark plug gaps** - see 'Section 04 General Specification'.
C. **Throttle levers** - see 'Throttle lever setting procedure'.
D. **Ignition timing** - static - see 'Section 86 Electrical'.

**CAUTION: IF THE ENGINE IS MISFIRING, IT SHOULD BE IMMEDIATELY SHUT DOWN AND THE CAUSE RECTIFIED. FAILURE TO DO SO WILL RESULT IN IRREPARABLE DAMAGE TO THE CATALYSTS.**

**NOTE:** If the previous checks and adjustments are satisfactory but the engine will not start the ignition and fuel injection electrical circuitry must be checked using the appropriate recommended equipment.

## RANGE ROVER

### Recommended Equipment -

Lucas 'Electronic Ignition Analyser'
Lucas Part Number - YWB 119.

Lucas 'E.F.I. Throttle Potentiometer Adjustment Gauge'
Lucas Part Number - YWB 121

Lucas Diagnostic Equipment
Lucas Part Number - 60600965 (complete kit)

Individual part numbers for the above kit are as follows:

Hand held test unit - Model 2HHT
Lucas Part Number - 84772

Interface unit - Model 2IU
Lucas Part Number - 84773

Serial link lead
Lucas Part Number - 54744753

Memory card
Lucas Part Number - 54744754

Operating manual
Lucas Part Number - XXB825

Plastic case
Lucas Part Number - 54744755

**NOTE:** The Lucas diagnostic equipment can be connected to the diagnostic plug located by the E.C.U.

Use in conjunction with the Lucas Operating Instruction Manuals.

If the above equipment is unavailable the tests can be carried out using a multi-meter, following the instructions given in the charts.

**CAUTION: Ensure the multi-meter is correctly set to volts or ohms, dependent upon which test is being undertaken.**

Carry out the following static checks before undertaking the continuity procedure:-

A. **Fuse 18** - in main fuse panel - is intact.
B. **Fuel** - ample fuel in fuel tank.
C. **Battery Condition** - state of charge.
D. **Air Leaks** - no unmetered air entering engine system.
E. **Electrical Connections** - dry, clean and secure.

## RANGE ROVER

### FUEL INJECTION SYSTEM

### CONTINUITY TEST PROCEDURE

The continuity procedure and instructions on the following pages must be followed precisely to prevent damage occurring to any of the fuel system components.

To enable the tests to be carried out when the 40 way multi-plug is connected to the E.C.U, it is necessary to remove the two screws securing the shroud to the plug to enable the multi-meter probes to be inserted into the back of the appropriate pin.

**CAUTION: Tests that require the plug to be removed from the E.C.U., must also have the meter probes inserted into the back of the plug. If the probes are inserted into the plug sockets, damage will occur to the sockets resulting in poor connections when the plug is reconnected.**

### TESTING

1. Remove the E.C.U., and harness plug from beneath the front right hand seat, access is gained through the rear opening of the seat base.
2. Remove the seal and two retaining screws from the front of the plug. Maneouver the plug shroud along the harness until there is enough clearance enabling meter probes to be inserted into the back of the plug.
3. There are 6 pin numbers, 1, 13, 14, 27, 28, 40 moulded onto the rear of the plug for pin position identification as shown in the illustration below. (for clarity the electrical leads have been omitted).

Pins 1 to 13 top row.
Pins 14 to 27 centre row.
Pins 28 to 40 bottom row.

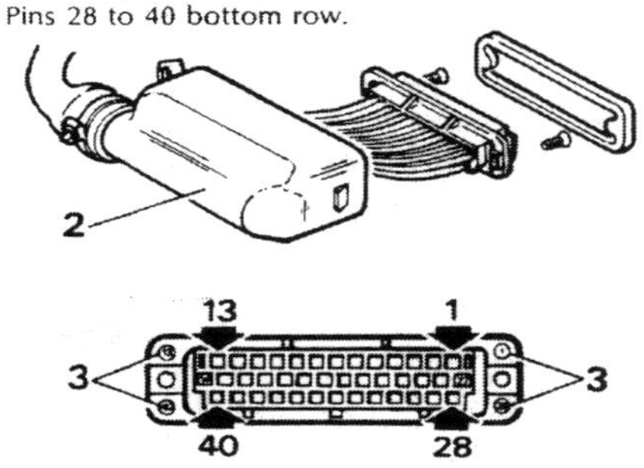

| PIN NOS. | CABLE COLOUR |
|---|---|
| 1. | Red/green |
| 2. | Brown/orange |
| 3. | Yellow |
| 4. | Black |
| 5. | Brown/purple |
| 6. | Yellow |
| 7. | Green/blue |
| 8. | Not used |
| 9. | White/light green |
| 10. | Black/Yellow |
| 11. | Yellow/white |
| 12. | Blue/red |
| 13. | Yellow/blue |
| 14. | Black |
| 15. | Brown |
| 16. | Blue/purple |
| 17. | Not used |
| 18. | White/pink |
| 19. | White/grey |
| 20. | Red |
| 21. | Yellow/blue |
| 22. | Blue/red |
| 23. | Blue |
| 24. | Blue |
| 25. | Red/black |
| 26. | Green/white |
| 27. | Black/grey |
| 28. | Blue/grey |
| 29. | Orange |
| 30. | Not used |
| 31. | Not used |
| 32. | Grey/white |
| 33. | Not used |
| 34. | Black/orange |
| 35. | Blue/green |
| 36. | Not used |
| 37. | White/yellow |
| 38. | Not used |
| 39. | White/black |
| 40. | Black |

The last colour denotes the wire tracer colour.

Continued

RR2320E

## RANGE ROVER

# 19 FUEL INJECTION SYSTEM

### TESTS - Using a Multi-Meter

The following continuity tests are intended as a guide to identifying where a fault may be within a circuit; reference should be made to the fuel injection circuit diagram for full circuit information.

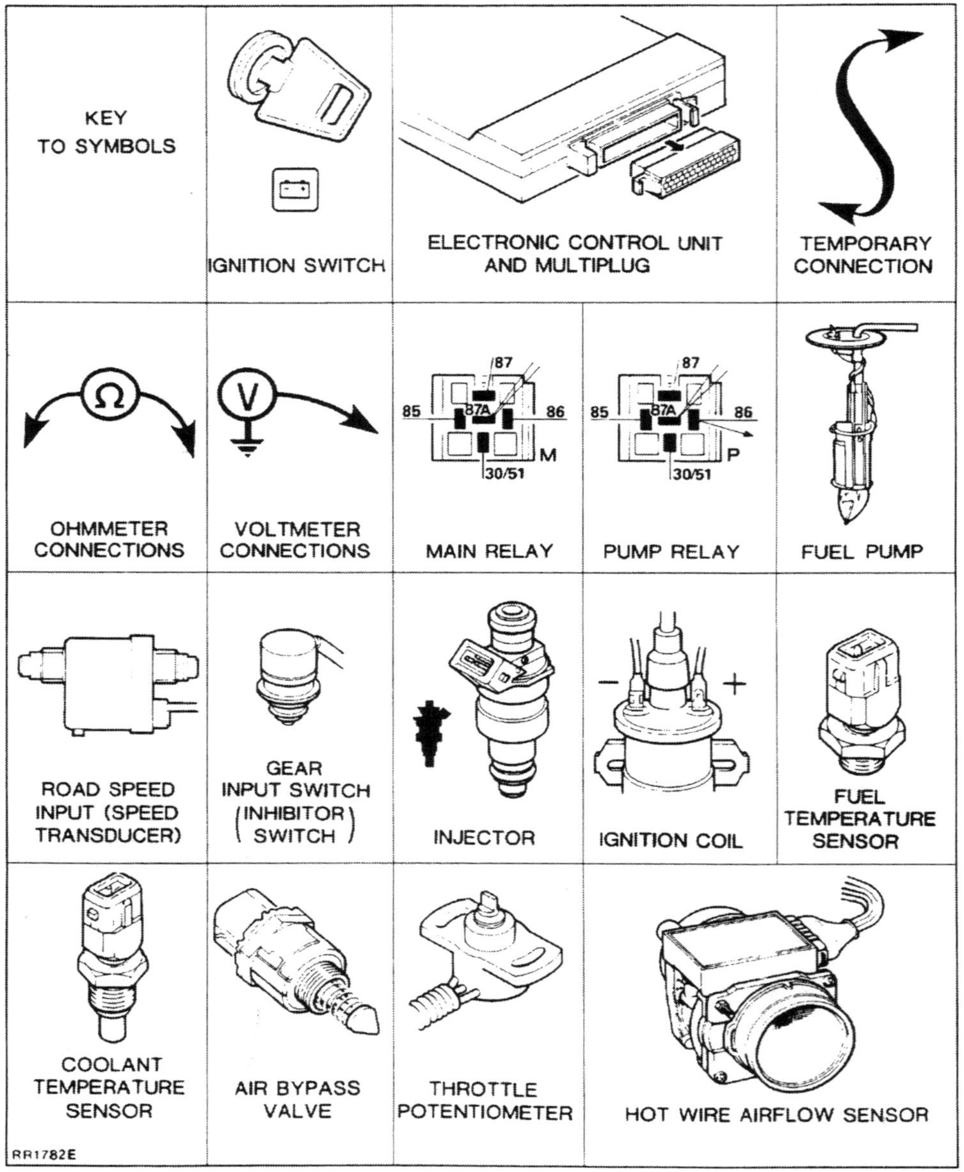

**KEY TO SYMBOLS**

| IGNITION SWITCH | ELECTRONIC CONTROL UNIT AND MULTIPLUG | PUMP RELAY | FUEL PUMP |
| --- | --- | --- | --- |
| VOLTMETER CONNECTIONS | MAIN RELAY | INJECTOR | FUEL TEMPERATURE SENSOR |
| OHMMETER CONNECTIONS | GEAR INPUT SWITCH (INHIBITOR SWITCH) | THROTTLE POTENTIOMETER | HOT WIRE AIRFLOW SENSOR |
| | ROAD SPEED INPUT (SPEED TRANSDUCER) | AIR BYPASS VALVE | |
| | COOLANT TEMPERATURE SENSOR | | |
| | | TEMPORARY CONNECTION | IGNITION COIL |

RR1782E

---

## RANGE ROVER

# 19 FUEL INJECTION SYSTEM

**NOTE:** All tests are carried out from the electronic control unit (ECU) harness multi-plug unless stated otherwise in the test procedure.

### TEST PROCEDURE

**1.** Check battery supply to ECU

**RESULTS** - Check cables and units shown in **bold**

Voltmeter reading of battery volts - (minimum battery voltage 10 volts) Proceed to Test 2
Voltmeter reading of zero volts
Check:-

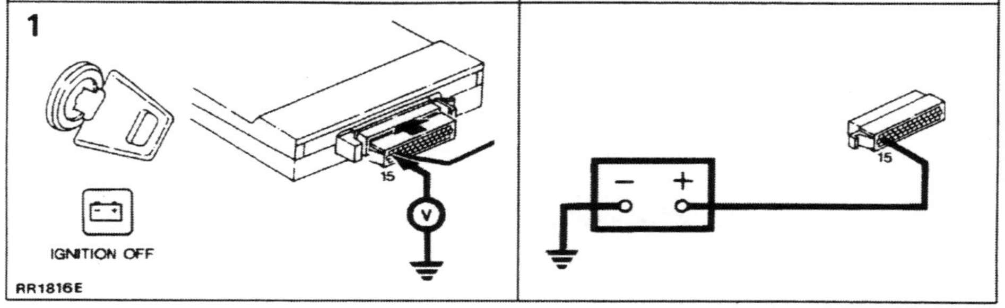

RR1816E

### TEST PROCEDURE

**2.** Check ignition supply to ECU

**RESULTS** - Check cables and units shown in **bold**

Voltmeter reading of battery volts - (minimum battery voltage 10 volts) Proceed to Test 3
Incorrect reading check:-

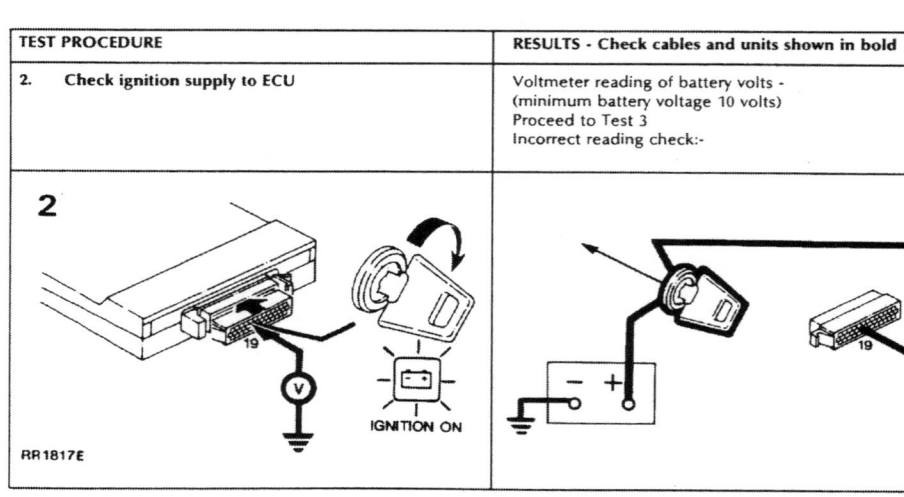

RR1817E

Continued

326

# FUEL INJECTION SYSTEM 19

## RANGE ROVER

| TEST PROCEDURE | RESULTS - Check cables and units shown in bold |
|---|---|
| 3. Check operation of Main relay | Voltmeter reading of battery volts - Proceed to Test 5 |
| | Voltmeter reading of zero volts - Proceed to Test 4 |

**3**

RR1818E

| TEST PROCEDURE | RESULTS - Check cables and units shown in bold |
|---|---|
| 4. Fault Diagnosis Main relay circuits | A. Voltmeter reading of battery volts - Check:- If OK Suspect ECU |
| | B. Voltmeter reading of zero volts Check:- |

**4**

RR1819E

---

# FUEL INJECTION SYSTEM 19

## RANGE ROVER

| TEST PROCEDURE | RESULTS - Check cables and units shown in bold |
|---|---|
| 5. Check operation of pump relay | Listen for audible 'click' from pump relay. If O.K - Proceed to Test 7 |
| | No audible 'click' from pump relay Check:- If OK proceed to Test 6. |

**5**

RR1820E

| TEST PROCEDURE | RESULTS - Check cables and units shown in bold |
|---|---|
| 6. Fault diagnosis Pump relay circuits | Voltmeter reading of battery volts - Suspect ECU |
| | Voltmeter reading of zero volts Check:- |

**6**

RR1821E

Continued

# FUEL INJECTION SYSTEM 19

## RANGE ROVER

| TEST PROCEDURE | RESULTS - Check cables and units shown in bold |
|---|---|
| 7. Check operation of Fuel pump<br><br>NOTE: It is not possible to place the multi-meter probes directly onto the pump terminals. A link lead attached to the pump is accessible behind the rear left hand wheel located between the chassis and stowage area floor panel.<br><br>KEY:<br>1. Harness plug - link lead<br>2. Fuse 18 | Voltmeter reading of battery volts - Pump operating - Proceed to Test 8<br><br>(A) Voltmeter reading of battery volts - Pump not operating Check:-<br><br>(B) Voltmeter reading of zero volts Check:- |

7

RR1822E

| TEST PROCEDURE | RESULTS - Check cables and units shown in bold |
|---|---|
| 8. Check engine speed signal Cable and resistor | Voltmeter reading of battery volts - Proceed to Test 9<br><br>Voltmeter reading of zero volts Check:- |

8

RR1823E

| TEST PROCEDURE | RESULTS - Check cables and units shown in bold |
|---|---|
| 9. Check injectors, Injector circuit<br><br>(Pin 13 left bank injectors 1,3,5,7). | Ohm-meter reading of 4-5 Ohms - Proceed to Test 10<br><br>Ohm-meter reading of 5-6 Ohms - Suspect 1 injector<br>Ohm-meter reading of 8-9 Ohms - Suspect 2 injectors<br>Ohm-meter reading of 16-17 Ohms - Suspect 3 injectors<br>Check for open circuit injector(s) or wiring faults.<br><br>Ohm-meter reading of Infinity Check: |

9

RR1824E

Continued

# FUEL INJECTION SYSTEM | 19

## RANGE ROVER

| TEST PROCEDURE | RESULTS - Check cables and units shown in bold |
|---|---|
| 10. Check injectors Injector circuit <br><br> (Pin 11 rightbank injectors 2,4,6,8) | Ohm-meter reading of 4-5 Ohms - Proceed to Test 11. <br><br> Ohm-meter reading of 5-6 Ohms - Suspect 1 injector <br> Ohm-meter reading of 8-9 Ohms - Suspect 2 injectors <br> Ohm-meter reading of 16-17 Ohms - Suspect 3 injectors <br> Check for open circuit injector(s) or wiring faults. <br><br> Ohm-meter reading of Infinity Check: |

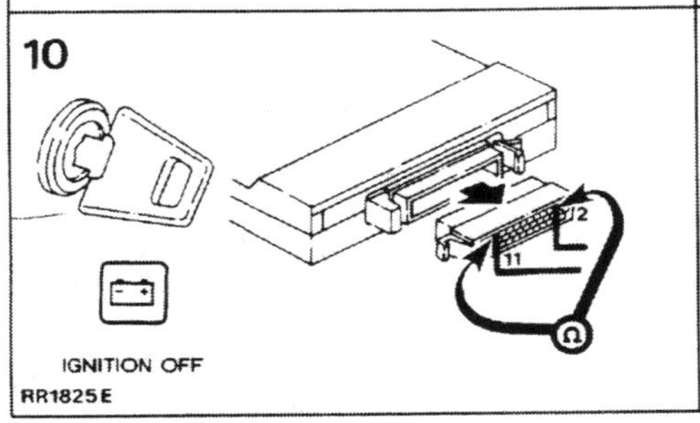

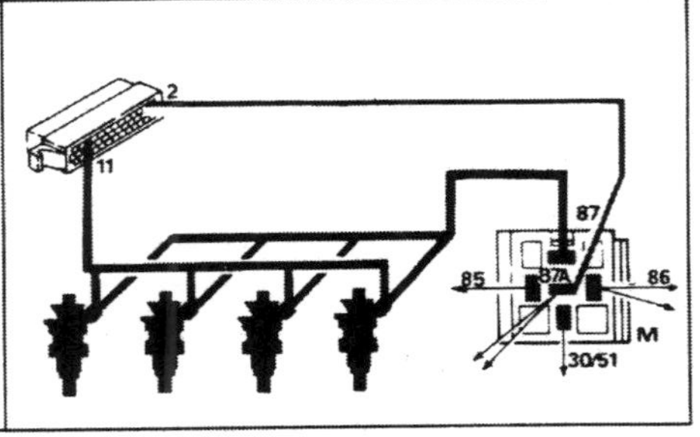

IGNITION OFF
RR1825E

| TEST PROCEDURE | RESULTS - Check cables and units shown in bold |
|---|---|
| 11. Check fuel temperature thermistor (sensor) | Correct reading-temperature to resistance <br> - Proceed to Test 12 <br> (Refer to Temperature Conversion Charts in Test 12 <br><br> Incorrect Ohm-meter reading Check |

IGNITION OFF
RR1826E

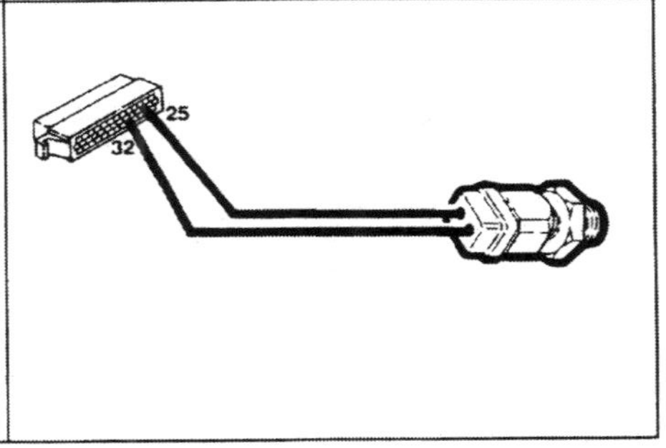

---

## RANGE ROVER

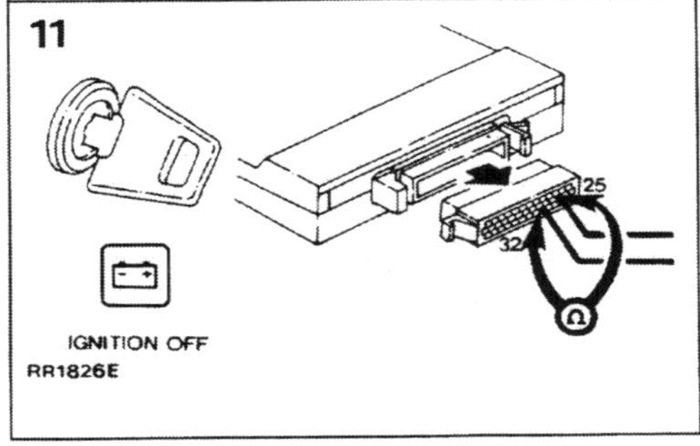

## FUEL INJECTION SYSTEM | 19

| TEST PROCEDURE | RESULTS - Check cables and units shown in bold |
|---|---|
| 12. Check coolant temperature thermistor (sensor) | Correct reading-Temperature to resistance <br> - Proceed to Test 13 <br> (Refer to Temperature Conversion Chart below). <br><br> Fuel and Coolant Temperature    Ohm-meter Reading Should be Ohms <br> °C    °F <br> -10°    14°    9100 - 9300 <br> 0°    32°    5700 - 5900 <br> 20°    68°    2400 - 2600 <br> 40°    104°    1100 - 1300 <br> 60°    140°    500 - 700 <br> 80°    176°    300 - 400 <br> 100°    212°    150 - 200 <br><br> Incorrect Ohm-meter reading Check:- |

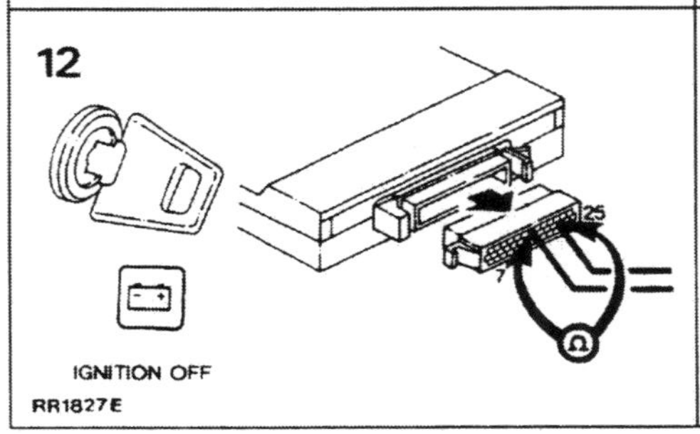

IGNITION OFF
RR1827E

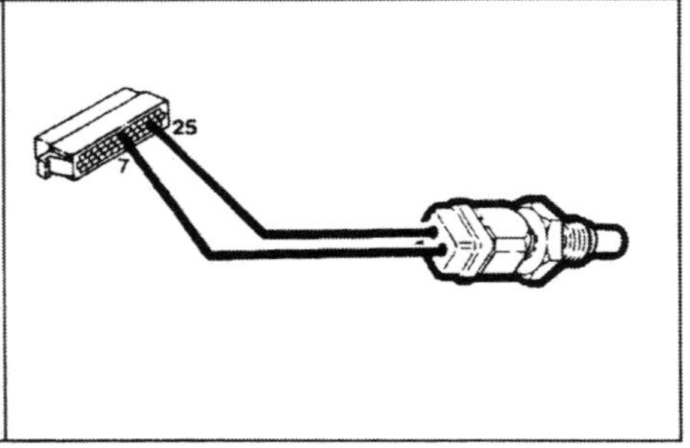

Continued

329

# FUEL INJECTION SYSTEM | 19

## RANGE ROVER

| TEST PROCEDURE | RESULTS - Check cables and units shown in bold |
|---|---|
| 13. Check air bypass valve - Part 1 | **Ohm-meter reading of 48-58 Ohms - Proceed to Test 14**<br><br>Incorrect reading Check:- |

13.

IGNITION OFF
RR1828E

| TEST PROCEDURE | RESULTS - Check cables and units shown in bold |
|---|---|
| 14. Check air bypass valve - Part 2 | **Ohm-meter reading of 48-58 Ohms - Proceed to Test 15**<br><br>Incorrect reading Check:- |

14.

IGNITION OFF
RR1829E

| TEST PROCEDURE | RESULTS - Check cables and units shown in bold |
|---|---|
| 15. Check throttle potentiometer - Part 1 | **Ohm-meter reading of 5000 Ohms - Proceed to Test 16**<br><br>Incorrect reading of Infinity Check:- |

15.

IGNITION OFF
RR1830E

| TEST PROCEDURE | RESULTS - Check cables and units shown in bold |
|---|---|
| 16. Check throttle potentiometer - Part 2 | **Correct voltmeter readings- Proceed to Test 17**<br><br>Throttle closed    Throttle open<br>0.29 )       ( 4.6 Volts<br>     ) smooth (<br>0.36 ) swing ( 5.0 Volts<br><br>Incorrect voltmeter readings Check:- |

16.

IGNITION ON
RR1831E

Continued

330

# FUEL INJECTION SYSTEM 19

| TEST PROCEDURE | RESULTS - Check cables and units shown in bold |
|---|---|
| 17. Check output of Airflow sensor | Voltmeter reading of 0.3-0.6 volts- Proceed to Test 18 |
| | Incorrect voltmeter reading Check:- |

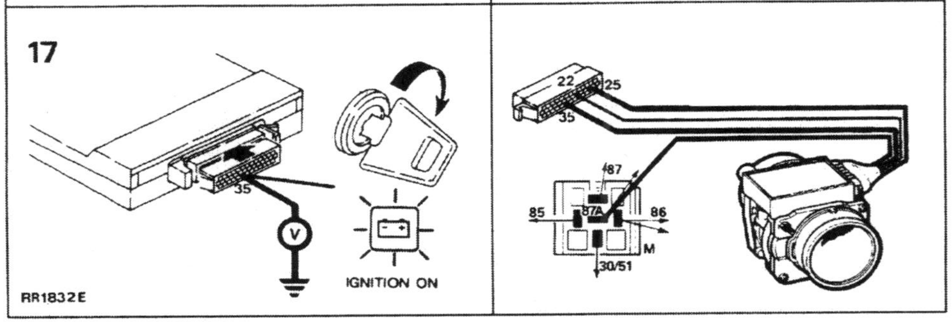

17.

RR1832E

**PRECAUTION:**

Depressurise the fuel system when fitting the fuel pressure gauge or disconnecting/replacing fuel system components.

**CAUTION:** Thoroughly clean the immediate area around the fuel filter and hose connections before disconnecting the fuel feed line from the filter. Failure to do so could cause foreign matter to be present in the fuel system which would be detrimental to the fuel system components.

**WARNING:** The spillage of fuel from the fuel filter is unavoidable when disconnecting the fuel feed line, ensure that all necessary precautions are taken to prevent fire and explosion due to fuel vapour and fuel seepage.

**DEPRESSURISING PROCEDURE**

a) Ignition off, pull pump relay off its terminal block.
b) Crank engine for a few seconds - engine may fire and run until fuel pressure is reduced.
c) Switch off the ignition.
d) Connect fuel pressure gauge in the fuel supply line between the fuel rail and the fuel filter, adjacent to the filter (see Test 18).
e) Reconnect the pump relay.

331

---

# FUEL INJECTION SYSTEM 19

| TEST PROCEDURE | RESULTS - Check cables and units shown in bold |
|---|---|
| 18. Check fuel system pressure Service tool 18G 1500<br><br>NOTE: Insert the pressure gauge in the fuel feed line immediately after the fuel line filter. The filter is located beneath the right hand rear wheel arch attached to the chassis | (A) Expected reading 2,4-2,6 kgf/cm² <br><br>(34.0-37.0 p.s.i.)<br><br>(B) Pressure drop-max 0.7 kgf/cm² (10 p.s.i.) in one minute |

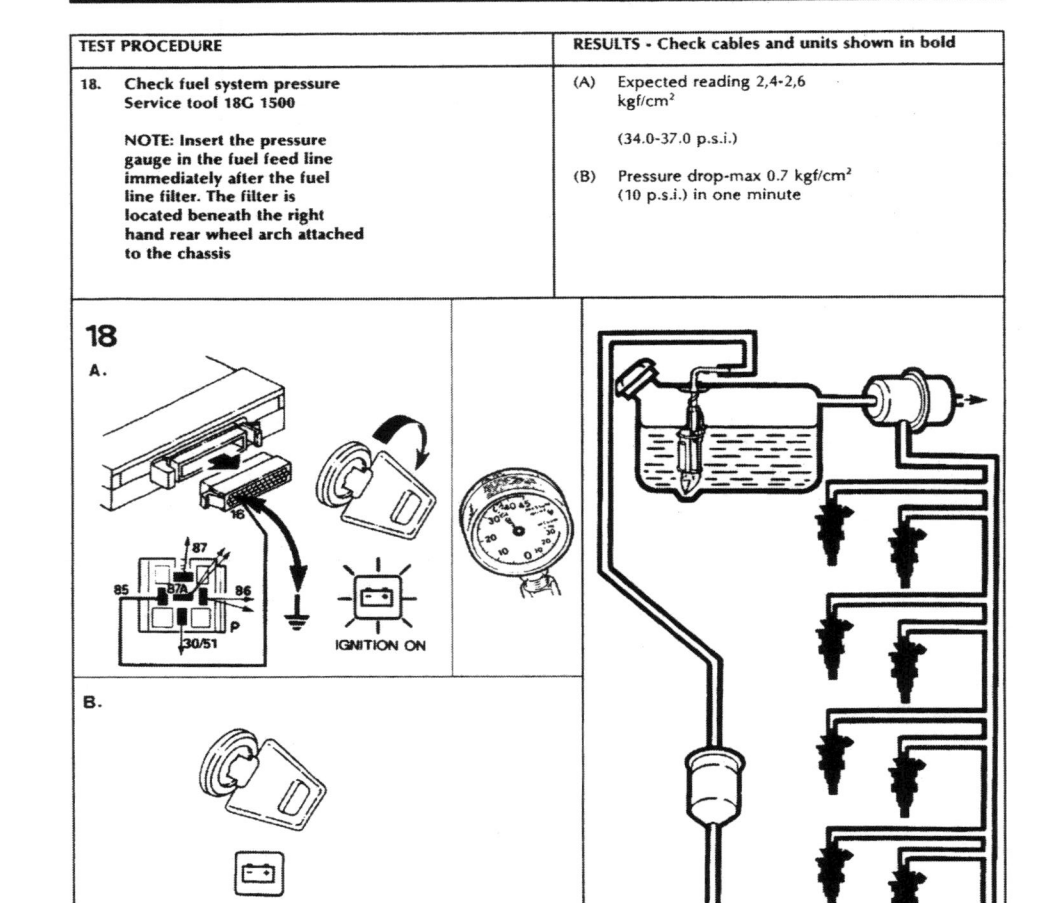

18.
A.

B.

RR1833E

Continued

## FUEL INJECTION SYSTEM | 19

**RANGE ROVER**

| TEST PROCEDURE | RESULTS - Check cables and units shown in bold |
|---|---|
| 19. Check for leaking injector<br><br>NOTE: Before removing any of the injectors, remove and examine the spark plugs, check for consistent colouration of plugs. A leaking injector will result in the appropriate spark plug being 'sooted up'.<br><br>Remove all injectors from manifold but do not disconnect from fuel rail | **WARNING: Ensure that all necessary precautions are taken to prevent fire and explosion.**<br><br>Replace any injector which leaks more than 2 drops of fuel per minute. |

19

RR1834E

---

## FUEL INJECTION SYSTEM | 19

**RANGE ROVER**

| TEST PROCEDURE | RESULTS - Check cables and units shown in bold |
|---|---|
| 20. Check for injector operation<br>Left bank injectors 1,3,5,7 | **WARNING: Ensure that all necessary precautions are taken to prevent fire and explosion.**<br><br>Repeat test for other injectors<br>Replace any injector which does not operate.<br>**NOTE: Fuel flow is 167cc minimum per minute per injector** |

20

RR1835E

| TEST PROCEDURE | RESULTS - Check cables and units shown in bold |
|---|---|
| 21. Right bank injectors 2,4,6,8 | **WARNING: Ensure that all necessary precautions are taken to prevent fire and explosion.**<br><br>Repeat test for other injectors<br>Replace any injector which does not operate<br>**NOTE: Fuel flow is 167cc minimum per minute per injector** |

21

RR1836E

Continued

# 19 FUEL INJECTION SYSTEM

## RANGE ROVER

### TEST PROCEDURE

22. Check gear switch input

| TEST PROCEDURE | RESULTS - Check cables and units shown in bold |
|---|---|
| 22. Check gear switch input | Voltmeter reading of zero volts - Neutral and park |
| | Voltmeter reading of 4.5-5.0 Volts - R.D.3.2.1 - Proceed to Test 23 |
| | Incorrect reading Check: |

RR1837E

---

# 19 FUEL INJECTION SYSTEM

## RANGE ROVER

### TEST PROCEDURE

23. Check road speed input

NOTE: Raise and rotate the left hand rear road wheel slowly

| TEST PROCEDURE | RESULTS - Check cables and units shown in bold |
|---|---|
| 23. Check road speed input | Voltmeter reading of 0 to 12V fluctuating 6 times per revolution - Proceed to Test 24 |
| | Incorrect reading Check: |

RR1838E

### TEST PROCEDURE

24. Check Lambda sensor heater coils

NOTE: Remove pump relay from its connector

| TEST PROCEDURE | RESULTS - Check cables and units shown in bold |
|---|---|
| 24. Check Lambda sensor heater coils | Ohm-meter reading of 2.65-3.35 Ohms Incorrect reading Check: |
| | NOTE: A reading of 5.3 to 6.7 Ohms indicates a faulty Lambda sensor |

RR1840E

After completing the tests with either the 'Diagnostic' equipment or multi-meter, re-test the vehicle to ensure the faults have been rectified. If faults still persist, recheck using the Lucas diagnostic equipment.

# FUEL INJECTION SYSTEM

## ENGINE TUNING PROCEDURE

Before carrying out 'Engine Tuning' on fuel injection vehicles, it is important that all other engine related setting procedures are undertaken first; air flow sensor to air cleaner correctly fitted, ignition and throttle potentiometer correctly set; all hoses correctly fitted and secured.

These checks should be carried out with the engine coolant temperature between 80° to 95°C (176° to 203°F).

### CHECK AND ADJUST IGNITION TIMING

1. Check that ignition timing is at 6° ± 1° BTDC.
2. Timing to be checked when engine speed is less than 800 rev/min using a stroboscopic lamp.
3. If adjustment is necessary, loosen the distributor clamp nut and rotate clockwise to retard or counter-clockwise to advance. When the required setting has been attained, tighten the clamp nut and re-check the setting.

**NOTE: Timing to be checked with vacuum hose connected.**

**IDLE SPEED is preset at the factory and should not normally require adjustment.**

**CAUTION:**

A. If engine fails to start within a maximum time of 12 seconds the cause must be rectified. Following rectification the engine must be run at 1500 rpm (no load) for 3 minutes to clear any accumulation of fuel in the catalysts.

B. If the engine is misfiring, it should be immediately shut down and the cause rectified.

**Failure to comply with A or B will result in irreparable damage to the catalysts.**

## AIR CLEANER

### Remove and Refit

#### Removing

1. Release the two clamps securing the air cleaner to the airflow sensor.
2. Release the two nuts and bolts securing the air cleaner to the left hand valance mounting bracket.
3. Detach the airflow sensor from the air cleaner, and lay carefully to one side.
4. Detach the air cleaner from the centre mounting bracket and withdraw from the engine compartment.
5. Remove the large 'O' ring from the outlet tube of the air cleaner, inspect for condition, fit a new 'O' ring if in poor condition.
6. Unclip the three catches securing the inlet tube to the air cleaner canister and remove the inlet tube.
7. Remove the nut and end plate securing the air cleaner element in position.
8. Withdraw the air cleaner element and discard.
9. Inspect the dump valve for condition and that it is clear of obstructions.

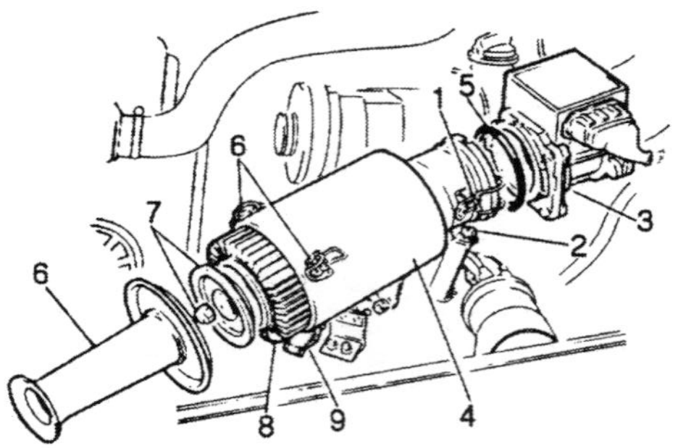

## AIR FLOW SENSOR

### Remove and refit

#### Removing

**NOTE: The air flow sensor is not a serviceable item. In the event of failure or damage the complete unit is to be replaced.**

1. Disconnect the battery negative terminal.
2. Release the large hose clamp at the rear of the air flow meter and disconnect the hose from the sensor.
3. Disconnect the multi-plug.
4. Release the two clips securing the air flow sensor to the air cleaner case detach sensor from the case and withdraw it from the engine compartment.

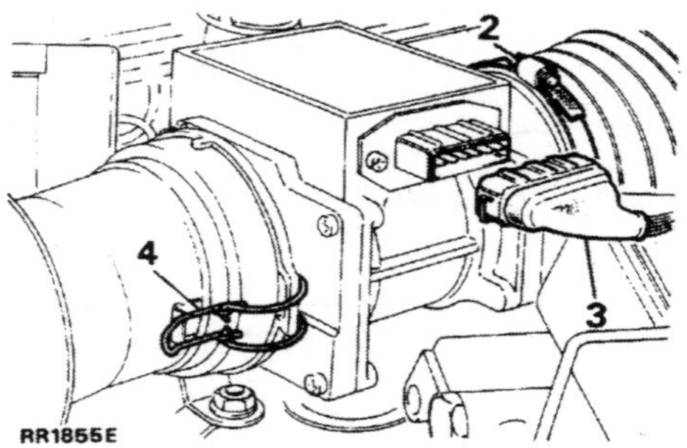

### Refitting

5. Reverse the removal procedure ensuring that the multi-plug is firmly reconnected to the air flow sensor and that the hose clamp at the rear of the sensor is securely tightened, to prevent un-metered air entering the engine.

### Refitting

10. Fit a new element and secure in position.
11. Refit the inlet tube to the air cleaner canister.
12. Refit the air cleaner to the mounting bracket and tighten the two nuts and bolts.
13. Clip the air flow sensor to the air cleaner.

# FUEL INJECTION SYSTEM 19

## THROTTLE POTENTIOMETER

### Remove

1. Disconnect the battery negative terminal.
2. Disconnect the electrical three-pin plug.
3. Remove the two screws securing the switch to the plenum chamber and carefully pull the switch off the throttle valve shaft.
4. Remove the old gasket.

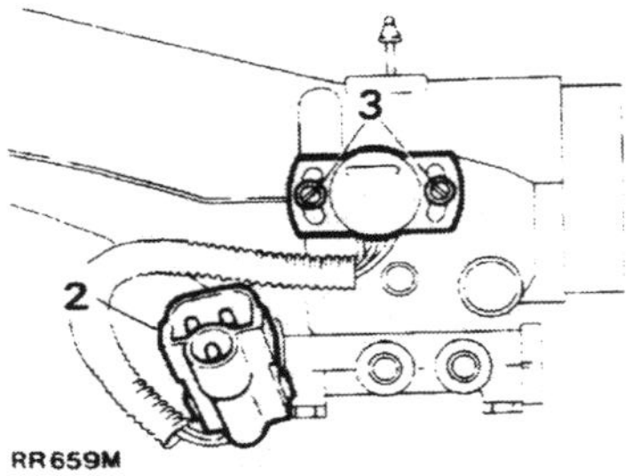

### Refit

5. Fit a new gasket between the throttle switch and plenum chamber.
6. Align the switch and shaft flats; slide the switch on to the throttle shaft and secure the switch to the plenum chamber.
7. The throttle potentiometer must be reset using a potentiometer adjustment gauge.

### Setting the potentiometer

**CAUTION: The throttle mechanism must not be operated while the potentiometer is loosely fitted, otherwise damage may be caused to the potentiometer wiper track.**

Equipment required:-

Lucas throttle potentiometer adjustment gauge - **Lucas Part Number YWB121.**

8. Loosen the potentiometer securing screws.
9. Disconnect the three-pin plug from the potentiometer electrical lead. Connect the adjustment gauge plug to the potentiometer.
10. Connect the two alligator clamps from the throttle potentiometer gauge to the appropriate battery terminals and select position 'R' on the gauge.
11. Rotate the potentiometer clockwise or counter-clockwise until only the middle lamp of the three indication lamps is illuminated.
12. Tighten the potentiometer securing screws.
13. Re-check the potentiometer setting by operating the throttle levers.
14. Disconnect the adjustment gauge from the potentiometer and battery terminals.
15. Re-connect the harness three-pin plug to the potentiometer.

**NOTE: If a potentiometer adjustment gauge is unavailable, the setting procedure can be carried out using a multi meter.**

**IF A MULTI METER IS USED TO CARRY OUT THIS CHECK-ENSURE IT IS SET TO VOLTS. A MULTI METER SETTING OTHER THAN VOLTS WILL RESULT IN DAMAGE TO THE POTENTIOMETER.**

### Setting the Potentiometer using a multi meter.

16. Loosen the potentiometer securing screws.
17. Switch on the ignition.
18. Connect the multi meter between the red and green leads at the potentiometer electrical plug.
19. Rotate the potentiometer clockwise or counter-clockwise, until the multi meter reads 325 ± 35 mV.
20. Tighten the potentiometer securing screws.
21. Re-check the multi-meter reading. Check also for a 'smooth swing' of the meter needle between minimum and maximum throttle opening between the voltage reading of 4.6 - 5.0 volts

**NOTE: After setting the potentiometer, lock and tamperproof each screw head by coating them with yellow paint.**

---

## SPEED TRANSDUCER

### Remove and refit

#### Removing

1. Place the vehicle on a hydraulic hoist and apply the parking brake.
2. Disconnect the battery negative terminal.
3. Raise the hoist and disconnect the speed transducer electrical plug.
4. Disconnect the speedometer cable from the transducer to the binnacle at the transducer.
5. Disconnect the speedometer cable from the transducer to the speedometer housing at the transducer.
6. Remove the single bolt securing the transducer to its mounting bracket and withdraw the unit from the vehicle.

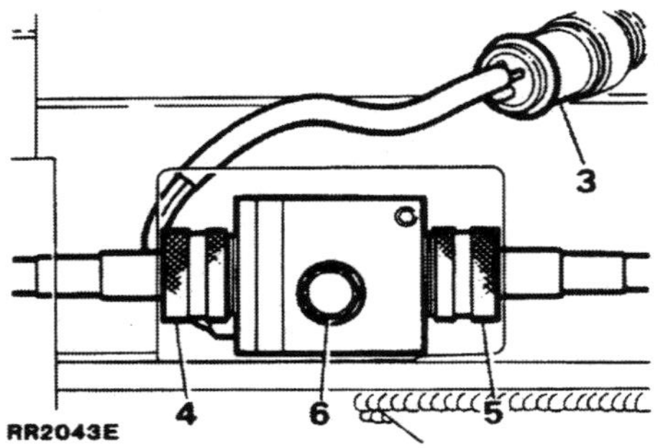

#### Refitting

7. Reverse the removal instructions.

---

## BY-PASS AIR VALVE (STEPPER MOTOR)

### Remove and refit

#### Removing

1. Disconnect the battery negative terminal.
2. Remove the multi-plug from the unit.
3. Unscrew the valve from its location at the rear of the plenum chamber.
4. Remove the captive washer.

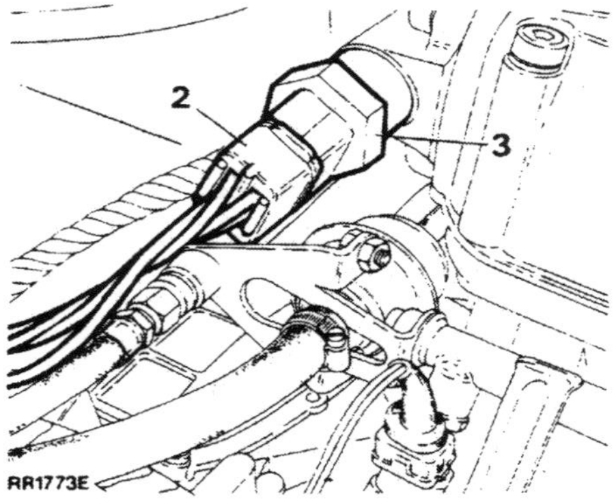

#### Refitting

5. Fit a NEW sealing washer.

**NOTE: If the same by-pass valve is being refitted clean any previous sealing compounds from the threads. Apply Loctite 241 to threads of the valve before reassembly.**

6. Tighten the valve to the specified torque (see Torque values-section 06).
7. Reverse the remaining removal instructions.

# FUEL INJECTION SYSTEM

## ELECTRONIC FUEL INJECTION RELAYS

Incorporated into the fuel injection electrical circuits are two relays. The relays are located beneath the front right-hand seat, adjacent to the E.C.U.

1. **fuel pump relay (mounted on a blue terminal block).**
2. **Main relay (mounted on a black terminal block).**
3. **Diagnostic plug.**

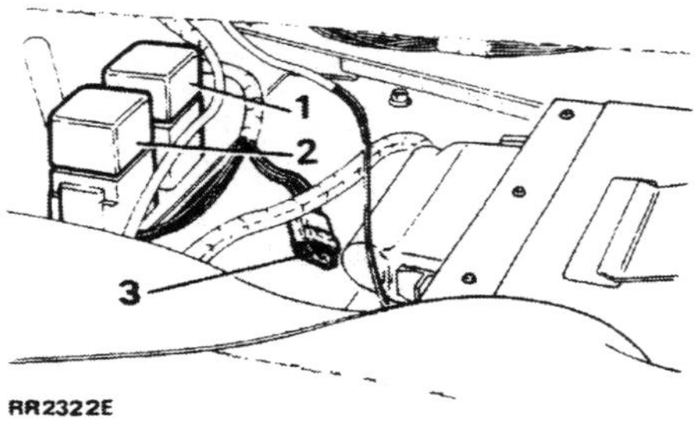

RR2322E

### Remove and refit

#### Removing

1. Disconnect the battery negative terminal.
2. Pull the relay(s) from the multi-plug(s).

#### Refitting

3. Reverse the removal procedure.

## ELECTRONIC CONTROL UNIT-ECU

**NOTE: The ECU is not itself a serviceable item, in the event of a unit failure, the ECU must be replaced.**

### Remove and refit

#### Removing

1. Disconnect the battery negative terminal.
2. The ECU is located under the front right hand seat and is accessible through the rear opening of the seat base when the seat is in its most forward position.
3. Release the E.C.U plug retaining clip.
4. Pull the rear of the multi-plug out of the ECU.
5. Maneouver the front of the plug (in the direction of the bold arrow) to release the hooked rear end of the plug from the retaining peg.
6. Release the screws securing the ECU to the mounting bracket.
7. Withdraw the ECU from the spring clip and remove it from the vehicle.

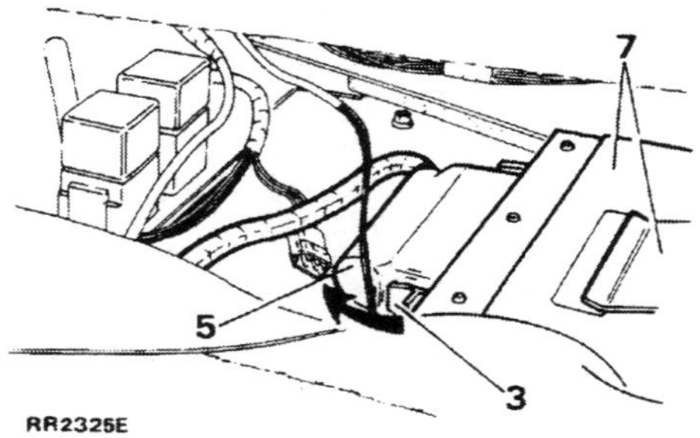

RR2325E

#### Refitting

8. Refit the E.C.U. securely in the spring clip and tighten the two screws.
9. Re-connect the E.C.U. harness plug, ensure the plug is firmly pushed into its location and that the retaining clip secures the plug in position.

## FUEL TEMPERATURE THERMISTOR (SENSOR)

### Remove and refit

#### Removing

**NOTE: No fuel leakage will occur when the thermistor is removed from the fuel rail therefore it is not necessary to depressurise the fuel system before removal.**

1. Disconnect the battery negative terminal.
2. Remove the electrical multi - plug from the thermistor.
3. Release the thermistor from the fuel feed rail.

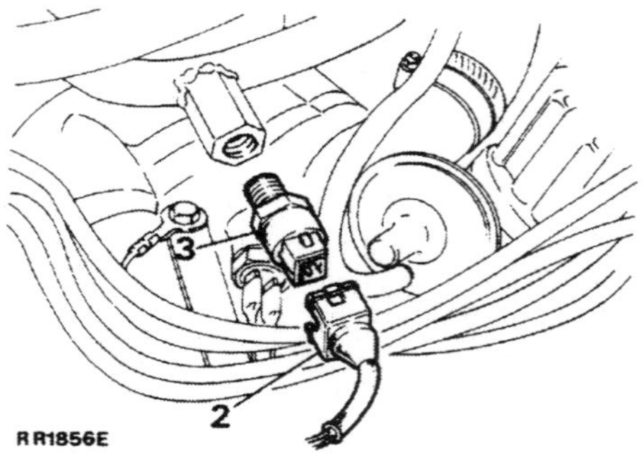

RR1856E

#### Refitting

4. Reverse the removal procedure, ensuring that the thermistor is tightened securely in the fuel rail.

# FUEL INJECTION SYSTEM 19

## RANGE ROVER

### COOLANT TEMPERATURE THERMISTOR (SENSOR)

**Remove and refit**

**Removing**

1. Remove the multi-plug from the thermistor.
2. Release the radiator bottom hose and partially drain the cooling system.
3. Refit the hose and tighten the clamp securely.
4. Remove the thermistor from the left hand front branch of the intake manifold.
5. Remove the copper washer.

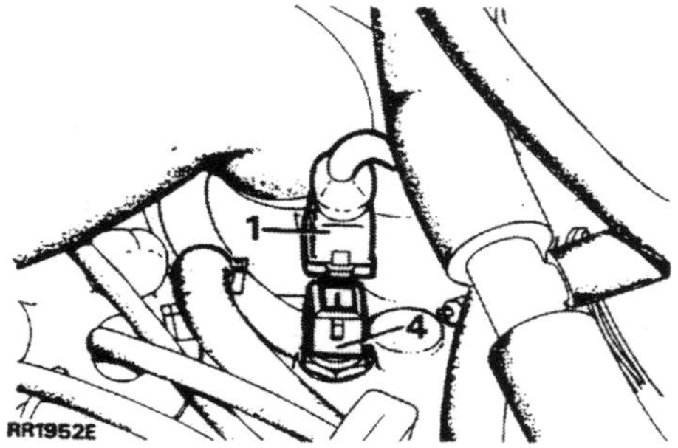

**Refitting**

6. Fit a NEW copper washer to the thermistor.
7. Fit the thermistor to the intake manifold and tighten securely.
8. Refill the cooling system.
9. Run the engine, check for water leaks around the coolant temperature thermistor.

### RESETTING THROTTLE LEVERS

NOTE: **The setting procedure outlined is applicable at minimum throttle condition only.**

1. Ensure that the throttle valve is retained at its 90° vertical setting by holding down the lever denoted by the bold arrow while adjusting the throttle operating levers.
2. Release the throttle operating lever securing screw and adjust the lever until contact is made with the top end of the slot in the throttle lever mounting bracket; retaining the lever in this position retighten the screw.
3. Lightly grease all throttle lever bearing surfaces and torsion spring with Admax 13 grease or a suitable equivalent.

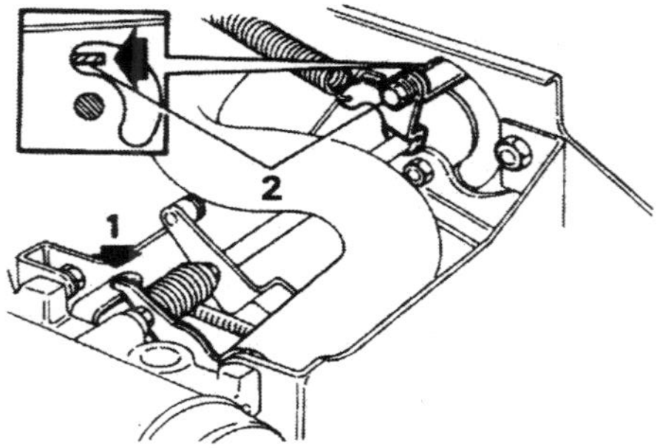

---

## RANGE ROVER

### THROTTLE CABLE

**Remove and refit**

**Removing**

1. Remove the cotter pin and clevis pin securing the cable to the lever.
2. Carefully pry the throttle cable adjustment nut out of the linkage mounting bracket.
3. Withdraw the cable from the mounting bracket.
4. Release the outer cable from the retaining clips within the engine compartment.
5. Remove the lower dash panel from beneath the steering column.
6. Disconnect the cable from the throttle pedal and release the cable locknut.
7. Feed the cable through the bulkhead grommet and into the engine compartment.

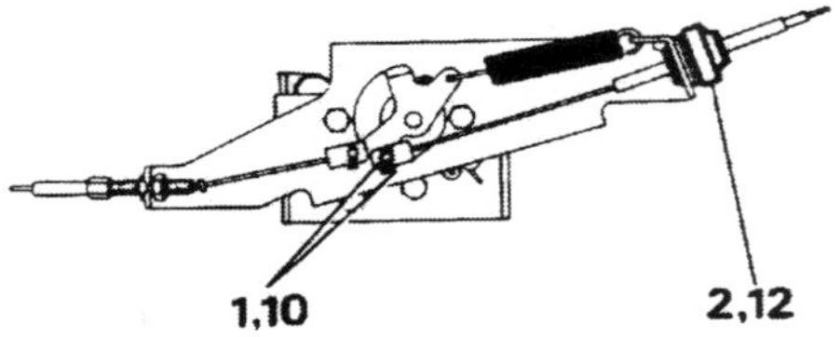

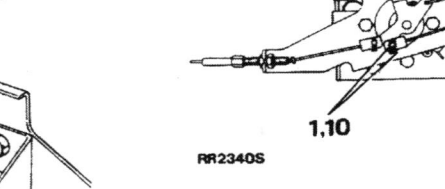

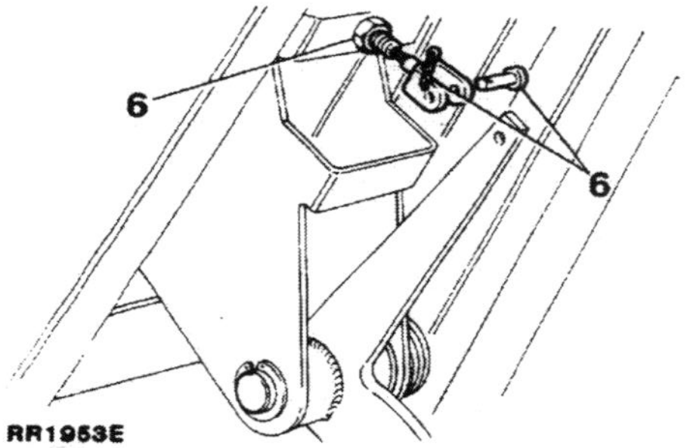

### FIT NEW THROTTLE CABLE

8. Feed the new cable from the engine compartment through the bulkhead grommet.
9. Connect the cable to the throttle pedal.
10. Connect the cable to the throttle linkage, fit a new cotter pin and secure in position.
11. Clip the outer cable adjustment nut into the mounting bracket.
12. Adjust the outer cable to give 1.57 mm (0.062 in) free play in the throttle cable and check the throttle operation.

### THROTTLE PEDAL

**Remove and refit**

**Remove**

1. Release the six screws securing the lower dash panel, lower the panel and disconnect the two electrical leads to the rheostat switch, detach the bulb check unit from the spring clip and remove the dash panel from the vehicle.
2. Remove the cotter pin and clevis pin securing the throttle cable to the throttle pedal.
3. Release the tension from the pedal return spring.
4. Remove the circlip from the pedal pivot pin.
5. Withdraw the pivot pin.

NOTE: **It may be necessary to remove the steering column fixings enabling the column to be lowered to gain access to the pedal pivot pin circlip.**

6. Withdraw the throttle pedal.

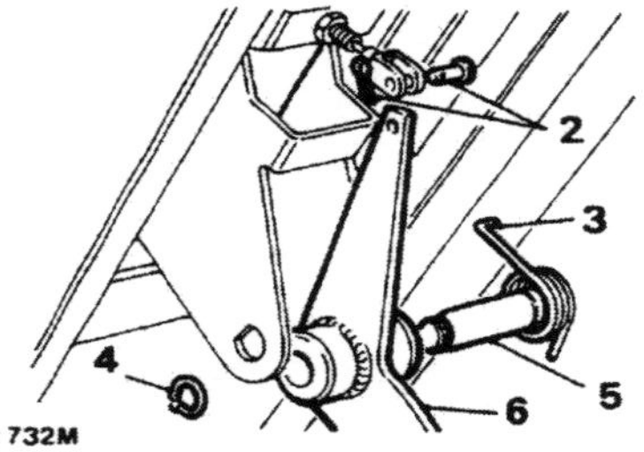

**Refitting**

7. Lightly grease the pivot pin and clevis pin before re-assembly.
8. Fit a NEW cotter pin to the clevis pin.
9. Reverse the remaining removal instructions.

# FUEL INJECTION SYSTEM

## PLENUM CHAMBER

### Remove and refit

#### Removing

1. Disconnect the battery negative terminal.
2. Release the radiator bottom hose and partially drain the cooling system, reconnect the hose to the radiator.
3. Release the two large hose clamps from the neck of the plenum chamber and outlet bore of the airflow sensor and remove the hose from its location.
4. Release the clamps and remove the two coolant hoses from the bottom of the plenum chamber inlet neck. Identify each hose to aid re-assembly.
5. Remove the distributor vacuum hose.
6. Remove the positive crankcase ventilation breather filter hose.
7. Remove the servo hose.
8. Disconnect the throttle potentiometer multi-plug.
9. Disconnect the multi-plug from the air by-pass valve.
10. Disconnect the small vacuum hose at the rear of the plenum chamber, located below the air by-pass valve.
11. Remove the hose from the air by-pass valve to plenum chamber to enable the small return spring located below the throttle levers to be unhooked.
12. Release the two throttle return springs.
13. Remove the two bolts (with spring washers) securing the throttle cable and kick-down cable anchor bracket to the throttle lever support bracket, lay the assembly to one side.
14. Remove the six socket head bolts (with plain washers) securing the plenum chamber to the ram housing.
15. Maneouver the plenum chamber and remove it from the ram housing.

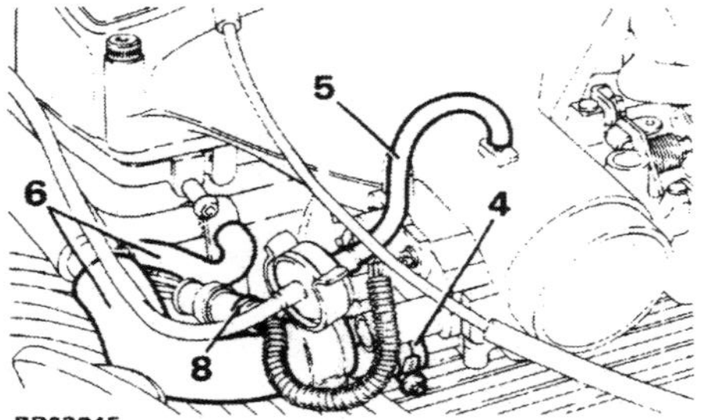

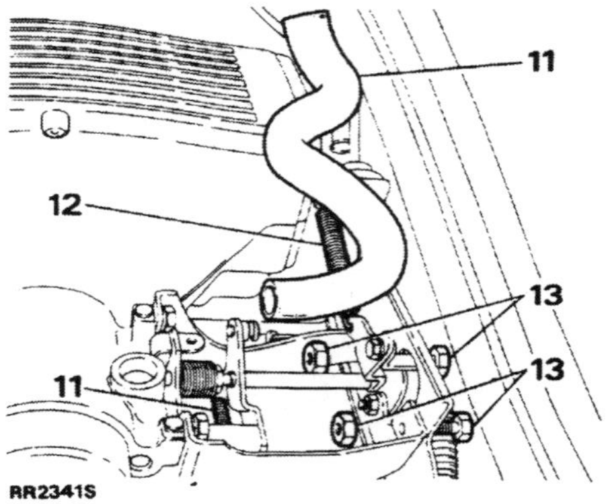

**NOTE: To prevent ingress of dirt into the ram tubes, place a protective cover over the ram tube openings.**

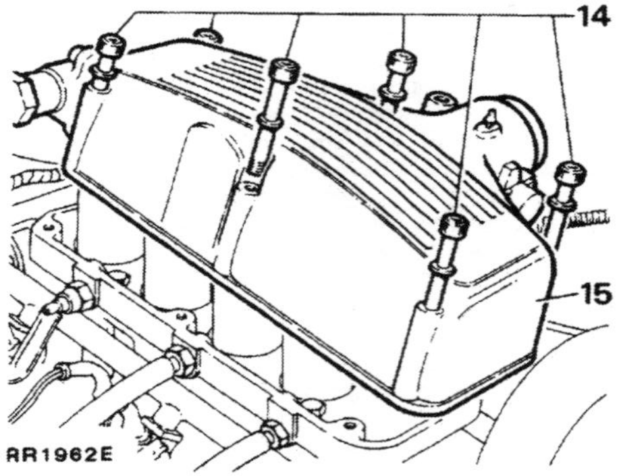

### Refitting

16. Ensure that all mating faces are free from any previous sealing compounds.
17. Coat the mating faces of the plenum chamber and ram housing with 'Hylomar' sealant.
18. Refit the plenum chamber and tighten the six bolts to the specified torque (see torque values-section 06).
19. When refitting the small return spring, item 11 in the removal procedure, it must be noted that the 'hooked' open end of the spring MUST face the plenum chamber as shown in illustration RR2292E below.

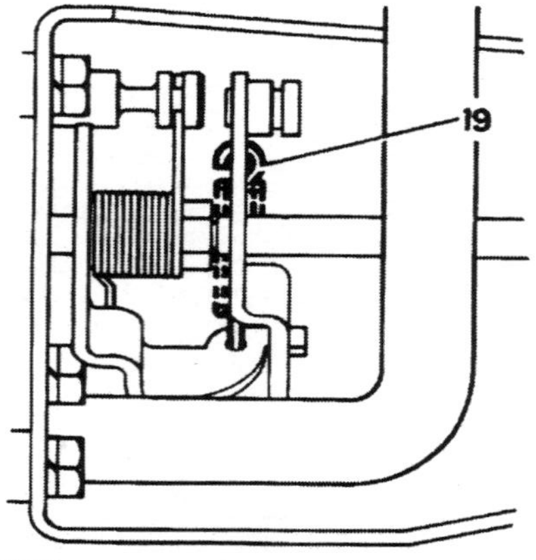

## RAM HOUSING

### Remove and refit

#### Removing

1. Disconnect the battery negative terminal.
2. Remove the plenum chamber (see Plenum Chamber remove and refit).
3. Release the hoses from around the outer edges of the ram housing.
4. Remove the six through bolts (with plain washers) securing the ram housing to the intake manifold.
5. Lift the ram housing off the intake manifold and remove it from the engine compartment.
6. Place a protective cover over the top of the intake manifold inlet bores to prevent ingress of dirt.

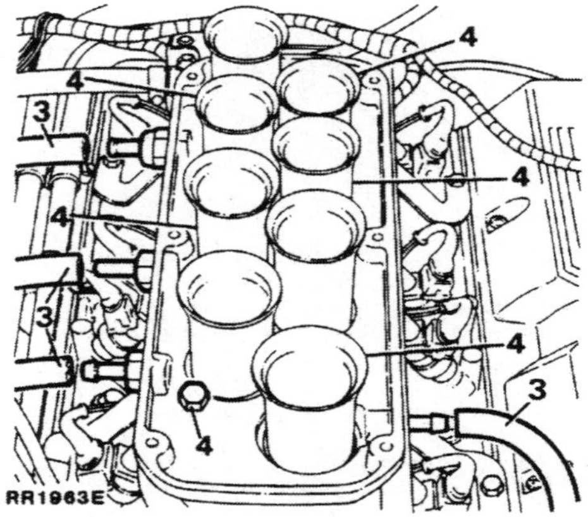

### Refitting

7. Ensure that all mating faces are clean and free from dirt and any previous sealing compounds.
8. Apply 'Hylomar' sealant to the intake manifold face before refitting the ram housing.
9. Fit the ram housing and retighten the bolts, working from the two centre bolts, diagonally towards the outer four bolts.
10. Tighten to the correct torque (See section 06-Torque values).
20. Reverse the remaining removal instructions.

**NOTE: Ensure that all hoses are connected securely to prevent un-metered air entering the engine.**

# FUEL INJECTION SYSTEM

## DEPRESSURISING THE FUEL SYSTEM

**WARNING:** Under normal operating conditions the fuel injection system is pressurised by a high pressure fuel pump, operating at 2.4 to 2.6 kgf/cm$^2$ (34 to 37 p.s.i.). When the engine is stationary this pressure is maintained within the system. To prevent pressurised fuel escaping and to avoid personal injury it is necessary to depressurise the fuel injection system before any service operations are carried out.

**NOTE:** If the vehicle has not been run there will still be a small amount of residual pressure in the fuel line. The depressurising procedure must still be carried out before disconnecting the component within the fuel system.

1. The fuel pump relay is located under the front right hand seat.
2. Pull the fuel pump relay off its multi-plug (see Electronic Fuel Injection Relays-Section 19, Page 28).
3. Start and run the engine.
4. When sufficient fuel has been used up causing the fuel line pressure to drop, the injectors will become inoperative, resulting in engine stall. Switch the ignition off.
5. Disconnect the battery negative terminal.

**NOTE:** Fuel at low pressure will remain in the system. To remove this low pressure fuel, place an absorbent cloth around the fuel feed hose at the fuel rail and release the fuel feed hose compression nut.

### Refitting

6. Refit the fuel feed hose.
7. Refit the fuel pump relay, reconnect the battery.
8. Crank the engine (engine will fire within approximately 6 to 8 seconds).

## FUEL PRESSURE REGULATOR

### Remove and refit

#### Removing

1. Depressurise the fuel system.
2. Disconnect the negative battery terminal.
3. Release the hose clamp securing the fuel return hose to the regulator and remove the hose.
4. Pull the vacuum hose from the rear of the regulator.
5. Remove the two nuts and bolts securing the regulator to the fuel rail, carefully ease the regulator fuel inlet pipe out of the fuel rail.

6. Withdraw the regulator from the engine compartment.

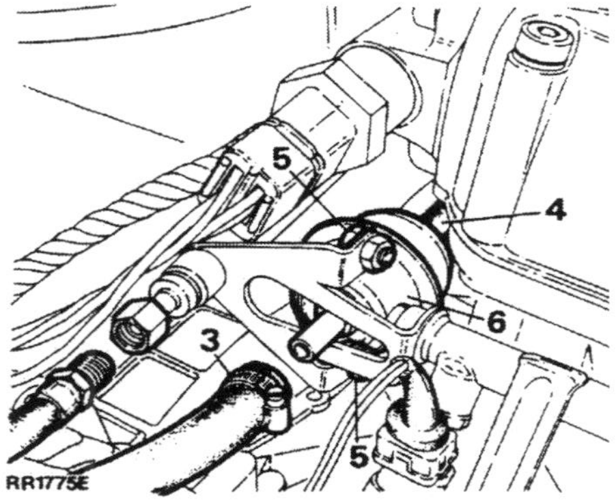

11. If necessary, remove the two nuts and bolts securing the regulator to the fuel rail, and carefully pull the regulator away from the rail.

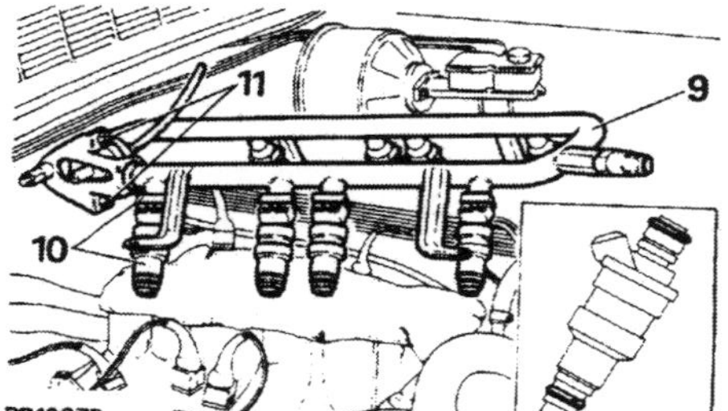

**NOTE: If the original regulator is being refitted, fit a NEW 'O' ring to the fuel inlet pipe.**

### Refitting

7. Lightly coat the 'O' ring with silicon grease 300 before fitting the regulator to the fuel rail.
8. Reverse the removal procedure.
9. Reconnect the battery and pressurise the fuel system and check that there are no fuel leaks around the regulator connections.

## FUEL SYSTEM HOSES

**CAUTION:** All fuel hoses are made up of two laminations, an armoured rubber outer sleeve and an inner viton core. If any of the fuel system hoses have been disconnected, it is imperative that the internal bore is inspected to ensure that the viton lining has not become separated from the armoured outer sleeve. If separation is evident, the hose must not be refitted. A new hose must be fitted.

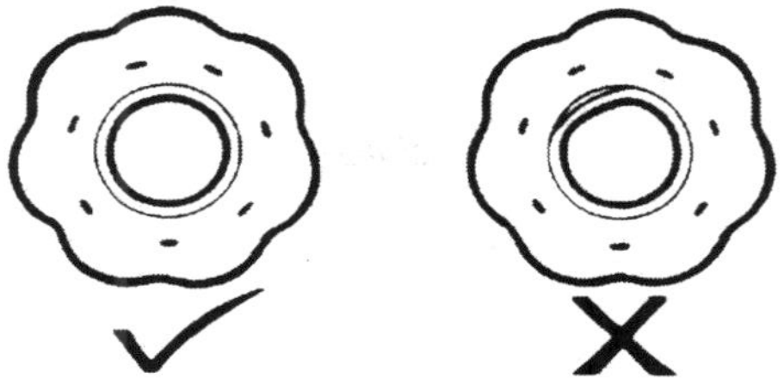

## FUEL RAIL-INJECTORS R/H AND L/H

### Remove and refit

#### Removing

1. Depressurise the fuel system.
2. Disconnect the negative battery terminal.
3. Remove the plenum chamber. (See Plenum Chamber, remove and refit).
4. Remove the ram housing. (See Ram Housing remove and refit).

**NOTE: Place a cloth over the ram tube openings to prevent ingress of dirt into the engine.**

5. Release the hose clamp and remove the fuel return hose from the pressure regulator.
6. Disconnect the multi-plug from the fuel temperature thermistor (sensor).
7. Disconnect the multi-plugs from the eight injectors.
8. Remove the five bolts securing the fuel rail support and heater pipe brackets to the intake manifold. Lay the heater pipes to one side.

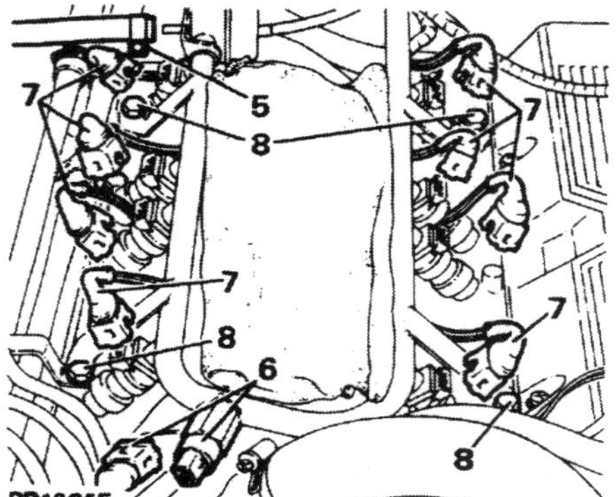

9. Remove the fuel rail, complete with injectors, from the intake manifold.
10. Remove the retaining clips securing the injectors to the fuel rail, ease the injectors from the rail.

### Refitting

12. Fit **NEW** 'O' rings, protective cap and supporting disc to the injectors, lightly coat the 'O' rings with silicon grease 300 and insert the injectors into the fuel rail, multi-plug connections facing outwards.
13. Refit the retaining clips.

**CAUTION: Care must be taken when refitting the fuel rail and injectors to the intake manifold to prevent damage occurring to the 'O' rings.**

14. Fit a NEW 'O' ring to the pressure regulator lightly coat the 'O' ring with silicon grease 300 and secure the regulator to the fuel rail.
15. Fit the fuel rail and heater pipe assemblies to the intake manifold, secure the rail and pipes in position with the five bolts.
16. Reverse the remaining removal instructions.
17. Pressurise the fuel system and check for fuel leaks around the injectors and pressure regulator.

# FUEL INJECTION SYSTEM 19

## INTAKE MANIFOLD

### Remove and refit

#### Removing

1. Depressurise the fuel system.
2. Disconnect the battery negative terminal.
3. Release the hose clamp and remove the radiator bottom hose to enable the cooling system to be partially drained, so that coolant level is below the thermostat housing, refit the hose and secure in position with the hose clamp.
4. Remove the plenum chamber (see Plenum Chamber, remove and refit).
5. Remove the ram housing (see ram housing remove and refit).

**CAUTION: Place a protective cover over the intake manifold openings to prevent ingress of dirt.**

6. Disconnect the electrical multi-plugs to the fuel temperature thermistor (sensor), coolant temperature thermistor (sensor) and injectors.
7. Remove the two nuts and bolts securing the pressure regulator to the fuel rail, ease the regulator out of the rail, seal the end of the fuel rail with suitable plastic plugs to prevent ingress of dirt.

**NOTE: The intake manifold can be removed from the cylinder block without removing the fuel rail and injectors.**

8. Disconnect the electrical leads from the air-conditioning engine coolant sensor located on the thermostat elbow.
9. Disconnect the electrical leads to the coolant temperature transmitter (sensor) located at the front of the intake manifold.
10. Remove the injector harnesses from behind the fuel rail and lay to one side.
11. Release the hose clamps securing the two heater hoses to the rigid heater pipes at the front of the right hand rocker cover.
12. Remove the two bolts securing the rigid heater pipes to the intake manifold and ease the pipes out of the hoses.

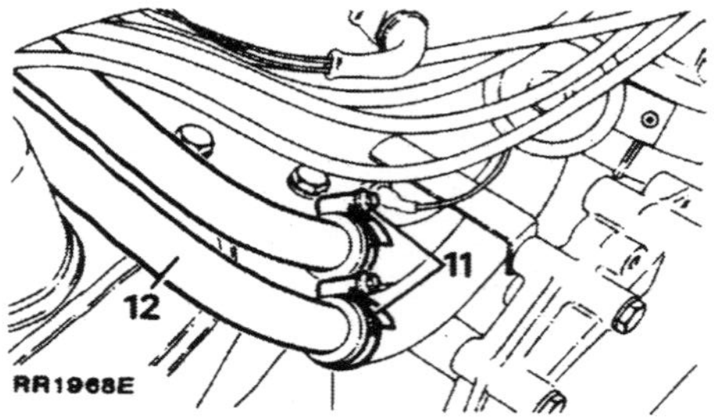

13. Lay the heater pipe assembly to one side.
14. Release the twelve bolts securing the intake manifold to the cylinder heads.
15. Lift the manifold off the cylinder heads and remove from the engine compartment.
16. Remove the two gasket clamps from the top of the cylinder block.
17. Lift off the gasket and remove the gasket seals.
18. Remove previous sealing compound from around the water passage openings of the cylinder heads.

#### Refitting

19. Locate the NEW seals in position with their ends engaged in the notches formed between the cylinder heads and block.

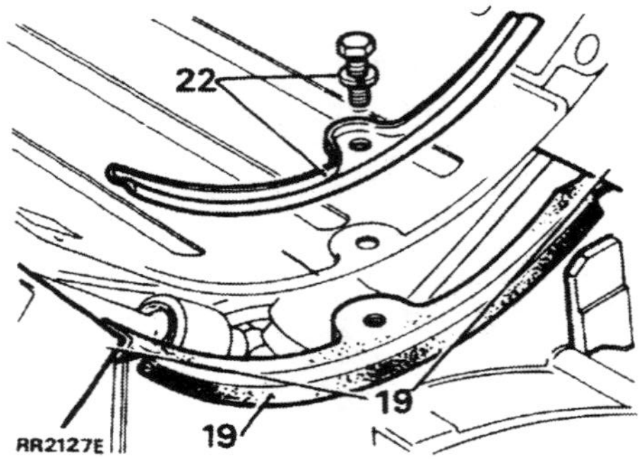

20. Lightly apply 'Hylomar' SQ32M sealant around the outside of the water passage openings on the cylinder heads, manifold gasket and intake manifold.

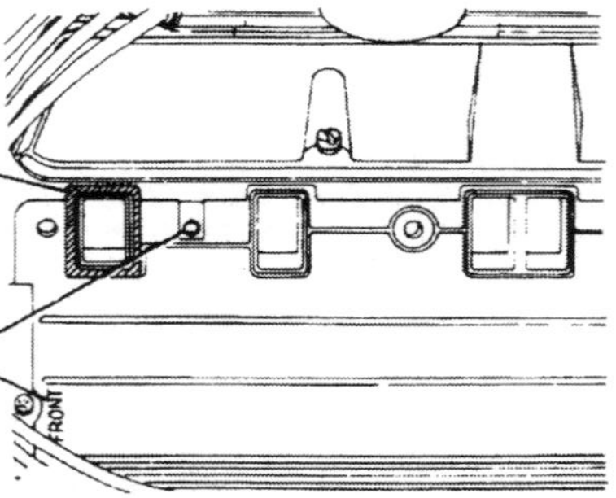

21. Fit the manifold gasket with the word 'FRONT' to the front and the open bolt hole to the front right hand side.
22. Fit the gasket clamps but **DO NOT** fully tighten the bolts at this stage.
23. Locate the intake manifold onto the cylinder heads, clean the threads of the manifold securing bolts.
24. Fit all manifold bolts and tighten them a little at a time, evenly, alternate sides working from the centre outwards.
25. Tighten to correct torque (see section 06 Torque values).
26. Tighten the gasket clamps to the correct torque (see section 06 Torque values).
27. Reverse remaining removal instructions.
28. Replenish the cooling system.
29. Start the engine, check for water and fuel leaks.

## FUEL LINE FILTER

### Remove and refit

#### Refitting

1. Depressurise the fuel system.
2. The fuel line filter is located on the right hand chassis side member forward of the fuel tank filler neck. Access to the filter is gained through the right hand rear wheel arch.
3. Thoroughly clean the immediate area around the hose connections to prevent ingress of foreign matter into the fuel system.
4. Loosen the two hose clamps nearest the filter to enable the hoses to be removed from the filter canister. Plug the end of the hoses to prevent ingress of dirt.
5. Release the securing bolt and bracket and remove the filter from the chassis side member.

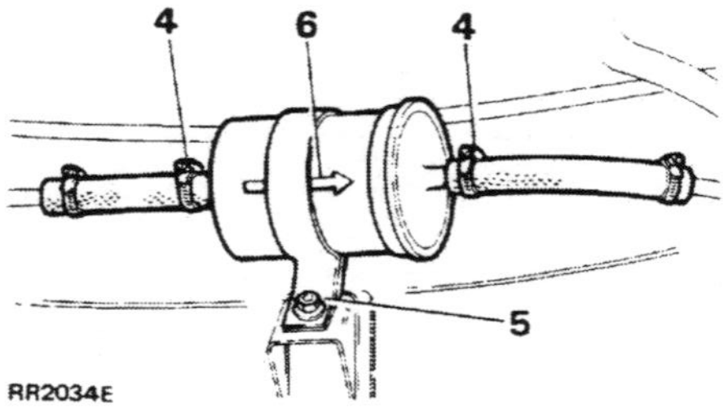

#### Refitting

6. Fit a new filter observing the direction of flow arrows stamped on the canister.
7. Start the engine and inspect for fuel leaks around the hose connections.

# FUEL INJECTION SYSTEM

## RANGE ROVER

### FUEL TANK

**Remove and refit**

**Removing**

**WARNING: Ensure that all the necessary precautions are taken against fuel spillage and fuel vapour to prevent fire or explosion.**

**CAUTION: Before disconnecting any part of the fuel system it is imperative that all dust, dirt and debris is removed from around the components to be removed to prevent ingress of foreign matter into the fuel system.**

1. Drive the vehicle onto a suitable hoist.
2. Depressurise the fuel system (see depressurising procedure-page 34).
3. Disconnect the battery negative terminal.
4. Disconnect the electrical leads to the fuel tank sender unit. Disconnect the fuel pump electrical multiplug, access to which is gained through the left hand rear wheel arch, the plug is located between the underside of the body and chassis side member.
5. Raise the hoist.
6. Remove the drain plug from the bottom of the fuel tank and drain the fuel into a suitable container that can be sealed afterwards, refit the drain plug. (Refer to Warning at start of procedure).

**From underneath the vehicle**

7. Disconnect the fuel hose from the inlet side of the fuel filter.
8. Disconnect the fuel return pipe to the fuel tank.
9. Remove the breather hose and three evaporative loss hoses from the fuel tank, seal all hose and pipe openings to prevent ingress of foreign matter.
10. Release the two large hose clamps, securing the inter-connecting hose to tank and filler tube, maneouver the hose up the outside of the filler tube to enable it to be withdrawn from the tank filler neck.
11. With assistance from a second person supporting the fuel tank, remove the four tank fixings.

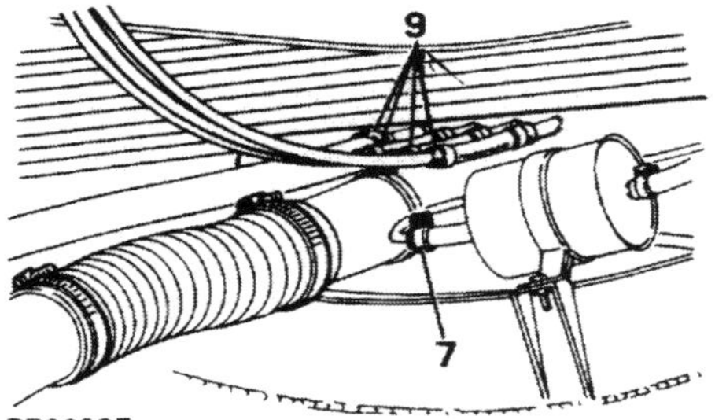

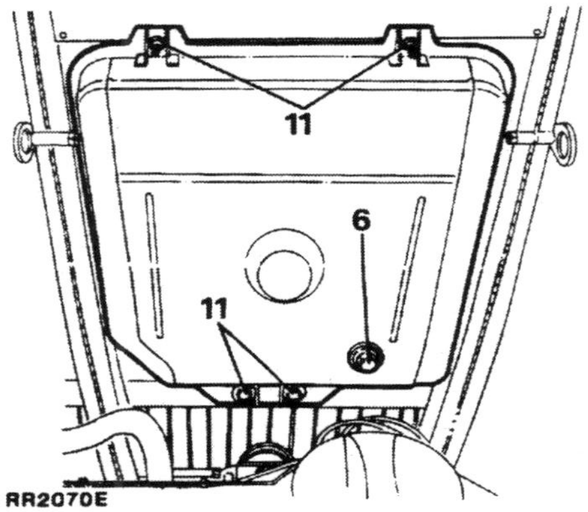

12. Tilt the left hand side of the tank downwards and maneouver it out of the chassis frame. Care should be taken to ensure that the fuel feed pipe to filter is not damaged when lowering the tank.
13. Place the tank in a safe area and ensure that all necessary precautions are undertaken to make all personnel within the vicinity aware that the tank will give off residual fuel fumes.
14. If necessary remove the fuel pump from the tank. (See Fuel Pump remove and refit).

**Refitting**

15. Refit the fuel tank to the chassis, taking care to relocate the fuel feed pipe grommets between the fuel tank and chassis.
16. Reverse the removal procedure, ensuring that all hose and pipe connections are securely tightened.
17. Re-pressurise the fuel system.
18. Inspect for fuel leaks from around the fuel filter, hose and pipe connections.

## RANGE ROVER

### FUEL PUMP

**Remove and refit**

**Removing**

**WARNING: Ensure that all necessary precautions are taken against fuel spillage and fuel vapour to prevent fire or explosion.**

1. Drive the vehicle onto a suitable hoist.
2. Depressurise the fuel pump system. (see depressurising procedure-page 34)
3. Disconnect the battery negative terminal.
4. Remove the fuel tank from the chassis frame. (see fuel tank remove and refit-page 38)
5. Place the tank in a safe area.
6. Disconnect the fuel supply hose from the pump.
7. Remove any previous sealant from the top of the pump flange.
8. Remove the five screws and withdraw the pump from the tank.

**Refitting**

9. Clean the immediate area around the pump opening in the fuel tank.
10. Fit a **NEW** pump seal.
11. Secure the pump to the tank and tighten the screws securely.
12. Liberally coat the heads of the screws and flange of the fuel pump with Sikaflex 221 flexible adhesive sealant.
13. Refit the tank to the vehicle.

**Continued**

# MANIFOLD AND EXHAUST SYSTEM [30]

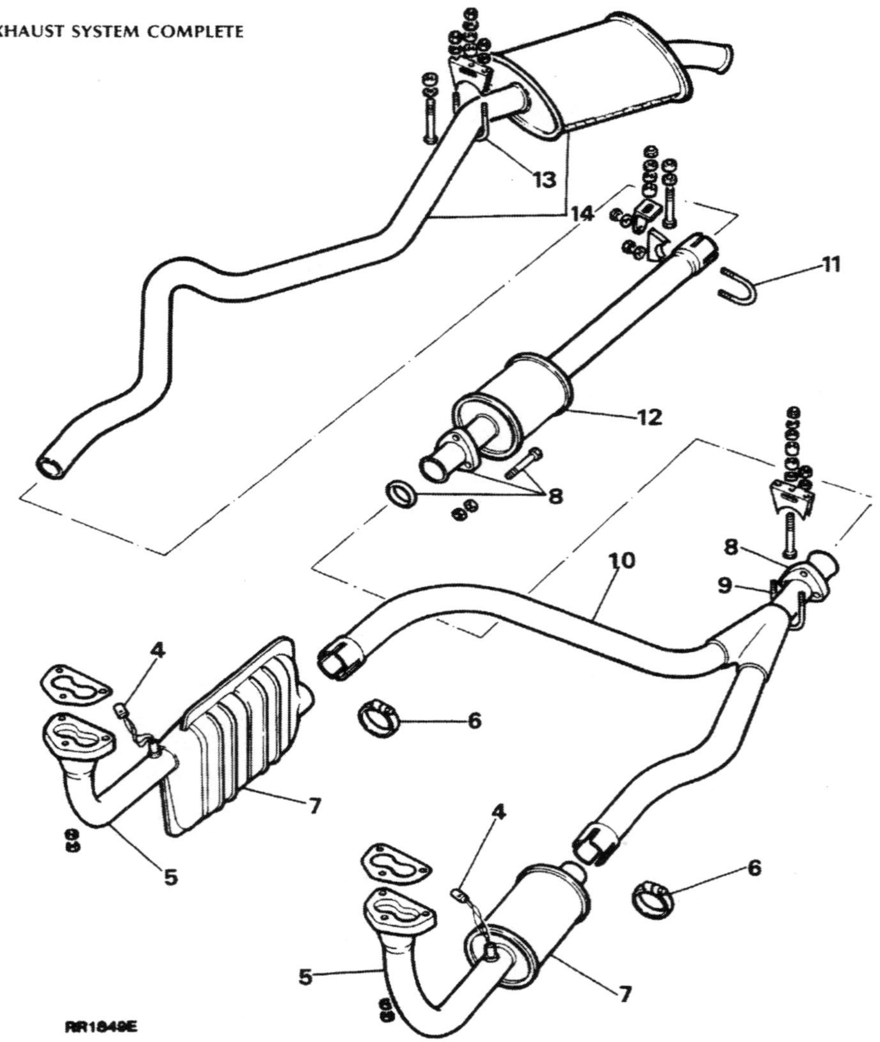

**RANGE ROVER**

## EXHAUST SYSTEM COMPLETE

## EXHAUST SYSTEM COMPLETE

**WARNING:** To prevent personal injury occurring from a hot exhaust system, DO NOT attempt to disconnect any of the components until ample time has elapsed to allow the exhaust system to cool.

**NOTE:** Ensure that no exhaust leaks are evident in either a new or an old exhaust system, as this will affect vehicle performance, and could contravene local emission regulations.

Continued

---

# FUEL INJECTION SYSTEM [19]

**RANGE ROVER**

## FUEL PIPES

**WARNING:** Depressurise fuel system before disconnecting any of the fuel pipes and ensure that all necessary precautions are taken against fuel spillage.

### KEY

1. Fuel feed hose to fuel rail.
2. Fuel return hose to fuel tank.
3. Rigid fuel feed pipe.
4. Rigid fuel return pipe.
5. Fuel filter.
6. Rigid fuel feed pipe to filter.
7. Breather hose.
8. In-tank fuel pump.
9. Fuel filler neck.
10. Fuel tank.

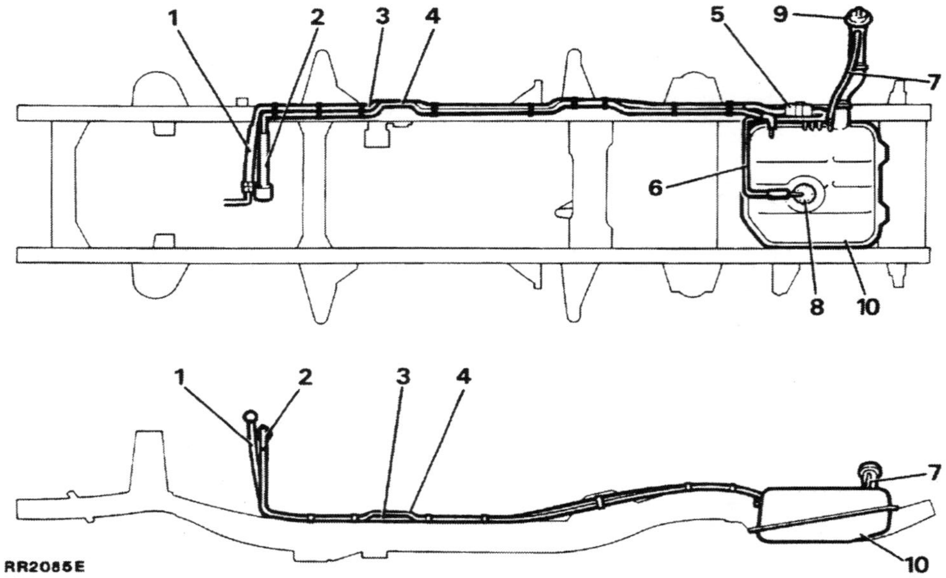

# MANIFOLD AND EXHAUST SYSTEM

## RANGE ROVER

### Remove and refit

#### EXHAUST MANIFOLD

**Remove and refit**

**Left hand**

**Right hand**

**Removing**

1. Raise the vehicle on a suitable hoist and apply the parking brake.
2. If fitted, remove the four fixings securing the grass shield to the mounting bracket at the centre catalyst.
3. Withdraw the grass shield from the bracket.

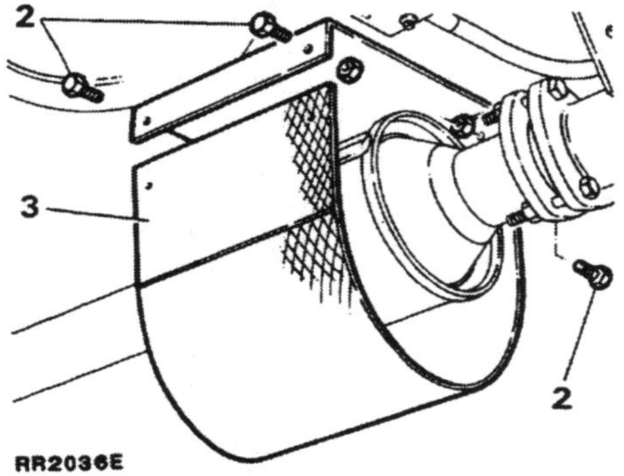

1. Disconnect the front exhaust pipe(s) from the manifold(s).
2. Tap back the bolt locking tabs and remove the eight bolts, lock tabs and washers.
3. Remove the manifold(s) and old gaskets.

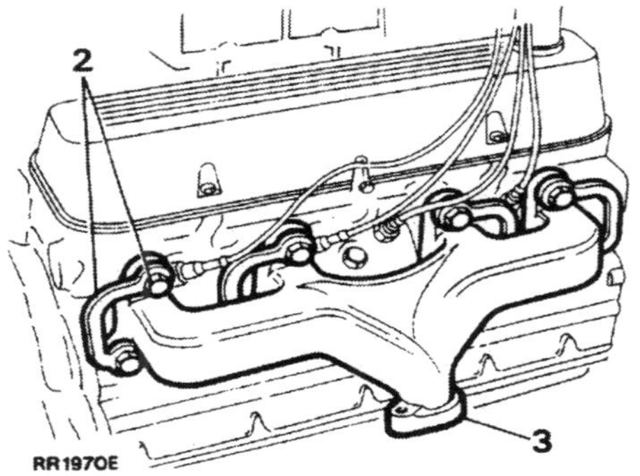

RR2036E

4. Disconnect the electrical plugs from the Lambda sensors.
5. Remove the nuts and disconnect the front pipe(s) from the manifold(s) and remove the gaskets.
6. Loosen the pinch bolt clamps securing the front pipe to the intermediate pipe.
7. Withdraw the front pipe(s) with catalysts.
8. Remove three bolts securing the intermediate pipe to the centre catalyst and withdraw the doughnut.
9. Remove the U-bolt from the pipe mounting bracket.
10. Withdraw the intermediate pipe.
11. Remove the U-bolt securing the centre catalyst to the main muffler.
12. Withdraw the catalyst.
13. Remove the U-bolt from the tail pipe mounting bracket.
14. Withdraw the tail pipe and rear muffler.

**Refitting**

NOTE: Apply Firegum Putty, Part No. 15608 to all exhaust system joints with the exception of the exhaust flange to manifold flange where new gaskets should be fitted.

15. Reverse the removal instructions.

RR1970E

**Refitting**

4. Ensure that the mating surfaces of the cylinder head and exhaust manifold are clean and smooth.
5. Coat the threads of each bolt with anti-seize compound.
6. Place the manifold and new gaskets in position on the cylinder head and fit the securing bolts, new lockplates and plain washers. The plain washers are fitted between the manifold and lockplates.
7. Evenly tighten the manifold bolts to the correct torque, see torque values-section 06, and bend over the lockplate tabs.
8. Reconnect the front exhaust pipe, using new exhaust flange gaskets.

#### INTAKE MANIFOLD

The removal and refit of the intake manifold is incorporated in the Fuel Injection System, Section 19.

---

# PROPELLER SHAFTS

## RANGE ROVER

### Propeller shafts

**Front propeller shaft**

The front propeller shaft fitted to catalytic exhaust vehicles is of the solid bar type.

NOTE: The front propeller shaft MUST be fitted with the sliding joint end of the shaft fitted to the drive flange at the front end of the transfer gearbox.

**Rear propeller shaft**

NOTE: The rear propeller shaft MUST be fitted with the sliding joint end of the shaft fitted to the brake drum at the rear end of the transfer gearbox.

# ELECTRICAL 86

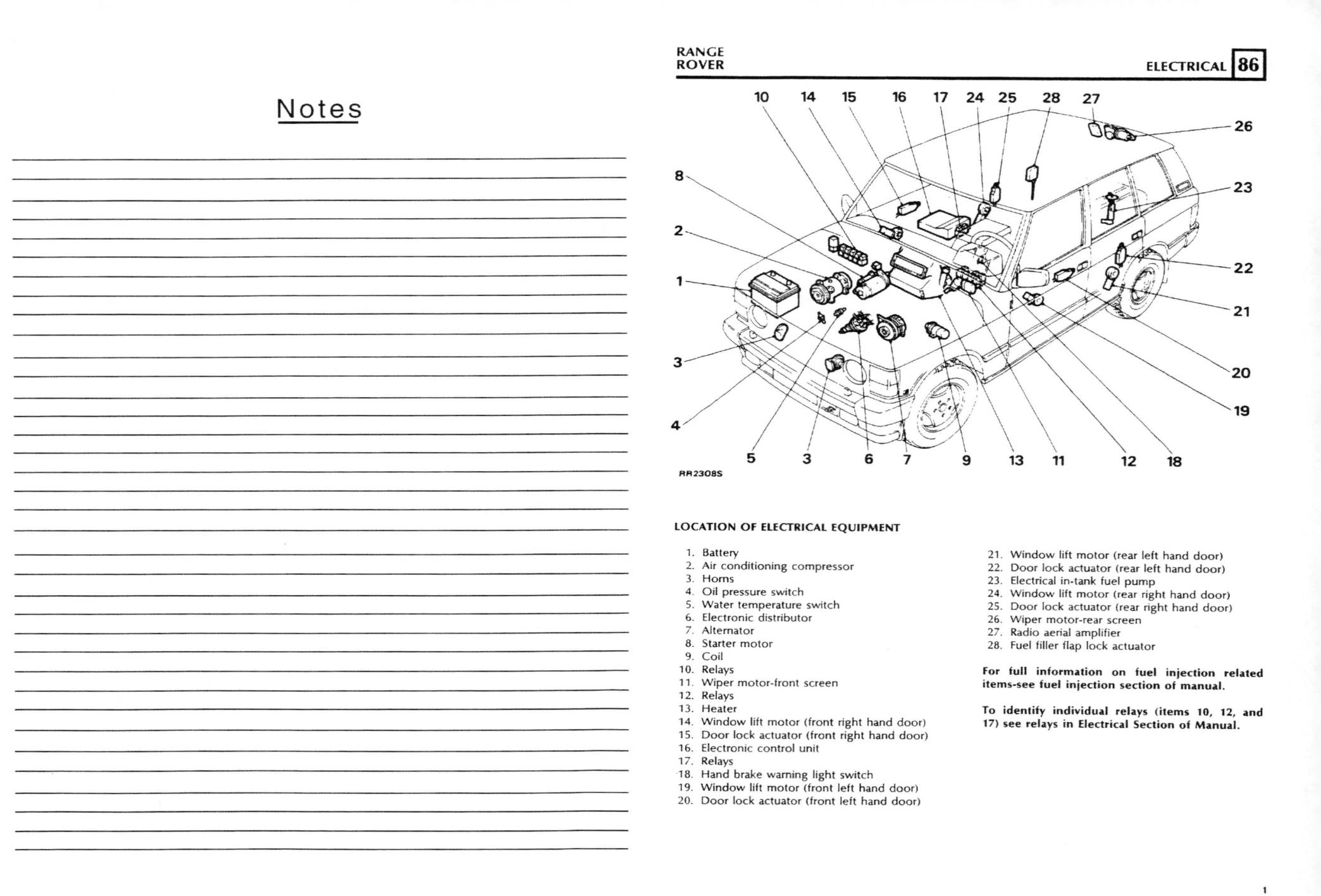

RANGE ROVER

## LOCATION OF ELECTRICAL EQUIPMENT

1. Battery
2. Air conditioning compressor
3. Horns
4. Oil pressure switch
5. Water temperature switch
6. Electronic distributor
7. Alternator
8. Starter motor
9. Coil
10. Relays
11. Wiper motor-front screen
12. Relays
13. Heater
14. Window lift motor (front right hand door)
15. Door lock actuator (front right hand door)
16. Electronic control unit
17. Relays
18. Hand brake warning light switch
19. Window lift motor (front left hand door)
20. Door lock actuator (front left hand door)
21. Window lift motor (rear left hand door)
22. Door lock actuator (rear left hand door)
23. Electrical in-tank fuel pump
24. Window lift motor (rear right hand door)
25. Door lock actuator (rear right hand door)
26. Wiper motor-rear screen
27. Radio aerial amplifier
28. Fuel filler flap lock actuator

**For full information on fuel injection related items-see fuel injection section of manual.**

**To identify individual relays (items 10, 12, and 17) see relays in Electrical Section of Manual.**

# Notes

# ELECTRICAL 86

## RANGE ROVER

### FAULT DIAGNOSIS

| SYMPTOM | POSSIBLE CAUSE | CURE |
|---|---|---|
| A-Battery in low state of charge | 1. Broken or loose connection in alternator circuit<br>2. Current voltage regulator not functioning correctly<br>3. Slip rings greasy or dirty<br>4. Brushes worn, not fitted correctly or wrong type<br>5. Fan belt broken | 1. Examine the charging and field circuit wiring. Tighten any loose connections, repair/replace broken leads. Examine the battery connection.<br>2. Check/fit new unit<br>3. Clean<br>4. Fit new brushes<br>5. Fit new belt |
| B-Battery overcharging leading to burnt out bulbs and frequent need for topping-up | 1. Current voltage regulator not functioning correctly | 1. Fit new unit |
| C-Lamps giving insufficient illumination | 1. Battery discharged<br>2. Bulbs discoloured through prolonged use<br>3. Fan belt broken | 1. Charge the battery from independent supply or by a long period of daylight running.<br>2. Fit new bulb<br>3. Fit new belt |
| D-Lamps light when switched on but fade out | 1. Battery discharged | 1. Charge the battery from an independent supply or by a long period of daylight running |
| E-Lights flicker | 1. Loose connection | 1. Tighten/clean |
| F-Failure of lights | 1. Battery discharged<br>2. Loose broken connection<br>3. Fan belt broken | 1. Charge the battery from an independent supply or by a long period of daylight running<br>2. Locate and rectify<br>3. Fit new belt |

## RANGE ROVER

### ELECTRICAL 86

| SYMPTOM | POSSIBLE CAUSE | CURE |
|---|---|---|
| G-Starter motor lacks power or fails to turn engine | 1. Stiff engine<br>2. Battery discharged<br>3. Broken or loose connection in starter circuit<br>4. Greasy or dirty slip rings.<br>5. Brushes worn, not fitted correctly or wrong type<br>6. Brushes sticking in holders or incorrectly tensioned.<br>7. Starter pinion jammed in mesh with flywheel | 1. Locate cause and remedy<br>2. Charge the battery either by a long period of daytime running or from independent electrical supply<br>3. Check and tighten all battery, starter and starter switch connections and check the cables connecting these units for damage<br>4. Clean<br>5. Fit new brushes<br>6. Rectify<br>7. Remove starter motor and investigate |
| H-Starter noisy | 1. Starter pinion or flywheel teeth chipped or damaged<br>2. Starter motor loose on engine<br>3. Armature shaft bearing | 1. Fit new components<br>2. Rectify, checking pinion and the flywheel for damage<br>3. Fit new bearing |
| J-Starter operates but does not crank the engine | 1. Pinion of starter does not engage with the flywheel | 1. Check operation of starter solenoid. If correct, remove starter motor and investigate |
| K-Starter pinion will not disengage from the flywheel when the engine is running | 1. Starter pinion jammed in mesh with the flywheel | 1. Remove starter motor and investigate |

345

# ELECTRICAL 86

## RANGE ROVER

| SYMPTOM | POSSIBLE CAUSE | CURE |
|---|---|---|
| L-Engine will not start | 1. The starter will not turn the engine due to a discharged battery | 1. Recharge battery by running the car for a long period during daylight or from an independent electrical supply |
| | 2. The starter will not turn due to incorrect gear selection. | 2. Select 'P' or 'N' |
| | 3. Sparking plugs faulty, dirty or incorrect plug gaps | 3. Rectify/ fit new plugs |
| | 4. Defective coil or distributor | 4. Carry out ignition checks. Fit a new coil or distributor |
| | 5. A fault in the low tension wiring circuit | 5. Examine all the ignition cables and check that the terminals are secure and not corroded. |
| | 6. Faulty amplifier | 6. Check/fit new component if necessary. |
| | 7. Air gap out of adjustment | 7. Adjust |
| | 8. Fuel system fault | 8. See Fuel System Section. |
| M-Engine misfires/ stalls | 1. Faulty sparking plugs | 1. Rectify |
| | 2. Air gap incorrectly set | 2. Adjust |
| | 3. Distributor cap cracked | 3. Fit new cap |
| | 4. Faulty pick-up or reluctor | 4. Fit new components |
| | 5. Excessive wear in distributor shaft brushes, etc. | 5. Fit a new components |
| | 6. Rotor arm and flash shield cracked or showing signs of tracking | 6. Fit new component |

## RANGE ROVER

# ELECTRICAL 86

| SYMPTOM | POSSIBLE CAUSE | CURE |
|---|---|---|
| N-Frequent recharging of the battery necessary | 1. Alternator inoperative | 1. Check the brushes, cables and connections or fit a new alternator |
| | 2. Loose or corroded connections | 2. Examine all connections especially the battery terminals and ground cables |
| | 3. Slipping fan belt | 3. Adjust |
| | 4. Voltage regulator faulty | 4. Fit new component |
| | 5. Excessive use of the starter motor | 5. In the hands of the operator, advise |
| | 6. Vehicle operation confined largely to night driving | 6. In the hands of the operator, advise |
| | 7. Abnormal accessory load | 7. Superfluous electrical fittings such as extra lamps, etc. |
| | 8. Internal discharge of the battery | 8. Fit new battery |
| P-Alternator not charging correctly | 1. Slipping fan belt | 1. Adjust |
| | 2. Voltage control not operating correctly | 2. Rectify/ fit new component |
| | 3. Greasy, charred or glazed slip rings | 3. Clean |
| | 4. Brushes worn, sticking or oily | 4. Rectify/fit new brushes |
| | 5. Shorted, open or burnt -out field coils | 5. Fit new field coils |
| Q-Alternator noisy | 1. Worn, damaged or defective bearings | 1. Fit new bearings |
| | 2. Cracked or damaged pulley | 2. Fit new pulley |
| | 3. Alternator out of alignment | 3. Rectify |
| | 4. Alternator loose in mounting | 4. Rectify |
| | 5. Excessive brush noise | 5. Check for rough or dirty slip rings, badly seating brushes, incorrect brush tension, loose brushes and loose field magnets. Rectify/fit new components |
| R-Poor performance of horns | 1. Low voltage due to discharged battery | 1. Recharge |
| | 2. Bad connections in wiring | 2. Carefully inspect all connections and horn push |
| | 3. Loose mounting nut | 3. Rectify |
| | 4. A faulty horn | 4. Fit new horn |

# RANGE ROVER

## ELECTRICAL EQUIPMENT

### DESCRIPTION

The electrical system is Negative ground, and it is most important to ensure correct polarity of the electrical connections at all times. Any incorrect connections made when reconnecting cables may cause irreparable damage to the semi-conductor devices used in the alternator and regulator. Incorrect polarity would also seriously damage any transistorised equipment such as radio and tachometer etc.

**Before carrying out any repairs or maintenance to an electrical component, always disconnect the battery negative lead.**

### ALTERNATOR - LUCAS A133/80

The alternator is a three phase, field sensed unit. The rotor and stator windings produce three phase alternating current, AC, which is rectified to direct current, DC. The electronic voltage regulator unit controls the alternator output voltage by high frequency switching of the rotor field circuit. Use only the correct Range Rover replacement fan belt. Occasionally check that the engine and alternator pulleys are accurately aligned.

It is essential that good electrical connections are maintained at all times. Of particular importance are those in the charging circuit (including those at the battery) which should be occasionally inspected to see that they are clean and tight. In this way any significant increase in circuit resistance can be prevented.

Do not disconnect battery cables while the engine is running or damage to the semi-conductor devices may occur. It is also inadvisable to break or make any connections in the alternator charging and control circuits while the engine is running.

The Model 15TR electronic voltage regulator employs micro-circuit techniques resulting in improved performance under difficult service conditions. The whole assembly is encapsulated in silicone rubber and housed in an aluminium heat sink, ensuring complete protection against the adverse effects of temperature, dust, and moisture etc.

The regulating voltage is set during manufacture to give the required regulating voltage range of 14.2 ± 0.2 volts, and no adjustment is necessary. The only maintenance needed is the occasional check on terminal connections and wiping with a clean dry cloth.

The alternator system provides for direct connection of a charge (ignition) indicator warning light, and eliminates the need for a field switching relay or warning light control unit. As the warning lamp is connected in the charging circuit, lamp failure will cause loss of charge. Lamp should be checked regularly and a spare carried.

When using rapid charge equipment to re-charge the battery, the battery must be disconnected from the vehicle.

---

| SYMPTOM | POSSIBLE CAUSE | CURE |
|---|---|---|
| S-Central door locking does not operate (on all doors) | 1. Battery discharged<br>2. Control unit in driver's door lock actuator faulty<br>3. Loose or broken connection in driver's door<br>4. Blown fuse | 1. Recharge<br>2. Fit new unit<br><br>3. Locate and rectify<br><br>4. Rectify |
| T-Central door locking does not operate (on one door only) | 1. Loose or broken connection<br>2. Lock actuator failure<br>3. Faulty lock<br>4. Mechanical linkages disconnected | 1. Locate and rectify<br><br>2. Fit new actuator<br>3. Rectify<br>4. Locate and rectify |
| U-Window lift will not operate | 1. Motor failure<br>2. Loose or broken connection<br>3. Faulty switch<br>4. Mechanical linkage faulty | 1. Fit new motor<br>2. Locate and rectify<br><br>3. Fit new switch<br>4. Rectify |
| V-Exterior mirrors fail to operate | 1. Loose or broken connection<br>2. Faulty switch<br>3. Mirror motor failure | 1. Locate and rectify<br><br>2. Fit new switch<br>3. Fit new motor |

347

# ELECTRICAL

## RANGE ROVER

### ALTERNATOR

**Remove and refit**

**Removing**

1. Disconnect battery ground lead
2. Disconnect leads from alternator.
3. Loosen alternator fixings, pivot alternator inwards and remove drive belt.
4. Remove three mounting bolts and lift the alternator clear of the engine.

**Refitting**

5. Fit the alternator and mounting bolts.

**NOTE: The fan guard is attached to the front fixing and the adjustment bracket bolt.**

6. Fit the drive belt and adjust the belt tension.
7. Tighten the mounting bolts and the adjustment bracket securing nut
8. Connect the wiring leads to the alternator.
9. Connect the battery.

### ALTERNATOR DRIVE BELT

**Adjust**

1. Loosen the alternator fixings and the adjustment link.
2. Pivot the alternator to give the required belt tension.
3. Belt tension should be 4 to 6mm (0.19 to 0.25 in) at the point indicated by the bold arrow.
4. Tighten the alternator fixing bolts and the adjustment link.

**NOTE: Check adjustment after running the engine at fast idle for 3 to 5 minutes, if a new belt has been fitted.**

### ALTERNATOR-LUCAS-TYPE A133/80

1. Cover
2. Regulator
3. Rectifier
4. Drive and bracket
5. Bearing assembly
6. Rotor
7. Slip ring end bearing
8. Slip rings
9. Slip ring end bracket
10. Stator
11. Brush box
12. Brushes
13. Through bolt
14. Suppressors

# ELECTRICAL

## ALTERNATOR-LUCAS-TYPE A133/80

### Overhaul

### Including Test (Bench)

**NOTE:** Alternator charging circuit-The ignition warning light is connected in series with the alternator field circuit. Bulb failure would prevent the alternator charging, except at very high engine speeds, therefore, the bulb should be checked before suspecting an alternator failure.

### Precautions

Battery polarity is **NEGATIVE GROUND**, which must be maintained at all times.

No separate control unit is fitted; instead a voltage regulator of micro-circuit construction is incorporated on the slip ring end bracket, inside the alternator cover.

Battery voltage is applied to the alternator output cable even when the ignition is switched off, the battery must be disconnected before commencing any work on the alternator. The battery must also be disconnected when repairs to the body structure are being carried out using electric welding equipment.

### Sequence of connections

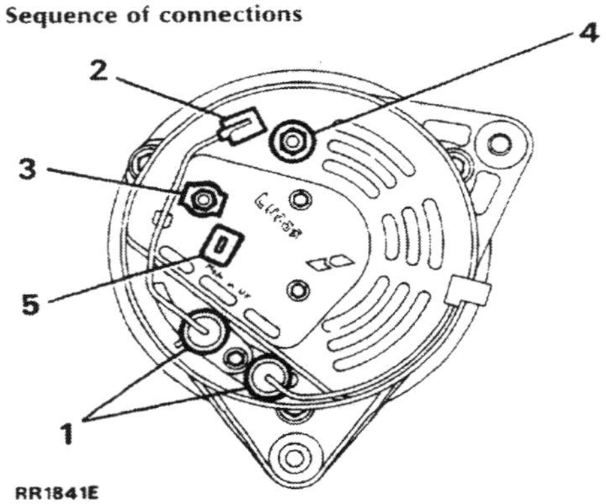

1. Suppression capacitors (two)
2. Positive suppression terminal
3. IND terminal
4. + output terminal
5. Sensing terminal

## ALTERNATOR TESTING

### Charging system check

1. Check the battery is in good condition, with an open circuit voltage of at leats 12.6 V. Recharge or fit a charged substitute battery to carry out test.
2. Check drive belt adjustment and condition. Rectify as necessary.
3. Check battery connections are clean and tight.
4. Check alternator connections are clean and tight.
5. Ensure that there is no continuous drain on battery due, for example, to interior, underhood or door edge lamps being left on.

### Alternator test

The following instructions refer to the use of suitable test equipment using a carbon pile rheostat.

6. Connect test equipment referring to the manufacturer's instructions.
7. Start engine and run at 3000 rev/min without accesory load.
8. Rotate the carbon pile load control (amps) to achieve the greatest output (amps) without allowing voltage to fall below 12.0 V. A reading of 80 amps, minus 10% to allow for EFI and Ignition loss, should be obtained.
9. Run engine at 3000 rev/min, switch selector to regulator test, read voltmeter. A reading of 13.6 to 14.4 V should be obtained.
10. Switch selector to diode/stator test, switch on headlamps to load alternator. Raise engine speed to 3000 rev/min, read voltmeter. The needle must be within the 'OK' range.

**Note: See also "Testing in position" - Section 86, page 13.**

### Testing-alternator removed

11. Withdraw the connectors from the alternator.
12. Remove the alternator.
13. Disconnect the suppressor and remove the alternator cover.

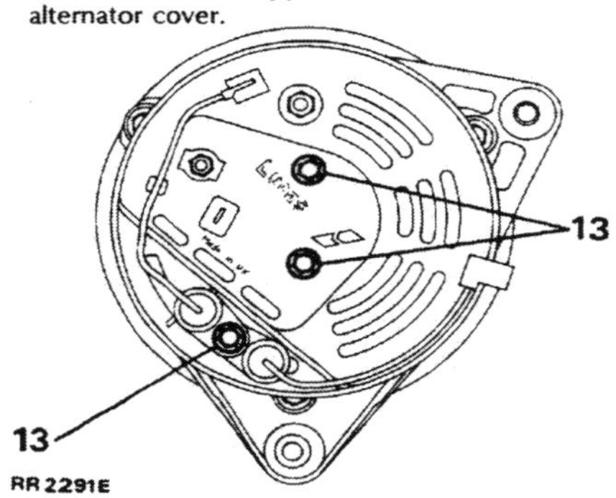

14. Disconnect the lead and remove the rectifier assembly.
15. Note the arrangement of the brush box connections and remove the screws securing the regulator to the brush box and withdraw. This screw also retains the inner brush mounting plate in position.
16. Remove the screw retaining the outer brush box in position and withdraw both brushes.
17. Check brushes for wear by measuring length of brush protruding beyond brush box moulding. If length is 10mm (0.4 in) or less, fit new brushes.
18. Check that brushes move freely in holders. If brush is sticking, clean with a mineral spirit moistened cloth or polish sides of brush with fine file.
19. Check brush spring pressure using push-type spring gauge. Gauge should register 136 to 279g (5 to 10 oz) when brush is pulled back until face is flush with housing. If reading is outside these limits, fit a new brush assembly.
20. Remove the two screws securing the brush box to the slip ring end bracket and lift off the brush box assembly.
21. Securely clamp alternator in a vice and release the stator winding cable ends from the rectifier by applying a hot soldering iron to the terminal tags of the rectifier. Pry out the cable ends when the solder melts.

## ELECTRICAL 86

22. Remove the two remaining screws securing the rectifier assembly to the slip ring end bracket and lift off the rectifier assembly. Further dismantling of the rectifier is not required.
23. Check the diodes. Connect the test equipment as shown and test each diode in turn, note whether lamp lights, then reverse test lead connections. The lamp should light in one direction only. Renew the rectifier assembly if a faulty diode is diagnosed.
24. Remove the slip ring end bracket bolts and lift off the bracket.
25. Connect a 12 volt battery and a 36 watt test lamp to two of the stator connections. Repeat the test replacing one of the two stator connections with the third. If test lamp fails to light in either test, fit a new stator.
26. Using a 110 volt a.c. supply and a 15 watt test lamp, test for insulation between any one of the three stator connections and stator laminations. If test lamp lights, fit a new stator.
27. Clean surfaces of slip rings using a solvent moistened cloth.
28. Inspect slip ring surfaces for signs of burning; remove burn marks using very fine sandpaper. On no account should emery cloth or similar abrasives be used, or any attempt made to machine the slip rings.
29. Note the position of the stator output leads in relation to the alternator fixing lugs, and lift the stator from the drive end bracket.
30. Connect an ohmmeter to the slip rings. A reading of 2.6 ohms should be recorded.
31. Using a 110 volt a.c. supply and a 15 watt test lamp, test for insulation between one of the slip rings and one of the rotor poles. If the test lamp lights, fit a new rotor.

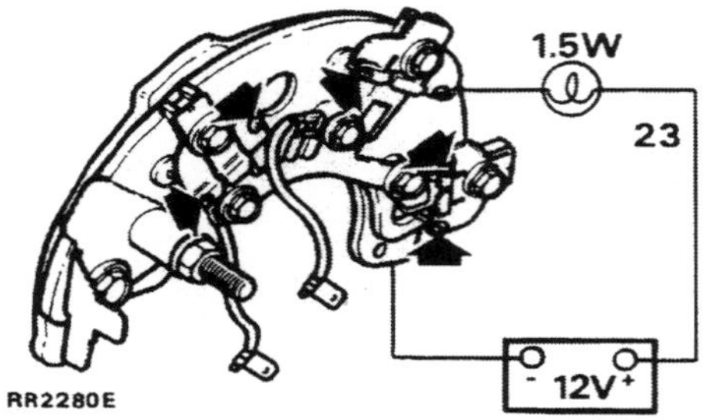

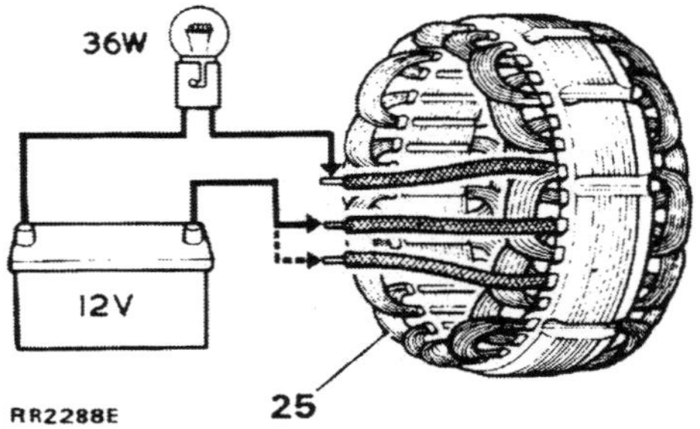

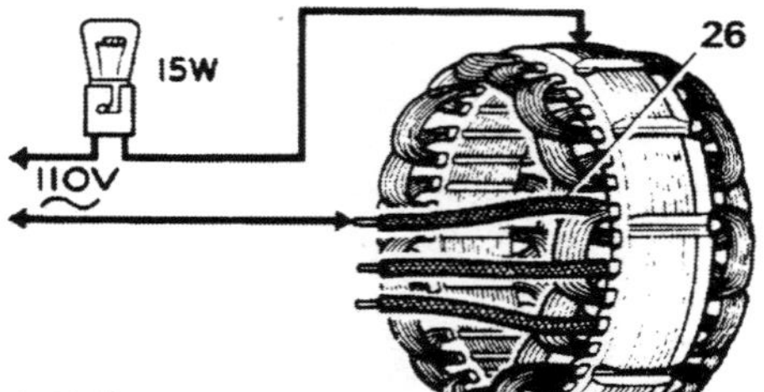

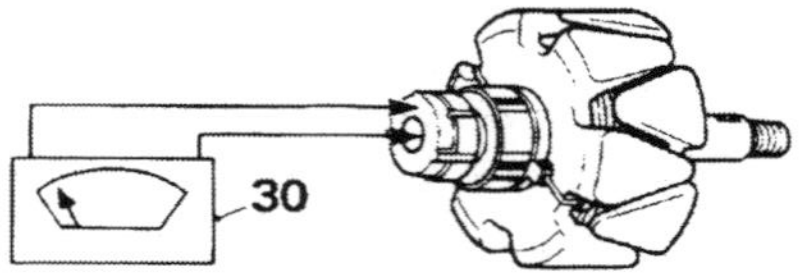

## ELECTRICAL 86

### Testing in position

### Charging circuit resistance test

1. Connect a low range voltmeter between the alternator terminal marked + and the positive terminal of the battery.
2. Switch on the headlamps and start the engine. Set the throttle to run at approximately 3000 rev/min. Note the voltmeter reading.
3. Transfer the voltmeter connections to the frame of the alternator and the negative terminal of the battery, and again note the voltmeter reading.

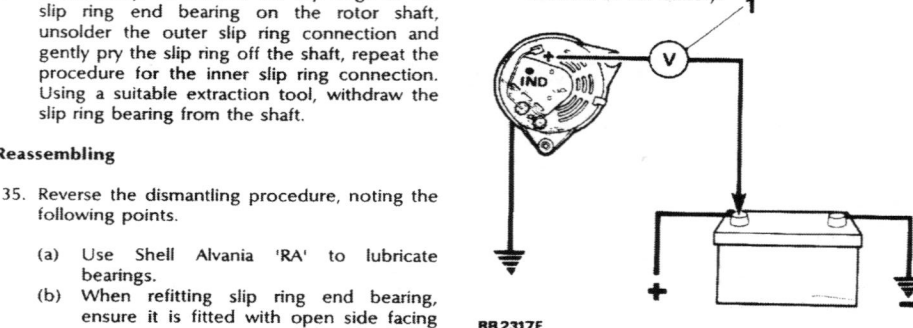

32. To separate the drive end bracket and rotor, remove the shaft nut, washers, woodruff key and spacers from the shaft.
33. Remove bearing retaining plate by removing the three screws. Using a press, drive the rotor shaft from the drive end bearing.
34. If necessary, to remove the slip rings or the slip ring end bearing on the rotor shaft, unsolder the outer slip ring connection and gently pry the slip ring off the shaft, repeat the procedure for the inner slip ring connection. Using a suitable extraction tool, withdraw the slip ring bearing from the shaft.

### Reassembling

35. Reverse the dismantling procedure, noting the following points.

    (a) Use Shell Alvania 'RA' to lubricate bearings.
    (b) When refitting slip ring end bearing, ensure it is fitted with open side facing rotor.
    (c) Use Fry's H.T.3 solder on slip ring field connections.
    (d) When refitting rotor to drive end bracket, support inner track of bearing. Do not use drive end bracket to support bearing when fitting rotor.
    (e) Tighten through-bolts evenly.
    (f) Fit brushes into housings before fitting brush moulding.
    (g) Tighten shaft nut to the correct torque, see Torque Values.
    (h) Refit regulator pack to brush moulding.

36. Reconnect the leads between the regulator, brush box and rectifier, as illustrated.
37. Refit the alternator.

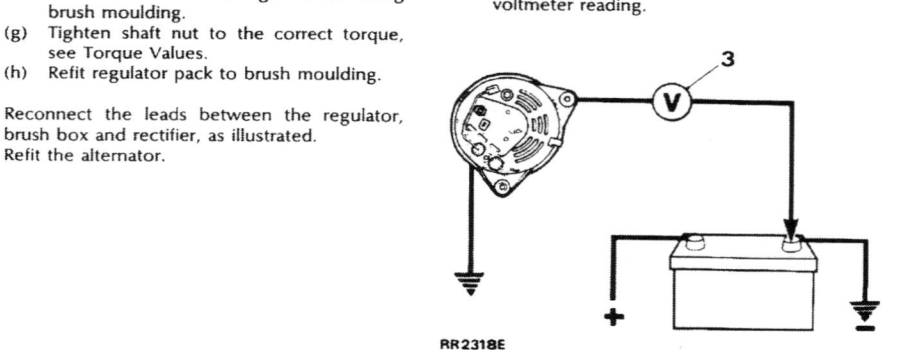

4. If the reading exceeds 0.5 volt on the positive side or 0.25 volt on the negative side, there is a high resistance in the charging circuit which must be traced and remedied.

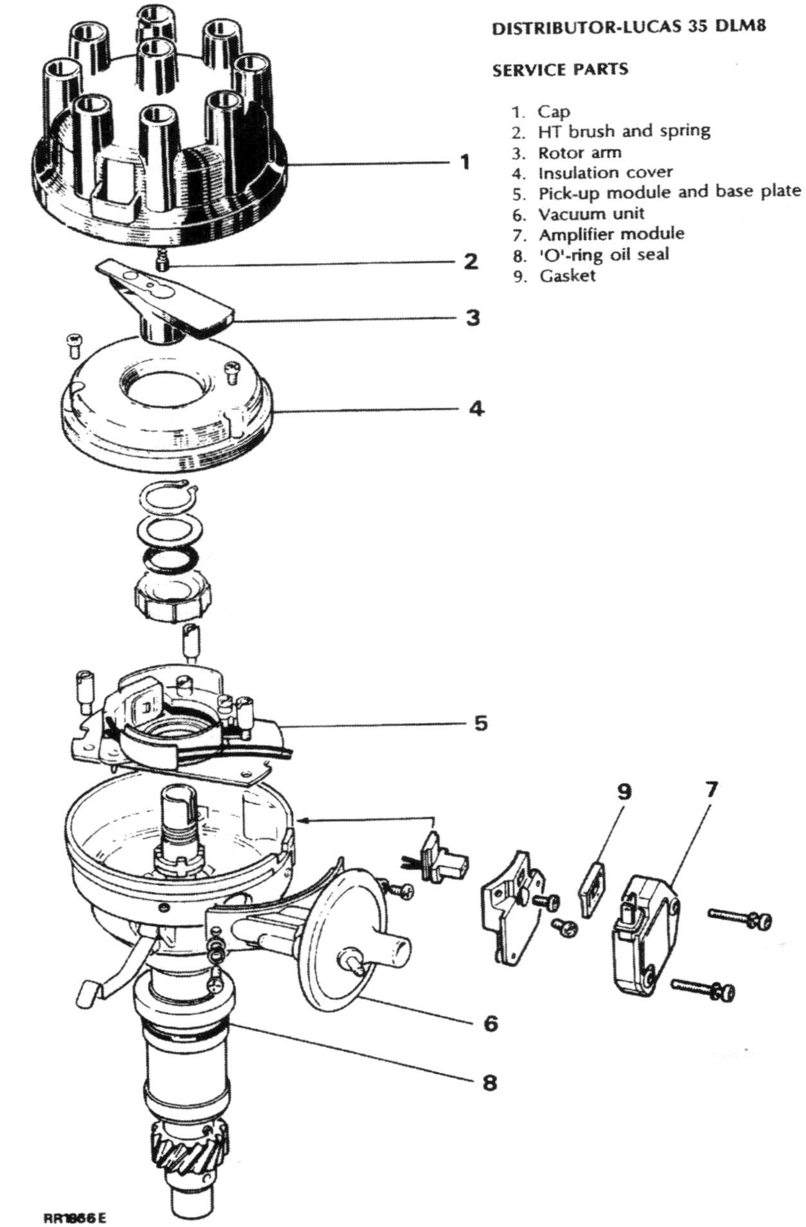

# ELECTRICAL 86

## RANGE ROVER

### DISTRIBUTOR-LUCAS 35 DLM8

#### SERVICE PARTS

1. Cap
2. HT brush and spring
3. Rotor arm
4. Insulation cover
5. Pick-up module and base plate assembly
6. Vacuum unit
7. Amplifier module
8. 'O'-ring oil seal
9. Gasket

---

## RANGE ROVER

### ELECTRONIC IGNITION

A Lucas 35DLM8 distributor is employed. This has a conventional vacuum advance unit and centrifugal automatic advance mechanism.

A pick-up module, in conjunction with a rotating timing reluctor inside the distributor body, generates timing signals. These are applied to an electronic ignition amplifier module mounted on the side of the distributor body.

**NOTE: The pick-up air gap is factory set. Do not adjust the gap unless the pick-up is being changed or the base plate has been moved. Use a non-ferrous feeler gauge to set the air gap.**

### DISTRIBUTOR

#### Remove and refit

#### Removing

1. Disconnect the battery negative lead.
2. Disconnect the vacuum hose.
3. Remove the distributor cap.
4. Disconnect low tension lead from the coil.
5. Mark distributor body in relation to centre line of rotor arm.
6. Add alignment marks to distributor and front cover.

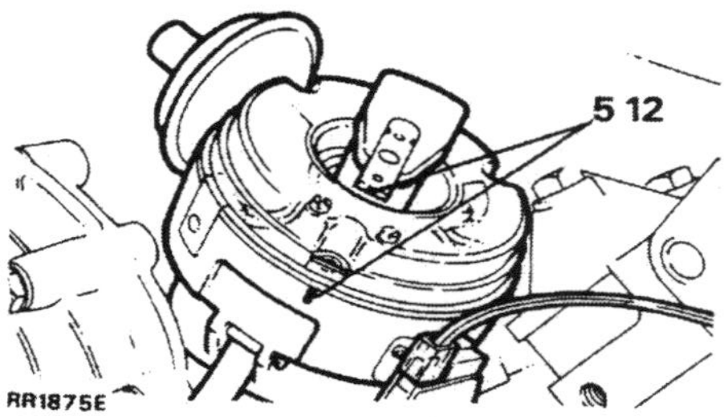

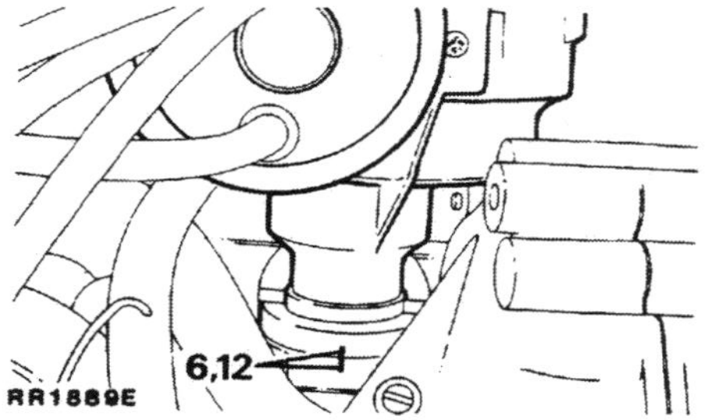

**NOTE: Marking distributor enables refitting in exact original position, but if engine is turned while distributor is removed, complete ignition timing procedure must be followed.**

7. Release the distributor clamp and remove the distributor.

#### Refitting

**NOTE: If a new distributor is being fitted, mark body in same relative position as distributor removed.**

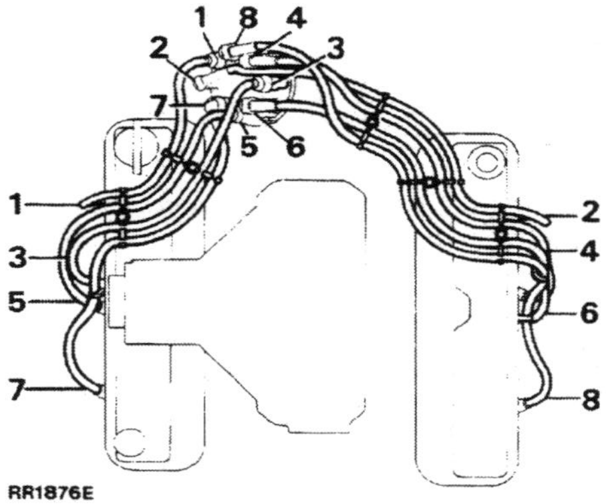

8. Leads for distributor cap should be connected as illustrated.
   Figures 1 to 8 inclusive indicate plug lead numbers.
   RH-Right hand side of engine, when viewed from the rear.
   LH-Left hand side of engine, when viewed from the rear.

Continued

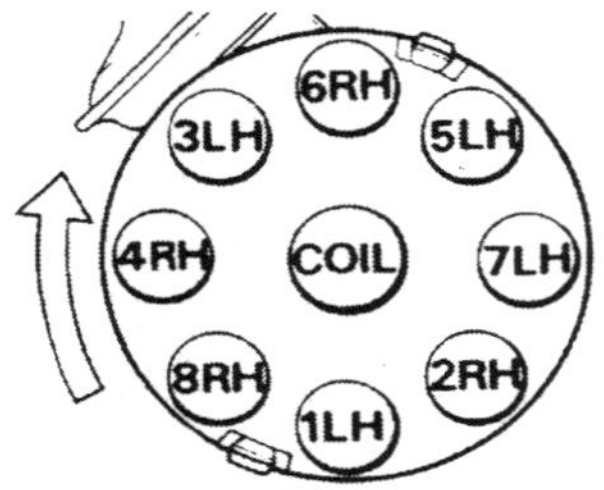

# ELECTRICAL 86

## RANGE ROVER

9. If engine has not been turned while distributor has been removed, proceed as follows (items 10 to 17). Alternatively proceed to instruction 18.
10. Fit new 'O' ring seal to distributor housing.
11. Turn distributor drive until centre line of rotor arm is 30° counter-clockwise from mark made on top edge of distributor body.
12. Fit distributor in accordance with alignment markings.

**NOTE: It may be necessary to align oil pump drive shaft to enable distributor drive shaft to engage in slot.**

13. Using suitable electronic equipment, set the ignition timing, see IGNITION TIMING-Adjust.
14. Fit clamp and bolt. Secure distributor in exact original position.
15. Connect vacuum hose to distributor and low tension lead to coil.
16. Fit distributor cap.
17. Reconnect battery.
18. If, with distributor removed, engine has been turned it will be necessary to carry out the following procedure.
19. Set engine-No. 1 piston to static ignition timing figure (see Engine Tuning Data- Section 05) on compression stroke.
20. Turn distributor drive until rotor arm is approximately 30° counter-clockwise from number one sparking plug lead position on cap.
21. Fit distributor to engine.
22. Check that centre line of rotor arm is now in line with number one sparking plug lead on cap. Reposition distributor if necessary.
23. If distributor does not seat correctly in front cover, oil pump drive is not engaged. Engage by lightly pressing down distributor while turning engine.
24. Fit clamp and bolt leaving both loose at this stage.
25. Set the ignition timing statically to within 2°-3° of T.D.C.
26. Connect the vacuum hose to the distributor.
27. Fit low tension lead to coil.
28. Fit distributor cap.
29. Reconnect the battery.
30. Using suitable electronic equipment set the ignition timing, see IGNITION TIMING-Adjust.

## DISTRIBUTOR-LUCAS 35DLM8

### Overhaul

### DISTRIBUTOR CAP

1. Unclip and remove the cap
2. Fit a new cap if known to be faulty.
3. Clean the cap and HT brush with a lint free cloth.

### ROTOR ARM

4. Pull rotor arm from shaft.
5. Fit a new rotor arm if known to be faulty.

### INSULATION COVER (Flash shield)

6. Remove cover, secured by three screws.
7. Fit a new cover if known to be faulty.

### VACUUM UNIT

8. Remove two screws from vacuum unit securing bracket, disengage vacuum unit connecting rod from pick-up base plate connecting peg, and withdraw vacuum unit from distributor body.

## FITTING PICK-UP AND BASE PLATE ASSEMBLY

18. Pick-up leads must be prevented from fouling the rotating reluctor. Both leads should be located in plastic guide as illustrated. Check during re-assembly.

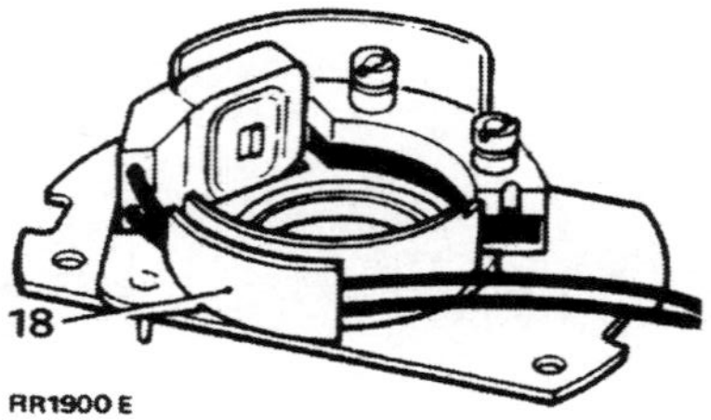

## AMPLIFIER MODULE

9. Remove two screws and withdraw the module.
10. Remove the gasket.
11. Remove two screws securing the cast heatsink and remove the heatsink.

**WARNING: The amplifier module is a sealed unit containing Beryllia. This substance is extremely dangerous if handled. Do not attempt to open or crush the module.**

## PICK-UP AND BASE PLATE ASSEMBLY

12. Use circlip pliers to remove the circlip retaining the reluctor on rotor shaft.
13. Remove the flat washer and then the 'O' ring recessed in the top of the reluctor.
14. Gently withdraw the reluctor from the shaft, taking care not to damage the teeth.

**NOTE: Coupling ring fitted beneath reluctor.**

15. Remove three support pillars and cable grommet. Lift out the pick-up and base plate assembly.

**NOTE: Do not disturb the two barrel nuts securing the pick-up module, otherwise the air gap will need re-adjustment.**

16. Fit a new pick-up and base plate assembly if module is known to be faulty, otherwise check pick-up winding resistance (2k-5k ohm).

## RE-ASSEMBLY

17. This is mainly a reversal of the dismantling procedure, noting the following points:

## LUBRICATION

**Apply clean engine oil:**

a. A spot into the rotor spindle before fitting rotor arm.

**Apply Omnilube 2 (or equivalent) grease.**

b. Auto advance mechanism.
c. Pick-up plate centre bearing.
d. Pre tilt spring and its rubbing area (pick-up and base plate assembly).
e. Vacuum unit connecting peg (pick-up and base plate assembly).
f. The connecting peg hole in vacuum unit connecting rod.

## REFITTING RELUCTOR

19. Slide reluctor as far as it will go on rotor shaft, then rotate reluctor until it engages with the coupling ring beneath the pick-up base plate. The distributor shaft, coupling ring and reluctor are 'keyed' and rotate together. Fit the 'O' ring, flat washer and retaining circlip.

## PICK-UP AIR GAP ADJUSTMENT

20. The air gap between the pick-up limb and reluctor teeth must be set within the specified limits, using a non-ferrous feeler gauge.
21. If adjustment is necessary, slacken the two barrel nuts to set the air gap. See Engine Tuning Data.

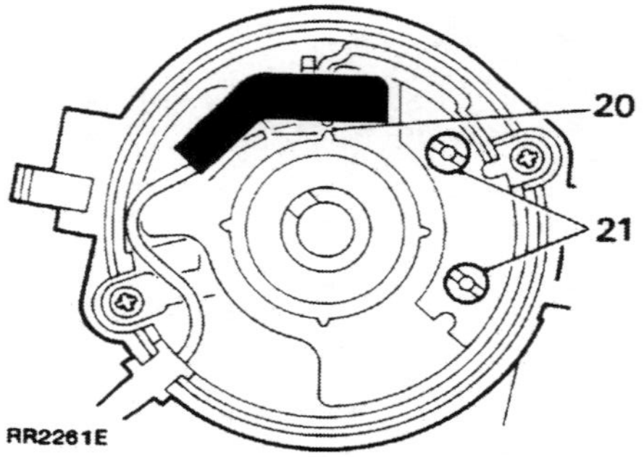

**Continued**

## ELECTRICAL

NOTE: When the original pick-up and base plate assembly has been refitted the air gap should be checked, and adjusted if necessary.

When fitting a new assembly the air gap will require adjusting to within the specified limits.

### AMPLIFIER MODULE

22. Before fitting the module, apply MS4 Silicone grease or equivalent heat-conducting compound to the amplifier module backplate, the seating face on distributor body and both faces of the heatsink casting.

### IGNITION COIL

### Remove and refit

**Removing**

1. Disconnect the battery negative terminal.
2. Disconnect the High Tension and Low Tension electrical leads from the ignition coil.

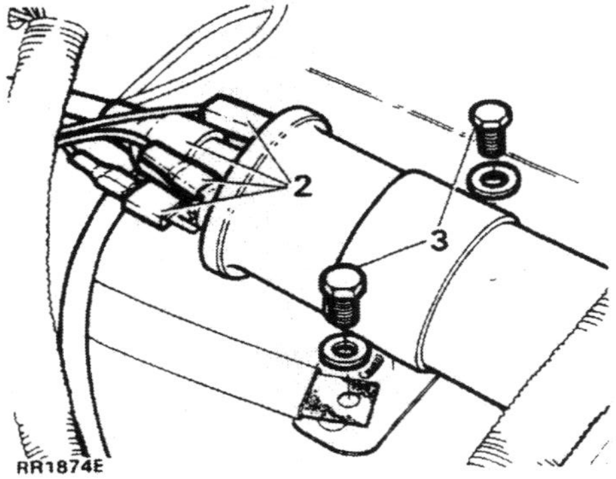

3. Remove the two bolts securing the coil to the valance.

NOTE: A ground strap is located under one of the bolts.

4. Remove the coil from the engine compartment.

**Refitting**

5. Reverse the removal instructions.

NOTE: Ensure that the bolting location for the ground strap is free from paint and grease. Coat the area around the bolt with Petroleum Jelly.

### IGNITION TIMING

**Adjust**

1. It is essential that the following procedures are adhered to. Inaccurate timing can lead to serious engine damage and additionally create failure to comply with emission regulations. If the engine is being checked in the vehicle, the air conditioning compressor must be disengaged.
2. On initial engine build, or if the distributor has been disturbed for any reason, the ignition timing must be set statically to 6° B.T.D.C. (This sequence is to give only an approximation in order that the engine may be started) ON NO ACCOUNT MUST THE ENGINE BE STARTED BEFORE THIS OPERATION IS CARRIED OUT.

**Equipment required**

Calibrated Tachometer
Stroboscopic lamp

3. Couple stroboscopic timing lamp and tachometer to engine following manufacturer's instructions.
4. Disconnect the vacuum hose from the distributor.
5. Start engine, with no load and not exceeding 3,000 rev/min run engine until normal operating temperature is reached. (Thermostat open). Check that the normal idling speed falls within the tolerance specified in the data section.
6. Idle speed for timing purposes must not exceed 800 rev/min.
7. With the distributor clamping bolt loosened turn distributor until the timing flash coincides with the timing pointer and the correct timing mark on the rim of the torsional vibration damper as shown in the engine tuning section.

8. Retighten the distributor clamping bolt securely. Recheck timing in the event that retightening has disturbed the distributor position.
9. Refit vacuum hose.
10. Disconnect stroboscopic timing lamp and tachometer from engine.

### LUCAS CONSTANT ENERGY IGNITION SYSTEM 35DLM8-PRELIMINARY CHECKS

Inspect battery cables and connections to ensure they are clean and tight. Check battery state of charge if in doubt as to its condition.

Inspect all LT connections to ensure that they are clean and tight. Check the HT leads are correctly positioned and not shorting to ground against any engine components. The wiring harness and individual cables should be firmly fastened to prevent chafing.

### PICK-UP AIR GAP

Check the air gap between pick-up limb and reluctor teeth, using a non-ferrous gauge, see 'Engine Tuning Data'.

NOTE: The gap is set initiallt at the factory and will only require adjusting if tampered with or when the pick-up module is replaced.

### TEST 1:

**HT Sparking**

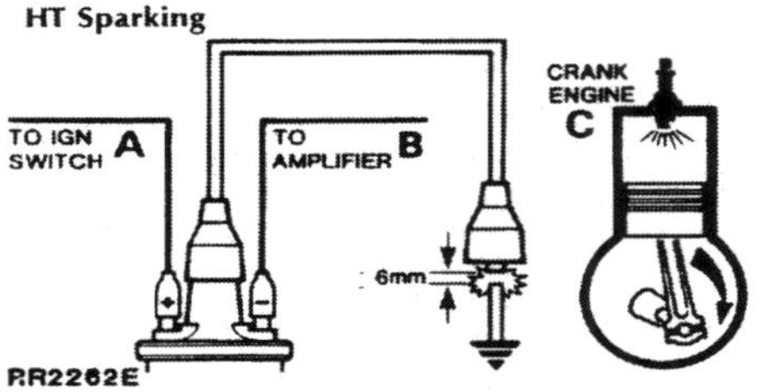

Remove coil/distributor HT lead from distributor cover and hold approximately 6mm (0.25 in) from the engine block, using suitable insulated pliers. Switch the ignition 'On' and operate the starter. Regular sparking indicates fault in HT distribution, plugs, timing or fuelling, proceed to Test 6. If no spark or weak spark occurs proceed to Test 2.

### Test 2:

**LT Voltage**

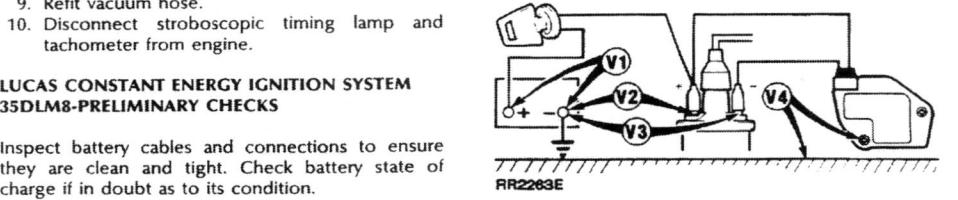

Swith the ignition 'On' - engine stationary.

(a) Connect voltmeter to points in the circuit indicated by V1 to V4 and make a note of the voltage readings.

(b) Compare voltages obtained with the specified values listed below:

**EXPECTED READINGS**

V1   More than 12 volts.
V2   1 volt maximum below volts at V1.
V3   1 volt maximum below volts at V1.
V4   0 volt - 0.1 volt.

(c) If all readings are correct proceed to Test 3.
(d) Check incorrect reading(s) with chart to identify area of possible faults, i.e. faults listed under heading SUSPECT and rectify.
(e) If coil and amplifier is suspected, disconnect LT lead at coil, repeat V3. If voltage is still incorrect, fit new coil. If voltage is now correct, check LT lead, if satisfactory fit new amplifier.
(f) If engine will not start proceed to Test 3.

Continued

# ELECTRICAL 86

| 1 | 2 | 3 | 4 | SUSPECT |
|---|---|---|---|---------|
| L | • | • | • | DISCHARGED BATTERY |
| • | L | L | • | IGN. SWITCH AND/OR WIRING |
| • | • | L | • | COIL OR AMPLIFIER |
| • | • | • | H | AMPLIFIER GROUND |

KEY

• Expected Voltage
H Voltage higher than expected
L Voltage lower than expected

### TEST 3:

**Amplifier Switching**

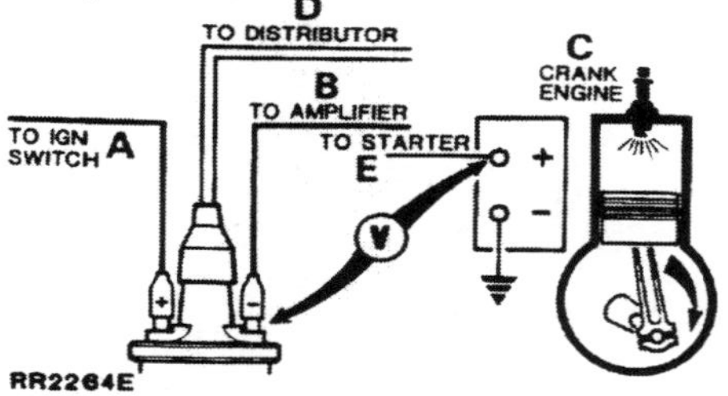

Connect the voltmeter between battery positive (+ve) terminal and H.T. coil negative (-ve) terminal, the voltmeter should register 0 volts.

Switch the ignition 'On' then crank the engine. The voltmeter reading should increase when cranking, in which case proceed to Test 5.

If there is no increase in voltage during cranking proceed to Test 4.

### TEST 4:

**Pick-up Coil Resistance**

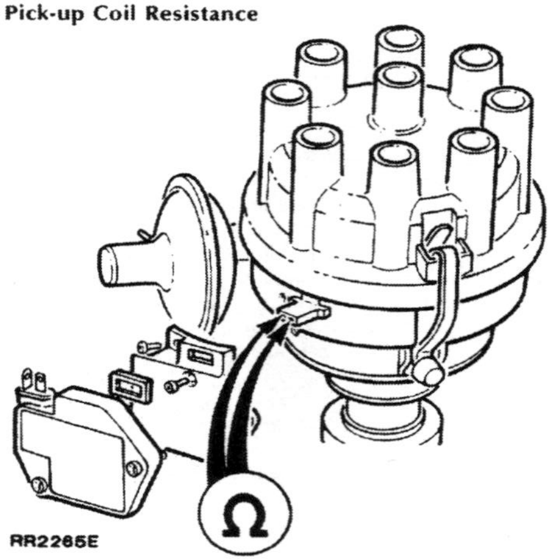

Remove the amplifier.

Connect the ohmmeter leads to the two pick-up terminals in the body of the distributor.

The ohmmeter should register between 2k and 5k ohm if pick-up is satisfactory. If ohmmeter reading is correct, check all connections between pick-up and amplifier, if satisfactory, fit new amplifier. If the engine still does not start carry out Test 5.

Change the pick-up if ohmmeter reading is incorrect. If the engine still does not start proceed to Test 5.

Continued

---

# ELECTRICAL 86

### TEST 5:

**Coil H.T. Sparking**

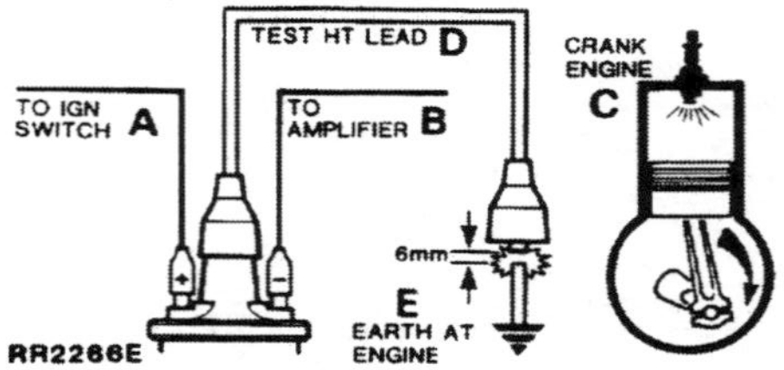

Remove existing coil/distributor H.T. lead and fit test H.T. lead to coil tower. Using suitable insulated pliers, hold free end about 6mm (0.25 in) from the engine block and crank the engine. There should be good H.T. sparking.

If weak or no sparking, fit new coil, repeat test.

H.T. sparking good, repeat test with original H.T. lead. If sparking is good carry out Test 6.

If weak or no sparking, fit new H.T. lead, if engine will not start carry out Test 6.

### TEST 6:

**Rotor Arm**

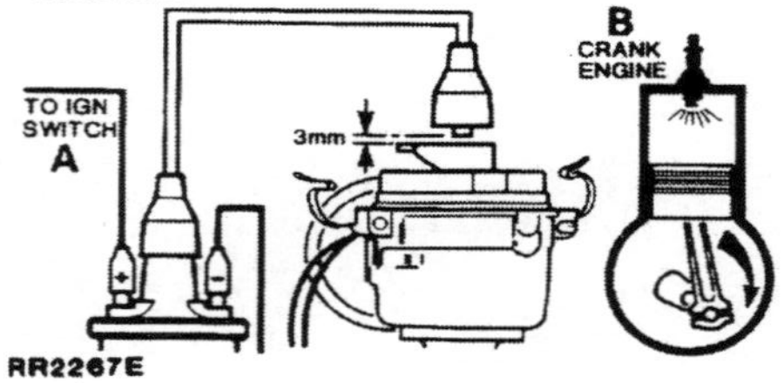

Remove distributor cover. Disconnect coil H.T. lead from cover, using insulated pliers hold about 3mm (0.13 in) above rotor arm electrode and crank the engine.

There should be no H.T. sparking between rotor and H.T. lead. If satisfactory carry out Test 7.

If H.T. sparking occurs, an earth fault on rotor arm is indicated. Fit new rotor arm. If engine will not start carry out Test 7.

### TEST 7:

**Visual and H.T. Cable Checks**

| Examine: | Should be: |
|----------|-----------|
| 1. Distributor Cover | Clean, dry, no tracking marks |
| 2. Coil Top | Clean, dry, no tracking marks. |
| 3. HT Cable Insulation | Must not be cracked, chafed or perished |
| 4. HT Cable Continuity | Must not be open circuit |
| 5. Sparking Plugs | Clean, dry, and set to correct gap |

NOTE:

1. Reluctor — Must not foul pick-up or leads
2. Rotor and Insulation Cover — Must not be cracked or show signs of tracking marks

## ELECTRICAL 86

### HEADLAMP ASSEMBLY/BULB REPLACEMENT

**Remove and refit**

**Removing**

1. Disconnect the battery negative lead.
2. Remove the radiator grille - see Body Section 76 in LSM 180 WS 1.
3. Remove three crosshead screws and the headlamp retaining rim.

4. DO NOT disturb the two adjusting screws.
5. Withdraw the headlamp unit and disconnect the wiring plug from the rear of the unit.

6. Remove the rubber dust cover.
7. To remove bulb: release the bulb retaining clips and withdraw the halogen bulb.
8. Remove three securing screws, prise away the grommet and withdraw the headlamp bowl.

**Refitting**

9. Reverse removal procedure ensuring that the quartz envelope of the bulb is not touched. Wipe the bulb gently using methylated spirits if contact does occur.

---

## ELECTRICAL 86

### INSTRUMENT BINNACLE WARNING LIGHT SYMBOLS

- Trailer connected-flashes with direction indicators (green)
- Direction indicator- left turn / right turn (green)
- Seat belt (red)
- Headlamp high beam on (blue)
- Engine oil pressure, low (red)
- Electronic fuel injection warning lamp (red) indicates failure of air flow sensor, throttle potentiometer, water temperature thermistor, or Lambda sensor
- Ignition on (red)
- Low coolant (red)
- Automatic gearbox oil temperature high (red)
- Fuel indicator, low (amber)
- Low wash fluid (amber)
- Transmission hand brake on (red)
- Brake pad wear (amber)
- Brake fluid level low (red)

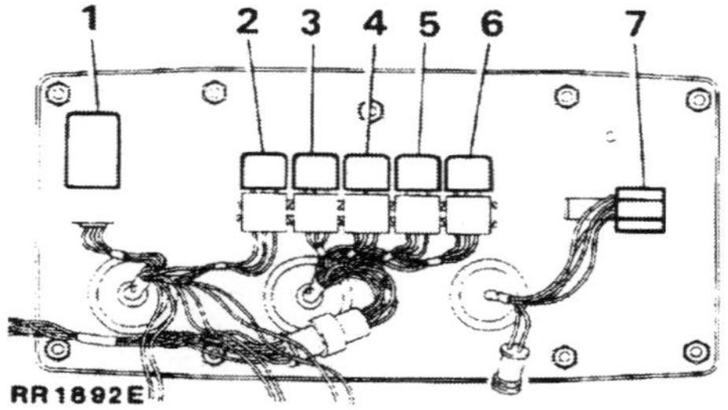

# ELECTRICAL 86

| Relay | Circuit Diagram Item Number | Colour |
|---|---|---|
| 1. Headlamp wash timer unit | 17. Main circuit diagram | Black |
| 2. Heated rear window | 67. Main circuit diagram | Natural |
| 3. Starter solenoid relay | 6. Main circuit diagram | Natural |
| 4. Compressor clutch | 11. Air conditioning diagram | Natural |
| 5. Condenser fan | 9. Air conditioning diagram | Natural |
| 6. Air conditioning/heater | 5. Air conditioning diagram | Natural |
| 7. Stowage position | Not used | ------ |
| 8. Rear wiper delay | 132. Main circuit diagram | Black |
| 9. Ignition load relay | 1. Main circuit diagram | Black |
| 10. Window lift relay | 63A. Main circuit diagram | Natural |
| 11. Auxiliary lamp relay | 86. Main circuit diagram | Red |
| 12. Front wiper delay | 14. Main circuit diagram | Yellow |
| 13. Voltage sensitive switch | 70. Main circuit diagram | Red |
| 14. Interior lamp delay | 99. Main circuit diagram | Blue |
| 15. Flasher/Hazard unit | 73. Main circuit diagram | Natural |
| 16. Main EFI relay | 22. EFI circuit diagram | Natural |
| 17. Fuel pump relay | 21. EFI circuit diagram | |

Closure panel viewed from the engine compartment, with protective cover removed.

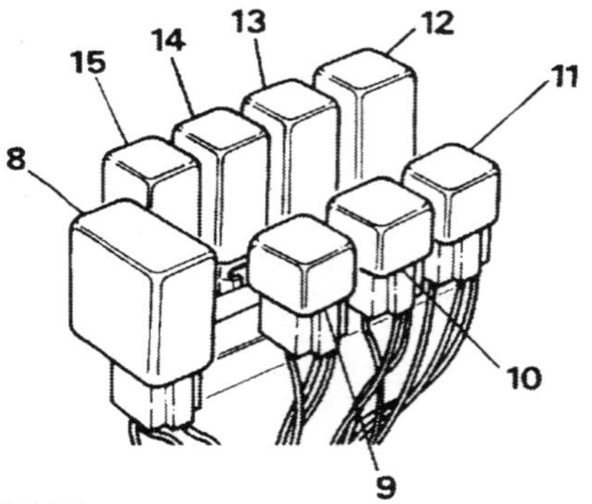

Steering column mounted relays viewed with the lower dash panel removed.

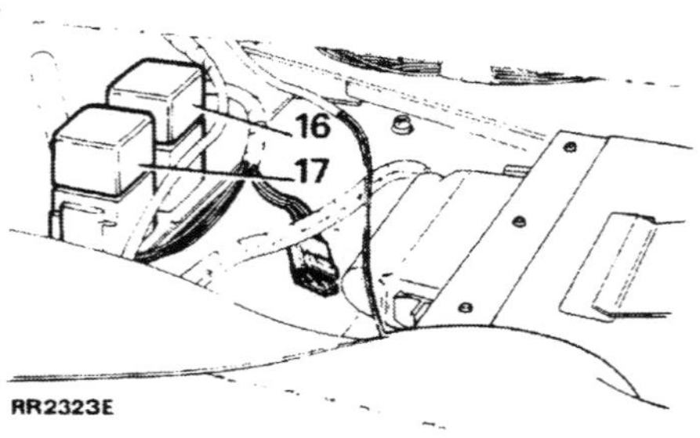

5. Pull the appropriate relay off its multi-plug.

### Refitting

6. Reverse the removal procedure.

## RELAYS-(Mounted on the steering column support bracket)

### Remove and refit

### Removal.

1. Disconnect the battery negative lead.
2. Remove the six screws securing the lower fascia panel.
3. Lower the dash panel, disconnect the electric leads from the dimming control switch and remove the fascia panel.
4. Locate the appropriate relay on the relay mounting bracket, carefully pull the relay off the multi-plug.

### Refitting

5. Reverse the removal procedure.

## RELAYS-(Floor mounted beneath front seat)

### Remove and refit

### Removing

1. Position seat to gain access to the required relay.
2. Disconnect the battery negative lead.
3. Carefully pull the relay off the multi-plug.

### Refitting

4. Reverse the removal procedure.

Main EFI (black terminal block) and fuel pump relays (blue terminal block) mounted beneath right hand front seat.

**NOTE: Refer to fuel injection section of manual for full information on E.F.I. relays.**

## RELAYS-(Mounted on the engine compartment closure panel).

### Remove and refit

### Removing

1. Lift the hood.
2. Disconnect the battery negative lead.
3. Remove the bolt securing the relay protective cover, located on the front of the engine compartment closure panel.
4. Remove the cover.

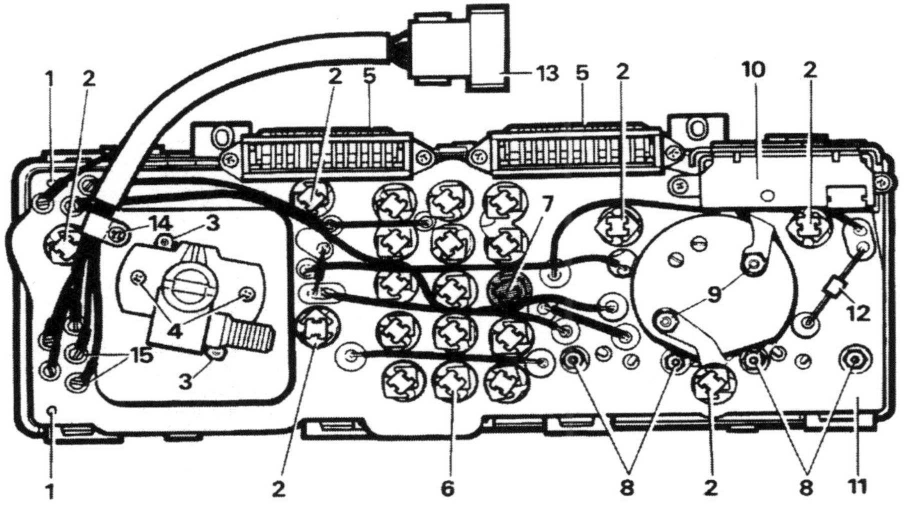

# ELECTRICAL 86

# RANGE ROVER

## Instrument case (back)

1. Locating pegs
2. Panel light bulbs
3. Speedometer securing screw
4. Speedometer drive securing screws
5. Harness connectors
6. Warning light bulb (14)
7. No charge warning light bulb (red holder)
8. Temperature and fuel gauge unit securing nuts
9. Tachometer securing nuts
10. Multi-function unit
11. Printed circuit
12. Pull-up resistor-high temperature gearbox oil
13. Single multi-plug
14. Single multi-plug securing screw
15. Single multi-plug wiring connecting screws (5)

## PRINTED CIRCUIT HARNESS CONNECTIONS

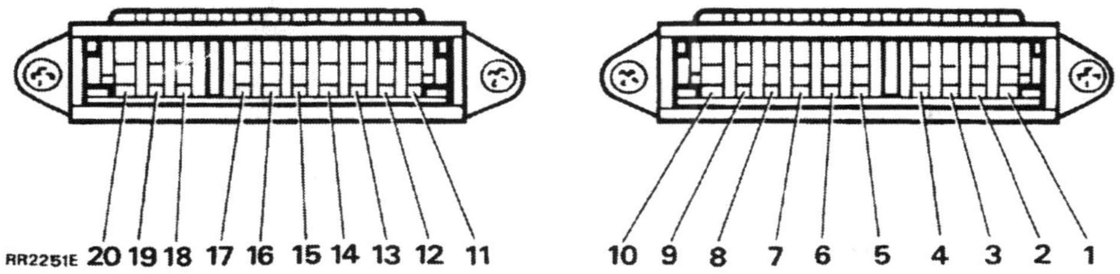

Sequence of connections looking towards the back of instrument case.

## CIRCUIT SERVED

1. Tacho signal
2. Ignition switch 12V+
3. Low coolant input
4. Ground-VE
5. Ignition warning light
6. Low oil level/pressure warning light
7. High beam warning light
8. Zero volts from dimmer
9. Trailer warning light
10. Direction indicators warning light
11. Seat belts warning light
12. E.F.I. warning light
13. Temperature warning light (automatic gearbox)
14. Low wash fluid warning light
15. Not used
16. 12V+ from dimmer
17. Brake fail warning light
18. Panel illumination bulbs (6 off)
19. Low fuel warning light
20. Low coolant warning light
21. Brake pad wear warning light
22. Hand brake warning light
23. Fuel tank unit and fuel gauge
24. Temperature gauge
25. Not used

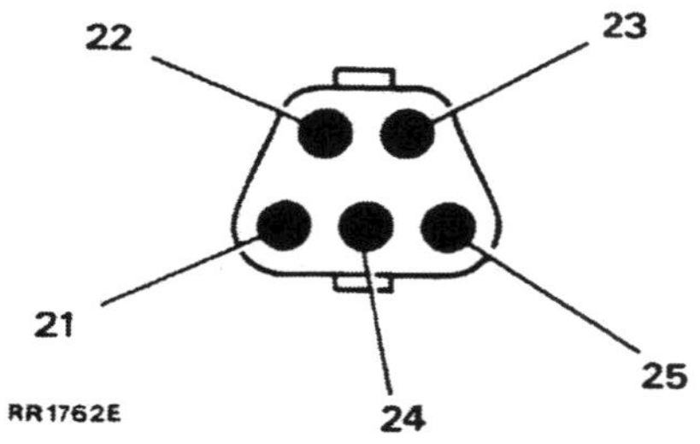

**NOTE:** The following 21 to 25 are connected at the single multi-plug located behind the binnacle

Sequence of pin connections viewed on the binnacle harness plug.

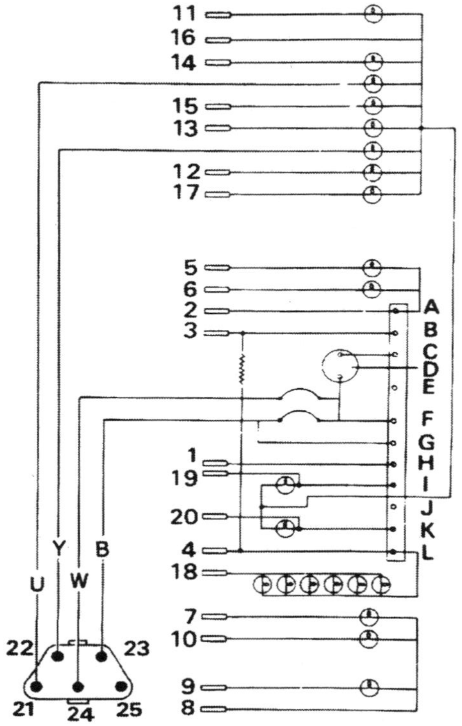

## MULTI-FUNCTION UNIT

A. 12V+ supply
B. Input to low coolant circuit
C. Tachometer drive
D. Tachometer
E. Spare
F. 10V+ stabilised
G. Input to fuel tank unit - stabilised
H. Tachometer signal
I. Low fuel warning light
J. Spare
K. Low coolant warning light
L. Ground

# ELECTRICAL 86

## RANGE ROVER

Window lifts and door locks
Circuit diagram

RR2282E

---

## RANGE ROVER

# ELECTRICAL 86

## ELECTRICAL EQUIPMENT-CIRCUIT DIAGRAMS

RR2245E

**ELECTRIC MIRRORS -**
**Circuit diagram - RR2245E**

1. Clinch
2. Main cable connections
   NK: (+) Battery feed - central locking
   WO: (+) Key position 1 - window lift
   B: (-) Ground
3. Switch unit-central door locking (drivers door)
4. Fuel flap actuator
5. Lock unit-central door locking (front passenger door)
6. Window lift motor L/H front
7. Window lift motor R/H front
8. Isolator switch

**WINDOW LIFTS AND DOOR LOCKS -**
**Circuit diagram - RR2282E**

1. Main cable connections
2. Clinches
3. Switch unit-central door locking (drivers door)
4. Fuel flap actuator
5. Lock unit-central door locking (front passenger door)
6. Mirror control switch
7. Ground - via main cable
8. Mirror heating elements - active with heated rear screen
9. Fuse 10 - heating elements
10. Fuse 13
11. Window lift switch L/H front
12. Window lift switch R/H front
13. Window lift switch L/H rear
14. Window lift switch R/H rear
15. Window lift motor L/H rear
16. Window lift motor R/H rear
17. Lock unit central door locking L/H rear door
18. Lock unit central door locking R/H rear door
19. Rocker switch operating levers

**CABLE COLOUR CODE**

| | | | | | |
|---|---|---|---|---|---|
| B | Black | N | Brown | R | Red | W | White |
| G | Green | O | Orange | S | Grey | Y | Yellow |
| K | Pink | P | Purple | U | Blue | | |

The last letter of a colour code denotes the tracer.

358

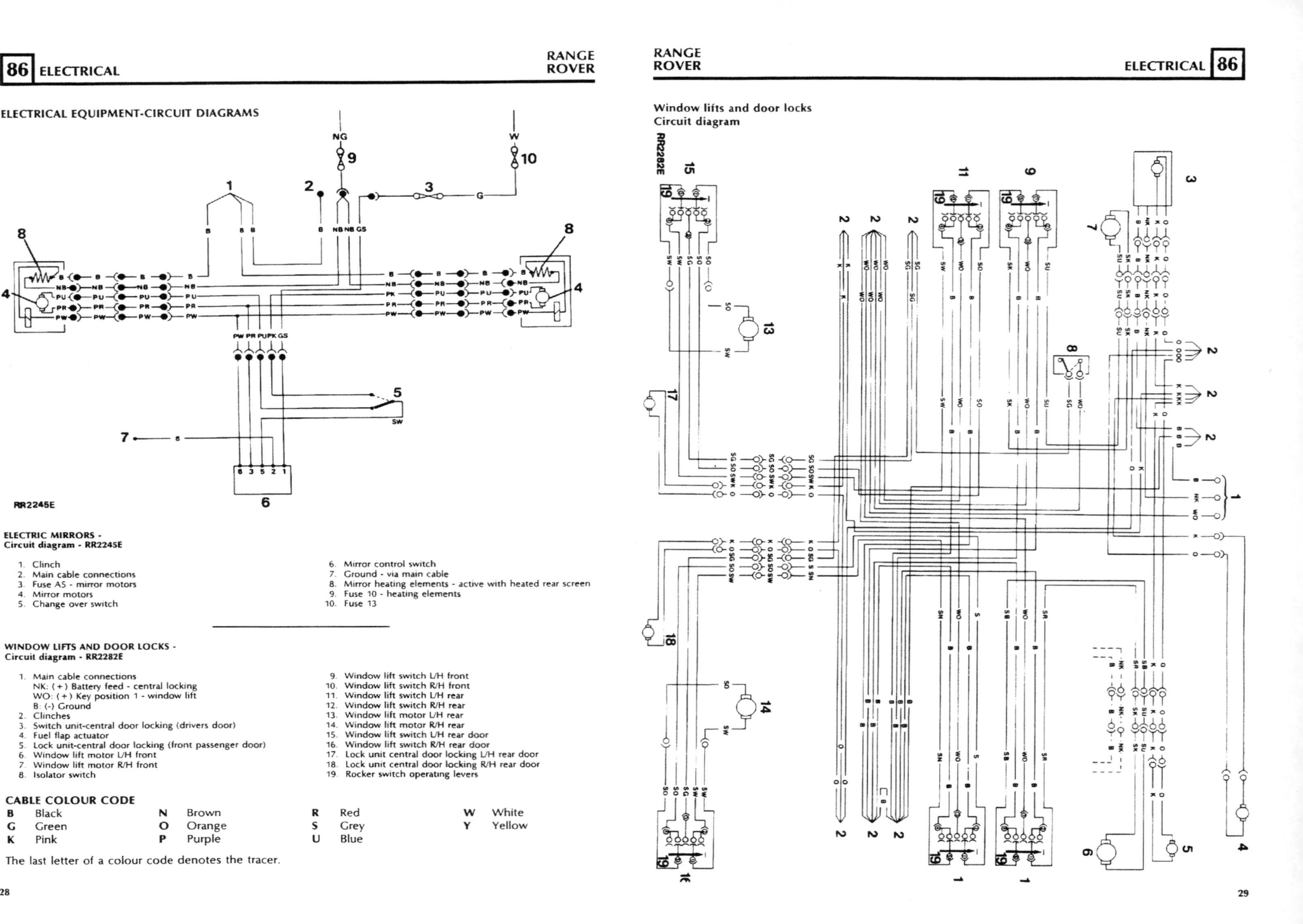

# ELECTRICAL 86

## MAIN CIRCUIT DIAGRAM
### Left hand steering - RR2334S and RR2335S

1. Ignition load relay
2. Battery
3. Terminal post
4. Starter solenoid
5. Starter motor
6. Starter relay
7. Starter inhibit switch (automatic)
8. Ignition switch
9. Tachometer
10. Ignition warning lamp
11. Alternator
12. Fuse 7
13. Front wipe/wash switch
14. Front wipe delay unit
15. Front wiper motor
16. Front wash pump
17. Headlamp wash timer unit (option)
18. Headlamp wash pump (option)
19. Main lighting switch
20. Fuse 6
21. Fuse 5
22. LH side lamp
23. LH tail lamp
24. Number plate lamp (2 off)
25. Main beam dip/flash switch
26. Radio illumination
27. RH side lamp
28. RH tail lamp
29. Rheostat
30. Fuse 3
31. Fuse 4
32. Fuse 1
33. Fuse 2
34. Rear fog switch
35. Fuse 12
36. Switch illumination (2 off)
37. Cigar lighter illumination (2 off)
38. Heater illumination (4 off)
39. Clock illumination
40. Automatic gear selector illumination (2 off)
41. Instrument illumination (6 off)
42. Rear fog warning lamp
43. LH rear fog
44. RH rear fog
45. LH dip beam
46. RH dip beam
47. LH main beam
48. RH main beam
49. Main beam warning lamp
50. Fuel gauge
51. Fuel gauge sender unit
52. Water temperature gauge
53. Water temperature sender unit
54. Fuse 11
55. Horn switch
56. RH horn
57. LH horn
58. Under bonnet illumination switch
59. Under bonnet light
60. Clock
61. Fuse 19
62. Fuse 20
63. Pick-up point central locking/window lift
63 (a). Window lift relay
64. Heated rear window relay
65. Fuse 9
66. Radio aerial amplifier
67. Heated rear screen
68. Heated rear screen switch
69. Heated rear screen warning lamp
70. Voltage sensitive switch
71. Fuse 13
72. Hazard switch
73. Flasher unit
74. Direction indicator switch
75. Hazard/indicator warning lamp
76. LH rear indicator lamp
77. LH front indicator lamp
78. LH side repeater lamp
79. RH side repeater lamp
80. RH front indicator lamp
81. RH rear indicator lamp
82. Trailer warning lamp
83. Fuse 15
84. Stop lamp switch
85. Reverse lamp switch
86. Auxiliary lamp relay
87. LH stop lamp
88. RH stop lamp
89. LH reverse lamp
90. RH reverse lamp
91. LH auxiliary lamp
92. RH auxiliary lamp
93. Auxiliary lamp switch
94. Fuse 17
95. Dash cigar lighter
96. Cubby box cigar lighter
97. LH interior lamp
98. RH interior lamp
99. Interior lamp delay unit
100. LH door edge lamp
101. RH door edge lamp
102. LH puddle lamp
103. RH puddle lamp
104. Interior lamp switch
105. LH rear door switch
106. RH rear door switch
107. Tailgate switch
108. LH front door switch
109. RH front door switch
110. Differential lock warning lamp
111. Differential lock switch
112. Oil pressure warning lamp
113. Oil pressure switch
114. Fuse 18
115. Speed transducer
116. Fuel pump
117. Ignition coil
118. Capacitor
119. Distributor
120. EFI Harness plug
121. Coil negative (engine RPM input to ECU)
122. Radio choke
123. Radio fuse
124. Radio
125. Four speakers
126 (a). Not used
127. Sunroof connection point
128. Automatic transmission oil temperature warning lamp
129. Automatic transmission oil temperature switch
130. Fuse 16
131. Rear wash wipe switch
132. Rear wipe delay unit
133. Rear wiper motor
134. Rear screen wash pump
135. Low screen wash fluid level warning lamp
136. Low screen wash switch
137. Low coolant switch
138. Multi-function unit in binnacle
139. Low coolant level warning lamp
140. Low fuel level warning lamp
141. E.F.I. warning lamp
142. Handbrake warning lamp
143. Handbrake warning switch
144. Brake fluid level warning switch
145. Brake fluid level warning lamp
146. Brake pad wear warning lamp
147. Brake pad wear sensors
148. Brake check unit
149. Split charge relay (option)
150. Split charge terminal post (option)
151. Heater/air conditioning connections
152. Fuse 8
153. Ignition load relay (+)
154. Battery feed (+)
155. Ignition auxiliary (+)
156. Ignition on (+)
157. Earth (-)
158. Warning lights common earth (-)
159. Warning lights supply (+)

## CABLE COLOUR CODE

| | | | | | |
|---|---|---|---|---|---|
| B | Black | N | Brown | R | Red |
| G | Green | O | Orange | S | Grey |
| K | Pink | P | Purple | U | Blue |
| | | W | White | Y | Yellow |

The last letter of a colour code denotes the tracer.

# Notes

## ELECTRICAL 86

### RANGE ROVER

### MAIN CIRCUIT DIAGRAM
Left hand steering - RR2334S and RR2335S

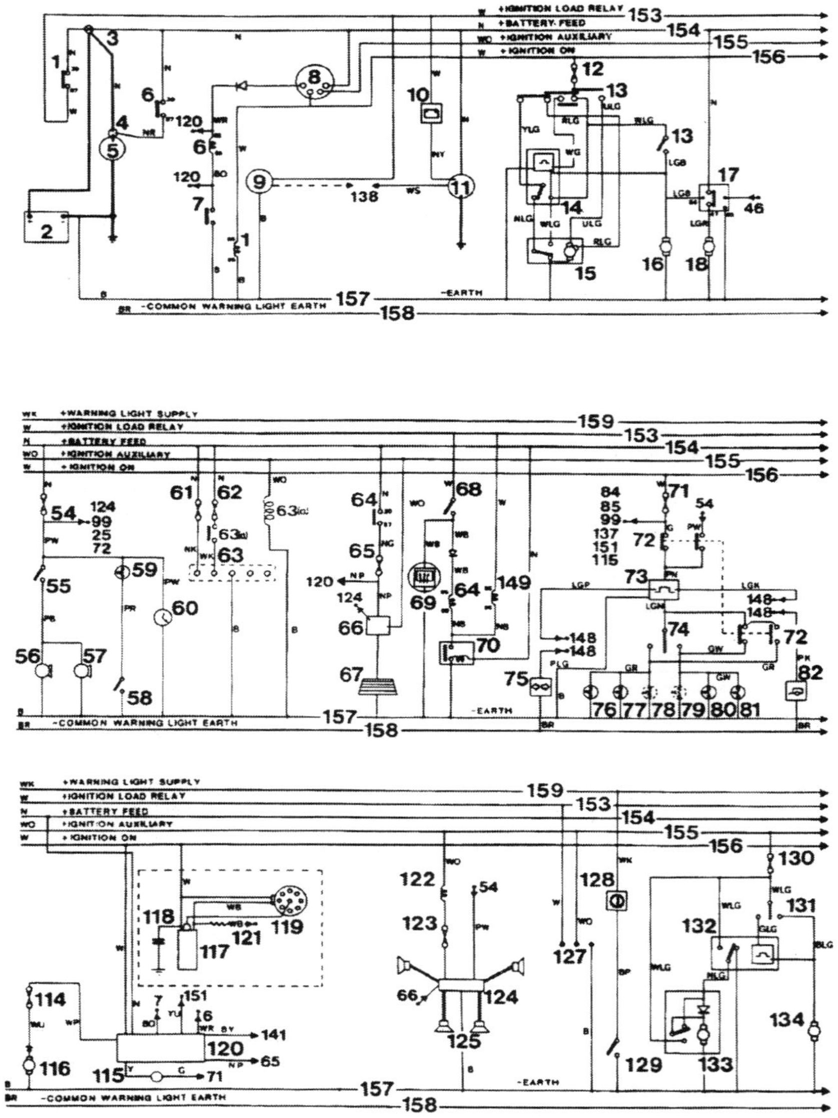

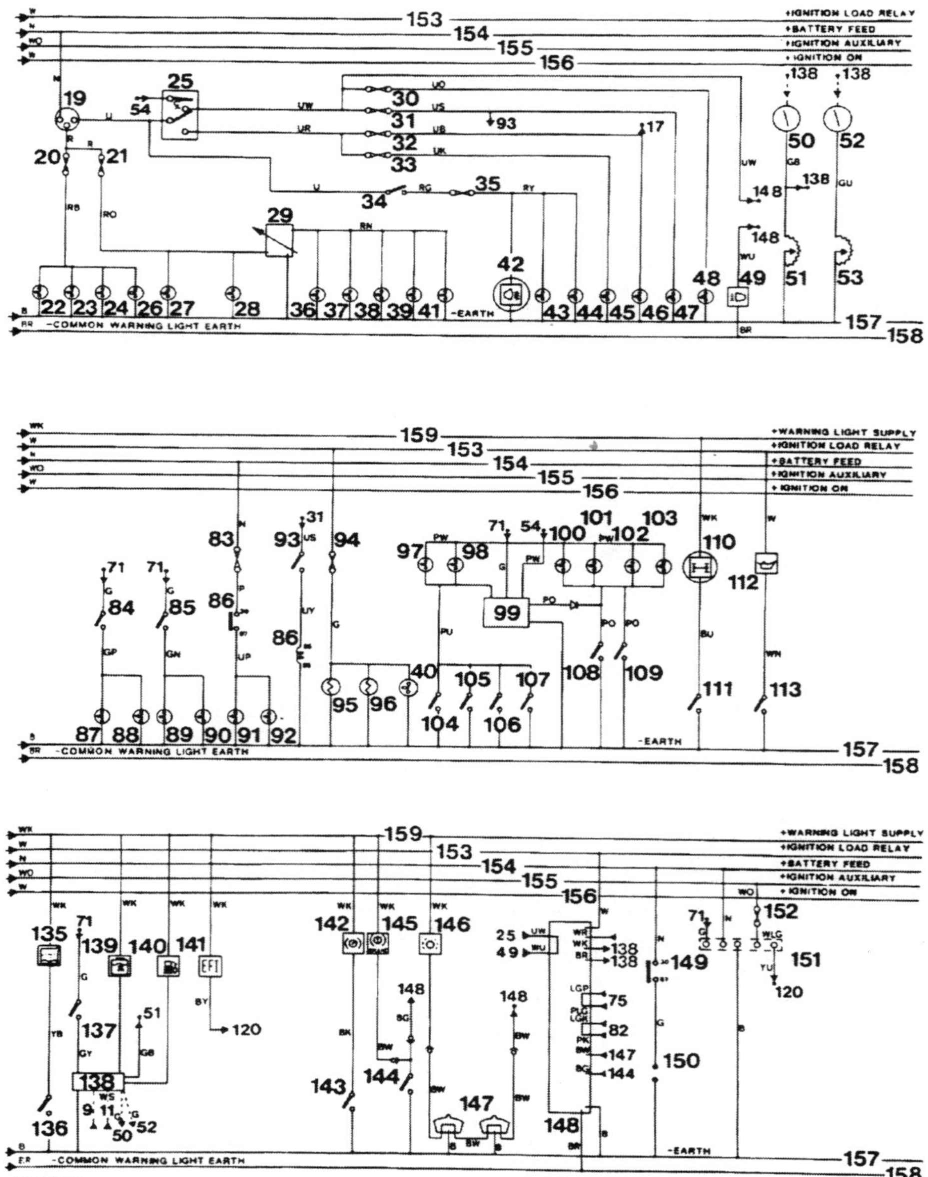

**Land Rover
Range Rover**

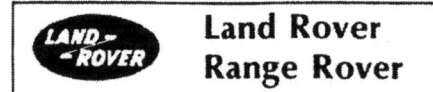

**Workshop Bulletin**

# Range Rover
## June 1987
## Lucas A127/65 AMP Alternator
## Workshop Bulletin
## Bulletin Number LSM180WB1

|  | Section | Page |
|---|---|---|
| Introduction | 01 | 362 |
| General Specification Data | 04 | 363 |
| Electrical | 86 | 363 |

Land Rover
Lode Lane,
Solihull
West Midlands
B92 8NW
England

Section

| 01 | **INTRODUCTION** | |
| --- | --- | --- |
| | Introduction | 1 |
| 04 | **GENERAL SPECIFICATION DATA** | |
| | Alternator | 2 |
| 06 | **TORQUE WRENCH SETTINGS** | |
| | Alternator | 2 |
| 86 | **ELECTRICAL** | |
| | Alternator | 3 |
| | Electric mirrors - circuit diagram | 7 |
| | Main circuit diagrams | 8 |

# INTRODUCTION 01

**RANGE ROVER**

**1987 MODEL YEAR IMPROVEMENTS**

**INTRODUCTION**

As part of the 1987 model year programme Land Rover Ltd is continually looking at ways to improve the quality and characteristics of its products. This bulletin contains servicing details of the Lucas A127/65 alternator now fitted to Range Rover models.

The A127 alternator incorporates a combined regulator and brush box assembly readily detachable from the rear of the alternator body.

This bulletin also includes the latest main circuit diagrams for both left and right hand drive models plus a revised electric mirror circuit diagram.

1

## ELECTRICAL 86

### 04 GENERAL SPECIFICATION DATA

**ALTERNATOR**

| | |
|---|---|
| Manufacturer | Lucas |
| Type | A127/65 |
| Polarity | Negative earth |
| Maximum operating speed | 15000 rev/min |
| Maximum DC output at 6000 rev/min | 65 amp |
| Stator windings | 3 phase |
| Stator winding resistance | 0.15 ohm/phase |
| Regulator type | 21TR |
| Regulator controlled voltage | 13.6-14.4V |
| Field winding rotor poles | 12 |
| Rotor winding resistance | 3.2 ohms |
| Brush length - new | 17 mm(0.67 in) )From |
| Renew regulator/brushbox assembly at | 5 mm(0.20 in) )moulding |
| Brush spring pressure |  |
| - flush with brushbox moulding | 1.3-2.7N(4.7-9.8oz) |

### 06 TORQUE WRENCH SETTINGS

**ALTERNATOR**

| | Nm | lbf ft |
|---|---|---|
| Alternator shaft nut | 50-70 | 37-52 |
| Alternator through bolts | 5.5 | 4.0 |
| Alternator rectifier screws | 3.5 | 2.5 |
| Regulator/brushbox screws | 2.5 | 2.0 |
| Terminal nut - main output | 4.0 | 3.0 |
| Terminal nut - phase | 4.0 | 3.0 |

### ALTERNATOR - LUCAS - A127-65

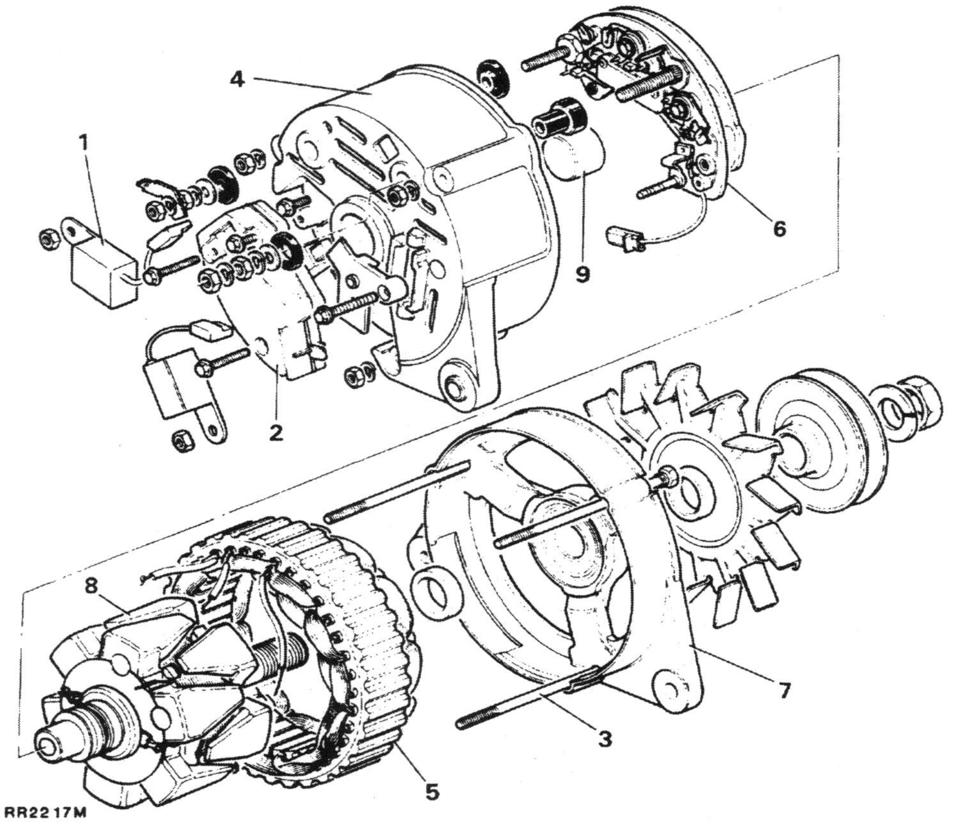

RR22 17M

**Service parts**

1. Suppression capacitors - 2 off
2. Regulator/Brushbox assembly
3. Through bolts
4. Slip ring end (SRE) bracket
5. Stator
6. Rectifier
7. Drive end bracket (DRE)/bearing assembly
8. Rotor
9. Slip ring end bearing

# ELECTRICAL

## ALTERNATOR - LUCAS - A127-65

### Description

The model A127 alternator is a three phase, field sensed unit, the rotor and stator windings generate three phase alternating current (AC) which is rectified to direct current (DC). The electronic voltage regulator unit controls the alternator output voltage by high frequency switching of the rotor field circuit.

It is essential that good electrical connections are maintained at all times. Of particular importance are those on the charging circuit (including those at the battery) which should be occasionally inspected to see that they are clean and tight. In this way any significant increase in circuit resistance can be prevented.

Do not disconnect battery cables while the engine is running or damages to the semi-conductor devices may occur. It is also inadvisable to break or make any connections in the alternator charging and control circuits while the engine is running.

The Model 21TR electronic voltage regulator employs micro-circuit techniques resulting in improved performance under difficult service conditions. The whole assembly is encapsulated in silicone rubber and housed in an aluminium heat sink, ensuring complete protection against the adverse effects of temperature, dust, and moisture etc.

The brushbox assembly is incorporated in the regulator unit.

Surge protection is incorporated in the regulator unit.

The regulating voltage is set during manufacture to give the required regulating voltage range of 13.6 to 14.4 volts, and no adjustment is necessary. The only maintenance needed is the occasional check on terminal connections and wiping with a clean dry cloth.

The alternator system provides for direct connection of a charge (ignition) indicator warning light, and eliminates the need for a field switching relay or warning light control unit. As the warning lamp is connected in the charging circuit, lamp failure will cause loss of charge. Lamp should be checked regularly and a spare carried.

When using rapid charge equipment to re-charge the battery, the battery must be disconnected from the vehicle.

### Dismantle, Overhaul, Test components

**NOTE: Alternator charging circuit - The ignition warning light is connected in series with the alternator field circuit. Bulb failure would prevent the alternator charging, except at very high engine speed, therefore, the bulb should be checked before suspecting an alternator failure.**

### Precautions

Battery polarity is NEGATIVE EARTH, which must be maintained at all times.

No separate control unit is fitted; instead a voltage regulator of micro-circuit construction is mounted on the slip ring end bracket.

Battery voltage is applied to the alternator output cable even when the ignition is switched off, the battery must be disconnected before commencing any work on the alternator. The battery must also be disconnected when repairs to the body structure are being done by arc welding.

### Dismantle

1. Withdraw the connectors from the alternator.
2. Remove the alternator
3. Remove the nuts from the through bolts, disconnect the connectors and remove two suppression capacitors.
4. Remove three screws, disconnect one lead and withdraw the regulator and brushbox assembly.

5. Check the brush lengths by measuring length of protrusion from moulding. If dimension A is 5mm (0.20in) or less, change the assembly.

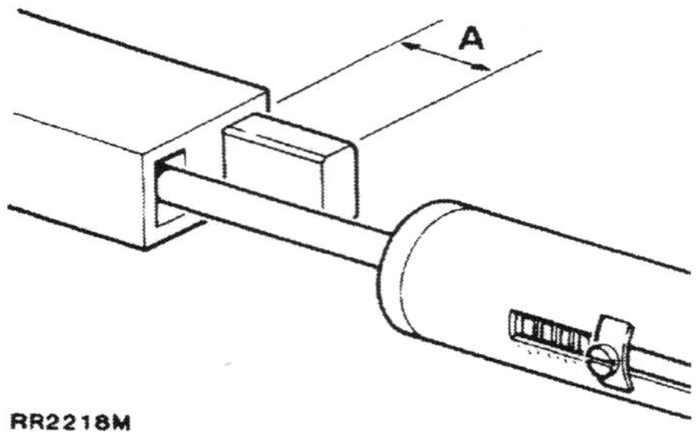

6. Check brush spring pressure using a push type spring gauge. Gauge should register 136 to 279 g (5 to 10 oz) when brush is pressed until face is flush with housing. If reading is outside these limits, renew assembly.
7. Mark the relative positions of the end brackets and stator. Remove through bolts and withdraw slip ring end bracket and stator assembly. Carefully tap the mounting lugs with a mallet if necessary.
8. Remove nuts, washers and insulators from stud terminals noting their position for reassembly. Remove two screws and insulation cover and withdraw rectifier and stator from slip ring end bracket.

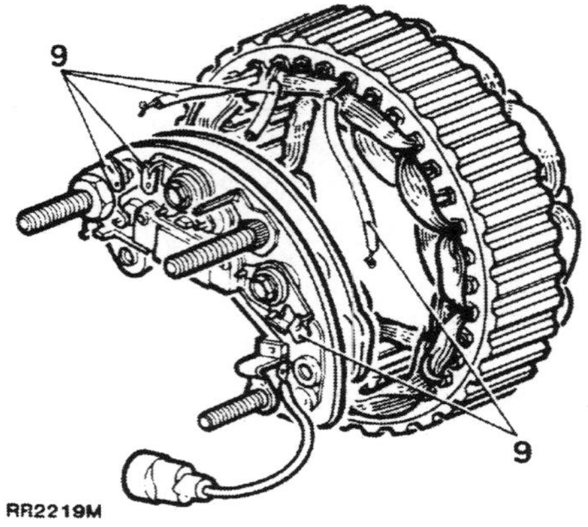

9. Release the stator winding cable ends from the rectifier by applying a hot soldering iron to the terminal tags of the rectifier. Prise out the cable ends when the solder melts, and separate the rectifier and stator. Further dismantling of the rectifier is not required.
10. Check the diodes. Connect the test equipment as shown and test each diode in turn, note whether lamp lights, then reverse test lead connections. The lamp should light in one direction only. Renew the rectifier assembly if a faulty diode is diagnosed.

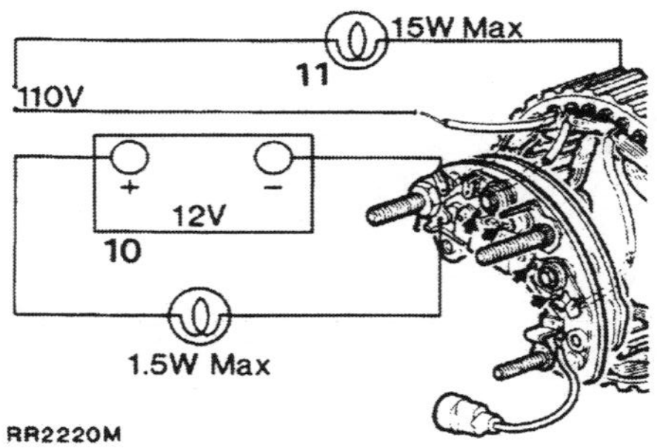

11. Visually inspect the stator windings for signs of damage or burning. Check the stator insulation using a suitable 110 volt test lamp. Connect the test leads to the laminated yoke and to each of the three leads in turn. If test lamp lights, fit a new stator.
12. Remove the fan and pulley assembly. Push out the rotor shaft from the bearing using a suitable press and remove the spacer from the shaft.

*Continued*

# ELECTRICAL

13. Clean surfaces of slip rings using a solvent moistened cloth.
14. Inspect slip ring surfaces for signs of burning, remove burn marks using extra fine glasspaper. On no account should emery cloth or similar abrasives be used, or any attempt made to machine the slip rings.
15. Check the insulation of the rotor field windings, using a suitable 110 volt test lamp connected between one of the slip rings and one of the rotor poles. If the test lamp lights, the rotor must be renewed.

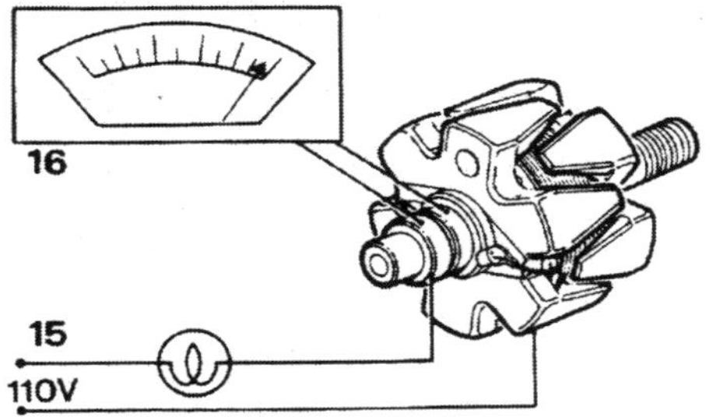

RR2221M

16. Check the resistance/continuity of the rotor field windings by connecting an ohmmeter to the slip rings as shown. A reading of 3.2 ohms should be obtained.
17. Check condition of bearings. If signs of rubbing between rotor poles and stator is evident, both bearings are excessively worn, and must be renewed.
18. Use a suitable press, applying pressure from the rear of the slip ring end bracket, to withdraw or refit the bearing. Note that the drive end bracket and bearing are only supplied as a unit.

## Reassembling

19. Fit the spacer and press the rotor into the drive end bracket. Insert the through bolts. Fit the spacer, fan, pulley and spring washer to the shaft. Fit and tighten the pulley nut to the correct torque.
20. Fit the stator and rectifier assembly to the drive end bracket in the position previously marked.
21. Fit the slip ring end bracket in the position previously marked, locating the regulator lead through its aperture. Fit the insulation cover and tighten two rectifier securing screws.
22. Fit and tighten the through bolt nuts evenly and to the correct torque.
23. Connect the lead to the regulator, locate the brushes on the slip rings and secure the regulator/brushbox assembly, taking care not to damage the brushes.
24. Fit both suppression capacitors.
25. Refit the alternator.

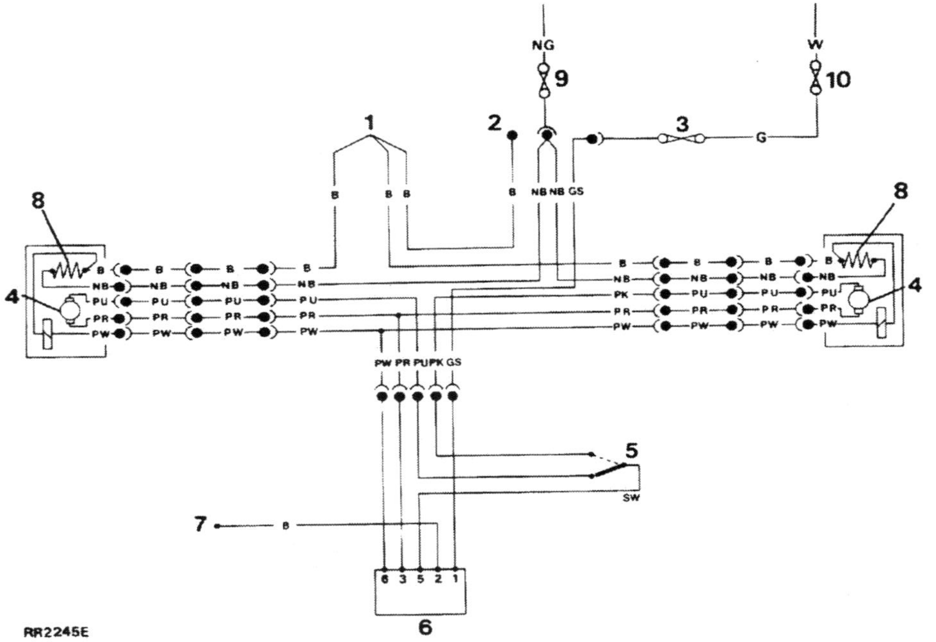

RR2245E

## ELECTRIC MIRRORS -
Circuit diagram

1. Clinch
2. Main cable connections
3. Fuse A5-mirror motors
4. Mirror motors
5. Change over switch
6. Mirror control switch
7. Earth-via main cable
8. Mirror heating elements-active with heated rear screen
9. Fuse 10-heating elements
10. Fuse 13

## CABLE COLOUR CODE

| | |
|---|---|
| B | Black |
| U | Blue |
| N | Brown |
| G | Green |
| O | Orange |
| P | Purple |
| R | Red |
| S | Slate |
| W | White |
| Y | Yellow |

The last letter of a colour code denotes the tracer.

# ELECTRICAL 86

## MAIN CIRCUIT DIAGRAM
### Right Hand Steering - RR2343M & RR2344M

1. Ignition load relay
2. Battery
3. Terminal post
4. Starter solenoid
5. Starter motor
6. Starter relay
7. Starter inhibit switch (Automatic)
8. Ignition switch
9. Tachometer
10. Voltage transformer(dim dip)
11. Ignition warning lamp
12. Alternator
13. Fuse 7
14. Front wipe/wash switch
15. Front wiper delay unit
16. Front wiper motor
17. Front wash switch
18. Front wash pump
19. Headlamp wash timer unit (option)
20. Headlamp wash pump (option)
21. Main lighting switch
22. Fuse 6
23. Fuse 5
24. LH side lamp
25. LH tail lamp
26. Number plate lamps (2 off)
27. Main beam dip flash switch
28. Radio illumination
29. RH side lamp
30. RH tail lamp
31. Rheostat
32. Fuse 3
33. Fuse 4
34. Fuse 1
35. Fuse 2
36. Rear fog switch
37. Fuse 12
38. Switch illumination (2 off)
39. Cigar lighter illumination (2 off)
40. Heater illumination (4 off)
41. Clock illumination
42. Automatic gear selector illumination (2 off)
43. Instrument illumination (6 off)
44. Rear fog warning lamp
45. LH rear fog

46. RH rear fog
47. LH dip beam
48. RH dip beam
49. LH main beam
50. RH main beam
51. Main beam warning lamp
52. Fuel gauge
53. Fuel gauge sender unit
54. Water temperature gauge
55. Water temperature sender unit
56. Fuse 11
57. Horn switch
58. RH horn
59. LH horn
60. Under bonnet illumination switch
61. Under bonnet light
62. Clock
63. Fuse 19
64. Fuse 20
65. Pick-up point central locking/window lift (option)
66. Heated rear window relay
67. Fuse 9
68. Radio aerial amplifier
69. Heated rear screen
70. Heated rear screen switch
71. Heated rear screen warning lamp
72. Voltage sensitive switch
73. Fuse 13
74. Hazard switch
75. Flasher unit
76. Direction indicator switch
77. Hazard/indicator warning lamp
78. LH rear indicator lamp
79. LH front indicator lamp
80. LH side repeater lamp
81. RH side repeater lamp
82. RH front indicator lamp
83. RH rear indicator lamp
84. Trailer warning lamp
85. Fuse 15
86. Stop lamp switch
87. Reverse lamp switch
88. Auxiliary lamp relay (option)
89. LH stop lamp
90. RH stop lamp

## MAIN CIRCUIT DIAGRAM
### Right Hand Steering - RR2343M & RR2344M

91. LH reverse lamp
92. RH reverse lamp
93. LH auxiliary lamp (option)
94. RH auxiliary lamp (option)
95. Auxiliary lamp switch (option)
96. Fuse 17
97. Dash cigar lighter
98. Cubby box cigar lighter
99. LH interior lamp
100. RH interior lamp
101. Interior lamp delay unit
102. LH door edge lamp
103. RH door edge lamp
104. LH puddle lamp
105. RH puddle lamp
106. Interior lamp switch
107. LH rear door switch
108. RH rear door switch
109. Tailgate switch
110. LH front door switch
111. RH front door switch
112. Differential lock warning lamp
113. Differential lock switch
114. Oil pressure warning lamp
115. Oil pressure switch
116. Fuse 18
117. Fuel cut off relay (carburetter models)
118. Fuel pump (petrol models)
119. Ignition coil
120. Capacitor
121. Distributor
122. Efi Harness plug
123. Fuel shut off solenoid (Diesel)

124. Radio choke
125. Radio fuse
126. Radio and four speakers
127. Ignition pick up points
128. Automatic transmission oil temperature warning lamp
129. Automatic transmission oil temperature switch
130. Fuse 16
131. Rear wash wipe switch
132. Rear wipe delay unit
133. Rear wiper motor
134. Rear screen wash pump
135. Low screen wash fluid level warning lamp
136. Low screen wash switch
137. Low coolant switch
138. Multi-function unit in binnacle
139. Low coolant level warning lamp
140. Low fuel level warning lamp
141. Cold start/diesel glow plug warning lamp
142. Cold start switch - carburetter
143. Glow plug timer (diesel)
144. Glow plugs (diesel)
145. Handbrake warning lamp
146. Brake fail warning lamp
147. Handbrake warning switch
148. Brake tail warning switch
149. Brake pad wear warning lamp
150. Brake pad wear sensors
151. Brake check relay
152. Split charge relay (option)
153. Split charge terminal post (option)
154. Heater/air conditioning connections
155. Fuse 8
156. Coil negative (engine RPM input to ECU)

## CABLE COLOUR CODE

| | |
|---|---|
| B | Black |
| U | Blue |
| N | Brown |
| G | Green |
| L | Light |
| O | Orange |
| K | Pink |
| P | Purple |
| R | Red |
| S | Slate |
| W | White |
| Y | Yellow |

## ELECTRICAL 86

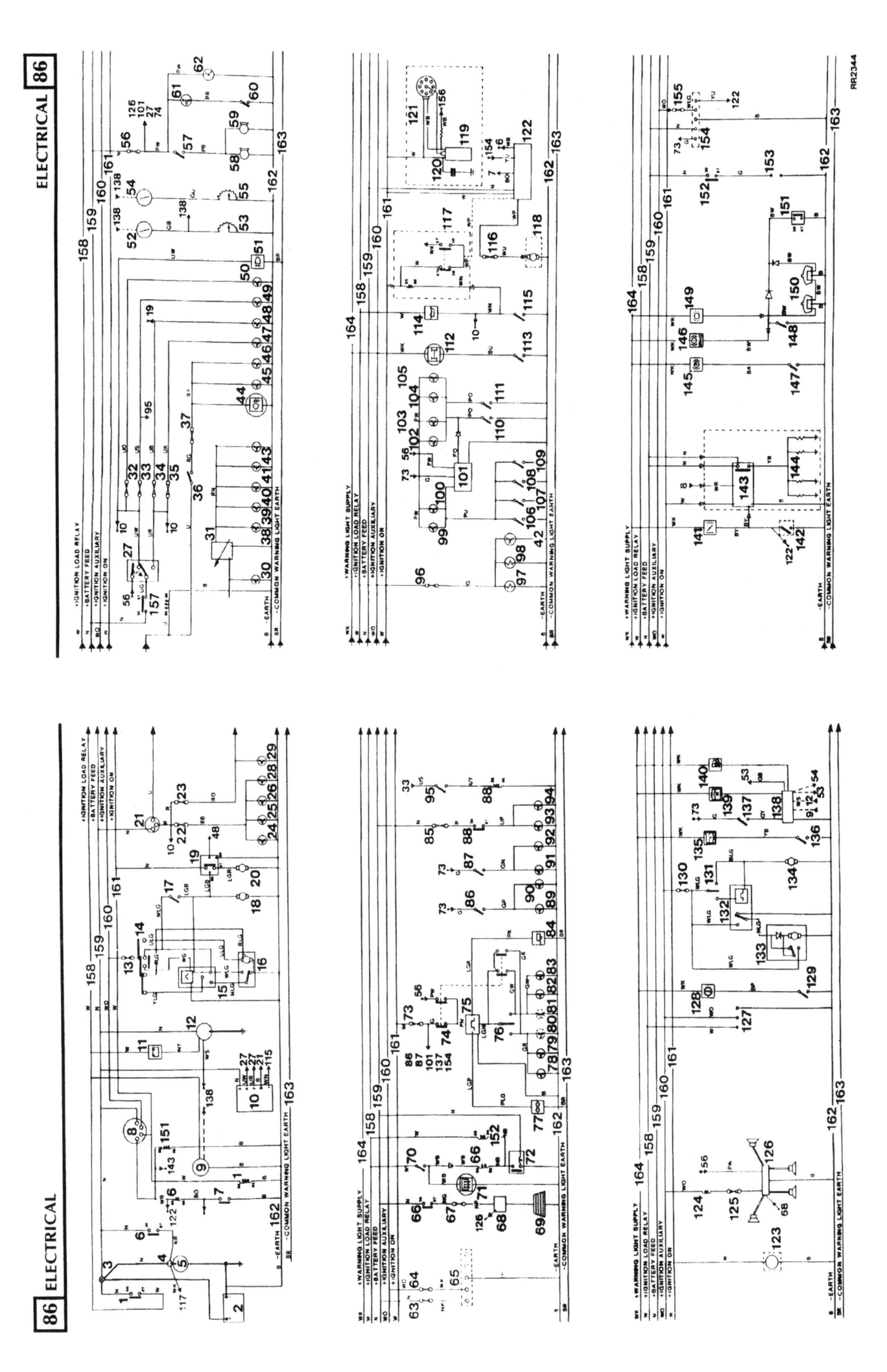

# ELECTRICAL 86

## MAIN CIRCUIT DIAGRAM
### Left-hand Steering - RR2345M & RR2346M

1. Ignition load relay
2. Battery
3. Terminal post
4. Starter solenoid
5. Starter motor
6. Starter relay
7. Starter inhibit switch (automatic)
8. Ignition switch
9. Tachometer
10. Ignition warning lamp
11. Alternator
12. Fuse 7
13. Front wipe/wash switch
14. Front wipe delay unit
15. Front wiper motor
16. Front wash pump
17. Headlamp wash timer unit (option)
18. Headlamp wash pump (option)
19. Main lighting switch
20. Fuse 6
21. Fuse 5
22. LH side lamp
23. LH tail lamp
24. Number plate lamp (2 off)
25. Main beam dip/flash switch
26. Radio illumination
27. RH side lamp
28. RH tail lamp
29. Rheostat
30. Fuse 3
31. Fuse 4
32. Fuse 1
33. Fuse 2
34. Rear fog switch
35. Fuse 12
36. Switch illumination (2 off)
37. Cigar lighter illumination (2 off)
38. Heater illumination (4 off)
39. Clock illumination
40. Automatic gear selector illumination (2 off)
41. Instrument illumination (6 off)
42. Rear fog warning lamp
43. LH rear fog
44. RH rear fog
45. LH dip beam
46. RH dip beam
47. LH main beam
48. RH main beam
49. Main beam warning lamp
50. Fuel gauge
51. Fuel gauge sender unit
52. Water temperature gauge
53. Water temperature sender unit
54. Fuse 11
55. Horn switch
56. RH horn
57. LH horn
58. Under bonnet illumination switch
59. Under bonnet light
60. Clock
61. Fuse 19
62. Fuse 20
63. Pick-up point central locking/window unit
64. Heated rear window relay
65. Fuse 9
66. Radio aerial amplifier
67. Heated rear screen
68. Heated rear screen switch
69. Heated rear screen warning lamp
70. Voltage sensitive switch
71. Fuse 13
72. Hazard switch
73. Flasher unit
74. Direction indicator switch
75. Hazard/indicator warning lamp
76. LH rear indicator lamp
77. LH front indicator lamp
78. LH side repeater lamp
79. RH side repeater lamp
80. RH front indicator lamp
81. RH rear indicator lamp
82. Trailer warning lamp
83. Fuse 15
84. Stop lamp switch
85. Reverse lamp switch
86. Auxiliary lamp relay
87. LH stop lamp
88. RH stop lamp
89. LH reverse lamp
90. RH reverse lamp

## MAIN CIRCUIT DIAGRAM
### Left-hand Steering - RR2345M & RR2346M

91. LH auxiliary lamp (option)
92. RH auxiliary lamp (option)
93. Auxiliary lamp switch (option)
94. Fuse 17
95. Dash cigar lighter
96. Cubby box cigar lighter
97. LH interior lamp
98. RH interior lamp
99. Interior lamp delay unit
100. LH door edge lamp
101. RH door edge lamp
102. LH puddle lamp
103. RH puddle lamp
104. Interior lamp switch
105. LH rear door switch
106. RH rear door switch
107. Tailgate switch
108. LH front door switch
109. RH front door switch
110. Differential lock warning lamp
111. Differential lock switch
112. Oil pressure warning lamp
113. Oil pressure switch
114. Fuse 18
115. Fuel shut-off relay - carburetter
116. Fuel pump - petrol models
117. Ignition coil
118. Capacitor
119. Distributor
120. EFI Harness plug
121. Fuel shut-off solenoid-Diesel
122. Radio choke
123. Radio
124. Radio fuse
125. Four speakers
126. Seat belt warning lamp
127. Speed transducer, Saudi only
128. Resistor
129. Audible warning unit
130. Transfer box neutral switch
131. Seat buckle switch
132. Overspeed monitor (Saudi only)
133. Overspeed buzzer (Saudi only)
134. Ignition pick up points
135. Automatic transmission oil temperature warning lamp
136. Automatic transmission oil temperature switch
137. Fuse 16
138. Rear wash wipe switch
139. Rear wipe delay unit
140. Rear wiper motor
141. Rear screen wash pump
142. Low screen wash fluid level warning lamp
143. Low screen wash switch
144. Low coolant switch
145. Multi-function unit in binnacle
146. Low coolant level warning lamp
147. Low fuel level warning lamp
148. Cold start Diesel glow plug warning lamp
149. Choke switch - carburetter
150. Glowplug timer/Diesel
151. Glowplugs Diesel
152. Handbrake warning lamp
153. Handbrake warning switch
154. Brake tail warning lamp
154a. Brake tail warning switch
155. Brake pad wear warning lamp
156. Brake pad wear sensors
157. Brake check unit
158. Split charge relay (option)
159. Split charge terminal post
160. Heater air conditioning connections
161. Fuse 8
162. Coil negative (engine RPM input to ECU)

## CABLE COLOUR CODE

| | |
|---|---|
| B | Black |
| U | Blue |
| N | Brown |
| G | Green |
| L | Light |
| O | Orange |
| K | Pink |
| P | Purple |
| R | Red |
| S | Slate |
| W | White |
| Y | Yellow |

# ELECTRICAL 86

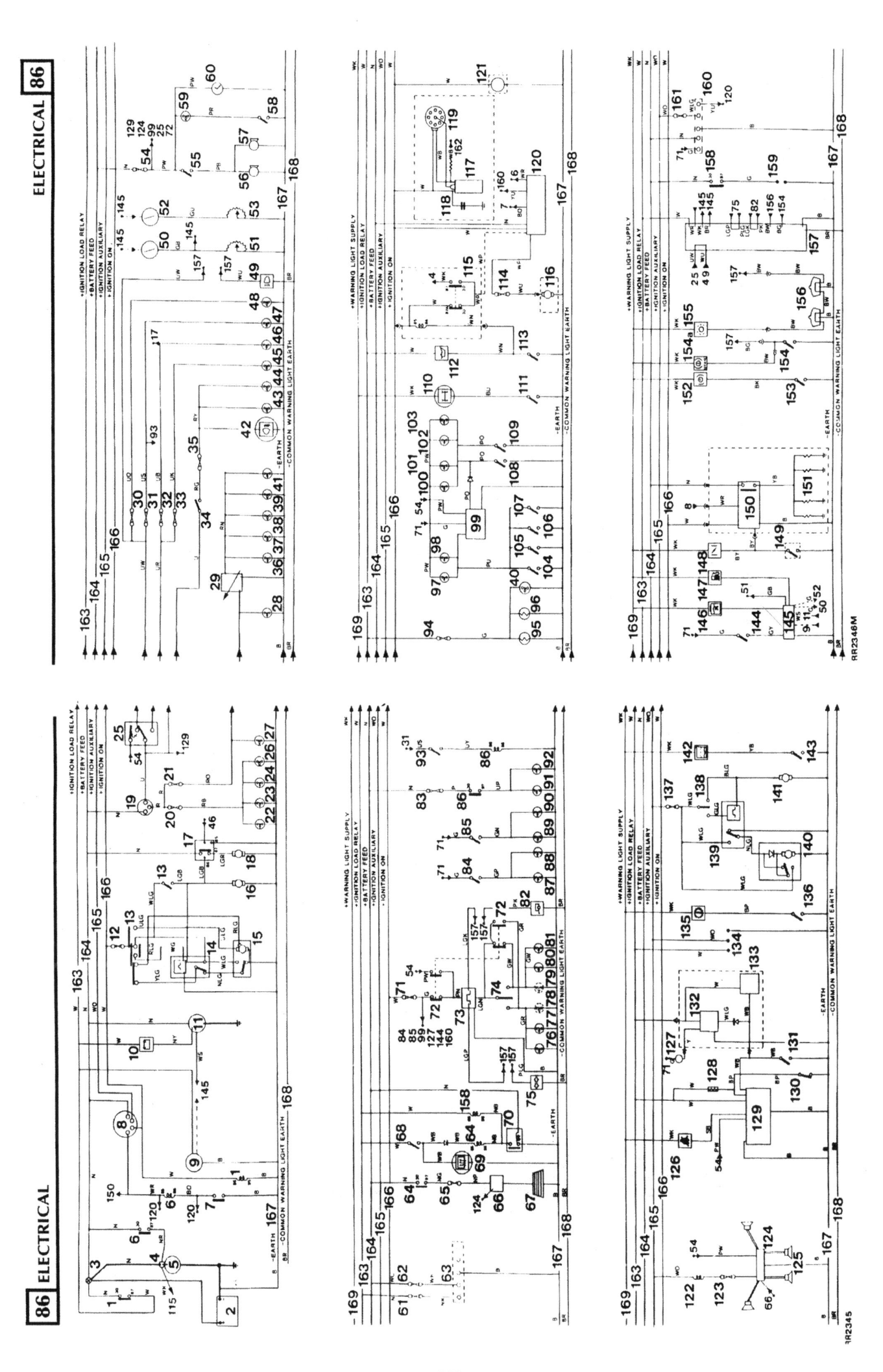

# Land Rover / Range Rover Workshop Bulletin

**Range Rover
June 1987
Lucas Girling-Type 115 Servo and Master Cylinder Assembly
Workshop Bulletin
Bulletin Number LSM180WB2**

|  | Section | Page |
|---|---|---|
| Introduction | 01 | 371 |
| General Specification Data | 04 | 372 |
| Maintenance | 10 | 372 |
| Brakes | 70 | 373 |

Land Rover
Lode Lane,
Solihull
West Midlands
B92 8NW
England

Section

| | | |
|---|---|---|
| **01** | **INTRODUCTION** | |
| | Introduction | 1 |
| **04** | **GENERAL SPECIFICATION DATA** | |
| | Servo and master cylinder | 2 |
| **06** | **TORQUE WRENCH SETTINGS** | |
| | Servo assembly | 2 |
| | Master cylinder | 2 |
| **09** | **LUBRICANTS FLUIDS AND CAPACITIES** | |
| | Brake fluid | 2 |
| **10** | **MAINTENANCE** | |
| | Requirements | 3 |
| | Top up brake fluid | 3 |
| **70** | **BRAKES** | |
| | Description | 5 |
| | Bleed | 10 |
| | Master cylinder - remove, overhaul and refit | 11 |
| | Servo - remove and refit | 16 |
| | Pedal assembly - remove, overhaul and refit | 18 |

# INTRODUCTION 01

## RANGE ROVER

## 1987 MODEL YEAR IMPROVEMENTS

### INTRODUCTION

As part of the 1987 model year programme Land Rover Ltd are continually looking at ways to improve the quality and characteristics of its products, the introduction of the following components to Range Rover will improve the brake system and also reduce service costs.

The braking system is uprated with the introduction of a new type-115 Lucas Girling servo and master cylinder assembly which gives improved brake pedal feel and response by reducing brake pedal effort. The effectiveness of the braking system is now controlled by brake pedal travel.

# Notes

# MAINTENANCE

## GENERAL SPECIFICATION DATA

### BRAKING SYSTEM
Servo/master cylinder

| | |
|---|---|
| Manufacturer | Lucas Girling |
| Servo type | LSC 115 |
| Boost ratio | 5.6:1 |
| Master cylinder type | 25.4mm AS/AS (steel tube construction) |
| Fluid displacement - Primary | 5.16 cm$^3$ |
|                - Secondary | 9.17 cm$^3$ |
| Nominal split | 36/64 |
| Unit weight | 3.75 Kg |
| Maximum bleed pressure | 2.76 bar (40 lbf/in$^2$) |

## TORQUE WRENCH SETTINGS

### BRAKING SYSTEM

| | Nm | lbf ft |
|---|---|---|
| Servo assembly to pedal box | 22 to 25 | 16 to 19 |
| Brake pipes to master cylinder | 9 to 11 | 7 to 8 |
| Master cylinder to servo | 21 to 29 | 15 to 22 |

## LUBRICANTS, FLUIDS AND CAPACITIES

| COMPONENT | SPECIFICATION | AMBIENT TEMPERATURE |
|---|---|---|
| Brake Reservoir | Brake fluid must have a minimum boiling point of 260°C (500°F) and comply with FMVSS/116/DOT4. | °C  -30 -20 -10  0  10  20  30  40  50<br>°F  -22  -4  +14 +32  50  68  86  104  122 |

### Approved fluids

Castrol/Girling     K624 DOT4
Automotive Products     429S Super DOT4

## MAINTENANCE SCHEDULE

The fitting of the new Servo and master cylinder does not necessitate the issue of additional Maintenance Schedules, therefore existing schedules should be used and the service intervals adhered to.

## CHECK AND TOP UP BRAKE FLUID RESERVOIR

1. Visually check the brake fluid level against the 'MIN' and 'MAX' level markings on the side of the reservoir.
2. If necessary top up with the recommended grade of fluid (refer to section 09 lubricants fluidsand capacities) **DO NOT OVERFILL**.
3. Release the cap with combined fluid level switch by rotating anti- clockwise. Withdraw the cap and switch, top up the reservoir.

**CAUTION : When topping up the reservoir care should be taken to ensure that the brake fluid does not come into contact with any of the vehicle paintwork. Should this occur, wash the affected area IMMEDIATELY with copious amounts of water.**

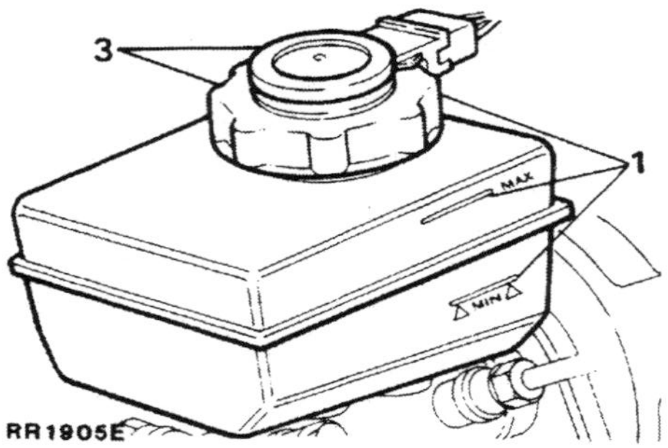

RR1805E

# BRAKES

## BRAKE SYSTEM

### Description

The hydraulic braking system fitted to Range Rovers is the dual line type, incorporating primary and secondary hydraulic circuits.

**NOTE: References made to primary or secondary do not imply main service brakes or emergency brakes but denote hydraulic line identification.**

The brake pedal is connected to a vacuum assisted mechanical servo which in turn operates a tandem master cylinder. The front disc brake calipers each house four pistons, the upper pistons are fed by the primary hydraulic circuit, the lower pistons by the secondary hydraulic circuit. The rear disc brake calipers each house two pistons, these are fed by the secondary hydraulic circuit via a pressure reducing valve.

A brake fluid level switch is incorporated into the reservoir cap assembly, the switch having detected either low or sudden fluid loss will immediately illuminate a warning light in the instrument binnacle.

The brake fluid reservoir is divided, the section closest to the servo feeds the primary circuit and the section furthest away from the servo feeds the secondary circuit. Under normal operating conditions both the primary and secondary circuits operate simultaneously on brake pedal application. In the event of a failure in the primary circuit the secondary circuit will still function and operate front and rear calipers. Alternatively, if the secondary circuit fails the primary circuit will still function and operate the upper pistons in the front calipers, allowances should be made and vehicle speed adjusted accordingly to allow for the lack of full braking efficiency.

If the servo should fail, both hydraulic circuits will still function but would require greater pedal effort due to the lack of vacuum assistance.

The hand operated transmission brake acts on a brake drum at the rear of the transfer gearbox and is completely independant of the hydraulic circuits.

Brake pad wear sensors are incorporated in the front and rear right hand side inboard brake pads. The sensor will illuminate a brake pad wear warning light in the instrument binnacle, when pad thickness has been reduced to approximately 3mm (0.118 in)

**CAUTION: THOROUGHLY CLEAN ALL BRAKE CALIPERS, PIPES AND FITTINGS BEFORE COMMENCING WORK ON ANY PART OF THE BRAKE SYSTEM. FAILURE TO DO SO COULD CAUSE FOREIGN MATTER TO ENTER THE SYSTEM AND CAUSE DAMAGE TO SEALS AND PISTONS WHICH WILL SERIOUSLY IMPAIR THE EFFICIENCY OF THE BRAKE SYSTEM.**

To ensure the brake system efficiency is not impaired the following warnings must be adhered to :-

**DO NOT** use any petroleum (gasoline) based cleaning fluids or any proprietary fluids containing petroleum (gasoline). The preferred fluids are clear methylated spirits or industrial alcohol.

**DO NOT** use brake fluid previously bled from the system.

**DO NOT** use old or stored brake fluid.

**ENSURE** that only new fluid is used and that it is taken from a sealed container.

**DO NOT** flush the brake system with any fluid other than the recommended brake fluid.

The brake system should be drained and flushed at the recommended service intervals.

Continued

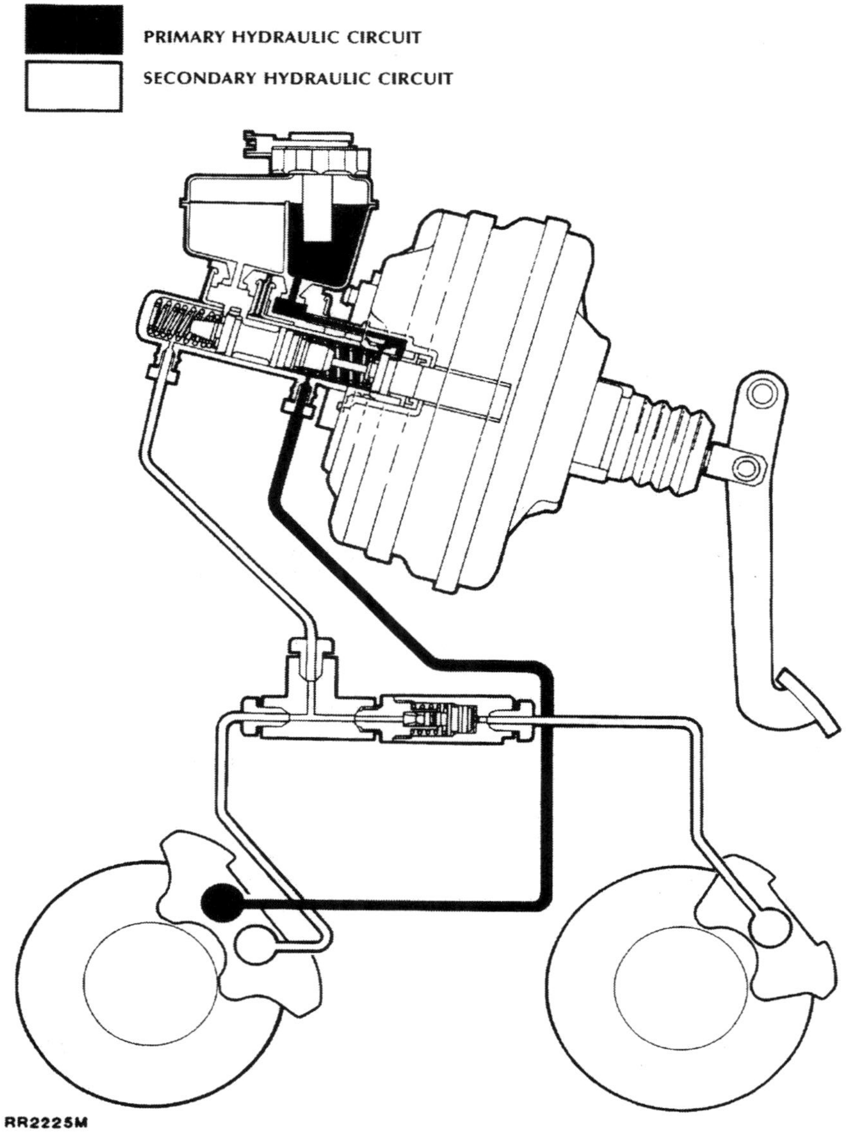

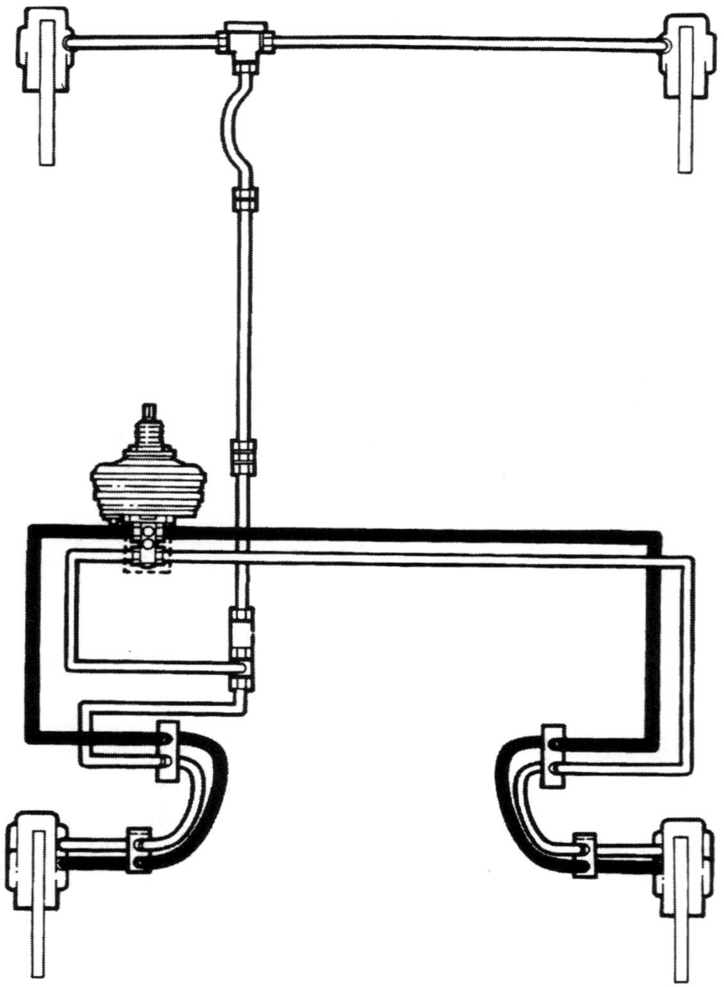

## BRAKES 70

**BRAKE PIPE LAYOUT**
Left hand drive

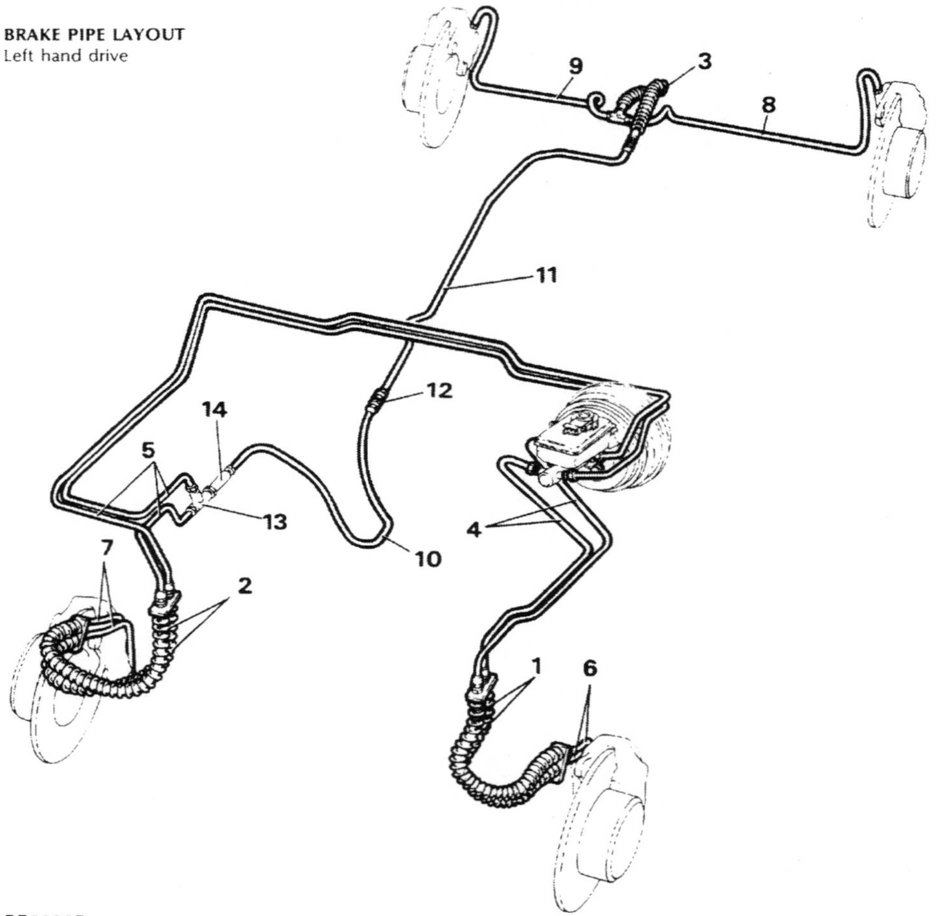

RR2222E

**HOSES**
1. Front left hand flexible hoses.
2. Front right hand flexible hoses.
3. Intermediate flexible hose.

**PIPES**
4. Feed to front left hand hose connector.
5. Feed to front right hand hose connector.
6. Feed to front left hand caliper.
7. Feed to front right hand caliper.
8. Feed to rear left hand caliper.
9. Feed to rear right hand caliper.
10. Feed to two way connector.
11. Feed to intermediate hose.
12. Two way connector.
13. Three way connector.
14. Pressure reducing valve.

## 70 BRAKES

IR2226M

Right hand drive brake circuit illustrated, left hand drive circuit is a mirror image of the right hand drive circuit.

374

## BRAKES

**Right hand drive**

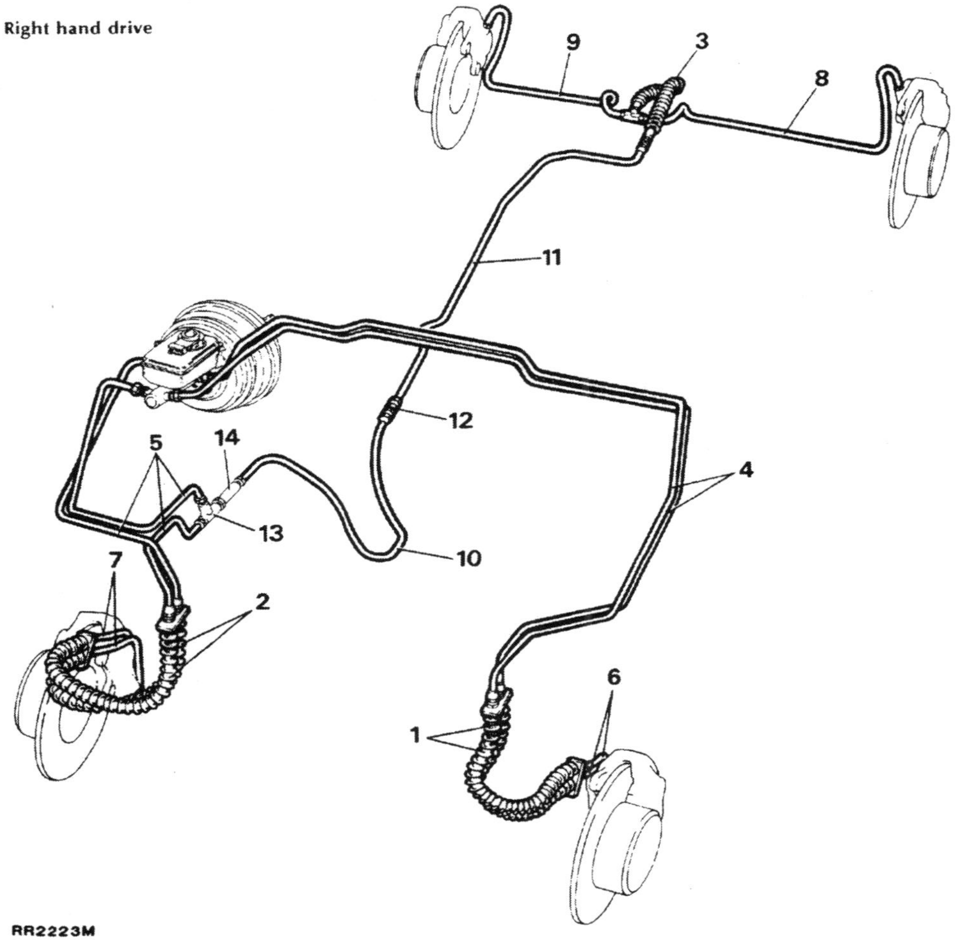

RR2223M

WARNING: Some components on the vehicle such as gaskets and friction surfaces (brake linings, clutch discs, or automatic transmission brake bands), may contain asbestos. Inhaling asbestos dust is dangerous to your health and the following essential precautions must be observed :-

* Work out of doors or in a well ventilated area and wear a protective mask.

* Dust found on the vehicle or produced during work on the vehicle should be removed by vacuuming or by using a well dampened cloth and not by blowing.

* Dust waste should be dampened, placed in a sealed container and marked to ensure safe disposal.

* If any cutting, drilling etc, is attempted on materials containing asbestos the item should be dampened and only hand tools or low speed power tools used.

---

6. Repeat 5 until fluid clear of air bubbles appears in the container, then keeping the pedal fully depressed, tighten the bleed screw.
7. Remove the bleed tube and replace the dust cap on the bleed screw.

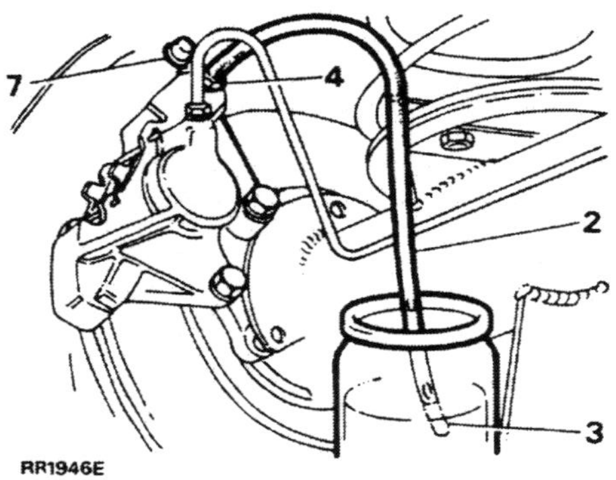

### Bleed

The hydraulic system comprises two completely independent circuits. The rear calipers and the lower pistons in the front calipers form the secondary circuit, while the upper pistons in the front calipers form the primary circuit. The following procedure covers bleeding the complete system, but it is permissible to bleed one circuit only if disconnections are limited to that circuit.

Bleeding will be assisted if the engine is run or a vacuum supply is connected to the servo.

**WARNING: IF THE ENGINE IS RUNNING DURING THE BRAKE BLEEDING PROCESS ENSURE THAT NEUTRAL GEAR IS SELECTED AND THAT THE HANDBRAKE IS APPLIED.**

When bleeding any part of the secondary circuit, almost full brake pedal travel is available. When bleeding the primary circuit only, brake pedal travel will be restricted to approximately half.

**NOTE: When bleeding the system commence with the caliper furthest from the master cylinder and bleed from the screw on the same side as the fluid inlet pipes, then close the screw and bleed from the screw on the opposite side of the same caliper. Tighten the bleed screws to the correct torque. See section 06**

### Bleeding

1. Fill the fluid reservoir with the correct fluid, see section 09 lubricants and fluids.

   **NOTE: The correct fluid level must be maintained throughout the procedure of bleeding.**

2. Connect a bleed tube to the bleed screw on the rear caliper furthest from the master cylinder.
3. Submerge the free end of the bleed tube in a container of clean brake fluid.
4. Slacken the bleed screw 1/2-3/4 turn.
5. Operate the brake pedal fully and allow to return.

   **NOTE: Allow at least five seconds to elapse with the foot right off the pedal to ensure that the pistons fully return before operating the pedal again.**

8. Repeat 1 to 7 for the other rear caliper.
9. Remove the front wheel on the side furthest from the master cylinder.
10. Connect a bleed tube to the primary bleed screw on the front caliper furthest from the master cylinder.
11. Connect a bleed tube to the secondary bleed screw on the same side of the caliper as the primary screw.

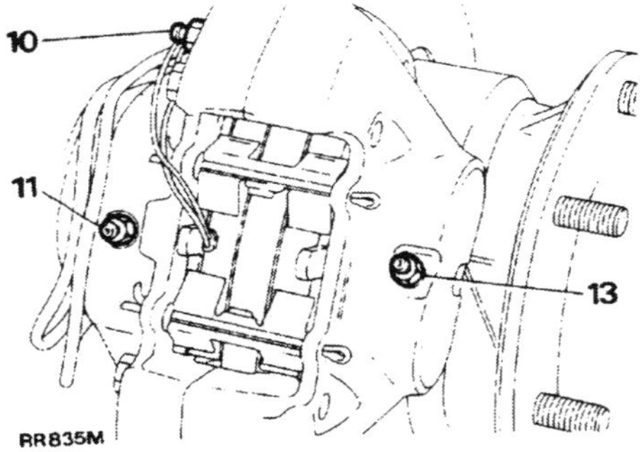

Continued

# BRAKES 70

12. Repeat 3 to 7 for the front caliper, bleeding from the two screws simultaneously.
13. Connect a bleed tube to the other screw on the front caliper furthest from the master cylinder.
14. Repeat 3 to 7 for the second secondary screw on the front caliper.
15. Refit the front wheel.
16. Repeat 9 to 15 for the front caliper nearest the master cylinder.

## MASTER CYLINDER- Lucas Girling -Type 25.4mm AS/AS

### Remove, overhaul and refit

#### Removing

1. Disconnect the battery negative terminal.
2. Place a suitable container under the master cylinder to catch any brake fluid which may seep from the cylinder when the brake pipes are disconnected from the outlet ports.
3. Thoroughly clean the immediate area around all outlet ports. Remove each of the brake pipes from the master cylinder in turn, sealing each pipe and outlet port with suitable plugs as they are disconnected, to prevent ingress of foreign matter and excessive fluid loss.
4. Disconnect the electrical plug from the low fluid switch located on the reservoir cap.
5. Remove the two nuts securing the master cylinder to the servo unit remove also the spring and plain washers.
6. Detach the master cylinder from the servo, remove the reservoir cap and drain the brake fluid into a suitable container.

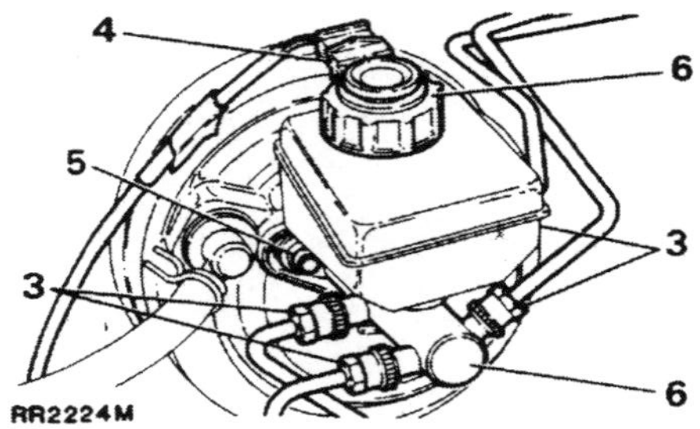

WARNING: Do not use brake fluid previously drained or bled from the system. Carefully dispose of unwanted fluid, if stored in a sealed container, ensure that the container is marked **USED BRAKE FLUID.**

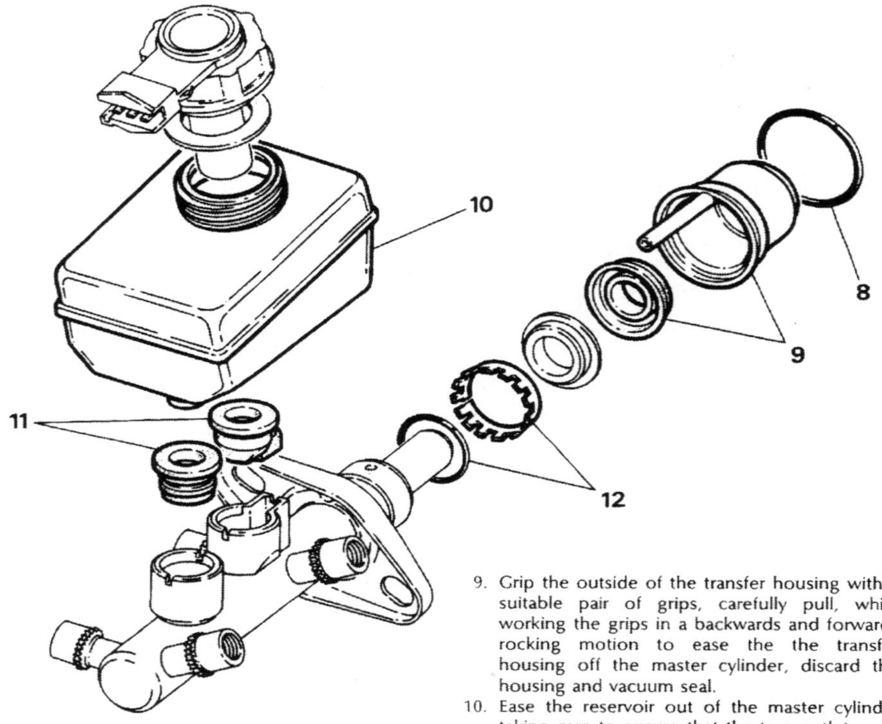

9. Grip the outside of the transfer housing with a suitable pair of grips, carefully pull, while working the grips in a backwards and forwards rocking motion to ease the the transfer housing off the master cylinder, discard the housing and vacuum seal.
10. Ease the reservoir out of the master cylinder taking care to ensure that the two outlet ports on the bottom of the reservoir do not become damaged during this process.

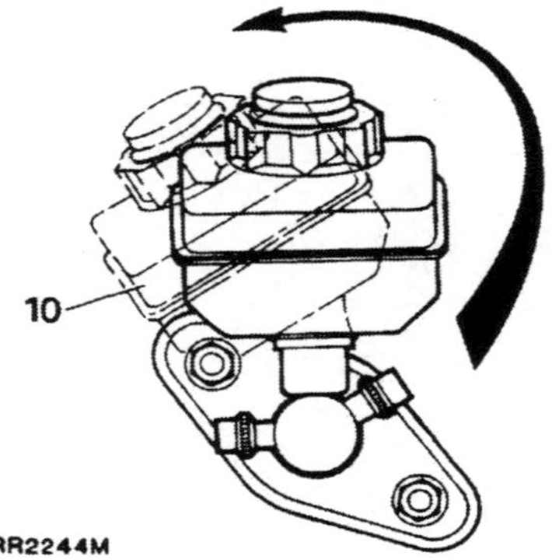

### Overhaul

WARNING: Use only clear methylated spirits or unused brake fluid to clean any part of the brake system. DO NOT use petrol, paraffin or other mineral based fluids.

7. Before commencing the overhaul procedure thoroughly clean the master cylinder and inspect the outer surfaces for damage and condition, renew the complete assembly if necessary.
8. Using two slave spacers, one either side of the master cylinder flange, clamp the flange in a suitable engineers vice. Remove the water ingress 'O' ring seal from the master cylinder to servo flange and discard.

Continued

# BRAKES 70

## KEY TO MASTER CYLINDER

1. Water ingress seal
2. Transfer housing
3. Vacuum seal
4. Guide ring
5. Retaining ring
6. 'O' ring seal
7. Primary plunger assembly
8. 'L' seal
9. Secondary plunger
10. Washer
11. Recuperating seal
12. Seal retainer
13. Springs (2 off)
14. Swirl tube
15. Master cylinder body
16. Reservoir seals
17. Reservoir
18. Low fluid level switch and cap

NOTE: Thoroughly check that no debris of any description is lodged in any of the fluid passageways and drillings. If debris is found, carefully remove, re-clean the cylinder and re-check.

Continued

---

11. Withdraw the two reservoir seals from the master cylinder inlet ports, the seals are different and should be noted for assembly, discard both of the seals.
12. Remove the retaining ring and 'O' ring seal from the machined outer surface of the master cylinder, discard both the seal and retaining ring.
13. Remove the guide ring from the mouth of the master cylinder which supports the primary plunger assembly and place to one side, this component is not part of the master cylinder service kit and is to be refitted on assembly of the unit.
14. Pull the primary plunger assembly out of the master cylinder.

NOTE: The primary plunger assembly cannot be stripped down any further and is serviced as a complete unit. Discard the assembly.

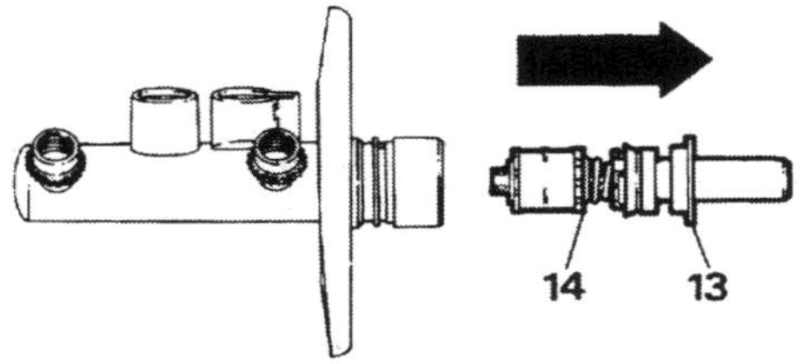

15. The secondary plunger assembly will remain at the bottom of the master cylinder bore, the plunger can be easily expelled by tapping the assembly on a piece of wood until the plunger appears at the cylinder mouth, carefully pull the plunger out of the master cylinder.
16. If the swirl tube was not expelled at the same time as the secondary plunger, repeat the above operation to expel it from the bottom of the master cylinder bore and discard.
17. Clean all parts with Girling cleaning fluid or unused brake fluid and place the cleaned parts onto a sheet of clean paper. Inspect the cylinder bores and plungers for signs of corrosion, ridges and score marks. Provided the working surfaces are in perfect condition, new seals from a Girling Service Kit may be used.

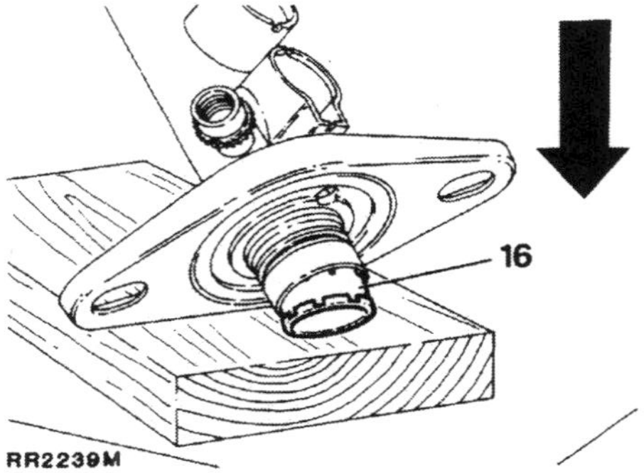

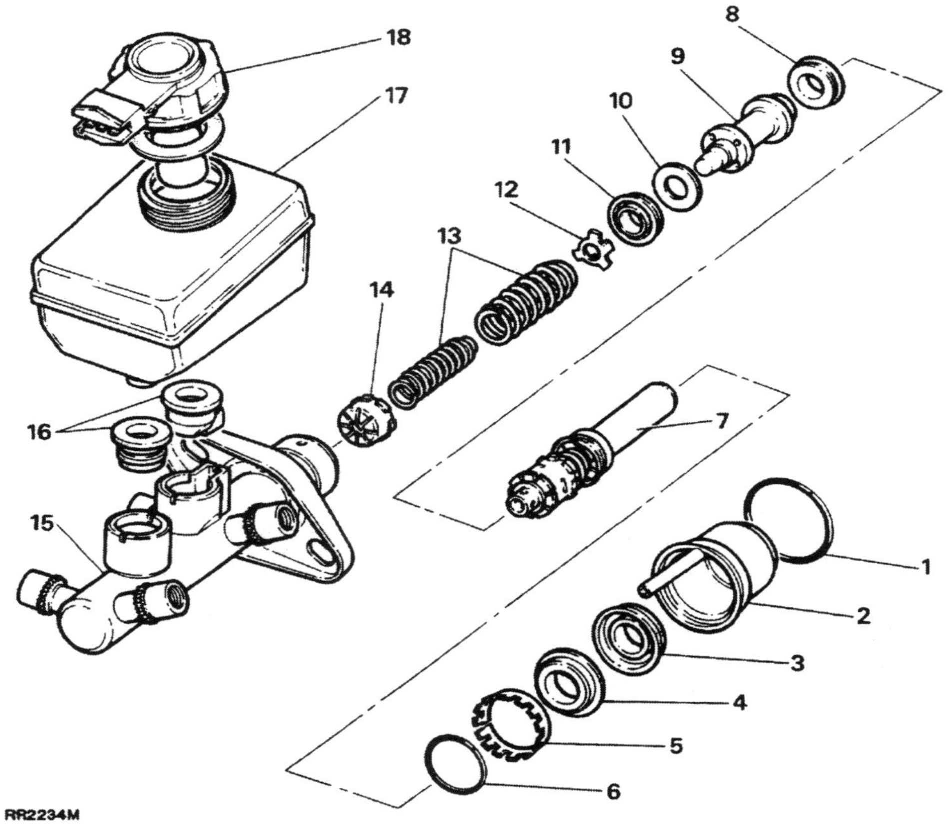

### Renewing secondary plunger seals

18. Remove from the following components from the secondary plunger and discard:

NOTE: A small screwdriver with the end rounded off and polished is required to remove the 'L' seal. DO NOT damage the secondary plunger.

(A) Springs
(B) Seal retainer
(C) Recuperating seal
(D) Washer
(E) 'L' seal

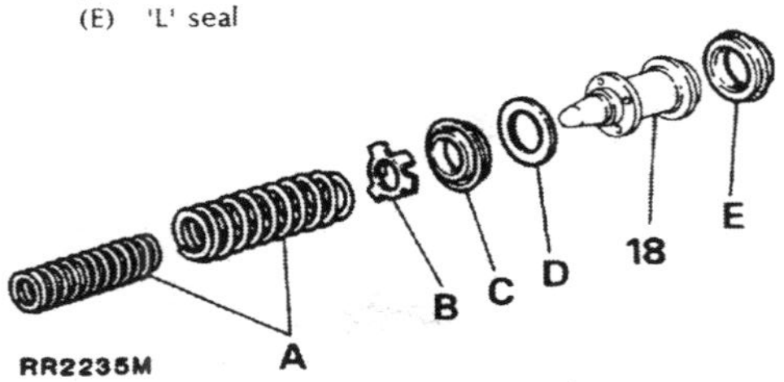

# BRAKES

19. Coat the new seals in unused brake fluid and firstly fit the 'L' seal to the plunger.
20. Fit the washer followed by the recuperating seal. Fit the seal retainer and springs, ensure the springs are correctly seated.

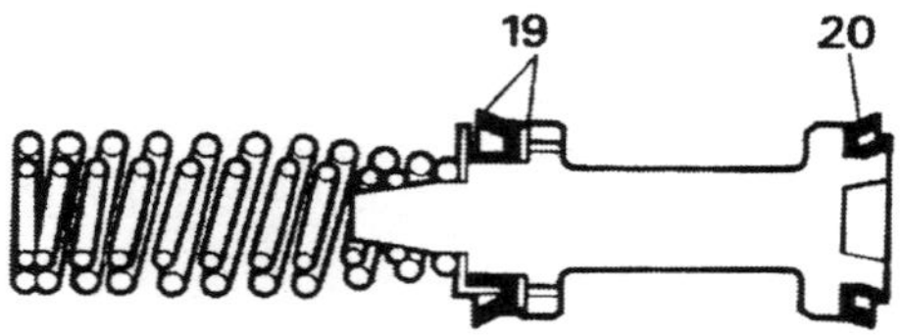

## ASSEMBLING THE MASTER CYLINDER

**CAUTION:** It is important that the following instructions are carried out precisely, otherwise damage could be caused to the new seals when inserting the plungers into the cylinder bore. Generous amounts of new brake fluid should be used to lubricate the the parts during assembly. Never use old fluid or any other form of cleaning and lubricating material. Cleanliness throughout is essential.

21. Fit the new swirl tube to the bottom of the cylinder bore.
22. Lubricate the secondary plunger and cylinder bore. Offer the plunger assembly to the cylinder until the recuperation seal is resting centrally in the mouth of the bore. Gently introduce the plunger with a circular rocking motion, as illustrated. Ensuring that the seal does not become trapped, ease the seal into the bore and slowly push the plunger down the bore in one continuous movement.

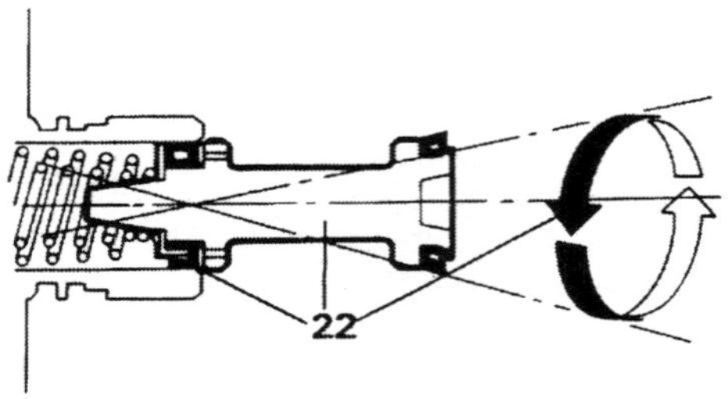

23. Fit the primary plunger assembly using the same method as for the secondary plunger, push the plunger down the bore.
24. Fit the original guide ring to support the primary plunger.
25. Coat a new 'O' ring with brake fluid and fit to its respective groove on the outer location surface of the master cylinder.

**NOTE:** The 'O' ring should not be rolled down the outer location surface of the master cylinder but should be slightly stretched and eased down the cylinder and into its groove. **DO NOT OVER STRETCH THE SEAL.**

26. Fit a new retaining ring on the outer surface of the master cylinder ensuring that the serrations of the ring are facing the mounting flange.
27. Fit the two new reservoir seals in their respective ports.
28. Fit a new vacuum seal to either the primary plunger or to the bottom of the transfer housing bore, open face of the seal towards the primary plunger guide ring.
29. Lubricate the vacuum seal with brake fluid, fit the transfer housing to the master cylinder, push the housing fully upto the cylinder mouting flange. **DO NOT ADJUST THE TRANSFER HOUSING AFTER FITTING.**
30. Lubricate a new water ingress seal with brake fluid, slightly stretch the seal and ease it down the housing until the seal is in the correct position between the housing and flange.
31. Roll the reservoir into the top of the master cylinder. (Reverse the operation described in instruction 10 illustration RR2244M).

32. Fit the master cylinder to the servo fit the plain and spring washers and secure in position with the two nuts. Tighten to the specified Torque- see section 06.
33. Fit the brake pipes to the master cylinder and tighten to the specified Torque- see section 06
34. Top-up the master cylinder with the correct grade of brake fluid (see section 09) bleed the brake systems.

**WARNING:** Do not use brake fluid previously drained or bled from the system. Carefully dispose of unwanted fluid, if stored in a sealed container, ensure that the container is marked **USED BRAKE FLUID.**

35. Fit the cap with combined low level fluid switch and refit the electrical plug. Re-connect the battery.

## SERVO ASSEMBLY

### Remove and refit

**NOTE:** Other than replacing the filter, non-return valve and gromet the servo is not a serviceable component, in the event of failure or damage fit a new unit.

### Removing

1. Disconnect the battery negative terminal.
2. Disconnect the electrical plug to the low fluid level switch.
3. Thoroughly clean the immediate area around all master cylinder outlet ports. Remove each of the brake pipes from the master cylinder in turn, sealing each pipe and outlet port as they are disconnected with suitable plugs, to prevent ingress of foreign matter and excessive fluid loss.
4. Disconnect the vacuum supply hose to the servo.
5. From inside the vehicle remove the lower fascia panel to gain access to the spring clip securing the servo push rod to the brake pedal assembly.
6. Release the spring clip and remove the clevis pin securing the servo push rod to the brake pedal.
7. Remove the two nuts and plain washers securing the servo to the bulkhead pedal box.

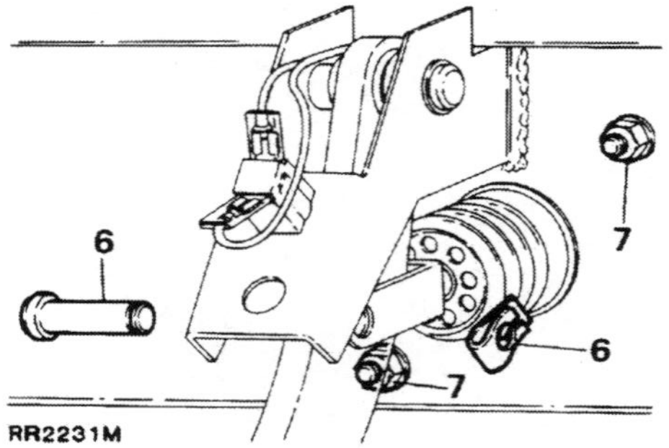

Continued

# BRAKES

8. From within the engine compartment withdraw the servo and master cylinder assembly.
9. Remove the cap from the reservoir and drain the brake fluid into a suitable container if the master cylinder can be kept horizontal it will not be necessary to drain the fluid.

**WARNING: Do not use brake fluid previously drained or bled from the system. Carefully dispose of unwanted fluid, it stored in a sealed container, ensure that the container is marked USED BRAKE FLUID.**

10. Detach the spacer from the servo mounting face of the servo, it is important that the spacer is fitted to the mounting face of the new servo to ensure that stringent pedal to servo operating dimensions are maintained.

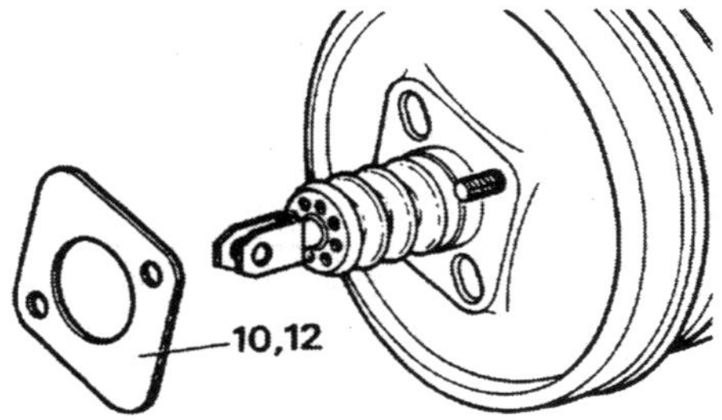

## Refitting

11. If a new servo is being fitted it will be necessary to remove the master cylinder from the existing servo and refitted to the new unit.
12. Fit the spacer previously removed from the old servo, to the new servo.
13. Fit the servo to the pedal box assembly.
14. From inside the vehicle lightly grease the brake pedal around the area that the servo push rod pivots.
15. Fit the push rod to the brake pedal and secure in position with the clevis pin and clip.
16. Fit the plain washers and secure the servo with the two nuts. Tighten the nuts to the specified torque-see section 06.
17. Refit the lower fascia panel.
18. Fit the master cylinder to the servo, refit the plain and spring washers and secure the master cylinder to the servo with the two nuts. Tighten the nuts to the specified torque-see section 06.
19. Remove the sealing plugs, fitted to the master cylinder outlet ports during the removal procedure and refit the brake pipes to their respective ports. Tighten the brake pipes to the correct torque-see section 06.
20. Refit the vacuum supply hose.
21. Fill the master cylinder to between the 'MAX' and 'MIN' level markings with the correct grade of brake fluid-see section 09.
22. Bleed the brake systems.

## PEDAL ASSEMBLY

### Remove, overhaul and refit

### Remove

1. Disconnect the battery negative terminal.
2. From inside the vehicle remove the lower fascia panel to gain access to the pedal assembly.
3. Release the spring clip and remove the clevis pin securing the servo push rod to the brake pedal.
4. Remove the circlip from the 'D' shaped end of the pedal shaft.
5. Remove the pedal shaft from the pedal assembly and withdraw the pedal from the pedal box.
6. Remove the return spring from the pedal.

### Overhaul

7. Remove the bushes from the pedal pivot tube.
8. Press new bushes into the pedal pivot tube. If necessary ream out the bushes to 15.87mm plus 0.05mm (.625 in plus .002in).
9. Lightly grease the bushes.

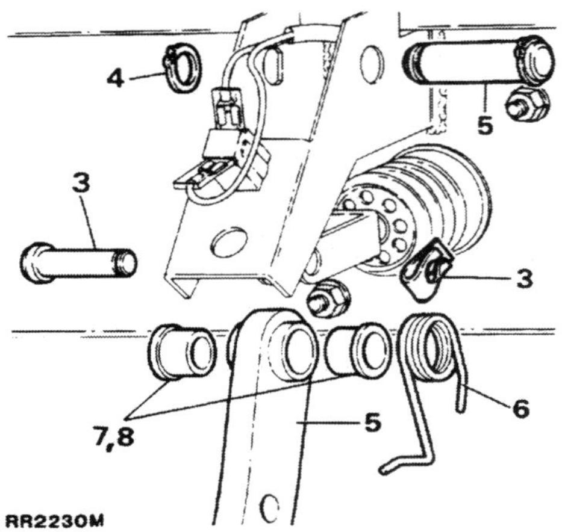

### Refitting

10. Fit the return spring to the pedal.
11. Fit the pedal to the pedal box assembly and refit the pedal shaft. Secure the pedal assembly in position with the circlip.
12. Lightly grease the servo push rod and secure in position with the clevis pin and clip.
13. Refit the lower fascia panel and reconnect the battery.

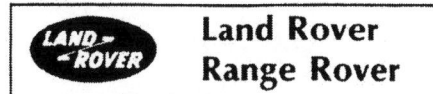

**Land Rover / Range Rover — Workshop Bulletin**

# Range Rover
## August 1987
## Tudor Webasto Electrically Operated Sunshine Roof
## Workshop Bulletin
## Bulletin Number LSM180WB3

|  | Section | Page |
|---|---|---|
| Introduction | 01 | 381 |
| Body | 76 | 382 |
| Electrical | 86 | 385 |

Land Rover
Lode Lane,
Solihull
West Midlands
B92 8NW
England

Section

## 01 INTRODUCTION

Introduction ................................................................ 1

## 76 BODY

Emergency operation ................................................ 2
Headlining - remove and refit ..................................... 6
Motor, micro-switch and stepper relay .......................... 2
Operation ................................................................ 2
Roof panel seals - remove and refit .............................. 1
Sunroof assembly .................................................... 3
Sunroof assembly - remove and refit ........................... 5
Sunroof blind assembly - remove and refit

## 86 ELECTRICAL

Relays .................................................................... 1
Fuse ....................................................................... 1
Circuit diagram ........................................................ 2

---

## INTRODUCTION 01

### RANGE ROVER

### 1987 MODEL YEAR IMPROVEMENTS

### INTRODUCTION

As part of the 1987 model year programme Land Rover Ltd is continually looking at ways to improve the quality and characteristics of its products. This bulletin contains details of the electrically operated sunshine roof which will further enhance the specification of the Range Rover. A further model improvement is the introduction of a one piece headlining.

# BODY 76

## ELECTRICALLY OPERATED SUNSHINE ROOF

RR2293E

**SUNROOF ASSEMBLY**

1. Roof panel
2. Blind
3. Insulation pad
4. Roof seals (front and rear)
5. Blind retaining clips (6 off)
6. Roof panel retaining screws (6 off)
7. Water channel
8. Water channel connectors
9. Support bracket-water channel
10. Slide shoe-water channel
11. Motor bracket/guide tube assembly
12. Operating motor
13. Motor retaining screws
14. Relay
15. Lower guide rails
16. Front guide rails
17. Slide mechanism
18. Rear guide
19. Pivot bracket
20. Slide shoe
21. Rear edge trim finisher
22. Wind deflector assembly
23. Wind deflector operating arms
24. Support bracket (6 off)

# Notes

# BODY

## OPERATION

The sunroof operates in a tilt and slide action controlled by the rocker switch adjacent to the interior roof lamp, with ignition switched 'ON'.

1. Depress front of switch to lift rear edge of sunroof into 'Tilt' position.
2. Depress rear edge of switch to close roof.
3. Depress rear of switch to slide sunroof into 'Open' position.
4. Depress front of switch to slide sunroof into 'Close' position.

**NOTE: The electric drive cuts out automatically in the closed, tilt and open positions. If the switch is operated in the wrong direction in one of these positions, press the switch once to reset and again to operate roof.**

## EMERGENCY OPERATION

If the sun roof fails in the open position carry out the following procedure:

1. Lower the interior lamp mounting panel by releasing two turnbuckles.
2. Remove the emergency handle from the vehicle tool kit.
3. Engage handle in motor drive spindle and turn to close roof.

## MAINTENANCE

At each service blow drain tubes out to ensure they are not blocked or kinked. Blow upwards to clear rear drain tubes which are clipped to the rear mud flap supports. Blow downwards to clear front drain tubes which run down each 'A' post and exit at back of engine bay.
Annually: Clean sunroof opening thoroughly.

**Note: Where the vehicle is operated in extremely dusty conditions more frequent cleaning is recommended.**

## OPERATING MOTOR, MICRO-SWITCH AND RELAY

### Remove and refit

#### Removing

1. Ensure sunroof is fully closed. Disconnect the battery negative lead.
2. Remove interior lamp mounting panel by loosening two turnbuckles to gain access to the motor.
3. Disconnect two wiring connectors.
4. Remove three securing screws and withdraw motor.
5. Remove securing screw and withdraw relay.
6. Remove micro-switch from motor by drilling out securing rivets, if required.

#### Refitting

7. Secure new micro-switch using suitable nuts and bolts to replace rivets. Tighten nuts and apply a spot of paint to threads.
8. Ensure motor is in the 'park' position, i.e. hole on driven gear aligned with drive spindle.
9. Reverse removal procedure.
10. Check operation of sunroof in all positions.

## SUNROOF ASSEMBLY

### Remove and refit

- including roof panel, sliding mechanism, wind deflector, motor mounting bracket and guide tubes.

#### Removing

**NOTE: Lower vehicle headlining ONLY if removing motor mounting bracket and guide tubes. Note that the headlining is secured around the roof opening using an adhesive tape. Remove edge trim and eight edge clips, carefully peel back headlining to remove.**

1. Partially open sunroof rearwards and carefully unclip blind from roof panel front. Slide the blind back fully.
2. Position sunroof in the tilt position and disconnect battery negative terminal.
3. Remove three roof panel fixing screws from each side and remove panel.
4. Remove motor securing screws and withdraw motor.
5. Remove the guide rail screws, seven each side.
6. Remove the pivot bracket and remove front guide rail.
7. Remove the slide and tilt mechanism complete with flexible drive cable from both sides. Do not strip these assemblies unless replacement parts are required.
8. Unclip both wind deflector operating arms from rear mounting brackets. Remove arms from deflector, if required.

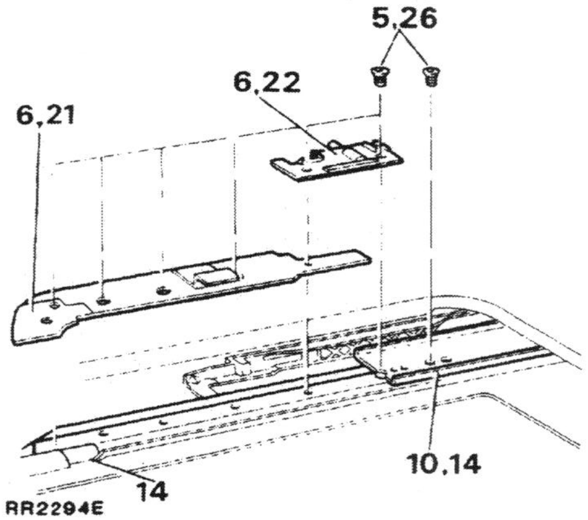

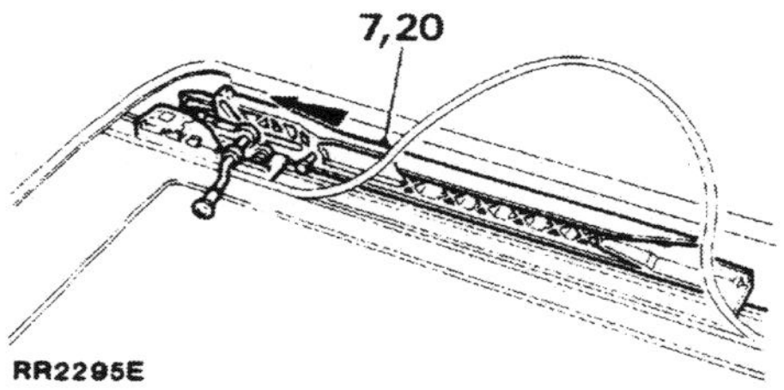

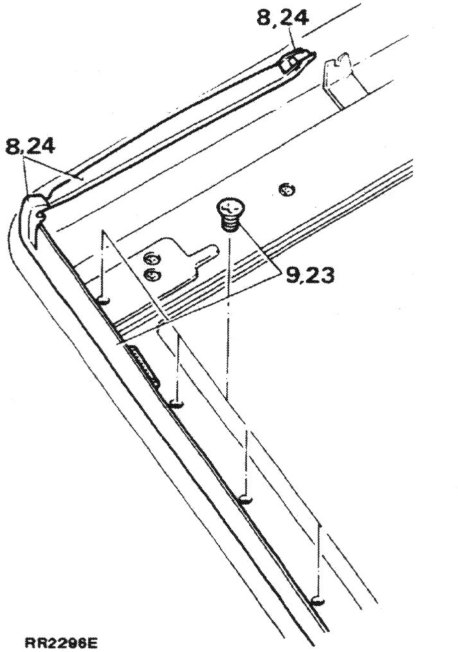

## ROOF PANEL SEALS

### Remove and refit

#### Removing

1. Position sunroof in the tilt position and disconnect battery negative terminal.
2. Unclip blind from roof panel front and slide the blind back fully.
3. Remove three roof panel fixing screws from each side and remove panel.
4. Place the roof panel on a suitable surface to avoid damage, and remove both seals.

#### Refitting

5. Position front seal to front edge of panel, ensuring that there is an equal length of seal each side of the centre point. Secure seal using a rubber mallet.
6. Position the rear seal immediately to rear of the front seal. Secure seal around panel edge ensuring a good fit around corners. Trim off excess seal at joint with front seal using suitable snips, ensuring that the joint is closed.
7. Refit the roof panel.

Continued

9. Remove seven fixing screws, and withdraw wind deflector.
10. Remove lower guide rails and rear edge finisher, seven screws, if required.
11. Remove fixing screws from guide tubes, two each side. Remove five fixing screws and withdraw motor mounting bracket.
12. Pull blind assembly forward and remove.

**Refitting**

**NOTE: During assembly lightly lubricate all sliding parts using a silicon spray.**

13. Position motor bracket and guide tube assembly. Fit and tighten the securing screws.
14. Position right hand guide rail in rear retaining bracket, push rearwards fully and locate leading edge under drive cable opening. Repeat operation for left hand guide rail.
15. Align fixing holes, and loosely fit screw in seventh hole from the front.
16. Position finisher to rear edge of sun roof opening and secure using seven screws.
17. Position blind assembly into outer guide runners, and push fully rearwards.
18. Lightly lubricate drive cables. Ensure that the slide and tilt mechanism is fully assembled
19. Push the cable fully into the right hand side guide tube. Loop remaining cable and enter the rear end into the right hand inner side runner.
20. Repeat operation 19. for left hand side. Push both assemblies rearwards to take up slack in the cables, and push a further 75 mm (3 in) to the rear.
21. Position both front guide rails, aligning with the four forward holes. Secure with screws, do not tighten.
22. Position both pivot brackets, fit but do not tighten fixing screws.
23. Fit the wind deflector and fully tighten fixings.
24. Position right hand operating arm in locating slot in deflector. Secure opposite end in frame bracket. Repeat for left hand side.
25. Pull right hand slide and tilt mechanism forward, align with pivot bracket and secure in position using setting key. Repeat operation for left hand side.

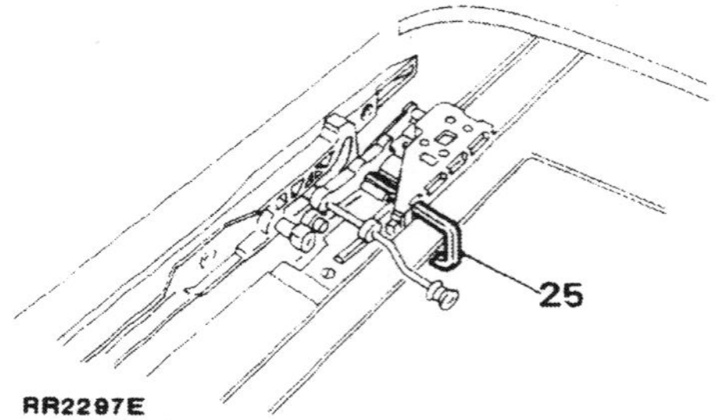

RR2297E

26. Tighten screws to guide rails, seven each side.
27. Fit and secure relay.
28. Ensure operating motor is in the park position i.e. hole on driven gear aligned with drive spindle. Fit and secure to mounting bracket.
29. Remove setting keys. Temporarily connect operating switch and reconnect wiring including battery.
30. Operate switch sequence to 'tilt' position.
31. Position roof panel into roof opening and secure with six fixing screws, do not tighten.
32. Put roof into 'closed' position and adjust roof profile. The panel profile should be 0.5 mm low at forward edge. 1 mm high at rear edge.
33. Tighten roof panel screws.
34. Tilt sunroof, pull blind forward and locate rear brackets in tilt mechanism. Align front six clips and push to secure.
35. Check operation of sunroof in all operating modes.

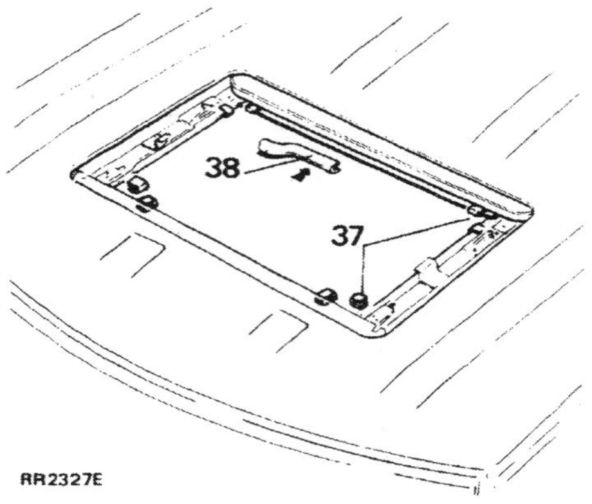

RR2327E

36. Refit vehicle headlining. The headlining is secured around the sunroof opening with '3M' adhesive tape.
37. Refit the eight edge clips as shown.
38. Finally fit the edging finisher.

4. Secure connecting arms to frame using suitable rivets before retrimming frame.
5. Retrim frame using a new blind cover. Inset shows section through frame indicating where adhesive is applied. Dimension 'A' should be radially constant.

**Refitting**

6. Refit blind assembly and reassemble sunroof.

**SUNROOF BLIND ASSEMBLY**

**Remove, retrim and refit**

**Removing and retrimming**

1. Remove sunroof blind assembly as detailed in Sunroof Assembly - remove and refit.
2. Remove trim covering from frame assembly. Note that it is not normally necessary to remove the three pads and insulation pad shown in illustration.
3. If required: remove water channel by unclipping connecting arms. Drill out rivets securing connecting arms to frame.

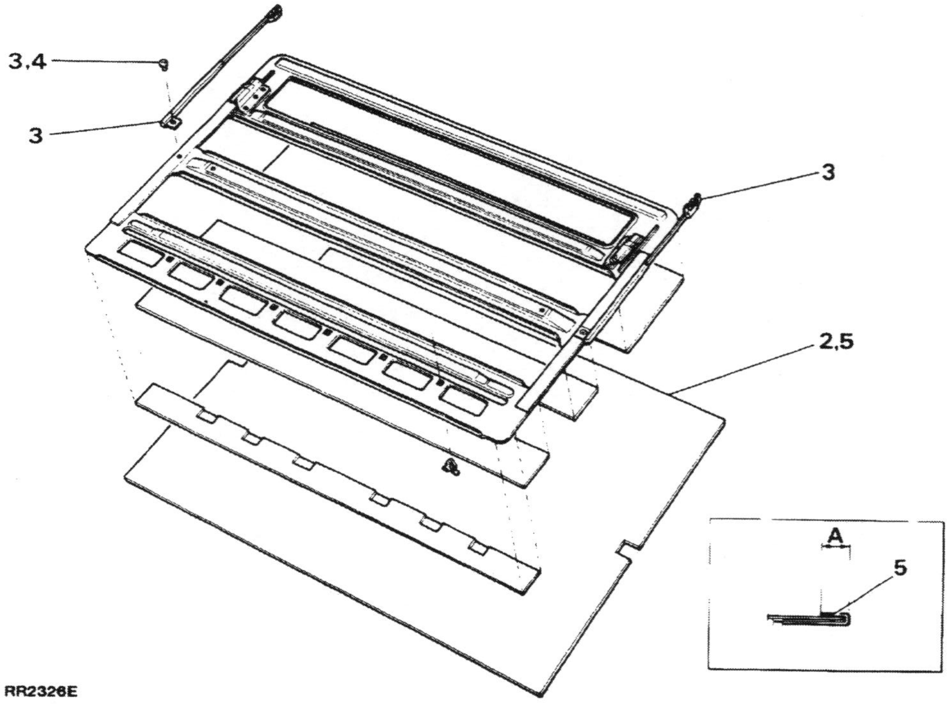

RR2326E

## ELECTRICAL 86

### RELAYS

**Stepper relay:** mounted on motor mounting bracket, accessible by lowering interior lamp mounting bracket.

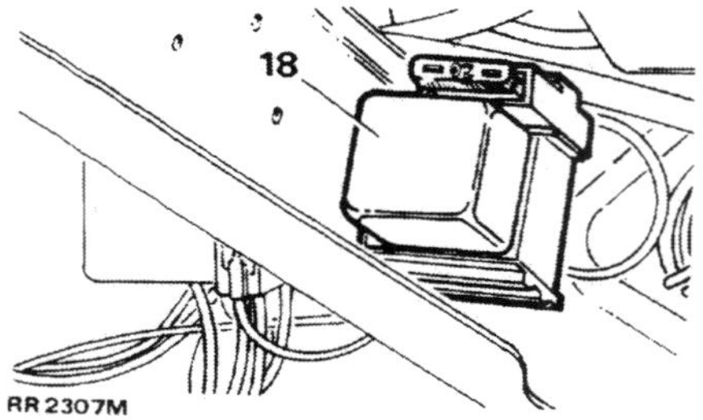

RR 2307M

**Auxiliary relay (18):** mounted on side of pedal box. Accessible by removing lower fascia panel.

**Fuse:** 20 amp located on side of auxiliary relay.

Continued

## 76 BODY

### HEADLINING (One piece)

**Remove and refit**

**Removing**

1. Remove spare wheel from the vehicle.
2. Remove the rear seat belt upper guide brackets and inertia reel assemblies.
3. Fold the rear seat backrest forward and recline the front seats as far as possible.
4. Disconnect the battery negative lead.
5. Remove the two roof lamp assemblies. (Lower and remove the interior lamp mounting panel if the vehicle has a sunroof)
6. Remove the rear view mirror and mounting bracket.
7. Remove the two sun visors and centre retaining bracket.
8. Remove the front and rear passenger grab handles. **Sunroof vehicles:** Remove edge trim and eight edge clips from roof opening, and carefully peel back headlining to remove.
9. With assistance support the front of the headlining, while removing the two plastic retaining clips above the rear quarter light glass.
10. While the headlining is still being supported, remove the two plastic retaining clips securing the rear end of the headlining, located adjacent to the upper tailgate hinges.
11. Pull the headlining forward to clear the rear quarter trim. Lower the headlining and disconnect the electrical leads from the roof mounted speakers.
12. Remove the headlining through the tailgate.

**CAUTION: To assist removal tilt the headlining at an angle. DO NOT flex the headlining as damage may occur.**

**Refitting**

8. Reverse the removal procedure. **Sunroof vehicles:** The headlining is secured around the roof opening with '3M' adhesive tape. Refit the eight edge clips and fit the edging finisher. See also Sunroof assembly, remove and refit.

385

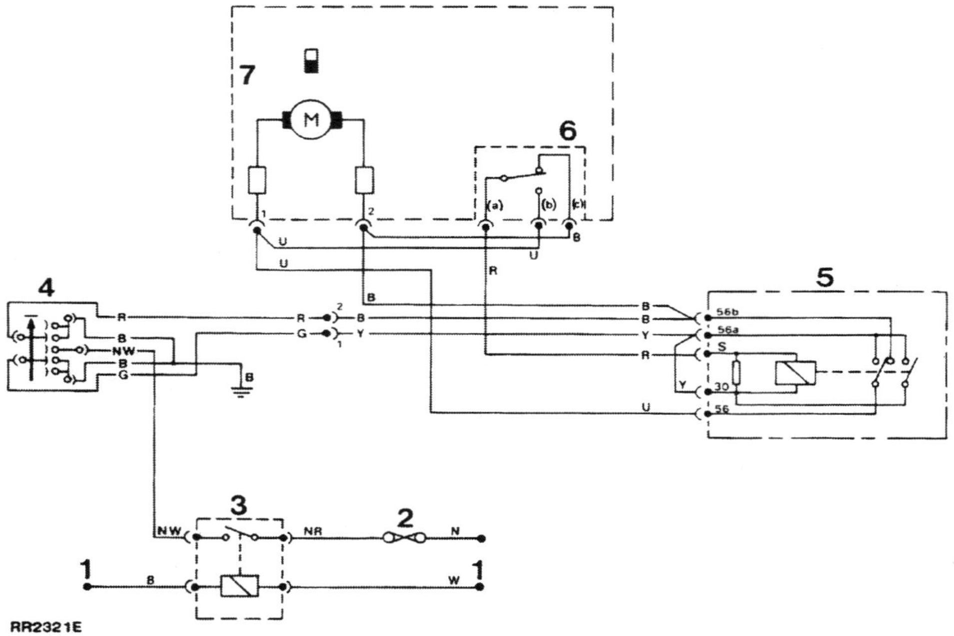

# ELECTRICAL

## Notes

**SUNSHINE ROOF - Circuit diagram**

RR2321E

1. Main harness connections
   Brown - live positive feed
   White - ignition positive feed
   Black - earth
2. Fuse
3. Auxiliary relay
4. Operating switch
5. Stepper relay
6. Micro-switch - motor switching:
   Contact (a) and (c)-CLOSED
   Contact (a) and (b)-OPEN/TILT
7. Drive motor

**CABLE COLOUR CODE**

- B Black
- U Blue
- N Brown
- G Green
- R Red
- W White
- Y Yellow

The last letter of a colour code denotes the tracer.

(08/87)

**Land Rover
Range Rover**

**Workshop Bulletin**

**Range Rover
March 1988
Vogue SE
Workshop Bulletin
Bulletin Number LSM180WB4**

|  | Section | Page |
|---|---|---|
| Introduction | 01 | 387 |
| General Specification Data | 04 | 388 |
| Torque Wrench Settings | 06 | 388 |
| Body | 76 | 388 |
| Electrical | 86 | 389 |

**Land Rover
Lode Lane,
Solihull
West Midlands
B92 8NW
England**

Section

## 01 INTRODUCTION
Introduction .................................................................................... 1

## 04 GENERAL SPECIFICATION
Alternator ....................................................................................... 2

## 06 TORQUE WRENCH SETTINGS
Alternator ....................................................................................... 2

## 76 BODY
Front seat-remove and refit ........................................................... 1

## 86 ELECTRICAL
Alternator - remove and refit ......................................................... 2
Alternator drive belt - adjust .......................................................... 2
Alternator - overhaul and testing .................................................. 4
Auxiliary fuse box ........................................................................... 12
Electrical equipment - description ................................................ 1
Front seat adjustment motors - remove and refit ........................ 8
Main circuit diagram - left hand drive - non catalyst vehicles ... 18
Main circuit diagram - left hand drive - catalyst vehicles .......... 22
Main circuit diagram - right hand drive ........................................ 14
Relays - identification - remove and refit ..................................... 11
Seat adjustment control switch - remove and refit ..................... 8
Seat adjustment circuit diagram ................................................... 9

# INTRODUCTION 01

## RANGE ROVER VOGUE SE

### INTRODUCTION

Land Rover Ltd introduce the addition of a Vogue SE version of Range Rover to complement the existing model range. The Vogue SE will be fitted with, as standard equipment;

- Automatic Transmission
- Air Conditioning
- Electric Sunroof
- Leather Seats - Electrically Operated
- Polished Burr Walnut door Cappings and Door Pulls
- Body Colour Paint to Alloy Wheels
- Coach Line
- Premium In-car Entertainment
- A133/80 Alternator

This bulletin contains servicing details for the Lucas A133/80 Alternator and Electrically operated seats, and includes up to date wiring diagrams for the complete Range Rover model range.

1

## 04 GENERAL SPECIFICATION DATA

### ELECTRICAL

System .................................................................... 12 volt, negative ground

**Alternator**

Manufacturer ........................................................... Lucas
Type ........................................................................ 133/80
Polarity .................................................................... Negative ground
Brush length
 New ...................................................................... 20 mm (0.78 in)
 Worn, minimum free protrusion
 from brush box ................................................... 10 mm (0.39 in)
Brush spring pressure flush with brush box face .... 136 to 279 g (5 to 10 oz)
Rectifier pack output rectification ........................... 6 diodes (3 positive side and 3 ground side)
Field winding supply rectification ........................... 3 diodes
Stator windings ....................................................... 3 phase-delta connected
Field winding rotor poles ........................................ 12
Maximum speed ...................................................... 16,000 rev/min
Winding resistance at 20°C .................................... 2.6 ohms
Control ..................................................................... Field voltage sensed regulation
Regulator-type ......................................................... 15 TR
 voltage .................................................................. 13.6 to 14.4 volts
Nominal output
 Condition ............................................................. Hot
 Alternator speed ................................................. 6000 rev/min
 Control voltage ................................................... 14 volt
 Amp ..................................................................... 80 amp

## 06 TORQUE WRENCH SETTINGS

### ELECTRICAL

| | Nm | lbf/ft |
|---|---|---|
| Alternator mounting bracket to cylinder head | 34 | 25 |
| Alternator to mounting bracket | 24 | 17 |
| Alternator to adjusting link | 24 | 17 |
| Alternator shaft nut | 27.2 to 47.5 | 20 to 35 |
| Alternator through bolts | 4.5 to 6.2 | 3.3 to 4.6 |
| Alternator rectifier bolts | 3.4 to 3.96 | 2.5 to 2.9 |

Charts below give torque values for all screws and bolts used-except for those that are specified.

| SIZE | METRIC | | SIZE | UNC | | UNF | |
|---|---|---|---|---|---|---|---|
| | Nm | ft lb | | Nm | ft lb | Nm | ft lb |
| M5 | 5-7 | 3.7-5.2 | 1/4 | 6.8-9.5 | 5-7 | 8.1-12.2 | 6-9 |
| M6 | 7-10 | 5.2-7.4 | 5/16 | 20.3-27.1 | 15-20 | 20.3-27.1 | 15-20 |
| M8 | 22-28 | 16.2-20.7 | 3/8 | 35.3-43.4 | 26-32 | 35.3-43.4 | 26-32 |
| M10 | 40-50 | 29.5-36.9 | 7/16 | 67.8-88.1 | 50-65 | 67.8-88.1 | 50-65 |
| M12 | 80-100 | 59.0-73.8 | 1/2 | 81.3-101.7 | 60-75 | 81.3-101.7 | 60-75 |
| M14 | 90-120 | 66.4-88.5 | 5/8 | 122.0-149.1 | 90-110 | 122.0-149.1 | 90-110 |
| M16 | 160-200 | 118.0-147.5 | | | | | |

2

# BODY

## FRONT SEAT

### Remove and refit

#### Removing

1. Remove the three fixings securing the seat cushion side trim panel and withdraw the panel. **Remove the bolt securing the seat belt to the side of the seat.**
2. Move the seat until it is in its most rearward position.
3. Remove the two fixings securing the front of the seat located in each seat slide channel.
4. Move the seat until it is in its most forward position.
5. Remove the four fixings securing the rear of the seat located inside each seat slide channel.
6. Disconnect the battery negative terminal.
7. Disconnect the electrical multi-plugs to the seat motors and seat control switch.

Note: Certain models may be fitted with a seat belt warning system depending upon territory requirements. If the system is fitted disconnect the electrical connector at the seat belt buckle.

8. Withdraw the seat from the vehicle.
9. If necessary the seat motors and operating switch can be removed. (Refer to Section 86 Electrical).
10. Remove the single screw securing the seat base side trim panel and withdraw the panel.
11. Remove the three screws securing the seat base front trim panel to the front footwell, remove the single screw securing the top of the front trim panel to the seat base located under the seat base cushion, and withdraw the panel.

#### Refitting

12. Reverse the removal instructions.
13. Arrange the electrical leads beneath the seat to ensure that they do not become trapped by the seat slide mechanism.

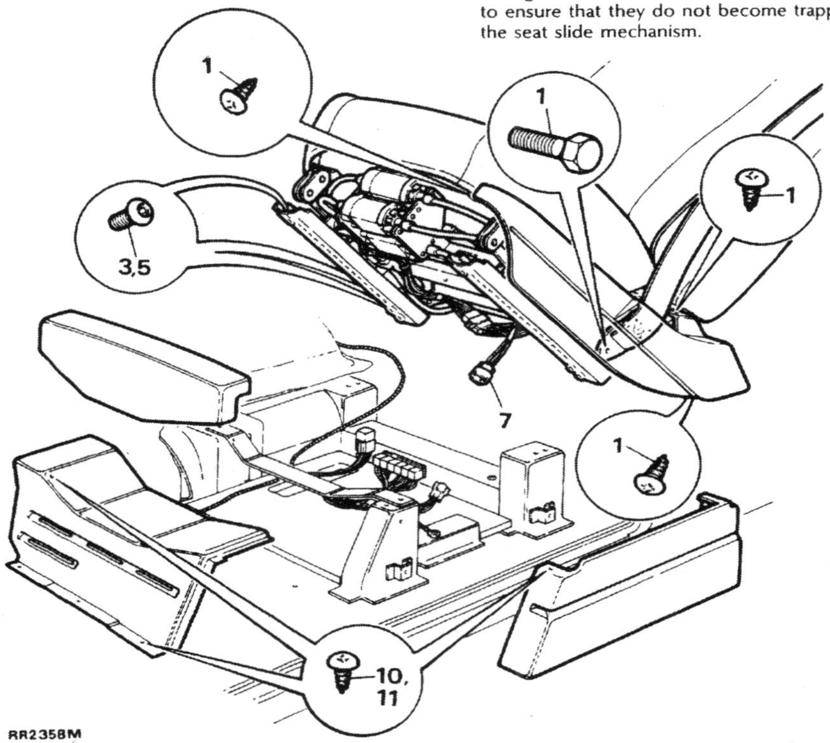

# ELECTRICAL

## ELECTRICAL EQUIPMENT

### DESCRIPTION

The electrical system is Negative ground, and it is most important to ensure correct polarity of the electrical connections at all times. Any incorrect connections made when reconnecting cables may cause irreparable damage to the semi-conductor devices used in the alternator and regulator. Incorrect polarity would also seriously damage any transistorized equipment such as radio and tachometer etc.

**WARNING: During battery removal or before carrying out any repairs or maintenance to electrical components always disconnect the battery negative lead first. If the positive lead is disconnected with the negative lead in place, accidental contact of the wrench to any grounded metal part could cause a severe spark, possibly resulting in personal injury. Upon installation of the battery the positive lead should be connected first.**

Do not disconnect battery cables while the engine is running or damage to the semi-conductor devices may occur. It is also inadvisable to break or make any connections in the alternator charging and control circuits while the engine is running.

The Model 15TR electronic voltage regulator employs micro-circuit techniques resulting in improved performance under difficult service conditions. The whole assembly is encapsulated in silicone rubber and housed in an aluminium heat sink, ensuring complete protection against the adverse effects of temperature, dust, and moisture etc.

The regulating voltage is set during manufacture to give the required regulating voltage range of 14.2 ± 0.2 volts, and no adjustment is necessary. The only maintenance needed is the occasional check on terminal connections and wiping with a clean dry cloth.

The alternator system provides for direct connection of a charge (ignition) indicator warning light, and eliminates the need for a field switching relay or warning light control unit. As the warning lamp is connected in the charging circuit, lamp failure will cause loss of charge. Lamp should be checked regularly and a spare carried.

When using rapid charge equipment to re-charge the battery, the battery must be disconnected from the vehicle.

### ALTERNATOR - LUCAS A133/80

The alternator is a three phase, field sensed unit. The rotor and stator windings produce three phase alternating current, AC, which is rectified to direct current, DC. The electronic voltage regulator unit controls the alternator output voltage by high frequency switching of the rotor field circuit. Use only the correct Range Rover replacement fan belt. Occasionally check that the engine and alternator pulleys are accurately aligned.

It is essential that good electrical connections are maintained at all times. Of particular importance are those in the charging circuit (including those at the battery) which should be occasionally inspected to see that they are clean and tight. In this way any significant increase in circuit resistance can be prevented.

# ELECTRICAL 86

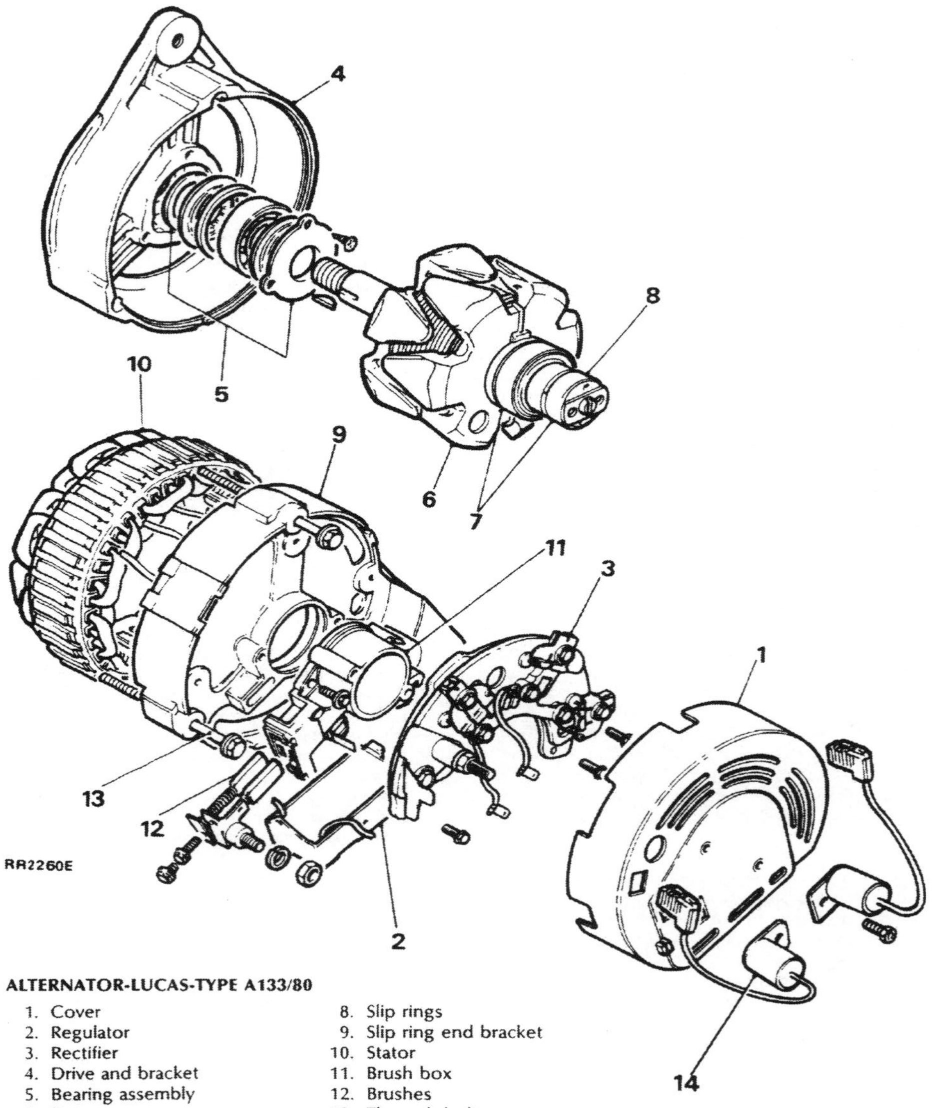

## ALTERNATOR-LUCAS-TYPE A133/80

1. Cover
2. Regulator
3. Rectifier
4. Drive and bracket
5. Bearing assembly
6. Rotor
7. Slip ring end bearing
8. Slip rings
9. Slip ring end bracket
10. Stator
11. Brush box
12. Brushes
13. Through bolt
14. Suppressors

## ALTERNATOR

### Remove and refit

#### Removing

1. Disconnect battery ground lead.
2. Disconnect leads from alternator. 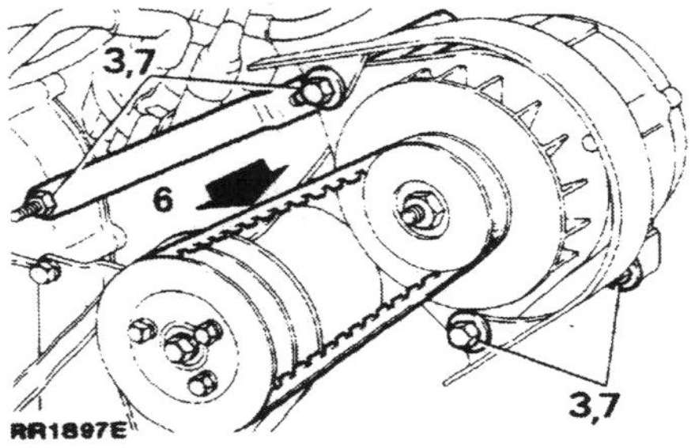
3. Loosen alternator fixings, pivot alternator inwards and remove drive belt.
4. Remove three mounting bolts and lift the alternator clear of the engine.

#### Refitting

5. Fit the alternator and mounting bolts.

**NOTE: The fan guard is attached to the front fixing and the adjustment bracket bolt.**

6. Fit the drive belt and adjust the belt tension.
7. Tighten the mounting bolts and the adjustment bracket securing nut.
8. Connect the wiring leads to the alternator.
9. Connect the battery.

## ALTERNATOR DRIVE BELT

### Adjust

1. Loosen the alternator fixings and the adjustment link.
2. Pivot the alternator to give the required belt tension.
3. Belt tension should be 4 to 6mm (0.19 to 0.25 in) at the point indicated by the bold arrow. 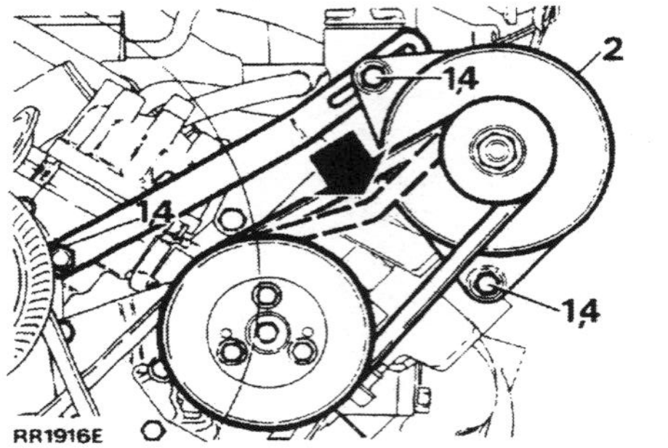
4. Tighten the alternator fixing bolts and the adjustment link.

**NOTE: Check adjustment after running engine at fast idle speed for 3 to 5 minutes if a new belt has been fitted.**

# ELECTRICAL 86

## ALTERNATOR-LUCAS-TYPE A133/80

### Overhaul

**Including Test (Bench)**

NOTE: Alternator charging circuit-The ignition warning light is connected in series with the alternator field circuit. Bulb failure would prevent the alternator charging, except at very high engine speeds, therefore, the bulb should be checked before suspecting an alternator failure.

### Precautions

Battery polarity is **NEGATIVE GROUND**, which must be maintained at all times.

No separate control unit is fitted; instead a voltage regulator of micro-circuit construction is incorporated on the slip ring end bracket, inside the alternator cover.

Battery voltage is applied to the alternator output cable even when the ignition is switched off, the battery must be disconnected before commencing any work on the alternator. The battery must also be disconnected when repairs to the body structure are being carried out using electric welding equipment.

### Sequence of connections

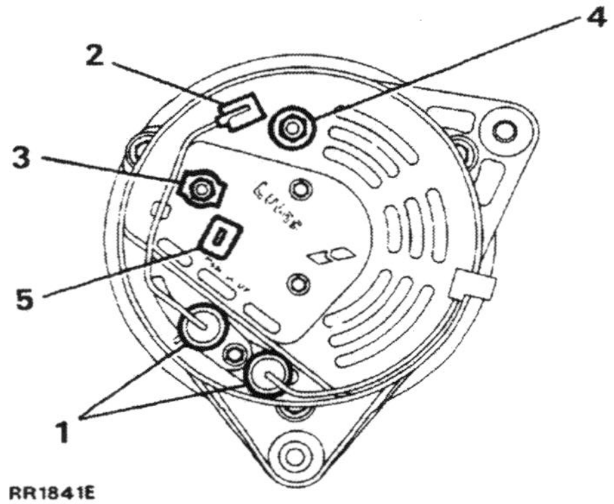

RR1841E

1. Suppression capacitors (two)
2. Positive suppression terminal
3. IND terminal
4. + output terminal
5. Sensing terminal

## ALTERNATOR TESTING

### Charging system check

1. Check the battery is in good condition, with an open circuit voltage of at least 12.6 V. Recharge or fit a charged substitute battery to carry out test.
2. Check drive belt adjustment and condition. Rectify as necessary.
3. Check battery connections are clean and tight.
4. Check alternator connections are clean and tight.
5. Ensure that there is no continuous drain on battery due, for example, to interior, underbonnet or door edge lamps being left on.

### Alternator test

The following instructions refer to the use of suitable test equipment using a carbon pile rheostat.

6. Connect test equipment referring to the manufacturer's instructions.
7. Start engine and run at 3000 rev/min without accesory load.
8. Rotate the carbon pile load control to achieve the greatest output (amps) without allowing voltage to fall below 12.0 V. A reading of 80 amps, minus 10% to allow for EFI and Ignition loss, should be obtained.
9. Run engine at 3000 rev/min, switch selector to regulator test, read voltmeter. A reading of 13.6 to 14.4 V should be obtained.
10. Switch selector to diode/stator test, switch on headlamps to load alternator. Raise engine speed to 3000 rev/min, read voltmeter. The needle must be within the 'OK' range.

NOTE: **See also charging circuit resistance test, page 7.**

### Testing-alternator removed

11. Withdraw the connectors from the alternator.
12. Remove the alternator.
13. Disconnect the suppressor and remove the alternator cover.

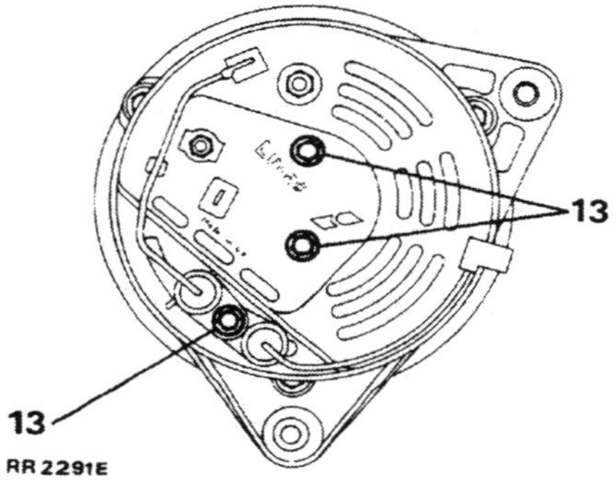

RR 2291E

14. Disconnect the lead and remove the rectifier assembly.
15. Note the arrangement of the brush box connections and remove the screws securing the regulator to the brush box and withdraw. This screw also retains the inner brush mounting plate in position.
16. Remove the screw retaining the outer brush box in position and withdraw both brushes.
17. Check brushes for wear by measuring length of brush protruding beyond brush box moulding. If length is 10mm (0.4 in) or less, fit new brushes.
18. Check that brushes move freely in holders. If brush is sticking, clean with a mineral spirit moistened cloth or polish sides of brush with fine file.
19. Check brush spring pressure using push-type spring gauge. Gauge should register 136 to 279g (5 to 10 oz) when brush is pulled back until face is flush with housing. If reading is outside these limits, fit a new brush assembly.
20. Remove the two screws securing the brush box to the slip ring end bracket and lift off the brush box assembly.
21. Securely clamp alternator in a vice and release the stator winding cable ends from the rectifier by applying a hot soldering iron to the terminal tags of the rectifier. Pry out the cable ends when the solder melts.

# ELECTRICAL 86

22. Remove the two remaining screws securing the rectifier assembly to the slip ring end bracket and lift off the rectifier assembly. Further dismantling of the rectifier is not required.
23. Check the diodes. Connect the test equipment as shown and test each diode in turn, note whether lamp lights, then reverse test lead connections. The lamp should light in one direction only. Renew the rectifier assembly if a faulty diode is diagnosed.
24. Remove the slip ring end bracket bolts and lift off the bracket.
25. Connect a 12 volt battery and a 36 watt test lamp to two of the stator connections. Repeat the test replacing one of the two stator connections with the third. If test lamp fails to light in either test, fit a new stator.
26. Using a 110 volt a.c. supply and a 15 watt test lamp, test for insulation between any one of the three stator connections and stator laminations. If test lamp lights, fit a new stator.
27. Clean surfaces of slip rings using a solvent moistened cloth.
28. Inspect slip ring surfaces for signs of burning; remove burn marks using very fine sandpaper. On no account should emery cloth or similar abrasives be used, or any attempt made to machine the slip rings.
29. Note the position of the stator output leads in relation to the alternator fixing lugs, and lift the stator from the drive end bracket.
30. Connect an ohmmeter to the slip rings. A reading of 2.6 ohms should be recorded.
31. Using a 110 volt a.c. supply and a 15 watt test lamp, test for insulation between one of the slip rings and one of the rotor poles. If the test lamp lights, fit a new rotor.
32. To separate the drive end bracket and rotor, remove the shaft nut, washers, woodruff key and spacers from the shaft.
33. Remove bearing retaining plate by removing the three screws. Using a press, drive the rotor shaft from the drive end bearing.
34. If necessary, to remove the slip rings or the slip ring end bearing on the rotor shaft, unsolder the outer slip ring connection and gently pry the slip ring off the shaft, repeat the procedure for the inner slip ring connection. Using a suitable extraction tool, withdraw the slip ring bearing from the shaft.

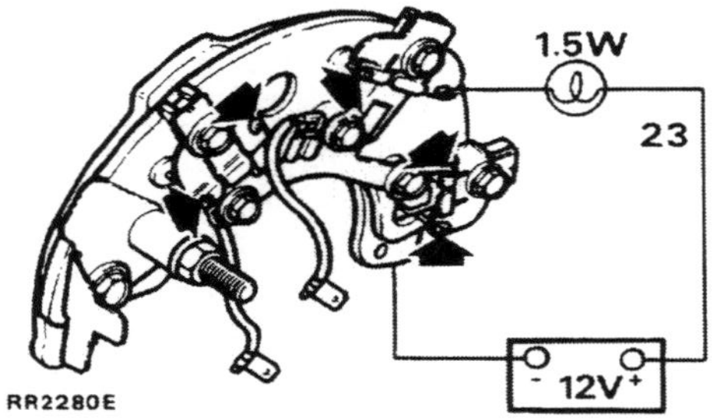

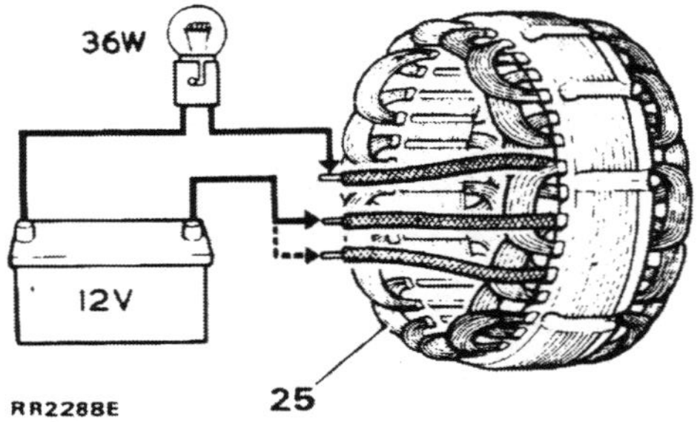

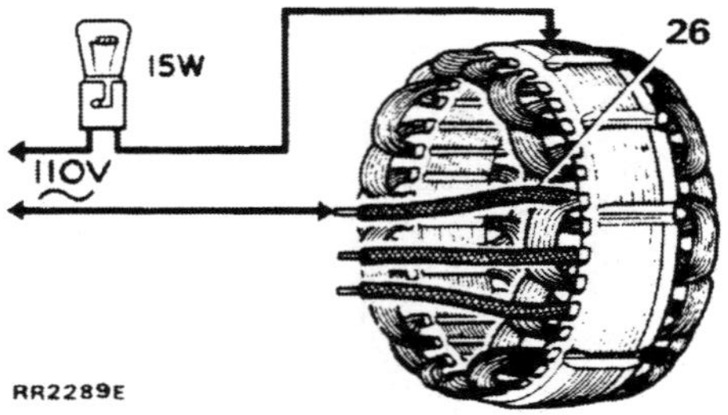

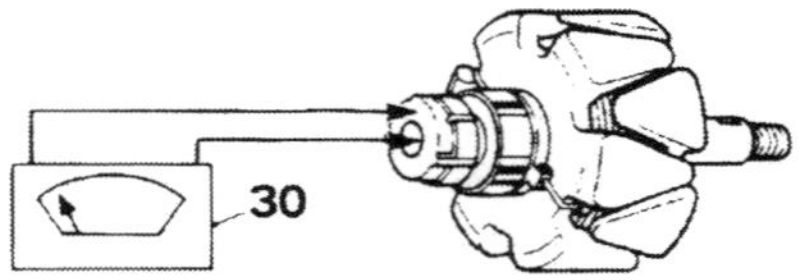

## Reassembling

35. Reverse the dismantling procedure, noting the following points.
    (a) Use Shell Alvania 'RA' to lubricate bearings.
    (b) When refitting slip ring end bearing, ensure it is fitted with open side facing rotor.
    (c) Use Fry's H.T.3 solder on slip ring field connections.
    (d) When refitting rotor to drive end bracket, support inner track of bearing. Do not use drive end bracket to support bearing when fitting rotor.
    (e) Tighten through-bolts evenly.
    (f) Fit brushes into housings before fitting brush moulding.
    (g) Tighten shaft nut to the correct torque, see Torque Values.
    (h) Refit regulator pack to brush moulding.
36. Reconnect the leads between the regulator, brush box and rectifier.
37. Refit the alternator.

## Testing in position

### Charging circuit resistance test.

1. Connect a low range voltmeter between the alternator terminal marked + and the positive terminal of the battery.
2. Switch on the headlamps and start the engine. Set the throttle to run at approximately 3000 rev/min. Note the voltmeter reading.
3. Transfer the voltmeter connections to the frame of the alternator and the negative terminal of the battery, and again note the voltmeter reading.
4. If the reading exceeds 0.5 volt on the positive side or 0.25 volt on the negative side, there is a high resistance in the charging circuit which must be traced and remedied.

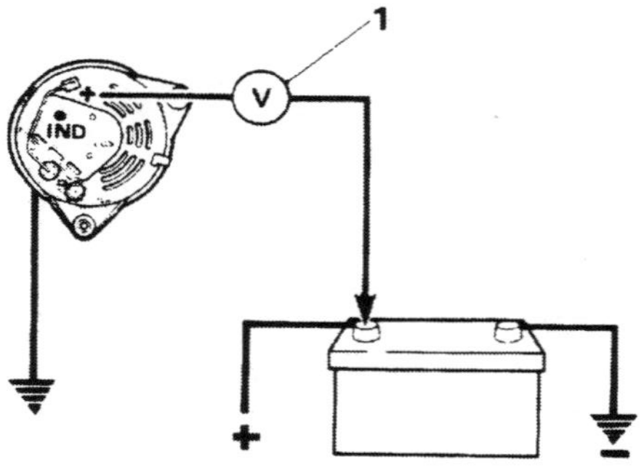

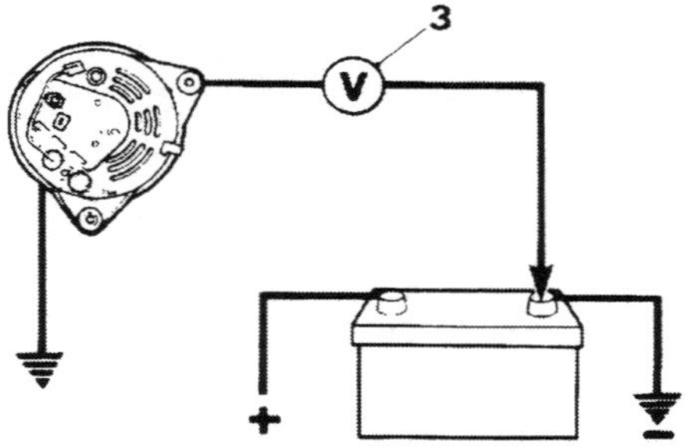

# ELECTRICAL 86

## FRONT SEAT ADJUSTMENT MOTORS

### Remove and refit

Four electric motors mounted beneath each front seat control the fore and aft movement, the cushion height front and rear, and the angle of recline of the seat. Adjustment is possible with either front door open, or with ignition switched ON.

### Removing

1. Position the seat to give access to the motors.
2. Disconnect the battery negative lead.
3. Remove the seat base trim.
4. Remove two securing screws from each side of the required motor.
5. Withdraw the motor from its mounting.
6. Disconnect the drive cables by unscrewing the ferrule.
7. Disconnect the wires from the multi-plug and remove the motor.

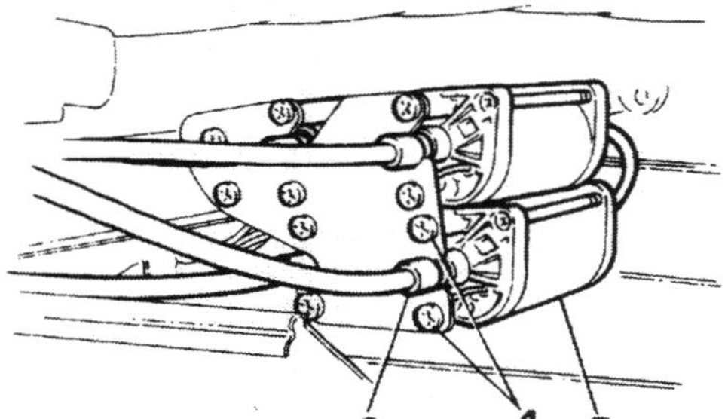

### Refitting

8. Reverse the removal procedure.
9. Check the seat adjustment for correct operation.

## SEAT ADJUSTMENT CONTROL SWITCH

### Remove and refit

### Removing

1. Disconnect the battery negative lead.
2. Pry the two finger tip controls from the top of the switch housing.
3. Removing the switch housing cover by lightly depressing the sides of the cover to disengage the clips.
4. Remove two crosshead screws and washers and lift the switch assembly to gain access to the two multiplugs.
5. Disconnect the multiplugs and withdraw the switch assembly.

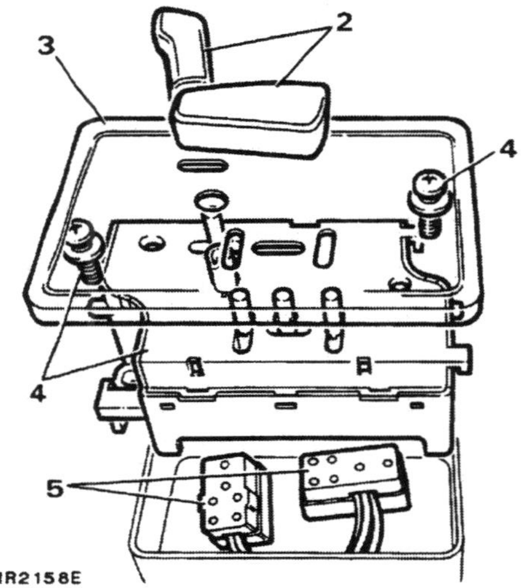

### Refitting

6. Reverse instructions 1 to 5.

NOTE: If switch housing removal is required it is necessary to remove the seat to gain access to the two securing screws- see Body Section 76.

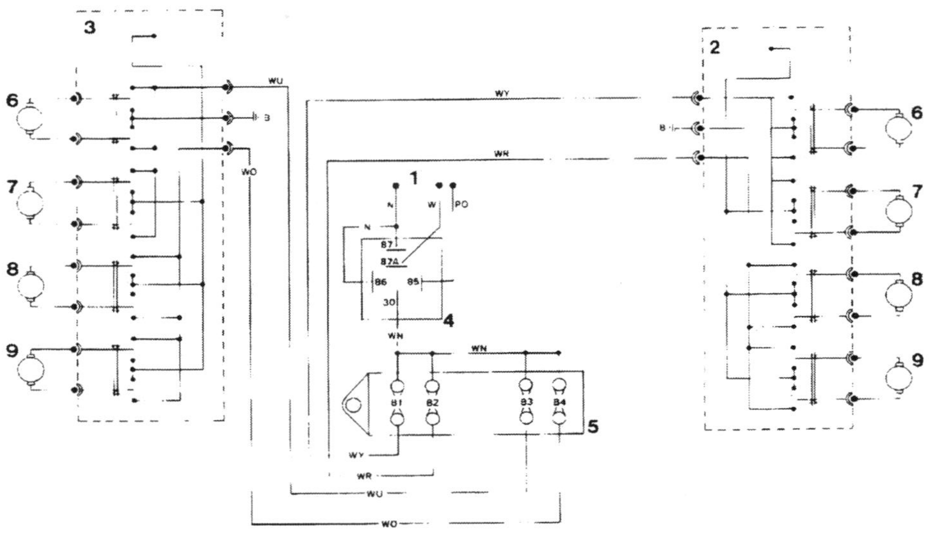

## SEAT ADJUSTMENT - Circuit diagram

1. Main connections - Item 126 on main circuit diagram left hand steer catalyst vehicles.
   - item 170 on main circuit diagram - left hand steer non - catalyst vehicles.
   - item 165 on main circuit diagram - right hand steer vehicles.

**Brown** - Live positive feed
**White** - Ignition positive feed
**Purple/Orange** - Door switch

2. Left hand seat control.
3. Right hand seat control.
4. Load control relay.
5. Auxiliary fuse box (B).
6. Seat recline motor.
7. Seat height (rear) motor.
8. Seat base adjust motor.
9. Seat height (front) motor.

### CABLE COLOUR CODE

B  Black
U  Blue
N  Brown
G  Green
O  Orange
P  Purple
R  Red
W  White
Y  Yellow

The last letter of a colour code denotes the tracer.

# ELECTRICAL 86

| Relay | Main Circuit Diagram Item Number |||
|---|---|---|---|
| | Right hand steer | Left hand steer Non-catalyst | Left hand steer Catalyst |
| 1. Headlamp wash timer unit | 19. M | 17. M | 17. M |
| 2. Heated rear window | 66. M | 64. M | 64. M |
| 3. Starter solenoid relay | 6. M | 6. M | 6.M |
| 4. Brake check relay | 151. A | Not fitted | Not fitted |
| 5. Headlamp relay | 157. A | Not fitted | Not fitted |
| 6. Compressor clutch | 11. A | 11. A | 11. A |
| 7. Condenser fan | 9. A | 9. A | 9. A |
| 8. Air conditioning/heater | 5. A | 5. A | 5. A |
| 9. Stowage position | Not used | Not used | Not used |
| 10. Rear wiper delay | 132. M | 139. M | 132. M |
| 11. Ignition load relay | 1. M | 1. M | 1. M |
| 12. Window lift relay (if fitted) | 65a. M | 63a. M | 63a. M |
| 13. Seat adjustment relay (if fitted) | 165a M | 170a. M | 126a. M |
| 14. Auxiliary lamp relay (if fitted) | 88. M | 86. M | 86. M |
| 15. Flasher/hazard unit | 74. M | 72. M | 72. M |
| 16. Interior lamp delay | 101. M | 99. M | 99. M |
| 17. Voltage sensitive switch (air conditioning) | 72. M | 70. M | 70. M |
| 18. Front wiper delay | 15. M | 14. M | 14. M |
| 19. Seat adjustment relay (load control - if fitted) | 4. S | 4. S | 4. S |
| 20. Main EFI relay | 10. E | 10. E | 22. E |
| 21. Fuel pump relay | 11. E | 11. E | 21. E |
| 22. Sunshine roof auxiliary relay (if fitted) | 3. SR | 3. SR | 3. SR |

M = Main circuit diagram
A = Air conditioning circuit diagram
S = Seat adjustment circuit diagram
E = EFI circuit diagram
SR = Sunroof circuit diagram

**NOTE: This relay chart applies to ALL petrol model Range Rovers. Vogue SE models have electric seat adjustment, window lift, air conditioning, auxiliary lamps and sunshine roof fitted as standard equipment. Saudi vehicles are fitted with two extra relays, an overspeed monitor and buzzer, located immediately below the instrument binnacle.**

Access to the two units is gained by removing the lower dash panel and steering column shroud.

## RELAYS-Identification

RR2376V shows left hand drive configuration of relays.

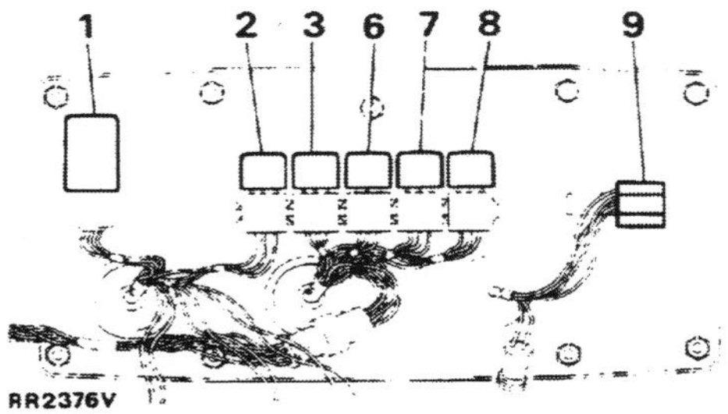

Closure panel viewed from the engine compartment, with protective cover removed.

RR2371V shows right hand drive configuration of relays.

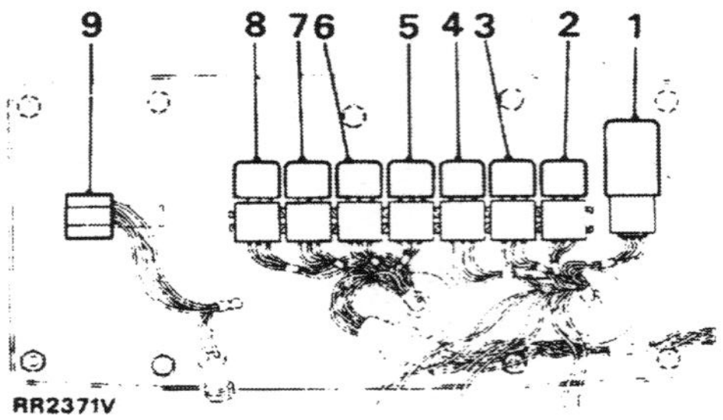

Closure panel viewed from the engine compartment, with protective cover removed.

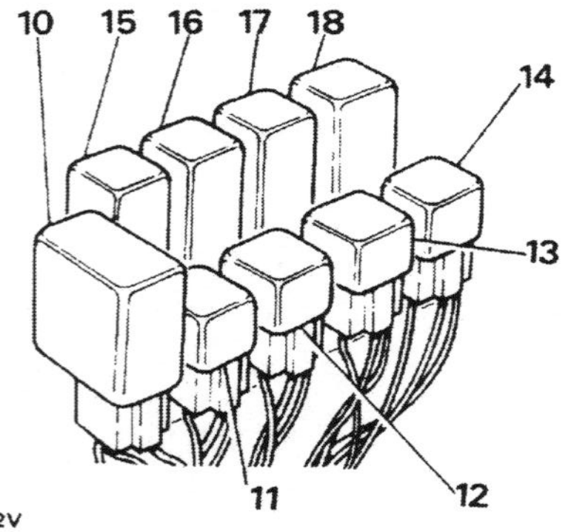

Main EFI (black terminal block) and fuel pump relays (blue terminal block) mounted beneath right hand front seat.

**NOTE: Refer to fuel injection section of manual for full information on E.F.I. relays.**

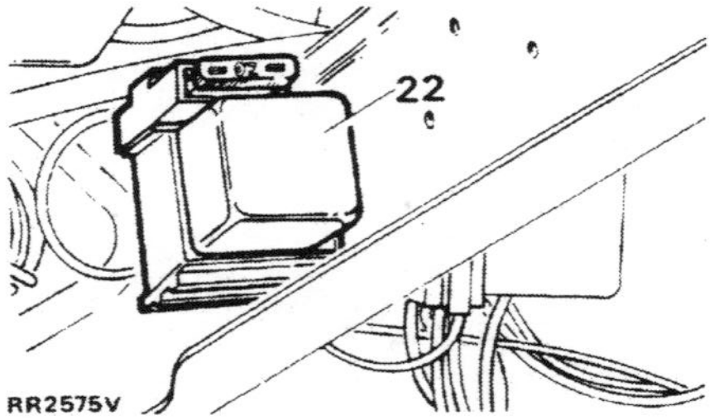

Steering column mounted relays viewed with the lower dash panel removed.

**NOTE: Left hand drive configuration of relays illustrated above. Unit 10, on right hand drive vehicles is located at the right hand side of the bank of relays.**

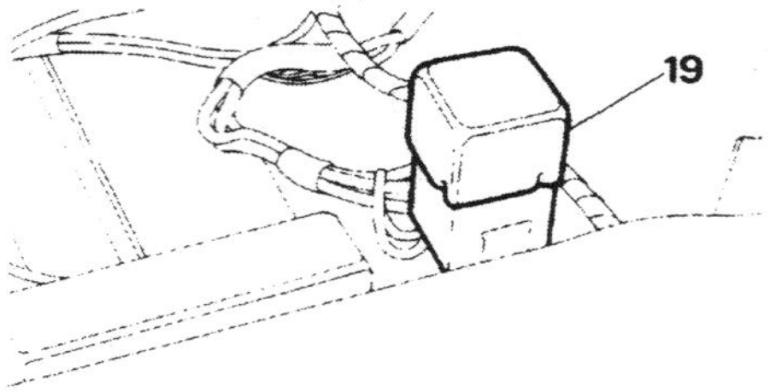

Sunshine roof auxiliary relay located on side of the steering column support bracket located behind the lower dash panel. (Left hand drive shown).

## RELAYS - (Mounted on the engine compartment closure panel).

### Remove and refit

**Removing**

1. Lift the hood.
2. Disconnect the battery negative lead.
3. Remove the bolt securing the relay protective cover, located on the front of the engine compartment closure panel.
4. Remove the cover.
5. Pull the appropriate relay off its multi-plug.

**Refitting**

6. Reverse the removal procedure.

Seat adjustment relay (load control) located beneath the left hand front seat adjacent to fuse box (B).

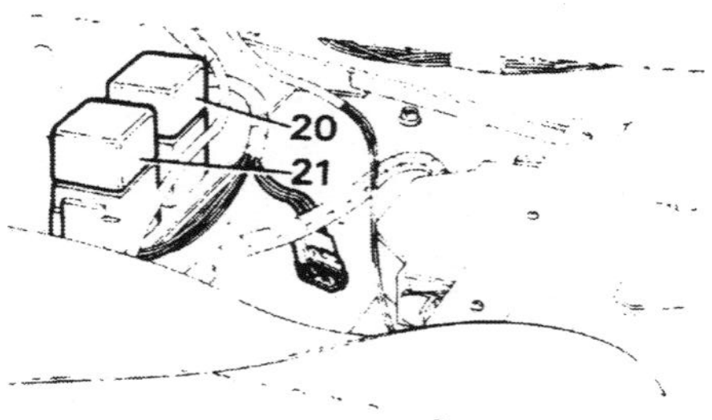

# ELECTRICAL 86

## RELAYS - (Mounted on the steering column support bracket)

### Remove and refit

**Removal.**

1. Disconnect the battery negative lead.
2. Remove the six screws securing the lower fascia panel.
3. Lower the dash panel, disconnect the electric leads from the dimming control switch and remove the fascia panel.
4. Locate the appropriate relay on the relay mounting bracket, carefully pull the relay off the multi-plug.

### Refitting

5. Reverse the removal procedure.

## AUXILIARY FUSE BOX

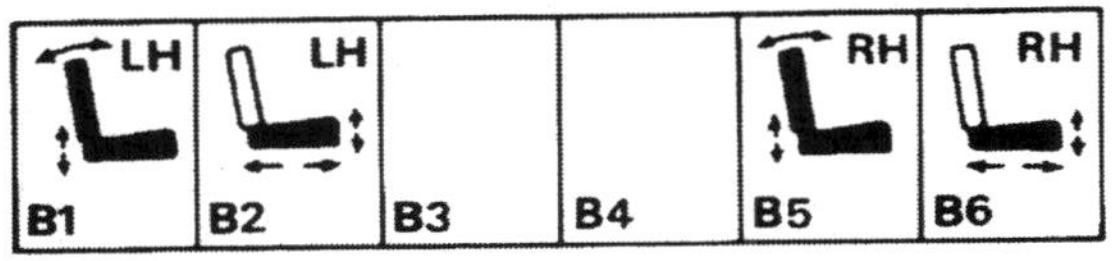

RR176OE

**AUXILIARY FUSE BOX (B)-Located under the front left-hand seat**

| FUSE NO | COLOUR CODE | FUSE VALUE | CIRCUIT SERVED |
|---|---|---|---|
| B1 | Green | 30 amp | Seat recline |
| B2 | Green | 30 amp | Seat base |
| B3 | ---- | ---- | Spare |
| B4 | ---- | ---- | Spare |
| B5 | Green | 30 amp | Seat recline |
| B6 | Green | 30 amp | Seat base |

## AUXILIARY FUSE BOX

### Remove and refit

**Removing**

1. Disconnect the battery negative lead.
2. Remove the clip-on fuse box cover.
3. Remove the fuses from auxiliary fuse box.
4. Remove the single screw securing the top auxiliary.
5. Remove the leads from the fuse box, by inserting a small screwdriver into each fuse socket to depress the small retaining tab on the back of the lucar connections, withdraw the leads from the rear of the fuse box.

## RELAYS - (Floor mounted beneath front seats)

### Remove and refit

**Removing**

1. Position seat to gain access to the required relay.
2. Disconnect the battery negative lead.
3. Carefully pull the relay off the multi-plug.

### Refitting

4. Reverse the removal procedure.

### Refitting

6. Reverse the removal instructions ensuring that all leads are refitted to the correct fuse socket (refer to circuit diagram).

**NOTE: When refitting the leads to the fuse box, the retaining tabs on the back of the lucar connectors must be in their raised position to prevent the leads being pushed out of the rear of the fuse box when the fuse is refitted.**

# Main Circuit Diagrams

Note: Right and left hand drive non-catalyst wiring diagrams cover the full Range-Rover model range, therefore it will be noted that references are made to Diesel Components on both circuit diagrams, these references should be ignored when using the diagrams for Vogue SE models.

# ELECTRICAL 86

## MAIN CIRCUIT DIAGRAM
### Right Hand Steering - RR2378M & RR2379M

1. Ignition load relay
2. Battery
3. Terminal post
4. Starter solenoid
5. Starter motor
6. Starter relay
7. Starter inhibit switch (Automatic)
8. Ignition switch
9. Tachometer
10. Voltage transformer(dim dip)
11. Ignition warning lamp
12. Alternator
13. Fuse 7
14. Front wipe/wash switch
15. Front wipe delay unit
16. Front wiper motor
17. Front wash switch
18. Front wash pump
19. Headlamp wash timer unit (option)
20. Headlamp wash pump (option)
21. Main lighting switch
22. Fuse 6
23. Fuse 5
24. LH side lamp
25. LH tail lamp
26. Number plate lamp(2 off)
27. Main beam dip/flash switch
28. Radio illumination
29. RH side lamp
30. RH tail lamp
31. Rheostat
32. Fuse 3
33. Fuse 4
34. Fuse 1
35. Fuse 2
36. Rear fog switch
37. Fuse 12
38. Switch illumination (2 off)
39. Cigar lighter illumination (2 off)
40. Heater illumination (4 off)
41. Clock illumination
42. Automatic gear selector illumination (2 off)
43. Instrument illumination (6 off)
44. Rear fog warning lamp
45. LH rear fog
46. RH rear fog
47. LH dip beam
48. RH dip beam
49. LH main beam
50. RH main beam
51. Main beam warning lamp
52. Fuel gauge
53. Fuel gauge sender unit
54. Water temperature gauge
55. Water temperature sender unit
56. Fuse 11
57. Horn switch
58. RH horn
59. LH horn
60. Under bonnet illumination switch
61. Under bonnet light
62. Clock
63. Fuse 19
64. Fuse 20
65. Pick-up point central locking/window lift (option)
65. (a) Window lift relay (option)
66. Heated rear window relay
67. Fuse 9
68. Radio aerial amplifier
69. Heated rear screen
70. Heated rear screen switch
71. Heated rear screen warning lamp
72. Voltage sensitive switch
73. Fuse 13
74. Hazard switch
75. Flasher unit
76. Direction indicator switch
77. Hazard/indicator warning lamp
78. LH rear indicator lamp
79. LH front indicator lamp
80. LH side repeater lamp
81. RH side repeater lamp
82. RH front indicator lamp
83. RH rear indicator lamp

## MAIN CIRCUIT DIAGRAM
### Right Hand Steering - RR2378M & RR2379M

84. Trailer warning lamp
85. Fuse 15
86. Stop lamp switch
87. Reverse lamp switch
88. Auxiliary lamp relay (option)
89. LH stop lamp
90. RH stop lamp
91. LH reverse lamp
92. RH reverse lamp
93. LH auxiliary lamp (option)
94. RH auxiliary lamp (option)
95. Auxiliary lamp switch (option)
96. Fuse 17
97. Dash cigar lighter
98. Cubby box cigar lighter
99. LH interior lamp
100. RH interior lamp
101. Interior lamp delay unit
102. LH door edge lamp
103. RH door edge lamp
104. LH puddle lamp
105. RH puddle lamp
106. Interior lamp switch
107. LH rear door switch
108. RH rear door switch
109. Tailgate switch
110. LH front door switch
111. RH front door switch
112. Differential lock warning lamp
113. Differential lock switch
114. Oil pressure warning lamp
115. Oil pressure switch
116. Fuse 18
117. Fuel cut off relay (carburetter models)
118. Fuel pump(petrol models)
119. Ignition coil
120. Capacitor
121. Distributor
122. EFI Harness plug
123. Fuel shut off solenoid (Diesel)
124. Radio choke
125. Radio fuse
126. Radio and four speakers
127. Sun roof (option) pick up points
128. Automatic transmission oil temperature warning lamp
129. Automatic transmission oil temperature switch
130. Fuse 16
131. Rear wash wipe switch
132. Rear wipe delay unit
133. Rear wiper motor
134. Rear screen wash pump
135. Low screen wash fluid level warning lamp
136. Low screen wash switch
137. Low coolant switch
138. Multi-function unit in binnacle
139. Low coolant level warning lamp
140. Low fuel level warning lamp
141. Cold start/diesel glow plug warning lamp
142. Cold start switch - carburetter
143. Glow plug timer (diesel)
144. Glow plugs (diesel)
145. Handbrake warning lamp
146. Brake fail warning lamp
147. Handbrake warning switch
148. Brake fail warning switch
149. Brake pad wear warning lamp
150. Brake pad wear sensors
151. Brake check relay
152. Split charge relay (option)
153. Split charge terminal post (option)
154. Heater/air conditioning connections
155. Fuse 8
156. Coil negative (engine RPM input to ECU)
157. Headlamp relay
158. Ignition load relay (+)
159. Battery feed (+)
160. Ignition auxiliary (+)
161. Ignition on (+)
162. Earth (-)
163. Fuse 14
164. Trailer pick up point
165. Electric seats pick up point (option)
165. (a) Electric seats relay (option)
166. Fuse 10
167. Electric mirrors pick up point (option)

### CABLE COLOUR CODE

| B | Black | N | Brown | R | Red | W | White |
|---|---|---|---|---|---|---|---|
| G | Green | O | Orange | S | Grey | Y | Yellow |
| K | Pink | P | Purple | U | Blue | | |

The last letter of a colour code denotes the tracer.

# ELECTRICAL 86

**MAIN CIRCUIT DIAGRAM**
Right hand steering - RR2378M & RR2379M

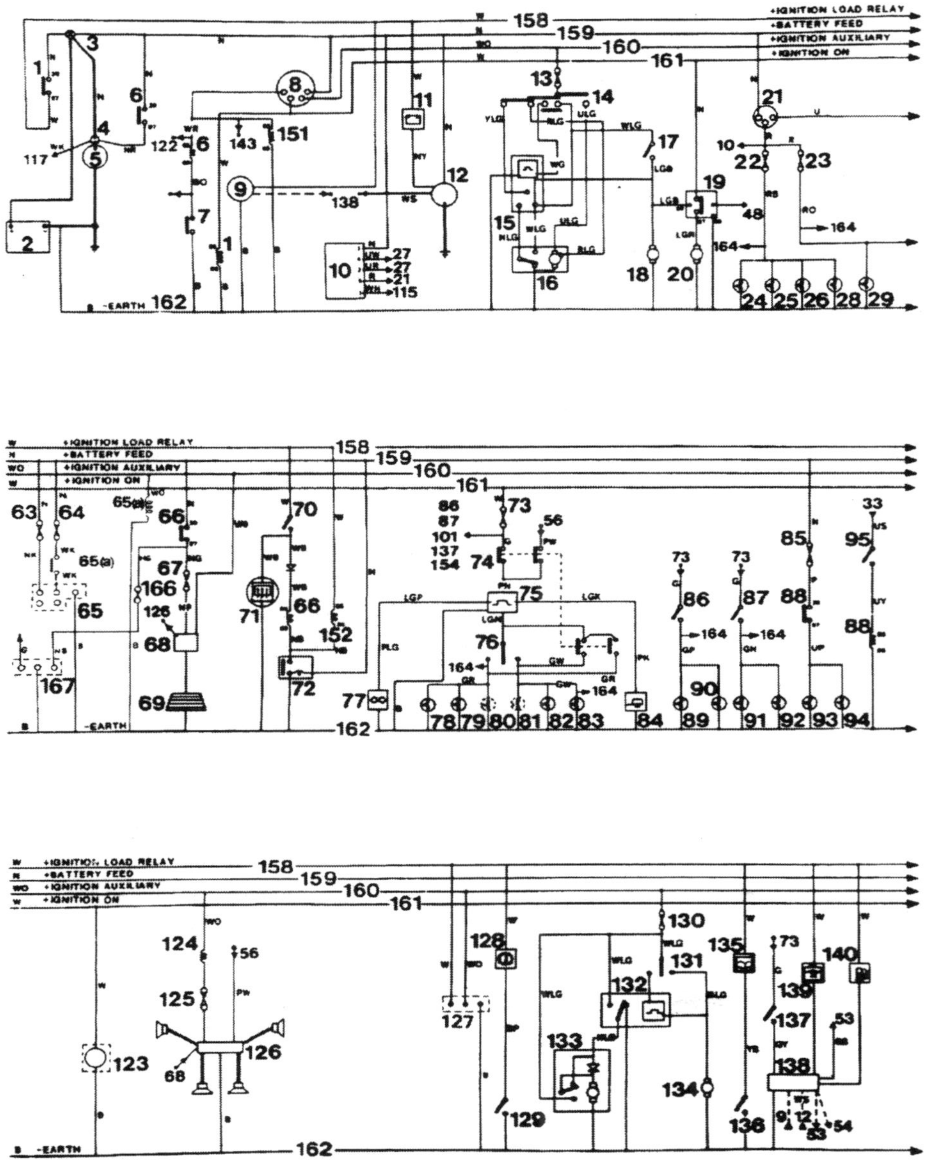

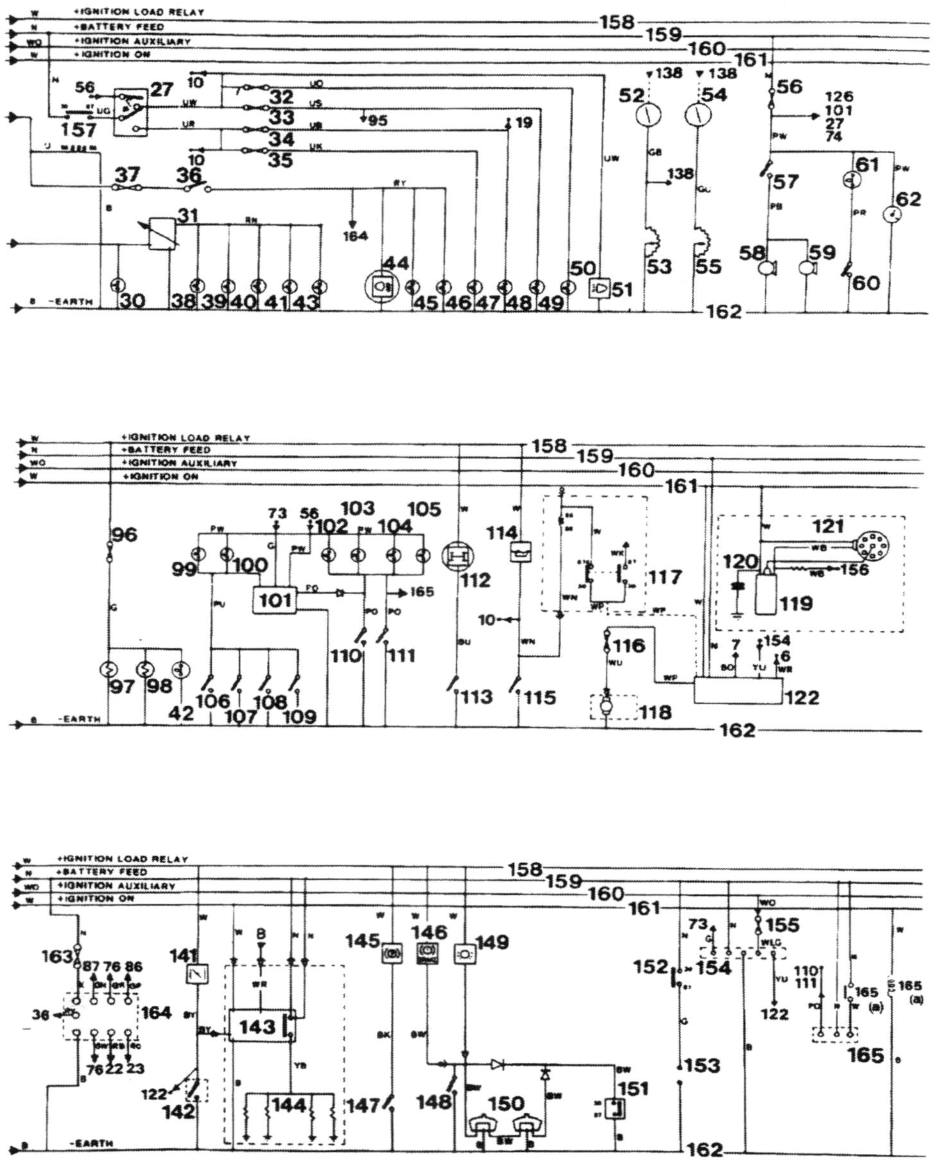

# ELECTRICAL 86

## MAIN CIRCUIT DIAGRAM - NON CATALYST VEHICLES
### Left-hand Steering - RR2380M & RR2381M

1. Ignition load relay
2. Battery
3. Terminal post
4. Starter solenoid
5. Starter motor
6. Starter relay
7. Starter inhibit switch (automatic)
8. Ignition switch
9. Tachometer
10. Ignition warning lamp
11. Alternator
12. Fuse 7
13. Front wipe/wash switch
14. Front wipe delay unit
15. Front wiper motor
16. Front wash pump
17. Headlamp wash timer unit (option)
18. Headlamp wash pump (option)
19. Main lighting switch
20. Fuse 6
21. Fuse 5
22. LH side lamp
23. LH tail lamp
24. Number plate lamp (2 off)
25. Main beam dip/flash switch
26. Radio illumination
27. RH side lamp
28. RH tail lamp
29. Rheostat
30. Fuse 3
31. Fuse 4
32. Fuse 1
33. Fuse 2
34. Rear fog switch
35. Fuse 12
36. Switch illumination (2 off)
37. Cigar lighter illumination (2 off)
38. Heater illumination (4 off)
39. Clock illumination
40. Automatic gear selector illumination (2 off)
41. Instrument illumination (6 off)
42. Rear fog warning lamp
43. LH rear fog
44. RH rear fog
45. LH dip beam
46. RH dip beam
47. LH main beam
48. RH main beam
49. Main beam warning lamp
50. Fuel gauge
51. Fuel gauge sender unit
52. Water temperature gauge
53. Water temperature sender unit
54. Fuse 11
55. Horn switch
56. RH horn
57. LH horn
58. Under bonnet illumination switch
59. Under bonnet light
60. Clock
61. Fuse 19
62. Fuse 20
63. Pick-up point central locking/window lift
63. (a). Window lift relay (option)
64. Heated rear window relay
65. Fuse 9
66. Radio aerial amplifier
67. Heated rear screen
68. Heated rear screen switch
69. Heated rear screen warning lamp
70. Voltage sensitive switch
71. Fuse 13
72. Hazard switch
73. Flasher unit
74. Direction indicator switch
75. Hazard/indicator warning lamp
76. LH rear indicator lamp
77. LH front indicator lamp
78. LH side repeater lamp
79. RH side repeater lamp
80. RH front indicator lamp
81. RH rear indicator lamp
82. Trailer warning lamp
83. Fuse 15

### CABLE COLOUR CODE

| | | | | | |
|---|---|---|---|---|---|
| B | Black | N | Brown | R | Red |
| G | Green | O | Orange | S | Grey |
| K | Pink | P | Purple | U | Blue |
| | | | | W | White |
| | | | | Y | Yellow |

The last letter of a colour code denotes the tracer.

# ELECTRICAL 86

## MAIN CIRCUIT DIAGRAM - NON CATALYST VEHICLES
### Left-hand Steering - RR2380M & RR2381M

84. Stop lamp switch
85. Reverse lamp switch
86. Auxiliary lamp relay
87. LH stop lamp
88. RH stop lamp
89. LH reverse lamp
90. RH reverse lamp
91. LH auxiliary lamp (option)
92. RH auxiliary lamp (option)
93. Auxiliary lamp switch (option)
94. Fuse 17
95. Dash cigar lighter
96. Cubby box cigar lighter
97. LH interior lamp
98. RH interior lamp
99. Interior lamp delay unit
100. LH door edge lamp
101. RH door edge lamp
102. LH puddle lamp
103. RH puddle lamp
104. Interior lamp switch
105. LH rear door switch
106. RH rear door switch
107. Tailgate switch
108. LH front door switch
109. RH front door switch
110. Differential lock warning lamp
111. Differential lock switch
112. Oil pressure warning lamp
113. Oil pressure switch
114. Fuse 18
115. Fuel shut-off relay - carburetter
116. Fuel pump - petrol models
117. Ignition coil
118. Capacitor
119. Distributor
120. EFI Harness plug
121. Fuel shut-off solenoid-Diesel
122. Radio choke
123. Radio fuse
124. Radio
125. Four speakers
126. Seat belt warning lamp
127. Speed transducer, Saudi only
128. Resistor
129. Audible warning unit
130. Transfer box neutral switch
131. Seat buckle switch
132. Overspeed monitor (Saudi only)
133. Overspeed buzzer (Saudi only)
134. Sun roof pick up point (option)
135. Automatic transmission oil temperature warning lamp
136. Automatic transmission oil temperature switch
137. Fuse 16
138. Rear wash wipe switch
139. Rear wipe delay unit
140. Rear wiper motor
141. Rear screen wash pump
142. Low screen wash fluid level warning lamp
143. Low screen wash switch
144. Low coolant switch
145. Multi-function unit in binnacle
146. Low coolant level warning lamp
147. Low fuel level warning lamp
148. Cold start/Diesel glow plug warning lamp
149. Choke switch - carburetter
150. Glowplug timer/Diesel
151. Glowplugs/Diesel
152. Handbrake/warning lamp
153. Handbrake warning switch
154. Brake fail warning lamp
154. (a).Brake fail warning switch
155. Brake pad wear warning lamp
156. Brake pad wear sensors
157. Brake check unit
158. Split charge relay (option)
159. Split charge terminal post
160. Heater/air conditioning connections
161. Fuse 8
162. Coil negative (engine RPM input to ECU)
163. Ignition load relay (+)
164. Battery feed (+)
165. Ignition auxiliary (+)
166. Ignition on (+)
167. Earth (-)
168. Warning lights common earth (-)
169. Warning lights supply (+)
170. Electric seats pick up point (option)
170. (a).Electric seats relay (option)
171. Fuse 14
172. Trailer pick up point
173. Fuse 10
174. Electric mirrors pick up point (option)

# ELECTRICAL 86

MAIN CIRCUIT DIAGRAM - NON CATALYST
Left hand steering - RR2380M & RR2381M

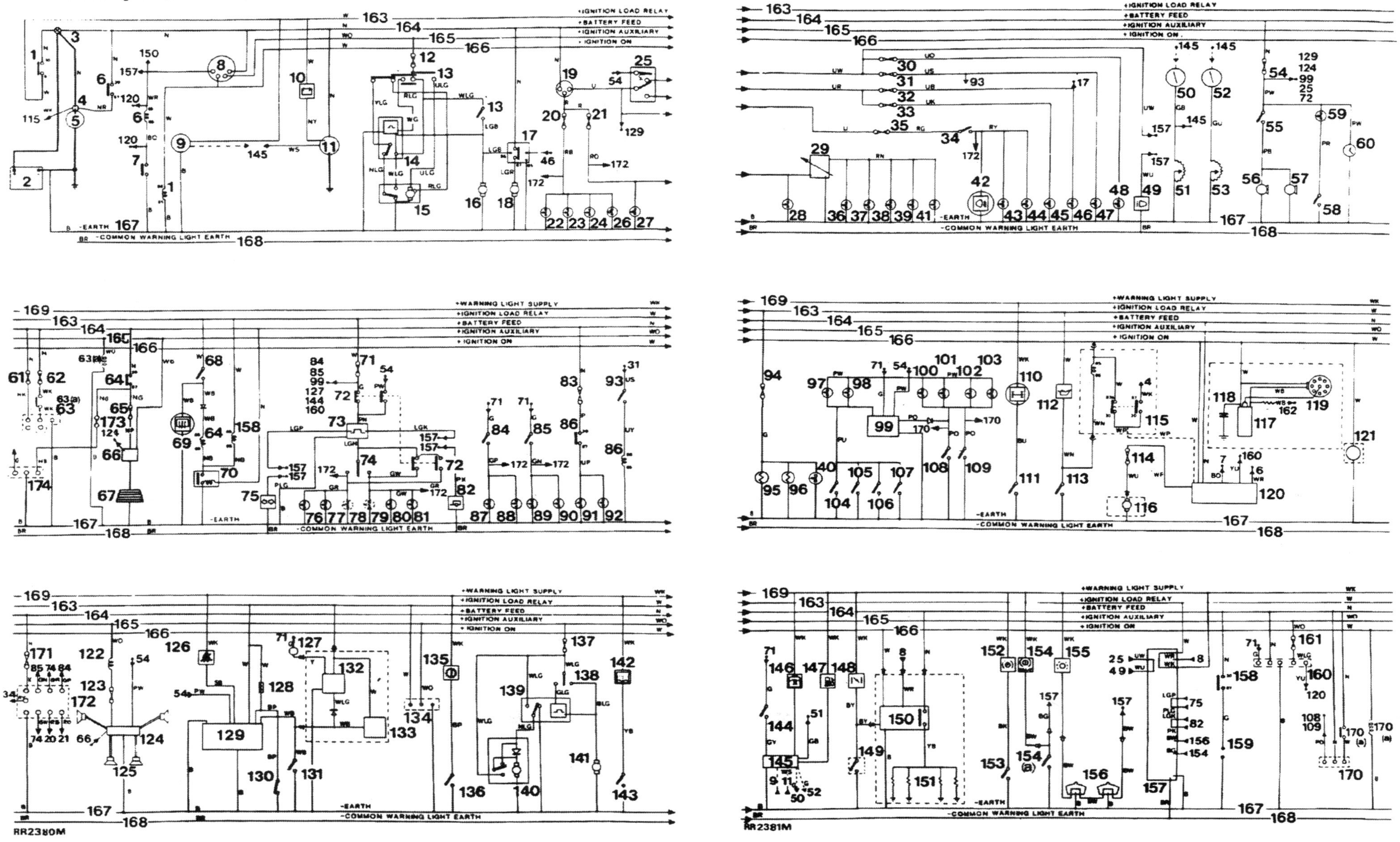

# ELECTRICAL 86

## MAIN CIRCUIT DIAGRAM - CATALYST VEHICLES
### Left hand steering - RR2382S and RR2383S

1. Ignition load relay
2. Battery
3. Terminal post
4. Starter solenoid
5. Starter motor
6. Starter relay
7. Starter inhibit switch (automatic)
8. Ignition switch
9. Tachometer
10. Ignition warning lamp
11. Alternator
12. Fuse 7
13. Front wipe/wash switch
14. Front wipe delay unit
15. Front wiper motor
16. Front wash pump
17. Headlamp wash timer unit (option)
18. Headlamp wash pump (option)
19. Main lighting switch
20. Fuse 6
21. Fuse 5
22. LH side lamp
23. LH tail lamp
24. Number plate lamp (2 off)
25. Main beam dip/flash switch
26. Radio illumination
27. RH side lamp
28. RH tail lamp
29. Rheostat
30. Fuse 3
31. Fuse 4
32. Fuse 1
33. Fuse 2
34. Rear fog switch
35. Fuse 12
36. Switch illumination (2 off)
37. Cigar lighter illumination (2 off)
38. Heater illumination (4 off)
39. Clock illumination
40. Automatic gear selector illumination (2 off)
41. Instrument illumination (6 off)
42. Rear fog warning lamp
43. LH rear fog
44. RH rear fog
45. LH dip beam
46. RH dip beam
47. LH main beam
48. RH main beam
49. Main beam warning lamp
50. Fuel gauge
51. Fuel gauge sender unit
52. Water temperature gauge
53. Water temperature sender unit
54. Fuse 11
55. Horn switch
56. RH horn
57. LH horn
58. Under bonnet illumination switch
59. Under bonnet light
60. Clock
61. Fuse 19
62. Fuse 20
63. Pick-up point central locking/window lift
63. (a). Window lift relay
64. Heated rear window relay
65. Fuse 9
66. Radio aerial amplifier
67. Heated rear screen
68. Heated rear screen switch
69. Heated rear screen warning lamp
70. Voltage sensitive switch
71. Fuse 13
72. Hazard switch
73. Flasher unit
74. Direction indicator switch
75. Hazard/indicator warning lamp
76. LH rear indicator lamp
77. LH front indicator lamp
78. LH side repeater lamp
79. RH side repeater lamp
80. RH front indicator lamp

## MAIN CIRCUIT DIAGRAM - CATALYST VEHICLES
### Left hand steering - RR2382S and RR2383S

81. RH rear indicator lamp
82. Trailer warning lamp
83. Fuse 15
84. Stop lamp switch
85. Reverse lamp switch
86. Auxiliary lamp relay
87. LH stop lamp
88. RH stop lamp
89. LH reverse lamp
90. RH reverse lamp
91. LH auxiliary lamp
92. RH auxiliary lamp
93. Auxiliary lamp switch
94. Fuse 17
95. Dash cigar lighter
96. Cubby box cigar lighter
97. LH interior lamp
98. RH interior lamp
99. Interior lamp delay unit
100. LH door edge lamp
101. RH door edge lamp
102. LH puddle lamp
103. RH puddle lamp
104. Interior lamp switch
105. LH rear door switch
106. RH rear door switch
107. Tailgate switch
108. LH front door switch
109. RH front door switch
110. Differential lock warning lamp
111. Differential lock switch
112. Oil pressure warning lamp
113. Oil pressure switch
114. Fuse 18
115. Speed transducer
116. Fuel pump
117. Ignition coil
118. Capacitor
119. Distributor
120. EFI Harness plug
121. Coil negative (engine RPM input to ECU)
122. Radio choke
123. Radio fuse
124. Radio
125. Four speakers
126. Electric seats pick up point (option)
126. (a). Electric seat relay (option)
127. Sunroof connection point (option)
128. Automatic transmission oil temperature warning lamp
129. Automatic transmission oil temperature switch
130. Fuse 16
131. Rear wash wipe switch
132. Rear wipe delay unit
133. Rear wiper motor
134. Rear screen wash pump
135. Low screen wash fluid level warning lamp
136. Low screen wash switch
137. Low coolant switch
138. Multi-function unit in binnacle
139. Low coolant level warning lamp
140. Low fuel level warning lamp
141. E.F.I. warning lamp
142. Handbrake warning lamp
143. Handbrake warning switch
144. Brake fluid level warning switch
145. Brake fluid level warning lamp
146. Brake pad wear warning lamp
147. Brake pad wear sensors
148. Brake check unit
149. Split charge relay (option)
150. Split charge terminal post (option)
151. Heater/air conditioning connections
152. Fuse 8
153. Ignition load relay (+)
154. Battery feed (+)
155. Ignition auxiliary (+)
156. Ignition on (+)
157. Earth (-)
158. Warning lights common earth (-)
159. Warning lights supply (+)
160. Fuse 14
161. Trailer pick up point
162. Fuse 10
163. Electric mirrors pick up point (option)

## CABLE COLOUR CODE

| B | Black | N | Brown | R | Red | W | White |
|---|---|---|---|---|---|---|---|
| G | Green | O | Orange | S | Grey | Y | Yellow |
| K | Pink | P | Purple | U | Blue | | |

The last letter of a colour code denotes the tracer.

22    400    23

## ELECTRICAL 86

MAIN CIRCUIT DIAGRAM - CATALYST VEHICLES
Left hand steering - RR2382S & RR2383S

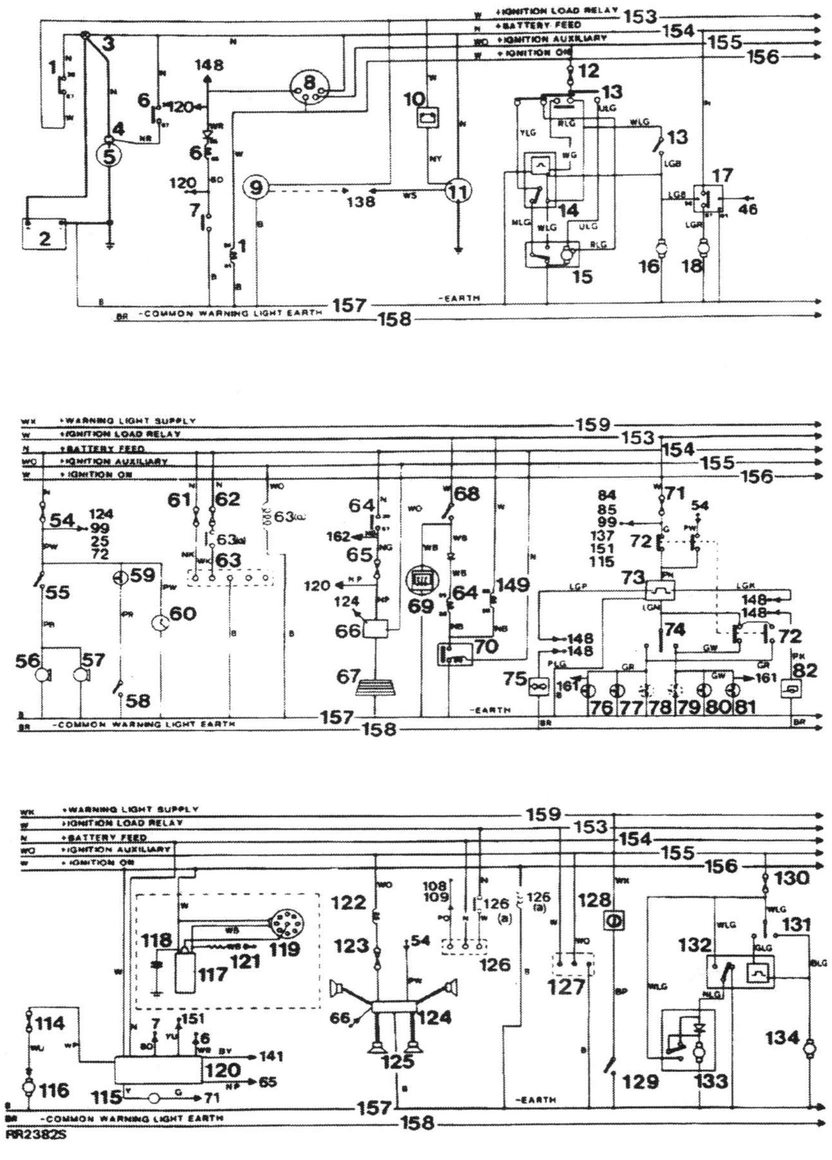

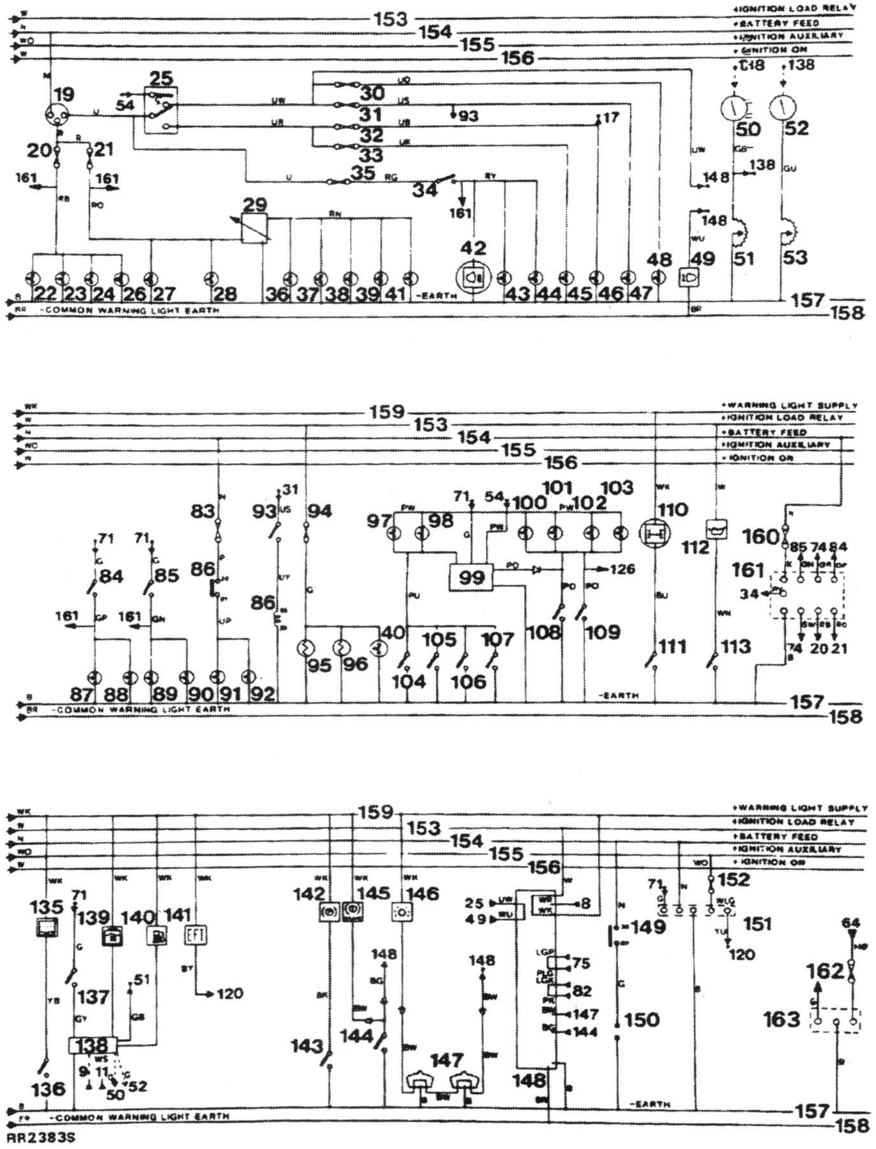

# Range Rover
# 1989 Model Year
# Workshop Manual Supplement
# Publication Number LSM180WS3

|  | Section | Page |
|---|---|---|
| Introduction | 01 | 404 |
| General Specification Data | 04 | 405 |
| Torque Wrench Settings | 06 | 406 |
| Lubricants and Fluids | 09 | 407 |
| Maintenance | 10 | 408 |
| Manual Gearbox and Transfer Box | 37 | 410 |
| Automatic Gearbox | 44 | 440 |
| Brakes | 70 | 443 |
| Body | 76 | 444 |
| Air Conditioning | 82 | 448 |
| Wipers & Washers | 84 | 449 |
| Electrical | 86 | 450 |

Land Rover
Lode Lane,
Solihull
West Midlands
B92 8NW
England

Section
Number

# 01 INTRODUCTION

Introduction .......................................................... 1
Used engine oil handling precaution ................................... 2
Disposing of used engine oil .......................................... 2
Identification number - Borg Warner transfer gearbox .................. 2

# 04 GENERAL SPECIFICATION DATA

Transfer and main gearbox ratios ...................................... 1
Automatic shift speed specification ................................... 2

# 06 TORQUE WRENCH SETTINGS

Transfer gearbox ...................................................... 1
LT77 Five speed manual gearbox ........................................ 1
Seat belt fixings ..................................................... 1

# 09 LUBRICANTS, FLUIDS AND CAPACITIES

Chart - Recommended lubricants ........................................ 1
Chart - Oil specification ............................................. 2
Transfer gearbox capacity ............................................. 2

# 10 MAINTENANCE

Top up transfer gearbox oil ........................................... 1
Renew transfer gearbox oil ............................................ 1
Adjust handbrake ...................................................... 2
Rolling road testing .................................................. 3
Towing ................................................................ 3

# 37 BORG WARNER TRANSFER GEARBOX

Viscous unit - In vehicle check ....................................... 1
In - situ operations
  - Renew speedometer drive pinion .................................... 1
  - Renew rear output shaft oil seal .................................. 1
  - Renew front output shaft oil seal ................................. 2
Transfer gearbox adaptor plate ........................................ 3
Transfer gearbox - Remove and refit ................................... 7
                  - Dismantle and overhaul ........................... 11

# 37 LT77 FIVE SPEED GEARBOX

Main and transfer gearbox adaptor plate ............................... 29
Main and transfer gearbox - remove and refit .......................... 31
LT77 main gearbox - Dismantle and overhaul ............................ 34

Notes

# INTRODUCTION  01

## INTRODUCTION - 1989 MODEL YEAR

A number of model improvements are introduced on all Range Rover models for the 1989 Model Year.

Specifications for individual vehicles may vary, but all will include some of the new features and options summarised below.

Borg Warner transfer gearbox incorporating viscous coupling
Revised LT77 five speed manual gearbox
Revised handbrake linkage
Rear asymmetric seat locking mechanism
Interior door trim panels
Central locking on upper tailgate
Revised heater distribution and controls
Single touch electric window lift (Drivers door only)
Heated screen washer jets
Variable delay windscreen wipers
Ignition override of headlamps

Service and repair information is included in this supplement, which should be used in conjunction with the main Range Rover Workshop Manual.

This Workshop Manual Supplement is designed to assist skilled technicians in the efficient repair and maintenance of Range Rover vehicles.

**Individuals who undertake their own repairs should have some skill and training, and limit repairs to components which could not affect the safety of the vehicle or its passengers. Any repairs required to safety critical items such as steering, brakes, or suspension should be carried out by a Range Rover Dealer. Repairs to such items should NEVER be attempted by untrained individuals.**

Continued

---

### 44  AUTOMATIC GEARBOX

| | |
|---|---|
| Main and transfer gearbox adaptor plate | 1 |
| Main and transfer gearbox - Remove and refit | 3 |

### 70  BRAKES

| | |
|---|---|
| Handbrake cable - Remove and refit | 1 |
| - Adjust | 1 |

### 76  BODY

| | |
|---|---|
| Front door interior trim panel - 4 door models | 1 |
| Front door lock and door release handles - 4 door models | 2 |
| Front door lock and door release handles - adjustment - 4 door models | 3 |
| Rear door interior trim panel - 4 door models | 4 |
| Rear door lock and door release handles - 4 door models | 5 |
| Rear door lock and door release handles - adjustment - 4 door models | 6 |
| Asymmetric split rear seat locking mechanism | 7 |

### 82  AIR CONDITIONING

| | |
|---|---|
| Circuit diagram | 1 |

### 84  WIPERS AND WASHERS

| | |
|---|---|
| Heated washer jets - remove and refit | 1 |
| Heated washer jets - thermostat - remove and refit | 2 |

### 86  ELECTRICAL

| | |
|---|---|
| Fuses | 1 |
| Relays - identification | 2 |
| Central locking tailgate actuator - remove and refit | 5 |
| Circuit diagrams | |
| - Central door locking | 6 |
| - Electric seat adjustment | 7 |
| - Electric window lift | 8 |
| - Main circuit diagram - right hand steering vehicles | 10 |
| - Main circuit diagram - left hand steering non catalyst vehicles | 14 |

# INTRODUCTION

## ENGINE OIL HANDLING PRECAUTIONS

Prolonged and repeated contact with mineral oil will result in the removal of natural fats from the skin, leading to dryness, irritation and dermatitis. In addition, used engine oil contains potentially harmful contaminants which may cause skin cancer. Adequate means of skin protection and washing facilities should be provided.

### WARNING:

1. Avoid prolonged and repeated contact with oils, particularly used engine oils.
2. Wear protective clothing, including impervious gloves where practicable.
3. Do not put oily rags in pockets.
4. Avoid contaminating clothes, particularly underwear, with oil.
5. Overalls must be cleaned regularly. Discard unwashable clothing and oil impregnated footwear.
6. First aid treatment must be obtained immediately for open cuts and wounds.
7. Use barrier creams, before each work period, to help the removal of oil from the skin.
8. Wash with soap and water to ensure all oil is removed (skin cleansers and nail brushes will help). Preparations containing lanolin replace the natural skin oils which have been removed.
9. Do not use petrol, kerosene, diesel fuel, gas oil, thinners or solvents for washing the skin.
10. If skin disorders develop, obtain medical advice as soon as possible.
11. Where practicable, degrease components prior to handling.
12. Where there is a risk of eye contact, eye protection must always be worn, for example, goggles or face shields; in addition an eye wash facility should be provided.

## DISPOSING OF USED OILS

### Enviromental protection precautions

It is illegal to pour used oil on to the ground, down sewers or drains, or into water courses.

The burning of used engine oil in small space heaters or boilers is not recommended unless emission control equipment is fitted.

Dispose of used oil through authorised waste disposal contractors to licensed waste disposal sites, or to the waste oil reclamation trade. If in doubt, contact the Local Authority for advice on disposal facilities.

## TRANSFER GEARBOX - BORG WARNER
### - 1989 MODEL YEAR

The gearbox serial number is stamped on a plate which is attached to the gearbox casing and is located between the filler/level and drain plugs adjacent to the rear output housing.

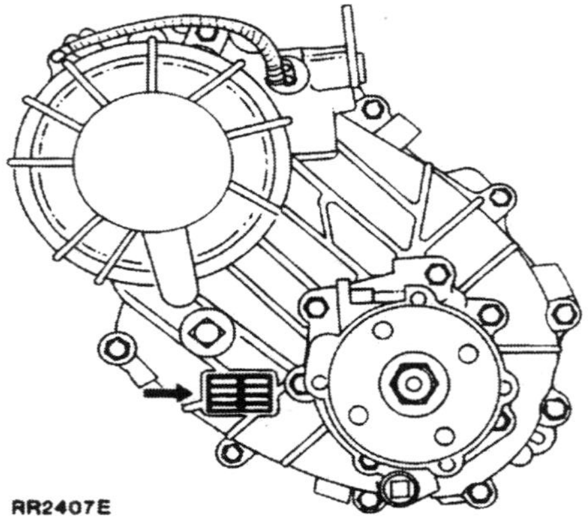

# GENERAL SPECIFICATION

## TRANSMISSION

### Borg Warner transfer gearbox

Type .................... Two speed reduction on main gearbox output, front and rear drive permanently engaged via a centre differential controlled by a Viscous unit giving a 50/50 nominal front and rear torque split.

### Transfer gearbox ratios

High .................... 1.206:1
Low .................... 3.244:1

### Automatic gearbox ratios

4th .................... 0.728:1
3rd .................... 1.000:1
2nd .................... 1.480:1
1st .................... 2.480:1
Reverse .................... 2.086:1

### Overall ratios (final drive):

| | High transfer | Low transfer |
|---|---|---|
| 4th | 3.11:1 | 8.36:1 |
| 3rd | 4.27:1 | 11.48:1 |
| 2nd | 6.32:1 | 17.00:1 |
| 1st | 10.59:1 | 28.50:1 |
| Reverse | 8.91:1 | 23.96:1 |

### Manual gearbox LT77 ratios

5th .................... 0.770:1
4th .................... 1.000:1
3rd .................... 1.397:1
2nd .................... 2.132:1
1st .................... 3.321:1
Reverse .................... 3.429:1

Diesel models - low 1st gear .................... 3.692:1

### Overall ratios (final drive):

| | High transfer | Low transfer |
|---|---|---|
| 5th | 3.29:1 | 8.84:1 |
| 4th | 4.27:1 | 11.48:1 |
| 3rd | 5.97:1 | 16.04:1 |
| 2nd | 9.10:1 | 24.48:1 |
| 1st | 14.18:1 | 38.14:1 |
| Reverse | 14.64:1 | 39.38:1 |
| Diesel models - low 1st gear | 15.76:1 | 42.40:1 |

RR2407E

## TORQUE WRENCH SETTINGS 06

### BORG WARNER TRANSFER GEARBOX

| | Nm | lb ft | lb in |
|---|---|---|---|
| Brake drum back plate to rear output housing | 65 - 80 | 48 - 59 | - |
| Brake drum to drive flange | 22 - 28 | 16 - 21 | - |
| Centre differential (front to rear) | 55 - 64 | 40 - 47 | - |
| Drive flanges to transfer gearbox | 203 - 244 | 150 - 180 | - |
| Driven gear to centre differential | 41 - 61 | 30 - 45 | - |
| Front cover to rear cover - main case | 30 - 49 | 22 - 36 | - |
| Front output housing to main case | 24 - 41 | 18 - 30 | - |
| Gearbox mounting brackets to chassis | 40 - 50 | 29 - 37 | - |
| Mounting bracket to gearbox | 92 - 112 | 68 - 83 | - |
| Neutral warning switch | 34 - 47 | 25 - 35 | - |
| Oil drain plug | 19 - 30 | 14 - 22 | - |
| Oil filler/level plug | 19 - 30 | 14 - 22 | - |
| Oil pump fixings | 4 - 8.5 | - | 35 - 75 |
| Propeller shafts to drive flanges | 41 - 52 | 30 - 38 | - |
| Rear output housing to main case | 30 - 49 | 22 - 36 | - |
| Selector lever shaft - Torx screw | 7 - 9 | 5 - 7 | 60 - 84 |
| Selector fork operating arm - Torx screw | 7 - 9 | 5 - 7 | 60 - 84 |

### LT77 MAIN GEARBOX (FIVE - SPEED)

| | Nm | lb ft. |
|---|---|---|
| Bottom cover to clutch housing | 7 - 10 | 5 - 7 |
| Oil pump body to extension case | 7 - 10 | 5 - 7 |
| Clip to clutch release lever | 7 - 10 | 5 - 7 |
| Attachment plate to gearcase | 7 - 10 | 5 - 7 |
| Extension case to gearcase | 22 - 28 | 16 - 21 |
| Pivot - clutch lever to bell housing | 22 - 28 | 16 - 21 |
| Guide clutch release sleeve | 22 - 28 | 16 - 21 |
| Slave cylinder to clutch housing | 22 - 28 | 16 - 21 |
| Front cover to gearcase | 22 - 28 | 16 - 21 |
| 5th support bracket | 22 - 28 | 16 - 21 |
| Clutch housing to gearbox | 65 - 80 | 48 - 59 |
| Oil drain plug | 40 - 47 | 30 - 35 |
| Oil filter plug | 65 - 80 | 48 - 59 |
| Breather | 14 - 16 | 10 - 12 |
| Oil level plug | 25 - 35 | 19 - 26 |
| Upper gear lever to lower gear lever | 22 - 28 | 16 - 21 |
| Upper gear lever to lower gear lever - pinch bolt | 22 - 28 | 16 - 21 |
| 5th layshaft gear retaining nut | 204 - 231 | 150 - 170 |
| Attachment plate to gear change housing | 7 - 10 | 5 - 7 |
| Gear change housing to extension case | 22 - 28 | 16 - 21 |
| Plunger housing to gear change housing | 22 - 28 | 16 - 21 |
| Adjustment plate to gear change housing | 22 - 28 | 16 - 21 |
| Cover to gear change housing | 7 - 10 | 5 - 7 |
| Bell housing to cylinder block bolts | 36 - 45 | 27 - 33 |
| Yoke to selector shaft | 22 - 28 | 16 - 21 |

### SEAT BELT FIXINGS

| | Nm | lb ft |
|---|---|---|
| Front and rear seat belt fixings (all) | 20.3 | 15 |

## GENERAL SPECIFICATION 04

**SHIFT SPEED SPECIFICATION**
Automatic ZF4HP22 Gearbox

| OPERATION | SELECTOR POSITON | VEHICLE SPEED APPROX MPH | VEHICLE SPEED APPROX KPH | ENGINE SPEED APPROX (RPM) |
|---|---|---|---|---|
| **KICKDOWN** | | | | |
| KD4 - 3 | D | 79 - 96 | 127 - 155 | |
| KD3 - 2 | 3(D) | 57 - 62 | 91 - 99 | |
| KD2 - 1 | 2(D,3) | 27 - 34 | 44 - 56 | |
| KD3 - 2 | D | N/A | N/A | |
| KD2 - 3 | D(3) | 60 - 63 | 96 - 104 | 4750 - 5200 |
| KD1 - 2 | D(3,2) | 34 - 40 | 56 - 64 | 4600 - 5250 |
| **FULL THROTTLE** | | | | |
| FT4 - 3 | D | 61 - 67 | 98 - 108 | |
| FT3 - 2 | 3(D) | 40 - 46 | 64 - 73 | |
| FT3 - 4 | D | 74 - 80 | 119 - 129 | 3980 - 4330 |
| FT2 - 3 | D(3) | 55 - 60 | 88 - 96 | 4350 - 4800 |
| FT1 - 2 | D(3,2) | 29 - 34 | 48 - 56 | 3950 - 4650 |
| **PART THROTTLE** | | | | |
| PT4 - 3 | D | 47 - 54 | 75 - 86 | |
| PT3 - 2 | D(3) | 29 - 37 | 48 - 59 | |
| PT2 - 1 | D(3,2) | 10 - 12 | 16 - 19 | |
| **LIGHT THROTTLE** | | | | |
| LT3 - 4 | D | 26 - 30 | 43 - 49 | 1430 - 1650 |
| LT2 - 3 | D(3) | 18 - 22 | 29 - 35 | 1420 - 1820 |
| LT1 - 2 | D(3,2) | 9 - 10 | 14 - 16 | 1180 - 1220 |
| **ZERO THROTTLE** | | | | |
| ZT4 - 3 | D | 19 - 25 | 31 - 41 | |
| ZT3 - 2 | D(3) | 12 - 15 | 19 - 24 | |
| ZT2 - 1 | D(3,2) | 6 - 7 | 10 - 11 | |
| **TORQUE CONVERTER** | | | | |
| Lock up (IN) | D | 51 - 54 | 81 - 86 | 1875 - 2000 |
| Unlock (OUT) | D | 49 - 52 | 78 - 83 | 1825 - 1930 |

NOTE: The speeds given in the above chart are approximate and only intended as a guide. Maximum shift changes should take place within these tolerance parameters.

# LUBRICANTS, FLUIDS AND CAPACITIES 09

## Recommended Lubricants and fluids - 1989 Model year

Use only the recommended grades of oil set out below.
These recommendations apply to climates where operational temperatures are above - 10°C (14°F)

| COMPONENTS | BP | CASTROL | DUCKHAM | ESSO | MOBIL | PETROFINA | SHELL | TEXACO |
|---|---|---|---|---|---|---|---|---|
| Petrol engine sump Carburetter Dashpots Oil can | BP Visco 2000 (15W/40) or BP Visco Nova (10W/40) | Castrol GTX (15W/50) or Castrolite (10W/40) | Duckhams 15W/50 Hypergrade Motor Oil | Esso Superlube plus (15W/40) | Mobil Super (10W/40) or Mobil 1 Rally Formula | Fina Supergrade Motor Oil (15W/40) or (10W/40) | Shell Super Motor Oil (15W/40) or (10W/40) | Havoline Motor Oil (15W/40) or Eurotex HD (10W/30) |
| Diesel engine sump ** | BP Vanellus C3 Extra (15W/40) | Castrol Turbomax (15W/40) | Duckhams Fleetmaster SHPD (15W/40) | Esso Super Diesel Oil TD (15W/40) | Mobil Delvac 1400 Super (15W/40) | Fina Kappa LDO (15W/40) | Shell Myrina (15W/40) | Texaco URSA Super TD (15W/40) |
|  | The following list of oils to MIL - L - 2104D or CCMC D2 or API Service levels CD or SE/CD are for emergency use only if the above oils are not available. They can be used for topping up without detriment, but if used for engine oil changing, they are limited to a maximum of 5,000 km (3,000 miles) between oil and filter changes | | | | | | | |
|  | BP Vanellus C3 Multigrade 15W/40) | Castrol Deusol RX Super (15W/40) | Duckhams Hypergrade (15W/50) | Esso Essolube XD - 3 plus (15W/40) | Mobil Delvac Super (15W/40) | Fina Dilano HPD (15W/40) | Shell Rimula X (15W/40) | Texaco URSA Super Plus (15W/40) |
| Automatic gearbox | BP Autran DX2D | Castrol TQ Dexron IID | Duckhams Fleetmatic CD or Duckhams D - Matic | Esso ATF Dexron IID | Mobil ATF 220D | Fina Dexron IID | Shell ATF Dexron IID | Texamatic Fluid 9226 |
| Manual gearbox | BP Autran G | Castrol TQF | Duckhams Q - Matic | Esso ATF Type G | Mobil ATF 210 | Fina Purfimatic 33G | Shell Donax TF | Texamatic Type G or Universal |
| Front and Rear differential Swivel pin housings and LT230 Transfer gear box | BP Gear Oil SAE 90EP | Castrol Hypoy SAE 90EP | Duckhams Hypoid 90 | Esso Gear Oil GX (85W/90) | Mobil Mobilube HD90 | Fina Pontonic MP SAE (80W/90) | Shell Spirax 90EP | Texaco Multigear Lubricant EP (85W/90) |
| Propeller shaft Front and Rear | BP Energrease L2 | Castrol LM Grease | Duckhams LB 10 | Esso Multi-purpose Grease H | Mobil Grease MP | Fina Marson HTL 2 | Shell Retinax A | Marfak All Purpose Grease |
| Power steering box and fluid Reservoir Borg Warner Transfer Gearbox | BP Autran DX2D or BP Autran G | Castrol TQ Dexron IID or Castrol TQF | Duckhams Fleetmatic CD or Duckhams Q - matic | Esso ATF Dexron IID or Esso ATF Type G | Mobil ATF 220D or Mobil ATF 210 | Fina Dexron IID or Fina purilimatic 33G | Shell ATF Dexron IID or Shell Donax TF | Texamatic Fluid 9226 or Texamatic Type G or 4291A Universal |
| Brake and clutch reservoirs | Brake fluids having a minimum boiling point of 260°C (500°F) and complying with FMVSS 116 DOT4 | | | | | | | |
| Lubrication nipples (hubs, ball joints etc.) | BP Energrease L2 | Castrol LM Grease | Duckhams LB 10 | Esso Multi-purpose Grease H | Mobil Grease MP | Fina Marson HTL 2 | Shell Retinax A | Marfak All Purpose Grease |
| Ball joint assembly Top Link | Dextragrease Super GP | | | | | | | |
| Seat slides Door lock striker | BP Energrease L2 | Castrol LM Grease | Duckhams LB 10 | Esso Multi-purpose Grease H | Mobil Grease MP | Fina Marson HTL 2 | Shell Retinax A | Marfak All purpose grease |

NLGI - 2 Multi - purpose Lithium - based Grease

** **Other approved oils include:** Agip Sigma Turbo, Aral OL P327, Autol Valve - SHP, Aviation Turbo, Caltex RPM Delo 450, Century SHPD, Chevron Delo 450 Multigrade, Divinol Multimax Extra, Ecubsol CD Plus, Elf Multiperformance, Esso Special Diesel, Fanal Indol X, Fuchs Titan Truck 1540, Gulf Superfleet Special, IP Taurus M, Total Rubia TIR, Valvoline Super HD LD.

# Notes

# MAINTENANCE 10

Check/top up transfer gearbox oil every - 6,000 miles (10,000 Km)
Renew transfer gearbox oil every - 24,000 miles (40,000 Km)

## TOP UP TRANSFER GEARBOX OIL
### - Borg Warner gearbox

**NOTE: The existing maintenance intervals for the LT230 are also applicable to the Borg Warner transfer gearbox.**

1. Before topping up the oil ensure that the vehicle is level, either on a hoist or on the ground.
2. Disconnect the battery negative terminal.
3. Clean the immediate area around the filler/level plug.
4. Remove the plug and fill the gearbox with the recommended grade of oil, until oil starts to seep from the filler/level hole.
5. Clean any previously applied sealant from the filler/level plug.
6. Apply Hylomar sealant to the threads of the plug and refit the plug. Tighten to the specified torque. 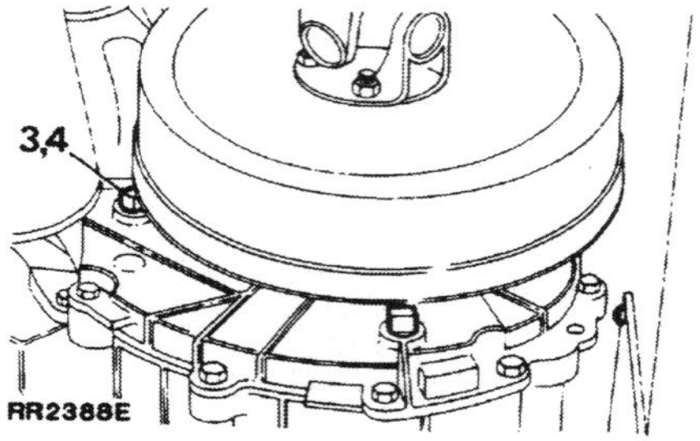
7. Wipe away any surplus oil.
8. Reconnect the battery.

## RENEW TRANSFER GEARBOX OIL
### - Borg Warner gearbox

**NOTE: The existing maintenance intervals for the LT230 are also applicable to the Borg Warner transfer gearbox.**

1. Before renewing the oil ensure that the vehicle is level, either on a hoist or on the ground.
2. Disconnect the battery negative terminal.
3. Clean the immediate area around the filler/level and drain plugs.

**WARNING: When draining the gearbox care should be taken to ensure that the oil is not hot as personal scalding could result.**

4. Place a container under the gearbox to drain the oil into.
5. Remove the filler/level plug to vent the gearbox and assist draining.
6. Remove the drain plug and allow the oil to drain. 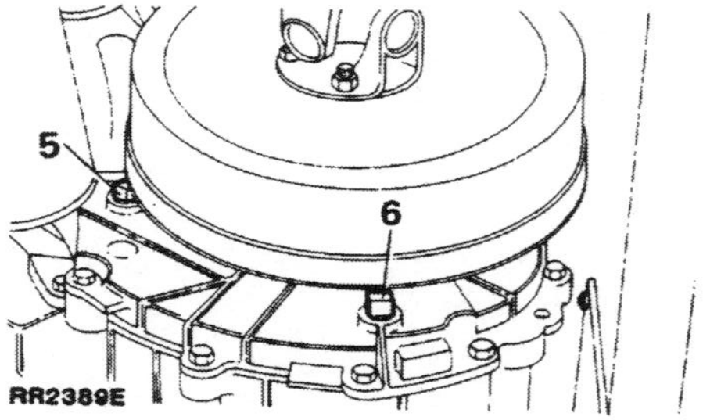
7. Thoroughly clean the drain plug threads prior to applying fresh 'Hylomar' sealant. Fit and tighten the plug to the specified torque.
8. Fill the gearbox with the correct quantity and grade of oil until oil seeps from the filler level hole. Wipe away any surplus oil.
9. Thoroughly clean the filler/level plug threads prior to applying fresh 'Hylomar' sealant. Fit and tighten the plug to the specified torque.
10. Reconnect the battery.

---

# 09 LUBRICANTS, FLUIDS AND CAPACITIES

## RECOMMENDED LUBRICANTS AND FLUIDS - ALL CLIMATES AND CONDITIONS - 1989 Model year

| COMPONENTS | SERVICE CLASSIFICATION Specification | SAE Classification |
|---|---|---|
| **Petrol models** | | |
| Engine sump Carburetter Dashpots Oil can | Oils must meet BLS.22.OL.07 or CCMC G3 or API service levels SF | 5W/30 5W/40 5W/50 10W/30 10W/40 10W/50 |
| | Oils must meet BLS.22.OL.02 or CCMC G1 or G2 or API service levels SE or SF | 15W/40 15W/50 20W/40 20W/50 25W/40 25W/50 |
| **Diesel models** engine sump | SHPD oils meeting CCMC D3 * Emergency only: Oils meeting MIL - L - 2104D or CCMCD2 or API CD | 10W/30 15W/40 |
| Main Gearbox Automatic | ATF Dexron IID | |
| Main Gearbox manual | ATF M2C33 (F or G) | |
| Transfer gearbox Final drive units Swivel pin housings | API GL4 or GL5 MIL - L - 2105 or MIL - L - 2105B | 90 EP 80W EP |
| Power steering | ATF M2C 336 or ATF Dexron IID | |
| Borg Warner Transfer Gearbox | | |

* Oils for emergency use only if the SHPD oils are not available. They can be used for topping up without detriment, but if used for engine oil changing, they are limited to a maximum of 5,000 km (3,000 miles) between oil and filter changes. (See * * on previous page)

## CAPACITIES
Transfer gearbox capacity - 2.1 Litres
- 3.7 UK Pints
- 4.4 US Pints

* Diesel Models - Engine Sump

# MAINTENANCE

## ADJUST HANDBRAKE

1. Set the vehicle on level ground and select 'P' in automatic gearbox or neutral in manual gearbox. Disconnect the battery negative terminal.
2. Chock the road wheels.
3. Fully release the handbrake lever.
4. From underneath the vehicle, rotate the adjuster on the brake drum back plate clockwise until the brake shoes are fully expanded against the brake drum.

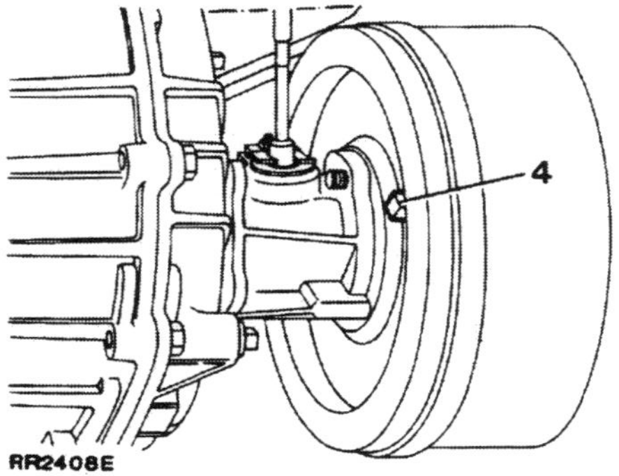

RR2408E

5. Back off the adjuster until the drum is free to rotate.
6. Release the four screws and remove the glove box liner.
7. Rotate the adjustment thumbwheel below the handbrake lever until the handbrake is fully operational on the third notch of the ratchet.

**NOTE: The handbrake adjustment thumbwheel must only be used for initial setting and to compensate for cable stretch, it must not be used to take up brake shoe wear, which must continue to be adjusted at the brake drum.**

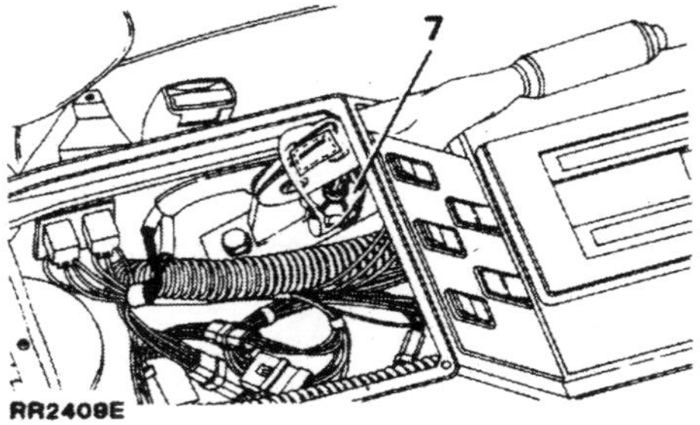

RR2409E

8. Operate the handbrake once or twice to settle the brake shoes, recheck that the handbrake is fully operational on the third notch of the ratchet. Readjust as necessary.
9. Refit the glove box liner.
10. Reconnect the battery and remove the wheel chocks.

## ROLLING ROAD TESTING OF PERMANENT FOUR WHEEL DRIVE VEHICLES

**NOTE: THIS INFORMATION APPLIES TO VEHICLES FITTED WITH BORG WARNER TRANSFER GEARBOX WITH VISCOUS COUPLING**

These vehicles are identified by the absence of the diff-lock position on the transfer gearbox lever.

### Viscous coupling

The front and rear axles cannot be driven independently due to the viscous coupling. This eliminates the need for the differential lock by progressively locking the centre differential automatically if any slip occurs at any wheel.

**WARNING: DO NOT attempt to drive individual wheels with the vehicle supported on floor jacks or stands.**

### Four wheel rolling roads

Provided that the front and rear rollers are rotating at identical speeds and that normal workshop safety standards are applied, there is no speed restriction during testing except for any that may apply to the tyres.

### Two wheel rolling roads

**IMPORTANT: Use a four wheel rolling road for brake testing if possible.**

If brake testing on a single axle rig is necessary it must be carried out with the propeller shaft to the rear axle removed, AND neutral selected in BOTH main gearbox and transfer gearbox. When checking brakes, run engine at idle speed to maintain servo vacuum.

If checking engine performance, the transfer box must be in high range and the propeller shaft to the stationary axle must be removed.

### TOWING

Note the towing procedure for previous models applies to vehicles fitted with Borg Warner transfer gearbox. The main gearbox and transfer gearbox must be in neutral when the vehicle is being towed.

# BORG WARNER TRANSFER GEARBOX [37]

## BORG WARNER TRANSFER GEARBOX

### VISCOUS UNIT (Front output housing) - In vehicle check

The viscous unit is located in the front output housing and its integrity can be checked while the unit is installed in the transfer gearbox as follows.

Remove either the front or rear propeller shaft from the vehicle to eliminate drive to one of the axles. If the viscous unit is operating effectively drive will be transferred to the axle that is still connected via the propeller shaft to the gearbox and the vehicle should remain driveable.

If the viscous unit has failed drive will not be transmitted to the axle.

A partially failed unit will be identified by excessively high engine revs and little vehicle movement when attempting to drive the vehicle.

### IN - SITU OPERATIONS

The following operations can be carried out with the gearbox in the vehicle which for ease of working, should be raised on a ramp or placed over a pit. Disconnect the battery negative terminal

**NOTE: The front and rear output housings can also be removed while the gearbox is in the vehicle. Reference should be made to the Overhaul procedure for the removal of these assemblies.**

### RENEW SPEEDOMETER DRIVE PINION

**NOTE: Driven gear identification:-**
**Non catalyst vehicles: RED**
**Catalyst vehicles : BLACK**

1. Remove the nut securing speedometer drive clamp and withdraw the cable. Prise the drive pinion assembly from the output housing.
2. Push in a new assembly, fit the speedometer cable and secure with the clamp and nut.

## RENEW REAR OUTPUT SHAFT OIL SEAL

**Service tool: 18G1422**

1. Disconnect the rear propeller shaft from the output drive flange and tie the shaft to one side.
2. Ensuring that the handbrake is applied to restrain the drive flange, release the drive flange nut.
3. Release the handbrake and remove the two screws which secure the brake drum and withdraw the drum.

**NOTE: While the brake drum is removed from the rear output housing the transmission brake assembly can be overhauled, the procedure for this operation is the same as the LT230, therefore reference should be made to Section 70 Brakes, of the main Workshop Manual.**

4. Remove the bottom two bolts which secure the oil catcher to the back plate and withdraw the oil catcher.
5. Remove the output shaft nut, steel washer, rubber seal and withdraw the flange.
6. Carefully tap the dust cover from the housing and prise out the oil seal.
7. Lubricate and carefully install the new seal using service tool 18G1422 with the spring side of the seal abuting the circlip.
8. Fit the dust cover.
9. If necessary release the circlip from the drive flange to allow new bolts to be installed.
10. Examine the flange for damage or wear particularly the seal running surface, if the surface is corroded or a groove has been worn by the previous seal discard the flange.
11. Lubricate the seal running surface of the flange. Fit the flange, if necessary fit a new rubber seal, steel washer and secure with a new nut. Do not tighten the nut at this stage.
12. Seal the oil catcher to the back plate using silicone rubber sealant and secure the assembly with the two bolts.
13. Fit the brake drum and secure with the two screws. Apply the handbrake to restrain the drum and tighten the new drive flange nut to the specified torque.
14. Reconnect the propeller shaft and secure with new nuts, tighten to the specified torque.

Continued

# BORG WARNER TRANSFER GEARBOX 37

## RENEW FRONT OUTPUT SHAFT OIL SEAL

**Service tool: 18G1422**

1. Disconnect the front propeller shaft from the flange and tie the shaft to one side.
2. Ensuring that the handbrake is applied to restrain the transmission release the drive flange nut.
3. Remove the output shaft nut, steel washer, rubber seal and withdraw the flange.
4. Prise out the oil seal.
5. Lubricate and carefully install the new seal using service tool 18G1422 with the spring side of the seal abutting the circlip.
6. Examine the flange for damage or wear, particularly the seal running surface, if the surface is corroded or a groove has been worn by the previous seal discard the flange.
7. Lubricate the seal running surface of the flange.
8. Fit the flange, if necessary fit a new rubber seal, steel washer and secure with a new nut, tighten to the specified torque.
9. Refit the propeller shaft and secure with new nuts tightened to the specified torque.

# BORG WARNER TRANSFER GEARBOX 37

## REMOVE TRANSFER GEARBOX

Adaptor plate for removing transfer gearbox

The transfer gearbox should be removed from underneath the vehicle, using a hydraulic transmission hoist. An adaptor plate for locating the transfer gearbox onto the hoist can be manufactured locally to the drawing below.

Material: Steel plate BS 1449 Grade 4 or 14.
Holes marked thus * to be drilled to fit hoist being used.

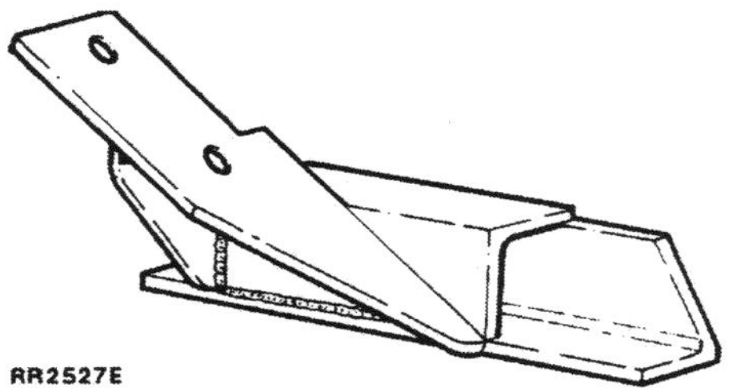

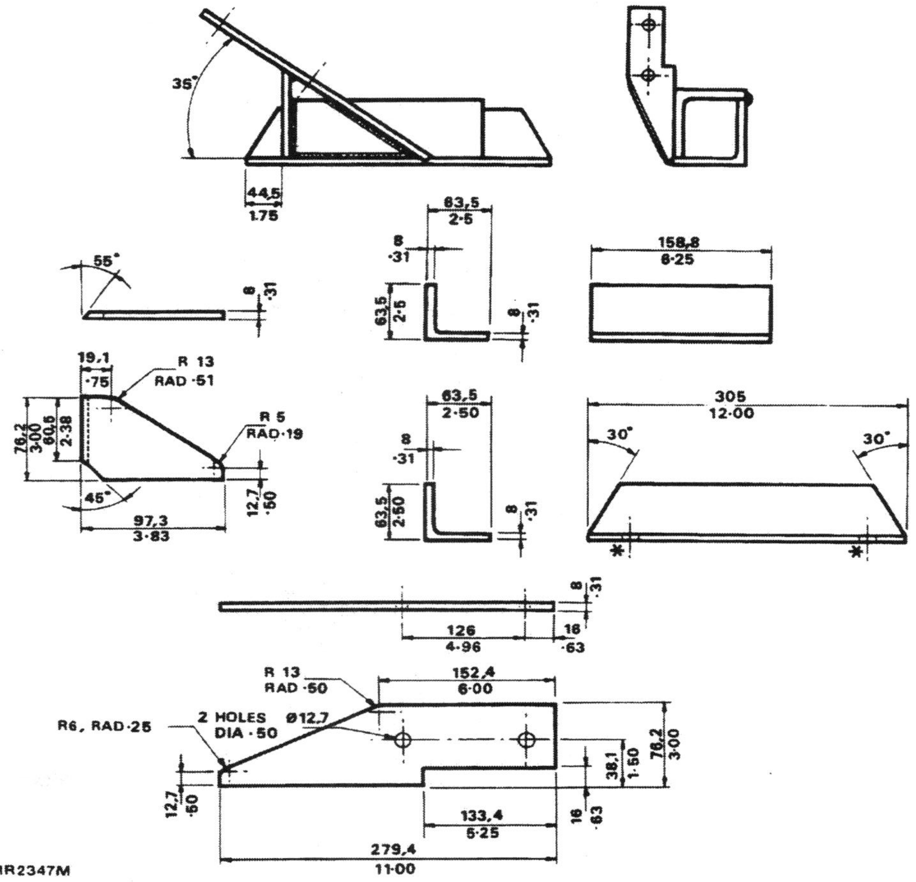

## BORG WARNER TRANSFER GEARBOX | 37

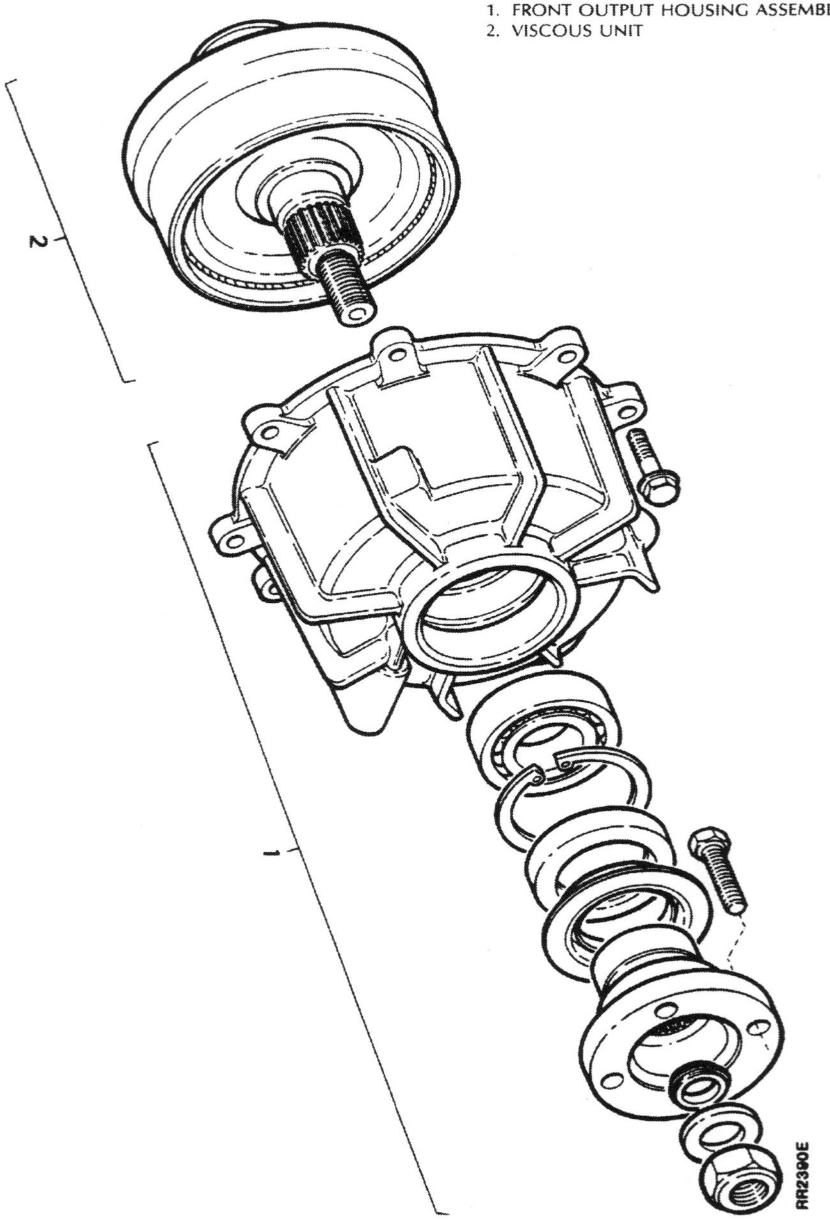

1. REAR OUTPUT HOUSING AND OUTPUT SHAFT ASSEMBLY
2. TRANSMISSION BRAKE DRUM ASSEMBLY
3. SPEEDOMETER DRIVE PINION ASSEMBLY

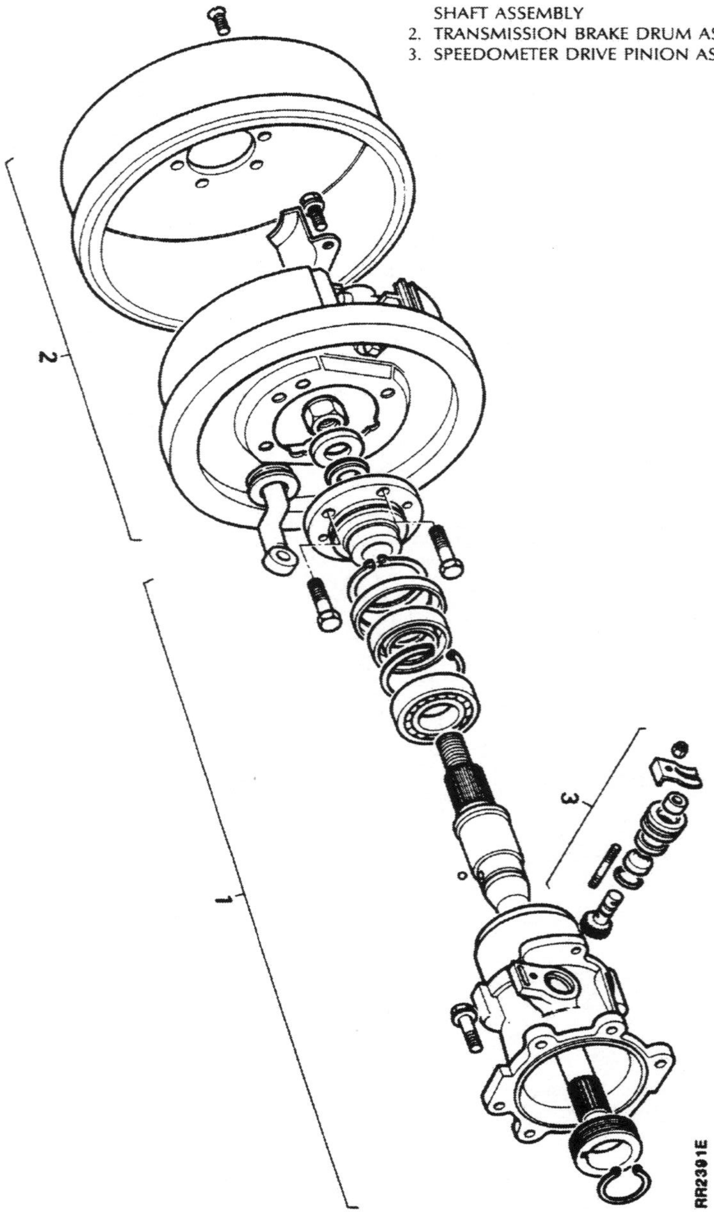

## 37 | BORG WARNER TRANSFER GEARBOX

1. FRONT OUTPUT HOUSING ASSEMBLY
2. VISCOUS UNIT

# BORG WARNER TRANSFER GEARBOX | 37

1. FRONT COVER - MAIN CASING
2. TRANSFER SPROCKET, CENTRE DIFFERENTIAL ASSEMBLY, DRIVEN SPROCKET AND CHAIN
3. BEARING CARRIER AND TRANSFER SHAFT
4. SELECTOR FORK AND LEVER ASSEMBLY
5. SELECTOR SLEEVE
6. OIL PICK - UP PIPE AND FILTER
7. PLANETARY SET (EPICYCLIC UNIT)
8. REAR COVER - MAIN CASE
9. END CAP

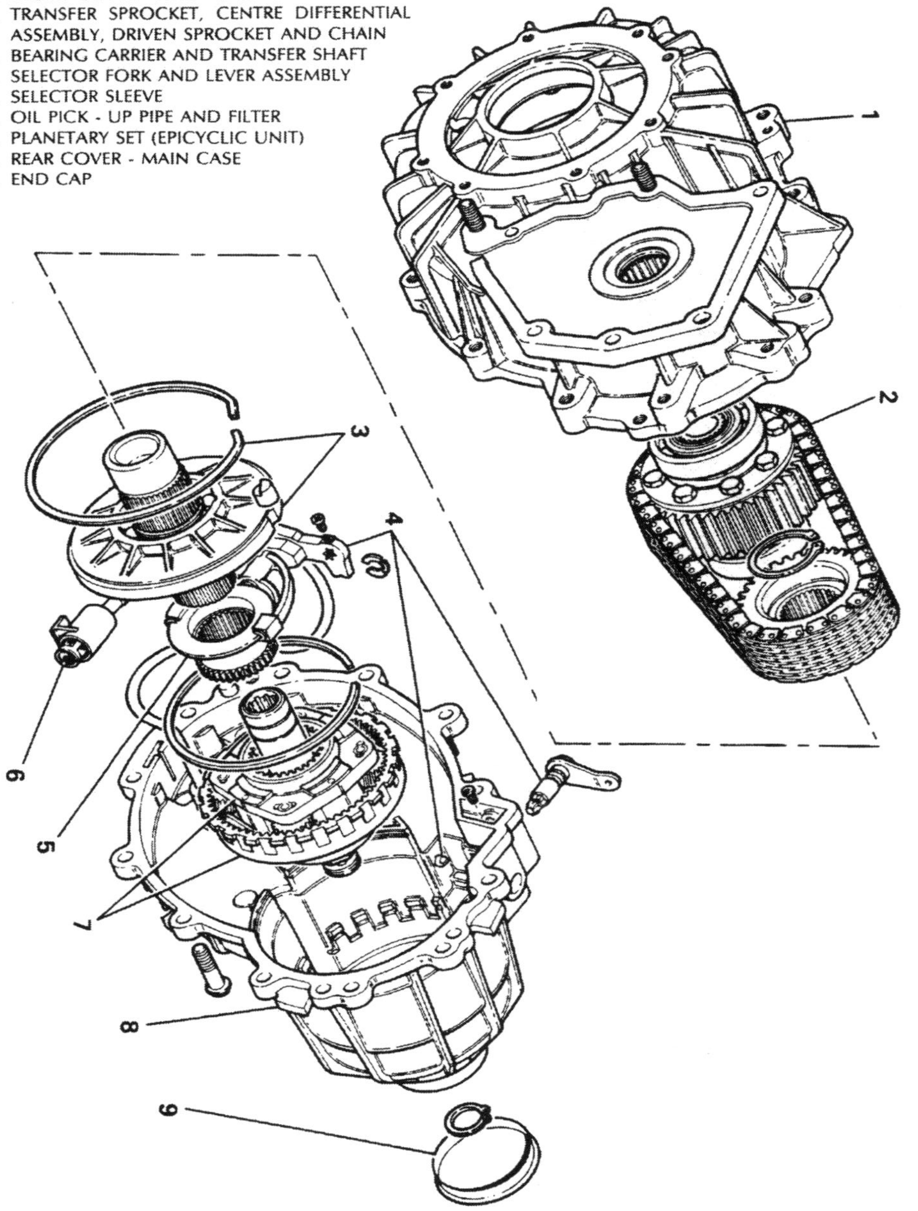

## REMOVE TRANSFER GEARBOX

NOTE: The following preparation work is necessary prior to the removal of the gearbox to avoid unnecessary damage to associated components.

WARNING: Where the use of a transmission hoist is necessary, it is ABSOLUTELY ESSENTIAL to follow the hoist manufacturer's instructions to ensure safe and effective use of the equipment.

### Preparation

### Outside the vehicle

1. Install the vehicle on a ramp and chock the wheels. Disconnect the battery negative terminal.
2. Fuel Injection models only - release the airflow meter to plenum chamber hose.
3. Carburetter models only - Remove the air intake elbows and withdraw the air cleaner from its location.
4. Release and remove the fan blade assembly noting that the fan blade has a left hand thread, removing the fan blade assembly will enable the engine to be tipped rearwards when the transmission is ready to be removed.

### Inside the vehicle

5. Remove the four screws securing the glove box liner to the glove box and lift out the liner. Detach the two relay mounting blocks from the clip on the side of the glove box.
6. Disconnect the electrical leads to the rear cigar lighter.
7. Carefully prise the window lift, switch panel away from the front of the glove box, manoeuvre the switch panel complete with switches back inside the glove box and allow to lie loose on the gearbox tunnel.
8. Select low range, unscrew and remove the transfer lever knob.
9. **Automatic gearbox models only** - Unclip the top cover of the main gearbox selector and remove the circlip, withdraw the detent button. Remove the circlip above the selector knob retention nut, remove the nut, serrated washer and withdraw the selector knob. **Manual gearbox models only** - Unscrew and remove the main gear lever knob.
10. Carefully prise the centre panel out of the floor mounted console **(Automatic models only)** - disconnect the electrical leads to the graphics panel on the underside of the centre panel) and remove it from the vehicle.

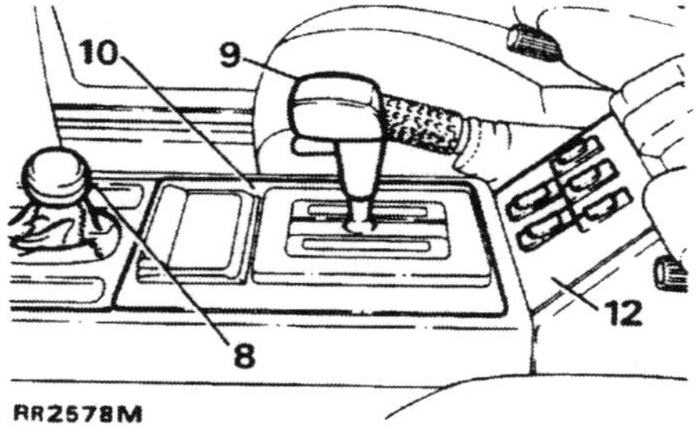

RR2578M

11. Release the two bolts and two screws securing the console assembly to the gearbox tunnel.

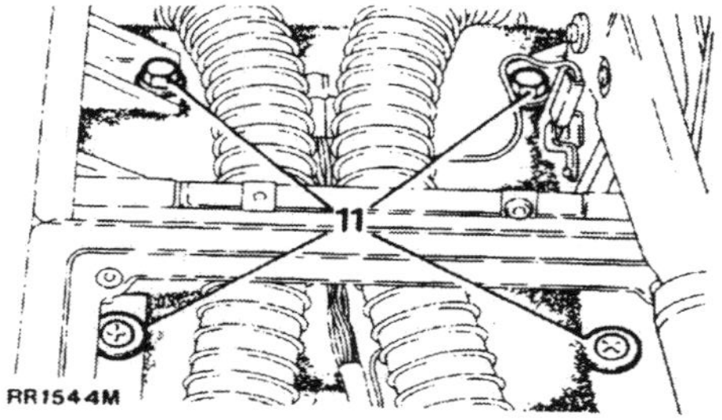

### Automatic version illustrated

RR1544M

12. Release the handbrake, pull the gaiter forward to gain access to the clevis pin. Remove the split pin, clevis pin and washer securing the handbrake cable to the handbrake lever.
13. Detach the console locating tab from the radio housing by easing the console slightly rearwards.
14. Carefully manoeuvre the glove box assembly rearwards (while raising the handbrake lever to its uppermost position) away from the radio housing and remove the glove box assembly from the vehicle.

# BORG WARNER TRANSFER GEARBOX

15. Remove the sound deadening material from the top of the gearbox tunnel. **Manual gearbox models only** - slacken the pinch bolt and remove the upper gear lever. Remove the four screws and detach the retaining plate securing the rubber gaiter to the top of the main gearbox.

### Underneath the vehicle

16. Raise the the vehicle on the ramp and remove the transfer gearbox drain plug and allow the oil to drain into a suitable container, meanwhile continue with the preparation operations.

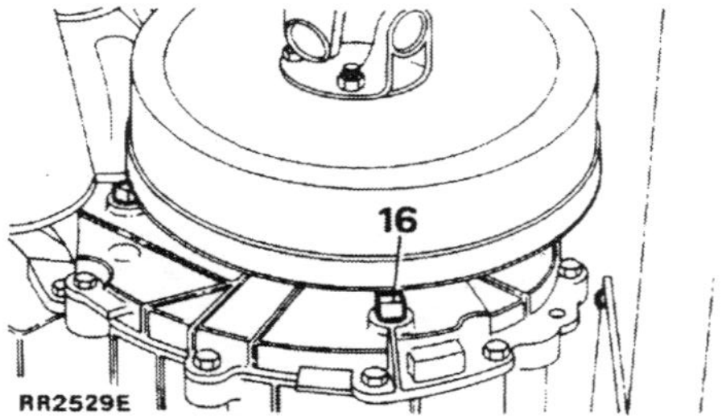

17. Refit the oil drain plug.
18. **Catalyst models only** - Disconnect the multi-plugs to the Lambda sensors.
19. Remove the crossmember from below the main gearbox.
20. Remove the front exhaust down pipes and intermediate pipe complete with centre silencer (or catalyst).

   **NOTE: Operation 19 will require the assistance of a second operator to support the exhaust system while the various fixings are released.**

21. Remove the underbody floor mounted centre silencer heat shield.
22. Mark each drive flange with an identification line to aid reassembly. Remove the nuts securing the front and rear propeller shafts to the transfer gearbox and tie both shafts clear of the working area.

23. Release the clamp and withdraw the speedometer cable from the rear output housing, also free the cable from the clip at the left hand side of the transfer gearbox. **Automatic models only** - Remove the three fixings securing the tie bar to the transmission.
24. Secure the adaptor plate to the gearbox hoist with two nuts and bolts.

   **NOTE: To ensure that the weight of the transfer gearbox is centralised on the hoist, fit the adaptor plate on the hoist platform so that the split line of the gearbox is aligned with the centre line of the ram.**

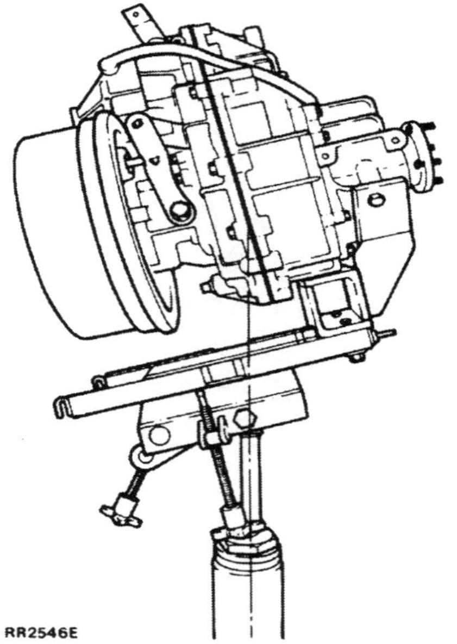

### Remove the transfer gearbox

26. Using a suitable hydraulic hoist, support the main gearbox.

27. Slacken the transfer gearbox right hand mounting rubber, upper nut and remove the lower nut. Remove the fixings and withdraw the transfer gearbox to chassis outer half of the mounting bracket.
28. Remove the inner half of the mounting bracket from the transfer gearbox.
29. Adjust the height of the hoist and place in position under the transfer gearbox so that the adaptor plate holes align with the transfer gearbox mounting bracket location.
30. Using the two short bolts previously removed from the right hand gearbox mounting bracket, secure the adaptor plate to the gearbox.

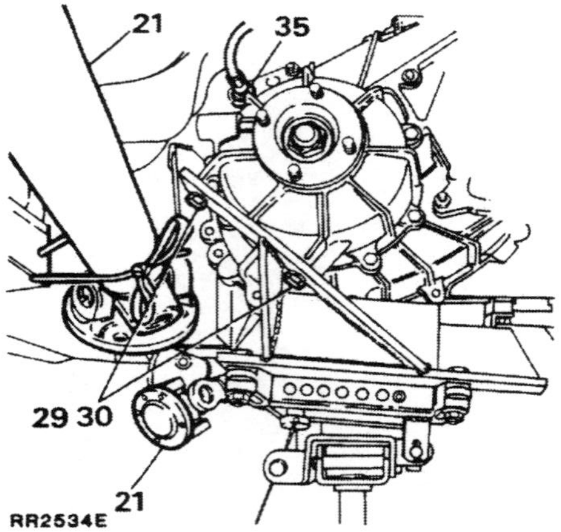

31. Remove the fixings securing the left hand mounting bracket to chassis.
32. Lower the front hoist to allow the transmission to be lowered.
33. Lower the transmission assembly until the top of the transfer gearbox clears the rear passenger footwell.
34. Remove the split pin, clevis pin and washer securing the handbrake cable to the brake drum actuating lever and disconnect the cable.
35. Remove the clip that secures the handbrake cable to the support bracket, feed the cable through the bracket and tie the cable to one side.

36. Remove the breather pipe from the top of the transfer gearbox.
37. Release the spring clip retaining the clevis pin, withdraw the clevis pin and clip assembly, remove the high/low rod from the transfer gearbox selector lever.

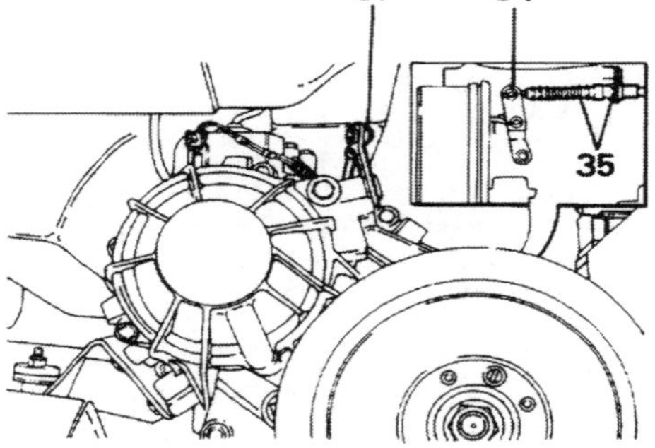

38. Support the main gearbox with the previously removed hoist before detaching the transfer gearbox.
39. Remove the upper and lower bolts and two nuts securing the transfer box to the main gearbox.
40. Manoeuvre the transfer gearbox rearwards to detach it from the main gearbox.
41. Thoroughly clean the exterior of the transfer gearbox before undertaking the overhaul procedure.

Continued

# BORG WARNER TRANSFER GEARBOX

## Transfer Gearbox - Refitting

42. Ensure that the joint faces of the transfer gearbox and main gearbox extension case are clean.

43. Lubricate the oil seal in the joint face of transfer gearbox, secure the transfer gearbox to the adaptor plate on the lifting hoist and raise the hoist until the input shaft enters the transfer gearbox.

**CAUTION: CARE MUST BE TAKEN DURING THIS OPERATION TO ENSURE THAT THE INPUT SHAFT SPLINES DO NOT DAMAGE THE OIL SEAL IN THE TRANSFER GEARBOX.**

During this process it may be necessary to rotate either one of the drive flanges to engage the input shaft splines.

44. Secure the transfer gearbox to the main gearbox by fitting the nuts to the two studs. Fit the remaining bolts noting that the longest bolt is fitted to the upper left hand fixing that locates the ring dowel and tighten all fixings to the correct torque.

## Transfer gearbox high/low link adjustment

45. Ensure that the selector lever at the gearbox is in the neutral position.

46. Set the transfer gearbox lever in a vertical position (at right angles to the centre line of the main gearbox), rotate the clevis on the end of the rod clockwise or counter clockwise which will shorten or lengthen the operating rod until the hole in the clevis aligns with the hole in the selector lever.

47. Fit the clevis pin and retaining clip assembly. Select high and low transfer to ensure full engagement is occurring. Repeat the above procedure if full engagement is not evident.

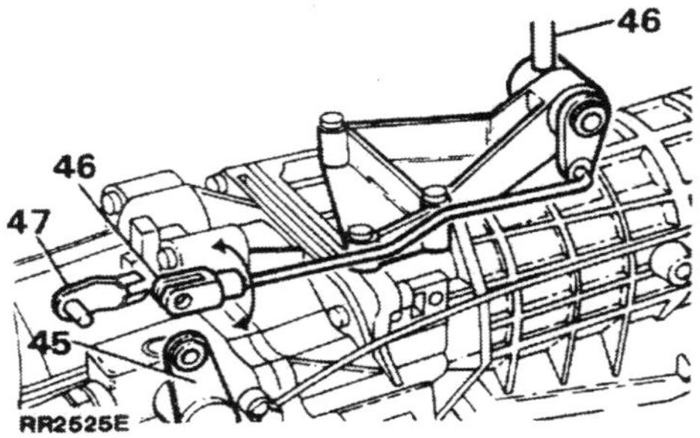

---

## BORG WARNER TRANSFER GEARBOX

### DISMANTLING, OVERHAUL AND REASSEMBLY

**Service Tools:**

18G1422 - Oil seal replacer
18G1205 - Adjustable flange holding wrench
18G134 - Bearing and oil seal replacer
LST550 - 6 - Input shaft oil seal replacer

**NOTE:** Before commencing the overhaul procedure thoroughly clean the exterior of the transfer gearbox. If the gearbox oil has not previously been drained, drain the oil into a suitable container.

### DISMANTLING

**NOTE:** Before commencing the dismantling procedure remove the brake drum assembly (refer to section 70 - Brakes, of the main Workshop Manual).

### Rear output housing

1. Remove the six bolts and withdraw the rear output housing complete with output shaft.

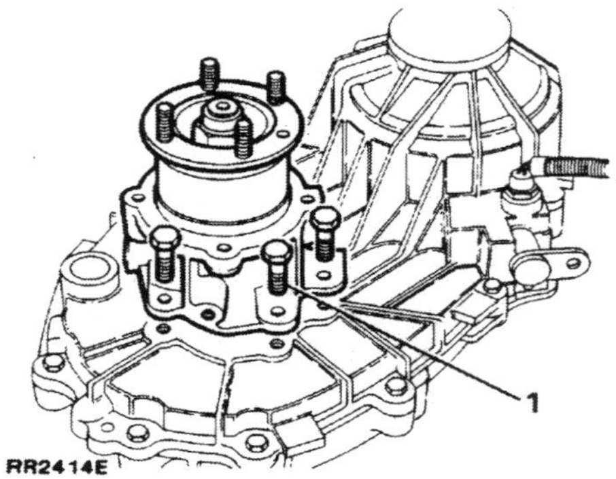

### Front output housing

**NOTE:** Invert the gearbox. Level up the assembly by placing wooden blocks under the transfer gearbox to main gearbox joint face.

2. Remove the eight bolts and withdraw the front output housing complete with viscous unit.

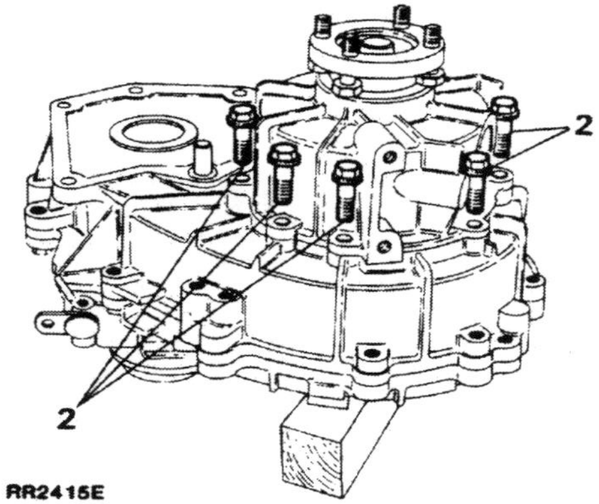

### Front cover - main casing

3. Remove the eleven bolts securing the front and rear cover (main case) together.

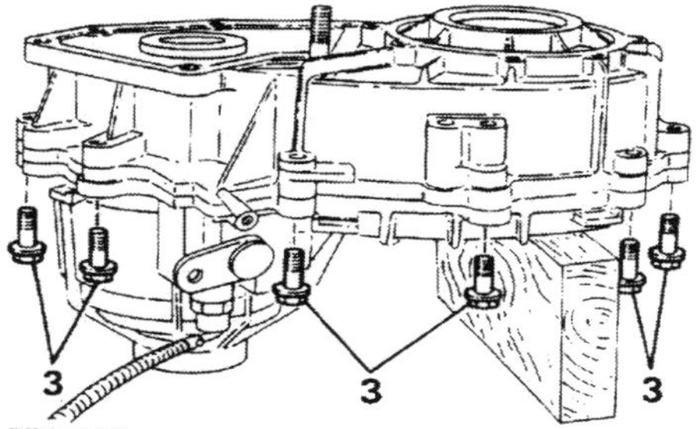

4. Clean any previous sealant from the threads of the bolts.

Continued

---

## Refitting (continued)

48. Complete the refitting procedure by reversing the removal sequence, noting the following important points.

49. After removing the lifting hoist and adaptor plate from the transfer gearbox, clean the threads of the bolts for the transfer gearbox and fit them together with the mounting bracket to the gearbox. Tighten to the specified torque.

50. Fit the three fixings which secure the right hand mounting bracket to the chassis. Tighten to the specified torque.

51. Fit the propeller shafts and tighten to the specified torque.

52. Remove the transfer gearbox combined oil filler and level plug, refill the transfer gearbox with the correct grade and quantity of oil until the oil starts to seep from the filler/level hole. Coat the plug with Hylomar sealant and refit the combined filler and level plug. Tighten to the specified torque, wipe away any surplus oil.

53. Check, and if necessary top - up the oil level in the main gearbox. Use the correct grade oil.

54. Check the operation of the handbrake and adjust as necessary.

# BORG WARNER TRANSFER GEARBOX

5. Using two levers between the cast lugs on the outer edges of the casing, to assist in separating the gearbox, carefully prise the front cover from the rear cover.

**CAUTION: DO NOT LEVER BETWEEN THE MATING FACES.**

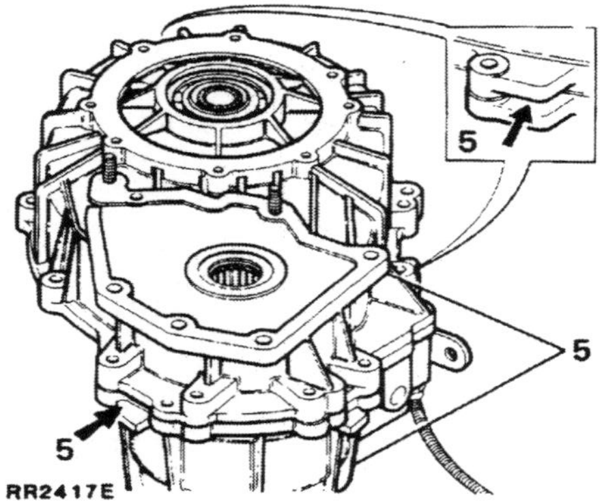

## Transfer sprocket, centre differential, assembly and chain

**NOTE: Before dismantling, mark one chain link and corresponding tooth on the transfer sprocket with an identification line. This is to ensure that the balance of the unit is maintained when reassembled with original components and that the chain is fitted the correct way up.**

6. Remove the circlip retaining the transfer sprocket to the transfer shaft.
7. Place two thin pieces of wood on the joint face to prevent damage and using two levers behind the differential assembly carefully lever the differential bearing from its bore while simultaneously easing the transfer sprocket off the transfer shaft to maintain alignment during removal.
8. Remove the transfer sprocket from the chain.
9. Remove the differential assembly from the chain.

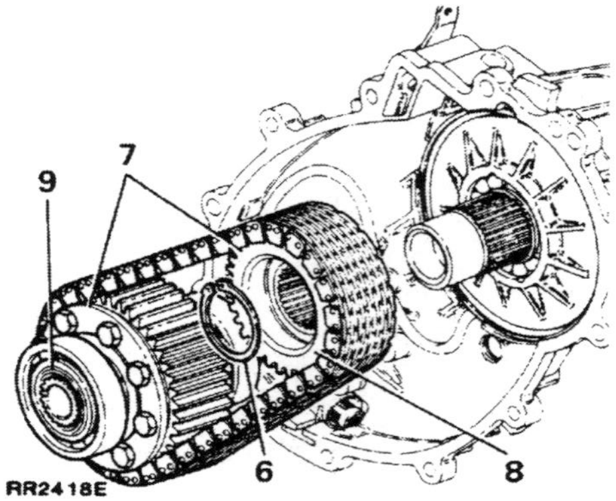

## Bearing carrier and transfer shaft

10. Insert a screw driver between the anti-rotation dowel and snap ring gently prise the snap ring out of the groove.
11. Withdraw the carrier complete with transfer shaft.
12. Withdraw the anti-rotation dowel.

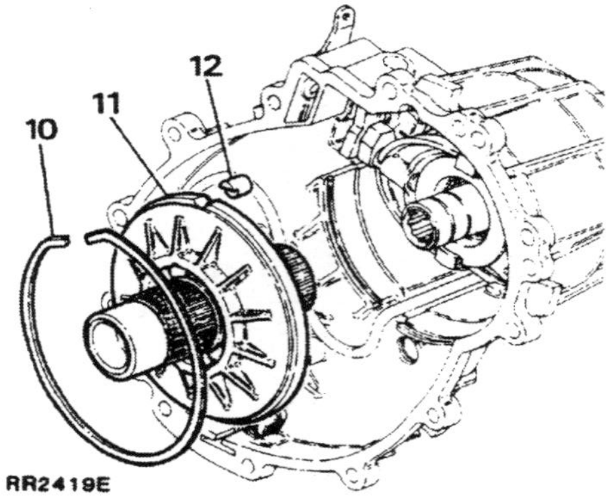

## Selector fork assembly

13. Using Torx bit 25 remove the screw securing the selector arm to the selector lever shaft.
14. Remove the retaining clip securing the selector fork arm to the selector lever shaft.

15. Using Torx bit 25 remove the screw retaining the selector lever.
16. Remove the selector lever shaft from the case and fork assembly.
17. Withdraw the selector fork assembly and selector sleeve.
18. Retrieve the selector plunger and spring from the rear cover.

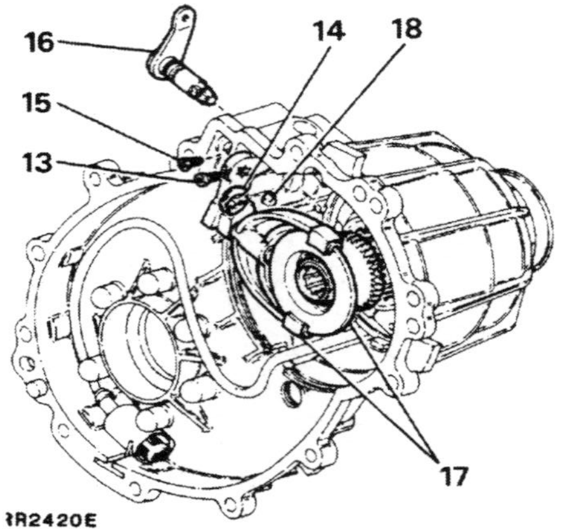

## Planetary set (Epicyclic unit)

19. Turn the case over and prise the end cap off the planetary set housing.
20. Remove the circlip retaining the sun gear shaft.
21. Turn the casing over and remove the large snap ring retaining the planetary set.
22. Withdraw the annulus and planetary assembly from the planetary set housing, complete with oil pump, feed pipe and filter.

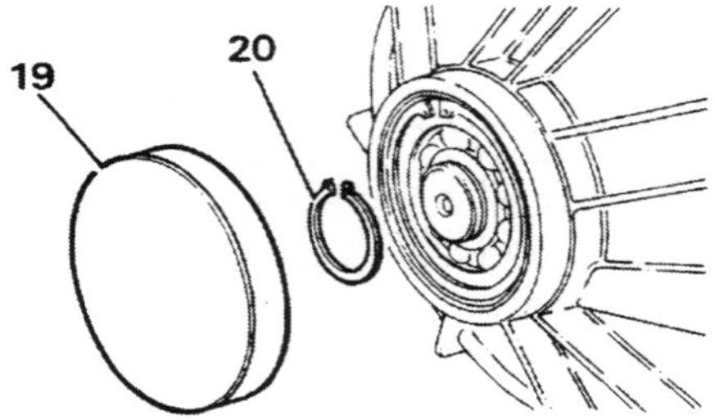

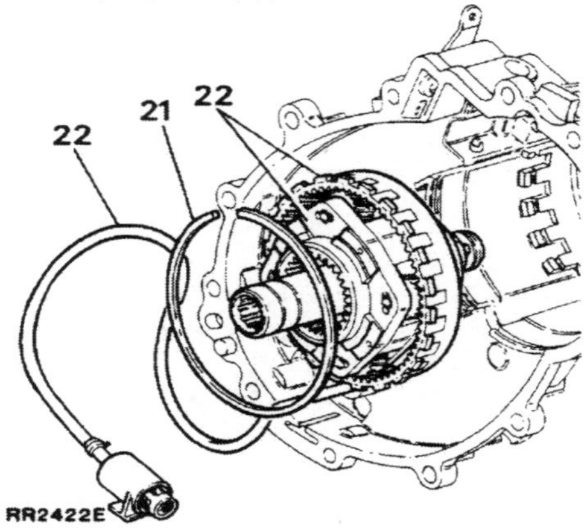

## DISMANTLE, INSPECTION AND OVERHAUL

### Rear cover main case
### - Dismantle and inspection

1. Remove the circlip retaining the bearing in the rear cover.
2. Drive or press the bearing from the cover and discard the bearing.

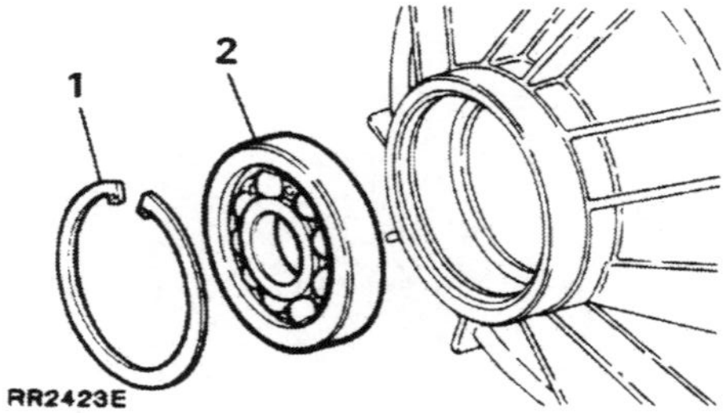

Continued

# BORG WARNER TRANSFER GEARBOX 37

3. Remove any previous sealant evident on the rear cover joint faces.
4. Using a suitable solvent thoroughly clean the cover.
5. Examine the cover for damage, cracks and porosity, renew if necessary.
6. Check the selector lever shaft bore, for ovality and wear. If worn renew the cover.

### Rear cover main case - Assemble

7. Drive or press a new bearing into the cover and secure in position with the circlip.
8. Place the cover aside until the gearbox is ready to be assembled.

### Front cover main case - Dismantle and inspection

9. Prise the input shaft oil seal from the front cover and discard the oil seal.
10. Remove the snap ring retaining the needle roller bearing, withdraw the bearing and discard.
11. Remove any previous sealant from the joint faces of the front cover.
12. Thoroughly clean the cover using a suitable solvent.
13. Examine the cover for damage, cracks and porosity, renew if necessary.
14. Check the inside edges of the case for witness marks which may indicate a chain that has stretched.

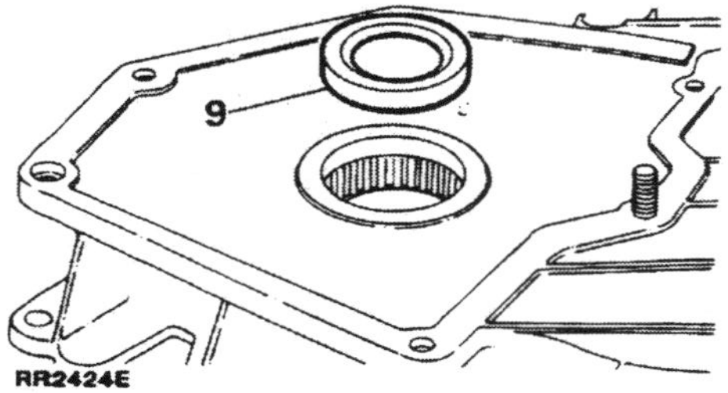

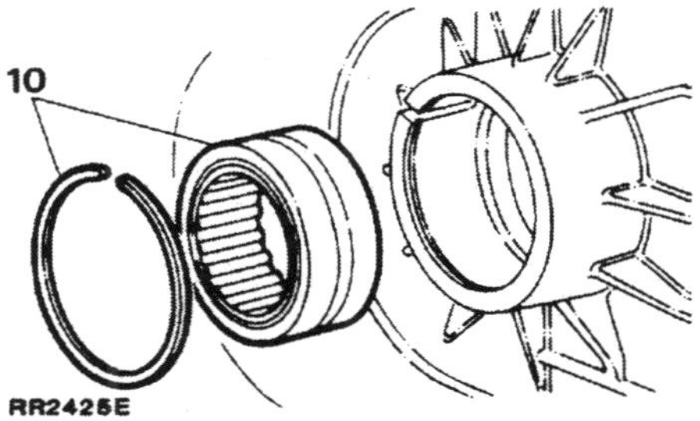

### Front cover main case - Assemble

15. Lubricate a new oil seal. Using service tool LST 550 - 6 in conjunction with bearing and oil seal replacer 18G134 fit the seal, open side of the seal leading, until the face of the seal is 1 mm (0.039 in) below the surface of the boss.
16. Lubricate a new needle roller bearing and drive or press the bearing into its recess until contact is made with the shoulder at the bottom of the bore.
17. Fit the snap ring to retain the bearing.
18. Place the cover aside until the gearbox is ready to be assembled.

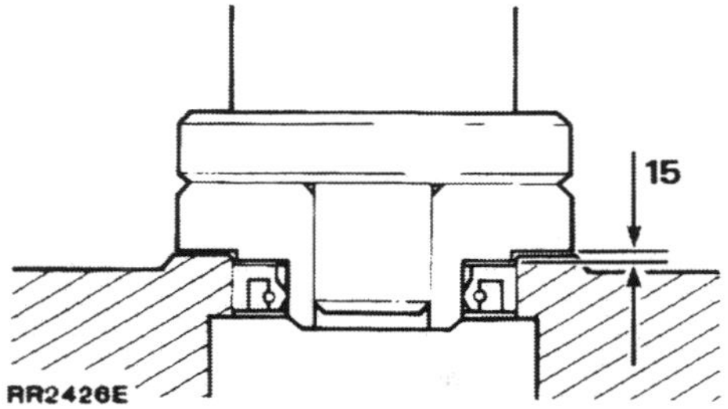

### Planetary set (Epicyclic unit) and oil pump - Dismantle and inspection

**NOTE: The Epicyclic unit and oil pump are serviced as a complete assembly, if after inspection either of the units is found to be worn a complete new assembly must be fitted.**

1. Remove the annulus from the planetary set.
2. Thoroughly clean all components using a suitable solvent.
3. Examine the helical teeth of the annulus for wear or damage. If damage is evident it will be necessary to renew both the annulus and planetary set.
4. Examine the planetary gears and high/low gear teeth for wear or damage. If damaged renew both the annulus and planetary set.
5. Check the end float of the four planet gears, between the end of the gear and planetary set carrier. End float of each planet gear should not exceed 0.83 mm (0.033 in) if any one of the planet gears is out of limits renew the planetary set assembly.
6. Check the end float of the sun gear to the planetary set carrier by supporting the body of the assembly on the top of a vice. Using a dial test indicator attached to a magnetic base, position the base on top of the assembly and zero the indicator on the end of the sun gear shaft, lift the shaft and check the end float. End float should not exceed 0.83 mm (0.033 in). Fit a new planetary set assembly if out of limits.

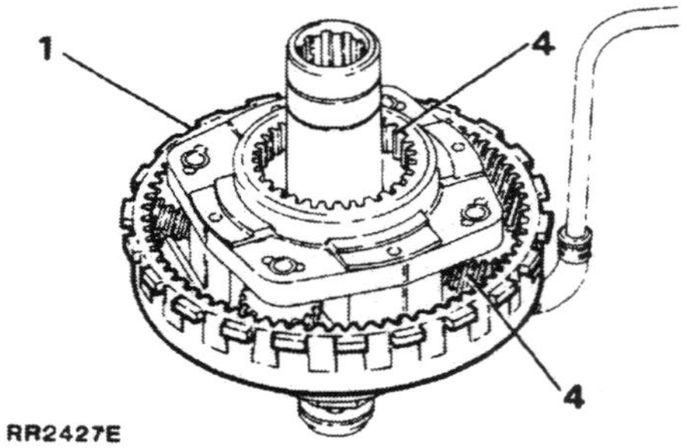

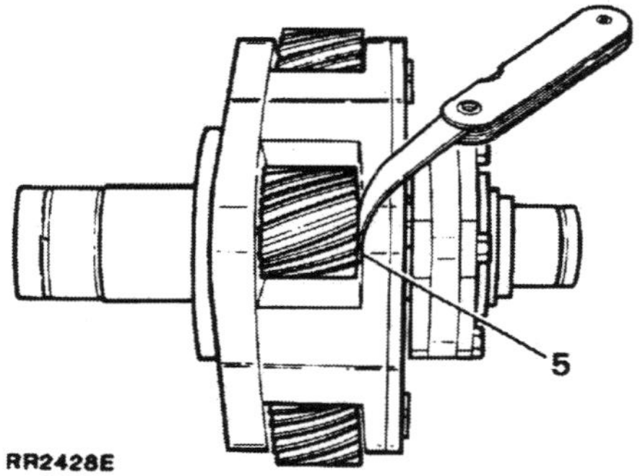

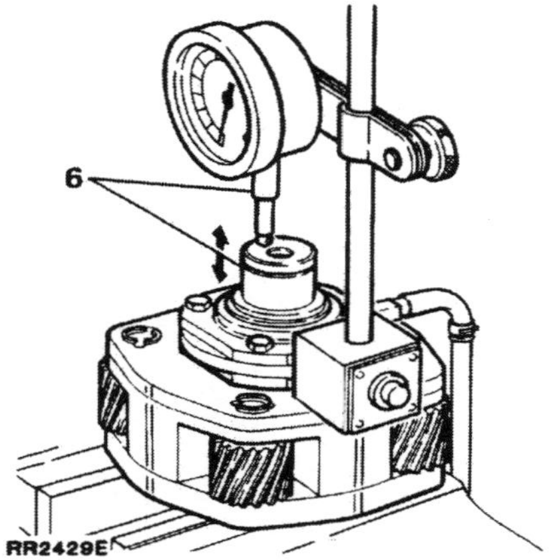

**NOTE: If the previous inspection instructions prove the assembly to be in an acceptable condition carry out the following examination of the oil pump.**

Continued

15

# BORG WARNER TRANSFER GEARBOX 37

## Oil pump and filter
### Dismantle and inspection

7. To aid re - assembly mark an identification line on the edges of the oil pump plates. Remove the four bolts securing the pump front and rear plates, separate the pump by removing the plungers, spring and bearing plate.

**NOTE: The front plate of the oil pump is stamped 'TOP', the centre bearing plate is stamped 'REAR' and the rear plate is stamped 'TOP REAR'. The fixing holes of the plates and body are also offset to ensure correct re - assembly of the pump.**

8. Depress the retaining clips, remove the oil pick - up pipe and rubber connection tubes. Examine the tubes and pipe for damage or fractures, renew as necessary.

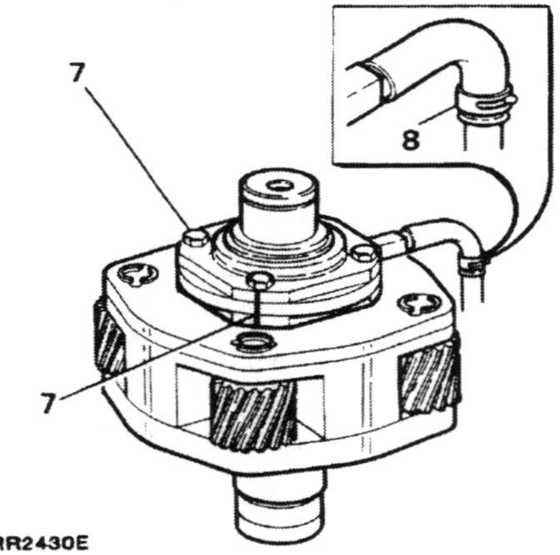

9. Clean the pump components and check for damage and wear, ie: blueing of the pump plungers, scoring of the centre bearing plate, if any wear is evident a new planetary set must be fitted, as the pump is part of the complete assembly.

10. Thoroughly clean the oil pick - up filter, examine the filter screen for damage and blockage, renew or clean the filter as necessary.

### Oil pump and filter - Assemble

11. Clean the sealant from the oil pump securing screws.
12. Prior to assembly lubricate the pump components with clean oil.
13. Fit the plate stamped 'TOP' to the sun gear shaft with the word 'TOP' facing the planetary assembly.
14. Fit the plungers and spring noting that the flats on the plungers must be uppermost to enable the 'TOP REAR' plate to be fitted.
15. Compress the plungers and fit the middle bearing plate with the word 'REAR' uppermost. Align the offset fixing holes and also noting the previously marked identification line.
16. Fit the top rear plate with the words 'TOP REAR' uppermost.
17. Apply Loctite 242 to the threads of the four screws and fit the screws, tighten to the specified torque.
18. Fit the rubber connection tube and oil pick - up pipe to the oil pump, fit the retaining clip. Note that the clip securing the tube to the pick - up pipe is positioned in front of the flare on the pipe.
19. Fit the rubber connection tube to the filter end of the pipe, fit the clip ensuring that the tube is clamped by the clip in front of the flare on the pipe.
20. Push the filter into the tube. The radial position of the filter to pipe at this stage is unimportant.

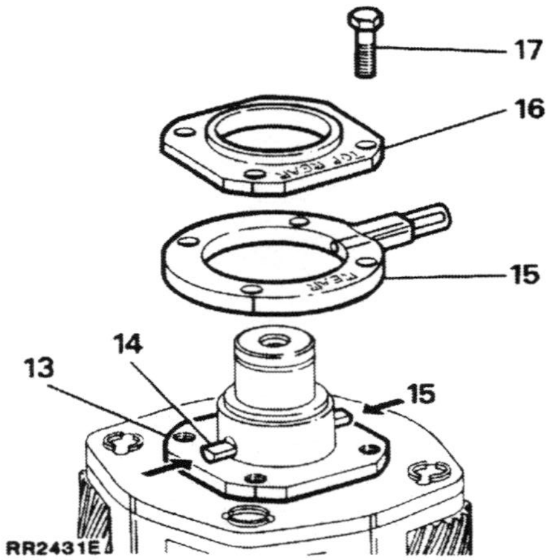

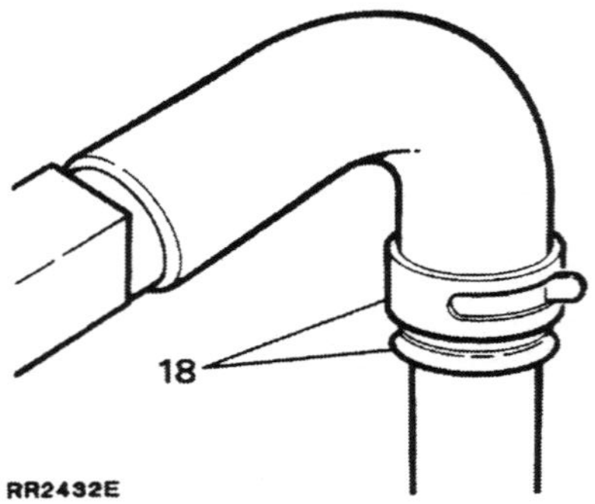

### Planetary set (Epicyclic unit) and annulus - Assemble

21. Lubricate the planetary set and annulus with clean oil.
22. Position the annulus around the planetary set, fit the assembly to the rear cover locating the oil pump inlet port in the groove at the bottom of the planetary set housing, the sun gear shaft in the bearing and the lugs on the outer edge of the annulus in the anti - rotation lugs. It may be necessary to tap the sun gear shaft into the bearing to enable the large ring gear snap ring to be fitted.
23. Fit the snap ring with the stepped ends adjacent to the selector shaft bore. 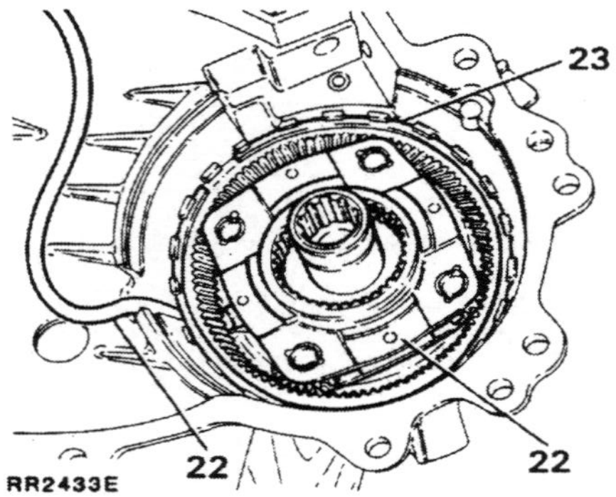
24. Turn the rear cover over and fit the circlip to retain the sun gear shaft.
25. Remove any previous sealant from the end cap. Apply Dow Corning 732 silicon sealant or a suitable equivalent to the inner edges of the cap, evenly tap the cap into position.
26. If necessary re - position the filter on the oil pick - up pipe until the lug on the filter can be pushed into the slot in the rear cover.

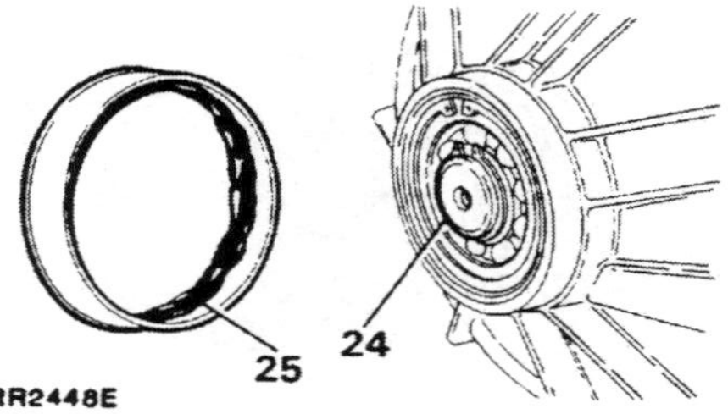

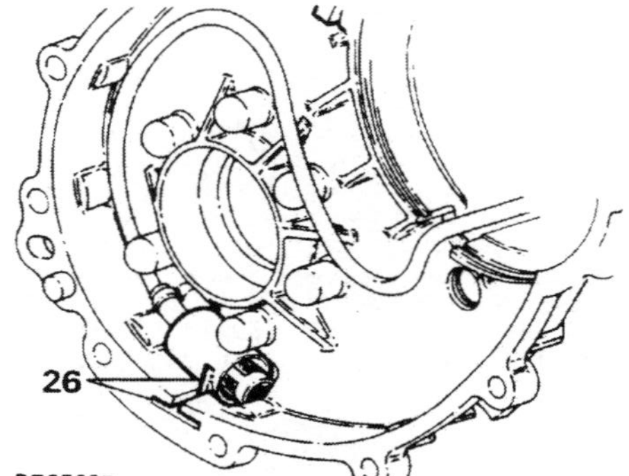

Continued

## BORG WARNER TRANSFER GEARBOX

### Selector fork
### - Dismantle and inspection

1. Remove the retaining clip and separate the fork from the arm.
2. Detach the two nylon slippers from the selector fork feet and discard.
3. Thoroughly clean all components.
4. Examine the fork, arm and pivot pin for wear.
5. Remove the 'O' ring and discard. Examine the shaft and lever for wear and damage, renew as necessary.
6. Examine the selector sleeve teeth and internal splines for damage and wear. Renew as necessary.

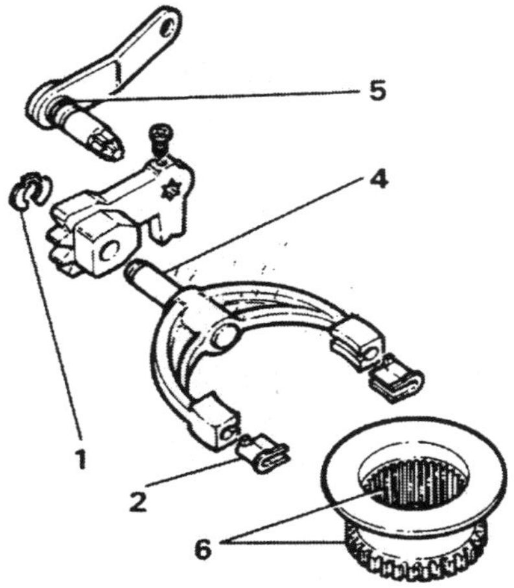

### Selector fork - Assemble

7. Fit new nylon slippers to the fork.
8. Assemble the fork to the selector arm and secure in position using a new retaining clip.
9. Lightly lubricate and fit the spring and selector plunger.

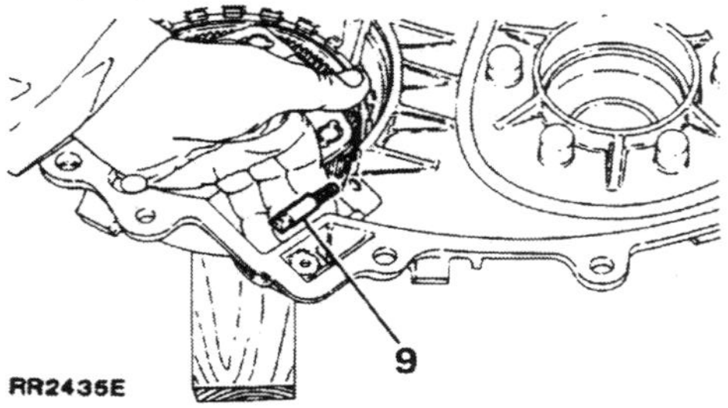

10. While compressing the plunger and spring, fit the selector fork, operating arm assembly and selector sleeve simultaneously.
11. Select neutral gear position at the operating arm.
12. Fit a new 'O' ring to the selector lever shaft. Lubricate the 'O' ring and fit the lever assembly to the rear cover, noting that when fully assembled the lever should lie parallel with the joint face of the rear cover.
13. Fit a new retaining clip to secure the selector fork operating arm to the selector lever shaft.
14. Remove any previous sealant from the Torx screw. Align the selector lever shaft groove to the retaining screw hole, apply a small amount of Loctite 242 to the screw threads and using Torx bit 25 fit and tighten the screw to the specified torque. Ensure that the screw locates in the groove of the shaft.
15. Clean any previous sealant from the Torx screw. Apply a small amount of Loctite 242 to the threads of the screw and fit to the selector fork operating arm, tighten using Torx bit 25 to the specified torque.

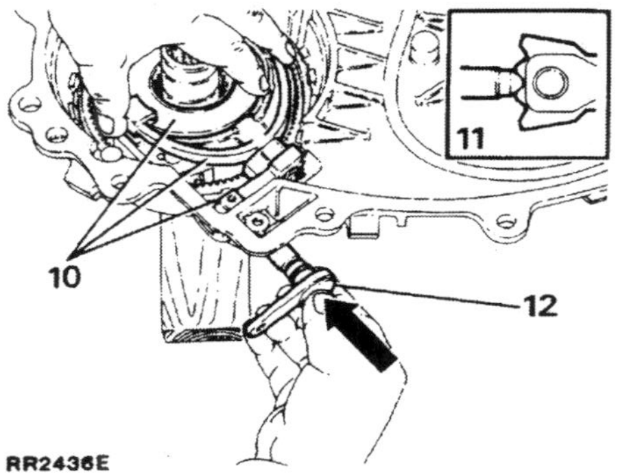

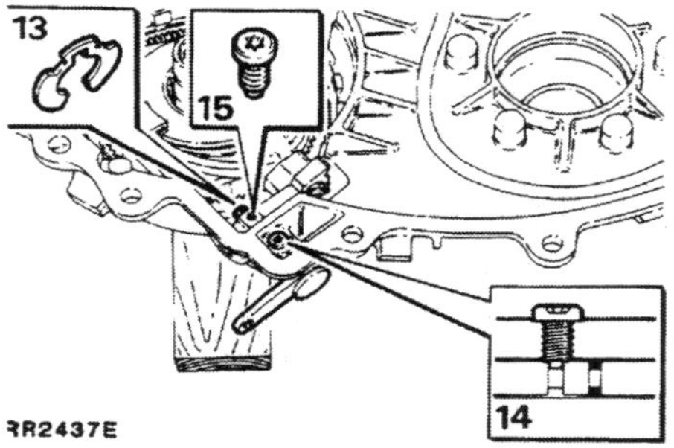

### Bearing carrier
### - Dismantle, inspection and assemble

1. Remove the circlip and drive or press the transfer shaft from the bearing.
2. Remove the circlip retaining the bearing in the carrier.
3. Drive or press the bearing from the carrier and discard the bearing.
4. Clean and examine the carrier for cracks and general condition. Renew as necessary.
5. Press or drive a new bearing into the carrier and secure with the circlip.

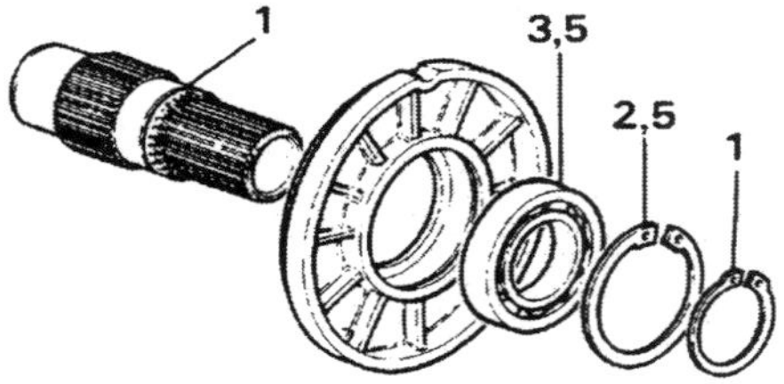

### Transfer shaft
### - Inspection and assemble

1. Clean the transfer shaft.
2. Visually examine the external splines for damage and wear, if worn fit a new the component.
3. Check the phosphor bronze bush for wear by measuring the internal diameter of the bush with internal calipers and a micrometer or with an internal micrometer. The bush diameter must not exceed 38.515 mm (1.516 in) fit a new transfer shaft if the bush has worn above the figure given.

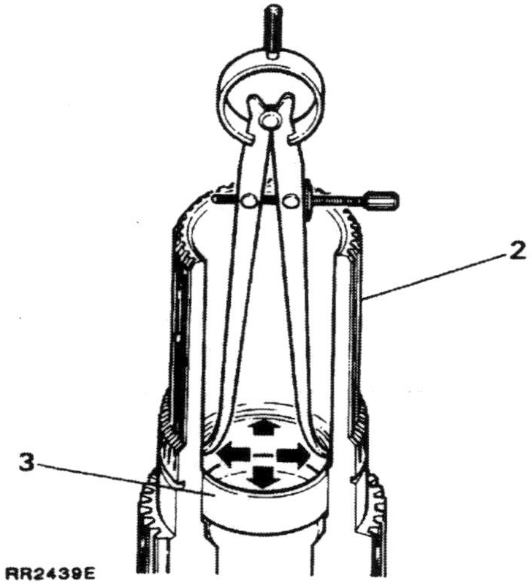

Continued

419

# BORG WARNER TRANSFER GEARBOX | 37

4. Drive or press the transfer shaft into the bearing in the carrier. Secure the shaft with the circlip.
5. Fit the carrier to the rear cover, fit the anti-rotation dowel and secure the assembly with the snap ring, noting that the open ends of the snap ring must be positioned by the cast relief in the bearing carrier upper face.

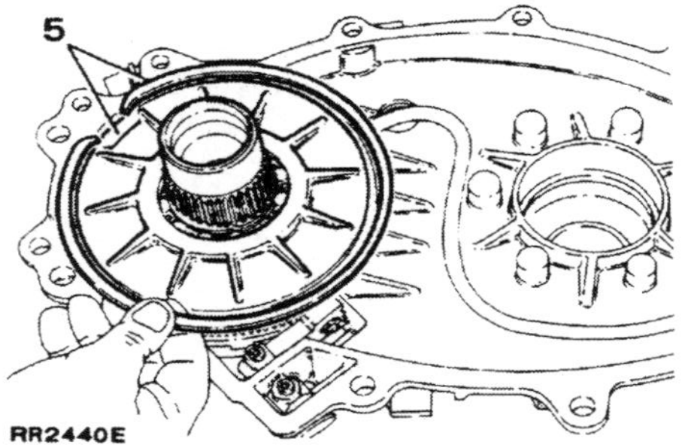

## Transfer sprocket - Inspection

1. Examine the sprocket teeth and splines for wear and damage, if either are evident discard the sprocket, otherwise clean and place to one side.

## Chain - Inspection

**NOTE: A stretched chain can be identified by either excessive noise when the gearbox is operational or by witness marks on the inside edges of the case. If either is evident, renew the chain.**

1. Using a suitable solvent thoroughly clean the chain.
2. Check the chain links for wear and damage, if necessary renew the chain.
3. Place the chain to one side.

## Centre differential and sprocket
## - Dismantle and inspection

1. Place the differential unit in a vice fitted with soft jaws. If the original components are to be refitted mark an identification line on the sprocket and differential unit.
2. Remove the bolts securing the sprocket to the differential.
3. Lift the differential assembly from the sprocket.
4. Examine the sprocket teeth for wear and damage, if either are evident renew the sprocket. Place the sprocket aside until the differential is ready to be assembled.

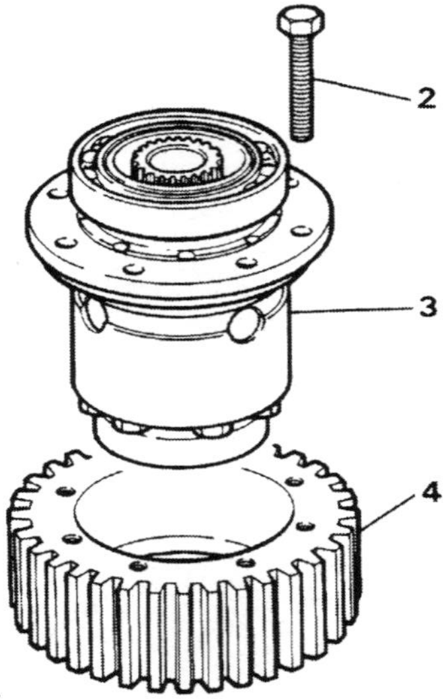

## Centre differential
## - Dismantle and inspection

1. Using a two legged puller, ease the bearings from the differential assembly and discard the bearings.

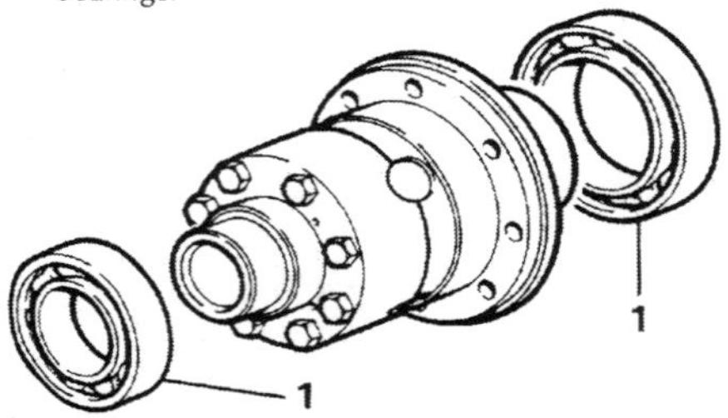

2. Secure the front half of the differential unit in a vice fitted with soft jaws, remove the eight retaining bolts securing the front and rear halves of the assembly together, lift off the rear part of the differential unit. Note the identification marks on the exterior of the differential unit.

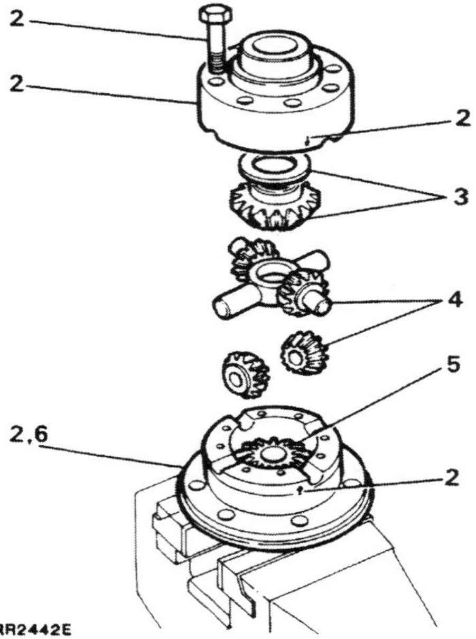

3. Remove the rear upper bevel gear and thrust washer.
4. Remove the pinion gears and dished washers along with the cross shaft.
5. Remove the front lower bevel gear and thrust washer from the front half of the differential unit.
6. Remove the front half of the differential unit from the vice and clean all components. Examine for wear or damage, renew if necessary.

## Differential pinions - rolling resistance

7. Using soft jaws secure the front half of the differential unit in the vice.
8. Fit the front bevel gear without the thrust washer. Lightly lubricate and fit the cross shaft, pinion gears and new dished washers.
9. Fit the rear bevel gear together with the thinnest thrust washer to the rear half of the differential. Assemble both halves of the differential noting the identification marks. Fit the bolts and tighten to the specified torque.

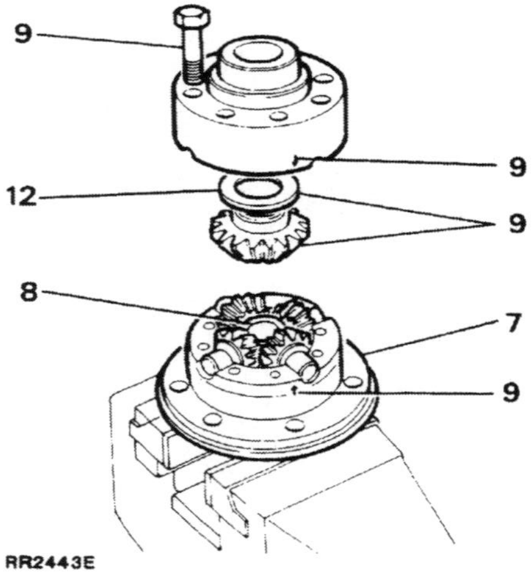

Continued

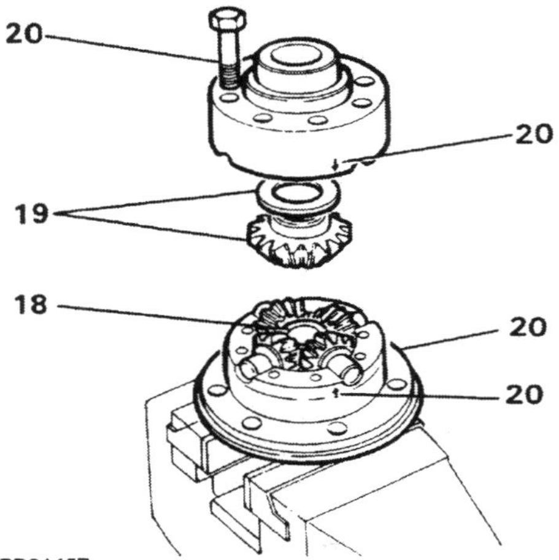

# BORG WARNER TRANSFER GEARBOX

10. Invert the differential unit in the vice, fit the front output housing to the differential, locating the viscous unit splines on the front bevel gear. Fit the drive flange to the viscous unit and place the brake drum on top of the drive flange, secure with the nut. Check that the gears are free to rotate.
11. Tie a length of string around the brake drum, attach a spring balance to the free end and carefully tension the string until a load to turn is achieved. Alternatively use a torque wrench applied to the drive flange nut. Rotate the brake drum slowly by hand to overcome the initial load when using either method.

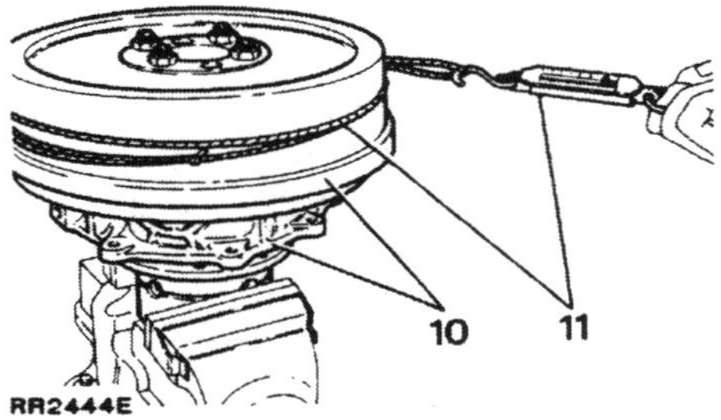

NOTE: Gears that have been run will rotate smoothly and will require a torque of 0.56 Nm (5 in lb), equivalent force using a spring balance 0.45 kg (1 lb). New gears will rotate with a notchy feel and will require a torque of not more than 2.26 Nm (20 in lb), equivalent force using a spring balance 7.72 kg (3.8 lb). Keep all components lubricated when carrying out these adjustments.

12. Change the thrust washer for a thicker one if the torque reading is too low and re - check the torque. Five thrust washers are available in 0.10mm steps ranging from 1.05 to 1.45mm.
13. Dismantle the unit when the rear bevel gear thrust washer has been selected.
14. Remove and retain the rear bevel gear and thrust washer combination.
15. Repeat the procedure to obtain the correct thrust washer for the front bevel gear, it is not necessary to fit the rear bevel gear when checking the front bevel gear rolling resistance.
16. When the thrust washer has been selected for the front bevel gear, again dismantle the differential unit and retain the thrust washer and front bevel gear combination.

## Centre differential - Assemble

17. Fit the thrust washer and front bevel gear into the front half of the differential unit.
18. Fit the pinion gears with dished washers to the cross shaft and fit the assembly to the differential unit.
19. Fit the thrust washer and rear bevel gear to the rear half of the differential unit.
20. Align both halves of the differential noting the identification marks. Secure both halves together with the eight bolts. Tighten the bolts to the specified torque.
21. Check the overall torque required to turn the differential, this should be approximately equal to both bevel gears added together.
22. Drive or press new bearings onto the differential, noting that the smaller of the two bearings is fitted to the rear half of the differential.

## Centre differential sprocket - Assemble

1. Fit the sprocket to the differential noting that the face of the sprocket with the relieved threads must contact the flange of the differential housing. Observe the previously marked identification lines if the original components are being refitted.
2. Fit new bolts and tighten evenly to the specified torque.

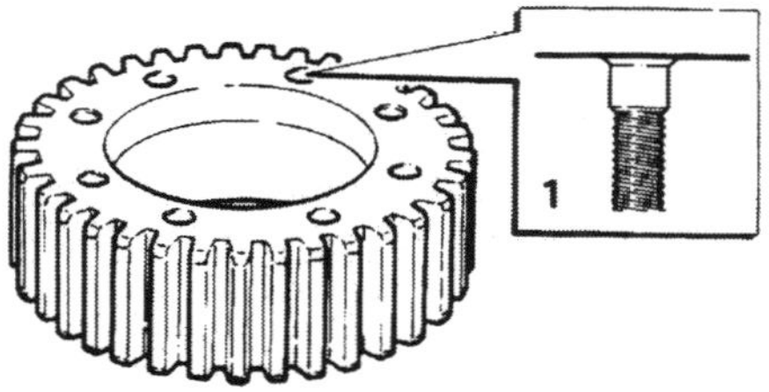

## Transfer sprocket, centre differential assembly and chain - Assemble

1. Place the differential assembly and transfer sprocket inside the chain. If the original components are being refitted observe the identification marks previously applied to the chain and transfer sprocket. Fit the complete assembly simultaneously.
2. Carefully tap the differential bearing into its bore while easing the transfer sprocket onto the transfer shaft.
3. Ensuring that the transfer sprocket is fully down, secure the sprocket to the transfer shaft with the circlip.

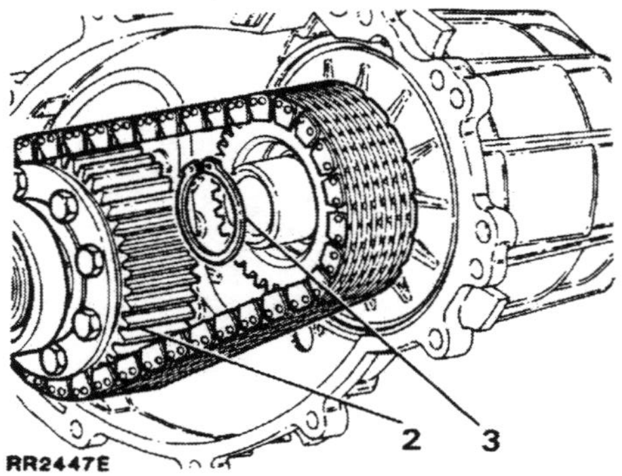

Continued

# BORG WARNER TRANSFER GEARBOX 37

## Front and rear cover - main casing - Assemble

1. Ensuring that the joint faces of the front and rear covers are clean, apply a bead of Dow Corning 732 or a suitable equivalent silicone sealant to the joint face of the rear cover and evenly spread the sealant over the face. Do not over apply the sealant.
2. Fit the front cover, secure with the eleven bolts, tightening evenly to the specified torque. Do not wipe away the surplus sealant which is forced out of the joint.

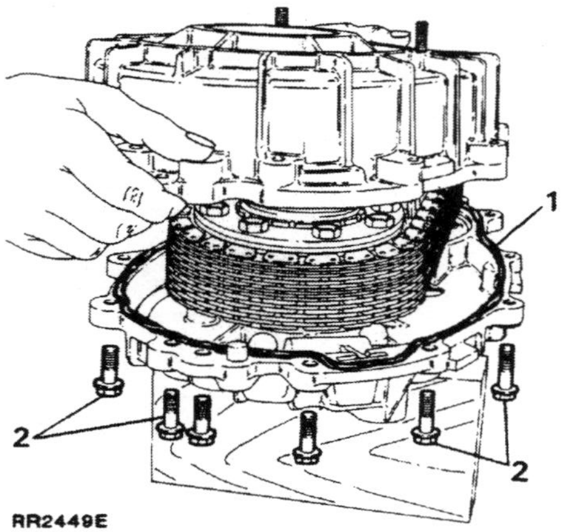

RR2449E

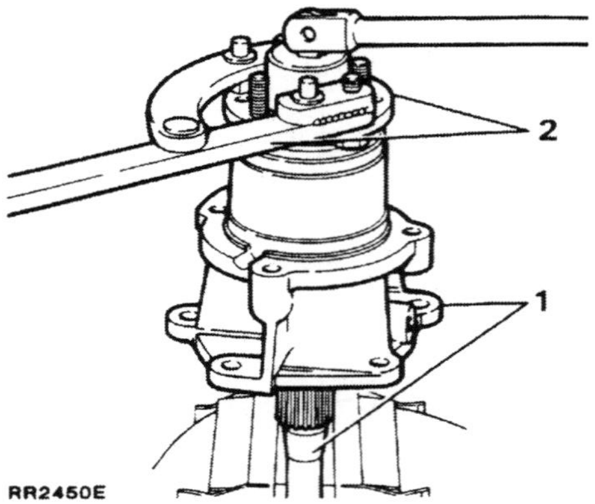

## Rear output housing
### - Dismantle and inspection

Service tools:
18G1422 - Oil seal replacer
18G1205 - Adjustable flange holding wrench

1. Support the rear output housing by the output shaft in a vice fitted with soft jaws.
2. Using service tool 18G1205 to restrain the drive flange, release and remove the nyloc nut and plain washer securing the drive flange to the output shaft, withdraw the rubber seal. Discard the nut and seal.

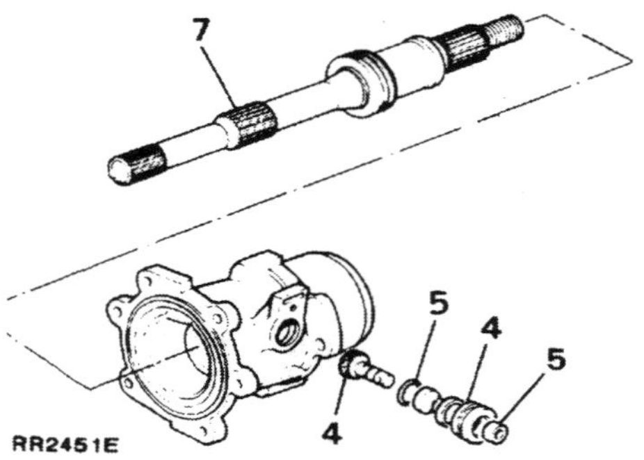

RR2451E

3. Remove the drive flange from the output shaft. Examine the flange for damage or wear particularly the seal running surface, if the surface is corroded or a groove has been worn by the previous seal discard the flange.
4. Prise the speedometer sleeve and driven gear from the housing. Examine the gear teeth for wear, if worn discard the gear.
5. Prise the oil seal from the sleeve and remove the 'O' ring, discard both the seal and 'O' ring.
6. Clean the sleeve and place to one side.
7. Drive or press the output shaft from the housing.

RR2450E

8. Clean and examine the splines and speedometer drive gear for wear or damage. The output shaft can be further dismantled if either the speedometer drive gear or output shaft is worn: remove the circlip and slide the gear from the shaft, retrieve the ball bearing from the indent in the shaft. Discard the worn component.

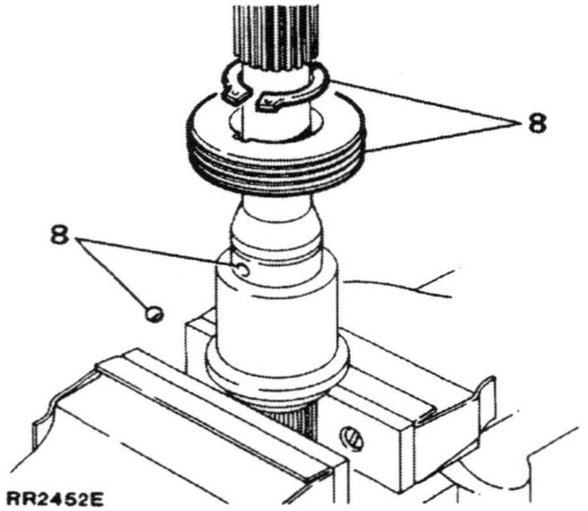

RR2452E

NOTE: While the output shaft is removed from the rear output housing, the shaft can be utilised for checking the rolling resistance of the viscous unit as follows.

## Viscous unit - rolling resistance
### Bench check

NOTE: Testing should be carried out in an ambient of 20 °C.

9. Secure the output shaft in a vice fitted with soft jaws, gripping the shaft on the drive flange splines.
10. With the viscous unit still installed in the front output housing place the assembly on the rear output shaft spline.
11. Apply a torque of 27 Nm (20lb ft) to the output flange nut, if no resistance to turn is felt, the viscous unit requires replacing.
12. If resistance to turn is felt, apply a torque of 20 Nm (20lb ft), to the output flange nut for 1 minute, this should result in a rotation of approximately 25° - 30°. If this does not occur, the unit requires replacing.

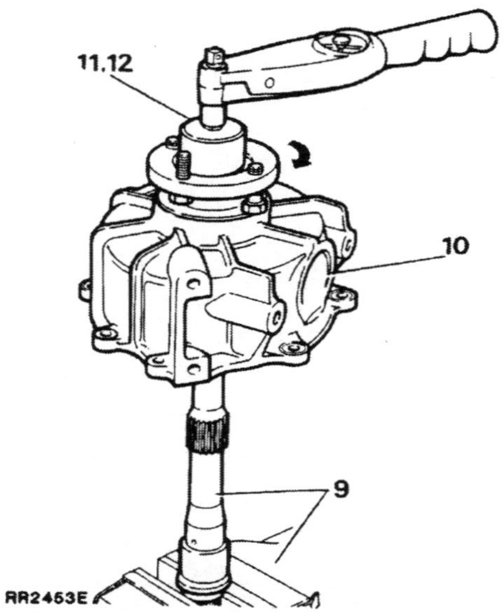

RR2463E

## Rear output housing (continued)

13. Lever off the dust shield.
14. Prise the oil seal from the housing and discard the seal.
15. Remove the circlip retaining the bearing.
16. Drive or press the bearing from the housing. Discard the bearing.
17. Remove any previous sealant from the housing joint face.

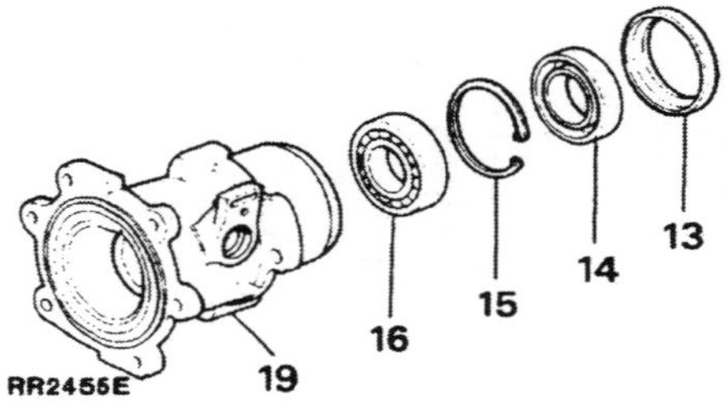

RR2456E

Continued

# BORG WARNER TRANSFER GEARBOX

18. Thoroughly clean all components with a suitable solvent.
19. Examine the housing for damage and wear. Renew as necessary.

### Rear output housing - Assemble

20. Drive or press a new bearing into the housing until the bearing contacts the shoulder.
21. Fit the circlip.
22. Lubricate a new oil seal. Using oil seal replacer 18G1422 fit the seal, lip side leading until it contacts the circlip.
23. Fit the dust shield.
24. Place the ball bearing in the indent on the output shaft, fit the speedometer drive gear to the shaft, secure together with the circlip.
25. Press or drive the output shaft into the housing until the shoulder of the shaft contacts the bearing.
26. Lubricate the oil seal bearing surface of the drive flange and fit the flange followed by a new rubber seal. Fit the steel washer and secure the flange to the shaft using a new nut. Tighten to the specified torque.
27. Lubricate a new speedometer sleeve oil seal, press the seal into the top of the sleeve.
28. Fit a new 'O' ring to the outside of the sleeve, push the driven gear spindle into the sleeve.
29. Lubricate the 'O' ring and push the sleeve and gear assembly into the housing. It may be necessary to rotate the output shaft to ensure that the driven gear engages with the drive gear on the shaft.
30. Apply Dow Corning 732 or a suitable equivalent silicone sealant to the rear output housing joint face on the main casing. Evenly spread the sealant on the face to ensure a good seal.
31. Fit the housing to the main casing and secure with the six bolts tightened to the specified torque.

## Front output housing
### - Dismantle and Inspection

Service tools:
18G1422 - Oil seal replacer.
18G1205 - Adjustable flange holding wrench

1. Support the viscous unit and front output housing in a vice fitted with soft jaws gripping on the two flats of the viscous unit.
2. Using service tool 18G1205 to restrain the drive flange, release and remove the nyloc nut and plain washer securing the drive flange to the output shaft, withdraw the rubber seal. Discard the nut and seal.
3. Remove the drive flange from the viscous unit. Examine the flange for damage or wear particularly the seal running surface, if the surface is corroded or a groove has been worn by the previous seal discard the flange.
4. If necessary the oil catcher can be carefully pressed from the drive flange, if either a new oil catcher or bolts are being fitted.
5. Carefully tap the viscous unit out of the housing. If the original unit is being refitted wipe clean with a clean cloth.

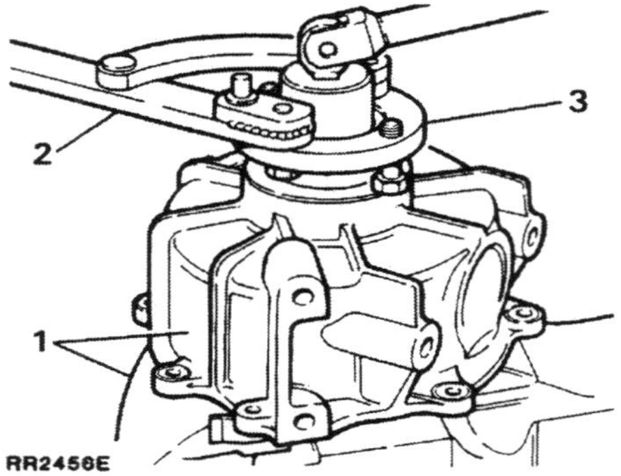

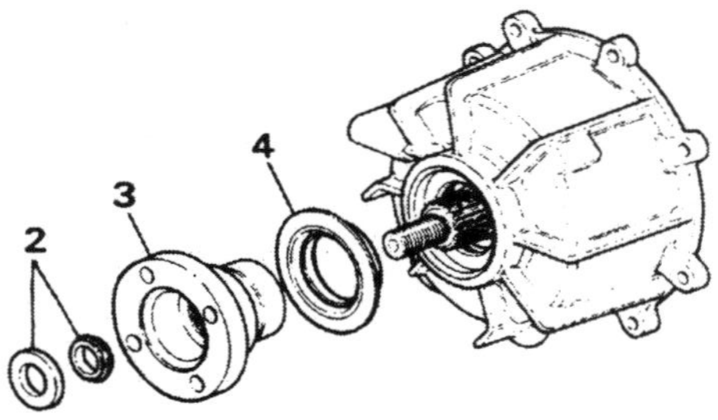

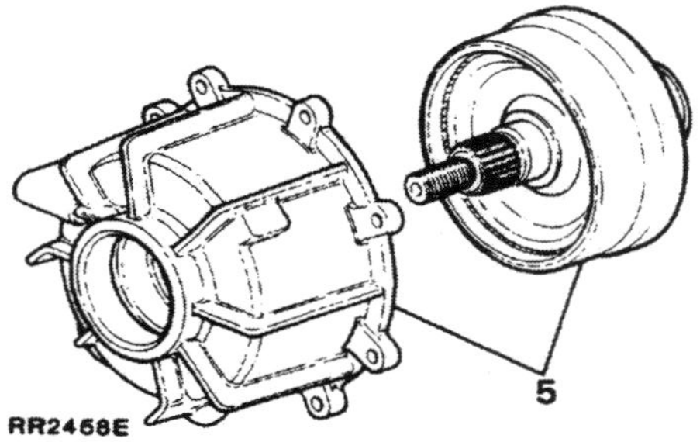

**NOTE: The viscous unit is a sealed assembly and cannot be further dismantled, a new unit should be fitted if the unit is damaged or if the torque to turn is out of limits.**

6. Prise the oil seal out from the front output housing and discard.
7. Remove the circlip retaining the bearing.
8. Drive or press the bearing from the housing and discard.
9. Clean the housing with a suitable solvent.
10. Remove any previous sealant from the joint face of the housing.
11. Examine the housing for damage and wear, renew the housing if necessary.

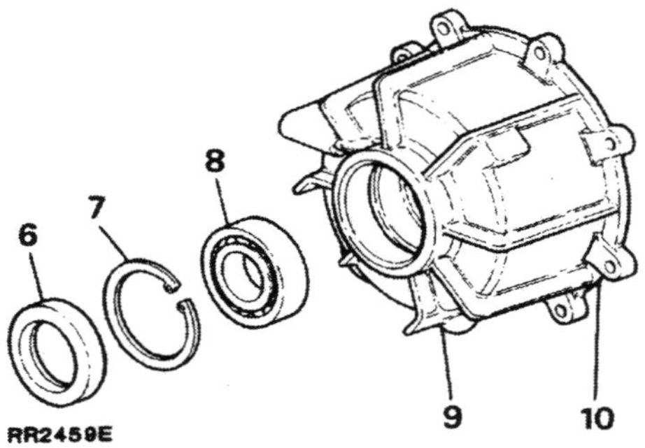

### Front output housing - Assemble

12. Drive or press a new bearing into the housing.
13. Fit the circlip to retain the bearing.
14. Lubricate a new oil seal. Using oil seal replacing tool 18C1422 fit the seal, lip side of the seal leading until it contacts the circlip.
15. Carefully tap the original or new viscous unit into the housing until contact is made with the face of the bearing.
16. Lubricate the lips of the seal and fit the flange followed by a new rubber seal; fit the steel washer and secure the flange with a new nut. Tighten to the specified torque.
17. Apply Dow Corning 732 or a suitable equivalent silicone sealant to the output housing joint face of the main casing. Evenly spread the sealant on the face to ensure a good seal.
18. Fit the housing to the main casing and secure in position with the eight bolts tightened evenly to the specified torque.
19. Refit the gearbox to the vehicle. (Refer to transfer gearbox remove and refit).

# LT77 FIVE SPEED GEARBOX 37

## LT77 FIVE SPEED GEARBOX AND BORG WARNER TRANSFER BOX

### Remove and refit

To assist in the removal of the transmission assembly it is necessary to manufacture an adaptor plate to use in conjunction with a transmission hoist.

**NOTE:** Four holes (A) to be countersunk on underside suit hoist.

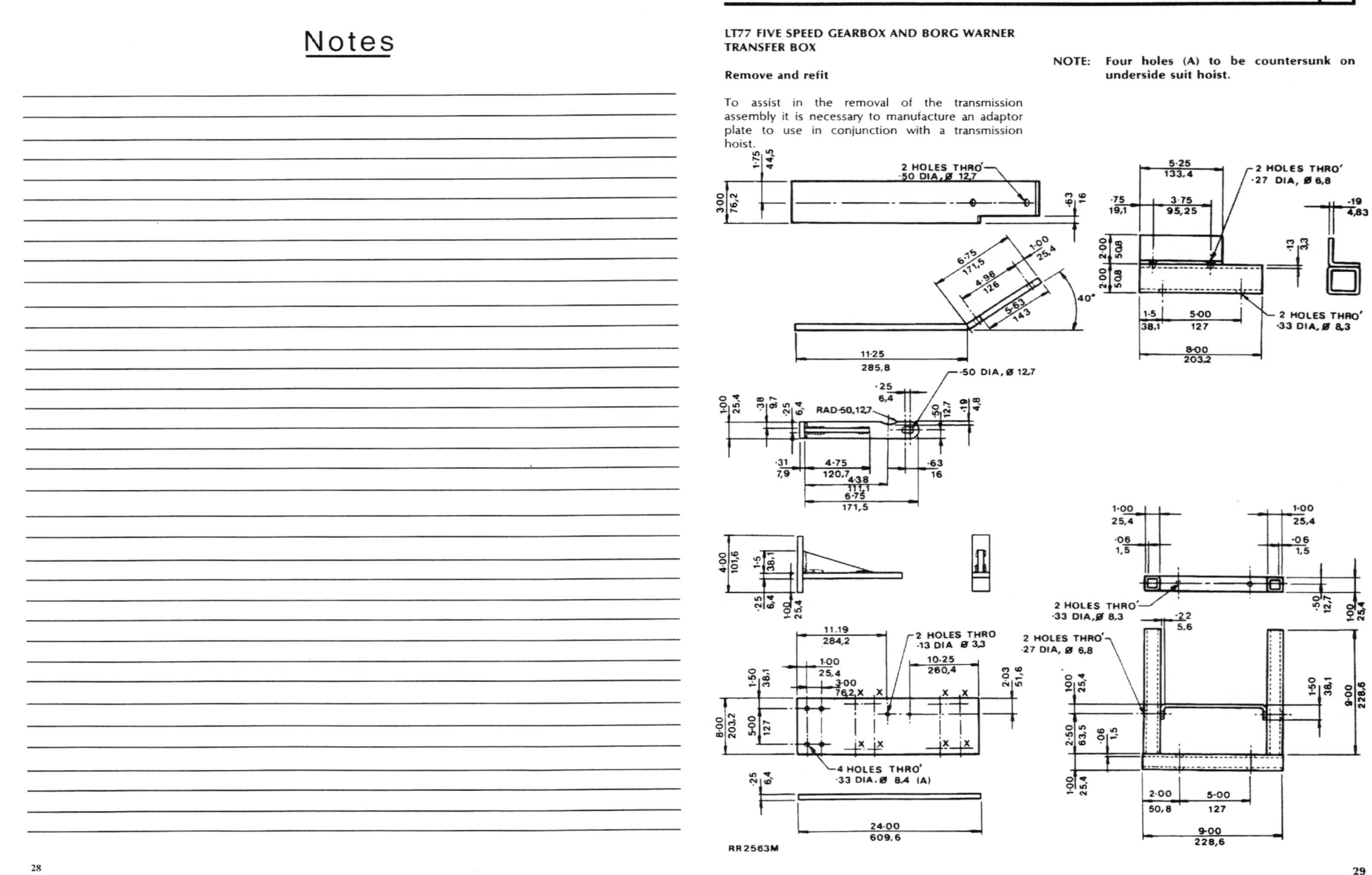

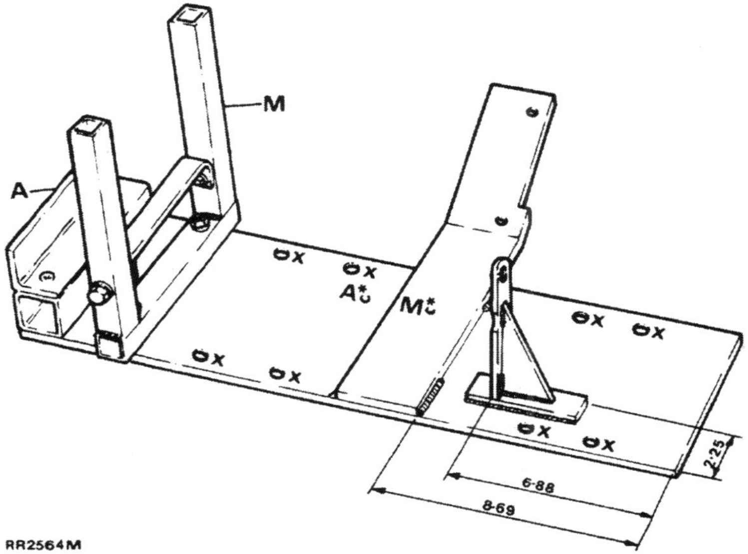

# LT77 FIVE SPEED GEARBOX

RR2564M

A: Automatic gearbox models
M: Manual gearbox models
A*: Centre of lifting hoist(automatic models)
M*: Centre of the lifting hoist (manual models)
X: Drill fixing holes to suit hoist table

Material: Steel plate BS1449 Grade 4 or 14.

## Preparation-under bonnet

1. Install the vehicle on a ramp and chock the wheels.
2. Disconnect the battery negative lead.
3. Release and remove the fan blade assembly. Note that the nut securing the viscous unit has a left hand thread, released by turning clockwise when viewed from the front.
4. **Fuel injection models** - Disconnect the airflow meter to plenum chamber hose.
5. **Carburetter models only** - Remove the air cleaner from its location.

## Inside vehicle

6. Remove the main gear lever knob, select low range and remove transfer gear lever knob.
7. Release four screws and remove the glove box liner.
8. Carefully prise the front of the glove box panel complete with switches back through the panel aperture and place on transmission tunnel.
9. Carefully prise the centre panel around the main gear lever out of the floor mounted console.
10. Release the two bolts and two screws securing the glove box/console assembly to the transmission tunnel.
11. Detach the two relay blocks from their mounting inside the glove box.
12. Disconnect the leads to the rear cigar lighter.
13. Release the handbrake, pull the gaiter forward to gain access to the clevis pin. Remove split pin, washer and clevis pin to release inner cable from handbrake lever.
14. Carefully manoeuvre the glove box assembly rearwards (while raising the handbrake lever to its uppermost position) away from the radio housing and remove the glove box assembly from the vehicle.
15. Remove the sound deadening pad from the top of the transmission tunnel.
16. Slacken the pinch bolt and remove the upper gear lever.
17. Remove the screws and detach the high low lever and main gearlever retaining plates.

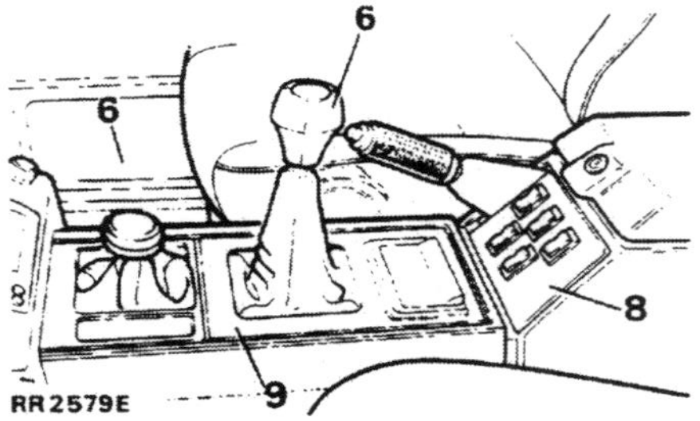

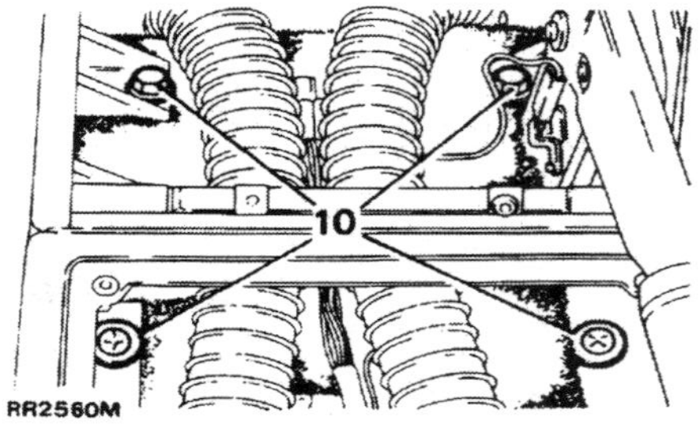

## Underneath vehicle

18. Raise the ramp.
19. Place a suitable container under the transmission, remove the transfer gearbox, main gearbox and extension housing drain plugs, allow the oil to drain and refit the plugs. Clean the filter on the extension housing plug before refitting.

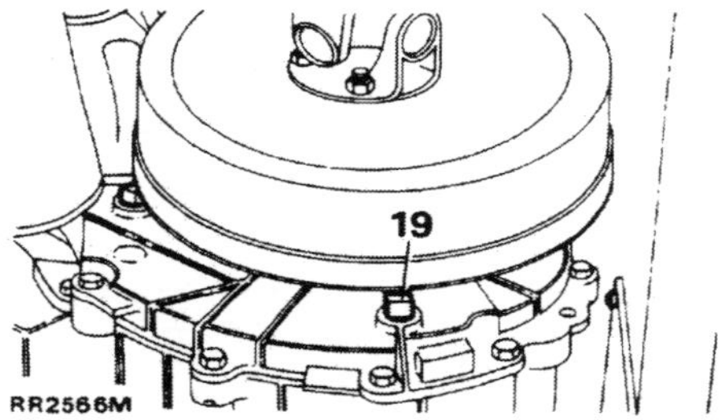

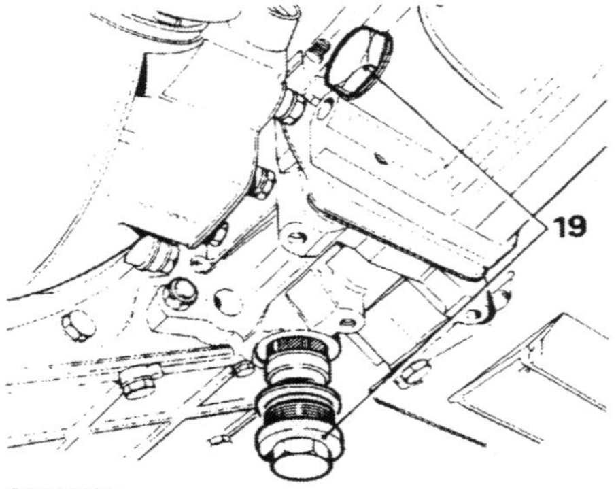

# LT77 FIVE SPEED GEARBOX

20. Remove the eight nuts and bolts securing the chassis cross member and remove the cross member. Use a suitable means of spreading the chassis, if necessary.
21. Remove the front exhaust downpipes and intermediate pipe with centre silencer.

   **NOTE: Operation 21 will require the assistance of a second operator to support the exhaust system while the fixings are released.**

22. Mark each drive flange for reassembly and disconnect the front and rear propeller shafts from the transfer box. Tie the shafts to one side.
23. Release the clamp and disconnect the speedometer cable from the rear output housing, also free the cable from its clip at the left hand side of the transfer gearbox. Tie the cable to one side.
24. Remove the two bolts and withdraw the clutch slave cylinder from the bell housing.

## Remove the transmission assembly

**WARNING: Where the use of a transmission hoist is necessary, it is ABSOLUTELY ESSENTIAL to follow the hoist manufacturer's instructions to ensure safe and effective use of the equipment.**

25. Position a suitable transmission hoist on the rear output housing or brake drum to support the weight of the transmission assembly.
26. Remove the fixings and withdraw the transfer gearbox mountings.
27. Secure the previously manufactured fixture to a transmission hoist, raise the hoist and position the fixture under the transfer box mounting points.
28. Secure the fixture to the transfer box mounting points using the original mounting bolts.
29. Remove the transmission hoist from the rear of the transfer box.
30. Carefully lower the transmission until the top of the transfer gearbox clears the rear passenger floor.
31. Disconnect the handbrake cable by removing the split pin, washer and clevis pin.
32. Remove the clip securing the handbrake outer cable to the support bracket, feed the cable through the bracket, and tie cable to one side.
33. Position the transmission hoist under the engine to support the weight while removing bellhousing bolts.
34. Remove the bolts from the bell housing.
35. Ensuring all connections to the engine and chassis are released, withdraw the transmission.

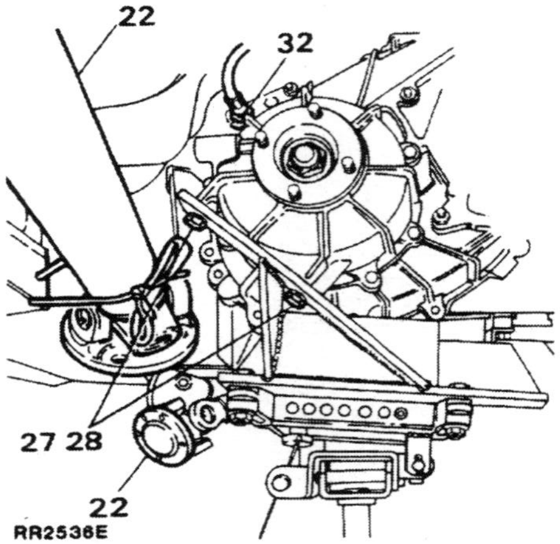

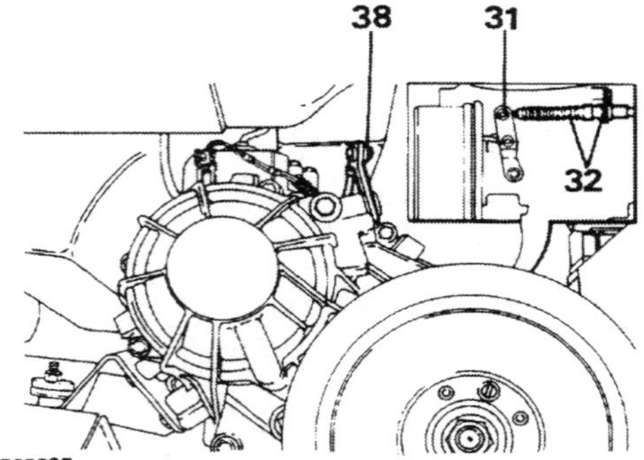

## Separating the transfer box from gearbox

36. Remove the transmission assembly from the hoist and cradle and install it safely on a bench.
37. Place a sling round the transfer box and attach to a hoist.
38. Detach the high low link from the transfer gearbox selector lever and remove the breather pipe.
39. Remove the upper and lower bolts and two nuts retaining the transfer box to the extension housing and withdraw the transfer box.

## Assembling transfer box to main gearbox

40. Stand the gearbox vertically on the bell housing face on two pieces of wood to prevent damage to the primary pinion which protrudes beyond the bell housing face. Lower the transfer gearbox onto the main gearbox, care should be taken to prevent any damage to seals. Secure the transfer gearbox to the main gearbox and tighten all bolts to the correct torque.
41. Refit the breather pipe and selector link.

## Transfer gearbox high/low link adjustment

42. Ensure that the selector lever at the transfer gearbox is in the neutral position.
43. Set the transfer gearbox lever in a vertical position (at right angles to the centre line of the main gearbox). Rotate the fork end of the rod until the holes align with the hole in the selector lever.
44. Fit the clevis pin and retaining clip. Select high and low transfer to ensure full engagement is obtained. Repeat the adjustment procedure if full engagement is not evident.

## Refitting

45. Fit the cradle to the transmission hoist and the transmission to the cradle. Smear Hylomar on bell housing mating face with engine.
46. Select any gear in the main and transfer gearbox to facilitate entry of the input shaft. Ensure that the clutch centre plate is in alignment.
47. Position and raise the hoist to line up with the engine, feed the handbrake cable through the aperture in the tunnel, ensure that any pipes or electrical leads do not become trapped.
48. Fit the transmission assembly to the engine and tighten the securing bolts.
49. Reverse the removal procedure noting the following points.
50. Tighten all fixings to the correct torque.
51. Check that the three drain plugs are tight and remove the main gearbox and transfer box filler level plugs. Fill both the main and transfer gearboxes with the recommended oil up to the level of the filler hole. Apply Hylomar sealant to the threads and fit the level plugs and wipe away any surplus oil.
53. Finally road test vehicle.

# LT77 FIVE SPEED GEARBOX

## LT77 FIVE SPEED GEARBOX

### OVERHAUL

The following overhaul procedure covers existing gearboxes and also the latest modified version. Improvements include increased capacity layshaft bearings and a new upper gearlever and gear knob. These modifications will improve durability and gearchange quality.

NOTE: Modified gearboxes are identified by the serial number prefix 'F'.

**Service Tools:**
18G705    - Puller - Bearing remover
18G705    - 1A - Adaptor for mainshaft
18G705    - 5 - Adaptor for layshaft
18G705    - 7 - Adaptor for layshaft - increased capacity bearings
18G1400   - Remover for synchromesh hub and gear cluster
18G1400   - 1 - Adaptor mainshaft fifth gear
MS47      - Hand press
18G47BA   - Adaptor, input shaft bearing
18G47BAX  - Conversion kit
18G284    - Impulse extractor
18G284AAH - Adaptor for input shaft pilot bearing track
18G1422   - Mainshaft rear oil seal replacer
18G1431   - Mainshaft fifth gear and oil seal collar replacer
18G1205   - Flange holder

### Locally manufactured tools

In addition to the service tools, the following items should be manufactured locally to facilitate overhauling the gearbox.

A. Dummy centre bearing, used for selection of first gear bush, material mild steel.

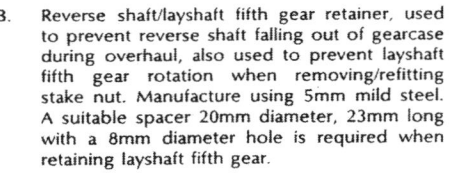

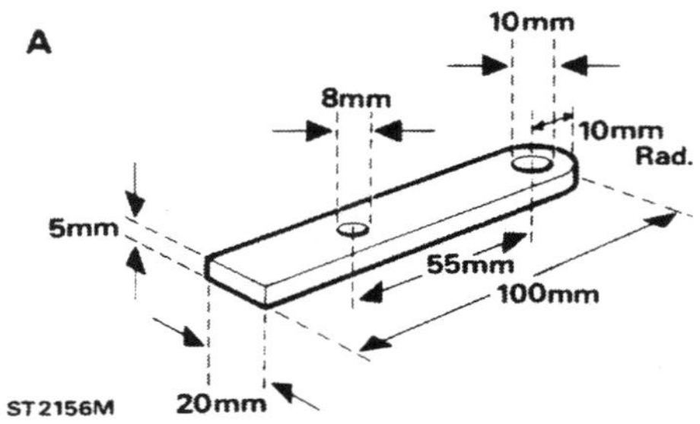

B. Reverse shaft/layshaft fifth gear retainer, used to prevent reverse shaft falling out of gearcase during overhaul, also used to prevent layshaft fifth gear rotation when removing/refitting stake nut. Manufacture using 5mm mild steel. A suitable spacer 20mm diameter, 23mm long with a 8mm diameter hole is required when retaining layshaft fifth gear.

C. Pilot studs, 4 off, threaded 8mm, used when separating centre plate from main gearcase.

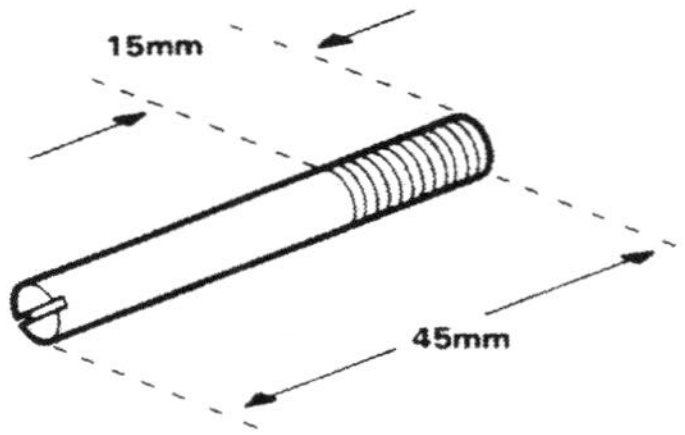

D. Layshaft support plate is fitted using two 8 x 25mm bolts and washers to the front of the gearbox case, it also supports the input shaft bearing outer track.

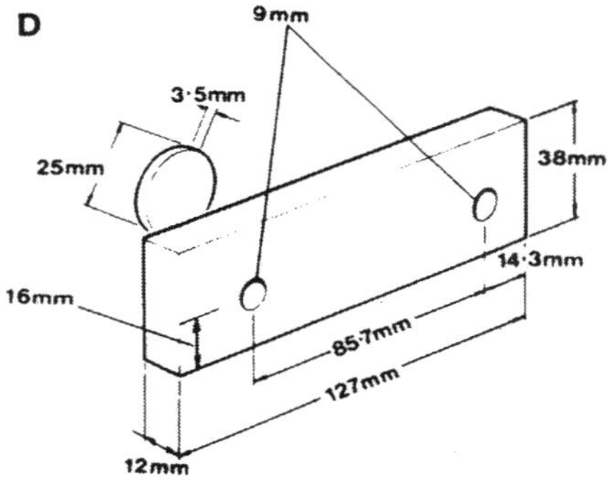

Continued

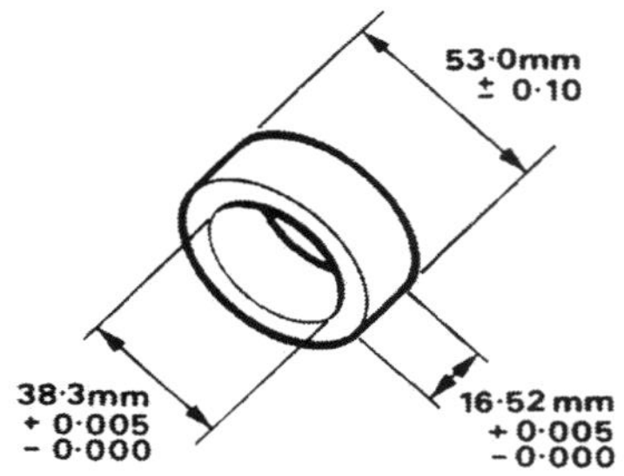

## LT77 FIVE SPEED GEARBOX

E. Gearbox workstand, securely locates gearbox unit during overhaul. Manufacture from 30mm x 30mm angle iron. Plan view of top is first scale. Gearbox security hole (A) is drilled 10mm through material. Four countersunk holes (B) are for gearbox location. Countersink using a 10mm drill. DO NOT drill through material.

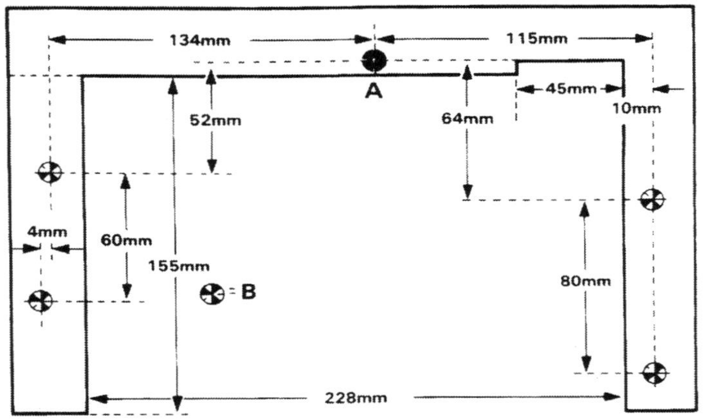

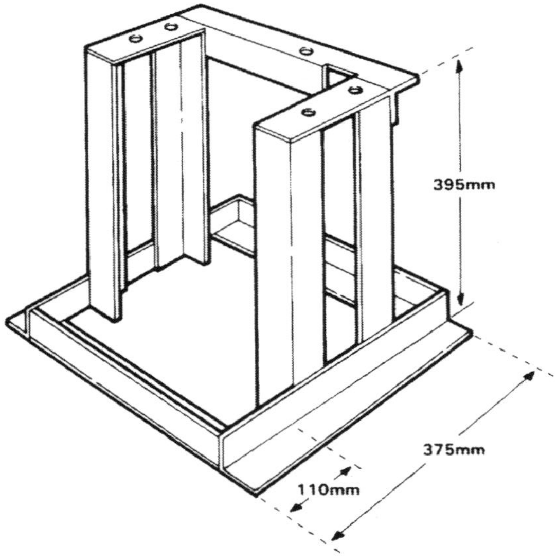

## LT77 FIVE SPEED GEARBOX

### Dismantle

1. Place gearbox on a bench with the transfer gearbox removed, ensuring the oil is first drained. Thoroughly clean the exterior of the gearbox case.
2. Remove the two pan head screws securing the gear change housing top cover. Raise the cover to give access to the gear change housing to extension case securing bolt adjacent to the reverse plunger assembly, and remove the bolt.
3. Remove the three remaining gear change housing to extension case securing bolts, remove the housing complete with transfer gear change assembly.
4. Remove the clutch release bearing plastic staple and remove the clutch release bearing.

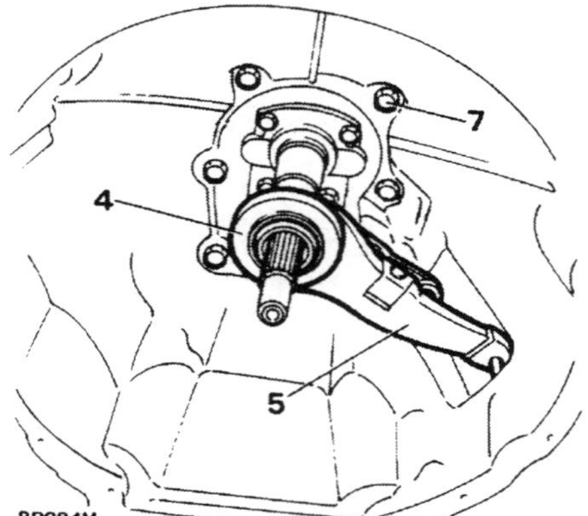

5. Release the single bolt and remove the spring clip from the clutch release lever. Pull the lever off the clutch release pivot.
6. If necessary: Remove the clutch release pivot and single bolt retaining the clutch release bearing guide. Withdraw the guide from the input shaft.

7. Remove the six bolts and washers securing the bell housing.
8. Carefully ease the bell housing off the dowels and withdraw it from the gearbox.
9. Remove two dowel tubes from the front of the casing and secure the gearbox unit to the workstand by one nut and bolt.

### Extension housing

1. Remove the circlip which retains the mainshaft oil seal collar located at the rear of the gearbox. Using tools 18G705 and 18G705 - 1A remove the oil seal collar.

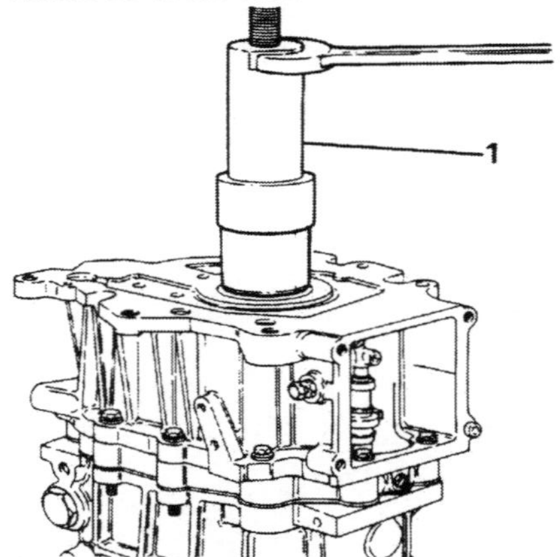

2. Remove the ten bolts and spring washers securing the fifth gear extension case to the gearcase. Carefully withdraw the extension case ensuring that the centre plate does not separate from the main case. Discard the gasket.

Continued

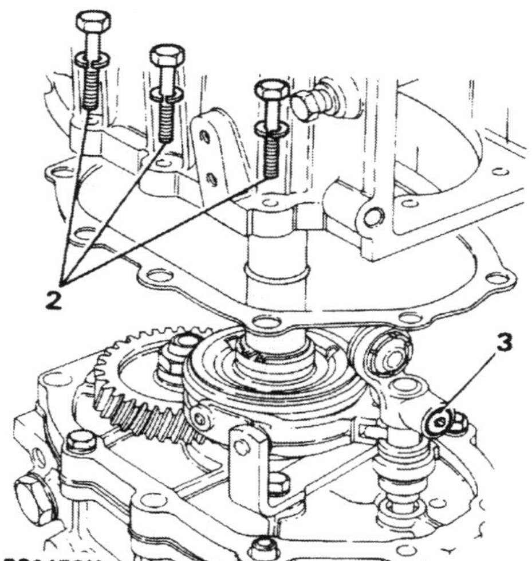

# LT77 FIVE SPEED GEARBOX

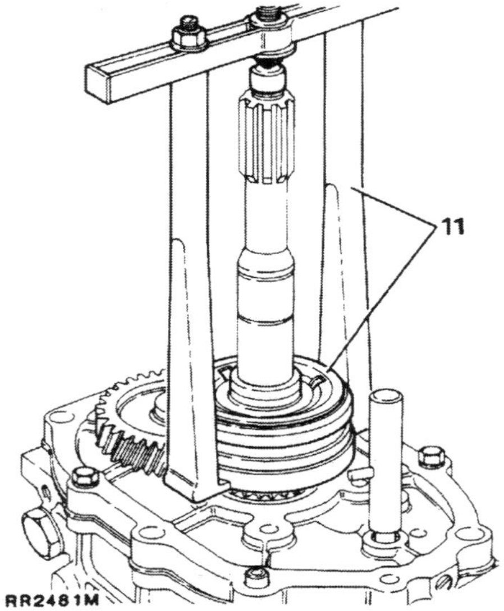

12. Using tools 18G705 and 18C705 - 1A, remove the layshaft fifth gear.

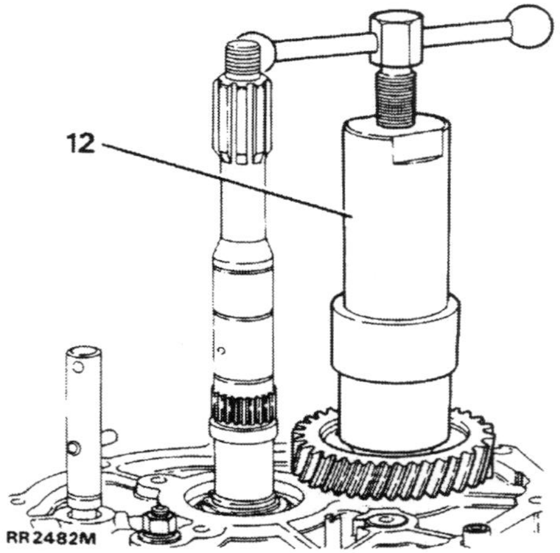

3. Remove the six bolts and spring washers from the front cover, withdraw the cover and discard the gasket.
4. Retrieve the input shaft and layshaft selective washers from the gearcase.
5. Remove the two bolts and washers and withdraw the retainer for the selector shaft front spool. Note that later models have an 'O' ring with a counterbore in the gearcase.

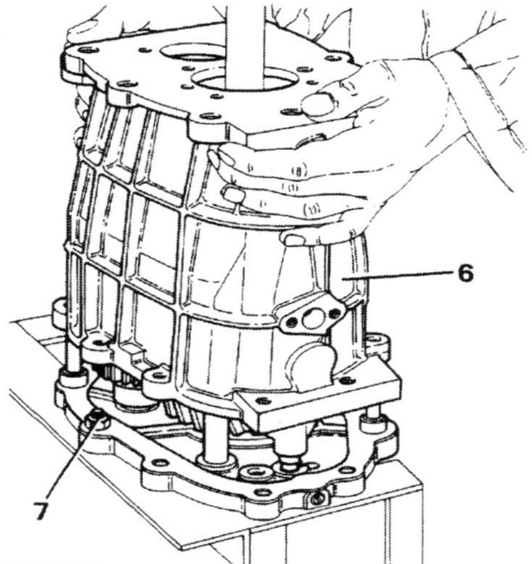

## Main gearbox case

1. Fit the reverse shaft retainer using one of the fifth gear bracket mounting bolts. Fit the four guide studs to the main gearbox case to locate in the workstand.
2. Release the gearbox from the workstand, invert the assembly and locate the four studs on the workstand.
6. Remove the slave bolts and carefully lift the gearcase, leaving the centre plate and gear assemblies in position.

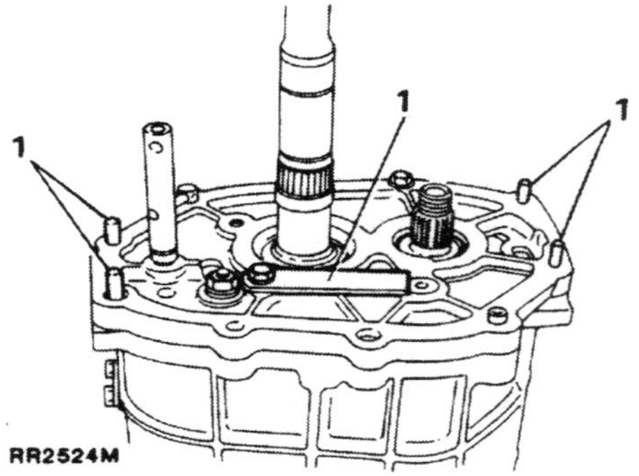

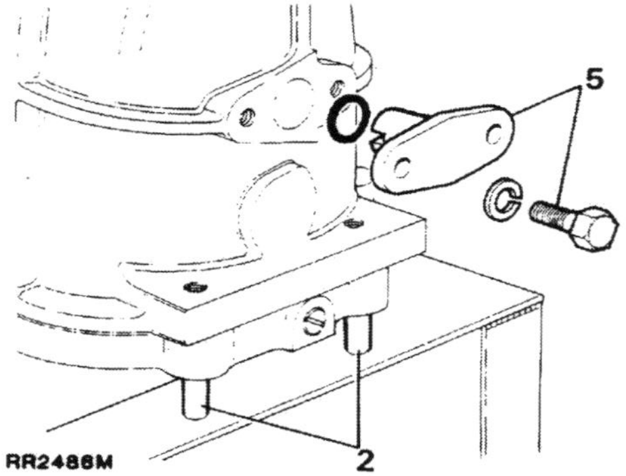

## LT77 FIVE SPEED GEARBOX

9. De-stake the nut securing fifth layshaft gear and remove the nut.
10. Release the circlip retaining the fifth gear synchromesh assembly to the mainshaft.
11. Using tools 18C1400 - 1 and 18C1400 withdraw the selective washer, fifth gear synchromesh hub and baulk ring, fifth gear (driven), spacer and split roller bearing from the mainshaft.

3. Slacken the socket head screw and remove the selector yoke from the selector shaft.

## Fifth gear

1. Fit two slave bolts (8 X 35 mm) to the casing to retain the centre plate to the main case.
2. Remove the oil seal collar 'O' ring from the mainshaft.
3. Withdraw the oil pump drive shaft.
4. Remove the two 'E' clips from the selector fork pivot pins retaining the 5th selector fork to its bracket and remove the pins, fork and pads.
5. Withdraw the fifth gear selector spool.
6. Remove the two bolts and spring washers and withdraw the fifth gear selector fork bracket.
7. Secure the flange holder 18C1205 to left hand side of gearcase.
8. Bolt the layshaft fifth gear retainer to the gearcase using the spacer. Insert a suitable length of bar, or a 10 mm bolt, through pierced hole in gear to prevent rotation.

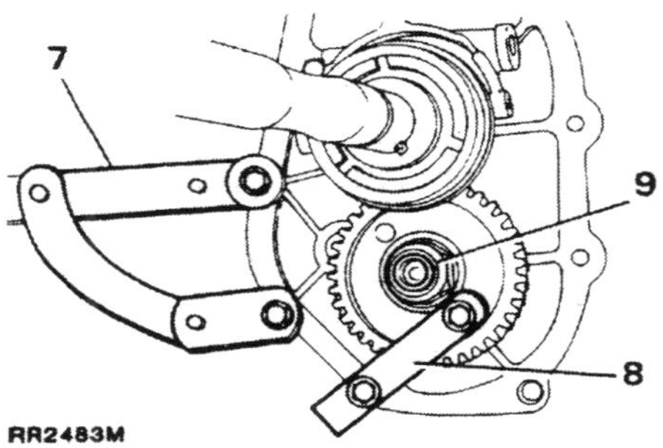

## LT77 FIVE SPEED GEARBOX | 37

7. Secure the centre plate to the workstand with a nut and bolt when the plate and casing have separated. Discard the gasket.

### Reverse shaft, layshaft and mainshaft

1. Release the reverse shaft retainer and remove the shaft.
2. Lift off the thrust washer, reverse gear and spacer from the centre plate.
3. Remove the pivot pin securing the reverse lever without removing the 'E' clip.
4. Remove the reverse lever and slipper pad.

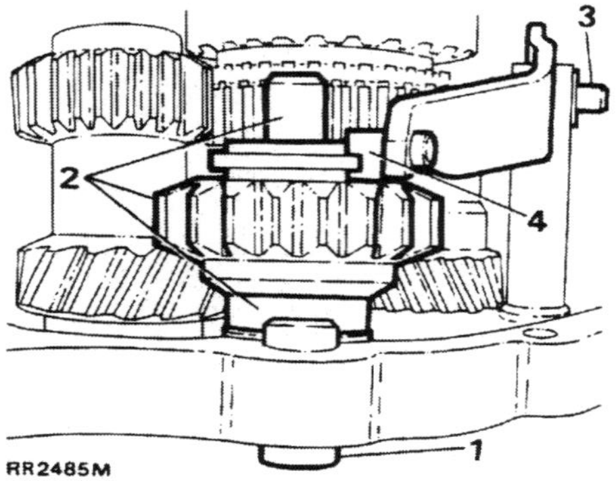

RR2485M

5. Remove the input shaft and fourth gear baulk ring.
6. Withdraw the selector plug, spring and detent ball from the centre plate.
7. Lift off the layshaft cluster by tilting it away from the mainshaft, simultaneously lifting the mainshaft slightly to clear the rear layshaft bearing.
8. Rotate the fifth gear selector shaft anti-clockwise (viewed from above) to align the fifth gear selector pin with the slot in the centre plate.

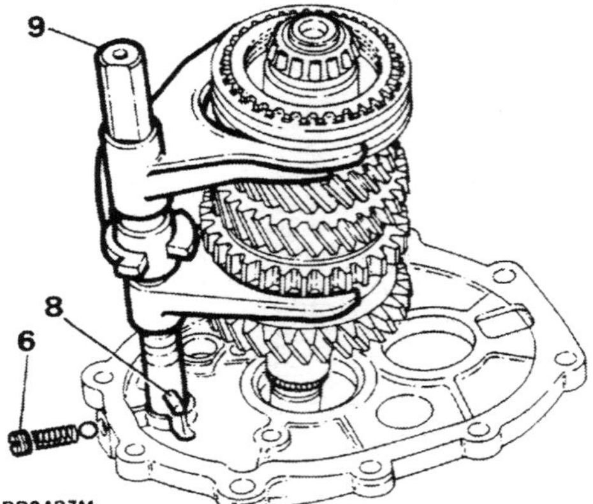

## LT77 FIVE SPEED GEARBOX | 37

### INSPECT AND PREPARE FOR REBUILDING

**NOTE: It is essential that all components are thoroughly cleaned and inspected before the rebuild is commenced. During the rebuild it is recommended that new bearings are fitted.**

### Main gearbox casing

1. Remove the mainshaft and layshaft bearing tracks from the main casing.
2. Remove the plastic oil trough from the front of the casing.
3. Clean gearcase thoroughly using a suitable solvent. Inspect case for cracks, stripped threads in the various bolt holes, and machined mating surfaces for burrs, nicks or any condition that would render the gearcase unfit for further service. If threads are stripped, install Helicoil, or equivalent inserts.
4. Insert a new plastic oil scoop inside the front of the casing, ensuring that the scoop side faces the top of the casing.

### Front cover

1. Remove and discard the oil seal from the front cover. Do not fit a new oil seal at this stage.

### Centre plate

1. Remove the layshaft and mainshaft bearing tracks from the centre plate. If required remove the reverse pivot post.
2. Inspect the bearing plate for damage and check the selector rail bore hole for wear.

Continued

### Mainshaft

13. Remove the centre bearing circlip.
14. Using press MS47 and two suitable metal bars to support first gear, remove the centre bearing, first gear bush, first gear and needle bearings and first gear baulk ring.

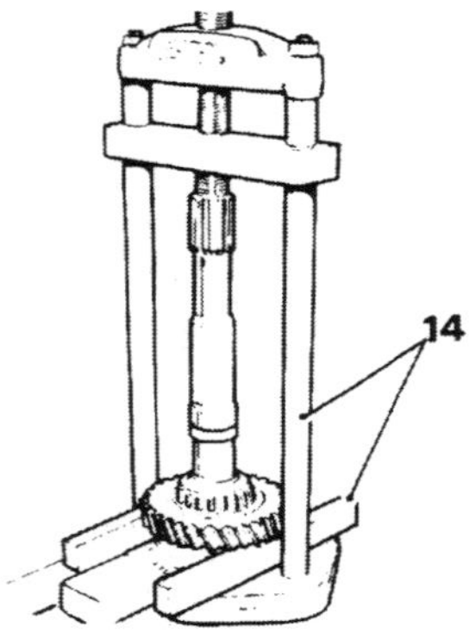

RR2489M

15. If a difficulty is experienced in removing the first and second gear synchromesh hub, support the second gear with the bars, and operate the press to release the first/second synchromesh unit, second gear, baulk ring and needle bearings.
16. Turn the mainshaft through 180° and repeat the operation using press MS47 and a suitable extension. Support third gear, press the mainshaft through the pilot bearing spacer, third and fourth synchromesh unit, third gear baulk ring, third gear and needle bearings.

### Layshaft

12. Using press 18C705 and tool 18C705 - 5 (18C705 - 7 if the increased capacity bearings are fitted) remove the layshaft bearings.

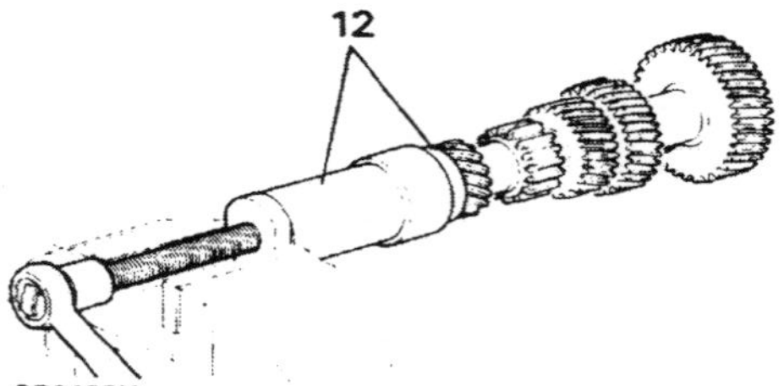

RR2488M

## LT77 FIVE SPEED GEARBOX 37

### Extension case

1. Examine the extension case for obvious signs of damage to threads and joint faces.
2. Remove the three oil pump housing bolts, spring washers and oil pump gears and housing. Inspect gears, renew if necessary.
3. Do not withdraw oil pick - up pipe.
4. Remove the plug, washer and filter.
5. Ensure that the oil pick - up pipe is free of contamination or blockage.
6. Invert casing and extract the oil seal.
7. Press out the ferrobestos bush from the casing.

**WARNING: This component contains asbestos. DO NOT use an air line when cleaning, as breathing asbestos dust is dangerous to your health. Use methylated spirit or denatured alcohol to clean asbestos components.**

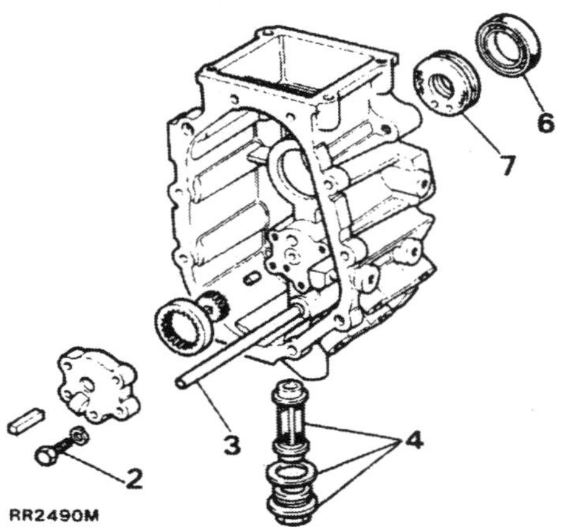

8. Press a new ferrobestos bush fully into position, ensuring the two drain holes are towards the bottom of the case.

**NOTE: If a new extension case is fitted, it is essential that a grub screw is securely fitted in the main oilway located in the rear of the case.**

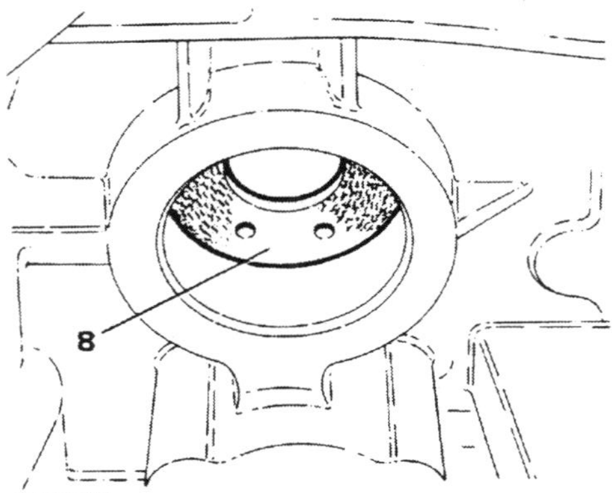

9. With the aid of tool 18C1422, fit a new oil seal to the rear of the extension case. Ensure the seal lips are towards the ferrobestos bush. Lubricate the seal lips with a suitable SAE 140 oil.
10. Assemble and fit the fibre oil pump gears to the oil pump cover, whilst ensuring the centre rotor square drive faces the layshaft.
11. Fit the three bolts and spring washers to secure the oil pump cover, and tighten to the correct torque.
12. Fit a new oil filter, fibre gasket and tighten plug to the correct torque.

### Gear change housing

**NOTE: The upper and lower gear levers are loctited together on vehicles produced from late 1985 onwards. If difficulty is experienced in separating these two components, a complete new assembly should be fitted.**

**LATER VEHICLES: To eliminate this problem a new upper gear lever, using a pinch bolt fixing to lower lever, is now fitted. The top of the lower gearlever is grooved to locate upper gear pinch bolt.**

1. Release the single bolt with plain washer securing the reverse plunger to the gear change housing.
2. Withdraw the plunger and retain the shims.

3. Remove the two bolts and washers anchoring the bias springs.
4. Release the springs from the register at the gear lever and remove them from the dowels. Restrain the springs using suitable grips whilst releasing bolts.
5. Remove the remaining two bolts securing the bias adjustment plate to the housing.
6. Remove the lower gear lever and bias plate from the housing.
7. Check the security of the spool guide bolts, located on the underside of the gear change housing.
8. Prise the seal from the bottom of the gear change housing.
9. Examine the gear change housing, ensuring that the cross pin location slots are not worn.
10. Inspect the condition of the bias springs renew if necessary.
11. Fit a new seal to the bottom of the housing. lips of the seal uppermost.
12. Lightly grease the lower gear lever ball with Shell Alvania R3.
13. Fit a new railko bush.
14. Fit the assembly to the housing locating the two pegs on the ball in the recesses of the ball seat.
15. Locate the bias adjustment plate, coat two of the retaining bolts with Hylomar PL32 or Loctite 290 and fit them forward of the gear lever. Do not tighten the bolts at this stage.
16. Fit the springs onto the posts, coat the remaining two bolts with Hylomar PL32 or Loctite 290 and fit to the housing to secure the springs in position. Do not tighten the bolts at this stage.
17. Carefully lever the free end of the springs around the rear of the gear lever until they are retained by the stop on the adjustment plate.
18. Do not fit the top cover at this stage.

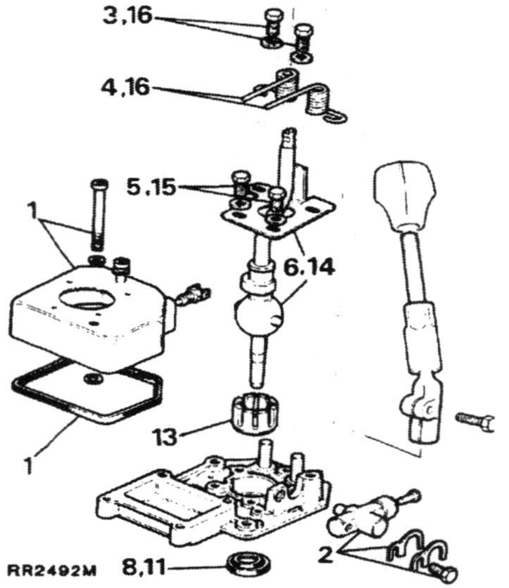

### Reverse gear plunger assembly

**NOTE: The plunger assembly is not a serviceable item. To check that the unit is operating correctly proceed as follows.**

1. Apply a load of between 45 to 55 kg (100 to 120 lb) to the plunger nose. If the plunger is operational within these limits the plunger is satisfactory. If the plunger operates outside the limits renew the plunger assembly.

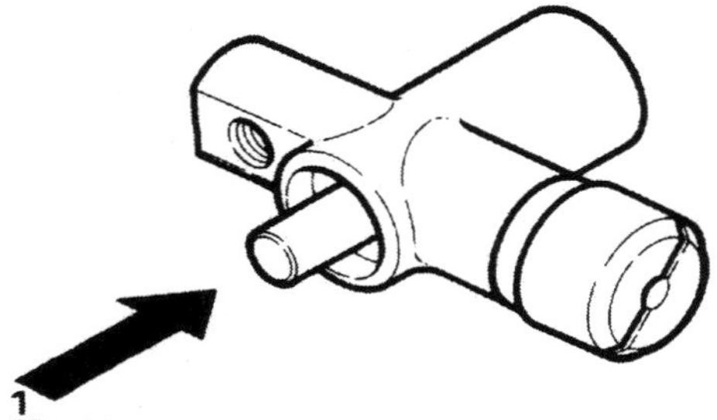

Continued

# LT77 FIVE SPEED GEARBOX

## Synchromesh assemblies

1. Mark the hub and sleeve to aid reassembly. Lever the backing plate off the fifth gear synchromesh assembly. Remove the slipper springs from the front and rear of each synchromesh assembly.
2. Withdraw the slippers and hub from the sleeve.
3. Inspect the springs and slippers for wear or breakage, fit new components where necessary. Inspect all the synchromesh components and fit new parts if there is evidence of chipped teeth or excess wear.

**NOTE: With the outer sleeve held, a push - through load applied to the outer face of the synchromesh hub should register 8,2 to 10 kgf m (18 to 22 lbf ft) to overcome the spring detent in either direction.**

4. Assemble the first and second synchromesh assembly by locating the shorter splines on inner member towards the second gear. Note that the outer member selector fork groove is to the rear of the gearbox

RR2494M

5. Refit the slippers and locate the slipper springs to each side of the assembly, ensuring that the hooked ends of both slipper springs are located in the same slipper, but running in opposite directions and fully located against the other two slippers.

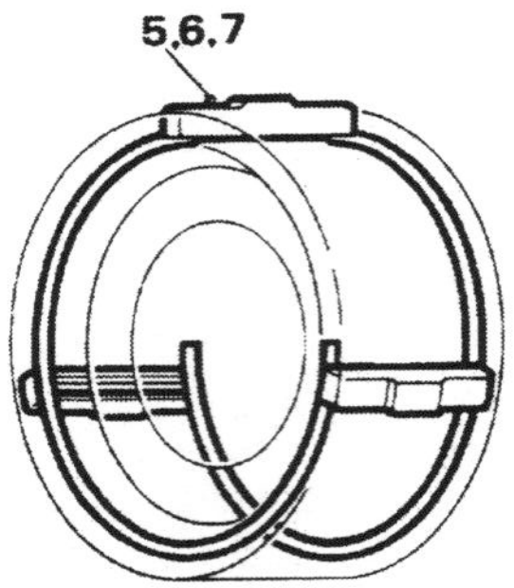

RR2495M

6. Assemble the third and fourth synchromesh assembly. Refit the slippers and locate the slipper springs to each side of the assembly, ensuring that the hooked ends of both slipper springs are located in the same slipper, but running in opposite directions and fully located against the other two slippers.
7. Assemble the fifth synchromesh hub assembly again ensuring the hooked ends of both slipper springs are located in the same slipper, but running in opposite directions. Fit the backplate onto the rear of the synchromesh hub assembly. Ensure the tag of the synchromesh hub assembly locates in the slot on the hub.
8. Check the wear between all the baulk rings and gears by pushing the baulk ring against the gear and measuring the gap between the gear and baulk ring. The minimum clearance (earlier vehicles) is 0,64 mm (0.025 in). Molybdenum coated baulk rings are fitted to later boxes which may be used on all gears on any LT77 gearbox, in which case the minimum clearance is 0,38mm (0.015 in). If required clearance is not met, fit new baulk rings.

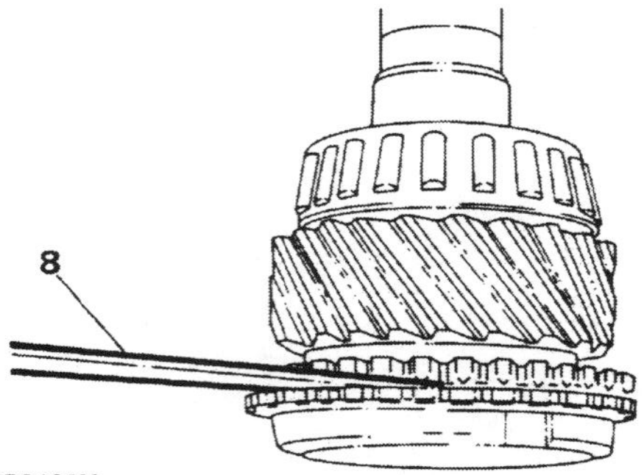

## Mainshaft

1. Examine each roller bearing surface for wear, and check the condition of the circlip grooves. Inspect the mainshaft splines, especially if any of the synchromesh units were found to be a loose fit during dismantling.
2. Ensure the oilways are free from sludge or contamination. Thoroughly clean with compressed air, observing the necessary safety requirements.
3. Check that the roll pins, pressed into the oil outlets to restrict oil flow to the bearings, are fitted below the bearing surface.

## Mainshaft end float checks

### Fifth gear

1. Fit the thrust washer, split roller bearing and fifth gear to the mainshaft, place a straight edge on the shoulder and using feeler gauges check the end float between gear and shoulder. End float should not be in excess of 0,20 mm (0.008 in) maximum. If end float is outside limit inspect the washer and gear faces for wear.

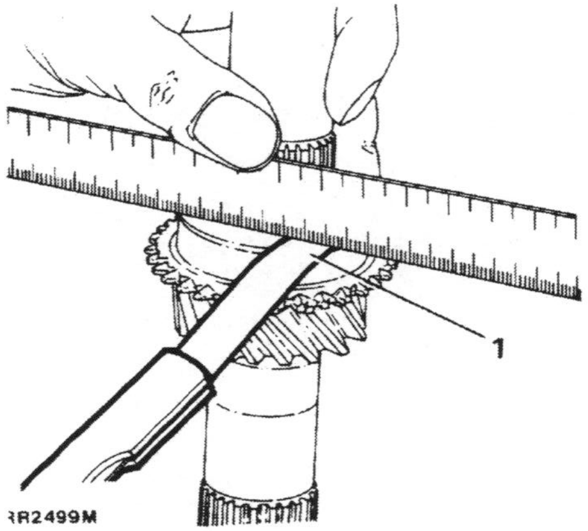

RR2496M

## Input shaft

1. With the aid of tools 18C284AAH and 18C284, extract the pilot bearing track.
2. Using tools MS47 and 18C47BA, remove the input shaft bearing.

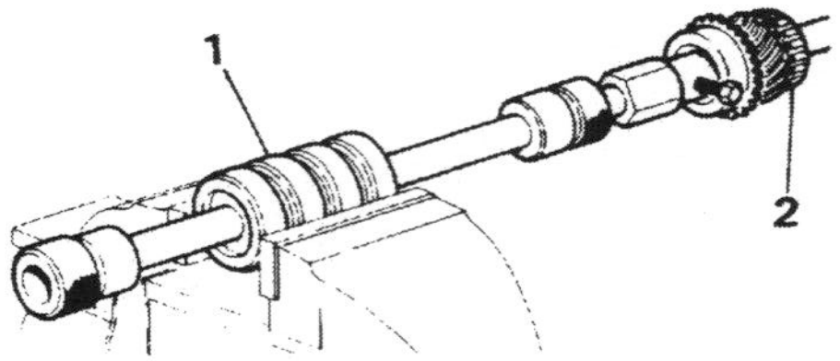

RR2497M

3. Inspect the shaft for wear or damage, and polish the oil seal track using fine emery cloth if required.
4. Using tool MS47 and a suitable tube, fit a new pilot bearing track to the input shaft.
5. Fit a new input shaft bearing using tools MS47, 18C47BA and 18C47BA - X.

Continued

## LT77 FIVE SPEED GEARBOX

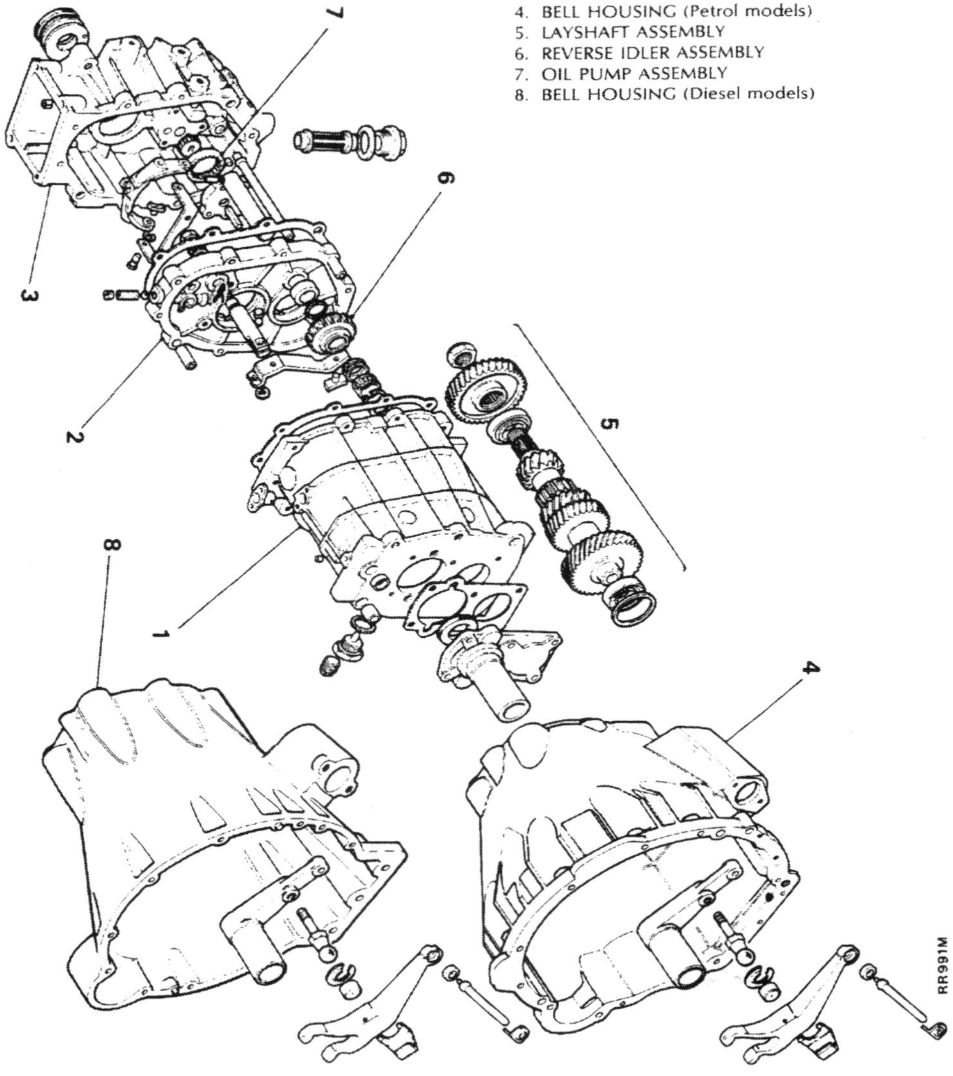

9. MAINSHAFT AND GEAR ASSEMBLY
10. GEAR SELECTOR ASSEMBLY
11. PRIMARY PINION ASSEMBLY

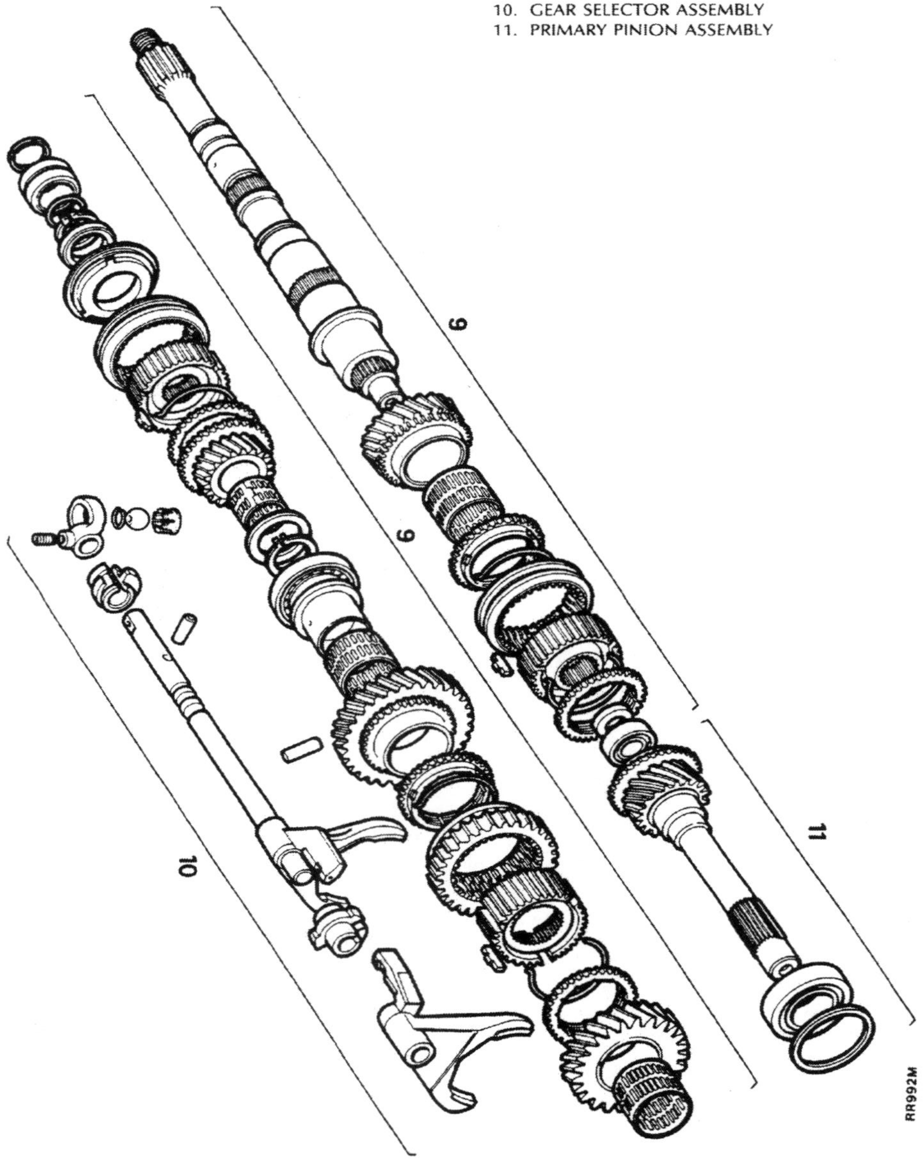

## LT77 FIVE SPEED GEARBOX

1. MAIN GEARCASE
2. CENTRE PLATE
3. EXTENSION CASE
4. BELL HOUSING (Petrol models)
5. LAYSHAFT ASSEMBLY
6. REVERSE IDLER ASSEMBLY
7. OIL PUMP ASSEMBLY
8. BELL HOUSING (Diesel models)

# LT77 FIVE SPEED GEARBOX 37

## First gear bush end float

**NOTE: It is essential to select first gear bush before checking first gear end float.**

5. Fit the first and second synchromesh hub assembly with the selector fork groove to the rear of the mainshaft.

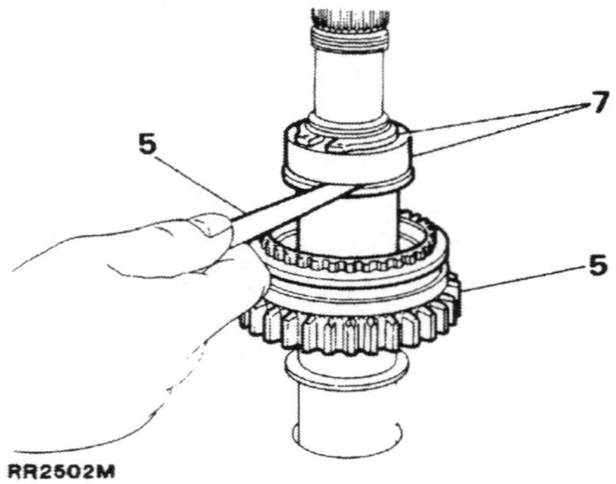

Continued

## Third gear

2. Refit the third gear and roller bearing to the mainshaft. Check third gear end float, by placing a staight edge on the mainshaft and checking the clearance between the gear and flange on the mainshaft, between the gear face and mainshaft flange. The end float should not be in excess of 0.20 mm (0.008 in) maximum. If end float is outside specified limit check flange and gear faces for wear.

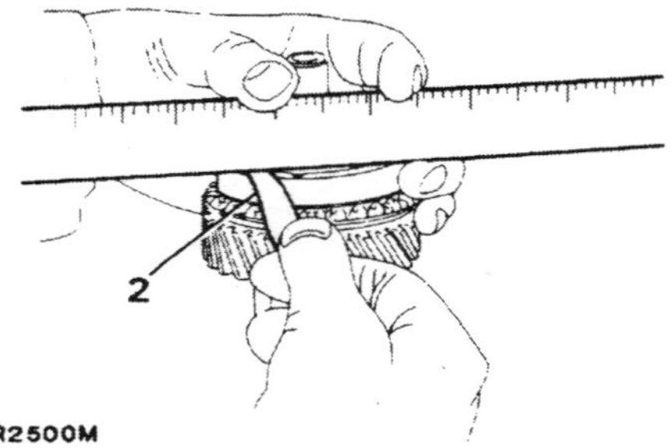

## Second gear

3. Lubricate the second gear needle bearing with a light oil and fit the bearing and second gear to the mainshaft.
4. Check the second gear end float, by placing a staight edge on the mainshaft and checking the clearance between the gear and flange on the mainshaft, end float should not be in excess of 0.20 mm (0.008 in) maximum. If end float is outside limit inspect the flange and gear faces for wear.

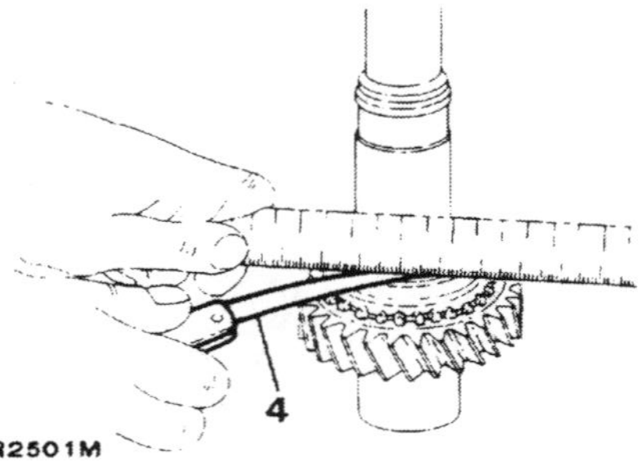

---

# LT77 FIVE SPEED GEARBOX

12. BIAS PLATE ASSEMBLY
13. UPPER GEAR LEVER ASSEMBLY
14. GEARBOX TOP COVER AND SEAL
15. LOWER GEAR LEVER AND BUSH
16. REVERSE PLUNGER AND SHIMS

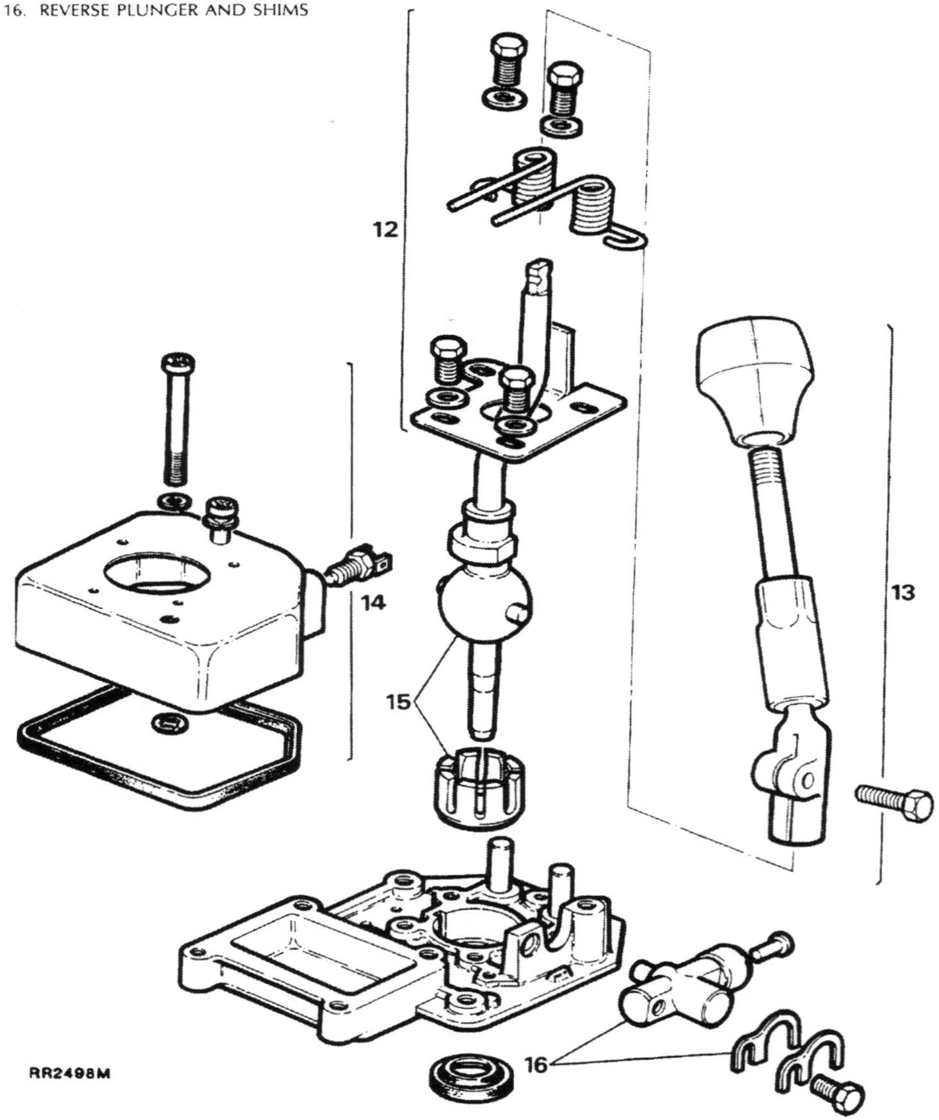

# LT77 FIVE SPEED GEARBOX

6. Manufacture a spacer to the dimensions provided in the illustration at the beginning of this section, this will represent a slave bearing.
7. Fit the first gear bush and slave bearing spacer and a new circlip to the mainshaft. When fitting the circlip, care must be taken to ensure it is not opened (stretched) beyond the minimum necessary to pass over the shaft.
8. Press the slave bearing spacer back against the circlip to allow the bush maximum end float. Measure the clearance between the rear of the first gear bush and front face of the slave bearing spacer with a feeler gauge. The clearance should be within 0.075 mm (0.003 in) maximum. The first gear bush is available with flanges of different thickness. Select a bush with a flange to give the required end float. The bush must be free to rotate easily with the required end float.
9. Remove the circlip, slave bearing spacer and first gear bush from the mainshaft.
10. First gear bushes are available in the following sizes:

| Part No. | Length (mm) |
| --- | --- |
| FRC5243 | 40,16 - 40,21 |
| FRC5244 | 40,21 - 40,26 |
| FRC5245 | 40,26 - 40,31 |
| FRC5246 | 40,31 - 40,36 |
| FRC5247 | 40,36 - 40,41 |

## First gear end float

11. Fit the selected first gear bush to the first gear, place the gear and bush on a clean, flat surface so that the gear is sitting firmly on the shoulder of the bush.
12. Place a straight - edge across the bush and measure the clearance between bush and gear, this should not be in excess of 0,20 mm (0.008 in) maximum.

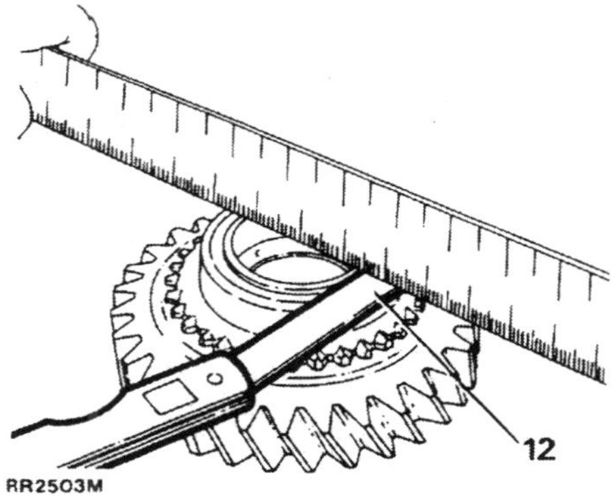

RR2503M

13. If the clearance is in excess of the maximum permissible limit, inspect gear faces for wear.

## Layshaft

1. Examine layshaft for excessive wear, or damaged teeth. Renew as necessary.
2. Fit new bearings fully onto the layshaft using MS47 and a suitable tube.

## Reverse gear and shaft

1. Remove the circlip from the reverse idler gear.
2. Remove both needle roller bearings and remaining circlip from the gear.
3. Check the condition of the reverse idler gear and its mating teeth on the layshaft and synchromesh outer unit.
4. Examine the reverse shaft for wear and renew if necessary.
5. Fit a new circlip to the rear of the reverse idler gear.
6. Lubricate with light oil and fit two new needle roller bearings. The needle roller bearings may be fitted either way round.
7. Fit a new circlip to the front of the reverse idler gear.

## Selectors

1. Check the selector rail for worn or loose pins. Note that the selector rail is only supplied complete with first and second selector fork, renew the assembly if necessary.

## Selector assembly

2. Place the first/second selector fork and rail assembly on a flat surface, locate the selector pin in the jaw of the fork.
3. Fit the front spool and third/fourth selector fork, engaging the spool in the jaw of the fork.

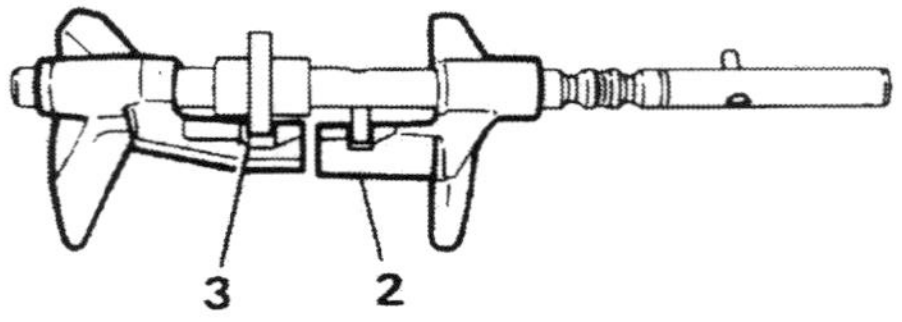

RR2504M

4. Slide the spool and fork towards the first/second selector until the slot in the spool locates over the selector pin and the spool remains engaged in the third/fourth selector fork jaw.

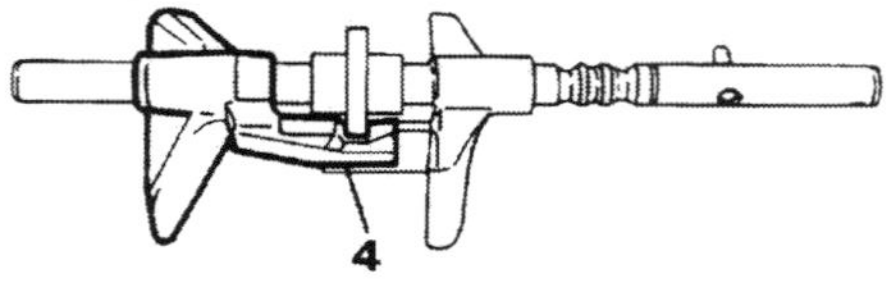

RR2505M

5. Examine the remaining selector components for wear, renew as necessary.
6. Examine the selector yoke components. If necessary remove the nylon seating by releasing the snap ring from the yoke.

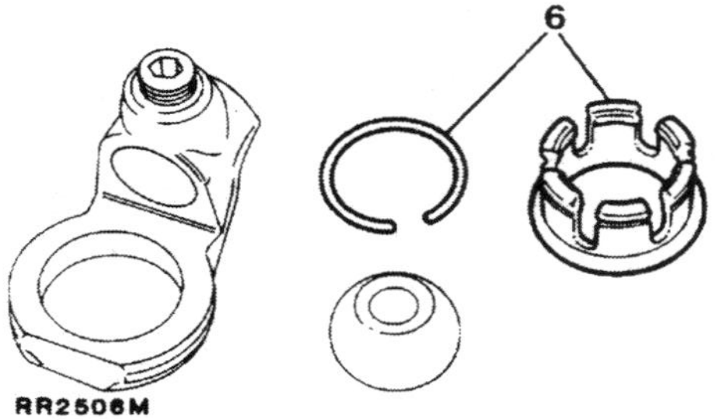

RR2506M

Continued

## LT77 FIVE SPEED GEARBOX

### ASSEMBLY

#### Mainshaft - rear end

**NOTE: Lubricate all the needle roller bearings with a light oil before assembly.**

1. Fit second gear and needle roller bearing to the mainshaft followed by the baulk ring and first/second synchromesh assembly. The synchromesh assembly may require gently tapping onto the splines using a plastic hammer.

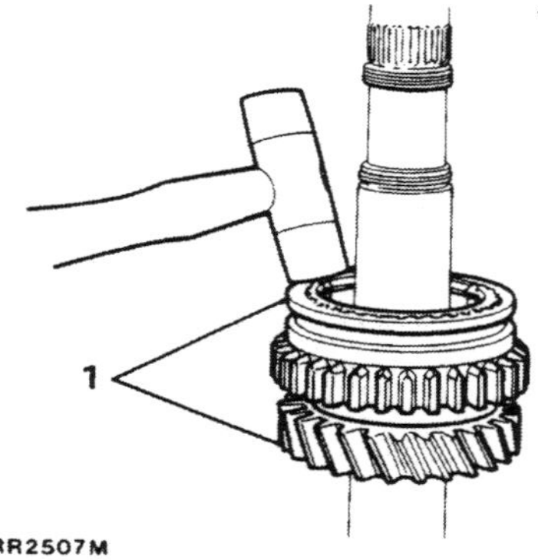

2. Fit the first gear baulk ring, first gear, needle roller bearing and selected bush to the mainshaft.

3. Using tools MS47, 18G47BA and 18G47BA - X refit the centre bearing to the mainshaft. Note that the larger spigot diameter of 18G47BA - X locates in the bearing cage. Ensure that the slots in the baulk ring align with the slipper blocks when pressing bearing.

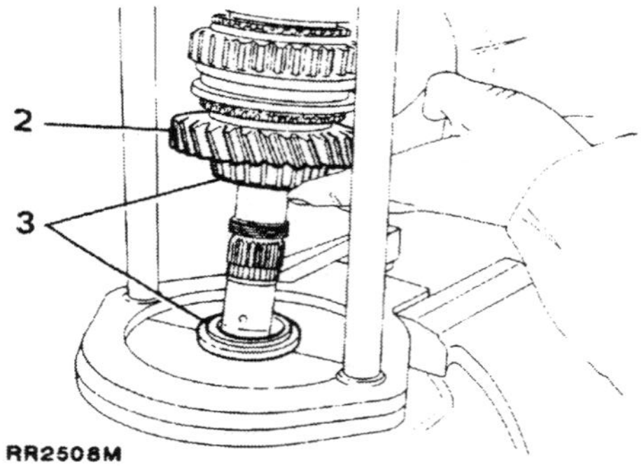

4. Fit the circlip, ensuring that the bearing, gear and circlip are in their rearmost position on the mainshaft. This is achieved by placing two bars under the 1st gear and carefully pressing the assembly rearwards on the mainshaft using press MS47. Check that there is no clearance between circlip and bearing, and that the bush is free to turn.

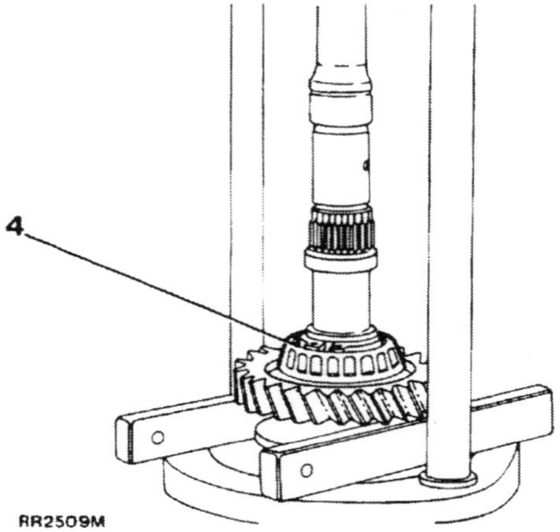

#### Mainshaft - front end

5. Invert the mainshaft, lubricate the third gear needle roller bearing with light oil, fit to the front end of the mainshaft.

6. Fit the third gear to the mainshaft; and locate the third gear baulk ring to the third gear.

7. Fit the third/fourth synchromesh assembly (with the raised centre boss of the synchromesh hub to the front of the gearbox) to the mainshaft.

8. Fit the spacer and press the pilot bearing to the front of the mainshaft.

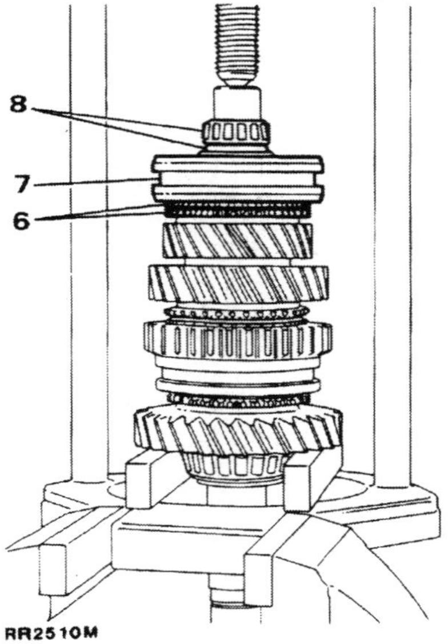

**WARNING: The mainshaft assembly MUST be supported on 1st gear when fitting the pilot bearing.**

#### Centre plate

1. Fit the centre plate to the workstand and secure with a nut and bolt.

2. Place the new mainshaft and layshaft bearing tracks in the centre plate.

3. Ensure both synchromesh units are in neutral. Fit the selector shaft assembly complete, engaging both selector forks in their respective synchromesh sleeves on the mainshaft.

4. Lubricate the selector shaft with a light oil. Engage the selector shaft and mainshaft assemblies in the centre plate, whilst rotating the fifth gear selector pin to align with the slot in the centre plate.

5. Fit the layshaft assembly to the centre plate, lifting the mainshaft assembly slightly to clear the rear layshaft bearing.

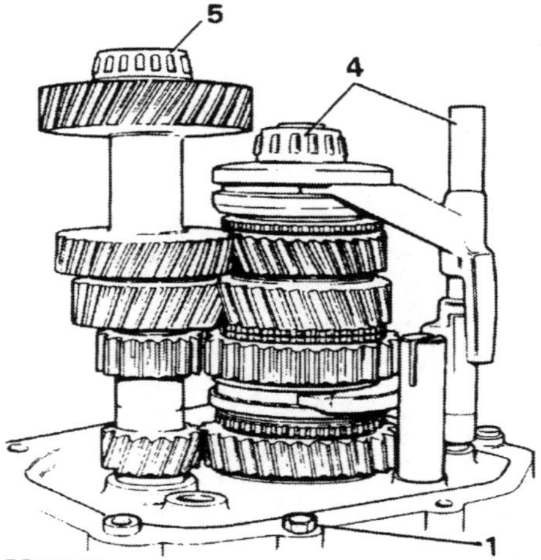

6. Rotate the selector shaft and spool to enable the reverse crossover lever forks to correctly align to the selector pin. Reposition the selector shaft and locate the lever into the slot of the reverse gear pivot shaft.

7. Insert pivot pin and fit a new circlip (earlier models only), ensuring that it is not opened beyond the minimum necessary to pass over the shaft.

Continued

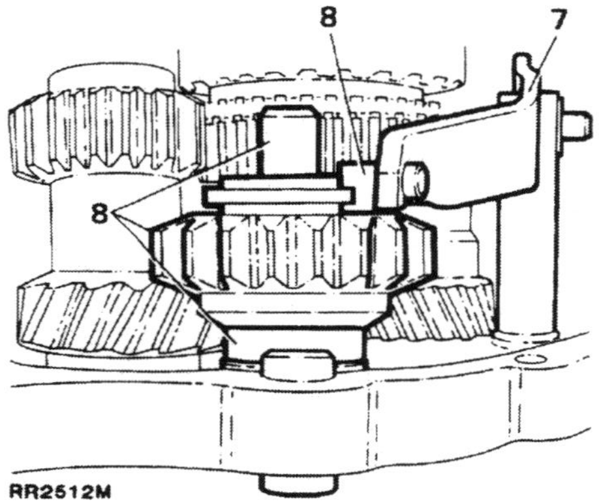

# LT77 FIVE SPEED GEARBOX

8. Fit the slipper pad to the reverse lever. Fit the reverse gear spacer and reverse gear assembly, locating the slipper pad lip to the reverse gear groove. Engage the reverse gear shaft from the underside of the centre plate, ensuring the roll pin is aligned with the slot in the centre plate casing. Secure the reverse shaft using the retaining plate.
9. Check that a running clearance exists between the slipper pad and lever, with a maximum clearance of 0,725mm (0.025in).
10. Fit the reverse gear thrust washer to the reverse gear shaft.
11. Locate the fourth gear baulk ring to the third/fourth synchromesh assembly.
12. Fit the input shaft to the mainshaft.
13. Release the bolt securing the centre plate to the workstand and fit a new gasket to the centre plate.

## Main gearbox casing

14. Ensure that the selector rail and spool are in the neutral position.
15. Lubricate the detent ball and spring with light oil, and fit to the top of centre plate. Fit and tighten the plug fully to ensure the selector rail does not move when the gearcase is fitted.
16. Fit the two guide studs into the casing, one each side.
17. Carefully lower the gearcase into position over the gear assemblies. DO NOT USE FORCE. Ensure the centre plate dowels and selector shaft are engaged in their respective locations.
18. Resecure the assembly to the workstand. Using two 8 X 35 mm slave bolts, with plain washers to prevent damage to the rear face of the centre plate, temporarily secure the gear case to the centre plate.
19. Fit the front spool retainer (with new 'O' ring if fitted) to the top of the gearcase using Hylomar PL32 on the joint face, if 'O' ring is not fitted. Smear Hylomar PL32 on the bolt threads, fit bolts and spring washers. Finally tighten to the correct torque.

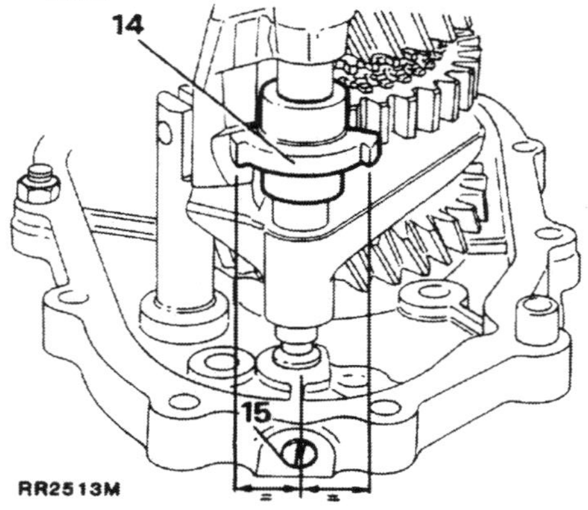

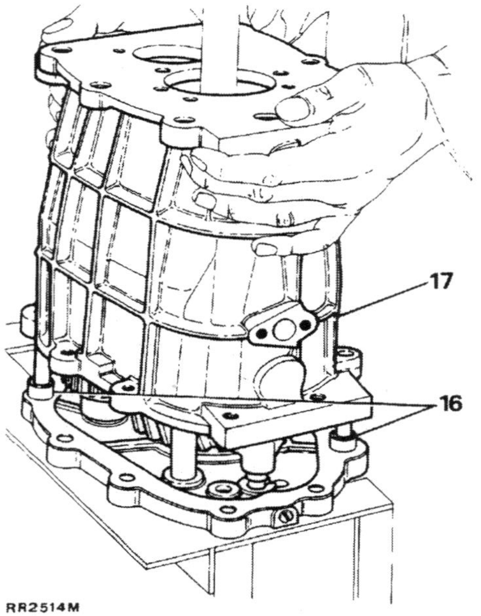

NOTE: Do not use force when fitting spool retainer. Provided the spool has not been disturbed the retainer will slide into position. If the selector rail is rotated or disturbed during assembly remove the main gearbox casing and dismantle the centre plate to rectify.

20. Remove the detent retaining plug, smear the threads with Loctite 290 or Hylomar PL32, and screw in until flush with case. Stake the plug to prevent rotation using a suitable centre punch.
21. Fit the layshaft and input shaft bearing outer tracks.
22. Fit the layshaft support plate using two 8 X 25 mm bolts and washers to the front of the gearbox, with the plain washer situated between the support plate and layshaft. The plate also retains the input shaft bearing outer track.
23. Remove the assembly from the workstand and invert, resecuring to the workstand. Remove the reverse retainer plate.

## Fifth gear

1. Press or drive the fifth gear onto the layshaft, using 18G1422, ensuring that the annular extraction groove is to the rear.
2. Fit a new 22 mm stake nut and tighten sufficiently to retain.
3. Fit the fifth speed thrust washer, roller bearing, fifth gear and baulk ring to the mainshaft.
4. Press fifth gear synchromesh hub assembly fully onto the shaft using tool 18G1431, ensuring that the slipper pads locate in the three slots in the baulk ring.

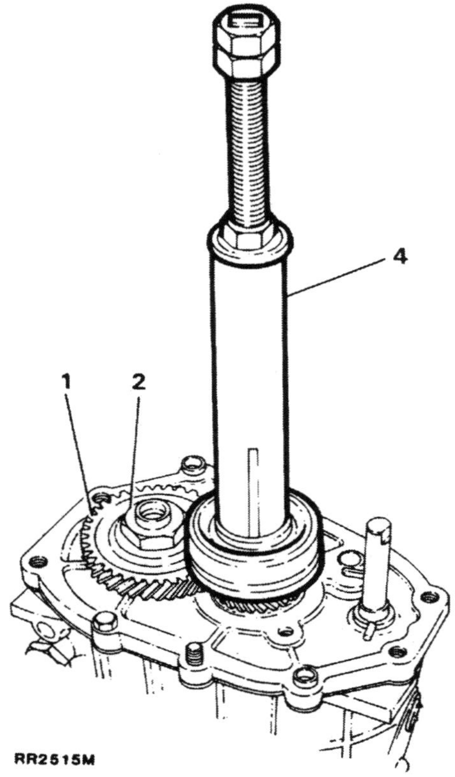

# LT77 FIVE SPEED GEARBOX 37

5. Fit a dummy selective washer which has an oversize bore for ease of fitting. It is recommended that the thinnest washer (FRC 5284) is used. Locate the circlip in its groove. Measure the clearance between the washer and circlip, which should be 0,005 to 0,055 mm,(0.0002 - 0.002in) maximum. The washers are available in ten sizes to obtain the correct clearance.

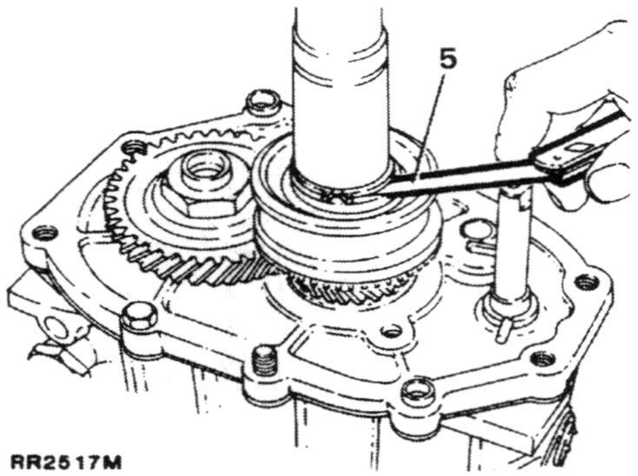

| Part No. | Thickness (mm) | Part No. | Thickness (mm) |
|---|---|---|---|
| FRC5284 | 5,10 | FRC5294 | 5,40 |
| FRC5286 | 5,16 | FRC5296 | 5,46 |
| FRC5288 | 5,22 | FRC5298 | 5,52 |
| FRC5290 | 5,28 | FRC5300 | 5,58 |
| FRC5292 | 5,34 | FRC5302 | 5,64 |

6. Fit the correct selective washer and new circlip.
7. Secure the flange holder 18C1205 to the gearcase.
8. Fit fifth gear retainer to case and fifth layshaft gear.
9. Tighten the staked nut securing layshaft fifth gear to the correct torque. Stake the nut using a suitable punch to secure. Remove flange holder and gear retainer.

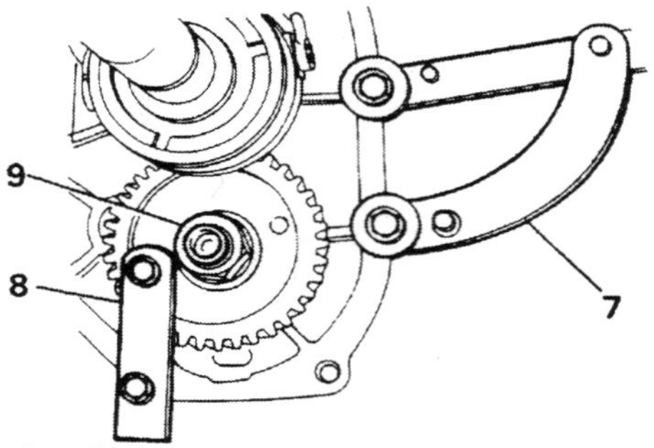

7. Using tool 18C1431 fit a NEW oil seal collar to the mainshaft, ensuring the collar is pushed on the shaft with sufficient clearance to allow the circlip to engage in its groove. The groove may be observed through the slot in the tool.

### Extension case

1. Remove the two bolts securing the centre plate and fit a new gasket to the joint face.
2. Fit the oil pump drive shaft to layshaft and align the pump gears.
3. Secure the selector yoke to the selector shaft with a NEW 10 mm grub screw tightened to the correct torque.

**NOTE: The NEW grub screw is encapsulated with loctite during manufacture.**

4. Carefully lower the extension casing into position. DO NOT use force, if difficulty is encountered, remove the case and align oil pump gears and drive shaft.
5. Remove the guide studs and fit the extension case bolts and spring washers, tighten to correct torque.
6. Cover the mainshaft splines with masking tape and fit a new oil seal collar 'O' ring. Remove the masking tape.

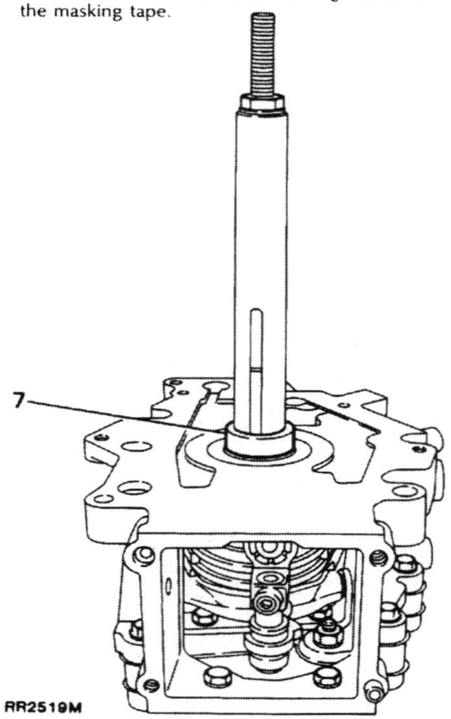

### Fifth gear selector fork assembly

1. Fit the fifth gear selector fork bracket to the centre plate with the bolts and spring washers and tighten to the correct torque.
2. Fit fifth gear spool to the selector shaft, the longer shoulder of the spool fitted towards the front of the gearbox.
3. Fit the bronze pads to the 5th gear selector fork (retain with vaseline), engage the fork, ensuring that the larger side of the selector jaw faces to the rear, with the spool, bracket and synchromesh sleeve. Insert the two dowel pins and secure in position with the two 'E' clips.

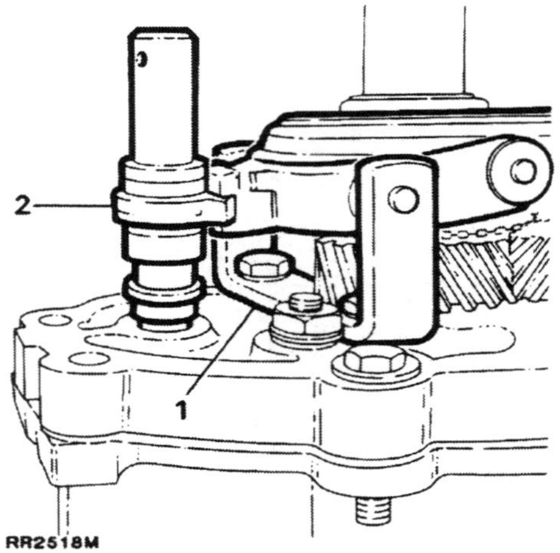

### Input/mainshaft bearing adjustment

1. Invert the assembly on the workstand, and remove the layshaft retaining plate.
2. Fit the original selective washer to the input shaft bearing and fit the front cover without the gasket. Do not fit the layshaft washer at this stage.
3. Retain the cover with four bolts, with plain washers only, around the input shaft. Secure evenly, finger tight.
4. Check the clearance between the cover and the gearcase using two feeler gauges as shown. If necessary change the selective washer to obtain a clearance of 0,20 to 0,26 mm (0.008 to 0.010 in). This will ensure the end float is within tolerance when the gasket is fitted and the cover tightened to the correct torque.

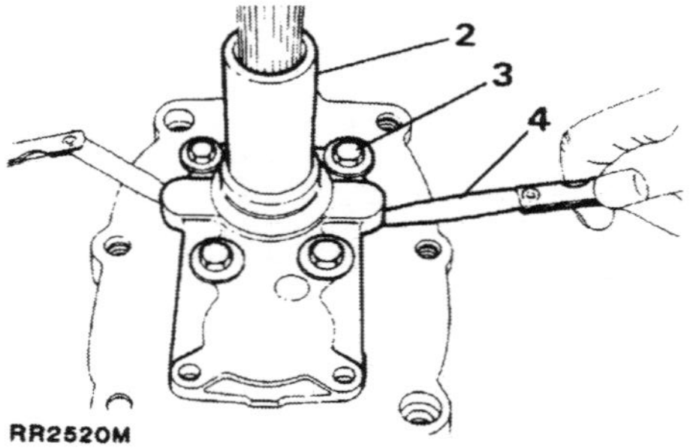

Continued

# LT77 FIVE SPEED GEARBOX

## Input/main shaft selective washers

| Part No. | Thickness (mm) | Part No. | Thickness (mm) |
|---|---|---|---|
| FRC4327 | 1,51 | FRC4349 | 2,17 |
| FRC4329 | 1,57 | FRC4351 | 2,23 |
| FRC4331 | 1,63 | FRC4353 | 2,29 |
| FRC4333 | 1,69 | FRC4355 | 2,35 |
| FRC4335 | 1,75 | FRC4357 | 2,41 |
| FRC4337 | 1,81 | FRC4359 | 2,47 |
| FRC4339 | 1,87 | FRC4361 | 2,53 |
| FRC4341 | 1,93 | FRC4363 | 2,59 |
| FRC4343 | 1,99 | FRC4365 | 2,65 |
| FRC4345 | 2,05 | FRC4367 | 2,71 |
| FRC4347 | 2,11 | FRC4369 | 2,77 |

5. Remove the cover and retain selective washer for final assembly.

## Layshaft bearing adjustment

1. The correct adjustment for the layshaft bearing is 0.025mm (0.001in) end float, and zero to 0,025 mm (0.001 in) preload. The following operation will ensure a preload figure within these tolerances. The input bearing selective washer must not be fitted during this operation.
2. Measure the thickness of a new cover gasket and the original selective washer.
3. Fit the original layshaft selective washer, and secure the cover (without gasket) with four bolts and plain washers, finger tight, around the layshaft.
4. Check the clearance using two sets of feeler gauges and select a washer to give a clearance equal to the thickness of the new gasket.

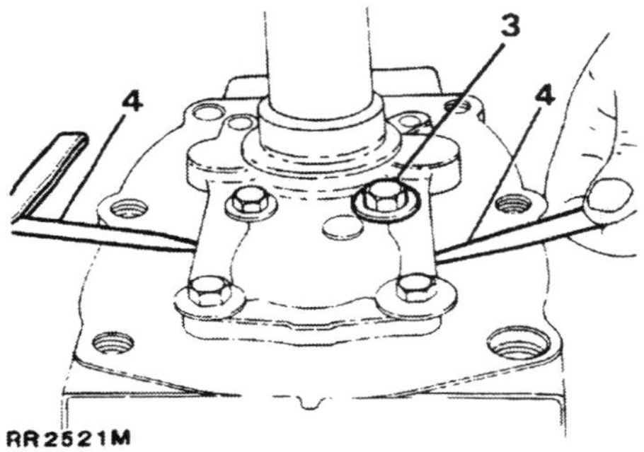

### Layshaft selective washers:

| Part No. | Thickness (mm) | Part No. | Thickness (mm) |
|---|---|---|---|
| TKC4633 | 1,69 | TKC4649 | 2,17 |
| TKC4635 | 1,75 | TKC4651 | 2,23 |
| TKC4637 | 1,81 | TKC4653 | 2,29 |
| TKC4639 | 1,87 | TKC4655 | 2,35 |
| TKC4641 | 1,93 | TKC4657 | 2,41 |
| TKC4643 | 1,99 | TKC4659 | 2,47 |
| TKC4645 | 2,05 | TKC4661 | 2,53 |
| TKC4647 | 2,11 | TKC4663 | 2,59 |

### Layshaft selective washers - increased capacity bearings:

| Part No. | Thickness (mm) | Part No. | Thickness (mm) |
|---|---|---|---|
| FTC0262 | 1,36 | FTC0280 | 1,90 |
| FTC0264 | 1,42 | FTC0282 | 1,96 |
| FTC0266 | 1,48 | FTC0284 | 2,02 |
| FTC0268 | 1,54 | FTC0286 | 2,08 |
| FTC0270 | 1,60 | FTC0288 | 2,14 |
| FTC0272 | 1,66 | FTC0290 | 2,20 |
| FTC0274 | 1,72 | FTC0292 | 2,26 |
| FTC0276 | 1,78 | FTC0294 | 2,32 |
| FTC0278 | 1,84 | FTC0296 | 2,38 |

5. Remove the front cover. Having ascertained the mainshaft and layshaft end float, fit the appropriate mainshaft and layshaft selective washers to the mainshaft and layshaft bearing tracks.
6. Fit a new oil seal to the front cover, ensuring the seal lips face towards the gearbox. Lubricate the seal lips with SAE 140 gear oil.
7. Mask the splines with masking tape to protect the oil seal, refit the front cover and remove the spline masking tape.
8. Refit the bolts and spring washers having applied Hylomar PL32 to the bolt threads. Tighten to the specified torque.

## Bell housing

1. Refit the dowels, and locate the bell housing on the dowels and fit the two long bolts (12 X 45 mm) with spring and plain washers to the dowel positions. The remaining four bolts (12 X 30 mm) are fitted with spring washers only. Tighten to the correct torque.
2. Slide the release bearing guide over the input shaft, and secure in position with the single bolt and clutch release pivot post, tighten to the correct torque.
3. Prior to assembly, lubricate the following items with a thin film of molybdenum disulphide grease.
   (a) Clutch release pivot post and locating socket in release lever.
   (b) Ball end of the clutch operating push rod.
   
   NOTE: Do not lubricate the bearing guide.

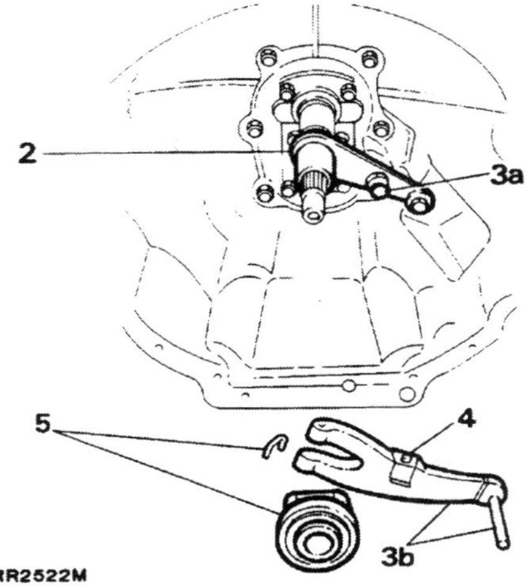

4. Locate the clutch release bearing lever on the pivot ball and secure in position with the spring clip, tighten the clip securing bolt to the correct torque.
5. Slide the clutch release bearing over the bearing guide. Fit a new plastic staple to clutch release lever and bearing to prevent the bearing sliding off the guide.

## Gear change housing

1. Remove the gearbox from the stand and place on the bench.
2. Refit the gearchange housing to the extension housing using a new gasket. Ensure that the gear lever engages with the selector yoke, and the spool and guide engage correctly.
3. Fit and tighten the securing bolts.

## Bias adjustment plate setting

4. Ensure that the bias plate is free to slide. Select fourth gear and load the gear lever fully to the right hand side of the gearbox.
5. Tighten the four adjustment plate bolts to the correct torque.
6. Check the adjustment by testing third and fourth gear selection. After adjustment the lever should lie centrally in the third and fourth plane, with no dog leg into gears. Equal side movement should exist in third and fourth gears.

Continued

# AUTOMATIC GEARBOX | 44

## ZF MAIN GEARBOX AND BORG WARNER TRANSFER GEARBOX - ADAPTOR PLATE

To assist in the removal of the transmission assembly from the vehicle it is necessary to locally manufacture an adaptor plate to use in conjunction with a transmission hoist.

**NOTE:** Four holes (A) to be countersunk on underside to suit hoist.

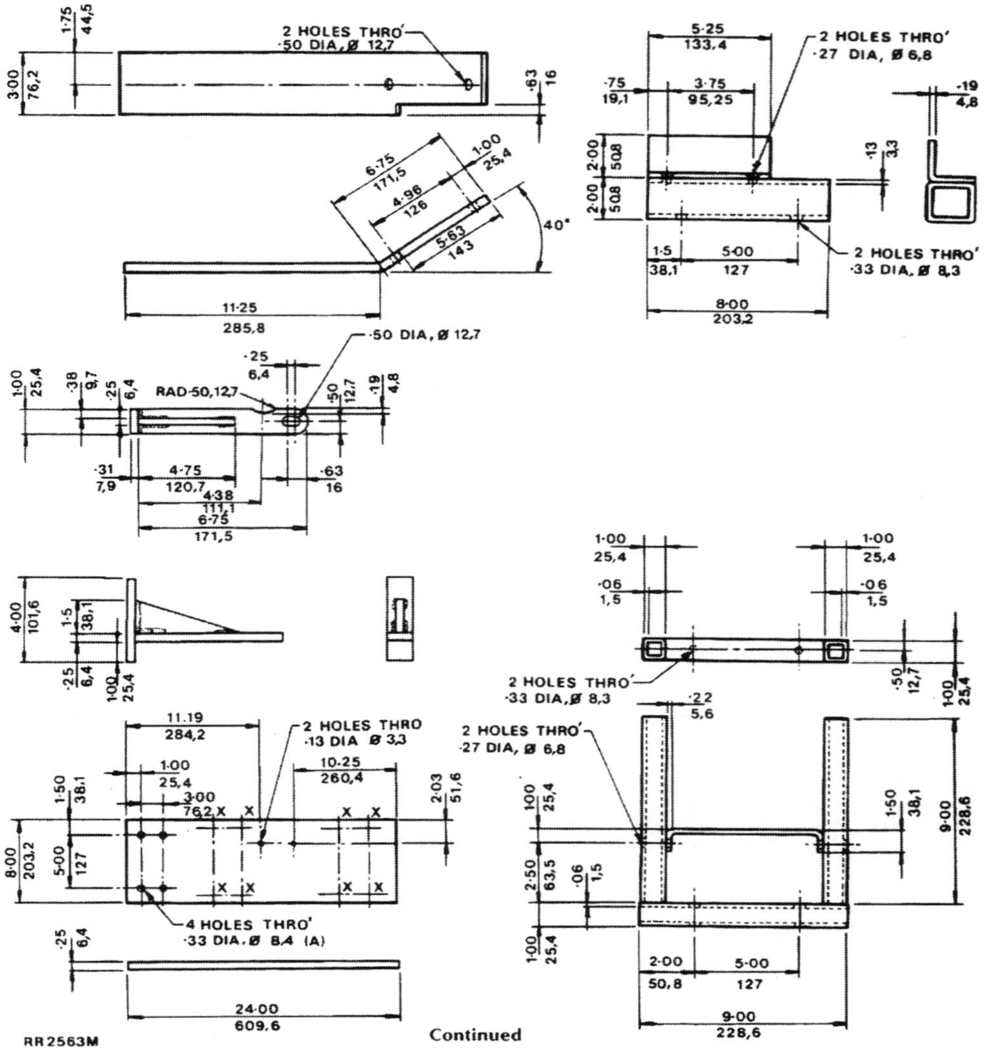

Continued

---

# LT77 FIVE SPEED GEARBOX | 37

### Reverse plunger assembly setting

7. Fit the assembly to the housing and secure in position with the single bolt and washer.
8. Select first gear. Using feeler gauges, check the gap between the reverse plunger nose and the side of the gear lever.
9. The required setting is 0,6 to 0,85 mm (0.024 to 0.034 in) clearance. Adjust the gap by adding or removing the shims behind the plunger assembly.
10. Ensure that the reverse light operating plunger is fitted to the reverse plunger. Fit the top cover and sealing rubber and tighten the securing screws. Refit and adjust reverse light switch.
11. Fit a new fibre washer to the oil drain plug, fit the plug to the gearbox and tighten to the correct torque.
12. Refill the gearbox with the correct quantity and grade of oil as specified in the 'Recommended Lubricants' section.
14. Refit the gearbox oil level plug, and tighten to the correct torque.
15. Refit the transfer gearbox and fit the assembly to the vehicle.

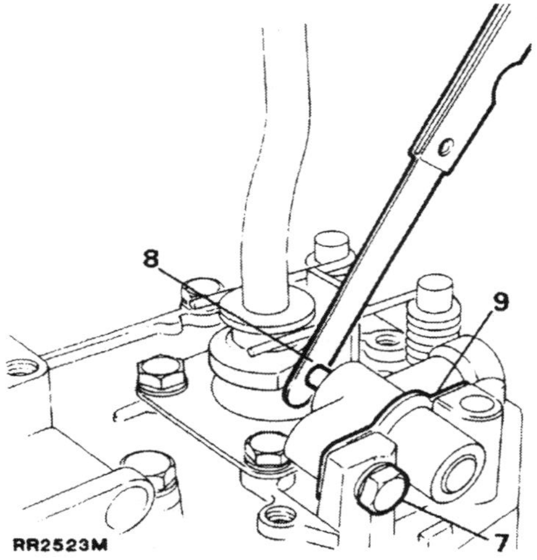

440

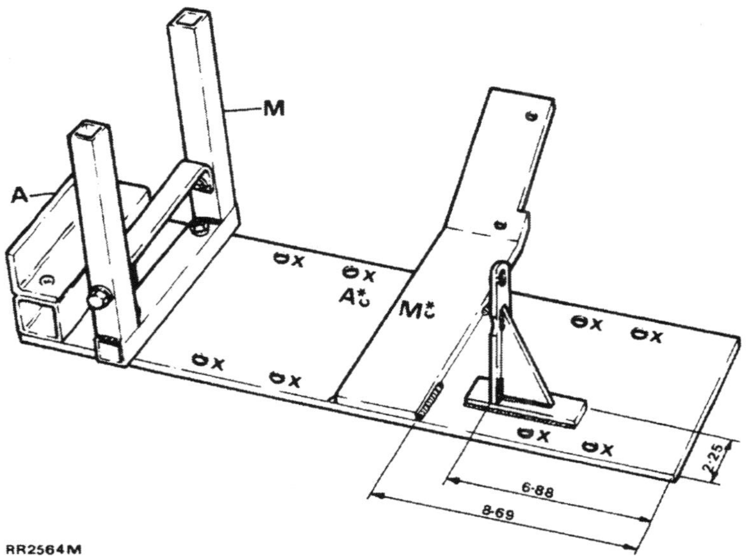

# AUTOMATIC GEARBOX 44

A: Automatic gearbox models
M: Manual gearbox models
A*: Centre of the lifting hoist (Automatic models)
M*: Centre of the lifting hoist (Manual models)
X: Drill fixing holes to suit hoist table

Material: Steel plate BS 1449 Grade 4 or 14

## ZF MAIN GEARBOX AND BORG WARNER TRANSFER GEARBOX

### Remove and refit

### Preparation - under bonnet

**WARNING: Where the use of a transmission hoist is necessary, it is ABSOLUTELY ESSENTIAL to follow the hoist manufacturer's instructions to ensure safe and effective use of the equipment.**

1. Install the vehicle on a hydraulic ramp and chock the road wheels.
2. Disconnect the battery negative terminal.
3. Release and remove the fan blade assembly. Note the assembly has a left hand thread.
4. **Fuel injection models** - Release the clamp and remove the air intake hose from the neck of the plenum chamber.
   **Carburetter models** - Remove the air cleaner and elbows.
5. Disconnect the kickdown cable from the throttle linkage.
6. Release the two gearbox breather pipes from the clip located on the lifting eye at the rear of the right hand cylinder head.
7. Remove the gearbox dipstick.

### Inside the vehicle

8. Select low range, unscrew and remove the transfer gearbox knob.
9. Unclip the top cover of the main gearbox selector and remove the circlip, withdraw the detent button. Remove the circlip above the selector knob retention nut, remove the nut, serrated washer and withdraw the selector knob.
10. Carefully prise the inset panel out of the floor mounted console, complete with gear selector illumination panel and ashtray. Disconnect the electrical multi - plug to the graphics panel, and remove the inset panel.
11. Release the four screws and remove the glove box liner.
12. Carefully prise the window lift switch panel from the front of the glove box. Push the panel complete with switches back through the panel opening and place on the gearbox tunnel.
13. Release the two bolts and two screws securing the glove box/console assembly to the gearbox tunnel.
14. Detach the two relays from the inner side of the glove box.
15. Disconnect the electrical leads to the rear cigar lighter.
16. Disconnect the handbrake cable from the handbrake lever. Raise the lever while simultaneously detaching the glove box/console assembly from the lower dash. Remove the assembly from the vehicle.
17. Remove the retaining clip and pull the handbrake adjustment thumb wheel from the outer sleeve. Push the inner sleeve to the underside of the vehicle.
18. Remove the sound deadening trim from the top of the gearbox tunnel.
19. Remove the screws and detach the retaining plate from around the transfer gearbox lever.

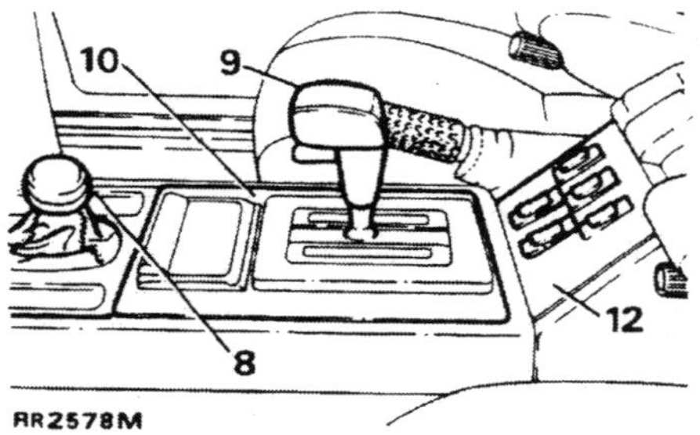

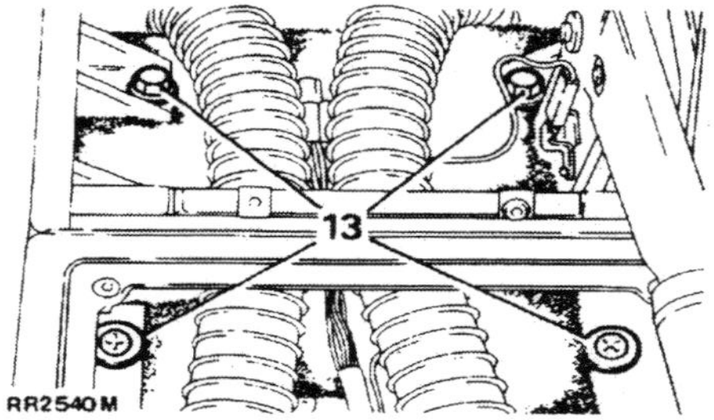

Continued

# AUTOMATIC GEARBOX

## Under the vehicle

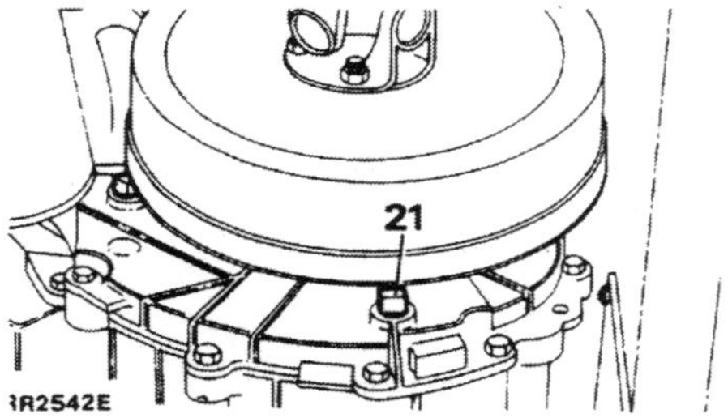

20. Raise the hydraulic ramp.
21. Remove the main and transfer gearbox oil drain plugs. Where applicable remove the filler plug to assist draining and drain the oil into suitable containers. While the oil is draining continue with the following operations.
22. **Catalyst models only** - Disconnect the multi-plugs to the Lambda sensors.
23. Remove the eight fixings securing the cross member. Note it may be necessary to spread the chassis to enable the cross member to be withdrawn.

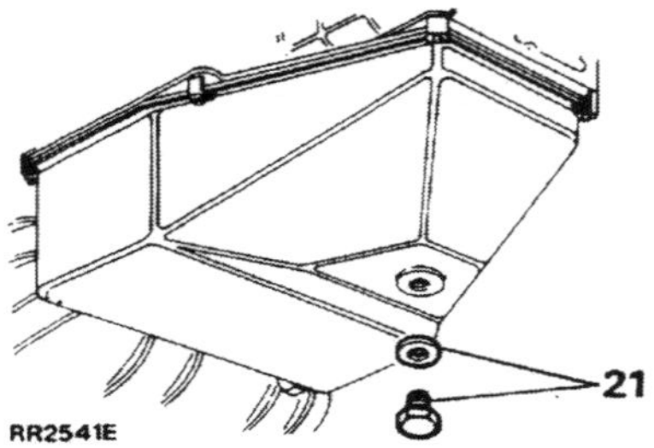

NOTE: The above operation will require the assistance of a second operator to support the exhaust system while the various fixings are released.

24. Remove the front exhaust down pipes and intermediate pipe complete with centre silencer (or catalyst).
25. Release the two clamps at the side of the engine sump that secure the two gearbox oil cooler feed and return pipes.
26. Place a suitable container below the gearbox, disconnect the oil cooler feed and return pipes from the bottom and side of the gearbox. Plug the pipes and openings to prevent ingress of foreign matter.
27. Disconnect the dipstick tube from the front of the gearbox oil pan.
28. Mark each propeller shaft drive flange at the transfer gearbox with an identification line to aid re - assembly, remove the fixings and disconnect the propeller shafts at the output flanges. Tie the shafts to one side.
29. Release the nut and disconnect the speedometer cable from the rear output housing, tie the cable to one side.
30. Disconnect the main gearbox selector cable and rod from the left side of the gearbox. Lay the cable aside.
31. Disconnect the main gearbox inhibitor switch multi - plug from the main harness.
32. If fitted disconnect the speed transducer multi - plug from the main harness.

## Remove the transmission assembly

33. Remove the nine front cover plate bolts from the bottom of the gearbox bellhousing, detach the cover plate to gain access to the four torque converter fixing bolts.
34. Rotate the engine using the crankshaft pulley until two of the access holes in the drive plate/ring gear assembly are visible through the bell housing bottom cover opening.
35. Remove the two bolts that are visible through the access holes, which secure the drive plate to the torque converter. Mark one of the access holes and a bolt hole in the converter with an identification line to aid re - assembly and to maintain original build setting.
36. Rotate the crankshaft 180° until the remaining access holes are visible, remove the remaining two bolts.
37. Position a suitable transmission floor jack on the rear output housing or brake drum to support the weight of the transmission assembly.
38. Remove the fixings and withdraw the transfer gearbox mountings.
39. Fit the previously manufactured fixture on a transmission hoist, raise the hoist and position the fixture and hoist under the transfer gearbox mounting points.
40. Using the original gearbox mounting bolts secure the fixture to the gearbox.
41. Remove the transmission floor jack from the rear of the transfer gearbox.
42. Carefully lower the transmission until the top of the transfer gearbox clears the rear passenger footwell.
43. Position the transmission floor jack under the engine to support the weight while the bellhousing bolts are removed.
44. Remove the bellhousing bolts noting that one of the bolts also secures the gearbox dipstick tube.
45. Withdraw the transmission assembly from the engine, ensuring that the torque convertor is removed with the gearbox and does not stay on engine.

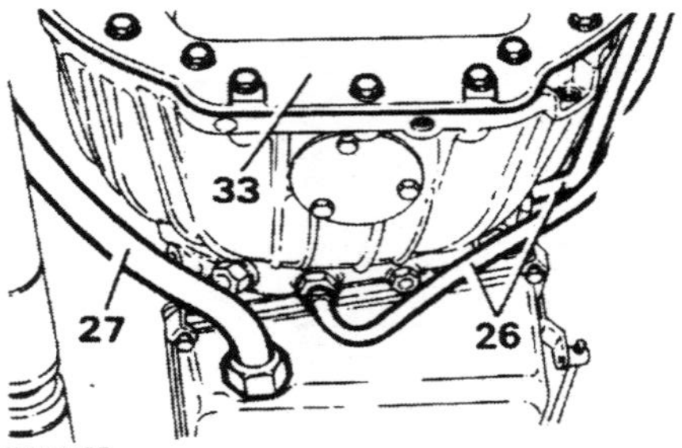

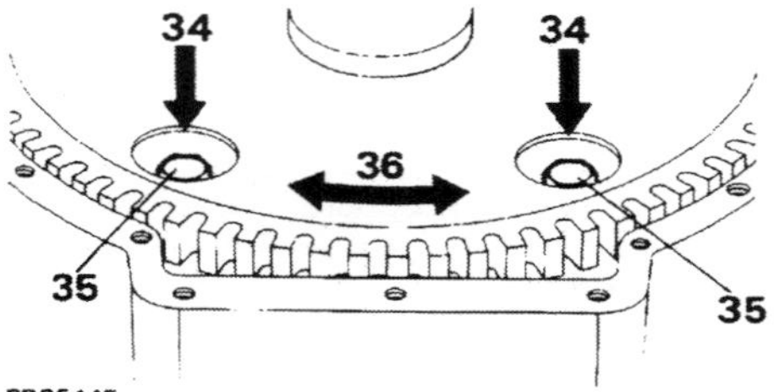

## Refitting

46. Refitting the gearbox is a reversal of the removal procedure noting the following points.
47. The flexible drive plate to torque converter bolts are to be coated with Loctite 270 prior to assembly.
48. Tighten all fixings to the specified torque. Refer to torque charts in main Workshop Manual.
49. New gaskets are to be fitted to the exhaust flanges, all joints other than those fitted with doughnuts, to be coated with 'Firegum Putty'. Check the system, if any leaks are evident reseal as necessary.

# BRAKES 70

## HANDBRAKE CABLE

### Remove and refit

#### Removing

#### Inside the vehicle

1. Set the vehicle on level ground or install on a hoist, select 'P' in automatic gearbox and neutral in manual gearbox. Disconnect the battery negative terminal.
2. Chock the road wheels.
3. Fully release the handbrake lever.
4. Remove the four retaining screws and lift out the glove box liner to gain access to the bottom of the handbrake lever.
5. Remove the split pin and withdraw the clevis pin and washer securing the cable to the handbrake lever.
6. Remove the clip from above the adjustment thumbwheel, push the inner and outer cable to the underside of the vehicle.
7. Pull the thumbwheel from the outer sleeve.

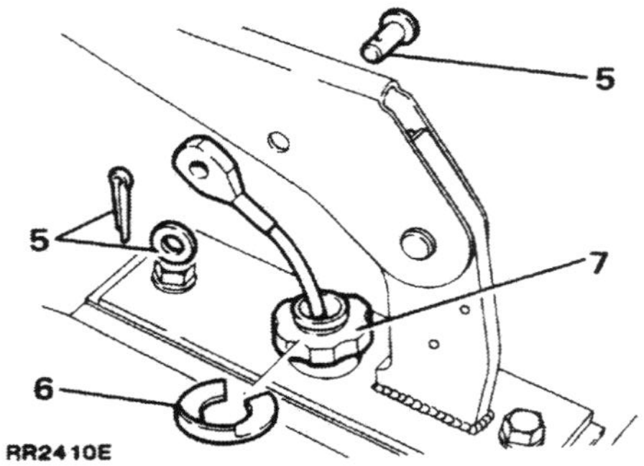

#### Underneath the vehicle

8. Remove the split pin and withdraw the clevis pin and washer.
9. Detach the retaining clip securing the outer cable to the support bracket located on the front cover of the transfer gearbox.
10. Release the cable from the 'P' clip located on the left hand side of the transfer gearbox.
11. Withdraw the cable.

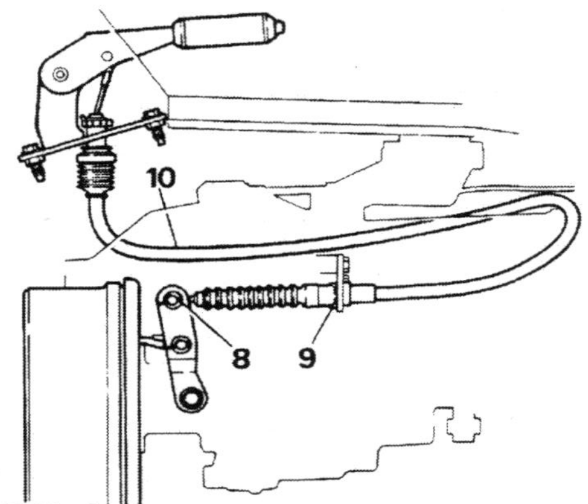

### Fit new cable

12. Reverse instructions 5 to 11 when fitting a new cable.

### Adjust handbrake

13. Ensure that the handbrake lever is fully released.
14. From underneath the vehicle, rotate the adjuster on the brake drum back plate clockwise until the brake shoes are fully expanded against the brake drum.

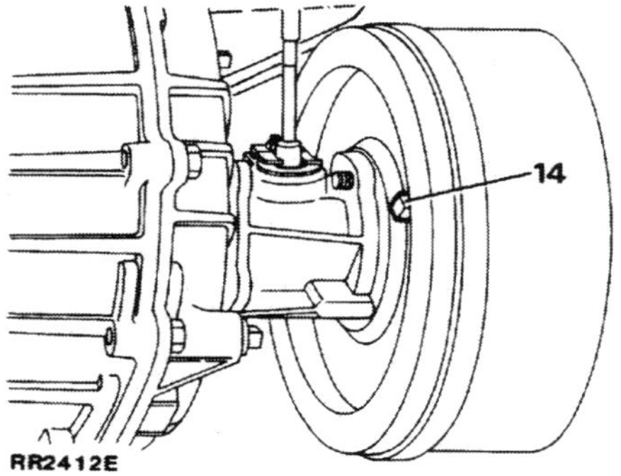

Continued

## Notes

## BRAKES

15. Back off the adjuster until the drum is free to rotate.
16. Rotate the adjustment thumbwheel below the handbrake lever until the parking brake is fully operational on the third notch of the ratchet.

    **NOTE: The handbrake adjustment thumbwheel must only be used for initial setting and to compensate for cable stretch, it must not be used to take up brake shoe wear, which must continue to be adjusted at the brake drum.**

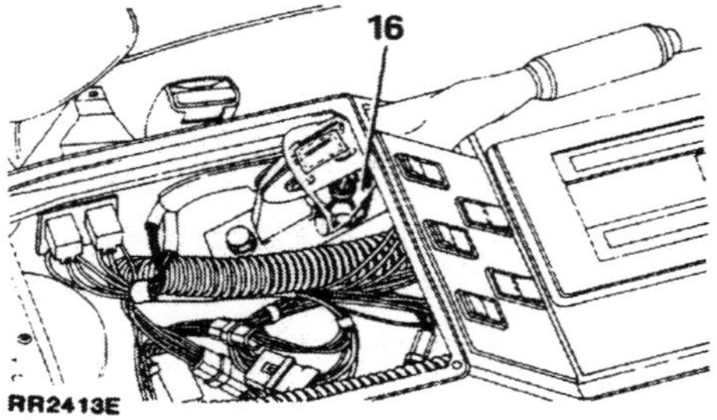

17. Operate the handbrake once or twice to settle the brake shoes, recheck that the handbrake is fully operational on the third notch of the ratchet. Re - adjust as necessary.
18. Refit the glove box liner.
19. Reconnect the battery and remove the wheel chocks.

## BODY

### FRONT DOOR - TRIM PANEL - 4 Door models

**Remove and refit**

**Removing**

1. Disconnect the battery negative terminal.
2. Remove the screw securing the handle bezel.
3. Remove the bezel.

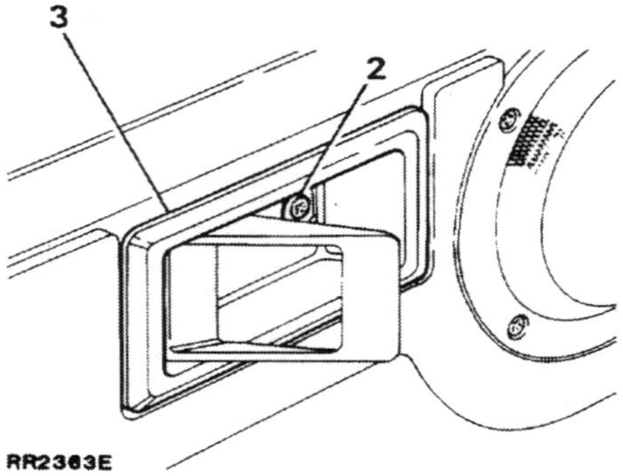

4. Carefully prise the sill locking button bezel from the trim panel.

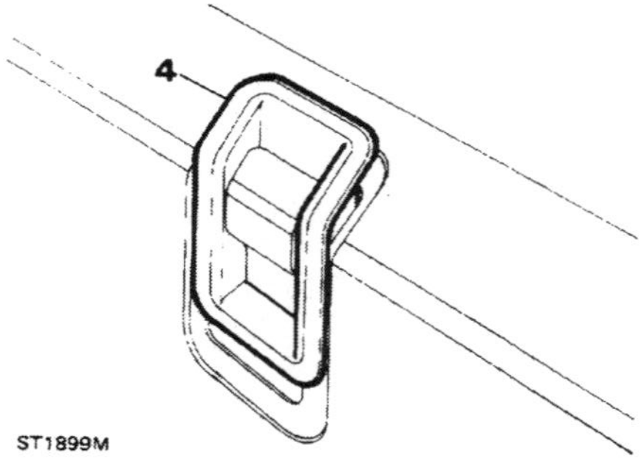

5. Remove the two finisher buttons from the bottom of the door pull pocket to reveal the securing screws.
6. Remove the screws and withdraw the pocket from the trim panel.

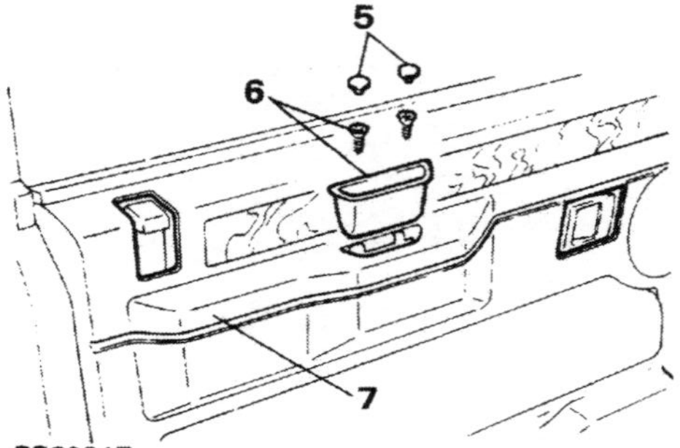

7. Using a screwdriver, carefully prise the trim panel away from the door.

   **NOTE: Support the trim panel while the speaker leads are disconnected.**

8. If necessary the stowage bin front panel can be removed by releasing the screws at the rear of the trim panel.

**Refitting**

9. Reverse the removal procedure.

Continued

## BODY 76

### FRONT DOOR LOCK, OUTSIDE AND INSIDE DOOR RELEASE HANDLES - 4 Door models

**Remove and refit**

**Removing**

1. Remove the interior door release handle bezel and sill locking button bezel.
2. Remove the two screws from the bottom of the door pull pocket remove the pocket and prise the door trim panel from the door. Disconnect the electrical leads to the door speaker and remove the plastic barrier sheet.
3. Remove the window lift motor. (Refer to electrical section in main Workshop Manual).
4. Remove the door glass and regulator. (Refer to door glass and regulator remove and refit in main Workshop Manual).
5. Remove the door lock actuator. (Refer to electrical section in main Workshop Manual).
6. Disconnect the control rod from the private key operated lock by releasing the metal clip at the bottom of the rod.
7. Disconnect the control rod from the outside door release handle by pulling it out of the plastic ferrule.
8. Disconnect the control rod connector between the inside door release handle and the door lock by releasing the metal clip and pulling one of the control rods out of the plastic connecting block. This is accessible through the small centre cut - out in the door panel. (The control rod also passes through a guide bracket on the inside of the inner door panel).
9. From inside the door panel push out the small pin which secures the quadrant to the inner door panel. Push the quadrant out of the panel.
10. Remove the two screws securing the sill locking button to the door. Manoeuvre the sill button and remove it from the control rod.
11. Release the door lock by removing the two countersunk screws from the door edge and the single screw with shakeproof washer on the inner door panel.

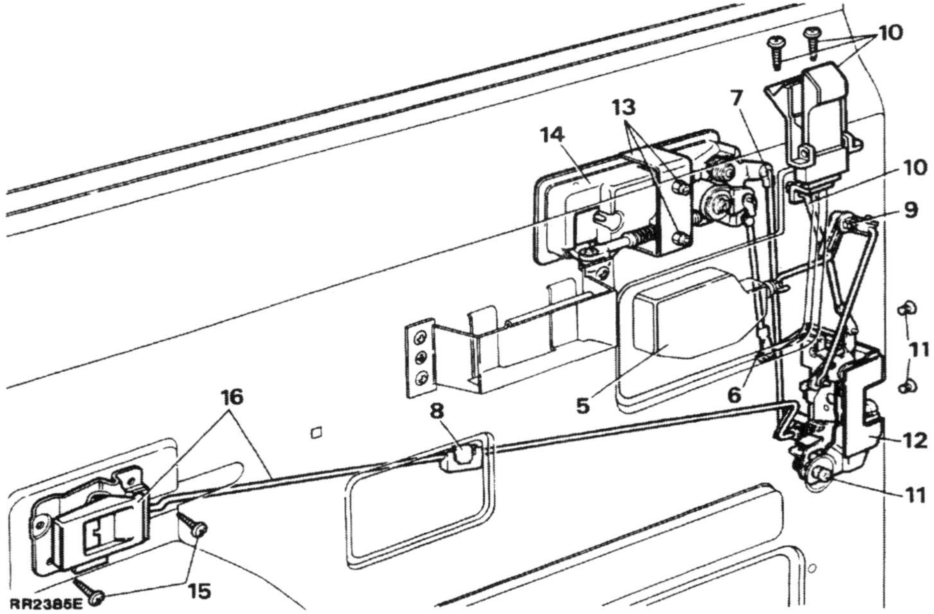

## BODY 76

12. Withdraw the lock through the lower rear cut - out on the inner door panel.

**NOTE: If necessary the following items can be removed.**

13. Remove the two nuts (with shakeproof washers) and retaining bracket securing the **outside release handle** to the outer door panel, accessible through the upper rear cut - out on the inner door panel.
14. Carefully detach the door release handle from the outer door panel.
15. Remove the two screws securing the **inside door release handle** to the inner door panel.
16. Withdraw the handle from its location with half of the connecting rod attached.

**Refitting**

17. Reverse the removal procedure items 1 to 16.

**NOTE: When refitting the door glass frame, ensure that it is positioned to suit the door opening before fully tightening the frame securing bolts.**

### ADJUSTMENT - FRONT DOOR LOCK AND HANDLE ASSEMBLY - 4 Door models

**Inside door release handle to lock**

1. Refit the inside door release handle bezel before any adjustment is made, allowing the handle to be set for the correct operating position.
2. Rotate the spring tensioned nyloc nut at the door lock clockwise or counter - clockwise to shorten or extend the operating length of the rod as required.

**Outside door release handle to lock**

3. Disconnect the connecting rod at the rear of the outer door release handle by releasing the rod from the plastic ferrule, rotate the rod clockwise or counter - clockwise to shorten or extend the operating length, refit the rod to the ferrule.

**NOTE: Door release should be effective before the total handle movement is exhausted to provide a small overthrow movement.**

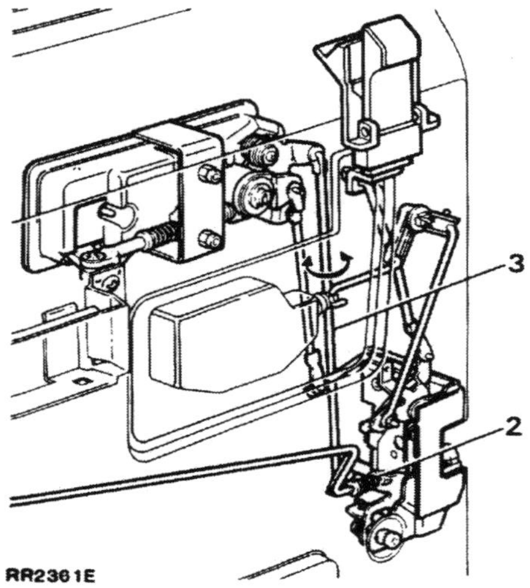

Continued

# BODY

## REAR DOOR - TRIM PANEL - 4 Door models

### Remove and refit

#### Removing

1. Disconnect the battery negative terminal.
2. Remove the screw securing the handle bezel.
3. Remove the bezel.
4. Prise the door locking button bezel from the trim panel.
5. Remove the two finisher buttons from the bottom of the door pull pocket to reveal the securing screws.
6. Remove the screws and withdraw the pocket from the trim panel.
7. Using a taped screwdriver, carefully prise the trim panel away from the door.
8. Disconnect the electrical plug from the window lift switch.
9. Remove the window lift switch by pushing from behind the trim panel.

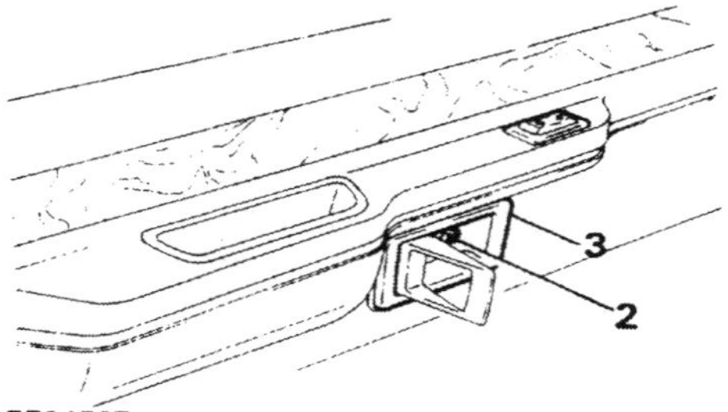

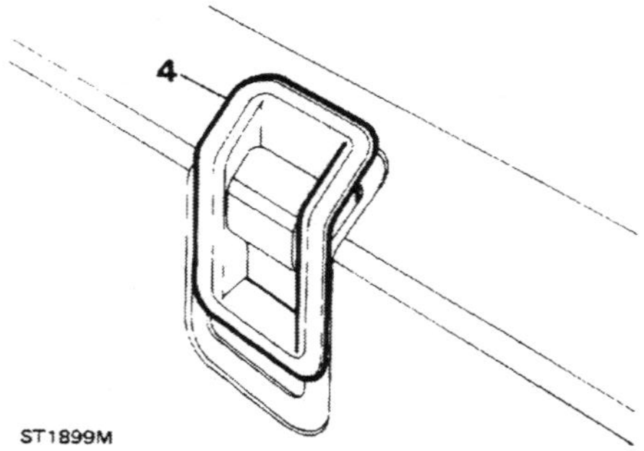

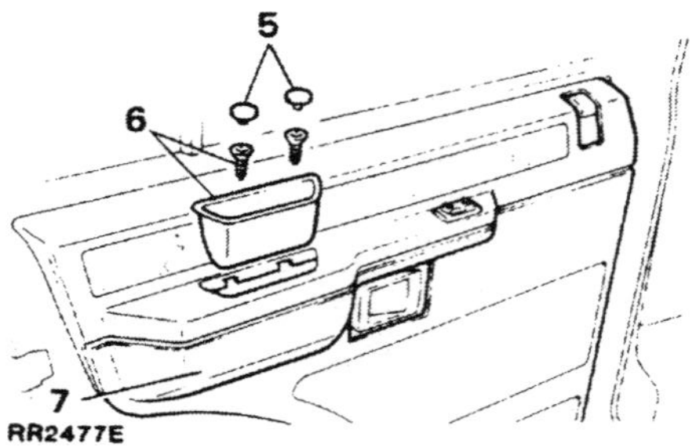

#### Refitting

10. Reverse the removal instructions.

## REAR DOOR LOCK, OUTSIDE AND INSIDE DOOR RELEASE HANDLES - 4 Door models

### Remove and refit

#### Removing

1. Ensure the window is fully closed position and disconnect the battery.
2. Remove the interior door handle bezel, prise the window lift switch from the trim panel and disconnect the electrical multi - plug.
3. Prise the sill button from the trim panel.
4. Remove the two screws from the bottom of the door pull pocket accessible after removing the two buttons and detach the trim panel from the door panel. Remove the plastic barrier sheet.
5. Disconnect the control rod from the inside door release handle by pulling the rod out of its location at the door lock.
6. Disconnect the sill locking control rod from the door lock by releasing the metal clip.
7. Disconnect the control rod from the outside door release handle by pulling it out of the plastic ferrule.
8. Release the door lock by removing the two countersunk screws from the door edge and the single screw (with shakeproof washer) on the inside of the door. Retrieve any spacing washers which may be fitted between the inner door panel and lock.
9. Withdraw the lock through the upper rear opening in the inner door panel.

**NOTE: If necessary the following items can also be removed.**

10. Remove the two nuts (with shakeproof washers) and retaining bracket securing the outside door release handle accessible through the upper rear cut - out on the inner door panel.
11. Carefully detach the outside door release handle from the outer door panel.
12. Remove the two screws (with plain washers) securing the inside door release handle to the inner door panel.
13. Withdraw the handle from its location with the connecting rod attached.

Continued

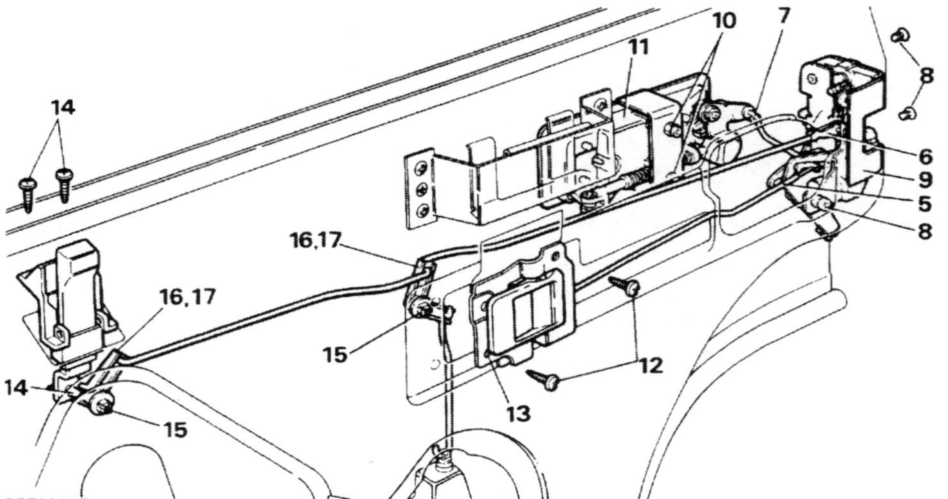

# BODY

14. Remove the two screws securing the sill locking button to the inner door panel and detach the sill button from the quadrant.

## Sill locking quadrants

15. Using a small screwdriver, or 3.175 mm diameter rod, press the plastic locking pins through the respective square inserts in the inner door panel, until they can be retrieved from inside the door.
16. Release the quadrants from the inner door panel and unhook the respective connecting rods.
17. Withdraw the quadrant from the inner door panel.

**NOTE: When refitting the quadrants the locking pins are entered into the square insert from outside and pressed in flush.**

## Refitting

18. Reverse the removal procedure items 1 to 17.

**NOTE: When refitting the door glass frame, ensure that it is positioned to suit the door opening before fully tightening the frame securing bolts.**

## ADJUSTMENT - REAR DOOR LOCK AND HANDLE ASSEMBLY - 4 Door models

### Outside door release handle to lock

1. Disconnect the short offset connecting rod at the rear of the door outer release handle, rotate the rod clockwise or counter - clockwise to shorten or extend the operating length of the rod.

**NOTE: Door release should be effective before the total handle movement is exhausted to provide a small overthrow movement.**

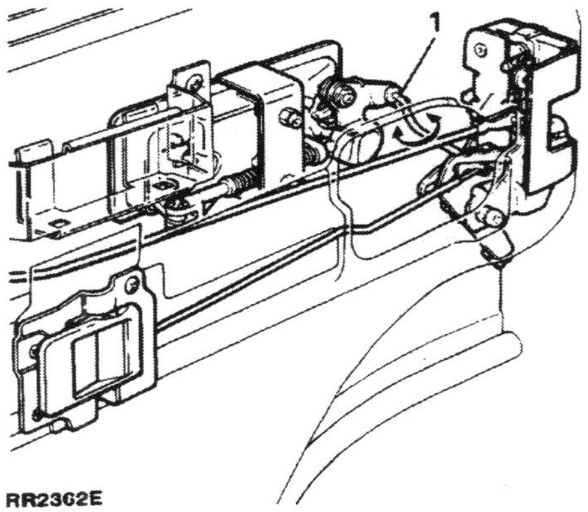

## ASYMMETRIC SPLIT REAR SEAT - LOCKING MECHANISM

**NOTE: 1989 Model Year vehicles have a revised seat locking mechanism which incorporates a push button release in place of a finger lift button. The revised latch and striker give the seat an improved positive location.**

### Remove and refit

#### Removing

1. Depress the seat release button and fold the seat back forward.
2. Unscrew and remove the seat release button.
3. Prise out the two trim buttons securing the trim covering to the latch tower.
4. Manoeuvre the trim covering from the tower.
5. Remove the three screws securing the latch to the tower, noting that access to the single screw is gained through the hole in the front of the tower.
6. Retrieve the latch from the opening at the rear of the tower, also if necessary retrieve the single screw.
7. The operating rod can be removed from the latch by releasing the rod at the plastic clip.

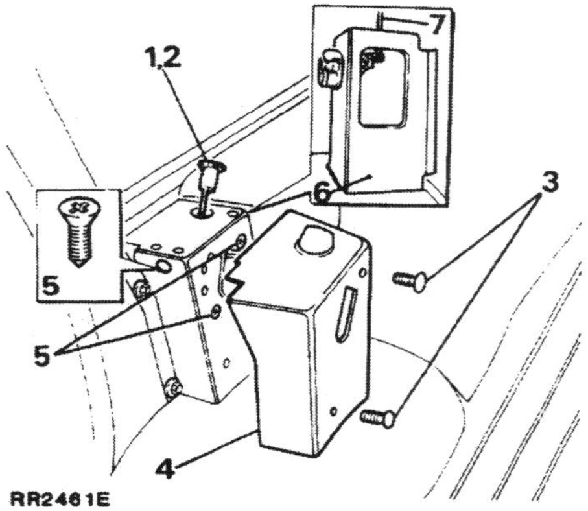

### Refitting

8. Refitting is a reversal of the removal procedure noting that when the seat release button is screwed onto the operating rod there must be a gap of 5 - 8 mm (0.196 - 0.312 inch) between the head of the button and lip of the trim covering after the button has been depressed.

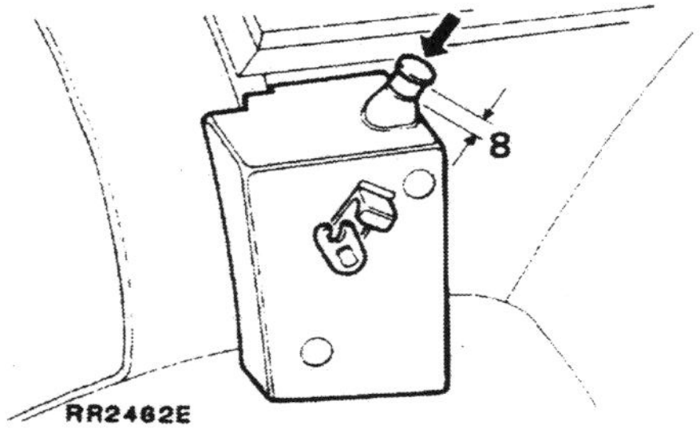

447

# AIR CONDITIONING 82

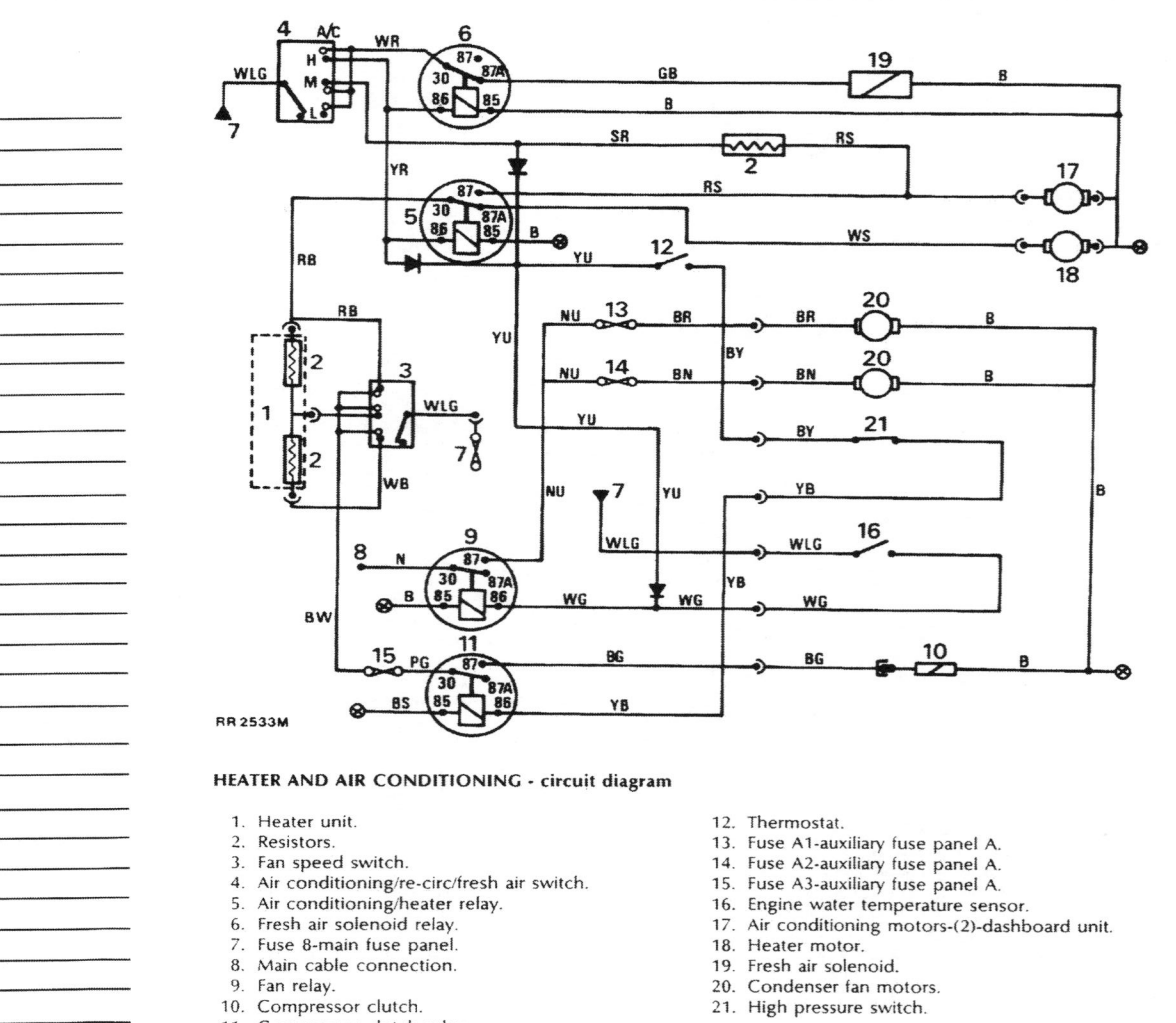

**HEATER AND AIR CONDITIONING - circuit diagram**

1. Heater unit.
2. Resistors.
3. Fan speed switch.
4. Air conditioning/re-circ/fresh air switch.
5. Air conditioning/heater relay.
6. Fresh air solenoid relay.
7. Fuse 8-main fuse panel.
8. Main cable connection.
9. Fan relay.
10. Compressor clutch.
11. Compressor clutch relay.
12. Thermostat.
13. Fuse A1-auxiliary fuse panel A.
14. Fuse A2-auxiliary fuse panel A.
15. Fuse A3-auxiliary fuse panel A.
16. Engine water temperature sensor.
17. Air conditioning motors-(2)-dashboard unit.
18. Heater motor.
19. Fresh air solenoid.
20. Condenser fan motors.
21. High pressure switch.

**Cable colour code**

| B | Black | N | Brown | R | Red | W | White |
| G | Green | O | Orange | S | Grey | Y | Yellow |
| L | Light | P | Purple | U | Blue | | |

The last letter of a colour code denotes the tracer.

## Notes

# WIPERS AND WASHERS 84

## HEATED WINDSCREEN WASHER JETS

1989 model year vehicles feature electrically heated windscreen washer jets. The operating thermostat fitted on the righthand headlamp mounting panel senses temerature and will operate the jet heaters at a temperature of 4°C ± 3° C.

## Washer jets

### Remove and refit

### Removing

1. Disconnect the battery negative lead.
2. Disconnect the electrical connection at the plug.
3. Withdraw the washer tube from the jet.
4. Push the jet upwards to remove from its mounting.
5. Remove the washer jet mounting from the bonnet, if necessary.

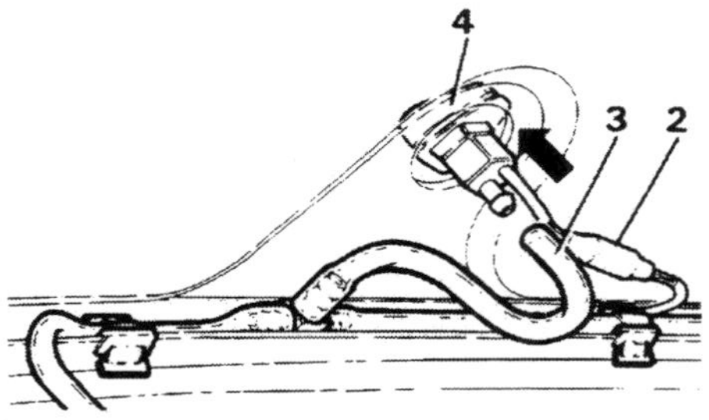

RR2549M

### Refitting

6. Reverse the removal procedure.

## Thermostat

The thermostat will operate (close) at 4°C ± 3° C. and re-open at 10°C ± 3° C.

### Remove

1. Disconnect the battery negative lead.
2. Remove the radiator grill.
3. Remove two screws securing the thermostat to the right hand headlamp mounting panel.
4. Withdraw the thermostat, disconnecting the electrical connector.

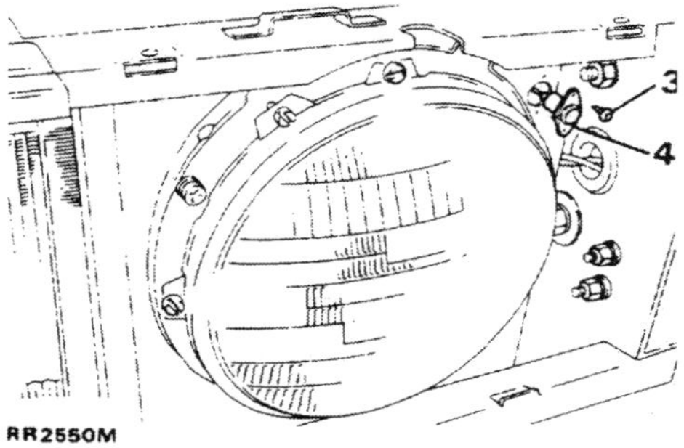

RR2550M

### Refitting

5. Reverse the removal procedure.

# Notes

## ELECTRICAL EQUIPMENT 86

**FUSE BOX**

RR2475M

| FUSE NO. | COLOUR CODE | FUSE VALUE | CIRCUIT SERVED | IGNITION KEY CONTROLLED |
|---|---|---|---|---|
| **MAIN FUSE PANEL** | | | | |
| 1 | Brown | 7.5 amp | RH headlamp low beam and power wash | |
| 2 | Brown | 7.5 amp | LH headlamp low beam | |
| 3 | Brown | 7.5 amp | RH headlamp high beam | |
| 4 | Brown | 7.5 amp | LH headlamp high beam, auxiliary lamp switch | |
| 5 | Tan | 5 amp | RH parking lights and instrument illumination | |
| 6 | Tan | 5 amp | LH parking lights and radio illumination | |
| 7 | Blue | 15 amp | Front wash/wiper motors, seat relay, window lift relay, antennae amplifier | Aux |
| 8 | Green | 30 amp | Heater/air con. motor | Aux |
| 9 | White | 25 amp | Heated rear screen | Ign |
| 10 | Green | 30 amp | Window lifts rear-option | Aux |
| 11 | Blue | 15 amp | Interior light delay, clock, radio, under bonnet illumination | Aux |
| 12 | Red | 10 amp | Rear fog guard (from dipped-headlamps) | |
| 13 | Blue | 15 amp | Direction indicators, stop lights, reverse lights, electric mirror pick up point, low coolant, heated jets, interior lamp delay, heater/air con relay | Ign |
| 14 | Yellow | 20 amp | Hazard lights, horn, headlamps flash | |
| 15 | Blue | 15 amp | Auxiliary driving lamps | Ign |
| 16 | Red | 10 amp | Rear wash/wipe motor, heated rear screen switch | Ign |
| 17 | Yellow | 20 amp | Cigar lighters (front and rear) | Ign |
| 18 | Red | 10 amp | Fuel pump | |
| 19 | Red | 10 amp | Central door locking-option | |
| 20 | Green | 30 amp | Electric window lifts front-option | Aux |

NOTE: Radio Cassette combination. An in-line 5 amp fuse is incorporated in the power input lead of the unit.

**AUXILIARY FUSE PANEL - (A)**

| A1 | Yellow | 20 amp | Air conditioning fan | IGN |
| A2 | Yellow | 20 amp | Air conditioning fan | IGN |
| A3 | Tan | 5 amp | Air conditioning compressor clutch | IGN |
| A4 | | | Spare | |
| A5 | Violet | 3 amp | Electric mirror motors | IGN |
| A6 | | | Spare | |

## Notes

# ELECTRICAL EQUIPMENT | 86

## AUXILIARY FUSE BOX

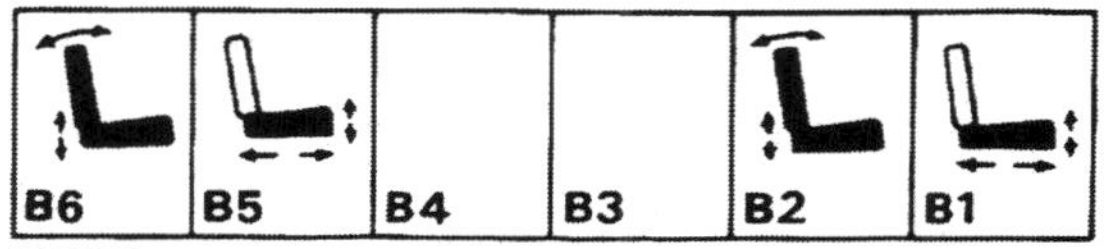

**AUXILIARY FUSE BOX (B)** - Located under the front left-hand seat

| FUSE NO | COLOUR CODE | FUSE VALUE | CIRCUIT SERVED |
|---|---|---|---|
| B1 | Green | 30 amp | Drivers seat base/height front |
| B2 | Green | 30 amp | Drivers seat recline/height rear |
| B3 | ---- | ---- | Spare |
| B4 | ---- | ---- | Spare |
| B5 | Green | 30 amp | Passengers seat base/height front |
| B6 | Green | 30 amp | Passengers seat recline/height rear |

RR2575E

---

# ELECTRICAL EQUIPMENT | 86

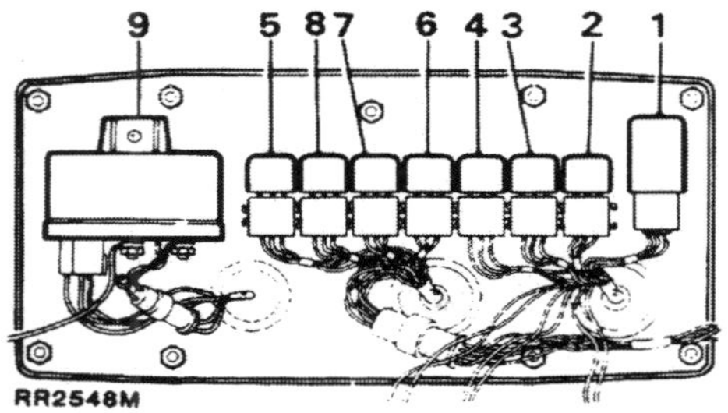

| Relay/delay/timer unit | Main Circuit Diagram Item Number | |
|---|---|---|
| | Right hand steer | Left hand steer Non-catalyst |
| 1. Headlamp wash timer unit | 19. M | 17. M |
| 2. Heated rear window relay | 66. M | 64. M |
| 3. Starter solenoid relay | 6. M | 6. M |
| 4. Brake check relay | 151. M | 157. M |
| 5. Fresh air solenoid relay | 6. A | 6. A |
| 6. Compressor clutch relay | 10. A | 10. A |
| 7. Condenser fan relay | 9. A | 9. A |
| 8. Air conditioning/heater | 5. A | 5. A |
| 9. Glow plug timer unit (Diesel models) | 143. M | 150. M |
| 10. Rear wiper delay | 132. M | 139. M |
| 11. Auxiliary lamp relay (if fitted) | 88. M | 86. M |
| 12. Ignition load relay | 1. M | 1. M |
| 13. Headlamp relay | 157. M | 26. M |
| 14. Heater/air con. relay | 163. M | 175. M |
| 15. Interior lamp delay | 101. M | 99. M |
| 16. Flasher/hazard unit | 75. M | 73. M |
| 17. Voltage sensitive switch (air conditioning) | 72. M | 70. M |
| 18. Front wiper delay | 15. M | 14. M |
| 19. Seat adjustment relay, two (if fitted) | 5/6. S | 5/6. S |
| 20. Main EFI relay | 10. E | 10. E |
| 21. Fuel pump relay | 11. E | 11. E |
| 22. Sunshine roof auxiliary relay (if fitted) | 3. SR | 3. SR |
| 23. Rear window lift relay (if fitted) | 13. W | 13. W |
| 24. Front window lift relay (if fitted) | 14. W | 14. W |
| 25. Window lift one touch unit (if fitted) | 1. W | 1. W |

M = Main circuit diagram
A = Air conditioning circuit diagram
S = Seat adjustment circuit diagram
E = EFI circuit diagram
SR = Sunroof circuit diagram
W = Window lift circuit diagram

**NOTE:** The brake check relay on left hand drive vehicles is situated inside the lower fascia panel.

**NOTE:** Refer to fuel injection section of manual for full information on E.F.I. relays.

Saudi vehicles are fitted with two extra relays, an overspeed monitor and buzzer, located immediately below the instrument binnacle.

Access to the two units is gained by removing the lower dash panel and steering column shroud.

---

## RELAYS-Identification

RR2547M shows left hand drive configuration of relays.

RR2548M shows right hand drive configuration of relays.

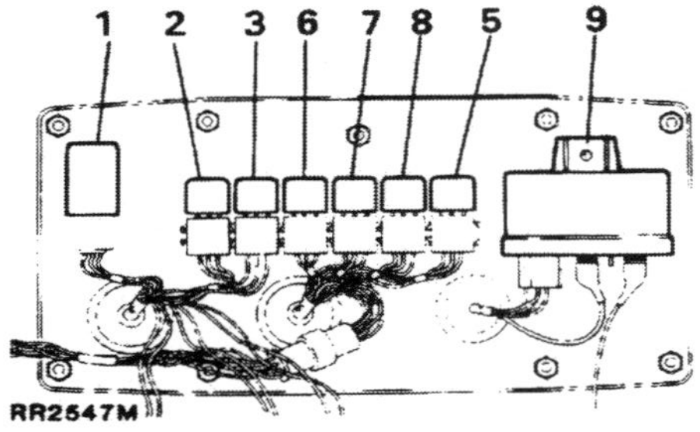

Closure panel viewed from the engine compartment, with protective cover removed.

451

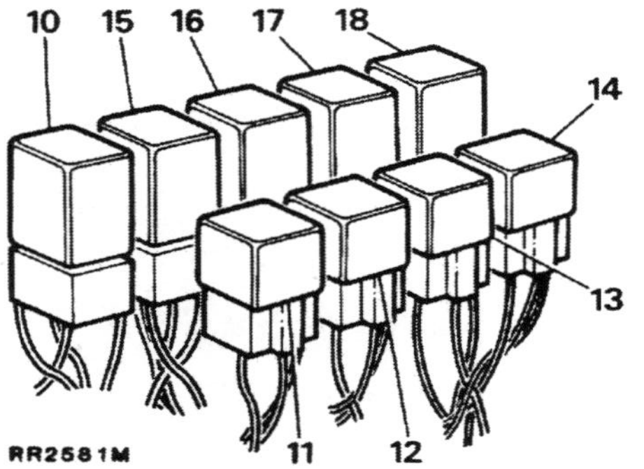


RR2581M

Steering column mounted relays viewed with the lower dash panel removed.

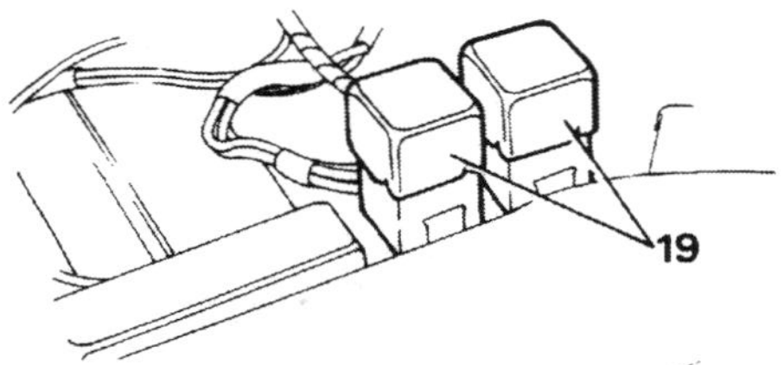


RR2554E

Seat adjustment relay (load control) located beneath the left hand front seat adjacent to fuse box (B).

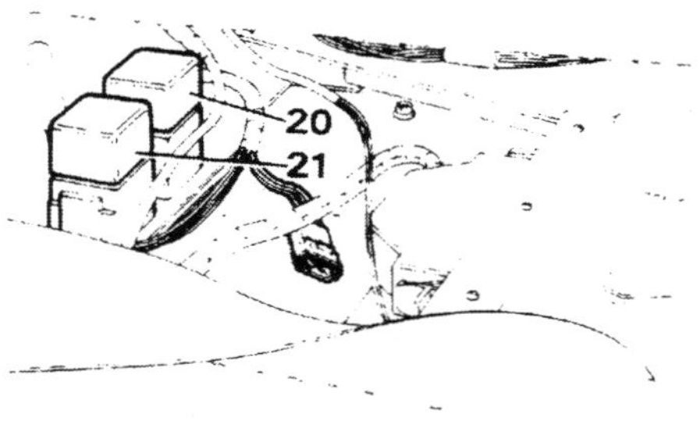


RR2374V

Main EFI (black terminal block) and fuel pump relays (blue terminal block) mounted beneath right hand front seat.

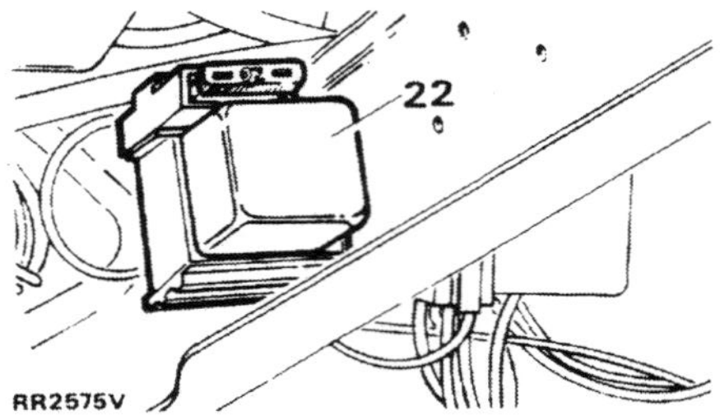


RR2575V

Sunshine roof auxiliary relay located on side of the steering column support bracket located behind the lower dash panel. (Left hand drive shown).

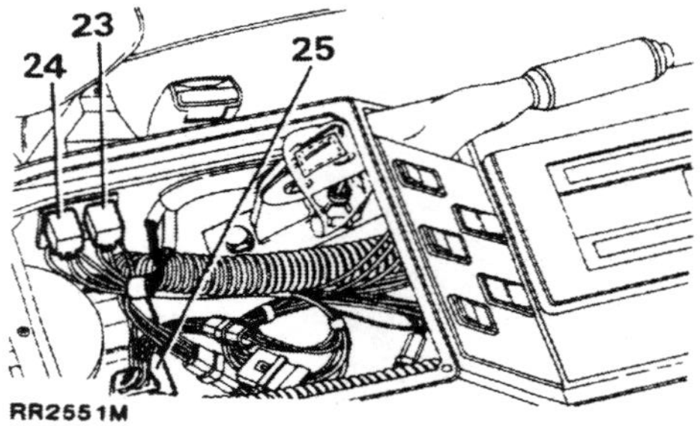


RR2551M

Front (black terminal block) and rear (blue terminal block) window relays. One touch control unit (25) is located inside the glove box, accessible by removing glove box liner.



## ELECTRICALLY OPERATED CENTRAL DOOR LOCKING SYSTEM

The optional central door locking system now includes an actuator unit to lock the upper tailgate.

Locking or unlocking the drivers door from outside by key operation, or from inside by sill knob automatically locks or unlocks all four doors, the upper tailgate and the fuel filler flap.

Front and rear passenger doors can be independently locked or unlocked from inside the vehicle by sill knob operation but can be overidden by further operation of the driver locking control.

On rear doors only a child safety lock is provided which can be mechanically pre-set to render the interior door handles inoperative.

Failure of an actuator will not affect the locking of the remaining three doors, tailgate or fuel filler flap. The door/tailgate with the inoperative actuator can still be locked or unlocked manually, but not the fuel filler flap.

It is also possible to override the tailgate central locking by use of the key.

**NOTE: The door lock actuator units contain non-serviceable parts. If a fault should occur replace the unit concerned with a new one.**

Before carrying out any maintenance work disconnect the battery.

## UPPER TAILGATE ACTUATOR UNIT

### Remove and refit

### Removing

1. Remove two screws and the trim covering to gain access to the actuator.
2. Disconnect the electrical connection.
3. Remove the two actuator retaining screws.
4. Manoeuvre the actuator assembly to detach the operating rod 'eye' from the actuator link to the lock.
5. Withdraw the tailgate actuator unit.

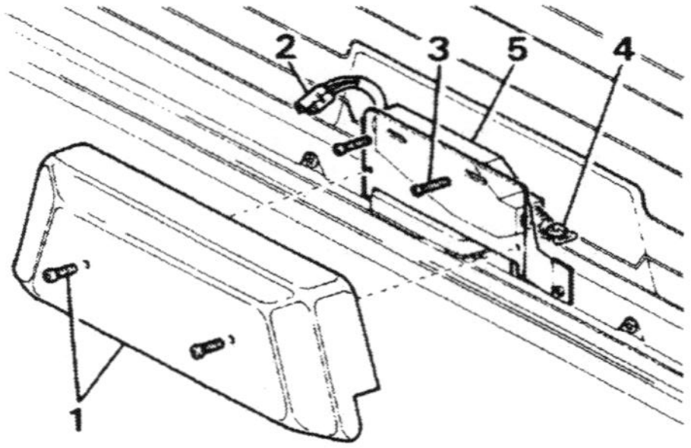


RR2570E

### Refitting

6. Reverse the removal procedure.
7. Check the operation of the central locking system.

# ELECTRICAL EQUIPMENT 86

## ELECTRIC SEAT ADJUSTMENT

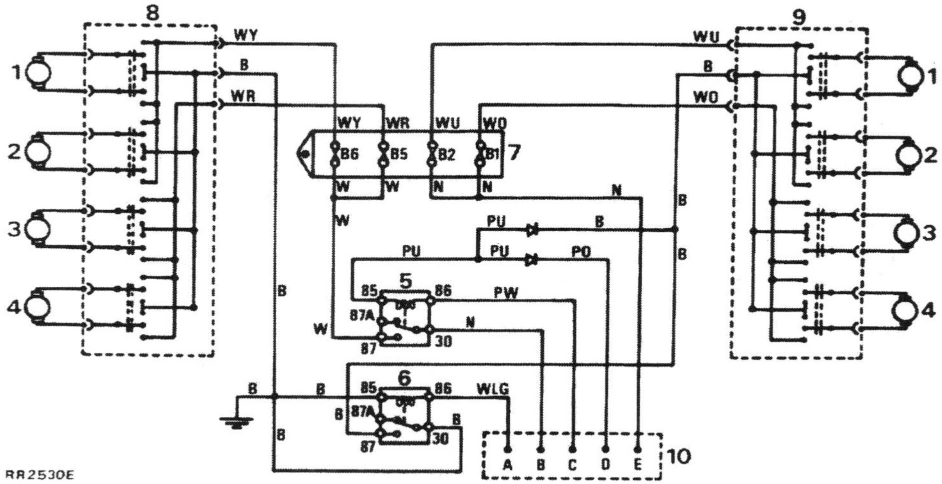

### Circuit diagram - RR2530E

1. Seat recline motor
2. Seat height (rear) motor
3. Seat base adjust motor
4. Seat height (front) motor
5. Load relay-from driver's door courtesy switch
6. Load relay-fused auxiliary feed controlled
7. Auxiliary fuse box (B)
8. Driver's seat control
9. Passenger's seat control
10. Main cable connections:
    - A: Fused auxiliary feed
    - B: Battery feed
    - C: Fused 12 volt
    - D: Courtesy switch earth
    - E: Battery feed

### Cable colour code

| B | Black | G | Green | P | Purple | Y | Yellow |
|---|---|---|---|---|---|---|---|
| U | Blue | S | Grey | R | Red | L | Light |
| N | Brown | O | Orange | W | White | | |

The last letter of a colour code denotes the tracer.

---

# ELECTRICAL EQUIPMENT 86

## ELECTRICAL EQUIPMENT - CIRCUIT DIAGRAMS
### - 1989 Model year

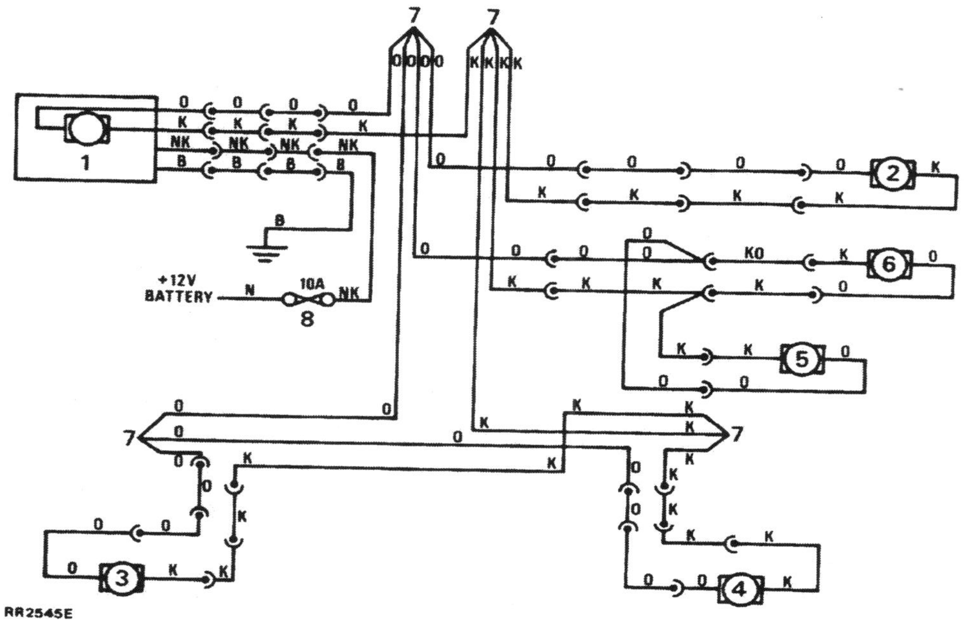

## CENTRAL DOOR LOCKING
### - Circuit diagram RR2545E

1. Switch/lock unit drivers door
2. Lock unit front passenger door
3. Lock unit-left hand rear door
4. Lock unit-right hand rear door
5. Fuel flap actuator
6. Lock unit-tailgate
7. Clinches
8. Fuse 19

### Cable colour code

| B | Black | L | Light | P | Purple |
|---|---|---|---|---|---|
| G | Green | N | Brown | R | Red |
| K | Pink | O | Orange | S | Grey |
| | | U | Blue | | |
| | | W | White | | |
| | | Y | Yellow | | |

The last letter of a colour code denotes the tracer.

453

# ELECTRICAL EQUIPMENT 86

## ELECTRIC WINDOW LIFT

### Circuit diagram - RR2531E

1. One touch control unit-drivers window
2. Window lift motor-drivers window
3. Window lift motor-front passengers side
4. Window lift motor LH rear
5. Window lift motor RH rear
6. Window lift switch drivers window
7. Window lift switch front passengers window
8. Window lift switch LH rear door
9. Window lift switch RH rear door
10. Isolator switch
11. Window lift switch in LH rear door
12. Window lift switch in RH rear door
13. Relay-rear windows
14. Relay-front windows
15. Clinches
16. Main cable fuses
    - a: Fuse 10
    - b: Fuse 20
    - c: Fuse 7

### Cable colour code

| | | | | | | |
|---|---|---|---|---|---|---|
| B | Black | L | Light | P | Purple | U | Blue |
| G | Green | N | Brown | R | Red | W | White |
| K | Pink | O | Orange | S | Grey | Y | Yellow |

The last letter of a colour code denotes the tracer.

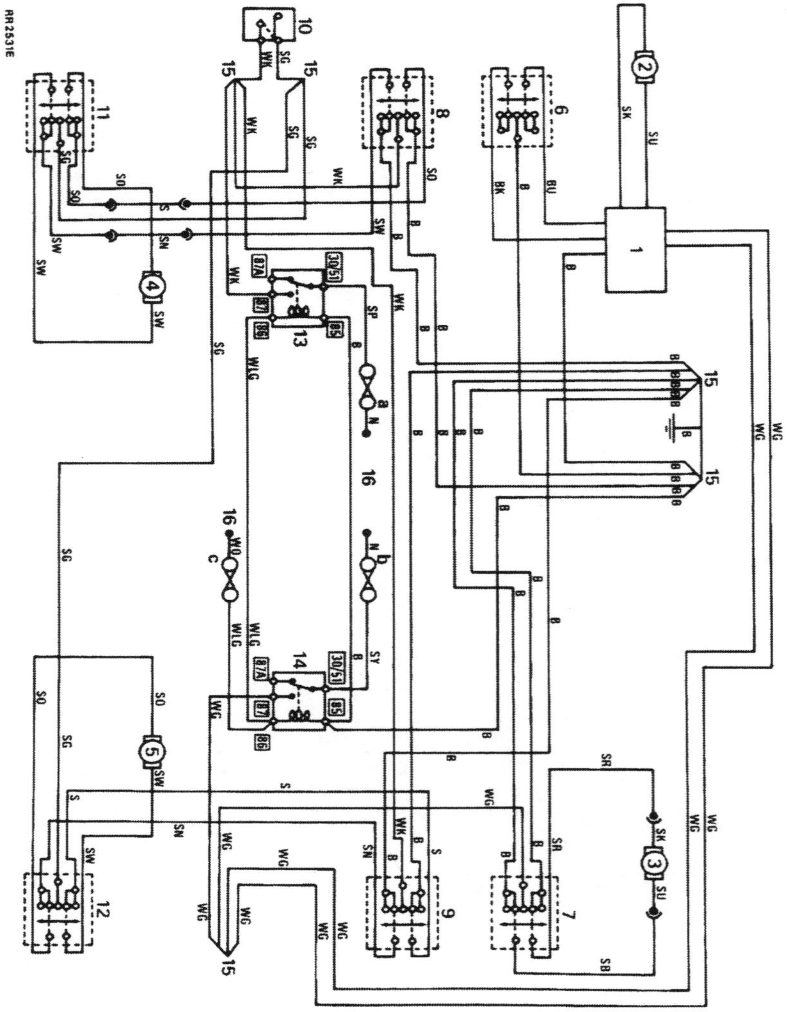

454

# ELECTRICAL EQUIPMENT 86

## MAIN CIRCUIT DIAGRAM
### Right Hand Steering - RR2571M & RR2572M

### Numerical key

|  |  |
|---|---|
| 1 | Ignition load relay |
| 2 | Battery |
| 3 | Terminal post |
| 4 | Starter solenoid |
| 5 | Starter motor |
| 6 | Starter relay |
| 7 | Starter inhibit switch (Automatic) |
| 8 | Ignition switch |
| 9 | Tachometer |
| 10 | Voltage transformer(dim dip) |
| 11 | Ignition warning lamp |
| 12 | Alternator |
| 13 | Fuse 7 |
| 14 | Front wipe/wash switch |
| 15 | Headlamp wash pump (option) |
| 16 | Front wipe delay unit |
| 17 | Front wiper motor |
| 18 | Front wash pump |
| 19 | Headlamp wash timer unit (option) |
| 20 | Headlamp wash pump (option) |
| 21 | Main lighting switch |
| 22 | Fuse 5 |
| 23 | Fuse 6 |
| 24 | LH side lamp |
| 25 | LH tail lamp |
| 26 | LH number plate lamp |
| 26b | RH number plate lamp |
| 27 | Main beam dip/flash switch |
| 28 | RH side lamp |
| 29 | RH tail lamp |
| 30 | Rheostat |
| 31 | Fuse 3 |
| 32 | Fuse 4 |
| 33 | Fuse 1 |
| 34 | Fuse 2 |
| 35 | Rear fog switch |
| 36 | LH rear fog |
| 37 | RH rear fog |
| 38 | Switch illumination (2 off) |
| 39 | Cigar lighter illumination (2 off) |
| 40 | Heater illumination (4 off) |
| 41 | Clock illumination |
| 42 | Automatic gear selector illumination (2 off) |
| 43 | Instrument illumination (6 off) |
| 43b | Column switch illumination |
| 44 | Rear fog warning lamp |
| 45 | LH rear fog |
| 46 | RH rear fog |
| 47 | LH dip beam |
| 48 | RH dip beam |
| 49 | LH main beam |
| 50 | RH main beam |
| 51 | Main beam warning lamp |
| 52 | Fuel gauge |
| 53 | Fuel gauge sender unit |
| 54 | Water temperature gauge |
| 55 | Water temperature sender unit |
| 56 | Fuse 11 |
| 57 | Horn switch |
| 58 | RH horn |
| 59 | LH horn |
| 60 | Under bonnet light switch |
| 61 | Under bonnet light |
| 62 | Clock |
| 63 | Fuse 19 |
| 64 | Fuse 20 |
| 65 | Pick-up point central locking/window lift (option) |
| 66 | Heated rear window relay |
| 67 | Fuse 9 |
| 68 | Radio aerial amplifier |
| 69 | Heated rear screen |
| 70 | Heated rear screen switch |
| 71 | Heated rear screen warning lamp |
| 72 | Voltage sensitive delay unit |
| 73 | Fuse 13 |
| 74 | Hazard switch |
| 75 | Flasher unit |
| 76 | Direction indicator switch |
| 77 | Hazard/indicator warning lamp |
| 78 | LH rear indicator lamp |
| 79 | LH front indicator lamp |
| 80 | LH side repeater lamp |
| 81 | RH side repeater lamp |
| 82 | RH front indicator lamp |
| 83 | RH rear indicator lamp |
| 84 | Trailer warning lamp |
| 85 | Stop lamp switch |
| 86 | Reverse lamp switch |
| 87 | Auxiliary lamp relay (option) |
| 88 | Auxiliary lamp switch (option) |
| 89 | LH stop lamp |
| 90 | RH stop lamp |
| 91 | LH reverse lamp |
| 92 | RH reverse lamp |
| 93 | LH auxiliary lamp (option) |
| 94 | RH auxiliary lamp (option) |
| 95 | Auxiliary lamp switch (option) |
| 96 | Fuse 17 |
| 97 | Dash cigar lighter |
| 98 | Cubby box cigar lighter |
| 99 | Front interior lamp |
| 100 | Rear interior lamp |
| 101 | Interior lamp delay unit |
| 102 | LH door edge lamp |
| 103 | RH door edge lamp |
| 104 | LH puddle lamp |
| 105 | RH puddle lamp |
| 106 | Interior lamp switch |
| 107 | LH rear door switch |
| 108 | RH rear door switch |
| 109 | Tailgate switch |
| 110 | LH front door switch |
| 111 | RH front door switch |
| 112 | Heated jets |
| 113 | Thermostat heated jets |
| 114 | Oil pressure warning switch |
| 115 | Oil pressure warning lamp |
| 116 | Fuse 18 |
| 117 | Fuel cut off relay (carburetter models) |
| 118 | Fuel pump (petrol models) |
| 119 | Ignition coil (petrol models) |
| 120 | Capacitor (petrol models) |
| 121 | Distributor (petrol models) |
| 122 | EFi harness plug |
| 123 | Fuel shut off solenoid (Diesel) |
| 124 | Not used |
| 125 | Radio fuse |
| 126 | Radio and four speakers |
|  | - LF-left hand front speaker |
|  | - LR-left hand rear speaker |
|  | - RF-right hand front speaker |
|  | - RR-right hand rear speaker |
| 127 | Sun roof pick up point (option) |
| 128 | Automatic transmission oil temperature warning lamp |
| 129 | Automatic transmission oil temperature switch |
| 130 | Fuse 16 |
| 131 | Rear wash wipe switch |
| 132 | Rear wipe delay unit |
| 133 | Rear wiper motor |
| 134 | Rear screen wash pump |
| 135 | Low screen wash fluid level warning lamp |
| 136 | Low screen wash switch |
| 137 | Low coolant switch |
| 138 | Multi-function unit in binnacle |
| 139 | Low coolant level warning lamp |
| 140 | Low fuel level warning lamp |
| 141 | Cold start/diesel glow plug warning lamp |
| 142 | Cold start switch (carburetter) |
| 143 | Glow plug timer (diesel) |
| 144 | Glow plugs (diesel) |
| 145 | Handbrake warning lamp |
| 146 | Brake tail warning lamp |
| 147 | Handbrake warning switch |
| 148 | Brake fluid level warning switch |
| 149 | Brake pad wear warning lamp |
| 150 | Brake pad wear sensors |
| 151 | Brake check relay |
| 152 | Split charge terminal post (option) |
| 153 | Split charge relay |
| 154 | Heater/air con relay |
| 155 | Fuse 8 |
| 156 | Coil negative (engine RPM input to ECU) |
| 157 | Headlamp relay |
| 158 | Ignition load relay ( + ) |
| 159 | Battery feed ( + ) |
| 160 | Ignition auxiliary ( + ) |
| 161 | Ignition on ( + ) |
| 162 | Earth (-) |
| 163 | Heater/air con relay |
| 164 | Trailer pick up point |
| 165 | Electric seats pick up point (option) |
| 166 | Electric mirrors pick up point (option) |
| 167 | Electric mirrors pick up point (option) |
| 168 | Alarm connection (dealer fit) |

### CABLE COLOUR CODE

| | | | | | | | |
|---|---|---|---|---|---|---|---|
| B | Black | L | Light | P | Purple | U | Blue |
| G | Green | N | Brown | R | Red | W | White |
| K | Pink | O | Orange | S | Grey | Y | Yellow |

The last letter of a colour code denotes the tracer.

---

# ELECTRICAL EQUIPMENT 86

## MAIN CIRCUIT DIAGRAM
### Right Hand Steering - RR2571M & RR2572M

### Alphabetical key

|  |  |
|---|---|
| Alarm connection (dealer fit) | 168 |
| Alternator | 12 |
| Automatic gear selector illumination (2 off) | 42 |
| Automatic transmission oil temperature switch | 129 |
| Automatic transmission oil temperature warning lamp | 128 |
| Auxiliary lamp relay (option) | 88 |
| Auxiliary lamp switch (option) | 95 |
| Battery | 2 |
| Battery feed ( + ) | 159 |
| Brake check relay | 151 |
| Brake fluid level warning switch | 148 |
| Brake pad wear sensors | 150 |
| Brake pad wear warning lamp | 149 |
| Brake tail warning lamp | 146 |
| Capacitor (petrol models) | 120 |
| Cigar lighter illumination (2 off) | 39 |
| Clock | 62 |
| Clock illumination | 41 |
| Coil negative (engine RPM input to ECU) | 156 |
| Cold start/diesel glow plug warning lamp | 141 |
| Cold start switch (carburetter) | 142 |
| Column switch illumination | 43b |
| Cubby box cigar lighter | 98 |
| Dash cigar lighter | 97 |
| Direction indicator switch | 76 |
| Distributor (petrol models) | 121 |
| Earth (-) | 162 |
| EFi harness plug | 122 |
| Electric mirrors pick up point (option) | 167 |
| Electric seats pick up point (option) | 165 |
| Flasher unit | 75 |
| Front interior lamp | 99 |
| Front wash pump | 18 |
| Front wipe delay unit | 16 |
| Front wiper motor | 17 |
| Front wipe/wash switch | 14 |
| Fuel cut off relay (carburetter models) | 117 |
| Fuse 1 | 34 |
| Fuse 2 | 35 |
| Fuse 3 | 32 |
| Fuse 4 | 33 |
| Fuse 5 | 23 |
| Fuse 6 | 22 |
| Fuse 7 | 13 |
| Fuse 8 | 155 |
| Fuse 9 | 67 |
| Fuse 10 | 56 |
| Fuse 11 | 155 |
| Fuse 13 | 73 |
| Fuse 14 | 28 |
| Fuse 15 | 85 |
| Fuse 16 | 130 |
| Fuse 17 | 96 |
| Fuse 18 | 116 |
| Fuse 19 | 63 |
| Fuse 20 | 64 |
| Fuel gauge | 52 |
| Fuel gauge sender unit | 53 |
| Fuel pump (petrol models) | 118 |
| Fuel shut off solenoid (Diesel) | 123 |
| Glow plugs (diesel) | 144 |
| Glow plug timer (diesel) | 143 |
| Handbrake warning lamp | 145 |
| Handbrake warning switch | 147 |
| Hazard/indicator warning lamp | 77 |
| Hazard switch | 74 |
| Headlamp relay | 157 |
| Headlamp wash pump (option) | 20 |
| Headlamp wash timer unit (option) | 19 |
| Heated jets | 112 |
| Heated rear screen | 69 |
| Heated rear screen switch | 70 |
| Heated rear screen warning lamp | 71 |
| Heater/air conditioning connections | 154 |
| Heater/air con relay | 163 |
| Heater illumination (4 off) | 40 |
| Horn switch | 57 |
| Ignition auxiliary ( + ) | 160 |
| Ignition coil (petrol models) | 119 |
| Instrument illumination (6 off) | 43 |
| Interior lamp delay unit | 101 |
| Interior lamp switch | 106 |
| Earth (-) | 1 |
| Ignition load relay | 158 |
| Ignition on ( + ) | 161 |
| Ignition switch | 8 |
| Electric seats pick up point (option) | 165 |
| Flasher unit | 75 |
| Front interior lamp | 99 |
| Front wash pump | 17 |
| Front wipe delay unit | 18 |
| Front wiper motor | 15 |
| Front wipe/wash switch | 16 |
| Fuel cut off relay (carburetter models) | 14 |
| Split charge terminal post (option) | 117 |
| Split charge relay | 34 |
| Heater/air con relay | 35 |
| Heater/air conditioning connections | 32 |
| Fuse 8 | 33 |
| Coil negative (engine RPM input to ECU) | 23 |
| Fuse 16 | 130 |
| LH horn | 59 |
| LH auxiliary lamp (option) | 93 |
| LH dip beam | 47 |
| LH door edge lamp | 102 |
| LH front door switch | 110 |
| LH front indicator lamp | 79 |
| LH horn | 59 |
| LH main beam | 49 |
| LH number plate lamp | 26 |
| LH puddle lamp | 103 |
| LH rear door switch | 107 |
| LH rear fog | 45 |
| LH rear indicator lamp | 78 |
| LH reverse lamp | 91 |
| LH side lamp | 24 |
| LH side repeater lamp | 80 |
| LH stop lamp | 89 |
| LH tail lamp | 25 |
| Low coolant level warning lamp | 139 |
| Low coolant switch | 137 |
| Low fuel level warning lamp | 140 |
| Low screen wash fluid level warning lamp | 135 |
| Low screen wash switch | 136 |
| Main beam dip/flash switch | 27 |
| Main beam warning lamp | 51 |
| Main lighting switch | 21 |
| Multi-function unit in binnacle | 138 |
| Oil pressure warning switch | 114 |
| Oil pressure warning lamp | 115 |
| Pick-up point central locking/window lift (option) | 65 |
| Radio aerial amplifier | 68 |
| Radio and four speakers | 126 |
| Radio fuse | 125 |
| Rear fog switch | 36 |
| Rear fog warning lamp | 44 |
| Rear interior lamp | 100 |
| Rear screen wash pump | 134 |
| Rear screen wash wipe switch | 131 |
| Rear wipe delay unit | 132 |
| Rear wiper motor | 133 |
| Reverse lamp switch | 87 |
| Rheostat | 31 |
| RH auxiliary lamp (option) | 94 |
| RH dip beam | 48 |
| RH door edge lamp | 104 |
| RH front indicator lamp | 82 |
| RH horn | 58 |
| RH main beam | 50 |
| RH number plate lamp | 26b |
| RH puddle lamp | 105 |
| RH rear door switch | 108 |
| RH rear fog | 46 |
| RH rear indicator lamp | 83 |
| RH reverse lamp | 92 |
| RH side lamp | 29 |
| RH side repeater lamp | 81 |
| RH stop lamp | 90 |
| RH tail lamp | 30 |
| Split charge relay | 153 |
| Split charge terminal post (option) | 152 |
| Starter inhibit switch (Automatic) | 7 |
| Starter motor | 5 |
| Starter relay | 6 |
| Starter solenoid | 4 |
| Stop lamp switch | 86 |
| Sun roof pick up point (option) | 127 |
| Switch illumination (2 off) | 38 |
| Tachometer | 9 |
| Tailgate switch | 109 |
| Terminal post | 3 |
| Thermostat heated jets | 113 |
| Trailer warning lamp | 84 |
| Under bonnet light | 61 |
| Under bonnet light switch | 60 |
| Voltage transformer(dim dip) | 10 |
| Voltage sensitive delay unit | 72 |
| Water temperature gauge | 54 |
| Water temperature sender unit | 55 |

# ELECTRICAL EQUIPMENT 86

**MAIN CIRCUIT DIAGRAM**
Right hand steering - RR2571M & RR2572M

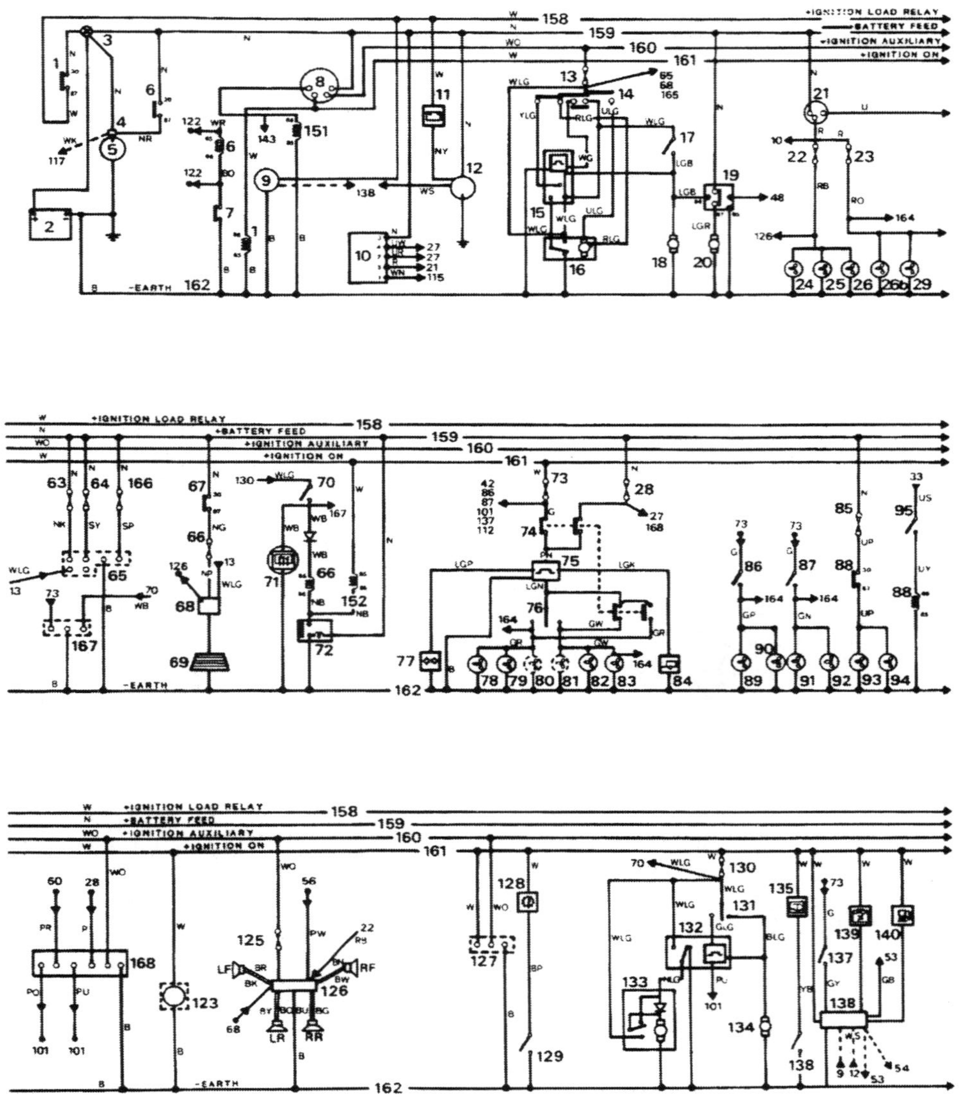

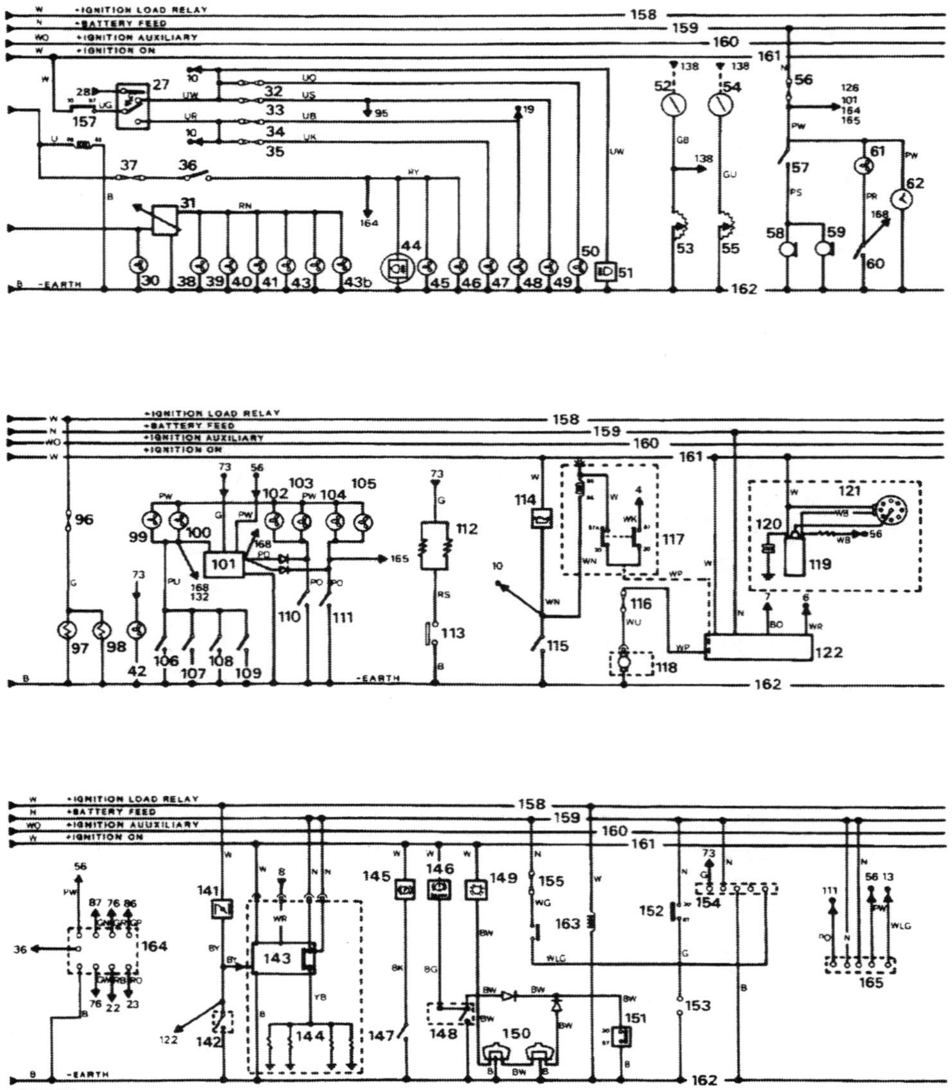

# ELECTRICAL EQUIPMENT 86

## MAIN CIRCUIT DIAGRAM - NON CATALYST VEHICLES
### Left hand Steering - RR2573M & RR2574M

### Numerical key

|  |  |
|---|---|
| 1 | Ignition load relay |
| 2 | Battery |
| 3 | Terminal post |
| 4 | Starter solenoid |
| 5 | Starter motor |
| 6 | Starter relay |
| 7 | Starter inhibit switch (automatic) |
| 8 | Ignition switch |
| 9 | Tachometer |
| 10 | Ignition warning lamp |
| 11 | Alternator |
| 12 | Fuse 7 |
| 13 | Front wipe/wash switch |
| 14 | Front wipe delay unit |
| 15 | Front wash pump |
| 16 | Front wiper motor |
| 17 | Headlamp wash timer unit (option) |
| 18 | Headlamp wash pump (option) |
| 19 | Main lighting switch |
| 20 | Fuse 6 |
| 21 | LH side lamp |
| 22 | RH side lamp |
| 23 | LH tail lamp |
| 24 | LH number plate lamp |
| 24 (a) | RH number plate lamp |
| 25 | Main beam dipflash switch |
| 26 | Headlamp relay |
| 27 | RH side lamp |
| 28 | RH tail lamp |
| 29 | Rheostat |
| 30 | Fuse 5 |
| 31 | Fuse 4 |
| 32 | Fuse 3 |
| 33 | Fuse 2 |
| 34 | Fuse 1 |
| 35 | Fuse log switch |
| 36 | Switch illumination (2 off) |
| 37 | Cigar lighter illumination (2 off) |
| 38 | Heater illumination (4 off) |
| 39 | Clock illumination |
| 40 | Automatic gear selector illumination (2 off) |
| 41 | Instrument illumination (6 off) |
| 41 (a) | Column switch illumination |
| 42 | Rear log warning lamp |
| 43 | LH rear log |
| 44 | RH rear log |
| 45 | LH dip beam |
| 46 | LH dip beam |
| 47 | LH main beam |
| 48 | Main beam warning lamp |
| 49 | RH main beam |
| 50 | Fuel gauge |
| 51 | Fuel gauge sender unit |
| 52 | Water temperature gauge |
| 53 | Water temperature sender unit |
| 54 | Fuse 11 |
| 55 | Horn switch |
| 56 | RH horn |
| 57 | LH horn |
| 58 | Under bonnet illumination switch |
| 59 | Under bonnet light |
| 60 | Clock |
| 61 | Fuse 19 |
| 62 | Fuse 20 |
| 63 | Pick-up point central locking/window lift |
| 64 | Heated rear window relay |
| 65 | Fuse 9 |
| 66 | Radio aerial amplifier |
| 67 | Heated rear screen |
| 68 | Heated rear screen switch |
| 69 | Heated rear screen warning lamp |
| 70 | Voltage sensitive switch |
| 71 | Fuse 13 |
| 72 | Hazard switch |
| 73 | Flasher unit |
| 74 | Direction indicator switch |
| 75 | Hazard/indicator warning lamp |
| 76 | LH rear indicator lamp |
| 77 | LH front indicator lamp |
| 78 | LH side repeater lamp |
| 79 | RH side repeater lamp |
| 80 | RH front indicator lamp |
| 81 | RH rear indicator lamp |
| 82 | Trailer warning lamp |
| 83 | Fuse 15 |
| 84 | Stop lamp switch |
| 85 | Reverse lamp switch |
| 86 | Auxiliary lamp switch |
| 87 | LH stop lamp |
| 88 | RH stop lamp |
| 89 | LH reverse lamp |
| 90 | RH reverse lamp |
| 91 | LH auxiliary lamp (option) |
| 92 | RH auxiliary lamp (option) |
| 93 | Fuse 17 |
| 94 | Cubby box cigar lighter |
| 95 | Dash cigar lighter |
| 96 | RH interior lamp |
| 97 | LH interior lamp |
| 98 | RH interior lamp |
| 99 | Interior lamp delay unit |
| 100 | LH door edge lamp |
| 101 | RH door edge lamp |
| 102 | LH puddle lamp |
| 103 | RH puddle lamp |
| 104 | Interior lamp switch |
| 105 | LH rear door switch |
| 106 | RH rear door switch |
| 107 | Tailgate switch |
| 108 | LH front door switch |
| 109 | RH front door switch |
| 110 | Heated jets |
| 111 | Thermostat-heated jets |
| 112 | Oil pressure warning lamp |
| 113 | Oil pressure switch |
| 114 | Fuse 18 |
| 115 | Fuel shut-off relay - carburetter |
| 116 | Fuel pump - petrol models |
| 117 | Ignition coil |
| 118 | Capacitor |

### CABLE COLOUR CODE

| | | | | | |
|---|---|---|---|---|---|
| B | Black | L | Light | P | Purple |
| G | Green | N | Brown | R | Red |
| K | Pink | O | Orange | S | Grey |

The last letter of a colour code denotes the tracer.

## MAIN CIRCUIT DIAGRAM - NON CATALYST VEHICLES
### Left Hand Steering - RR2573M & RR2574M

|  |  |
|---|---|
| 119 | Distributor |
| 120 | EFI Harness plug |
| 121 | Fuel shut-off solenoid-Diesel |
| 122 | Trailer pick up point |
| 123 | Radio fuse |
| 124 | Radio |
| 125 | Four speakers |
| 126 | Alarm pick up point |
| 127 | Speed transducer, (Saudi only) |
| 131 | Seat buckle switch |
| 132 | Overspeed monitor (Saudi only) |
| 133 | Overspeed buzzer (Saudi only) |
| 134 | Sun roof pick up point (option) |
| 135 | Automatic transmission oil temperature warning lamp |
| 136 | Automatic transmission oil temperature switch |
| 137 | Fuse 16 |
| 138 | Rear wash wipe switch |
| 139 | Rear wipe delay unit |
| 140 | Rear wiper motor |
| 141 | Rear screen wash pump |
| 142 | Low screen wash fluid level warning lamp |
| 143 | Low screen wash switch |
| 144 | Low coolant switch |
| 145 | Multi-function unit in binnacle |
| 146 | Low coolant level warning lamp |
| 147 | Low fuel level warning lamp |
| 148 | Cold start/Diesel glow plug warning lamp |
| 149 | Choke switch - carburetter |
| 150 | Glowplug timer/Diesel |
| 151 | Glowplugs/Diesel |
| 152 | Handbrake/warning switch |
| 153 | Handbrake warning lamp |
| 154 | Brake fluid level warning lamp |
| 154 (a) | Brake fluid level warning switch |
| 155 | Brake pad wear warning lamp |
| 156 | Brake pad wear sensors |
| 157 | Brake check unit |
| 158 | Split charge terminal post |
| 159 | Split charge terminal post |
| 160 | Heater/air conditioning connections |
| 161 | Fuse 8 |
| 162 | Coil negative (engine RPM input to ECU) |
| 163 | Ignition load relay (+) |
| 164 | Battery feed (+) |
| 165 | Ignition on (+) |
| 166 | Ignition switch |
| 167 | Earth (-) |
| 168 | Warning lights common earth (-) |
| 169 | Warning lights supply (+) |
| 170 | Electric seats pick up point (option) |
| 171 | Fuse 14 |
| 173 | Fuse 10 |
| 174 | Electric mirrors pick up point (option) |
| 175 | Heater aircon relay |

### Alphabetical key

|  |  |
|---|---|
| Main lighting switch | 19 |
| Multi-function unit in binnacle | 145 |
| Oil pressure switch | 113 |
| Oil pressure warning lamp | 112 |
| Overspeed buzzer (Saudi only) | 133 |
| Overspeed monitor (Saudi only) | 132 |
| Pick-up point central locking/window lift | 63 |
| RH auxiliary lamp (option) | 92 |
| RH dip beam | 46 |
| RH door edge lamp | 102 |
| RH front door switch | 109 |
| RH front indicator lamp | 80 |
| RH horn | 56 |
| RH interior lamp | 98 |
| RH main beam | 48 |
| RH number plate lamp | 24 (a) |
| RH puddle lamp | 103 |
| RH rear door switch | 106 |
| RH rear indicator lamp | 81 |
| RH rear log | 44 |
| RH reverse lamp | 90 |
| RH side lamp | 27 |
| RH side repeater lamp | 79 |
| RH stop lamp | 88 |
| RH tail lamp | 28 |
| Radio | 124 |
| Radio aerial amplifier | 66 |
| Radio fuse | 123 |
| Rear log warning lamp | 42 |
| Rear screen wash pump | 141 |
| Rear wash wipe switch | 138 |
| Rear wipe delay unit | 139 |
| Rear wiper motor | 140 |
| Split charge relay (option) | 127 |
| Split charge terminal post | 158 |
| Starter inhibit switch (automatic) | 159 |
| Starter motor | 7 |
| Starter relay | 5 |
| Starter solenoid | 6 |
| Stop lamp switch | 4 |
| Sun roof pick up point (option) | 84 |
| Switch illumination (2 off) | 134 |
| Tachometer | 36 |
| Tailgate switch | 9 |
| Terminal post | 107 |
| Thermostat-heated jets | 3 |
| Trailer pick up point | 111 |
| Trailer warning lamp | 122 |
| Under bonnet illumination switch | 82 |
| Under bonnet light | 58 |
| Voltage sensitive switch | 59 |
| Warning lights common earth (-) | 70 |
| Warning lights supply (+) | 168 |
| Water temperature gauge | 169 |
| Water temperature sender unit | 52 |
| | 53 |

|  |  |
|---|---|
| Fuse 15 | 83 |
| Fuse 16 | 137 |
| Fuse 17 | 114 |
| Fuse 18 | 61 |
| Fuse 19 | 62 |
| Fuse 20 | 150 |
| Glowplug timer/Diesel | 151 |
| Glowplugs/Diesel | 86 |
| Auxiliary lamp relay | 93 |
| Auxiliary lamp switch (option) | 2 |
| Battery | 164 |
| Battery feed (+) | 75 |
| Brake check unit | 26 |
| Brake fluid level warning lamp | 157 |
| Brake fluid level warning switch | 154 (a) |
| Brake pad wear sensors | 156 |
| Brake pad wear warning lamp | 155 |
| Capacitor | 118 |
| Choke switch - carburetter | 149 |
| Cigar lighter illumination (2 off) | 37 |
| Clock | 60 |
| Clock illumination | 39 |
| Coil negative (engine RPM input to ECU) | 162 |
| Cold start/Diesel glow plug warning lamp | 148 |
| Column switch illumination | 41 (a) |
| Cubby box cigar lighter | 96 |
| Dash cigar lighter | 95 |
| Direction indicator switch | 74 |
| Distributor | 119 |
| EFI Harness plug | 120 |
| Earth (-) | 167 |
| Electric mirrors pick up point (option) | 174 |
| Electric seats pick up point (option) | 170 |
| Flasher unit | 73 |
| Four speakers | 125 |
| Front wash pump | 16 |
| Front wipe delay unit | 14 |
| Front wipewash switch | 13 |
| Front wiper motor | 15 |
| Fuel gauge | 50 |
| Fuel gauge sender unit | 51 |
| Fuel pump - petrol models | 116 |
| Fuel shut-off solenoid-Diesel | 121 |
| Fuse 8 | 161 |
| Fuse 9 | 65 |
| Fuse 10 | 173 |
| Fuse 11 | 54 |
| Fuse 12 | 35 |
| Fuse 13 | 171 |
| Fuse 14 | |
| Fuse 1 | 32 |
| Fuse 2 | 33 |
| Fuse 3 | 30 |
| Fuse 4 | 31 |
| Fuse 5 | 20 |
| Fuse 6 | 12 |
| Fuse 7 | 65 |
| Fuse 8 | 173 |
| Fuse 9 | 54 |
| Fuse 10 | 143 |
| Fuse 12 | 25 |
| Fuse 13 | 49 |
| Fuse 14 | 171 |

| | | | | |
|---|---|---|---|---|
| | | U | Blue | |
| | | W | White | |
| | | Y | Yellow | |

Alarm pick up point 126
Alternator 11
Automatic gear selector illumination (2 off) 40
Fuel shut-off solenoid-Diesel 135
Radio fuse 136

Heated rear screen 67
Heated rear screen switch 68
Heated rear screen warning lamp 69
Heater air/con relay 64
Heater illumination (4 off) 175
Heater/air conditioning connections 38
Horn switch 160
Ignition coil 55
Ignition load relay 117
Ignition load relay (+) 1
Ignition on (+) 163
Ignition switch 166
Ignition warning lamp 8
Instrument illumination (6 off) 10
Interior lamp delay unit 41
Interior lamp switch 99
LH auxiliary lamp (option) 91
LH dip beam 45
LH door edge lamp 100
LH front door switch 108
LH front indicator lamp 77
LH horn 57
LH interior lamp 97
LH main beam 47
LH number plate lamp 24
LH puddle lamp 101
LH rear door switch 105
LH rear indicator lamp 76
LH rear log 43
LH reverse lamp 89
LH side lamp 22
LH side repeater lamp 78
LH stop lamp 87
LH tail lamp 23
Low coolant level warning lamp 146
Low coolant switch 144
Low fuel level warning lamp 147
Low screen wash fluid level warning lamp 142
Low screen wash switch 143
Main beam dipflash switch 25
Main beam warning lamp 49

457

# ELECTRICAL EQUIPMENT 86

**MAIN CIRCUIT DIAGRAM - NON CATALYST**
Left hand steering - RR2573M & RR2574M

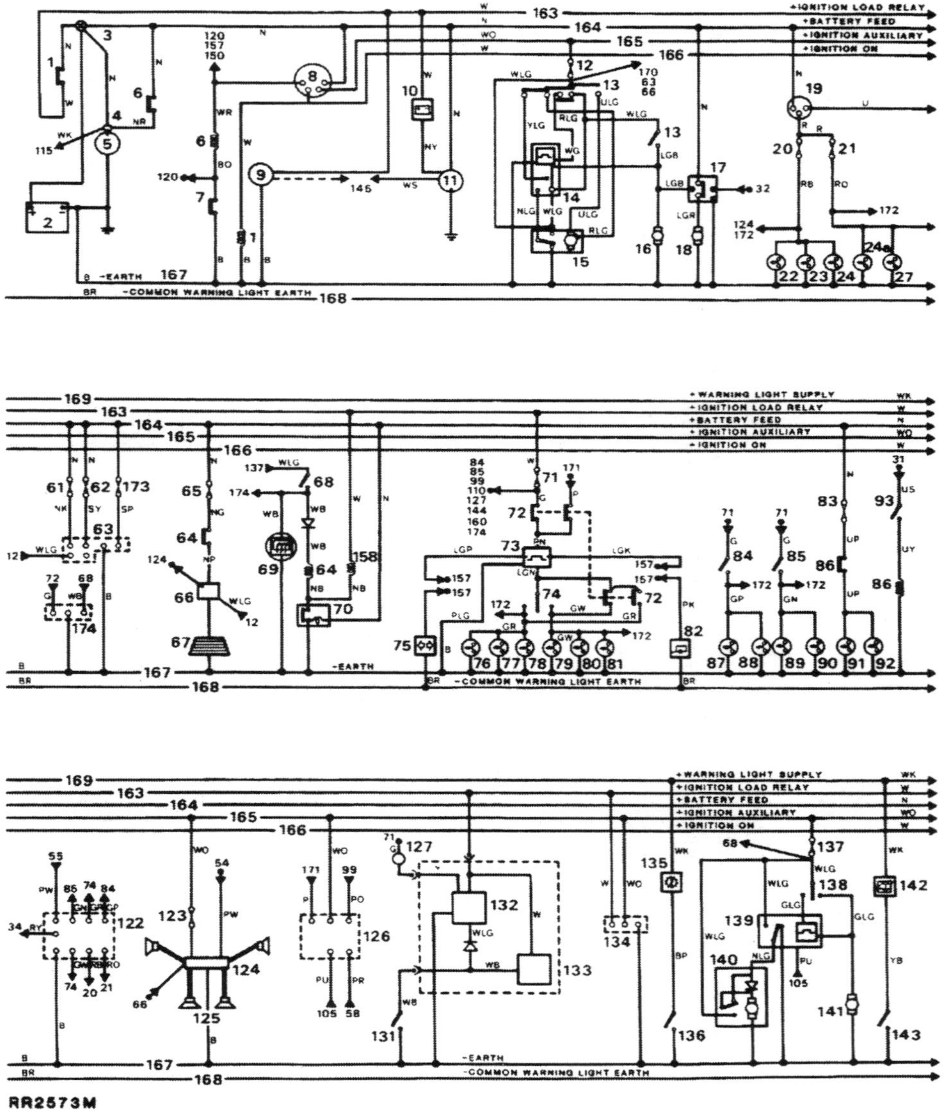

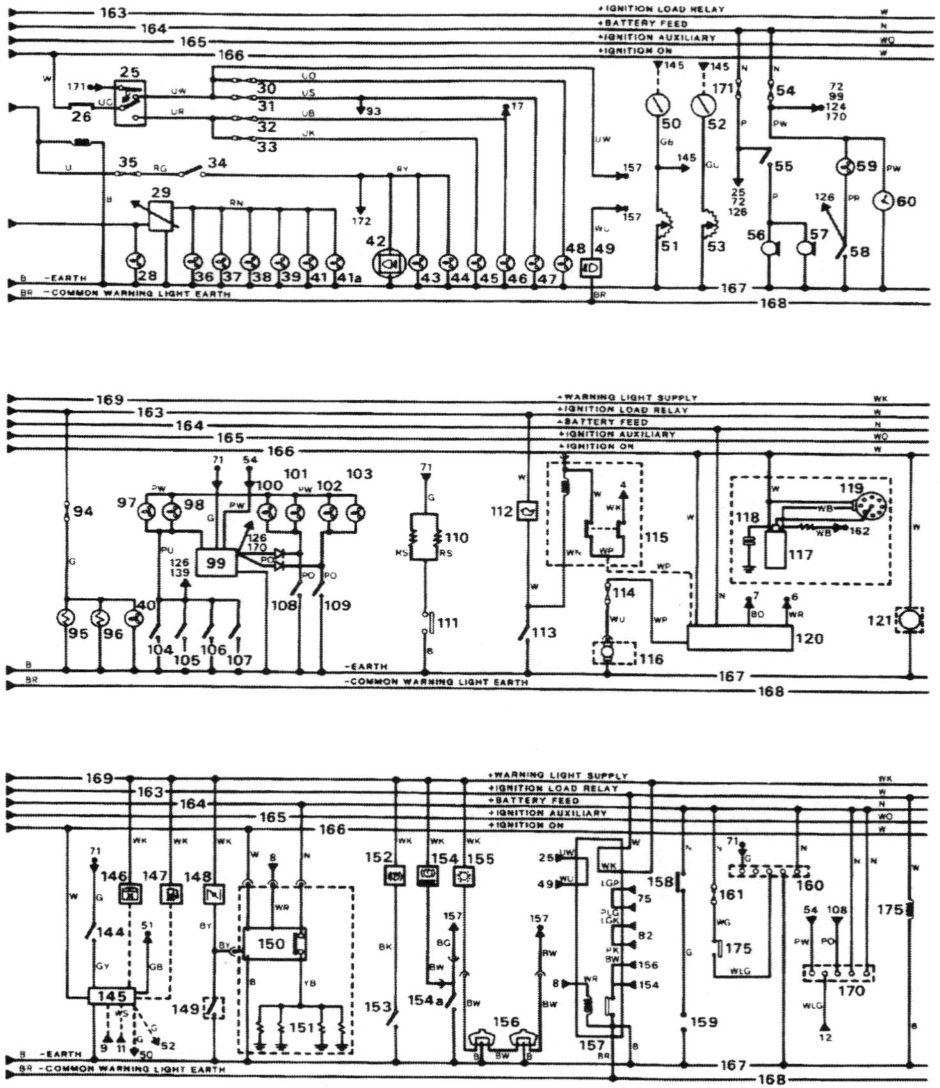

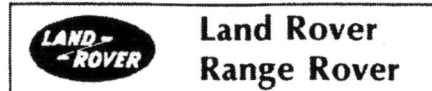

**Land Rover / Range Rover** — **Workshop Bulletin**

# Range Rover - Four Door
# Concealed Door Hinges
# 1989 Model Year
# Workshop Manual Supplement
# Bulletin Number LSM180WB8

|  | Section | Page |
|---|---|---|
| Introduction | 01 | 460 |
| Torque Wrench Settings | 06 | 460 |
| Lubricants and Fluids | 09 | 461 |
| Maintenance | 10 | 462 |
| Body | 76 | 463 |

Land Rover
Lode Lane,
Solihull
West Midlands
B92 8NW
England

| Section Number | | |
|---|---|---|
| **01** | **INTRODUCTION** | |
| | -Introduction | 1 |
| **06** | **TORQUE WRENCH SETTINGS** | |
| | -Body | 1 |
| **09** | **RECOMMENDED LUBRICANTS, FLUIDS AND CAPACITIES** | |
| | -Recommended lubricants and fluids - 1989 Model Year | 1 |
| **10** | **MAINTENANCE** | |
| | -Drive belts - Adjust or Renew | 1 |
| **76** | **CHASSIS AND BODY** | |
| | -Front door - remove, refit and adjust | 1 |
| | -Rear passenger door - remove refit and adjust | 2 |
| | -Decker panel - remove and refit | 3 |

## RANGE ROVER

# INTRODUCTION  **01**

## INTRODUCTION

A model enhancement for 1989 model year four door Range Rovers is the introduction of concealed door hinges.
This bulletin contains service details of the new door hinges and the decker panel.
This bulletin also contains a revised list of recommended lubricants deleting reference to LT230 Transfer gearbox. Note that the Borg Warner Transfer Gearbox is fitted to ALL 1989 model year Range Rovers.
In Section 10 Maintance, the correct procedure for tensioning all drive belts is given. If a new drive belt is fitted the engine must be run as instructed before retensioning the belt.

## RANGE ROVER

# TORQUE WRENCH SETTINGS  **06**

### BODY

| | Nm | lbf ft |
|---|---|---|
| Front and rear seat belt fixings (ALL) | 20.3 | 15 |
| Front door hinges to door and body | 25 | 19 |
| Rear passenger door hinges to door and body | 25 | 19 |

## Notes

# RANGE ROVER

# LUBRICANTS, FLUIDS AND CAPACITIES 09

## Recommended Lubricants and Fluids - 1989 Model year

Use only the recommended grades of oil set out below.
These recommendations apply to climates where operational temperatures are above -10°C (14°F).

| COMPONENTS | BP | CASTROL | DUCKHAM | ESSO | MOBIL | PETROFINA | SHELL | TEXACO |
|---|---|---|---|---|---|---|---|---|
| Petrol engine sump Carburetter Dashpots Oil can | BP Visco 2000 (15W/40) or BP Visco Nova (10W/40) | Castrol GTX (15W/50) or Castrolite (10W/40) | Duckhams 15W/50 Hypergrade Motor Oil | Esso Superlube plus (15W/40) | Mobil Super (10W/40) or Mobil 1 Rally Formula | Fina Supergrade Motor Oil (15W/40) or (10W/40) | Shell Super Motor Oil (15W/40) or (10W/40) | Havoline Motor Oil (15W/40) or Eurotex HD (10W/30) |
| Diesel engine sump ** | BP Vanellus C3 Extra (15W/40) | Castrol Turbomax (15W/40) | Duckhams Fleetmaster SHPD (15W/40) | Esso Super Diesel Oil TD (15W/40) | Mobil Delvac 1400 Super (15W/40) | Fina Kappa LDO (15W/40) | Shell Myrina (15W/40) | Texaco URSA Super TD (15W/40) |
|  | The following list of oils to MIL - L - 2104D or CC/MC D2 or API Service levels CD or SE/CD are for emergency use only if the above oils are not available. They can be used for topping up without detriment, but if used for engine oil changing, they are limited to a maximum of 5,000 km (3,000 miles) between oil and filter changes. ||||||||
|  | BP Vanellus C3 Multigrade 15W/40) | Castrol Deusol RX Super (15W/40) | Duckhams Hypergrade (15W/50) | Esso Essolube XD -3 plus (15W/40) | Mobil Delvac Super (15W/40) | Fina Dilano HPD (15W/40) | Shell Rimula X (15W/40) | Texaco URSA Super Plus (15W/40) |
| Automatic gearbox | BP Autran DX2D | Castrol TQ Dexron IID | Duckhams Fleetmatic CD or Duckhams D - Matic | Esso ATF Dexron IID | Mobil ATF 220D | Fina Dexron IID | Shell ATF Dexron IID | Texamatic Fluid 9226 |
| Manual gearbox | BP Autran G | Castrol TQF | Duckhams Q - Matic | Esso ATF Type G | Mobil ATF 210 | Fina Purfimatic 33G | Shell Donax TF | Texamatic Type G or Universal |
| Front and Rear differential Swivel pin housings | BP Gear Oil SAE 90EP | Castrol Hypoy SAE 90EP | Duckhams Hypoid 90 | Esso Gear Oil GX (85W/90) | Mobil Mobilube HD90 | Fina Pontonic MP SAE (80W/90) | Shell Spirax 90EP | Texaco Multigear Lubricant EP (85W/90) |
| Propeller shaft Front and Rear | BP Energrease L2 | Castrol LM Grease | Duckhams LB 10 | Esso Multi - purpose Grease H | Mobil Grease MP | Fina Marson HTL 2 | Shell Retinax A | Marfak All Purpose Grease |
| Power steering box and fluid Reservoir Borg Warner Transfer Gearbox | BP Autran DX2D or BP Autran G | Castrol TQ Dexron IID or Castrol TQF | Duckhams Fleetmatic CD or Duckhams Q - matic | Esso ATF Dexron IID or Esso ATF Type G | Mobil ATF 220D or Mobil ATF 210 | Fina Dexron IID or Fina purflimatic 33G | Shell ATF Dexron IID or Shell Donax TF | Texamatic Fluid 9226 or Texamatic Type G or 4291A Universal |
| Brake and clutch reservoirs | Brake fluids having a minimum boiling point of 260°C (500°F) and complying with FMVSS 116 DOT4 ||||||||
| Lubrication nipples (hubs, ball joints etc.) | BP Energrease L2 | Castrol LM Grease | Duckhams LB 10 | Esso Multi - purpose Grease H | Mobil Grease MP | Fina Marson HTL 2 | Shell Retinax A | Marfak All Purpose Grease |
| Ball joint assembly Top Link | Dextragrease Super GP ||||||||
| Seat slides Door lock striker | BP Energrease L2 | Castrol LM Grease | Duckhams LB 10 | Esso Multi - purpose Grease H | Mobil Grease MP | Fina Marson HTL 2 | Shell Retinax A | Marfak All purpose grease |

NLGI - 2 Multi - purpose Lithium - based Grease

** Other approved oils include: Agip Sigma Turbo, Aral Oil P327, Autol Valve - SHP, Aviation Turbo, Caltex RPM Delo 450, Century SHPD, Chevron Delo 450 Multigrade, Devinol Multimax Extra, Ecubsol CD Plus, Elf Multiperformance, Esso Special Diesel, Fanal Indol X, Fuchs Titan Truck 1540, Gulf Superfleet Special, IP Taurus M, Total Rubia TIR, Valvoline Super HD LD, Veedol Turbostar.

## Notes

# MAINTENANCE 10

## RANGE ROVER

### DRIVE BELTS - adjust or renew

**Illustration A** RR606M

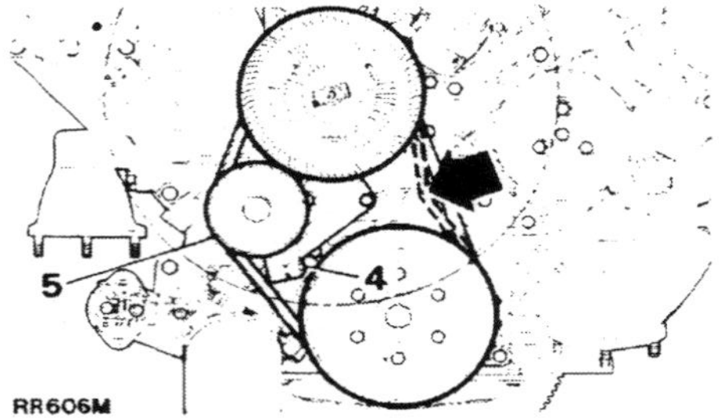

**Illustration B** RR607M

**Illustration C** RR608M

**Illustration D** RR2695M
4mm / 6mm
0·19in / 0·25in

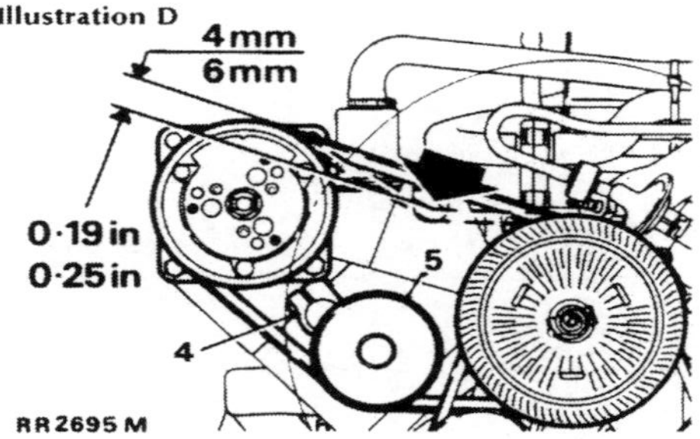

### Check driving belts, adjust or renew

1. Examine the following belts for wear and condition and renew if necessary:
   (A) Crankshaft-Jockey Pulley-Water Pump
   (B) Crankshaft-Steering Pump
   (C) Steering Pump-Alternator
   (D) Compressor-Jockey Pulley-Water Pump

2. Each belt should be sufficiently tight to drive the appropriate auxiliary without undue load on the bearings.

3. Slacken the bolts securing the unit to its mounting bracket.

4. Slacken the appropriate pivot bolt or jockey wheel and the fixing at the adjustment link where applicable.

5. Pivot the unit inwards or outwards as necessary and adjust until the correct belt tension is obtained.

6. Belt tension should be approximately 4 to 6 mm (0.19 to 0.25 in) at the points denoted by the bold arrows.

7. Tighten all unit adjusting bolts. Check adjustment again.

   **CAUTION:** When fitting a new drive belt, tension the belt as described above. Reconnect the battery and start and run the engine for 3 to 5 minutes at fast idle, after which time the belt must be re-checked, retension the belt if necessary.

---

## 09 LUBRICANTS, FLUIDS AND CAPACITIES

### RANGE ROVER

### RECOMMENDED LUBRICANTS AND FLUIDS - ALL CLIMATES AND CONDITIONS - 1989 Model year

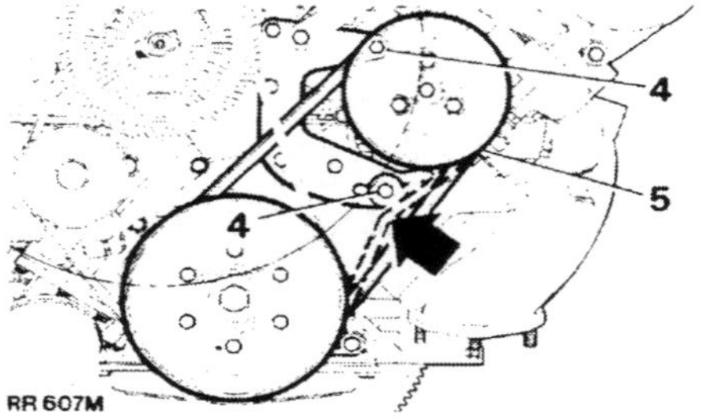

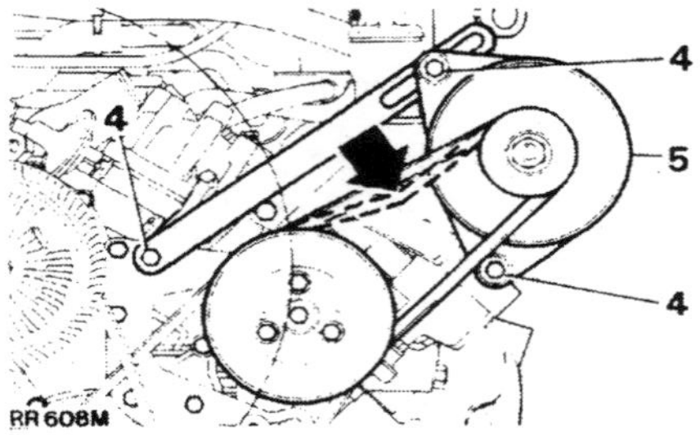

| COMPONENTS | SERVICE CLASSIFICATION | SAE Classification | AMBIENT TEMPERATURE °C (-30 -20 -10 0 10 20 30 40 50) |
|---|---|---|---|
| **Petrol models** Engine sump Carburetter | Oils must meet BLS.22.OL.07 or CCMC G3 or API service levels SF | 5W/30 5W/40 5W/50 10W/30 10W/40 10W/50 | |
| Dashpots Oil can | | | |
| | Oils must meet BLS.22.OL.02 or CCMC G1 or G2 or API service levels SE or SF | 15W/40 15W/50 20W/40 20W/50 25W/40 25W/50 | |
| **Diesel models** engine sump | SHPD oils meeting CCMC D3 | 10W/30 15W/40 | |
| | *Emergency only: Oils meeting MIL - L - 2104D or CCMCD2 or API CD | | |
| Main Gearbox Automatic | ATF Dexron IID | | |
| Main Gearbox manual | ATF M2C33 (F or G) | | |
| Final drive units Swivel pin housings | API GL4 or GL5 MIL - L - 2105 or MIL - L - 2105B | 90 EP 80W EP | |
| Power steering | ATF M2C 336 or ATF Dexron IID | | |
| Borg Warner Transfer Gearbox | | | |

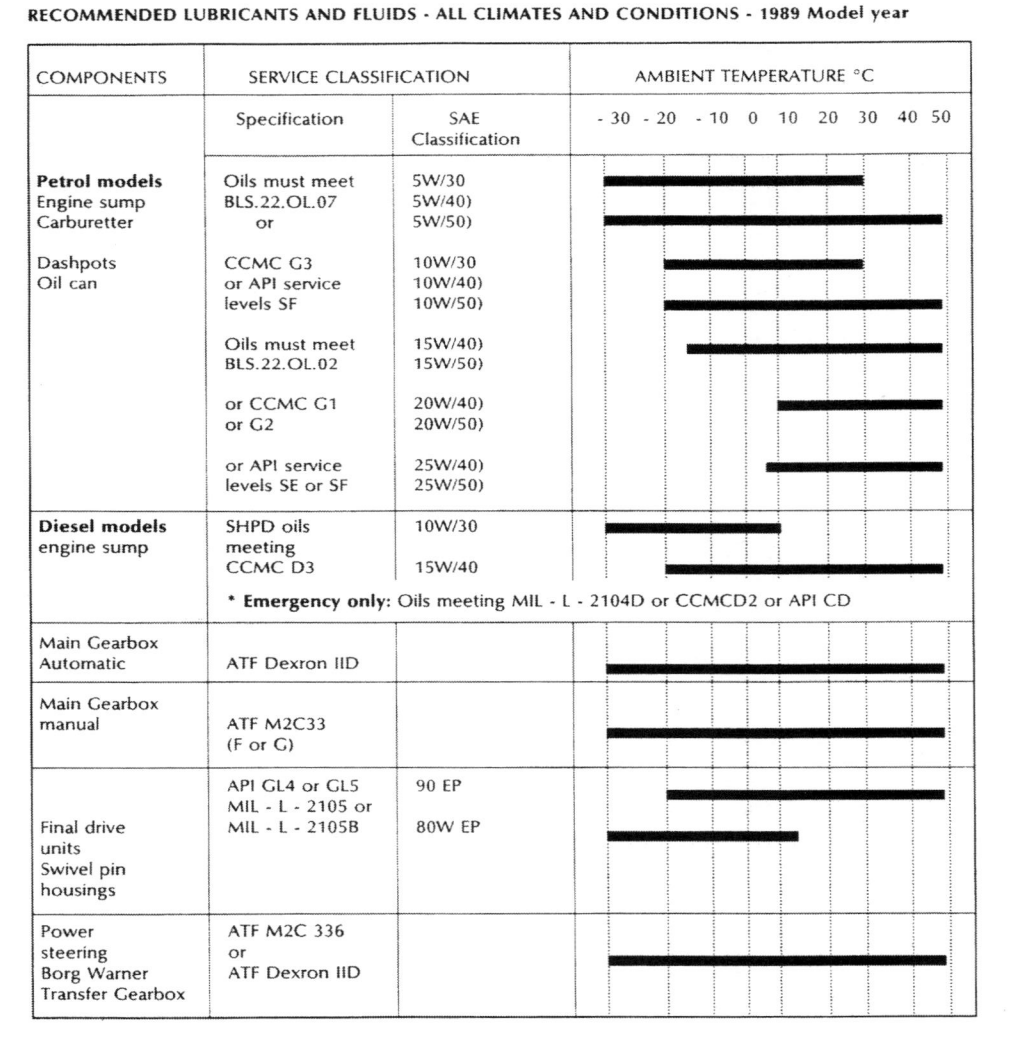

* Diesel Models - Engine Sump
Oils for emergency use only if the SHPD oils are not available. They can be used for topping up without detriment, but if used for engine oil changing, they are limited to a maximum of 5,000 km (3,000 miles) between oil and filter changes. (See * * on previous page)

### CAPACITIES

Transfer gearbox capacity - 2.1 Litres
                        - 3.7 UK Pints
                        - 4.4 US Pints

462

RANGE ROVER

# CHASSIS AND BODY 76

## FRONT DOOR

### Remove, refit and adjust.

**Removing**

1. Disconnect the battery negative lead.
2. Open the door to be removed.
3. Remove the trim panel from side of footwell by carefully levering under the trim and prising out the two plastic clips.
4. Locate and disconnect all door wiring plugs.
5. Disengage the grommets either side of 'A' post and feed wiring out.
6. Drive out the roll pin from the door check link.
7. Remove 'C' clips from grooves in hinge pins.

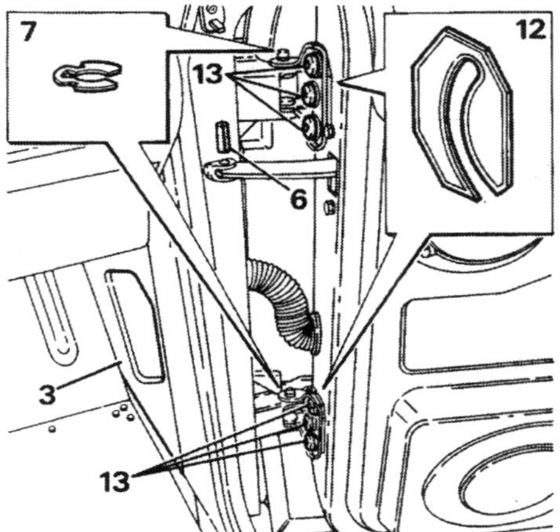

**WARNING: Instruction 8. MUST BE carried out with assistance.**

8. Carefully lift the opened door off the hinge pins.

**Refitting**

9. Reverse the removal procedure. Renew the 'C' clips if worn or distorted.
10. With door fully open reconnect wiring plugs ensuring they are located above the trim panel.
11. Check the operation of the door and lock. If necessary, adjust the door and striker plate.

**Adjusting**

12. Adjust the door by means of shims between the hinge and door to move the door forward or rearward in the opening.
13. Loosen the six Torx screws securing the hinges to the door to adjust the door up and down or in and out of the opening. Retighten the screws to the specified torque settings.
14. The door lock striker can be adjusted by loosening the striker and moving it in the appropriate direction or adding and subtracting spacing washers between the striker and 'B' post.
15. Note: If it is necessary to remove hinges from 'A' post they should be refitted in exactly the same position using the same thickness of shims.

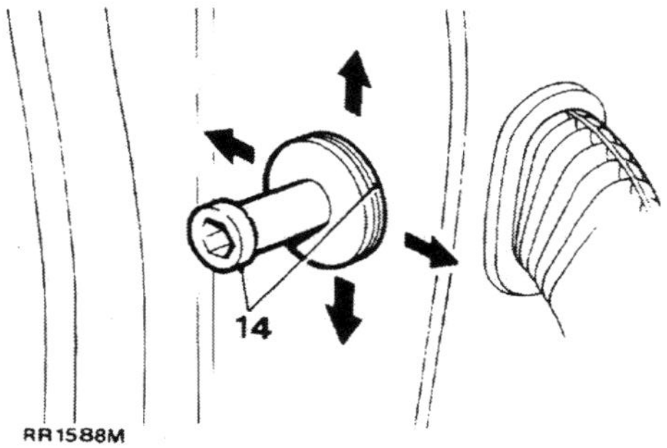

# CHASSIS AND BODY 76

## RANGE ROVER

### REAR PASSENGER DOOR

**Remove, refit and adjust.**

**Removing**

1. Disconnect the battery negative lead.
2. Remove wiring grommet from the 'B' post.
3. Withdraw the door wiring plugs from the 'B' post and disconnect them.
4. Remove the two bolts securing the check strap to 'B' post.
5. Remove 'C' clips from grooves in the hinge pins.

**WARNING: Instruction 6. MUST BE carried out with assitance.**

6. Carefully lift the opened door off the hinge pins.

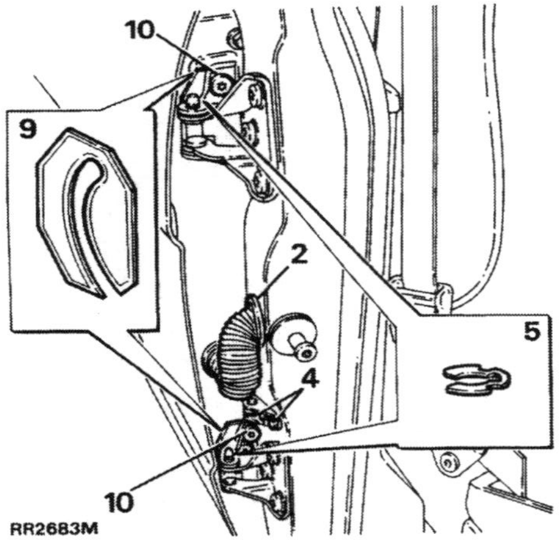

**Adjusting**

9. Adjust the door by means of shims between the hinge and door to move the door forward or rearward in the opening.
10. Loosen the six Torx screws securing the hinges to the door to adjust the door up and down or in and out of the opening. Retighten the screws to the specified torque settings.
11. Adjustment to the door striker is identical to front doors.
12. Note: If it is necessary to remove hinges from 'B' post they should be refitted in exactly the same position using the same thickness of shims.

**Refitting**

7. Reverse the removal procedure. Renew 'C' clips if worn or distorted.
8. Check the operation of the door and lock. If necessary, adjust the door and striker plate.

---

# CHASSIS AND BODY 76

## RANGE ROVER

6. With assistance place a tube over each of the hinges and lower to enable the decker panel to be fed over the hinges, gradually return the hinges to their upright position.

**WARNING: Gradually let the torsion bar spring tension return the hinges to their upright position to prevent the possibility of personal injury or damage to the vehicle.**

**Refitting**

7. Reverse the decker panel removal instruction.
8. Using a soft blunt implement ease the windscreen rubber up onto the top of the decker panel.

### DECKER PANEL

**Remove and Refit**

**Removing**

1. Remove bonnet.
2. Remove the wiper arms and two nuts securing the wheel boxes to the decker panel and remove the two sealing rubbers.
3. Remove the two cross-head screws retaining the panel to the 'A' post mounting brackets located above the front door hinges.

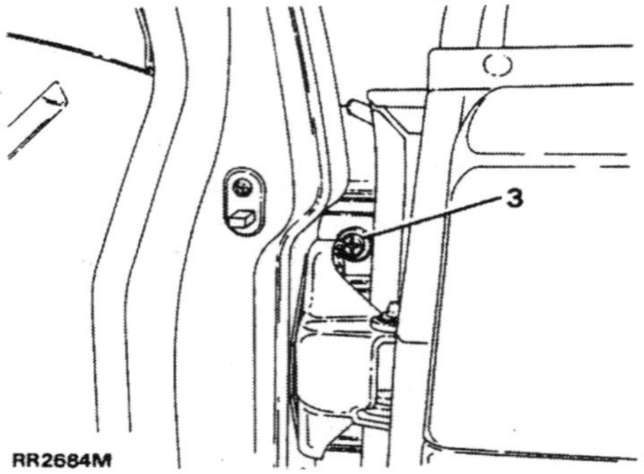

4. Remove the nine cross-head screws securing the front of the decker panel.
5. Remove the four bolts with spring and plain washers securing decker panel to front wings accessible from the front of the decker panel.

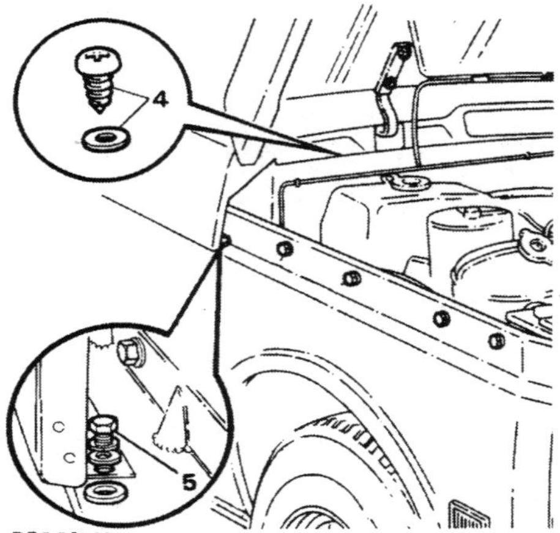

# Range Rover
## 2.4 & 2.5 Turbo Diesel Engines
## Workshop Manual Supplement
## Publication Number LSM180WS8 (edition 2)

|  | Section | Page |
|---|---|---|
| Introduction | 01 | 467 |
| General Specification Data | 04 | 468 |
| Engine Tuning Data | 05 | 474 |
| Torque Wrench Settings | 06 | 475 |
| Lubricants and Fluids | 09 | 476 |
| Maintenance | 10 | 478 |
| Service Tools | 99 | 483 |
| VW Diesel Engine | 12 | 484 |
| Fuel System | 19 | 511 |
| Cooling Systems | 26 | 514 |
| Electrical | 86 | 515 |

Land Rover
Lode Lane,
Solihull
West Midlands
B92 8NW
England

Section
Number

## 01 INTRODUCTION
- Introduction ... 1

## 04 GENERAL SPECIFICATION DATA
- 2.5 Litre Engine ... 1
- 2.4 Litre Engine ... 7

## 05 ENGINE TUNING DATA
- 2.5 Litre Engine ... 1
- 2.4 Litre Engine ... 2

## 06 TORQUE WRENCH SETTINGS
- Engine ... 1
- Electrical ... 1

## 09 RECOMMENDED LUBRICANTS AND FLUIDS
- Recommended lubricants and fluids ... 1
- Recommended lubricants and fluids - all climates and conditions ... 3
- Anti-freeze ... 4

## 10 MAINTENANCE SCHEDULES AND OPERATIONS
- Maintenance Schedule ... 1
- Engine Coolant ... 2
- Check oil level ... 3
- Engine, oil refill and filter renewal ... 3

**Fuel system maintenance**
- Main fuel filter ... 4
- Fuel tank breather ... 4
- Fuel sedimenter ... 5

## 99 SERVICE TOOLS
- Special tools for engine overhaul ... 1

## Notes

# RANGE ROVER

## INTRODUCTION

This supplement supersedes publication No's. LSM227 WS, LSM180 WS4 and combines the existing 2.4 litre Diesel Engine with the introduction of the new 2.5 Litre Diesel Engine for the Range Rover. It should be used in conjunction with the existing Range Rover Workshop Manual publication No. LSM180 WM.

### SYNTHETIC RUBBER

Many 0-ring seals, flexible pipes and other similar items which appear to be natural rubber are made of synthetic materials called Fluoroelastomers. Under normal operating conditions this material is safe, and does not present a health hazard. However, if the material is damaged by fire or excessive heat, it can break down and produce highly corrosive Hydrofluoric acid which can cause serious burns on contact with skin. Should the material be in a burnt or overheated condition handle only with seamless industrial gloves. Decontaminate and dispose of the gloves immediately after use.
If skin contact does occur, remove any contaminated clothing immediately and obtain medical assistance without delay. In the meantime, wash the affected area with copious amounts of cold water or limewater for fifteen to sixty minutes.

### RECOMMENDED SEALANTS

A number of branded products are recommended in this manual for use during maintenance and repair work.
These items include: **HYLOMAR GASKET AND JOINTING COMPOUND** and **HYLOSIL RTV SILICON COMPOUND**.
They should be available locally from garage equipment suppliers. If there is any problem obtaining supplies, contact one of the following companies for advice and the address of the nearest supplier.

### COPYRIGHT

© Rover Group Limited 1991

All rights reserved. No part of this publication may be produced, stored in a retrieval system or transmitted in any form, electronic, mechanical, recording or other means without prior written permission of Land Rover.

**MARSTON LUBRICANTS LTD.**
Hylo House,
Cale lane,
New Springs,
Wigan WN2 1JR,

0942 824242

---

Section Number

## 12 VM DIESEL ENGINE

- Engine fault diagnosis — 1
- Engine removal and refit — 4
- Cylinder heads remove and refit — 6
- Retorque cylinder heads — 10
- Liner protrusion check — 10
- Head gasket select — 11
- Engine external components — 12
- Engine internal components — 16
- Engine dismantling, overhaul and reassembly — 18
- Inspection and overhaul of components — 24
- Assembling engine — 41

## 19 FUEL SYSTEMS

- Turbocharger description — 1
- Turbocharger check — 2
- Turbocharger boost pressure — 3
- Waste gate valve — 3
- Turbocharger 'end float' check — 4
- Air filter check — 4
- Turbocharger faultfinding — 5
- Injection pump timing — 6

## 26 COOLING SYSTEM

- Fit coolant system kit — 1
- Viscous fan - check operation — 2

## 86 ELECTRICAL EQUIPMENT

- Circuit Diagram 1986 Model Year — 3

# RANGE ROVER

## GENERAL SPECIFICATION DATA 04

### MODEL: DIESEL RANGE ROVER 2.5 LITRE ENGINE.

| | | |
|---|---|---|
| Type | 95 A VM type HR 4924 HI | |
| Number of cylinders | 4 | |
| Bore | 92 mm | 3.62 in |
| Stroke | 94 mm | 3.7 in |
| Capacity | 2500 cm$^3$ | 152.32 in$^3$ |
| Injection order | 1 - 3 - 4 - 2 | |
| Compression ratio | 22.5 : 1 (± 0.5) | |

**Crankshaft**

| | | |
|---|---|---|
| Front main journal diameter | 62,995 to 63,010 mm | 2.4801 to 2.4807 in |
| Clearance in main bearing | 0,05 to 0,115 mm | 0.0019 to 0.0045 in |
| Minimum regrind diameter | 62,495 mm | 2.4604 in |
| Central main journal diameter | 63,005 to 63,020 mm | 2.4805 to 2.4811 in |
| Clearance in main bearing | 0,03 to 0,088 mm | 0.0012 to 0.0034 in |
| Minimum regrind diameter | 62,52 mm | 2.4614 in |
| Rear main journal diameter | 69,985 to 70,00 mm | 2.7551 to 2.7559 in |
| Clearance in main bearing | 0,040 to 0,070 mm | 0.0015 to 0.0027 in |
| Minimum regrind diameter | 69,485 mm | 2.7354 in |
| Crankpin journal diameter | 53,94 to 53,955 mm | 2.123 to 2.124 in |
| Clearance in big end bearing | 0,022 to 0,076 mm | 0.0008 to 0.0030 in |
| Minimum regrind diameter | 53,44 mm | 2.104 in |
| End float | 0,153 to 0,304 mm | 0.006 to 0.0119 in |
| Adjustment | Thrust washers | |
| Thrust washers available | 2,311 to 2,362 | 0.090 to 0.093 in |
| | 2,411 to 2,462 mm | 0.095 to 0.097 in |
| | 2,511 to 2,562 mm | 0.099 to 0.101 in |

**Thrust spacer**

| | | |
|---|---|---|
| Thickness | 7,9 to 8,1 mm | 0.311 to 0.319 in |
| Diameter | 89,96 to 90 mm | 3.542 to 3.543 in |

**Main bearings**

Standard
Internal diameter:

| | | |
|---|---|---|
| Front | 63,060 to 63,11 mm | 2.4872 to 2. 4845 in |
| Centre | 63,050 to 63,09 mm | 2.4823 to 2.4838 in |
| Rear | 70,040 to 70,055 mm | 2.7574 to 2.7580 in |

Bearing undersizes:
0.25 mm (0.01 in) and 0.5 mm (0.02 in) less than the dimensions given.

**Main bearing carriers**

Internal diameter:

| | | |
|---|---|---|
| Front | 67,025 to 67,050 mm | 2.639 to 2.640 in |
| Centre | 66,67 to 66,687 mm | 2.624 to 2.625 in |
| Rear | 75,005 to 75,030 mm | 2.953 to 2.954 in |
| Piston oil jet opening pressure | 1,5 to 2,0 kg/cm$^2$ | 22 to 29 lb/in$^2$ |

## Notes

# GENERAL SPECIFICATION DATA

## 2.5 LITRE ENGINE CONTINUED

### Liners
Internal diameter:
- White .................................................. 92,000 to 92,010 mm — 3.6220 to 3.6224 in
- Standard
- Red .................................................... 92,010 to 92,020 mm — 3.6224 to 3.6228 in
- Standard
- Protrusion .......................................... 0,01 to 0,06 mm — 0.0004 to 0.002 in
- Adjustment ........................................ Shims
- Shims available .................................. 0,15 mm — 0.006 in
- 0,20 mm — 0.008 in
- 0,23 mm — 0.009 in
- Maximum ovality ................................ 0,100 mm — 0.004 in
- Maximum taper .................................. 0,100 mm — 0.004 in

### Cylinder heads
Minimum thickness ................................ 89,95 to 90,05 mm — 3.541 to 3.545 in
Gaskets
Free thickness | Identity
- Number STC 654 ..... No notch ........ 1,51 to 1,59 mm — 0.059 to 0.062 in
- Number STC 656 ..... 1 notch .......... 1,75 to 1,83 mm — 0.069 to 0.072 in
- Number STC 655 ..... 2 notches ...... 1,65 to 1,73 mm — 0.065 to 0.068 in

Fitted thickness
- Number STC 654 .............................. 1,42 mm ± 0,04 — 0.056 in ± 0.001575
- Number STC 656 .............................. 1,62 mm ± 0,04 — 0.064 in ± 0.001575
- Number STC 655 .............................. 1,52 mm ± 0,04 — 0.059 in ± 0.001575

### End plates
Height .................................................. 91,26 to 91,34 mm — 3.593 to 3.596 in

### Connecting rods
Weights (connecting rod complete with small end bush, big-end cap and big-end bolts, but without the big-end shell).
- Letter Code
  - L ...................................................... 1156 to 1172 gr
- Fully machined balanced

### Pistons
Skirt diameter:
(measured at approximately 15 mm (0.6 in) above the bottom of the skirt).
- Class A .............................................. 91,92 to 91,93mm — 3.6188 to 3.6192 in
- Class B .............................................. 91,93 to 91,94mm — 3.6192 to 3.6196 in
- Piston skirt wear limit ......................... 0,05 mm — 0.0019 in
- Maximum ovality of gudgeon pin bore ... 0,05mm — 0.0019 in
- Piston clearance.
- Top of piston to cylinder head ............ 0,95 to 1,04mm — 0.0374 to 0.0409 in
- Piston protrusion above crankcase ..... 0,38 to 0,47mm — 0.0149 to 0.0185 in
  - Fit gasket 1,42
  - 0,58 to 0,67mm — 0.0228 to 0.0263 in
  - Fit gasket 1,62
  - 0,48 to 0,57mm — 0.0189 to 0.0224 in
  - Fit gasket 1,52
- Maximum piston to liner clearance ...... 0,15mm — 0.006 in

## 2.5 LITRE ENGINE CONTINUED

### Small end bush
Internal diameter:
- Minimum ........................................... 30,030 mm — 1.1823 in
- Maximum .......................................... 30,045 mm — 1.1828 in
- Wear limit between bush and gudgeon pin .......... 0,100 mm — 0. 004 in

### Big-end bearings
Standard
- Internal diameter ............................... 53,977 to 54,016 mm — 2.125 to 2.126 in
- Bearing undersizes:
  - 0.25 mm (0.01 in) and 0.5 mm (0.02 in) less than the dimensions given.

### Piston rings
Clearance in groove:
- Top ................................................... 0,080 to 0,130 mm — 0.0031 to 0.0051 in
- Second ............................................. 0,070 to 0,102 mm — 0.0027 to 0.004 in
- Oil control ........................................ 0,040 to 0,072 mm — 0.0015 to 0.0028 in

Fitted gap:
- Top ................................................... 0.25 to 0,50 mm — 0.0098 to 0.0196 in
- Second ............................................. 0.25 to 0.45 mm — 0.0098 to 0.0177 in
- Oil control ........................................ 0.25 to 0.58 mm — 0.0098 to 0.0228 in

### Gudgeon Pins
Type .................................................... Fully floating
Diameter .............................................. 29,990 to 29,996 mm — 1.180 to 1.181 in
Clearance in connecting rod
Wear limit between gudgeon pin and connecting rod bush ... 0,034 to 0,055 mm — 0.0013 to 0.0022 in
                                                      0,100 mm — 0.004 in

### Camshaft
Journal diameter: Front ......................... 53,495 to 53,51 mm — 2.1061 to 2.1067 in
- Bearing clearance ............................. 0,030 to 0,095 mm — 0.0012 to 0.0037 in
- Centre ............................................... 53,45 to 53,47 mm — 2.1043 to 2.1051 in
- Bearing clearance ............................. 0,07 to 0,14 mm — 0.0027 to 0.0055 in
- Rear .................................................. 53,48 to 53,50 mm — 2.1055 to 2.1063 in
- Bearing clearance ............................. 0,04 to 0,11 mm — 0.0016 to 0.0043 in

Cam lobe minimum dimensions:

Inlet (A)
- (c) ..................................................... 38,5 mm — 1.516 in
- (d) .................................................... 45,7 mm — 1.799 in

Exhaust (B)
- (c) ..................................................... 37,5 mm — 1.476 in
- (d) .................................................... 45,14 mm — 1.777 in

Thrust plate thickness ........................... 3,95 to 4,05 mm — 0.155 to 0.159 in

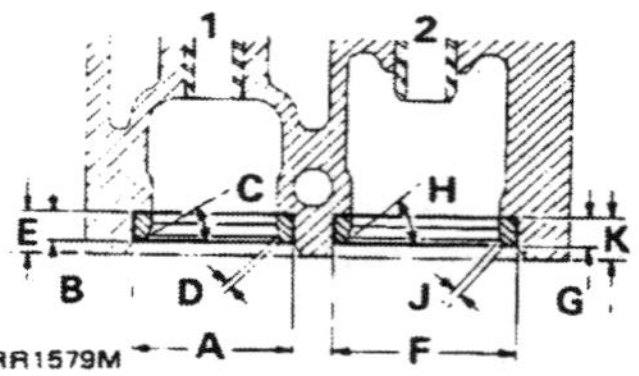

# GENERAL SPECIFICATION DATA

## 2.5 LITRE ENGINE CONTINUED

**Tappets**
Outside diameter .................................................. 14,965 to 14,985 mm / 0.589 to 0.590 in

**Rocker gear**
Shaft diameter ..................................................... 21,979 to 22,00 mm / 0.865 to 0.866 in
Bush internal diameter ........................................ 22,020 to 22,041 mm / 0.867 to 0.868 in
Assembly clearance ............................................ 0,020 to 0,062 mm / 0.0008 to 0.0024 in
Wear limit between bush and shaft ................... 0,2 mm / 0.008 in

**Valves**
Face angle:
Inlet ...................................................................... 55° 30'
Exhaust ................................................................ 45° 30'
Head diameter:
Inlet ...................................................................... 40,05 to 40,25 mm / 1.576 to 1.584 in
Exhaust ................................................................ 33,80 to 34,00 mm / 1.331 to 1.338 in
Head stand down:
Inlet ...................................................................... 0,80 to 1,20 mm / 0.0315 to 0.0472 in
Exhaust ................................................................ 0,79 to 1,19 mm / 0.0311 to 0.468 in
Stem diameter:
Inlet ...................................................................... 7,940 to 7,960 mm / 0.312 to 0.313 in
Exhaust ................................................................ 7,920 to 7,940 mm / 0.311 to 0.312 in
Clearance in guide:
Inlet ...................................................................... 0,040 to 0,075 mm / 0.0016 to 0.0029 in
Exhaust ................................................................ 0,060 to 0,095 mm / 0.0024 to 0.0037 in

**Valve guides**
Inside diameter ................................................... 8 to 8,015 mm / 0.314 to 0.315 in
Fitted height (above spring
plate counterbore) ............................................... 13,5 to 14 mm / 0.531 to 0.551 in

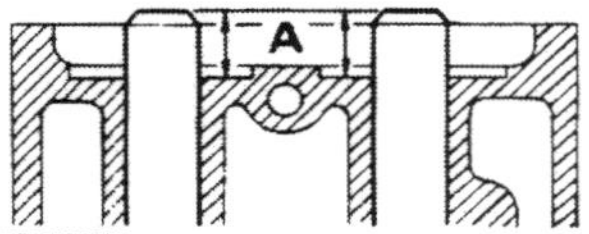

**Valve seat inserts**
Machining dimensions
Exhaust (1)
A ........................................................................... 36,066 to 36,050 mm / 1.4199 to 1.4193 in
B ........................................................................... 7,00 to 7,05 mm / 0.275 to 0.277 in
C ........................................................................... 44° 30'
D ........................................................................... 1,65 to 2,05 mm / 0.065 to 0.080 in
E ........................................................................... 10,15 to 10,25 mm / 0.399 to 0.403 in
Inlet (2)
F ........................................................................... 42,070 to 42,086 mm / 1.6536 to 1.6569 in
G ........................................................................... 7,14 to 7,19 mm / 0.281 to 0.283 in
H ........................................................................... 34° 30'
J ........................................................................... 1,8 to 2,2 mm / 0.071 to 0.086 in
K ........................................................................... 10,3 to 10,4 mm / 0.405 to 0.409 in

---

# GENERAL SPECIFICATION DATA

## 2.5 LITRE ENGINE CONTINUED

**Valve springs**
Free length .......................................................... 44,65 mm / 1.76 in
Fitted length ........................................................ 38,6 mm / 1.52 in
Load at fitted length ........................................... 34 ± 3% Kg / 75 ± 3% lbf.
Load at top of lift ................................................ 92,5 ± 3% Kg / 204 ± 3% lbf.
Number of coils ................................................... 5,33

**Valve timing**
Rocker clearance: Timing
Inlet ...................................................................... 0,30 mm / 0.012 in
Exhaust ................................................................ 0,30 mm / 0.012 in
Inlet valve:
Opens ................................................................... 22° ± 5° B.T.D.C.
Closes .................................................................. 48° ± 5° A.B.D.C.
Exhaust valve:
Opens ................................................................... 60° ± 5° B.B.D.C.
Closes .................................................................. 24° ± 5° A.T.D.C.

**Lubrication**
System pressure with oil at 90-100° C
at 4,000 rev/min. ................................................. 3,5 to 5,0 kgf/cm² / 50 to 70 lbf/in².
Pressure relief valve opens ................................ 6.38 kgf/cm² / 91 lbf/in².
Pressure relief valve spring
- free length ........................................................ 57,5 mm / 2.26 in.
Oil pump:
Outer rotor end float ........................................... 0,04 to 0,087 mm / 0.0015 to 0.0034 in.
Inner rotor end float ............................................ 0,04 to 0,087 mm / 0.0015 to 0.0034 in.
Outer rotor to body
diametrical clearance ......................................... 0,130 to 0,230 mm / 0.005 to 0.009 in.
Rotor body to drive gear
clearance (pump not fitted) ............................... 0,15 to 0,25 mm / 0.0059 to 0.0098 in

**COOLING SYSTEM**
Thermostat .......................................................... 80°C ± 2°C
Pressure cap ....................................................... 1,05 kgf cm² / 15 lb f/in²

**DRIVE BELT TENSIONING**
Installed drive belts using a recognised driving belt
tension gauge to be :-
Air conditioning compressor .............................. 450N / 95 lbf
Power steering pump .......................................... 400N / 90 lbf
Alternator/water pump ........................................ 490N / 110 lbf

# GENERAL SPECIFICATION DATA

## FUEL SYSTEM

| | | |
|---|---|---|
| Fuel lift pump | mechanical, driven by camshaft | |
| Turbo charger: | | |
| Shaft radial clearance | 0,35 mm | 0.0137 in |
| Shaft axial clearance | 0,10 mm | 0.0039 in |
| Waste gate valve: | | |
| Opening pressure | 0.9 kgf cm² | 13 lbf/in² |

## CLUTCH

| | | |
|---|---|---|
| Make and type | Valeo, diaphragm | |
| Diameter | 235mm | 9.25 in |

## GEARBOX

| | |
|---|---|
| Model | LT77 (manual) |
| Type | Five speed, single helical constant mesh with synchromesh on all forward gears |

## TRANSFER GEARBOX

| | |
|---|---|
| Model | BORG WARNER |
| Type | 13-61 with viscous controlled unit |

## STARTER MOTOR

| | |
|---|---|
| Make and type | BOSCH 0.001. 362.092 |

## ALTERNATOR

| | |
|---|---|
| Make and type | Magnetti Marelli A127 - 65A |
| On Diesel Vogue Range Rover | Magnetti Marelli A133 - 80A |

---

# GENERAL SPECIFICATION DATA

## MODEL: 2.4 LITRE DIESEL RANGE ROVER ENGINE

| | | |
|---|---|---|
| Type | 11A VM type HR 492 HI | |
| Number of cylinders | 4 | |
| Bore | 92 mm | 3.62 in |
| Stroke | 90 mm | 3.54 in |
| Capacity | 2393 cm³ | 146.03 in³ |
| Injection order | 1 - 3 - 4 - 2 | |
| Compression ratio | 21.5 : 1 (± 0.5) | |

### Crankshaft

| | | |
|---|---|---|
| Front main journal diameter | 62,98 to 63 mm | 2.4795 to 2.4803 in |
| Clearance in main bearing | 0,06 to 0,13 mm | 0.0023 to 0.005 in |
| Minimum regrind diameter | 62,48 mm | 2.4498 in |
| Central main journal diameter | 62,98 to 63 mm | 2.4795 to 2.4803 in |
| Clearance in main bearing | 0,05 to 0,113 mm | 0.0019 to 0.0044 in |
| Minimum regrind diameter | 62,48 mm | 2.4498 in |
| Rear main journal diameter | 69,98 to 70 mm | 2.7551 to 2.7559 in |
| Clearance in main bearing | 0,06 to 0,105 mm | 0.0023 to 0.0041 in |
| Minimum regrind diameter | 69,48 mm | 2.7354 in |
| Crankpin journal diameter | 53,92 to 53,94 mm | 2.1228 to 2.1236 in |
| Clearance in big end bearing | 0,035 to 0,094 mm | 0.0014 to 0.0037 in |
| Minimum regrind diameter | 53,42 mm | 2.1032 in |
| End float | 0,12 to 0,323 mm | 0.005 to 0.0127 in |
| Adjustment | Thrust washers | |
| Thrust washers available | 2,311 to 2,362 mm | 0.090 to 0.093 in |
|  | 2,411 to 2,462 mm | 0.095 to 0.097 in |
|  | 2,511 to 2,562 mm | 0.099 to 0.101 in |

### Thrust spacer

| | | |
|---|---|---|
| Thickness | 7,9 to 8,1 mm | 0.311 to 0.319 in |
| Diameter | 89,96 to 90 mm | 3.542 to 3.543 in |

### Main bearings

| | | |
|---|---|---|
| Standard | | |
| Internal diameter: | | |
| Front | 63,060 to 63,11 mm | 2.4872 to 2. 4845 in |
| Centre | 63,050 to 63,09 mm | 2.4823 to 2. 4838 in |
| Rear | 70,060 to 70,085 mm | 2.7582 to 2.7592 in |
| Bearing undersizes: | | |
| 0,25 mm (0.01 in) and 0.5 mm (0.02 in) less than the dimensions given. | | |

### Main bearing carriers

| | | |
|---|---|---|
| Internal diameter: | | |
| Front/centre | 66,67 to 66,687 mm | 2.624 to 2.625 in |
| Rear | 75,005 to 75,030 mm | 2.953 to 2.954 in |
| Piston oil jet opening pressure | 1,5 to 2,0 kg/cm² | 22 to 29 lb/in² |

# GENERAL SPECIFICATION DATA

## 2.4 LITRE ENGINE CONTINUED

**Liners**
Internal diameter:
- White ........................................... 92,000 to 92,010 mm ... 3.6220 to 3.6224 in
- Standard ....................................... 92,010 to 92,020 mm ... 3.6224 to 3.6228 in
- Red
- Standard ....................................... 0 to 0,05 mm ........... 0 to 0.002 in
- Protrusion

Adjustment ...................................... Shims
Shims available ................................. 0,15 mm ................ 0.006 in
                                                 0,20 mm ................ 0.008 in
                                                 0,23 mm ................ 0.009 in
                                                 0,100 mm ............... 0.004 in

Maximum ovality ................................. 0,100 mm ............... 0.004 in
Maximum taper ................................... 0,100 mm ............... 0.004 in

**Cylinder heads**
Minimum thickness ............................... 89,95 to 90,05 mm ...... 3.541 to 3.545 in
Gaskets
Free thickness                  Identity
- Number STC 654 ................ No notch ....... 1,60 mm ................ 0.063 in
- Number STC 656 ................ 1 notch ........ 1,80 mm ................ 0.071 in
- Number STC 655 ................ 2 notches ...... 1,70 mm ................ 0.067 in
Fitted thickness
- Number STC 654 ................................. 1,42 mm ................ 0.056 in
- Number STC 656 ................................. 1,62 mm ................ 0.064 in
- Number STC 655 ................................. 1,52 mm ................ 0.059 in

**End plates**
Height .......................................... 91,26 to 91,34 mm ...... 3.593 to 3.596 in

**Connecting rods**
Weights (connecting rod complete with small end bush, big-end cap and big end bolts, but without the big-end shell).

Letter Code
- A ............................................. 1100 to 1109 gr ........ 38.80 to 39.12 oz
- B ............................................. 1110 to 1119 gr ........ 39.15 to 39.47 oz
- C ............................................. 1120 to 1129 gr ........ 39.51 to 39.82 oz
- D ............................................. 1130 to 1139 gr ........ 39.86 to 40.17 oz
- E ............................................. 1140 to 1149 gr ........ 40.21 to 40.53 oz
- F ............................................. 1150 to 1159 gr ........ 40.56 to 40.88 oz
- G ............................................. 1160 to 1169 gr ........ 40.92 to 41.23 oz
- H ............................................. 1170 to 1179 gr ........ 41.27 to 41.58 oz
- I ............................................. 1180 to 1189 gr ........ 41.62 to 41.94 oz

**Small end bush**
Internal diameter:
- Minimum ....................................... 30,030 mm .............. 1.1823 in
- Maximum ....................................... 30,045 mm .............. 1.1828 in
Wear limit between bush and gudgeon pin ......... 0,100 mm ............... 0.004 in

**Big-end bearings**
Standard
Internal diameter .............................. 53,975 to 54,014 mm .... 2.125 to 2.126 in
Bearing undersizes:
0,25 mm (0.01 in) and 0,5 mm (0.02 in) less than the dimensions given.

---

## 2.4 LITRE ENGINE CONTINUED

**Pistons**
Skirt diameter:
(measured at approximately 15 mm (0.6 in) above the bottom of the skirt).
- Class A ....................................... 91,965 to 91,975 mm ... 3.6207 to 3.6211 in
- Class B ....................................... 91,975 to 91,985 mm ... 3.6211 to 3.6214 in
Piston skirt wear limit ......................... 0,05 mm ................ 0.0019 in
Maximum ovality of gudgeon pin bore ............. 0,05 mm ................ 0.0019 in
Piston clearance.
- Top of piston to cylinder head ................ 0.85 to 0.94 mm ........ 0.0335 to 0.0370 in
- Piston protrusion above crankcase ............. 0.48 to 0.57 mm ........ 0.0189 to 0.0224 in
                                                 Fit gasket 1.42
- Piston protrusion above crankcase ............. 0.68 to 0.77 mm ........ 0.0268 to 0.0303 in
                                                 Fit gasket 1.62
- Piston protrusion above crankcase ............. 0.58 to 0.67 mm ........ 0.0228 to 0.0263 in
                                                 Fit gasket 1.52
Maximum piston to liner clearance ............... 0,15 mm ................ 0.006 in

**Piston rings**
Clearance in groove:
- Top ........................................... 0,080 to 0,130 mm ...... 0.0031 to 0.0051 in
- Second ........................................ 0,070 to 0,102 mm ...... 0.0027 to 0.004 in
- Oil control ................................... 0,030 to 0,062 mm ...... 0.0012 to 0.0024 in
Fitted gap:
- Top ........................................... 0.40 to 0.65 mm ........ 0.0157 to 0.0256 in
- Second ........................................ 0.25 to 0.45 mm ........ 0.0098 to 0.0177 in
- Oil control ................................... 0.25 to 0.58 mm ........ 0.0098 to 0.0228 in

**Gudgeon Pins**
- Type .......................................... Fully floating
- Diameter ...................................... 29,990 to 29,996 mm .... 1.180 to 1.181 in
- Clearance in connecting rod .................... 0,034 to 0,055 mm ...... 0.0013 to 0.0022 in
- Wear limit between gudgeon pin and connecting rod bush ................ 0,100 mm ............... 0.004 in

**Camshaft**
- Journal diameter .............................. 53.48 to 53.50 mm ...... 2.105 to 2.106 in
- Clearance in bearings ......................... 0,040 to 0,11 mm ....... 0.0016 to 0.0043 in

Cam lobe minimum dimensions:

Inlet (A) ....................................... 38,5 mm ................ 1.516 in
      (d) ....................................... 45,7 mm ................ 1.799 in
Exhaust (B) ..................................... 37,5 mm ................ 1.476 in
        (d) ..................................... 45,14 mm ............... 1.777 in
Thrust plate thickness .......................... 3,95 to 4,05 mm ........ 0.155 to 0.159 in

# GENERAL SPECIFICATION DATA

## 2.4 LITRE ENGINE CONTINUED

**Tappets**
Outside diameter ................................................. 14,965 to 14,985 mm / 0.589 to 0.590 in

**Rocker gear**
Shaft diameter ..................................................... 21,979 to 22,00 mm / 0.865 to 0.866 in
Bush internal diameter ....................................... 22,020 to 22,041 mm / 0.867 to 0.868 in
Assembly clearance ........................................... 0,020 to 0,062 mm / 0.0008 to 0.0024 in
Wear limit between bush and shaft ................... 0,2 mm / 0.008 in

**Valves**
Face angle:
Inlet ..................................................................... 55° 30'
Exhaust .............................................................. 45° 30'
Head diameter:
Inlet ..................................................................... 40,05 to 40,25 mm / 1.576 to 1.584 in
Exhaust .............................................................. 33,80 to 34,00 mm / 1.331 to 1.338 in
Head stand down:
Inlet ..................................................................... 0,80 to 1,20 mm / 0.0315 to 0.0472 in
Exhaust .............................................................. 0,79 to 1,19 mm / 0.0311 to 0.0468 in
Stem diameter:
Inlet ..................................................................... 7,940 to 7,960 mm / 0.312 to 0.313 in
Exhaust .............................................................. 7,920 to 7,940 mm / 0.311 to 0.312 in
Clearance in guide:
Inlet ..................................................................... 0,040 to 0,075 mm / 0.0016 to 0.0029 in
Exhaust .............................................................. 0,060 to 0,095 mm / 0.0024 to 0.0037 in

**Valve guides**
Inside diameter .................................................. 8 to 8,015 mm / 0.314 to 0.315 in
Fitted height (above spring plate counterbore) .. 13,5 to 14 mm / 0.531 to 0.551 in

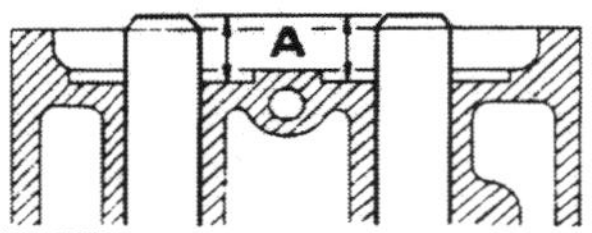

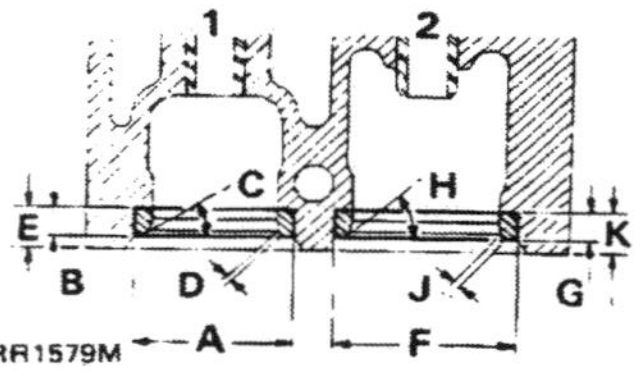

**Valve seat inserts**
Machining dimensions
Exhaust (1)
A ......................................................................... 36,066 to 36,050 mm / 1.4199 to 1.4193 in
B ......................................................................... 7,00 to 7,05 mm / 0.275 to 0.277 in
C ......................................................................... 44° 30'
D ......................................................................... 1,70 to 1,80 mm / 0.067 to 0.071 in
E ......................................................................... 10,00 to 10,10 mm / 0.393 to 0.397 in
Inlet (2)
F ......................................................................... 42,070 to 42,086 mm / 1.6536 to 1.6669 in
G ......................................................................... 7,14 to 7,19 mm / 0.281 to 0.283 in
H ......................................................................... 34° 30'
J ......................................................................... 1,9 to 2,0 mm / 0.075 to 0.079 in
K ......................................................................... 10,25 to 10,35 mm / 0.403 to 0.407 in

## 2.4 LITRE ENGINE CONTINUED

**Valve springs**
Free length ......................................................... 44,65 mm / 1.76 in
Fitted length ....................................................... 38,6 mm / 1.52 in
Load at fitted length .......................................... 34 ± 3% Kg / 75 ± 3% lbf.
Load at top of lift ............................................... 92,5 ± 3% Kg / 204 ± 3% lbf.
Number of coils ................................................. 5.33

**Valve timing**
Rocker clearance: Timing
Inlet ..................................................................... 0,30 mm / 0.012 in
Exhaust .............................................................. 0,30 mm / 0.012 in
Inlet valve:
Opens ................................................................. 22° ± 5° B.T.D.C.
Closes ................................................................ 48° ± 5° A.B.D.C.
Exhaust valve:
Opens ................................................................. 60° ± 5° B.B.D.C.
Closes ................................................................ 24° ± 5° A.T.D.C.

**Lubrication**
System pressure with oil at 90 - 100°C
at 4,000 rev/min. ............................................... 3,5 to 5,0 kgf/cm$^2$ / 50 to 70 lbf/in$^2$
Pressure relief valve opens .............................. 4 to 4,5 kgf/cm$^2$ / 57 to 64 lbf/in$^2$
Pressure relief valve spring
- free length ....................................................... 57,5 mm / 2.26 in.
Oil pump:
Outer rotor end float .......................................... 0,081 to 0,097 mm / 0.003 to 0.004 in.
Inner rotor end float ........................................... 0,081 to 0,097 mm / 0.003 to 0.004 in.
Outer rotor to body
diametrical clearance ........................................ 0,130 to 0,230 mm / 0.005 to 0.009 in.
Rotor body to drive gear
clearance ........................................................... 0,050 to 0,070 mm / 0.0 02 to 0.003 in.

## COOLING SYSTEM

Thermostat ......................................................... 83°C ± 2°C
Pressure cap ..................................................... 1,05 kgf cm$^2$ / 15 lb f/in$^2$

## DRIVE BELT TENSIONING

On 'V' type installed drive belts using a recognised driving belt tension gauge to be :-
On 12,7 mm wide belts ..................................... 450N / 95 lbf

**"In field" Tensioning - No gauge available**

Deflection of belt run between longest belt centres to be:- ........................ 0.5 mm per 25 mm of belt run

# GENERAL SPECIFICATION DATA — 04

## FUEL SYSTEM

Fuel lift pump ..................................... mechanical, driven by camshaft
Turbo charger:
Shaft radial clearance ........................ 0,42 mm ........................ 0.016 in
Shaft axial clearance ......................... 0,15 mm ........................ 0.006 in
Waste gate valve:
Opening pressure .............................. 0,9 kgf/cm² .................... 13 lbf/in²

## CLUTCH

Make and type ................................... Valeo, diaphragm
Diameter ........................................... 235,0 mm ....................... 9.25 in

---

# ENGINE TUNING DATA — 05

## ENGINE TUNING DATA

**Model: Diesel Range Rover** ......................... 1990 MODEL YEAR 2.5 LITRE ENGINE

### Engine
Type ................................................. 95A VM Type HR 4924 HI
Capacity .......................................... 2500 cm³ ....................... 152.32 in³
Compression pressure ..................... 24 to 26 kgf/cm² ........... 340 to 370 lbf/in²
Injection order ................................. 1 - 3 - 4 - 2
Idling speed at running temperature .... 750 - 800 rev/min
Idling speed cold start temperature ..... 1000 - 1100 rev/min
Maximum light running speed ........... 4700 to 4730 rev/min
Maximum governed road speed ........ 4200 rev/min
Valve rocker clearances (cold)
Inlet ................................................. 0,30 mm ........................ 0.012 in
Exhaust ............................................ 0,30 mm ........................ 0.012 in

### Fuel injection pump
Make and type ................................ Bosch Rotary VE 4 10F 2100 L269
Injection pump timing ...................... 3° -0 + 1° B.T.D.C.

### Injectors
Make and type ................................ Bosch KBE 58 S 4/4
Nozzle type ..................................... DNO SD 263 or SDV 4011379
Opening pressure ........................... 150 +8/-0 BAR

### Heater plugs
Make and type ................................ Bosch 0.250.201.012
Nominal voltage .............................. 11 volts

# 05 ENGINE TUNING DATA

## RANGE ROVER

### ENGINE TUNING DATA

**Model: Diesel Range Rover** ............... **1986 MODEL YEAR 2.4 LITRE ENGINE**

**Engine**
| | |
|---|---|
| Type | 11A VM Type HR 492 HI |
| Capacity | 2393 cm³ — 146.03 in³ |
| Compression pressure at crank speed 150 rev/min | 32 to 35 kgf/cm² — 450 to 500 lbf/in² |
| Injection order | 1 - 3 - 4 - 2 |
| Idling speed at running temperature | 750 - 800 rev/min |
| Maximum light running speed | 4700 to 4730 rev/min |
| Maximum governed road speed | 4200 rev/min |
| Valve rocker clearances (cold) | |
| Inlet | 0.30 mm — 0.012 in |
| Exhaust | 0.30 mm — 0.012 in |

**Fuel injection pump**
| | |
|---|---|
| Make and type | Bosch Rotary VE L 168-1 |
| Injection pump timing | 3° B.T.D.C. |

**Injectors**
| | |
|---|---|
| Make and type | Bosch KBE 58 S 4/4 |
| Nozzle type | DNO SD 263 |
| Opening pressure | 150 + 8/-0 BAR |

**Heater plugs**
| | |
|---|---|
| Make and type | Bosch 0.250.201.012 |
| Nominal voltage | 11 volts |

# 06 TORQUE WRENCH SETTINGS

## RANGE ROVER

### TORQUE WRENCH SETTINGS

**ENGINE** — Nm
- Camshaft screws — 24
- Connecting rod bolts — 81 *
- Crankshaft pulley nut — 152
- Cylinder head bolts — SEE SPECIAL PROCEDURE
- Cylinder head oil pipe unions — 8
- Engine coolant rail bolts — 8
- Engine mountings — 49
- Engine sump bolts — 11
- Engine sump pan bolts — 11
- Exhaust manifold nuts — 32
- Exhaust pipe flange bolts — 27
- Flywheel bolts — 108
- Flywheel housing bolts — 49
- Fuel line unions — 19
- Heater plugs — 23
- Idler gear screws 2.4 litre engine — 27
- Injection pump mounting nut — 31
- Injection pump gear nut — 88
- Injector nut — 27
- Inlet manifold nuts — 32
- Main bearing carrier bolts — 42
- Oil drain plugs — 79
- Oil filter base — 38
- Oil pump screws — 27
- Oil thermostat — 74
- Rear main bearing carrier nuts — 27
- Rocker cover nuts — 9
- Rocker shaft pedestal nuts — 108 *
- Timing cover screws — 12
- Turbo charger to manifold nuts — 26
- Vacuum pump nuts — 21
- Vacuum pump screws 2.5 Litre Engine — 28
- Valve gear oil pipe unions — 8
- Water pump screws — 24

**ELECTRICAL** — Nm
- Alternator tie rod — 49
- Alternator bracket to crankcase — 54
- Alternator pulley nut — 54
- Alternator bottom fixing — 54
- Starter motor to flywheel housing — 68

\* Apply Molyguard to threads before fitting.

# RECOMMENDED LUBRICANTS AND FLUIDS

## RECOMMENDED LUBRICANTS AND FLUIDS

Use only the recommended grades of oil set out below.
These recommendations apply to climates where operational temperatures are above -10°C.

| Component | Recommended Oils |
|---|---|
| Petrol engine sump Oil can | BP Visco 2000 plus<br>Castrol Syntron X<br>Castrol GTX -2<br>Castrolite TXT<br>Duckhams Hypergrade Motor Oil<br>Esso Superlube EX2<br>Mobil Super Duckhams QXR<br>Mobil 1 Rally Formula Esso Vitra<br>Fina Supergrade Motor Oil<br>Fina First<br>Shell Super Motor Oil<br>Shell Gemini<br>Havoline X1<br>Havoline multigrade<br>UK only - Land Rover Parts 15W/40 |
| Diesel engine sump ** | BP Vanellus C3 Extra (15W/40)<br>Castrol Turbomax (15W/40)<br>Duckhams Fleetmaster SHPD (15W/40)<br>Esso Super Diesel Oil TD (15W/40)<br>Mobil Delvac 1400 Super (15W/40)<br>Fina Kappa LDO (15W/40)<br>Shell Myrina (15W/40)<br>Texaco URSA Super TD (15W/40)<br>UK only - Land Rover Parts SHPD<br><br>The following list of oils to MIL - L - 2104D or CCMC D2 or API Service levels CD are for emergency use only if the above oils are not available. They can be used for topping up without detriment, but if used for engine oil changing, they are limited to a maximum of 5,000 km (3,000 miles) between oil and filter changes.<br><br>BP Vanellus C3 Multigrade (15W/40)<br>Castrol RX Super (15W/40)<br>Duckhams Hypergrade (15W/50)<br>Esso Essolube XD - 3 plus (15W/40)<br>Mobil Delvac Super (15W/40)<br>Fina Dilano HPD (15W/40)<br>Shell Rimula X (15W/40)<br>Texaco URSA Super Plus (15W/40) |
| Automatic gearbox | BP Autran DX2D<br>Castrol TQ Dexron IID<br>Duckhams Fleetmatic CD<br>Duckhams D - Matic<br>Esso ATF Dexron IID<br>Mobil ATF 220D<br>Fina Dexron IID<br>Shell ATF Dexron IID<br>Texamatic Fluid 9226<br>UK only - Land Rover Parts ATF Dexron II |
| Manual gearbox | BP Autran G<br>Castrol TQF<br>Duckhams Q - Matic<br>Esso ATF Type G<br>Mobil ATF 210<br>Fina Purfimatic 33G<br>Shell Donax TF<br>Texamatic Type G or Universal<br>UK only - Land Rover Parts ATF Type 'G' |
| Front and Rear differential Swivel pin housings | BP Gear Oil SAE 90EP<br>Castrol Hypoy SAE 90EP<br>Duckhams Hypoid 90<br>Esso Gear Oil GX (85W/90)<br>Mobil Mobilube HD90<br>Fina Pontonic MP SAE (80W/90)<br>Shell Spirax 90EP<br>Texaco Multigear Lubricant EP (85W/90)<br>UK only - Land Rover Parts EP90 |

**Other approved oils include:** Agip Sigma Turbo, Aral OL P327, Autol Valve - SHP, Aviation Turbo, Caltex RPM Delo 450, Century Centurion, Chevron Delo 450 Multigrade, Divinol Multimax Extra, Ecubsol CD Plus, Elf Multiperformance 4D, Esso Special Diesel, Fanal Indol X, Fuchs Titan Truck 1540, Gulf Superfleet Special, IP Taurus M, Total Rubia TIR XLD, Valvoline Super HD 4D LD, Veedol Turbostar, Gulf Superfleet (GB), Silkolene Turbolene D, Kuwait Q8 T700.

## Notes

# RECOMMENDED LUBRICANTS AND FLUIDS

| | | |
|---|---|---|
| Propeller shaft Front and Rear | BP Energrease L2<br>Castrol LM Grease<br>Duckhams LB 10<br>Esso Multi - purpose Grease H | Mobil Grease MP<br>Fina Marson HTL 2<br>Shell Retinax A<br>Marfak All Purpose Grease |
| Power steering box and fluid Reservoir Transfer Gearbox | BP Autran DX2D<br>BP Autran G<br>Castrol TQ Dexron IID<br>Castrol TQF<br>Duckhams Fleetmatic CD<br>Duckhams Q - matic<br>Esso ATF Dexron IID | Fina Dexron IID<br>Fina Purfimatic 33G<br>Shell ATF Dexron IID<br>Shell Donax TF<br>Texamatic Fluid 9226<br>Texamatic Type G or 4291A Universal<br>UK only - Land Rover Parts ATF Dexron II |
| | Esso ATF Type G<br>Mobil ATF 220D<br>Mobil ATF 210 | or Type G |
| Brake and clutch reservoirs | Brake fluids having a minimum boiling point of 260°C (500°F) and complying with FMVSS 116 DOT4 | |
| Lubrication nipples (hubs, ball joints etc.) | BP Energrease L2<br>Castrol LM Grease<br>Duckhams LB 10<br>Esso Multi - purpose Grease H | Mobil Grease MP<br>Fina Marson HTL 2<br>Shell Retinax A<br>Marfak All Purpose Grease |
| Ball joint assembly Top Link | BPL21M<br>Castrol M53<br>Shell Retinax AM | Duckhams LBM10<br>Esso MP<br>Mobil Supergrease |
| Seat slides Door lock striker | BP Energrease L2<br>Castrol LM Grease<br>Duckhams LB 10<br>Esso Multi - purpose Grease H<br>Mobil Grease MP | Fina Marson HTL 2<br>Shell Retinax A<br>Marfak All purpose grease<br>NLGI - 2 Multi - purpose Lithium - based Grease |

# RECOMMENDED LUBRICANTS AND FLUIDS

Recommended lubricants and fluids - All climates and conditions

| COMPONENT | SPECIFICATION | VISCOSITY | AMBIENT TEMPERATURE °C |
|---|---|---|---|
| **Petrol models**<br>Engine sump<br>Oil can | Oils must meet:<br>RES.22.OL.G-4<br>or<br>CCMC G-4<br>API service level SG | 5W/30<br>5W/40<br>5W/50<br>10W/30<br>10W/40<br>10W/50<br>15W/40<br>15W/50<br>20W/40<br>20W/50<br>25W/40<br>25W/50 | |
| **Diesel models**<br>Engine sump | RES 22 OLD-5<br>CCMC D-5<br>API CE | 15W/40<br>10W/30 | |
| * Emergency use: | MIL - L - 2104D,<br>CCMCD2 or<br>API CD | 10W/30 | |
| Main Gearbox Automatic | ATF Dexron IID | | |
| Main Gearbox manual | ATF M2C33<br>(F or G) | | |
| Final drive units<br>Swivel pin housings | API GL4 or GL5<br>MIL - L - 2105 or<br>MIL - L - 21-05B | 90 EP<br>80W EP | |
| Power steering<br>Borg Warner<br>Transfer Box | ATF M2C 336<br>or<br>ATF Dexron IID | | |

* Diesel Models - Engine Sump

Oils for emergency use only if the SHPD oils are not available. They can be used for topping up without detriment, but if used for engine oil changing, they are limited to a maximum of 5,000 km (3,000 miles) between oil and filter changes. (See previous page)

# MAINTENANCE SCHEDULES AND OPERATIONS 10

## SUPPLEMENTARY MAINTENANCE SCHEDULE FOR DIESEL - RANGE ROVER

The following supplementary schedule should be used together with the schedule in the main Workshop Manual, for the complete maintenance of Range Rover Diesel models.

The maintenance intervals in this schedule are for European highway driving conditions, for change intervals of engine oil and all filters, under severe abnormal operating conditions, consult your nearest Land Rover Dealer.

**Every 500 km (250 miles)**

- Check engine oil level

**After first 1,500 km (1000 miles)**

- Tighten inlet manifold, exhaust manifold and turbo-charger bolts
  (See **Section 06** for torque wrench settings)
- Change engine oil and filter
- Check engine coolant level
- Check drive belt tension
- General check for fluid leaks
- Check tappet clearance

**Every 10,000 km (6,000 miles)**

- Change engine oil and oil filter
- Drain sedimenter
- Change fuel filter
- Check for fluid leaks
- Check drive belt tension

**Every 20,000 km (12,000 miles)**

- Clean lift pump filter
- Clean fuel sedimenter
- Clean fuel tank breather pipe
- Change air filter element
- Check engine cold idle speed

**Every 40,000 km (24,000 miles)**

- Check tappet clearance
- Check glow plug operation (continuity)
- Remove diesel injectors, spray test and refit

**Every 80,000 km (48,000 miles)**

- Remove intercooler element and flush out using 'GENKLENE' produced by ICI Ltd

**Every 96,000 km (60,000 miles)**

- Check turbo-charger impeller shaft axial and radial clearance
  (See **Section 04** General Specification Data)
- Check wastegate operation

### SPECIAL MAINTENANCE INSTRUCTION

**First 40,000Km (24,000 miles) only**

**NOTE: These instructions must be carried out at the first 40,000 Km (24,000 miles) service. The use of new type gasket eliminates the need to retorque head bolts at 1,500 Km (1,000 miles).**

1. Centre bolts, starting with bolt A: Without slackening bolts, tighten each bolt in sequence through 10 - 15°.
2. Side bolts: Without slackening bolts cheque that torque of each bolt is 85 - 90 Nm. first M1 then M2.

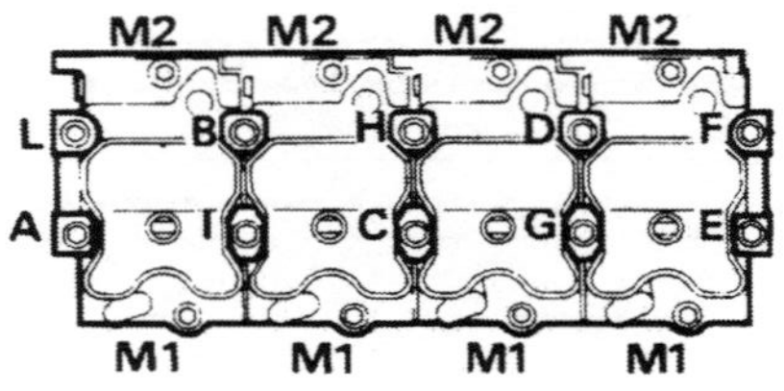

---

# 09 RECOMMENDED LUBRICANTS AND FLUIDS

| | |
|---|---|
| Engine cooling system | Use an ethylene glycol based anti-freeze (containing no methanol) with non-phosphate corrosion inhibitors suitable for use in aluminium engines to ensure the protection of the cooling system against frost and corrosion in all seasons. Use one part anti-freeze to one part water for protection down to -36°C. **IMPORTANT: Coolant solution must not fall below proportions one part anti-freeze to three parts water, i.e. minimum 25% anti-freeze in coolant otherwise damage to engine is liable to occur.** |
| Battery lugs. Earthing surfaces where paint has been removed | Petroleum jelly. **NOTE: Do not use Silicone Grease** |
| Air Conditioning System Refrigerant | **METHYLCHLORIDE REFRIGERANTS MUST NOT BE USED** Use only with refrigerant 12. This includes 'Freon 12' and 'Arcton 12' |
| Compressor Oil | Shell Clavus 68   BP Energol LPT68   Sunisco 4GS Texaco Capella E Wax/Free 68. Castrol Icematic 99 |
| ABS Sensor bush-rear | Silicone grease: Staborags NBU - Wabco 830 502.0634 Wacker chemie 704 - Wabco 830 502.0164 Kluber GL301 |

## ANTI-FREEZE

| ENGINE TYPE | MIXTURE STRENGTH | PERCENTAGE CONCENTRATION | PROTECTION LOWER TEMPERATURE LIMIT |
|---|---|---|---|
| V8 (aluminium) Diesel VM | One part anti-freeze One part water | 50% | |
| **Complete protection** Vehicle may be driven away immediately from cold | | | - 36°C |
| **Safe limit protection** Coolant in mushy state. Engine may be started and driven away after warm-up period | | | - 41°C |
| **Lower protection** Prevents frost damage to cylinder head, block and radiator. Thaw out before starting engine | | | - 47°C |

# MAINTENANCE SCHEDULES AND OPERATIONS

## RANGE ROVER

### ENGINE COOLANT

The level of coolant in the expansion tank should be checked daily or weekly dependent on the operating conditions.

The expansion tank is located in the engine compartment and:-

On 2.4 litre engines is fitted with a spring loaded filler cap. (1) Fig. RR1154.

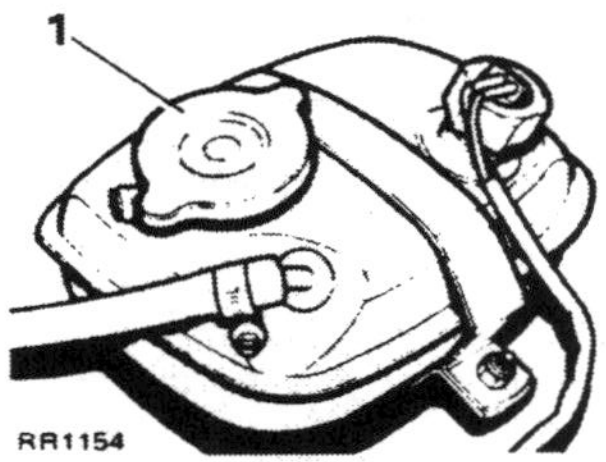

On 2.5 litre engines is fitted with a plastic filler cap and combined coolant level sensor. (2) Fig. RR2729M

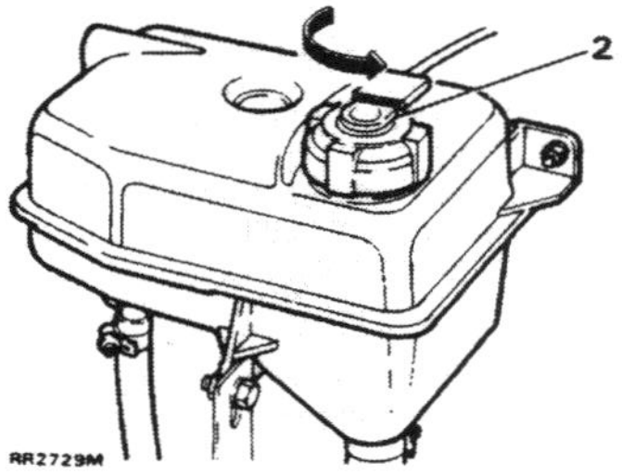

**WARNING: Do not remove the expansion tank filler cap when the engine is hot, because the cooling system is pressurised and personal scalding could result.**

When removing the filler cap, first turn it anti-clockwise a quarter of a turn and allow all pressure to escape, before turning further in the same direction to lift off.

With a cold engine the expansion tank should be approximately half full.

When replacing the filler cap, it is important that it is tightened down fully. Failure to tighten the filler cap properly may result in water loss, with possible damage to the engine through overheating.

### Frost precautions and engine protection.

To prevent corrosion of the aluminium alloy engine parts it is imperative that the cooling system is filled with the specified strength solution of clean water and the correct type of anti-freeze, winter and summer.

The cooling system should be drained and flushed out and refilled with anti-freeze every 40,000 km (24,000 miles) or sooner where the purity of the water is questionable.

After the second winter the system should be drained and thoroughly flushed by using a hose inserted in the radiator filler orifice.

**NOTE: Whenever the cooling system has been drained and refilled, the vehicle should be run for approximately 20 minutes to ensure that the thermostat is open. Recheck the coolant level top up as necessary.**

---

## MAINTENANCE SCHEDULES AND OPERATIONS

## RANGE ROVER

### ENGINE OIL LEVEL CHECKING AND TOPPING UP - Fig. RR1155

Withdraw the dipstick (1) and wipe the blade clean.

Re-insert the dipstick fully, then withdraw it and check the oil level indication, which must be between the 'MAX' (top) and 'MIN' (bottom) mark.

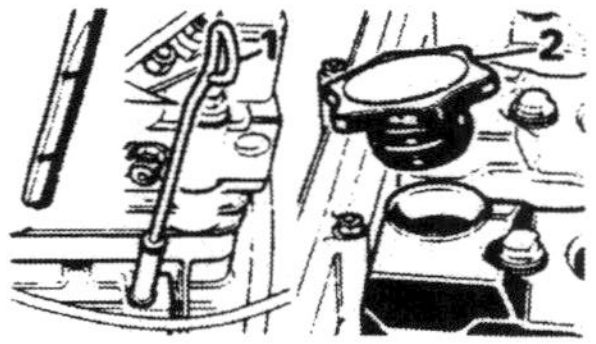

To top-up, remove the filler cap (2) and top-up the engine with new oil, then repeat the checking and topping-up procedure until the oil level is correct. Do not overfill. Do not forget to replace the filler cap.

### Oil draining and refilling - Fig RR1156

The oil should be drained after a run when the engine is warm. The oil filter can be renewed while the oil is draining.

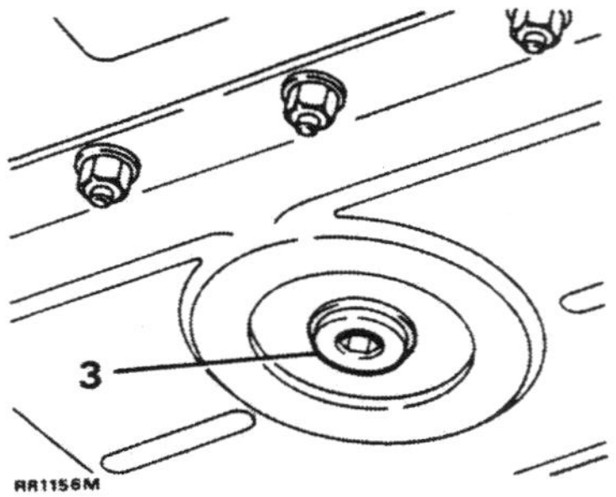

Place a container under the engine that has a capacity of at least 7 litres (12 pints)

Unscrew the drain plug (3) and drain the oil. Clean the drain plug; use a new sealing washer if necessary and refit the drain plug.

Fill the engine with the correct quantity of new oil and recheck the level.

### ENGINE OIL REFILL AND FILTER RENEWAL

Following any drain and refill of the engine oil or renewal of the engine oil filter cartridge the engine must be run at idle speed for a short period to allow oil pressure to build up in the turbo-charger.

**CAUTION: Serious damage to the turbo-charger will result if the engine is run above idling speed before oil pressure is restored.**

### Oil filter cartridge renewal - Fig. RR1157

Slacken the clip and disconnect the air intake hose from the turbo-charger.

Clean the area around the filter head, and place a container beneath the engine.

Unscrew the oil filter cartridge (1) and discard it.

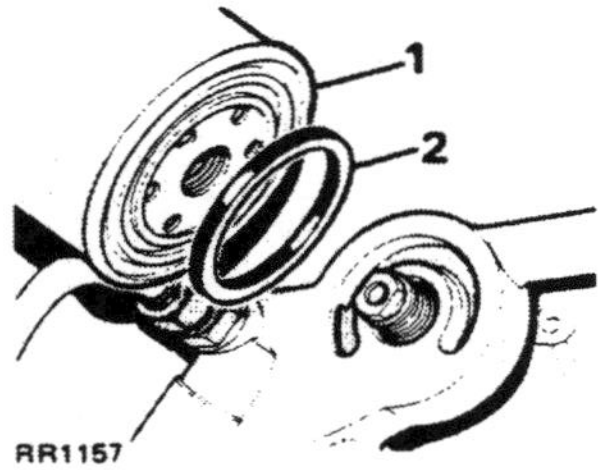

Wet the seal (2) of the new oil filter with engine oil.

Screw the new filter into position, using hand force only.

Check the engine oil level.

Refit the air intake hose to the turbo-charger and tighten the clip.

Start the engine and check for leaks.

Stop the engine, wait a few minutes, then check the oil level and top-up if necessary.

# MAINTENANCE SCHEDULES AND OPERATIONS

## MAIN FUEL FILTER - Fig. RR1161

### Draining off water and sediment

It is essential that any water and sediment in the fuel filter is drained off, as water in the fuel can result in damage to the injection pump.

Hold a small receptacle beneath the drain cock.

Unscrew the drain cock (1) at the bottom of the filter half a turn.

Drain off water and sediment.

Immediately fuel starts to flow from the drain cock tighten the drain cock.

**NOTE: Any delay in tightening the drain cock when the fuel starts to flow could possibly mean bleeding the fuel system.**

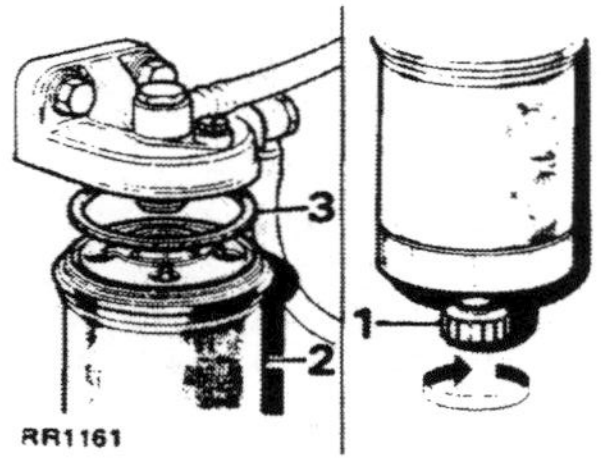

### Renewing the fuel filter element

Clean the area around the filter head, and place a container beneath the filter.

Unscrew the filter (2) - a quantity of fuel will be released - and discard the filter. A hexagon is formed on the base of the filter for unscrewing it with a spanner.

Wet the seal (3) of the new filter with fuel.

Screw the new filter into position and tighten with a spanner.

Ensure that the drain cock at the bottom of the filter is screwed up tight.

## FUEL SEDIMENTER

The sedimenter is attached to the left-hand side of the chassis frame near the fuel tank, and increases the working life of the fuel filter by the larger droplets of water and larger particles of foreign matter from the fuel.

Drain off water as follows:

### Drain off water - Fig. RR1159

Slacken off drain plug (1) and allow water to run out. When pure diesel fuel is emitted, tighten the drain plug.

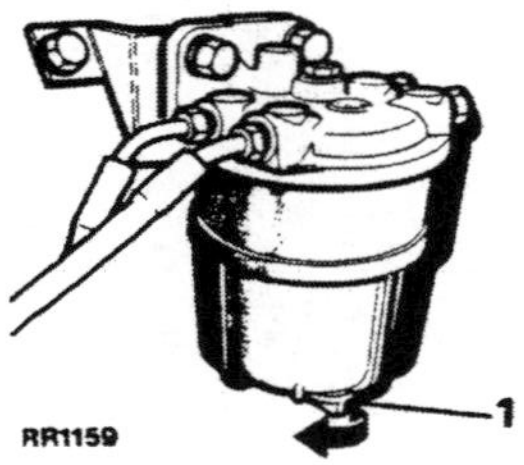

Support the sedimenter bowl (1), unscrew the bolt (2) on the top of the unit and remove the bowl.

Remove the sedimenter element (3) and clean all parts in kerosene. Fit new seals (4) and reassemble the sedimenter.

Slacken off the drain plug, when pure diesel fuel runs out, tighten plug. Start the engine and check the sedimenter for leaks.

### Clean element - Fig. RR1160

If fuel is used from dubious storage facilities, the sedimenter should be removed and cleaned as circumstances require or as specified in the maintenance schedule.

Disconnect the fuel inlet pipe from the sedimenter and raise pipe above the the level of the fuel tank and support in this position to prevent fuel draining from the tank.

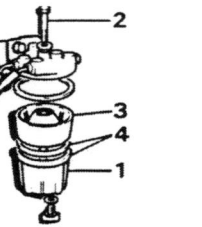

## CLEANING FUEL TANK BREATHER PIPE - Fig. RR1168

The fuel tank breather pipe must be cleaned regularly to prevent diesel oil residue and road dust causing blockage. The pipe is located underneath the vehicle and runs down the body panel joint, to the rear of the fuel tank filler neck.

Clean the pipe at the intervals specified in the maintenance schedule, or more frequently if operating in dusty or muddy conditions.

Wipe clean the end of the breather pipe (1) and use a stout piece of wire to clear the inside.

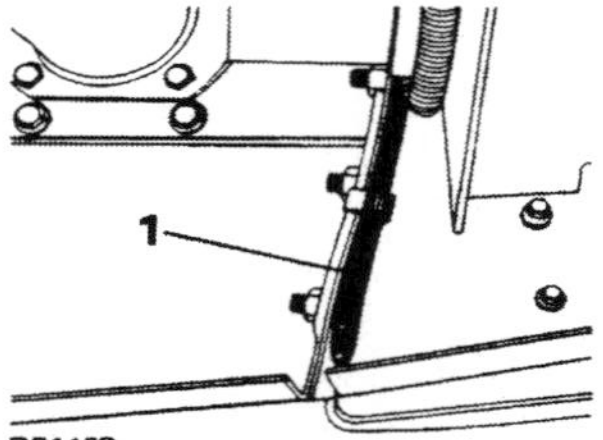

# MAINTENANCE SCHEDULES AND OPERATIONS

## RENEW AIR CLEANER ELEMENT - Figs. RR1158/RR1171

Disconnect the hose (1) from the air cleaner. Release the retaining strap (2) and lift up the air cleaner assembly.

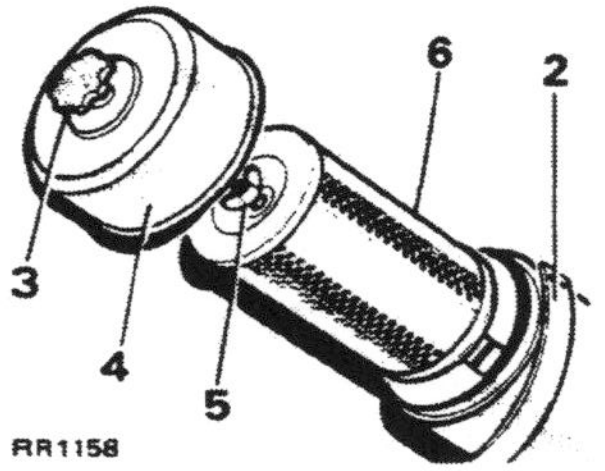

Unscrew the knob (3) and remove the end cover (4) from the air cleaner casing. Unscrew the wing nut (5), discard the element (6) and wipe clean the casing and cover.

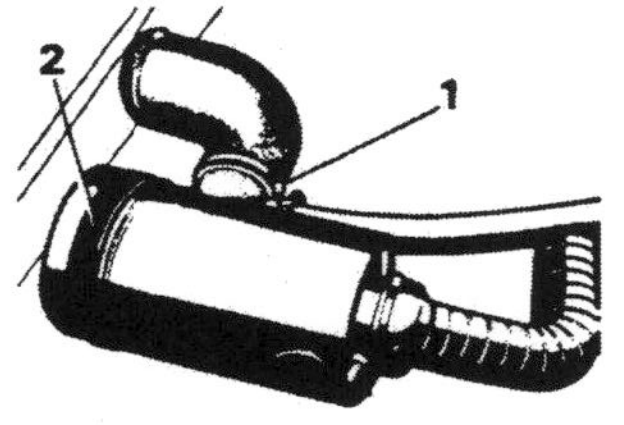

## CHECK AIR CLEANER DUMP VALVE - Fig. RR1169

Squeeze open the dump valve (7) and check that the interior is clean. Also, check that the rubber is flexible and in good condition. If necessary, remove the dump valve to clean the interior. Fit a new valve if the original is in poor condition

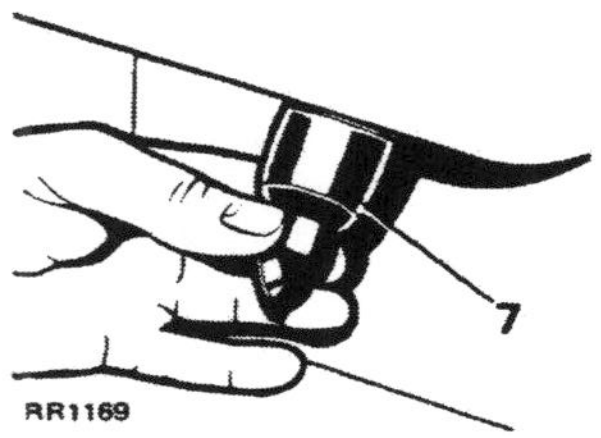

Fit a new element, rubber seal end first, and reassemble the air cleaner.

---

# MAINTENANCE SCHEDULES AND OPERATIONS

## TAPPET ADJUSTMENT - Fig. RR1164

The correct clearance is: inlet and exhaust 0,30 mm (0.012 in) engine cold.

### Remove rocker cover

Unscrew the centre retaining bolts and remove the rocker covers for each cylinder, taking care not to lose the seals from the top of the rocker cover.

### Check and adjust the tappets

Turn the engine over until number 1 valve (counting from front of engine) is fully open.

Using a 0,30 mm (0.012 in) feeler gauge (1) check the clearance between the valve tip and rocker pad of number 7 valve.

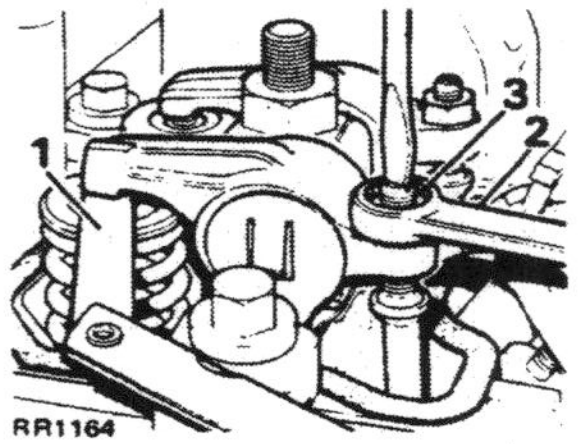

Adjust the clearance by slackening the lock nut (3) and turning the tappet adjusting screw clockwise to reduce clearance and anti-clockwise to increase clearance. Recheck the clearance after tightening the lock nut.

Continue to check and adjust the remaining tappets in the following sequence:

With No.1 valve fully open adjust No.7 valve.
With No.8 valve fully open adjust No.2 valve.
With No.5 valve fully open adjust No.3 valve.
With No.4 valve fully open adjust No.6 valve.
With No.7 valve fully open adjust No.1 valve.
With No.2 valve fully open adjust No.8 valve.
With No.3 valve fully open adjust No.5 valve.
With No.6 valve fully open adjust No.4 valve.

### Refitting the rocker covers

Clean the rocker cover gasket seating face.

Inspect the rocker cover gaskets; renew if damaged.

Position the rocker cover with the oil filler cap on No.1 cylinder, and the rocker cover with the breather pipe to No.3 cylinder

Check that the collars and seals are located on the top of the rocker covers, then fit the rocker covers and tighten the retaining nuts.

## INJECTORS - Fig. RR1165

To locate a faulty injector, slacken the feed pipe union nut on the suspected injector and run the engine slowly. If there is no change in engine performance or if a faulty condition, such as a smoky exhaust, has disappeared, it can be assumed that the injector is faulty and a replacement injector should be fitted.

Unscrew the retaining nut and remove the rocker cover adjacent to the injector to be removed.

Disconnect the fuel leak-off pipe (2) from the injector.

Unscrew the mounting nut (3), and remove the mounting clamp, injector (4) and sealing washer.

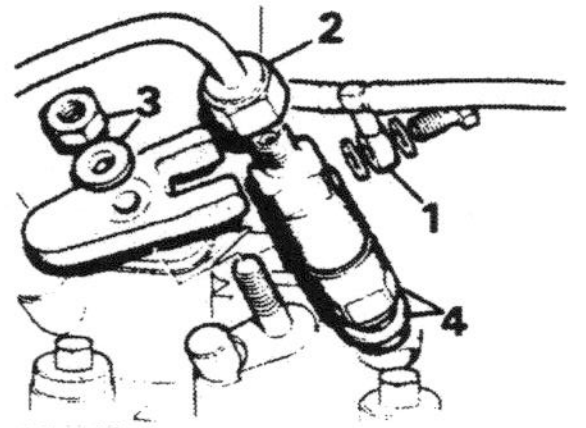

Before fitting an injector fit a new sealing washer.

Fit the injector, its mounting clamp and tighten the injector retaining nut to a torque of 1,7 kg/m.

Refit the high pressure feed pipe and leak-off pipe.

Refit the rocker cover; renew gasket if it is damaged; check that the collars and seals are located on top of the rocker cover before fitting and tightening the rocker cover.

NOTE: Fit the rocker cover with the oil filler cap on No.1 cylinder and the rocker cover with the breather pipe to No. 2 and 3 cylinders.

## MAINTENANCE SCHEDULES AND OPERATIONS

### CHECK DRIVE BELTS - adjust or renew

Right-hand steering - Fig. RR1162

Left-hand steering - Fig. RR1163

**WARNING: Disconnect the battery to prevent any possibility of the starter motor being operated.**

The procedure for checking and adjusting the drive belts for the alternator (1), power steering pump (2) and the optional, air conditioning compressor (3) is similar. Examine all belts for wear and renew if necessary.

**NOTE: Any marks on the outside of the air conditioning drive belt, caused by belt slipper bracket, can be ignored.**

Check the tension of each drive belt, the belts should fit within the following dimensions, when checked at mid-point between the pulleys on the longest side of the belt.

Using a recognised drive belt tensioning gauge the tensions to be:-

On 'V' type drive belts:-
12.7mm wide belts ............ 450N ...... 95 lbf

On poly 'V' drive belts 2.5 Litre Engines:-
Power steering pump ............ 400N ...... 90 lbf
Alternator/water pump ............ 490N ...... 110 lbf

**"In field" Tensioning-No gauge available**

Using normal hand pressure to check deflection, the belt should be tensioned to give a deflection of 0,5 mm per 25 mm of belt run between belt centres.

If any of the drive belts require adjustment, slacken the applicable pivot bolt (4) and the adjusting bracket nut and screw (5), pull the driven unit away from the engine until the belt is tight. Tighten the adjusting bracket then tighten the pivot bolt. Check the belt tension and readjust if necessary.

**CAUTION: When fitting a new drive belt,tension the belt as described above. Reconnect the battery, start and run the engine for 3 to 5 minutes at fast idle, after which time the belt must be re-checked. If necessary retension the belt.**

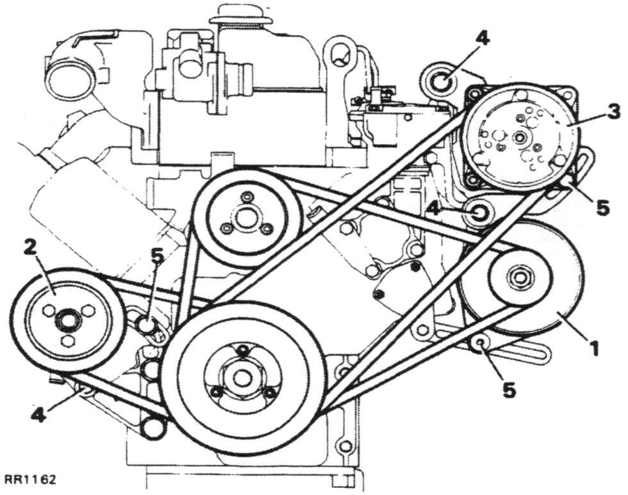

LEFT HAND STEERING

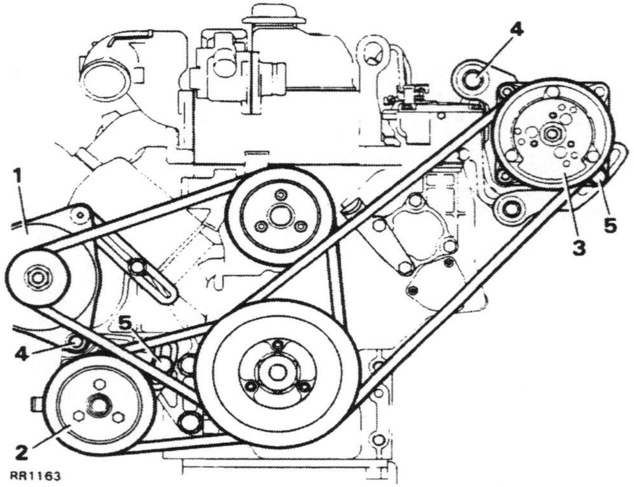

RIGHT HAND STEERING

## SERVICE TOOLS

### RANGE ROVER

18G.1368 — (no label)

18G.1367-1A — (no label)

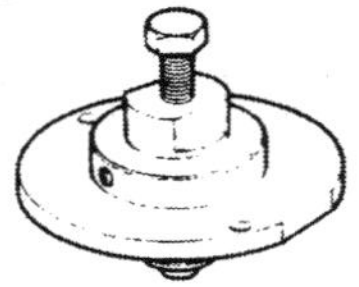

Remover and holder injection pump drive gear 2.4 engine
18G.1377

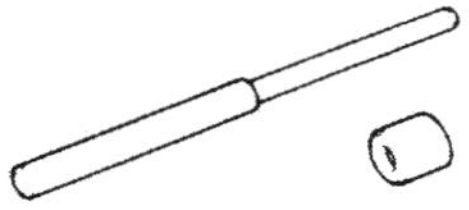

Remover/replacer valve guide
18G.1370B

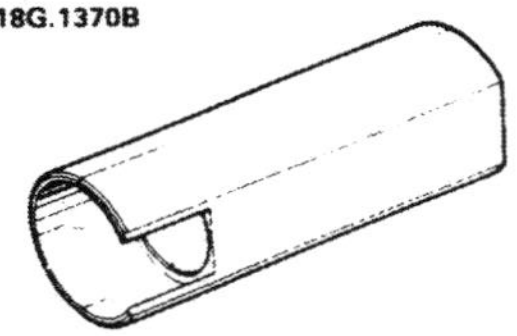

Remover/replacer sleeve crankshaft
18G.1372BX

Thrust pad

18G.1367A — (no label)

Remover crankshaft pulley
18G.1374

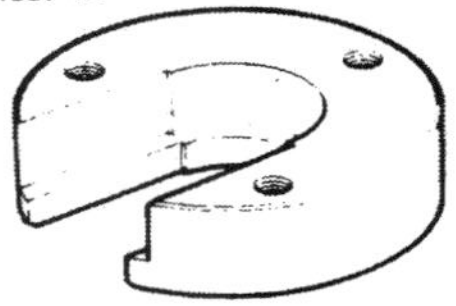

Adaptor crankshaft gear remover
18G.1375

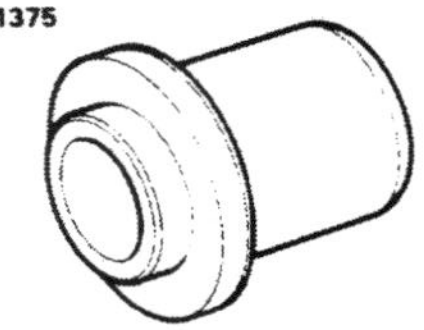

Replacer timing cover oil seal
18G.1369A

Timing gauge

18G.1372B

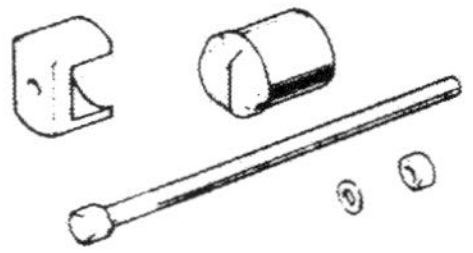

Remover/replacer crankshaft bearings

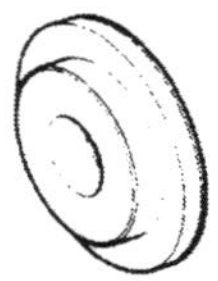

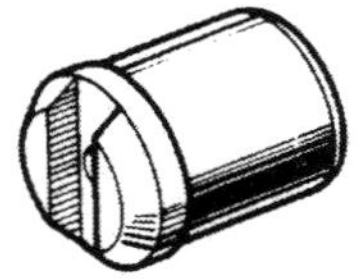

Replacer crankshaft rear oil seal
18G.1378B

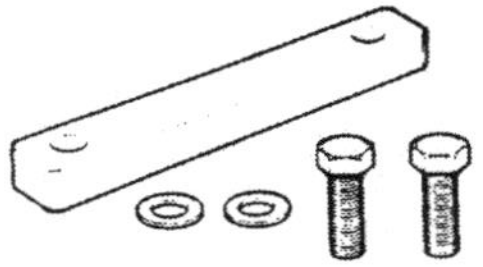

Retainer beam and gauge block cylinder liner
18G.1371

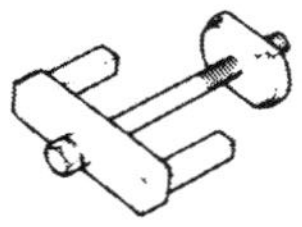

Remover cylinder liner

---

## MAINTENANCE SCHEDULES AND OPERATIONS

### RANGE ROVER

**CHECK COLD START ADJUSTMENT AND COLD IDLE SPEED 2.4 LITRE ENGINE**

NOTE: It is important that these checks are carried-out when the engine is cold.

**Cold start adjustment**

1. Check dimension 'A' which should be 3mm to 4mm.
   If adjustment is required slacken the cable clamp nut and move the clamp forward or rearward as necessary to achieve the correct dimensions and tighten the clamp nut.

**Cold idle speed**

2. The cold idle speed should be between 1000 and 1100 r.p.m. If adjustment is required slacken the nut and move the lever stop (3) to increase or decrease the speed accordingly and tighten the nut.

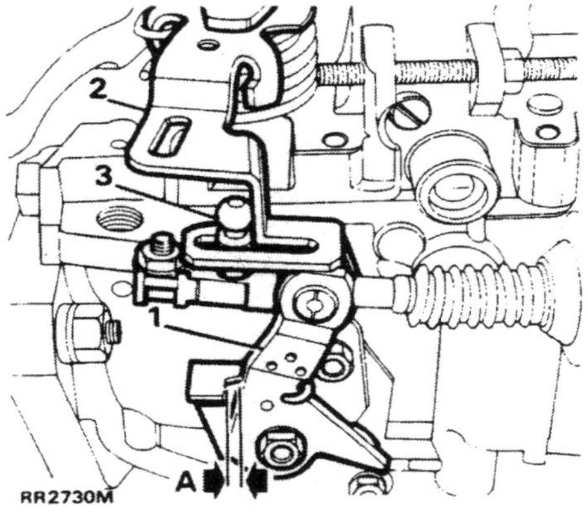

**CHECK AND ADJUST FAST IDLE SETTING PROCEDURE 2.5 LITRE ENGINE.**

NOTE: It is important that these checks are carried-out when the engine is warm - above 40°C.

**Fast idle adjustment**

1. Dimension 'A' should be 4.5 mm. Adjust the lever inserting a 4.5mm distance piece into the gap to hold this dimension.
2. Move the accelerator lever to achieve engine speed 1000 to 1100 r.p.m. (no load).
3. Release and move the lever stop until it rests against the stop tab on the accelerator lever. Retighten the lever stop. Remove the distance piece.

483

# VM DIESEL ENGINE

## DIESEL ENGINE FAULT DIAGNOSIS

### SYMPTOMS

**Engine will not start**
Start with check No. 1 and proceed as directed

**Engine lacks power** (ensure that the vehicle is not overloaded)
Start with check No. 34 and proceed as directed

**Incorrect idling**
Start with check No. 26 and proceed as directed

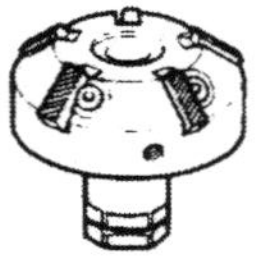

**Excessive exhaust**
Start with check No. 17 and proceed as directed

**Engine misfires**
Start with check No. 29 and proceed as directed

---

## SERVICE TOOLS

**18G.1373**
Remove/replacer front and rear crankshaft bearings

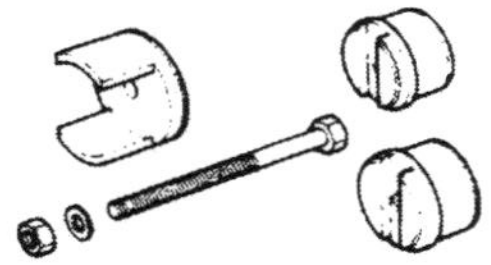

**MS.76B**
Basic handle set valve seat cutter

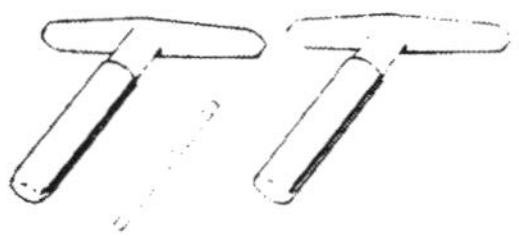

**MS.150-8**
Dia. 7.9 mm–8.5 mm
Adjustable pilot

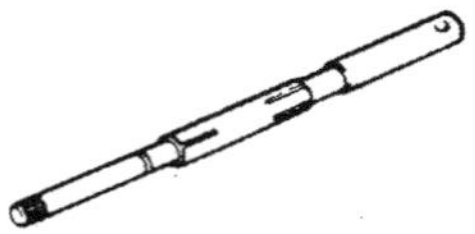

**MS.621**
Dia. Range 28.5 mm–44 mm 15° & 45°
Adjustable valve seat cutter

**LST 122**
Angle gauge

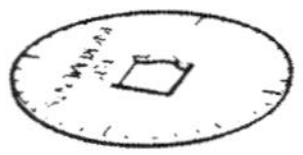

**MS.107**
Adaptor timing injector fuel pump

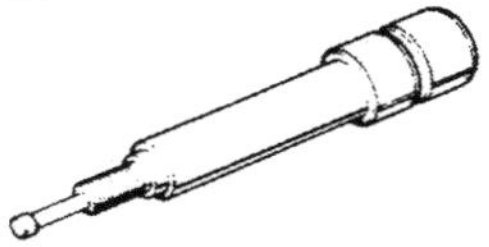

**MS.690**
Dia. Range 52 mm–42.5 mm 35°

**18G.79**
Clutch centralising tool

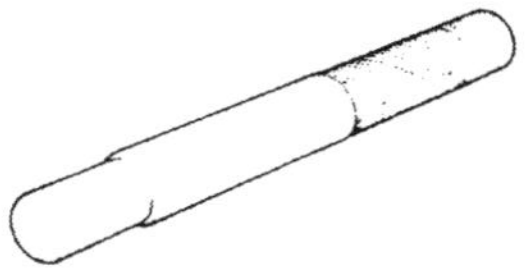

**LST-139**
Adjustable cutter

Remover and holder injection pump drive gear 2.5 engine.

# VM DIESEL ENGINE 12

## RANGE ROVER

### ENGINE FAULT DIAGNOSIS

**CHECK:**

1. Does the starter motor turn the engine?
   - YES: Check 2
   - NO: Check 4

2. Does the starter turn the engine at normal starting speed?
   - YES: Check 6
   - NO: Check 3

3. Is the engine oil of the correct grade?
   - YES: Check 4
   - NO: Change the oil

4. Is the battery charged and in good condition?
   - YES: Check 5
   - NO: Charge or renew the battery as necessary

5. Are all the cables and connections in the starter and solenoid circuit satisfactory?
   - YES: Suspect faulty starter or solenoid
   - NO: Repair as necessary

6. Are the heater plugs operating?
   - YES: Check 8
   - NO: Check 7

7. Is the heater plug electrical circuit satisfactory?
   - YES: Check the heater plug
   - NO: Repair the circuit

8. Does the manual cold start advance operate correctly?
   - YES: Check 9
   - NO: Renew cold start device

9. Is fuel reaching the injectors?
   - YES: Check 17
   - NO: Check 10

10. Is the fuel cut-off solenoid working?
    - YES: Check 12
    - NO: Check 11

11. Is the solenoid electrical circuit satisfactory?
    - YES: Suspect faulty solenoid
    - NO: Repair as necessary

12. Is there a supply of clean fuel in the tank?
    - YES: Check 13
    - NO: Fill the tank and bleed the system

13. Are there leaks at fuel pipes or connections?
    - YES: Repair the leaks and bleed the system
    - NO: Check 14

14. Is there a blockage in the fuel system?
    - YES: Clear the blockage or renew the filter system
    - NO: Check 15

15. Is the fuel lift pump operating?
    - YES: Check 16
    - NO: Renew the lift pump

16. Does the fuel system require bleeding?
    - YES: Bleed the fuel system
    - NO: Suspect faulty injection pump

17. Are the injector pipes connected in the correct firing order?
    - YES: Check 18
    - NO: Correct the firing order

18. Are the correct injectors fitted?
    - YES: Check 19
    - NO: Fit correct injectors

## RANGE ROVER

### ENGINE FAULT DIAGNOSIS

**CHECK:**

19. Are the injectors fitted correctly
    - YES: Check 20
    - NO: Rectify the error

20. Is the injection timing correct?
    - YES: Check 21
    - NO: Re-set the timing

21. Is the air cleaner or trunking blocked?
    - YES: Clear the blockage
    - NO: Check 22

22. Is the injector spray pattern, opening pressure and test performance satisfactory?
    - YES: Check 23
    - NO: Clean or renew injectors as necessary

23. Are valve clearances correct?
    - YES: Check 24
    - NO: Adjust the valve clearances

24. Are the cylinder compression pressures satisfactory?
    - YES: Check 25
    - NO: Locate and correct the fault

25. Is the injection pump delivery correct?
    - YES: Suspect faulty turbo-charger
    - NO: Adjust or renew the injection pump

26. Does the throttle cable operate correctly
    - YES: Check 27
    - NO: Repair or renew the throttle cable

27. Does the throttle cable have at least 1.5 mm (1/16") free play?
    - YES: Check 28
    - NO: Adjust the throttle cable

28. Is the idle speed screw setting correct?
    - YES: Check 29
    - NO: Adjust the engine idle speed

29. Is the fuel tank air vent restricted?
    - YES: Clear the restriction
    - NO: Check 30

30. Are there leaks at the fuel pipes or connections?
    - YES: Repair the leaks and bleed the system
    - NO: Check 31

31. Is there a blockage in the fuel system?
    - YES: Clear the blockage and bleed the system
    - NO: Check 32

32. Is the lift pump operating correctly?
    - YES: Check 33
    - NO: Renew the lift pump

33. Does the fuel system require bleeding?
    - YES: Bleed the system
    - NO: Check 17

34. Are the brakes binding?
    - YES: Adjust the brakes
    - NO: Check 35

35. Is the throttle cable transmitting full travel to the throttle lever?
    - YES: Check 17
    - NO: Adjust the throttle cable

485

# VM DIESEL ENGINE

## REMOVING AND REFITTING ENGINE

The procedure for engine remove and refit is similar to the petrol engine. The major component differences are highlighted in the following procedure. All instructions refer to both 2.4 and 2.5 engines unless otherwise stated.

**CAUTION: Seal all pipe ends against the ingress of dirt after disconnecting oil, fuel, fluid, vacuum or air conditioning pipelines.**

### Removing

1. Remove the radiator and intercooler unit. The radiator unit has a built in engine oil cooler, access to the lower union is possible when the fan cowl is removed.
2. Remove the air cleaner assembly and connecting hoses.
3. Remove the power steering pump outlet and inlet hoses.
4. Disconnect the engine harness multiplug and, if the vehicle has air conditioning, the wiring to the compressor clutch.
5. (Air conditioning vehicles only) turn the high and low pressure compressor service valves to the OFF position (fully clockwise). Depressurise the compressor and remove the high and low pressure hoses.
6. Disconnect the heater return hose at the water pump and draw it clear.
7. Remove the heater inlet hose at the bulkhead connection.
8. Remove the split pin securing the inner throttle cable to the fuel injection pump.
9. Depress the tags on the outer cable adjusting screw to release the cable from the mounting bracket.
10. Disconnect the vacuum pipe from the vacuum pump.
11. Disconnect the glow plug feed wire.
12. Remove the main fuel line at the fuel pump, retaining the washers.
13. Remove the inlet and outlet fuel lines at the filter assembly.
14. Remove the spill return pipe union at the fuel injection pump.
15. Remove the exhaust manifold heat shield.
16. Release the exhaust flange nuts and disconnect the exhaust down pipe.
17. Remove the starter motor heat shield, wiring connections and fixings to the bell housing. Leave the starter motor attached to the engine block.
18. Remove one centre engine mounting nut from each side.
19. Remove the fixings securing the bell housing to the engine.
20. Attach a suitable lifting chain and hoist to the engine lifting hooks.
21. Raise the engine clear of the mountings and support the gearbox.
22. Remove the right hand engine mounting. 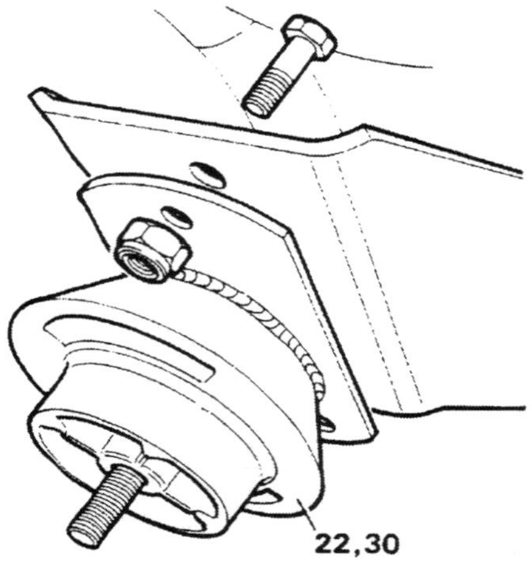
23. Remove the centre bolt from the left hand engine mounting.
24. Withdraw the engine from the gearbox and release the gearbox and transfer box breather pipes from their securing clip.
25. Lift the engine clear of the vehicle.

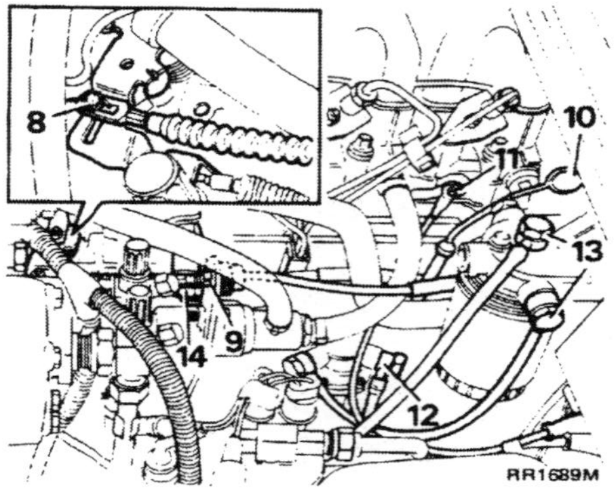

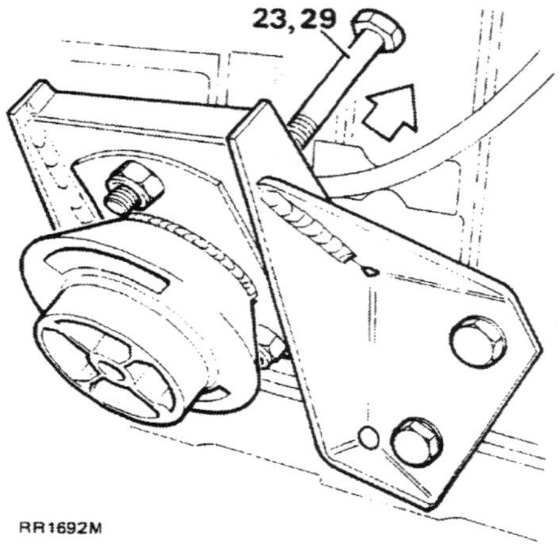

### Refitting.

Before refitting the engine

Smear the splines of the primary pinion, the clutch centre and withdrawal unit abutment faces with molybdenum disulphide grease, Rocol MTS.1000. Smear the engine to gearbox joint faces with Hylomar jointing compound.

26. Attach a lifting chain and hoist to the engine lifting hooks.
27. Lower the engine into the engine bay and locate the gearbox and transfer box breather pipes in their securing clip.
28. Locate the primary pinion into the clutch and secure the engine to the bell housing with at least two bolts.
29. Fit the left hand centre engine mounting bolt.
30. Fit the right hand engine mounting and centre bolt.
31. Lower the engine on to the mountings.
32. Secure the fixings at both front engine mountings.
33. Remove the lifting equipment and the gearbox support.
34. Reverse instructions 1 to 17.
35. Prime the fuel system.

# VM DIESEL ENGINE

## CYLINDER HEADS

**NOTE:** Before removing cylinder heads check alignment of heads, evidence of head gasket or manifold gasket blowing and evidence of water leaks.

### Remove and refit

### Removing

1. Disconnect battery negative lead.
2. Remove expansion tank filler cap. Drain coolant.
3. Disconnect breather hoses from rocker covers.
4. Disconnect brake servo hose.
   Disconnect air conditioning temperature switch.
5. Disconnect by-pass hose and top hose at thermostat housing.
6. Disconnect cold start hose at water rail.
7. Disconnect bleed hose at water rail.
8. Disconnect vacuum pipe from inlet manifold.
9. Remove intercooler pipe.
10. Remove fuel feed pipes from injector. Remove all injectors, with spill pipe, lay aside. Retain injector dowels.
11. Remove heater plug feed wire.
12. Remove cold start hose from cylinder head.
13. Disconnect temperature sensor connector.
14. Remove rocker covers.
15. Remove rocker assemblies. Remove push rods. inspect.
16. Remove eight bolts securing water rail. Lay water rail aside on heater hose.
17. Remove oil feed banjo bolts from cylinder heads.
18. Remove exhaust heat shield.
19. Remove turbocharger oil feed pipe
20. Remove four nuts securing turbocharger to exhaust manifold.
21. Remove outer cylinder head bolts.
22. Remove centre cylinder head bolts.
23. Lay cylinder head oil feed pipe against bulkhead.
24. Remove cylinder heads complete with manifolds.
25. Remove inlet and exhaust manifolds, discard gaskets. Inspect cylinder heads

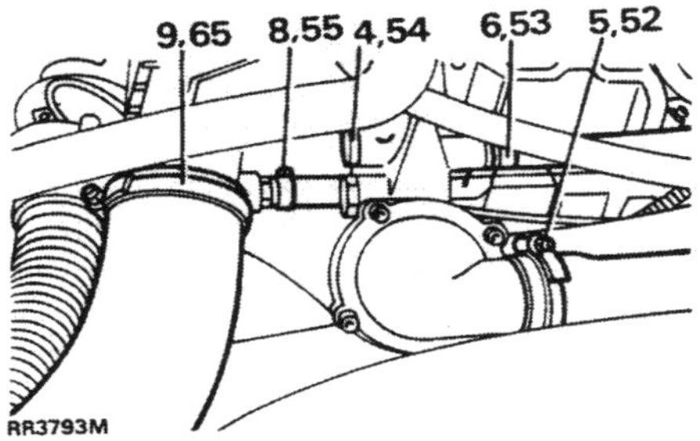

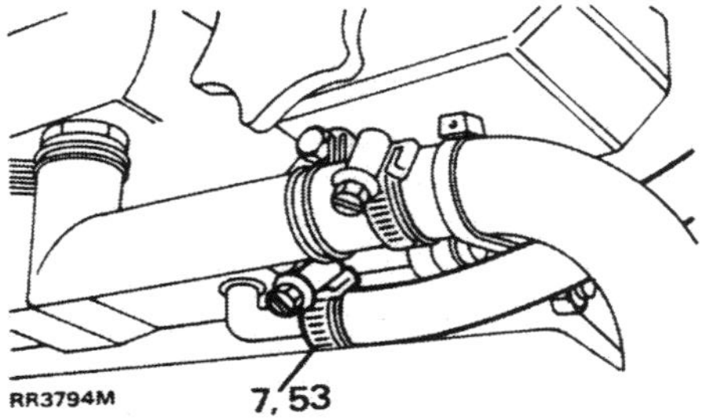

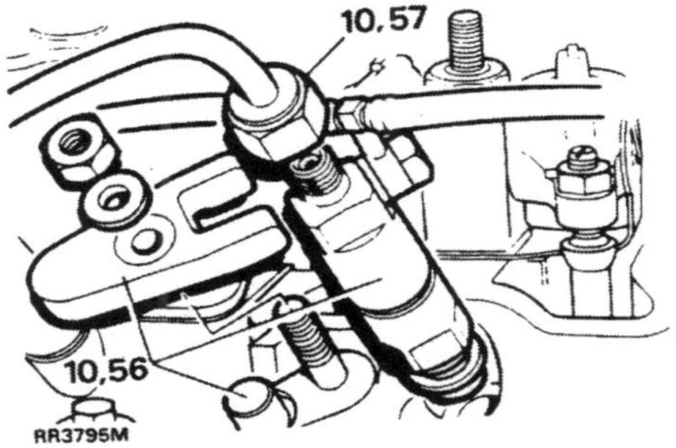

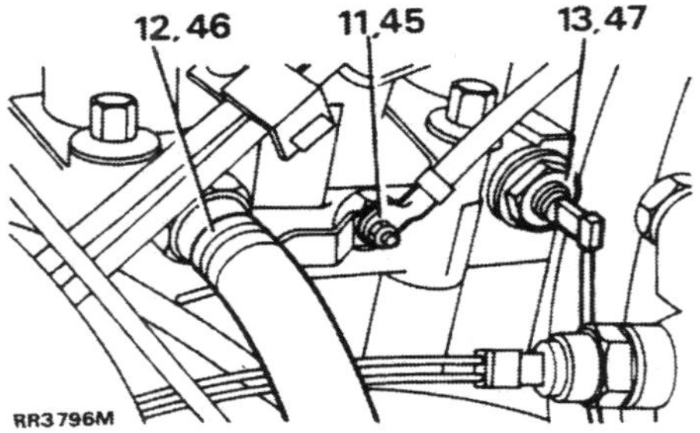

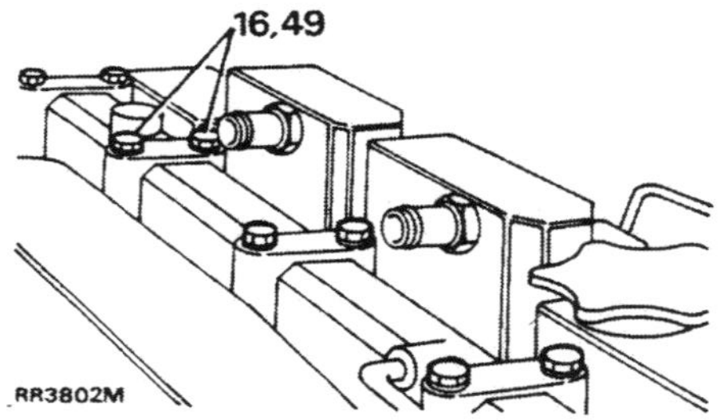

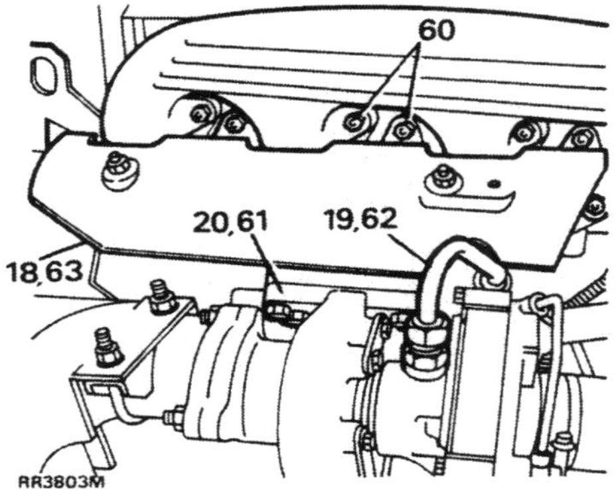

### INSPECT CYLINDER HEADS

Inspect cylinder heads, using the checks below. Any head that fails one or more check must be replaced with a new component, retaining those heads which pass all the checks.

a) Minimum width - 109 mm.
b) Height - 90 mm ± 0.05mm.
c) Inspect for cracks across valve bridge.
d) Distortion of mating faces.
e) Indentation of upper face caused by clamps.
f) Coolant leakage.
g) Measure end plate height - 91.26 to 91.43 mm.

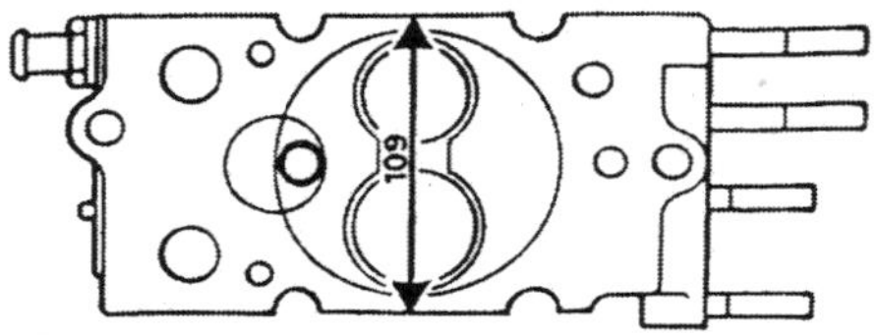

If components pass above checks they may be refitted, using latest gasket and new centre bolts.

26. If new heads are being fitted, remove heater plugs, oil feed dowels, coolant adaptor and temperature transmitter. Fit these items to new cylinder heads. Fit new injector shrouds. Using an airline, check rocker oil feed drilling is free of obstruction.
27. Inspect gaskets, attempt to determine area of failure. Remove old gaskets from block. Thoroughly clean all traces of old gasket material from face of block. Check liner protrusion, see **LINER PROTRUSION CHECK.**

**WARNING: Failure to clean block face thoroughly could lead to head gasket failure.**

28. Remove oil filter, catching any oil spillage.
29. Remove fan assembly, left hand thread.

### Refitting

30. Fit inlet manifold loosely. Fit exhaust manifold loosely, fitting lifting eye.
31. Thoroughly clean face of new cylinder heads.
32. Determine thickness of head gasket required - see **HEAD GASKET SELECT.** Fit gaskets to cylinder block correctly.
33. Fit cylinder head assembly to block, locating studs to turbocharger.
34. Align head assembly with gaskets.
35. Ensure head side holes align with gasket and holes in block. Gaps between heads should be parallel, see RR3809M.

**NOTE: 2.4 litre models, where bulkhead clearance is limited, fit number 8 push rod into cylinder head before fitting head assembly. To ensure push rod does NOT protrude below face of cylinder head, tape it in place.**

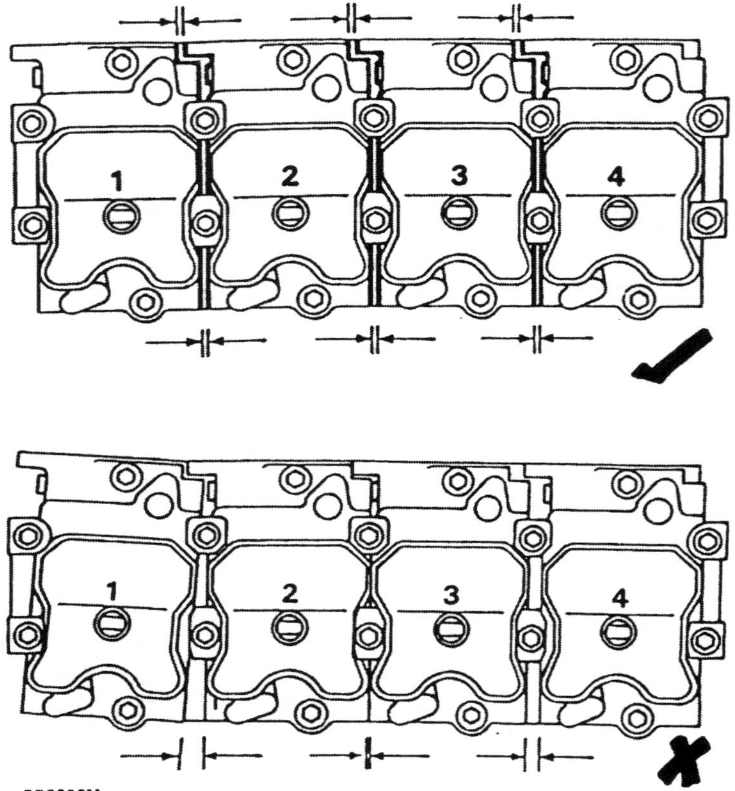

# VM DIESEL ENGINE

## RANGE ROVER

36. Lubricate side bolts (without washers) with engine oil, fit loosely.
37. Ensure inlet manifold is fitted square to cylinder heads to bring heads into alignment.
38. Lubricate threads and underside of central bolt heads with Molybdenum Disulphide. Fit centre bolts loosely, with end plates at front and rear. Align oil feed pipe.
39. Partially tighten centre bolts, holding end plates flush with cylinder heads. Ensure gasket positions are square and have not moved.

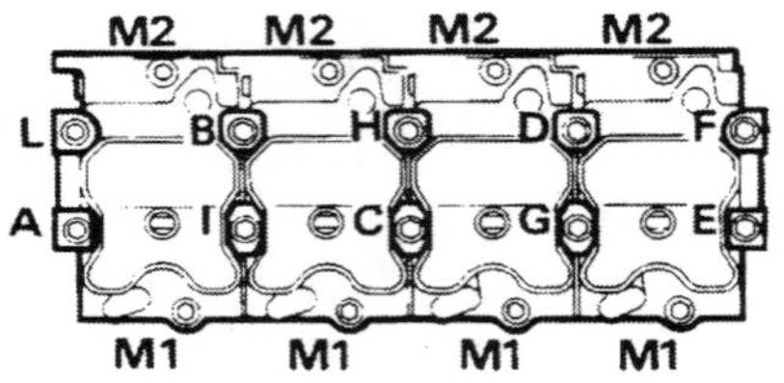

40. Centre bolts:
    a) Torque centre bolts to 30 Nm in sequence shown, starting from bolt A. REPEAT procedure for each bolt.
    b) Tighten each bolt through an angle of 70°, in sequence.
    c) Tighten each bolt an additional 70° in sequence.

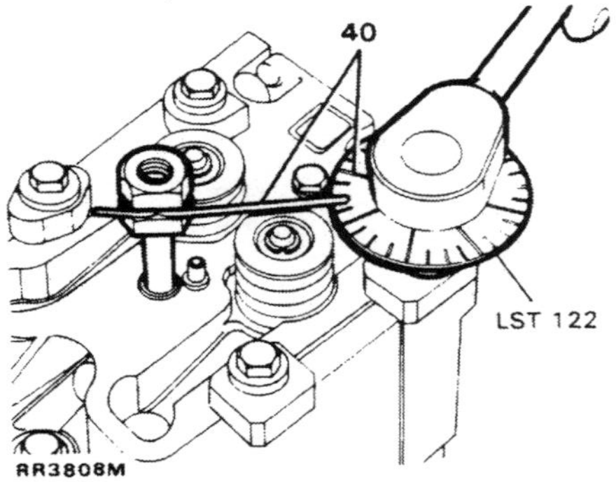

41. Torque side bolts to 80 Nm. in the sequence bolts M1 then bolts M2.
42. Fit oil feed pipe, tighten bolts to 8 Nm.
43. Fit push rods and rocker assemblies, tighten single fixing to 108 Nm.

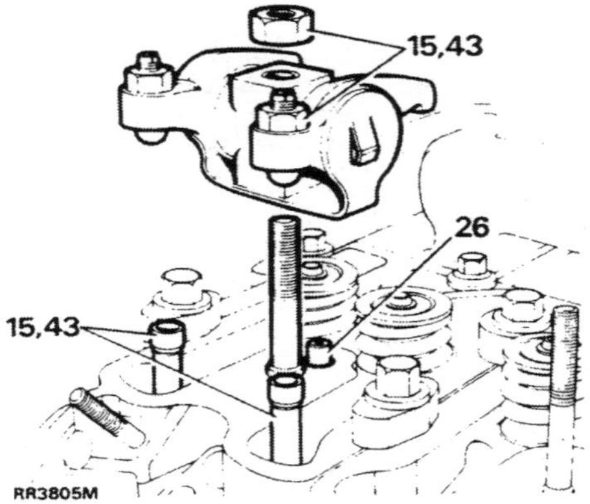

44. Adjust tappets.
45. Fit heater plug feed wire, tighten connector plates.
46. Fit cold start hose to cylinder head.
47. Connect temperature sensor connector.
48. Fit fan assembly.
49. Fit two rear bolts to water rail. Fit water rail attaching by pass hose. Tighten eight bolts to 8 Nm.
50. Tighten by pass hose clip.
51. Fit cold start hose to water rail, tighten clip.
52. Fit top hose, tighten clip.
53. Fit bleed hose to water rail, tighten clip.
54. Connect air conditioning temperature switch.
55. Connect vacuum pipe from inlet manifold, tighten clip.
56. Fit injectors with copper sealing washers. Locate dowels, tighten clamp nuts to 26 Nm.
57. Fit fuel supply pipes to injectors, tighten to 19 Nm.
58. Fit rocker covers, tighten to 9 Nm.
59. Fit breather pipes to rocker covers. Connect brake servo hose.
60. Tighten exhaust and inlet manifold nuts to 32 Nm.
61. Tighten four turbocharger to manifold nuts to 25 Nm.
62. Fit and tighten turbocharger oil feed pipe.
63. Fit exhaust heat shield
64. Fit oil filter.
65. Fit intercooler pipe, tighten clip.

# VM DIESEL ENGINE

## Coolant refill

66. Remove coolant pipe from top of radiator.
67. Fill system through expansion tank until radiator is full.
68. Refit pipe to radiator and tighten.
69. Start engine, run until operating temperature is achieved, top up expansion tank as necessary.
70. Fit expansion tank filler cap, run engine for twenty minutes.

## Retorque cylinder heads

71. Allow engine to cool completely.
72. Drain coolant.
73. Remove rocker covers.
74. Remove water rail.
75. Centre bolts, starting with bolt A:
    a) Loosen bolt, torque to 30 Nm.
    b) Tighten bolt through an angle of 120°.

NOTE: 120° may be achieved by tightening through 60°, immediately followed by a further 60°. The total 120° MUST BE ACHIEVED BEFORE proceeding to next bolt.

c) Repeat for each bolt in sequence shown.
76. Retorque outer bolts to 90 Nm without loosening, first M1 then M2.
77. Fit rocker assemblies, tighten single fixing to 108 Nm.
78. Adjust tappets.
79. Fit water rail using new gaskets.
80. Fit hoses, tighten clips.
81. Check top up oil.
82. Refill cooling system. Run engine until operating temperature is reached, top up if necessary.

## RETORQUE CYLINDER HEADS

First 40,000Km (24,000 miles) only

NOTE: These instructions must be carried out at the first 40,000 Km (24,000 miles) service OR 40,000 Km (24,000 miles) AFTER the above procedure has been carried out. The use of new type gasket eliminates the need to retorque head bolts at 1,500 Km (1,000 miles).

1. Centre bolts, without slackening bolts, start with bolt A, tighten each bolt in sequence through 10 - 15°.
2. Side bolts: Without slackening bolts cheque that torque of each bolt is 85 - 90 Nm, first M1 then M2.

## LINER PROTRUSION CHECK

1. Ensure face is clean
   Correct reading:
   2.4 Litre and 2.5 Litre - 0,00 to 0,06 mm
2. To obtain the correct liner protrusion, attach special tool 18G 1378 B as illustrated, to the cylinder block and tighten the bolts to 30 Nm. Fit a dial test indicator so that the stylus rests in a loaded condition on the external rim of the liner and set the dial to zero. Slide the stylus across to the cylinder block and note the reading.
   Repeat the above procedure to the remaining cylinders. 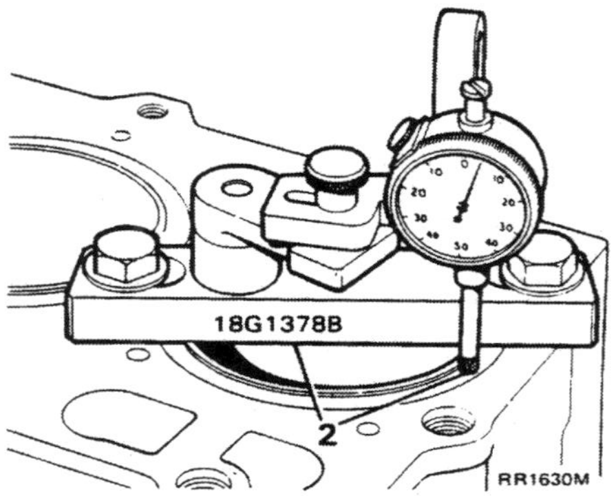
3. Remove the liners and add shims as required to achieve the protrusion.

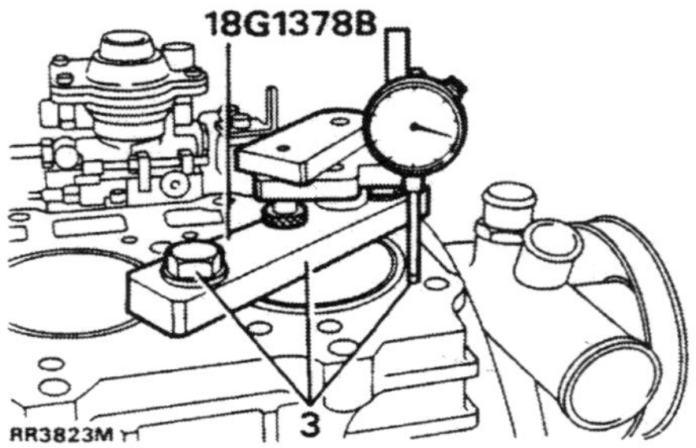

3. Fit special tool 18G 1378B to the cylinder block and tighten the bolts to 30 Nm Attach the dial test indicator to the tool, as illustrated, and position the stylus, in a loaded condition, on the cylinder block and zero the gauge. Slide the indicator over so that the stylus rests on the piston crown and note the reading.

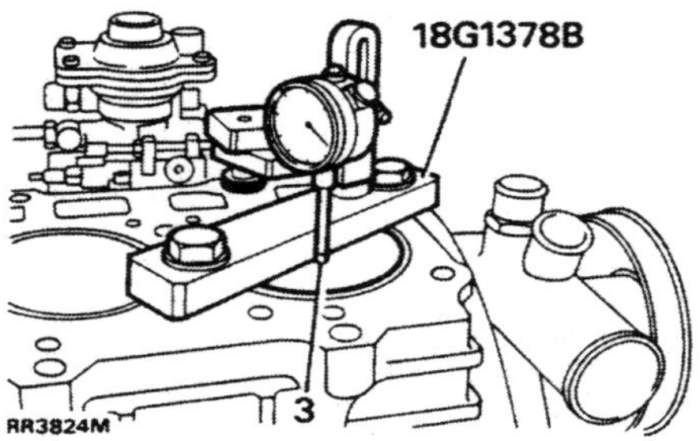

## HEAD GASKET SELECT

1. Before fitting the cylinder heads it is necessary to determine the thickness of gasket that must be used to achieve the correct clearance between each piston crown and cylinder head. Three thicknesses of gasket are available, see table below. The following procedure should be used to determine which size to fit. However, only one thickness of gasket must be used on all four cylinders, this being the one for the cylinder which calls for the thickest gasket.

| Identification | Part no. | Fitted thickness |
|---|---|---|
| No notch | STC 654 | 142mm ± 0.04 |
| One notch | STC 656 | 162mm ± 0.04 |
| Two notches | STC 655 | 152mm ± 0.04 |

2. Turn the crankshaft to bring number one piston to T.D.C.

Example
Required piston clearance .......... 0,85 to 0,94 mm
Plus measured height ................ 0,60   0,60 mm

Thickness of gasket required ...... 1,45 to 1,54 mm

The nearest compressed thickness of gasket available is 1.52 mm part number STC655

4. Repeat the above instruction on the remaining cylinders. The thickest gasket required is the one which must be fitted to all cylinders.

# VM DIESEL ENGINE

## ENGINE EXTERNAL COMPONENTS 2.4 LITRE ENGINE

1. Inlet manifold
2. Engine coolant rail
3. Turbo-charger
4. Engine lifting eye
5. Heat shield
6. Exhaust manifold
7. Sealing rings
8. Cylinder head bolt and spacer block
9. Oil filler cap
10. Rocker cover
11. Valve gear oil feed pipe
12. Dipstick
13. Injector pipes
14. Vacuum pump
15. Oil filter element
16. Cylinder head end-plate
17. Cylinder head
18. Injector dowel
19. Heater plug
20. Heater plug copper link
21. Turbo-charger support bracket
22. Coolant thermostat
23. Oil thermostat
24. Vacuum pump gear
25. Retaining clamp - vacuum pump
26. Oil filter base adapter
27. Cylinder block
28. Fuel injection pump
29. 'O' ring
30. Flywheel housing
31. Flywheel
32. Water pump
33. Crankshaft pulley
34. Timing cover
35. Oil pressure relief valve
36. Sump
37. Sump pan
38. Fuel lift pump
39. Cold start device

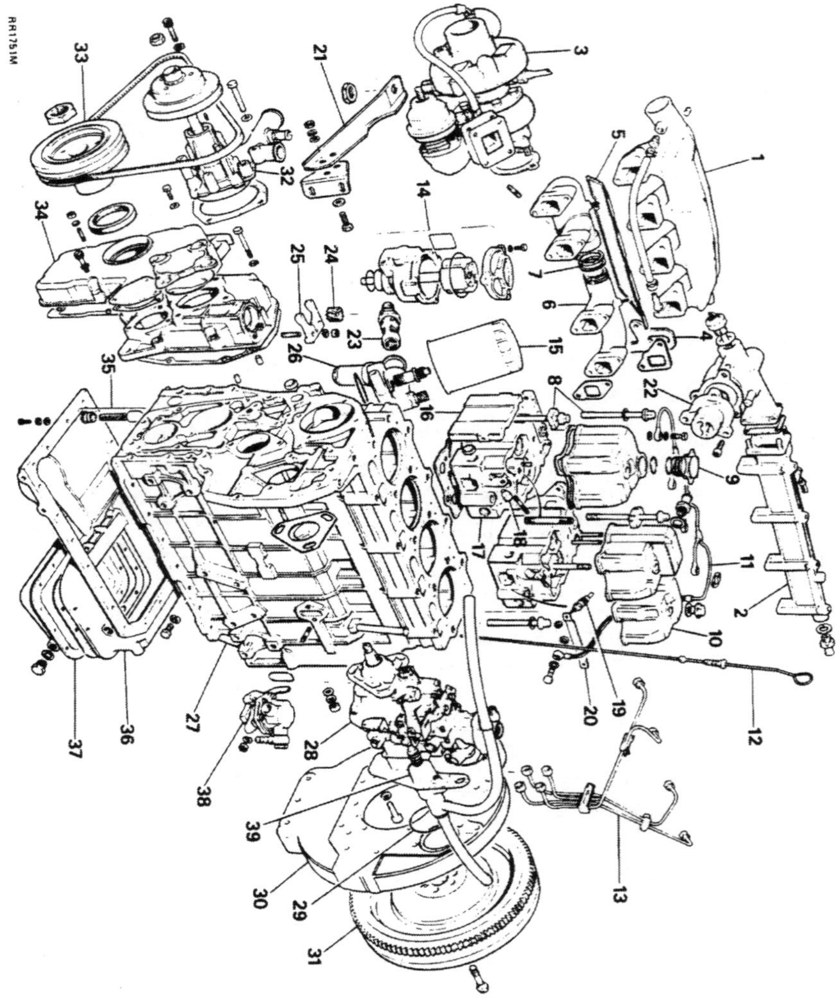

# VM DIESEL ENGINE

## ENGINE EXTERNAL COMPONENTS 2.5 LITRE ENGINE

1. Inlet manifold
2. Engine coolant rail
3. Turbo-charger
4. Engine lifting eye
5. Heat shield
6. Exhaust manifold
7. Sealing rings
8. Cylinder head bolt and spacer block
9. Oil filler cap
10. Rocker cover
11. Valve gear oil feed pipe
12. Dipstick
13. Injector pipes
14. Screw plug
15. Oil filter element
16. Cylinder head end-plate
17. Cylinder head
18. Injector dowel
19. Heater plug
20. Heater plug copper link
21. Turbo-charger support bracket
22. Coolant thermostat
23. Oil thermostat
24. Support bracket
25. Bolt
26. Oil filter base adapter
27. Cylinder block
28. Fuel injection pump
29. 'O' ring
30. Flywheel housing
31. Flywheel
32. Water pump
33. Crankshaft pulley
34. Timing cover
35. Oil pressure relief valve
36. Sump
37. Sump bolt
38. Fuel lift pump
39. Cold start device

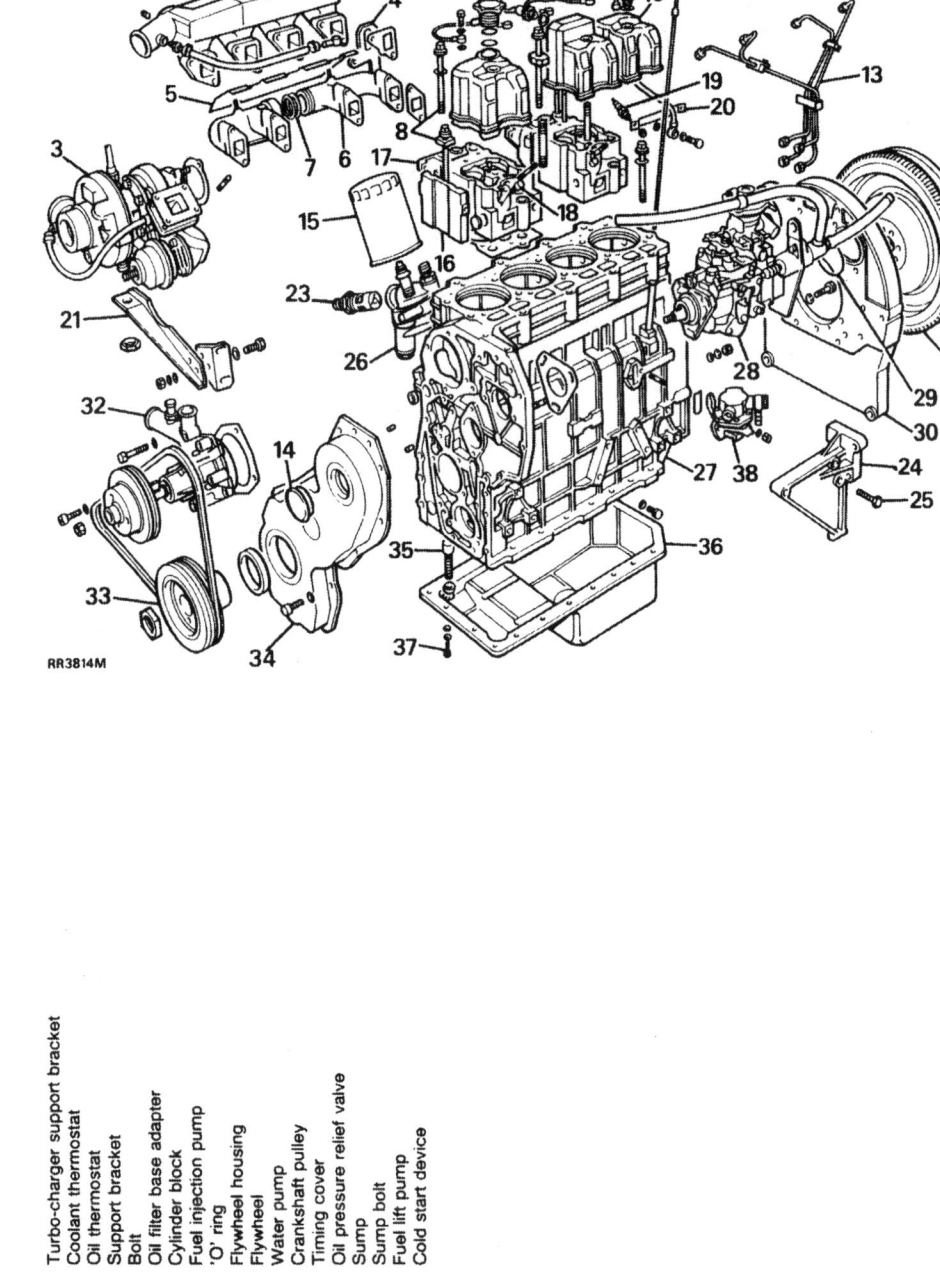

# VM DIESEL ENGINE

## KEY TO ENGINE INTERNAL PARTS

1. Liner
2. Shim
3. 'O' ring seals
4. Compression rings
5. Oil control ring
6. Valve spring cap
7. Valve guide
8. Rocker shaft bush
9. Inlet rocker arm
10. Rocker shaft (pedestal)
11. Exhaust rocker arm
12. Push rod
13. Injector
14. Snap ring
15. Gudgeon pin
16. Valve clearance adjusting screw
17. Piston
18. Inlet valve
19. Exhaust valve
20. Spring clip
21. Tappet
22. Oil pressure switch
23. Small end bush
24. Thrust plate
25. Connecting rod
26. Camshaft bearings
27. Connecting rod bearing shell
28. Carrier location and lubrication shaft
29. Central main bearing carrier
30. Central main bearing shell
31. Camshaft
32. Gear retaining nut (injection pump)
33. Idler gear
34. Crankshaft gear
35. 'O' ring seal
36. Oil pump assembly
37. Camshaft gear
38. Fuel injection pump gear
39. 'O' ring seal
40. Front main bearing
41. Crankshaft
42. 'O' ring seal
43. Oil pick-up pipe and strainer
44. Thrust washer halves
45. Rear main bearing
46. Rear main bearing carrier
47. 'O' ring seal
48. Crankshaft thrust spacer and 'O' ring seal
49. Socket headed screw
50. Crankshaft rear oil seal

## KEY TO 2.5 LITRE DIESEL ENGINE VARIATIONS INSET A

32. Retaining nut and washer
33. Vacuum pump and split gear assembly
34. Crankshaft gear
35. 'O'-ring seal
36. Oil pump assembly
37. Camshaft gear
38. Fuel injection pump gear

# VM DIESEL ENGINE

## DISMANTLING, OVERHAUL AND REASSEMBLY

### Special Tools

| | |
|---|---|
| 18G 29 | Valve lapping tool |
| 18G 55A | Piston ring compressor |
| 18G 79 | Clutch centralising tool |
| 18G 106A | Valve spring compressor |
| 18G 257 | Circlip pliers (large) |
| 18G 284 | Circlip pliers (small) |
| 18G 1004 | Impulse extractor |
| 18G 284-10 | Adaptor remover injector |
| 18G 1367 | Remover crankshaft pulley |
| 18G 1367-1A | Adaptor remover gear |
| 18G 1368 | Remover and holder injection pump drive gear 2.4 engine |
| LST - 139 | Remover and holder injection pump drive gear 2.5 engine |
| 18G 1369A | Timing marker |
| 18G 1370B | Remover replacer sleeve crankshaft |
| 18G 1371 | Remover replacer cylinder liner |
| 18G 1372B | Remover replacer camshaft bearings |
| 18G 1373 | Remover replacer crankshaft front and rear main bearings |
| 18G 1374 | Replacer crankshaft rear oil seal |
| 18G 1375 | Replacer timing cover oil seal |
| 18G 1377 | Remover replacer valve guides |
| 18G 1378B | Retainer cylinder liner |
| MS 70 | Oil filter wrench |
| MS 76 | Basic handle set |
| MS 107 | Timing adaptor fuel injection pump |
| MS 150-7 | Expandable pilot |
| MS 150-8 | Expandable pilot |
| MS 621 | Adjustable valve seat cutter |
| MS 690 | Adjustable valve seat cutter |
| LST 122 | Cylinder head bolt angle gauge |

## DISMANTLING

### Removing ancilliary equipment

**NOTE: All instructions refer to both 2.4 and 2.5 litre diesel engines unless otherwise stated.**

Remove the engine from the vehicle. Clean the exterior and in the interests of safety and efficient working, secure the engine to a recognised engine stand and drain the oil from the sump. Before commencing make a careful note of the position of brackets, clips, harnesses, pipes, hoses, filters and other miscellaneous items to facilitate re-assembly.

1. Remove the alternator and mounting bracket.
2. Remove the starter motor.
3. Remove the power steering pump.
4. Remove the Air Conditioning Compressor and mounting brackets.
5. Remove the oil filter cartridge.
6. Remove the oil drain pipe from the turbo-charger and engine block union.
7. Remove the two socket headed bolts securing the turbo-charger support bracket to the cylinder block.
8. Disconnect the oil feed to the turbo-charger.
9. Remove the four bolts and release the turbo-charger from the exhaust manifold.
10. Remove the inlet and exhaust manifolds.
11. Remove the special nuts and lift off the four rocker covers and joint washers.
12. Release the six bolts and remove the coolant rail, complete with thermostat housing.
13. Disconnect the injector pipes from the injectors and injector pump.
14. Release the clamp nuts and remove the injectors complete with spill rail and collect the four dowels. 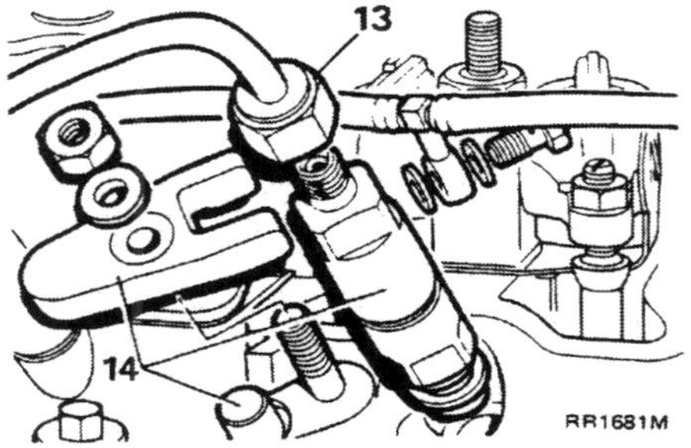
15. Turn the crankshaft in a clockwise direction to bring number one piston to T.D.C. on the firing stroke with number four cylinder valves on the "rock." This condition is necessary for removal of the injection pump at a later stage.
16. Remove the four single nuts and lift off each rocker assembly keeping them identified with their respective cylinder heads. Check that the oil feed dowels are in position in the heads and not inside the pedestals. Remove the push-rods.
17. Remove all cylinder head bolts and clamps. Lift of each cylinder head and gasket and number it according to the bore from which it was removed.
18. Withdraw the four bolts and remove the water pump complete with pulley.

**NOTE: Instructions 19 and 20 refer to 2.4 Litre engines only.**

19. Remove the vacuum pump oil feed pipe.
20. Remove the two nuts and clamp plate and withdraw the vacuum pump.

### Remove injection pump 2.4 litre engines

1. Remove the injection pump drive gear access plate from front cover.
2. Remove the injection pump drive gear retaining nut. 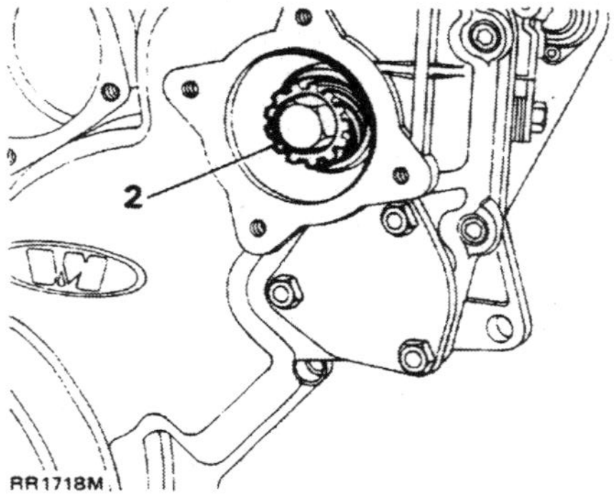 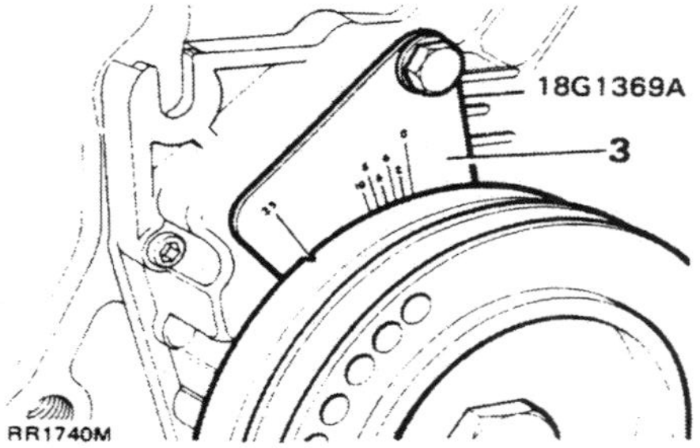
3. Fit the timing marker 18G 1369A to the front cover. The 'O' on the scale should line-up with the groove in the crankshaft pulley.
4. Turn the crankshaft anti-clockwise, beyond the 25° mark on the scale, to remove backlash, and then turn it clockwise until the groove in the crankshaft pulley is aligned with the 25° B.T.D.C. mark. The key on the injection pump shaft should now be at the 11 o'clock position.
5. Fit special tool 18G 1368 to the front cover. Lock the flywheel, and slacken the three nuts that secure the injection pump flange to the engine block.
6. Turn the centre bolt of the tool until the gear releases from the taper. Remove the tool, and pump retaining nuts and withdraw the pump complete with cold start device. 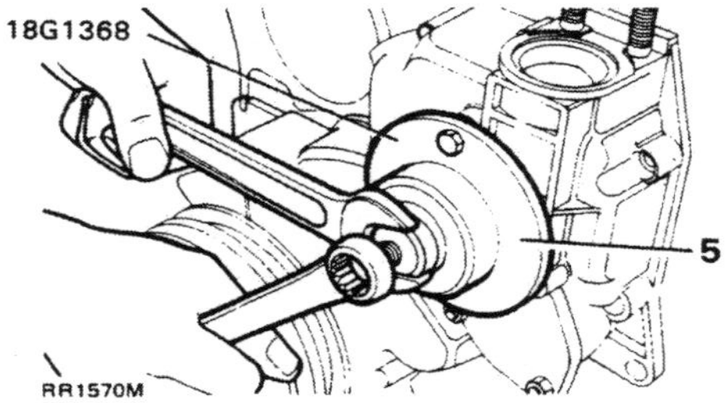
7. Lock the flywheel and remove the crankshaft pulley securing nut.
8. Using special tool 18G 1367A withdraw the crankshaft pulley. 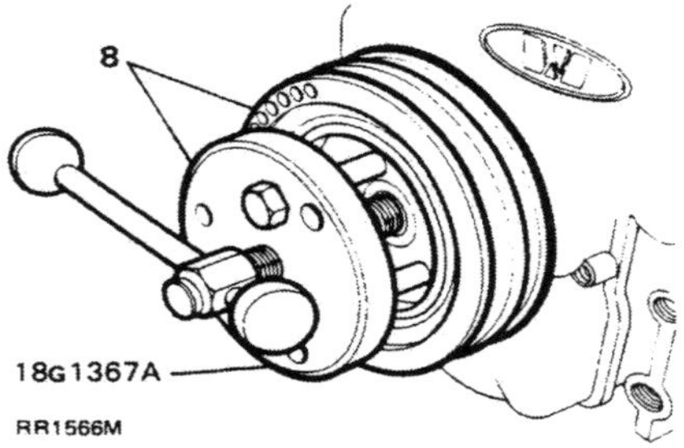
9. Remove the timing cover and retrieve the injector pump drive gear.

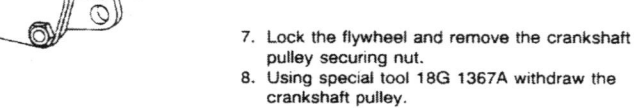

## VM DIESEL ENGINE 12

### Remove injection pump 2.5 litre engines

1. Using the pegged component, item 6 of special tool LST - 139 remove the injection pump drive gear access plate from front cover.
2. Remove the injection pump drive gear retaining nut and washer.

**CAUTION: Ensure the washer is removed to prevent it from dropping inside the timing cover.**

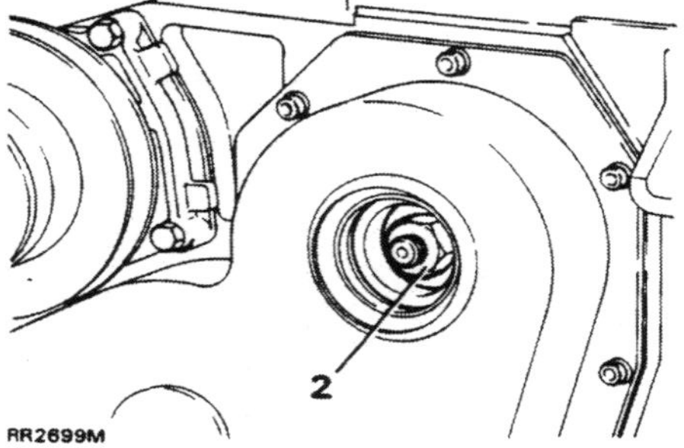

3. Remove bolt from timing cover, using a 47 mm tube spacer and M6 x 55 mm bolt, fit timing gauge 18G 1369A to front cover. The 'O' on the scale should line-up with the groove in the crankshaft pulley.
4. Turn the crankshaft anti-clockwise, beyond the 25° mark on the scale, to remove backlash, and then turn it clockwise until the groove in the crankshaft pulley is aligned with the 25° B.T.D.C. mark. The key on the injection pump shaft should now be at the 11 o'clock position viewed from the front of the engine.
5. Fit item 5 of special tool LST - 139 flush to the front cover. Lock the flywheel, and slacken the three nuts that secure the injection pump flange to the engine block.
6. Fit item 6 into the injection pump drive gear. Ensure a flush fit against item 5 and the centre bolt is fully retracted.
7. Turn the centre bolt of the tool until the gear releases from the taper. Retain the gear on the tool. Remove the injection pump retaining nuts and withdraw the pump complete with cold start device.
8. Lock the flywheel and remove the crankshaft pulley securing nut.
9. Using special tool 18G 1367A withdraw the crankshaft pulley.
10. Remove the timing cover with injection pump drive gear.
11. Remove the special tool from the timing cover and injection pump drive gear.
12. Reassemble the special tool LST - 139.

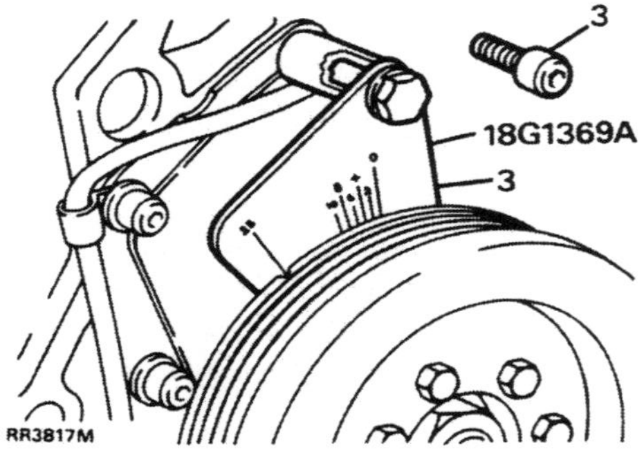

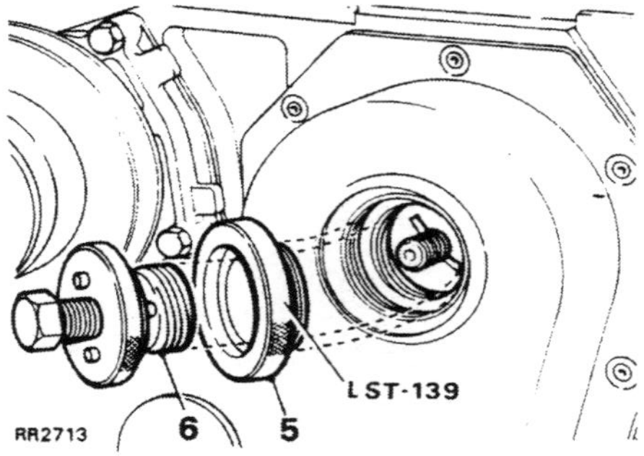

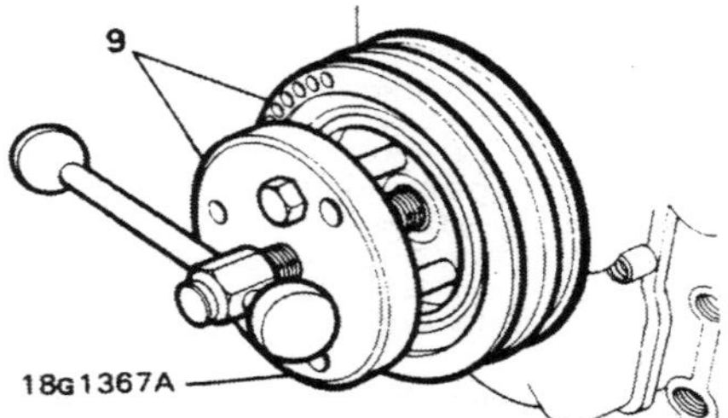

### Remove remaining components both 2.4 & 2.5 engines

1. Remove the clutch pressure plate and centre plate.
2. Remove the three bolts and withdraw the spigot bearing plate.
3. Remove the six bolts and lift off the flywheel.
4. To assist with the removal of the flywheel fit two 8 mm bolts approximately 100 mm (4.0") long into the clutch retaining bolt holes, diametrically opposite, and lift the flywheel from the engine.
5. Remove the nine bolts and six nuts and withdraw the flywheel housing.
6. Remove the single socket-headed screw and withdraw the crankshaft thrust plate and outer thrust washer halves.
7. Remove the oil filter adaptor housing.
8. Remove the sump oil pan.
9. Remove the twenty-one screws and remove the sump.
10. Remove the three bolts and remove the oil pump pickup pipe and strainer and 'O' ring.

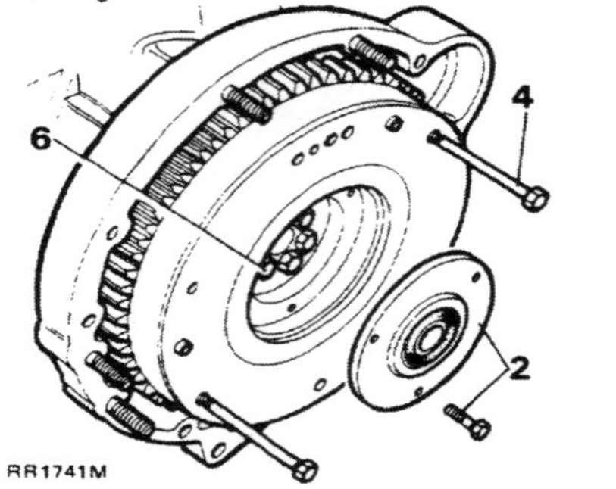

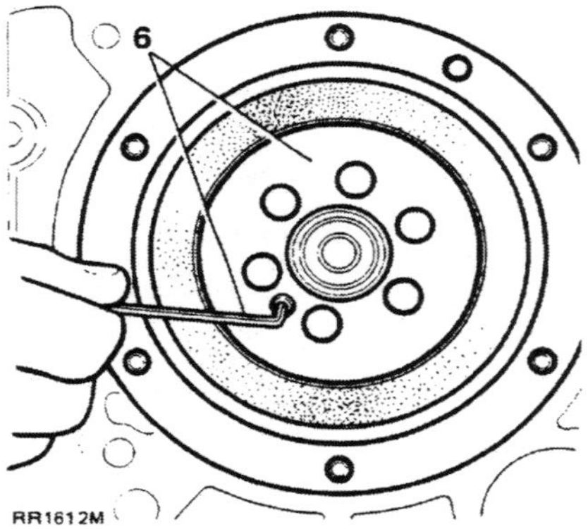

**NOTE: Before performing the next instruction mark the top of each piston with the number of the bore commencing at the front of the engine. Unlike most engines the connecting rods are not numbered relative to the bores.**

11. Turn the crankshaft to bring numbers one and four connecting rod caps to an accessible position. Remove each cap and lower bearing shell, in turn, and push the connecting rod and piston up the bore and withdraw from the top. Immediately refit the cap to the connecting rod with the number on the same side. Repeat the procedure for numbers two and three connecting rod assemblies.
12. Using a suitable piece of timber drift-out the rear main bearing carrier assembly complete with bearing shells and oil seal.
13. Remove the three screws retaining the oil pump to the crankcase and withdraw the pump complete with drive gear.
14. Position the cylinder block horizontal with the crankcase uppermost and remove the two screws securing the camshaft retaining plate to the cylinder block and carefully withdraw the camshaft complete with gear. It is necessary to have the cylinder block inverted so that the tappets will not drop, and foul the cams.

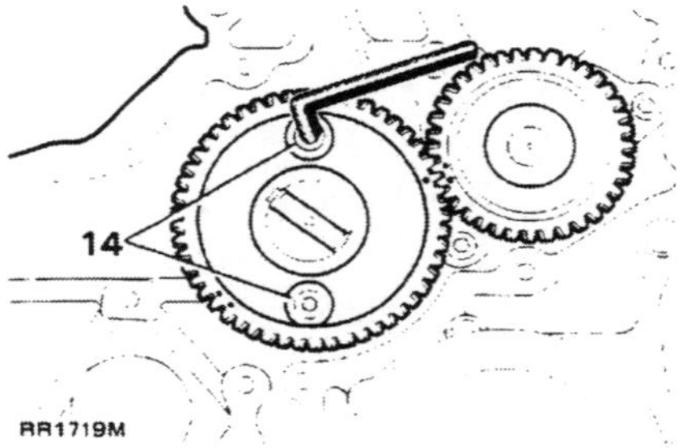

# VM DIESEL ENGINE

15. Remove the three screws and :-
    - On 2.4 Litre engines remove the idler gear.
    - On 2.5 Litre engines remove the vacuum pump and gear assembly.

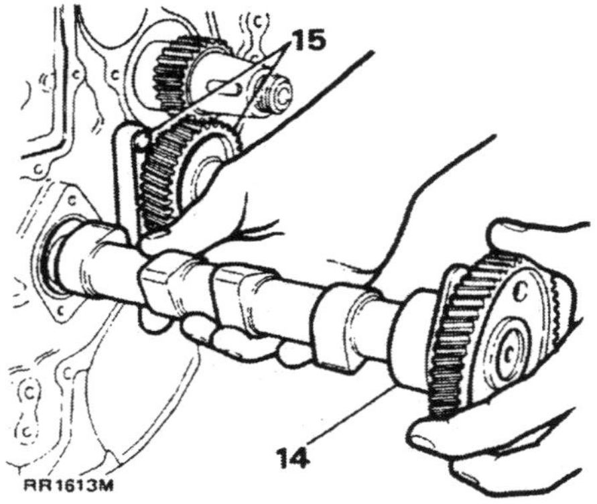

16. Mark for re-assembly and remove from the left hand side of the cylinder block the three main bearing oil feed and carrier location shafts, and identify for re-assembly. Remove the oil pressure switch.

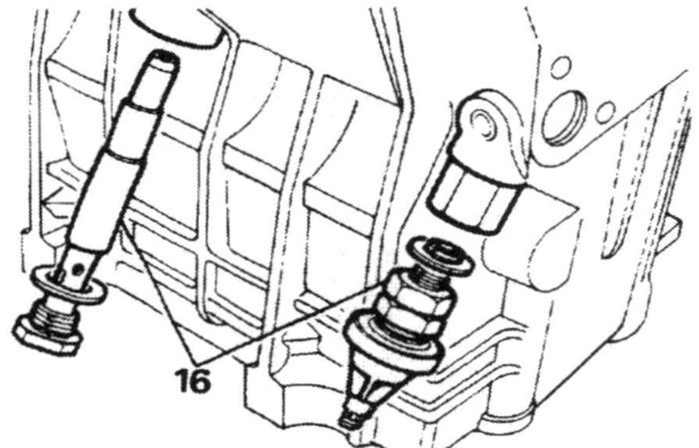

17. To remove the crankshaft and main bearing carrier assembly from the crankcase, slide special tool 18G 1370B over the crankshaft gear, as illustrated, and with assistance withdraw the complete assembly rearwards.

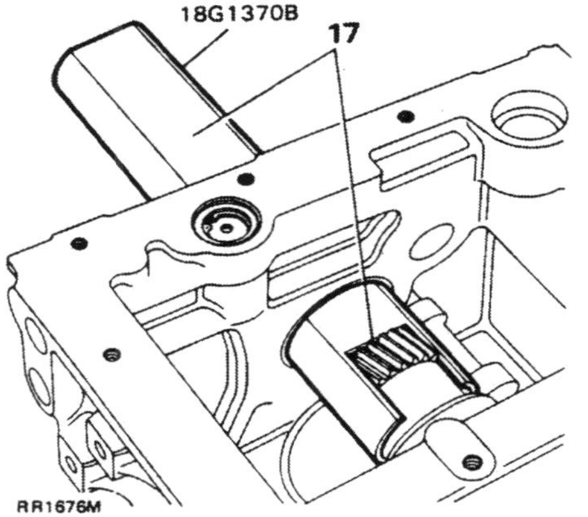

18. Should difficulty be experienced in removing the complete assembly as described above, slide the assembly rearwards sufficiently to gain access to the main bearing carrier bolts. Mark the carriers for assembly and remove the bolts, two for each carrier.

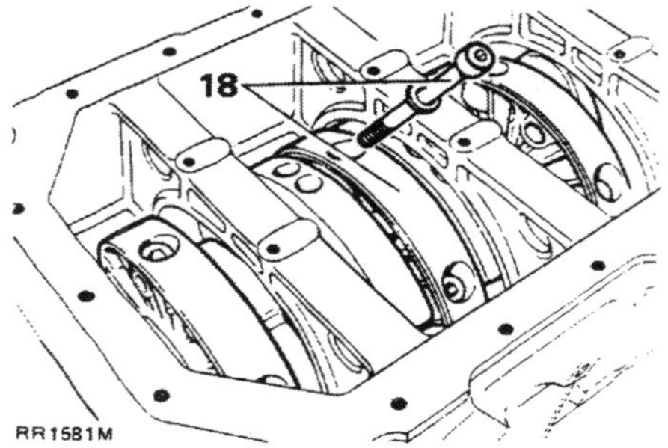

19. Separate the two halves of each carrier, remove from the crankshaft and temporarily re-assemble the carriers. Withdraw the crankshaft through the rear of the crankcase.

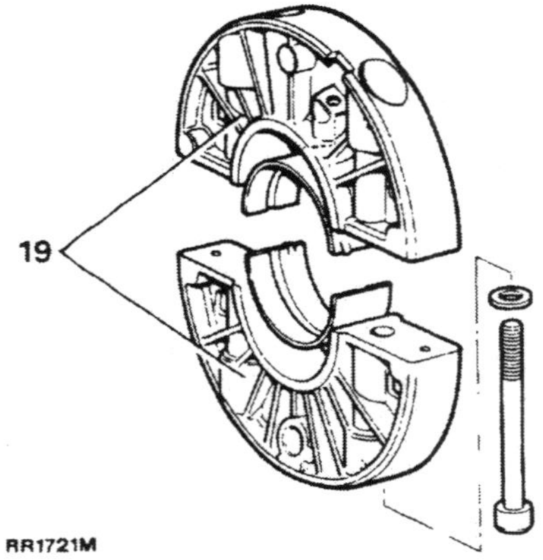

20. Remove the cam followers and identify for possible re-assembly to their original locations.

21. If after inspection it is necessary to renew the cylinder liners then they should be removed as follows: position special tool 18G 1371 as illustrated and turn the centre bolt clockwise to withdraw each liner from the cylinder block. Each liner is fitted with three red 'O' rings; the lower one for oil sealing and the others for coolant sealing. The shim under the lip is for achieving the correct protrusion of the liner above the cylinder block face.

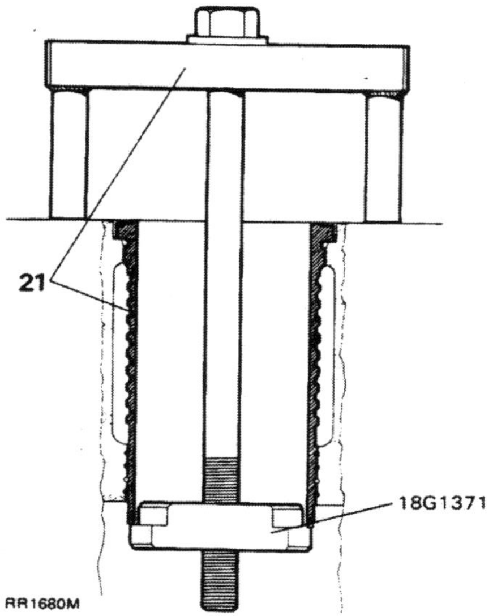

22. Remove the pressure relief valve assembly by removing the circlip which will release the cap, spring and relief valve.

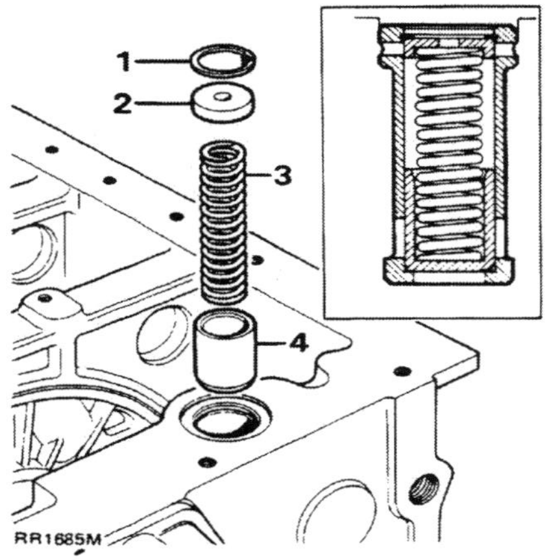

1 Circlip
2 Cap
3 Spring
4 Plunger

## VM DIESEL ENGINE 12

### INSPECTION AND OVERHAUL OF COMPONENTS

#### Cylinder head assemblies

Ensure that the marks made when the cylinder heads were removed are maintained and that during the following instructions the various parts of the cylinder heads are similarly identified.

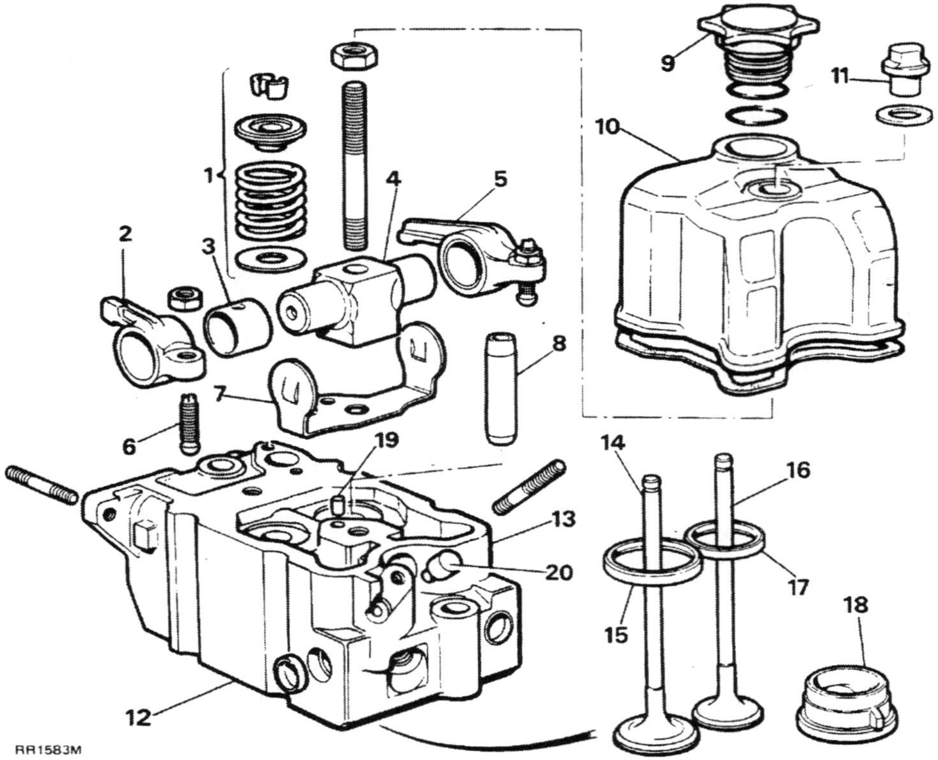

**Key to cylinder head and associated components.**

1. Valve spring, cap seat and cotters.
2. Inlet valve rocker.
3. Bush.
4. Rocker shaft. (Pedestal)
5. Exhaust valve rocker.
6. Rocker adjusting screw.
7. Spring clip.
8. Valve guide.
9. Oil filler cap.
10. Rocker cover.
11. Rocker cover nut.
12. Cylinder head.
13. Injector locating dowel.
14. Inlet valve.
15. Inlet valve seat.
16. Exhaust valve.
17. Exhaust valve seat.
18. Pre-combustion chamber. (Hot plug.)
19. Pedestal lubrication dowel.
20. Injector clamp dowel

---

### Valves

1. Clean the valves and renew any that are bent, have worn stems, or are burnt and damaged. Valves that are satisfactory for further service can be refaced. This operation should be carried out using a valve grinding machine. Only the minimum of material should be removed from the valve face to avoid thinning of the valve edge which must be not less than 1.30 mm (dimension A). Check the valves against the dimensions given in the data section. In addition dimensions B should be as follows:-

   D. Inlet valve .................... 2,73 to 3,44 mm
   E. Exhaust valve ................ 2,45 to 3,02 mm

   Angle C
   Of inlet valve D .................. 55° 30'
   Of exhaust valve E ............. 45° 30'

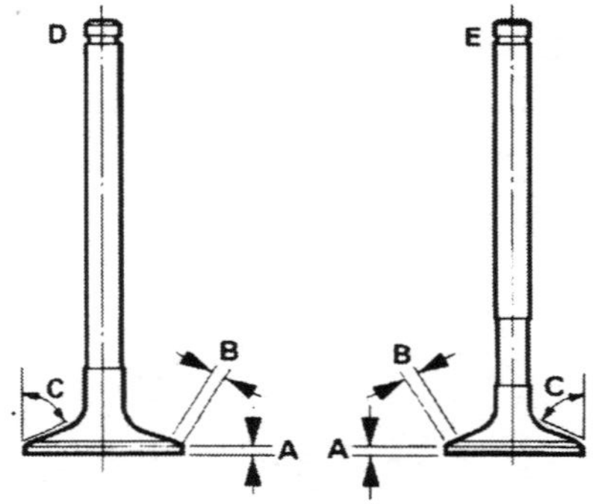

2. Using valve spring compressor 18G 106A or suitable alternative remove the collets, spring cups, springs and valves.

3. Remove the rocker arm pedestal stud and manifold studs.

4. Degrease and remove carbon deposits from the cylinder heads. Examine the cylinder head mating face for cracks pitting and distortion. Renew if necessary.

**CAUTION: The cylinder heads are plated therefore the face must not be machined.**

4. Cracked or burned hot plugs can be removed by heating the cylinder head uniformly in an oven to 150°C. Tap out the hot plug using a thin drift inserted through the injector hole. Clean-out the hot plug pocket in the cylinder head.

5. Measure the depth of seat (D) and the new hot plug height (B) to establish they meet the fitted tolerance detailed below. If necessary machine the outer face of the hot plug to suit.

6. To fit the new hot-plug cool in liquid nitrogen whilst maintaining the cylinder head at the above temperature fit the hot plug. Ensure that the small pip on the side of the hot plug locates in the groove in the side of the pocket. Allow the cylinder head to cool slowly.

| | | |
|---|---|---|
| Hot plug diameter | A | 30,380 to 30,395 mm |
| Hot plug height | B | 23,350 to 23,440 mm |
| Hot plug seat dia. | C | 30,340 to 30,370 mm |
| Depth of seat | D | 23,570 to 23,730 mm |

Maximum protrusion above cylinder head  0.02 mm
Maximum depth below cylinder head  0.03 mm

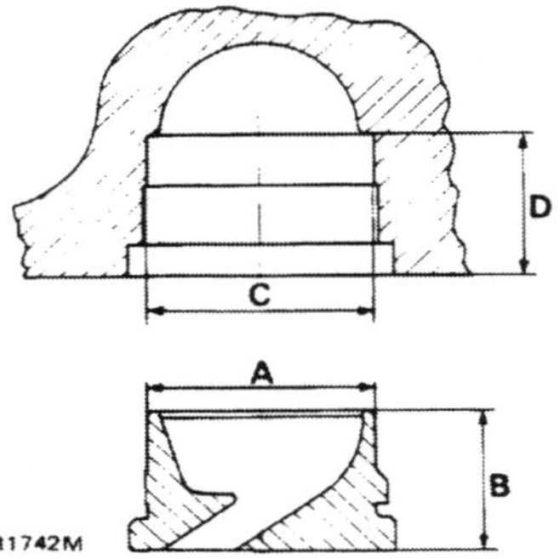

# VM DIESEL ENGINE

## Valve guides

1. Visually examine the guides for damage, cracks, scores and seizure marks. Insert the appropriate serviceable or new valve in the guides and check that the stem-to-guide clearance is within the tolerance given in the data.
2. To renew valve guides, heat the cylinder head to a temperature of between 80°C and 90°C and using special tool 18G 1377 without height gauge 18G 1377/2 press the guides out through the top of the cylinder head.
3. Whilst maintaining the above temperature and using the same tool, but with height gauge 18G 1377/2 drive-in new guides from the top of the cylinder head to the distance determined by the gauge or to dimensions in the data.

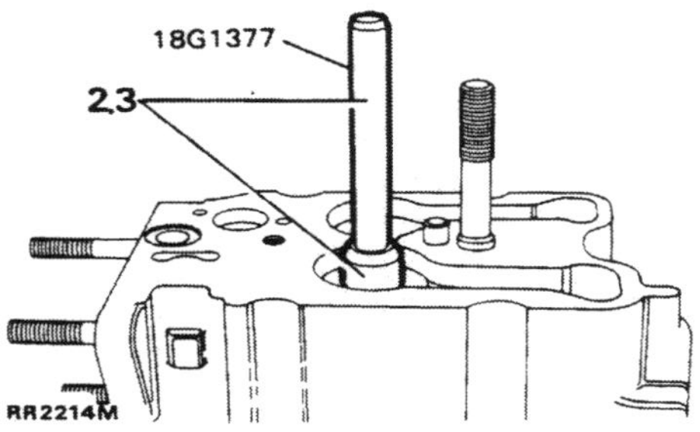

## Valve seat inserts

1. Examine the valve seat inserts for damage, wear and cracks. the seats can be restored provided they are not abnormally wide due to refacing operations. If the seat cutting operation, however, excessively lowers the valve recess or if the seat cannot be narrowed to within the limits given in the data, the insert should be renewed.
2. To recut an inlet valve seat use an expandable pilot M.S. 150-8 loosely assemble the collet, expander and nut. Ensure that the chamfered end of the expander is towards the collet. Insert the assembled pilot into the valve guide from the combustion face side of the cylinder head until the shoulder contacts the valve guide and the whole of the collet is inside the valve guide. Expand the collet in the guide by turning the tommy bar clockwise whilst holding the knurled nut.
8. Perform the above instructions to recut an EXHAUST valve seat using cutter MS 621 until the seat width is in accordance with dimension D in data. Check that the valve head recess is within the data limits.
9. To remove either an inlet or exhaust valve seat, hold the cylinder head firmly in a vice, wear protective goggles and grind the old insert away until thin enough to be cracked and prised out. Take care not to damage the insert pocket. Remove any burrs and swarf from the pocket. Failure to do this could cause the new insert to crack when being fitted.
10. Heat the cylinder head, uniformly in an oven, to a temperature of 150°C cool the new seat insert by dipping into liquid Nitrogen. This will enable the seat to be positioned without the use of pressure. Allow the cylinder head to cool naturally to avoid distortion.

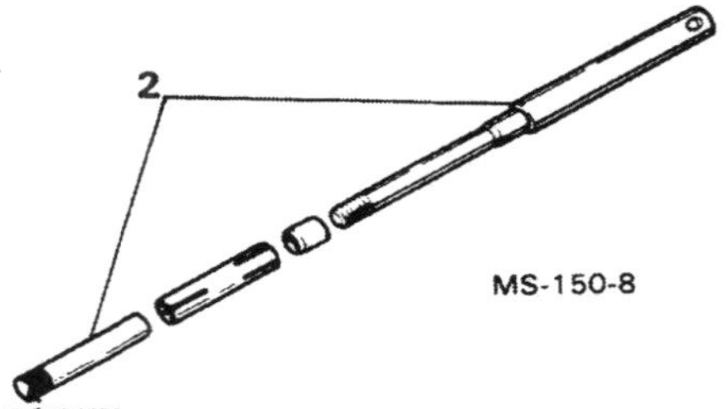

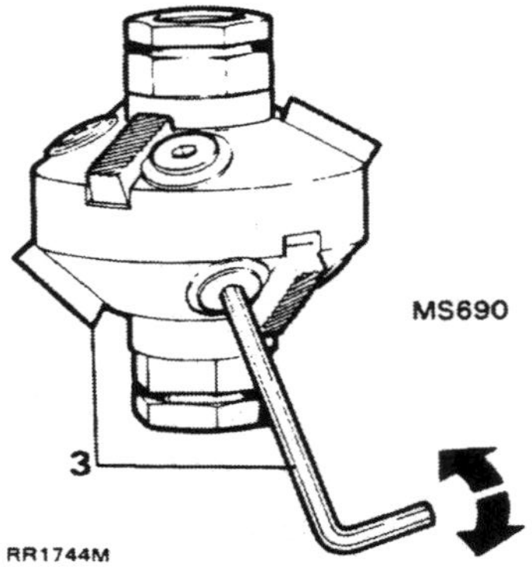

## Lapping in valves

1. To ensure a gas tight seal between the valve face and the valve seat it is necessary to lap-in the appropriate valve to its seat. It is essential to keep the valve identified with its seat once the lapping in operation has been completed.
2. Unless the faces to be lapped are in poor condition it should only be necessary to use fine valve lapping paste. Smear a small quantity of paste on the valve face and lubricate the valve stem with engine oil.
3. Insert the valve in the appropriate guide and using a suction type valve lapping tool employ a light reciprocating action while occasionally lifting the valve off its seat and turning it so that the valve returns to a different position on the seat.
3. Select cutter MS 690 and ensure that the cutter blades are correctly fitted to the cutter head with the angled end of the blade downwards facing the work, as illustrated. Check that the cutter blades are adjusted so that the middle of the blade contacts the area of material to be cut. Use the key provided in the hand set MS 76.
4. Fit the wrench to the cutter head, apply it to the seat to be refaced and turn clockwise using only very light pressure. Continue cutting until the width of the seat is in accordance with the dimension J in data.
5. To check the effectiveness of the cutting operation use engineer's blue or a feeler gauge made from cellophane.
6. Smear a quantity of engineer's blue round the valve seat and revolve a properly ground valve against the seat. A continuous fine line should appear round the valve. If there is a a gap of not more than 12 mm it can be corrected by lapping.
7. Alternatively, insert a strip of cellophane between the valve and seat, hold the valve down by the stem and slowly pull out the cellophane. If there is a drag the seal is satisfactory at that spot. Repeat this in at least eight places. Lapping in will correct a small open spot.

# VM DIESEL ENGINE 12

## RANGE ROVER

4. Continue the operation until a continuous matt grey band round the valve face is obtained. To check that the lapping operation is successful, wipe off the valve paste from the valve and seat and make a series of pencil lines across the valve face. Insert the valve into the guide and while pressing the valve onto the seat revolve the valve a quarter turn a few times. If all the pencil lines are cut through no further lapping is required.
5. Wash all traces of grinding paste from the valves and cylinder head seats.

### Valve springs.

1. Examine the valve springs for damage and overheating and discard any that are visually faulty.
2. New and used valve springs, in the interests of uniformity, should be subjected to load and height tests as shown in the table and diagram below.
The amount of distortion D must not exceed 2,0 mm (0.078 in).

| | Test load (Kg) | | Height (mm) | Condition |
|---|---|---|---|---|
| A | 0.00 | H1 | 43.20 | Free height |
| B | 33-35 | H2 | 37.00 | Closed valve |
| C | 88-94 | H3 | 26.61 | Open valve |

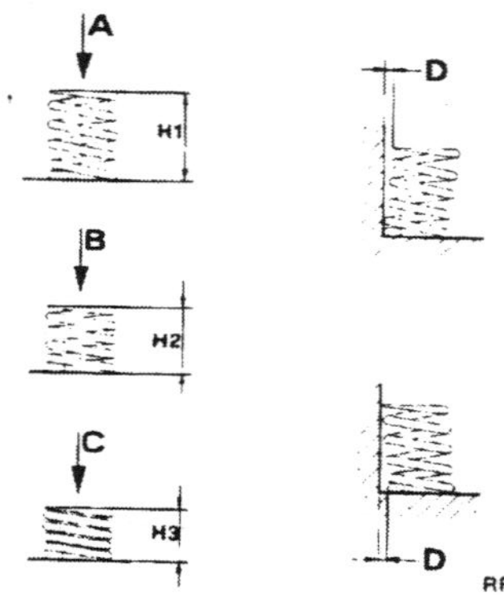

### Assembling the cylinder head

1. Assemble the valves to their respective positions in the cylinder head. Fit the spring plates, springs and cups and secure the assembly with the split collets using valve spring compressor 18G 106A or equivalent.
2. Using feeler gauges check the inlet and exhaust valve head stand down i.e clearance of valve heads below cylinder head combustion face, see data.
3. When renewing the cylinder head water jacket plugs secure them in position with Loctite 601
4. Renew the manifold retaining studs and when fitting a new pedestal stud secure it with Loctite 270.
5. Fit the rocker pedestal location and lubrication dowel into each cylinder head and ensure that the oil hole is clear. Place the heads to one side ready for assembly to the cylinder block at a later stage.

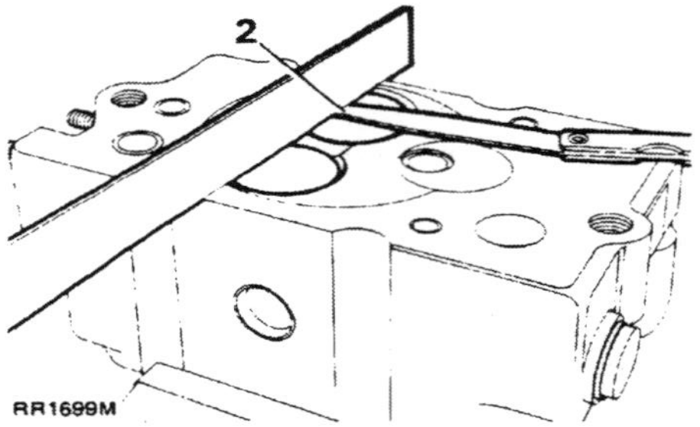

### Rocker assembly and push rods.

1. Remove the spring clip and slide the rockers from the shaft.
2. Clean and examine the rocker shafts and check for ovality, overall wear taper, and surface condition. Compare the dimensions with those given in data.
3. Examine the rockers and renew any that have worn rocker pads. It is not permissible to grind a pad in an attempt to restore a rocker.
4. Examine the rocker adjusting screws and renew any that are worn.
5. Check the internal dimensions of the bushes against the figures in data. If necessary renew the bushes ensuring that the oil hole in the bush aligns with the hole in the rocker arm. Check that the rocker arm to shaft clearance is within the figures in data.
6. Assemble the rocker-arms to the shaft noting that they are handed and that when assembled the pad ends point inwards. Retain the assembly with the spring clip and place to one side for fitting to the cylinder head at a later stage.
7. Examine the push rods and discard any that are bent or have worn or pitted ends.

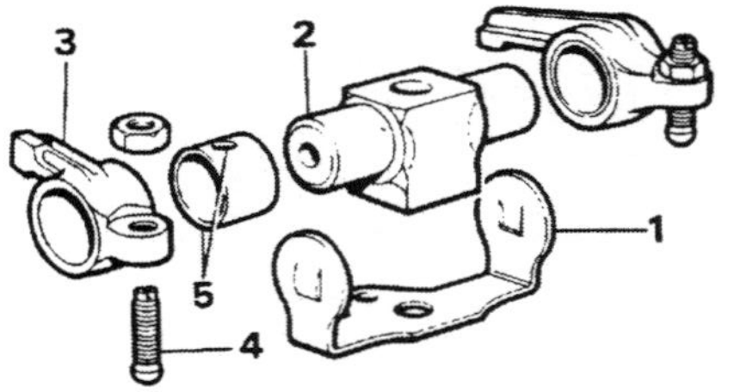

### Cylinder block

1. Clean the cylinder block with kerosene or suitable solvent and blow dry with compressed air all oil passages and water ways. Carry out a careful visual examination checking for cracks and damage.
2. Measure the cylinder liner bores for ovality, taper and general wear using any suitable equipment. An inside micrometer is best for checking ovality and a cylinder gauge for taper.
3. Check the ovality of each bore by taking measurements at the top of the cylinder just below the ridge at two points diametrically opposite. The difference between the two figures is the ovality of the top of the bore. Similar measurements should be made approximately 50 mm (2.0 in) up from the bottom of the bore so that the overall ovality may be determined.
4. The taper of each cylinder is determined by taking measurements at the top and bottom of each bore at right angles to the gudgeon pin line. The difference between the two measurements is the taper.
5. To establish maximum overall bore wear, take measurements at as many points possible down the bores at right angles to the gudgeon pin line. The largest recorded figure is the maximum wear and should be compared with the original diameter of the cylinder liner. (See **Section 04** General specification data).
6. If the cylinder bores are excessively worn outside the limits the cylinder liners must be renewed. See ENGINE ASSEMBLY.
7. Alternatively, if the overall wear, taper and ovality are well within the acceptable limits and the original pistons are serviceable new piston rings may be fitted. It is important however, that the bores are deglazed, with a hone, to give a cross-hatched finish to provide a seating for the new rings. It is vital to thoroughly wash the bores afterwards to remove all traces of abrasive material.

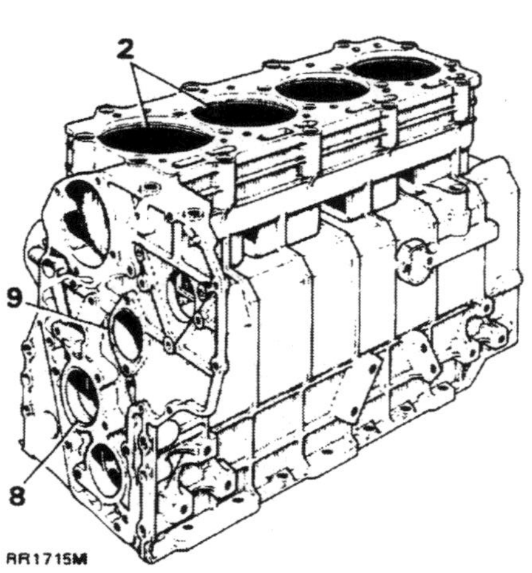

## VM DIESEL ENGINE

8. Using an inside micrometer check the front main bearings for general condition, overall wear, taper and ovality. If outside the limits given in data remove the bearing. Use special tool 18G 1373 to renew the bearing, see ENGINE ASSEMBLY.

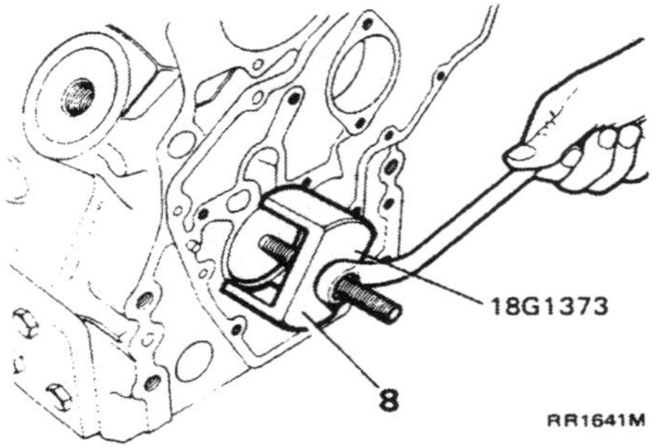

9. Measure the internal diameter of each camshaft bearing at several points using an internal micrometer. A comparison of the bearing diameters with those of the respective camshaft journals will give the amount of clearance. The bearings should be renewed if the clearance is excessive or if they are scored or pitted. Use special tool 18G 1372B as illustrated, to remove the bearings.

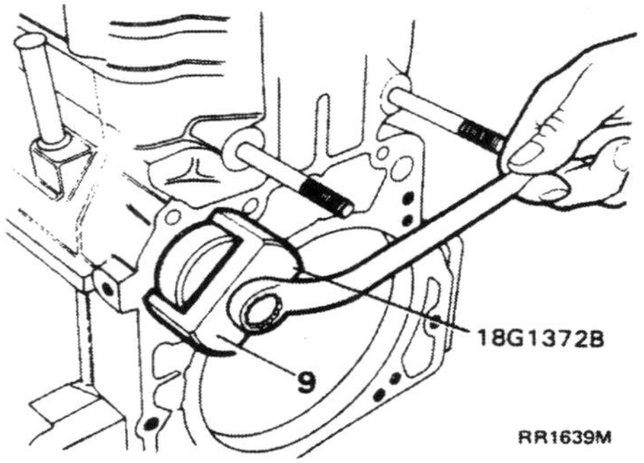

### Crankshaft

1. Identify for reassembly and remove the main bearing carriers from the crankshaft.
2. Degrease the crankshaft and clear out the oil ways, which can become clogged after long service.
3. Mount the crankshaft on "V" blocks and examine visually, the crankpins and main bearing journals, for obvious wear, scores, grooves and overheating.
4. With a micrometer, measure and note the ovality and taper of each main bearing journal and crankpin as follows:
   **Ovality** - Take two readings at right-angles to each other at various intervals.
   **Taper** - Take two readings parallel to each other at both ends of the main bearing journal and crankpin.

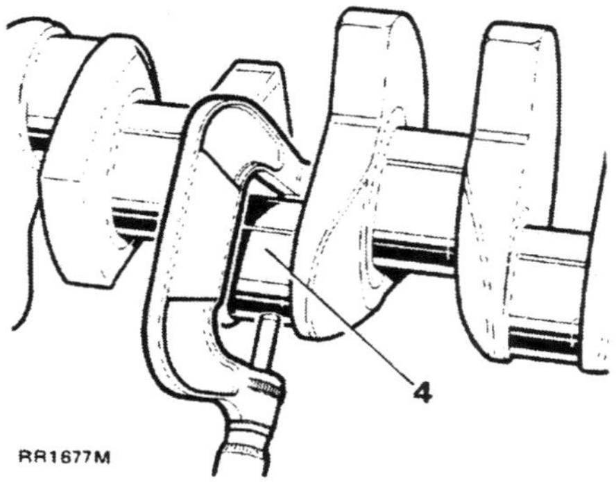

8. To fit a new gear, heat in an oven to 180°C to 200°C, and press-on to the shaft up to the shoulder. Fit a new key for the crankshaft pulley.

### Crankshaft carriers

1. Assemble the three main bearing carriers with the bearings fitted and tighten to the correct torque.
2. Using an internal micrometer check the internal diameters of the bearings against the figures in data and renew if necessary or in any event if the crankshaft is being reground.
3. Remove the bearings from the carriers, reassemble and tighten bolts to correct torque. With an internal micrometer check the carrier bore against the figures in data, and for excessive ovality.
4. Check that the piston oil jets in the carriers open at the correct pressure and renew if necessary. Drift the old jet out through the carrier bore, apply a thin coat of Loctite AVX Special around the new jet before fitting.

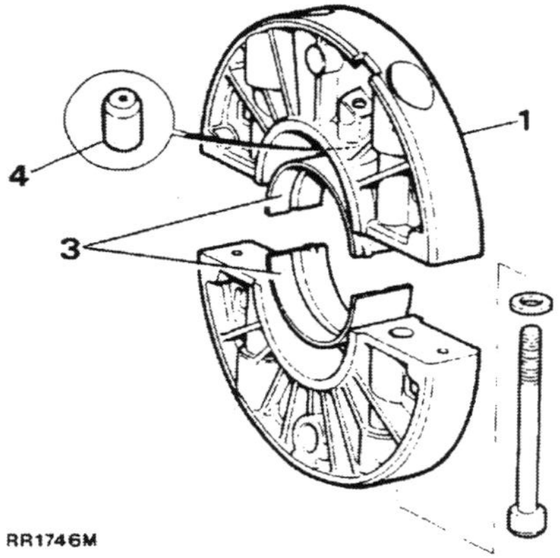

5. If the overall wear exceeds 0.01 mm (0.004 ins) for both main bearing journals and crankpins regrind and fit undersize bearings. When regrinding do not remove any material from thrust faces.
6. After grinding it is important to restore the journal fillet radii as illustrated.

A = 2.7 to 3.00mm
B = 2.5mm

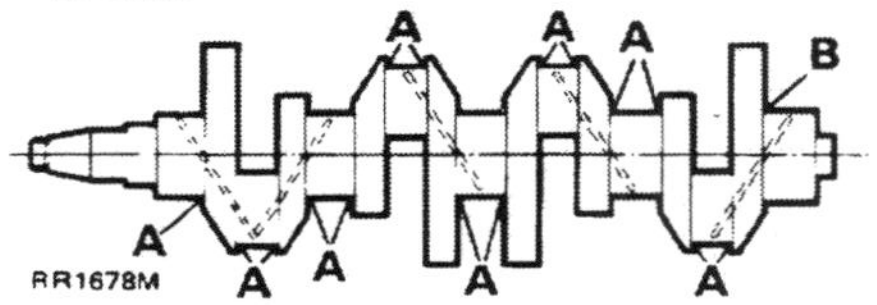

7. Examine the timing gear teeth and if worn remove the gear with special tool 18G 1367-1A and 18G 1367-A

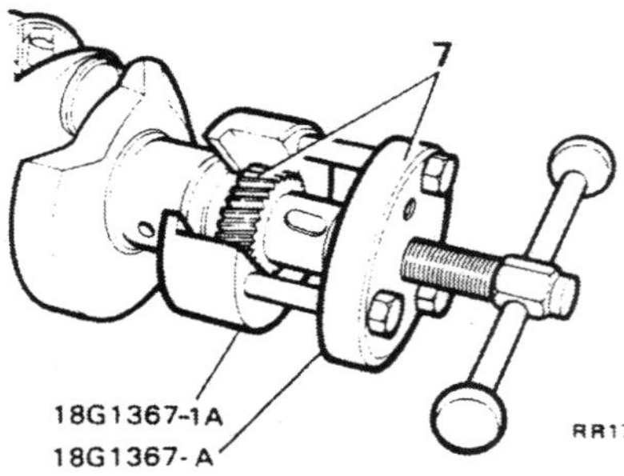

# VM DIESEL ENGINE

## Rear main bearing carrier

1. Extract the oil seal taking care not to damage the carrier bore.
2. Using an internal micrometer check the bearing dimensions against the figures in data.
3. If required remove the bearing using special tool 18G 1373, as illustrated.
4. Check the carrier bearing bore for wear against the figures in data.
5. With special tool 18G 1374 fit a new oil seal to the rear carrier, lipside leading.

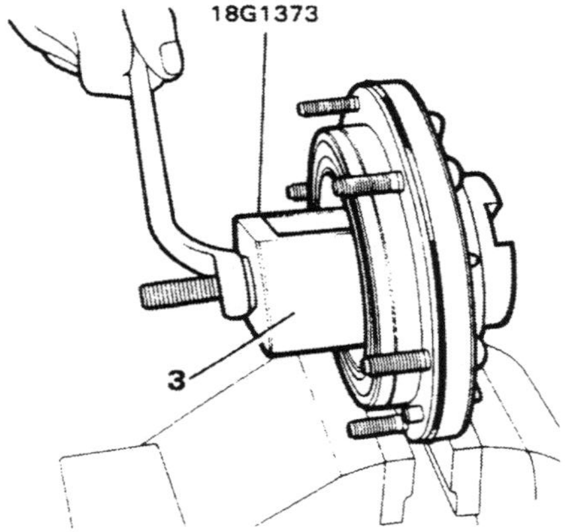

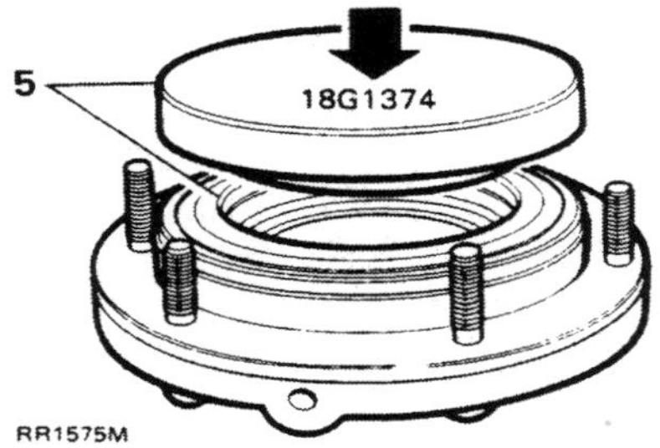

## Thrust spacer

1. Examine the spacer thrust face for damage, scratches, cracks and seizure marks. Ensure that outer diameter on which the seal runs is free from imperfections.
2. With micrometers check the thickness A and the diameter B at four diametrically opposite points and compare with the figures in data.

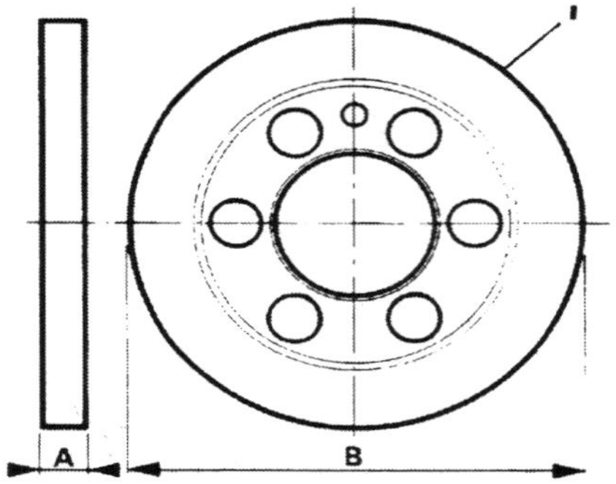

## Camshaft

1. Carry-out a visual examination of the cam lobes and bearing surfaces. If these are worn, scored or cracked the shaft should be renewed.
2. If visually satisfactory, carry out the dimensional checks detailed in the data section to the cams and bearing journals.
3. Check the camshaft for straightness, by mounting between centres and checking with a dial test gauge on the centre bearing journal. The shaft may be straightened under a press if the bend exceeds 0.05 mm (0.002 in). This work, however, should be entrusted to a specialist.
4. Examine the gear teeth and if worn or damaged press the shaft from the gear, together with the thrust plate.
5. Before fitting a new gear, check the thrust plate thickness at the four points illustrated. Renew the plate if the dimensions do not conform to the limits in data.
6. Heat the new gear in an oven to 180°C to 200°C, fit the thrust plate and press the gear onto the shaft until the gear is hard against the shoulder. If, when the gear has cooled the thrust plate turns freely on the shaft the camshaft end-float will be correct when fitted.

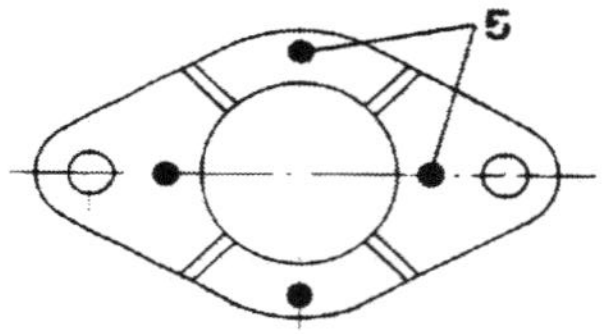

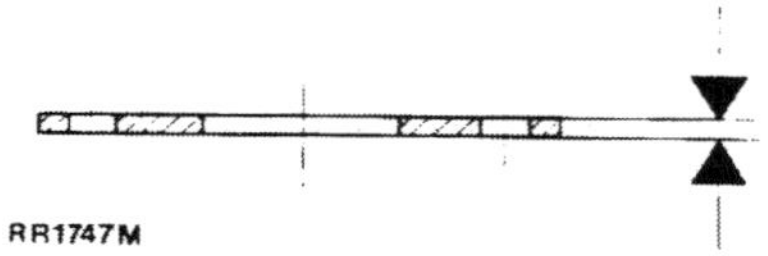

## Cam followers (Tappets)

1. Examine the cam followers and discard any that are worn, pitted or scored on the cam contact face. Check also the cups in which the push rods seat.
2. Check the stem diameter for general wear, ovality and taper. Take measurements at several points round the circumference and along the length of the stem.

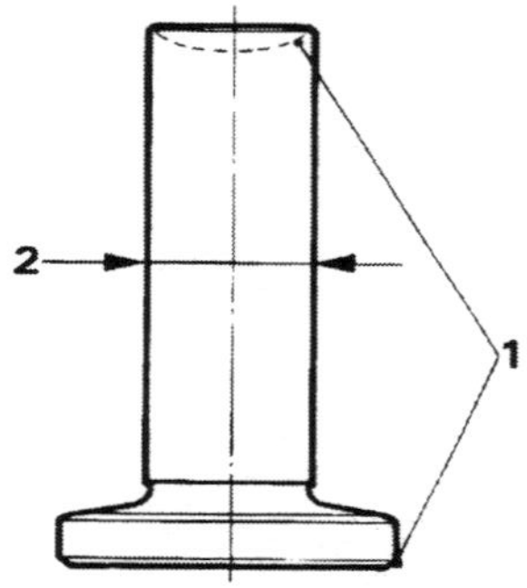

## Connecting rods and pistons

1. Whilst keeping each piston and connecting rod identified for possible refitting, separate the pistons from the rods and remove the piston rings. Degrease and decarbonise the pistons and rings ready for examination. Likewise prepare the connecting rods for inspection.

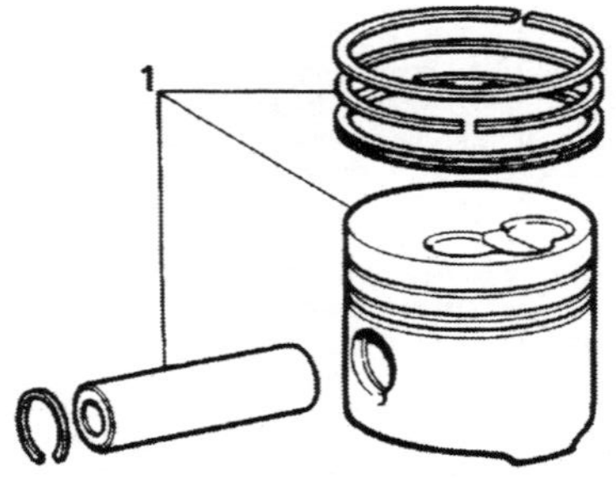

## Flywheel

1. Examine the flywheel clutch face for cracks, grooves and signs of over-heating. If excessive damage is evident renew or reface the fly-wheel.

## Flywheel face run-out

2. The above check should be carried out during engine assembly. See fitting flywheel.

# VM DIESEL ENGINE

## Pistons and rings

1. Examine the pistons for scores, cracks signs of overheating and general wear.
2. If visually satisfactory measure the piston skirt at right angles to the gudgeon pin 15 mm above the bottom of the piston skirt. If the wear is in excess of the maximum permitted in data and the piston to liner clearance is in excess of 0,15 mm (0.006 in) new pistons and liners must be fitted.
3. Check the gudgeon pins for wear, scores, pitting and signs of overheating. Check the gudgeon pin bore for ovality.
4. Examine the piston rings for damage, wear and cracks. Fit the rings to the pistons as illustrated and using a feeler gauge check the side clearance in the grooves.
5. To check the piston ring fitted gap insert the ring squarely into the bottom of the bore at the lowest point of piston travel. To ensure squareness push the ring down the bore with a piston. Using an appropriate feeler gauge check the gaps of all the rings in turn. The correct gaps are given in data. If any gap is less than that specified, remove the ring and file the ends square whilst holding the ring in a filing jig or vice.
6. The previous instruction should also be carried out when new pistons and rings are fitted to new liners but the rings may be inserted squarely in any position in the bore.

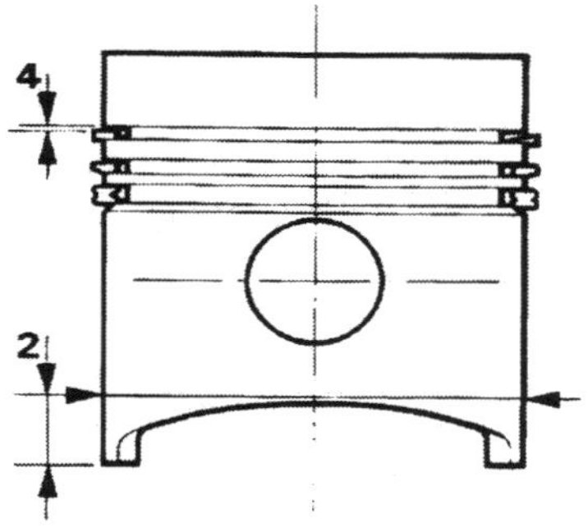

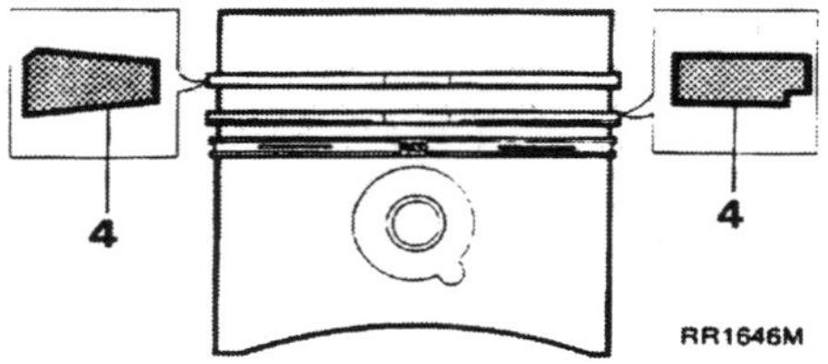

NOTE: **The difference in weight between the four pistons must not exceed 5 grams. When renewing pistons and liners they should all belong to the same classification A or B.**

## Connecting rods

1. Examine the connecting rods and caps for cracks using a recognised crack testing process.
2. Assemble the cap and rod and tighten to the correct torque. Check the crank pin bore using an inside micrometer and three different points. the bore must be 57,563 to 57,582 mm. Renew rods if the tolerance exceeds 0,02 mm.
3. Examine the connecting rod shells and discard if worn, scored or show signs of overheating. Assemble the rods, caps and shells and tighten to the correct torque. Check the internal diameter against the figures in data.
4. Inspect the small end bush for wear against the figures in data. Check that the wear limit between bush and gudgeon pin does not exceed 0,100 mm (0.004 in). When renewing the bush ensure that the oil hole aligns with the connecting rod hole.
5. Check the rod for bend and twist, taking measurements at approximately 100 mm from the centre of the rod using a recognised alignment gauge. Twist or bend must not exceed 0,5 mm (0.019 in).
6. If it is necessary to renew connecting rods check that the weight difference between them does not exceed 10 grams, see letter code in data (2.4 Litre engines only).
7. Slightly warm the pistons and assemble to the connecting rods ensuring that the recess in the piston crown is on the same side as the number on the connecting rod big end. Insert the gudgeon pins and secure with the circlips.
8. Fit the connecting rod bearing shells ensuring that the tags locate in the cutouts.

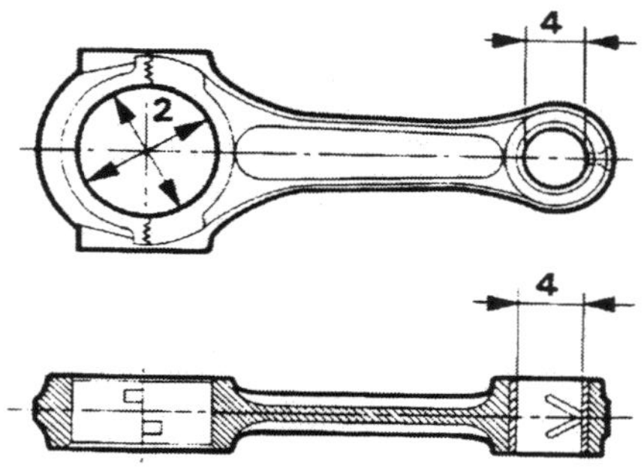

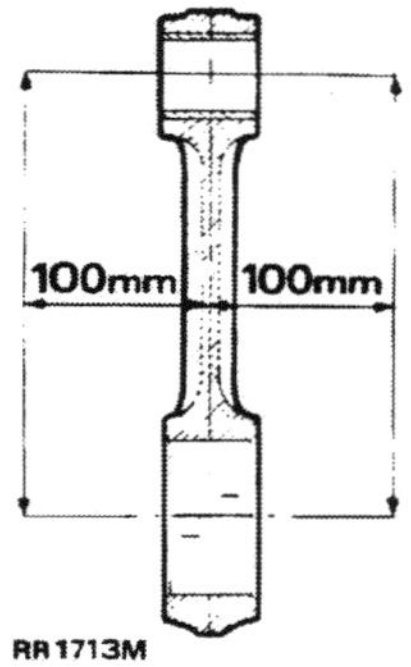

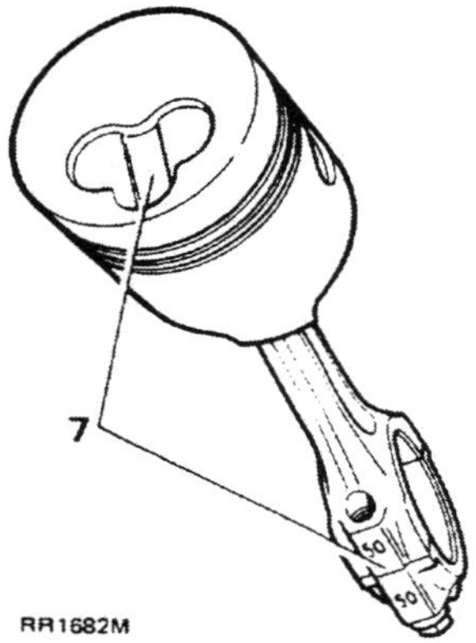

# VM DIESEL ENGINE

## Oil pump

NOTE: **The oil pump is only supplied as an assembly complete with drive gear.**

1. Dismantle the oil pump and clean with kerosene or solvent. Examine the rotors and body for wear and pitting.
2. Assemble the oil pump noting that the chamfered side of the outer rotor is fitted downwards towards the drive gear.
3. Check, with a feeler gauge, the clearance between the inner and outer rotor A.
4. Check the clearance between the pump body and outer rotor B and compare the figures in data.
5. Examine the gear teeth for wear, chips and pitting.

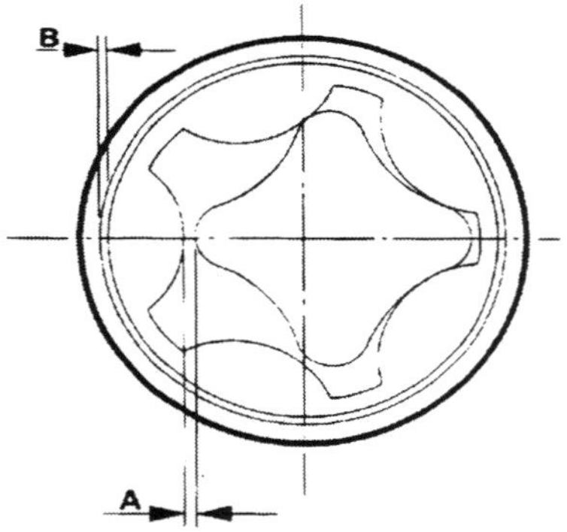

## Oil pressure relief valve

1. Examine the plunger for scores and pitting. If necessary the valve plunger may be lapped to its seat, to restore efficiency, using fine valve grinding compound. Make sure that all trace of the compound is removed before assembling valve to the crankcase.
2. Check the free length of the spring against the figure in data.

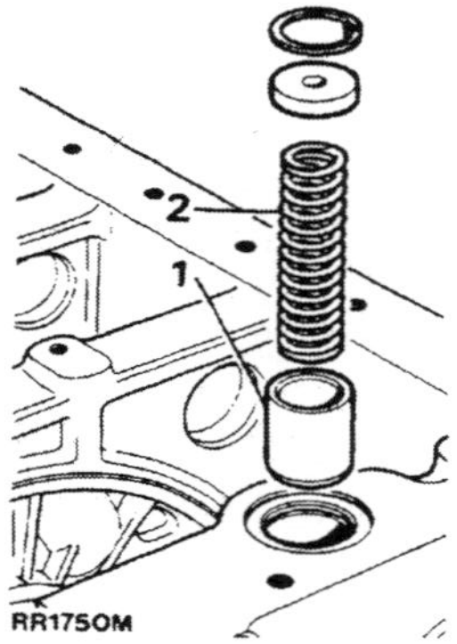

## Idler gear assembly 2.4 litre engine only

1. Check the idler gear for wear and damage and for wear in the bushes. Check that the lubrication hole at the back of the mounting plate is clear. If the gear is unserviceable the complete unit should be renewed.

## Vacuum pump and gear assembly 2.5 litre engine only

1. Inspect the gear for wear and damage. Check the vanes for wear. Examine the vacuum pump housing for scouring or damage.
If the unit is worn or damaged the complete assembly should be renewed.

## Injection pump drive gear

1. Check the injection pump gear (and combined vacuum pump gear on 2.4 litre engines) for damage, wear and pitting. Examine the bore and keyways for wear. Renew if any gear is unsatisfactory.

# VM DIESEL ENGINE

## Vacuum pump 2.4 litre engines only

1. Remove the three screws and withdraw the top cover and "O" ring seal.
2. Check the rotor and vanes for wear.
3. Examine the drive gear for wear.

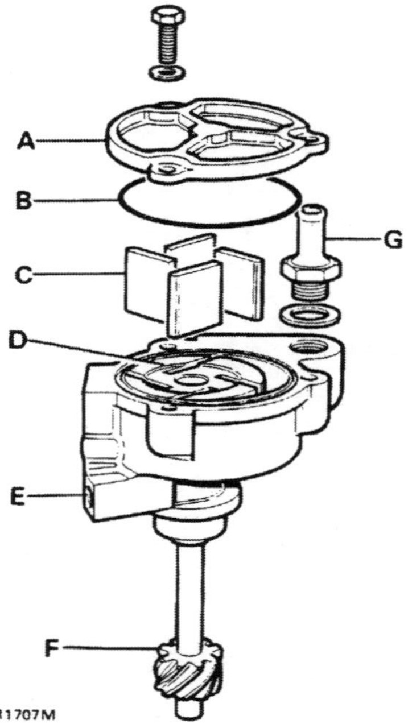

A   Cover
B   'O' ring
C   Vanes
D   Rotor
E   Lubrication port
F   Drive gear
G   Vacuum hose adapter and non return valve

## Oil filter adaptor housing

1. This housing contains a by-pass valve which opens to maintain oil circulation when a difference in pressure exists between the filter base outlet to the oil cooler and the main oil gallery due to a restriction in the oil cooling system. A thermostat which opens at 80°C, to allow oil to pass to the oil cooler is also incorporated in the housing.

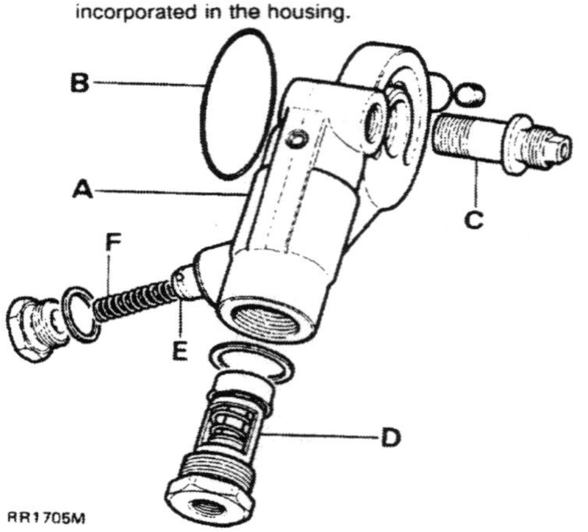

A   Adaptor housing.
B   'O' ring
C   Oil filter and adaptor housing union screws
D   Thermostat
E   By-pass plunger
F   By-pass plunger spring

2. Remove the thermostat and check the opening temperature. Place the thermostat in vessel containing water and a thermometer. Apply heat and observe the temperature at which the thermostat opens. Refit or renew as necessary, using a new sealing washer.
3. Remove the by-pass valve plug and remove the spring and plunger. Check the plunger for scores and pitting. Refit or renew as necessary using a new sealing washer.

## VM DIESEL ENGINE

4. Check the rotor end float by placing a straight edge across the pump body and with a feeler gauge measure the clearance between the machined outer diameter and straight edge. The end float should be 0,07 to 0,14 mm (0.002 to 0.004 in).
5. Fit the vanes, noting that the round edge must face outwards.
6. Check operation of vacuum non return valve.
7. Fit a new "O" ring seal and secure the cover with the three screws.

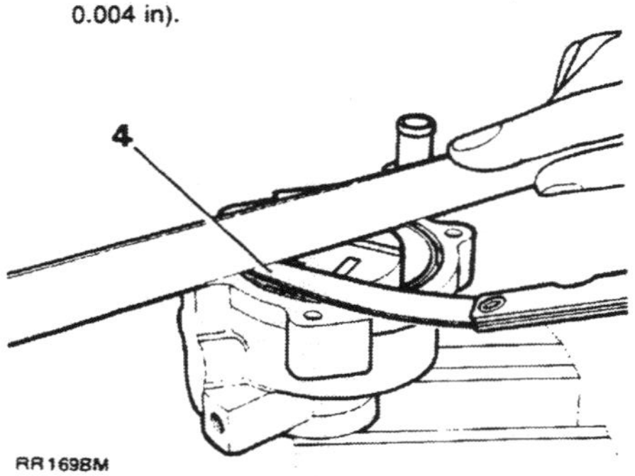

### Water pump

1. Since the water pump is not serviceable the complete assembly should be renewed if the impeller is worn and corroded or if there is excessive end float or side movement in the impeller shaft.

### Fuel lift pump

1. Mark the relationship of the pump cover to the body to facilitate reassembly.
2. Remove the six retaining screws and lift-off the cover.
3. Remove the valve plate.
4. Press down on the diaphragm and twist to release the diaphragm from the body.
5. Remove the diaphragm spring.
6. Clean and examine all parts. The diaphragm can be renewed if faulty.

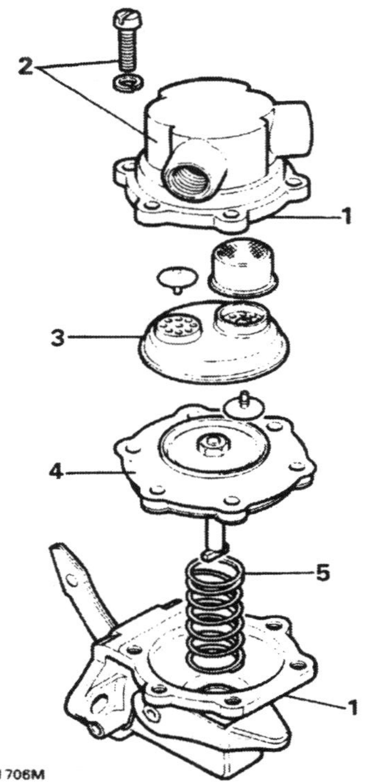

7. Reassemble the pump reversing the above procedure.

### Thermostat and housing 2.4 litre engine only

1. Remove the three socket headed screws and withdraw the thermostat and body from the water rail.

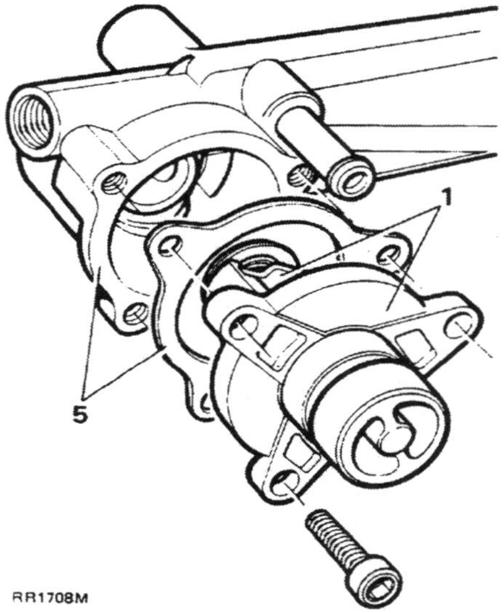

2. Hold the body in a vice and press down upon the two "ears" of the thermostat and twist to release it from the body.

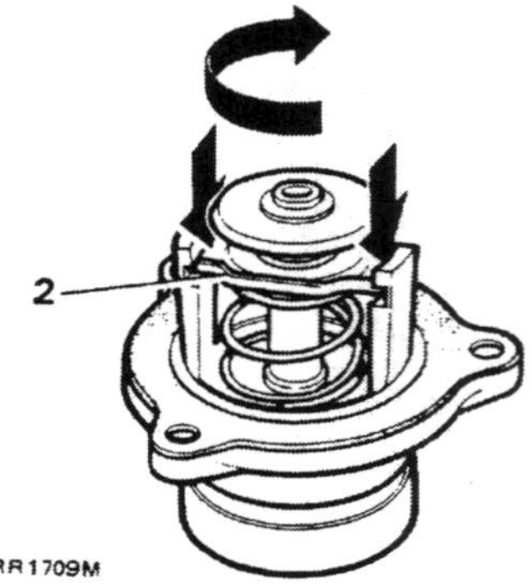

3. To test the thermostat, note the opening temperature stamped on the end of the thermostat and place it in a vessel containing water and a thermometer. Apply heat and observe the temperature at which the thermostat opens. Renew if necessary.
4. Fit the thermostat to the body, reversing the removal instructions.
5. Using a new joint washer, fit the thermostat and body to the water rail.

### Thermostat and housing 2.5 litre engine only

1. Remove the four socket headed screws and lift the outlet elbow clear to remove the thermostat with its fitted seal from the thermostat housing.

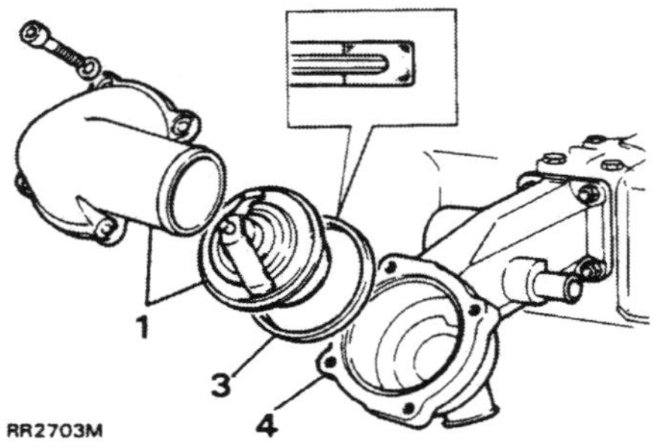

2. To test the thermostat, remove the seal (3) note the opening temperature stamped on the end of the thermostat and place it in a vessel containing water and a thermometer. Apply heat and observe the temperature at which the thermostat opens. Renew if necessary.
3. Fit a new joint seal onto the edge of the thermostat ensuring it is fitted evenly.
4. Refit the thermostat ensuring location of the seal into the recess of the thermostat housing.
5. Refit the outlet elbow and tighten the screws evenly.

# VM DIESEL ENGINE

## Inlet and exhaust manifold

1. Examine the manifold for damage and cracks.
2. Check the mating faces with the cylinder head for distortion by mounting on a surface plate and checking with feeler gauges. If necessary, the flange faces may be machined to restore maximum surface contact with the cylinder head.
3. The exhaust manifold is manufactured in two sections and piston ring type seals are used to provide a flexible gas tight seal. Renew the rings if cracked and assemble the two sections using Vaseline on the rings to facilitate assembly.

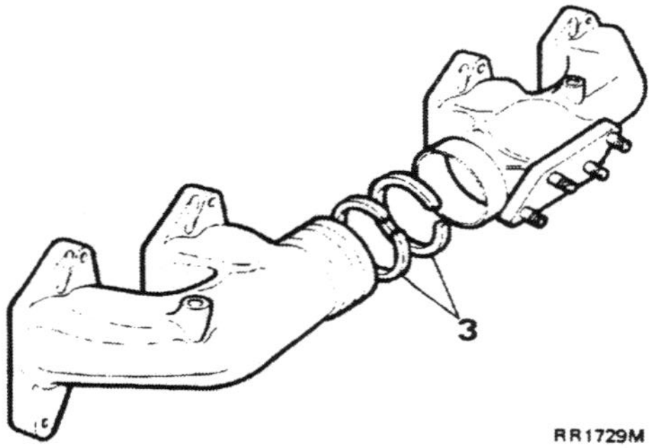

## ASSEMBLING ENGINE

### Fitting cylinder liners

1. Clean the liners and the cylinder block areas of contact. Fit the liners without 'O' rings. The liners should drop into position under their own weight, if not, further cleaning is necessary.
2. To obtain the correct liner protrusion, attach special tool 18G 1378,B as illustrated, to the cylinder block and tighten the bolts to 30 Nm (22 lbf/ft). Fit a dial test indicator so that the stylus rests in a loaded condition on the external rim of the liner and set the dial to zero. Slide the stylus across to the cylinder block and note the reading.
   Repeat the above procedure to the remaining cylinders.
3. Remove the liners and add shims as required to achieve the protrusion given in the data Section 04.
4. Remove the liners and fit three new 'O' rings. Apply molybdenum disulphide grease, such as 'Marston's Molycote' to the 'O' ring contact area in the cylinder block.
5. Apply 'Loctite 275' to areas A and B. Avoid any sealant contacting the shim and face C.
   An 'O' ring is fitted to top of liner, on later 2.5 engines or when fitting new liners to 2.4 and early 2.5 engines. Fit 'O' ring seal to position D and apply Loctite to area A only. Avoid sealant contacting the shim, face C and 'O' ring.
6. Fit the liners to the cylinder block and hold them in using the cylinder head spacers and slave bolts, tighten the bolts to 30 Nm (22 lb/ft), leave the spacers and slave bolts in position for approximately two hours until the Loctite is set.

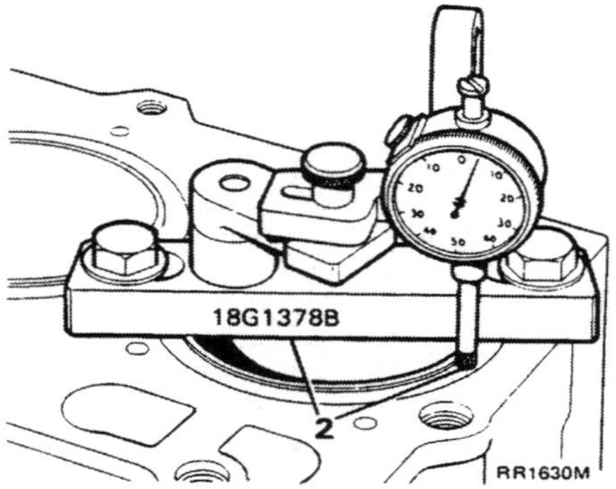

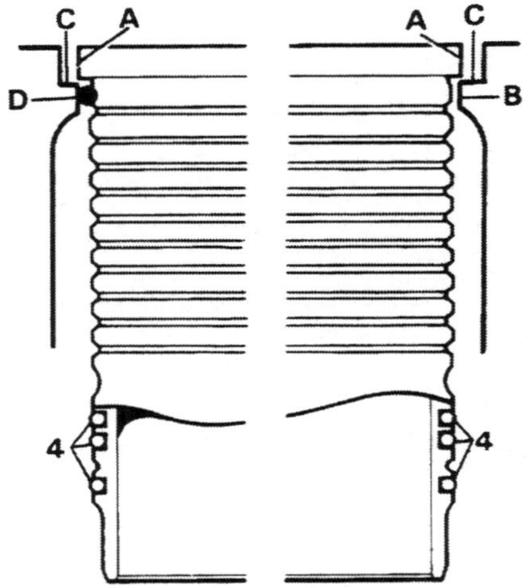

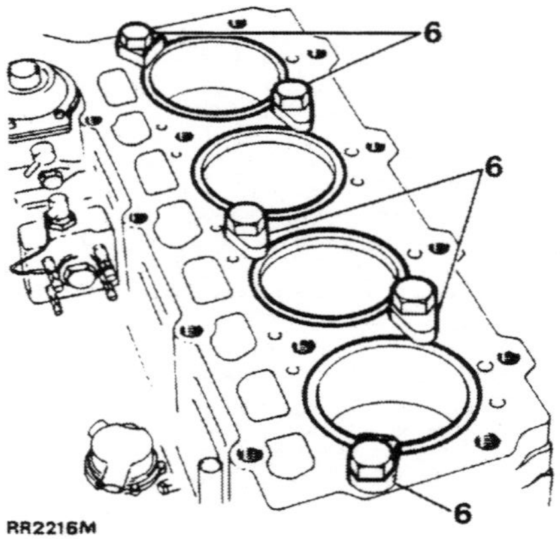

# VM DIESEL ENGINE

## Fitting front and rear main bearings.

1. Use special tool 18G 1373 to refit the front main bearings to the cylinder block, ensuring that the oil hole in bearing aligns with oil hole in the bearing bore.
2. Use the same tool 18G 1373 to refit the rear main bearings to the carrier assembly, ensuring that the oil holes in bearing and carrier align.

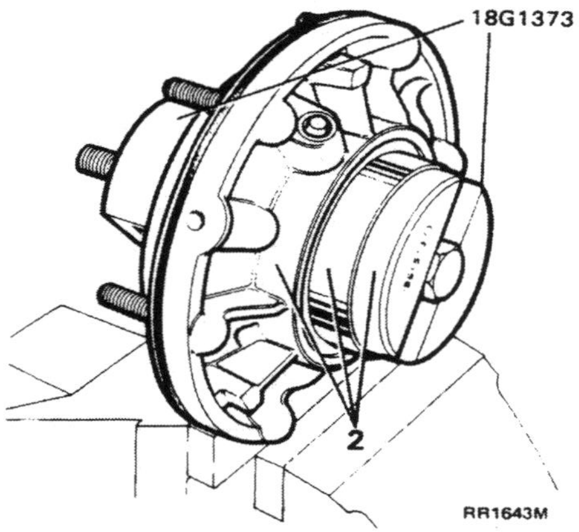

## Fitting camshaft and followers

1. Invert the cylinder block and smear the cam followers with clean engine oil and fit them to their original locations in the cylinder block.

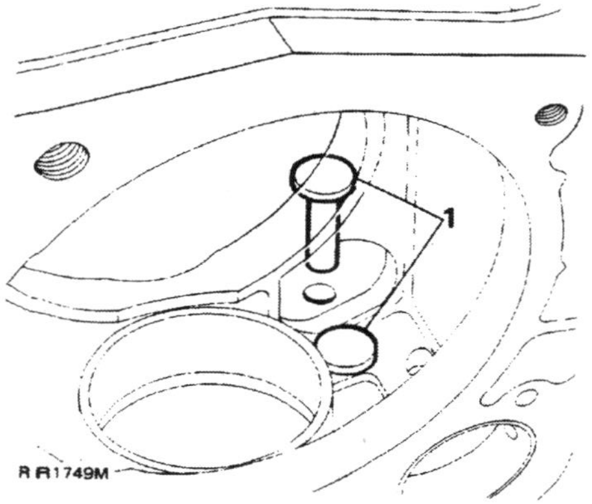

## Camshaft bearings

1. Renew the camshaft bearings in the cylinder block using special tool 18G 1372.
2. Each bearing shell has two oil holes and it is essential that these align exactly with the corresponding oil drillings in the cylinder block. The illustration shows the camshaft rear bearing being fitted.

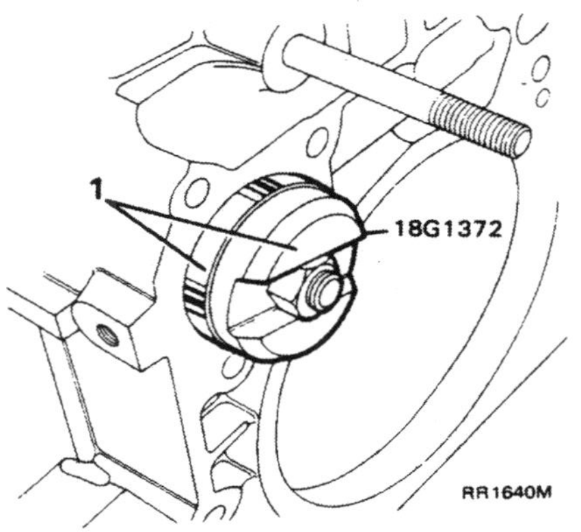

## Fitting crankshaft and carrier assembly.

1. Fit new main bearing shells to each of the carrier halves.
2. Assemble the carriers to the crankshaft journals, ensuring that the same carriers are fitted to their original locations and that the piston jet cut-a-way is towards the front of the crankshaft. Secure each carrier with the two bolts tightening evenly to the correct torque. Check that the oil jet is in position.
3. Slide special tool 18G 1370 over the crankshaft gear and, if necessary, with assistance insert the crankshaft and carrier assembly into the crankcase in the same manner as for removal.
4. Align the holes in the lower carriers, as illustrated, with the centre of the crankcase webs.
5. Secure each carrier assembly to the crankcase with the appropriate oil feed and carrier location shaft. Ensure that the shafts are fitted to their original locations with new washers.
   The correct locations are as follows:-
   Front carrier shaft - Oil feed to vacuum pump.
   Centre carrier shaft - Oil feed to turbo-charger.
   Rear carrier shaft - Blank
   Tighten the shafts to the correct torque.
6. Fit the oil pressure switch.

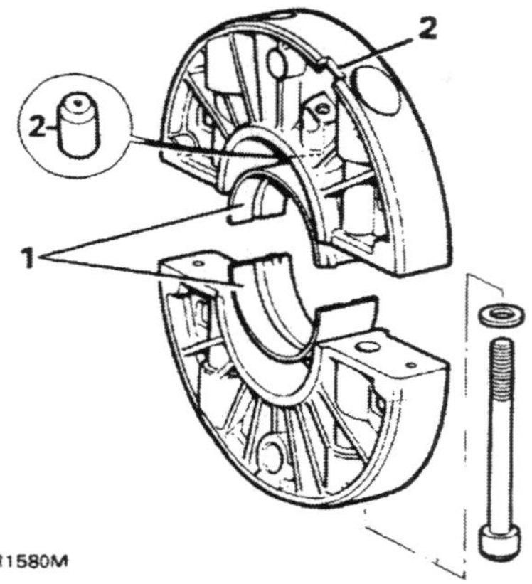

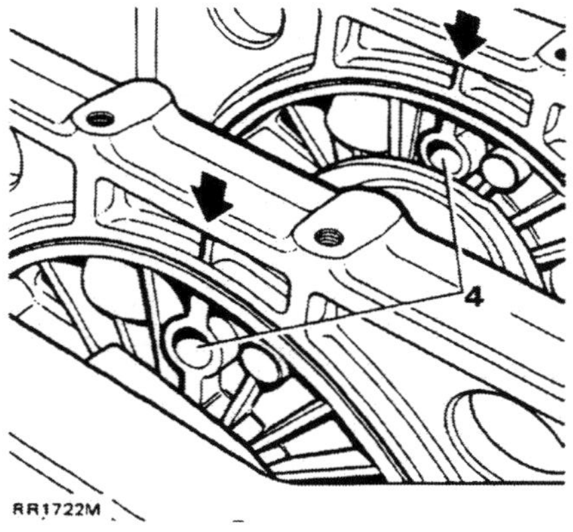

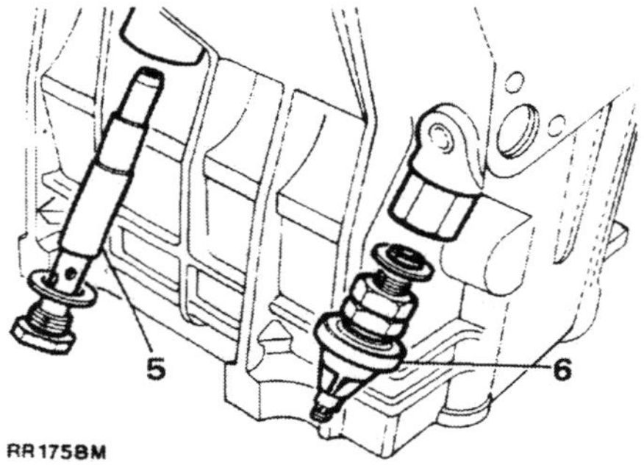

2. Smear the camshaft journals with clean engine oil and and carefully insert the camshaft complete with thrust plate and gear. Temporarily secure the camshaft to the cylinder block with the two screws.

## VM DIESEL ENGINE

**Fitting rear main carrier assembly.**

1. Fit a new 'O' ring seal to the rear main carrier.

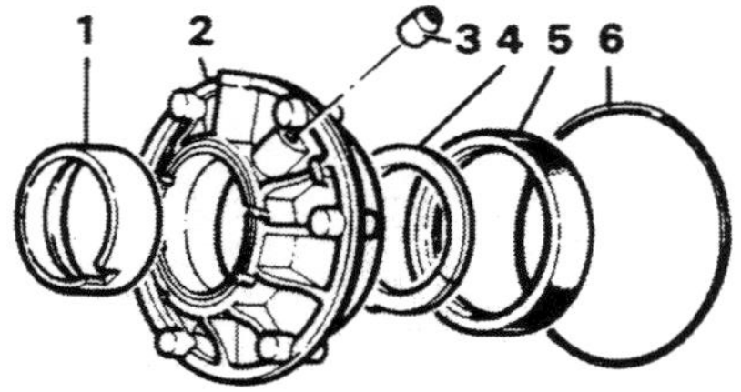

1. Bearing
2. Carrier
3. Oil jet
4. Outer thrust washers
5. Oil seal
6. 'O' ring seal

2. Fit new outer thrust halves to the oil seal side with the oil grooves outwards. Ensure that both halves are of the same thickness value and that the thrust with the tag locates in the keyway in the carrier. Hold the thrusts in position with Vaseline.

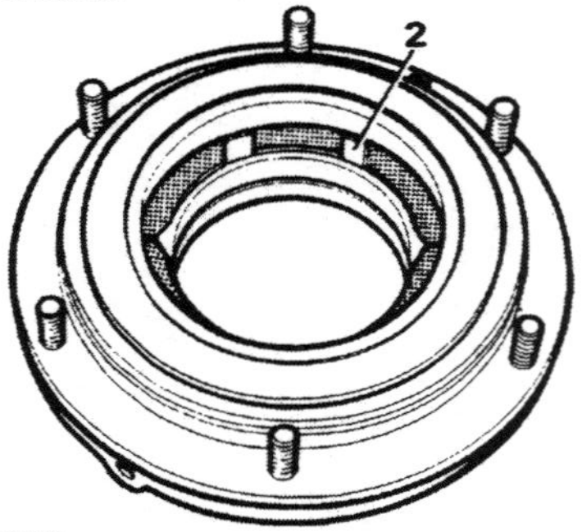

## RANGE ROVER

3. With the cylinder block still in the inverted position, lubricate the oil seal with clean engine oil and fit the carrier assembly to the crankcase. Ensure that the oil hole in the crankcase is aligned with the oil hole in the carrier as illustrated.

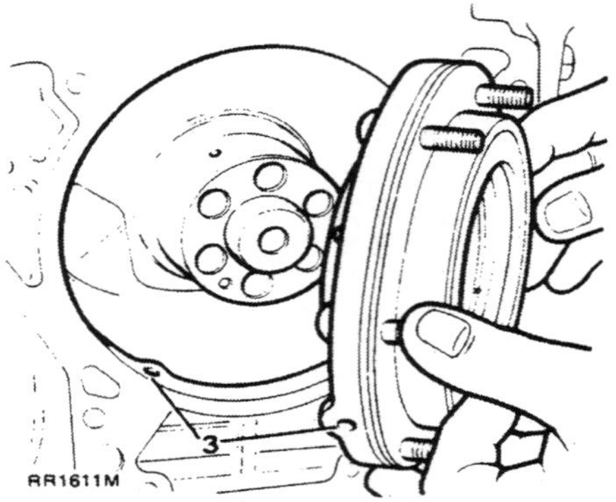

4. When correctly aligned the dowel in the carrier must be at the 1 o'clock position. Final alignment will be achieved when the flywheel housing is fitted.

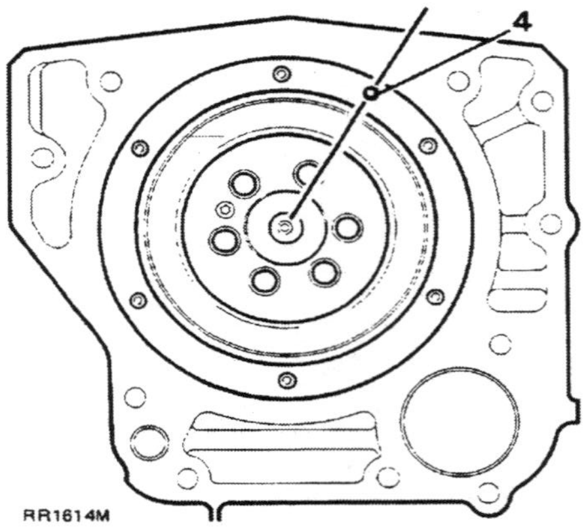

5. Fit a new 'O' ring seal to the rear of the flywheel housing.

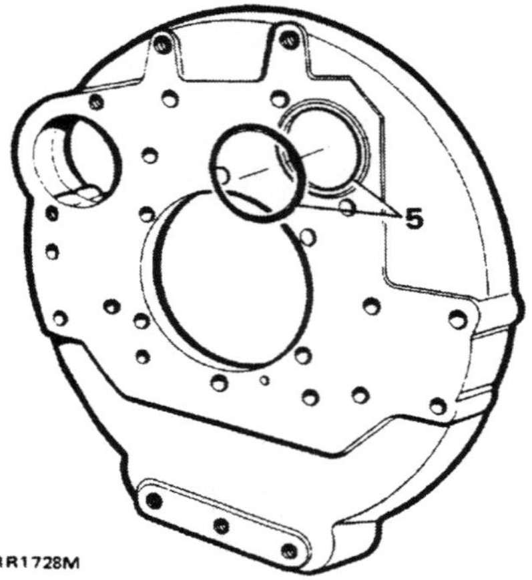

6. Fit the flywheel housing and secure with the nine bolts, tightening evenly to the correct torque. Fit and evenly tighten, to the correct torque, the six carrier retaining nuts.

7. Fit the thrust spacer and a new 'O' ring seal and secure with the socket headed screw.

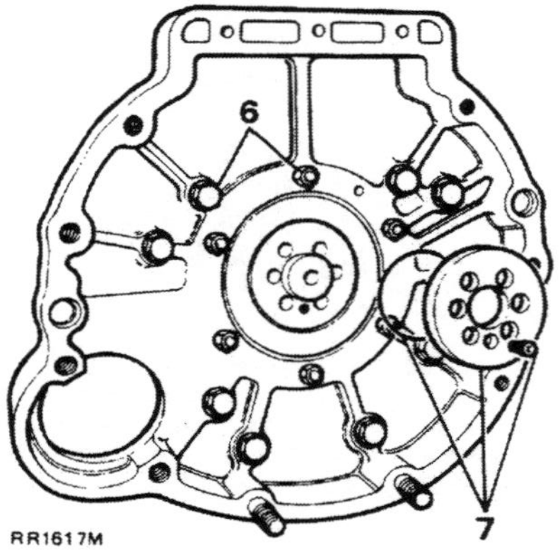

## VM DIESEL ENGINE

## RANGE ROVER

8. To check the crankshaft end-float, insert two flywheel bolts in the crankshaft using spacers equivalent to thickness of the flywheel and tighten to the correct torque. Mount a dial test indicator with the stylus resting, in a loaded condition, on the thrust spacer. Lever the crankshaft back and forth and note the reading. Adjust the end-float, if necessary, by substituting with washers of an appropriate thickness, see data section for available washers.

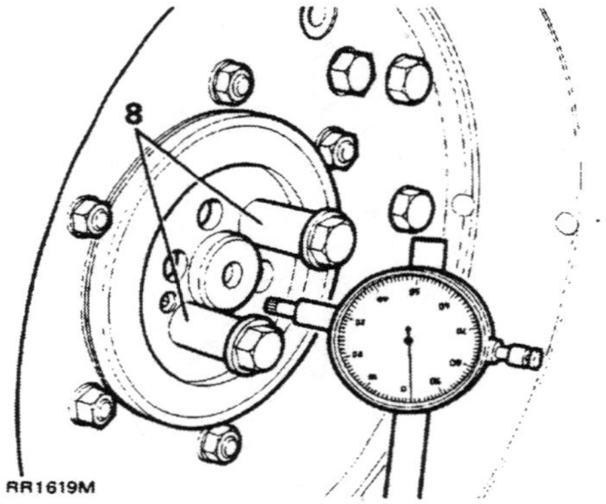

# VM DIESEL ENGINE

## Fitting flywheel

1. Fit the flywheel using the same method as for removal. Fit and evenly tighten the six retaining bolts to the correct torque.

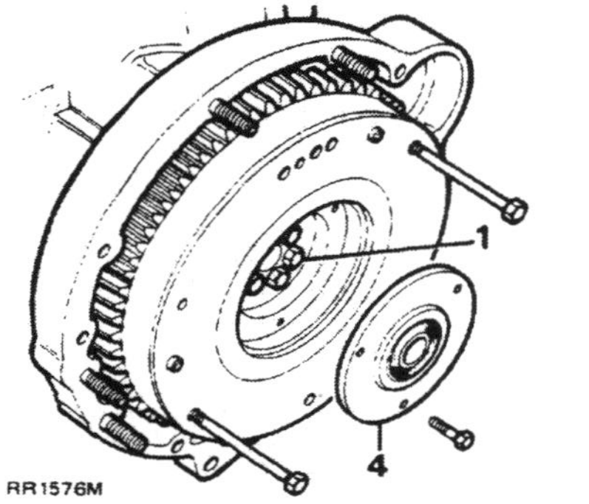

## Checking flywheel face run-out

2. Mount a dial test indicator on the flywheel housing with the stylus positioned in a loaded condition on the flywheel face and zero the gauge.
3. Turn the flywheel and take readings every 90°. The difference between the highest and lowest readings taken at all four points should not exceed 0,10 mm (0.004 in) which is the maximum permissible run-out.
4. Fit the spigot bearing and plate and secure with the three bolts.

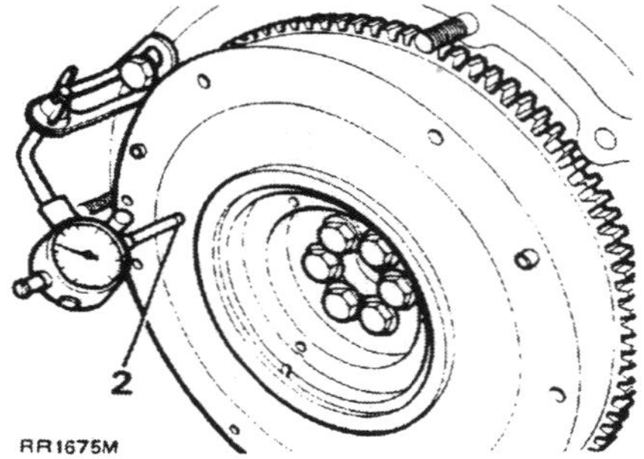

## Fitting idler gear and oil pump 2.4 litre engine.

1. Whilst maintaining the cylinder block in the inverted position, remove the two socket headed screws and partially withdraw the camshaft.
2. Fit the idler gear assembly with the three socket headed screws and tighten evenly.
3. Turn the crankshaft and idler gear until the dots align, as illustrated, with the single dot on the idler gear between the two dots on the crankshaft gear.
4. Refit the camshaft and align the gears so that the single dot on the camshaft gear is between the two dots on the idler gear, as illustrated. Fit and tighten the two camshaft retaining screws.
5. Using a new 'O' ring seal fit the oil pump assembly and secure with the three socket headed screws tightening evenly to the correct torque.

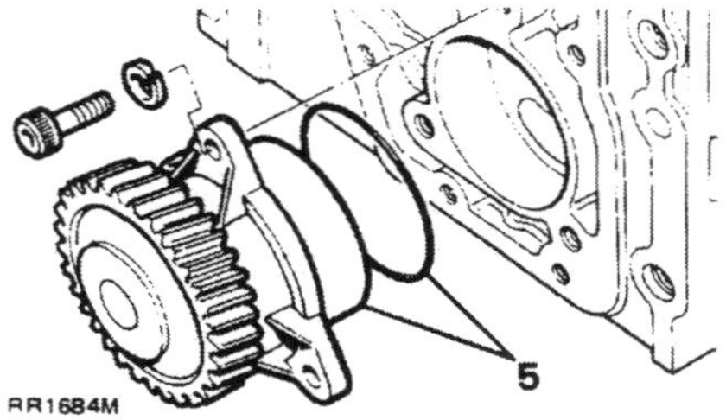

## VM DIESEL ENGINE

4. Refit the camshaft and align the gears so that the single dot on the camshaft gear is between the two dots on the vacuum pump gear, as illustrated. Fit and tighten the two camshaft retaining screws.
5. Fit the oil pump assembly and secure with the three socket headed screws tightening evenly to the correct torque.

## Fitting oil pressure relief valve

1. Clean the valve seating in the crankcase and fit the relief valve, spring and cap and secure with the circlip using 18G 257 or suitable alternative pliers.

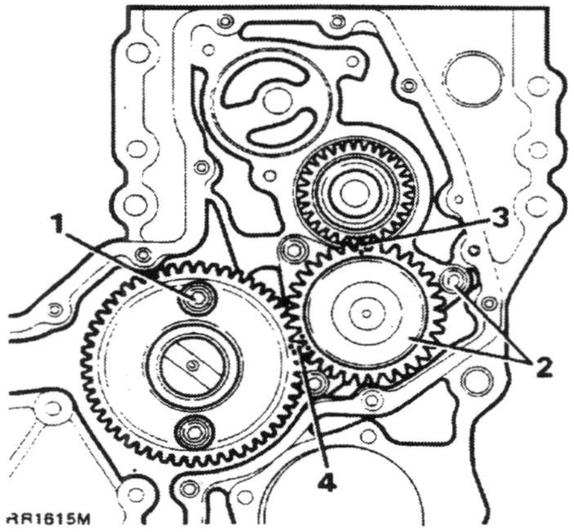

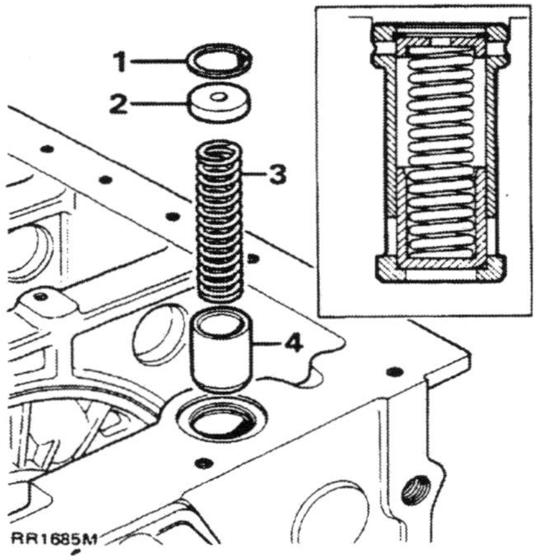

1 Circlip
2 Cap
3 Spring
4 Plunger

## Fitting vacuum and oil pumps 2.5 litre engine.

1. Whilst maintaining the cylinder block in the inverted position, remove the two socket headed screws and partially withdraw the camshaft.
2. Using a new 'O' ring seal offer the vacuum pump and gear assembly into its location.
3. Turn the crankshaft and vacuum pump gear until the dots align, as illustrated, with the single dot on the vacuum pump gear between the two dots on the crankshaft gear. Fully house the vacuum pump, tightening the three socket headed screws to the correct torque.

NOTE: The screw with the smaller diameter head should be fitted closest to the camshaft gear.

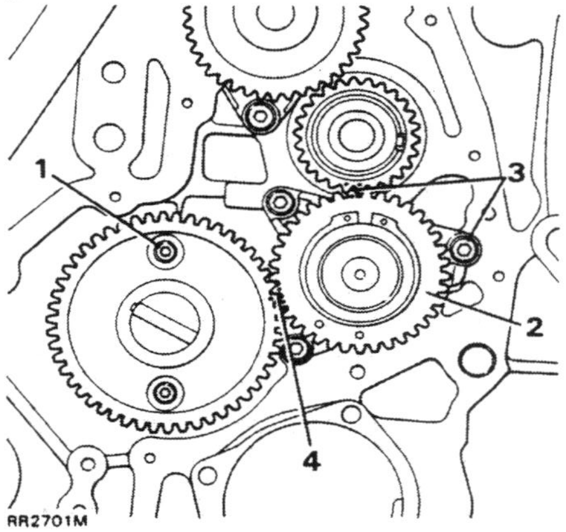

## VM DIESEL ENGINE

### Fitting connecting rods and pistons

1. If the original pistons and connecting rods are being refitted ensure that they are returned to their original locations.
2. Turn the cylinder block over to an upright position.
3. Turn the crankshaft to bring numbers one and four crankpins to the B.D.C position.
4. Stagger the piston ring gaps as follows :—

   A    Compression ring gap 30° to the right of the combustion chamber recess.

   B    Scraper ring gap on the opposite side of the combustion chamber recess.

   C    oil control rings gap 30° to the left of the combustion chamber recess.

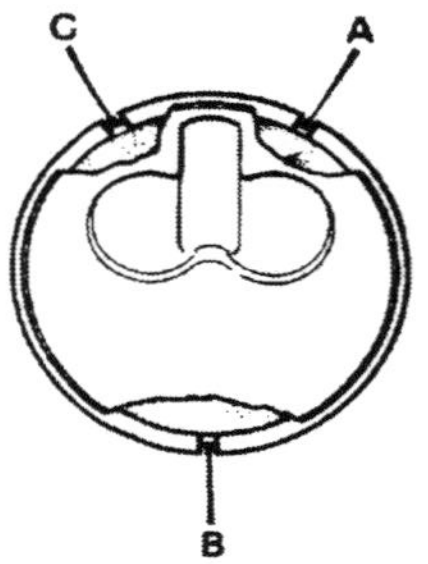

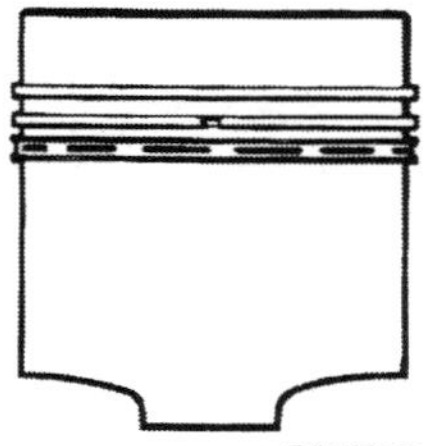

5. Check that the recess area in the piston crown is on the same side as the figures on the connecting rod. Fit the connecting rod bearing shells. Using piston ring compressor 18G 55A or a suitable alternative, insert number one and number four pistons into the cylinder bores ensuring that the recess area in the piston crown is toward the camshaft side of the engine. Tap the pistons into position in the bores.

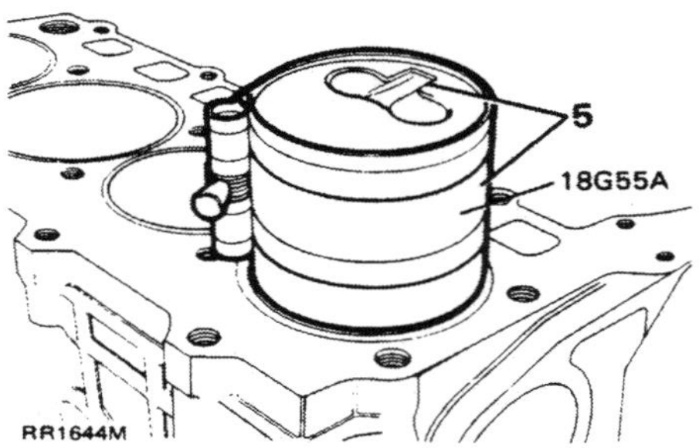

6. Turn the cylinder block over and fit the connecting rod caps so that the figures are on the same side. Apply 'Molyguard' to the threads of the NEW bolts and tighten to the correct torque.
7. Repeat the above instructions to fit number two and three pistons.

### Fitting oil strainer and sump.

1. Fit a new 'O' ring seal to the oil pick-up pipe and insert into the crankcase. Secure the strainer end of the pipe to the crankcase with two bolts. See items 42 and 43 on illustration of engine internal components.
2. Clean the sump and crankcase mating faces and apply 'Loctite 518' to both surfaces. Secure the sump with the twenty-one bolts tightening evenly to the correct torque.
3. Apply 'Hylosil RTV' to the oil pan and sump mating faces and secure the pan to the sump with eighteen nuts and evenly tighten to the correct torque. Tighten the drain plug to the correct torque.

### Fitting cylinder heads.

1. The fitting of the cylinder heads requires a precise sequence of instructions to be carried out. It includes - checking cylinder liner protrusion - selecting head gasket thickness and tightening the head bolts in the correct order.

   For these details see: **CYLINDER HEADS** remove and refit. Section 12 pages 6 to 11.

### Fitting and timing fuel injector pump.

1. Temporarily fit the timing cover and crankshaft pulley and turn the crankshaft until the T.D.C. mark on the cover aligns with the groove in the crankshaft pulley so that number one piston is at T.D.C. on the compression stroke, with number four valves 'rocking'.
2. Attach the special timing gauge 18G 1369A to the timing cover and turn the crankshaft anti-clockwise until the pulley groove aligns with the 25° B.T.D.C. mark on the scale.
3. Remove the pulley and timing cover and mesh the injection pump and camshaft gears so the tooth marked '4' is offset from the two camshaft teeth marked with dots, also the two keyways positioned exactly as illustrated.
4. Whilst holding the gear in this position fit the injection pump with a new joint washer ensure the key on the shaft is at the 11 o'clock position viewed from front of engine. Secure the three nuts, finger tight only. Fully tighten the injection pump gear retaining nut to the correct torque.
5. Release the screw on the cold start cable and turn the trunnion 90° until the lever is fully released.

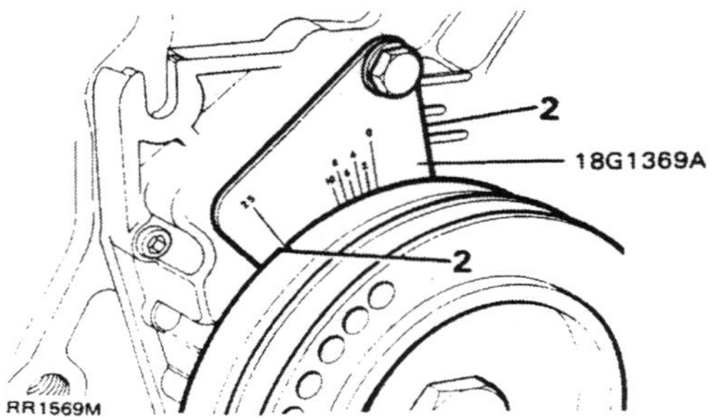

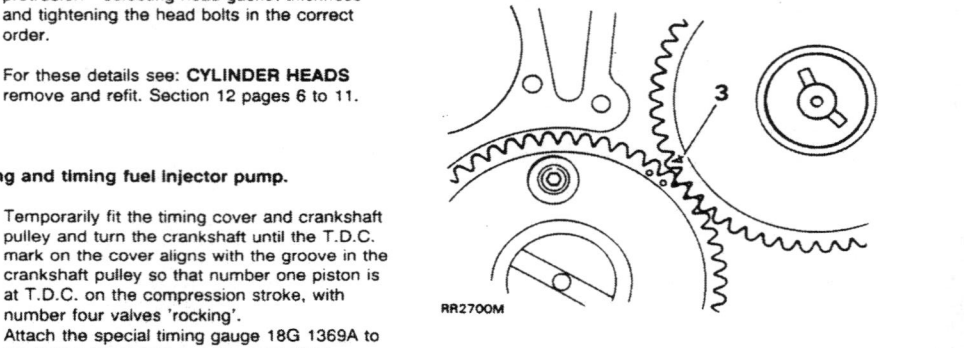

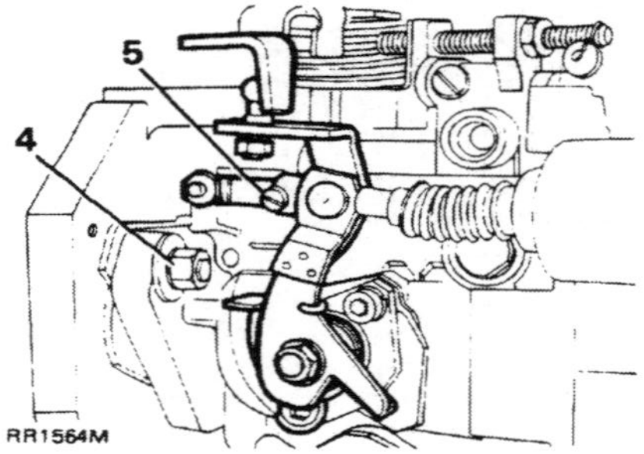

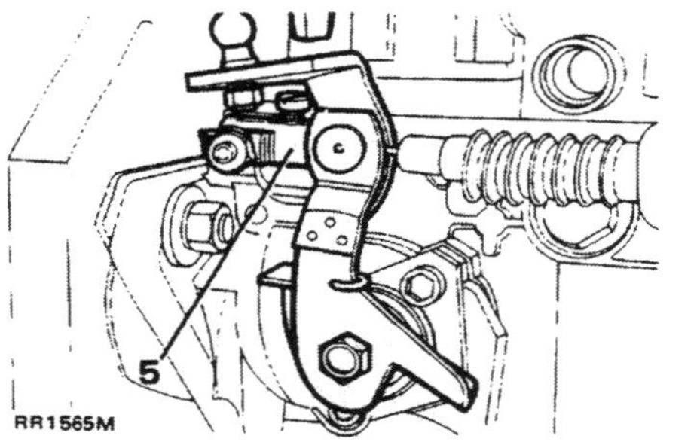

## VM DIESEL ENGINE

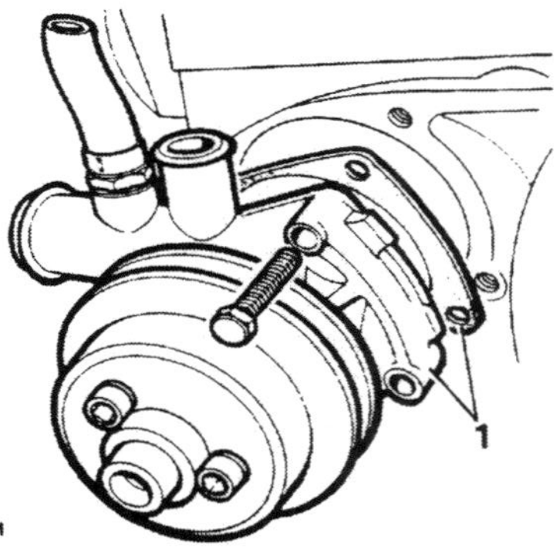

### Fitting valve rocker assemblies

1. Check that the oil feed dowels are in position in each cylinder head.
2. Fit the push rods ensuring that the ball-end locates correctly in the cam follower cup.
3. Slacken-off the tappet adjusting screws. Fit the valve rocker assemblies to the cylinder head over the oil feed dowels and locate the tappet adjusting screws in the push rod cups. Secure with the single nut and tighten to the correct torque.
4. Adjust the inlet and exhaust valve tappet clearances to 0.30 mm (0.012 in) in the following manner and sequence. The feeler gauge should be a sliding fit between the rocker and valve tip. Slacken the rocker adjusting screw locknut and turn the screw clockwise to decrease or anti-clockwise to increase the clearance. When correct hold the screw against rotation and tighten the locknut. Two sequences may be used to adjust the clearances.

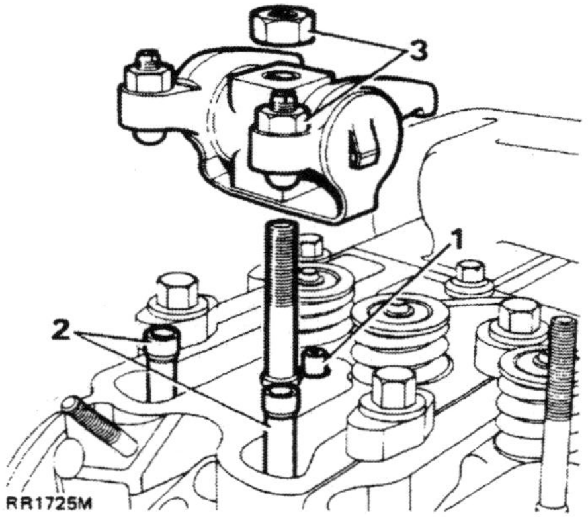

15. Fit the pump drive gear cover plate using a new 'O' ring seal. Secure with the four bolts and tighten evenly.
16. Fit the crankshaft pulley and tighten the nut to the correct torque.

### Fitting vacuum pump 2.4 litre engine

Fit the vacuum pump with a new 'O' ring seal and secure with the clamp and two nuts and tighten to the correct torque. Check that the backlash between the vacuum pump drive gear and worm drive does not exceed 0.200 mm (0.008 in).

### Fitting water pump

1. Using a new joint washer fit the water pump and pulley assembly and secure with the four bolts, tightening evenly to the correct torque.

---

10. Move the cold start lever rearward to the normal running position prior to instruction 5, and tighten the screw. **See Maintenance Section 10** for cold start adjustment.
11. Turn the crankshaft until the T.D.C. mark on the cover or timing gauge aligns with the pulley groove.
12. Remove the pulley and timing cover and fit a new joint washer and 'O' ring seal to the crankshaft.

**NOTE: Hylosil RTV is used in place of a joint washer on 2.5 Litre engines.**

13. At the same time check that the timing marks on the gear train all align, as illustration.
14. Fit the timing cover and secure with the twelve socket-headed screws and one bolt and tighten evenly to the correct torque. Using special tool 18G 1375 drive in a new timing cover seal, cavity side leading.

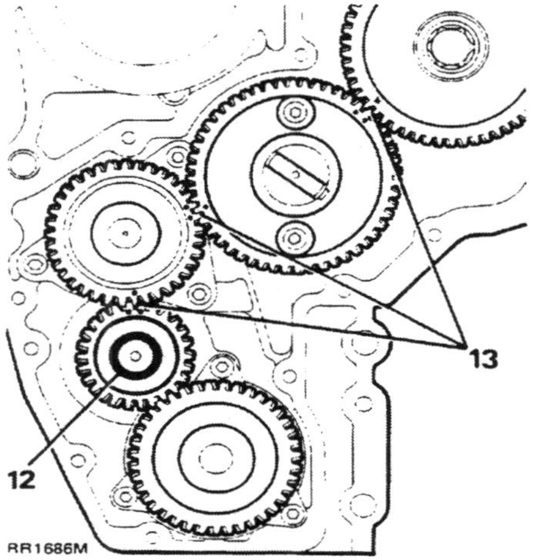

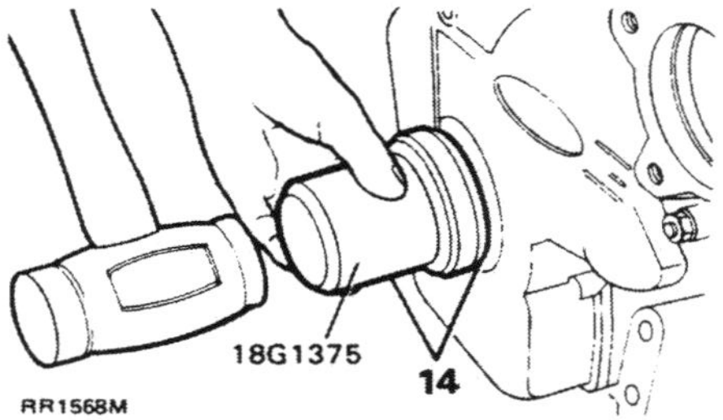

6. Fit the special tool MS107 and dial test indicator to the rear of the pump.
7. Fit the timing cover and scale and crankshaft pulley and turn the crankshaft to T.D.C. Then turn crankshaft anti-clockwise until the indicator needle stops and zero the indicator. The groove on the pulley should now be approximately aligned with the 25° B.T.D.C. mark.
8. Turn the crankshaft clockwise so that the pulley groove is aligned with the 3° B.T.D.C. mark. Turn the injector pump body, clockwise or anti-clockwise as necessary until the indicator reads 50 (0.5 mm).
9. Tighten the pump body retaining nuts and check that the dial reads 68 (0.68 mm).

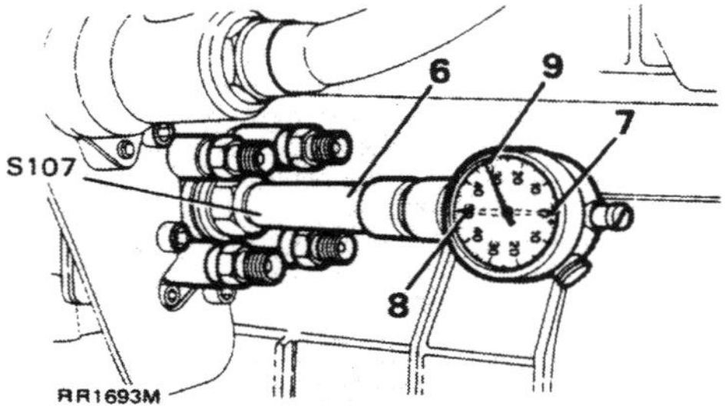

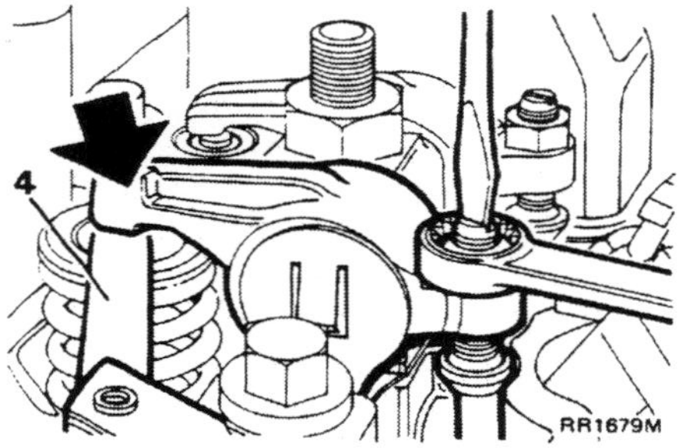

## VM DIESEL ENGINE 12

### RANGE ROVER

Sequence A

With No.1 valve fully open adjust No.7 valve.
With No.8 valve fully open adjust No.2 valve.
With No.5 valve fully open adjust No.3 valve.
With No.4 valve fully open adjust No.6 valve.
With No.7 valve fully open adjust No.1 valve.
With No.2 valve fully open adjust No.8 valve.
With No.3 valve fully open adjust No.5 valve.
With No.6 valve fully open adjust No.4 valve.

Sequence B

Rotate the crankshaft until the valves of number four cylinder are rocking then adjust the clearance of number one valve. Adjust the remaining valve clearances in the following order:-

Adjust:-
Valves of No. 3 cyl with No. 2 valves rocking
Valves of No. 4 cyl with No. 1 valves rocking
Valves of No. 2 cyl with No. 3 valves rocking

### Fitting injectors and pipes

1. Fit the sealing washer the the injector to the cylinder head.
2. Locate the dowel and clamp and tighten the nut to the correct torque.
3. Fit the remaining injectors and spill rail using a new washer both sides of the banjo unions. 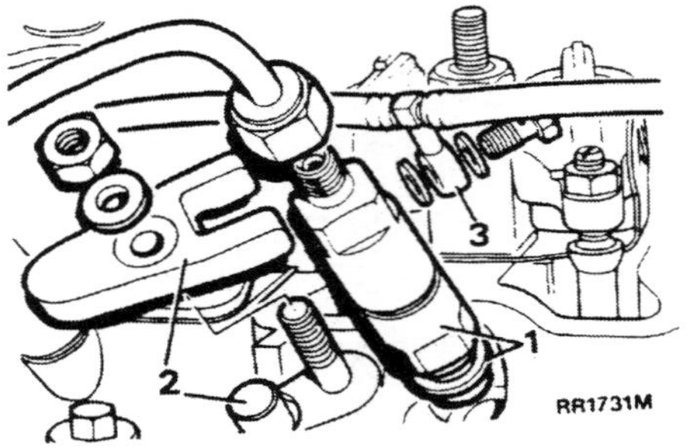
4. Fit the heater plugs and three connecting terminal bars.
5. Fit the supply pipes to the injectors and injector pump. Do not overtighten the union nuts.

### Fit rocker covers and coolant rail

1. Using new gaskets fit the rocker covers noting that the tallest covers are fitted to numbers two and three cylinders and the oil filler cap to number one cylinder. Tighten the special nuts to the correct torque.
2. Fit the engine coolant rail complete with thermostat housing to the cylinder heads using new gaskets. Tighten the eight bolts evenly to the correct torque.
3. Fit the water hose from the injector pump cold start device to number three cylinder head rocker cover and the hose from the thermostat housing to cold start device.
4. Fit the by-pass hose between thermostat housing and and water pump.

### Fit fuel lift pump

1. Using a new gasket fit the fuel lift pump to the cylinder block. Ensure that the actuating lever rides on top of the cam.

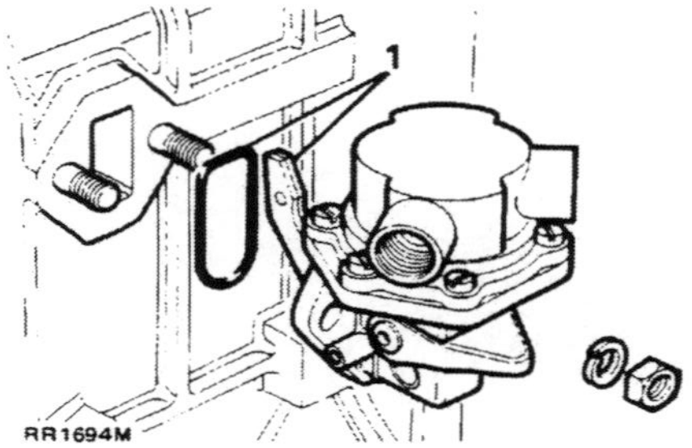

### Fit oil filter adaptor

1. Fit the oil filter adaptor, using a new 'O' ring seal, to the cylinder block. Ensure that the adaptor is fitted, as illustrated, with the elongated cavity on the side facing the cylinder block at the bottom. Secure with the union screw to the correct torque. 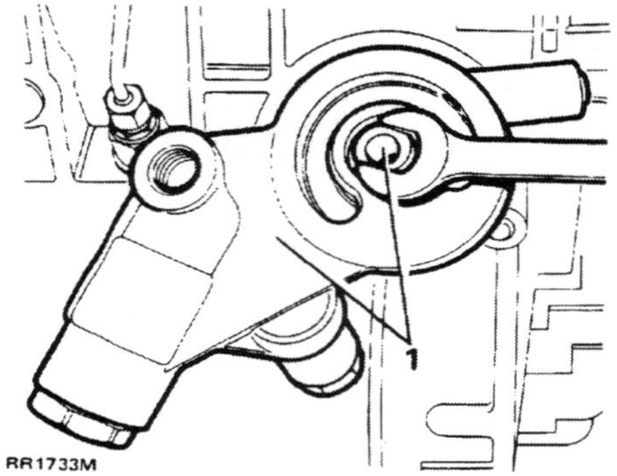
2. Smear the oil filter canister seal with clean engine oil and screw the canister on to the adaptor until contact then turn a further half turn by hand only. See maintenance **Section 10**.
3. Connect the oil feed pipe to the front main bearing carrier adaptor union and the banjo hose end to the vacuum pump.

### Fit the turbocharger

1. Fit the turbocharger support bracket to the cylinder block attachment bracket.
2. Also fit the starter motor heat shield rear support bracket which shares a common fixing point on the cylinder block.

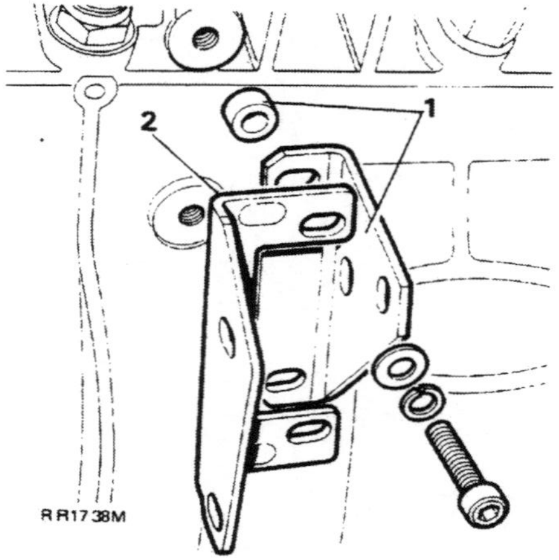

## 12 VM DIESEL ENGINE

3. Fit the oil feed hose to the centre union on the cylinder block.
4. Fit the oil return hose to the crankcase union.
5. Fit a new gasket to the exhaust manifold and fit the turbo-charger and tighten the four nuts evenly to the correct torque.
6. Connect the oil feed and oil return pipes to the turbocharger.

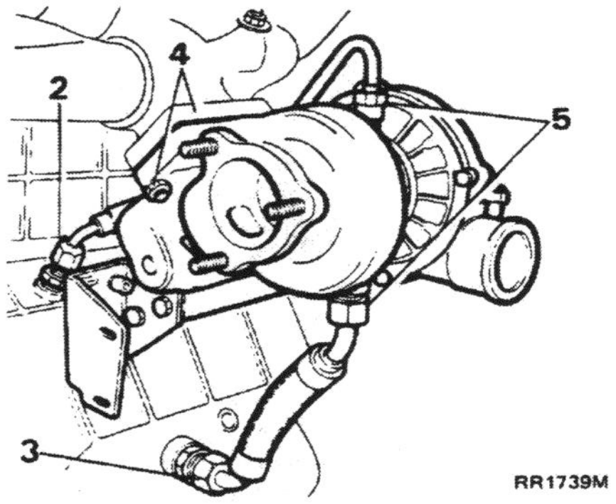

RR1739M

7. Fit the heat shield to the exhaust manifold.

### Fit power steering pump

1. Fit the power steering pump and support bracket to the engine and fit the drive belt. Adjust the drive belt tension. **See Maintenance Section 10.** To tension the belt move the pump away from the engine and tighten the pivot and adjusting bolts.

### Fit the alternator

1. Right hand steer vehicles have the alternator mounted on the left side of the engine. On left hand steer vehicles the alternator is mounted on the right hand side.
2. Fit the alternator and drive belt. Adjust the belt tension. **See Maintenance Section 10.** To tension the belt, lever the alternator away from the engine and tighten the pivot and adjusting nuts and bolts. Do not apply pressure to the stator or slip ring end of the alternator, whilst tensioning, or damage could result.

### Fit the air conditioning compressor

1. Fit the mounting bracket to the cylinder block and attach the compressor, noting that on R.H.S. vehicles the compressor and alternator share a common pivot belt. Fit and tension the drive belt. **See Maintenance Section 10.** Pivot the compressor anti-clockwise and tighten the pivot and adjusting nuts and bolts.

**CAUTION: When fitting a new drive belt, tension the belt as described above. Start and run the engine for 3 to 5 minutes at fast idle, after which time the belt must be re-checked. If necessary retension the belt.**

### Fitting starter motor

1. Fit the starter motor to the flywheel housing and secure with either two bolts or two nuts. Also attach the heat shield to the lower fixing, together with the earth strap.
2. Secure the rear-end of the starter motor to the rear support bracket, fitted earlier, with two bolts and attach the rear of the heat shield to the top bolt.

### Fitting clutch

1. Clean the flywheel and clutch assembly faces.
2. Place the clutch centre friction plate in position on the flywheel with the flat side towards the flywheel.
3. Fit the clutch assembly and loosely secure with the six bolts.
4. Centralise the centre plate using special tool 18G 79 or a spare primary shaft and tighten the six bolts evenly to the correct torque.
5. Smear the splines of the centre plate with a Molybdenum disulphide grease.

## 19 DIESEL FUEL SYSTEM

### TURBOCHARGER DESCRIPTION

A turbocharger is a simple but efficient means of increasing engine power. It consists of an exhaust gas driven air compressor that delivers high volumes of air into the combustion chamber, which may increase the engine's power output by up to 30%.

The turbocharger is fed by the main gallery oil pressure which lubricates and stabilises the fully floating bearings. When in operation the turbine shaft usually revolves between 1,000 and 130,000 rev/min. Therefore it is extremely important that the recommended change periods for oil and air filters are adhered to.

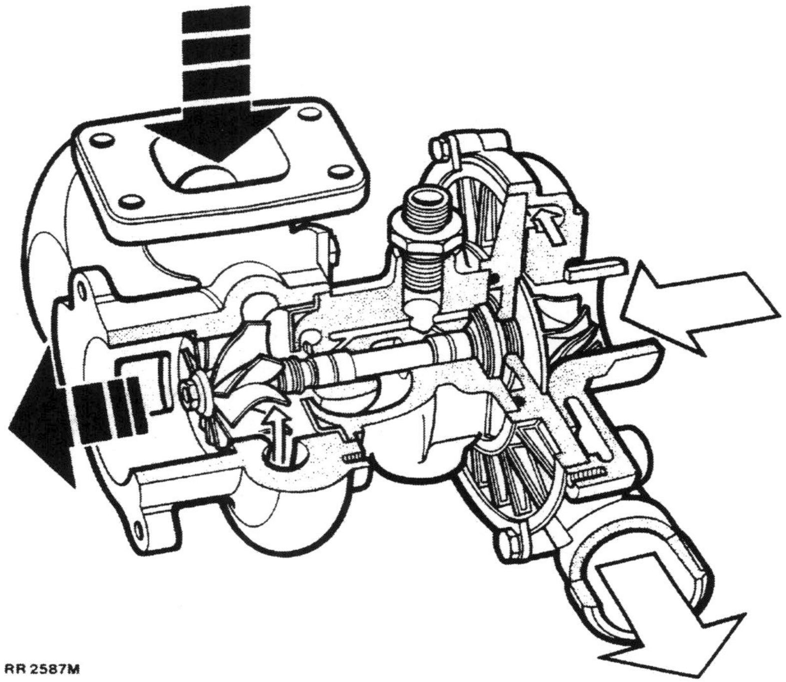

RR 2587M

# DIESEL FUEL SYSTEM

## TURBOCHARGER - CHECK

If the turbocharger unit is suspected of being faulty, the following simple test may be carried out. The assistance of a second operative is required to carry out this operation.

1. Open the bonnet.
2. Start the engine and allow it to idle.
3. Depress the turbocharger to intercooler feed pipe with one hand, the air pressure increase in the pipe may be detected as the second operative increases the engine revs.

NOTE: Although the above test indicates the operation of the turbocharger. It does not indicate it's efficiency.

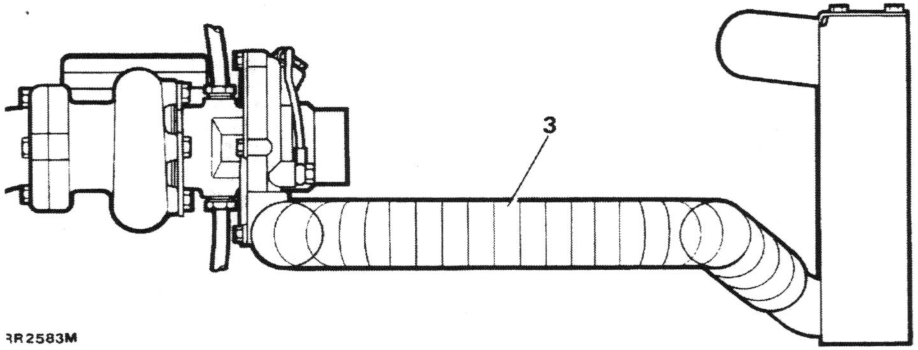

## TURBOCHARGER BOOST PRESSURE

Service tools:
**18G.1116-1 Pressure test adaptor**

### Check

1. Remove the grub screw, located in the inlet manifold.

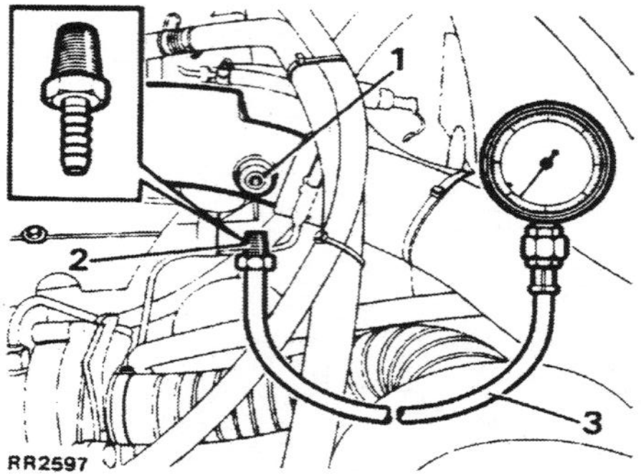

2. Insert adaptor 18G.1116-1 into the grub screw orifice
3. Attach a suitable pressure gauge, with sufficient length of tube to reach from the inlet manifold to the cab of the vehicle.
4. Drive the vehicle in 3rd gear at 3800 rev/min to give a satisfactory reading of 0.9kg/cm.

## WASTE GATE VALVE

The turbocharger waste gate diverts exhaust gas flow to by-pass the turbine when the boost pressure is higher than 0.9kg/cm.

### Adjust

The boost pressure may be adjusted by loosening the lock nut and turning the screw marked 'A' in the diagram. Turn the screw clockwise to increase the spring load on the valve and consequently increase the boost pressure. Unscrew to decrease both spring load and boost pressure.

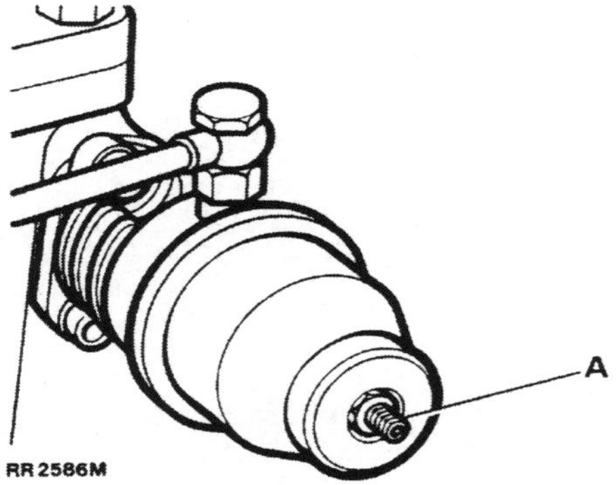

NOTE: There is a small hole located in the waste gate housing. To ensure efficient operation of the waste gate diaphragm it is neccessary to clean this hole. A small piece of sturdy wire, or a similar object, is a suitable tool for this operation. Take care not to insert the wire too far in to the waste gate housing, as the diaphragm is made of a heat resistant rubber and is subsequently easily damaged.

# 19 DIESEL FUEL SYSTEM

## TURBOCHARGER 'END FLOAT' CHECK

Use a dial test gauge and the set up shown in the diagram. Set the gauge to zero on the turbine wheel, and by moving the shaft in a linear motion the end play may be established. The maximum allowable end play is 0,15mm (0.006in).

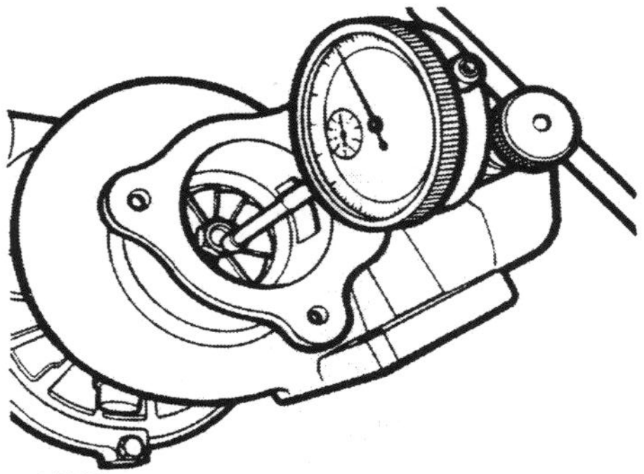

RR2584M

### Radial clearance

Push the turbine wheel to the extreme side position and set the dial test gauge to zero, on the indicator, as shown in the diagram. Check the side clearance of the turbine shaft by observing total radial movement of the turbine wheel. Maximum side clearance allowable is 0,42mm (0.016in).

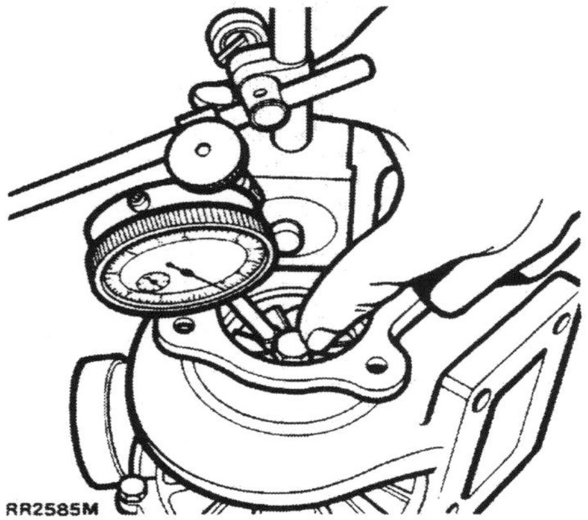

RR2585M

# RANGE ROVER

## AIR FILTER CHECK

To ensure that the correct volume of air is supplied to the turbocharger unit, the air filter should be checked for cleanliness. Firstly remove the filter box from the securing brackets. Then remove the air filter from it's housing. A visual inspection of the filter will verify it's condition. Fit a new air filter if there are any signs of oil contamination or blockage of any description.

## TURBOCHARGER FAULTFINDING

This workshop bulletin has looked at the principles on which the turbocharger operates, and the fundamentals of maintenance. It is now necessary to identify symptoms and probable causes of a suspect turbocharger. As the exhaust gas drives the turbocharger, it is capable of speeds up to 130,000 rev/min and temperatures of up to 650°C. In order to ensure that the bearings are lubricated and cooled, the turbocharger is connected to the normal engine lubrication system. Obviously very high quality seals must be used in the turbocharger, to prevent lubricating oil entering the inlet or exhaust system.

Should oil leak past a seal into the exhaust system, dense pale blue smoke will be emitted continuously. If however the oil leaks into the inlet system it will be burnt at a higher temperature and produce a darker shade of blue smoke. The engine speed may also be permanently higher than normal, as the engine will burn the oil as extra fuel and an excessive oil leak in to the inlet system may even cause the engine to accelerate, however for this to occur the operator would have to ignore all earlier signs of impending trouble.

Blockage of the large oil drain pipe from the turbocharger, though very unlikely, is certainly the worst condition as oil under pressure would be forced past the seals and into both inlet and exhaust systems. If grey or black smoke is being emitted, the turbocharger may be partly blocked or the shaft may not be perfectly free to spin. This will cause a restriction in the air inlet and result in grey/black exhaust smoke, which usually increases at higher engine speeds.

**CAUTION: If the driver is in the habit of accelerating the engine before switching the engine off, the turbine will continue to spin after the engine has come to rest and the lubrication to the turbine bearings ceased. It is therefore possible that this practice will cause damage or seizure of the bearings.**

# DIESEL FUEL SYSTEM 19

### Other symptoms of turbocharger faults

A change in the normal noise level is usually the first indication of a fault in the turbocharger or its hose connections. A higher pitched sound usually indicates a possible air leak into the suction side of the compressor and the inlet system, or an escape of compressed air between the compressor and the inlet manifold. Obviously an excessive escape of compressed air from the manifold is not only noisy but will also cause some loss of power.

Slight leaks from the turbine housing or exhaust manifold, whilst noisy, are easily detected and have little effect on the power output. Cyclic sounds (e.g. a continuous rubbing noise) are an indication of a restriction to the compressor or that the compressor wheel is coated in dirt.

If the waste gate valve sticks in the open position the engine will be down on power.

## COOLING SYSTEM 26

### Refitting

8. Fit new low coolant sensor (in kit) to new expansion tank.
9. Fit new expansion tank and tighten pinch bolt.
10. Fit radiator hose, tighten clip

**Bleed pipe (Y piece) - see RR3806M**

NOTE: Passage of coolant along hose A causes air to be extracted from radiator along hose B.

11. Remove existing bleed pipe from vehicle.
12. Fit new bleed pipe from kit. Place pipe assembly on vehicle. Fit shortest hose to expansion tank, tighten clip.

CAUTION: It is essential that the new bleed pipe is installed correctly. Interchanging position of hoses A and B will render the air bleed function of the Y piece inoperative. The shorter of the two long hoses, identified by a blue plastic tag, MUST be fitted to the engine.

13. Identify hose A and B. Fit correctly to engine and radiator, tighten clips.
14. Fit bleed pipe to front left hand inner wing, by the M8 screw washer and nut, using one of two existing holes.

## RANGE ROVER

### FIT COOLANT SYSTEM KIT

- 2.4 VM diesel.

**Part No. RTC 6863**

This kit should be used where the engine is prone to overheating, or requires new cylinder heads/gaskets. Prior to fitting the kit, it is essential to check if a non-factory air conditioning system has been installed. If such a system is fitted, check that a 16 fin/inch radiator, part number BTP 1742 has also been fitted.

### Expansion tank

### Removing

WARNING: Do not remove expansion tank filler cap when engine is hot. The system is pressurised and personal scalding could result.

1. Remove expansion tank filler cap. Turn expansion cap a quarter turn anti-clockwise, allow pressure to escape, continue turning in same direction to remove cap.
2. Drain cooling system.
3. Disconnect hose to radiator.
4. Disconnect overflow pipe.
5. Disconnect wiring to low coolant sensor.
6. Remove pinch bolt.
7. Remove expansion tank.

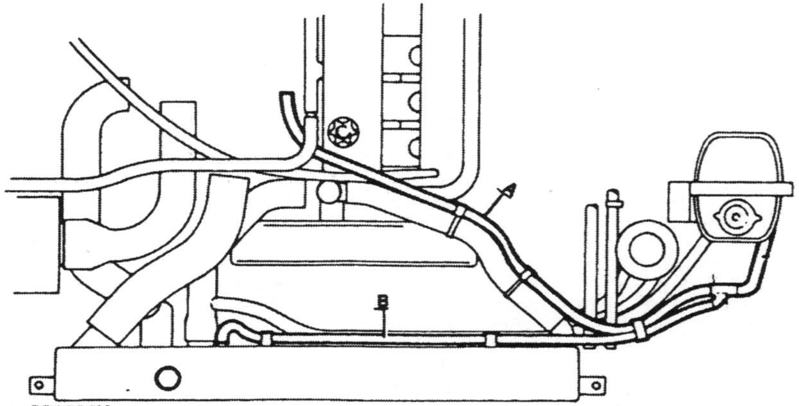

## RANGE ROVER

## 19 DIESEL FUEL SYSTEM

### INJECTION PUMP TIMING

### Check and adjust

**Service tools:**

18G.1376 Timing adaptor
18G.1369A Timing gauge

When it has been established that the injection pump requires a timing check the following procedure should be followed.

1. Attach the timing gauge 18G.1369A to the engine front cover. Rotate the engine until the mark on the crankshaft pulley lines up with the top dead centre TDC mark on the timing gauge.
2. Release the cable tensioner on the cold start mechanism to ensure an accurate result.
3. Remove the blanking plug from the rear of the pump assembly and insert adaptor 18G. 1376. Attach a dial test gauge to allow a reading to be obtained.
4. Rotate the engine again until the pulley mark is lined up with the 25° BTDC mark. The dial on the test gauge should then be zeroed. Ensure at this stage that there is sufficient pressure being applied to the stylus to give a deflection on the indicator.
5. The engine should be turned again to 3° BTDC and the needle deflection noted. A deflection of 50 (0.5mm) should be read.
6. To adjust the dimension, slacken the three locking nuts, which secure the pump to the front cover. Then rotate the injection pump assembly until a correct reading has been established, retighten the locking nuts. As a double check the deflection on the dial test gauge should be a further 16-18 (0,16-0,18mm) when the engine is turned to TDC.
7. When adjustment is complete remove the service tools and refit the blanking plug to injection pump and the engine front cover bolt.
8. Finally retension the cable on the cold start mechanism and tighten the retaining screw.

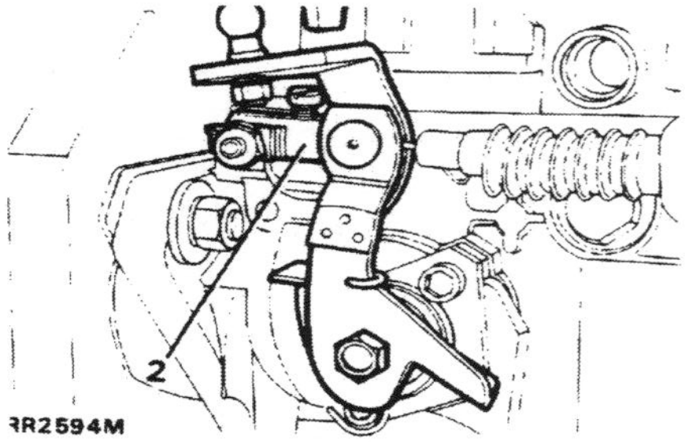

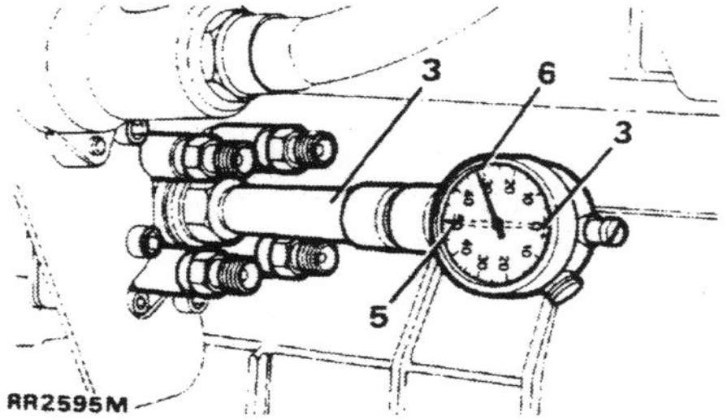

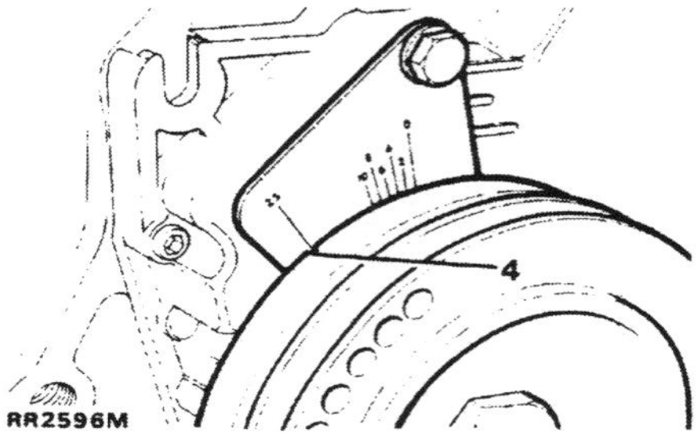

514

# ELECTRICAL 86

RANGE
ROVER

NOTE: The circuit diagram RR1166 is for 2.4 litre, 1986 Model Year Diesel Range Rovers.
For later circuit diagrams refer to main workshop manual or supplements.

| Supplements | Publication No. |
|---|---|
| 1987 Model Year | LSM180 WS1 ed2 |
| 1988 Model Year | LSM180 WS2 |
| 1989 Model Year | LSM180 WS3 |
| 1990 Model Year | LSM180 WS5 ed2 |
| 1991 Model Year | LSM180 WS6 |

---

# 26 COOLING SYSTEM

RANGE ROVER

### Coolant temperature sensor

15. Remove coolant temperature sensor from thermostat. Blank off hole using blank plug supplied.
16. Remove blank plug from No. 4 cylinder head. Fit adaptor from kit, applying Loctite Superfast 572.
17. Fit coolant temperature sensor to adaptor.
18. Re-route sensor wiring. Ensure that cables are protected where necessary with PVC electrical tape to prevent chafing.

### Cooling system fill

NOTE: If cooling system is dry, fill radiator with correct quantity and solution of coolant before carrying out this fill procedure.

WARNING: Do not remove radiator filler plug unless system is cold and expansion tank filler cap is first removed. The system is pressurised and personal scalding could result.

19. With system cold remove header tank filler cap.
20. Remove radiator filler plug.
21. Start engine run at 1500 rev/min while carrying out instructions 22 to 24.
22. Add coolant to radiator until full.
23. Refit radiator filler plug.
24. Add coolant until it is within 25 mm of bottom of filler neck. Disregard level plate.
25. Refit header tank filler cap.

### VISCOUS FAN - CHECK

When investigating instances of engine or cylinder head overheating on the VM 2.4 litre diesel engine, it is important that you check the operation of the viscous coupling to ensure it is functioning correctly. The following procedure should be used.

NOTE: When an engine is cold, for instance on the first start-up of the day, some noise will be evident from the fan. This noise is normal and is evident that the unit functioning correctly. After a few minutes the fan noise will reduce.

1. Remove the viscous coupling assembly from the drive shaft

NOTE: The hexagonal coupling on the input shaft is a left hand thread.

2. Examine the assembly for general damage and especially for fluid leakage from either the valve on the front of the unit or from the rear in the area of the input shaft/hexagonal coupling. The viscous fluid is normally dark grey in appearance, although if the leak is fresh and the fluid has not been contaminated it will appear transparent.
3. Inspect the bimetallic spring, which operates the valve on the front of the viscous assembly to ensure that it is not damaged and is properly secured. The spring is fixed to a pressed steel bracket by silicon rubber adhesive and the bracket is in turn rivetted to the aluminium housing of the viscous coupling.
4. Ensure that the input shaft rotates smoothly with no evidence of tight spots or grating, and also that a significant degree of constant resistance is felt during rotation - the shaft must not run too freely.

NOTE: If there is evidence of a fault with the viscous fan coupling it must be replaced. Do not attempt to dismantle or overhaul the unit.

515

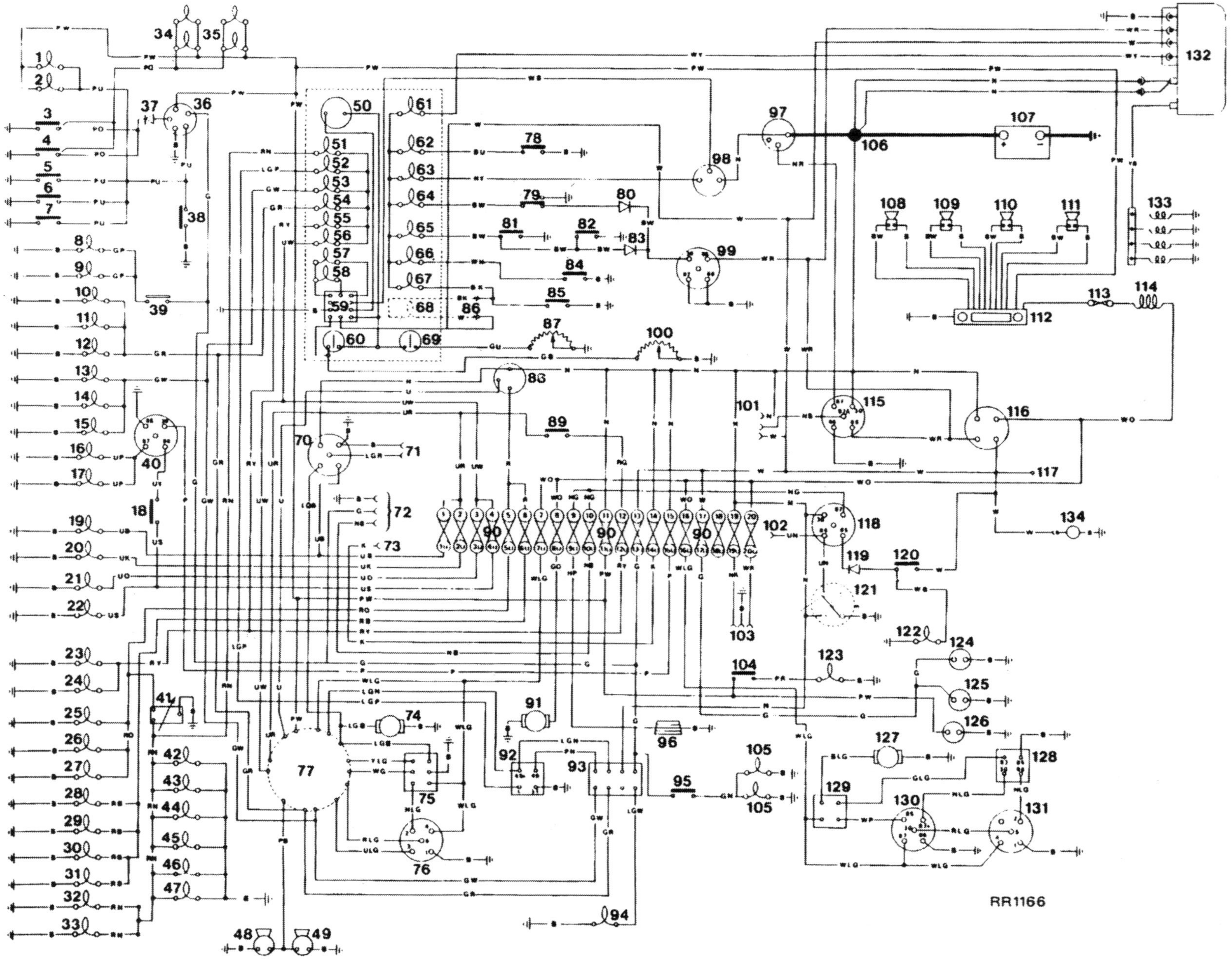

ELECTRICAL 86

RANGE ROVER

RANGE ROVER

86 ELECTRICAL

RR1166

# ELECTRICAL 86

## RANGE ROVER

## KEY TO CIRCUIT DIAGRAM - Fig. RR1166

1. Front interior lamp
2. Rear interior lamp
3. LH front door switch
4. RH front door switch
5. Tailgate switch
6. LH rear door switch
7. RH rear door switch
8. LH stop lamp
9. RH stop lamp
10. LH front indicator lamp
11. LH rear indicator lamp
12. LH side repeater lamp
13. RH front indicator lamp
14. RH rear indicator lamp
15. RH side repeater lamp
16. LH auxiliary driving lamp
17. RH auxiliary driving lamp
18. Auxiliary driving lamp switch
19. RH headlamp dip
20. LH headlamp dip
21. RH headlamp main
22. LH headlamp main
23. RH rear fog lamp
24. LH rear fog lamp
25. RH number plate lamp
26. RH side lamp
27. RH tail lamp
28. LH number plate lamp
29. LH side lamp
30. LH tail lamp
31. Radio illumination
32. Switch illumination
33. Switch illumination
34. LH door lamps
35. RH door lamps
36. Interior lamp delay
37. Diode
38. Interior lamp switch
39. Stop lamp switch
40. Auxiliary lamps relay
41. Rheostat
42. Front cigar lighter illumination
43. Clock illumination
44. Heater illumination
45. Heater illumination
46. Heater illumination
47. Heater illumination
48. LH Horn
49. RH Horn
50. Tachometer
51. Instrument illumination (6 bulbs)
52. Trailer warning light
53. RH indicator warning light
54. LH indicator warning light
55. Rear fog warning light
56. Headlamp warning light
57. Not used
58. Low fuel warning light
59. Multifunction unit in binnacle
60. Fuel indicator gauge
61. Cold start warning light (carburetter versions only)
62. Differential lock warning light
63. Ignition warning light
64. Brake failure warning light
65. Brake pad wear warning light
66. Oil pressure warning light
67. Park brake warning light
68. Park brake warning light (Australia)
69. Water temperature gauge
70. Headlamp washer timer (option)
71. Headlamp wash pump (option)
72. Heated electric mirrors (option)
73. Trailer socket (option)
74. Front screen wash
75. Front wiper delay
76. Wiper motor
77. Steering column switches
78. Differential lock switch
79. Brake failure switch
80. Diode
81. Front brake pad wear
82. Rear brake pad wear
83. Diode
84. Oil pressure switch
85. Park brake switch
86. Pick up point - park brake warning light (Australia)
87. Water temperature transducer
88. Light switch
89. Rear fog lamp switch
90. Main fuse box
91. Heater motor and switch unit
92. Flasher unit
93. Hazard switch
94. Hazard warning lamp
95. Reverse lamp switch
96. Heated rear screen
97. Starter solenoid
98. Alternator
99. Brake failure warning lamp check relay
100. Fuel tank unit
101. Air conditioning (option)
102. Split charge relay (option)
103. Electric windows and central door locking (option)
104. Under bonnet illumination switch
105. Reverse lamps
106. Terminal post
107. Battery
108. LH rear speaker (option)
109. RH rear speaker (option)
110. LH front speaker
111. RH front speaker
112. Radio (option)
113. Radio fuse
114. Radio choke
115. Starter solenoid relay(on)
116. Ignition heat start switch
117. Split charge relay (option)
118. Heated rear windows relay
119. Diode
120. Heated rear window switch
121. Voltage switch (option)
122. Heated rear window warning lamp
123. Bonnet lamp
124. Cigar lighter (dash)
125. Cigar lighter (cubby box)
126. Clock
127. Rear screen wash motor
128. Rear wiper delay
129. Rear wash wipe switch
130. Rear wiper relay
131. Rear wiper motor
132. Timer for glow plugs
133. Glow plugs
134. Fuel shut off solenoid

## KEY TO CABLE COLOURS

B . Black
K . Pink
N . Brown
P . Purple
G . Green
L . Light
O . Orange
R . Red
U . Blue
Y . Yellow
S . Slate
W . White

517

# BROOKLANDS BOOKS LAND ROVER PUBLICATIONS

| Title | Ref | ISBN |
|---|---|---|
| Land Rover Series 1 Workshop Manual | 4291 | 9781783181841 |
| Land Rover Series 1 1948-1953 Parts Catalogue | 4051 | 9781855201194 |
| Land Rover Series 1 1954-1958 Parts Catalogue | 4107 | 9781855201071 |
| Land Rover Series 1 1948-1958 Instruction Manual | 4277 | 9781855207912 |
| Land Rover Series 1 & II Diesel Instruction Manual | 4343 | 9781855201286 |
| Land Rover Series II & IIA Workshop Manual | AKM8159 | 9781783180295 |
| Land Rover Series II & Early IIA Bonnetted Control Parts Catalogue | 605957 | 9781855202382 |
| Land Rover Series IIA Bonnetted Control Parts Catalogue | RTC9840CC | 9781855202757 |
| Land Rover Series IIA, III & 109 V8 Optional Equipment Parts Catalogue | RTC9842CE | 9781855202870 |
| Land Rover Series IIA/IIB Instruction Manual | LSM64IM | 9781855201231 |
| Land Rover Series 2A and 3 88 Parts Catalogue Supplement (USA Spec) | 606494 | 9781783180264 |
| Land Rover Series III Workshop Manual | AKM3648 | 9781855201088 |
| Land Rover Series III Workshop Manual V8 Supplement (edn. 2) | AKM8022 | 9781855203105 |
| Land Rover Series III 88, 109 & 109 V8 Parts Catalogue | RTC9841CE | 9781855202139 |
| Land Rover Series III Owners Manual (handbook) 1971-1978 | 607324B | 9781855201293 |
| Land Rover Series III Owners Manual (handbook) 1979-1985 | AKM8155 | 9781855202269 |
| Military Land Rover (Lightweight) Series III Parts Catalogue | 61278 | 9781855201545 |
| Military Land Rover Series III (L.W.B.) User Handbook | 608179 | 9781855208919 |
| Military Land Rover (Lightweight) Series III User Manual | 608180 | 9781855200159 |
| Land Rover 90/110 & Defender Workshop Manual 1983-1992 | SLR621ENWM | 9781855202504 |
| Land Rover Defender Workshop Manual 1993-1995 | LDAWMEN93 | 9781783181711 |
| Land Rover Defender 300 Tdi & Supplements Workshop Manual 1996-1998 | LRL0097ENGBB | 9781855205048 |
| Land Rover Defender Td5 Workshop Manual & Supplements 1999-2006 | LRL0410BB | 9781855206977 |
| Land Rover Defender Electrical Manual Td5 1999-06 & 300Tdi 2002-2006 | LRD5EHBB | 9781855206984 |
| Land Rover 110 Parts Catalogue 1983-1986 | RTC9863CE | 9781855202887 |
| Land Rover Defender Parts Catalogue 1987-2006 | STC9021CC | 9781855207127 |
| Land Rover 90 • 110 Handbook 1983-1990 MY | LSM0054 | 9781855204560 |
| Land Rover Defender 90 • 110 • 130 Handbook 1991 MY - Feb. 1994 | LHAHBEN93 | 9781855206502 |
| Land Rover Defender 90 • 110 • 130 Handbook Mar. 1994 - 1998 MY | LRL0087ENG/2 | 9781855206519 |
| Military Land Rover 90/110 All Variants (Excluding APV & SAS) User Manual | 2320-D-122-201 | 9781855208926 |
| Military Land Rover 90 & 110 2.5 Diesel Engine Versions User Handbook | SLR989WDHB | 9781783180233 |
| Military Land Rover Defender XD - Wolf Workshop Manual | 2320D128 - 302 522 523 524 | 9781783180257 |
| Military Land Rover Defender XD - Wolf Parts Catalogue | 2320D128711 | 9781783180240 |
| Land Rover Discovery Workshop Manual 1990-1994 (Petrol 3.5, 3.9, Mpi & Diesel 200 Tdi) | SJR900ENWM | 9781855203129 |
| Land Rover Discovery Workshop Manual 1995-1998 (Petrol 2.0 Mpi, 3.9, 4.0 V8 & Diesel 300 Tdi) | LRL0079BB | 9781855207332 |
| Land Rover Discovery Series II Workshop Manual 1999-2003 (Petrol 4.0 V8 & Diesel Td5 2.5) | VDR100090/6 | 9781855208681 |
| Land Rover Discovery 3 2004-2009 TDI Diesel Workshop Manual | | 9781783182183 |
| Land Rover Discovery 3 2004-2009 V8 4.4L Petrol Workshop Manual | | 9781783182176 |
| Land Rover Discovery 3 2004-2009 V6 4.0 Petrol Workshop Manual | | 9781783182169 |
| Land Rover Discovery 3 2004-2009 Chassis and Body Workshop Manual | | 9781783182152 |
| Land Rover Discovery Parts Catalogue 1989-1998 (2.0 Mpi, 3.5, 3.9 V8 & 200 Tdi & 300 Tdi) | RTC9947CF | 9781855206144 |
| Land Rover Discovery Parts Catalogue 1999-2003 (Petrol 4.0 V8 & Diesel Td5 2.5) | STC9049CA | 9781855208858 |
| Land Rover Discovery Owners Handbook 1990-1991 (Petrol 3.5 V8 & Diesel 200 Tdi) | SJR820ENHB90 | 9781855202849 |
| Land Rover Discovery Series II Handbook 1999-2004 MY (Petrol 4.0 V8 & Td5 Diesel) | LRL0459BB | 9781855208438 |
| Land Rover Freelander Workshop Manual 1998-2000 (Petrol 1.8 and Diesel 2.0) | LRL0144 | 9781855206151 |
| Land Rover Freelander Workshop Manual 2001-2003 ON (Petrol 1.8L, 2.5L & Diesel Td4 2.0) | LRL0350ENG/4 | 9781855208742 |
| Land Rover 101 1 Tonne Forward Control Repair Operation Manual | RTC9120 | 9781855201392 |
| Land Rover 101 1 Tonne Forward Control Parts Catalogue | 608294B | 9781855201385 |
| Land Rover 101 1 Tonne Forward Control User Manual | 608239 | 9781855201439 |
| Range Rover Workshop Manual 1970-1985 (Petrol 3.5) | AKM3630 | 9781855201224 |
| Range Rover Workshop Manual 1986-1989 (Petrol 3.5 & Diesel 2.4 Turbo VM) | SRR660ENWM & LSM180WM | 9781783180707 |
| Range Rover Workshop Manual 1990-1994 (Petrol 3.9 V8, 4.2 V6 & Diesel 2.5 200 Turbo Tdi) | LHAWMENA02 | 9781783180691 |
| Range Rover Workshop Manual 1995-2001 (Petrol 4.0, 4.6 V8 & BMW 2.5 Diesel) | LRL0326ENG | 9781855207462 |
| Range Rover Workshop Manual 2002-2005 (BMW Petrol 4.4 & BMW 3.0 Diesel) | LRL0477 | 9781855209046 |
| Range Rover Electrical Manual 2002-2005 UK version (Petrol 4.4 & 3.0 Diesel) | RR02KEMBB | 9781855209053 |
| Range Rover Electrical Manual 2002-2005 USA version (BMW Petrol 4.4) | RR02AEMBB | 9781855209060 |
| Range Rover Parts Catalogue 1970-1985 (Petrol 3.5) | RTC9846CH | 9781855202528 |
| Range Rover Parts Catalogue 1986-1991 (Petrol 3.5, 3.9 & Diesel 2.4 & 2.5 Turbo VM) | RTC9908CB | 9781855202931 |
| Range Rover Parts Catalogue 1992-1994 MY & 95 MY Classic (Petrol 3.9, 4.2 & Diesel 2.5 Turbo VM, 200 Tdi & 300 Tdi) | RTC9961CB | 9781855206137 |
| Range Rover Parts Catalogue 1995-2001 MY (Petrol 4.0, 4.6 & BMW 2.5 Diesel) | RTC9970CE | 9781855206168 |
| Range Rover Owners Handbook 1970-1980 (Petrol 3.5) | 606917 | 9781855201736 |
| Range Rover Owners Handbook 1981-1982 (Petrol 3.5) | AKM8139 | 9781855202795 |
| Range Rover Owners Handbook 1983-1985 (Petrol 3.5) | LSM0001HB | 9781855202801 |
| Range Rover Owners Handbook 1986-1987 (Petrol 3.5 & Diesel 2.4 Turbo VM) | LSM129HB | 9781855202900 |
| Range Rover Owners Handbook 1988-1989 (Petrol 3.5 & Diesel 2.4 Turbo VM) | SRR600EN | 9781855202917 |

**Engine Overhaul Manuals for Land Rover & Range Rover**

| Title | Ref | ISBN |
|---|---|---|
| Land Rover 300 Tdi Engine, R380 Manual Gearbox & LT230T Transfer Gearbox Overhaul Manuals | LRL003, 070 & 081 | 9781855205215 |
| Land Rover Petrol Engine V8 3.5, 3.9, 4.0, 4.2 & 4.6 Overhaul Manual | LRL004 & 164 | 9781855205284 |
| Land Rover/Range Rover Driving Techniques | LR369 | 9781855202863 |
| Working in the Wild - Manual for Africa | SMR684MI | 9781855202856 |
| Winching in Safety - Complete guide to winching Land Rovers & Range Rovers | SMR699MI | 9781855202986 |

**Workshop Manual Owners Edition**

| Title | ISBN |
|---|---|
| Land Rover 2 / 2A / 3 Owners Workshop Manual 1959-1983 | 9780713625127 |
| Land Rover 90 • 110 Workshop Manual 1983-1995 MY Owners Edition 1983-1995 | 9781855203112 |
| Land Rover Discovery Workshop Manual Owners Edition 1990-1998 | 9781855207660 |

www.brooklandsbooks.com